S0-AKY-189

METAL AND METALLOID AMIDES

Syntheses, Structures, and Physical and Chemical Properties

First published in 1980 by

ELLIS HORWOOD LIMITED

Market Cross House, Cooper Street, Chichester, West Sussex, PO19 1EB, England

The publisher's colophon is reproduced from James Gillison's drawing of the ancient Market Cross, Chichester.

Distributors:

Australia, New Zealand, South-east Asia:
Jacaranda-Wiley Ltd., Jacaranda Press,
JOHN WILEY & SONS INC.,
G.P.O. Box 859, Brisbane, Queensland 40001, Australia.

Canada:
JOHN WILEY & SONS CANADA LIMITED
22 Worcester Road, Rexdale, Ontario, Canada.

Europe, Africa:
JOHN WILEY & SONS LIMITED
Baffins Lane, Chichester, West Sussex, England.

North and South America and the rest of the world:
Halsted Press, a division of
JOHN WILEY & SONS
605 Third Avenue, New York, N.Y. 10016, U.S.A.

British Library Cataloguing in Publication Data

Metal & Metalloid Amides:
Syntheses, Structures, and Physical and Chemical Properties.–
(Ellis Horwood Series in Chemical Science)
1. Organometallic compounds 2. Amides
I. Lappert, Michael Franz
547'.05 QD411 79-40253

ISBN 0-85312-098-6 (Ellis Horwood Ltd., Publishers)
ISBN 0-470-26573-6 (Halsted Press)

Typeset in Press Roman by Ellis Horwood Ltd.
Printed in Great Britain by W. & J. Mackay Ltd., Chatham, Kent.

METAL AND METALLOID AMIDES

SYNTHESES, STRUCTURES, AND PHYSICAL AND CHEMICAL PROPERTIES

M. F. LAPPERT, F.R.S.
University of Sussex

P. P. POWER
Stanford University, California

A. R. SANGER
Alberta Research Council, Canada

R. C. SRIVASTAVA
University of of Lucknow, India

ELLIS HORWOOD LIMITED
Publishers Chichester

Halsted Press: a division of
JOHN WILEY & SONS
New York - Chichester - Brisbane - Toronto

ELLIS HORWOOD SERIES IN CHEMICAL SCIENCE

KINETICS AND MECHANISMS OF POLYMERIZATION REACTIONS: Applications and Physico-chemical Systematics
P. E. M. ALLEN, University of Adelaide, Australia and
C. R. PATRICK, University of Birmingham
ELECTRON SPIN RESONANCE
N. M. ATHERTON, University of Sheffield
METAL IONS IN SOLUTION
J. BURGESS, University of Leicester
ORGANOMETALLIC CHEMISTRY – A Guide to Structure and Reactivity
D. J. CARDIN and R. J. NORTON, Trinity College, University of Dublin
STRUCTURES AND APPROXIMATIONS FOR ELECTRONS IN MOLECULES
D. B. COOK, University of Sheffield
LIQUID CRYSTALS AND PLASTIC CRYSTALS
Volume I, Preparation, Constitution, and Applications
Volume II, Physico-Chemical Properties & Methods of Investigation
Edited by G. W. GRAY, University of Hull and
P. A. WINSOR, Shell Research Ltd.
POLYMERS AND THEIR PROPERTIES: A Treatise on Physical Principles and Structure
J. W. S. HEARLE, University of Manchester Institute of Science and Technology
BIOCHEMISTRY OF ALCOHOL AND ALCOHOLISM
L. J. KRICKA and P. M. S. CLARK, University of Birmingham
STRUCTURE AND BONDING IN SOLID STATE CHEMISTRY
M. F. C. LADD, University of Surrey
METAL AND METALLOID AMIDES: Syntheses, Structures, and Physical and Chemical Properties
M. F. LAPPERT, University of Sussex
A. R. SANGER, Alberta Research Council, Canada
R. C. SRIVASTAVA, University of Lucknow, India
P. P. POWER, Stanford University, California
BIOSYNTHESIS OF NATURAL PRODUCTS
P. MANITTO, University of Milan
ADSORPTION
J. OSCIK, Head of Institute of Chemistry, Marie Curie Sladowska, Poland
CHEMISTRY OF INTERFACES
G. D. PARFITT, Tioxide International Limited and
M. J. JAYCOCK, Loughborough University of Technology
METALS IN BIOLOGICAL SYSTEMS: Function and Mechanism
R. H. PRINCE, University Chemical Laboratories, Cambridge
APPLIED ELECTROCHEMISTRY: Electrolytic Production Processes
A. Schmidt
CHLOROFLUOROCARBONS IN THE ENVIRONMENT
Edited by T. M. SUGDEN, C.B.E., F.R.S., Master of Trinity Hall, Cambridge and
T. F. West, former Editor-in-Chief, Society of Chemical Industry
HANDBOOK OF ENZYME BIOTECHNOLOGY
Edited by A. WISEMAN, Department of Biochemistry, University of Surrey

Contents

Authors' Preface

Metal and metalloid amides are compounds in which an NH_2^-, NHR^-, or, most importantly, the NR_2^- ligand is attached to a metal or metalloid M. They are interesting both in their own right and as reagents, and progress has been extensive especially in the last two decades. Their chemistry covers a wide spectrum, ranging from theoretical to biological. In the case of the latter, important molecules which are metal amides include chlorophyll, haemin, and Vitamin B_{12} coenzyme. With regard to the former, some relevant problems are the nature of the M–N bond and the stabilisation of unusual metal or metalloid oxidation states by means of ligands such as $\bar{N}(SiMe_3)_2$.

This treatise on metal and metalloid amides has no forerunner. In it we have attempted to provide a detailed survey of the subject but, inevitably, in a work of this size some restrictions were necessary. Firstly, the chemistry of compounds having macrocyclic ligands, such as those of biological significance is only dealt with briefly. Secondly, the amides of the non-metals (H, C, N, O, Se, Te and the halogens) are considered to be outside our scope. Finally, although amides of phosphorus are discussed at some length (Chapter 6 for synthesis, structures, and physical properties, and Chapters 10–18 for selected chemical properties), they have not been treated with the same degree of completeness as we have tried to achieve for the amides of the other elements. With these reservations then, this volume represents a summary of the knowledge currently available in the major areas of research on metal and metalloid amides and, although our own approach to the field must undoubtedly be reflected, we trust that we have not been unduly selective. Indeed, we have attempted to provide a comprehensive literature coverage and hence, not surprisingly, have cited about three thousand original references up to the end of 1978. There is considerable advantage in our broad spectrum treatment because it enables common patterns to be revealed which might otherwise be submerged as isolated facts. As we state in Chapter 1, one of the principal methodologies of inorganic chemistry is exploration through the Periodic Table, and amides, which form volatile compounds with virtually all the elements, often in various oxidation states, are almost uniquely placed for such a study of periodicity, which involves examination of trends.

The introductory chapter provides a brief outline of the subject and explains the arrangement of the book; it also deals with questions of classification and nomenclature. The remaining chapters fall into two sections: Part 1 (Chapters 2-9) deals with the preparation, structures, and physical properties of the amides, using the Periodic Table as the basis for subdivision; Part 2 covers the chemical properties of amides, including their application in organic synthesis and catalysis, with separate chapters devoted to different reaction types. For Part 1, references are found, listed alphabetically, at the end of each chapter, and for Part 2, at the end of Chapter 18. There is an extensive Author Index which enables the reader to locate an author's citation, by-passing if necessary the chapter bibliographies. The Subject Index is more modest in size, partly because each chapter is prefaced by a detailed list of contents.

Tables form an integral feature of this work, and in the main have one of two functions. In Chapters 2-5 and 7-9, individual amides of the elements are listed, together with appropriate references and notes on the methods of preparation, melting and boiling points, and a comment as to which physical methods have been used in their study. In Part 2, individual reactions on specific amides, listed according to the position of M in the Periodic Table, are cited with relevant bibliography.

This book was planned more than ten years ago, and a first incomplete draft was produced in 1971. It was then set aside and not taken up again until 1976. One of us (M.F.L) has been closely involved in research on metal and metalloid amides for more than twenty years and he wishes to thank his able collaborators for their many contributions and for nourishing his growing interest in this field. We are grateful to a number of friends and colleagues for helpful comments, especially Norman Billingham, Harold Goldwhite, Heinz Nöth, and David Smith. We owe a considerable debt to Paula Keilthy for typing the manuscript, to Lorna Lappert for proof-reading, and to the staff of Ellis Horwood (especially Clive Horwood, James Gillison, and Mick Wasley) for their friendly assistance throughout.

To

Louise Gross-Lappert

1

Metal and Metalloid Amides: Introduction

A CLASSIFICATION AND NOMENCLATURE

A metal or metalloid amide is a compound which contains one or more $\bar{N}H_2$ ligand(s) or a simple derivative [such as $\bar{N}HMe$, $\bar{N}Me_2$, $\bar{N}Ph_2$, or $\bar{N}(SiMe_3)_2$] attached to a metal or metalloid M. With a few exceptions (some ionic alkali metal amides, Chapter 2), these are thus molecular compounds which, in the simplest instance, have the structural unit (I) [R and R′ are the same or different and each is H, alkyl, alkenyl, alkynyl, aryl, or $M'R_3$ (M' = Si, Ge, or Sn)].

$$M-\ddot{N}\begin{matrix} \diagup R \\ \diagdown R' \end{matrix}$$

(I)

The amides represent one of the most prolific of ligands. Stable compounds are found for almost all the elements and may be homoleptic, i.e., $M(NRR')_n$, or heteroleptic, i.e., mixed ligand complexes such as Me_3SnNMe_2 or $[Ti(\eta\text{-}C_5H_5)_2(NMe_2)_2]$.*

For a main group metal amide to be thermally stable in an inert atmosphere at ambient temperature, it must generally have a closed shell configuration, although a few kinetically stable subvalent compounds (the diamagnetic Ge^{II}, Sn^{II}, or Pb^{II} and paramagnetic Ge^{III}, Sn^{III}, P^{II}, or As^{II}) have recently been obtained using bulky amido ligands (Chapters 5-7). Transition metal amides of d^0-d^{10} configurations are known (Chapter 8), but are somewhat sparse for the late transition metals, especially of the 4d- and 5d-series.

*This distinction was first proposed for the alkyls and other hydrocarbyls of the elements [4].

Metal or metalloid amides may be mono-, bi-, oligo-, or poly-nuclear. This complexity arises because in the monomeric structure (I) there is in principle both a donor ($\ddot{N}$) and an acceptor (M) site, and hence species other than monomer are derived from it by autocomplexation. Such aggregation is characteristic of the *s*- and *d*- block and of the main group 3 elements, although steric effects, especially for the light Be, Mg, and B (where it is particularly significant because of the higher co-ordination number of boron), may supervene. It is commonly held that for amides of these elements in particular, but also Si, there is the alternative possibility of intra- (II), rather than inter- (III), molecular donor-acceptor pairing. This is variously referred to as back-co-ordination or π-bonding; $(p \leftarrow p)\pi$- for amides of group 1-3 elements or $(d \leftarrow p)\pi$-bonding for amides of

$$\mathrm{M{-}\ddot{N}(R)(R')} \longleftrightarrow \mathrm{\bar{M}{=}\overset{+}{N}(R)(R')}$$

(II)

$$\mathrm{M{-}\overset{+}{N}(R)(R'){-}\bar{M}'} \longleftrightarrow \mathrm{\bar{M}{-}\overset{+}{N}(R)(R'){-}M'}$$

(III)

Si, Ge, Sn, or *d*-block elements, and is sometimes designated as $M \leftleftharpoons NR_2$. Such theories have had a considerable impact on the development of the chemistry of metal and metalloid amides, but there is increasing scepticism about their importance, especially of $(d \leftarrow p)\pi$-bonding.

An alternative method of defining the monomer/higher aggregate problem is to recognise that the amido ligand $\bar{N}RR'$ may function not only in a terminal (I), but also a bridging (III) fashion. This procedure has an advantage because it leads to the recognition of different types of bridges: (a) homo- [M = M′, in (III)] or hetero- ($M \neq M'$) metallic; (b) μ-$\bar{N}RR'$ or μ,μ'-$(\bar{N}RR')_2$ between adjacent metal centres, as in $[Li\{N(SiMe_3)_2\}]_3$ and $[Cu\{N(SiMe_3)_2\}]_4$ on the one hand, or $(Cl_2BNMe_2)_2$ and $[Ti(NMe_2)_3]_2$ on the other (Chapters 2, 4, and 8); and (c) where μ-$\bar{N}RR'$ is found together with a μ-$\bar{X}$ bridge, as in $B_2H_5(NMe_2)$ or $Al_2Me_5(NPh_2)$ (X = H or Me) (Chapter 4).

For the purpose of this book we shall restrict the definition of 'metal or metalloid amide' to those compounds in which the nitrogen is in a three-co-ordinate environment.* Consequently complexes having the structural unit (III) are excluded. In Tables of compounds or reactions, and in most equations a di-, oligo-, or polymeric amide will be written as monomer, e.g., ethyl(diethylamido)-beryllium has the dimeric structure (IV), but will generally appear as $EtBeNEt_2$ (Chapter 3).

*Or an oligomer or polymer of an amide; such compounds, having structure (III), are treated only briefly, except for Al and Ga where they represent the main body of amide chemistry.

$$\mathrm{EtBe}\begin{matrix}\overset{\mathrm{Et_2}}{\mathrm{N}}\\ \\ \underset{\mathrm{Et_2}}{\mathrm{N}}\end{matrix}\mathrm{BeEt}$$

(IV)

On the other hand, amides may still be cyclic, e.g., $\overline{RBN(R')CH_2CH_2CH_2}$, as well as acyclic, and mono-, bi-, or poly-nuclear (within the limitation of three-co-ordination for nitrogen; the last two may alternatively be regarded as μ-imido complexes), as in $(RBNR')_{2,3,\text{ or }4}$.

The vast majority of metal or metalloid amides consists of neutral compounds, but there are isolated cases of cationic or anionic complexes, especially for amides of group 3 elements, e.g., $[B(bipy)(NMe_2)_2]^+$ (bipy = 2,2′-bipyridyl) or $[Al(NC_4H_4)_4]^-$ ($\bar{N}C_4H_4$ is the *N*-pyrrolyl anion) (Chapter 4).

Compounds are named variously as metal or metalloid amides, amidometals, or aminometallanes. For example, Me_3SiNMe_2 may be called trimethylsilicon dimethylamide, trimethylsilyldimethylamine, dimethylamido(trimethyl)silicon, or dimethylamino(trimethyl)silane. The aminometallane or metalyl nomenclature is commonly used only for elements which form stable hydrides, whence amino-boranes, -alanes, -silanes, -germanes, -phosphines, and -arsines (but not -stannane, -titanane, etc.).

B OBJECTIVES, SCOPE, AND ORGANISATION OF SUBJECT MATTER

Our principal aim is to provide a comprehensive overview of the field of metal and metalloid amide chemistry. Although selected topics (e.g., those concerning amides of transition metals or of insertion reactions of amides) have been surveyed before, the only previous attempt to give an account of amides of a wide range of metals has focussed specifically on the bis(trimethylsilyl)-amides and related complexes [11]. There is considerable advantage, in our view, in the broad spectrum coverage which we propose, especially because it enables common patterns to be revealed which otherwise might be submerged as isolated facts. Indeed one of the principal methodologies of inorganic chemistry is exploration through the Periodic Table; and amides, which form volatile compounds with virtually all the elements, are almost uniquely placed for such a study of periodicity, which involves examination of trends.

The range of elements covered is shown in Table 1. Of the non-radioactive elements, only the non-metals H, C, N, O, S, Se, Te, the halogens and the rare gases are excluded, while amides of Sr, Ba, Os, Ir, Pd, Tb, Dy, or Er have not yet been reported.* The exclusion of H, C, N, O, and the halogens is largely

*Or the actinides, except Th or U.

on the grounds that the amides of these elements are very different from those of the metals and metalloids and are in any event (e.g., the amines) well-understood and documented. There would have been a better case for the inclusion of amides of S, Se, and Te (which are barely known); sulphur amides have clear similarities with analogous compounds of B, Si, and especially P.* Phosphorus seemed to us a marginal case, and in the event the coverage here (Chapter 6) is somewhat more limited than for the remainder of the metals and metalloids, and data on individual amides of phosphorus are not tabulated. Xenon amides $FXeN(SO_2F)_2$ [12b] and $[XeN(SO_2F)_2][AsF_6]$ [5a] have been reported.

Tables form an integral feature of the book, and in the main have one of two functions. In Chapters 2-5 and 7-9, individual amides of the elements are listed, together with appropriate references, and notes on the method(s) of preparation, melting and/or boiling points, and a comment as to which physical method(s) (such as infrared or ^1H nuclear magnetic resonance spectroscopy, or X-ray diffraction) have been used in their study. In Chapters 10-18, individual reactions of specific amides are tabulated together with literature citations.

The main text is divided into two parts. The first, Chapters 2-9, deals with the preparation, physical properties, and structures of the metal amides, using the Periodic Table as the basis for subdivision. The second, Chapters 10-18, covers the chemical properties of the metal amides, including their application in synthesis and catalysis, with separate chapters devoted to different reaction types; to avoid duplication, references relevant to Chapters 10-18 are collected at the end of Chapter 18, whereas for Chapters 1-9 they are found at the end of each chapter.

As stated in Section A, we are restricting detailed consideration to compounds which have the structural unit (I), M-NRR′, but the parent compounds, derived from the unit $M\text{-}NH_2$, are only discussed briefly. Ionic metal nitrides, such as Li_3N, are excluded, but covalent analogues such as $N(SiH_3)_3$ or $N(BO_2C_6H_4\text{-}o)_3$ are covered. Thus amides included here may also have the unit (V) or (VI), and hence we regard our terms of reference as mono-, bis-, or tris-metallo- substituted ammonia; *N*-, or *N,N*-metallo- substituted primary amines; or *N*-metallo- substituted secondary amines. In order to keep this work within acceptable length, we place certain limitations on coverage of the amide chemistry of the most prolific elements B, Si, and P; for these, cyclic or polymeric compounds, including borazenes, cyclosilazanes, polysilazanes, and phos-

*The chemistry of sulphur amides is long-established and compounds such as RSO_2NH_2 are generally covered in textbooks of organic chemistry. A number of amidoselenium heterocycles, e.g., SeNHCH=X=CH—Y (X = CH or N, Y = N or CH) are known, and the compounds $OSe(NRR')_2$ [R = H = R′, or R = H and R′ = Me, or NRR′ = $N(H)CH_2$] are rather unstable [11a]. A recent report on Se amides concerns $Me_3SiOSe(O)NR_2$ [1]. As for tellurium, only the Te^{VI} amides, such as F_5TeNMe_2 or $F_5Te(NHSiMe_3)$, are well-established [7a, 15a], but $Te[N(SiMe_3)_2]_2$ [from $TeCl_4$ and $LiN(SiMe_3)_2.OEt_2$] is moderately thermally stable, m.p. 61–63° C [12a].

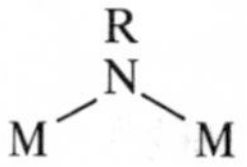

(V)

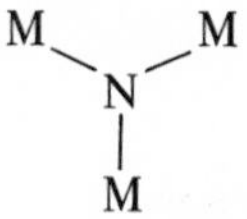

(VI)

phazenes, are not discussed at length. Finally, we draw attention to the fact that metallo- derivatives of tetrapyrrolic and related ligands are metal amides. They include many important natural products, such as haemin (a porphin), (VII), chlorophyll (a dihydroporphin or chlorin), (VIII), and vitamin B_{12} coenzyme (a corrin) (IX) (R = X); and metallo-phthalocyanines, some of which, e.g., the Cu^{II} complex (XI), are useful pigments. By and large, the macrocyclic ligand may be held to form the equatorial template for metal binding and to provide a suitable electronic and steric framework for important functions of the complex to be exercised by axial binding to the metal centre. Such compounds will be excluded.

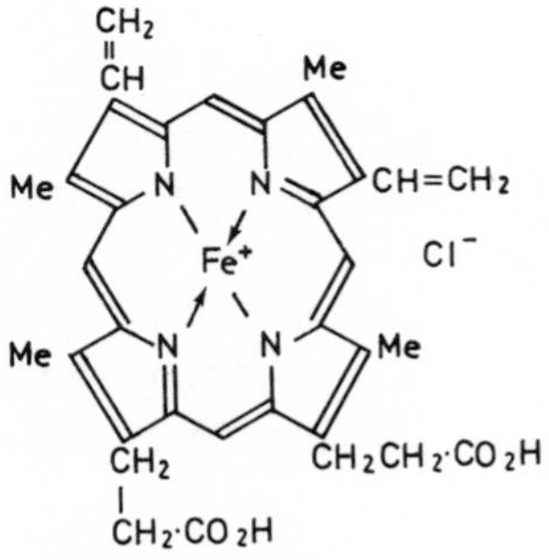

(VII)

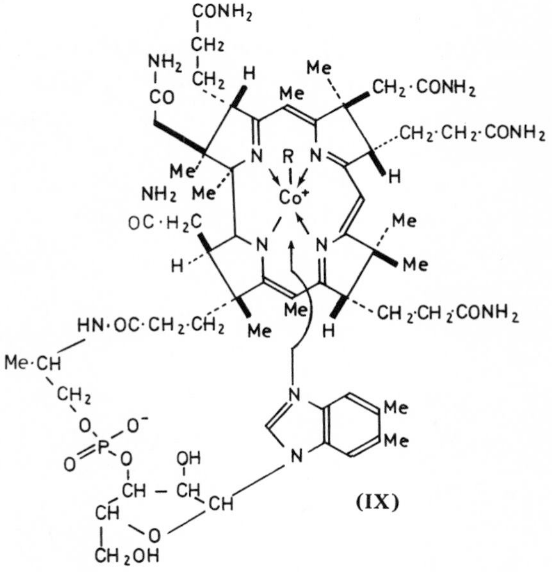

(IX)

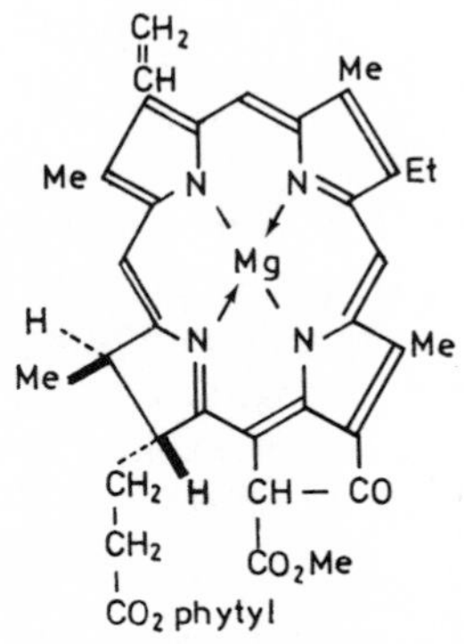

(VIII)

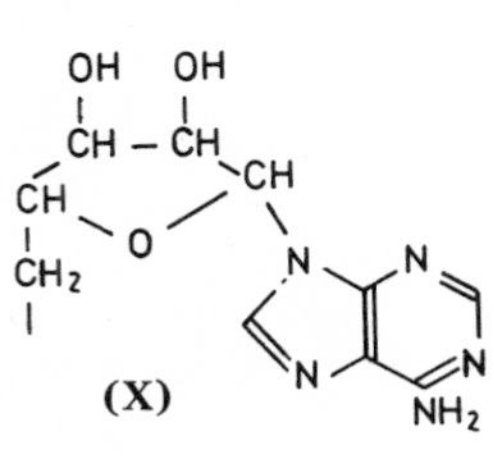

(X)

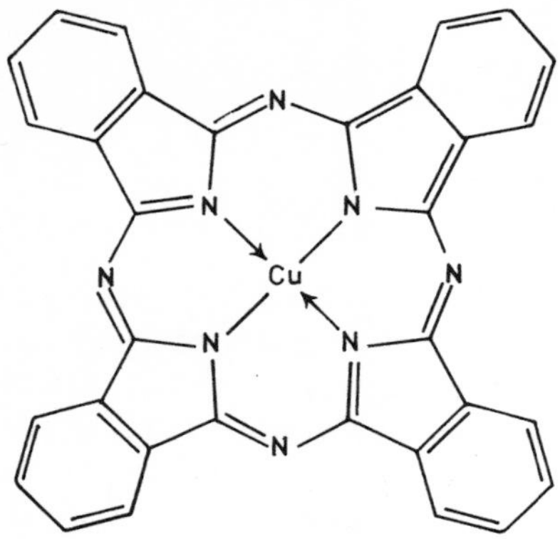

(XI)

The literature has been scanned to the end of 1978, but references to work after May 1978 were inserted at the proof stage and comments are accordingly somewhat cursory.

C DEVELOPMENTS AND PERSPECTIVES

The first metal amide, $Zn(NEt_2)_2$, was prepared by Frankland in 1856 [7].* This may be misleading as an indicator of subsequent development. A better guide is provided by Table 1 which, in the form of a fragment of the Periodic Table, shows the date of the first report of a well-characterised monomeric amide for each element. Apart from an isolated pre-World War II excursion into titanium chemistry, $[Ti(NPh_2)_4]$ was obtained in 1935 [5], transition metal amides were not discovered until the late 1950's, while the lanthanide analogues date from the 1970's, as do amides of Mg, Ga, In, and Tl, although the first three were known in oligomeric form for a substantially longer period.

Another purpose of Table 1 is to show for each element the number of literature citations (these are taken from the Tables of compounds at the ends of Chapters 2-5 and 7-9, or the bibliography of Chapter 6, and should be regarded not only as approximations, but also as minimum values). It is clear that boron, silicon, and phosphorus have attracted the greatest attention. The reasons are diverse, but include the ready accessibility and low cost of the starting materials, the relative simplicity of manipulative techniques (although boron compounds, in particular, are readily hydrolysed by atmospheric moisture), and the wide interest in bonding between close neighbours among the *p*-block elements (see, e.g., Ebsworth [6]). This last aspect was first given explicit recognition by E. Wiberg, who noted that an aminoborane was isoelectronic and isosteric with an olefin.

*$(HO)_2(O)PNH_2$ dates from 1857 [15]; $NaNH_2$ and KNH_2 were prepared earlier (see Chapter 2).

Table 1

Metal and metalloid dialkyl- or bis(trimethylsilyl)- amides: A historical survey showing elements which form amides, dates of first preparations,[a] and numbers of publications[b]

Li[c] 16	Be 1965,23											B 1933,484		
Na[d] 14	Mg 1972,56											Al 1961,184	Si 1887,463	P 1857,275
K[d] 4	Ca[e] 3	Sc 1972,4	Ti 1935,88	V 1957,29	Cr 1961,21	Mn 1964,8	Fe 1963,19	Co 1964,15	Ni 1964,14	Cu 1964,26	Zn 1856,31	Ga 1971,25	Ge 1927,78	As 1896,82
Rb[d] 4		Y 1972,3	Zr 1959,18	Nb 1957,11	Mo 1957,9				Pd 1[f]	Ag 1964,9	Cd 1965,7	In 1971,12	Sn 1961,166	Sb 1964,14
Cs[d] 4		* †	Hf 1965,6	Ta 1959,11	W 1962,20	Re 1972,2			Pt 1962,7	Au 1970,2	Hg 1907,52	Tl 1971,14	Pb 1959,28	Bi 1966,5
		*La 1972,4	Ce 1972,2	Pr 1972,4	Nd 1972,3	Sm 1972,4	Eu 1972,5	Gd 1972,2		Ho 1972,2		Yb 1972,4	Lu 1972,4	
		†	Th 1969,4		U 1956,9									

a Dates refer to first preparations of a well-characterised monomeric amide (for details see Chapters 2–9).

b These refer to references cited in Tables at the ends of Chapters 2–5 and 7–9, or the bibliography of Chapter 6.

c The only well-authenticated molecular compound is a trimer, $[LiN(SiMe_3)_2]_3$ (Chapter 2).

d Ionic compounds only are known and very few are well-characterised in the solid state (Chapter 2).

e No clear case of a monomeric amide (Chapter 3).

f No date indicated; the sole reference to a Pd–N< compound (Table, Chapter 8) is *not* to an amide but to a more complex derivative $[Pd\{N(CONHC_6H_{11})C(NH_2):NH\}_2]$

The amides of the more electropositive elements are generally significantly moisture sensitive, and often rather severe forcing conditions are required for their synthesis. Whereas amides of B, Si, and P continue to attract much attention [for example cyclic BN (Nöth, et al.) or SiN (Wannagat, et al.) compounds, and PN compounds of unusual stereochemistry (Niecke, Pohl, and Schmutzler, and their co-workers)], there is much activity elsewhere, notably with *d*- and *f*-block elements, aluminium [e.g., the poly(iminoalanes) especially by Cucinella and by J. D. Smith et al.], and tin. Significant early contributions to the problem of autocomplexation, especially for amides of Be, Mg, B, Al, and Zn, were made by Coates, et al.

By far the most important synthetic route to a transition or inner-transition metal amide involves a metathetic exchange between the metal halide (usually chloride) and the lithium (or sodium) amide, which was pioneered by Bradley and Thomas [2]. It is with the same elements that the $\bar{N}(SiMe_3)_2$ ligand first gained prominence, initially due to the studies of Wannagat and co-workers. It has also been used, as to a lesser extent have other bulky amides [e.g., $\bar{N}(CMe_3)SiMe_3$, $\bar{N}Pr^i_2$, $\bar{N}Bu^t_2$, or $\bar{N}\overline{C(Me_2)(CH_2)_3C}Me_2$], to stabilise complexes of unusually low co-ordination number, especially of *d*- and *f*-block elements (cf., Bradley and Chisholm [1a]) but also of main group elements, e.g., the monomeric $Be[N(SiMe_3)_2]_2$ and recently by this means of the open shell compounds $M[N(SiMe_3)_2]_2$ (M = Ge, Sn, Pb, P, or As) and $M'[N(SiMe_3)_2]_3$ (M′ = Ge or Sn) (Lappert and co-workers) [3, 8, 10] (for the Sn^{II} amide, see also ref. [14]).

Considering that SiN compounds have been known for about 150 years, and the first silicon amide $Cl_2Si(NHPh)_2$ since Harden's preparation in 1887 [9], it is surprising that tin(IV) amide chemistry did not develop until the 1960's. The compound $Me_3Sn{-}NMe_2$ has been widely used as a model to study metal-amide reactions, because of the high polarity $\overset{\delta+}{M}{-}\overset{\delta-}{N}$ and weakness of the SnN bond (Jones and Lappert [12]). Consequently the range of reactions is similar to that of Grignard reagents, and involves polar substrates. Thus, we find:

(i) insertion reactions (Chapter 10),
(ii) reactions with protic compounds (Chapter 11), including
(iii) metal hydrides (Chapter 12) and
(iv) substrates which may be deprotonated (Chapter 14),
(v) metathetical exchange reactions (e.g., NR_2/Cl) (Chapter 13),
(vi) reactions with Lewis bases (Chapter 15),
(vii) reactions with Lewis acids (Chapter 16),
(viii) systems in which an amide behaves as a polymerisation initiator (Chapter 17),
(ix) and miscellaneous reactions, including molecular rearrangements, oxidative addition, reductive elimination, disproportionation, M—N homolysis,or elimination (Chapter 18).

Reactivity, in general, depends on M–N bond polarity and bond strength; for example, the relatively non-polar (σ- and π-bond polarity may be compensating) and strong SiN bonds make for relatively unreactive Si amides.

Among the physicochemical studies, the most far reaching have been those of electron or single crystal X-ray diffraction. In the last 10–15 years, this has been the single most influential development. We describe the results of about 110 X-ray investigations, including 42 for amides of transition or lanthanide metals, many by Hursthouse, et al.;* Sheldrick and co-workers have contributed notably to main group metal amide structural chemistry by X-ray and electron diffraction. The latter technique is used for gaseous molecules (20 results are cited here) and (with a single microwave study) has also had a major impact, not least on discussions of bonding (see Section D).

Other experimental data relevant to bonding relate to spectroscopic and thermochemical investigations. Earlier studies placed much emphasis on vibrational spectroscopy and force constant calculations; however, this appears to have had relatively little lasting impact, no doubt due to the complexities of the mixing of vibrational states and of approximations inherent in the calculations. However, for amides of the lighter elements such as boron (Becher, Bürger, Goubeau, and their co-workers) some diagnostic features have emerged. $^{14}N(I = 1)$ or $^{15}N(I = ½)$ nuclear magnetic resonance or ^{14}N nuclear quadrupole resonance has not yet made much impression but, with the onset of a new generation of FT nmr spectrometers capable of examining ^{15}N in natural abundance, useful information, especially correlations between metal-^{15}N coupling and bonding character, may emerge.

Thermochemical data are still scant (Table 2), but with an ever widening range of homoleptic amides $M(NRR')_n$ available, the approximations necessary to derive mean thermochemical bond energy terms $\bar{E}(M-N)$ from standard heats of formation (usually derived from heats of hydrolysis or alcoholysis) become more acceptable. The original measurements are due to Skinner and co-workers in the mid-1950's on aminoboranes.

Table 2

Metal-Nitrogen bond energy terms in metal or metalloid amides, from ΔH_f° data [13]

Bond	$\bar{E}(M-N)$ (Range, kcal mol^{-1})	Bond	$\bar{E}(M-N)$ (Range, kcal mol^{-1})
B-N	100-110	Ti-N	70-85
Si-N	70-95	Zr-N	*ca.* 90
Ge-N	60-70	Hf-N	*ca.* 95
Sn-N	40-60	Zn-N	45-50
Pb-N	*ca.* 40	Cd-N	*ca.* 30
		Hg-N	*ca.* 25

*See also Chisholm and Cotton, and their co-workers.

He(I) photoelectron spectroscopy of volatile small molecules, especially when used in conjunction with molecular orbital calculations, has made a useful contribution, providing information on occupied valence orbital energies and especially first ionisation potentials. These are relevant to discussions of relative basic character of series of metal amides (cf., the introductory section of Chapter 16) (e.g., see Bock et al., for amides of B, and Chapters 5 or 8 for amides of Si, Ge, Sn, Pb, or transition elements).

The barrier to M–N rotation in metal amides is often quite low unless unusual steric constraints are present. However, for dialkylamides of boron or phosphorus they are on the nuclear magnetic resonance energy scale, and hence dynamic ^{1}H, or less often ^{13}C, nmr at various temperatures is the method of choice. The problem was first recognised in 1961 in the context of boron chemistry (Chapter 4). Initially, interpretations laid much emphasis on electronic effects; e.g., B–N $(p \leftarrow p)\pi$-bonding in a boron amide would significantly raise the barrier in a way which the cylindrically-symmetrical $(d \leftarrow p)\pi$-bond of say Si-N would not, or $\ddot{P}$-$\ddot{N}$ lone-pair repulsions in an aminophosphine could also cause a relatively high barrier. It has gradually become evident that steric effects probably have the more dominant role.

Several more general statements about the amido ligand are in order. The chemistry of metal amides may be compared with the isoelectronic alkyls, alkoxides, and fluorides. Their chemical behaviour certainly has much in common with those of alkyls and alkoxides, but little with fluorides. Bond strengths and bond polarity $(\overset{\delta+}{M}-\overset{\delta-}{X})$ increase monotonically in the sequence M-R < M-NRR′ < M-OR < M-F (see Chapter 8). This parallels electronegativity C<N<O<F. The amido ligand may thus be regarded as a relatively hard ligand, but not to the extent of $\bar{O}R$ or $\bar{F}$, and not surprisingly is found for all oxidation states of the metal including the highest. Taking chromium as an example, it is interesting that the highest oxidation state (apart from complexes of O^{2-}) is +4 and is found in an alkyl $[Cr(CH_2SiMe_3)_4]$, alkoxide $[Cr(OBu^t)_4]$, or amide $[Cr(NEt_2)_4]$, whereas the highest fluoride is CrF_3. Substitution of a silyl- group at N makes the amido ligand more polarisable; consequently the softer $\bar{N}(SiMe_3)_2$ is found even in low metal oxidation states, e.g., note the Ni^{I} complex in Table 4. Differences between amides and phosphides are much more substantial; thus the latter are only rarely found as terminal ligands and then generally when P carries bulky substituents such as Bu^t.

In principle, a metal amide may be primary, secondary, or tertiary, with units $M\text{-}NH_2$, M-NHR, or $M\text{-}NR_2$, respectively. In practice, for covalent compounds the primary is exceedingly, and secondary moderately, rare. The reason is that unless there is some steric hindrance making for kinetic stabilisation, condensation with loss of NH_3 or H_2NR takes place to yield compound (V) or (VI).

D STRUCTURAL DATA AND THEIR SIGNIFICANCE

X-ray and electron diffraction data on metal or metalloid amides show that the nitrogen atom in a terminal amido ligand is almost invariably in a planar or almost planar three-co-ordinate environment. In addition, the M–N distance is usually rather short. For instance, in the case of SiN compounds, a typical Si–N interatomic distance is ca. 1.73 Å, while the 'single bond distance' is often calculated as 1.80–1.84 Å. These two observations are often interpreted as a manifestation of significant π-bonding: $(d \leftarrow p)\pi$ in, for example, a Si or a *d*- or *f*-block metal amide, (XII), or $(p \leftarrow p)\pi$ in, for example, $>$BN$<$, (XIII). The suggestion that planarity at nitrogen is important to maximise upon π-bonding has been questioned, especially in the context of SiN chemistry (see Chapter 5); certainly inversion barriers at nitrogen are remarkably low.

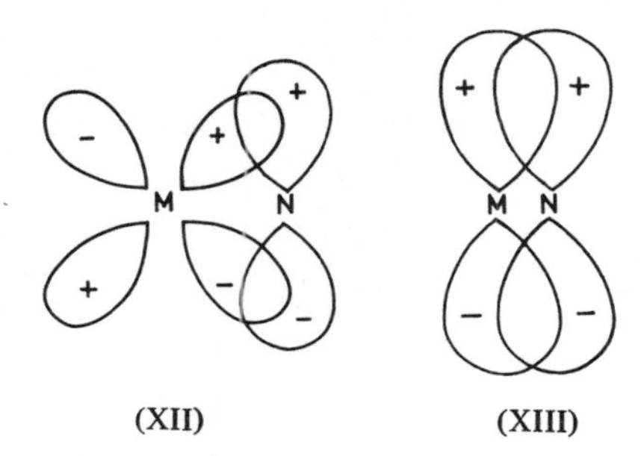

(XII) (XIII)

In discussions of M–N π-bonding in the literature, much emphasis has been placed on the fact that these bonds are invariably shorter than the sum of the covalent radii. These radii are estimated from homonuclear bond lengths where available and from heteronuclear bond lengths. Considering the dihalogens, the covalent single bond radii are taken to be one half of the internuclear distance. Similarly, the covalent radius of carbon, derived from the C–C distance in diamond, is assumed to be 0.77 Å. In a heteronuclear molecule AB, if the covalent radius of either A or B is known, then measurement of the AB bond length leads to the radius of the other partner.

By this means the covalent radius of nitrogen might be estimated from the N–N single bond (however the bond may be untypically long* due to lone-pair-lone-pair repulsion). This affords a value of 0.73 Å; a value of 0.70 Å derived from H_2NMe is generally preferred.

Marked deviation has also been found in other systems, especially where there is a large disparity in electronegativity between the bonding atoms. For example, in SiF_4 the Si–F bond length is 1.54 Å, whereas the calculated distance is 1.81 Å; likewise, in HF the calculated and measured bond lengths are 1.08 and 0.92 Å, respectively.

*E.g., in N_2H_4.

The M–N bond lengths derived from covalent (r_{cov}) and the Bragg-Slater ($r_{B\text{-}S}$) radii [16] are listed in Table 3 and compared with experimental data for monomeric amides.

Table 3

Selected covalent and Bragg-Slater radii and corresponding calculated and experimental (for amides with 3-co-ordinate nitrogen) M–N bond distances

M	Covalent radius approximation		Bragg-Slater approximation		Experimental bond length (Å)
	Radius (Å)	Bond length (Å)	Radius (Å)	Bond length (Å)	
Li	1.34	2.04	1.45	2.10	
Be	1.25	1.95	1.05	1.70	1.56–1.65
B	0.90	1.60	0.85	1.50	1.38–1.46
C	0.77	1.47	0.70	1.35	ca. 1.45 in amines
Na	1.54	2.24	1.80	2.45	2.35
Mg	1.45	2.15	1.50	2.15	2.20
Al	1.30	2.00	1.25	1.90	ca. 1.96
Si	1.18	1.88	1.10	1.75	ca. 1.74
P	1.10	1.80	1.00	1.65	ca. 1.63–1.68
K	1.96	2.66	2.20	2.85	2.70
Sc			1.60	2.25	2.05
Ti		1.981	1.40	2.05	1.85–2.2
V		1.948	1.35	2.00	1.91
Cr		1.938	1.40	2.05	1.80–2.10
Mn	1.39	2.09	1.40	2.05	1.99
Fe	1.25	1.95	1.40	2.05	ca. 1.92
Co	1.26	1.96	1.35	2.00	ca. 1.93
Ni	1.21 tet. 1.16 sq.	1.91 1.86	1.35	2.00	ca. 1.87
Cu			1.35	2.00	
Zn	1.2	1.9	1.35	2.00	2.07
Ga	1.2	1.9	1.30	2.00	1.97
Ge	1.22	1.92	1.25	1.90	1.81–1.83
As	1.22	1.92	1.15	1.80	1.76–1.88
Y			1.80	2.45	
Nb			1.45	2.10	1.90–2.06
Mo			1.45	2.10	1.97
In			1.55	2.20	2.25
Sn	1.4	2.10	1.45	2.05	2.09
W					1.90–2.02
Pt	1.7–1.8	2.4–2.5		1.35	
U	1.9	2.60			2.21–2.37

From Table 3 it is clear that many factors as well as π-bonding, including electronegativity, co-ordination number, and hybridisation of M σ-orbitals, in addition to steric effects, are important. Even when the same amido ligand is examined for low co-ordination number complexes, curious irregularities are noted. This is illustrated in Table 4 (for details, see Chapters 2-5 and 8). There is no obvious correlation between ligand parameters and MN ionic character; the K and Na amides are certainly ionic, the others are molecular compounds, with the possible exception of $Eu[N(SiMe_3)_2]_3$.

Table 4
Comparison of selected structural parameters for the $\bar{N}(SiMe_3)_2$ ligand[a]

Compound	Si-N Distance (Å)	Si-C Distance (Å)	Si-N-Si Angle (°)
$K[N\{Si(CH_3)_3\}_2].2C_4H_8O_2$	1.64(1)	1.90(3)	136.2(1.2)
$Eu[N\{Si(CH_3)_3\}_2]_3$	1.68	–	129.4
$Na[N\{Si(CH_3)_3\}_2]$	1.690(5)	1.866(5)	125.6(1)
$Cr[N\{Si(CH_3)_3\}_2]_2.2C_4H_8O$	1.69(2)	–	–
$Ni[N\{Si(CH_3)_3\}_2].2P(C_6H_5)_3$	1.70(1)	–	126(1)
$Co[N\{Si(CH_3)_3\}_2]_2.P(C_6H_5)_3$	1.706(9)	–	125(1)
$Cr[N\{Si(CH_3)_3\}_2]_3.NO$	1.72(3)	–	–
$Be[N\{Si(CH_3)_3\}_2]_2$	1.722(7)	1.876(4)	129.2(7)
$Sc[N\{Si(CH_3)_3\}_2]_3$	1.73	–	121
$Fe[N\{Si(CH_3)_3\}_2]_3$	1.73(3)	1.886(12)	121.2(4)
$H[N\{Si(CH_3)_3\}_2]$	1.735(12)	1.867(4)	125.5(1.8)
$Al[N\{Si(CH_3)_3\}_2]_3$	1.75(2)	1.90(2)	118.0(1.5)

a Reproduced, with permission, from Grüning, R., and Atwood, J. L., *J. Organometallic Chem.*, 1977, **137**, 101, (for refs., see Chapters 2-4 and 8; see also $Sn[N(SiMe_3)_2]_2$ in Chapter 5).

E SYNTHETIC METHODS

Although the first metal amide was prepared by Frankland in 1856 by the elimination of alkane [Eq. (1)], the method has not been widely employed

$$ZnEt_2 + 2HNR_2 \longrightarrow Zn(NR_2)_2 + 2C_2H_6 \quad (1)$$

except for Be, Mg, and the heavier group 3B metal amides, and Cd, because only these elements have readily available (contrast with *d*- or *f*-block elements) alkyls which are susceptible to reaction with protic reagents.

The ammonium halide elimination, illustrated by Eq. (2) (L_n represents the sum of ligands other than one Cl), was thus the first method of metal

$$L_nM\text{-}Cl + 2HNR_2 \longrightarrow L_nMNR_2 + [R_2NH_2]Cl \quad (2)$$

amide synthesis to be widely applied. It is particularly well suited for the synthesis of amides of B, Si, P, and As, as well as of Ti; the metal or metalloid chloride clearly has to be covalent and susceptible to nucleophilic attack. The main advantage of the method is that the starting materials are easily accessible and the procedure is relatively straightforward, provided moisture is excluded from the system.

Reaction (2) is often inapplicable to the synthesis of an amide of a heavier *p*-block or transition element. This is due partly to the greater ionic character of such a chloride and to the formation by the latter of a stable donor-acceptor complex with an amine. The difference is illustrated by a comparison of the reaction of a primary or secondary amine with $SiCl_4$ and $SnCl_4$ [12]. The former in most instances affords, depending on the stoicheiometry, an amino(chloro)-silane, whereas the latter yields a thermally-stable complex of the formula $SnCl_4.(HNRR')_n$ (n is usually 2).

In recent years (starting ca. 1960 [2]) the method of transmetallation, exemplified by Eq. (3) (M′ is usually Li or Na) has played an increasingly important role in the synthesis of metal amides. The lithium amide is synthesised *in situ*, usually as shown in Eq. (4); the commercial availability of LiBu has

$$L_nMX + M'NRR' \longrightarrow L_nMNRR' + M'X \quad (3)$$

$$HNRR' + LiBu \longrightarrow LiNRR' + C_4H_{10} \quad (4)$$

contributed to the utility of the technique, but in any case the combination of reactions (3) and (4) has had a profound impact on the progress of the chemistry of amides of the metals. Other advantages of transmetallation are that the yields are usually high, and it is the most applicable procedure for the controlled synthesis of a heteroleptic metal amide.

Less widely applied, and related to the alkane elimination of Eq. (1), is the hydrogen elimination route, Eq. (5). This method has proved very useful in the

$$L_nMH + HNR'_2 \longrightarrow L_nMNR'_2 + H_2 \quad (5)$$

synthesis of Be, Mg, B, Al, Ga, and Zn amides. The advantages, like those of Eq. (1), are the generally quantitative yield and the ease of purification, because the amide is the sole non-volatile product; however, few metal hydrides, except those of B or Al, are readily obtained.

Transamination, Eq. (6), has often been used to synthesise metal amides. The reaction is usually controlled by volatility, the more volatile amine being

$$L_nMNR_2 + HNR'_2 \longrightarrow L_nMNR'_2 + HNR_2 \quad (6)$$

eliminated, although steric factors also play a role. The yields are generally quantitative and purification is simple, since a volatile by-product (e.g., $HNMe_2$) is released. The method is particularly advantageous for the synthesis of a compound containing two different amido ligands, or of primary amidometal derivatives M–NHR from M–NMe_2.

F REFERENCES

[1] Barashenkov, G. G., and Derkach, N. Ya., *J. Gen. Chem. U.S.S.R.*, 1978, **48**, 1012.
[1a] Bradley, D. C., and Chisholm, M. H., *Accounts Chem. Res.*, 1976, **9**, 273.
[2] Bradley, D. C., and Thomas, I. M., *Proc. Chem. Soc.*, 1959, 225.
[3] Cotton, J. D., Cundy, C. S., Harris, D. H., Hudson, A., Lappert, M. F., and Lednor, P. W., *J.C.S. Chem. Comm.*, 1974, 651.
[4] Davidson, P. J., Lappert, M. F., and Pearce, R., *Accounts Chem. Res.*, 1974, **7**, 209.
[5] Dermer, O. C., and Fernelius, W. C., *Z. Anorg. Chem.*, 1935, **221**, 83.
[5a] DesMarteau, D. D., *J. Amer. Chem. Soc.*, 1978, **100**, 6270.
[6] Ebsworth, E. A. V., in *Organometallic Compounds of the Group IV Elements*, Vol. 1, Part I, p. 1, Marcel Dekker, 1968.
[7] Frankland, E., *Proc. Roy. Soc.*, 1856–7, **8**, 502.
[7a] Fraser, G. W., Peacock, R. D., and Watkins, P. M., *J. Chem. Soc. (A)*, 1971, 1125.
[8] Gynane, M. J. S., Hudson, A., Lappert, M. F., Power, P. P., and Goldwhite, H., *J. C. S. Chem. Comm.*, 1976, 623.
[9] Harden, A., *J. Chem. Soc.*, 1887, **51**, 40.
[10] Harris, D. H., and Lappert, M. F., *J. C. S. Chem. Comm.*, 1974, 895.
[11] Harris, D. H., and Lappert, M. F., *J. Organometallic Chem. Library*, 1976, **2**, 13.
[11a] Hopf, G., and Paetzold, R., *Z. Anorg. Chem.*, 1973, **401**, 179.
[12] Jones, K., and Lappert, M. F., in *Organotin Compounds*, ed. Sawyer, A.K., Marcel Dekker, 1971, Vol. 2, p. 509.
[12a] Lappert, M. F., and Martin, T. R., unpublished results, 1978.
[12b] LeBlond, R. D., and DesMarteau, D. D., *J.C.S. Chem. Comm.*, 1974, 555.
[13] Pedley, J. B., and Rylance, J., *Sussex–N. P. L. computer analysed thermochemical data, organic and organometallic compounds*, University of Sussex, 1977.
[14] Schaeffer, C. D., and Zuckerman, J. J., *J. Amer. Chem. Soc.*, 1974, **96**, 7160.
[15] Schiff, H., *Ann.*, 1857, **101**, 299.
[15a] Seppelt, K., *Inorg. Chem.*, 1973, **12**, 2837.
[16] Slater, J. C., *J. Chem. Phys.*, 1964, **41**, 3199.

2

Amides of the Alkali Metals: Synthesis, Physical Properties, Structures, and an Outline of their Chemical Behaviour

A INTRODUCTION

Amides of sodium and potassium (MNH_2, M = Na or K), the first examples of metal amides, were synthesised in the early part of the nineteenth century [20, 21]. Since then numerous studies have been made with these and other alkali metal amides and the earlier results have been reviewed [3, 4]. Such a compound is generally prepared by passage of gaseous ammonia over the molten metal [Eq. (1)] (in a vessel which is not attacked by the molten alkali metal, e.g., of copper [14, 23, 61]) or by dissolving the metal in liquid ammonia usually in the presence of a catalyst which accelerates the formation of the metal amide in liquid ammonia [Eq. (2)]. The alkali metals are such strong bases that

$$\underset{\text{(molten)}}{M} + \underset{\text{(gas)}}{NH_3} \longrightarrow MNH_2 + \tfrac{1}{2}H_2 \quad (1)$$

$$M + \underset{\text{(liquid)}}{n\,NH_3} \longrightarrow \text{(Blue solution)} \xrightarrow{\text{catalyst}} MNH_2 + \tfrac{1}{2}H_2 \quad (2)$$

they liberate hydrogen from ammonia, however only under forcing conditions. With sodium and gaseous ammonia, a temperature of 300–350°C is required, but in liquid ammonia the reaction proceeds quite readily at ambient temperature in presence of, for example, $Fe(NO_3)_3$ [10]. The nature of the dark-blue solution formed by dissolving the alkali metal in liquid ammonia has caused much speculation; evaporation of ammonia from a freshly-prepared solution allows the metal to be recovered. The blue solution is used as a reducing agent, for example in the Birch reduction of an aromatic compound [19e].

Alternative methods of preparation of the simple alkali metal amides are illustrated in Eqs. (3) [77d] or (4)-(6) [3, 4]. The essential feature in the liquid ammonia procedures of Eqs. (4) and (5) is the use of an organic unsaturated substrate as an H_2-acceptor; naphthalene ($C_{10}H_8$) may also be used, whence the co-product with sodamide is dekalin ($C_{10}H_{12}$) [3, 4]. Perhaps the catalytic role of a transition metal compound, such as $Fe(NO_3)_3$, is to function similarly. Another technique is to electrolyse an alkali metal salt (Cl^-, NO_3^-, or NO_2^-) in liquid ammonia.

$$M_2O + NH_3 \longrightarrow MNH_2 + MOH \quad (3)$$

$$C_6H_5CH{=}CH_2 + 2Na + 2NH_3 \longrightarrow C_6H_5CH_2CH_3 + 2NaNH_2 \quad (4)$$

$$3C_6H_5C{\equiv}CH + 4Na + 2NH_3 \longrightarrow 2C_6H_5C{\equiv}CNa + C_6H_5C_2H_5 + 2NaNH_2 \quad (5)$$

$$\tfrac{1}{2}(Et_3Ge)_2 + Li \longrightarrow LiGeEt_3 \xrightarrow{NH_3} HGeEt_3 + LiNH_2 \quad (6)$$

The alkali metal amides are colourless or translucent, crystalline solids, which melt to clear liquids often tinged slightly yellow or green due to impurity, e.g., of catalyst. The standard heats of formation (kcal mol^{-1}) of the solid amides MNH_2 are: Na, −28.4; K, −28.3; Rb, −25.7; and Cs, −25.4 [40a]. Heating the amide produces successively the imide M_2NH (e.g., for Li at 450° C) and nitride M_3N [10].

In this book, our objective is to describe the chemistry of metal amides which are distinct and well-defined molecular compounds. For this reason, emphasis is on dialkylamides or related compounds, e.g., on metal complexes having $\bar{N}R_2$ or $\bar{N}(SiMe_3)_2$ as ligands. For the case of lithium, one well-characterised amide $[Li\{N(SiMe_3)_2\}]_3$, a cyclic trimer (Fig. 1 [48], see Section C), is known; but even the relatively 'soft' bissilylamido ligand favours an ionic structure in the sodium (Fig. 2) [30c] or potassium (Fig. 3) [19b] complexes. Accordingly, the alkali metal amides present us with an atypical situation compared with that found for other groups in the Periodic Table. Hence the scope of the present chapter differs from that of Chapters 3-9, in that an outline of chemical reactions is also provided here, largely because the alkali metal amides are often prepared *in situ* and, without isolation, are then used for further syntheses. They certainly have an important role in metal amide chemistry as amide transfer reagents.

Reviews on purely inorganic aspects are found in refs. [10, 40a, 74a]. A brief account of $\bar{N}(SiMe_3)_2$ derivatives appears in ref. [75a], and a more detailed survey elsewhere [31]. The alkali metal (Li, Na, or K) bis(trimethylsilyl)amides were the first examples of metal complexes of this ligand, and are synthesised by various procedures (Scheme 1) [75a]; they are by far the most important precursors for all other metal, organometal, or metalloid bis(trimethylsilyl)amides [31]. An early review deals with silylamides of P and As [63].

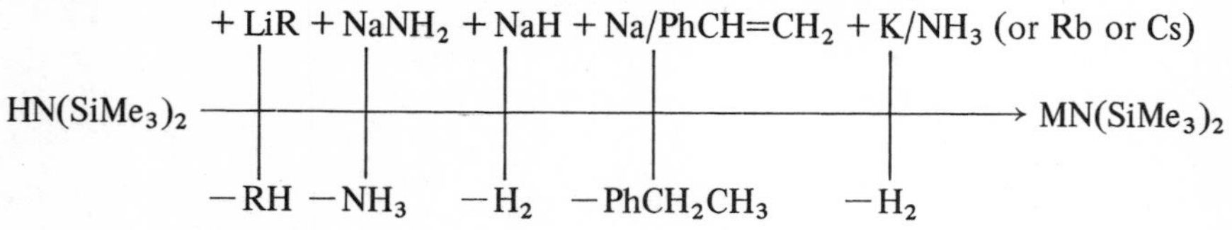

$$HN(SiMe_3)_2 \xrightarrow[-RH\;\;-NH_3\;\;-H_2\;\;-PhCH_2CH_3\;\;-H_2]{+LiR\;\;+NaNH_2\;\;+NaH\;\;+Na/PhCH{=}CH_2\;\;+K/NH_3\text{ (or Rb or Cs)}} MN(SiMe_3)_2$$

Scheme 1 Routes to $MN(SiMe_3)_2$ (M = Li, Na, K, Rb, or Cs; R = Me, Bu, or Ph) [75a].

B SYNTHESIS

One of the important reactions of an alkali metal amide, MNH_2, is with certain primary or secondary amines. Lower aliphatic amines do not react with $NaNH_2$ or KNH_2 at ambient temperature [6]. However, fused $NaNH_2$ attacks a primary aliphatic amine with the formation of cyanamide and a mixture of hydrogen and hydrocarbons [68]. A recent patent report, however, suggests that a primary aliphatic amine can be metallated by an alkali metal amide in toluene at 50–100°C [e.g., Eq. (7)] [38]. An aromatic amine, on the other hand, is readily metallated by an alkali metal amide in liquid ammonia or benzene to produce the corresponding arylamide [e.g., Eq. (8)] [41, 44, 69, 77e, 80]. ^{15}N-Labelled sodium bis(trimethylsilyl)amide has been prepared in this fashion [Eq. (9) and Scheme 1] [77].

$$NaNH_2 + H_2NBu \xrightarrow{PhMe} NaNHBu + NH_3 \qquad (7)$$

$$MNH_2 + H_2NPh \xrightarrow{C_6H_6} MNHPh + NH_3 \qquad (8)$$

$$NaNH_2 + H^{15}N(SiMe_3)_2 \longrightarrow Na^{15}N(SiMe_3)_2 + NH_3 \qquad (9)$$

The alkyl-, aryl-, or silyl-substituted alkali metal amides, which are important precursors for most other metal or metalloidal amides (Chapters 3–9), are in general more conveniently prepared by interaction of a metal alkyl or aryl and a primary or secondary amine at low temperature, as illustrated in Eqs. (10) [84] or (11) [31]. Although the lithium derivatives are now prepared from the commercially available n-butyl-lithium [e.g., Eq. (11)] [1, 13a], use of methyl- or phenyl-lithium has also frequently been made, e.g., refs. [1, 25]. An alternative to this procedure, which is useful for the preparation of derivatives of those alkali metals whose alkyl or aryl compounds are unstable or not easily available, and is also economical for large scale syntheses, is to treat the

$$RM + HNR'R'' \longrightarrow MNR'R'' + RH \qquad (10)$$

$$LiBu + HN(SiMe_3)(CMe_3) \longrightarrow LiN(SiMe_3)(CMe_3) + BuH \qquad (11)$$

alkali metal with a primary or secondary amine in the presence of an unsaturated organic compound (e.g., styrene, butadiene, or naphthalene) which acts as a hydrogen-acceptor, as in Eqs. (12) or (13) [19, 19c, 82, 83]. Metallation of a secondary amine by the alkali metal (Na or K) in liquid ammonia [Eq. (14)]

$$2M + 2HNMePh + PhCH{=}CH_2 \xrightarrow{\text{Li or Na}} 2MNMePh + PhCH_2CH_3 \qquad (12)$$

$$2Na + H_2NC_6H_{11} + PhCl \longrightarrow NaNHC_6H_{11} + NaCl + PhH \qquad (13)$$

[56] or the metal hydride in toluene, [e.g., Eqs. (15) or (16)] has also been reported [19a, 77b].

$$2Na + 2HNRPh \xrightarrow{NH_3} 2NaNRPh + H_2 \qquad (14)$$

$$2K + 2HN(SiMe_3)_2 \xrightarrow{NH_3} 2KN(SiMe_3)_2 + H_2 \qquad (15)$$

$$MH + HNMePh \xrightarrow{\text{Na or K}} MNMePh + H_2 \qquad (16)$$

$LiN(SiMe_3)_2$ is a by-product in the synthesis of $M[CH(SiMe_3)_2]_2$ (M = Ge [17a] or Pb [45c]) from $M[N(SiMe_3)_2]_2$ and $Li[CH(SiMe_3)_2]$.

C CHARACTERISATION AND STRUCTURES

The alkyl- and aryl- amides of the alkali metals are generally stable, colourless or yellowish white solids and only sparingly soluble in hydrocarbon solvents. The colourless, crystalline disilyl- substituted derivatives are, however, readily soluble in aromatic hydrocarbons. The Li, Na, or K compounds $MN(SiMe_3)_2$ are dimeric in benzene solution [13, 74a, 77a, 77b, 77c]. Lithium bis-(trimethylsilyl)amide shows (^1H or ^7Li variable temperature nmr) a monomer-dimer equilibrium in this and other solvents [42], with some decomposition. All these amides are soluble in liquid ammonia, and the solutions are highly conducting [13, 45], but benzene solutions or melts of $MN(SiMe_3)_2$ (M = Li, Na, or K) are poor conductors [13]. The lithium or sodium compounds are distillable.

A single crystal X-ray structure determination of bis(trimethylsilyl)-amidolithium reveals that the compound is a trimer, Fig. 1, in the solid state with a planar $(LiN)_3$ six-membered ring [48]; a more accurate diffractometer study has become available [59a]. $Li[\overline{NCMe_2(CH_2)_3CMe_2}]$ is a cyclotetramer [45e].

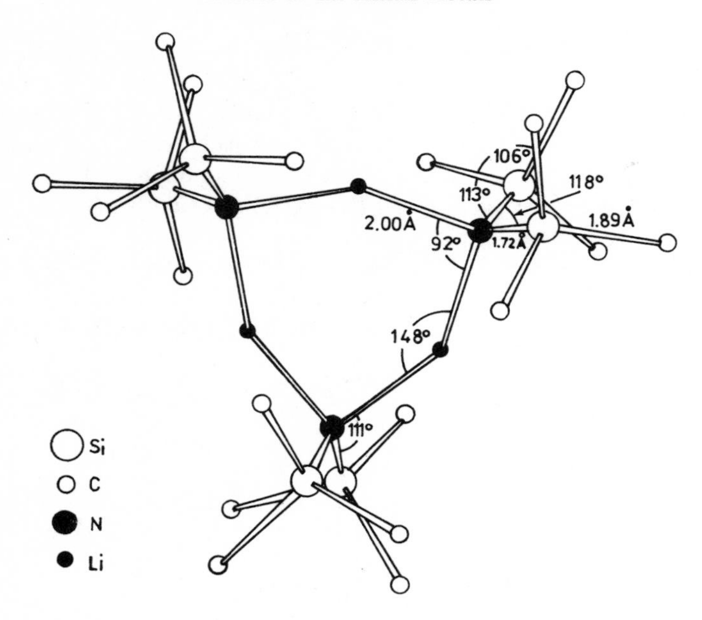

Fig. 1 – Crystal structure of $[LiN(SiMe_3)_2]_3$. *[Reproduced, with permission, from Mootz, D., Zinnius, A., and Böttcher, B.,* Angew. Chem. Internat. Edn., *1969,* **8**, *378.]* See also ref. [59a].

The sodium analogue exists as an infinite chain of cations and anions in the solid state, Fig. 2, as demonstrated by single crystal X-ray diffraction [30c]. The methyl groups have configurations such that the angle of rotation of the tri-methylsilyl moiety about the Si-N bond is 30° from the eclipsed position; presumably because this minimises methyl-methyl repulsion, the closest C····C approach thus being 3.60 Å, (I).

3.60 Å

(I)

The co-ordination of the potassium ion and the anion geometry of crystal-line $KN(SiMe_3)_2.2C_4H_8O_2$, as revealed by a single crystal X-ray study, is illus-trated in Fig. 3 [19b]; the oxygen atom positions are well-defined, but the carbon atoms of the solvent 1,4-dioxane are disordered. The five-co-ordination of K^+ is unusual. The shortest non-bonded C–C contact of the type shown in (I) is ca. 3 Å.

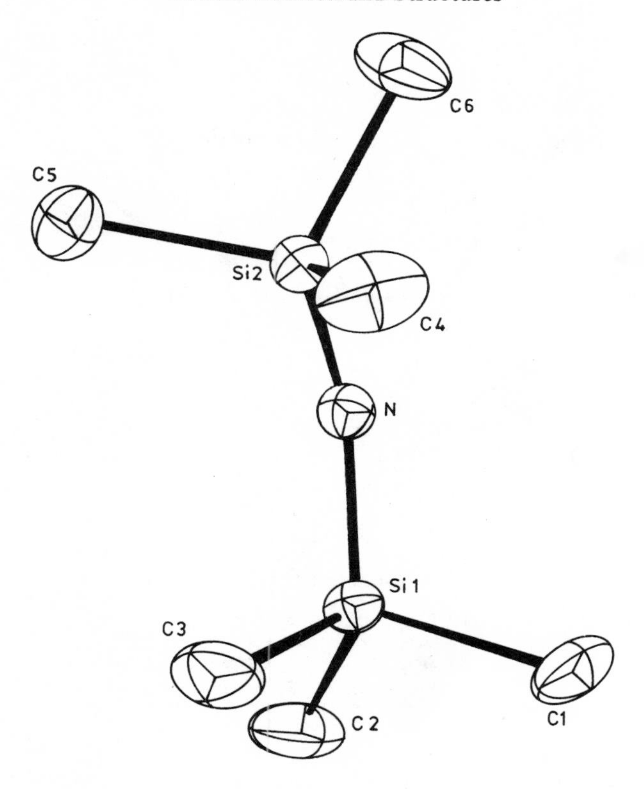

Bond Distances (Å)

N -Si1	1.694(2)	N -Si2	1.687(2)
Si1-C1	1.865(4)	Si2-C4	1.856(4)
Si1-C2	1.865(4)	Si2-C5	1.868(4)
Si1-C3	1.869(4)	Si2-C6	1.870(4)
Na^{1a}-N	2.352(2)	Na^{11}-N	2.358(3)

Bond Angles (°)

N -Si1-C1	112.4(2)	N -Si2-C4	116.7(2)
N -Si1-C2	111.2(1)	N -Si2-C5	110.0(1)
N -Si1 -C3	115.9(2)	N -Si2-C6	112.0(2)
C1-Si1-C2	105.1(2)	C4-Si2-C5	105.4(2)
C1-Si1-C3	106.2(2)	C4-Si2-C6	105.7(2)
C2-Si1-C3	105.1(2)	C5-Si2-C6	106.4(2)
Si1-N-Si2	125.6(1)	N-Na-N^{111}	150.2(1)
		Na^{1}-N-Na^{11}	102.0(1)

Fig. 2 – Crystal structure, and atom numbering scheme in the anion, of NaN-$(SiMe_3)_2$. *[Reproduced, with permission, from Grüning, R., and Atwood, J. L.,* J. Organometallic Chem., *1977,* **137**, *101.]*

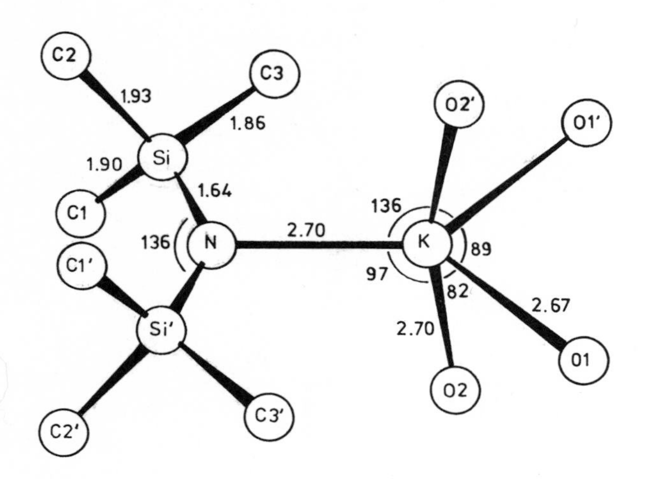

Bond Lengths (Å)

Si-N	1.64(1)	K-N	2.70(2)
Si-C(1)	1.90(2)	K-O(1)	2.67(2)
Si-C(2)	1.93(2)	K-O(2)	2.70(2)
Si-C(3)	1.86(2)		

Bond Angles (°)

Si—N-Si′	136.2(1.2)	O(1)-K-O(2)	81.9(1.0)
C(1)-Si-N	113.5(1.2)	O(1)-K-O(2′)	88.4(1.1)
C(2)-Si-N	115.7(1.0)	O(1)-K-O(1′)	89.1(1.0)
C(3)-Si-N	109.4(1.1)	O(2)-K-O(2′)	116.4(1.2)
C(1)-Si-C(2)	104.6(1.3)	N—K-O(1)	135.5(0.9)
C(2)-Si-C(3)	103.1(1.3)	N—K-O(2)	96.8(1.0)
C(3)-Si-C(1)	109.9(1.3)		

Non-bonded distances less than 4.0 Å

Si^{II}——Si	3.04	Si^{III}——K	3.64
Si^{II}——C(1)	3.94	K^{IV}——C(3)	3.36
Si^{II}——C(2)	3.94	N^{I}——C(1)	2.97
$C(3)^{V}$—O(1)	3.93	N^{I}——C(2)	3.03
$O(2)^{VI}$—O(1)	2.71	N^{I}——C(3)	2.86
$O(1)^{I}$—O(2)	3.52	$C(1)^{I}$—C(2)	3.03
$O(2)^{VII}$—O(1)	3.75	$C(1)^{I}$—C(3)	3.08
$O(1)^{VII}$—O(1)	3.75	$C(2)^{I}$—C(3)	2.97
		$C(1)^{II}$—C(2)	3.90

Fig. 3 – Anion geometry and co-ordination of potassium in $KN(SiMe_3)_2.2C_4H_8O_2$ *[Reproduced, with permission, from Domingos, A. M., and Sheldrick, G. M.,* Acta Cryst., *1974,* **30B**, *517.]*

Compared with the structures of other metal or metalloid $\bar{N}(SiMe_3)_2$ derivatives (Chapters 3-9), it is interesting that the only two ionic compounds (Na^+ and K^+ derivatives,* Figs. 2 and 3) have a short Si-N distance [1.64(1) (K) and 1.690(5) (Na) Å] and a wide SiNSi angle [136.2(1.2)° (K) and 125.6(1)° (Na)]. The former parameter has been proposed as a more reliable criterion of degree of ionic character [30c], whence in the sodium amide the Na····N bond is assigned as having significant covalent character. Reference may be made to Chapter 1, Section D.

D APPLICATIONS IN SYNTHESIS OR CATALYSIS

1. Introduction

In most of their reactions alkali metal amides behave as strong nucleophiles, but they can also exhibit electrophilic character, as shown by the formation of donor-acceptor complexes by the silyl derivatives with bases such as NH_3, C_5H_5N, Et_2O, T.H.F., or $HN(SiMe_3)_2$ [13, 75]; see also the dioxane complex of Fig. 3. Reactions of MNH_2 [3, 4] and $MN(SiMe_3)_2$ [31] are discussed elsewhere. There are substantial applications in organic chemistry [19e], as briefly summarised in Chapter 14. Sodamide reacts with ketones [47] and catalyses the polymerisation of styrene [61a]; $CsNH_2$ reacts with oxygen [59].

2. Amide transfer reagents

One of the most important and extensively used reactions of alkali metal amides is with halides in the sense of Eq. (17) (M = alkali metal, X = halide, M′ = main group or transition element, and R and R′ are H, alkyl, aryl, or substituted silyl). Such reactions are exemplified by Eqs. (18) [14a], (19) [52], and (20) [63], and are dealt with at length in Chapters 3-9.

$$n\text{MNRR}' + \text{M}'\text{X}_n \longrightarrow n\text{MX} + \text{M}'(\text{NRR}')_n \qquad (17)$$

$$[\text{WCl}_4(\text{OEt}_2)_2] + \text{LiNMeEt} \longrightarrow [\text{W}_2(\text{NMeEt})_6] + \text{LiCl} \qquad (18)$$

$$\text{PhBCl}_2 + \text{LiN(CMe}_3)(\text{SiMe}_3) \longrightarrow \text{PhClBN(CMe}_3)(\text{SiMe}_3) + \text{LiCl} \qquad (19)$$

$$\text{Me}_2\text{MCl} + \text{NaN(SiMe}_3)_2 \xrightarrow{(\text{M = P, As, or Sb})} \text{Me}_2\text{MN(SiMe}_3)_2 + \text{NaCl} \qquad (20)$$

Halogenobenzenes (except fluorobenzene) react readily with KNH_2 in liquid ammonia at −33°C to give a mixture of aniline and diphenylamine, with smaller quantities of triphenylamine and *p*-aminobiphenyl [5]. On the other hand, $NaNH_2$ reacts with bromobenzene in presence of a primary or secondary aliphatic alkylamine to yield the corresponding alkylarylamine, e.g., Eq. (21)

*And possibly Eu^{3+} [22].

[8, 11]. Alkyl- or aryl-substituted amides react more smoothly and give better results with either an alkyl or an aryl halide, e.g., Eqs. (22) or (23) (R = Et or Bu, or $NR_2 = NC_5H_{10}$; X = Cl, Br, or I; R′ = Bu, Oct, 2-ethylhexyl, Ph, or tolyl) [34, 56, 57, 58]. The reactivity of the halogenobenzenes towards piperidinato-

$$NaNH_2 + PhBr + HNMe_2 \longrightarrow Me_2NPh + NaBr + NH_3 \qquad (21)$$

$$NaNHPh + EtI \xrightarrow{-40^\circ C} HNEtPh + NaI \qquad (22)$$

$$LiNR_2 + R'X \longrightarrow R_2NR' + LiX \qquad (23)$$

lithium decreases in the sequence $F \gg Br > Cl > I$. A reaction mechanism involving two step aryne liberation from the halogenoarene and its subsequent reaction with the free amine, Eq. (24), has been proposed [8, 11, 35–37, 62].

$$R'C_6H_4X + LiNR_2 \underset{+HNR_2}{\overset{-HNR_2}{\rightleftharpoons}} R'C_6H_3(Li)X \xrightarrow{-LiX} R'C_6H_3 \text{ (aryne)} \xrightarrow{HNR_2} R'C_6H_4NR_2 + R'C_6H_4NR_2 \qquad (24)$$

An aryl halide in which the halogen atom is *ortho* to an ether, sulphide, or an amine group reacts with $NaNH_2$ in liquid ammonia, or $LiNR_2$ in ether, to give a rearranged product in which the entering amino group is *meta* [e.g., Eqs. (25), (26)] [2, 24, 25, 27, 28, 30]. With *p*-bromoanisole, a *p*-halogenotoluene, or *p*-bromophenyltrimethylsilane, a lithium dialkylamide yields a

$$o\text{-}ClC_6H_4OMe + NaNH_2 \xrightarrow{NH_3} m\text{-}H_2NC_6H_4OMe + NaCl \qquad (25)$$

$$o\text{-}BrC_6H_4NMe_2 + LiNEt_2 \xrightarrow{Et_2O} m\text{-}Et_2NC_6H_4NMe_2 + LiBr \qquad (26)$$

mixture of *p*- and *m*-isomers, Eqs. (27) (Y = OMe and R = Et, or Y = Me_3Si and R = Me), (28) (X = F, Cl, or Br) [26, 29, 36]. A 1- or 2- naphthyl halide similarly gives the same isomeric mixture of 1- and 2- piperidinonaphthalene with $LiNC_5H_{10}$, Eq. (29) [62]. 2-Bromopyridine or 2-chloroquinoline, however, gives no isomerisation, e.g., Eq. (30) (X = Cl, Br, or I) [33].

$$p\text{-}YC_6H_4Br + LiNR_2 \xrightarrow[-LiBr]{ether} p\text{-} + m\text{-}YC_6H_4NR_2 \qquad (27)$$

$$p\text{-}MeC_6H_4X + LiNC_5H_{10} \xrightarrow{-LiX} p\text{-} + m\text{-}MeC_6H_4NC_5H_{10} \qquad (28)$$

$$1\text{- or }2\text{-}C_{10}H_7X + LiNC_5H_{10} \xrightarrow[Et_2O]{HNC_5H_{10}} 1\text{-} + 2\text{-}C_{10}H_7NC_5H_{10} \quad (29)$$

$$2\text{-}BrC_5H_4N + NaNMePh \longrightarrow 2\text{-}(PhMeN)C_5H_4N \quad (30)$$

Vinyl halides undergo dehydrohalogenation in presence of an alkali metal amide producing an alkyne, as in Eq. (31) or (32). A saturated or unsaturated polyhalide, on the other hand, yields an alkynylamine (ynamine) in presence of a Li or Na amide, as in Eqs. (33), (34), or (35) [3, 4, 18, 53, 72, 73].

$$-CH{=}CHX + NaNH_2 \longrightarrow -C{\equiv}CH + NaX + NH_3 \quad (31)$$

$$PhCH{=}CHBr + NaNH_2 \longrightarrow PhC{\equiv}CH + NaBr + NH_3 \quad (32)$$

$$Cl_2C{=}CH(SR') + 2LiNR_2 \longrightarrow R_2NC{\equiv}C(SR') + 2LiCl + HNR_2 \quad (33)$$

$$FHC{=}CCl_2 \xrightarrow[-HNR_2]{LiNR_2} (F)(Li)C{=}CCl_2 \xrightarrow{-LiCl} (FC{\equiv}CCl) \xrightarrow{LiNR_2} R_2NC{\equiv}CCl$$

$$\xrightarrow{HNR_2} (R_2N)_2C{=}CHCl \xrightarrow{LiNR_2} R_2NC{\equiv}CNR_2 \quad (34)$$

$$Bu^tCHBr.CHBrF \xrightarrow{LiNEt_2} Bu^tC{\equiv}CNEt_2 \quad (35)$$

Condensation reactions involving dehydrohalogenation, as exemplified by Eq. (36) proceed readily in presence of LiNRR′ [58].

$$C_6H_{10}(OH)(H) + X\text{-}CH{=}CHR'' \xrightarrow{LiNRR'} C_6H_{10}(OH)(CH{=}CHR'') \quad (36)$$

We shall next consider reactions with halides of silicon (see also Chapter 5), nitrogen, phosphorus (Chapter 6), and sulphur, or of a dihalogen or a pseudo-halogen halide; the majority of these are metathetical X/NRR′ exchanges, but we shall mention only those in which more complicated pathways are found (see also Chapter 18). An example of a simple metathetical exchange is the formation of *N*-silylpyrrole from H_3SiBr and KNC_4H_4 [30b].

Sodamide with Me_3SiCl affords $HN(SiMe_3)_2$, $NaN(SiMe_3)_2$, and/or $N(SiMe_3)_3$, depending upon the conditions [76]. The reaction with $LiN(CMe_3)$-$(SiMe_3)$ and Me_3SiCl is solvent dependent, Eq. (37) [13a, 50, 75].

$$\mathrm{LiN(CMe_3)(SiMe_3) + Me_3SiCl} \begin{cases} \xrightarrow[-\mathrm{LiCl}]{\mathrm{Et_2O}} \mathrm{N(CMe_3)(SiMe_3)_2} \\ \xrightarrow[-\mathrm{LiCl}]{\mathrm{T.H.F.}} \mathrm{Me_3SiCH_2SiMe_2NH(CMe_3)} \end{cases} \quad (37)$$

The first examples of disilyltriazenes, $RN{=}N{-}N(SiMe_3)_2$ (R = alkyl or aryl), were obtained according to Eq. (38) [78a].

$$\mathrm{MN(SiMe_3)_2 + ArN_2Cl \xrightarrow[Et_2O]{-20^\circ C} ArN{=}N{-}N(SiMe_3)_2 + MCl} \quad (38)$$

The reaction of a phosphorus or sulphur halide with a silyl-substituted amide is often complicated due to the elimination, not only of HCl, but also halogenosilane. Whereas mixing PCl_3 with $NaN(SiMe_3)_2$ leads to a P-N-P-N polymer [75], the product with the lithium analogue depends upon the reaction conditions, Eq. (39) [67]. The equimolar reaction of $LiN(CMe_3)(SiMe_3)$ and

$$\mathrm{2LiN(SiMe_3)_2 + PCl_3 \xrightarrow{25^\circ C} PCl[N(SiMe_3)_2]_2 + 2LiCl}$$

$$\mathrm{\xrightarrow[-Me_3SiCl]{150^\circ C} (Me_3Si)_2N{-}P{=}N(SiMe_3)} \quad (39)$$

PCl_3, on the other hand, gives mainly the (rare) diazadiphosphetidine of Eq. (40) [64]. Changing the stoicheiometry affords an exceptionally stable phosphazene containing two-co-ordinate phosphorus, Eq. (41) [65]. A related phosphazene $(Me_3Si)_2N{-}P{=}N(CMe_3)$ is obtained according to Eq. (42) [66].

$$\mathrm{2LiN(CMe_3)(SiMe_3) + 2PCl_3 \longrightarrow \begin{matrix} Me_3C{-}N{-}P{-}Cl \\ | \quad\ | \\ Cl{-}P{-}N{-}CMe_3 \end{matrix} + 2LiCl + 2Me_3SiCl} \quad (40)$$

$$\mathrm{2LiN(CMe_3)(SiMe_3) + PBr_3 \xrightarrow[(ii)\ 130^\circ C]{(i)\ -78^\circ C} (Me_3C)(Me_3Si)N{-}P{=}N(CMe_3) + 2LiBr + Me_3SiBr} \quad (41)$$

$$\mathrm{PCl_3 \xrightarrow[\substack{(ii)\ LiN(CMe_3)(SiMe_3),\ -78^\circ C \\ (iii)\ heat}]{(i)\ LiN(SiMe_3)_2,\ 0^\circ C} (Me_3Si)_2N{-}P{=}N(CMe_3) + 2LiCl + Me_3SiCl} \quad (42)$$

Unlike POF_3, which reacts normally with $LiN(SiMe_3)_2$, $POCl_3$ gives a trimethylsiloxyphosphinimine, Eq. (43), formed presumably from $POCl_2$-$[N(SiMe_3)_2]$ by a 1,3-trimethylsilyl shift [13d]. Bismuth oxyiodide and MNH_2 (M = Na or K) in liquid ammonia yield $Bi(O)NH_2$ [77d].

$$LiN(SiMe_3)_2 \begin{cases} \xrightarrow[Et_2O,\ 0^\circ C]{POF_3} POF_2[N(SiMe_3)_2] + LiF \\ \xrightarrow[Et_2O,\ 36^\circ C]{POCl_3} (Me_3SiO)Cl_2P{=}N(SiMe_3) + LiCl \end{cases} \quad (43)$$

In case of a sulphur substrate containing an S=O bond, loss of bistrimethylsiloxane may also occur, e.g., Eq. (44) [45f]. This is the preferred route for the synthesis of tris(*N*-trimethylsilylimido)sulphur, $S(=NSiMe_3)_3$ rather than Eq. (45) [19d, 30a].

$$3NaN(SiMe_3)_2 + SOF_4 \xrightarrow{-40^\circ C} S(=NSiMe_3)_3 + 3NaF + Me_3SiF + (Me_3Si)_2O \quad (44)$$

$$3LiN(SiMe_3)_2 + 2NSF_3 \longrightarrow S(=NSiMe_3)_3 + SF_2(=NSiMe_3)_2 + Me_3SiF + LiF \quad (45)$$

Di-iodine has been reported to form the salt Na_2NI_3 when treated with $NaNH_2$ in liquid ammonia [39]. With $NaN(SiMe_3)_2$, however, both Cl_2 and I_2 give the corresponding halogenoamine, Eq. (46) (X = Cl or I) [74, 78].

$$NaN(SiMe_3)_2 + X_2 \longrightarrow XN(SiMe_3)_2 + NaX \quad (46)$$

Bromocyanogen reacts with $NaNH_2$ to give sodium dicyanamide, Eq. (47). Aryl dicyanamides are prepared by a related reaction, Eq. (48) [7].

$$3NaNH_2 + 2BrCN \longrightarrow NaN(CN)_2 + 2NH_3 + 2NaBr \quad (47)$$

$$KN(CN)R + BrCN \longrightarrow RN(CN)_2 + KBr \quad (48)$$

Formation of a silyl-substituted carbodi-imide from BrCN and $NaN(SiMe_3)_2$ is interesting and probably involves rearrangement of the expected product $(Me_3Si)_2NCN$, Eq. (49) [9].

$$NaN(SiMe_3)_2 + BrCN \longrightarrow Me_3Si{-}N{=}C{=}N{-}SiMe_3 + NaBr \quad (49)$$

3. Metallation

In these reactions, the metal amide functions as a strong, but somewhat sterically-hindered, nucleophile and hence in the main as a proton abstractor (see Chapter 14). An example for an acidic hydrocarbon is shown in Eq. (50) ($X^1 = H = X^2$, X^3 = Cl, Br, or I; or $X^1 = SiMe_3$ and $X^2 = X^3$ = Br) [46a,46b] and Eq. (51) [13b, 54a, 68a].

$$HCX^1X^2X^3 + MN(SiMe_3)_2 \longrightarrow M(CX^1X^2X^3) + HN(SiMe_3)_2 \quad (50)$$

$$RC{\equiv}CH + LiNMe_2 \longrightarrow RC{\equiv}CLi + HNMe_2 \quad (51)$$

Enolisable ketones can be metallated by MNR_2 and often give isomeric products, as illustrated in Eq. (52) [15, 16, 47, 60]. Non-enolisable ketones,

$$MNR_2 + R'CH{=}COR'' \xrightarrow{-HNR_2} (R'\bar{C}H\text{-}COR'' \rightleftharpoons R'CH{=}CR''\text{-}O^-)Na^+$$

$$\downarrow -NaX,\ R'''X$$

$$R'R'''CH\text{-}COR'' + R'CH{=}CR''\text{-}OR''' \quad (52)$$

such as Ph_2CO, may undergo substitution [Eq. (53)] [45b, 75] or reduction [Eq. (54)] [79].

$$NaN(SiMe_3)_2 + Ph_2CO \longrightarrow Ph_2C{=}NSiMe_3 + NaOSiMe_3 \quad (53)$$

$$LiN(R)CH_2R' + Ph_2CO \xrightarrow{H_2O} Ph_2CHOH + RN{=}CHR' \quad (54)$$

Condensation of esters may be effected by LiNRR′, as in Eq. (55) [30d].

$$2R''CH_2COOR''' + LiNRR' \longrightarrow R''CH_2CO\text{-}CH(R'')COOR''' \quad (55)$$

Metallation of an aliphatic nitrile of type R_2CHCN (R = H, alkyl, or aryl) produces the corresponding salt $R_2C(M)CN$, which is a useful reactive intermediate, as in Eqs. (56)-(58) [13c, 17, 32, 45a, 45d, 84, 85] (M = Li and R = alkyl, R′ = R″ = Me, $R''' = CH_2{=}CHCH_2$; or M = Na and R = H, R′ = Ph, R″ = alkyl, and $R''' = ClCH_2CH_2$).

$$NaN(SiMe_3)_2 + CH_3CN \xrightarrow{-HN(SiMe_3)_2} NaCH_2CN \xrightarrow[-NaCN]{\text{cyclohexene}} \text{bicyclo[4.1.0]heptane} \quad (56)$$

$$LiNR_2 + RCH_2CN \xrightarrow{-HNR_2} LiCHRCN \xrightarrow{RCH_2C{\equiv}N} RCH_2C({=}NLi)CH(R)CN$$

$$\xrightarrow[-LiCl]{Me_3SiCl} Me_3SiNHC(CRH_2){=}CRCN \quad (57)$$

$$MNR_2 + R'R''CHCN \xrightarrow{-HNR_2} R'R''C(M)CN \xrightarrow[R'''Cl]{-MCl} R'R''C(CN)R''' \quad (58)$$

Tosylhydrazine is metallated at room temperature by $MN(SiMe_3)_2$; the tosylhydrazide thus formed was used for an important synthesis, Eq. (59) (M = Li, Na, or K; Tos = p-$MeC_6H_4SO_2$) [77f].

$$MN(SiMe_3)_2 + TosNH\text{-}NH_2 \xrightarrow[25°C]{C_6H_6} TosN(M)NH_2 + HN(SiMe_3)_2$$

$$TosN(M)NH_2 \xrightarrow[<10^{-4}\,mmHg]{heat} HN{=}NH + MTos \quad (59)$$

4. Insertion

A lithium dialkylamide adds across the C=C double bond of a vinylacetylene [Eqs. (60) (n = 1, 2) or (61)] [54, 55]. Insertion of benzonitrile into the Li-N bond of $LiN(SiMe_3)_2$ in diethyl ether is the only reported insertion reaction involving an alkali metal bis(trimethylsilyl)amide [Eq. (62)] [61b]. In petroleum as solvent $LiN(SiMe_3)_2$ does not react, and $LiNMe_2$ causes trimerisation of PhCN [45d, 61b]; $NaNH_2$ reacts according to Eq. (63) [43].

$$HC{\equiv}C(CH_2)_n CH{=}CH_2 \xrightarrow[(ii)\ H^+]{(i)\ LiNR_2} HC{\equiv}C(CH_2)_{n+2}NR_2 \quad (60)$$

$$Me_3SiC{\equiv}CCH{=}CH_2 \xrightarrow[(ii)\ H^+]{(i)\ LiNR_2} Me_3SiC{\equiv}CCH_2CH_2NR_2 + Me_3SiCH{=}C{=}CHCH_2NR_2 \quad (61)$$

$$ArC{\equiv}N + LiNR_2 \xrightarrow{Et_2O} LiN{=}C(Ar)NR_2 + (ArCN)_3 \quad (62)$$

$$PhC{\equiv}N + NaNH_2 \longrightarrow PhC({=}NH)NHNa \xrightarrow{H_2O} PhC({=}NH)NH_2 \quad (63)$$

The reaction of $NaNH_2$ with carbon monoxide gives Na(HCONH) [40, 51]. With $NaN(SiMe_3)_2$, however, CO yields NaCN and $(Me_3Si)_2O$. The expected product, $Na[CON(SiMe_3)_2]$, if formed, decomposes [77c]. Metal carbonyls, except $[Fe(CO)_5]$ [Eq. (64)] [77c], react similarly, as illustrated in Eq. (65) [10a, 42a, 77e]. On the other hand, $LiNEt_2$ adds to $[Cr(CO)_6]$ to afford a carbenemetal complex, Eq. (66) [19f, 19g].

$$MN(SiMe_3)_2 + [Fe(CO)_5] \xrightarrow[Li\ or\ Na]{-10\ to\ 25°C} [Fe(CO)_4(CNSiMe_3)] + MOSiMe_3 \quad (64)$$

$$NaN(SiMe_3)_2 + [M(CO)_n] \xrightarrow[C_6H_6]{25°C} Na[M(CO)_{n-1}(CN)] + (Me_3Si)_2O \quad (65)$$

$$LiNEt_2 + [Cr(CO)_6] \xrightarrow{Et_2O} \left[Cr(CO)_5C\begin{matrix} OLi.OEt_2 \\ NR_2 \end{matrix}\right]$$

$$\xrightarrow{[Et_3O][BF_4]} \left[Cr(CO)_5C\begin{matrix} OEt \\ NR_2 \end{matrix}\right] \quad (66)$$

Lithium diethylamide catalyses the insertion of an epoxide into a Sn–N bond Eq. (67) [71a].

$$\underset{\text{(epoxide, O bridging)}}{RCH\text{-}CH_2} + LiNEt_2 \longrightarrow \underset{|\atop OLi}{RCHCH_2NEt_2}$$

$$\xrightarrow{Bu_3SnNEt_2} \underset{|\atop OSnBu_3}{RCHCH_2NEt_2} + LiNEt_2 \quad (67)$$

Diphenylamidosodium reacts with carbon bisulphide to yield sodium diphenylthiocarbamate [Eq. (68)] [41]. With $NaN(SiMe_3)_2$, however, the addition product, if formed, decomposes as shown in Eq. (69) [75].

$$NaNPh_2 + CS_2 \longrightarrow NaSC(S)NPh_2 \quad (68)$$

$$NaN(SiMe_3)_2 + CS_2 \longrightarrow NaSCN + (Me_3Si)_2S \quad (69)$$

5. Polymerisation initiation

In a broader context, of metal amides generally, this topic is the subject matter of Chapter 17. Alkali metal amides possess the ability to initiate polymerisation of some vinyl monomers. The mechanisms are either of anionic polymerisation or a closely related process. Accordingly, the vinylic monomer is restricted to one which is relatively electron-deficient. Thus, acrylonitrile or methacrylonitrile is polymerised at –85°C in presence of $NaNH_2$ in liquid ammonia or $LiNPh_2$ in ether, respectively, to give high molecular weight polymers [10b, 61a]. Polymerisation of styrene may be initiated by the KNH_2/liquid ammonia system, and proceeds via an anionic mechanism which involves the opening of the C=C bond by the amide ion, thereby forming a 2-aminoalkyl carbanion which propagates the chain [33a]. Conjugated dienes are readily polymerised in presence of $LiNR_2$ [41a, 43a, 49, 52a]. The lithium dialkyl-amide- initiated polymerisation of butadiene in hexane or benzene yields polymer containing 85–89% 1,4 units, of which 40–54% have the *trans*-configuration [41a, 52a].

E TABLE OF COMPOUNDS

Table 1
Amides and substituted amides of the alkali metals

Compound	Method of preparation[a]	M.P. (°C)	Comments	References
$LiNH_2$	A	380–400	Gives Li_2NH at 450°C; Li_3N at higher temp	[70]
$NaNH_2$	A,C	210	Colourless solid	[20,21,46,81]
KNH_2	A,C	338	Distils rapidly at red heat with slight decomposition Colourless solid	[81]
$RbNH_2$	A	309	Distils unchanged 400°C Colourless solid	[71]
$CsNH_2$	A	362	Colourless solid	[59]
LiNHPh	A,B	–	Colourless solid	[19,82,83]
$LiNHC_{10}H_7$	A	–	Colourless solid	[19,82,83]
$LiNMe_2$	B	–	Colourless crystals	[29]
$LiNEt_2$	B	–	Colourless crystals	[26]
$LiNPr_2$	B	–	Colourless crystals	[30d]
$LiNBu_2$	B	–	Colourless solid	[54,55]
$LiNC_5H_{10}$	B		Colourless crystals	[34]
$LiNC_4H_8O$	B	–	Colourless crystals	[24,27,30]
$LiN(C_6H_{11})_2$	B	–	Colourless crystals	[19,82,83]
$LiNPh_2$	A,B,C	–	Colourless crystals	[19,82,83]
LiNMePh	C	–	Colourless crystals	[19,82,83]
LiNEtPh	A,C	–	Colourless solid	[19,82,83]
$NaNHPr^i$	A	–	Colourless solid	[19,38,82,83]
$NaNHBu^s$	A	–	Colourless solid	[19,82,83]
$NaNHC_6H_{11}$	C	–	Colourless solid	[19,82,83]
NaNHPh	A,C	–	Glassy pale-yellow solid	[56,69]
$NaNHC_{10}H_7$	A,C	–	Colourless solid	[69]
$NaNHC_6H_4Me$-*p*	A	–	Colourless solid	[69]
$NaNHC_6H_4Me$-*o*	A	–		[56]
$NaNMe_2.2BH_3$				[1a]
$NaNEt_2$	A	–		[19,82,83]
$NaNPh_2$	A,C	–		[56,57,69]
NaNMePh	C	–		[19,82,83]

Table 1 *(C'td.)*

Compound	Method of preparation[a]	M.P. (°C)	Comments	References
$LiN(SiMe_3)_2$	B	87	b.p. 155°C/1 mm, X-ray	[17a, 45c, 48, 59a,75a,77a,77b]
$NaN(SiMe_3)_2$	A	165–7 172	b.p. 170°C/2 mm, X-ray	[30c,74a,75a, 77a,77b]
$KN(SiMe_3)_2$	A	165		[13]
$KN(SiMe_3)_2$.2dioxan	–	–	X-ray	[12,19b,22]
$KNEt_2.BH_3$				[1a]
$RbN(SiMe_3)_2$	A	178		[13,75a]
$CsN(SiMe_3)_2$	A	195		[13,75a]
$LiN(CMe_3)(SiMe_3).Et_2O$	A	–		[13a]
$LiN(SiMe_3)_2.Et_2O$	B		Colourless crystals	[31]
$LiN(GeMe_3)_2.Et_2O$	B		Colourless crystals	[31]
$\overline{LiNCMe_2(CH_2)_3CMe_2}$	B		Colourless crystals, X-ray	[45e]

a This is denoted by a letter assigned to the following reactions:
A = metallation by a metal amide, hydride, or metal itself;
B = metallation by a metal alkyl or aryl;
C = metallation by a metal in presence of an organic substrate which is reduced by the metal or forms a metal–carbon bond.

F REFERENCES

[1] Amonoo-Neizer, E. H., Shaw, R. A., Skovlin, D. O., and Smith, B. C., *Inorg. Synth.*, 1966, **8**, 19.

[1a] Balulescu, C. R., and Keller, P. C., *Inorg. Chem.*, 1978, **17**, 3707; Keller, P. C., *Inorg. Chem.*, 1971, **10**, 2256; Keller, P. C., *Synth. Inorg. Metal-Org. Chem.*, 1973, **3**, 307.

[2] Benkeser, R. A., and De Boer, C. E., *J. Org. Chem.*, 1956, **21**, 285, 365.

[3] Bergstrom, F. W., and Fernelius, W. C., *Chem. Rev.*, 1933, **12**, 43.

[4] Bergstrom, F. W., and Fernelius, W. C., *Chem. Rev.*, 1937, **20**, 413.

[5] Bergstrom, F. W., Wright, R. E., Chandler, C., and Gilkey, W. A., *J. Org. Chem.*, 1936, **1**, 170.

[6] Bergstrom, F. W., and Wood, D., unpublished work cited in ref. 3.

[7] Biechler, J., *Compt. Rend.*, 1935, **200C**, 141.

[8] Biehl, E. R., Smith, S. M., and Reeves, P. C., *J. Org. Chem.*, 1971, **36**, 1841.

[9] Birkofer, L., Ritter, A., and Richter, P., *Tetrahedron Letters*, 1962, **5**, 195.

[10] Brauer, G., *Handbook of Preparative Inorganic Chemistry*, 2nd. ed., Vol. 1, p. 464 et seq., Academic Press, 1963.

[10a] Brunner, H., *Chem. Ber.*, 1969, **102**, 305.

[10b] Bullitt, O. H., U.S. Patent 2,608,555; *Chem. Abstr.*, 1953, **47**, 1430.
[11] Bunnett, J. F., and Brotherton, T. K., *J. Org. Chem.*, 1957, **22**, 832.
[12] Bürger, H., *Angew. Chem. Internat. Edn.*, 1973, **12**, 474.
[13] Bürger, H., and Seyffert, H., *Angew. Chem. Internat. Edn.*, 1964, **3**, 646.
[13a] Bush, R. P., Lloyd, N. C., and Pearce, C. A., *J. Chem. Soc. (A)*, 1969, 253, 257, 808.
[13b] Campbell, K. N., and Campbell, B. K., *Proc. Indiana Acad. Sci.*, 1940, **50**, 123.
[13c] Cason, J., Sumrell, G., and Mitchell, R. S., *J. Org. Chem.*, 1950, **15**, 850.
[13d] Czieslik, G., Flaskerud, G., Hoefer, R., and Glemser, O., *Chem. Ber.*, 1973, **106**, 399.
[14] Chambers, R. W., and Scherer, P. C., *J. Amer. Chem. Soc.*, 1926, **48**, 1054.
[14a] Chisholm, M. H., and Extine, M.W., *J. Amer. Chem. Soc.*, 1975, **97**, 1623.
[15] Cornubert, R., and Humeau, R., *Bull. Soc. Chim. Fr.*, 1931, **49**, 1492.
[16] Cornubert, R., Humeau, R., Le Bihan, H., and Maurel, A., *Bull. Soc. Chim. Fr.*, 1931, **49**, 1260.
[17] Cuvigny, T., and Normant, H., *Compt. Rend.*, 1969, **268C**, 1380.
[17a] Davidson, P. J., Harris, D. H., and Lappert, M. F., *J.C.S. Dalton*, 1976, 2286.
[18] Delavarenne, S. Y., and Viehe, H. G., *Tetrahedron Letters*, 1969, 4761.
[19] De Pree, D. O., U.S. Patent, 1957, 2,799,705; *Chem. Abstr.*, 1958, **52**, 1202.
[19a] Deutsche Gold und Silber-Scheideanstalt, Brit. Patent, 293,040; *Chem. Abstr.*, 1929, **23**, 1418.
[19b] Domingos, A. M., and Sheldrick, G. M., *Acta Cryst.*, 1974, **30B**, 517.
[19c] Ethyl Corporation, Brit. Patent, 742,790; *Chem. Abstr.*, 1956, **50**, 16826i.
[19d] Feser, M. F., Glemser, O., von Halasz, S. P., and Saran, H., *Inorg. Nucl. Chem. Letters*, 1972, **8**, 321.
[19e] Fieser, F., and Fieser, M., *Reagents for Organic Synthesis*, Vol. 1-6, Wiley, 1967–1976.
[19f] Fischer, E. O., and Kollmeier, H. J., *Angew. Chem. Internat. Edn.*, 1970, **9**, 309.
[19g] Fischer, E. O., Winkler, E., Kreiter, C. G., Huttner, G., and Krieg, B., *Angew. Chem. Internat. Edn.*, 1971, **10**, 922.
[20] Gay-Lussac, J. L., and Thenard, L. J., *Ann. Physik.*, 1809, **32**, 30, 34.
[21] Gay-Lussac, J. L., and Thenard, L. J., *Physicochimiques (I)*, Paris, 1811, 337, 341, 354, 356.
[22] Ghotra, J. S., Hursthouse, M. B., and Welch, A. J., *J.C.S. Chem. Comm.*, 1973, 669.
[23] Gilbert, H. N., Scott, N. D., Zimmerli, W. F., and Hansley, V. L., *Ind. Eng. Chem.*, 1933, **25**, 735.

[24] Gilman, H., and Avakian, S., *J. Amer. Chem. Soc.,* 1945, **67**, 349.
[25] Gilman. H., Crounse, N. N., Massie, S. P., Benkeser, R. A., and Spatz, S. M., *J. Amer. Chem. Soc.,* 1945, **67**, 2106.
[26] Gilman, H., and Kyle, R. H., *J. Amer. Chem. Soc.,* 1948, **70**, 3945.
[27] Gilman, H., and Kyle, R. H., *J. Amer. Chem. Soc.,* 1952, **74**, 3027.
[28] Gilman, H., Kyle, R. H., and Benkeser, R. A., *J. Amer. Chem. Soc.,* 1946, **68**, 143.
[29] Gilman, H., Melvin, H. W., and Goodman, J. J., *J. Amer. Chem. Soc.,* 1954, **76**, 3219.
[30] Gilman, H., and Nobis, J. F., *J. Amer. Chem. Soc.,* 1945, **67**, 1479.
[30a] Glemser, O., and Wegener, J., *Angew. Chem. Internat. Edn.,* 1970, **9**, 309.
[30b] Glidewell, C., and Rankin, D. W. H., *J. Chem. Soc. (A),* 1970, 279.
[30c] Grüning, R., and Atwood, J. L., *J. Organometallic Chem.,* 1977, **137**, 101.
[30d] Hammell, M., and Levine, R., *J. Org. Chem.,* 1950, **15**, 162.
[31] Harris, D. H., and Lappert, M. F., *J. Organometallic Chem. Library,* 1976, **2**, 13.
[32] Hastings, R., and Cloke, J. B., *J. Amer. Chem. Soc.,* 1934, **56**, 2136.
[33] Hauser, C. R., and Weiss, M. J., *J. Org. Chem.,* 1949, **14**, 310.
[33a] Higginson, W. C. E., and Wooding, N. S., *J. Chem. Soc.,* 1952, 1178.
[34] Horning, C. H., and Bergstrom, F. W., *J. Amer. Chem. Soc.,* 1945, **67**, 2110.
[35] Huisgen, R., and Sauer, J., *Angew. Chem.,* 1957, **69**, 390.
[36] Huisgen, R., and Sauer, J., *Chem. Ber.,* 1958, **91**, 1453.
[37] Huisgen, R., and Sauer, J., *Chem. Ber.,* 1959, **92**, 192.
[38] Humphreys, D. D., U.S. Patent, 1954, 2,685,604; *Chem. Abstr.,* 1955, **49**, 10358e.
[39] Hurd, C. D., Sweet, A. D., and Thomas, C. L., *J. Amer. Chem. Soc.,* 1933, **55**, 335.
[40] Jacobson, R. A., and Carothers, W. H., *J. Amer. Chem. Soc.,* 1933, **55**, 1622.
[40a] Jolly, W. C., *The Inorganic Chemistry of Nitrogen,* W. A. Benjamin, 1964, p. 35.
[41] Jones, P. C., U.S. Patents, 1935, 2,046,875 and 2,046,876; *Chem. Abstr.,* 1936, **30**, 5592.
[41a] Kibler, R. W., Bozzacco, F. A., and Forman, L. E., U.S. Patent 2,849,432; *Chem. Abstr.,* 1959, **53**, 10844i.
[42] Kimura, B. Y., and Brown, T. L., *J. Organometallic Chem.,* 1971, **26**, 57.
[42a] King, R. B., *Inorg. Chem.,* 1967, **6**, 25.
[43] Kirsanov, A. V., and Poliakowa, I., *Bull. Soc. Chim. Fr.,* 1936, **3**, 1600.
[43a] Kodomari, M., Sawa, S., Morozumi, K., and Ohkita, T., *Nippon Kagaku Kaishi,* 1976, **2**, 301.

[44] Kraus, C. A., and Bien, P. B., *J. Amer. Chem. Soc.*, 1933, **55**, 3609.
[45] Kraus, C. A., and Hawes, W. W., *J. Amer. Chem. Soc.*, 1933, **55**, 2776.
[45a] Krüger, C., *J. Organometallic Chem.*, 1967, **9**, 125.
[45b] Krüger, C., Rochow, E. G., and Wannagat, U., *Chem. Ber.*, 1963, **96**, 2132, 2138.
[45c] Lappert, M. F., and Miles, S. J., unpublished results, 1977.
[45d] Lappert, M. F., and Sanger, A. R., unpublished results, 1969.
[45e] Lappert, M. F., Slade, M. J., Atwood, J. L., and Shakir, R., unpublished results, 1978.
[45f] Lidy, W., Sundermeyer, W., and Verbeek, W., *Z. Anorg. Chem.*, 1974, **406**, 228.
[46] McGee, J. M., *J. Amer. Chem. Soc.*, 1921, **43**, 586.
[46a] Martel, B., and Aly, E., *J. Organometallic Chem.*, 1971, **29**, 61.
[46b] Martel, B., and Hiriart, J. M., *Tetrahedron Letters*, 1971, 2737.
[47] Merling, G., Chrzesciuski, O., and Pfeffer, O., U.S. Patent, 1,169, 341; *Chem. Abstr.*, 1916, **10**, 952.
[48] Mootz, D., Zinnius, A., and Böttcher, B., *Angew. Chem. Internat. Edn.*, 1969, **8**, 378.
[49] Mueller, E., and Marwede, G., Ger. Patent, 2,355, 941 (1975); *Chem. Abstr.*, 1975, **83**, 133050.
[50] Murray, M., Schirawski, G., and Wannagat, U., *J.C.S. Dalton*, 1972, 911.
[51] Nast, R., and Dilly, P., *Angew. Chem. Internat. Edn.*, 1967, **6**, 357.
[52] Neilson, R. H., and Wells, R. L., *Inorg. Chem.*, 1974, **13**, 480.
[52a] Nikolaev, N. I., Geller, N. M., Dolgoplask, B. A., Zgonnik, V. N., and Kropachev, V. A., *Vysokomol. Soedin.*, 1963, **5**, 811; *Chem. Abstr.*, 1963, **59**, 10324a.
[53] Ozanne, I. R., and Marvel, C. S., *J. Amer. Chem. Soc.*, 1931, **52**, 5267.
[54] Petrov, A. A., and Kromer, V. A., *Dokl. Akad. Nauk S.S.S.R.*, 1959, **126**, 1278.
[54a] Petrov, A. A., and Kromer, V. A., *Zh. Obshch. Khim.*, 1964, **34**, 1868.
[55] Petrov. A. A., Kromer, V. A., and Stadnichak, T. V., *Zh. Obshch. Khim.*, 1960, **30**, 3890.
[56] Picon, M., *Compt. Rend.*, 1930, **175C**, 5267.
[57] Puterbaugh, W. H., and Hauser, C. R., *J. Org. Chem.*, 1959, **24**, 416.
[58] Reisodorf, D., and Normant, H., *Compt. Rend.*, 1969, **268C**, 959.
[59] Rengade, E., *Compt. Rend.*, 1905, **140C**, 1183.
[59a] Rogers, R. D., Atwood, J. L., and Grüning, R., *J. Organometallic Chem.*, 1978, **157**, 229.
[60] Rühlmann, K., and Kührt, G., *Angew. Chem. Internat. Edn.*, 1968, **7**, 809.
[61] Rupe, H., Seiberth, M., and Kussmaul, W., *Helv. Chim. Acta*, 1920, **3**, 50.
[61a] Sanderson, J. J., and Hauser, C. R., *J. Amer. Chem. Soc.*, 1949, **71**, 1595.
[61b] Sanger, A. R., *Inorg. Nucl. Chem. Letters*, 1973, **9**, 351.
[62] Sauer, J., Huisgen, R., and Hauser, A., *Chem. Ber.*, 1958, **98**, 1461.

[63] Scherer, O. J., *Angew. Chem. Internat. Edn.*, 1969, **8**, 861.
[64] Scherer, O. J., and Klusmann, P., *Angew. Chem. Internat. Edn.*, 1969, **8**, 752.
[65] Scherer, O. J., and Kuhn, N., *Angew. Chem. Internat. Edn.*, 1974, **13**, 811.
[66] Scherer, O. J., and Kuhn, N., *J. Organometallic Chem.*, 1974, **82**, C3.
[67] Scherer, O. J., and Kuhn, N., *Chem. Ber.*, 1974, **107**, 2123.
[68] Schurman, I., and Fernelius, W. C., *J. Amer. Chem. Soc.*, 1930, **52**, 2425.
[68a] Stadnichak, T. V., Kromer, V. A., and Petrov, A. A., *Zh. Obshch. Khim.*, 1964, **34**, 3284.
[69] Titherley, A. K. W., *J. Chem. Soc.*, 1897, **71**, 460.
[70] Titherley, A. K. W., *J. Chem. Soc.*, 1894, **65**, 504.
[71] Titherley, A. K. W., *J. Chem. Soc.*, 1897, **71**, 469.
[71a] Tzschach, A., and Reiss, E., *J. Organometallic Chem.*, 1967, **8**, 255.
[72] Vaughn, T. H., Vogt, R. R., and Nieuwland, J. A., *J. Amer. Chem. Soc.*, 1934, **56**, 2120.
[73] Viehe, H. G., *Angew. Chem. Internat. Edn.*, 1967, **6**, 767.
[74] Wannagat, U., *Angew. Chem.*, 1963, **75**, 173.
[74a] Wannagat, U., *Angew. Chem.*, 1964, **76**, 234.
[75] Wannagat, U., *Advan. Inorg. Chem. Radiochem.*, 1964, **6**, 225.
[75a] Wannagat, U., *Pure Appl. Chem.*, 1969, **19**, 329.
[76] Wannagat, U., Bürger, H., Geymayer, P., and Torper, G., *Monatsh.*, 1964, **95**, 39.
[77] Wannagat, U., Bürger, H., Peach, M. E., Hensen, K., and Lebert, K. H., *Z. Anorg. Chem.*, 1965, **336**, 129.
[77a] Wannagat, U., and Niederprüm, H., *Angew. Chem.*, 1959, **71**, 574.
[77b] Wannagat, U., and Niederprüm, H., *Chem. Ber.*, 1961, **94**, 1540.
[77c] Wannagat, U., and Seyffert, H., *Angew. Chem. Internat. Edn.*, 1965, **4**, 438.
[77d] Watt, G. W., and Fernelius, W. C., *J. Amer. Chem. Soc.*, 1939, **61**, 1692.
[77e] White, G. F., Morrison, A. B., and Anderson, E. G. E., *J. Amer. Chem. Soc.*, 1924, **46**, 961.
[77f] Wiberg, N., Fischer, G., and Bachhuber, H., *Angew. Chem. Internat. Edn.*, 1972, **11**, 829.
[78] Wiberg, N., and Gieren, A., *Angew. Chem.*, 1962, **74**, 942.
[78a] Wiberg, N., and Pracht, H. J., *Chem. Ber.*, 1972, **105**, 1377, 1392, 1399; *J. Organometallic Chem.*, 1972, **40**, 289.
[79] Wittig, G., and Haeusler, G., *Ann.*, 1971, **746**, 185.
[80] Wohl, A., and Lange, M., *Chem. Ber.*, 1907, **40**, 4728.
[81] Wohler, L., and Stang-Lund, F., *Z. Elektrochem.*, 1918, **24**, 261.
[82] Ziegler, K., Ger. Patent, 1935, 615,468; *Chem. Abstr.*, 1935, **29**, 6250.
[83] Ziegler, K., U.S. Patent, 2,141,058; *Chem. Abstr.*, 1939, **33**, 2538.
[84] Ziegler, K., Eberle, H., and Ohlinger, H., *Ann.*, 1933, **504**, 94.
[85] Ziegler, K., and Ohlinger, H., *Ann.*, 1932, **495**, 84.

3
Amides of Beryllium, Magnesium, and Calcium: Synthesis, Physical Properties, and Structures

A BERYLLIUM AMIDES

1. Introduction

There are no reviews dealing specifically with amides of beryllium, but the topic is discussed in more general monographs or review articles [17c, 21a, 27a, 28, 56a].

Few amides and imides of beryllium have been prepared and characterised. $[Be(NH_2)_2]_n$, a white crystalline material, is prepared from Be and NH_3 by a high pressure procedure at 130–370°C. Its properties are partially dependent on the mode of preparation [36b]. Heating to ca. 230°C *in vacuo* affords the imide $(BeNH)_n$, which decomposes to the nitride Be_3N_2 at 250°C. The latter compound may also be obtained from BeR_2 (R = Me or Et) by treatment with NH_3 and heating to 50°C. $R_2Be.NH_3$ and $(RBeNH_2)_n$ are isolable intermediates [43a]; the compounds react readily with moisture, liberating ammonia, although steric effects may stabilise the compounds in this respect [25]. Reaction of beryllium with an alkali metal in liquid ammonia yields $M[Be(NH_2)_3]$ (M = Na, K, Rb, or Cs) [9a, 54a] (Fig. 3); thermolysis affords ammonia and either $M_2[Be_2(NH)_3]$ (M = Na or K) or $M[Be(NH_2)NH]$ (M = K, Rb, or Cs).

Reactions of beryllium–nitrogen bonded compounds have not been studied in any depth. With trimethylamine-alane, $H_3Al.NMe_3$, trimethylalane, or an alkylaluminium hydride, bis(dimethylamido)beryllium acts as a Lewis base, but exchange of ligands may also occur [e.g., Eq. (1)] [50] (see Chapter 4, Table 15). Imidoberyllium compounds $[Be(N{=}CR_2)_2]_2$ and $[ClBe(N{=}CR_2)]_2$ have also been studied [56b].

$$2Be(NMe_2)_2 + H_2(Me)Al.NMe_3 \longrightarrow H(Me)AlNMe_2.2HBeNMe_2 \qquad (1)$$

A list of amides is shown in Table 1.

2. Synthesis

Except for reports of insertion reactions, to give beryllium amides [Eq. (2)] [8] or hydrazides [Eq. (3)] [30a], preparation of these compounds generally involves alkane [Eqs. (4), (5)] [7, 8, 18–22, 25, 31, 41], or salt [Eq. (6)] [13a], elimination.

$$EtBeH + 3NC_5H_5 \longrightarrow EtBeNC_5H_6.2py \qquad (2)$$

$$Me_2Be + 2PhN{:}NPh \longrightarrow Be(NPhNMePh)_2 \qquad (3)$$

$$Me_2Be + HNPr^i_2 \longrightarrow MeBeNPr^i_2 + CH_4 \qquad (4)$$

$$MeBeNMe_2 + HNMe_2 \longrightarrow Be(NMe_2)_2 + CH_4 \qquad (5)$$

$$BeCl_2 + 2NaN(SiMe_3)_2 \longrightarrow Be[N(SiMe_3)_2]_2 + 2NaCl \qquad (6)$$

3. Physical properties and structures

Co-ordination saturation for beryllium is four [28]. However, in its amides beryllium may also be two- (see Fig. 1) or three-co-ordinate if bulky groups are present. The higher co-ordination numbers are achieved by intra- [e.g., (I)] [25] or inter- [e.g., (III)] [18] molecular donor (N) – acceptor (Be) dative bonding. A monomeric beryllium amide is shown in (I) ($R = Bu^t$) [18]. However, when a less bulky group is attached to beryllium, then the dimeric structure (II) ($R = Pr$, Pr^i, or Ph) is found [24a]. $Be[N(SiMe_3)_2]_2$ is monomeric in solution [13a] and the vapour [17a, 17b]. Electron diffraction details are in Fig. 1. The NBeN skeleton is linear, with D_{2d} symmetry for the $Si_2NBeNSi_2$ framework, which may indicate $N \rightleftharpoons Be \leftrightharpoons N$ pseudo-allene bonding [17a, 17b], and extended π-bonding $\ddot{N}$–Si and $\ddot{N}$–Be. The structural assignments for (I)-(III) were made largely on the basis of the determination of molecular weights in benzene solution; nevertheless, a structure similar to (II) has been firmly established for a magnesium analogue (see Section B.3) [42a]. It is interesting that when the smaller Be atom of (I) is replaced by the larger Mg or Zn (Chapter 9), the resulting compounds are dimeric.

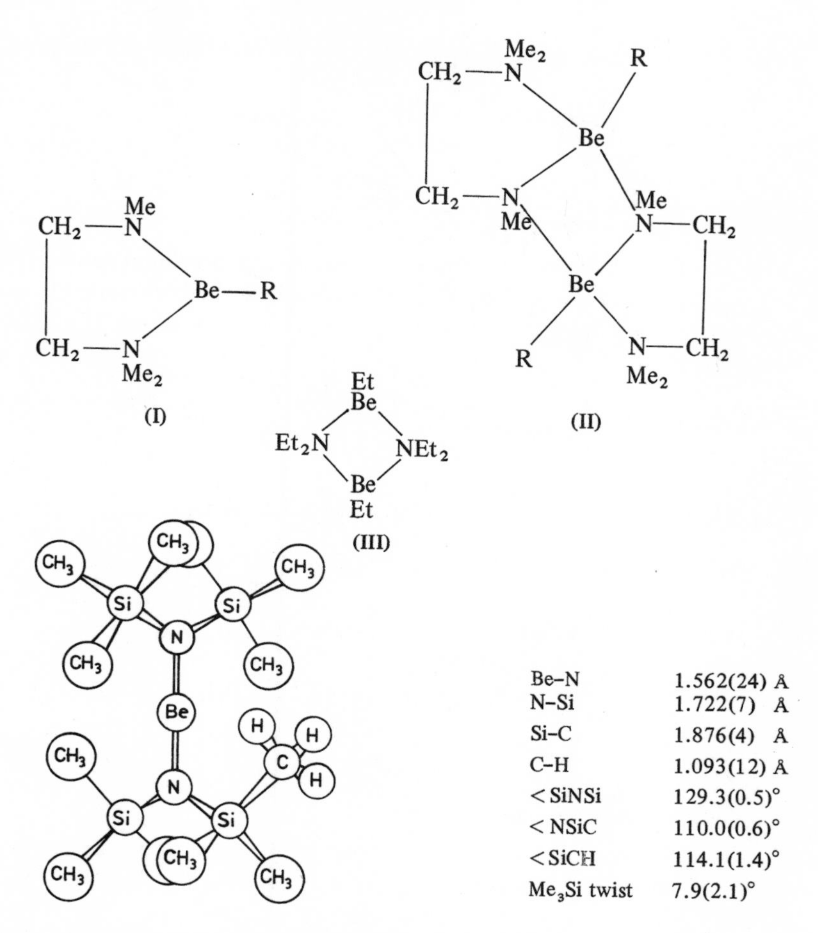

Be–N	1.562(24) Å
N–Si	1.722(7) Å
Si–C	1.876(4) Å
C–H	1.093(12) Å
< SiNSi	129.3(0.5)°
< NSiC	110.0(0.6)°
< SiCH	114.1(1.4)°
Me_3Si twist	7.9(2.1)°

Fig. 1 – Molecular structure of $Be[N(SiMe_3)_2]_2$. *[Reproduced, with permission, from Clark, A. H., and Haaland, A.,* Chem. Comm., *1969, 912;* Acta Chem. Scand., *1970,* **24**, 3024.]

The bis(dimethylamide), $Be(NMe_2)_2$, is a trimer in benzene solution [19]. It was originally proposed to be a six-membered ring compound with alternating Be and N atoms. This was ruled out by the ^1H nmr spectrum, which showed the presence of two different Me environments in the ratio 2:1 [29]. The problem was resolved by an X-ray single crystal analysis, which showed it to be an open bridged chain of D_{2d} symmetry, containing three- and four-co-ordinate Be and N atoms, as illustrated in Fig. 2 [2]. Additionally, $J(^{13}C-^{1}H)$ of 138 Hz for the terminal NMe_2 groups was regarded as evidence for Me-N π-bonding [2] following a relationship proposed earlier [26d, 44a] for a methyl group attached to three-co-ordinate nitrogen.

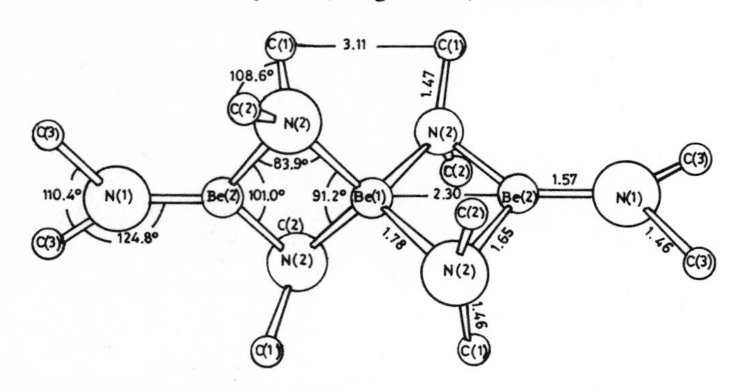

Fig. 2 – Molecular structure of $[Be(NMe_2)_2]_3$. *[Reproduced, with permission, from Atwood, J. L., and Stucky, G. D.,* J. Amer. Chem. Soc., *1969*, **91**, *4426.]* For clarity, H atoms are omitted.

The structure of $K[Be(NH_2)_3]$ has been determined by X-ray diffraction [31a]. The molecule is monomeric with approximate D_{3h} symmetry, Fig. 3.

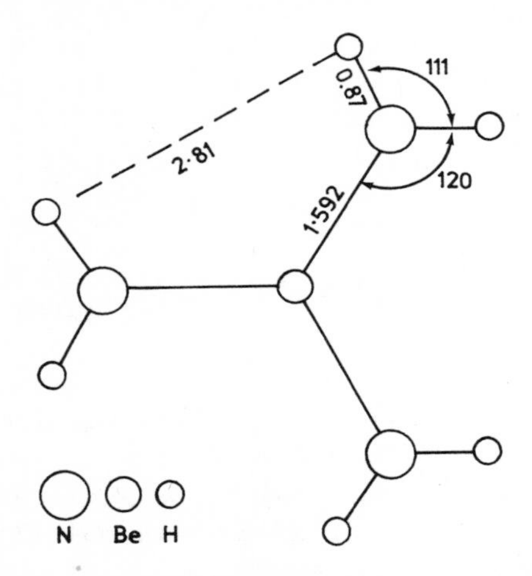

Fig. 3 – Mean dimensions in the $[Be(NH_2)_3]^-$ anion. *[Reproduced, with permission, from Guemas, L., Drew, M. G. B., and Goulter, J. E.,* J.C.S. Chem. Comm., *1972, 916.]*

The varying state of aggregation of some organoberyllium amides $(RBeNR'_2)_n$ has been examined, with the following results [18].

Trimers:	$(MeBeNMe_2)_3$	$(MeBeNEt_2)_3$	$(PhBeNMe_2)_3$
	$(EtBeNMe_2)_3$	$(Pr^iBeNMe_2)_3$	
Dimers:	$(MeBeNPr_2)_2$	$(MeBeNPh_2)_2$	$(EtBeNEt_2)_2$
	$(PhBeNPh_2)_2$	$(EtBeNPh_2)_2$	$[MeBeN(CH_2Ph)Ph]_2$

The formation of variously a monomer (I), dimers $(RBeNR'_2)_2$ or trimers $[Be(NMe_2)_2]_3$ and $(RBeNR'_2)_3$, is largely explained in terms of steric effects. These are increasingly more severe in the series trimers<dimers<monomer. The absence of higher oligomers or an infinite polymer, as in $(BeMe_2)_n$, is at first sight surprising; however, it becomes clearer from a closer examination of the structure of $[Be(NMe_2)_2]_3$, as shown in Fig. 2 [2]. The Me–Me non-bonded distance for the bridging NMe_2 groups in this compound is 3.11 Å, which is extremely short and is near the limit for non-bonding interaction. It is only achieved by a considerable distortion of these groups away from the centre of the molecule; this would not be possible if it were an infinite polymer. $(MeBeNMe_2)_n$ is a glass at room temperature and therefore may be polymeric, despite being a trimer in solution [9]. The trimers have been assumed to have $(BeN)_3$ six-membered rings [18]; however, by analogy with $[Be(NMe_2)_2]_3$, it is possible that they are open-chain compounds, similar to the structure shown in Fig. 2. There is ^1H nmr spectroscopic evidence in favour of the six-membered ring for $(MeBeNMe_2)_3$, which is assumed to undergo rapid intermolecular interconversion [9].

As to the bonding in these compounds, for four-co-ordinate BeN pairs, there is no particular problem. For three-co-ordinate pairs, there is the possibility of $(p\text{-}p)\pi$-bonding ($N\rightleftharpoons Be$). Fig. 2 shows that the C_2N angle at the terminal N's of $[Be(NMe_2)_2]_3$ is considerably smaller than the 120° expected for sp^2 N; however, the $BeNC_2$ arrangement is essentially co-planar. The small C_2N angle probably minimises non-bonding interactions.

B MAGNESIUM

1. Introduction

The composition of Grignard reagents, and the mechanism of their reactions including those with amines, has been reviewed [1]. A summary of the reactions of certain magnesium–nitrogen compounds up to 1962 is available [51]; this relates to 'magnesylamines' or amidomagnesium halides $XMgNR_2$ or $(XMg)_2NR$.

Magnesium amide $[Mg(NH_2)_2]_n$ (Fig. 5) and imide $(MgNH)_n$ have been obtained from Mg and NH_3, in procedures similar to those for the Be analogues [36c]. Double amides of magnesium with the alkali metals, $M_2[Mg(NH_2)_4]$ (M = Na, K, Rb, or Cs) were obtained from Mg and liquid ammonia solutions of M. Like the simple amide, they decompose thermally *in vacuo* to corresponding imides $M_2[Mg(NH)_2]$ and NH_3, and yield Mg_3N_2 and M on further heating.

2. Synthesis

Since 1903 [44] a primary or secondary amine has been used to displace an alkane from (i) a dialkylmagnesium, to yield the bis(amido)magnesium [e.g., Eq. (7)] [24] or the alkyl(amido)magnesium [e.g., Eq. (8)]; or (ii) from a

Grignard reagent [e.g., Eq. (9)] [42]. An iodide or bromide is more reactive than a chloride for formation of the halogeno(amido)magnesium [51]. Another example involves MeMgX or $MgEt_2$ with $HN(SiMe_3)_2$ [57a].

$$Et_2Mg + 2HNMe_2 \longrightarrow Mg(NMe_2)_2 + 2C_2H_6 \quad (7)$$

$$Et_2Mg + HNPh_2 \longrightarrow EtMgNPh_2 + C_2H_6 \quad (8)$$

$$EtMgBr + HNEt_2 \longrightarrow BrMgNEt_2 + C_2H_6 \quad (9)$$

Reaction of $MgBu^s_2$ and $HNPr^i_2$ affords $Bu^sMgNPr^i_2$ which reacts with H_2 at 25°C and 3000 p.s.i., or $LiAlH_4$, to form $HMgNPr^i_2$ [3a]. $EtMgNPr^i_2$ and H_2 at 110°C gave $MgH_2 + C_2H_6 + HNPr^i_2$. $HMgNBu_2$ is obtained either from $EtMgNBu_2$ and $LiAlH_4$ or $BrMgNBu_2$ and KH in T.H.F. $HMgNPh_2$ is prepared by hydrogenation of Bu^sMgNPh_2 or, preferably, from $BrMgNPh_2$ and KH in T.H.F. $HMgNEt_2$ or $HMgN(Me)CH_2CH_2NMe_2$ is formed by hydrogenation of Bu^sMgNEt_2 or $EtMgN(Me)CH_2CH_2NMe_2$. A series of T.H.F.-soluble compounds $HMgNR_2$ has been obtained from active magnesium hydride with either $Mg(NR_2)_2$ in T.H.F. or HNR_2, and also from $LiNR_2$ and ClMgH [1a, 3a]; the bulkier amides tend to be dimeric (ebullioscopy, T.H.F.) at high dilution.

Insertion reactions have been employed to prepare more complex Mg-N compounds, as in Eq. (10)] [5]. Others are obtained by insertion of an imine or carbodi-imide into a Mg-C bond, e.g., Eqs. (11), (12) [14, 15, 24].

$$RCONH_2 + 2R'MgX \longrightarrow RR'C(OMgX)NHMgX + R'H \quad (10)$$

$$Et_2Mg + PhCH{:}NPh \longrightarrow EtMgNPhCHEtPh \quad (11)$$

$$MeMgI + PhN{:}C{:}NPh \longrightarrow IMgNPhCMe{:}NPh \quad (12)$$

Bis(anilido)magnesium is obtained in high yield by the reaction of aniline with hot magnesium [Eq. (13)] [57]. Magnesium, Al, and H_2NBu^t in T.H.F. under H_2 pressure affords $[\{(T.H.F.)Mg\}(HAlNBu^t)_3]$ [26a].

$$Mg + 2H_2NPh \longrightarrow Mg(NHPh)_2 + H_2 \quad (13)$$

The reaction of 2-(dimethylamino)chloroethane with magnesium surprisingly gives chloro(dimethylamido)magnesium [Eq. (14)] [32]; the reaction probably proceeds via an intermediate (IV) rather than a Grignard reagent [32].

$$Me_2NCH_2CH_2Cl + Mg \longrightarrow ClMgNMe_2 + C_2H_4 \quad (14)$$

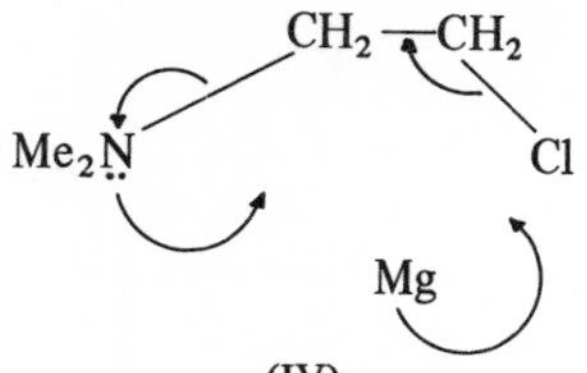

$$Me_2\ddot{N}-CH_2-CH_2-Cl \quad Mg$$

(IV)

The reactions of a 'magnesylamine' with (i) a halide to yield an *N*- or *C*-substituted product [Eqs. (15), (16)], (ii) a nitrile to yield an amidine [Eq. (17)],* (iii) a compound having a carbonyl group to yield either an insertion adduct [Eq. (18)], or the oxygen-abstracted derivative [Eq. (19)], or (iv) a carboxylic acid or related reagent, have been discussed in the review [51] mentioned in Section B. 1. The use of a 'magnesylamine' in condensation reactions for the synthesis of heterocycles has been discussed [51], as has the role of magnesium amides as polymerisation catalysts [28] (Chapter 17).

$$XMgNR_2 + R'X \longrightarrow R_2NR' + MgX_2 \tag{15}$$

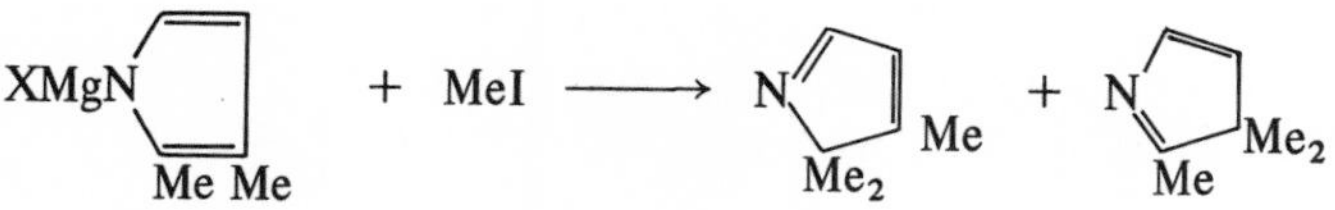

$$XMgN(C_4H_2Me_2) + MeI \longrightarrow N(C_4HMe_2)Me + N(C_4HMe_2)Me \tag{16}$$

(Structures: XMgN-pyrrolyl ring bearing Me Me; products: ring with Me, Me_2; ring with Me_2, Me)

$$RCN + XMgNR'R'' \longrightarrow XMgN{:}CRNR'R'' \tag{17}$$

$$BrMgNPhCO_2Et \xrightarrow[2)\ H_2O]{1)\ PhCHO} PhCH(OH)NPhCO_2Et \tag{18}$$

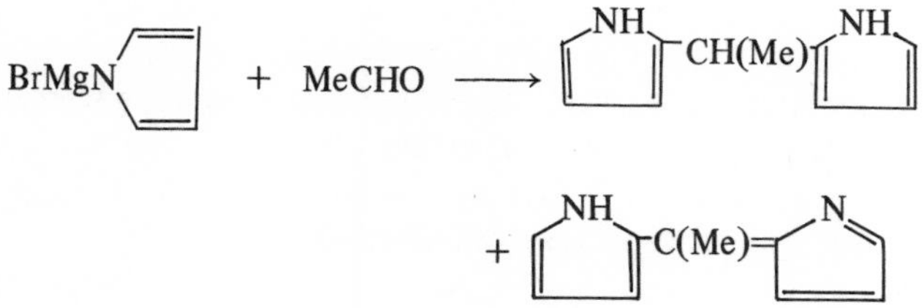

$$BrMgN(C_4H_4) + MeCHO \longrightarrow (C_4H_3NH)CH(Me)(C_4H_3NH) + (C_4H_3NH)C(Me){=}(C_4H_3N) \tag{19}$$

An attempt to prepare a model haemoglobin precursor prompted the formation of an indolyliron(III) chloride [Eq. (20)] [46]. A list of amides is in Table 2.

*Or a guanidine [11].

$$2BrMgN(C_8H_6) + FeCl_3 \longrightarrow ClFe(NC_8H_6)_2 + 2MgBrCl \quad (20)$$

3 Physical properties and structures

Studies of the solubilities and molecular weights of magnesium amides have shown that they are polymeric, unless bulky ligands are bonded to magnesium or nitrogen [24]. Complexation, for example with tetrahydrofuran (T.H.F.), may reduce the degree of polymerisation. For example, the dimeric $[(EtMgNR_2)_2(T.H.F.)]_2$ and monomeric $EtMgNPh_2(T.H.F.)_2$ have been characterised [24]. The dimeric $[Mg(NPr^i_2)Pr^i]_2$ was believed to have the structure (V), rather than those in which bridging was not by NPr^i_2 but by either one or both Pr^i groups [24]. The compound $Mg[N(SiMe_3)_2]_2$ forms adducts $Mg[N(SiMe_3)_2]_2.MgX_2.nOEt_2$ ($n = 2$ for X = Cl or Br, or = 4 for X = I), $Mg[N(SiMe_3)_2]_2.OEt_2$, and $Mg[N(SiMe_3)_2]_2.3$dioxane [57a], as well as $Mg[N(SiMe_3)_2]_2.Pr^iC(:O)OR$ (R = Me or Bu^t) [39a], which loses $HN(SiMe_3)_2$ with concomitant formation of (VI). The preference for amide, rather than alkyl,

(V)

(VI)

group bridging is certainly shown from the X-ray single crystal examination of $[MeMgNMe(CH_2)_2NMe_2]_2$, as illustrated in Fig. 4 [42a]. The exceptionally small $(Me_2)NMgNMe$ angle of 83.7° (average) shows that the local symmetry of each magnesium atom is only very approximately tetrahedral.

The hydridomagnesium amides $HMgNR_2$ are highly associated in tetrahydrofuran, suggestive of $\bar{N}R_2$ and $\bar{H}$ bridging between successive Mg atoms [3a].

The structure of $Mg(NH_2)_2$ has been solved by single crystal X-ray crystallography, Fig. 5 [36a], as has that of $[\{(T.H.F.)Mg\}(HAlNBu^t)_3]$ [26a]; the latter is related to that of a calcium analogue, Fig. 6 (see Section C) [26c], and has a magnesium atom taking the place of one Al atom of a cubic $(AlN)_4$ skeleton as found in $(HAlNBu^t)_4$ (see Chapter 4, Section D.4). The M–N ionic character increases progressively in the series Al<Mg<Ca; the mean Mg–N distance is 2.090(4) Å.

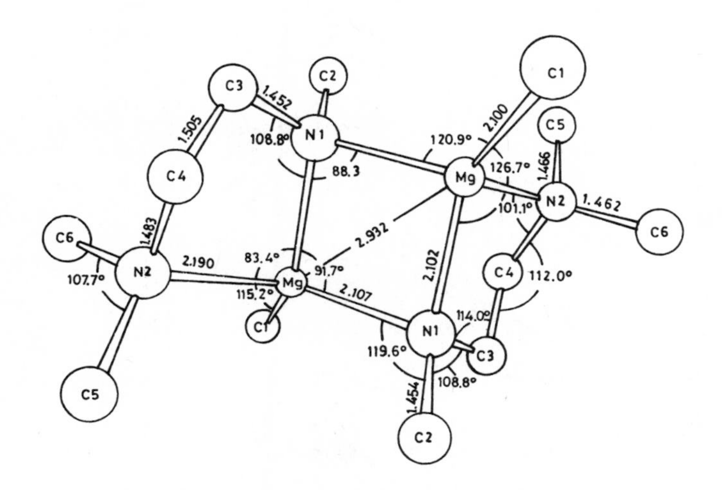

Fig. 4 – Molecular structure of $[MeMgNMe(CH_2)_2NMe_2]_2$. *[Reproduced, with permission, from Magnuson, V. R., and Stucky, G. D.,* Inorg. Chem., *1969,* **8**, *1427.]* For clarity, H atoms are omitted.

C CALCIUM

Amides of calcium have been only sparsely investigated (see Table 3).

The preparation of calcium amides has been reported in a comparative study of the reactions of triphenylmethylcalcium chloride with alcohols, thiols, amines, phosphines, arsines, or selenols [43]; the products of the reactions with amines [Eq. (21)] are polymeric.

$$Ph_3CCaCl + HNEt_2 \longrightarrow Ph_3CH + \frac{1}{n}[CaCl(NEt_2)]_n \qquad (21)$$

Reaction of either metallic Ca, Al, and H_2NBu^t in T.H.F. under hydrogen pressure or $Ca[AlH_4]_2$.T.H.F., $H_3Al.NMe_3$, and H_2NBu^t yields $[\{(T.H.F.)_3Ca\}(HAlNBu^t)_3]$ [26a]. The single crystal X-ray analysis of this compound, like a Mg analogue (Section B.3), has been solved [26c] and reveals the distorted cubane structure, Fig. 6, related to that of $(HAlNBu^t)_4$ (see Chapter 4, Section D.4). The mean Ca–N bond length, 2.490(2) Å corresponds fairly closely to the sum of the covalent radii for the two atoms.

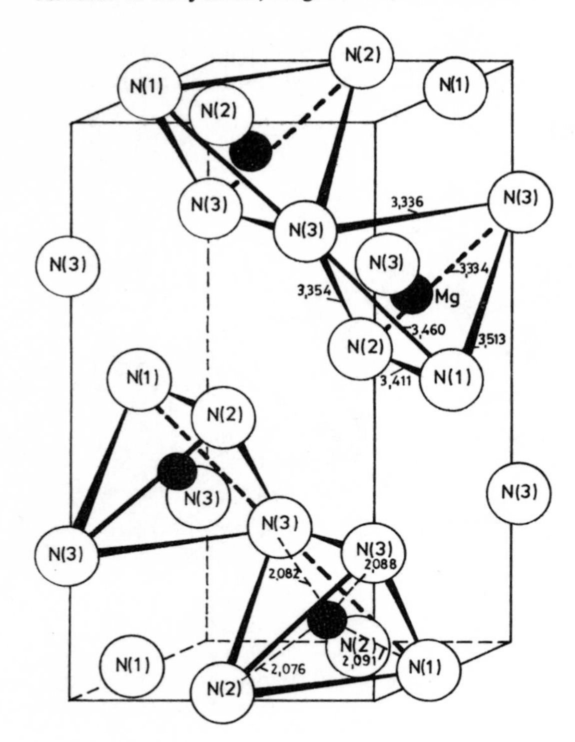

Mg–Mg	4 Mg:	3.496	3.513	3.513	3.607	(2)
Mg–N	1 N(1):	2,091	(4)			
	1 N(2):	2.076	(3)			
	2 N(3):	2.082	2.088	(5)		
N(1)–N	2·2 N(2):	3.411	3.947	(4)		
	2·4 N(3):	3.460	3.513	3.753	3.813	(6)
N(2)–N	2·2 N(1):	3.411	3.947	(4)		
	2·4 N(3):	3.334	3.354	3.882	3.901	(5)
N(3)–N	4 N(1):	3.460	3.513	3.753	3.813	(6)
	4 N(2):	3.334	3.354	3.882	3.901	(5)
	4 N(3):	3.336	3.336	3.815	4.190	(7)
Mg–H	2 H(1):	2.41	2.62	(6)		
	2 H(2):	2.42	2.44	(6)		
	2 H(3):	2.42	2.56	(6)		
	2 H(4):	2.54	2.63	(6)		
N(1)–H	2 H(1):	0.93	(6)			
N(2)–H	2 H(2):	0.87	(6)			
N(3)–H	1 H(3):	0.89	(6)			
	1 H(4):	0.94	(6)			
H(1)–H(1)	N(1):	1.52	(9)			
H(2)–H(2)	N(2):	1.57	(9)			
H(3)–H(4)	N(3):	1.44	(9)			

Fig. 5 – The unit cell, and some bond lengths of $Mg(NH_2)_2$. *[Reproduced, with permission, from Jacobs, H.,* Z. Anorg. Chem., *1971,* **382**, *97].*

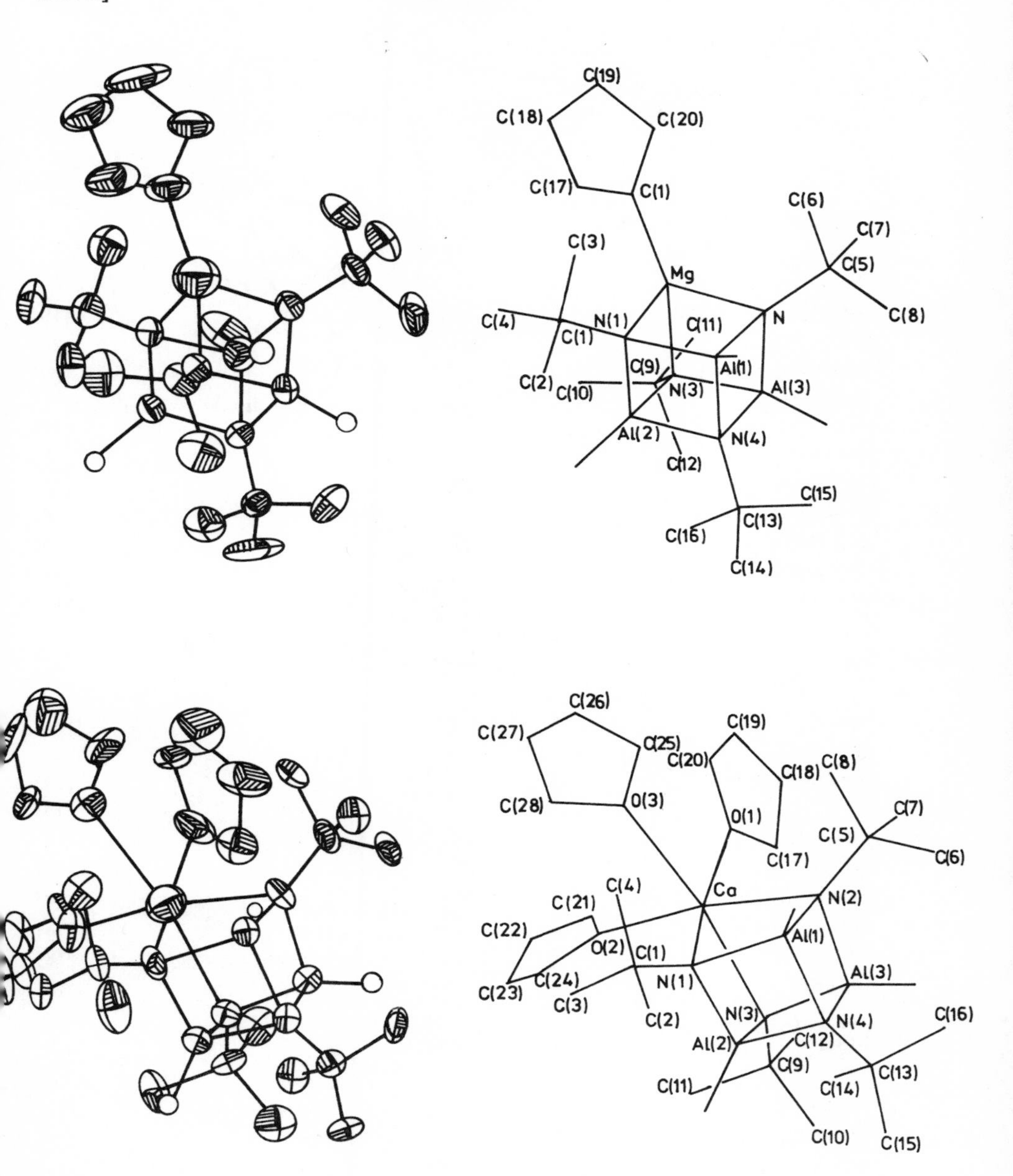

Fig. 6 – Perspective view of the molecular structures of [{(T.H.F.)Mg}(HAlNBu^t)$_3$] and [{(T.H.F.)$_3$Ca}(HAlNBu^t)$_3$], T.H.F. *[Reproduced, with permission, from Del Piero, G., Cesari, M., and Mazzei, A.,* J. Organometallic Chem., *1977,* **137,** *265.]* [Mg-N (av), 2.090(4) Å, Al-N (av), 1.913(4) Å; and Ca-N (av), 2.490(2) Å; Al-N (av), 1.908(11).]

D TABLES OF COMPOUNDS

Table 1† Beryllium Amides

Compound	Method	B.p. (°C/mm.) (m.p., °C)	Comments	References
$HBeNMe_2$	E			[19]
$HBeNMe_2 \cdot H_2AlNMe_2$	A		nmr	[50]
$HBeNMe_2 \cdot HAlMeNMe_2$	A		nmr	[50]
$HBeNMe_2 \cdot Me_2AlNMe_2$	A		nmr	[50]
MeBeNHMe	E	(110)d		[20]
$MeBe_2(NMe)NHMe$	E			[20]
$MeBeNMe_2$	E	(55)	nmr, trimer	[9, 20, 21]
			ir, nmr	[18]
$MeBeNMe_2 \cdot py$	E	(225)d	ir	[18]
$MeBeNMe_2 \cdot 2py$	E	(175)d	ir	[18]
$MeBeNMe_2 \cdot bipy$	E			[18]
$MeBeNMe(CH_2)_2NMe_2$	E	(116–8)	dimer	[22]
$MeBeNEt_2$	E		trimer	[18]
$MeBeNPr_2$	E		dimer, ir, nmr	[18]
MeBeN(piperidino ring)	E	d(40)²		[31]
MeBeN(morpholino ring, O)	E	d(40)		[31]
$MeBeNPhCH_2Ph$	E		dimer	[18]
	D		nmr	[8]
$MeBeNPh_2$	E	(141)d	trimer, ir, nmr	[18]
$MeBeNPh_2 \cdot py$	E	(softens 158)	ir	[18]
$MeBeNPh_2 \cdot 2py$	E	(83–5)d	ir	[18]
$MeBeNPh_2 \cdot bipy$	E	(110)d	uv	[18]
$EtBeNMe_2$	E		dimer, ir, nmr	[18]
$EtBeNEt_2$	E		dimer, ir, nmr	[18]
EtBeN(dihydropyridyl ring)·2py	D	(109–11)		[8]
$EtBeNPh_2$	E	(154)d	dimer, ir, nmr	[18]
$EtBeNPh_2 \cdot 2py$	E	(85)d	ir	[18]
$EtBeNPh_2 \cdot bipy$	E	(135)d	uv	[18]
Pr^iBeNMe_2	E	liquid	trimer, ir, nmr	[19]
$Bu^tBeNHMe.Bu^tBeNMe_2$	E			[18]
$PrBeNMe(CH_2)_2NMe_2$	E	d(60)	dimer	[24a]
$Pr^iBeNMe(CH_2)_2NMe_2$	E	(165–7)	dimer	[24a]

Table 1 *(C'td.)*

Compound	Method	B.p. (°C/mm.) (m.p., °C)	Comments	References
$Bu^tBeNMe(CH_2)_2NMe_2$	E	45/1	ir, monomer	[18]
$PhBeNMe_2$	E	(154)	trimer	[18, 25]
$PhBeNMe(CH_2)_2NMe_2$	E	(255)	dimer	[25]
$PhBeNPh_2$	E	(260–2)	dimer	[18, 25]
$[Be(NHCH_2CH_2NH)]_n$	E		polymer	[22]
$Be(NMe_2)_2$	E	175 (88–90)	trimer, ir, nmr	[19, 29]
			X-ray	[2]
$Be(NMe_2)_2 \cdot bipy$	E			[18]
$Be(NMe_2)_2 \cdot AlMe_3$	E		nmr	[50]
$Be(NMe_2)_2 \cdot 2AlMe_3$	E		nmr	[50]
$Be[NMe(CH_2)_2NMe_2]_2$	E	(86)	ir	[7]
$[BeNMe(CH_2)_2NMe]_n$	E		polymer	[22]
Be(N⟨piperidine ring⟩)$_2$				[31]
Be(N⟨morpholine ring⟩O)$_2$				[31]
$Be(NPh_2)_2$				[41]
$[BeN(CH_2)_2NMe_2]_n$	E		polymer	[22]
$Be[N(SiMe_3)_2]_2$	A	110/3	n_D^{20} 1.4369, d_4^{20} 0.8249	[13a]
		(−3 to −1)	ir, nmr, Raman, ed	[17a]
			X-ray	[17b]
$Be(NH_2)_2$				[36b]
$KBe(NH_2)_3$			X-ray	[31a]
$MBe(NH_2)_3$ (M = Na, K, Rb, Cs)				[54a]

FOOTNOTES TO TABLE 1: *(these are rather extensive because they are also used in conjunction with other tables of compounds, Chapters 4–9).*

1. M.p.s with decomposition denoted '(110)d' means compound melts at 110°C with decomposition; 'd(40)' means compound decomposes at 40°C.

The *Empirical Formulae* are tabulated in the column headed 'Compound', and the state of molecular aggregation is indicated under 'Comments'.

The letter in the column headed 'Method' refers to the method of preparation as described in footnote 2. The following abbreviations are used to indicate that a physical property is recorded in the reference cited:

Spectra: ir (infra-red); uv (ultra-violet); nmr (nuclear magnetic resonance); esr (electron spin resonance); MCD (magnetic circular dichroism); Möss-bauer, pes (photoelectron), Raman, or mass spectra are so denoted.

mag.:	magnetic properties	$[M]_D^0$:	rotatory dispersion
μ :	dipole moment	subl. :	sublimation temperature
μ_{eff} :	magnetic moment	X-ray :	X-ray diffraction analysis (single crystal)
$[\alpha]_D^0$:	specific rotation		

ed :	electron diffraction	Me :	methyl
ΔH_f^0 :	standard heat of formation	Et :	ethyl
en :	ethylenediamine	Pr :	propyl
pn :	propylenediamine	Bu :	butyl
pic :	picoline	Pent :	pentyl
lut :	lutidine	Hex :	hexyl
quin :	quinoline	Oct :	octyl
isoquin :	isoquinoline	Ph :	phenyl
py :	pyridine	Np :	naphthyl
bipy :	2,2′-bipyridyl	Ac :	acetyl
THF :	tetrahydrofuran	C_3H_5 :	allyl
DMSO :	dimethyl sulphoxide	C_2H_3 :	vinyl
		C_5H_5 :	cyclopentadienyl (or Cp)

2. The methods of preparation of amido-derivatives of metals or metalloids are shown in A–K.

A Transmetallation, which may be salt elimination (Eq. 1) or metathetical exchange (Eq. 2).

$$n\text{LMX} + \text{M}'(\text{NRR}')_n \longrightarrow n\text{LMNRR}' + \text{M}'\text{X}_n \quad (1)$$

$$\text{LMX} + \text{L}'\text{M}'\text{NRR}' \longrightarrow \text{LMNRR}' + \text{L}'\text{M}'\text{X} \quad (2)$$

(X = halogen, oxygen, alkoxide, or alkyl).

B Transamination (Eq. 3).

$$\text{LMNRR}' + \text{HNR}''\text{R}''' \longrightarrow \text{LMNR}''\text{R}''' + \text{HNRR}' \quad (3)$$

C Reactions of amines with metal-halides, -oxides, -sulphides, or -oxyanions (e.g., nitrate, sulphate) (Eq. 4).

$$\text{LMX} + 2\text{HNRR}' \longrightarrow \text{LMNRR}' + \text{HNRR}'.\text{HX} \quad (4)$$

(X = halogen, oxygen, sulphur, or oxyanion).

D Insertion reactions (Eq. 5).

$$\text{LMX} + \text{Y=Z} \longrightarrow \text{LMY–ZX} \quad (5)$$

E Alkane elimination (Eq. 6).

$$\text{LMR}'' + \text{HNRR}' \longrightarrow \text{LMNRR}' + \text{R}''\text{H} \quad (6)$$

F Hydrogen elimination (Eq. 7).

$$\text{LMH} + \text{HNRR}' \longrightarrow \text{LMNRR}' + \text{H}_2 \quad (7)$$

G Elimination of alcohol or thiol (Eq. 8).

$$\text{LMXR}'' + \text{HNRR}' \longrightarrow \text{LMNRR}' + \text{R}''\text{XH} \quad (8)$$

(X = oxygen or sulphur).

H Oxidative addition (e.g. Eq. 9).

$$(\text{Ph}_3\text{P})_4\text{Pt} + \text{HNR}_2 \longrightarrow trans\text{-}(\text{Ph}_3\text{P})_2\text{Pt(H)NR}_2 + 2\text{Ph}_3\text{P} \quad (9)$$

I Disproportionation, redistribution, or scrambling reactions.

J Alkyl halide elimination (Eq. 10).

$$\text{LMR}'' + \text{XNRR}' \longrightarrow \text{LMNRR}' + \text{R}''\text{X} \quad (10)$$

K Cleavage of metal-nitrogen bonds (e.g., Eq. 11).

$$\text{Zr(NMe}_2)_4 + 2\text{C}_5\text{H}_6 \longrightarrow (\eta\text{-C}_5\text{H}_5)_2\text{Zr(NMe}_2)_2 + 2\text{HNMe}_2 \quad (11)$$

Table 2† Magnesium Amides

Compound	Method	M.p.(°C)	Comments	References
$Mg(NH_2)_2$			X-ray	[36a, 36c]
$HMgNR_2$	A, E, I			
[R = Et, Pr, Pr^i, Bu, Bu^s, Ph; NR_2 = $NMePr^i$, $\overline{N(CH_2)_5}$,			ir, nmr, X-ray	
$\overline{NCMe_2(CH_2)_3CMe_2}$, $N(Bu^t)SiMe_3$]			mol. wt.	[1a, 3a]
$HMgN(Me)(CH_2)_2NMe_2$	E		ir	[3a]
$MeMgNMe_2$	E			[24]
$MeMgNMe(CH_2)_2NMe_2$	E		dimer	[23]
			X-ray	[42a]
$EtMgNHBu$	E			[34]
$EtMgNHBu^i$	E			[34]
$EtMgNHCH_2Ph$	E			[34]
$EtMgNHPh$	E			[34]
$EtMgNEt_2$				[38]
.THF	E		nmr	[24]
$EtMgNPr^i_2$.THF	E		nmr	[24]
EtMgNPhCHEtPh	D	115–25 d	nmr, dimer	[24]
$EtMgNPh_2$	E	d 210	nmr	[24, 37, 38]
.2THF	E	110–4 d	nmr	[24]
EtMgN  (ring)	E			[38]
EtMgN  (ring)	E			[37, 38]
EtMgN 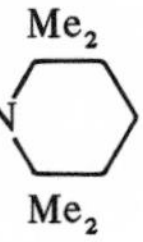 (ring, Me_2, Me_2)	E	151–3 d	nmr	[24]
EtMgN 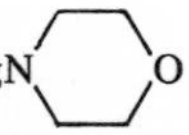 (ring) O	E			[37, 38]
$Pr^iMgNPr^i_2$	E	120 d	nmr	[24]
Pr^iMgNPh_2	E	170–5 d	nmr, polymer	[24]
$.2OEt_2$	E	62–110	nmr	[24]
$Mg(NHPh)_2$	E			[56]
	F			[57]
$Mg(NMe_2)_2$	E	> 360	nmr	[24]
$Mg(NBu_2)_2$	A			[26b]
$Mg(NPh_2)_2$	E			[36]

Table 2 *(C'td.)*

Compound	Method	References
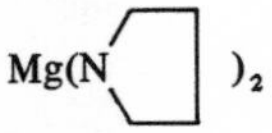 $Mg(N\ \ )_2$		[37]
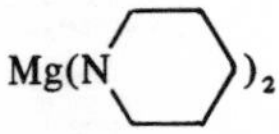 $Mg(N\ \ )_2$		[37, 38]
$BrMgNHPr^i$		[39]
BrMgNHBu	E	[33]
BrMgNHCMeEtOMgBr	B	[5]
BrMgNHCMePhOMgBr	B	[5]
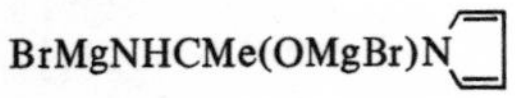 BrMgNHCMe(OMgBr)N	B	[55]
$BrMgNHCEt_2OMgBr$	B	[5]
BrMgNHCEtPrOMgBr	B	[5]
$BrMgNHCEtBu^iOMgBr$	B	[5]
BrMgNHCEtPhOMgBr	B	[5]
$BrMgNHCH_2Ph$	E	[36]
BrMgNHPh	E	[17, 26, 27, 33, 53]
o-$BrMgNHC_6H_4Me$	E	[53]
m-$BrMgNHC_6H_4Me$	E	[53]
p-$BrMgNHC_6H_4Me$	E	[53]
$BrMgNHC_6H_3Me_2$-2,4	E	[53]
p-$BrMgNHC_6H_4Cl$	E	[53]
p-$BrMgNHC_6H_4Br$	E	[53]
p-$BrMgNHC_6H_4OMe$	E	[53]
p-$BrMgNHC_6H_4OEt$	E	[53]
1-BrMgNHNp	E	[53]
2-BrMgNHNp	E	[53]
$BrMgNMe_2$	D	[32]
BrMgNMePh	E	[17, 26, 33]
$BrMgNEt_2$	E	[16, 42, 58]
BrMgNEtPh	E	[17]
$BrMgNPr_2$	E	[42]
$BrMgNPr^i_2$	E	[30]
$BrMgNBu_2$	E	[30, 33, 34, 42]
BrMgNBuPh		[59]
$BrMgNPent_2$	E	[42]
$BrMgNHex_2$	E	[42]
$BrMgN(C_7H_{15})_2$	E	[42]

Table 2 *(C'td.)*

Compound	Method	References
$BrMgN(C_8H_{17})_2$	E	[42]
$BrMgN(CH_2Ph)_2$	E	[42]
$BrMgNPhCH_2Ph$	E	[60]
$BrMgNPh(CH_2)_3NEt_2$	E	[33]
$BrMgN(C_6H_{11})_2$		[30]
$BrMgNPh_2$		[42]
BrMgN		[59]
BrMgN	E	[4, 31]
Me BrMgN Me	E	[48]
Me BrMgN Me	E	[35]
BrMgN		[59]
CO BrMgN CEt BrMgO	E	[6]
IMgNHEt		[58]
$IMgNHCH_2Ph$	E	[12]
IMgNHPh	E	[42, 44]
$IMgNHC_6H_4Me$-*o*	E	[13]
$(IMgNH)_2C_6H_4$-*o*	E	[3]
$IMgNMe_2$		[58]
IMgNMePh	E	[44]
$IMgNEt_2$	E	[40, 58]

Table 2 *(C'td.)*

Compound	Method	References
IMgNPhCHMePh	D	[14]
$IMgNPhCHPh_2$	D	[14]
IMgN 	E	[45]
IMgN  .2py		[44]
IMgN 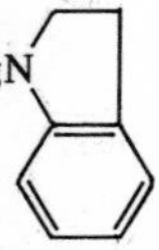	E	[45, 46, 47]
Me IMgN 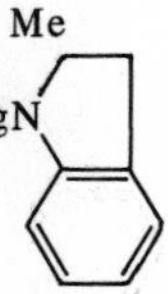	E	[47]
Me IMgN 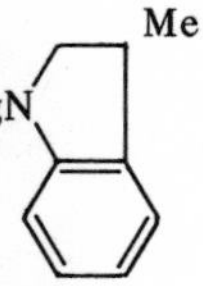	E	[46, 47]
Et IMgN 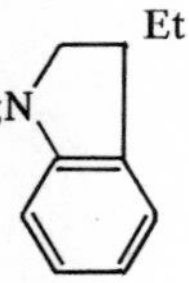	E	[47]
IMgN 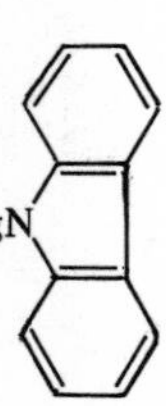	E	[45, 46]

Table 2 *(C'td.)*

Compound	Method	M.p. (°C)	Comments	References
$(BrMg)_2NPr$	E			[49]
$(BrMg)_2NPh$	E			[52, 54]
$(BrMg)_2NC_6H_4Me$-*o*	E			[52]
$(BrMg)_2NNp$-2	E			[52]
BrMgNHCOMe	B			[55]
$BrMgNHCO_2Et$	E			[10]
$BrMgNPhCO_2Et$	E			[10]
BrMgNPhCPh:NH	D			[15]
BrMgNPhC(1-Np):NH	D			[15]
BrMgNPhC(1-Np):NPh	D			[15]
IMgNPhCMe:NPh	D			[15]
$IMgNPhC(CH_2Ph)$:NPh	D			[15]
IMgNPhCPh:NPh	D			[15]
$Mg[N(SiMe_3)_2]_2.MgX_2.nEt_2O$				
(X = Cl, *n* = 2		(106)	ir, nmr	[57a]
X = Br, *n* = 2		(98)	ir, nmr	[57a]
X = I, *n* = 4)		(60)	ir, nmr	[57a]
$Mg[N(SiMe_3)_2].Et_2O$	E	(39)	ir, nmr, mass	[57a]
	E	(116)	ir, nmr, mass	[57a]
$R''MgN(SiMe_3)_2.Pr^iC(O)OR'$				
R' R''				
Me Br			ir, nmr	[39a]
Me $N(SiMe_3)_2$			ir, nmr	[39a]
Bu^t Br			ir, nmr	[39a]
Bu^t $N(SiMe_3)_2$			ir, nmr	[39a]
Pr^i O=C $[(Me_3Si)_2N]_2Mg$ CMe_2 O=C Pr^i			ir, nmr	[39a]
CMe_2 // O–C $(Me_3Si)_2NMg$ CMe_2 O=C OMe			ir, nmr	[39a]
$[\{(THF)Mg\}(HAlNBu^t)_3]$	F		ir, nmr, X-ray	[26a, 26c]

† For explanation of abbreviations, see footnotes to Table 1, p. 57.

Table 3† Calcium Amides

Compound	Method	Comments	References
$Ca(NBu_2)_2$	A		[26a]
$ClCaNEt_2$	E	polymer	[43]
$ClCaNPh_2$.THF	E	polymer	[43]
ClCaN(piperidine ring)	E	polymer	[43]
ClCaN(morpholine ring)O.THF	E		[43]
$[\{(THF)_3Ca\}(HAlNBu^t)_3]$	E	ir, nmr, X-ray	[26a, 26c]

† For explanation of abbreviations, see footnotes to Table 1, p. 57.

E REFERENCES

[1] Ashby, E. C., *Quart. Rev.*, 1967, **21**, 259.
[1a] Ashby, E. C., and Goel, A. B., *Inorg. Chem.*, 1978, **17**, 1862.
[2] Atwood, J. L., and Stucky, G. D., *Chem. Comm.*, 1967, 1169; *J. Amer. Chem. Soc.*, 1969, **91**, 4426; 1971, **92**, 1107.
[3] Bassett, H. L., and Thomas, C. R., *J. Chem. Soc.*, 1954, 1188.
[3a] Beach, R. G., and Ashby, E. C., *Inorg. Chem.*, 1971, **10**, 906.
[4] Beau, G. P., *J. Org. Chem.*, 1967, **32**, 228.
[5] Béis, C., *Compt. Rend.*, 1903, **137C**, 575.
[6] Béis, C., *Compt. Rend.*, 1904, **138C**, 987.
[7] Bell, N. A., *J. Chem. Soc. (A)*, 1966, 542.
[8] Bell, N. A., and Coates, G. E., *J. Chem. Soc. (A)*, 1966, 1069.
[9] Bell, N. A., Coates, G. E., and Emsley, J. W., *J. Chem. Soc. (A)*, 1966, 49.
[9a] Bergstrom, F. W., *J. Amer. Chem. Soc.*, 1928, **50**, 652.
[10] Binaghi, R., *Gazz. Chim. Ital.*, 1927, **57**, 676.
[11] Birtwell, S., Curd, F. H. S., and Rose, F. L., *J. Chem. Soc.*, 1949, 2556.
[12] Blazević, K., Houghton, R. P., and Williams, C. S., *J. Chem. Soc. (C)*, 1968, 1704.
[13] Bodroux, F., and Taboury, F., *Compt. Rend.*, 1907, **144C**, 1437; *Bull. Soc. Chim. Fr.*, 1907, **1**, 911.
[13a] Bürger, H., Forker, C., and Goubeau, J., *Monatsh.*, 1965, **96**, 597.
[14] Busch, M., *Ber.*, 1904, **37**, 2691.

[15] Busch, M., and Hobein, R., *Ber.*, 1908, **40**, 4296.
[16] Buyalla, B., *Rev. real, acad. cien., Madrid,* 1911, **9**, 635; *Chem. Abstr.,* 1911, **5**, 3802.
[17] Chelinsiev, V. V., and Pataraya, A. V., *J. Gen. Chem. U.S.S.R.,* 1941, **11**, 461.
[17a] Clark, A. H., and Haaland, A., *Chem. Comm.,* 1969, 912.
[17b] Clark, A. H., and Haaland, A., *Acta Chim. Scand.,* 1970, **24**, 3024.
[17c] Coates, G. E., *Rec. Chem. Progr.,* 1967, **28**, 3.
[18] Coates, G. E., and Fishwick, A. H., *J. Chem. Soc. (A),* 1967, 1199.
[19] Coates, G. E., and Glockling, F., *J. Chem. Soc.,* 1954, 22.
[20] Coates, G. E., Glockling, F., and Huck, N. D., *J. Chem. Soc.,* 1952, 4496.
[21] Coates, G. E., Glockling, F., and Huck, N. D., *J. Chem. Soc.,* 1952, 4512.
[21a] Coates, G. E., Green, M. L. H., and Wade, K., *Organometallic Compounds,* Methuen, Vol. 1, Ch. II, 1967.
[22] Coates, G. E., and Green, S. I. E., *Proc. Chem. Soc.,* 1961, 376; *J. Chem. Soc.,* 1962, 3340.
[23] Coates, G. E., and Heslop, J. A., *J. Chem. Soc. (A),* 1966, 26.
[24] Coates, G. E., and Ridley, D., *J. Chem. Soc. (A),* 1967, 56.
[24a] Coates, G. E., and Roberts, P. D., *J. Chem. Soc. (A),* 1968, 2651.
[25] Coates, G. E., and Tranah, M., *J. Chem. Soc. (A),* 1967, 236.
[26] Colonge, J., *Compt. Rend.,* 1933, **196C**, 1414.
[26a] Cucinella, S., Dozzi, G., Perego, G., and Mazzei, A., *J. Organometallic Chem.,* 1973, **63**, 17; 1977, **137**, 257.
[26b] Cuvigny, T., and Normant, H., *Compt. Rend.,* 1969, **268C,** 834.
[26c] Del Piero, G., Cesari, M., and Mazzei, A., *J. Organometallic Chem.,* 1977, **137**, 265.
[26d] Drago, R. S., and Matwiyoff, N. A., *J. Organometallic Chem.,* 1965, **3**, 62.
[27] Durand, J. F., and Naves, R., *Compt. Rend.,* 1925, **180C**, 521.
[27a] Everest, D. A., *The Chemistry of Beryllium,* Elsevier, 1964.
[28] Fetter, N. R., *Organometallic Chem. Rev. (A),* 1968, **3**, 1.
[29] Fetter, N. R., and Peters, F. M., *Can. J. Chem.,* 1965, **43**, 1884.
[30] Frostick, F. C., and Hauser, C. R., *J. Amer. Chem. Soc.,* 1949, **71**, 1350.
[30a] Gilman, H., and Schulze, F., *Rec. Trav. Chim.,* 1929, **48**, 1129.
[31] Godnev, T. N., and Naryshkin, N. A., *Ber.,* 1925, **58B**, 2703.
[31a] Guemas, L., Drew, M. G. B., and Goulter, J. E., *J. C. S. Chem. Comm.,* 1972, 916.
[32] Gurien, H., *J. Org. Chem.,* 1963, **28**, 878.
[33] Hauser, C. R., and Weiss, M. J., *J. Org. Chem.,* 1949, **14**, 310.
[34] Houghton, R. P., and Williams, C. S., *Tetrahedron Letters,* 1967, **40**, 3929.

[35] Ingraffia, F., *Gazz. Chim. Ital.*, 1933, **63**, 584.
[36] Issleib, K., and Deyling, H. J., *Ber.*, 1964, **97**, 946.
[36a] Jacobs, H., *Z. Anorg. Chem.*, 1971, **382**, 97.
[36b] Jacobs, H., and Juza, R., *Z. Anorg. Chem.*, 1970, **370**, 248.
[36c] Jacobs, H., and Juza, R., *Z. Anorg. Chem.*, 1970, **370**, 254.
[37] Joh, Y., Yoshihara, T., Kotake, Y., Ide, F., and Nakatsuka, K., *J. Polymer Sci., (B)*, 1966, **4**, 673.
[38] Joh, Y., Yoshihara, T., Kotake, Y., Imai, Y., and Kurihara, S., *J. Polymer Sci., (A)*, 1967, **5**, 2503.
[39] Kamio, K., Kojima, S., and Daimon, H., *Kogyo Kagaku Zasshi*, 1963, **66**, 246; *Chem. Abstr.*, 1963, **59**, 10241.
[39a] Lochmann, L., and Šorm, M., *Coll. Czech. Chem. Comm.*, 1973, **38**, 3449.
[40] Longi, P., Bassi, I. W., Greco, F., and Cambini, M., *Tetrahedron Letters*, 1964, 995.
[41] Longi, P., Mazzanti, G., and Bernardini, F., *Gazz. Chim. Ital.*, 1960, **90**, 180.
[42] Lorz, E., and Baltzly, R., *J. Amer. Chem. Soc.*, 1948, **70**, 1904.
[42a] Magnusson, V. R., and Stucky, G. D., *Inorg. Chem.*, 1969, **8**, 1427.
[43] Masthoff, R., Krieg, G., and Vieroth, C., *Z. Anorg. Chem.*, 1969, **364**, 316.
[43a] Masthoff, R., and Vieroth, C., *Z. Chem.*, 1965, **5**, 142.
[44] Meunier, L., *Compt. Rend.*, 1903, **136C**, 758.
[44a] Niedzielski, R. J., Drago, R. S., and Middaugh, R. L., *J. Amer. Chem. Soc.*, 1964, **86**, 1694.
[45] Oddo, B., *Gazz. Chim. Ital.*, 1911, **41**, 222, 248, 255.
[46] Oddo, B., *Gazz. Chim. Ital.*, 1914, **44**, 268, 482.
[47] Oddo, B., *Mem. Accad. Lincei.*, 1923, **14**, 510; *Chem. Abstr.*, 1925, **19**, 2492.
[48] Oddo, B., and Acuto, G., *Gazz. Chim. Ital.*, 1936, **66**, 380.
[49] Okado, H., Kamio, K., and Kojima, S., Jap. Patent 20,499 (1963).
[50] Peters, F. M., and Fetter, N. R., *J. Organometallic Chem.*, 1965, **4**, 181.
[51] Petyunin, P. A., *Russ. Chem. Rev.*, 1962, **31**, 100.
[52] Petyunin, P. A., Berdinskii, I. S., and Shklyaev, V. S., *J. Gen. Chem. U.S.S.R.*, 1954, **24**, 181.
[53] Petyunin, P. A., and Tetyueva, L. A., *J. Gen. Chem. U.S.S.R.*, 1957, **27**, 545.
[54] Petyunin, P. A., and Tetyueva, L. A., *J. Gen. Chem. U.S.S.R.*, 1958, **28**, 1105.
[54a] Rouxel, J., and Brisseau, L., *Bull. Soc. Chim. Fr.*, 1971, 2000.
[55] Sanna, A., *Gazz. Chim. Ital.*, 1934, **64**, 857.
[56] Schlenk, W., *Ber.*, 1931, **64B**, 736.
[56a] Schubert, J., *Chimia*, 1959, **13**, 321.
[56b] Summerford, C., Wade, K., and Wyatt, B. K., *Chem. Comm.*, 1969, 61;

J. Chem. Soc. (A), 1970, 2016.
[57] Terent'ev, A., *Bull. Soc. Chim. Fr.,* 1924, **35**, 1164.
[57a] Wannagat, U., Autzen, H., Kuckertz, H., and Wismar, H. -J., *Z. Anorg. Chem.,* 1972, **394**, 254.
[58] Wawzonek, S., and Nagler, R. C., *J. Amer. Chem. Soc.,* 1955, **77**, 1796.
[59] Wellcome Foundation Ltd., Brit. Patent 619,659 (1949).
[60] Wittig, G., and Dieter, H., *Chem. Ber.,* 1964, **97**, 3541.

4

Amides of Boron, Aluminium, and the Group 3B Metals: Synthesis, Physical Properties, and Structures

A INTRODUCTION TO BORON-NITROGEN CHEMISTRY

During the last twenty years, boron chemistry has been among the most active areas of chemistry, and boron-nitrogen compounds have been at the forefront of attention, together with topics such as hydrides, carboranes, and organoboron compounds. There is now definite evidence of a slackening of interest and the general impression, especially in B—N chemistry, is that the subject is well-established and that the ground rules are adequately understood. As an illustration, one can point to the twenty or so complete X-ray or electron-diffraction studies on BN compounds other than of borazines or Lewis acid-base adducts (see Section C.4). It is also appropriate to indicate that the chemistry of boron amides has been studied as thoroughly as that of any other element [see also Si (Chapter 5) and P (Chapter 6)]. With both B and N as elements of the first short period, problems of bonding do not involve possible complications of *d*-orbital participation, and molecular orbital calculations may be carried out fairly readily (e.g., refs. [11, 12, 14, 50, 157, 213, 341]) (see C.4).

Boron-nitrogen chemistry has been well-served by reviews. The subject generally is discussed in a book [287]. An early important source book of data is ref. [373]. More recently there are detailed compilations of Gmelin with bibliography up to the end of 1973 [143c, 143d] or 1975 [143a, 143b,* 143f]: (i) of compounds with isolated trigonal boron atoms (the aminoboranes and BN-heterocycles) [143c]: (ii) of boron-pyrazole derivatives and spectroscopic studies (mass, vibrational, photoelectron, and ^{11}B and ^{14}N n.m.r.) [143d]; (iii) of BN compounds having four-co-ordinate boron such as the μ-aminodiborane derivatives $R_2NB_2H_5$ (e.g., R = H or Me) and the boronium salts {e.g., $[B(bipy)_2]^{3+}$, $[B(Ph)py_3]^{2+}$, or $[BPh_2py_2]^{+}$} (see also Chapter 15) [143e]; and (iv) amine-boranes (e.g., $Cl_3B.NMe_3$) [143f]. Other general articles are found in refs. [83, 135, 209, 211, 245, 301, 302a, 349]. Specialised topics include borazines [243, 246, 366], other BN-heterocycles [127], heteroaromatic compounds [104, 212, 235], BN oligomers and polymers [211], thermochemistry [125], infrared spectra [237], and diboron derivatives [183].

The isoelectronic relationship of BN and CC atomic pairs [399] has provided the impetus for much research and rationalisation [211]; such compounds are also isostructural. Thus, corresponding to (i) the alkanes, there are the Lewis acid-base adducts (e.g., $F_3B.NH_3$), (ii) the alkenes, there are the aminoboranes or boron amides (the topic discussed here), (iii) the aromatics or other cyclo-(polyacetylenes), there are the borazines [e.g., (I)], and BN heteroaromatics [e.g., (II)], and (iv) the alkynes, there are the boron imides [e.g., PhB=Nmesityl].

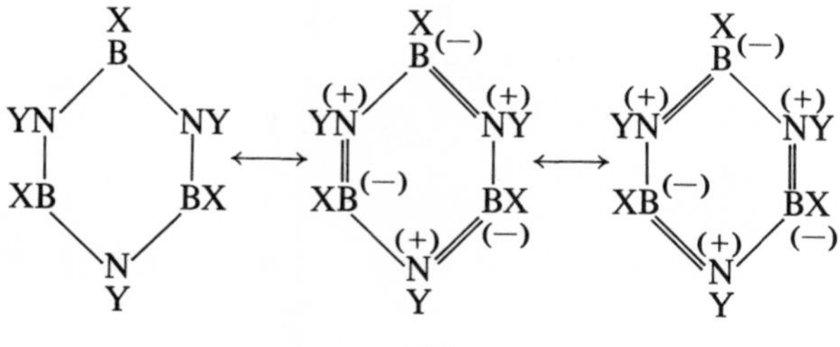

(I)

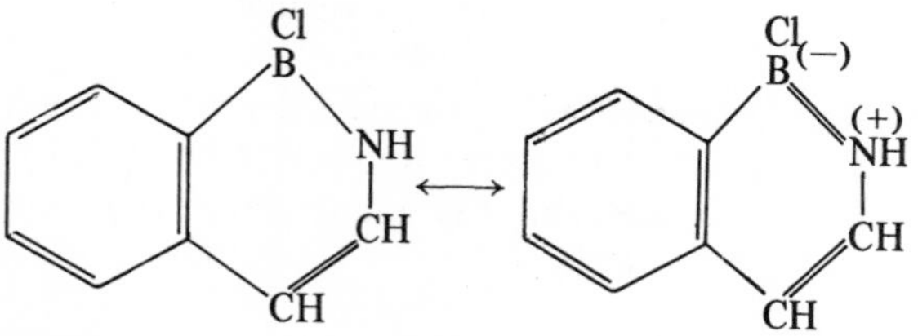

(II)

* Polymeric BN compounds.

The existence of BN compounds of class (iv) as stable entities though documented (cf. ref. [336]) is increasingly questionable. However some two-co-ordinate boron compounds having amide ligands are somewhat more firmly established, e.g., $[Fe(BNMe_2)(CO)_4]$ [358a, 359] (see Chapter 16), and these, together with other metalloboron compounds, are discussed elsewhere [357]. Aminoboranes, like olefins, are in principle capable of oligomerisation or polymerisation [211, 229], because the monomers have potential donor (N) and acceptor (B) sites (see Section A.2): these may be regarded as related to the cycloalkanes or polyalkenes. Until 1962 the borazines were the only cyclic BN compounds with three-co-ordinate B and N. However, four- [215, 359, 374a] [the diazaboretanes or *cyclo*-1,3-diaza-2,4-diboranes, (III)] and eight- [386] [the borazocines or *cyclo*-1,3,5,7-tetra-aza-2,4,6,8-tetraboranes,† (IV)] membered ring compounds have since been characterised and their formation rather than that of the borazines is dependent *inter alia* on steric effects. Such effects are particularly well documented also in other areas of BN chemistry, no doubt because of the small size of the two core atoms (B,N). An example of a particularly bulky boron amide is dimesityl(diphenylamino)borane [146]. Compounds of types (i), (iii), and (iv) will not be discussed extensively (but see Section C.4).

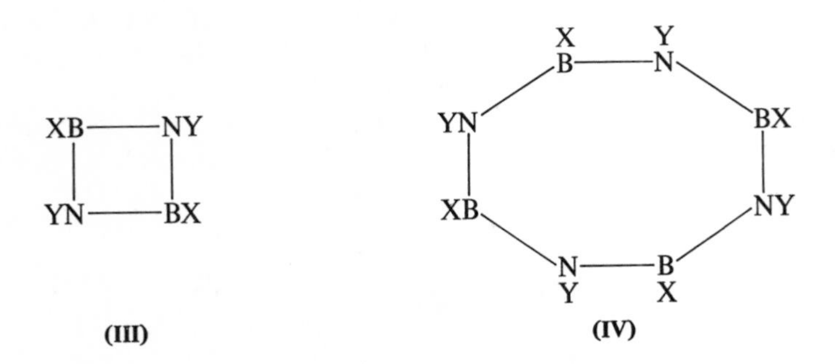

Other BN compounds which fall outside the scope of the present work are (a) the diborylimines, of which class one derivative $Ph_2B{-}N{=}N{-}BPh_2$ [409] has been characterised, (b) the hydrazinoboranes (see ref. [143d]), (c) the aldimides, such as $(Bu_2B{-}N{=}CHBu^t)_{1\ or\ 2}$ [113], (d) the ketimides, such as $(R_2B{-}N{=}CR'_2)_{1\ or\ 2}$ [378], (e) boron-pyrazole derivatives including the 1-pyrazolylborates [384, 385a], (f) amino-borazines (cf., ref. [243]) or boroxines [91] [e.g., $(MeNHBO)_3$], (g) various BN heterocycles, (h) acylamino-(amido-), ureido-, or guanidino-boranes (but see Table 12), (i) polyboranes or carborane derivatives (but see Table 13), and (j) diboryl- or triboryl-amines (but see Table 14).

† E.g., $(MeNBF)_4$ [371a].

Compounds of class (e) have gained some prominence, especially the poly-(1-pyrazolyl)borates as ligands in transition metal chemistry [143d, 365a]. The amines (abbreviated as $[H_nBpz_{4-n}]^-$) are accessible by reactions according to Eq. (1) or a simple variant. Di(1-pyrazolyl)borates are bidentate, e.g., in the square planar (M = Ni or Cu) or tetrahedral (M = Mn, Fe, Co, or Zn) M^{2+} com-

$$BH_4^- + 2HN\!-\!N\ (\text{pyrazole}) \xrightarrow{ca.\ 100^\circ C} [H_2B(N\!-\!N)_2]^- \xrightarrow[\text{pyrazole}]{180^\circ C}$$

$$[H\!-\!B(N\!-\!N)_3]^- \xrightarrow[\text{pyrazole}]{220^\circ C} [B(N\!-\!N)_4]^- \qquad (1)$$

plexes shown in (V). Tri(1-pyrazolyl)borates may act as bi-, e.g., (VI), or tri-, e.g., (VII), dentate ligands.

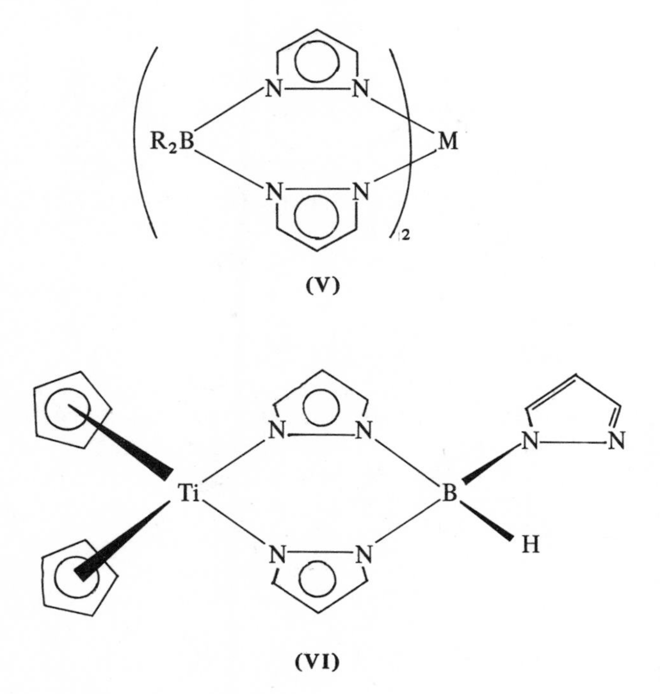

(V)

(VI)

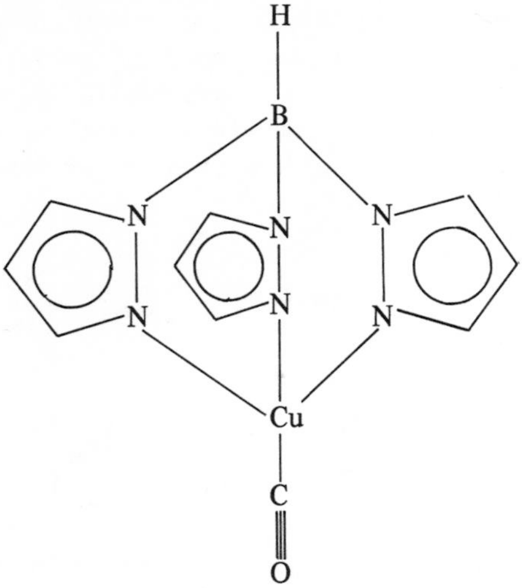

(VII)

A very large number of BN heterocyclic compounds (class g) is now known. The following formulae illustrate the range, each being an endocyclic amino-borane: (VIII) [374a], (IX) [374a], (X) (X = O or S) [374a], (XI) [203a], (XII) [327a], (XIII) [86a], (XIV) [385], (XV) [315a], (XVI) [295a, 296b, 302a, 315a], (XVII) [18c], (XVIII) [296a], (XIX) [392], (XX) [327a], (XXI) [132a], (XXII) [328a], (XXIII) [328a], (XXIV) [296a], (XXV) [276a], (XXVI) [328], (XXVII) [95a], and (XXVIII) [48b].

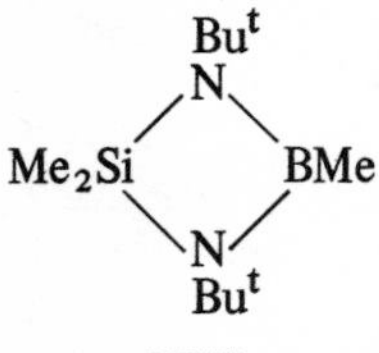

(VIII) 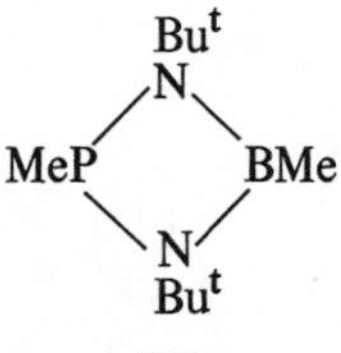 (IX) 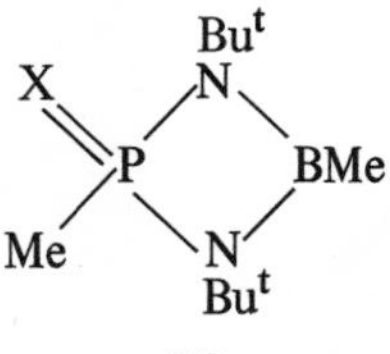 (X) 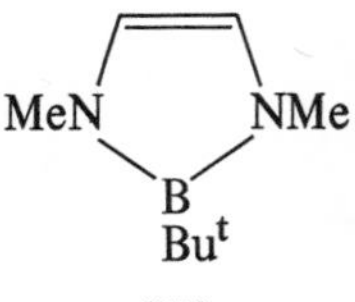 (XI)

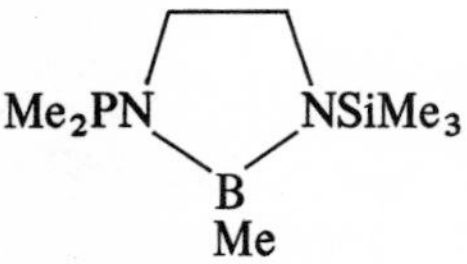

(XII) 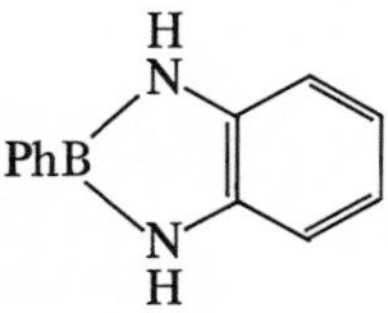 (XIII) 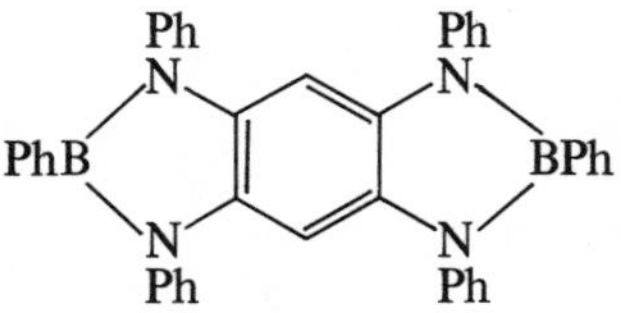 (XIV)

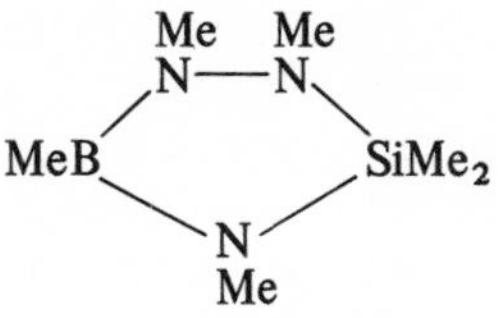

(XV) 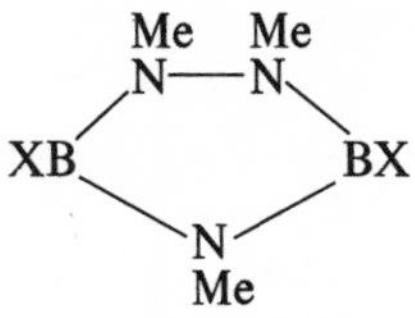 (XVI) 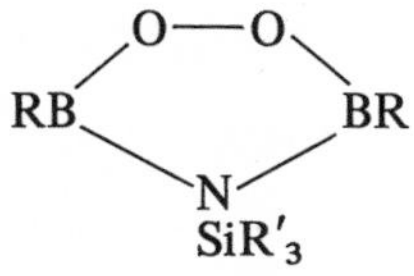 (XVII)

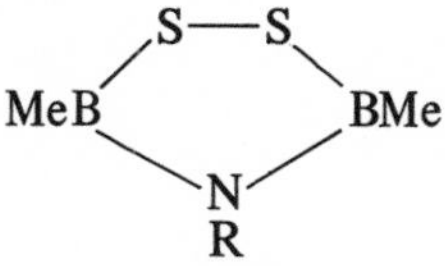

(XVIII) 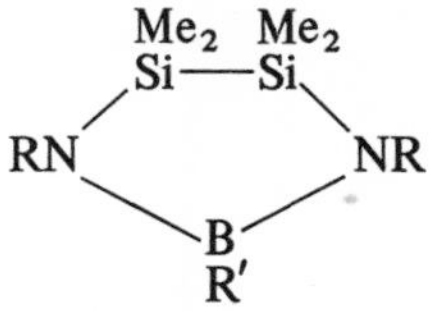 (XIX) 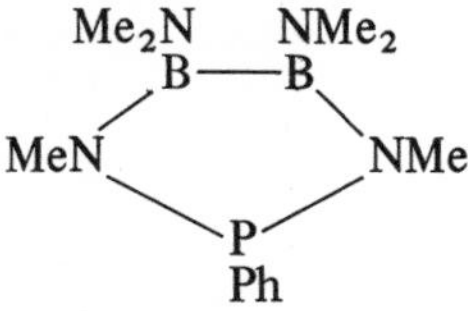 (XX)

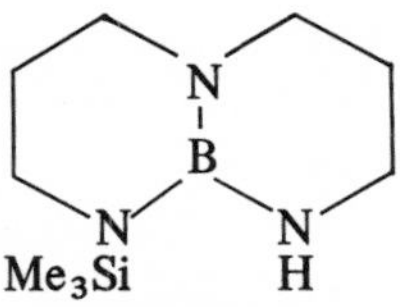

(XXI) 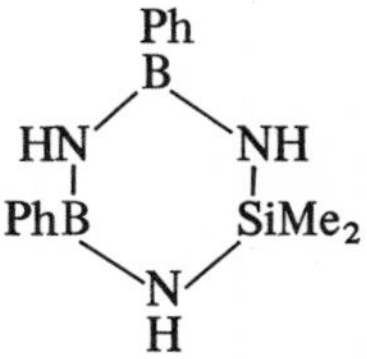 (XXII) 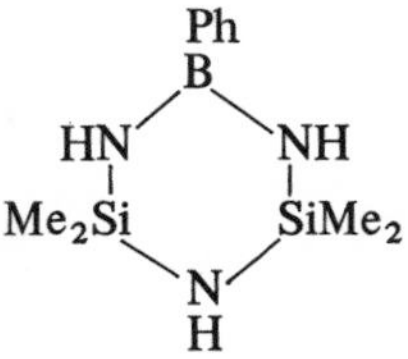 (XXIII) 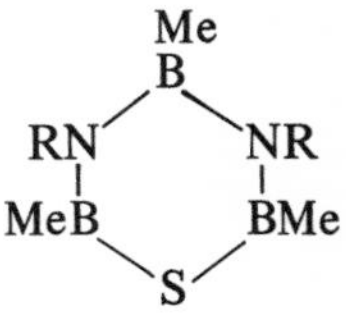 (XXIV)

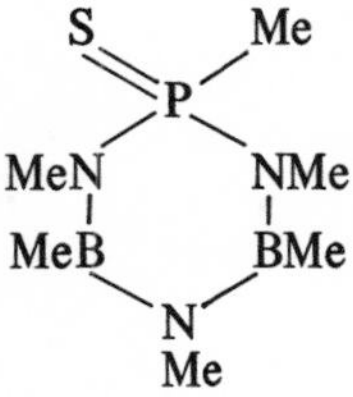

(XXV) 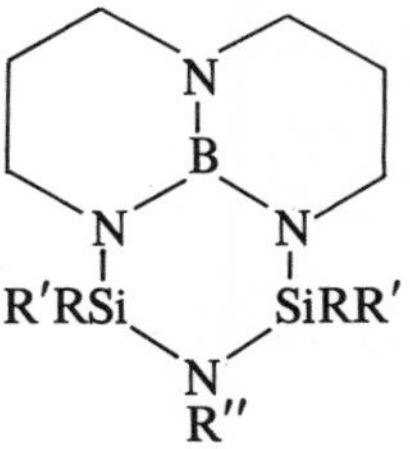 (XXVI) 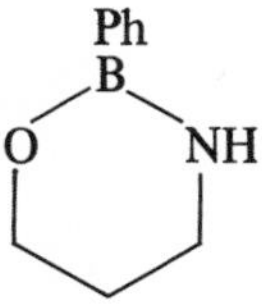 (XXVII) 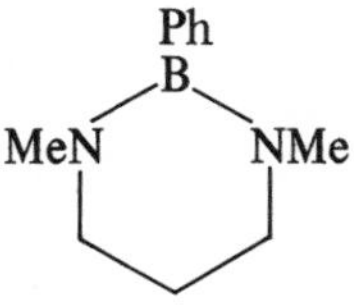 (XXVIII)

Boron, in its nitrogen chemistry, has the oxidation state +3, but a paramagnetic +2 compound $(Me_2N)_2B.py$ is known [204]*. As noted already, the co-ordination number for boron and nitrogen is 3 or 4 (but see the two-co-ordinate $FeBNR_2$ compound). The latter are generally neutral {e.g., $F_3B.NH_3$ or more unusually $H_2B(NMe_2)_2Al(BH_4)_2$ and $H_2B(NMe_2)BH_2NMe_2BH_3$ [192, 193]} or positively charged (the boronium ions); however, there are examples of anions such as $[B(N\;\;)_4]^-$ [130d], which are derived from $[BH_4]^-$. As $B(NR_2)_3$ is only a feeble Lewis acid (Chapter 15), it is not surprising that so few tetra-co-ordinate anions exist; steric and electronic factors are unfavourable.

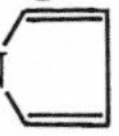

*For B^I, see Table 9; $Me_2B.bipy$ is diamagnetic [204].

Data on the various boron amides are summarised in Tables 4–13. Their chemistry has been widely studied (see Chapters 10–18). Until 1959, tris-amides were known only in the $B(NRR')_3$ form, but derivatives of primary amines such as $B(NHR)_3$ were then characterised [6]. Their reactions, and those of related mono-$X_2B(NHR)$ and bis-$XB(NHR)_2$ amides, are not discussed separately. Their principal unique properties depend on the possibility of hydride displacement, especially metallation by reaction, for example, with LiBu; however, B–N cleavage is a competing process [217], or condensation [6] to form successively a diborylamine (XXIX) and borazine, diazaboretane, or borazocine (XXX) [6, 216] [Eq. (2)].

$$B(NHR)_3 \xrightarrow{-H_2NR} (RHN)_2B{-}N(R){-}B(NHR)_2 \; \textbf{(XXIX)} \xrightarrow{-B(NHR)_3} [RHNB{=}NR] \; \text{(unstable)} \longrightarrow (RHNB{-}NR)_{2,\,3,\,\text{or}\,4} \; \textbf{(XXX)} \qquad (2)$$

The first triborylamine (XXXI) was prepared according to Eq. (3) [222].

$$o\text{-}C_6H_4O_2BCl + HN(SiMe_3)_2 \longrightarrow HN(BO_2C_6H_4\text{-}o)_2 \xrightarrow{o\text{-}C_6H_4O_2BCl.NEt_3} N(BO_2C_6H_4\text{-}o)_3 \; \textbf{(XXXI)} \qquad (3)$$

Its stability was attributed to steric hindrance to disproportionation. Thus, the hypothetical $N(BF_2)_3$ would probably decompose with elimination of BF_3 as a result of a 1,3- shift such as that shown in Eq. (2). Consequently the presently known di- or tri-borylamines (Table 14) are either sterically constrained or are molecules such as $HN(BPh_2)_2$ or $N(BPh_2)_3$ in which a nucleophilic 1,3- shift is energetically unfavourable (no lone pair on the incipient Ph migrating group). Similarly $MeN(BH_2)_2$ is unstable [417]. The low temperature molecules $HB(NH_2)_2$ (with a structural analysis by rotational spectroscopy) [39a] and H_2BNH_2 [234] have been reported, but the simplest trisaminoborane is still $B(NHMe)_3$ [6], although hindered unsymmetrical species having a single NH_2 group, such as $B[N(SiMe_3)_2](NMe_2)NH_2$, are stable [154c].

Also listed in Table 13 are rare examples of polyboranes having B—B linkages including $B_6(NMe_2)_8$ generally formed by a Wurtz-type coupling, e.g., from $ClB(NMe_2)_2$ and K [172]. It is generally assumed that they are stabilised by amido ligands, both for electronic ($\ddot{N}$—B π-bonding) and steric reasons. An unusual dinuclear boron compound is $CH_2[B(NMe_2)_2]_2$ made in low yield from Li_2CH_2 and $ClB(NMe_2)_2$ [203].

The significance of steric effects in BN chemistry may be demonstrated by reference to some reactions of boron amides. For example, in transamination displacement orders generally follow the sequences $NH_2 > NHR > NR_2$ and $NMe_2 > NBu_2 > NHBu^t > NPr^i_2$ [7, 8, 121, 214]. While most aminoboranes are readily hydrolysed or alcoholysed, others such as $B(NMePh)_3$ or $B(NBu^i_2)_3$ are remarkably stable [5, 7, 8, 112]. Highly hindered compounds, such as $B(NPr^i_2)_3$, are not, or not readily, accessible [217]. Mixed compounds, e.g., $B(NHPr^i)(NHBu^t)(NPr^i_2)$ [225], are stable to disproportionation [unlike most mixed borates, $B(OR)(OR')_2$ or mixed halides such as BF_2Cl] because steric constraints make the activation energy for redistribution unfavourable, boron having to become four-co-ordinate in the transition state.

It will be noted (see C.1) that autocomplexation of Group 3 metal amides is rather general unless steric effects supervene. Thus the amido ligand NR_2^- is capable of behaving not only as a terminal, but also as a bridging ligand, as in $[Al(NMe_2)_3]_2$. Boron is unique among these elements in only rarely forming μ-bis(amido)-bridged complexes. A singly bridging amide is found in compounds such as $B_2H_5NMe_2$ [191]. This difference from Al, Ga, and In may be steric in origin. Dimers (XXXII), trimers (XXXIII), or polymers $(H_2BNH_2)_n$ are well-established for several types of aminoboranes [212], providing that the groups pendant to B and N are small. Recent examples of related systems are (XXXIV) [63], $[H_2BN(R)CHX]_3$ (RNCX = MeNCO, PhNCO, or MeNCS, but not PhNCS) [274], and (XXXV) [168a].

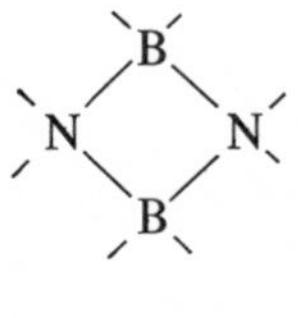

(XXXII) 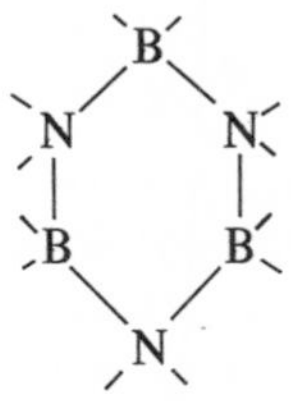 **(XXXIII)** 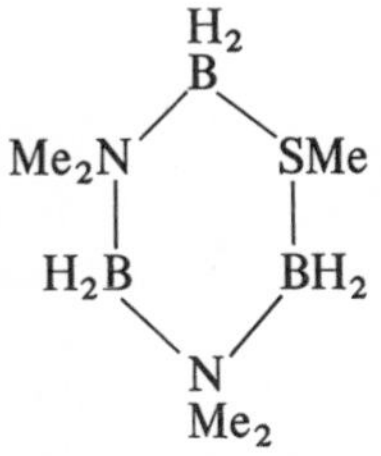 **(XXXIV)** 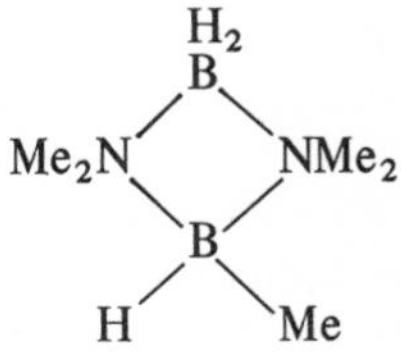 **(XXXV)**

The analogy drawn above between an aminoborane and an olefin poses the question of the existence of *cis*- *trans*-isomers due to restricted rotation about a B–N bond. This problem has been examined by dynamic n.m.r. measurements (Section C.3). An interesting example of isomer interconversion is shown in Eq. (4) with the less hindered *trans*-isomer thermodynamically favoured [163].

$$\mathrm{(PhCH_2)(Ph)B{-}N(Me)(CH_2Ph)} \underset{> -22^\circ\,C}{\overset{h\nu,\ -22^\circ\,C}{\rightleftharpoons}} \mathrm{(PhCH_2)(Ph)B{-}N(CH_2Ph)(Me)} \quad (4)$$

B PREPARATION OF AMINOBORANES

1. Displacement reactions of halogenoboranes

Boron halides or amine-boranes react with excess of amine to cleave boron-halogen bonds, e.g., Eq. (5) [189] or (6) [6].

$$\mathrm{Cl_3B.NH_2Ph + 5H_2NPh \longrightarrow B(NHPh)_3 + 3[PhNH_3]Cl} \quad (5)$$

$$\mathrm{BCl_3 + 6H_2NMe \longrightarrow B(NHMe)_3 + 3[MeNH_3]Cl} \quad (6)$$

When the above reactions are not performed at low temperatures, the products condense to yield the corresponding borazine and free amine. Attempts to prepare mixed amido derivatives generally yield symmetrical products (disproportionation may be obviated by steric hindrance [5, 8, 42, 225], see above).

Steric hindrance may prevent complete displacement of halide, e.g., Eq. (7) [5, 8].

$$\mathrm{BCl_3 + (excess)\ HNPr^i_2 \longrightarrow ClB(NPr^i_2)_2 + 2[Pr^i_2NH_2]Cl} \quad (7)$$

Examples of salt elimination are provided by Eqs. (8) [332], (9) [102], and (10) (R = Me or Et) [371].

$$\mathrm{R_2BCl + LiNMeBR_2 \longrightarrow (R_2B)_2NMe + LiCl} \quad (8)$$

$$\mathrm{BF_3 + 3H_2NEt + 3Li \longrightarrow B(NHEt)_3 + 3LiF + {}^3\!/_2\,H_2} \quad (9)$$

$$\mathrm{[NR_4][BF_4] + Al \xrightarrow{heat} F_3B.NR_3 \xrightarrow{Al,\ heat} F_2BNR_2} \quad (10)$$

Displacement of F^- is generally more difficult than of Cl^-, and in some cases aminoboranes have only been detected as intermediates in a reaction sequence [e.g., Eq. (11)], using microwave spectroscopy (for F_2BNH_2 [228]), mass spectrometry (for F_2BNH_2 [347]), or photoelectron spectroscopy (for F_2BNRR', R = H or Me = R′ or R = H and R′ = Me [203c]). Boron trifluoride has also been treated with $H(R)NSiMe_3$ [117] or $HN(SiMe_3)_2$ [119] when HF and Me_3SiF elimination are competing processes. Similarly, there is mass spectrometric evidence for the transient existence of Cl_2BNH_2 in a gas-phase

$$\mathrm{F_3B + NH_nMe_{3-n} \rightleftharpoons F_3B.NH_nMe_{3-n} \rightarrow F_2BNH_{n-1}Me_{2-n} \rightarrow [FBNH(or\ Me)]_3} \quad (11)$$

Cl_3B/NH_3 flow system [205]. In place of the gaseous boron trifluoride, the liquid diethyl etherate may conveniently be used, e.g., with LiNRR′ [340, 345].

Amides of Si or Sn are convenient substrates for preparing aminoboranes (see Chapter 13); e.g., Me_3SnNMe_2 for Cl_2BNMe_2 (using BCl_3) or $B(NMe_2)_3$ (using $F_3B.OEt_2$) [133]; or $N(MMe_3)_3$ (M = Si [18a] or Sn [375, 375a, 397]) for di- or tri-borylamines. Elimination of $PbBr_2$ has been used, Eq. (12) [327b];

$$Br_2BNMe_2 + \text{[cyclic } S_2Pb\text{]} \longrightarrow \text{[cyclic } S_2\text{]}BNMe_2 + PbBr_2 \quad (12)$$

other examples of metal halide elimination from a halogeno(amino)borane include synthesis of $XB(NMe_2)_2$ from $ClB(NMe_2)_2$ and AgX (X = NCO or NCS [221a], or CN [241a]) and $R(Et)BNMe_2$ from RMgBr (R = $CH_3CH{:}CHCH_2$) and $Cl(Et)BNMe_2$ [164]. A reverse lithium chloride elimination was pioneered in borazine chemistry [390a]. A more recent example relates to the conversion of Me_2BNHEt to $Me_2BN(SiMe_3)Et$ by treatment with successively LiBu and Me_3SiCl [132a].

2. Transamination

Transamination reactions of aminoboranes, e.g., Eq. (13) [281], are subject to steric effects (see above).

$$Et_2BNMe_2 + HN\text{[ring]} \longrightarrow Et_2BN\text{[ring]} + HNMe_2 \quad (13)$$

With aminoborazines, either of two reactions may take place. For *B*-tris-(dimethylamino)-*N*-trimethylborazine, exocyclic transamination is found, Eq. (14) [373], but with an isopropylamino analogue, ring-opening and subsequent displacement of isopropylamine is observed, Eq. (15) [121].

$$(Me_2NBNMe)_3 + 3HNR_2 \longrightarrow (R_2NBNMe)_3 + 3HNMe_2 \quad (14)$$

$$(Pr^iNHBNPr^i)_3 + 9HN\text{[ring]} \longrightarrow 3B(N\text{[ring]})_3 + 6H_2NPr^i \quad (15)$$

Hindered unsymmetrical aminoboranes have been prepared by transamination, e.g., Eq. (16) [42].

$$B(NMe_2)_3 + H_2NPh \longrightarrow PhNHB(NMe_2)_2 + HNMe_2 \quad (16)$$

3. Displacement reactions of hydrido-, alkyl-, alkoxy-, hydroxy-, or alkylthioboranes

The principal type of reaction considered is an elimination such as that illustrated in Eq. (17) (X = R, OR, OH, SR, or H).

$$X{-}B\lt + HNRR' \longrightarrow \gt B{-}NRR' + HX \qquad (17)$$

Hydrogen elimination occurs when an appropriate amineborane is heated, e.g., Eq. (18) [401], or when an amine reacts with a borane or amineborane containing a BH bond, e.g., Eq. (19) [198]. Other examples are provided by Eqs. (20) [198] and (21) [198].

$$2H_3B.NHMe_2 \longrightarrow (H_2BNMe_2)_2 + 2H_2 \qquad (18)$$

$$H_3B.NEt_3 + HNR_2 \longrightarrow H_2BNR_2 + H_2 + NEt_3 \qquad (19)$$

$$H_2BNR_2 + HNR'_2 \longrightarrow HB(NR_2)NR'_2 + H_2 \qquad (20)$$

$$3H_3B.NEt_3 + 3H_2NR \longrightarrow (RNBH)_3 + 3H_2 + 3NEt_3 \qquad (21)$$

Using diborane and a simple primary or secondary amine affords H_2BNRR' (e.g., [375a]), but a 1,2-diamine, 2-aminoethanol, or 2-amino-ethanethiol yields the heterocycle (XXXVI) (R = H or alkyl, X = NR, O, or S) [281b]. The diborane may be made *in situ*, Eq. (22) [10]. An interesting variant employs an amide of lithium [194] or sodium [191a]; the latter is particularly useful for obtaining μ-$R_2NB_2H_5$, Eq. (23).

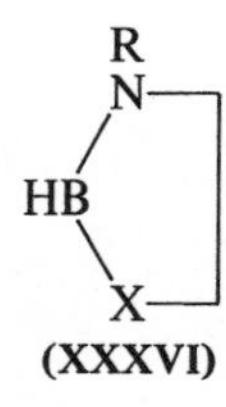

(XXXVI)

$$B(OPh)_3 + Al + \tfrac{3}{2}H_2 + HNR_2 \xrightarrow{100\text{–}150^\circ C} H_2BNR_2 \longrightarrow HB(NR_2)_2 \longrightarrow B(NR_2)_3 \qquad (22)$$

$$H_3B.NMe_2Na + B_2H_6 \longrightarrow \mu\text{-}Me_2NB_2H_5 + Na[BH_4] \qquad (23)$$

A $>$B–C bond is not cleaved as readily as $>$Al–C. Consequently reactions between a trialkylborane and a primary or secondary amine, with alkane elimination, require rather severe conditions. However, the more acid carboxylic acid amides react readily, as in Eq. (24) [381]; $R_2BNHC(O)Bu^t$

$$Et_3B + HN\text{(cyclic lactam, C=O)} \longrightarrow Et_2BN\text{(cyclic lactam, C=O)} + C_2H_6 \quad (24)$$

has been made similarly from BR_3 (R = Pr or Bu^i) [114].

The position of equilibrium of Eq. (25) generally lies well to the left [46, 139, 189, 211]. By using a chelating amine, as in Eq. (26), or by displacing a volatile alcohol with an involatile amine, the equilibrium may be displaced

$$B(OR)_3 + 3H_2NR' \rightleftharpoons B(NHR')_3 + 3ROH \quad (25)$$

to the right; alternatively the sodium amide may be used, as for $BuB(NHBu)_2$ or $B(NHBu)_3$ from $XB(OMe)_2$ (X = Bu or MeO), Na, and H_2NBu [147]. Benzeneboronic acid, $PhB(OH)_2$ [227], or its di-isopropyl ester [46], react with *o*-phenylenediamine to eliminate water or isopropanol, respectively, to yield the BN_2 heterocycle (XXXVII). The displacement of an alcohol by an amine is more difficult for a secondary amine than for a primary amine [248].

$$PhB(OR)_2 + C_6H_4(NH_2)_2 \longrightarrow PhB(NH)_2C_6H_4 + 2ROH \quad (26)$$

(XXXVII)
(R = H or Pr^i)

Alkylthioboranes have likewise been used to prepare aminoboranes [e.g., Eq. (27)] [363]; however, as a reaction type, this is also subject to equilibria [93], and the reverse reaction is favoured in presence of HCl [45]. For chelate compounds the procedure is particularly effective, e.g., using $B(SR)_3$ and $HXCH(R')CRR''NH_2$ (X = O or S) [95].

$$B(SEt)_3 + 3H_2NBu \longrightarrow B(NHBu)_3 + 3EtSH \quad (27)$$

4. Disproportionation or metathetical exchange

Many disproportionation reactions of unsymmetrical halogeno-, alkoxy-, alkylthio-, or alkyl-(amino)boranes have been reported (see Section A). An attempt to prepare bis(dibutylamino)fluoroborane, and distillation of the reaction mixture yielded the disproportionation products, [Eq. (28)] [112].

$$2FB(NBu_2)_2 \longrightarrow B(NBu_2)_3 + F_2BNBu_2 \quad (28)$$

However, other derivatives can be distilled unchanged; for example, $FB(NMe_2)_2$ [64], $ClB(NPr_2)_2$ [5], or $BrB(NMe_2)_2$ [44]. Anilinobis(ethylthio)borane disproportionates slowly at room temperature, and rapidly when warmed [Eq. (29)] [363].

$$3(EtS)_2BNHPh \longrightarrow B(NHPh)_3 + 2B(SEt)_3 \quad (29)$$

Metathetical exchange has been employed to prepare aminoboranes from various boron compounds, e.g., Eqs. (30) or (31) [348] (see Chapter 13). Et_2BNEt_2 is obtained in high yield from BEt_3 and $B(NEt_2)_3$ in presence of $H_3B.$(T.H.F.) as catalyst [159a].

$$2R_3B + Al(NMe_2)_3 \longrightarrow 2R_2BNMe_2 + R_2AlNMe_2 \quad (30)$$

$$PhB(OEt)_2 + Al(NMe_2)_3 \longrightarrow PhB(NMe_2)_2 + (EtO)_2AlNMe_2 \quad (31)$$

5. Insertion reactions

Insertion reactions (see also Chapter 10) have yielded compounds containing boron-nitrogen bonds. The insertion reactions of isocyanates into B—N, [e.g., Eq. (32)] [93], B—Cl [e.g., Eq. (33)] [221], B—C [e.g., Eqs. (33), (34)] [93, 221], or B—O bonds [e.g., Eq. (35)] [93] have been reported. A review is available [221]. Ring expansion is possible by this method [e.g., Eq. (36)] [130].

$$Pr_2BNHPh + PhNCO \longrightarrow Pr_2BNPhCONHPh \quad (32)$$

$$PhBCl_2 + 2PhNCO \longrightarrow PhCONPhB(Cl)NPhCOCl \quad (33)$$

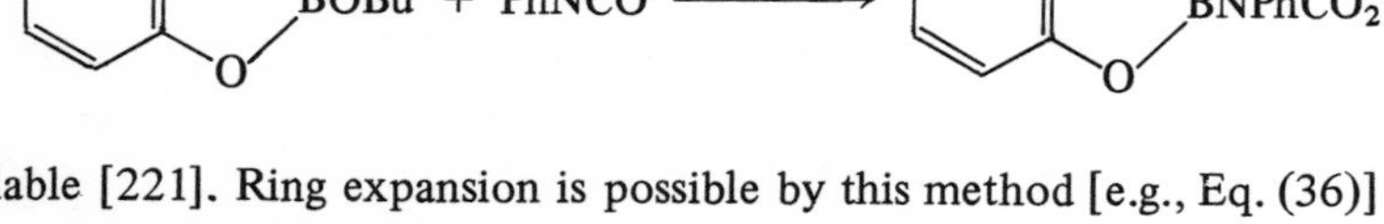

$$C_6H_4(O)_2BBu + PhNCO \longrightarrow C_6H_4(O)_2BNPhCOBu \quad (34)$$

$$C_6H_4(O)_2BOBu + PhNCO \longrightarrow C_6H_4(O)_2BNPhCO_2Bu \quad (35)$$

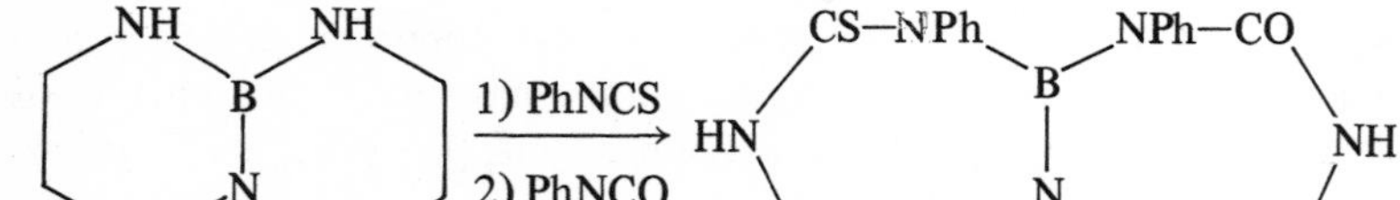

$$\text{(bicyclic NH–B(–N)–NH)} \xrightarrow[\text{2) PhNCO}]{\text{1) PhNCS}} \text{HN–CS–NPh–B(–N)–NPh–CO–NH (ring-expanded)} \quad (36)$$

Carbodi-imides have been inserted into B—Cl or B—C bonds [e.g., Eq. (37)] [187].

$$PhBCl_2 + 2RN{:}C{:}NR \longrightarrow RN{:}CPhNRB(Cl)NRCCl{:}NR \qquad (37)$$

More recent variants are shown in Eqs. (38) [273], (39) [239], (40) ($M = Li$ or Na) [3, 4], (41) [242], (42) ($X{=}O$ or S) [186a], and (43) [411].

$$B_2H_6 + 2PhN = CHAr \longrightarrow 2H_2BN(Ph)CH_2Ar \xrightarrow{heat} HB[N(Ph)CH_2Ar]_2 \qquad (38)$$

$$BR_3 + MeN{=}CHPh \longrightarrow R_2BN(Me)CHRPh \qquad (39)$$

$$M[R_3B\text{------}HBH_3] + 6Ar'N{=}CHAr \longrightarrow 2B[N(Ar')CH_2Ar]_3 \qquad (40)$$

$$B(NMe_2)_3 + 2COS \longrightarrow [Me_2NC(S)O]_2BNMe_2 \qquad (41)$$

$$XCNB(NMe_2)_2 + p\text{-TolNCX} \longrightarrow XNCB(NMe_2)[N(Tol\text{-}p)C(NTol\text{-}p)NMe_2] \qquad (42)$$

$$CH_2{=}CH(CH_2)_3NH_2 + BH_2Cl \longrightarrow \text{cyclic } \overset{+}{N}H_2\text{—}\overset{-}{B}HCl \longrightarrow \text{cyclic } NH\text{—}BCl \qquad (43)$$

6. Miscellaneous methods

A convenient synthesis of secondary amines with retention of configuration via hydroboration probably involves the insertion of a nitrene into a B—C bond, as illustrated in Eq. (44) [54]; dialkylchloroboranes or BR_3 may also be employed, but less effectively, and the aminoborane is not normally isolated.

$$RBCl_2 + R'N_3 \rightleftharpoons Cl\overset{R}{\underset{Cl}{B^-}}\text{—}\underset{N_2^+}{NR'} \longrightarrow Cl\overset{R}{\underset{Cl}{B}}\text{—}NR' + N_2 \xrightarrow{NaOH} HNRR' \qquad (44)$$

Another reaction used in amine or alkyl chloride synthesis, and based on hydroboration, involves $ClNR'_2$ as reagent, Eq. (45) [98].

$$BR_3 \xrightarrow{ClNR'_2} \begin{cases} \xrightarrow{(a)} NRR'_2 + R_2BCl \\ \xrightarrow{(b)} RCl + R_2BNR'_2 \end{cases} \qquad (45)$$

The pathway leading to amine, Eq. 45(a), follows a polar mechanism and is dominant in the presence of the free radical inhibitor galvinoxyl, whereas the reaction of Eq. 45(b) involves a radical chain, e.g., Eq. (46), and is enhanced by irradiation and the presence of $\dot{N}Me_2$ from $Me_2NNNNMe_2$.

$$\left.\begin{array}{l} \dot{N}Me_2 + BBu_3 \longrightarrow Bu_2BNMe_2 + \dot{B}u \\ \dot{B}u + ClNMe_2 \longrightarrow BuCl + \dot{N}Me_2 \end{array}\right\} \quad (46)$$

An insertion reaction of a different type involves carbon vapour (obtained using graphite electrodes) and H_2BNMe_2 [168a]; $Me(H)BNMe_2$ is believed to be the principal product and is isolated as its complex (XXXV) with the starting borane, whereas Me_2BNMe_2, $H_3B.NHMe_2$, and $O[B(H)NMe_2]_2$ are also obtained in trace quantities.

Excess BCl_3 and NCl_3 yield Cl_2BNCl_2 (with Cl_2) and $(ClBNCl)_3$ [160]. Several silylaminoboranes have been prepared from $Bu^tMe_2SiN(R)Li$ (R = H, Me, or $SiMe_3$) and various chloroboranes [37a].

C PHYSICAL PROPERTIES AND STRUCTURES OF AMINOBORANES

1. Oligomerisation and polymerisation

For various aminoboranes, dimeric (XXXII), as well as monomeric, forms are known. Cl_2BNMe_2 was the first of these to be studied [406, 407]. When initially formed it is monomeric, but dimerises slowly at room temperature, or more quickly if heated. The dimer is reconverted to the monomer by sublimation [51]. The tendency for dimerisation to occur with higher homologues is related to steric factors [277]. Me_2BNH_2 is dimeric at low temperature [404], but dissociates above 30°C. Ph_2BNH_2 exists only as the dimer [84]. Monomeric Me_2BNHMe dimerises upon cooling [403]. Ph_2BNHPh slowly changes from monomer into dimer at room temperature [260]. $MeHBNMe_2$ is almost wholly monomeric in the vapour; in the liquid, the dimer rapidly forms, and the dimer is a solid at room temperature [65]. H_2BNMe_2 is dimeric under ambient conditions, but dissociates into a monomer when heated *in vacuo* [20, 69, 161, 401]. Dimeric $(H_2BNMe_2)_2$ is converted into the trimer (see crystal data, C.4) by reaction with B_5H_9 at 100–110°C [70, 73]. Compounds o-$C_6H_4O_2BNRR'$ are dimeric, unreactive solids when sterically possible (RR' = Me, H; Et, H; Pr^i, H; Ph, H; or Me_2), while monomers (RR' = Bu^t, H; Et_2; or Bu_2) are reactive liquids [218]. H_2BNHMe exists as trimer, isosteric with 1, 3, 5-trimethylcyclohexane [33] and has been characterised in two distinct conformations [353]. The trimer was shown to be thermodynamically more stable than the polymer [53]; intermediates are important in determining the degree of oligomerisation [53, 55]. H_2BNH_2 is usually a polymer, but a trimer [97] is known; as are other oligomers [37]. For the equilibrium of Eq. (47), thermodynamic data were obtained from ^{11}B n.m.r. measurements [331]. Rates of dimerisation have similarly been examined [218]. Aminoboranes and aminodifluoroboranes dimerise rapidly, while the dimerisation of X_2BNMe_2 (X = Cl or Br) is rather slow [218]. In the gas phase [16] heavier halides (X = Cl, Br, or I) are present only as monomers, while for the fluoride the equilibrium still persists. X-ray data are also available for a dimer [173]. In all these compounds bridging is through nitrogen

$$2X_2BNR_2 \rightleftharpoons (X_2BNR_2)_2 \qquad (47)$$

or, exceptionally, through hydrogen. The compound $Et_2B\text{-}NH(C_5H_4N\text{-}2)$, obtained from $2\text{-}NH_2\text{-}C_5H_4N/BEt_3$ [116a, 154b], appears to exist in two forms (^{13}C n.m.r.), with the intramolecular 4-membered ring favoured at low temperature [154a].

2. Thermochemistry

Heats of hydrolysis, which afford standard heats of formation [125], have been measured for the series $Cl_nB(NMe_2)_{3-n}$ (n = 0, 1, 2, and 3) [367]. These suggest that the mean thermochemical bond energy term $\bar{E}$(B—N) is ca. 105 kcal mol^{-1}. However, Cl_2BNMe_2 (monomer) is more stable, and so to a lesser extent is $ClB(NMe_2)_2$, than would be the case if $\bar{E}$(B—N) and $\bar{E}$(B—Cl) are taken from data on $B(NMe_2)_3$ and BCl_3, respectively. A possible interpretation is in terms of rough constancy of B—N and B—Cl σ-bond strengths within the series, with BN(p-p) π-bonding more important than BCl.

Combustion calorimetry of $B(NHBu)_3$ and $BuB(NHBu)_2$ has yielded ΔH_f°(l) data: -141.7 ± 1.6 and -120.3 ± 3.1 kcal mol^{-1}, respectively [147]. Mass spectrometry (determination of first ionisation potential and appearance potential of fragment ions) yielded ΔH_f° (g) data (kcal mol^{-1}) for H_2NBF_2 (−225) and H_2NBH_2 (−75) [347], at variance for the latter with earlier results [234]. Computer processed data are in Table 1 [340a].

Table 1
Some thermochemical data for aminoboranes [340a]

Compound (state)	ΔH_f° (kcal mol^{-1})	Compound (state)	ΔH_f° (kcal mol^{-1})
Cl_2BNMe_2 (l)	−438.6	$B(NEt_2)_3$ (l)	−463.4
Cl_2BNMe_2 (g)	−401.3	Bu_2BNH_2 (l)	−398.6
$ClB(NMe_2)_2$ (l)	−376.6	Bu_2BNHBu (l)	−452.4
$ClB(NMe_2)_2$ (g)	−334.8	$Bu(MeO)BNEt_2$ (l)	−611.4
$B(NMe_2)_3$ (l)	−292.3	$Bu(MeO)BNEt_2$ (g)	−553.2
$B(NMe_2)_3$ (g)	−245.4	μ-$Me_2NB_2H_5$ (l)	−146.3
		μ-$Me_2NB_2H_5$ (g)	−117.5

3. Dipole moments, spectroscopy, and mass spectrometry

Me_2BNMe_2 has the low dipole moment of 1.40 ± 0.03 D [21]. Owing to the large electronegativity difference between B and N, this implies that there is a significant π-bond moment $(\overset{-}{B}-\overset{+}{N})$ to counterbalance the σ-moment $(\overset{+}{B}-\overset{-}{N})$ [84] and consistent with that view the small dipole moments in Ph_2BNMe_2 $(N \rightarrow B)$ and Ph_2BNPh_2 $(N \leftarrow B)$ are in opposite directions.

Bibliography on BN vibrational spectra is in ref. [9]; later data are in ref. [237]. In monomeric aminoboranes ν(BN) is found as a very strong absorption between 1400–1550 cm^{-1}. In the early literature, stretching frequencies (e.g., [9, 21, 250]) or simple valence force field force constants (e.g., [150]) were frequently interpreted in terms of bond strengths (assignments are assisted by the presence in significant natural abundance of two isotopes ^{10}B and ^{11}B of boron). While conclusions seem to be chemically sensible (e.g., see ref. [9]), more recent work has shown that there is significant mixing of BN and CN modes [24].

The ^{10}B and ^{11}B nuclei both have nuclear spins and there is considerable information, mainly ^{11}B chemical shift data $(I = 3/2)$ on BN compounds [275]. A notable feature is the large spread of chemical shifts and the distinction between three- and four-co-ordinate boron compounds (the latter are, of course, more shielded and resonances are at higher field), which is useful, for example, in distinguishing between aminoboranes and their polymers. A recent book is devoted to ^{11}B n.m.r. spectroscopy [333d].

There has been a good deal of routine study using ^{1}H, ^{11}B, ^{13}C, ^{19}F, and even ^{14}N n.m.r., as well as ^{15}N, ^{29}Si, ^{31}P, ^{35}Cl, ^{77}Se, ^{117}Sn, ^{119}Sn, and ^{207}Pb nuclei for appropriate compounds. More recent citations are in refs. [18a, 18c, 29, 48b, 101, 130d, 154b, 188a, 296a, 302a, 329, 333a, 333b, 333c, 375a, 413a, 413b, and 413c]. However, the most notable investigations have concerned measurements of rotational barriers about BN bonds in aminoboranes. The earliest measurements of ^{1}H spectra as a function of temperature were made in 1961 on $PhB(Cl)NMe_2$ [17] and $Me_2BNMePh$ [351]. At room temperature, there is restricted rotation about the BN bond and for unsymmetrical compounds such as PhB(Cl)NMePh [18], there is evidence for the coexistence of *cis-/trans-*isomers in solution. The barriers to rotation vary from ca. 10 to ca. 25 kcal mol^{-1}. This has been taken to imply the existence of significant $(p\text{-}p)\pi$ $\ddot{N}\overset{\curvearrowright}{-}B$) bonding, but there can be little doubt that steric effects play a major role in determining the magnitude of the activation energies. Other papers have provided additional data [48a, 96a, 96b, 110, 154c, 186, 225, 279, 361, 370, 398]. Some data (S.I. units) from ref. [18] are shown in Table 2 [II, III, and IV refer to observations based on respectively the ratio of (i) the central minimum to either maximum intensity, (ii) the band width at half height to the band maximum separation at infinitely slow exchange, and (iii) the maximum height of the band to that value which obtains at infinitely fast exchange].

Table 2*

(a) Arrhenius parameters and 'coalescence' temperatures

Compound	Activation energy, E (kJ mol^{-1})			Pre-exponential factor, A (sec^{-1})			'Coalescence' temperature (°K)
Method	II	III	IV	II	III	IV	
$PhB(Cl)NMe_2$	85	89	91	9×10^{12}	5×10^{13}	8×10^{13}	383
$PhB(Br)NMe_2$	96	92	94	3×10^{14}	1×10^{14}	3×10^{14}	385
$PhB(OMe)NMe_2$	51	69	55	4×10^{10}	2×10^{13}	2×10^{11}	319
$PhB(Cl)N(Me)Et$	51	56	55	4×10^{9}	3×10^{9}	9×10^{9}	371
$AnB(Cl)NMe_2$[b]	86	92	99	2×10^{13}	2×10^{13}	2×10^{14}	366

(b) Transition state parameters[a]

Compound	Activation enthalpy, $\Delta H^{\ddagger}$ (kJ mol^{-1})			Activation entropy, $\Delta S^{\ddagger}$ (J° K^{-1} mol^{-1})		
Method	II	III	IV	II	III	IV
$PhB(Cl)NMe_2$	83	84	87	−4	12	10
$PhB(Br)NMe_2$	94	88	86	28	3	3
$PhB(OMe)NMe_2$	50	69	54	−28	−68	−35
$PhB(Cl)N(Me)Et$	46	54	51	−96	−73	−65
$AnB(Cl)NMe_2$[b]	81	87	100	3	13	48

[a] Assuming a transmission coefficient of unity. [b] An = anisyl (*p*-methoxyphenyl).

* N.m.r. data for rotational barriers *[Reproduced, with permission, from Barfield, P. A., Lappert, M. F., and Lee, J.,* Trans. Faraday Soc., *1968,* **64,** *2571.]*

Variable temperature ^{13}C n.m.r. measurements on $PhB(Cl)NBu^s_2$ provide a value of 15.6 kcal mol^{-1} for the BN rotational barrier [48a].

Bis(amino)boranes $XB(NR_2)_2$ often have a somewhat lower torsional barrier [17, 110, 186] unless steric effects are especially pronounced as, for example, in $PhB(NMe_2)[N(SiMe_3)_2]$ (19.8 kcal mol^{-1}) [279] {cf. $PhB(NMe_2)[N(MMe_3)_2]$: M = C, 18.8; Ge, 14.5; or Sn, 10.8 kcal mol^{-1} [279]}. This lowering has been rationalised by supposing a decrease in π-bond order in the bis(amino)borane. However, the barrier for a trisaminoborane $B(NMe_2)_2[N(Me)Ph]$ is also high and this was attributed to the N(Me)Ph being forced to be nearly perpendicular to the BN_3 plane [110]. A series of BN_3 molecules, $B(NPr^i_2)_2(NRR')$, has been shown to possess helically chiral configurations, (XXXVIII), which are stereochemically rigid at low temperatures, and n.m.r. line shape analysis indicates that these molecules enantiomerise by correlated B–N rotations through transition states in which the substituents attached to two of the nitrogens are in the plane of the BN_3 unit while those on the third nitrogen are perpendicular to this plane [96a, 96b]. When R = R′ molecular symmetry only allows the determination of the barrier to the exchange (8-9 kcal mol^{-1}) in which one of the NPr^i_2 groups is coplanar with, while the other is perpendicular to, the BN_3 plane. In the series when R = H and R′ = alkyl, the $\Delta G^{\ddagger}$ for the three possible non-degenerate routes to enantiomerisation are 13–15, 8–9, and $\leqslant$ 5 kcal mol^{-1}, respectively. In principle there are four processes, the transition states for which are shown in (XXXIX) (zero-amine flip), (XL) (one-amine flip), (XLI) (two-amine flip), and (XLII) (three-amine flip) [96a, 96b]. In $PhB(X)(NPr^i_2)$ there are two rotamers of equal abundance at 25°C when X = Cl or NPr^i_2, whereas in $ClB(X)(NPr^i_2)$ there is only one for X = Cl or NEt_2 [225].

Mass spectra of boron-nitrogen compounds have provided a useful analytical tool, e.g., for cyclic compounds [202]. Species derived from aminoboranes often have a high abundance of doubly-charged [109, 222] and even triply-charged [222] ions. This has been taken as evidence of high BN bond strengths [109]. Acyclic or cyclic phenylboranes containing at least one BN bond give peaks assignable to the boratropylium or boracyclopentadienyl cations [96]. Mass spectra have provided first ionisation potentials (i.p.'s) for $B(NMe_2)_3$ (7.57 eV) and derivatives $X_nB(NMe_2)_{3-n}$ (X = F, Cl, Br, I, OMe, OEt, OPr, Et, or SMe) [15, 213, 219] and the data have been analysed in molecular orbital terms [213]. It was concluded that π-bonding is significant in BN chemistry. The first i.p.'s of H_2BNH_2 and F_2BNH_2 are at 11.0 ± 0.1 and 12.4 ± 0.4 eV respectively and corresponding BN bond dissociation energies are 8.1 and 7.6 eV [347].

The electronic spectra of gaseous $ClB(NMe_2)_2$ and $B(NMe_2)_3$ have been examined; $n \rightarrow \pi^*$ u.v. transitions are at ca. 200 nm [219] (see also ref. [130a]).

Photoelectron (p.e.) spectra of the series $X_nB(NMe_2)_{3-n}$ (n = 0, 1, 2, 3; X = F, Cl, Br, I) have been recorded [36, 196]. The two highest occupied levels were assigned [36] and variations were attributed to geometric factors; a more detailed analysis takes BN twisting into account [196]. The p.e. spectrum of

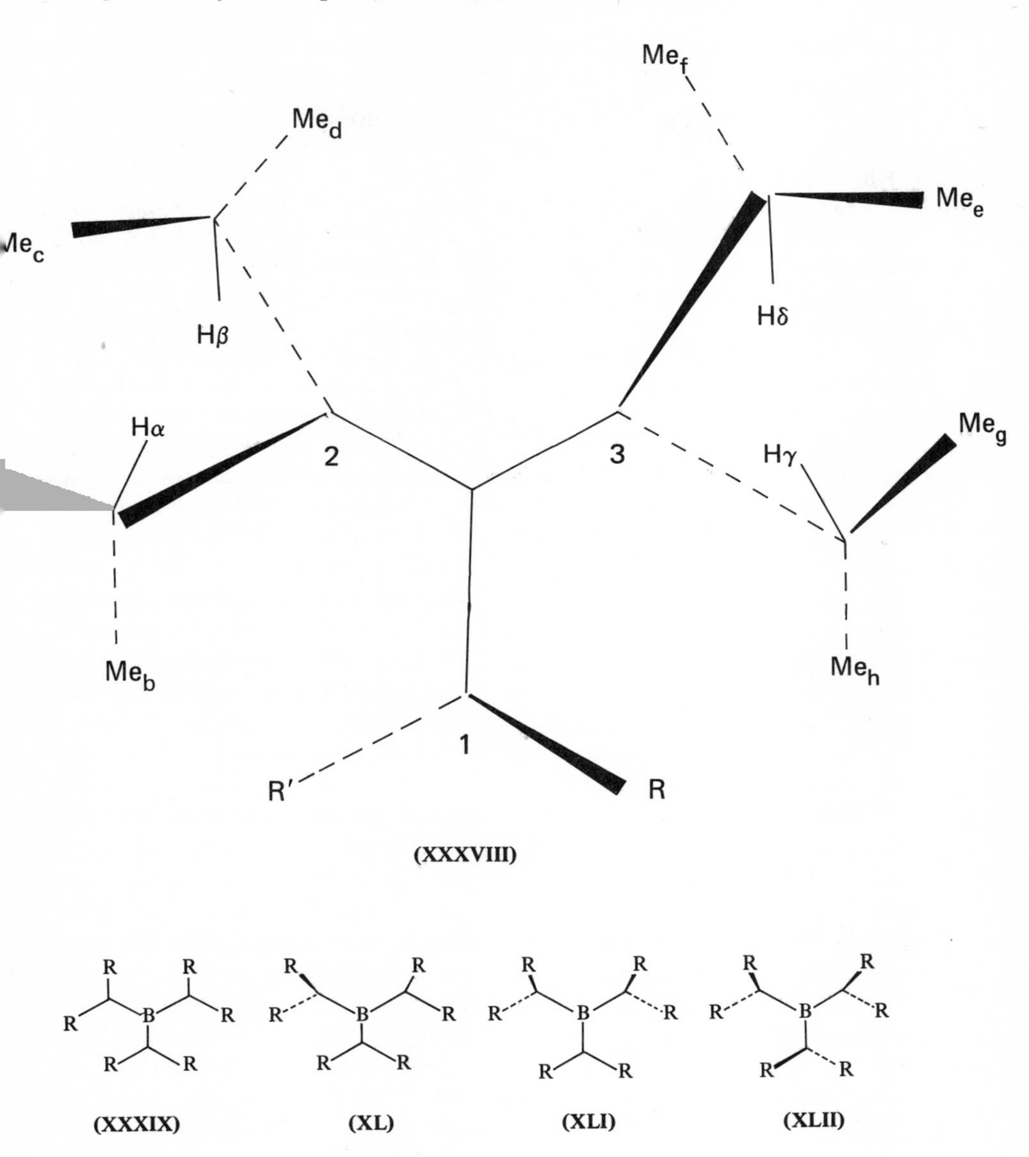

$B_2(NMe_2)_4$ has been recorded, assigned, and compared with that of the isostructural $C_2(NMe_2)_4$ [77]. Other data are in refs. [18d, 203a, 203b, 203c] (see also Chapter 16) and a detailed discussion is in ref. [143d] (pp. 170–196).

The n.q.r. parameters for polycrystalline samples of Me_2BNMe_2 and X_2BNMe_2 (X = F, Cl, or Br) have been determined [410].

4. Structures of BN compounds

Ab initio molecular orbital calculations at various levels of sophistication have been carried out on $H_3B.NH_3$ [11], H_2BNH_2 [11, 157], the hypothetical $(HBNH)_2$ [14], $H_2BNH_2.NH_3$ [12], the unknown $H_2NBHBHNH_2$ [50], and $HB(NH_2)(CH{=}CH_2)$ [50].

Apart from data on various borazines {see ref. [167] for a recent study, which gives the B–N bond length in $(HBNH)_3$ by electron diffraction as 1.435 ± 0.002 Å}, results of X-ray electron diffraction or microwave spectroscopy are available for twenty-one aminoboranes including cyclic derivatives (Table 3). These may be compared with the following information (X-ray data) on compounds having B and N in four-co-ordinate environments. The molecular structure of

Table 3
Structural data on aminoboranes

Compound	Method	Reference	Compound	Method	Reference
$HB(NH_2)_2$	rotational	[39a]			
$(HBNH)_3$	X-ray	[167]	$(o\text{-}C_6H_4O_2B)_3N$	X-ray	[57]
$(CH_2BNMe_2)_3$	X-ray	[175]	$\overline{(CH_2)_3NH}{-}B{-}\underline{N(CH_2)_3NH}$	X-ray	[58]
Cl_2BNMe_2	ed	[81]	Cl_2BNPh_2	X-ray	[415]
$F_2BN(SiH_3)_2$	ed	[346]	F_2BNH_2	microw.	[228]
$B(NMe_2)_3$	ed	[80]	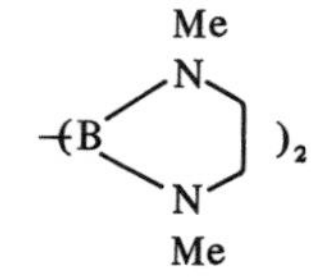 $-(B(NMe)_2)_2$	X-ray	[130b]
$[Me_3SiNBN(SiMe_3)_2]_2$	X-ray	[175]	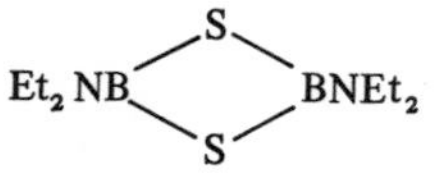 $Et_2NB(S)_2BNEt_2$	X-ray	[71a]
Me_2BNMe_2	X-ray	[59]	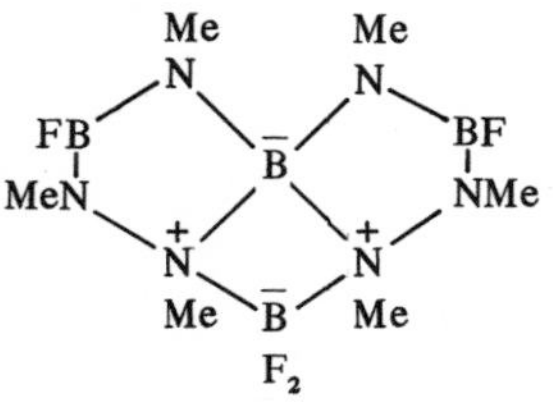 Me N, Me N, FB–MeN, $\overline{B}$, BF–NMe, $\overset{+}{N}$ Me, $\overline{B}F_2$, $\overset{+}{N}$ Me	X-ray	[130c]
$Ar(Cl)BNMeAr'$ ($Ar = Ph$, $Ar' = 4\text{-}BrC_6H_4$; $Ar = 2\text{-}MeC_6H_4$, $Ar' = 2\text{-}Me, 4\text{-}BrC_6H_3$)	X-ray	[418]	$(Bu^t_2P)_2BNPr^i_2$	X-ray	[19a]

$(H_2BNMe_2)_3$ is that of a six-membered ring of alternating B and N atoms in a chair configuration with B–N bond lengths of 1.61 ± 0.04 Å [383], while that of $(Cl_2BNMe_2)_2$ is of a planar four-membered ring of alternating B and N atoms with B–N bond lengths of 1.59 ± 0.01 Å [173]. An exceptionally long $B(sp^3)$–$N(sp^3)$ bond length of 1.66 Å has been recorded for the cage compound of Fig. 1, and the $B(sp^2)$–$N(sp^2)$ distance of 1.49 Å is also very long [380]. The geometry of the hydrazinoborane $B(NHNMe_2)_3$ is revealed by X-ray results [330], with a planar BN_6 unit and a B–N bond length of 1.420 ± 0.14 Å. Gaseous μ-$Me_2NB_2H_5$, not surprisingly, has a very long BN bond [1.544(10) Å], as shown by microwave spectroscopy [85].

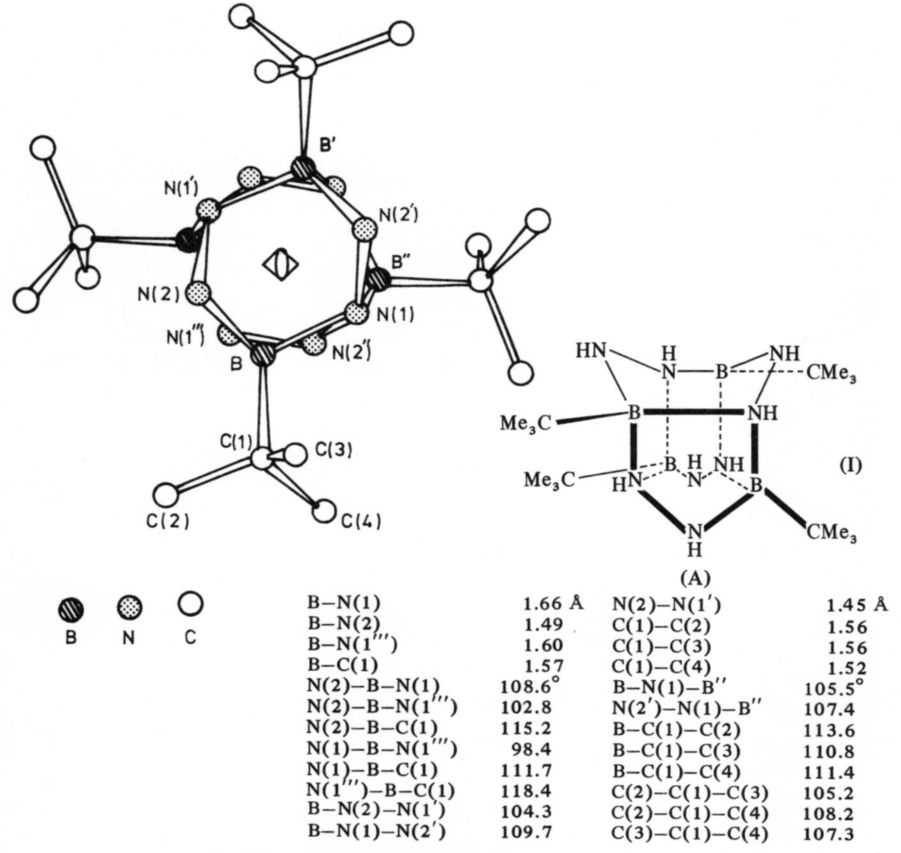

B–N(1)	1.66 Å	N(2)–N(1′)	1.45 Å
B–N(2)	1.49	C(1)–C(2)	1.56
B–N(1‴)	1.60	C(1)–C(3)	1.56
B–C(1)	1.57	C(1)–C(4)	1.52
N(2)–B–N(1)	108.6°	B–N(1)–B″	105.5°
N(2)–B–N(1‴)	102.8	N(2′)–N(1)–B″	107.4
N(2)–B–C(1)	115.2	B–C(1)–C(2)	113.6
N(1)–B–N(1‴)	98.4	B–C(1)–C(3)	110.8
N(1)–B–C(1)	111.7	B–C(1)–C(4)	111.4
N(1‴)–B–C(1)	118.4	C(2)–C(1)–C(3)	105.2
B–N(2)–N(1′)	104.3	C(2)–C(1)–C(4)	108.2
B–N(1)–N(2′)	109.7	C(3)–C(1)–C(4)	107.3

(a) View of the molecule, looking along the c-axis.
(b) Bond distances (± 0.01 Å) and angles (± 0.8°).

Fig. 1 – Molecular structure of the cage compound (A). *[Reproduced, with permission, from Thomas, P. C., and Paul, I. C.,* Chem. Comm., *1968, 1130.]* For clarity, H atoms are omitted.

For aminoboranes (apart from the above cage compound), B–N bond lengths vary from 1.379 ± 0.06 Å in Cl_2BNMe_2 (electron diffraction) [81] to 1.485 ± 0.022 Å in $F_2BN(SiH_3)_2$ (electron diffraction) [346]. However, these are extreme values and the remainder fall in the range 1.41–1.45 Å. The geometry around each B and N atom corresponds approximately to that expected on the basis of sp^2-hybridisation of the σ-orbitals of each atom. This suggests that $(p\text{-}p)\pi$ bonding $(\ddot{\mathrm{N}}\overset{\curvearrowright}{-}\mathrm{B})$ is significant and trends within series lend support to that view. Thus, the average BN distances (electron diffraction) in $B(NMe_2)_3$ at 1.431 ± 0.06 Å [80] are appreciably longer than the 1.379 ± 0.06 Å in Cl_2BNMe_2, which could be explained in terms of significant BN and less important BCl π-bonding. Other factors, namely variations in σ-bond strengths and non-bonding interactions, are obviously also important and it is certainly not necessary to call upon π-bonding arguments. The extremely long BN bond length in $F_2BN(SiH_3)_2$, quoted above, would require in π-bonding terms the assumption that FB π-bonding is unusually important, which is at variance with other data [213].

The structure of tris-1,3,5-dimethylamino-1,3,5-triboracyclohexane (X-ray) is illustrated in Fig. 2 [175], hexakis(trimethylsilyl)-2,4-diamino-1,3,2,4-diazaboretane (X-ray) in Fig. 3 [174], dimethylcyclotetrazenoborane (electron diffraction) in Fig. 4 [79], dimethylaminodimethylborane (low temperature X-ray) in Fig. 5 [59], trisdimethylaminoborane (electron diffraction) in Fig. 6 [80], disilylaminodifluoroborane (electron diffraction) in Fig. 7 [346], tris-(*o*-phenylenedioxyboryl)amine (X-ray) in Fig. 8 [57], 1,8,10,9-triazaboradecalin (X-ray) in Fig. 9 [58], and dichlorodimethylaminoborane (electron diffraction) in Fig. 10 [81]. In the substituted cyclobutadiene analogue (Fig. 3), the shortness

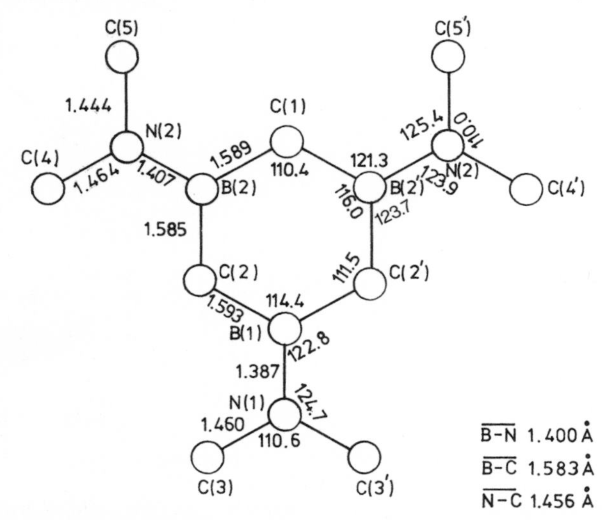

Fig. 2 – Molecular structure of $(CH_2BNMe_2)_3$. *[Reproduced, with permission, from Hess, H.,* Acta Cryst., *1969,* **25B**, *2334.]* For clarity, H atoms are omitted.

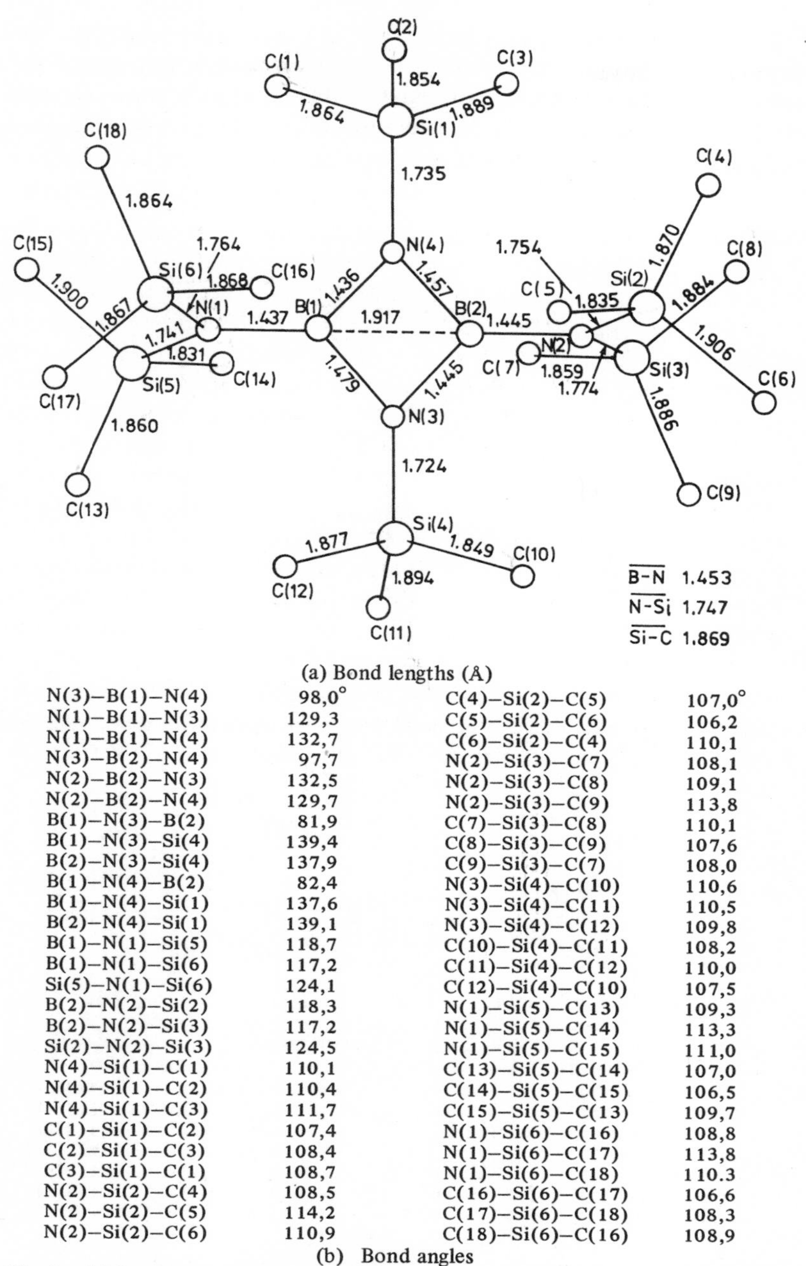

(a) Bond lengths (Å)

N(3)–B(1)–N(4)	98,0°	C(4)–Si(2)–C(5)	107,0°
N(1)–B(1)–N(3)	129,3	C(5)–Si(2)–C(6)	106,2
N(1)–B(1)–N(4)	132,7	C(6)–Si(2)–C(4)	110,1
N(3)–B(2)–N(4)	97,7	N(2)–Si(3)–C(7)	108,1
N(2)–B(2)–N(3)	132,5	N(2)–Si(3)–C(8)	109,1
N(2)–B(2)–N(4)	129,7	N(2)–Si(3)–C(9)	113,8
B(1)–N(3)–B(2)	81,9	C(7)–Si(3)–C(8)	110,1
B(1)–N(3)–Si(4)	139,4	C(8)–Si(3)–C(9)	107,6
B(2)–N(3)–Si(4)	137,9	C(9)–Si(3)–C(7)	108,0
B(1)–N(4)–B(2)	82,4	N(3)–Si(4)–C(10)	110,6
B(1)–N(4)–Si(1)	137,6	N(3)–Si(4)–C(11)	110,5
B(2)–N(4)–Si(1)	139,1	N(3)–Si(4)–C(12)	109,8
B(1)–N(1)–Si(5)	118,7	C(10)–Si(4)–C(11)	108,2
B(1)–N(1)–Si(6)	117,2	C(11)–Si(4)–C(12)	110,0
Si(5)–N(1)–Si(6)	124,1	C(12)–Si(4)–C(10)	107,5
B(2)–N(2)–Si(2)	118,3	N(1)–Si(5)–C(13)	109,3
B(2)–N(2)–Si(3)	117,2	N(1)–Si(5)–C(14)	113,3
Si(2)–N(2)–Si(3)	124,5	N(1)–Si(5)–C(15)	111,0
N(4)–Si(1)–C(1)	110,1	C(13)–Si(5)–C(14)	107,0
N(4)–Si(1)–C(2)	110,4	C(14)–Si(5)–C(15)	106,5
N(4)–Si(1)–C(3)	111,7	C(15)–Si(5)–C(13)	109,7
C(1)–Si(1)–C(2)	107,4	N(1)–Si(6)–C(16)	108,8
C(2)–Si(1)–C(3)	108,4	N(1)–Si(6)–C(17)	113,8
C(3)–Si(1)–C(1)	108,7	N(1)–Si(6)–C(18)	110.3
N(2)–Si(2)–C(4)	108,5	C(16)–Si(6)–C(17)	106,6
N(2)–Si(2)–C(5)	114,2	C(17)–Si(6)–C(18)	108,3
N(2)–Si(2)–C(6)	110,9	C(18)–Si(6)–C(16)	108,9

(b) Bond angles

Fig. 3 – Molecular structure of $[Me_3SiNBN(SiMe_3)_2]_2$. *[Reproduced, with permission, from Hess, H.,* Acta Cryst., *1969,* **25B**, *2342.]* For clarity, H atoms are omitted.

of the exocyclic B–N bond lengths (1.44 Å) is somewhat surprising since the substituents on boron and nitrogen are arranged nearly perpendicular to one another. The structure (X-ray) of $Ph_2C{=}NB(mesityl)_2$ has been determined, with B–N 1.40 Å, C=N 1.31 Å, and $B\widehat{N}C$ 170° [60]; the molecule thus has pseudo-allene geometry, implying $N{\rightleftharpoons}B$ bonding.

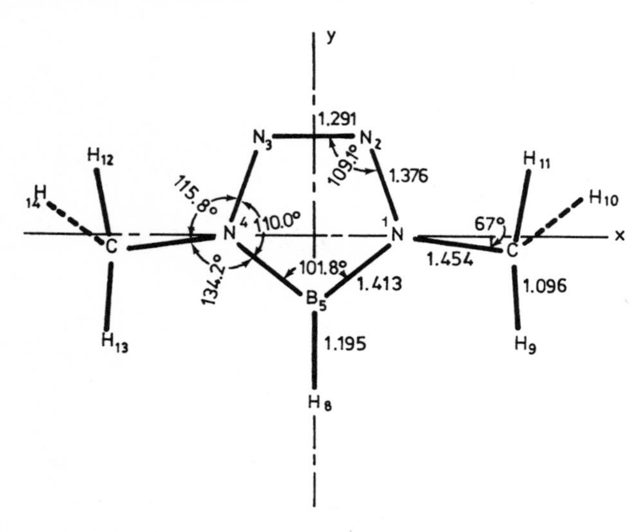

Fig. 4 – Molecular structure of Me_2N_4BH. *[Reproduced, with permission, from Chang, C. H., Porter, R. F., and Bauer, S. H.,* Inorg. Chem., *1969,* 8, *1677.]*

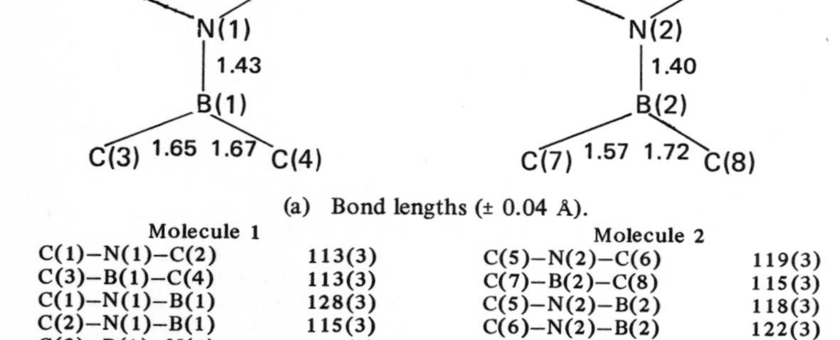

(a) Bond lengths (± 0.04 Å).

Molecule 1		Molecule 2	
C(1)–N(1)–C(2)	113(3)	C(5)–N(2)–C(6)	119(3)
C(3)–B(1)–C(4)	113(3)	C(7)–B(2)–C(8)	115(3)
C(1)–N(1)–B(1)	128(3)	C(5)–N(2)–B(2)	118(3)
C(2)–N(1)–B(1)	115(3)	C(6)–N(2)–B(2)	122(3)
C(3)–B(1)–N(1)	123(3)	C(7)–B(2)–N(2)	122(3)
C(4)–B(1)–N(1)	124(3)	C(8)–B(2)–N(2)	122(3)

(b) Bond angles (± 3°).

Fig. 5 – Molecular structure of Me_2BNMe_2. *[Reproduced, with permission, from Bullen, G. J., and Clark, N. H.,* J. Chem. Soc. (A), *1970, 992.]* For clarity, H atoms are omitted.

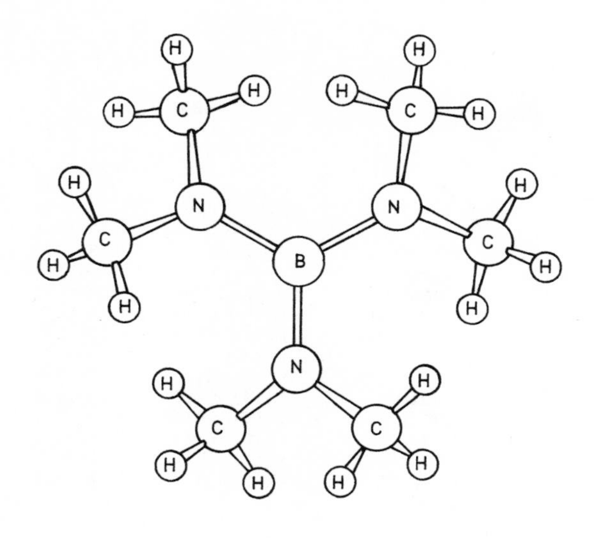

B–N	1.431 (12) Å
N–C	1.454 (7) Å
C–H	1.109 (12) Å
∠BNC	123.9° (1.0°)
∠NCH	111.1° (1.9°)
$-N(CH_3)_2$ twist	32.8° (2.6°)
$-CH_3$ twist	10.0° (4.1°)

Fig. 6 – Molecular structure of $B(NMe_2)_3$. *[Reproduced, with permission, from Clark, A. H., and Anderson, G. A.,* Chem. Comm., *1969, 1082.]*

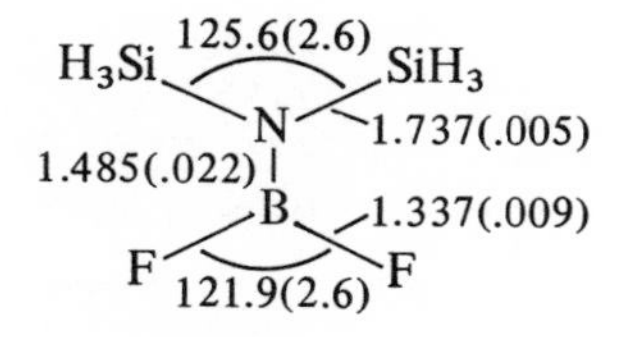

Fig. 7 – Molecular structure of $F_2BN(SiH_3)_2$. *[Reproduced, with permission, from Robiette, A. G., Sheldrick, G. M., Sheldrick, W. S., Beagley, B., Cruickshank, D. W. J., Monaghan, J. J., Aylett, B. J., and Ellis, I. A.,* Chem. Comm., *1968, 909.]*

B(1)–N	1.440(11)	C(1)–C(2)	1.369(13)
B(2)–N	1.438(11)	C(2)–C(3)	1.400(12)
B(3)–N	1.437(14)	C(3)–C(4)	1.385(12)
		C(4)–C(5)	1.411(14)
		C(5)–C(6)	1.376(10)
B(1)–O(1)	1.385(13)	C(6)–C(1)	1.377(13)
B(1)–O(2)	1.377(11)	C(7)–C(8)	1.386(10)
B(2)–O(3)	1.392(12)	C(8)–C(9)	1.407(14)
B(2)–O(4)	1.366(10)	C(9)–C(10)	1.385(15)
B(3)–O(5)	1.394(10)	C(10)–C(11)	1.391(11)
B(3)–O(6)	1.373(10)	C(11)–C(12)	1.394(13)
		C(12)–C(7)	1.328(12)
		C(13)–C(14)	1.397(13)
C(1)–O(1)	1.391(9)	C(14)–C(15)	1.374(13)
C(6)–O(2)	1.375(10)	C(15)–C(16)	1.385(11)
C(12)–O(3)	1.379(9)	C(16)–C(17)	1.413(13)
C(7)–O(4)	1.410(10)	C(17)–C(18)	1.360(12)
C(13)–O(5)	1.385(11)	C(18)–C(13)	1.386(10)
C(18)–O(6)	1.395(10)		

(a) Bond lengths.

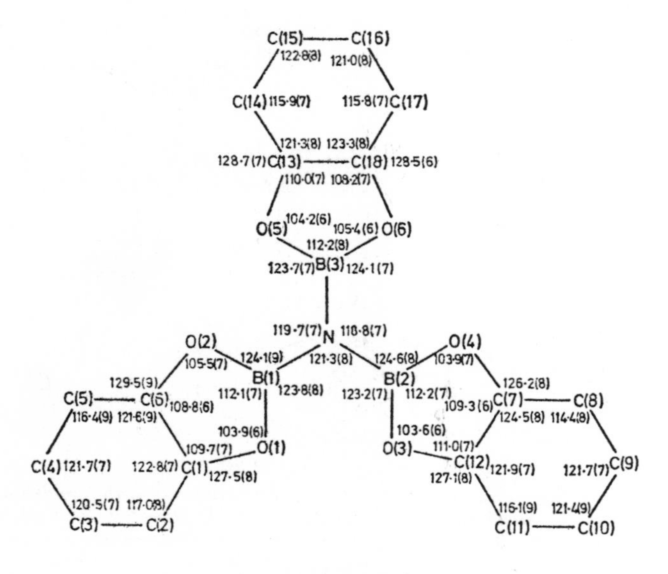

(b) Bond angles (°).

Fig. 8 – Molecular structure of $(o\text{-}C_6H_4O_2B)_3N$. *[Reproduced, with permission, from Bullen, G. J., and Clark, N. H.,* J. Chem. Soc. (A), *1970, 2213.]* For clarity, H atoms are omitted.

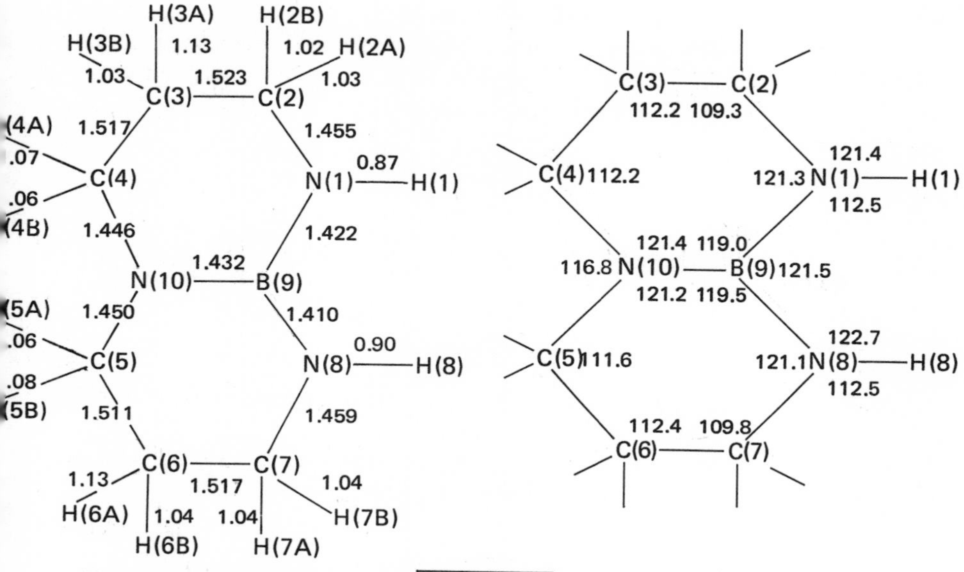

Fig. 9 – Molecular structure of $(CH_2)_3NH{-}B{-}N(CH_2)_3NH$. *[Reproduced, with permission, from Bullen, G. J., and Clark, N. H.,* J. Chem. Soc. (A), *1969, 58.]* For clarity, H atoms are omitted.

Parameter	r_g	l_g
C–H	1.119 ± 0.006	0.079 ± 0.008
B–N	1.379 ± 0.006	0.035 ± 0.015
C–N	1.465 ± 0.004	0.038 ± 0.005
Cl–B	1.770 ± 0.004	0.048 ± 0.003
∠out-of-plane[b]	0.0 ± 3	
∠Cl–B–N	122.1 ± 0.3	
∠H–C–N	109.7 ± 1	
∠C–N–B	123.2 ± 0.4	
C···C	2.450	0.067 ± 0.010
B···C[c]	2.502	0.067 ± 0.010
Cl···Cl	3.000	0.065 ± 0.005
N···Cl	2.762	0.056 ± 0.008
$(C\cdots Cl)_{short}$[d]	3.134	0.066 ± 0.008
$(C\cdots Cl)_{long}$[d]	4.144	0.068 ± 0.010

[a]Distances in ångströms. Angles in degrees. Estimated uncertainties include the effects of known systematic errors and random errors inferred from least-squares analyses according to L.S. Bartell in "Physical Methods in Chemistry," A. Weissberger and B.W. Rossiter, Ed., 4th ed, Interscience, New York, N.Y., 1971. The interaction between random and systematic errors as expressed in terms of eq 27 in this reference has not, however, been included. Hence, the standard errors listed are probably overoptimistic, particularly for the angles. [b]Angle between the planes C–N–B and Cl_2BN. [c]$l_g(B\cdots C)$ constrained to equal $l_g(C\cdots C)$. [d]$l_g(C\cdots Cl)_{short}$ and $l_g(C\cdots Cl)_{long}$ constrained to equal $l_g(N\cdots Cl)$ plus 0.010 and 0.012 Å, respectively.

Fig. 10 – Molecular parameters for Cl_2BNMe_2. *[Reproduced, with permission, from Clippard, F. B., and Bartell, L. S.,* Inorg. Chem., *1970,* **9**, *2439.]*

Dichlorodiphenylaminoborane (Fig. 11, X-ray) has a nearly planar Cl_2BNC_2 skeleton with the Ph rings tilted out of the plane; the B–N bond length, 1.380 Å, [415] is very similar to that in Cl_2BNMe_2 (see above), but significantly shorter than the 1.402 Å in gaseous F_2BNH_2 (Fig. 12) [228]. Three further X-ray structures:

Me Me Me Me Me
N N–N B N B
$(B$ N $)_2$ [130b], Me_2NB $BNMe_2$ [185a], and S N S [329a]
N–N B N B
Me Me Me Me Me

are in Figs. 13-15, respectively; in Et_2NB(S)(S)$BNEt_2$ the nitrogen co-ordination is trigonal planar and set at $< 10°$ to the ring plane with mean BN and BS bond lengths of 1.38(5) and 1.84(4) Å, respectively [71a].

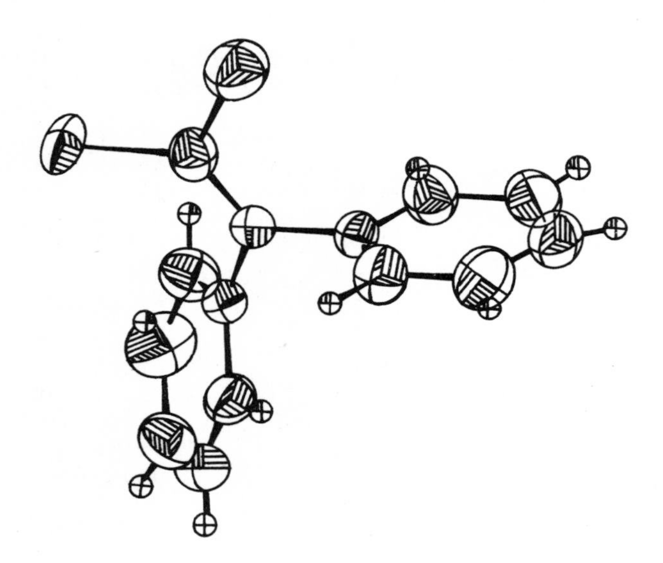

Fig. 11 – Molecular structure of Cl_2BNPh_2. *[Reproduced, with permission, from Zettler, F., and Hess, H.,* Chem. Ber., *1975,* **108,** *2269.]*

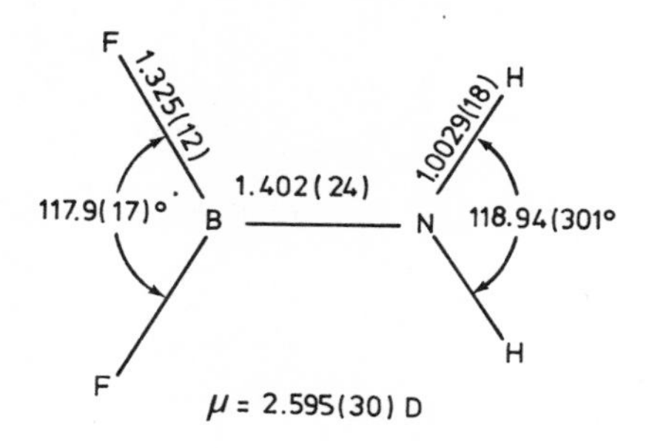

Fig. 12 – Molecular structure and electric dipole moment for F_2BNH_2. *[Reproduced, with permission, from Lovas, F. J., and Johnson, D. B.,* J. Chem. Phys., *1973,* **59**, *2347.]*

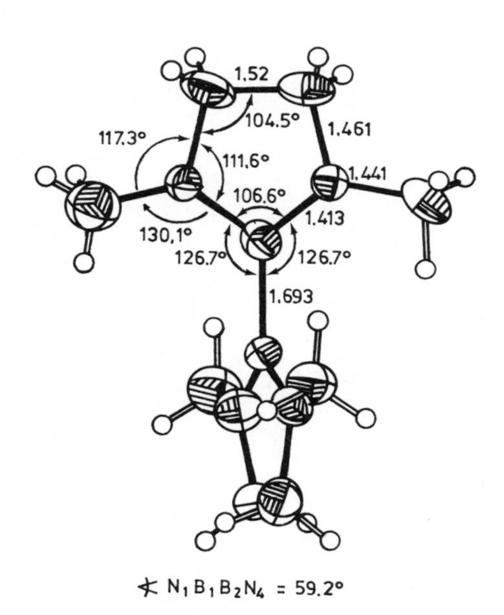

Fig. 13 – Molecular structure of ($B(NMe)_2C_2H_4$ ring)$_2$. *[Reproduced, with permission, from Fussstetter, H., Huffman, J. C., Nöth, H., and Schaeffer, R.,* Z. Naturforsch., *1976,* **31b**, *1441.]*

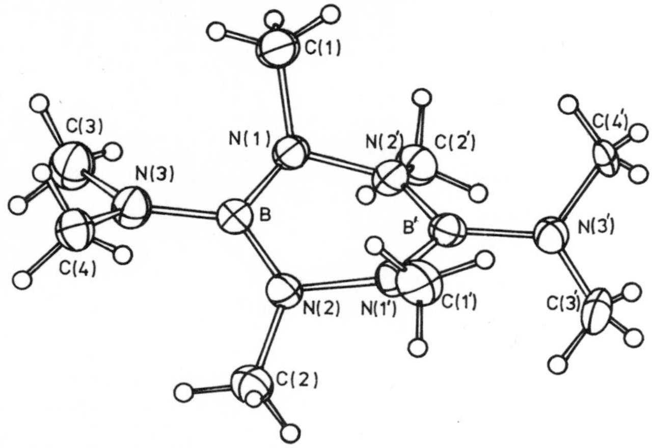

Bond distance	[Å]	Bond angles			[°]				
B–N(1)	1.422(3)	N(1)	B	N(2)	117.5(2)	N(2)	C(2)	H(6)	113.8
B–N(2)	1.432(3)	N(1)	B	N(3)	121.9(2)	H(4)	C(2)	H(5)	109.4
B–N(3)	1.430(3)	N(2)	B	N(3)	120.5(2)	H(4)	C(2)	H(6)	105.4
N(1)–N(2)	1.431(2)	B	N(1)	N(2)	118.5(2)	H(5)	C(2)	H(6)	104.6
N(1)–C(1)	1.448(3)	B	N(1)	C(1)	126.6(2)	N(3)	C(3)	H(7)	111.3
N(2)–C(2)	1.443(3)	N(2)	N(1′)	C(1′)	113.1(2)	N(3)	C(3)	H(8)	109.1
N(3)–C(3)	1.455(3)	B	N(2)	N(1′)	117.5(2)	N(3)	C(3)	H(9)	110.3
N(3)–C(4)	1.444(3)	B	N(2)	C(2)	124.8(2)	H(7)	C(3)	H(8)	108.5
C(1)–H(1)	0.978(23)	N(1)	N(2)	C(2)	112.6(2)	H(7)	C(3)	H(9)	108.0
C(1)–H(2)	0.982(26)	B	N(3)	C(3)	123.0(2)	H(8)	C(3)	H(9)	109.6
C(1)–H(3)	0.984(26)	B	N(3)	C(4)	123.9(2)	N(3)	C(4)	H(10)	109.8
C(2)–H(4)	0.965(25)	C(3)	N(3)	C(4)	112.7(2)	N(3)	C(4)	H(11)	112.6
C(2)–H(5)	0.986(23)	N(1)	C(1)	H(1)	111.6(12)	N(3)	C(4)	H(12)	113.4
C(2)–H(6)	1.012(27)	N(1)	C(1)	H(2)	110.2(12)	H(10)	C(4)	H(11)	107.3
C(3)–H(7)	1.037(25)	N(1)	C(1)	H(3)	109.2(14)	H(10)	C(4)	H(12)	107.6
C(3)–H(8)	1.012(22)	H(1)	C(1)	H(2)	107.0(18)	H(11)	C(4)	H(12)	105.8
C(3)–H(9)	0.994(25)	H(1)	C(1)	H(3)	110.3(19)				
C(4)–H(10)	0.979(24)	H(2)	C(1)	H(3)	108.4(20)				
C(4)–H(11)	0.979(24)	N(2)	C(2)	H(4)	110.9(14)				
C(4)–H(12)	1.024(26)	N(2)	C(2)	H(5)	112.3(13)				

Dihedral angles [°]

C(1)	N(1)	B	N(2)	— 146.2	N(1)	N(2′)	B′	N(3′)	— 164.9
C(1)	N(1)	B	N(3)	34.5	B	N(1)	N(2′)	C(2′)	122.3
C(1)	N(1)	N(2′)	C(2′)	— 72.1	B	N(1)	N(2′)	B′	— 33.7
C(1)	N(1)	N(2′)	B′	132.0	N(2)	N(1′)	B′	N(2′)	17.3
N(1)	B	N(2)	C(2)	— 136.9	N(2)	N(1′)	B′	N(3′)	— 161.9
N(1)	B	N(2)	N(1′)	15.9	N(2)	B	N(3)	C(4)	27.7
N(1)	B	N(3)	C(3)	34.2	N(2)	B	N(3)	C(3)	— 145.0
N(1)	B	N(3)	C(4)	— 153.1	N(3)	B	N(2)	C(2)	42.4

Fig. 14 – Molecular structure of $Me_2NB\langle(N(Me)\!-\!N(Me))_2\rangle BNMe_2$. *[Reproduced, with permission, from Huffman, J. C., Fussstetter, H., and Nöth, H.,* Z. Naturforsch., *1976,* **31b**, *289.]*

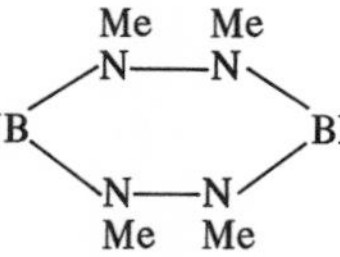

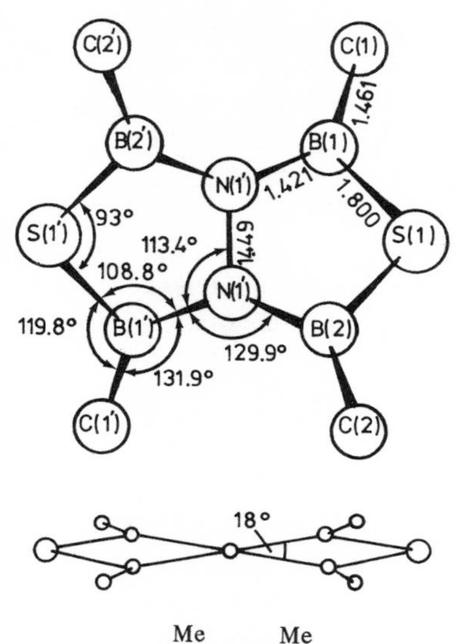

Fig. 15 – Molecular structure of
Me Me
B N B
S | S.
B N B
Me Me
[Reproduced, with permission, from Nöth., and Ullmann, R., Chem. Ber., *1975,* **108**, *3125.]*

D ALUMINIUM AMIDES*

1. Introduction

Compounds with Al–N bonds have been reviewed in the context of organo-aluminium chemistry [16, 52, 71a], while selected aspects have been discussed elsewhere [18f, 54, 107]. The subject has blossomed during the past decade or so. There are considerable experimental difficulties due to the extreme hydrolytic sensitivity of many of the compounds. The foundations are now firmly laid, especially with the availability of much structural data (see Section D.4). In the equations and tables, empirical formulae (i.e., of monomers) will usually be written; this is not intended to have structural implications.

By contrast to B–N compounds, there are only few examples of molecules in which the metal and N are both three-co-ordinate. These include the sterically hindered aminoalanes, such as $Al(NPr^i_2)_3$ and $Al[N(SiMe_3)_2]_3$. Other aminoalanes are dimers, trimers, oligomers, or polymers (see Section D.4), formed as a result of intermolecular dative bonding between monomer molecules, which have donor (N) as well as acceptor (Al) sites. A common route to Al–N compounds is shown in Eq. (48); this illustrates the point that simple Al–N compounds fall into various classes: the amine adducts, the aminoalanes, the type RAlNR', and

*Numbers in [] refer to the bibliography at the end of the Chapter, p. 225.

aluminium nitride, as well as (where appropriate) their oligomers or polymers. That monomeric aminoalanes are far less commonly found than corresponding aminoboranes (see Section A) is largely attributable to the smaller size of boron compared with aluminium. Elimination reactions of the type shown in Eq. (48) are progressively more ready in the series $B < Al > Ga$ [7].

With reference to Eq. (48), the subsequent treatment deals successively with the two reagent classes aminoalanes and poly(*N*-alkylimino)alanes, but the amine adducts are outside our terms of reference.

$$R_3Al + H_2NR' \longrightarrow \underset{\text{Alane-amine adducts}}{R_3Al.NH_2R'} \xrightarrow{-RH} \underset{\substack{\text{Aminoalanes} \\ (n = \text{usually 2 or 3})}}{(R_2AlNHR')_n}$$

$$\xrightarrow{-RH} \underset{\substack{\text{Poly(}N\text{-alkylimino)alanes} \\ (n = \geqslant 4)}}{(RAlNR')_n} \xrightarrow[\text{(if } R'=H)]{-RH} (AlN)_n \qquad (48)$$

In the past ten years there has been considerable interest in the polyimino-alanes of the general formula $(RAlNR')_n$ [Eq. (48)], which in many cases possess a cage-like structure [18i, 19a, 19c, 62a]. These compounds are strictly beyond the scope of this book. Nevertheless a brief discussion and a comprehensive series of references to these compounds is included, e.g., $\lbrack(H)AlN(Ph)\rbrack_n$ or $\lbrack H_2AlNPh_2\rbrack_n$ from $Li[AlH_4]$ and respectively $H_2NPh.HCl$ or $HNPh_2.HCl$ [53a].

Outside our present cover (but see Section D.4) are the ketimido derivatives,* which may be monomeric as in $Al(N{=}CR_2)_3$ (R = Ph or Bu^t) or dimeric as in $(Me_2AlN{=}CMeBu^t)_2$ [108], $(Cl_2AlN{=}CH_2)_2$ [100, 101], $[(MeN{=}CCl)_2$-$Al{-}N(Me){=}C(Cl)]_2$ [70], or $(Et_2AlN{=}CArNMe_2)_2$ [38].† Somewhat related to these compounds are phosphine-imido derivatives $(Me_2M{-}N{=}PR_3)_2$ (M = Al, Ga, or In; R = Me, or Et), $Me_2M(NSiMe_3)_2PPh_2$, $[Me_2M(N{=}PR_3)_2SiMe_2]^+$, and (LXIII) [94, 113]. Also recorded are a few examples of hydrazinoalanes [15, 22]. Data are in Tables 15-20.

The puckered 8-membered ring (LXIV) may describe the structure of $[Me_2AlOC(Ph){:}NPh]_2$ obtained from $Me_3Al/PhCONHPh$ [47]; similar structures may serve for the insertion products of PhNCX (X = O or S) into Al–X′ (X′ = Cl, Br, or Et) bonds [42], and for $(Me_2MONSMe_2)_2$ (M = Al or Ga) (from Me_3M/Me_2SNOH) [92]. However, the products from $R_3Al/R'NHCOR''_2$ are trimers in the crystalline form [44].

Al–N compounds have found applications as polymerisation initiators [18a, 18e, 102, 115], e.g., of isoprene by $(HAlNR)_n$-$TiCl_4$ [69], or C_2H_4 [121] (Chapter 17), reducing agents [13], and synthetic intermediates [5b, 6b, 41a, 73b, 78].

* See refs. [43, 58, 98].
† Or $(Ph_2AlN{=}CPhC_6H_4Br)_2.2C_6H_6$ [63].

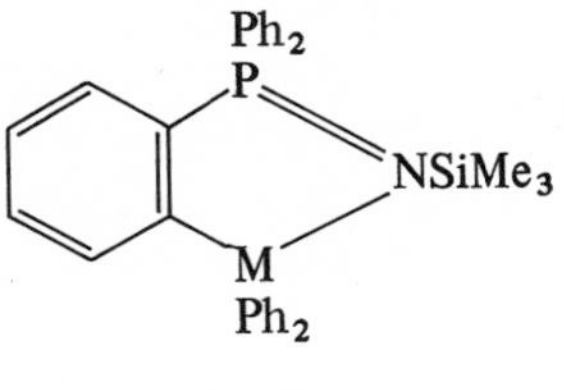

Ph$_2$ P NSiMe$_3$ M Ph$_2$

(LXIII)
(M = Al, Ga, or In)

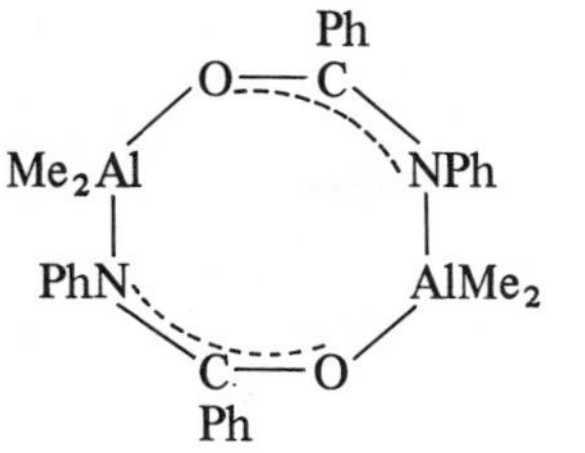

Ph O=C Me$_2$Al NPh PhN AlMe$_2$ C=O Ph

(LXIV)

Aminoalanes have typical properties of metal amides, as discussed in Chapters 10–18. Not considered there are the properties of primary amine derivatives. Compounds having $>Al-N(H)<$ [Eq. (50)], or $(H)>Al-N<$ [Eq. (49)] frameworks are additionally characterised by their H^- displacement reactions [e.g., Eq. (49)] [20] or elimination reactions [e.g., Eq. (50)] [55]. Displacement of one NRR′ on Al by another (transamination) is essentially governed by steric effects, with the order $H_2N > HMeN > HBu^tN > Me_2N$ for reactions of $(Et_2AlNRR')_n$ [31]. In $Al[N(SiMe_3)_2]_3$, the Al–N bonds are hydrolytically cleaved in preference to the Si–N bonds [79].

$$H_2AlNMe_2 + 2HCl \longrightarrow Cl_2AlNMe_2 + 2H_2 \qquad (49)$$

$$Et_2AlCl \xrightarrow{HNMe_2} Et_2AlCl.NH_2Me \xrightarrow[-C_2H_6]{heat} \tfrac{1}{2}[Et(Cl)AlNHMe]_2 \xrightarrow[-C_2H_6]{210\text{–}220^\circ C} (ClAlNMe)_n \qquad (50)$$

2. Synthesis of aminoalanes

Aminoalanes were first prepared in 1925 by the reaction of an amine with an alkylaluminium sesqui-iodide, Eq. (51) [57].

$$2Al \xrightarrow{3RI} R_3Al_2I_3 \xrightarrow{HNRR'/Et_2O} I_3Al_2(NRR')_3.OEt_2 + 3RH \qquad (51)$$

The principal method of synthesis of aminoalanes is by alkane elimination from an ammonia- or amine-alane adduct, illustrated in Eqs. (52) [18], (53) [19], (54) [31], and (55) [55]. The loss of the first alkane molecule requires mild conditions. Where the initial product contains the $>Al-N(H)-$

$$Et_3Al.NH_3 \xrightarrow{25^\circ C} Et_2AlNH_2 + C_2H_6 \qquad (52)$$

$$Me_3Al.NHMe_2 \xrightarrow{100^\circ C} Me_2AlNMe_2 + CH_4 \qquad (53)$$

$$Et_3Al.NH_2Bu^t \xrightarrow{100°C} Et_2AlNHBu^t + C_2H_6 \quad (54)$$

$$EtAlCl_2.NH_2Me \xrightarrow{130°C} Cl_2AlNHMe + C_2H_6 \quad (55)$$

linkage, further elimination may take place at elevated temperature to afford the polymeric imido-derivative Eq. (56) [31]. In general, use of an aryl- rather

$$4Et_2AlNHBu^t \xrightarrow{270°C} (EtAlNBu^t)_4 + 4C_2H_6 \quad (56)$$

than alkyl-amine makes loss of alkane more facile, e.g., Eq. (57) [26]. This may be due to the greater Brønsted acidity of the former. For complexes $Et_2(X)Al.NHMe_2$ or $Et_2(X)Al.NH_2Bu^t$, ease of C_2H_6 elimination follows the sequence $X = Et < Cl < Br < I$ [29a], and decomposition of $Et_2(Cl)Al.NHMe_2$ follows zeroth order kinetics suggestive of surface catalysis. Use of Ph_3Al and a diamine yields benzene [14].

$$Et_3Al + H_2NPh \xrightarrow{-78°C} Et_2AlNHPh + C_2H_6 \quad (57)$$

Hydrogen elimination is widely employed to prepare aminoalanes, e.g., Eq. (58) [5], (59) [85], and (60) [67]. This reaction proceeds with greater

$$Me_2AlH + HNMe_2 \longrightarrow Me_2AlNMe_2 + H_2 \quad (58)$$

$$H_3Al.NMe_3 + 3HNMe_2 \longrightarrow Al(NMe_2)_3 + 3H_2 + NMe_3 \quad (59)$$

$$H_2AlCl.NMe_3 + H\overline{N(CH_2)_5} \longrightarrow HAl(Cl)\overline{N(CH_2)_5} + H_2 + NMe_3 \quad (60)$$

facility than the alkane elimination. The yield of H_2 is quantitative, a feature often used as a means of estimating the H-Al content of an organo-aluminium compound, e.g., using aqueous ammonia [122]. In the reaction between H_3Al and amines, the formation of poly(iminoalane) can be monitored by measurement of electrical conductivity [19e].

Salt elimination [e.g., Eq. (61)] [36, 89], metathetical exchange [e.g., Eq. (62)] [85], or insertion reactions [e.g., Eq. (63)] [75] {or + Bu^i_2AlH/RCN [116a]} have also been used, as has transamination [e.g., Eq. (64)] [31, 85]. An interesting 'direct' method for preparing the alanes $H_{3-n}Al(NEt_2)_n$ ($n = 1$, 2, or 3), using moderate reaction conditions (e.g., 110°C at 4000 p.s.i. of H_2 in benzene), is illustrated in Eq. (65) [5, 5a].

$$LiAlH_4 + [Me_3NH]Cl \longrightarrow H_2AlNMe_2 + LiCl + H_2 + CH_4 \quad (61)$$

$$2R_3B + Al(NMe_2)_3 \longrightarrow 2R_2BNMe_2 + R_2AlNMe_2 \quad (62)$$

$$Bu^i_2AlH + 2N(\text{isoquinoline}) \longrightarrow Bu^i_2AlN(\text{dihydroisoquinoline}) \cdot N(\text{isoquinoline}) \quad (63)$$

$$[Al(NMe_2)_3]_2 + 2HNPr^i_2 \longrightarrow [(Me_2N)_2AlNPr^i_2]_2 + 2HNMe_2 \quad (64)$$

$$Al + \frac{3n}{2}H_2 + nHNEt_2 \longrightarrow H_{3-n}Al(NEt_2)_n + nH_2 \quad (65)$$

Alkyl groups from aluminium and nitrogen may couple as in Eq. (66) [117].

$$Et_3Al + CH_2{:}CHCH_2NEt_2 \longrightarrow Et_2AlNEt_2 + CH_2{:}CHCH_2Et \quad (66)$$

The cleavage of one Al–N bond in compounds containing more than one such bond has been employed to prepare mixed halogeno(amino)alanes [e.g., Eq. (67)] [26].

$$EtAl(NHPh)_2 + 2HCl \longrightarrow EtAlClNHPh + [PhNH_3]Cl \quad (67)$$

Redistribution reactions have been used to synthesise mixed derivatives [e.g., Eq. (68)] [41b].

$$2Et_2AlNMePh + AlCl_3 \longrightarrow 2EtAlClNMePh + Et_2AlCl \quad (68)$$

Other redistribution reactions include that between $H_3Al.2NMe_3$ and $B_2(NMe_2)_4$ to give various boron compounds and $Al_3B_3(NMe_2)_7H$ [34], Al_2Me_6 and $B_2(NMe_2)_4$ to give $Al_3Me_3(NMe_2)$ [95], $Al_4B(NMe_2)_3Me_6$ [94a], or $B_6(NMe_2)_{12}$-Al_6Me_{12} [12a], and Al_2Cl_6 and $B_2(NMe_2)_4$ to give a cage compound (X-ray) [106]). Further redistribution reactions between B_2H_6 and an aminoalane are described elsewhere [49, 50, 51]. From Al_2Me_6 and $(Me_2NBH_2)_2$, an interesting compound $Al_4Me_8(NMe_2)_2H_2$ is among the products [27], whereas excess of $HB(NMe_2)_2$ and Al_2Me_6 yields the product of Eq. (69) [31a].

$$Me_3Al + HB(NMe_2)_2 \longrightarrow H_2B\left\langle\begin{matrix} N(Me_2){-}Al(Me_2) \\ N(Me_2){-}Al(Me_2) \end{matrix}\right\rangle NMe_2 \quad (69)$$

The reactions of an amine with an aluminium oxy (or *S*) halide yields an amino-oxyalane or an *S*- analogue [e.g., Eq. (70)] [32].

$$Al(S)Cl + 2H_2NMe \longrightarrow SAlNHMe + [MeNH_3]Cl \quad (70)$$

Insertion reactions of isocyanates or isothiocyanates into an Al-C [e.g., Eq. (71)] [81] or Al-Br [e.g., Eq. (72)] [42] bond have yielded Al-N compounds.

$$Et_2AlCl + BuNCS \longrightarrow EtAl(Cl)[N(Bu)CSEt] \quad (71)$$

$$AlBr_3 + MeNCO \longrightarrow Br_2AlN(Me)COBr \quad (72)$$

An Al^I compound is thought to serve as a ligand in the dimeric complex $[\{Fe(Br)_2(CO)_3(AlNMe_2)\}_2]$ (see Chapter 16) and is formed from Br_2AlNMe_2 and $[Fe(CO)_5]$ [78k]; it reacts with a base, e.g., T.H.F. or PBu_3, to afford the monomeric $[Fe(Br)_2(CO)_3(AlNMe_2)base]$. Cleavage of bridging amido groups is indeed a characteristic reaction, as in Eq. (73) [82].

$$Me_2Al(\mu\text{-}NPh_2)_2AlMe_2 + 2L \xrightarrow[\text{or } SMe_2]{L = NMe_3,\ OEt_2,} 2Me_2AlNPh_2.L \quad (73)$$

3. Synthesis and properties of the poly(*N*-alkyliminoalanes)

Study of the title compounds has constituted one of the main growth areas in aluminium-nitrogen chemistry in recent years. These compounds were originally synthesised by Wiberg and May [110, 111], via reaction of AlH_3 with methylamine in ether, yielding $(HAlNMe)_n$. A linear polymeric structure was proposed [20a] for the compounds $(RAlNR')_n$ synthesised via reaction of an aluminium alkyl with a primary amine or the reaction of $LiAlH_4$ with $EtNH_2.HCl$.

The main synthetic routes are summarised by Eqs. (74)-(76). A comprehensive study of each method has been carried out [18i]; the product type is dependent on the method of preparation and the structure is largely determined by the nature of the alkyl groups. For example, in $(RAlNR')_n$ the degree of

$$AlH_3.NMe_3 + H_2NR \xrightarrow{Et_2O} \frac{1}{n}(HAlNR)_n + 2H_2 + NMe_3 \quad (74)$$

$$LiAlH_4 + RNH_2.HCl \xrightarrow{Et_2O} \frac{1}{n}(HAlNR)_n + 3H_2 + LiCl \quad (75)$$

$$AlR_3.NMe_3 + H_2NR' \xrightarrow{\text{heat}} \frac{1}{n}(RAlNR')_n + 2RH \qquad (76)$$

oligomerisation (n) varies as follows: when (i) $R = H$ and $R' = Bu^t$, $n = 4$, (ii) $R = H$ or Me and $R' = Pr^i$, $n = 4$ or 6, (iii) $R = Pr$ or Bu and $R' = H$, $n = 6$, (iv) $R = R' = Me$, $n = 7$ or 8, and (v) $R = H$ and $R' = Et$, $n = 8$ (for further details, see Table 16).

New methods of synthesis of the poly(*N*-iminoalanes) have recently been reported: (i) the interaction of aluminium metal and hydrogen with an amine, Eq. (77) [18j] leads to a high yield of products, (ii) the reaction of $LiAlH_4$ with a primary amine H_2NR (R = Et, Pr, Pr^i, Bu, Bu^s, Bu^i, Bu^t, or C_6H_{11}) [Eq. (78)] [18d].

$$nAl + nH_2NR \xrightarrow{H_2} \frac{1}{n}(HAlNR)_n + \frac{n}{2}H_2 \qquad (77)$$

$$LiAlH_4 + H_2NR \longrightarrow \frac{1}{n}(HAlNR)_n + LiH + 2H_2 \qquad (78)$$

The polymeric Al–N bonded compounds have also been shown to act as reducing agents [26a] and polymerisation catalysts [69].

4. Physical properties and structures

Whereas most tris(amino)boranes are monomeric, some tris(amino)alanes, e.g., $Al(NMe_2)_3$ and $(Me_2N)_2AlNPr^i_2$, are reported to be dimeric. However, $Al(NMe_2)_3$ has also been stated to exist as a monomer (ebullioscopy in OEt_2) [111], as is $Al(NPh_2)_3$. Several mono- or bis-(amino)alanes have been described as dimers, and $HAl(NMe_2)_2$ has variously been reported as a mixture of monomer and dimer [111], or of dimer and trimer [85]. An adduct, $HAl(NMe_2)_2.2Me_2NBH_2$, has been described [87]. Adducts of mono(amino)alanes with T.H.F. [116], ether [13, 116], and $TiCl_4$ or VCl_4 [64] are known. Aluminium is also four-co-ordinate in the ionic compounds $Li[Al(NR_2)_4]$ [46, 85] and $Li[AlH(NR_2)_3]$ [12, 59, 74], and in chelated aminoalanes $X_2\overline{AlN(R')(CH_2)_2N}R_2$ which at lower temperatures exist as *cis*- (preferred)/*trans*- mixtures [8].

The X-ray crystal structure of $[H_2AlNMe_2]_3$ is illustrated in Fig. 16 [97]. The dimethylaminoalane molecule is similar to a cyclohexane six-membered ring constructed from alternate Al and NMe_2 groups. The co-ordination at Al is distorted tetrahedral.

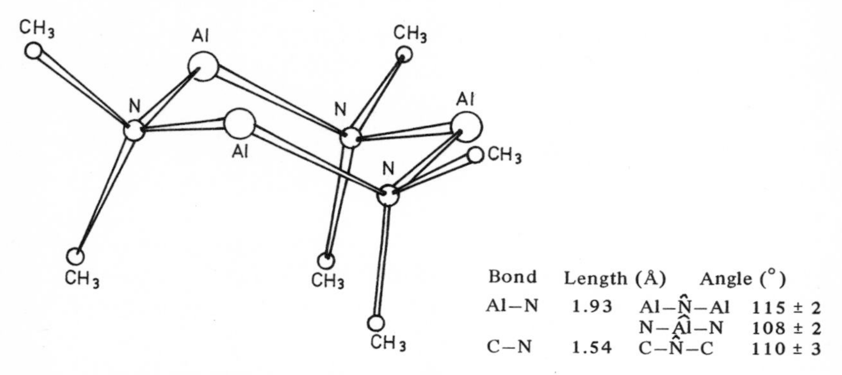

Bond	Length (Å)	Angle	(°)
Al–N	1.93	Al–$\hat{N}$–Al	115 ± 2
		N–$\hat{Al}$–N	108 ± 2
C–N	1.54	C–$\hat{N}$–C	110 ± 3

Fig. 16 – X-ray crystal structure of $[H_2AlNMe_2]_2$. *[Reproduced, with permission, from Semenenko, K. N., Lobkovskii, E. B., and Dorosinskii, A. L.,* J. Struct. Chem., *1972,* **13,** *696.]*

Stereoisomers of $(Me_2AlNHMe)_3$ have been characterised by ^1H n.m.r., as well as by X-ray crystallography (Fig. 17) [30]. The metastable monoclinic form (LXVI), which transforms to the stable rhombohedral form (LXV) (m.p. 110°C) at 20°C, has both axial and equatorial Me(NH), unlike the stable form. The mean molecular dimensions of (LXV), with the six-membered $(AlN)_3$ ring in chair conformation and equatorial *N*–Me groups, are Al–N, 1.953 ± 0.005 Å; Al–C, 1.980 ± 0.005 Å; N–C, 1.504 ± 0.010 Å; N–$\hat{Al}$–N, 102.1 ± 0.4°; C–$\hat{Al}$–C, 116.9 ± 0.4 Å; C–$\hat{Al}$–N, 109.4 ± 0.3°; Al–$\hat{N}$–Al, 122.3 ± 0.4°, and Al–$\hat{N}$–C, 108.6 ± 0.4°. The bond lengths and valency angles for (LXVI) are in good agreement with those in (LXV). $[Me_2AlN(Me)Ph]_2$ exists in benzene as an equilibrium mixture of *cis-* (preferred) and *trans-* dimer [108a, 108b]; n.m.r. provides ΔH and ΔS.

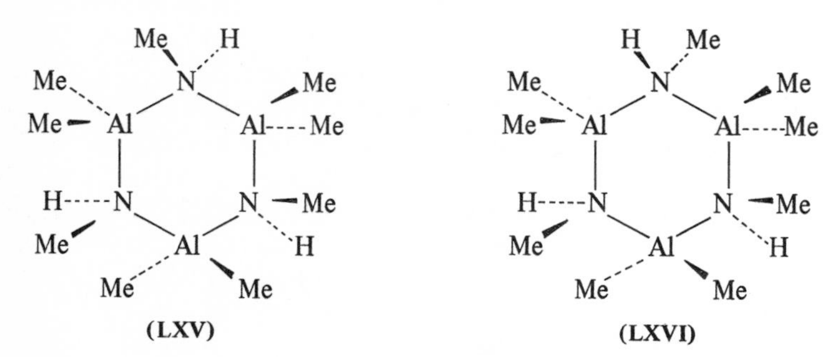

Fig. 17 – The chair (LXV) and skew-boat (LXVI) conformations for two isomers of $(Me_2AlNHMe)_3$. Torsion angles (°) about the Al–N bonds: (LXV), 53, –53, 53, –53, 53, –53; (LXVI), 35, –61, 20, 31, –62, 33. *[Reproduced, with permission, from Gosling, K., McLaughlin, G. M., Sim, G. A., and Smith, J. D.,* Chem. Comm., *1970, 1617; and McLaughlin, G. M., Sim, G. A., and Smith, J. D.,* J. C. S. Dalton, *1972, 2197.]*

Details of the crystal structures of the related ethyleneimino(dimethyl)alane trimer (chair conformation) [6] are in Fig. 18, and of dimethylamino(dimethyl)-alane dimer [35] in Fig. 19, which may be compared (Fig. 20) with the electron diffraction study of $(Cl_2AlNMe_2)_2$ [6a]. The compound $[(H_4B)_2Al\overline{N(CH_2)_2}]_2$ is a symmetrical dimer in the crystal (X-ray), with Al–N = 1.9 Å [97a]. The solid

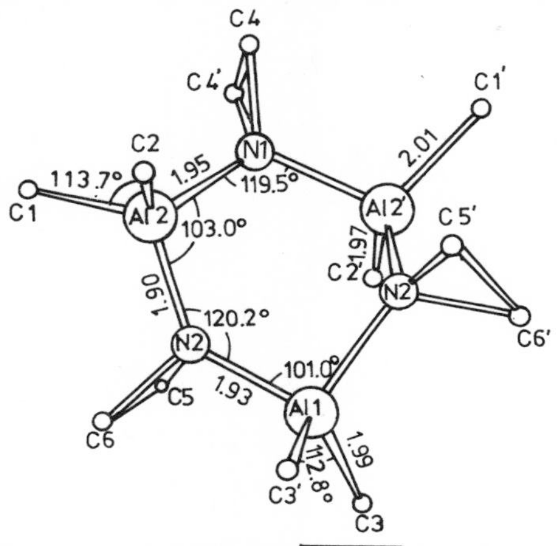

Fig. 18 – Molecular structure of $[Me_2Al\overline{N(CH_2)_2}]_3$. *[Reproduced, with permission, from Atwood, J. L., and Stucky, G. D.,* J. Amer. Chem. Soc., *1970,* **92,** *285.]* For clarity, H atoms are omitted.

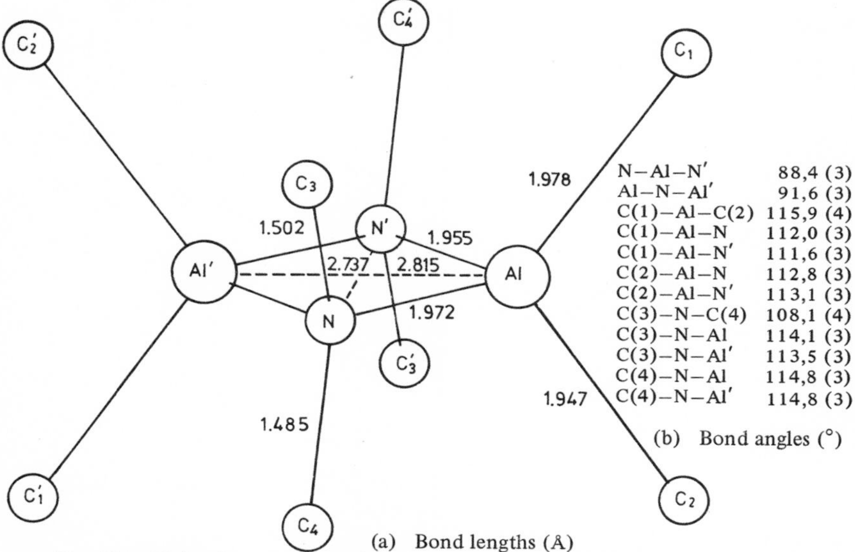

Fig. 19 – Molecular structure of $(Me_2AlNMe_2)_2$. For clarity, H atoms are omitted (see also ref. [60]). *[Reproduced, with permission, from Hess, H., Hinderer, A., and Steinhauser, S.,* Z. Anorg. Chem., *1970,* **377,** *1.]*

state (X-ray) structure of $(Cl_2AlNMe_2)_2$ [1a] is essentially identical to that of the vapour (Fig. 20) and is similar to that found for the crystalline $(X_2AlNMe_2)_2$ (X = Br or I) [1b].

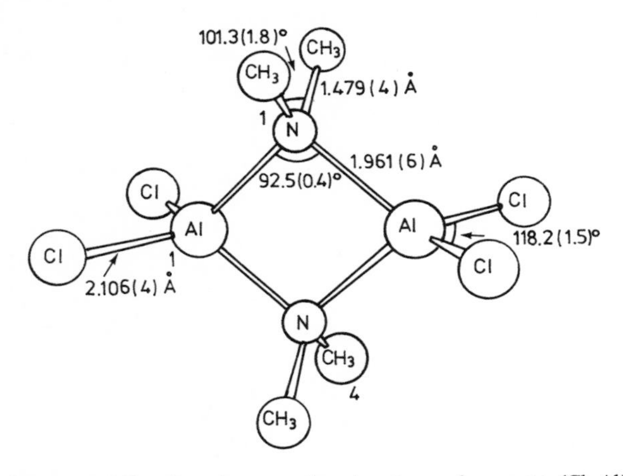

Fig. 20 – Electron diffraction data on the structure of gaseous $(Cl_2AlNMe_2)_2$. *[Reproduced, with permission, from Barthe, T. C., Haaland, A., and Novak, D. P.,* Acta Chim. Scand., *1975,* **29A,** *273.]*

Details of crystal structure determinations for other compounds are available, including the very interesting planar, monomeric tris[bis(trimethylsilyl)amino]-alane [99] (Fig. 21) which is isostructural with the Fe^{III} compound (see Chapter 8).

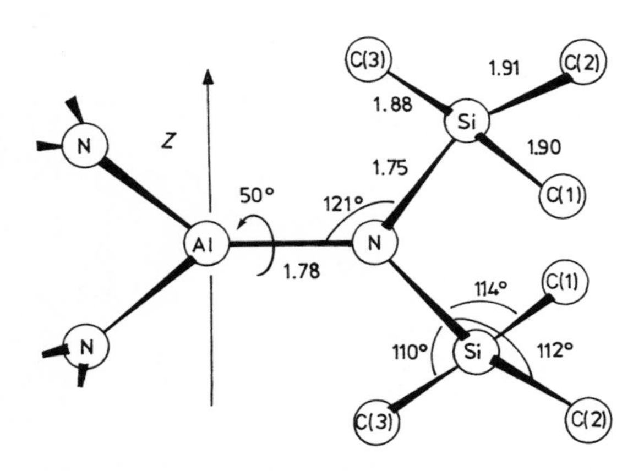

Fig. 21 – Molecular structure of $Al[N(SiMe_3)_2]_3$. For clarity, H atoms are omitted. *[Reproduced, with permission, from Sheldrick, G. M., and Sheldrick, W. S.,* J. Chem. Soc. (A), *1969, 2279.]*

The structure depicted in Fig. 22 provides a rare (by group 3 metal standards, although better known for Al) example of a molecule having NPh_2 and Me as bridging groups [61]; this arrangement is probably favoured here because $\bar{N}Ph_2$ is a relatively weak base (compared, say, with $\bar{N}Me_2$) and comparable with Me^-. Intramolecular Me exchange is a first-order process with an activation energy of 13.7 ± 0.45 kcal mol^{-1} at 290 K and probably involves a single bridge opening [82a], while with $GaMe_3$ the exchange is first-order in the Al compound and independent of $GaMe_3$ concentration.

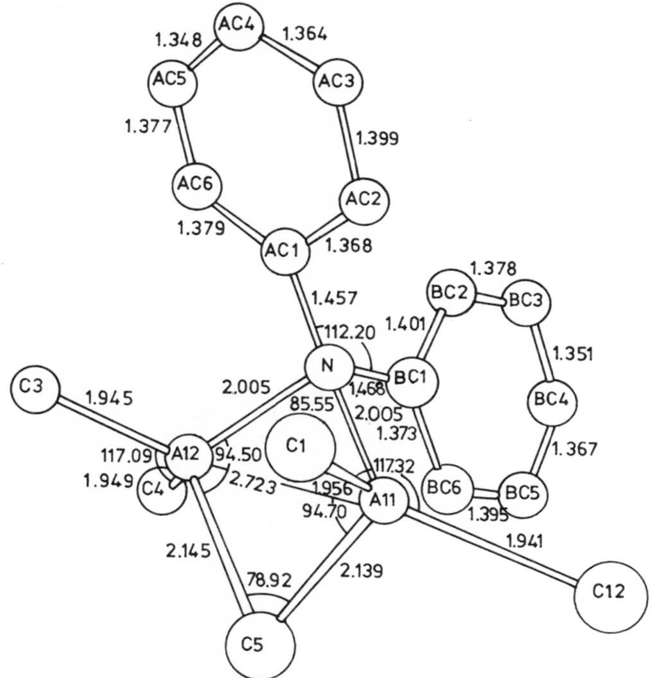

Fig. 22 – Molecular structure of $Me_5Al_2NPh_2$. For clarity, H atoms are omitted. *[Reproduced, with permission, from Magnusson, V. R., and Stucky, G. D.,* J. Amer. Chem. Soc., *1969,* **91,** *2544.]*

An aluminium oxamide $Me_2Al(CONMe)_2AlMe_2$ shows, by ^1H or ^{13}C n.m.r., as well as X-ray diffraction, interesting tautomerism, cf., Ga analogues (Section E.2) [23a].

One of the most interesting features of AlN chemistry has been the synthesis of several examples of molecules containing an aluminium-nitrogen cage framework. The simplest of these molecules are the tetrameric iminoalanes $(RAlNR')_4$, which, by X-ray crystallography, are seen to possess the cubane skeleton. The first reported structure of this type is illustrated in Fig. 23 [62a]. The faces of the Al_4N_4 cube show little significant departure from planarity. The Al–N distance is 1.914 Å and unremarkable.

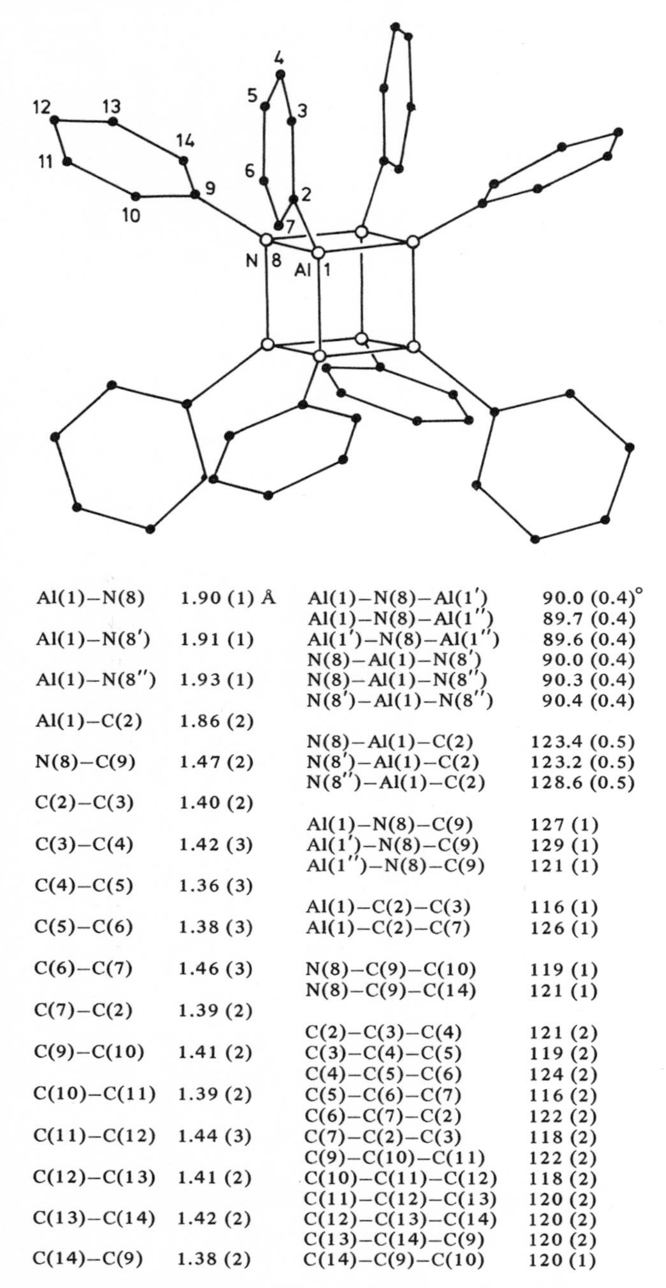

Al(1)–N(8)	1.90 (1) Å	Al(1)–N(8)–Al(1′)	90.0 (0.4)°
		Al(1)–N(8)–Al(1″)	89.7 (0.4)
Al(1)–N(8′)	1.91 (1)	Al(1′)–N(8)–Al(1″)	89.6 (0.4)
		N(8)–Al(1)–N(8′)	90.0 (0.4)
Al(1)–N(8″)	1.93 (1)	N(8)–Al(1)–N(8″)	90.3 (0.4)
		N(8′)–Al(1)–N(8″)	90.4 (0.4)
Al(1)–C(2)	1.86 (2)		
		N(8)–Al(1)–C(2)	123.4 (0.5)
N(8)–C(9)	1.47 (2)	N(8′)–Al(1)–C(2)	123.2 (0.5)
		N(8″)–Al(1)–C(2)	128.6 (0.5)
C(2)–C(3)	1.40 (2)		
		Al(1)–N(8)–C(9)	127 (1)
C(3)–C(4)	1.42 (3)	Al(1′)–N(8)–C(9)	129 (1)
		Al(1″)–N(8)–C(9)	121 (1)
C(4)–C(5)	1.36 (3)		
		Al(1)–C(2)–C(3)	116 (1)
C(5)–C(6)	1.38 (3)	Al(1)–C(2)–C(7)	126 (1)
C(6)–C(7)	1.46 (3)	N(8)–C(9)–C(10)	119 (1)
		N(8)–C(9)–C(14)	121 (1)
C(7)–C(2)	1.39 (2)		
		C(2)–C(3)–C(4)	121 (2)
C(9)–C(10)	1.41 (2)	C(3)–C(4)–C(5)	119 (2)
		C(4)–C(5)–C(6)	124 (2)
C(10)–C(11)	1.39 (2)	C(5)–C(6)–C(7)	116 (2)
		C(6)–C(7)–C(2)	122 (2)
C(11)–C(12)	1.44 (3)	C(7)–C(2)–C(3)	118 (2)
		C(9)–C(10)–C(11)	122 (2)
C(12)–C(13)	1.41 (2)	C(10)–C(11)–C(12)	118 (2)
		C(11)–C(12)–C(13)	120 (2)
C(13)–C(14)	1.42 (2)	C(12)–C(13)–C(14)	120 (2)
		C(13)–C(14)–C(9)	120 (2)
C(14)–C(9)	1.38 (2)	C(14)–C(9)–C(10)	120 (1)

Fig. 23 – X-ray crystal structure of $Al_4N_4Ph_8$. *[Reproduced, with permission, from McDonald, T. R. R., and McDonald, W. S.,* Acta Cryst., *1972,* **B28**, *1619.]*

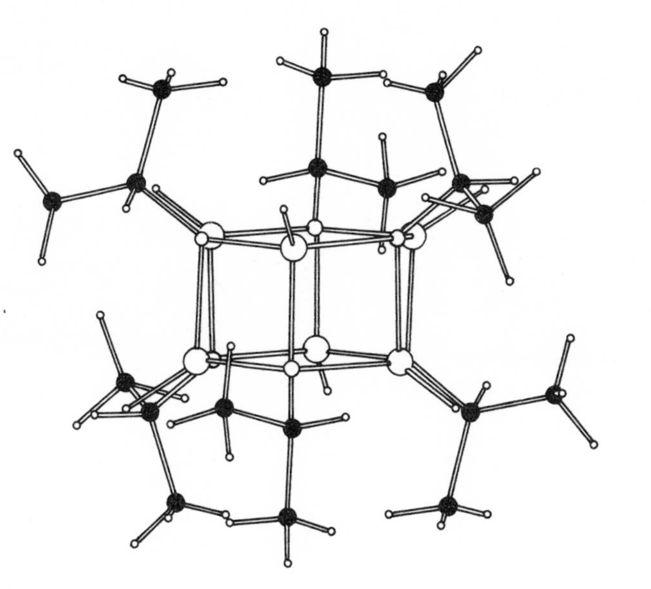

Unit cell has 4 molecules, 2 × 2 conformers arising from rotation around N–C of Pr^i group. Each ring is almost planar, with endocyclic Al–N = 1.898 Å and Al–N between rings = 1.956 Å; Al–H = 1.48 Å; N–C = 1.542 Å.

Fig. 24 – X-ray crystal and molecular structure of $(HAlNPr^i)_6$. *[Reproduced, with permission, from Cesari, M., Perego, G., Del Piero, G., Cucinella, S., and Cernia, E.,* J. Organometallic Chem., *1974,* **78**, *203.]*

More recent examples have been the tetramers $(HAlNPr^i)_4$ [19a], $(MeAlNPr^i)_4$ [19a], $(ClAlNBu^t)_4$ [18g], or $(ClAlNPr^i)_4$. The $(AlN)_4$ cubane framework is essentially identical with that in $Al_4N_4(Ph)_8$. The mean Al–N bond length 1.913(2) Å is normal. ^1H n.m.r. reveals that $[HAlN\{CH(Me)Ph\}]_4$ exists in both a closed (cubane-type) and open $Ph(Me)CH{-}NH{-}[HAlN\{CH(Me)Ph\}]_3AlH_2$ form [12b].

X-ray crystallographic studies on the hexamers $(RAlNR')_6$ confirm a cage structure. For example, $(HAlNPr^i)_6$ [18i] has the geometry illustrated in Fig. 24. The molecule consists of a hexagonal cage formed from two almost planar six-membered rings. The unit cell consists of four molecules, one of which has a different conformation from the other three; this arises from different rotation angles around the N–C bond. The compound $(HAlNPr)_6$ [19c] has a similar structure, and so has $(ClAlNPr^i)_6$ [18g] as shown by X-ray analysis [19c]; this is also found in $(Me_{0.83}H_{0.17}AlNPr^i)_6(MeAlNPr^i)_6$ [19d].

The structure (X-ray) of the crystalline heptamer $(MeAlNMe)_7$ has been reported [38a, 38b], Fig. 25, and consists of a framework of three six-membered $(AlN)_3$ rings, Al(1)N(3)Al(5)N(6)Al(4)N(1), Al(2)N(4)Al(6)N(7)Al(5)N(3), and Al(3)N(1)Al(4)N(5)Al(6)N(4), with a boat conformation and six four-membered $(AlN)_2$ rings. In the six-membered rings, the mean Al–N bond length is 1.96 Å and angles $Al{-}\widehat{N}{-}Al$ (120°) and $N{-}\widehat{Al}{-}N$ (110.7°) are similar to those in $(Me_2AlNHMe)_3$, while the four-membered rings are like those of $(Me_2AlNMe_2)_2$.

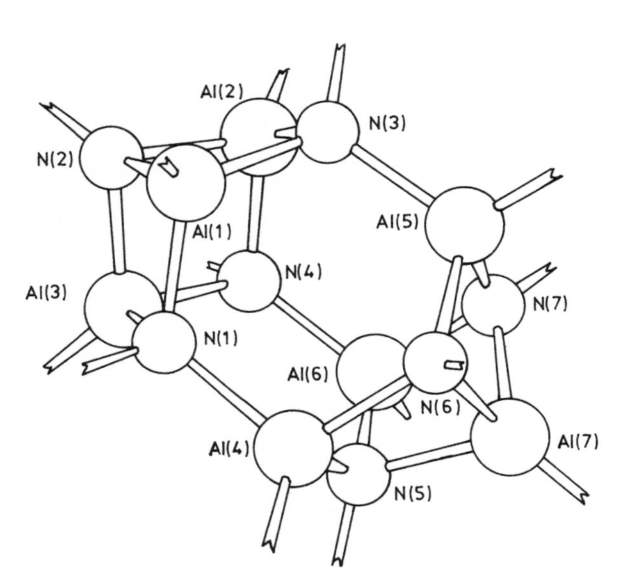

Fig. 25 – Molecular structure of $(MeAlNMe)_7$. *[Reproduced, with permission, from Hitchcock, P. B., Smith, J. D., and Thomas, K. M.,* J. C. S. Dalton, *1976, 1434.]*

Further examples of structural studies on cage compounds are provided by $Al_4Cl_4(NMe_2)_4(NMe)_2$ [106], $(HAlNPr^i)_6.AlH_3$ [78a], $[H(HAlNPr^i)_5AlH_2]$.-LiH/Et_2O [12d], $[(HAlNPr)_2(H_2AlNHPr^i)_3]$ [78f] (a pentamer) and the octamer $(HAlNPr)_8$ illustrated by Fig. 26 [19c]. The molecular framework may be derived from a combination of the hexamer cage with a square ring of a dimer. The mean bond distances are Al—N, 1.916 Å, N—C, 1.508 Å, and Al—H, 1.516 Å.

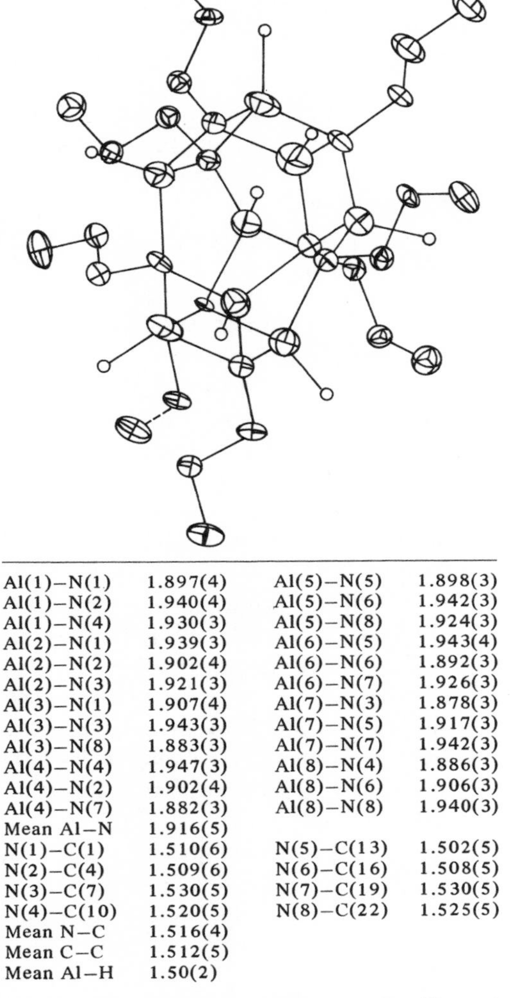

Al(1)–N(1)	1.897(4)	Al(5)–N(5)	1.898(3)
Al(1)–N(2)	1.940(4)	Al(5)–N(6)	1.942(3)
Al(1)–N(4)	1.930(3)	Al(5)–N(8)	1.924(3)
Al(2)–N(1)	1.939(3)	Al(6)–N(5)	1.943(4)
Al(2)–N(2)	1.902(4)	Al(6)–N(6)	1.892(3)
Al(2)–N(3)	1.921(3)	Al(6)–N(7)	1.926(3)
Al(3)–N(1)	1.907(4)	Al(7)–N(3)	1.878(3)
Al(3)–N(3)	1.943(3)	Al(7)–N(5)	1.917(3)
Al(3)–N(8)	1.883(3)	Al(7)–N(7)	1.942(3)
Al(4)–N(4)	1.947(3)	Al(8)–N(4)	1.886(3)
Al(4)–N(2)	1.902(4)	Al(8)–N(6)	1.906(3)
Al(4)–N(7)	1.882(3)	Al(8)–N(8)	1.940(3)
Mean Al–N	1.916(5)		
N(1)–C(1)	1.510(6)	N(5)–C(13)	1.502(5)
N(2)–C(4)	1.509(6)	N(6)–C(16)	1.508(5)
N(3)–C(7)	1.530(5)	N(7)–C(19)	1.530(5)
N(4)–C(10)	1.520(5)	N(8)–C(22)	1.525(5)
Mean N–C	1.516(4)		
Mean C–C	1.512(5)		
Mean Al–H	1.50(2)		

Fig. 26 – Crystal and molecular structure of $(HAlNPr)_8$. *[Reproduced, with permission, from Del Piero, G., Cesari, M., Perego, G., Cucinella, S., and Cernia, E.,* J. Organometallic Chem., *1977,* **129**, *289.]*

$[(MeAlNMe)_6(Me_2GaNHMe)_2]$ is isostructural with the all-Ga analogue (Fig. 29) [2b], and diethylaluminium-*N,N'*-dimethylacetamidine, $[Et_2Al(NMe)_2CMe]_2$, is similarly related to the Ga compound (Fig. 30) [34a], with Al–N = 1.925 Å.

An $(AlN)_4$-type cubane structure is found in the compounds $[(T.H.F.)Mg(HAlNBu^t)_3]$ and $[\{(T.H.F.)_3Ca\}(HAlNBu^t)_3]$ [18b], as shown by X-ray analysis (see Chapter 3).

E GALLIUM, INDIUM, AND THALLIUM†

1. Introduction and synthesis

Compared with aluminium the number of publications involving gallium, indium, or thallium amides has been small. The chemistry of gallium has been surveyed [14, 36], and the organometallic chemistry of these elements is discussed in ref. [9]. Apart from the thallium(I) compounds, the oxidation state +3 derivatives appear to resemble aluminium analogues (see Section D). The relative expense of these elements, compared with Al, probably accounts for the sparse attention that they have received. Some phosphine-imido-compounds [34, 43] were mentioned in connection with Al derivatives (see Section D), as was $(Me_2GaONSMe_2)$ [21]. Azomethine derivatives, the ketimide $(R_2GaN{=}CPh_2)_2$ (R = Me, Et, or Ph) [20] and the aldimide $(Et_2GaN{=}CHR)_2$ (R = Bu^t or Ph) [19], have been described. An indium carboxamide, $Et_2In[N(Ph)C(O)Et]$ is made by insertion of PhNCO into Et_3In [41a, 42]; excess of PhNCO probably affords $Et_2In[N(Ph)C(O)]_3Et$ or a related compound because hydrolysis yields the biuret $H[N(Ph)C(O)]_3H$. Pyrazolyl derivatives, such as M^{2+} complexes of $[Me_2M(pz)_2]^-$ (M = Ga or In), are discussed in refs. [3, 4. 18b].

The first amido-thallium compound was reported in 1915 [11].

Gallium, indium, or thallium halides form complexes with bases such as ethylenediamine [22, 25, 42] and piperidine [16], but the complexes of ethylenediamine with dimethylthallium halides are unstable [25]. Thermal stabilities of the complexes, with regard to loss of amine, have been studied [22]. The cleavage of metal-carbon bonds via reaction with amines has not been reported. For example, amine complexes of dimethylthallium halides [25] do not eliminate methane with formation of a Tl–N bond. Trimethylamine adducts of GaH_3 and GaD_3 have been described [15].

The main synthetic routes to gallium, indium, or thallium(III) amides involve hydrogen, Eq. (79) [39], or alkane, Eq. (80) [35], elimination. The thallium(I) derivatives of acid amides or phthalimide were prepared by transamination or alcohol elimination. However, $LiNMe_2$ or $NaN(SiMe_3)_2$ with $(GaCl_3)_2$ have been used in the synthesis of $[Ga(NMe_2)_3]_2$, $[ClGa(NMe_2)_2]_2$, $Li[Ga(NMe_2)_4]$ [30], or $Ga[N(SiMe_3)_2]_3$ [2a] respectively, as in Eq. (81).

†Numbers in [ ] refer to bibliography at the end of this Chapter, p. 232.

$$H_3Ga.NR_3 + H\overline{N(CH_2)_x} \longrightarrow \frac{1}{n}[H_2Ga\overline{N(CH_2)_x}]_n + H_2 + NR_3 \quad (79)$$

$$R_3M + H\overline{N(CH_2)_5} \longrightarrow \frac{1}{2}[R_2M\overline{N(CH_2)_5}]_2 + RH \quad (80)$$

(M = Ga or In; R = Me, Et, or Bu^i)

$$GaCl_3 + LiNMe_2 \longrightarrow [ClGa(NMe_2)_2]_2 \longrightarrow [Ga(NMe_2)_3]_2$$
$$\longrightarrow Li[Ga(NMe_2)_4] \quad (81)$$

Reaction of $H_3Ga.NMe_3$ with H_2NMe leads to the primary amine adduct, which loses H_2 even below room temperature to yield $(H_2GaNMe_2)_n$ [37], and is thus more reactive than the Al analogue. The related $(D_2GaNHMe)_n$ was similarly obtained [37]. The compounds are trimers in benzene and ^1H n.m.r. shows the co-existence in solution of two isomers, which are believed to be *cis*- and *trans*- forms of the $(GaN)_3$ ring in chair conformation.

The effect of the substituents at nitrogen on the degree of association or the structure of Ga amides has been investigated [38, 39]. For example, it has been found that in $[H_2GaN(H)R]_n$ when R is a non-branching alkyl substituent $n = 3$, but in the presence of chain branching $n = 2$. Similarly in the complex $[H_2N\overline{Ga(CH_2)_x}]_n$, when $x = 2$ or 3, $n = 3$, and when $x = 4$ or 5, a dimer-trimer equilibrium is obtained. As with aluminium, when R is very bulky, association is prevented by steric hindrance; thus the compounds $M[N(SiMe_3)_2]_3$ (M = Ga, In, or Tl) are monomers in solution [5, 24]. On the other hand, $[Ga(NMe_2)_3]_2$ and $[ClGa(NMe_2)_2]_2$ are dimers in solution; ^1H n.m.r. spectra show distinct terminal and bridging Me_2N ligands in the former, whereas the latter dissociates in solution with μ-$(Me_2N)_2$, μ-(Me_2N,Cl), and μ-(Cl_2) bridges [30].

2. Structures

The molecular structure of aziridinylgallane trimer shows the conformation of Fig. 27 [17], with torsion angles 59–61° and mean dimensions: Ga–N 1.97(2), N–C 1.54(4), C–C 1.55(5) Å; and angles N–$\widehat{Ga}$–N 100(1), Ga–$\widehat{N}$–Ga 121(1), and Ga–$\widehat{N}$–C 116(1)°, with angles in the three-membered rings close to 60°. The $(GaN)_3$ ring is in the chair conformation, as is the aluminium analogue $[H_2Al\overline{N(CH_2)_2}]_3$ (see Section D.4). The Ga–N bond distance, 1.97 Å, is also very close to the Al–N bond length in aminoalanes, as is expected, since the two group 3 metal atoms have almost identical covalent radii. Similar Ga–N bond distances are reported in the pyrazolylgallate complexes [31, 32].

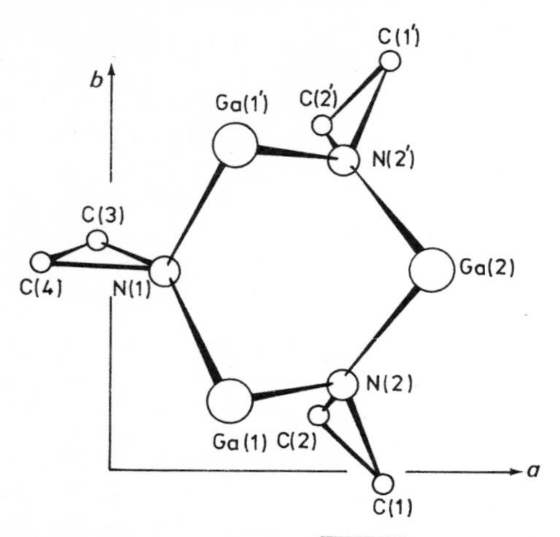

Fig. 27 – Molecular structure of $H_2Ga[\overline{N(CH_2)_2}]_3$. For clarity, H atoms are omitted. *[Reproduced, with permission, from Harris, W., Storr, A., and Trotter, J.,* J. Chem. Comm., *1971, 1101.*

The structure of the dimer $[Me_2InNMe_2]_2$ synthesised by alkane elimination, has also been determined [26]. The molecule consists of a planar four-membered $(InN)_2$ ring, and is illustrated in Fig. 28. The In–N bond length is greater than the sum of the covalent radii of the elements and is the longest $M–N_{amino}$ bond length known. The N–In–N angle (85.7°) is smaller than in the corresponding aluminium compound $[Me_2AlNMe_2]_2$ (88.4° Section D.4) and this is what is expected on the basis of electronegativity. The In–C and C–N bond lengths are within the normal range.

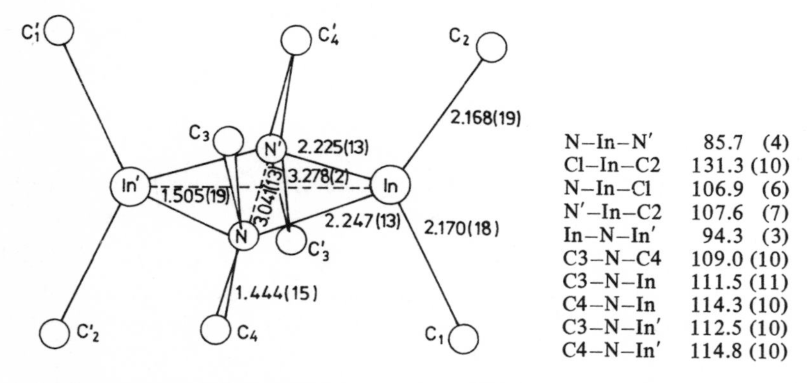

N–In–N'	85.7 (4)
Cl–In–C2	131.3 (10)
N–In–Cl	106.9 (6)
N'–In–C2	107.6 (7)
In–N–In'	94.3 (3)
C3–N–C4	109.0 (10)
C3–N–In	111.5 (11)
C4–N–In	114.3 (10)
C3–N–In'	112.5 (10)
C4–N–In'	114.8 (10)

Fig. 28 – The crystal and molecular structure of $[Me_2InNMe_2]_2$. *[Reproduced, with permission, from Mertz, K., Schwartz, W., Eberwein, B., Weidlein, J., Hess, H., and Hausen, H. D.,* Z. Anorg. Chem., *1977,* **429**, *102.*

A significant extension to the chemistry of the heavier group 3 metal amides has been the synthesis and structure of the first example of a poly-iminogallane of the formula $[(MeGaNMe)_6(Me_2GaNHMe)_2]$ [1a]. The structure of this compound is depicted in Fig. 29. The average Ga–N distance is ca. 1.96 Å, in close agreement with that found in $[H_2GaNMe_2]_3$. The elongated thermal ellipsoids of N_4 and $N_{4'}$ indicate disorder. Two isomers (one with both N_4 and $N_{4'}$ above the plane defined by $Ga_2N_2Ga_{2'}N_{2'}$ and the other with N_4 above the plane and $N_{4'}$ below) thus crystallise together; these isomers may be detected in solution by n.m.r.

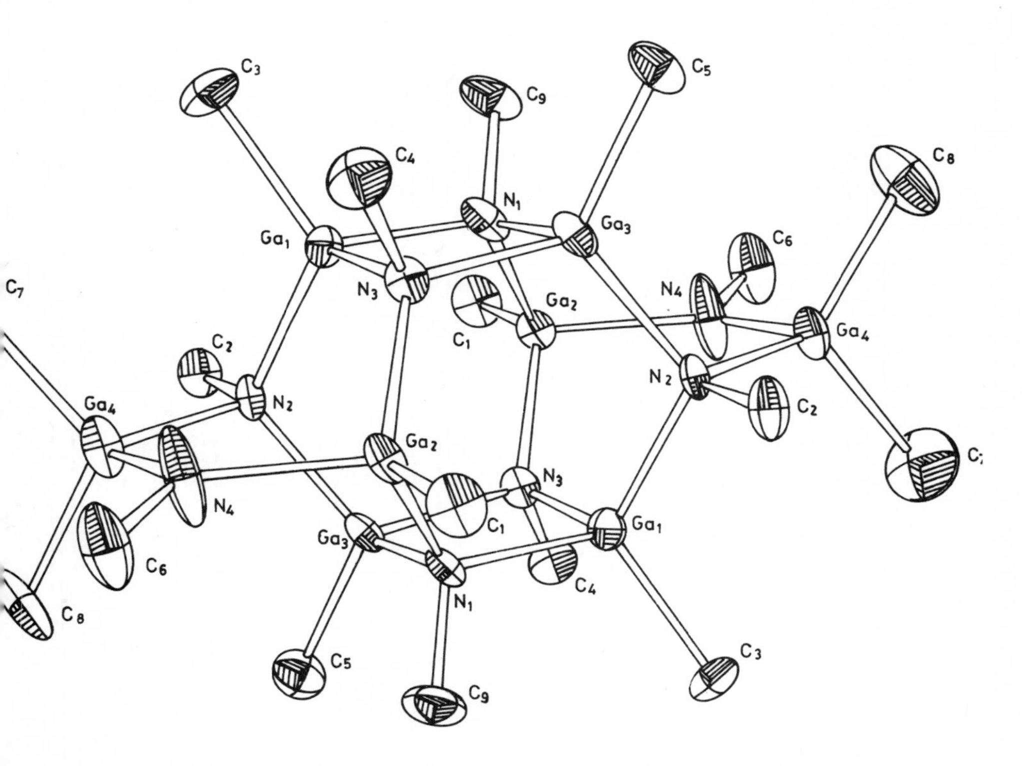

Fig. 29 – Crystal and molecular structure of $[(MeGaNMe)_6(Me_2GaNHMe)_2]$. *[From Amirkhalili, S., Hitchcock, P. B., and Smith, J. D., unpublished results, 1978.]*

Some oxamides $Me_2M(CONR_2)_2MMe_2$ (M = Al and R = Me; M = Ga and R = H or Me) are obtained from Me_3M and $(CONHR)_2$ [10a]; ^1H and ^{13}C n.m.r., as well as X-ray diffraction demonstrate the presence of tautomeric structures (LXVII) and (LXVIII).

(LXVII) (LXVIII)

The crystal and molecular structure of dimethylgallium-*N*,*N*′-dimethylacetamidine, as well as its isostructural Al compound, has been determined by X-ray crystallography, Fig. 30 [18a]; the metal atoms are tetrahedrally coordinated.

Single crystal X-ray data are now available for $Tl[N(SiMe_3)_2]_3$ [1]. It is isomorphous with the Al and Fe analogues: Tl–N, 2.089 Å; Si–N, 1.738 Å; Si–C, 1.889 Å; $Si\widehat{N}Si$, 122.6°; $N\widehat{Si}C$, 111.8°; $C\widehat{Si}C$, 107.1°, and a propellor twist of the SiN_2-groups versus the TlN_3 plane of 49.1°.

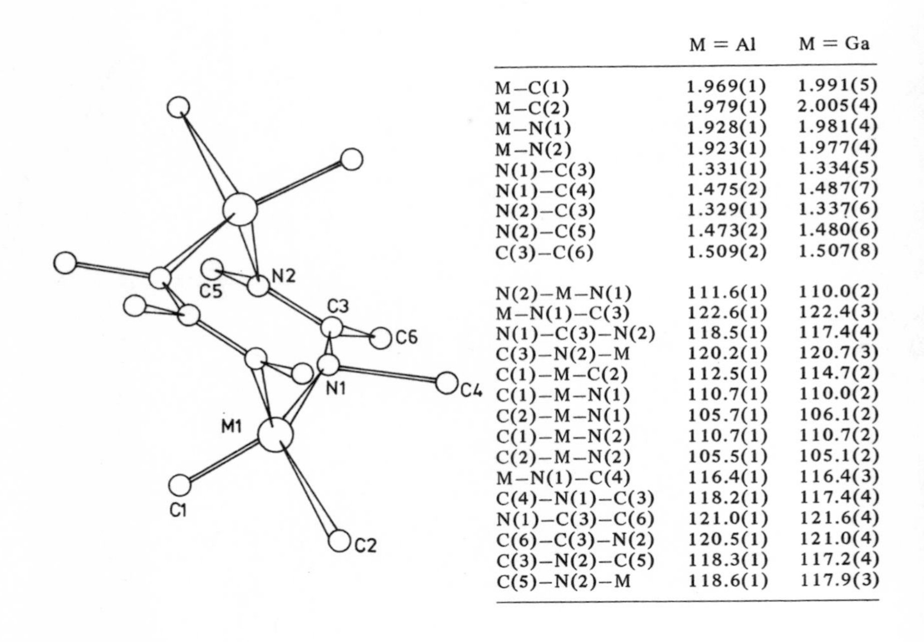

	M = Al	M = Ga
M–C(1)	1.969(1)	1.991(5)
M–C(2)	1.979(1)	2.005(4)
M–N(1)	1.928(1)	1.981(4)
M–N(2)	1.923(1)	1.977(4)
N(1)–C(3)	1.331(1)	1.334(5)
N(1)–C(4)	1.475(2)	1.487(7)
N(2)–C(3)	1.329(1)	1.337(6)
N(2)–C(5)	1.473(2)	1.480(6)
C(3)–C(6)	1.509(2)	1.507(8)
N(2)–M–N(1)	111.6(1)	110.0(2)
M–N(1)–C(3)	122.6(1)	122.4(3)
N(1)–C(3)–N(2)	118.5(1)	117.4(4)
C(3)–N(2)–M	120.2(1)	120.7(3)
C(1)–M–C(2)	112.5(1)	114.7(2)
C(1)–M–N(1)	110.7(1)	110.0(2)
C(2)–M–N(1)	105.7(1)	106.1(2)
C(1)–M–N(2)	110.7(1)	110.7(2)
C(2)–M–N(2)	105.5(1)	105.1(2)
M–N(1)–C(4)	116.4(1)	116.4(3)
C(4)–N(1)–C(3)	118.2(1)	117.4(4)
N(1)–C(3)–C(6)	121.0(1)	121.6(4)
C(6)–C(3)–N(2)	120.5(1)	121.0(4)
C(3)–N(2)–C(5)	118.3(1)	117.2(4)
C(5)–N(2)–M	118.6(1)	117.9(3)

Fig. 30 – Crystal and molecular structure of $[Me_2M(NMe)_2CMe]_2$ (M = Al or Ga). *[Reproduced, with permission, from Hausen, H. D., Gerstner, F., and Schwarz, W.,* J. Organometallic Chem., *1978,* **145,** *277.]*

F TABLES OF COMPOUNDS

Table 4† Mono(amino)boranes ($\supset BN \subset$ and $RR'NB_2H_5$)

Compound	Method	B.p., (°/mm.) (m.p., °C)	Comments	References
H_2BNH_2	F		ir, mass, IP	[234]
H_2BNHMe (trimer)	F		ir, nmr, mass	[20, 51, 53, 353]
			X-ray	[33]
(dimer)	A			[185]
H_2BNHEt (*cis*-trimer)	F	(197)	ir, nmr, mass	[53]
(*trans*-trimer)		(53)		
H_2BNHPr (*cis*-trimer)	F	(189)	ir, nmr, mass	[53]
(*trans*-trimer)		(31.5)		
H_2BNHPr	F	65–68/0.4	ir, nmr	[281b]
H_2BNHPr^i (dimer)	F	(105)	ir	[53]
	A			[185]
$H_2BNHPr^i.H_2BNMe_2$				[52]
$2H_2BNHPr^i.H_2BNMe_2$				[52]
H_2BNHBu (*cis*-trimer)	F	(179)	ir, nmr, mass	[53]
(*trans*-trimer)		(45)		
H_2BNHBu^t (dimer)	F	(24–6)	ir	[53]
(*cis*-trimer)	F	(120–1)	ir, nmr, mass	[53]
(*trans*-trimer)		(177)		
$(H_2BNHCH_2)_2$	F	(119–24)d		[195]
H_2BNMe_2 (dimer)	F	76–7/760 (73)		[343, 352]
	A			[298]
	F	(73.5)		[401, 404]
	F	(74)		[416]
(monomer)			pes, uv	[36, 130a]
H_2BNMe_2 (dimer)	F	(69–71)	ir, nmr	[281b]
	F	95/760 (74)		[402]
	F	25/8 (74)		[354]
	F	(74.3–4.5)		[62]
	F	23/9.1		[68]
		(74.5–5)	ir, nmr	[70, 78]
(trimer)			X-ray	[383]
D_2BNMe_2 (dimer)	F	(74.3–4.5)		[62]
$H_2BNMeCH_2CH_2Ph$	E	(75.5)	ir, mass	[198]
$H_2BNMeCH_2Ph$	E		ir	[198]
$H_2BNMeC_6H_{11}$	E	60–70/11 (104–6)	ir, mass	[198]
$H_2BNMePh$	E	(65–6)	ir, nmr, mass	[198]
H_2BNEt_2 (dimer)		(41–3)		[287]
	F	(42–4)		[252, 268]
	F	(47)		[416]
	E	120–0.5/160 (43)	ir, nmr	[198]
H_2BNEt_2 (dimer)	F	(39–41)	ir, nmr	[281b]
$H_2BNPr^i_2$	E	101/760	ir, nmr	[198]
H_2BNBu_2	E	70–4/14 (27–30)	ir	[198]

Table 4 *(c'td.)*

Compound	Method	B.p., (°/mm.) (m.p., °C)	Comments	References
$H_2BNBu^s_2$	E	51–3/16	mass	[198]
$H_2BNPent_2$ (dimer)	G	50–2/1		[287]
		(38–9)		[268]
$H_2B\overline{N(CH_2)_n}$ (dimer) ($n = 2$–5)	F		ir, nmr	[375b]
$H_2BN(C_6H_{11})_2$	E	66–7/0.2 (>90)	ir, nmr, mass	[198]
H_2BNPh_2	F	(128–30)	ir, nmr	[281b]
$H_2BN(Ph)CH_2Ph$ (monomer)	D			[273]
H_2BN (four-membered ring) (dimer)	F	186/760 (extr.) (51.2–1.7)	ir, nmr	[66, 375b]
H_2BN (five-membered ring) (dimer)	F	194/760 (extr.) (33.8–4.2)		[66, 375b]
	E	95–7/114 (35)	nmr, mass	[198, 375b]
H_2BN (six-membered ring)	F	202/760 (extr.) (106–10)		[66, 375b]
	E	(107–9)	ir, mass	[198, 375b]
H_2BN (unsaturated ring) $.NEt_3$	F	$65/10^{-4}$	ir, nmr, mass	[198]
$H\overline{BNMe(CH_2)_4}$	I	(106)	ir, nmr	[411]
(bicyclic structure: H, B, Me, N; B, H, N, Me) or (Me N–B ring–chain–NMe–BH_2)	I	(83)	ir, nmr	[411]
$MeBHNMe_2$	A	42 (extr.) (–136.4–6.2)		[65]
$EtBHNMeCH_2CH_2Ph$	I	104–5/10		[198]
$EtBHNMeCH_2Ph$	I	85–8/9		[198]
$BuBHNMe_2$		118–20/719	d_4^{20} 0.789, n_4^{20} 1.4334	[306]
	A	114	ir, nmr	[348]
$BuBHNEt_2$		68/46		[170]
Bu^iBHNEt_2	F	52/21	nmr	[170]
Bu^sBHNMe_2	A	49/91	ir, nmr	[348]

Table 4 *(c'td.)*

Compound	Method	B.p., (°/mm.) (m.p., °C)	Comments	References
Bu^sBHNEt_2	F	54/27	nmr	[170]
$Bu^tBHNHMe$	F	42/10	nmr	[170]
Bu^tBHNMe_2	F	48–50/80		[170]
	A	40/81	ir, nmr	[348]
Bu^tBHNEt_2	F	68/46	nmr	[170]
$Bu^tBHNPr^i_2$	E	74/31	nmr	[170]
Bu^tBHNBu_2	E	77/3.3	nmr	[170]
$Bu^tBHNBu^i_2$	E	54/3	nmr	[170]
HB(–CH_2—NPh–)(–NPh—CH_2–)BH (ring)	D			[413]
$PhBHNEt_2$	F			[252]
	F	60–5/2.5		[257]
Me_2BNHMe			ir, nmr, Raman	[27, 78]
dimer	A		nmr	[403]
				[332]
$Me_2BNHC_2H_3$			nmr	[361]
Me_2BNHPr^i	C		nmr	[302]
Me_2BNHBu	B	114/722		[297]
$Me_2BNHCH_2CH_2NH_2$	B	28–30/20		[297]
$(Me_2BNHCH_2)_2$	B	25–7/22		[297]
$Me_2BNHCH_2CH_2CHCH_2$	D	95	ir, nmr	[411]
Me_2BNHPh			ir, nmr	[78, 130d]
	A	77/12		[297]
Me_2BNMe_2			nmr, pes	[23, 27, 36]
			ir, nmr, Raman	[78, 150, 333c, 413a]
			X-ray	[59]
		49–51		[123, 124]
			uv	[130a]
$.NMe_3$				[103]
$.Cl_2BNMe_2$	I	(63–5)		[374]
$.Ni(C_3H_5)_2$			ir, nmr, mass	[356]
$Me_2BNMeC_6H_{11}$	C	70/11		[281]
$Me_2BNMeCH_2Ph$	C	64–5/5		[281]
$Me_2BNMePh$	A	63/12	ir, Raman	[13]
			dnmr	[23, 351]
			μ	[26]
			ir, nmr	[78]
	A		ir, nmr, uv	[369]
		51/4		[283]
$Me_2BNMeNp$-1		89/3		[35]
Me_2BNEt_2	B	90–2/712		[38]
$Me_2BNEtPh$		51/4		[18]
Me_2BNBu_2	B	68/10		[38]
$Me_2BN(C_6H_{11})_2$		259		[283]
Me_2BN (ring, Me, Me)	C	53–6/20	ir, nmr	[333a]

Table 4 *(c'td.)*

Compound	Method	B.p., (°/mm.) (m.p., °C)	Comments	References
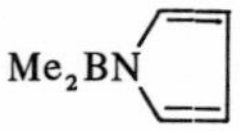 Me_2BN	E	98/757	ir, nmr, mass	[31, 333a]
.py	E	(72)	ir, nmr, mass	[31]
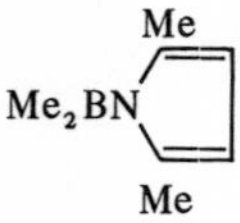 Me Me_2BN Me	C	32/1	ir, nmr	[333a]
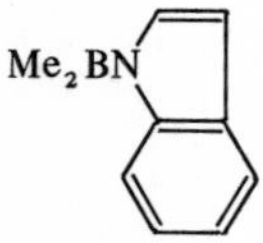 Me_2BN	C	66–7/1	ir, nmr	[333a]
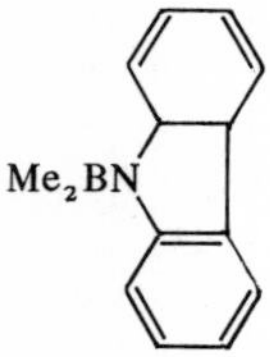 Me_2BN	C	(179–181)	ir, nmr	[333a]
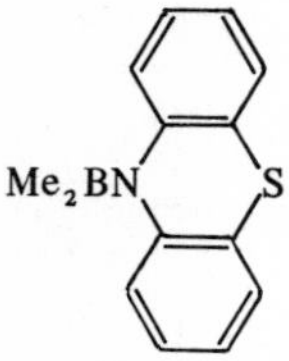 Me_2BN S	C	131–5/1	ir, nmr	[333a]
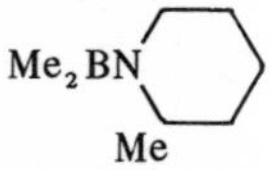 Me_2BN Me	C	38/3		[281]
Me_2BNPh_2			nmr	[23]
	A		μ	[26]
$MeBBuNMe_2$		33–5/1	d_4^{20} 0.738, n_D^{20} 1.4209	[307]
$PhCH_2BMeNMeCH_2Ph$			ir, nmr	[78]
$MeB(C_2H_3)NMe_2$			ir, nmr	[78]
	A		ir, nmr, uv	[369]
$MePhBNMe_2$			nmr	[18, 23]
			ir, nmr	[78]
	C			[286]

Table 4 *(c'td.)*

Compound	Method	B.p., (°/mm.) (m.p., °C)	Comments	References
MePhBNMePh	B		ir, Raman	[13]
			nmr	[23]
			ir, nmr	[78]
		98/3		[286]
MePhBN (ring)	C	75/3		[286]
Et_2BNHMe	A		nmr	[332]
$Et_2BNHCH_2CH_2Ph$	F	66/0.05	ir	[198]
$Et_2BNHC_6H_4Ph$-2	F	96–8/0.01	ir, nmr, mass	[198]
Et_2BNMe_2	K			[1]
$Et_2BNMeCH_2CH_2Ph$	F	115–8/9	ir, nmr, mass	[198]
$Et_2BNMeCH_2Ph$	F	108–10/9	ir, nmr, mass	[198]
		68–70/0.1		
Et_2BNEt_2	F	88–91/100		[198]
	I	154		[159a, 286]
$Et_2BNEtPh$	F	158–65/6		[284]
		168/6		[283]
$Et_2BNBu^s_2$	A	212–4		[283]
Et_2BN (ring)	B	53/3		[281]
Et_2BN (ring, O)	E		nmr	[381]
Et_2BN (ring)	F	68–9/18	ir, nmr, mass	[198]
	E	68/15	ir, nmr, mass	[31]
.py	F	(70–2)	nmr	[198, 333a]
	E	$44\text{–}5/10^{-4}$	ir, nmr, mass	[31]
.4-pic	E	$50\text{–}3/10^{-4}$	ir, nmr, mass	[31]
Et_2BN (ring, Me)	E	73/9	ir, nmr, mass	[31]
Et_2BN (ring, Pr)	E	103–10/12	ir, nmr, mass	[31]

Table 4 *(c'td.)*

Compound	Method	B.p., (°/mm.) (m.p., °C)	Comments	References
Et_2BN	E	$80-3/10^{-3}$	ir, nmr, mass	[31]
Et_2BN	E	$93-6/10^{-4}$ (35)	ir, nmr, mass	[31]
Et_2BN =N	B, I	(165–75)	ir, nmr, mass	[276b]
$Et_2BNHC_5H_4N$-2	E		ir, nmr	[116a, 154a, 154b]
Et_2BN Me Me	Li/THF	$48/10^{-3}$	nmr	[199]
Et_2B N, BEt_2 N, Me Me Me Me	Li/THF	(93)	nmr	[199]
Et_2BNPh_2	C	124/6 (23–4)		[286]
NH—CEt_2, EtB, BEt, CEt_2—NH	D	(65)	ir, nmr	[178]
NPh—CEt_2, EtB, BEt, CEt_2—NPh	D	(204)	ir, nmr	[177]
EtPhBNMeBu	C	73–89/3		[286]
$EtPhBNEt_2$	C	98/4		[286]
EtPhBNEtPh	C	106–114/3		[286]
$Et(CH_2CHCH_2)BNMe_2$	A	(57–60)	ir, nmr	[163]
$Et(CH_2CHCHMe)BNMe_2$	A, I	(62–4)	ir, nmr	[163]

Table 4 *(c'td.)*

Compound	Method	B.p., (°/mm.) (m.p., °C)	Comments	References
1-Me-2-RB-piperidine-type ring (N–Me, RB in six-membered ring)	F		ir, nmr	[411]
(R = Et		154/760		
Bu		107/0.27		
Hex)		70-3/0.01		
Pr_2BNMe_2	A	42/9	ir, nmr	[348]
		57-9/12		[298]
Pr_2BNEt_2	E	119-22/96, 92-3/34	d_4^{20} 0.7702, n_D^{20} 1.4323	[249]
Pr_2BNBu^i(cyclohexenyl)	G	109-11/2	n_D^{20} 1.5654	[116]
Pr_2BN(pyrrolidinyl)	F	91/15		[198]
Pr_2BNPh(cyclohexenyl)	G	119-22/1.5	d_4^{20} 0.930, n_D^{20} 1.5162	[115]
Pr_2BN(piperidinyl)	F	106/15		[198]
Pr_2BN(pyrrolyl)	F	93-5/14	nmr, mass	[198]
	E	93-5/14	ir, nmr, mass	[31]
.py	E	45-7/10^{-4} (67)	ir, nmr, mass	[31]
.4-pic	E	53-6/10^{-4}	ir, nmr, mass	[31]
Pr_2BN(7-azaindolyl)	E	(79)	ir, nmr, mass	[31]
R_2B(NH=CR″)(NR′)-cyclohexenyl chelate				
(R = Pr:				
R′ = Bu^i R″ = Ph	D	174-8/0.3	n_D^{20} 1.5615, ir, nmr	[116]
R′ = Ph R″ = Me	D	175-8/1.5 (106-8)	ir, nmr	[116]

Table 4 *(c'td.)*

Compound	Method	B.p., (°/mm.) (m.p., °C)	Comments	References
R' = Ph R'' = Pr	D	175–6/0.8	n_D^{20} 1.5686, ir, nmr	[116]
R = Bu:				
R' = Pent R'' = Me	D	174–5/1.5	n_D^{20} 1.5232, ir, nmr	[116]
R' = Ph R'' = Me	D	185–9/1 (63–6)	ir, nmr	[116]
R' = Ph R'' = Ph)	D	210–4/0.5 (51–4)	ir, nmr	[116]
Bu_2BNHMe	G	64–5/8	d_4^{20} 0.7706, n_D^{20} 1.4293	[251]
Bu_2BNHEt	G	73–4/8	d_4^{20} 0.7682, n_D^{20} 1.4279	[250]
			MCD	[225]
Bu_2BNHBu			MCD	[225]
$Bu_2BNH(CH_2)_3NHBBu_2$	B	134–5/2	ir	[291]
Bu_2BNHPh	G	136–7/7	n_D^{20} 1.4995	[249, 270]
Bu_2BNMe_2	A	110/18		[141]
	D			[324]
		79–81/9	n_D^{20} 1.4332	[363]
	A, I	77/9	ir, nmr	[348]
	A	78–80/10		[298]
	J			[94]
$Bu_2BNMeCH_2CH_2Ph$	D	100–2/0.3		[198]
	I		ir, mass	[200]
$Bu_2BNMeCH_2Ph$	D	~87–92/0.4		[198]
Bu_2BNEt_2	D	52–5/1.5	d_4^{20} 0.781, n_D^{20} 1.4377	[252]
			MCD	[225]
Bu_2BNBu_2			MCD	[225]
$(Bu_2B)_2NMe$	A, G			[332]
$Bu_2BNPent$(cyclohexenyl) 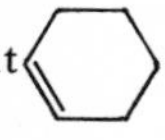	G	137–8/2	n_D^{20} 1.4690, ir, nmr	[116]
Bu_2BNPh(cyclohexenyl) 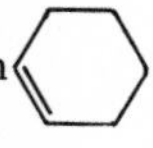	G	120–2/1	n_D^{20} 1.5102, ir, nmr	[116]
$(C_3H_5)_2BN(Me)CHPhC_3H_5$	D	84–6/0.01	ir, nmr	[239]
$Br(CH_2)_4(Bu)BNHBu$	I	100/2.5	ir, nmr	[388]
$Br(CH_2)_4(Bu)BNEt_2$	I	105–6/30	ir, nmr	[388]
$Br(CH_2)_4(Bu)BNHMe$	I		ir, nmr	[388]
$Et_2N(CH_2)_4(Bu)BNEt_2$	I		ir, nmr	[388]
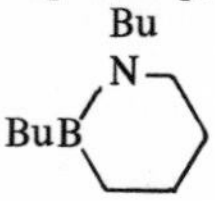 BuB(ring)N–Bu	C		ir, nmr	[388]
 (bicyclic B–N)	F	144–55/760	pes	[36]

Table 4 *(c'td.)*

Compound	Method	B.p., (°/mm.) (m.p., °C)	Comments	References
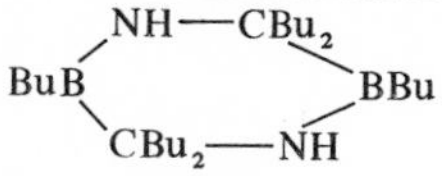 NH—CBu_2 BuB BBu CBu_2—NH	D	(71)	ir, nmr	[178]
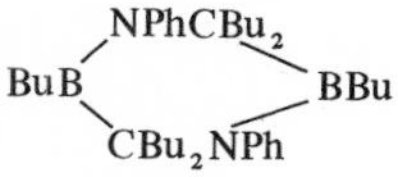 $NPhCBu_2$ BuB BBu CBu_2NPh	D	(186)	ir, nmr	[177]
NH— BuB (ring)	I			[276]
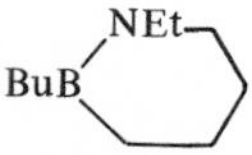 NEt— BuB (ring)	I	70–0.5/4	d_4^{20} 0.8833, n_D^{20} 1.4548	[272]
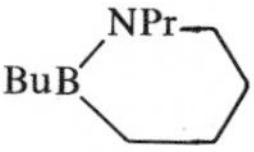 NPr— BuB (ring)	I	60.5/2.5	d_4^{20} 0.8320, n_D^{20} 1.4560	[272]
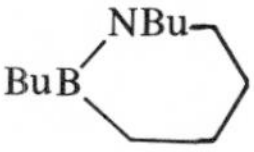 NBu— BuB (ring)	I	60/2	d_4^{20} 0.8315, n_D^{20} 1.4568	[272]
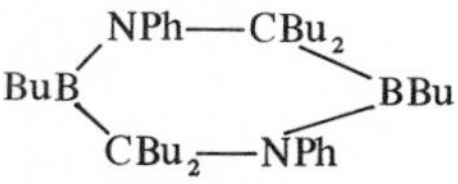 NPh—CBu_2 BuB BBu CBu_2—NPh	D	(186–9)		[177] [412]
NPh—$BCBu_3$ BuB | CBu_2—NPh	I	(132–3)		[412]
$BuB(C_2H_3)NMe_2$	A		ir, nmr, uv	[369]
$MeCCB(Bu)NMe_2$	A	93/15	ir, nmr, uv	[369]
$Pent_2BNHBu$			MCD	[225]
$Pent_2BNEt_2$	I	82/3		[281]
$Pent_2BNPent_2$	C	124/8		[294]
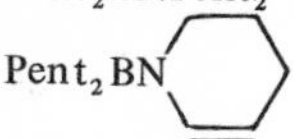 $Pent_2BN$ (ring)	B	93–4/2		[281]
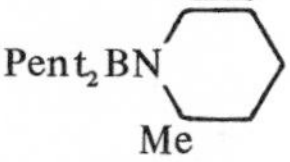 $Pent_2BN$ (ring) Me	B	95/2		[281]
Hex_2BNHBu			MCD	[225]
$HexB(Ph)NMeC_6H_{11}$		102–4/3		[281]
Oct_2BNEt_2	D	150–2/2	d_4^{20} 0.8072, n_D^{20} 1.4502	[252]
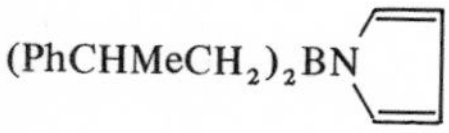 $(PhCHMeCH_2)_2BN$ (ring)	E	90–5/10^{-4}	ir, nmr, mass	[31]

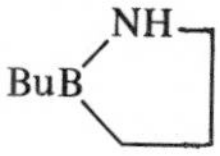

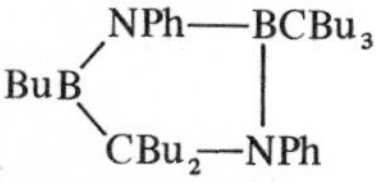

Table 4 *(c'td.)*

Compound	Method	B.p., (°/mm.) (m.p., °C)	Comments	References
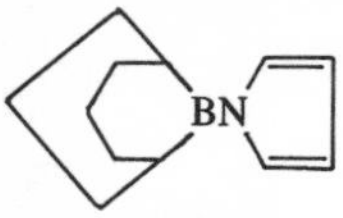 BN	E	72–4/0.1	ir, nmr, mass	[31]
$C_6H_{11}B(Ph)NMeC_6H_{11}$	C	102–24/3		[281]
$PhCH_2B(Ph)NMeCH_2Ph$	C	170–80/3		[281]
$PhCH_2B(Ph)NPhCH_2Ph$	C	170–80/3		[281]
B—N , B—N	B		nmr, pes, mass	[97,100, 105]
$(C_2H_3)_2BNMe_2$	A		ir, nmr	[413a]
$(C_2H_3)B(Me)NMe_2$	A		ir, nmr	[413a]
(X)$_2$ BNMe$_2$	A			[413a]
(X = O				
S		68–71/1	ir, nmr	
NMe)		85–7/1	ir, nmr	
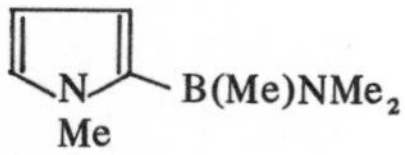 N Me B(Me)NMe$_2$	A	42–3/1	ir, nmr	[413a]
 Me BNRR′	F			[201]
(R = H = R′		86–8/0.1		
R = H, R′ = Et		35–7/8		
R = H, R′ = $PhCH_2$		84.5–5/0.1		
R = H, R′ = Ph		123–4/10		
R = Et = R′)		54/12		
$(C_2H_3)_2BNEt_2$	A	85/90		[105]
$(C_2H_3)BPhNMe_2$			ir, nmr	[78]
	A		ir, nmr, uv	[369]
$(CH_2CMe)_2BNEt_2$	A	56–60/15		[38]
$(MeCHCH)_2BNEt_2$	A	75/20		[38]
$(C_3H_5)_2BNHEt$	E	63–4.5/26	d_4^{20} 0.8394, n_D^{20} 1.4686	[269]
$(C_3H_5)_2BNHPh$	E	77–8/2	d_4^{20} 0.9215, n_D^{20} 1.5342	[269]
$(C_3H_5)_2BNEt_2$	E	79–80/20	d_4^{20} 0.8025, n_D^{20} 1.4562	[269]
$(HCC)_2BNMe_2$	A	41/30	ir, nmr	[413c]
(dimer)		(101–3)	ir, nmr	[413c]
$(HCC)_2BNEt_2$	A	54–6/18	ir, nmr	[413c]

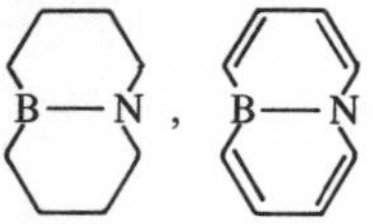

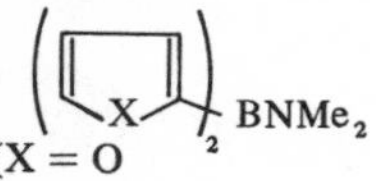

Table 4 *(c'td.)*

Compound	Method	B.p., (°/mm.) (m.p., °C)	Comments	References
$(MeCC)_2BNMeBu^t$	A	90/1.5	ir, nmr, uv	[369, 370]
		130/0.5 (70)	ir, nmr, uv	[369, 370]
$(MeCC)_2BNMe_2$	A	29–31/0.1	ir, nmr	[413c]
$(MeCC)_2BNMePh$	A	127/0.7	ir, nmr, uv	[369, 370, 413c]
$(MeCC)_2BNEt_2$	A	90/1	ir, nmr, uv	[369, 370]
$MeCCB(Ph)NMe_2$	A	112/3	ir, nmr, uv	[369, 370]
$(CH_2CMeCC)_2BNMePh$	A	138/0.8	ir, nmr, uv	[369, 370]
$(CH_2CMeCC)_2BNEt_2$	A	103/1	ir, nmr, uv	[369, 370]
$PhCCB(Ph)NMe_2$	A		ir, nmr, uv	[369]
Ph_2BNHMe	C			[56]
Ph_2BNHEt	F	124–8/3	n_D^{20} 1.5752	[256]
$Ph_2BNPr^i_2$	C		nmr	[340]
	C	124/4 (86–7)		[281]
Ph_2BNEt_2	C		mass	[95]
Ph_2BNHBu^i	F			[256]
Ph_2BNHBu^t	C	108/0.05	n_D^{20} 1.5531, ir	[93]
Ph_2BNHPh	G	202–6/1 (56–8)		[248]
	C	134–5/0.04 (57–9)		[260]
Ph_2BNMe_2			nmr	[23]
	A	80–2/0.002		[134]
	A	96–8/10^{-3}	ir, nmr	[348]
		102–4/0.05	ir, μ	[84]
	F	121–2/2.5	d_4^{20} 0.9988, n_D^{20} 1.5803	[256]
	C	(104–2)		[280]
$Ph_2BNMePh$	B		ir, Raman	[13]
			nmr	[23]
Ph_2BNEt_2		161/11		[284]
	F	128–32/2.5 (39–39.5)		[252]
		159–62/11 (36–7)	d_4^{20} 0.9953	[259]
		104/0.1	n_D^{20} 1.5606	[217]
$Ph_2BNEtPh$		158–64/5–6 (27)		[283]
Ph_2BN	C		ir, nmr	[333a]
Ph_2BN	C	200–3/14 (65–6)		[259]
Ph_2BNPh_2	C	(255–6)	ir, μ	[84]
	C, G	(196–7)		[261]
$Ph_2BN(C_6H_4Me\text{-}p)_2$	C	(132)	ir, μ	[84]
$Ph_2BN(1\text{-}Np)_2$	C	(174–5)		[261]
$Ph_2BN(H)(4\text{-pyridyl})$	B	(145–50)	ir, nmr	[87]

Table 4 *(c'td.)*

Compound	Method	B.p., (°/mm.) (m.p., °C)	Comments	References
$Ph_2BN(H)$(3-pyridyl)	B	(208–10)	ir, nmr	[87]
$Ph_2BN(H)$(2-pyridyl)	B	(145–50)	ir, nmr	[87]
p-$BrC_6H_4B(Ph)NPr^i_2$	C	152–4/3 (64–6)		[281]
$PhB(CCl_3)NMe_2$				[165]
$Ph(PhCH_2)BNMePh$-*cis*	I		nmr	[163]
-*trans*	I		nmr	[163]
(*p*-$MeC_6H_4)_2BNMe_2$	A	110–2/0.04–0.05	ir, μ	[84]
(*p*-$MeC_6H_4)_2BNEtPh$	A	160–80/3 (169)		[283]
(*p*-$MeC_6H_4)_2BNPh_2$	C	(72–3)	ir, μ	[84]
(*p*-$BrC_6H_4)_2BNMe_2$		(39–40)	ir, μ	[84]
(*p*-$BrC_6H_4)_2BNPr^i_2$	C	195–6/3 (132–4)		[281]
$(Mesityl)_2BNRR'$	A			
(R R'				
H Ph		(150–1)	ir, nmr, mass	[146]
H 2, 6-$Me_2C_6H_3$		(198–9)		[146]
H 1-Np		(147–8)		[146]
Me Ph		(136–7)		[146]
Me 1-Np		(138–40)		[146]
Ph Ph)		(219–22)		[146]
$Ph_2BN(Ph)$–(2,5-dioxo-cyclohexa-3,6-diene-1,4-diyl)–$N(Ph)BPh_2$	E	(303)	nmr, uv	[385]
1-Np_2BNHMe	C	(104–6)		[260]
1-Np_2BNHEt	C	(89–91)		[256]
1-Np_2BNHBu^i	C	240–5/1		[260]
1-Np_2BNHBu^t	C	229–31/2		[256]
1-Np_2BNHPh	C	(125–7)		[260]
1-Np_2BNEt_2	C	(176–7)		[256]
	C	(178–9)		
1-Np_2BNPh_2	C	(163–4)		[261]
$EtNHB_2H_5$	F	(–96.2)		[66]
$Me_2NB_2H_5$	F	(–54.6)		[68]
	I			[7]
	A			[298]
$Me_2NB_2H_5$	A		ir, nmr	[191a]
			microwave	[85]
$Me_2NB_2D_5$	F	(–57.0)		[62]
$RNHB_2H_5$	A			
(R = Pr			ir, nmr	[194]
Pr^i			ir, nmr	[194]
Bu			ir, nmr	[194]
Bu^s			ir, nmr	[194]
Bu^t			ir, nmr	[194]
Hex)			ir, nmr	[194]
$Pr_2NB_2H_5$	A		ir, nmr	[194]

Table 4 *(c'td.)*

Compound	Method	B.p., (°/mm.) (m.p., °C)	Comments	References
$Pr^i_2NB_2H_5$	A		ir, nmr	[194]
$[R\{NC(CH_2)_n\}N]BMe_2$				
n R				
1 CH_2CN		84/0.05	ir, nmr, mass	[240]
1 Me		67/25	ir, nmr, mass	[240]
2 Me		65/4	ir, nmr, mass	[240]
1 Bu		104/22	ir, nmr, mass	[240]
$[Me\{NC(CH_2)_2\}N]BBu_2$		$80/10^{-3}$	ir, nmr, mass	[240]
$\overline{Me_2NB(SiPh_2)}_4$	A	(261–3)	ir, nmr, Raman, uv	[171]

† Footnote to Table 4: for explanation of abbreviations, see footnotes to Table 1, p. 57.

Table 5† Bis(amino)boranes [RB(N⟨)$_2$]

Compound	Method	B.p. (°C/mm.) (m.p., °C)	Comments	References
$HB(NH_2)_2$	F		ir, nmr, mass, rotat.	[39a]
$HB(NHEt)_2$	F	0/6.3, 35.3/40.9		[355]
$HB(NHBu)_2$	B, G	46–8/2	d_4^{20} 0.7879, n_D^{20} 1.4306	[253. 256]
$HB(NHBu^i)_2$	B, G	59–61/9	d_4^{20} 0.7794, n_D^{20} 1.4263	[253, 256]
HB⟨NH—NH⟩ (ring)	B	97–9/760		[293]
$HB(NHC_3H_5)_2$	B, G	62–7/30	d_4^{20} 0.8181, n_D^{20} 1.4520	[253, 256]
$HB(NHPh)_2$	B	(103–10)		[255]
	F	(112–4)	ir, nmr, mass	[198]
$PhNHBHNEt_2$	G	80–2/1.5	d_4^{20} 0.9234, n_D^{20} 1.5295	[255]
$(PhNHBH)_2NPh$	F	(77–9)	ir, nmr	[198]
$(PhNHBNPh)_2$	F	(201–2)	ir	[198]
PhNHB(N⟨ring⟩)$_2$	F	$73/10^{-3}$	ir, mass	[198]
o-$MeC_6H_4NHBHNEt_2$	G	87–9/1.5	d_4^{20} 0.9145, n_D^{20} 1.5228	[255]

Table 5 *(c'td.)*

Compound	Method	B.p. (°C/mm.) (m.p., °C)	Comments	References
HB(NH)(NH) benzo-fused ring (structure)	B	(79–80)		[181]
	F	(80–5)	ir	[154]
HB(NH–)(NH–) naphtho-fused ring (structure)	C	(97–9)	ir	[75]
	F	(100.5)	ir	[154]
$HB(NMe_2)_2$	$(Me_2N)_2BCl/LiH$	106–8/720	d_4^{20} 0.7667, n_D^{20} 1.4232	[305]
		108/760 (extr.)		
	F	109/760 (extr.) (–45)		[400]
	K	107/760 (extr.) (–55–6)		[32]
	I	105.7/760 (extr.)		[69]
		(–57.5)	nmr	[130d]
	A			[298]
			ir, Raman, uv, pes	[36, 151, 130a]
$HB(NMeCH_2CH_2Ph)_2$	E+I	135–45/0.3	ir	[198]
$HB(NMeCH_2Ph)_2$	E+I	114–5/0.2	ir, nmr	[198]
	I		ir	[200]
$HB(NMePh)_2$	B, G	123–7/1.5	n_D^{20} 1.6037	[257]
	G	125–8/1.5	d_4^{20} 1.0331, n_D^{20} 1.6035	[255]
	E+I	99–100/0.15	ir, mass	[198]
$HB(NEt_2)_2$	B	26–8/2	n_D^{20} 1.4384	[305]
	G	29–34/2, 62–4/16	d_4^{20} 0.7870, n_D^{20} 1.4330	[255]
	E+I	58.7–9/13	d_4^{20} 0.7908, n_D^{20} 1.4354, ir, nmr	[198]
$HB(NPr_2)_2$	B	84–6/4	d_4^{20} 0.801, n_D^{20} 1.4428	[305]
$HB(NBu_2)_2$	B	125–6/3.5	d_4^{20} 0.8067, n_D^{20} 1.4481	[305]
	E+I	84–94/0.5	ir, mass	[198]
$HB(NPent_2)_2$	B, G	122–5/1.5	d_4^{20} 0.8147, n_D^{20} 1.4504	[257]
	G	123–8/1.5	n_D^{20} 1.4500	[258]
$HB[N(C_6H_{11})_2]_2$	E+I	(138–42)		[198]
$HB(NPhCH_2Ph)_2$		188/0.01	ir	[198]
$HB[N(C_3H_5)_2]_2$	B, G	77–81/4	d_4^{20} 0.8455,	

Table 5 *(c'td.)*

Compound	Method	B.p. (°C/mm.) (m.p., °C)	Comments	References
			n_D^{20} 1.4739	[257]
	G	69–74/2.5	n_D^{20} 1.4759	[258]
HB(N)$_2$	E+I	105–7/15	ir, nmr, mass	[198]
HB(N)$_2$	B, G	70–1/1.5	d_4^{20} 0.9247, n_D^{20} 1.4968	[257]
	E+I	116–9/12	ir, mass	[198]
$HB(NPh_2)_2$	F	(subl. 188/0.01) (140–2)	ir, nmr	[198]
$HB(NPhNp\text{-}1)_2$	F	(182–4)	ir	[198]
HB(NMe)(NMe)			nmr	[130d]
$MeB(NHMe)_2$	C	52/0.02	ir, nmr	[304]
MeB(NH)(NH)	E		ir	[291]
	B			[297]
MeB(NH)(NH)			ir	[102]
	B	36–8/20		[293]
			nmr	[295]
$(N,N)D_2$			nmr	[295]
MeB(NH)(NH)			ir, Raman	[292]
	B	39–42/6		[293]
MeB(NH)(NEt)	B	39–41/1.5		[395]
$MeB(NMe_2)_2$			nmr	[130d, 333a, 333d]
			ir, Raman	[151]
	C	29–32/15	d_4^{20} 0.799, n_D^{20} 1.4360	[306]
	I		uv, pes	[36, 82, 130a]
$.Fe(CO)_3$			ir, nmr, mass	[256]
$HN(BMeNMe_2)_2$	A	77–8/13	n_D^{20} 1.4455, nmr	[332]
$HB[N(Ph)CH_2Ph]_2$	I		ir, nmr	[273]
$HB[N(CH_2Ph)(Np\text{-}2)]_2$	I		ir, nmr	[273]

Table 5 *(c'td.)*

Compound	Method	B.p. (°C/mm.) (m.p., °C)	Comments	References
NMe MeB NMe	B	30–1/2	 nmr	[395] [295]
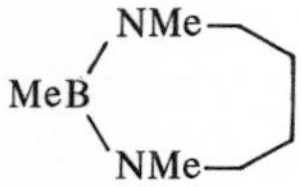 NMe MeB NMe	B	34–5/2		[395]
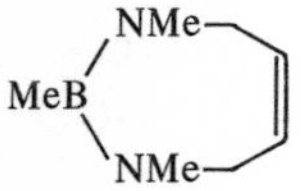 NMe MeB NMe	B	43–4/2		[395]
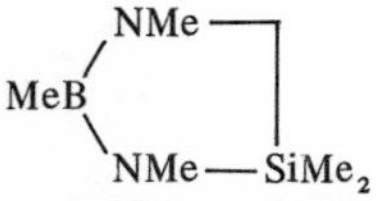 NMe MeB NMe—$SiMe_2$	A	151/760	nmr	[131]
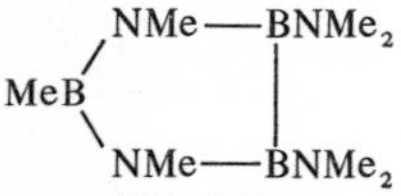 NMe—$BNMe_2$ MeB NMe—$BNMe_2$	B	54–5/~10^{-4}	ir, nmr	[304]
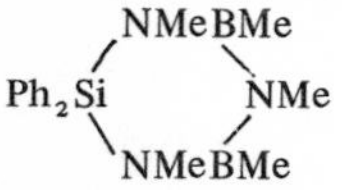 NMeBMe Ph_2Si NMe NMeBMe	A	(143–5)	nmr	[132]
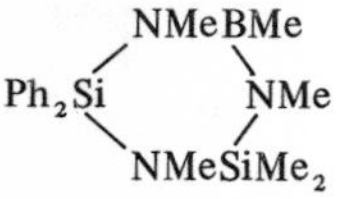 NMeBMe Ph_2Si NMe $NMeSiMe_2$	A		nmr	[132]
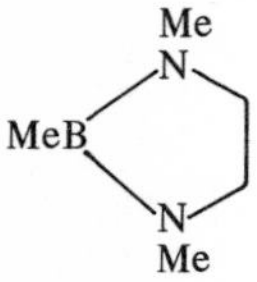 Me N MeB N Me	C	75/130	ir, nmr pes	[130d, 333a] [203a]
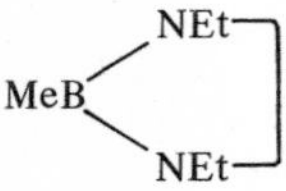 NEt MeB NEt	B	33–4/2		[395]
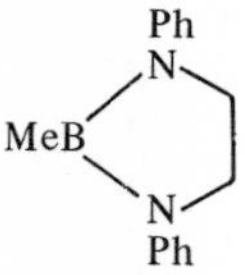 Ph N MeB N Ph	C	(97–8)	ir, nmr	[294a]
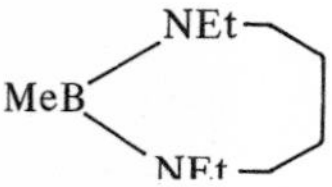 NEt MeB NEt	B	43–5/2		[395]

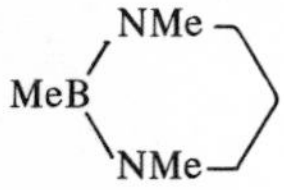

Table 5 *(c'td.)*

Compound	Method	B.p. (°C/mm.) (m.p., °C)	Comments	References
MeB(NHPr^i)$_2$	C		nmr	[302]
MeB[N(H)CH$_2$CH$_2$CHCH$_2$]$_2$	D	48–52/0.01	ir, nmr	[239]
CH$_2$[B(NMe$_2$)$_2$]$_2$	A		ir, nmr	[203]
EtB(NH)(NH) (ring)	B	44–5/13		[293]
			ir, Raman	[292]
			nmr	[295]
MeB(N)$_2$	C	88–90/1	ir, nmr	[333a]
MeB(N)$_2$	C	67/10^{-2}	ir, nmr	[333a]
MeB(N)$_2$ (Me, Me)	C	(50–5)	ir, nmr	[333a]
MeB(NMe$_2$)N	C	36/1	ir, nmr	[333a]
MeB (N-Bu^t, N-Bu^t)	C			[358b]
MeB (N-Ph, N-Ph)	C	112–5/0.05	ir, nmr	[394a]
.Cr(CO)$_3$			ir, nmr	[394a]
EtB(N)$_2$	A	73/0.1	ir, nmr, mass	[31]
EtB(N)$_2$ (Me)	A	83–90/0.1	ir, nmr, mass	[31]
EtB(N)$_2$ (Me, Me)	A	94–7/0.1	ir, nmr, mass	[31]

Table 5 *(c'td.)*

Compound	Method	B.p. (°C/mm.) (m.p., °C)	Comments	References
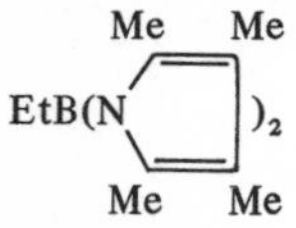 EtB(N(C(Me)=C(Me))₂ ring)₂ [structure: Me Me / EtB(N ...)₂ / Me Me]	A	129–30/0.1	ir, nmr, mass	[31]
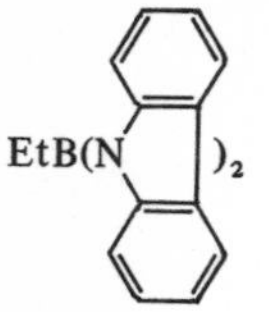 EtB(N-carbazolyl)₂ [structure]	E	(185)	ir, nmr, mass	[31]
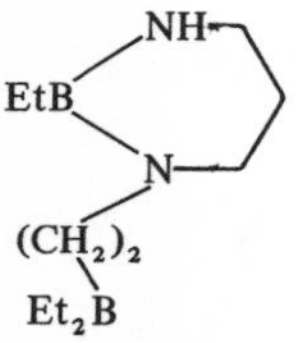 EtB(NH–(CH₂)₃–N(CH₂)₂BEt₂) cyclic [structure]	B+E	95–7/1	ir	[291]
$EtB(NMe_2)_2$	C	33–5/10	d_4^{20} 0.802, n_D^{20} 1.4388	[306]
$PrB(NHEt)_2$	G	66–7/12	d_4^{20} 0.7908, n_D^{20} 1.4272	[264]
			MCD	[225]
$PrB(NHBu)_2$	G	110–1/11	d_4^{20} 0.8041, n_D^{20} 1.4381	[264]
			MCD	[226]
$PrB(NHPh)_2$	G	162–3/1	n_D^{20} 1.5837	[248]
	B, G	133–4/0.35	n_D^{20} 1.5820	[264]
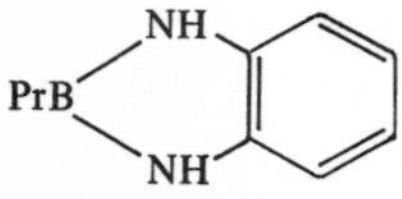 PrB(NH)₂C₆H₄ [structure]	B, C	(102–3)		[169]
	G	(92–4)	uv	[326]
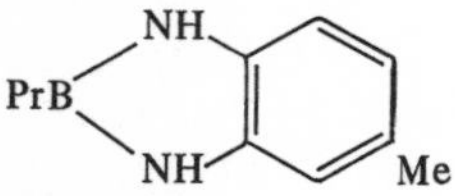 PrB(NH)₂C₆H₃Me [structure]	G	(99–101)	uv	[334]
$PrB(NMe_2)_2$	C	45–8/11	d_4^{20} 0.805, n_D^{20} 1.4397	[306]
$PrB(NEt_2)_2$	C	97–7.5/18	d_4^{20} 0.8120, n_D^{20} 1.4450	[265]
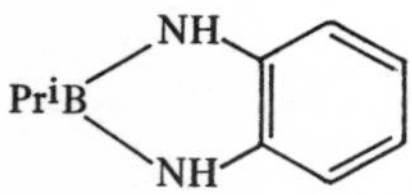 PrⁱB(NH)₂C₆H₄ [structure]	B, C	(124–6)		[169]
$BuB(NHMe)_2$	C	48/0.1	ir, nmr	[304]

Table 5 *(c'td.)*

Compound	Method	B.p. (°C/mm.) (m.p., °C)	Comments	References
$BuB(NHEt)_2$	G	74/14	d_4^{20} 0.7990,	
			n_D^{20} 1.4308	[264]
			MCD	[225]
$BuB(NHBu)_2$	G+I			[272]
			ΔH_f	[147]
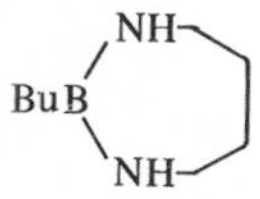 NH BuB NH	B	64/1		[293]
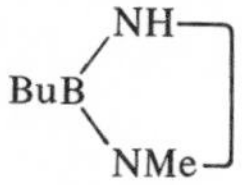 NH BuB NMe	B	33–5/2		[395]
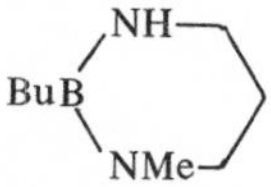 NH BuB NMe	B	44–6/2		[395]
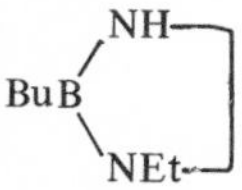 NH BuB NEt	B	45–7/2		[395]
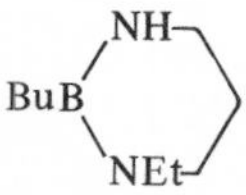 NH BuB NEt				[395]
$H_2N(CH_2)_3NHB(Bu)NBu_2$	B		ir	[291]
$BuB(NHPh)_2$	G	169–71/1	n_D^{20} 1.5750	[248, 270]
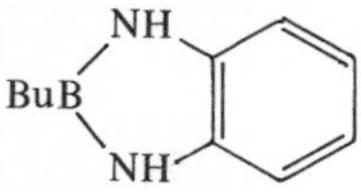 NH BuB NH	B, C	(66–7)		[169]
	G	(86–6.5)		[227]
$BuB(NMe_2)_2$	C	67–9/12	d_4^{20} 0.808,	
			n_D^{20} 1.4399	[306]
	A	50–1/5		[348]
	J			[99]
			nmr	[110]
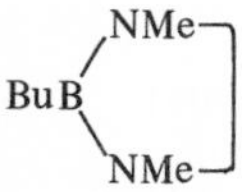 NMe BuB NMe	B	31–3/2		[395]
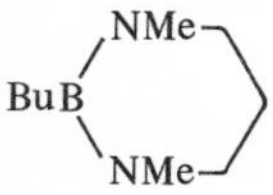 NMe BuB NMe	B	48–51/2		[395]

Table 5 *(c'td.)*

Compound	Method	B.p. (°C/mm.) (m.p., °C)	Comments	References
BuB(NMe—)(NMe—) [cyclic, –(CH₂)₄– bridge]	B	52–4/2		[395]
BuB(NMe—)(NMe—) [cyclic, bridge with C=C]	B	59–61/2		[395]
BuB(NMe—$BNMe_2$)(NMe—$BNMe_2$) [cyclic, B–B bond]	B	92–3/h.v.	ir, nmr	[304]
$BuB(NEt_2)_2$	F	77/0.3	nmr MCD	[170] [226]
BuB(NEt—)(NEt—) [cyclic, –CH_2CH_2– bridge]	B	45–7/2		[395]
BuB(NEt—)(NEt—) [cyclic, –(CH₂)₃– bridge]	B	65–8/2		[395]
Bu^sB(NH—)(NH—) [cyclic, o-phenylene bridge]	B, C	(61–2)		[169]
$Bu^sB(NMe_2)_2$	A	51/7	ir, nmr	[348]
$Bu^tB(NMe_2)_2$	A, I	66/20	ir, nmr	[348]
Bu^tB(NPh—)(NPh—) [cyclic, –CH_2CH_2– bridge]	D	35–7		[176]
$PentB(NEt_2)_2$			MCD	[225]
$Pent^iB(NHEt)_2$	G	53–4/3	n_D^{20} 1.4332	[264]
$Pent^iB(NHBu)_2$	G	93.5–4/2	d_4^{20} 0.8078, n_D^{20} 1.4422	[264]
$Pent^iB(NHPh)_2$	G, B	173–3.5/2	n_D^{20} 1.5700	[264]
Bu_3CB(NPh—)(NPh—) [cyclic, –CH_2CH_2– bridge]	D	(64–5)		[176]
$(PhCH_2)_3CB$(NPh—)(NPh—) [cyclic, –CH_2CH_2– bridge]	D	(97–9)		[176]

Table 5 *(c'td.)*

Compound	Method	B.p. (°C/mm.) (m.p., °C)	Comments	References
$(CH_2)_3[B(NMe_2)_2]_2$	G	97/2	d_4^{20} 0.8841, n_D^{20} 1.4775	[266]
$(CH_2)_3[B(NEt_2)_2]_2$	G	133–4/1	d_4^{20} 0.9202, n_D^{20} 1.4973	[266]
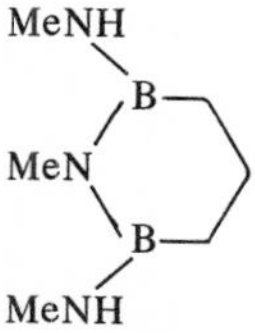 MeNH, MeN, MeNH (1,3-bis(methylamino)-2-methyl-1,3,2-diazaborinane ring structure: B–(CH₂)₃–B, bridged by MeN)	G	87–91/7	d_4^{20} 0.9328, n_D^{20} 1.4895	[266]
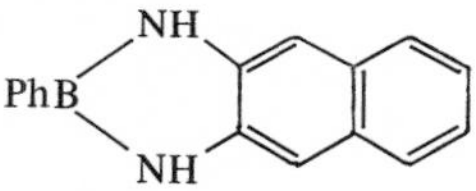 PhB(NH)₂ fused to naphthalene (structure)	G	(327–8)	uv	[334]
$PhB(NMe_2)_2$			nmr	[18]
			mass	[109]
	C	59/3		[259, 280]
	A	61/2.5	ir, nmr	[110, 348]
$.SnCl_4$		(146–150)	ir	[223]
$(PhBNMe_2)_2NH$	A		nmr	[332]
	B	158–60/3		[312]
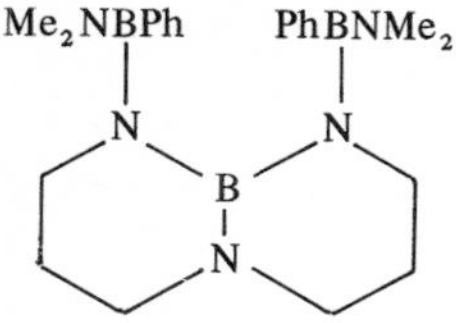 Me_2NBPh, $PhBNMe_2$ (structure: two N in six-membered rings, B–N)	A	205–8/1 (100–2)		[130]
$PhB(NMe_2)NHSiMe_3$	B	82–3/3		[188]
$PhB(NMe_2)NHSiEt_3$	B	119–22/3		[188]
$PhB(NMe_2)NRR'$ ($R = SiMe_3$; $R' = SiMe_3$, Bu^t, Pr^i, Et, Me, H; $R = GeMe_3$; $R' = GeMe_3$, Bu^t; $R = SnMe_3$; $R' = SnMe_3$, Me)	A		dnmr	[279]
$PhB(NH_2)N(SiMe_3)_2$			dnmr	[279]
PhB(NMe)₂ ring (structure: PhB, NMe, NMe)			mass	[109]
	B	73/3		[280]
			nmr	[295]
		(210)		[243a]

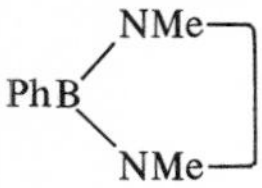

Table 5 *(c'td.)*

Compound	Method	B.p. (°C/mm.) (m.p., °C)	Comments	References
PhB(NMe)CH=CH(NMe) (ring: Me–N, N–Me)		250–1/745	ir, nmr	[243a]
$PhCH_2(Me)NB$ (benzo-fused ring, N–Me)		120–2/0.2		[201]
PhB(NMe–)₂ ring (NMe, NMe)	B	71–4/2		[395]
			nmr	[48b, 295]
PhB(NMe–)₂ ring, Me	B		nmr	[295]
PhB(NMe–)₂ ring	B	85–6/2		[395]
$PhB(NMePh)_2$	C	155/0.05	nmr, mass	[17]
PhB($NMeSiMe_2$)₂NMe ring	A	(116–8)	nmr	[132]
$PhB(NEt_2)_2$	C	70/2		[259,280]
	F	100/3	n_D^{20} 1.4992	[256]
$PhB(NHPr^i)_2$	C		nmr	[302]
$PhB(NMe_2)NHBu^t$	C	64/0.3	d_4^{20} 1.195, dnmr n_D^{20} 1.4983	[225]
$PhB(NEt_2)NHBu^t$	C	60/0.2	d_4^{20} 1.165, dnmr n_D^{20} 1.4896	[225]
$PhB(NPr^i_2)NHR'$ (R′ =				
Et	C	104/0.01	d_4^{20} 1.178, dnmr n_D^{20} 1.4952	[225]
Pr^i	C	74/0.05 (28)	dnmr	[225]
Bu^t)	C	70/0.05 (30)	dnmr	[225]
$PhB(NPr^i_2)_2.SnCl_4$		(104–5)	ir	[223]

Table 5 *(c'td.)*

Compound	Method	B.p. (°C/mm.) (m.p., °C)	Comments	References
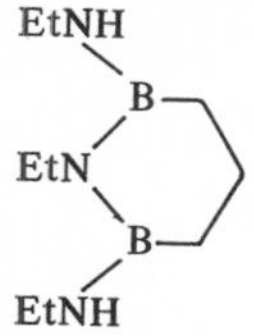 EtNH B EtN B EtNH	G	120–1/15, 62–6/1	d_4^{20} 0.8892, n_D^{20} 1.4740	[266]
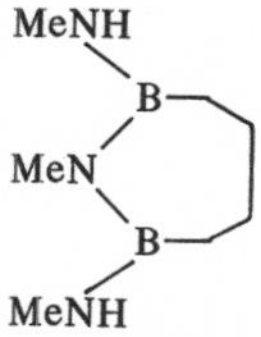 MeNH B MeN B MeNH	G			[271]
$C_2H_3B(NMe_2)_3$	A		ir, nmr	[413a]
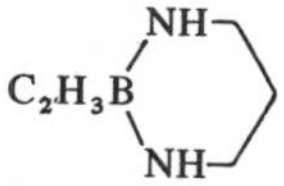 NH C_2H_3B NH	B	41/7	ir, Raman	[292] [248]
$CH_2CMeB(NMe_2)_2$	C	42.5/17		[38]
$MeCHCHB(NMe_2)_2$	A	96–7/15		[38]
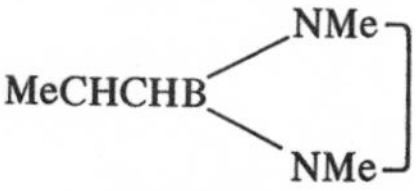 NMe MeCHCHB NMe		58–63/18		[38]
$MeCHCHB(NEt_2)_2$	C	58/22		[38]
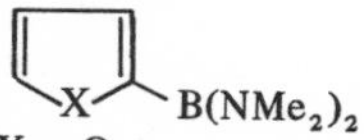 X B$(NMe_2)_2$	A		ir, nmr	[413a]
(X = O,		38–42/1		
S,		68–71/1		
NMe)		51–3/1		
$MeCCB(NMe_2)_2$	A	72/15	ir, nmr, uv	[368, 413c]
$MeCCB(NPh_2)_2$	A	96/0.5	ir, nmr, uv	[368]
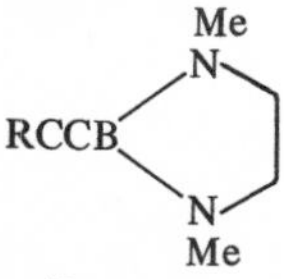 Me N RCCB N Me	A		ir, nmr	[413c]
(R = Me		31–3/1		
Ph)		(50–1)		
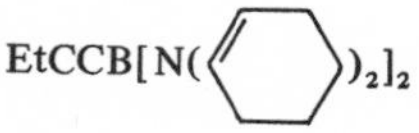 EtCCB[N( )$_2$]$_2$	A	120/1	ir, nmr, uv	[368]

Table 5 *(c'td.)*

Compound	Method	B.p. (°C/mm.) (m.p., °C)	Comments	References
$EtCCB[N(CMeCH_2)_2]_2$	A	72/1.5	ir, nmr, uv	[368]
$EtCCB(NPh_2)_2$	A	145/3	ir, nmr, uv	[368]
$PhB(NHMe)_2$	C	106–7/16	d_4^{20} 0.9645, n_D^{20} 1.5275	[61]
$PhB(NHEt)_2$	C	106–8/10	d_4^{20} 0.9248, n_D^{20} 1.5055	[61]
	C	97–8/3	d_4^{20} 0.924, n_D^{20} 1.5120	[247]
	B	77–9/2	n_D^{20} 1.5098	[256]
$PhB(NHPr)_2$	C	134–5/11	d_4^{20} 0.9097, n_D^{20} 1.5007	[61]
$PhB(NHPr^i)_2$	C	109–10/10	d_4^{20} 0.8870, n_D^{20} 1.4901	[61]
$PhB(NHBu)_2$	C	118–9/0.5	d_4^{20} 0.8997, n_D^{20} 1.5954	[61]
$PhB(NHBu^i)_2$	C		n_D^{20} 1.4929, ir, nmr	[362]
	C	80–1/0.1	d_4^{20} 0.8949, n_D^{20} 1.4930	[61]
	F	96–8/2	d_4^{20} 0.8892, n_D^{20} 1.4930	[256]
$PhB(NHBu^s)_2$	C	80–1/0.3	d_4^{20} 0.8862, n_D^{20} 1.4848	[61]
$PhB(NHBu^t)_2$	C	71–2/0.1	d_4^{20} 0.8795, n_D^{20} 1.4837	[61]
H N PhB N H (ring)			mass	[96]
H N — Me PhB N H (ring)			mass	[96]
NH PhB NMe (ring)	B	66–8/2		[395]
			nmr	[295]
NH PhB NEt (ring)	B	70–2/2		[395]

Table 5 *(c'td.)*

Compound	Method	B.p. (°C/mm.) (m.p., °C)	Comments	References
PhB(NH–)(NH–) (ring)	B	94–5/1 (50–2)	ir, Raman, mass	[96, 292] [293]
			nmr	[295]
$(N,N)D_2$			nmr	[295]
PhB(NH–)(NMe–) (ring)	B	74–6/2		[395]
PhB(NH–)(NEt–) (ring)	B	81–3/2		[18b, 395]
PhB(NH–)(N–) (ring) $(CH_2)_3NH_2$	B, I	121–3/1	ir	[291]
PhB(NH–)(NH–) (ring)	B	102–4/1 (27–9)	ir, Raman	[292] [293]
$PhB[N(C_3H_5)_2]_2$	B	101–2/2	d_4^{20} 0.9461, n_D^{20} 1.5290	[256]
$PhB(NHPh)_2$	C	(83.5–5.5)		[247]
	G	(84–6)		[248]
	G	178–80/0.6 (84–6)		[260, 263]
$PhB(NHC_6H_4Me\text{-}4)_2$	G	(85–7)		[248]
$PhB(NHC_6H_4Br\text{-}2)_2$	C		ir, nmr	[34]
$PhB(NHC_6H_3Br_2\text{-}2,6)_2$	C		ir, nmr	[34]
(PhB(NH)(NH)C_6H_3–)_2	B	(320–1)	ir, nmr	[86a]
(PhB(NH)(NH)C_6H_3–)_2O	B	(294–6)	ir, nmr	[86a]

Table 5 *(c'td.)*

Compound	Method	B.p. (°C/mm.) (m.p., °C)	Comments	References
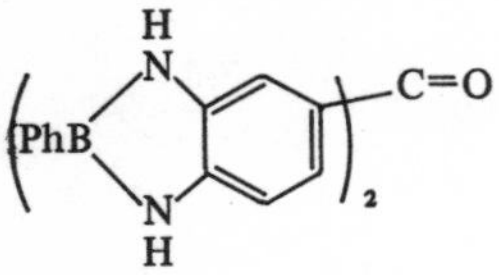 H, N, PhB, N, H, C=O, 2	B	(294–6)	ir, nmr	[86a]
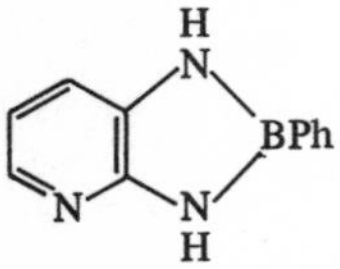 H, N, BPh, N, N, H	B	(221–3)	ir, nmr	[86a]
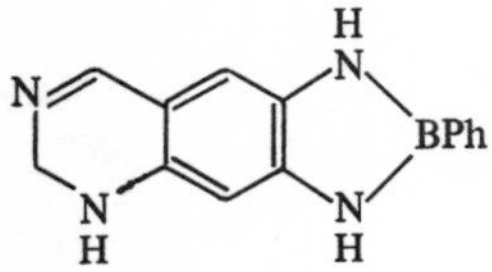 N, N, H, H, N, BPh, N, H	B	(353–5)	ir, nmr	[86a]
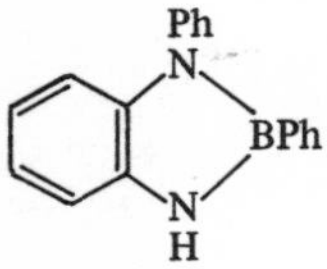 Ph, N, BPh, N, H	B	(114–6)	ir, nmr	[86a]
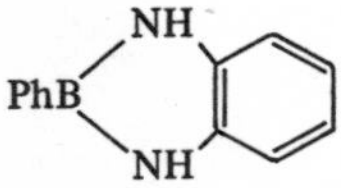 NH, PhB, NH	G			[46]
		(92.5–3.5)	ir	[75]
	C	(204–6)	uv	[108]
	G	(215–6)		[227]
	G	(212–4)	ir, uv	[334]
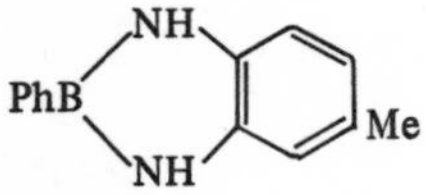 NH, PhB, NH, Me	G	(224–5)	uv	[334]
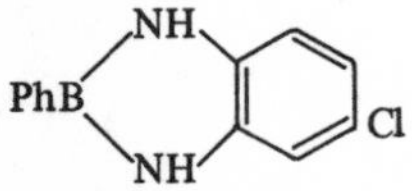 NH, PhB, NH, Cl	G	(183–4)	uv	[334]
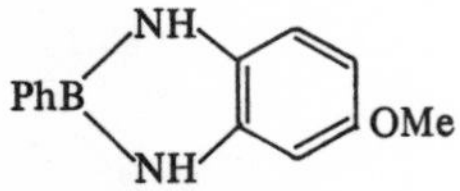 NH, PhB, NH, OMe	G	(138–40)	uv	[334]
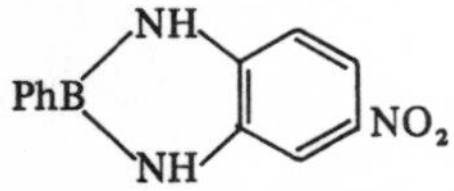 NH, PhB, NH, NO_2	G	(203–4)	uv	[334]

Table 5 *(c'td.)*

Compound	Method	B.p. (°C/mm.) (m.p., °C)	Comments	References
PhB, NH, NH, CO_2H	G	(209–10)	uv	[334]
PhB, NEt, NEt	B	95/8		[280]
			nmr	[295]
PhB, NEt, NEt	B	112–4/3		[395]
$PhB(NBu_2)_2$		118–9/0.5		[61]
PhB, NBu, NBu	B	100–5/0.2	d_4^{20} 0.9289, n_D^{20} 1.5047	[93]
$PhB(N\quad)_2$	B	156–60/4 (63–5)		[256]
O, C, H N, MeN, B, R, C, O, N Me, N H	G		ir, nmr, uv	[76]
(R = H		(262)		
o-Me		(258)		
m-Me		(291)		
p-Me		(294)		
o-OMe		(286)		
m-OMe		(293)		
p-OMe		(315)		
p-Br)		(290)		
p-$MeC_6H_4B(NHPh)_2$	G	163–5/0.04 (100–2)		[259, 263]
o-$Me_2NCH_2C_6H_4B(NMe_2)_2$	A			[128]
$PhCH_2NMeB$, MeN	I		ir	[200]
p-ClC_6H_4B, NH, NH	G	(219–21)	uv	[334]

Table 5 *(C'td.)*

Compound	Method	B.p., (°/mm.) (m.p., °C)	Comments	References
p-$BrC_6H_4B(NH)_2C_6H_4$ (1,2-phenylenediamino ring)	G	(232–3)		[227]
p-$MeOC_6H_4B(NH)_2C_6H_4$ (1,2-phenylenediamino ring)	G	(258–8.5)		[227]
	G	(242–3)		[334]
$pMeOC_6H_4B(NMe_2)_2$	C	82/0.3	n_D^{20} 1.5256, nmr, mass	[17]
p-$O_2NC_6H_4B(NH)_2C_6H_4$ (1,2-phenylenediamino ring)	G	(218–9)	uv	[334]
p-$HO_2CC_6H_4B(NH)_2C_6H_4$ (1,2-phenylenediamino ring)	G	(281–2)	uv	[334]
$CH_2{=}C(Me)C_6H_4B(NH)_2C_6H_4$ (1,2-phenylenediamino ring)	G	(237–8)		[227]
1-$NpB(NHBu^i)_2$	F	156–8/2.5	d_4^{20} 0.9845, n_D^{20} 1.5478	[256]
	F	156–63/2.5	n_D^{20} 1.5480	[256]
1-$NpB(NH)_2C_6H_4$ (1,2-phenylenediamino ring)	G	(149–50)		[227]
Me_2NB—$BNMe_2$ bridged via N,N of bicyclic $B(N)_3$ (B—N ring system)	A	105/~10^{-4}	ir, nmr	[303, 304]
$Me_2NB{-}NMe{-}BMe{-}NMe{-}BMe{-}NMe{-}BNMe_2$ (ring: $Me_2NB{-}NMe{-}BMe$, NMe, $Me_2NB{-}NMe{-}BMe$)	B	80–3/~10^{-4}	ir, nmr	[304]
$Me_2NB{-}NMe{-}BBu{-}NMe{-}BBu{-}NMe{-}BNMe_2$ (ring: $Me_2NB{-}NMe{-}BBu$, NMe, $Me_2NB{-}NMe{-}BBu$)	B	125/~10^{-4}	ir, nmr	[304]

Table 5 *(c'td.)*

Compound	Method	B.p., (°/mm.) (m.p., °C)	Comments	References
$Me_2NB{-}NMe$, BBu, $ClB{-}NMe$ (ring structure)	I			[304]
$Me_2NB{-}BCl$ linked through N, N to a bicyclic B–N ring system (structure)	I		ir, nmr	[304]
$(CF_3)_2C{=}NB(NMe_2)_2$		70–73/99	ir, nmr, mass	[119a]

† Footnote to Table 5: for explanation of abbreviations, see footnotes to Table 1, p. 57.

Table 6† Tris(amino)boranes [B(N<)$_3$]

Compound	Method	B.p. (°C/mm) (m.p., °C)	Comments	References
$B(NHMe)_3$	C	25/15–20		[148, 216, 231]
	B	41/12	d_4^{20} 0.8871, n_D^{20} 1.4465	[8]
	C	59/33	d_4^{20} 0.8871, n_D^{20} 1.4465	[6]
	B, C	70–80/48	n_D^{20} 1.4432	[121, 122]
			ir	[9, 281a]
$B(NHR)_3$ (R = Me, Et, Pr, Pr^i, Bu^s, Bu^t)			nmr	[101, 339]
$MeNHB(NPr^i_2)_2$	C	82/23	d_4^{20} 0.8494, n_D^{20} 1.4562	[5]
			dnmr	[96a, 96b]
$MeNHB(NHSiMe_3)N(SiMe_3)_2$	C	81–2/2	ir, nmr, mass	[397]
$B(NHEt)_3$	C	62–3/10	d_4^{20} 0.833, n_D^{20} 1.4380	[6, 210]
			MCD	[226]
			ir	[9]
$EtNHB(NHSiMe_3)N(SiMe_3)_2$	C	85–6/2	ir, nmr, mass	[397]
$EtNHB(NPr^i_2)_2$	C	62/0.05	d_4^{20} 0.8525, n_D^{20} 1.4566	[5]
			dnmr	[96a, 96b]
EtNHB(N-piperidyl)$_2$ (ring structure)	C			[230]
$B(NHPr)_3$		76/0.5	n_D^{20} 1.4409	[232]
$PrNHB(NPr_2)_2$	C	95/0.2	d_4^{20} 0.8375, n_D^{20} 1.4523	[5]
$PrNHB(NPr^i_2)_2$	C	92/0.9	d_4^{20} 0.8456, n_D^{20} 1.4554	[5]

Table 6 *(c'td.)*

Compound	Method	B.p. (°C/mm) (m.p., °C)	Comments	References
$B(NHPr^i)_3$	C	37/1	d_4^{20} 0.8000, n_D^{20} 1.4267	[6]
		81–2/18	n_D^{20} 1.4238	[233]
	C	85–7/21		[231, 233]
			ir	[9]
$Pr^iNHB(NHC_3H_5)_2$				[230]
$Pr^iNHB(NPr^i_2)_2$	C	62/0.2	d_4^{20} 0.8492, n_D^{20} 1.4565	[5]
			dnmr	[96a, 96b]
$B(NHBu)_3$	B, C	84/0.005	d_4^{20} 0.8347, n_D^{20} 1.4462	[6, 8]
	G	100–2/0.06	d_4^{20} 0.8378, n_D^{20} 1.4482	[363]
			ΔH_f	[147]
	G	105–6/1		[374]
	B, C	105–6/1	n_D^{20} 1.4438	[121, 122]
			MCD	[226]
			ir	[9]
$BuNHB(NPr^i_2)_2$	C	108/0.5	d_4^{20} 0.8438, n_D^{20} 1.4562	[5]
$B(NHBu^i)_3$		90–2/0.3	n_D^{20} 1.4398	[232]
$B(NHBu^s)_3$	C	49/0.1	d_4^{20} 0.8186, n_D^{20} 1.4355	[6]
	B, C	134–6/47	n_D^{20} 1.4324	[121, 122]
			ir	[9]
$B(NHBu^t)_3$	C	42/0.05	d_4^{20} 0.7971, n_D^{20} 1.4272	[6]
		95–8/20		[233]
		97–7.5/18	n_D^{20} 1.4253	[232]
	C	101–7/24		[148, 231]
			ir	[9]
$Bu^tNHB(NHSiMe_3)N(SiMe_3)_2$	C	96–7/2	ir, nmr, mass	[397]
$Bu^tNHB(NPr^i_2)_2$	C	88–91/1	d_4^{20} 0.8620, n_D^{20} 1.4616	[8]
$B(NHPent)_3$	G			[374]
$B(NHHex)_3$	B, C	150/0.1d	n_D^{20} 1.4549	[121, 122]
$B(NHC_6H_{11})_3$	G			[374]
$B(NHC_{18}H_{35})_3$	C	104–5/1 (65–71)		[122]
$B(NHCH_2Ph)_3$	C		d_4^{20} 1.0479, n_D^{20} 1.5805	[6]
			ir	[9]
B(NH-cyclopentyl)$_3$				[47]
B(NH-cyclohexyl)$_3$				[47]
			ir	[9]
B(NH-tetrahydronaphthyl)$_3$	G			[47, 374]

Table 6 *(c'td.)*

Compound	Method	B.p. (°C/mm) (m.p., °C)	Comments	References
H N Me_2NB N H		48/5	ir, nmr	[72]
Me N Me_2NB N Me		33–4/2	pes	[203a, 282]
NH Me_2NB NH	B		ir	[291]
		50/2	ir, nmr	[72]
R^2 R^3 N—N R^1B BR^4 N R^5	C, B		ir, nmr	[296]
(R^1 R^2 R^3 R^4 R^5				
Me Me Me Me Me		74/34		
Me Me Me Me H		74/60		
Me H Me Me Me		72/41		
Me H Me Me H)		74/75		
Me Me B N B MeN N NMe B N B Me Me	C, B	(117)	ir, nmr	[296]
NH NH B N	B	62/1		[293]
		(38–41)		
	B	68/2	ir	[291]
		(38–41)	X-ray	[57, 58]
NH NLi B N	E			[130]
$B(NHC_3H_5)_3$				[230]
$B(NHPh)_3$		(150–8)		[232]
	G	(163–7)		[363]
	B	(164–7)		[121]

Table 6 *(c'td.)*

Compound	Method	B.p. (°C/mm) (m.p., °C)	Comments	References
	C	(166–9)		[189]
		(168)		[208]
		(168–70)		[40,47]
	A	(170)		[112]
	F	(170–2)	ir	[9,185]
$(PhNH)_2BN$ 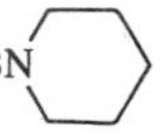 (piperidine ring)	F	(170–2)	ir	[198]
$PhNHB(NMe_2)_2$	B	72–5/0.1	n_D^{20} 1.5353	[42]
$[PhNHBNPh]_2$	I	(201–2)	ir	[198]
$B(NHC_6H_4Me\text{-}p)_3$	G	(157–60)		[248]
	C	(165–6)		[197]
$B(NHC_6H_4Cl\text{-}p)_3$		(207)		[208]
	A	(210)		[112]
$B(NHC_6H_4OMe\text{-}p)_3$	C	(124)d		[197]
$B(NH\text{-}3,5\text{-}C_6H_3Cl_2)_3$		(193)		[208]
$B(NH\text{-}2,4,6\text{-}C_6H_2Br_3)_3$	A	(225)		[112]
$B(NHNp\text{-}1)_3$	A	(160–90)d		[112]
$(H_2N)_2BNHNp\text{-}1$				[230]
$B(NHNp\text{-}2)_3$	A	(257)d		[112]
Me_2NB 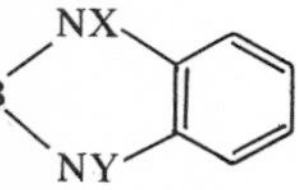 (NX, NY ring fused to benzene)	B		nmr, mass	[32,146b]
(X = H = Y		72–4/720		
		(220)d		
X = H, Y = $SiMe_3$		48–53/5 × 10^{-3}		
X = $SiMe_3$ = Y)				
B(NH-(pyridyl ring, N, Me))$_3$ 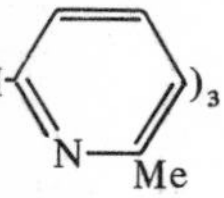	C	(92–6)	nmr	[278]
$B(NR_2)_3$				
(R = Me, Et, Pr, or Pr^i)			nmr	[101,339]
$B(NMe_2)_3$		38–9/10	d_4^{20} 0.84	[74]
		(−10)		
	C	39/10	d_4^{20} 0.8380, n_D^{20} 1.4462	[8]
	G	43/12		[374]
		48–50/20	d_4^{20} 0.834, n_D^{20} 1.4450	[19]
		(−16)		
	A	60/30		[134]
	A	146–7/760	ir, nmr	[9,24,112, 333c,348]
		147–8/760	n_D^{20} 1.4453	[41,233]
	J			[99]

Table 6 *(c'td.)*

Compound	Method	B.p. (°C/mm) (m.p., °C)	Comments	References
			1st I.P., pes	[15, 36, 196, 213]
	C	147.5/760 (extr.) (–40)		[406]
			mass	[109]
	I	152 ± 2/760 (extr.) (–16.5–16.1)		[69]
			electron diffraction	[80]
(H_{18} and D_{18})			ir	[238]
			pes	[36, 196]
$.HgCl_2$		d(120)	ir	[223]
$.VOCl_3$		d(80)	ir	[223]
$.W(CO)_3$		d(>60)	ir, nmr	[358]
$(Me_2NBNH)_3$				[137]
$(Me_2N)_2BNHSiMe_3$	B	83–91/3		[188]
$Me_2NB(NHSiMe_3)N(SiMe_3)_2$	C	93–4/2	ir, nmr, mass	[397]
$(Me_2N)_2BNMePh$	B	68–9/0.1	n_D^{20} 1.5338	[42]
			dnmr	[110]
Pr^i_2N-B(Y)$NHBu^t$ (Y	C			
$NHBu^t$		74/0.02	n_D^{20} 1.4445, d_4^{20} 1.083, ir, dnmr	[225]
$NHPr^i$		86/0.4	n_D^{20} 1.4327, d_4^{20} 1.078, ir, dnmr	[225]
NEt_2		48/0.2	n_D^{20} 1.4416, d_4^{20} 0.954, ir, dnmr	[225]
NMe_2)		60/0.1	n_D^{20} 1.4376, d_4^{20} 1.082, ir, dnmr	[225]
$(Me_2NBN$-[cyclohexenyl ring]$)_2$	I	130–50/10^{-3}	ir, uv	[337]
$(Me_2N)_2BNPr_2$	B	32/0.05	n_D^{20} 1.4485	[42]
$Me_2NB(NPr_2)_2$	B	75/0.2	n_D^{20} 1.4530	[42]
$Me_2NB(NPr^i_2)_2$	C	73/0.7	d_4^{20} 0.8523, n_D^{20} 1.4589	[5]
$(Me_2N)_2BN$-[ring with Me, Me]	C	36/1	dnmr	[96a, 96b]
			ir, nmr	[333a]

Table 6 *(c'td.)*

Compound	Method	B.p. (°C/mm) (m.p., °C)	Comments	References
$B(NMeC_{18}H_{37})_3$	A	(155)		[112]
$B(NMePh)_3$	A	(210)		[112, 208]
	B	(211–2)		[121]
	C	(214–6)		[8]
			ir	[9]
$B(NEt_2)_3$	C	50–3/0.4	d_4^{20} 0.826, n_D^{20} 1.4450	[139]
	C	52/0.2	d_4^{20} 0.8240, n_D^{20} 1.4460	[5]
	A	95/11, 220/760		[112]
			nmr	[342]
			MCD	[226]
			ir	[9]
$Et_2NB(NHSiMe_3)N(SiMe_3)_2$	C	105–7/0.2	ir, nmr, mass	[397]
$(Et_2N)_2BNPh$–(cyclohex-1-enyl)	K	153–4/1	n_D^{20} 1.5322, ir, nmr	[116]
$Et_2NB(NPr^i_2)_2$	C	110/4.5		[5]
		(76–8)	dnmr	[96a, 96b]
$(Et_2NBNPh)_2$	I	150–80/10^{-3}		[337]
$(Et_2NBNEt)_2$	I	125–35/10^{-3}		[337]
$B(NEtCH_2Ph)_3$	C			[120]
$B(NEtPh)_3$	A	(164–5)		[112]
$B(NPr_2)_3$	C	101/0.15	d_4^{20} 0.8462, n_D^{20} 1.4576	[5]
		117–9/21		[233]
		121–3/0.6		[226]
	A	123–4/32		[112]
			MCD	[226]
$B(NBu_2)_3$	C	136–42/0.1	d_4^{20} 0.8399, n_D^{20} 1.4578	[139]
	C	139/0.2	d_4^{20} 0.8409, n_D^{20} 1.4584	[5]
		148–54/0.5		[233]
	A	198/11		[112]
			MCD	[226]
			ir	[9]
$B(NBu^i_2)_3$	C	141/0.2		[5]
		(122–9.5)		
$B[N(CH_2Ph)_2]_3$	C	(266–8)		[5]
(bicyclic structure: N, B, N, N, $SiMe_2$)	A	96–7/8	nmr	[131]

Table 6 *(c'td.)*

Compound	Method	B.p. (°C/mm) (m.p., °C)	Comments	References
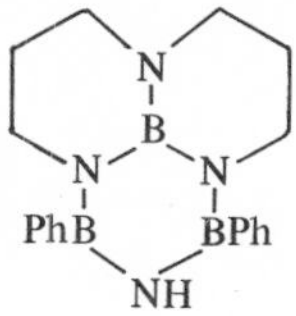 N, B, N, N, PhB, BPh, NH	B	228–30/3		[130]
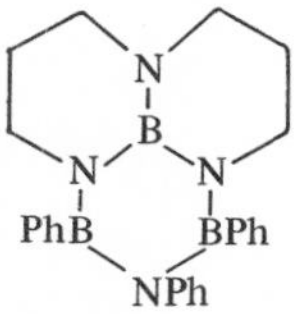 N, B, N, N, PhB, BPh, NPh	B	(259–60)		[130]
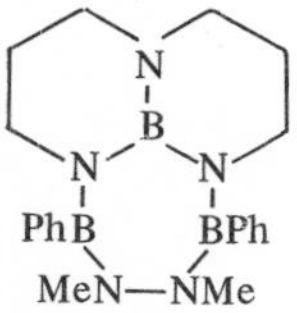 N, B, N, N, PhB, BPh, MeN—NMe	B	(106–8)		[130]
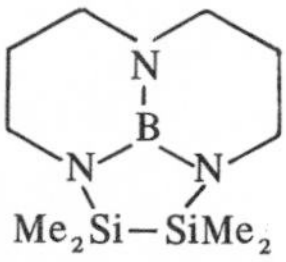 N, B, N, N, $Me_2Si-SiMe_2$	A	70/0.8 (58)	nmr	[131]
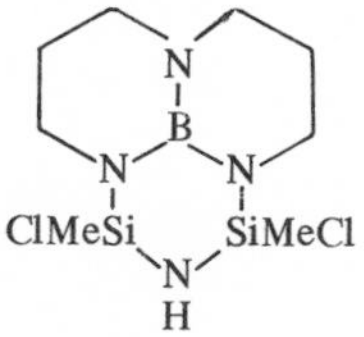 N, B, N, N, ClMeSi, SiMeCl, N, H	A		nmr	[132]
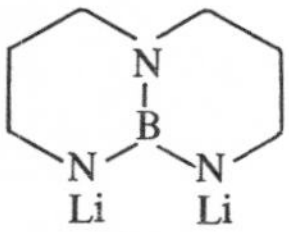 N, B, N, N, Li, Li	E			[130,304]
$B(NPh_2)_3$	B	(247–8)		[121]
$B[N(Ph)CH_2Ph]_3$	D	(201–1.5)	ir, nmr	[3, 4]
$B[N(CH_2Ph)(C_6H_4Me\text{-}p)]_3$	D	(191.5–2.5)	ir, nmr	[3, 4]
$B[N(Ph)(CH_2C_6H_4Cl\text{-}p)]_3$	D	(181–2)	ir, nmr	[3, 4]
$B[N(CH_2CH_2Ph)_2]_3$	A	(269–271)	ir, nmr	[345]
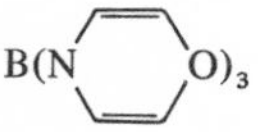 $B(N\ \ O)_3$	A	(72–5)	ir, nmr	[345]

Table 6 *(c'td.)*

Compound	Method	B.p. (°C/mm) (m.p., °C)	Comments	References
$B[N(CH_2CH_2CHMe_2)_2]_3$	A	134–6/0.025	ir, nmr	[345]
B(N)$_3$	C	67–9/22		[333a]
B(N)$_3$		164/13 (49)	ir, nmr	[112,333a]
B(N)$_3$	C	(208–210)	ir, nmr	[333a]
B(N)$_3$	C	(70.5)		[8]
		137–49/0.3 (71–4)		[233]
	B	142–9/1		[121]
			ir	[9]
B(N)$_2NH_2$	C			[8]
B(N)$_3$	B	(212)	ir, nmr, mass	[140,333a]
	E	(212)	ir, nmr, mass	[31]
	F			[198]
Me B(N)$_3$ Me	C	106–12/10^{-3}	ir, nmr	[333a]
B(N)$_3$	A	(348)		[112]
B(N)$_3$	A	(338)		[112]

Table 6 *(c'td.)*

Compound	Method	B.p., (°/mm.) (m.p., °C)	Comments	References
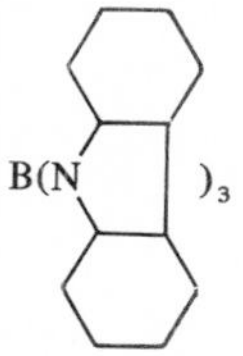 $B(N$ [perhydrocarbazolyl ring]$)_3$	A	(325)		[112]
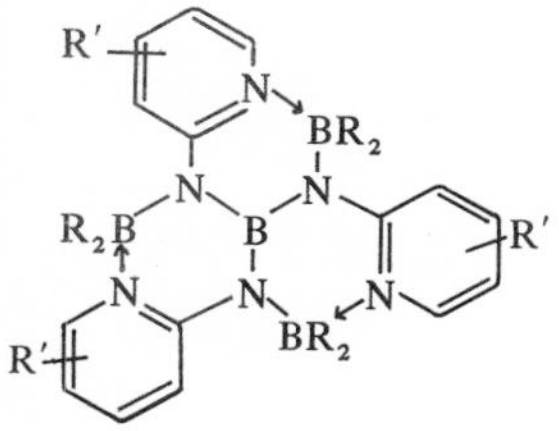 [cyclic structure: R'-pyridyl, $N \rightarrow BR_2$, R_2B, B, N, BR_2, R']	A+C			
(R = Me R' = H		(241.5–2)		[394]
R = Me R' = 4-Me		(198–9)	nmr	[278]
R = Et R' = H		(193.5–4)		[394]
R = Et R' = 5-Cl		(168–9)	nmr	[278]
R = Bu R' = H		(143–3.5)		[394]
R = Bu R' = 4-Bu		(166–9)	nmr	[278]
R = Bu R' = pyrimidyl)		(105–6)	nmr	[278]
$B(NMe_2)_2[N(CH_2CN)_2]$		$59/10^{-3}$ (79)	ir, nmr, mass	[240]
$B(NMe_2)_2[N(Me)CH_2CN]$		$67/10^{-3}$	ir, nmr, mass	[240]
$B(NMe_2)_2[N(Bu^t)CN]$		$83/10^{-3}$	ir, nmr, mass	[240]
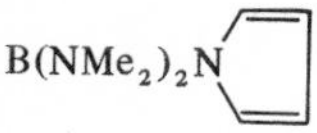 $B(NMe_2)_2N$ [pyrrolyl ring]	C	42–3/1	ir, nmr	[333a]
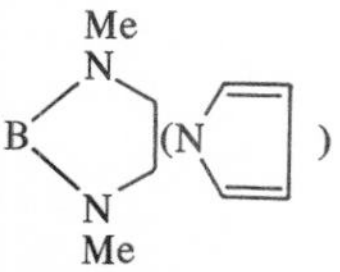 [B(NMe)CH$_2$CH$_2$(NMe) ring] B(N [pyrrolyl ring])	C	75–6/1	ir, nmr	[333a]
$B(NHSiMe_2Bu^t)_3$	A	$133/10^{-2}$	ir, nmr	[37a]
$B(NMe_2)[N(SiMe_3)SiMe_2Bu^t]NH_2$	A	$92/10^{-2}$	ir, nmr	[37a]

† Footnote to Table 6: for explanation of abbreviations, see footnotes to Table 1, p. 57.

Table 7† Aminohalogeno- or pseudohalogeno-boranes [(RBXN⊂) or (X_2BN⊂)]

Compound	Method	B.p. (°C/mm) (m.p., °C)	Comments	References
$HBClNMe_2$	K			[305]
dimer		(95–7)		[305]
$HBClNEt_2$	K	20–4/25, 112/760		[305]
dimer		(118–20)		
$HBClNPr_2$	K	41–1.5/10	n_D^{20} 1.4305	[305]
dimer		(117–9)		[305]
$HBClNBu_2$	K	52/3, 63–5/55	n_D^{20} 1.4410	[305]
$HBBrNMe_2$ dimer		(122–4)		[364, 390]
$MeBFNMe_2$		58		[405]
MeBClNHMe	K			[403]
$MeBClNMe_2$	I			[123, 159]
	K	94–6/720	d_4^{20} 0.922, n_D^{20} 1.4180	[306]
dimer		(100–1)		[306]
$MeBClNMe_2.Me_2BNMe_2$	I	(106–8)		[374]
MeBClNMePh	I			[26]
	A		ir, nmr, mass	[369]
$MeBBrNMe_2$	K	27–9/1	d_4^{20} 1.072, n_D^{20} 1.4537	[306]
	I	42/22	d_4^{20} 1.21, n_D^{20} 1.4602	[307]
dimer		(115–7)		[307]
$EtBClNMe_2$	C	20/1		[43, 286]
	K	116–9/720	n_D^{20} 1.4283	[306]
	I	116–8/720	d_4^{20} 0.924, n_D^{20} 1.4319	[307]
dimer		(83–5)		[307]
$PrBClNMe_2$	K	26–9/9		[306]
	I	139–41	d_4^{20} 0.918, n_D^{20} 1.4343	[307]
$BuBClNMe_2$	K	42–5/9	n_D^{20} 1.4359	[306]
	I	46–9/10	d_4^{20} 0.910, n_D^{20} 1.4357	[307]
$BuBINMe_2$	K	37–9/1	d_4^{20} 1.273, n_D^{20} 1.4900	[306]
$Bu^tBClNMe_2$	A	68	ir, nmr, uv	[369]
$C_2H_3BBrNMe_2$	A		ir, nmr, uv	[290]
$.Fe(CO)_3$			ir, nmr, mass	[356]
$PhBFNMe_2$	C	62/3	n_D^{20} 1.5058, nmr, mass	[17]
			nmr	[18]
$PhBClNMe_2$	C	82/0.2	n_D^{20} 1.5284, nmr, mass	[17]
			nmr, dnmr	[17, 18]
	A		ir, nmr, uv	[369]
	C	51–2/2	n_D^{20} 1.5320	[43, 286]
PhBClNMeEt			nmr	[18]
$PhBClNMePr^i$	C	90/0.2	n_D^{20} 1.5193, nmr, mass	[17]
			nmr	[18]
PhBClNMeBu	C	70–2/1		[286]
$PhBClNMeC_6H_{11}$	C	108/3		[281]
$PhBClNMeCH_2Ph$	C	123–7/3		[281]
PhBClNMePh			nmr	[18]
	C	103–10/3		[286]

Table 7 *(c'td.)*

Compound	Method	B.p. (°C/mm) (m.p., °C)	Comments	References
$PhBClNEt_2$	C	67/2		[286]
$PhBClNEtPh$	C	105–28/2		[286]
$PhBClNPr^i_2$	C	94/4		[286]
$PhBClNPr_2$			nmr	[48b]
PhBClN(piperidino ring)	C	82/2		[286]
			dnmr	[48a]
$PhBBrNMe_2$	C	63/0.8	n^{20}_D 1.5554, nmr, mass	[17]
			nmr	[18]
$PhBBrNMePh$	C	123/0.3 (88)	n^{20}_D 1.5990, nmr, mass	[17]
$PhBClNBu^s_2$	C		dnmr	[48a]
$PhBClN(R)SiMe_3$ ($R = SiMe_3, Bu^t$)			dnmr	[279]
$PhBN_3NEt_2$		57–9/11	ir	[337]
$(N_3)_2BNMe_2$		53–5/11	ir	[337]
$(N_3)_2BNEt_2$		75–7/11	ir	[337]
$NCB(Ph)NMe_2$		90–2/0.5	ir, nmr, mass	[241a]
$[NCB(Me)NMe_2]_{1-3}$		60/30 (155–160)	ir, nmr, mass	[241a]
$(OCN)_2BNPr^i_2$		74/2	n^{20}_D 1.4529, d^{20}_4 0.9113, ir	[221a]
p-$MeOC_6H_4B(Cl)NMe_2$	C	95/0.4	n^{20}_D 1.5438, nmr, mass	[17]
			nmr	[18]
F_2BNH_2			pes, microwave	[203c, 228]
F_2BNHMe	C	46		[405]
			pes	[203c]
F_2BNHBu^t	C	(140)		[117,155]
F_2BNMe_2 dimer	C	(152–7)		[365]
		(160–4)		[377]
		(165)		[299]
		(165–8)		[19,158]
	I	(132/760) (subl., extr.) (165–8)		[22, 64]
	C	(167–9) (subl. 90/2)		[49]
	A	(174.5–5)		[333, 371]
	I		ir	[64]
			1st I.P.	[213]
			uv, pes (monomer)	[36, 130a, 196, 203c, 213]
			nqr	[410]
			nmr	[413b]
F_2BNEt_2		(63)		[299]
	I	(65)		[389]

Table 7 *(c'td.)*

Compound	Method	B.p. (°C/mm) (m.p., °C)	Comments	References
	C	(65.5)		[139]
	A		ir, nmr, X-ray	[156,371]
F_2BNPr_2	C	99/45		[244]
$F_2BNPr^i_2$		60–78/10	n_D^{20} 1.4258	[217]
F_2BNBu_2		(35)		[299]
F_2BN(piperidino)		(162)		[299]
ArBClN(Me)Ar′ (Ar = Ph, Ar′ = 4-BrC_6H_4; Ar = 2-MeC_6H_4; Ar′ = 2-Me, 4-Br-C_6H_3)			X-ray	[418]
$F_2BNRSiMe_3$	C			
(R = Me		73	ir, nmr	[117]
Et		33/43	ir, nmr	[117]
Pr		43/40	ir, nmr	[117]
Bu		38/39	ir, nmr	[117]
Pr^i)		41/44	ir, nmr	[117]
Cl_2BNH_2	C		ir, mass	[205]
Cl_2BNHPr^i		(128–9)		[232]
Cl_2BNHBu^t dimer	C			[155, 386]
$Cl_2BNHHex$	C		ir	[141]
Cl_2BNH-(cyclohexyl)	C	(142–3)		[140]
Cl_2BNHCH_2Ph	C		ir	[140, 141]
Cl_2BNHPh	C		ir	[141]
$Cl_2BNHPh.py$	C		ir	[141]
$Cl_2BNHPh.H_2NPh$	C			[189]
$Cl_2BNHC_6H_4Ph$-*o*	C			[107]
$Cl_2BNHC_6H_4(C_6H_4Cl$-4)-2	C			[106]
$Cl_2BNHC_6H_3$(2-Ph)3-Cl	C			[106]
$Cl_2BNHC_6H_3$(2-Ph)6-Cl	C			[106]
$Cl_2BNHC_6H_4Me$-*p*	C		ir	[141]
$Cl_2BNHC_6H_4Cl$-*p*	C		ir	[141]
$Cl_2BNHC_6H_4Br$-*p*	C		ir	[141]
$Cl_2BNHC_6H_4OMe$-*p*	C		ir	[141]
Cl_2BNMe_2	C	(—43)	n_D^{20} 1.1394	[49, 55, 406]
		112/760 (extr.)		[149, 402, 407]
		(—46) 29/32 (—46)	d_4^{20} 1.150, n_D^{20} 1.4351	[19]
	K	35/35		[311]

Table 7 *(c'td.)*

Compound	Method	B.p. (°C/mm) (m.p., °C)	Comments	References
	A	44–6/80		[288]
	C	52–4/92		[367]
	I		ir	[16, 22, 24]
	A	51–3/90	μ	[26]
			1st I.P.	[15, 213]
	K	(–42.5)	nmr	[18b, 206, 413b]
			Raman	[112]
			ed	[81]
			uv, pes	[36, 130a, 196]
$Cl_2BNMe_2.2py$		70/0.1 (subl.)		[224]
$Cl_2BNMe_2.2p\text{-}MeC_5H_4N$		(85)		[224]
$2Cl_2BNMe_2.TiCl_4$			nmr	[206]
Cl_2BNMe_2 dimer	K	(135.5)	nmr	[206]
	A	(138–41)		[184]
	A	(142)		[19, 153, 159]
	I	(142.5)	μ	[55]
	C	(142)		[406, 407]
			nqr	[410]
			X-ray	[173]
$Cl_2BNMeEt$	C	94–8/7		[283, 294]
$Cl_2BNMeBu^t$	A	65–8/760	ir, nmr, uv	[369]
$Cl_2BNMeCH_2Ph$	C	68/2		[281]
$Cl_2BNMeC_6H_{11}$				[39]
$Cl_2BNMePh$	I	100/20	d_4^{20} 1.1976, n_D^{20} 1.5288	[8]
	C	55/1.5	ir, Raman	[13]
			μ	[26]
$Cl_2BNMeNp\text{-}1$	C	124/4 (70)		[283]
Cl_2BNEt_2	C	148 (–23)	d_4^{20} 1.0564	[335]
		150/760 (–25.5)		[149]
	C	(–58)	Raman	[153]
	C	48–9/18		[136]
		72.8/61.9		[39]
		75–5.5/63–4		[382]
	C	140–4d		[244]
		144		[140]
		148		[285]
$Cl_2BNEt_2.3py$		(82–6)		[159]
$Cl_2BNEtPh$	C	67/3		[283, 294]
Cl_2BNPr_2	C	99/45		[244]
$Cl_2BNPr^i_2$	C	61–3/13	n_D^{20} 1.4492	[136]

Table 7 *(c'td.)*

Compound	Method	B.p. (°C/mm) (m.p., °C)	Comments	References
	C	83/30		[281]
$Cl_2BNPr^i_2.py$		60/10 (subl.)		[224]
$Cl_2BNPr^i_2.p\text{-}MeC_5H_4N$		(74)		[224]
Cl_2BNBu_2	I	41/0.1	n_D^{20} 1.4500	[137]
$Cl_2BNBu^i_2$	C	92–5/17		[244]
$Cl_2BNBu^s_2$	C	62/6		[283]
$Cl_2BNPent_2$	C	121/10		[294]
			MCD	[226]
$Cl_2BN(C_6H_{11})_2$ (cyclohexyl)	C	114–6/4		[283]
$Cl_2BNC_4H_8$ (pyrrolidino)	C	70–4/17		[277]
dimer	C	(131–2)		[277]
$Cl_2BNC_5H_{10}$ (piperidino)	C	82–3/20		[277]
dimer	C	(105)		[277]
$Cl_2BNC_4H_8O$ (morpholino)	C	88–90/25		[277]
dimer	C	(103)		[277]
Cl_2BNPh_2	C	172/18		[136]
		(65–8)		
	I	(64–7)		[137]
	C		μ	[26]
			X-ray	[415]
Cl_2BNRR'	C			[54]
(R R'				
Bu Bu				
Bu C_6H_{11}				
Bu Ph				
Bu^i C_6H_{11}				
Bu^s C_6H_{11}				
Cp Bu				
Cp C_6H_{11}				
Cp Ph				
C_6H_{11} C_6H_{11}				
2-Me-1-C_5H_{11} C_6H_{11}				
3-C_6H_{13} C_6H_{11}				
C_5H_9 Bu				
C_5H_9 C_6H_{11}				
Bu Norbornyl				
Ph Ph)				

Table 7 *(c'td.)*

Compound	Method	B.p. (°C/mm) (m.p., °C)	Comments	References
$Br_2BN(H)(C_6H_4CN)$	C	CN at 3-position, (104–106)		[241]
		CN at 4-position, (> 200)		[241]
Br_2BNMe_2	F	163/760 (extr.) (–21.5)		[401]
	C	45.5–7/10, 162–3/772	d_4^{20} 1.755, n_D^{20} 1.4965	[364]
	A	49–51/1		[298]
	I	161/760		[408]
			1st I.P.	[15, 213]
	I		ir	[16]
	I	54/20	nmr	[391]
dimer	F	(152)		[149, 401]
			Raman	[112]
			nqr	[410]
			uv, pes (monomer)	[36, 130a, 196]
Br_2BNEt_2	C	76.2/20, 194/760 (extr.) (–7.5)		[149]
Br_2BN (ring)	C	97–100/15		[277]
dimer	C	(157–8)		[277]
Br_2BN (ring)	C	105–6/14		[277]
dimer	C	(130.5–1.5)		[277]
Br_2BNPh_2	I	(78)		[21]
I_2BNMe_2	I		ir	[16]
			1st I.P.	[213]
I_2BNEt_2				[322]
$FB(NMe_2)_2$	I	106 (–44.3)		[64]
	A		nmr, uv	[118, 130a, 130d]
			pes	[36, 196]
FB (ring: N–Me, N–Me)			nmr	[130d]
FB (ring: NMe, NMe)	K	46/22	nmr	[391]

Table 7 *(c'td.)*

Compound	Method	B.p. (°C/mm) (m.p., °C)	Comments	References
FB(NMe)(NMe) ring with Me (NMe / FB / NMe / Me)	K	49/16	nmr	[391]
FB(NEt)(NEt) ring (NEt / FB / NEt)	K	55/11	nmr	[391]
$FB(NEt_2)_2$	A	54/14	nmr	[340]
$FB(NPr_2)_2$	A	72/1	nmr	[340]
$FB(NBu_2)_2$	A	105/0.7	nmr	[340]
$FBNR_2N(SiMe_3)_2$	A		nmr	[118]
(R = Me		25/0.2		
Et		49/0.2		
Pr		60/0.2		
Bu)		79/0.2		
$ClB(NHBu^t)_2$	C,I	72–4/10	n_D^{20} 1.4330	[215]
$(Bu^tNHBCl)_2NBu^t$		82–6/0.02		[215]
ClB(NH)(NH) ring with Me (NH / ClB / NH / Me)	C	192/0.03 (50–5)		[8]
	B	(74–7)		[181]
ClB(NH)(NH) fused to benzene ring (NH / ClB / NH)	B	(320)d		[181]
$ClB(NR_2)_2$ (R = Me, Et, Pr^i)			nmr	[101]
$ClB(NPr^i_2)NHBu^t$	C	40/0.6	n_D^{20} 1.438, d_4^{20} 1.166, dnmr	[225]
$ClB(NPr^i_2)NEt_2$	C	50–52/0.2	n_D^{20} 1.454, d_4^{20} 1.199, dnmr	[225]
$ClB(NMe_2)_2$	C	38/10		[305]
	A	43/12		[298]
		50–2/25	n_D^{20} 1.4491	[44]
	K	51/20	d_4^{20} 0.832	[311]
		145/738		[406]
		145–8/760		[367]
			ir, Raman	[24, 151]
	K		nmr	[18b, 206]
	—D			[324]
			1st I.P.	[15, 213]
			uv, pes	[36, 130a, 196]
$ClB(NMe_2)N(SiMe_3)_2$			nmr	[398]

Table 7 *(c'td.)*

Compound	Method	B.p. (°C/mm) (m.p., °C)	Comments	References
$2ClB(NMe_2)_2.3TiCl_4$		(−37.5)	nmr	[206]
			ir	[207]
ClB(NMe)(NMe) (ring: NMe—NMe bridged)	C	72/45	nmr	[18b, 391]
			pes	[203a]
ClB(NMe)(NMe) (ring: NMe—NMe, curved bridge)	C	64/10		[50]
	C	100/40	nmr	[391]
$ClB(NEt_2)_2$	C	83–7/16	n_D^{20} 1.4549	[139]
	C	74–6/9		[305]
			MCD	[226]
			ir	[9]
ClB(NEt)(NEt) (ring: NEt—NEt bridged)	C	65–6/10		[50]
ClB(NEt)(NEt) (ring: NEt—NEt, curved bridge)	C	66/10	nmr	[391]
$ClB(NPr_2)_2$	C	86/0.15	d_4^{20} 0.9051, n_D^{20} 1.4558	[5]
$ClB(NPr^i_2)_2$	C	55/0.1	d_4^{20} 0.8997, n_D^{20} 1.4537	[8]
		65/0.13		[233]
	C	55/0.08	d_4^{20} 0.9031, n_D^{20} 1.4560	[5]
ClB(NPr^i)(NPr^i) (ring: NPr^i—NPr^i bridged)	C	64/3.5		[50]
$ClB(NBu^s_2)_2$	C	121/0.15	d_4^{20} 0.9212, n_D^{20} 1.4556	[5]
$ClB[N(C_6H_{11})_2]_2$	C	(116)		[5]
$ClB(NPh_2)_2$	C	192/0.06		[5]
$ClB(Ph)N(H)CH_2CH_2CHCH_2$	A	58–64/10^{-3}	nmr	[411]
$BrB(NMe_2)_2$		20–8/0.5		[44]
	A	36–8/3		
			1st I.P.	[15, 213]
			pes	[36, 196]
$[BrB(NMe_2)_2]_2(TiBr_4)_3$			ir	[379a]
$[Br_2B(NMe_2)]_2.TiBr_3$			ir, uv, mag	[130a, 379a]
BrB(NMe)(NMe) (ring: NMe—NMe bridged)	K	81/44	nmr	[391]

Table 7 *(c'td.)*

Compound	Method	B.p. (°C/mm) (m.p., °C)	Comments	References
NMe BrB NMe	C, K	80/10	nmr	[391]
BrB(Me)NMeCH$_2$CH$_2$CHCH$_2$	A	70–71/0.17	nmr	[411]
BrB[N(H)C$_6$H$_4$CN-4]$_2$	C	(>200)		[241]
Me N—N XB ‖ N—N Me			ir, nmr, uv, mass	[179]
(X = Cl	C	(12.3)		
Br)		(11.5)		
IB(NMe$_2$)$_2$				[321]
			1st I.P.	[213]
			pes	[36, 196]
NMe IB NMe	K	86/24	nmr	[391]
NMe IB NMe	C	59/3	nmr	[391]
N$_3$B(NMe$_2$)$_2$		40–4/11	ir	[337]
N$_3$B(NEt$_2$)$_2$		50–3/11	ir	[337]
NCB(NMe$_2$)$_2$			ir, Raman	[151]
NCB(NR$_2$)$_2$				[241a]
(R = Et		119–122/14	ir, nmr, mass	
Bu)		120–130/10^{-3}	ir, nmr, mass	
Me N NCB N Me		92–3/28–30	ir, nmr, mass	[241a]
Me N—N NCB ‖ N—N Me			ir, nmr, uv, mass	[179]
OCNB(NHBu^t)$_2$		72/0.1	n_D^{20} 1.4376, ir	[221a]

Table 7 *(c'td.)*

Compound	Method	B.p. (°C/mm) (m.p., °C)	Comments	References
$OCNB(NR_2)_2$				[221a]
(R = Me		30/0.3	n_D^{20} 1.4499, ir	
Et		61/0.1	n_D^{20} 1.4475, d_4^{20} 0.8678, ir	
Pr^i)		73–5/0.005	n_D^{20} 1.4502, d_4^{20} 0.8933, ir	
OCNB ring: $-N(Pr)-CH_2-CH_2-N(Pr)-$ (structure)		61/0.9	n_D^{20} 1.4575, d_4^{20} 0.9264, ir	[221a]
$OCNB(NPr^i_2)X$				[221a]
(X = $NHBu^t$		41–3/0.01	n_D^{20} 1.4453, ir	
NMePh		80/0.1	n_D^{20} 1.5051, d_4^{20} 1.081, ir	
OBu		90/0.25	n_D^{20} 1.4529, d_4^{20} 0.9694, ir	
OC_6H_{11})		32/0.05	n_D^{20} 1.4213, d_4^{20} 0.8722, ir	
$SCNB(NR_2)_2$				[221a]
(R = Me		50/0.4	n_D^{20} 1.5231, d_4^{20} 0.9941, ir	
Et)		65/0.1	n_D^{20} 1.5032, d_4^{20} 0.943, ir	
SCNB ring: $-N(Me)-N=N-N(Me)-$ (structure)			ir, nmr, uv, mass	[179]
SeCNB ring: $-N(Me)-N=N-N(Me)-$ (structure)		(40)	ir, nmr, uv, mass	[179]

† Footnote to Table 7: for explanation of abbreviations, see footnotes to Table 1, p. 57.

Table 8† Alkoxy(amino)boranes

Compound	Method	B.p. (°C/mm.) (m.p., °C)	Comments	References
HB(N(Me)CH₂CH₂O) (ring)	C	(67–74)		[281b]
ROB(N(Me)CH₂CH₂CH₂) (ring) (R = H, Me)	E		ir, nmr	[281b]
EtOB(Me)NMePh	B		ir, Raman	[13]
			nmr	[25]
MeOB(Ph)NMe$_2$	C	70/4.5	n_D^{20} 1.4966, nmr, mass	[17]
			nmr	[18]
EtOB(Ph)NEt$_2$		102–4/10	n_D^{20} 1.4835	[217]
Bu^{i}OB(Ph)NHEt	C	86–7/2	d_4^{20} 0.9057, n_D^{20} 1.4831	[262]
Bu^{i}OB(Ph)NMe$_2$	C	68–70/1		[262]
Bu^{i}OB(Ph)NEt$_2$	C	92–4/3	d_4^{20} 0.8931, n_D^{20} 1.4822	[262]
PhB(NH–C$_6$H$_4$–O) (ring)	E			[387]
	G	(101–3)		[46, 376]
	C	(105–6)	uv	[108]
PhB(N(H)CMe$_2$CH$_2$O) (ring)	C	84/0.9	ir, mass	[91, 95a]
PhB(OCH$_2$CH$_2$CH$_2$N(H)) (ring)	C	85/0.1	ir, mass	[95a, 96]
			nmr	[48b]
PhB(OCH$_2$CH$_2$N(H)) (ring)	C	64/0.15	ir, mass	[95a, 96]
PhB(OCH(Me)CH$_2$N(H)) (ring)	C	82/0.4	ir, mass	[95a, 96]
PhB(O(CH$_2$)$_n$–C$_5$H$_9$N) (ring, piperidine) (n = 2			nmr	[48b, 94a]
n = 1)				[94a]

Table 8 *(c'td.)*

Compound	Method	B.p. (°C/mm.) (m.p.,°C)	Comments	References
PhB, O, N, H, Me (ring)	C	70/0.15	ir, mass	[95a]
PhB, O, N, H, Ph (ring)	C	130/0.1	ir, nmr	[95a]
PhB, O, N, CH_2Ph (ring)	C	128/0.1	ir, mass	[95a, 96]
PhB, O, N, CH_2Ph, Me (ring)	C	122/0.1	ir, mass	[95a]
PhB, O, N, CH_2Ph, Me (ring)	C	130/0.1	ir, mass	[95a]
PhB, O, N, Me (ring)	C	103/5	ir, mass	[94a, 95]
EtSB, N, Me, O (ring)	C			[91]
o-$MeC_6H_4B(OBu^i)NEt_2$	C	122.5–3/8	d_4^{20} 0.8877, n_D^{20} 1.4808	[262]
o-$MeC_6H_4B(OBu^i)NHEt$	G	93–5/3	d_4^{20} 0.9042, n_D^{20} 1.4847	[263]
p-$MeC_6H_4B(OBu^i)NHEt$	G	110–2/2	d_4^{20} 0.9059, n_D^{20} 1.4891	[259, 263]
1-$NpB(OBu^i)NHEt$	G	182–5/9	d_4^{20} 0.9750, n_D^{20} 1.5470	[259, 263]
1-$NpB(OBu^i)NEt_2$	C	127–8/2	d_4^{20} 0.9502, n_D^{20} 1.5378	[262]
$(EtO)_2BN$ (pyrrole ring)		75–7/8	mass	[198]
EtOB—NH, O, CO (fused benzene ring)	G	120–40/3 (subl.)		[57, 236, 403]

Table 8 *(c'td.)*

Compound	Method	B.p. (°C/mm.) (m.p.,°C)	Comments	References
$(BuO)_2BNHEt$	C	49–51/0.1	d_4^{20} 0.846, n_D^{20} 1.4178	[7, 210]
			MCD	[226]
			ir	[9]
MeOB(NH)₂C₆H₄ (cyclic: MeOB bonded to two NH groups of an o-phenylene ring)	G		nmr	[146b]
ROB(NMe)₂ (cyclic: ROB bonded to two NMe groups linked by a CH_2CH_2 bridge)				
(R = Me	C	45/22	ir, nmr	[130d, 333a]
OBu^t	C	80/22		[333a]
OPh)	C	125/22		[333a]
$(PhO)_2BNEt_2$	A	120/0.5	ir, mass	[10]
$(PhO)_2BNPr^i_2$	A	125/0.5	ir, mass	[10]
$(BuO)_2BNEt_2$	C	65–8/0.45	d_4^{20} 0.856, n_D^{20} 1.4233	[140]
			MCD	[226]
			ir	[9]
$(BuO)_2BNBu_2$	C	107–10/0.2	n_D^{20} 1.4309	[140]
			MCD	[226]
			ir	[9]
$(CH_2O)_2BNHMe$ (cyclic)	B		ir, nmr	[89]
$(CH_2O)_2BNHEt$ (cyclic)	B	(86–8)	ir, nmr	[89]
$(CH_2O)_2BNHBu$ (cyclic)	B	(66–9)	ir, nmr	[89]
$(CH_2O)_2BNHBu^i$ (cyclic)	B	(73–5)	ir, nmr	[89]
$(CH_2O)_2BNHBu^t$ (cyclic)	B	liq.	ir, nmr	[89]

Table 8 *(c'td.)*

Compound	Method	B.p. (°C/mm.) (m.p.,°C)	Comments	References
O BNHPh O	B	(90–2)	ir, nmr	[89]
O $BNMe_2$ O	C		ir, nmr	[89, 94, 333a]
	C	55–62/60		[86]
		35/10	d_4^{20} 0.993, n_D^{20} 1.4198	[19]
O $BNEt_2$ O	C		ir, nmr	[89, 94, 333a]
	C	36/0.4	n_D^{20} 1.4284	[35]
O BNR_2 O ($R_2 = Pr_2$; Pr^i_2; Bu_2; Ph, Me; Ph, Et)	C		nmr	[94]
O BN O	C	67–9/1	ir, nmr	[333a]
Me O BNHMe O	C	38/10		[86]
Me O $BNMe_2$ O	C	45/20		[86]
O BN O	C	90–1/23	ir, nmr	[333a]
Me O BNMePh O	C			[86]
Me O $BNPr^i_2$ O	C			[86]
Me O $BN(C_6H_{11})_2$ O	C			[86]

Table 8 *(c'td.)*

Compound	Method	B.p. (°C/mm.) (m.p.,°C)	Comments	References
Me O O BN	C			[86]
O O $BNHEt$	C	24/0.05	n_D^{20} 1.4340	[126]
O O $BNMe_2$	C	24/0.1	n_D^{25} 1.4330	[126]
O O $BNEt_2$	C	33–4/0.01	n_D^{25} 1.4365	[126]
O O $BNPr^i_2$	C			[86]
O O $BN(C_6H_{11})_2$	C			[86]
Me Me O O $BNHPr^i$	C			[86]
O O $BNHMe$ oligomer	B,C,K	(300)	ir	[218]
O O $BNHEt$ dimer	B,C,K	(270)	ir	[218]
O O $BNHPr$	B		ir	[218]
O O $BNHPr^i$	B	liq.	ir, nmr	[89]
monomer/dimer	B,C,K	52/0.2	ir	[218]

Table 8 *(c'td.)*

Compound	Method	B.p. (°C/mm.) (m.p.,°C)	Comments	References
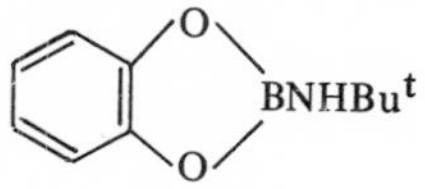 O, O, BNHBu^t	B,C,K	68/0.3	ir	[218]
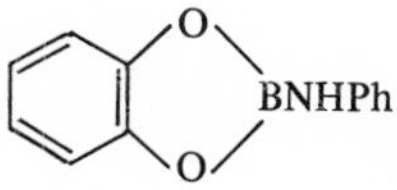 O, O, BNHPh dimer	B,C,K	(266)	ir	[218]
	—D	(268)	ir	[93]
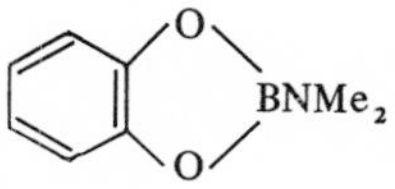 O, O, BNMe$_2$ dimer	B,C,K	54/0.3	ir	[218]
		(64)		
	—D	(60–2)		[180,324]
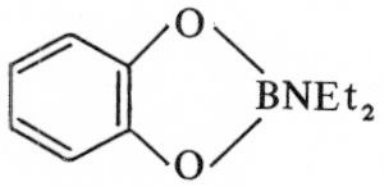 O, O, BNEt$_2$	B		ir	[218]
	C	72/0.1	n_D^{20} 1.5061	[138]
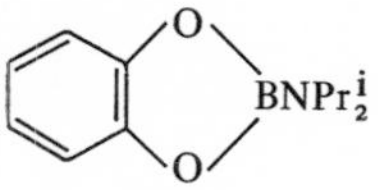 O, O, BNPr$^i_2$	B,C,K	(52)	ir	[218]
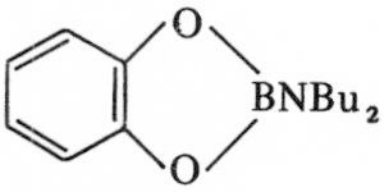 O, O, BNBu$_2$	C	108/0.005	n_D^{20} 1.4971	[138]
$(MeNHBO)_3$	C			[6]
$(Me_2NBO)_3$	I	221/752.4		[152]
		(64)		
		90/0.9		[323]
		(63–4)		
$(Et_2NBO)_3$		90–1/0.2	n_D^{20} 1.4338	[90]
$(Pr^n_2NBO)_3$		105/0.2	n_D^{20} 1.4703	[90]
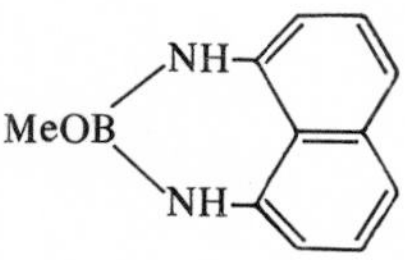 MeOB(NH–, NH–)		115.2–120/6	ir	[75]
$BuOB(NHEt)_2$	C	38–9/0.2	d_4^{20} 0.8391, n_D^{20} 1.4280	[7]
			MCD	[226]
			ir	[9]
$BuOB(NEt_2)_2$	C	68–8.5/0.3	d_4^{20} 0.8419, n_D^{20} 1.4348	[139]
			MCD	[226]
			ir	[9]

Table 8 *(c'td.)*

Compound	Method	B.p. (°C/mm.) (m.p.,°C)	Comments	References
$BuOB(NBu_2)_2$	C	124–31/0.1	n_D^{20} 1.4470	[139]
			MCD	[226]
			ir	[9]
$[CH_2OB(NEt_2)_2]_2$		67/0.3	n_D^{21} 1.4348	[35]
(cyclic NH–O)$BNH(CH_2)_2OB$(cyclic NH–O)				[232]
Me_2C—O, MeCH—O (ring) $BNHPr^i$	K			[232]

†Footnote to Table 8: for explanation of abbreviations, see footnotes to Table 1, p. 57.

Table 9† Amino(thioalkyl)boranes

Compound	Method	B.p. (°C/mm.) (m.p., °C)	d_4^{20}	n_D^{20}	References
$MeSB(Me)N(SiMe_3)C_6H_4NHSiMe_3$-2	G		(nmr, mass)		[146b]
$MeSBHNMe_2$	F	0/13			[314]
$MeSBHNEt_2$	F	0/3			[314]
		(−122)			
	F	59–61/20	0.8436	1.4626	[267, 268]
$EtSBHNEt_2$	F	65–7/19	0.8562	1.4616	[257]
	F	48–9/7		1.4618	[267, 268]
$EtSBHNPent_2$	F	78–80/1	0.8530	1.4608	[267, 268]
$PrSBHNMe_2$	F	48–50/13	0.8706	1.4701	[253, 254]
$PrSBHNEt_2$	F	79–81/17	0.848	1.4628	[252]
$PrSBHNPent^i_2$	G	87–92/1.5		1.4633	[258]
	F	110–4/3.5	0.8422	1.4640	[254]
$BuSBHNMe_2$	F	55–7/7	0.8666	1.4699	[254]
$BuSBHNEt_2$	F	51.5–2.5/2		1.4640	[267, 268]
BuSBHN(ring)	F	73–4/1.5	0.9170	1.4944	[254]
$Bu^tSBHNEt_2$	F	44–6/1		1.4640	[255]
	F	49–51/1.5	0.849	1.4636	[252]
$PhSBHNEt_2$	F	82–4/1.5	0.9736	1.5470	[255]
$EtSBPrNEt_2$	C, G	91–2/10	0.8688	1.4720	[265]
$BuSBPrNEt_2$	G	96–7.5/4	0.8628	1.4704	[264]
$BuSBBuNEt_2$	G	102–3/3	0.8620	1.4702	[264]

Table 9 *(c'td.)*

Compound	Method	B.p. (°C/mm.) (m.p., °C)	d_4^{20}	n_D^{20}	References
$BuSBPent^iNEt_2$	G	97–8/2	0.8572	1.4688	[264]
	C	102–3/3		1.4690	[265]
$BuSBBu^iNEt_2$	G	104–6.5/4	0.8613	1.4700	[264]
BSEt NMe BSEt	G	124–6/3	1.0175	1.5440	[265]
$PrSBPhNMe_2$	F, G	86–90/1.5	0.9665	1.5372	[256]
$PrSBPhNEt_2$	F, G	103–5/2	0.9454	1.5248	[256]
$BuSBPhNEt_2$	F, G	111–4/2	0.9398	1.5228	[256]
S PhB NH	B, K	(152–5)	(ir)		[289]
	C	(154–6)	(uv)		[108]
H N PhB S	C		(mass)		[96]
$p\text{-}MeC_6H_4B(SBu)NEt_2$	F, G	103–7/1	0.9328	1.5216	[256]
$(EtS)_2BNHEt$	G	51–4/0.07	0.9744	1.5122	[363]
$(EtS)_2BNHPh$	G		1.0773	1.5882	[363]
$(EtS)_2BNMe_2$	G	68–72/0.1	0.9826	1.5147	[363]
$(EtS)_2BNEt_2$	C	68–9/0.3			[144]
S $BNMe_2$ S	A, C	50/0.1	(ir, nmr)		[86, 327b, 333a]
S BN S	C	71–3/1	(ir, nmr)		[333a]
S BN S	C	(60–3)	(ir, nmr)		[333a]
$MeSB(NMe_2)_2$	C		(ir, nmr)		[333c]
Me N MeSB N Me	C	93/22	(ir, nmr)		[333a]
			(pes)		[203a]
$(Me_2NBS)_3$	C	(118)			[408]
$EtSB(NHEt)_2$	G	73–6/3	0.9219	1.4805	[363]
$EtSB(NMe_2)_2$	G	52–5/1	0.9037	1.4822	[363]

Table 9 *(c'td.)*

Compound	Method	B.p. (°C/mm.) (m.p., °C)	d_4^{20}	n_D^{20}	References
$EtSB(NEt_2)_2$	C	57–8/0.2			[144]
Me_2 H_2 N—B H_2B NMe_2 S—B Me H_2	E	(33.3–33.5)	(ir, nmr)		[63]

† Footnote to Table 9: for explanation of abbreviations, see footnotes to Table 1, p. 57.

Table 10† Miscellaneous compounds [(BXYZ) and $R_2NB.L$]

Compound X	Y	Z	Method	B.p. (°C/mm.) m.p.	Comments	References
Cl	OMe	NMe_2		41/18.5		[300]
Cl	OBu	NMe_2	I	64–5/20		[8, 300]
					ir	[9]
Cl	OBu	NEt_2	C	76–8/14	n_D^{20} 1.4275	[140]
					ir	[9]
Cl	OBu	NBu_2			MCD	[226]
Cl	SEt	NEt_2	C	40–2/0.1		[145]
Cl	SEt	NEtPh	C	95–100/0.2		[145]
H	PEt_2	NEt_2		$36/10^{-3}$		[323]
Cl	PEt_2	NEt_2		$45–8/10^{-3}$	n_D^{20} 1.4803	[320, 322]
OC_6H_{11}	PEt_2	NEt_2		89–94/h.v.	n_D^{20} 1.4831	[322]
PEt_2	PEt_2	NEt_2		$78–91/10^{-3}$		[320]
PEt_2	NMe_2	NMe_2		134–6/53	nmr	[146a, 317, 320]
Bu	SiH_3	NMe_2	A	22/12		[2]
Bu	$SiPh_3$	NMe_2	A	155–65/760d	nmr	[310]
Cl	$SiMe_3$	NMe_2	A	42/9	nmr	[310]
Cl	$SiPh_3$	NMe_2	A	150–60/h.v. (135)	nmr	[310]
Cl	Cl	$N(SiCl_3)_2$	A			[393]
Cl	Cl	$N(SiMe_3)_2$	A		ir	[71]
Cl	Cl	$N(SnMe_3)_2$	A			[309]
Cl	Cl	$N(Me)SiCl_3$	A		nmr, pes	[18c]
Cl	Cl	$N(Me)SiMe_3$	A		nmr	[18c]
Cl	Ph	$N(Me)SiMe_3$	A		nmr	[18c]
Br	Me	$N(Me)SiMe_3$	A		nmr	[18c]
Br	Br	$N(Me)SiMe_2Br$			nmr, pes	[18c]
Me	Me	$N(Me)PMe_2$	A	60/45	ir, nmr	[327a]
Me	Me	$N(Me)PPh_2$	A	$135–7/10^{-2}$	ir, nmr	[327a]
Br	Br	$N(Me)PPh_2$	A	$141/10^{-2}$	ir, nmr	[327a]

Table 10 *(C'ntd.)*

Compound X	Y	Z	Method	B.p. (°C/mm.) m.p.	Comments	References
Bu	Bu	N(Me)PPh_2	A	$153/10^{-2}$	ir, nmr	[327a]
Ph	Ph	N(Me)PPh_2	A	(83–6)	ir, nmr	[327a]
MeB ring: $N(PMe_2)$–CH_2CH_2–$N(SiMe_3)$ (cyclic)			A	$42/10^{-1}$	ir, nmr	[327a]
Me	MeO	NMe_2	C	87–9/725	ir, nmr	[333a, 333c]
Me	Bu^tO	NMe_2	C		ir, nmr	[333a]
Me	PhO	NMe_2	C	70–1/1	ir, nmr	[333a]
Me	MeS	NMe_2	C	39/12	ir, nmr	[333a]
Cl	$Mn(CO)_4PPh_3$	NEt_2	A	(135–40)	ir, nmr	[319]
$Mn(CO)_4PPh_3$	$PEt_2Ni(CO)_4$	NEt_2	A		ir, nmr	[319]
$[Mn(CO)_4PPh_3]_2$		NEt_2	A	(120)	ir, nmr	[319]
$Mn(CO)_4(PPh_3)$HN–C_6H_4–NH_2 (o-phenylene)			A	(120–5)	ir, nmr	[319]
$Mn(CO)_5$	NMe_2	NMe_2	A	(100)	ir, nmr	[318]
Cl	$N(SiMe_3)_2$	$N(SiMe_3)_2$	A		ir	[71]
Cl	$N(SnMe_3)_2$	$N(SnMe_3)_2$	A			[309]
$Mn(CO)_4PPh_3$	NMe_2	NMe_2	A	(125)	ir, nmr	[318]
$SiPh_3$	$SiPh_3$	NMe_2	A	(150–5)	nmr	[310]
SiH_3	NMe_2	NMe_2	A	23/8		[2]
SiH_3	NEt_2	NEt_2	A	21/2		[2]
$SiMe_3$	NMe_2	NMe_2	A	65/9	d_4^{20} 0.830, n_D^{20} 1.4661, nmr	[310]
Ph	NMe_2	NBu^tMMe_3	A			
(M = Si				70–1/0.01	ir, nmr	[397a]
Ge)				67–8/0.01	ir, nmr	[397a]
Ph	NBu^tMMe_3	NBu^tMMe_3	A			
(M = Si				59/0.01	ir, nmr	[397a]
Ge)				77–8/0.01	ir, nmr	[397a]
Ph	NMe_2	$N(SnMe_3)_2$	A	85/0.01	ir, nmr	[397a]
NMe_2	NMe_2	$N(SnMe_3)_2$	A	61–3/0.01	ir, nmr	[397a]
PEt_2	PEt_2	NMe_2	A			[379]
PMe_2	NMe_2	NMe_2	A	85/0.23	ir, nmr	[30]
$AsEt_2$	NMe_2	NMe_2	A	90–91/7	ir, nmr	[30]
$SbEt_2$	NMe_2	NMe_2	A		ir, nmr	[30]
$AsEt_2$	$AsEt_2$	NEt_2	A	$92–4/10^{-3}$	ir, nmr	[30]
$AsEt_2$	Br	NEt_2	A	$54–6/10^{-3}$	ir, nmr	[30]
$AsPh_2$	NMe_2	NMe_2	A	65–8/1	ir, nmr	[146a]
F	NMe_2	$N(SiMe_3)_2$	A	25/0.2	ir, nmr	[118]

Table 10 *(C'ntd.)*

Compound X	Y	Z	Method	B.p. (°C/mm.) m.p.	Comments	References
F	NEt_2	$N(SiMe_3)_2$	A	49/0.2	ir, nmr	[118]
F	NPr_2	$N(SiMe_3)_2$	A	60/0.2	ir, nmr	[118]
F	NBu_2	$N(SiMe_3)_2$	A	79/0.2	ir, nmr	[118]
F	$NHSiMe_3$	$N(SiMe_3)_2$	A	41/0.2	ir, nmr	[118]
F	$NMeSiMe_3$	$N(SiMe_3)_2$	A	100/12	ir, nmr	[118]
F	$NEtSiMe_3$	$N(SiMe_3)_2$	A	66/0.2	ir, nmr	[118]
F	$N(SiMe_3)_2$	$N(SiMe_3)_2$	A	41/0.2	ir, nmr	[119,130d]
F	F	$NMe(SiMe_3)$	C	(73)	ir, nmr	[117]
F	F	$NEt(SiMe_3)$	C	33/43	ir, nmr	[117]
F	F	$NPr(SiMe_3)$	C	43/40	ir, nmr	[117]
F	F	$NBu(SiMe_3)$	C	38/39	ir, nmr	[117]
F	F	$NPr^i(SiMe_3)$	C	41/44	ir, nmr	[177]
Cl	Cl	NCl_2				[160]
Cl	NMe_2	$N(SiMe_3)_2$	A	(39–41)	ir, nmr	[338]
Cl	$OC(S)NMe_2$	NMe_2	D	(94)	ir, nmr	[242]
Cl	Ph	$N(SiMe_3)_2$	A	66–9/0.01	ir, nmr	[397a]
Me	Me	$N(H)P(S)Me_2$	A	45–50/0.01	ir, nmr	[327]
Me	Me	$N(Me)P(S)Me_2$	A	89/0.01	ir, nmr	[327]
Me	Me	$N(Me)P(S)Ph_2$	A	174–7/0.01	ir, nmr	[327]
Me	Me	$N(Me)P(S)(NMe_2)_2$	A	78/0.01	ir, nmr	[327]
Me	Me	$N(Me)P(S)(NEt_2)_2$	A	101–3/0.01	ir, nmr	[327]
Me	Me	$N(Me)P(S)(CH_2NMe_2)_2$	A	88–9/0.01	ir, nmr	[327]
$SiPh_3$	NHPh	NHPh	K		ir, nmr	[310]
$SiPh_3$	NMe_2	NMe_2	A	150–60/10^{-3}	d_4^{20} 1.060, n_D^{20} 1.6048, nmr	[310]
$SiPh_3$	$NMe(CH_2)_2NMe$		A	(71–3)	nmr	[162]
$B_3Al_3(NMe_2)_7H_5$					ir, nmr	[162]
$H_2B(NMe_2)_2AlH(BH_4)$					ir, nmr	[194]
$H_2B(NMe_2)_2Al(BH_4)_2$					ir, nmr	[193]
$Me_4Al(NMe_2)_2(Me_2BNMe_2)$					ir, nmr	[200]
Cl	NMe_2	$N(SiMe_3)_2$	A		nmr	[398]
$N(SiMe_3)_2$	$N(SiMe_3)_2$	$N(SiMe_3)_2$	A		ir	[71]
$N(SiMe_3)_2$	$N(SnMe_3)_2$	$N(SnMe_3)_2$				[309]
$N(SiMe_3)_2$	NMe_2	NH_2	C	56–8/2	ir, nmr, dnmr	[154c]
$N(SiMe_3)_2$	NEt_2	NH_2	C	86/2	ir, nmr, dnmr	[154c]
$N(SiMe_3)_2$	NPr^i_2	NH_2	C	78/1	ir, nmr, dnmr	[154c]
Cl	$SnMe_3$	NMe_2			nmr	[130d]
OMe	$SnMe_3$	NMe_2			nmr	[130d]
SMe	$SnMe_3$	NMe_2			nmr	[130d]
NMe_2	$SnMe_3$	NMe_2			nmr	[130d]
$SnMe_3$	$SnMe_3$	NMe_2			nmr	[130d]
$Me_2NBFe(CO)_4$					ir, nmr	[190, 358a, 359]

Table 10 *(C'ntd.)*

Compound X	Y	Z	Method	B.p. (°C/mm.) m.p.	Comments	References
$Et_2NBFe(CO)_4$					ir, nmr	[190, 358a, 359]
$Me_2NB.bipy$ (monomer)			A	80/hv	nmr, uv	[204]
$B(N\langle\text{pyrrolyl}\rangle)_4^-$					nmr	[130d]
1,3-dimethyl-1,3,2-diazaborolidine $(CH_2)_2(NMe)_2BX$ (X = $SiMe_3$, $SnMe_3$, $PbMe_3$, PMe_2, SeMe)					nmr	[130d]
$B(NMe_2)_3Al_4Me_3$						[360]

† Footnote to Table 10: for explanation of abbreviations, see footnotes to Table 1, p. 57.

Table 11† Metal- or metalloid-substituted aminoboranes [$\rangle$BNRM]

Compound	Method	B.p. (°C/mm.) (m.p., °C)	Comments	References
$FB[N(SiMe_3)_2]_2$	A	63/0.5 (18)	ir, nmr	[350]
$ClB[N(H)SiMe_3][N(SiMe_3)_2]$	C	75–7/2.2	ir	[396]
$RB[N(Me)SiMe_3]_2$	I		nmr	[325]
(R = Me				
F		71.5/9.5 (−20)		
Cl		48/0.5 (−24)		
Ph)		$73\text{–}5/10^{-3}$		
$H_3SiNMeB_2H_5$		(−39)		[67]
$MeB(NMe\text{—}SiMe_2)_2$ (cyclic: MeB bonded to two NMe, NMe—$SiMe_2$—$SiMe_2$—NMe)	A	74–5/10	nmr	[131]
$MeB(NMeLi)_2$	E		ir, nmr	[304]
$BuB(NMeLi)_2$	E		ir, nmr	[304]
$F_2BNMeSiH_3$	A			[377]
$F_2BN(SiMe_3)_2$		21/5.2 (−37)	nmr	[130d, 350]
$Cl_2BN(SiMe_3)_2$	A	82/11 (6)	ir, nmr, d_4^{20} 1.011, n_D^{20} 1.4554	[142, 143, 396]
$B(NMe_2)_2N(SiMe_3)_2$	C		ir, nmr	[338, 397a]

Table 11 *(C'ntd.)*

Compound		Method	B.p. (°C/mm.) (m.p., °C)	Comments	References
Me N BN(H)$SiMe_3$ N Me		C		ir, nmr	[327]
R′ N RB $SiMe_2$ I $SiMe_2$ N R′		A		ir, nmr	[392]
(R	R′				
NMe_2	Me		(25)		
NMe_2	Et		(7)		
$N(SiMe_3)_2$	Me		(57)		
$N(SiMe_3)_2$	Et		(84)		
Ph	Me		(49) 60/0.1		
Ph	Et		(41) 72/0.1		
p-tolyl	Me		(123)		
p-tolyl	Et		(57)		
NMe_2	Me		(23) $54/2 \times 10^{-2}$		
NMe_2	Et		(7) 81/0.1		
$N(SiMe_3)_2$	Me		(57) 73/0.3		
$N(SiMe_3)_2$	Et		(84) 89/0.25		
EtO	Me		50/0.1		
EtO	Et)		$56/3 \times 10^{-2}$		
BRR′[N(Me)$SiMe_3$]		A		nmr	[325]
(R = Cl, R′ = NEt_2			48–50/0.3		
R = Me = R′					
R = NMe_2 = R′			65–6/8		
R = Ph = R′)					
N I B N N R R′		A		ir, nmr	[132a]
(R	R′				
H	$SiMe_3$		61/1		
$SiMe_3$	$SiMe_3$				
H	$SnMe_3$		75–9/1		
$SiMe_3$	Cl		115/1		
$SiCl_3$	H)		63–4/1		
(RR′ = > $SiMe_2$, $Me_2Si—SiMe_2$)			125/0.01		

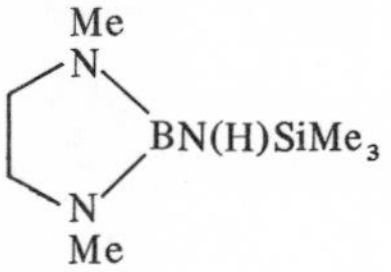

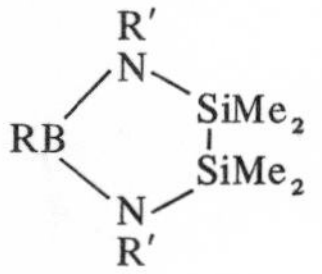

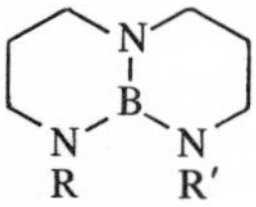

Table 11 *(C'ntd.)*

Compound	Method	B.p. (°C/mm.) (m.p., °C)	Comments	References
R′ R″$_2$ N—Si RB NR‴ N—Si R′ R″$_2$	A		ir, nmr	[328]
(R R′ R″ R‴				
Ph Me Me Me		(56)		
Ph Me Me H		63/0.1		
Ph Ph Me Me		(123–4)		
Me Et Me Me				
Ph Bu^s Me H)		40–3/0.01		
Me N Me_2Si BMe N Me	A			[328]
$Me_2Si(NMeBR–NMe_2)_2$	A			
(R = Me		96/11	ir, nmr	[328]
Bu)		76–9/0.01	ir, nmr	[328]
$Me_2BN(Me)B[N(Me)SiMe_3]_2$	A	$50/10^{-3}$		[326]
$Me_2BN(Et)SiMe_3$	A	79–81/85	ir, nmr	[132a]
$Ph_2BN(Me)SiMe_3$	A	85–7/0.01	ir, nmr	[132a]
$PhBuBN(Me)SiMe_3$	A			[132a]
$Me_2BN(Et)SnMe_3$	A	93/15	ir, nmr	[132a]
$Ph_2BN(Me)SnMe_3$	A	118/1	ir, nmr	[132a]
$Ph_2BN(Me)SnMe_2Cl$	A	$120/10^{-2}$	ir, nmr	[132a]
Ph B—NMe Ph_2BN | B—NMe Ph	A			[315]
$H_2BN(Me)SiMe_3$	A	116/755	ir, nmr	[132a]
$B_2H_5N(Me)SiMe_3$	A	115/760	ir, nmr	[132a]
Me-(cyclohexyl)-$BN(SnMe_3)_2$	A	117–9/0.01	ir, nmr	[375]
N B N N R R Si Si R′ N R′ R″	A		ir, nmr	[392]

Table 11 *(C'ntd.)*

Compound	Method	B.p. (°C/mm.) (m.p., °C)	Comments	References
(R R′ R″				
Ph Ph H		(201)		
Me Me H		(70)		
Me Me Me		(62)		
Me Me Me		(115)		
Me Cl Me		(152)		
Cl Cl H				
Me Cl H)				
Me N—$B(NMe_2)_2$ Me_2Si N—$B(NMe_2)_2$ Me	A	(44–6)	ir, nmr	[392]
Me N—$BNMe_2$ Me_2Si NCH_2Ph N—$BNMe_2$ Me	A	(32)	ir, nmr	[392]
$B[N(Me)SiMe_3][N(Me)BMe_2]_2$	A	39/0.01	ir, nmr	[326]
$Me(Br)BN(Me)SiMe_3$	A		ir, nmr	[18d]
$Ph(Cl)BN(Me)SiMe_3$	A	$66/10^{-3}$	ir, nmr	[18d]
$Cl_2BN(Me)SiMe_2Cl$	A	24/2	ir, nmr	[18d]
$Ph(Cl)BN(Me)SiMe_2Cl$	A	64/2	ir, nmr	[18d]
$Me_2BN(Me)SiMe_2Br$	A	32/2	ir, nmr	[18d]
$Br_2BN(Me)SiMeBr_2$	A	52/2	ir, nmr	[18d]
$Ph(Br)BN(Me)SiMeBr_2$	A	75–8/2	ir, nmr	[18d]
Et N—$SiMe_2$ EtOB N—$SiMe_2$ Et	A	56/0.3	ir, nmr	[392]
Me N R_2AsB N Me				
(R = Me		91–2/12	ir, nmr	[146a]
Ph)		113/0.1	ir, nmr	[146a]
$XB[N(SiMe_3)_2]_2$	A		ir, nmr	[142,143]
(X = F		76/0.1	d_4^{20} 0.922, n_D^{20} 1.4548	
Cl		91/0.1		
OPh)		$82/3 \times 10^{-2}$	n_D^{20} 1.4858	

Table 11 *(C'ntd.)*

Compound	Method	B.p. (°C/mm.) (m.p., °C)	Comments	References
(cyclic OCH_2CH_2O)$BN(R)SiMe_3$	A		ir, nmr	[372]
(R = H		33/0.2		
$SiMe_3$		65/0.2		
(cyclic $OCH(Me)CH_2CH(Me)O$)$BN(H)SiMe_3$	A	60/1	ir, nmr	[372]
$R''BN(R')P(S)R_2$			ir, dnmr	[316]
(e.g., R = Me = R'', R' = H)		(78.5)		

† Footnote to Table 11: for explanation of abbreviations, see footnotes to Table 1, p. 57.

Table 12† Acylamido-, ureido-, and guanidino-boranes, and related compounds

Compound	Method	B.p. (°C/mm.) (m.p.,°C)	Comments	References
$XB(NMe_2)[N(Z)C(Y)NMe_2]$				
(X = NCS, Y = O, Z = Ph;			ir, nmr	[186a]
X = NCO, Y = NTol-*p* = Z)			ir, nmr	[186a]
$H_2B[N(R)C(X)H]$			mass, trimer	[274]
(R, X = Me, O; Ph, O; Me, S)				
$Et_2BNHCOMe$ dimer	E	98/0.7		[381]
$Pr_2BNHCOMe$	E	83/2		[208]
$R''_2BN(R')COR$	C		ir, nmr	[235a]
(*e.g.*, R = CF_3, R' = Ph, R'' = Me)				
$Pr_2BNHCONHPh$	E	(220)		[208]
$(Pr_2BNH)_2CS$	E			[208]
$R_2BNHCOBu^t$				
(R = Pr		52–3/1.0	ir, nmr, monomer–dimer	[114]
Bu^i)		53–4/0.5	ir, nmr, monomer	[114]
$Ph_2BNPhCONHBu^t$	D	(77–9)	ir	[93]
$Ph_2BN(p\text{-}MeC_6H_4)C(NEt_2):NC_6H_4Me\text{-}p$	D	(129)	ir	[187]
$Ph_2BN(p\text{-}MeC_6H_4)C(OMe):NC_6H_4Me\text{-}p$	D	(194)	ir	[187]
$Ph_2BN(p\text{-}MeC_6H_4)C(SBu):NC_6H_4Me\text{-}p$	D	(35)	ir	[187]
PrB(–NH–C_6H_4–CO–NH–) (cyclic, fused benzene ring)	E	(146)		[208]

Table 12 *(C'ntd.)*

Compound	Method	B.p. (°C/mm.) (m.p., °C)	Comments	References
PhB, NH, $PhCH_2N$, CO (cyclic, fused benzene ring)		(190–2)		[129]
PhB, NH, RN, CO (cyclic, fused benzene ring) (R = 2,6-$Me_2C_6H_3$)		(265–7)		[129]
PhB, NH, RN, CO, NHCOMe (cyclic, fused benzene ring) (R = 2,6-$Me_2C_6H_3$)				[129]
PhB, NH, RN, CO, $NAcCH_2NMe_2$ (cyclic, fused benzene ring) (R = 2,6-$Me_2C_6H_3$)				[129]
1-NpB, NH, $PhCH_2N$, CO (cyclic, fused benzene ring)		(218–20)		[129]
Pr^i_2NB, NR—C:NR, NBu^t, NR—C:NR (cyclic) (R = p-MeC_6H_4)	D	(187)	ir	[187]
$PhB(NPhCOPh)_2$	D	(118–20)	ir	[220]
$PhB(NPhCONHBu^t)_2$	D	(158–60)	ir	[93]
$PhB(NPhCSNHBu^t)_2$	D	(79–82)	ir	[93]
$PhB(NPhCONHCH_2)_2$	D	(232)	ir	[93]
$PhB(NPhCONBuCH_2)_2$	D	156–60/0.2	n_D^{20} 1.5430, ir	[93]
$PhB(NRCPh{:}NR)_2$ (R = p-MeC_6H_4)	D	(166–8)	ir	[187]
$RB(NRCOR)_2$ (R = p-MeC_6H_4)	D	(96)	ir	[220]
$-\!\!($NBuB, NMe, BNBuCONH–C_6H_4–CH_2–C_6H_4–NHCO$)_n$; ring: MeN, NMe, BMe	D	(238–42)		[344]
NH, B, N, NPhCS, NH (bicyclic)	D	(110–2)	ir	[130]

Table 12 *(C'ntd.)*

Compound	Method	B.p. (°C/mm.) (m.p., °C)	Comments	References
CONPh NPhCO HN B NH N	D	(175)		[130]
CONPh NPhCS NH B NH N	D	(115)d	ir	[130]
NRCO PhNHB NR NMeCO	D	(212–4)		[344]
(R = $-C_6H_4-$CH$_2$$-C_6H_4-$NCO)				
ClB(NPhCOPh)$_2$	D	(190)	ir	[220]
PhCONPhB(Cl)NPhCOCl	D	(197)	ir	[220]
PhCSNPhB(Cl)NPhCSCl	D	(120–3)	ir	[220]
ClB(NPhCOCl)$_2$	D	(140)	ir	[220]
ClB(NRCOCl)$_2$ (R = 1-Np)	D	(164)	ir	[220]
ClB(NRCPh:NR)$_2$ (R = *p*-MeC$_6$H$_4$)	D	(206–7)	ir	[187]
ClB(NRCCl:NR)(NRCPh:NR) (R = *p*-MeC$_6$H$_4$)	D	(138–40)	ir	[187]
ClB(NRCCl:NR)$_2$ (R = *p*-MeC$_6$H$_4$)	D	(164–6)	ir	[187]
BrB(NRCBr:NR)$_2$ (R = *p*-MeC$_6$H$_4$)	D	(199–203)	ir	[187]
B(NPhCONMe$_2$)$_3$	D	(129)		[180]
B(NPhCSNMe$_2$)$_3$	D	(182–4)		[180]
B(NPhCONHBu^t)$_3$	D	(125–130)		[92]
Et$_2$NBClNRC(OMe):NR (R = *p*-MeC$_6$H$_4$)	D	(212)	ir	[187]
(MeO)$_2$BNRCCl:NR (R = *p*-MeC$_6$H$_4$)	D	(110)d	ir	[187]
C_6H_4(O)(O)BNPhCOBu	D	(181)	ir	[93]
C_6H_4(O)(O)BNPhCSNHBu^t	D	(268)	ir	[93]

Table 12 *(C'ntd.)*

Compound	Method	B.p. (°C/mm.) (m.p., °C)	Comments	References
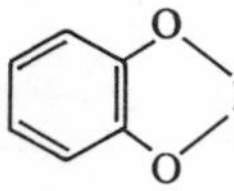 O BNPhCONEt$_2$ O	D	(204)	ir	[93]
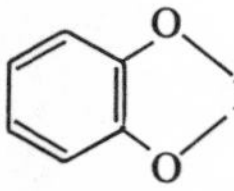 O BNPhCONRC(NEt$_2$):NR O (R = *p*-MeC$_6$H$_4$)	D	(242)	ir	[187]
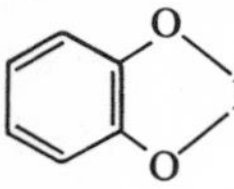 O BNPhCOOBu O	D	(58)	ir	[93]
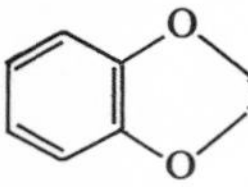 (O B)$_2$NCONHBu^t O	D	(198)	ir	[93]
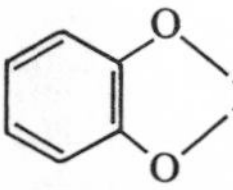 O BNRC(NEt$_2$):NR O (R = *p*-MeC$_6$H$_4$)	D	(79)	ir	[187]
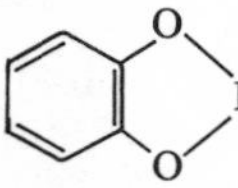 O BNRC(SBu):NR O (R = *p*-MeC$_6$H$_4$)	D	(109)	ir	[187]
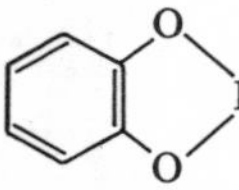 O BNRC(NR)NRC(NR)NEt$_2$ O (R = *p*-MeC$_6$H$_4$)	D	(93–7)	ir	[187]
PhB(NRCOSEt)$_2$ (R = *o*-MeC$_6$H$_4$, *p*-MeOC$_6$H$_4$, *o*-ClC$_6$H$_4$, *p*-ClC$_6$H$_4$, 2,4-Cl$_2$C$_6$H$_3$)				[88]
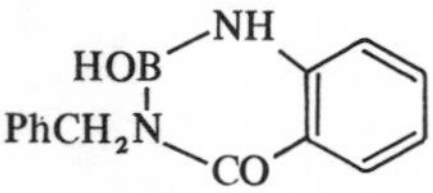 NH HOB PhCH$_2$N CO		(315)		[129]
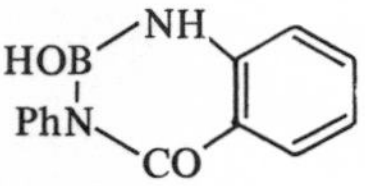 NH HOB PhN CO				[129]
MeOB(Ph)NRC(NR)OMe (R = *p*-MeC$_6$H$_4$)	D	(192)d	ir	[187]

Table 12 *(C'ntd.)*

Compound	Method	B.p. (°C/mm.) (m.p., °C)	Comments	References
PhB, O, 1-NpN, CO (ring fused to benzene)				[129]
BuSB(Ph)NRC(NR)NEt$_2$ (R = p-MeC$_6$H$_4$)	D	(51–4)	ir	[187]
o-ClC$_6$H$_4$B, S, PhN, CO (ring fused to benzene)				[220]
MeOBClNRCCl:NR (R = p-MeC$_6$H$_4$)	D	(93)d	ir	[187]
R$_2$N—B, N(Ph), CO, NR′, CO, PhN (ring)				
(R = Et, R′ = Et)		(65–8)	ir	[92]
(R = Pr^i, R′ = Bu^t)		(142–5)	ir	[92]
Bu^{t}NH—B, O, CO, NBu^t, CO, O (ring)		(118–120)	ir	[92]
[CH$_3$C(O)O]$_2$BN(H)Me dimer		(70–80)	ir, nmr	[168]
[CH$_3$C(O)O]$_2$BNMe$_2$ monomer		30/0.7	ir, nmr	[168]
Me$_2$NB[OC(S)NMe$_2$]$_2$		110	ir, nmr	[241]

† Footnote to Table 12: for explanation of abbreviations, see footnotes to Table 1, p. 57.

Table 13† Derivatives of polyboranes or carboranes

Compound	Method	B.p. (°C/mm.) (m.p., °C)	Comments	References
(EtBNMe$_2$)$_2$	B, Cl/Na	70–3/8	n_D^{20} 1.4440	[43]
(BuBNHMe)$_2$	C		ir, nmr	[304]
(PhBNMe$_2$)$_2$	B, Cl/Na	(101–3)		[43]
[B(NHMe)$_2$]$_2$	B	25/0.5		[44]
(MeNHBNMe$_2$)$_2$	C	25.8/0.5	ir, nmr	[304]
[(HexNH)$_2$B]$_2$	B		n_D^{20} 1.4606	[44]
(NH, NH ring B)$_2$	B	74–5/0.6		[50]

Table 13 *(C'ntd.)*

Compound	Method	B.p. (°C/mm.) (m.p., °C)	Comments	References
$[B(NHPh)_2]_2$	B	(180–200)		[44]
$[B(NMe_2)_2]_2$	B, Cl/Na	55–7/2		[44, 48, 166, 176, 182, 312]
	C		ir, Raman	[28]
			pes	[28]
$(B\langle^{NMe-}_{NMe-}]\,)_2$ (ring: B, NMe, NMe)	B	85/5		[50]
		(43–4)	X-ray	[130b]
$(B\langle^{NMe-}_{NMe-}\rangle)_2$ (larger ring: B, NMe, NMe)	B			[50]
$[B(NEt_2)_2]_2$	B, Cl/Na	84/0.1 (−65)		[312]
$(B\langle^{NEt-}_{NEt-}]\,)_2$	B	80/0.1		[50]
$(B\langle^{NPr^i-}_{NPr^i-}]\,)_2$	B	(130–4)		[50]
$[B(NBu_2)_2]_2$	B	170–83/0.55	n_D^{20} 1.4667	[44]
$[B(NMe_2)(OR)]_2$ (R = Me, Et)	A		ir, nmr	[333a]
$(Me_2N)_3B_2Cl$	I	45–6/1		[317]
	K			[313]
$(Me_2N)_3B_2Br$	I	51/0.5		[172, 317]
$(Me_2NBCl)_2$	I	57–60/1		[317]
	K			[313]
$(Me_2NBBr)_2$	I	76–7/1.5		[317]
$(Et_2NBPEt_2)_2$	B, Cl/Na	160–70/10^{-3}		[320]
$B_2H_4MeNMe_2$		(−94.5 to −95.3)	nmr, mass	[110, 111]
$B_3(NMe_2)_5$		85–8/0.8 (43)	ir, nmr, mass	[172]
$B_3(NMe_2)_4NEt_2$		85–6/0.1 (18)	n_D^{20} 1.4761, ir, nmr, mass	[172]
$B_3(NEt_2)_5$		(75–8)	ir, nmr, mass	[172]
$B_4(NMe_2)_6$		99–103/0.1 (98–100)	ir, nmr, mass	[172]
$B_5(NMe_2)_7$		119–24/0.1	ir, nmr, mass	[172]
$B_6(NMe_2)_8$		127–32/10^{-3}	n_D^{20} 1.4685, ir nmr, mass	[172]

Table 13 *(C'ntd.)*

Compound	Method	B.p. (°C/mm.) (m.p., °C)	Comments	References
$C_2B_{10}H_9$-3-R				[414]
(R = HNMe		(36–7)		
NHEt				
$NHCH_2Ph$		(49–50)		
.HCl		(232–5)		
$NHCH_2COOEt$		(119.5–20)		
$NHCH_2CONMe_2$.HCl		(119–201)		
NMeCHO)		(80–1)		
1-$MeC_2B_{10}H_8$-3-R				[414]
(R = NHMe		(39–40)		
NHEt				
.HCl		(228–31)		
$NHCH_2Ph$		(77–9)		
NHCHMePh		(61–3)		
.HCl		(229–31)		
$NHCHPh_2$		(72–3)		
NMeCHO)		(77.5–8.5)		
(B<[N(R)–N=N–N(R)])$_2$ (cyclic: B, N(R), N, N, N(R))	C			[179]
$B_2(Cl)(NMe_2)_3$	I		ir, nmr	[166]
$B_2(Cl)_2(NMe_2)_2$	I		ir, nmr	[166]
(B<[X–CHR'–CR''R'''–N])$_3$ (cyclic)	C			
(R' R'' R''' X				
H H H S		(190)	ir, nmr	[96]
H H H O		(150)	ir, nmr	[96]
Me H H O		(75)	ir, nmr	[96]
H H Me O		(102)	ir, nmr	[96]
H Me Me O)		(191)	ir, nmr	[96]
(B<[O–CH$_2$CH$_2$CH$_2$–N])$_3$ (cyclic)	C	(171)	ir, nmr	[96]
(B(SiMe$_3$)(NMe$_2$))$_2$			nmr	[130d]
$[B(NMe_2)_2]_2CH_2$	A		ir, nmr	[203]
$[Me_2BN(Me)+_2PPh$	A	115/0.1	ir, nmr	[327a]
$[B(NEt_2)_2]_2C_2$	A	106–7/1	ir, nmr	[413c]

Table 13 *(C'ntd.)*

Compound	Method	B.p. (°C/mm.) (m.p., °C)	Comments	References
(B[N(Me)CH₂CH₂N(Me)])₂C₂ — ring: B, N(Me), C, C, N(Me)	A	(102–6)	ir, nmr	[413c]
$[B(NMe_2)Me]_2$-(2,5-thienylene)	A	95/1	ir, nmr	[413c]

† Footnote to Table 13: for explanation of abbreviations, see footnotes to Table 1, p. 57.

Table 14† Di- and tri-borylamines [(⊂B)$_2$N- and (⊂B)$_3$N]

Compound	Method	B.p. (°/mm) (m.p., °C)	Comments	References
RN(B-cyclopentyl ring with Me)$_2$	F			[201]
(R = Et		121–3/11.5	d^{20} 0.8618, n_D^{20} 1.4762	
CH_2Ph		137–8/0.4	d_4^{20} 0.9304, n_D^{20} 1.5229	
Ph)		98.5–100/0.3	d_4^{20} 0.9337, n_D^{20} 1.5195	
$HN(BR_2)_2$ (R = Me, Et)			ir, nmr	[333c]
HN(B-cyclohexyl ring with Me)$_2$	A	43–5/0.1	nmr	[327]
$HN(BPh_2)_2$	A	(104–117)	nmr	[188a]
HN(B[N(Me)CH₂CH₂CH₂N(Me)])$_2$	C	59–60/10^{-2}	nmr	[327]
HN(B[O-C₆H₄-O])$_2$	A	subl. 200/2		[222]

Table 14 *(C'ntd.)*

Compound	Method	B.p. (°/mm) (m.p., °C)	Comments	References
MeN[BR{N(Me)SiMe$_3$}]$_2$	I		nmr	[325]
(R = Me		68–73/0.27		
Ph)		(94)		
MeN(BMe$_2$)$_2$	A		nmr	[188a, 413b]
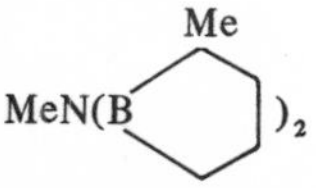 Me MeN(B)$_2$	A	61–3/10^{-2}	nmr	[327]
MeN(BMe$_2$)(BPh$_2$)	A		nmr	[188a]
MeN(BPh$_2$)$_2$	A	(135–9)	ir, nmr	[18d, 188a]
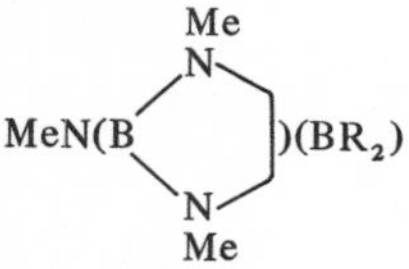 Me N MeN(B)(BR$_2$) N Me	C		nmr	[327]
(BR$_2$ = BMe$_2$		29–31/10^{-2}		
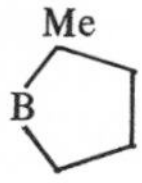 Me B		82–4/10^{-2}		
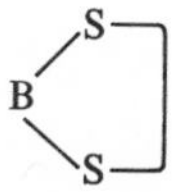 S B S		(73)		
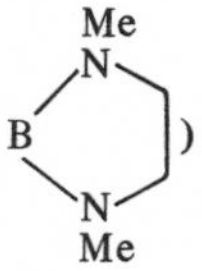 Me N B) N Me		(35)		
MeN[B(NHMe)$_2$]$_2$				[281a]
Bu^{t}N[B(NHBu^t)$_2$]$_2$	I	98–100/10^{-2} (52–4)	ir, nmr	[215]
Bu^{t}N[B(NHBu^t)$_2$][B(Cl)NHBu^t]	I	82–5/10^{-2}	ir, nmr	[215]
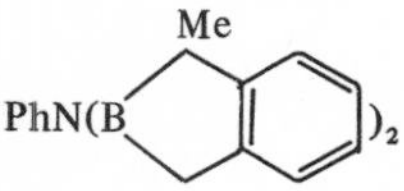 Me PhN(B)$_2$	I	(107–8)		[201]
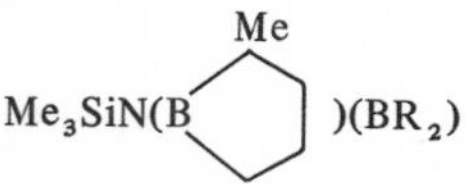 Me Me$_3$SiN(B)(BR$_2$)			nmr	[327]
(BR$_2$ = BMe$_2$		78–84/10^{-2}		

Table 14 *(C'ntd.)*

Compound	Method	B.p. (°/mm) (m.p., °C)	Comments	References
Me $Me_3SiN(B\ \ )_2$		115–120/0.1		
$Me_3SiN[B(Br)Me]_2$	A	36/1	nmr	[18a]
$Me_3SiN[B(Me)NMe_2]_2$	A	42–5/1	nmr	[18a]
$Me_3SiN[B(Me)SMe]_2$	C	54/1	nmr	[18a]
$Me_3SiN[B(Me)OMe]_2$	C	20/1	nmr	[18a]
$Me_3SiN[B(X)Me][B(Me)NMe_2]$	A			
(X = Cl		34/1	nmr	[18a]
Br)		44/1	nmr	[18a]
$Me_3SiN[B(Me)NMe_2][B(Me)SMe]$	C	50/1	nmr	[18a]
$Me_3SiN[B(Me)OMe][B(Me)NMe_2]$	A	28/1	nmr	[18a]
$Me_3SiN[B(Br)Me][B(Me)OMe]$	C	33–5/0.9	nmr	[18a]
$Me_3SiN[B(Me)OMe][B(Me)SMe]$	A	34/1	nmr	[18a]
$Me_3SiN[BBr(Me)][B(Me)SMe]$	C, I	55–6/2.5	nmr	[18a]
$Me_3SiN(BBr_2)[B(Me)NMe_2]$	C, I	58–60/0.1	nmr	[18a]
$Me_2BrSiN[B(Br)Me]_2$	I		nmr	[18a]
Me N $Me_2NN(B\ \ )_2$ N Me				[79a]
$Me_3SnN(BMe_2)_2$	A		nmr	[375a]
Me $Me_3SnN(B\ \ )_2$	A	110–2/1	nmr	[375, 375a]
$Me_3SnN(B\ \ Me)_2$	A, I	89–91/10^{-2}	nmr	[375]
$N(BMe_2)_3$	A	132/726	nmr	[130d, 375a]
Me $N(B\ \ Me)_3$	A, I	73–4/10^{-2}	nmr	[375]
Me $N(B\ \ )_3$	A	100–2/1	nmr	[375a]

Table 14 *(C'ntd.)*

Compound	Method	B.p. (°/mm) (m.p., °C)	Comments	References
Me $N(BMe_2)_2(B$ $)$	A	$40/10^{-2}$	nmr	[375a]
Me $N(BMe_2)(B$ $)_2$	A	$87\text{–}90/10^{-2}$	nmr	[375a]
$N(B$ O O $)_3$	C	subl. 260/0.1	mass X-ray	[222] [57]
$HN(B$ O O $)_2$	C		ir, nmr	[372]
$Me_3SiN(B$ O O $)_2$	C	$70/5 \times 10^{-2}$	ir, nmr	[372]
$HN(B$ O Me_2 O Me_2 $)_2$	C	60/1	ir, nmr	[372]
$EtB[(CH_2)_3NHBEt_2]_2$	B		ir, nmr	[119b]
$HN[(CH_2)_3NHBEt_2]_2$	B		ir, nmr	[119b]

† Footnote to Table 14: for explanation of abbreviations, see footnotes to Table 1, p. 57.

Table 15† Aluminium amides

Compound		Method	B.p. (°C/mm.) (m.p., °C)	Comments	References
$H_2AlNHPr^i$	dimer	A, F	(250)d	ir, mass	[78a]
$H_2AlNHBu^t$	dimer	F		mass	[78a]
$D_2AlNDBu^t$	dimer	F		mass	[78a]
H_2AlNMe_2	dimer	F	(89–90)		[2, 22]
			(89–91)		[23]
			(93–4)		[20]
	trimer	A	(89)		[85, 86]
		A, F	(89–90)	nmr	[88, 89]
				mass	[114]
				X-ray	[97]

Table 15 *(C'ntd.)*

Compound		Method	B.p. (°C/mm.) (m.p., °C)	Comments	References
.THF				ir	[116]
$.TiCl_4$					[64]
$.VCl_4$					[64]
$H_2AlNMe_2.(HBeNMe_2)_2$					[78i]
$H_5Al_2NMe_2$				ir	[116]
$.OEt_2$				ir	[116]
$HAl(NMe_2)BH_4$		A	(69)	nmr	[88]
$Al(NMe_2)(BH_4)_2$		A	(156)	nmr	[88]
$[H(Me_2N)Al]_2NMe$		C		ir, nmr	[78c]
$H_2AlNMePh.NEt_3$		F			[67, 68]
H_2AlNEt_2	dimer	A, F	(42)		[89]
		F, direct		ir, nmr	[2, 4, 5]
$H_2AlNPr^i_2$					[5a, 13]
	dimer	A, F	subl. 90 (130–1)		[85]
$H_2AlN(C_6H_{11})_2$	dimer	F			[67, 68]
$H_2AlNPh_2.OEt_2$					[13]
$.NEt_3$		F	(75–7)		[67, 68]
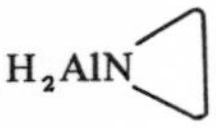 H_2AlN		F			[67, 68, 103a]
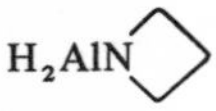 H_2AlN					[103a]
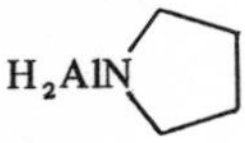 H_2AlN	trimer	F			[5a, 67, 68, 103a]
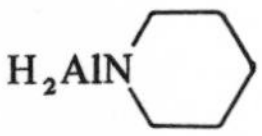 H_2AlN		A, F	subl. 80–130d		[5a, 53, 67, 68, 103a]
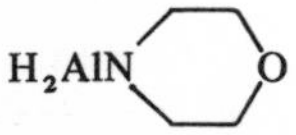 H_2AlN O		F	subl. 140/0.05		[67, 68]
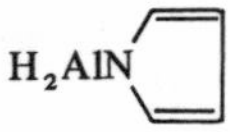 H_2AlN		F	(47)		[67]
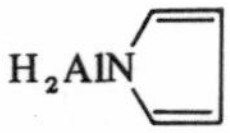 H_2AlN $.NMe_3$					[68]

Table 15 *(C'ntd.)*

Compound		Method	B.p. (°C/mm.) (m.p., °C)	Comments	References
H_2AlN (carbazolyl)		F	(180)d		[67, 68]
$Me_2AlNHMe$	trimer(A)	E	(110)	X-ray, nmr	[30]
	trimer(B)	E	(transforms 20)	X-ray, nmr	[30]
	cis	E	(113–4)	X-ray	[2]
	trans	E	(109–111)	X-ray	[2]
$Me_2AlNHEt$	trimer	E	72–3/10	ir, mass	[2]
Me_2AlNMe_2		F	(154–6)	ir	[11, 21a]
	dimer	A	(137–9)	ir, nmr	[1,7,15,106a]
		E	(154–6)		[19]
		E	(151–3)		[22, 23]
				X-ray	[2, 35]
				pes	[53b]
$Me_2AlNMe_2.Be(NMe_2)_2$					[78i]
$[Al_2HMe_4(NMe_2)]$	dimer			X-ray	[27]
Me_2AlNEt_2	dimer				[11]
$Me_2Al\overline{N(CH_2)_2}$	trimer			X-ray	[6]
		E	(193d)	ir, nmr, mass	[103]
$R_2Al\overline{NCH_2(CH_2)_x}$		E			
(R = Me, x = 2			(114–6)	ir, nmr, mass, mol. wt.	[103]
3			(56–8)	ir, nmr, mass, mol. wt.	[103]
4)			(64–6)	ir, nmr, mass, mol. wt.	[97b, 103]
(R = Et, Bu^i; x = 1)				mol. wt.	[103]
Me_2AlNPh_2	dimer	E			[17]
		E, C	(68–72)	ir, nmr	[48, 112]
$Me_2AlNPh_2.MMe_3$	(M = Al or Ga)			dnmr	[82a]
				X-ray (M = Al)	[61]
(Me_2AlNPh_2).base					
(base = NMe_3				ΔH_f	[82]
OEt_2				ΔH_f	[82]
SMe_2)				ΔH_f	[82]

Table 15 *(C'ntd.)*

Compound		Method	B.p. (°C/mm.) (m.p., °C)	Comments	References
Me_2Al ring: $N(Ph)$–$(CH_2)_2$–NMe_2 (and other $X_2AlN(R')(CH_2)_2NR_2$ species; also the *cis/trans* dimers)		E		ir, nmr	[8]
$Me_2AlN(Me)Ph$	dimer			nmr	[108a, 108b]
$Me_2AlNHSiMe_3$	dimer				[93]
$Al_2Me_5N(SiMe_3)_2$		C	(175–8)	nmr	[112]
$Me_2AlN(SiEt_3)_2$		C	75–6/10^{-2}	ir, nmr	[112]
$Me_2AlN(R)SiMe_3$		E			
(R = Me			(58–61)	ir, nmr	[90]
Ph)			(77–80)	ir, nmr	[90]
Me_2AlN (ring with N)		E	(83.5–84.5)	ir, nmr, mass	[3]
$MeBu^iAlNEt_2$		E	115–6/1.5		[117]
Et_2AlNH_2	trimer				[18]
$Et_2AlNHMe$	trimer	B		mass	[31]
$Et_2AlNHMe$	trimer				
	cis	E, I		ir, nmr, Raman	[2]
	trans	E, I		ir, nmr, Raman	[2]
$Et_2AlNHBu^t$				ir, mass	[109]
	dimer	B		mass	[31]
$Et_2AlNHPh$	dimer	E	(58)		[26]
Et_2AlNMe_2	dimer	E	90/5 × 10^{-3}		[22, 23]
			(5–6)	mass	[109, 121]
				ir, nmr	[37]
$.Et_2AlX$ (X = Cl, Br, I)				mass	[24]
$Et_2AlN(Me)C_6H_{11}$			145/5 × 10^{-4}		[121]
Et_2AlN (aziridine ring)		E		ir, nmr, Raman	[103]
Et_2AlN (piperidine ring)		E	95–100/10^{-3}	ir, nmr	[97b, 121]
Et_2AlN (pyrrole ring)	dimer			ir, nmr, mass	[3]

Table 15 *(C'ntd.)*

Compound	Method	B.p. (°C/mm.) (m.p., °C)	Comments	References
$(Et_2AlNMeCH_2)_2$	F			[109a,121]
$.OEt_2$	F			[109]
$Et_2AlN(C_6H_{11})_2$		(*ca.* 165)		[121]
Et_2AlN ring: $COCH_2$–CH_2–CH_2–CH_2–CH_2 (ring closing on N)				[121]
Et_2Al ring: NMe–NMe (cyclic)	F	164/0.3 (57–62)		[109]
$Et_2AlNMePh$	F	190/0.005d		[75,121]
Et_2AlNEt_2 polymer	F			[109]
	E	141-5/14		[117]
$.Et_2AlX$ (X = Br, OEt, SMe)			mass	[24]
$Et_2AlNEt_2.N$(isoquinoline)			uv	[76]
Et_2AlNPh_2				[104]
$Et_2Al[N(R)CH_2Ph]$	F			[41]
(R =				
CH_2Ph		(90)	ir, nmr	
Ph)		(112–115)		
$Et_2AlN(Ph)SiMe_3$	E	(81–4)	ir, nmr	[90]
$Et_2AlN(SiMe_3)_2$	E		ir, nmr	[59a,120]
$Et_2AlN(GeMe_3)_2$	E	241/3		[71]
Et_2AlNF_2	D			[91]
$Et_2AlN(Ph)$–C_6H_4–$EtAl$–$N(Ph)$–C_6H_4–$EtAl$–NPh_2		(223)	ir, nmr, Raman	[40]
EtAl–N(R)–CH_2 (benzo-fused ring)	E		ir, nmr	[41]
(R =				
CH_2Ph		(194–6)		
Ph)		(186–8)		

Table 15 *(C'ntd.)*

Compound		Method	B.p. (°C/mm.) (m.p., °C)	Comments	References
$(\triangleright)_2$AlN$\triangleleft$		E		ir, nmr, Raman	[73a]
Pr_2AlNMe_2		I			[86]
Bu_2AlNMe_2	dimer	F			[84]
		I			[86]
Bu_2AlNPh_2		E, F	(85–6)		[75]
$Bu^i_2AlNMePh$		F	d(110–4)		[75]
	.PhNHMe	F		uv	[76]
$Bu^i_2AlNEt(C_3H_5)$		E	149/1		[118]
Bu^{i}Al NEt (ring)		E	131/1		[118]
$Bu^i_2AlN(CH_2Ph)_2.PhCH_2N$:CHPh		D			[75]
$Bu^i_2AlNPhCH_2Ph$		D, F	(102–5)		[75]
	.PhN:CHPh	D	d(85)	ir	[75]
				uv	[76]
	$.PhNHCH_2Ph$	D	d(110–5)		[75]
$Bu^i_2AlN(C_6H_4Me$-*p*$)CH_2Ph.$*p*$-MeC_6H_4N$:CHPh		D	d(126)		[75]
$Bu^i_2AlN(1\text{-}Np)CH_2Ph.1$-NpN:CHPh		D	(40)		[75]
				uv	[76]
$Bu^i_2AlN(C_6H_4Me$-*p*$)CH_2C_6H_4Me$-*p*		D			[75]
$Bu^i_2AlN(1\text{-}Np)CH_2Np$-1		D			[75]
	.1-NpN:CHNp-1	D			[75]
$Bu^i_2AlN(SiMe_3)_2$		E	107–9/5	ir, nmr	[59a, 120]
$Bu^i_2AlNR'CHR''(o\text{-}C_6H_4R''')$ (R' R'' R'''		F		ir, nmr	[41]
Et H H			(96–8)		
Ph H H			(106–10)		
Ph H Me			(138–40)		
Ph Et H			oil		
Ph Ph H)			oil		
Bu^{i}Al, N(R'), CHR'', R''' (ring structure)		E			

Table 15 *(C'ntd.)*

Compound	Method	B.p. (°C/mm.) (m.p., °C)	Comments	References
(R′ R″ R‴			ir, nmr	[41]
Et H H		subl. 175–80/10^{-3}		
Ph H H		(165–8)		
Ph H Me		(179–80)		
Ph Et H		(235)d		
Ph Ph H)		(217–9)		
$Bu^i_2AlNPh_2$	F	d(80–5)		[75]
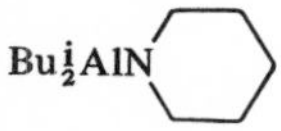 Bu^i_2AlN	E + I			[119]
	E		ir, nmr, Raman	[97b]
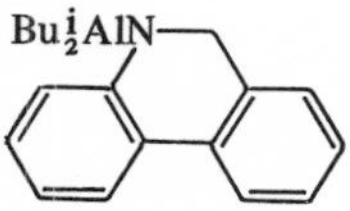 Bu^i_2AlN	D, F	d(162–5)		[75]
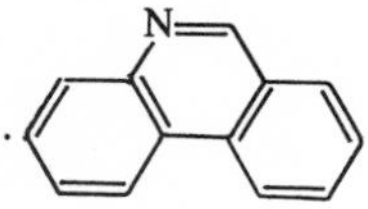 · N=	D	(118)		[75]
			uv	[76]
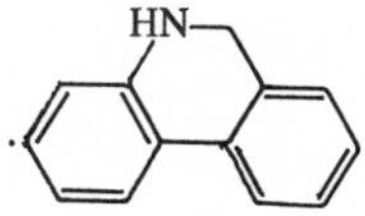 · HN	D	d(135–40)		[75]
			uv	[76]
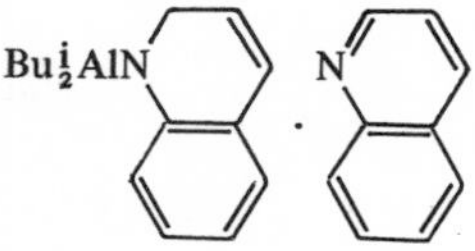 Bu^i_2AlN · N	D		uv	[76]
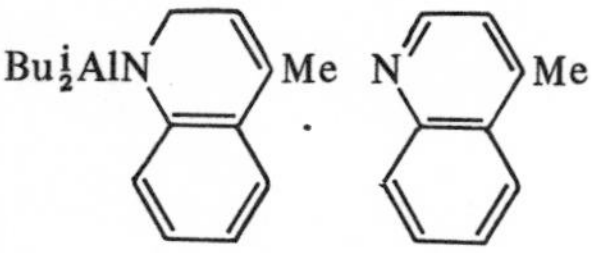 Bu^i_2AlN Me · N Me	D		uv	[76]
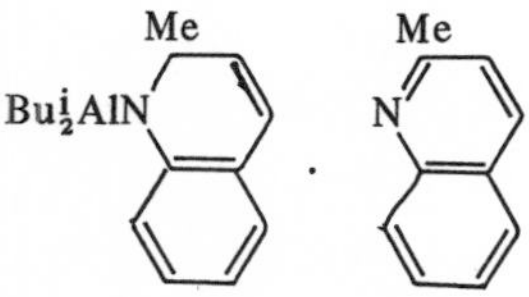 Me Bu^i_2AlN · Me N	D		uv	[76]

Table 15 *(C'ntd.)*

Compound		Method	B.p. (°C/mm.) (m.p., °C)	Comments	References
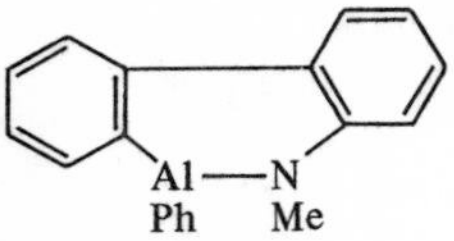 Al—N Ph Me		E			[21]
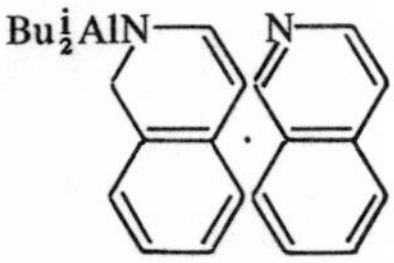 Bu^i_2AlN · N		D		uv	[76]
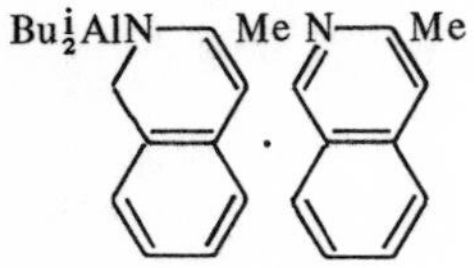 Bu^i_2AlN Me · N Me		D		uv	[76]
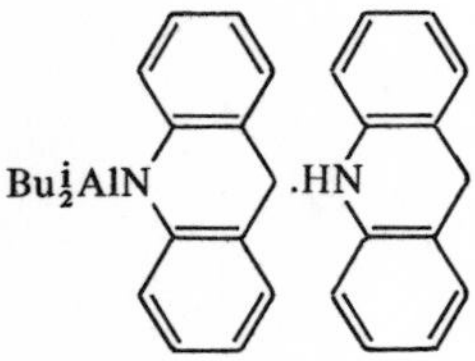 Bu^i_2AlN .HN		D	d(192)	uv	[75, 76]
$Ph_2AlNHMe$	polymer	E	subl. 125–30/ ~10^{-4}	ir	[56]
Ph_2AlNMe_2	polymer	E	subl. 190–200/ ~10^{-4}	ir	[56]
	dimer	F			[84]
		I			[86]
$(BH_4)_2AlN(CH_2)_2$	dimer			X-ray	[97a]
$(BH_4)_2AlNMe_2$			subl./vac 158	ir	[49]
$(BH_4)_2Al(NMe_2)_2BH_2$			(39–41)	ir, nmr mass	[50, 51]
$(BH_4)HAl(NMe_2)_2BH_2$					[50]
$H_2B(NMe_2)_3Al_2Me_4$				ir, nmr	[31a]
$B(NMe_2)_3Al_4Me_6$				ir, nmr	[94a]
$B_6(NMe_2)_{12}Al_6Me_{12}$				ir, nmr	[2a]
$HAl(NMe_2)_2$	mono-dimer		(63)		[111]
	di-/trimer	E	(62)		[85, 89]
$.2H_2BNMe_2$			(120)	ir, nmr	[87]
		F			[5]
Me_2AlNMe_2				nmr	[106a]
$Me_2AlNHPr^i$	dimer	E		ir, nmr, mass	[78d]

Table 15 *(C'ntd.)*

Compound		Method	B.p. (°C/mm) (m.p., °C)	Comments	References
$Et_2AlNHPr^i$	trimer	E		ir, nmr, mass	[78d]
$(Me_3SiCH_2)_2AlNHMe$	dimer	E		ir, nmr, mass	[78e]
$HAl(NMe_2)NEt_2$		F			[68]
$HAl(NMe_2)NPr^i_2$	dimer	B			[85]
$HAl(NMe_2)NPh_2$		F			[68]
$HAl(NEt_2)_2$		F, direct		ir, nmr	[4, 5, 5a]
$HAl(NPr^i_2)_2$					[5a, 13]
$HAl[\overline{N(CH_2)_n}]_2$ ($n = 4, 5$)		F			[5a]
$HAl(NPh_2)_2$					[13]
$HAl(NPh_2)N$⬡	dimer	F			[67, 68]
$HAl(N$⬡$)_2$	dimer	F			[67, 68]
$MeAl[N(SiMe_3)_2]_2$		I	67–8/0.5	ir, nmr	[112]
$EtAl[N(SiMe_3)_2]_2$		E			[59a]
$Bu^iAl[N(SiMe_3)_2]_2$		E	153/6		[59a]
$EtAl(NPh_2)_2$					[104]
$Bu^iAl(NPhCH_2Ph)_2$		I			[75]
$EtAl\langle NMe—NMe \rangle$ (ring)		F	164/0.3 (57–62)	ir, nmr	[109]
$(OC)_4Fe—C(—O—Al(NMe_2)_2)(—NMe_2)$		I		ir, nmr	[78j]
$Al(NHPh)_3$		F			[25]
$Al(NMe_2)_3$		E			[59]
	dimer	F	(87–9)		[85]
			(88–9)		[20]
	monomer		(87–8)		[111]
$Al(NR_2)_3$ ($R = Me, Et$)		F		ir, nmr	[5a]
$Li[Al(NMe_2)_4]$					[26b, 85]
$(Me_2N)_2AlNPr^i_2$	dimer	B			[85]
$Al(NEt_2)_3$		F		ir, nmr	[4]
$Al(NPr^i_2)_3$	monomer	E	(56–9)	nmr	[18j, 85]
$Li[Al(NBu_2)_4]$					[46]

Table 15 *(C'ntd.)*

Compound	Method	B.p. (°C/mm) (m.p., °C)	Comments	References
$Al(NPh_2)_3$	E			[59]
$Li[Al(NPh_2)_3H]$				[12]
$.OEt_2$				[59, 74]
$M[Al(NH_2)_4]$ (M = Na, K)	E		X-ray	[10, 72]
$Li[AlH(NEt_2)_3]$	E		nmr	[5, 12, 59]
$Li[AlEt(NBu_2)_3]$				[115]
$Li[Al(NEt_2)_4]$	E		nmr	[18]
$Al[N(SiMe_3)_2]_3$	A, C	(188)		[79, 120]
			X-ray	[99]
$Me_3Al_3(NMe_2)_2$	I		X-ray	[61]
$H_4Al_3(NPr^i_2).(T.H.F.)_2$	F		ir, nmr	[18a]

† Footnote to Table 15: for explanation of abbreviations, see footnotes to Table 1, p. 57.

Table 16† **The poly(*n*-alkyliminoalanes)**

Compound	Method	B.p. (°C/mm) m.p. (°C)	Comments	References
$(XAlNBu^t)_2$ (X = Me, Cl)			nmr	[12c]
$(HAlNPr^i)_4$	A, I		ir, nmr, mass, X-ray	[12c, 18a, 18e, 19a, 19e]
$(RAlNPr^i)_4$ (R = Me, Et)	E		ir, nmr, mass, X-ray	[18a, 19a]
$(ClAlNPr^i)_4$	A, B, F, I		ir, nmr, mass	[12c, 18g, 78a, 78b, 78c]
$(HAlNBu^s)_4$			nmr	[12c, 18e]
$(HAlNBu^t)_4$	A, B, I	(182)	ir, nmr, mass	[12c, 18d, 18e, 18g, 78a, 78b, 78c]
$[HAlN\{CH(Me)Ph\}]_4$			nmr	[12b, 12c]
$Ph(Me)CH\text{-}NH\text{-}[HAl\{CH(Me)Ph\}]_3\text{-}AlH_2$			nmr	[12b]
$(DAlNBu^t)_4$	A, I		ir, nmr, mass	[78a, 78b, 78c]
$(ClAlNBu^t)_4$	F, I	(270)	ir, nmr, mass	[12c, 18g, 78a]
$(BrAlNBu^t)_4$	I	subl. 140–150	ir, nmr, mass	[78a]
$HCl_3(AlNBu^t)_4$	I		ir, nmr, mass	[78a]
$HBr_3(AlNBu^t)_4$	I		ir, nmr, mass	[78a]
$H_2I_2(AlNBu^t)_4$	I		ir, nmr, mass	[78a]
$HI_3(AlNBu^t)_4$	I		ir, nmr, mass	[78a]
$[Al_4Cl_4(NMe_2)_4(NMe)_2]$	C, I		ir, nmr	[106]
$[(Pr^iHN)AlNPr^i]_4$	F, I		ir, nmr,	[18j]
$[PhAlNPh]_4$	E		ir, nmr, X-ray	[45, 62, 62a]
$(XAlNPr^i)_4$ (X = Me, Et, Cl)			nmr	[12c]

Table 16 *(C'ntd.)*

Compound	Method	B.p. (°C/mm) m.p. (°C)	Comments	References
$(EtAlNPh)_4$	F, I	subl. 250/10^{-3}		[26]
$(EtAlNBu^t)_4$	F, I	subl. 130–160/10^{-3}	ir, mass	[31]
$[H(HAlNPr^i)_5.AlH_2].LiH/Et_2O$	C, F, I		ir, nmr, X-ray	[12d]
$[(HAlNPr^i)_2(H_2AlNHPr^i)_3]$	F		ir, nmr, X-ray	[78f]
$(HAlNPr)_6$	C, F, I		ir, nmr, X-ray	[12c, 18i, 19e]
$(HAlNPr^i)_6$	C, F, I		ir, nmr, X-ray	[12c, 18g, 19e, 52a]
$(HAlNPr^i)_6.AlH_3$	C, F, I		ir, nmr, X-ray	[78g]
$(HAlNBu^s)_6$	C, F		ir, nmr	[12c, 18d, 18h, 18i, 18j]
$(HAlNBu^i)_6$	C, F		ir, nmr	[18d, 18f, 18h, 18i]
$(ClAlNPr^i)_6$	F		ir, nmr, X-ray	[18g, 19d]
$[HAlN\{CH(Me)Ph\}]_6$			nmr	[12b]
$(HAlNC_6H_{11})_6$	C, F		ir, nmr	[18d, 18h, 18i, 18j]
$(HAlNPr^i)_6(MeAlNPr^i)_6$	C, F, I		ir, nmr	[18h]
$(HAlNPr^i)_5(MeAlNPr)_5.AlMe_3$			ir, nmr	[18h]
$(Me_{0.83}H_{0.17}AlNPr^i)_6(MeAlNPr^i)_6$			X-ray	[19d]
$(MeAlNMe)_7$	F, I	subl. 105–115/10^{-3}	ir, nmr, mass, X-ray	[38a]
$(EtAlNMe)_7$	F, I	(353–6)	ir, nmr	[31, 38b]
$(HAlNEt)_6(H_2AlNHEt)_2$	A, I	(155)d	ir, nmr, mass	[78a]
$(HAlNEt)_8$	A, I	(193)	ir, nmr, mass	[78a]
$(HAlNEt)_{15,16}$	A, I		ir, nmr, mass	[78a]
$H_4Cl_4(AlNEt)_8$		subl. 220/10^{-6}	ir, nmr, mass	[78a]
$(ClAlNEt)_4(Cl_2AlNHEt)_2$	A, F, I		ir, nmr, mass	[78a]
$(HAlNPr)_8$	B, F, I		ir, nmr, mass, X-ray	[12c, 18c, 18g, 19c]
$(MeAlNMe)_6(Me_2AlNHMe)_2$	E, I	subl. 135–145/10^{-3}	ir, nmr, mass, X-ray	[2b]
$(MeAlNMe)_8$	E, I	subl. 115–125/10^{-3}	ir, nmr, mass	[2b]
$(MeAlNMe)_6(Me_2GaNHMe)_2$	E, I		ir, nmr, X-ray	[2b]
$(MeAlNPr^i)_5(HAlNPr^i)$				
$HAl[\{EtN(CH_2)_2NEt\}AlH_2]_2$			ir, nmr	[78h]
$(H_{0.85}Cl_{0.15}(Al[\{EtN(CH_2)_2NEt\}AlH(H_{0.7}Cl_{0.3})]_2$			ir, nmr, X-ray	[78h]
$[(THF)M(HAlNBu^t)_3]$ (M = Mg, Ca)	E, I		ir, nmr, X-ray	[18b, 19b]
$[Me_2Al(NMe)_2CMe]_2$			X-ray	[34a]

† Footnote to Table 16: for explanation of abbreviations, see footnotes to Table 1, p. 57.

Table 17† Halogeno(amino)alanes

Compound		Method	B.p. (°C/mm.) (m.p., °C)	Comments	References
$HAlClN(H)Bu^t$	dimer	I		ir, mass	[78a]
$HAlClNMe_2$					[13, 65]
			(83)		[84]
$HAlClNBu_2$					[66]
$HAlClNPh_2.NEt_3$		F			[68]
HAlClN 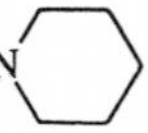 (ring structure)	dimer	F			[67, 68]
$HAlBrNMe_2$					[13, 53]
$.VCl_4$					[64]
MeAlClN 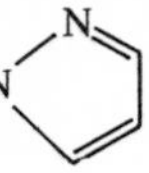 (ring structure containing N)			(153–5)	ir, nmr	[3]
EtAlClNHMe	dimer	E	(91)		[55]
EtAlClNHPh		E, K	(58–61)	X-ray	[26]
EtAlClNMePh		I	(98–100)		[41b]
$EtAlClNEt_2$		I	110–5/0.1		[41b]
$EtAlClNPh_2$		I	(117–8)		[39]
$EtAlBrNEt_2$		I	118–21/0.1		[41b]
$EtAlINEt_2$		I	$137/10^{-3}$ (98–101)		[41b]
EtAlClN 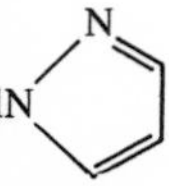 (ring structure containing N)			(65–66)	ir, nmr	[3]
$EtAlBrNMe_2$	dimer	E, I	(123–4)	ir, nmr	[28]
$EtAlINMe_2$	dimer	E, I	(149.5–151.5)	ir, nmr	[29]
$EtAlBrNHBu^t$ (two isomers)	dimer	C, I			[9]
$Cl_2AlNHPr^i$	dimer	I		ir, nmr, mass	[78a]
$Cl_2AlNHBu^t$	dimer	I	subl. $80/10^{-4}$	ir, nmr, mass	[78a]
$Cl_2AlNHPh$	dimer		(170)		[26]
Cl_2AlNMe_2	dimer	K	(148)		[26]
				ed	[6a]
				X-ray	[1a]
			(151)		[84]
		I	(151)		[85]
		F	(237–8) subl. $104–5/10^{-4}$		[20]
Cl_2AlNEt_2		I	(128–30)		[41b]
Cl_2AlNPh_2	dimer		(138)		[26]
Cl_2AlN 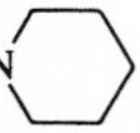 (ring structure)		I	(190–3)		[41b]

Table 17 *(C'ntd.)*

Compound		Method	B.p. (°C/mm.) (m.p., °C)	Comments	References
$Cl_2AlN(C_6H_{11})_2$		I	(84–7)		[39]
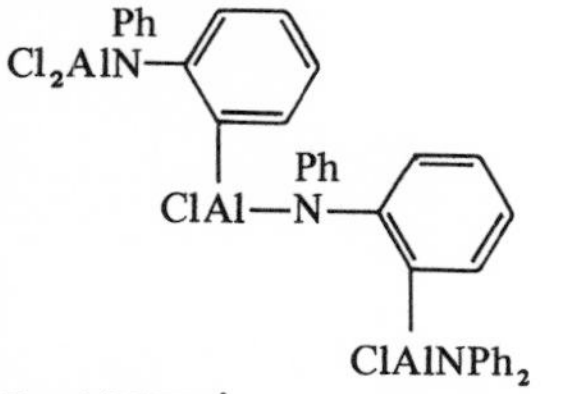 Ph, Cl_2AlN—(ring); Ph, ClAl—N—(ring); $ClAlNPh_2$		C, I	260	ir, nmr, Raman	[40]
$Br_2AlNHPr^i$	dimer	I		ir, nmr, mass	[78a]
Br_2AlNEt_2		I	(150–2)		[41b]
X_2AlNMe_2 (X = Br, I)	dimer			X-ray	[1b]
I_2AlNEt_2		I	(225)d		[41b]
$H_2NAlClNHPh$		K			[26]
$Me_2NAlClNHPh$	dimer	K	(92–4)		[26]
$ClAl(NMe_2)_2$	dimer	C	(55–7)		[85]
$.Me_2NH$		B	(132–4)		[26]
$.PhNH_2$		B	(72–3)		[26]
$[\{X(Me_2N)Al\}_2NMe]_2$ (X = Cl, Br, I)		C, F		ir, nmr	[78c]
$(EtAlCl)_2NPh$					[73]
$Al_2I_3(NHEt)_3.OEt_2$		E			[57]
$Al_2I_3(NHPh)_3.OEt_2$		E			[57]
$Al_2I_3(NMe_2)_3.OEt_2$		E			[57]
$Al_4Cl_4(NMe_2)_4(NMe)_2$		A		ir, mass	[77]
				X-ray	[106]
$(Cl_2Al)_2NBu$					[69]
$(Cl_2AlNBu)_3Al$					[69]
ClAlNMe	polymer	E			[55]

† Footnote to Table 17: for explanation of abbreviations, see footnotes to Table 1, p. 57.

Table 18† Alkoxy- and alkylthio-aminoalanes

Compound	Method	References
$(EtO)_2AlNMe_2$	K	[86]
AlONHMe	C	[32, 33, 83]
$AlONMe_2$	C	[33]
	K	[86]
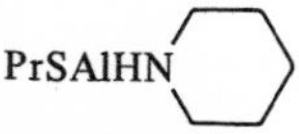 PrSAlHN(ring)	F	[68]
AlSNHMe	C	[32]
$.2MeNH_2$	C	[33, 83]
$AlSNMe_2.2Me_2NH$	C	[33]

† Footnote to Table 18: for explanation of abbreviations, see footnotes to Table 1, p. 57.

Table 19† Amido-, ureido-, and guanidino-alanes, and related compounds

Compound	Method	B.p. (°C/mm.) (m.p., °C)	Comments	References
$(R_2Al)_2(CONR)_2$	E			
(R = Me		(118)	ir	[96]
Et)		$90/10^{-4}$	ir	[96]
$Et_2AlNMeCSEt$	D			[80]
$Et_2AlNMeCONMe_2$	D	102–4/0.15	ir, nmr	[37]
$Et_2AlNMeCSNMe_2$	D			[37]
$Et_2AlNMeCOSEt$	D	120–3/0.3	ir, nmr	[37]
$Et_2AlNEtCOEt$	D	d > 150		[80]
$Et_2AlNBu^tCONMe_2$	D	75–8/0.04	ir, nmr	[37]
$Et_2AlN(C_6H_{11})COEt$	D			[80]
$Et_2AlN(C_6H_{11})CONMe_2$	D			[37]
$Et_2AlNPhCOEt$	D	(100–35)	ir, polymer	[42]
$Et_2AlNPhCSEt$	D	(117)	ir, polymer	[42]
$Et_2AlNPhCONMe_2$	D	112–5/0.07	ir, nmr	[37]
$Et_2AlNPhCSNMe_2$	D			[37]
$Et_2AlNPhCOSEt$	D	(95–100)d	ir, nmr	[37]
$Et_2AlN(C_6Cl_5)COEt$	D			[80]
$Bu^i_2AlN(C_3H_5)CSBu^i$	D			[80]
EtAlClNBuCSEt	D			[81]
$EtAlClN(C_3H_5)CSEt$	D			[81]
EtAlClNPhCOEt	D	(102)d	ir, polymer	[42]
EtAlClNPhCSEt	D	(126)d	ir, polymer	[42]
EtAlINPhCSEt	D			[81]
PhAlClNBuCOPh	D			[81]
$Cl_2AlNBuCOPh$	D			[81]
$Cl_2AlN(C_{14}H_{29})COEt$	D	d(130)		[81]
$Cl_2AlNPhCOEt$	D		ir, polymer	[42]
$Cl_2AlNPhCOBu^s$	D			[81]
$Br_2AlNMeCOBr$	D		ir	[42]
$Br_2AlNMeCSBr$	D		ir	[42]
$Br_2AlN(C_6H_{11})COEt$	D			[81]
$Br_2AlNPhCOPr$	D		ir	[42]
$Br_2AlNPhCONEt_2$	D		polymer	[42]
$Br_2AlNPhCOBr$	D		ir	[42]
$ClAl(NPhCOEt)_2$	D	(98)d	ir, polymer	[42]

†Footnote to Table 19: for explanation of abbreviations, see footnotes to Table 1, p. 57.

Table 20† N-Alkali metal-substituted aminoalanes

Compound	Method	References
$Et_2AlNBuNa$	E	[105]
Et_2AlNCH_2PhLi	E	[105]
$Et_2AlNPhLi$	E	[105]
$Et_2AlNPhK$	E	[105]
$Bu^i_2AlNPhLi$	E	[105]
$Ph_2AlNBuLi$	E	[105]
$Ph_2AlNBuNa$	E	[105]

† Footnote to Table 20: for explanation of abbreviations, see footnotes to Table 1, p. 57.

Table 21† Gallium, indium, and thallium amides

Compound		Method	B.p. (°C/mm.) (m.p., °C)	Comments	References
$H_2GaNHMe$	trimer	F		nmr	[37, 40]
H_2GaNHR (R = Et, Pr, Pr^i, Bu, Bu^s, Bu^t)		F		ir, nmr, mass	[38]
$H_2Ga\overline{N(CH_2)_2}$		F		X-ray	[17, 41]
$H_2Ga\overline{N(CH_2)_n}$ (n = 2–5)		F		ir, nmr	[39]
H_2GaN 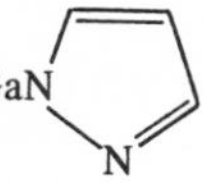 (pyrazolyl ring)		F	(44–5)	ir, nmr	[1b]
D_2GaN  (pyrrolyl ring)		F		ir, nmr	[1b]
H_2GaNMe_2	(^1H, ^2H)	F		ir, nmr	[13]
$Me_2GaNHMe$	dimer	E	226/760 (extr.)		[7]
	oligomer			nmr	[37]
Me_2GaNMe_2	dimer	F	(157–61)	ir	[2]
		E	212/760 (extr.) (97.2–7.5)		[7]
		E			[8]
Me_2GaNMe_2	dimer	F		ir	[2]
	oligomer	F		nmr	[37]
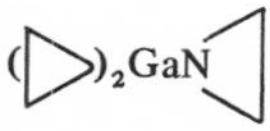 (cyclopropyl)$_2GaN$ (aziridinyl)		E			[29]

Table 21 *(C'ntd.)*

Compound	Method	B.p. (°C/mm.) (m.p., °C)	Comments	References
$Me_2Ga\overline{N(CH_2)_n}$ trimer	E			
(n = 2		(183–5)	ir, nmr, mass, mol. wt.	[18,41]
3		(128–31)	ir, nmr, mass, mol. wt.	[18,41]
4		(67–9)	ir, nmr, mass, mol. wt.	[18,41]
5)		(99–100)	ir, nmr, mass, mol. wt.	[18,35,41]
		(96)	ir, nmr	[35]
$[Me_2Ga(NMe)_2CMe]_2$			X-ray	[18a]
Me_2GaN (pyrazolyl)	E	(50–5)		[1b,32]
Me_2GaN (3-methylpyrazolyl, Me)	E		dimer, X-ray	[32]
Me_2GaN (indazolyl)	E		dimer, X-ray	[32]
Me_2GaN (dimethylpyrazolyl, Me, Me)	E		dimer, X-ray	[32]
$HO[Ga(Me_2)-N]_2$ (bridging dimethylpyrazolyl, Me, Me) $.2N_2C_5H_8$	E		X-ray	[32]
$Me_2In\overline{N(CH_2)_n}$	E			
(n = 2		(138–41)	ir, nmr, mass, mol. wt.	[18,41]
3		(102–4)	ir, nmr, mass, mol. wt.	[18,41]
4		(62–5)	ir, nmr, mass, mol. wt.	[18,41]
5)		(127–135)	ir, nmr, mass, mol. wt.	[18,41]
$Me_2Tl[N(SiMe_3)_2]$	A	$90/10^{-3}$	ir, nmr	[24]

Table 21 *(C'ntd.)*

Compound		Method	B.p. (°C/mm.) (m.p., °C)	Comments	References
Me_2GaNPh_2		E			[8]
Me_2InNMe_2		F	(174–5)	ir	[2]
	dimer	E			[10]
		E		X-ray, ir, nmr	[26]
Et_2NGaN (ring)		E		ir, nmr	[35]
$Et_2Ga\overline{N(CH_2)_2}$		E		ir, nmr, mass, mol. wt.	[18, 41]
Et_2InNEt_2					[33]
Et_2InN (ring)	dimer	E		ir, nmr	[35]
Bu^i_2GaN (ring)	dimer	E	(59)	ir, nmr	[35]
$ClGa(NMe_2)_2$	dimer	A	(60–1)	ir, nmr	[30]
$Ga(NMe_2)_3$	dimer	A	(90–2)	ir, nmr	[30]
$Ga[N(SiMe_3)_2]_3$		A	(187)	ir, nmr, pes	[2a, 5, 30]
$In[N(SiMe_3)_2]_3$		A	(168)	ir, nmr, pes	[2a, 5]
$Tl[N(SiMe_3)_2]_3$		A	(152–3)d	ir, nmr	[24]
				X-ray	[1]
$Na[Ga(NH_2)_4]$				X-ray	[27]
$Li[Ga(NMe_2)_4]$		A			[27]
$Na_2[Ga(NH_2)_5]$					[28]
$[R_2Ga][C(O)NMe]_2$		E			
(R = Me			(90–2)	ir, nmr	[34a]
CD_3			(90–2)	ir, nmr	
Et)			(7)	ir, nmr	
$[R_2In][C(O)NMe]_2$		E			
(R = Me			(160–3)	ir, nmr	[34a]
Et)			(131–2)	ir, nmr	
Me_2In—N (ring)—C(=O)R (R = H, Me)		E			[6]

Table 21 *(C'ntd.)*

Compound		Method	B.p. (°C/mm.) (m.p., °C)	Comments	References
$In(NC_4H_3COR)_3$ (pyrrolyl, 2-COR) (R = H, Me)		E			[6]
$Et_2InNPhCOEt$	dimer	D			[41a, 42]
Me_2TlNMe_2	dimer	E			[10]
TlNHCOMe		B			[11]
$.2NH_3$		B			[11]
TlNPhCOMe		B			[11]
$TlN(CO)_2C_6H_4$ (phthalimido)		G			[12]
$Li[Me_3GaNHEt]$		F + Li			[23]
$(MeGaNMe)_6(Me_2GaNHMe)_2$		E, I	subl. 140–150/10^{-3}	ir, nmr, mass X-ray	[1a]

† Footnote to Table 21: for explanation of abbreviations, see footnotes to Table 1, p. 57.

G REFERENCES

1. Boron

[1] Abeler, G., Bayrhuber, H., and Nöth, H., *Chem. Ber.*, 1969, **102**, 2249.
[2] Amberger, E., and Romer, R., *Z. Anorg. Chem.*, 1966, **345**, 1.
[3] Aronovich, P. M., Bochareva, N. M., and Mikhailov, B. M., *Zh. Obshch. Khim.*, 1971, **41**, 1526.
[4] Aronovich, P. M., Bogdanov, V. S., and Mikhailov, B. M., *Izv. Akad. Nauk SSSR Ser. Khim.*, 1970, 1682.
[5] Aubrey, D. W., Gerrard, W., and Mooney, E. F., *J. Chem. Soc.*, 1962, 1786.
[6] Aubrey, D. W., and Lappert, M. F., *J. Chem. Soc.*, 1959, 2927; Lappert, M. F., *Proc. Chem. Soc.*, 1959, 59.
[7] Aubrey, D. W., and Lappert, M. F., *Proc. Chem. Soc.*, 1960, 148.
[8] Aubrey, D. W., Lappert, M. F., and Majumdar, M. K., *J. Chem. Soc.*, 1962, 4088.
[9] Aubrey, D. W., Lappert, M. F., and Pyszora, H., *J. Chem. Soc.*, 1960, 5239.

[10] Ashby, E. C., and Kovar, R. A., *Inorg. Chem.*, 1971, **10**, 900.
[11] Aslangul, C., Daudel, R., Gallais, F., and Veillard, A., *Theor. Chem. Acta,* 1971, **23**, 211.
[12] Aslangul, C., Daudel, R., Gallais, F., and Veillard, A., *Compt. Rend.,* 1972, **33C**, 275.
[13] Baechle, H. T., and Becher, H. J., *Spectrochim. Acta,* 1965, **21**, 579.
[14] Bairt, N. C., *Inorg. Chem.*, 1973, **12**, 473.
[15] Baldwin, J. C., Lappert, M. F., Pedley, J. B., Riley, P. N. K., and Sedgwick, R. D., *Inorg. Nucl. Chem. Letters,* 1965, **1**, 57.
[16] Banister, A. J., Greenwood, N. N., Straughan, B. P., and Walker, J., *J. Chem. Soc.,* 1964, 995.
[17] Barfield, P. A., Lappert, M. F., and Lee, J., *J. Chem. Soc. (A),* 1968, 554; *Proc. Chem. Soc.,* 1961, 421.
[18] Barfield, P. A., Lappert, M. F., and Lee, J., *Trans. Faraday Soc.,* 1968, **64**, 2571.
[18a] Barlos, K., Christl, H., and Nöth, H., *Ann.,* 1976, 2272.
[18b] Barlos, K., Kroner, J., Nöth, H., and Wrackmeyer, B., *Chem. Ber.,* 1977, **110**, 2774.
[18c] Barlos, K., Nölle, D., and Nöth, H., *Z. Naturforsch.,* 1977, **32b**, 1095.
[18d] Barlos, K., and Nöth, H., *Chem. Ber.,* 1977, **110**, 2790.
[19] Bastin, E. L., *et al.,* quoted in ref. 373, Shell Development Co., Potential Chemical Warfare Agents, Final Report, Contract No. CML-4564, April 30th, 1954.
[19a] Baudler, M., Marx, A., and Hahn, J., *Z. Naturforsch.,* 1978, **33b**, 335.
[20] Beachley, O. T., *Inorg. Chem.,* 1967, **6**, 870.
[21] Becher, H. J., *Z. Anorg. Chem.,* 1952, **270**, 273; 1956, **287**, 285.
[22] Becher, H. J., *Z. Anorg. Chem.,* 1956, **288**, 235.
[23] Becher, H. J., and Baechle, H. T., *Chem. Ber.,* 1965, **98**, 2159.
[24] Becher, H. J., and Baechle, H. T., *Z. Phys. Chem.,* 1966, **48**, 359.
[25] Becher, H. J., and Baechle, H. T., *Spectrochim. Acta,* 1965, **21**, 579.
[26] Becher, H. J., and Diehl, H., *Chem. Ber.,* 1965, **98**, 526.
[27] Becher, H. J., and Goubeau, J., *Z. Anorg. Chem.,* 1952, **268**, 133.
[28] Becher, H. J., Nowodny, W., Nöth, H., and Meister, W., *Z. Anorg. Chem.,* 1962, **314**, 226.
[29] Beck, W., Becker, W., Nöth, H., and Wrackmeyer, B., *Chem. Ber.,* 1972, **105**, 2883.
[30] Becker, W., and Nöth, H., *Chem. Ber.,* 1972, **105**, 1962.
[31] Bellut, H., and Köster, R., *Ann.,* 1970, **738**, 86.
[32] Beyer, H., Niedenzu, K., and Dawson, J. W., *J. Org. Chem.,* 1962, **27**, 4701.
[33] Bissot, T. C., and Parry, R. W., *J. Amer. Chem. Soc.,* 1955, **77**, 3481.
[34] Blackborow, J. R., and Lockhart, J. C., *J. Chem. Soc. (A),* 1971, 1343.
[35] Blau, J. A., Gerrard, W., and Lappert, M. F., *J. Chem. Soc.,* 1960, 667;

idem, Mountfield, B. A., and Pyszora, H., *ibid.*, p. 380.
[36] Bock, H., and Fuss, W., *Chem. Ber.*, 1971, **104**, 1687.
[37] Böddeker, K. W., Shore, S. G., and Bunting, R. K., *J. Amer. Chem. Soc.*, 1966, **88**, 4396.
[37a] Bowser, J. R., Neilson, R. H., and Wells, R. L., *Inorg. Chem.*, 1978, **17**, 1882.
[38] Braum, J., and Normant, H., *Bull. Soc. Chim. Fr.*, 1966, 2557.
[39] Brey, W. S., Fuller, M. E., Ryschkewitsch, G. E., and Marshall, A. S., *Adv. Chem. Ser.*, 1964, **42**, 100.
[39a] Briggs, T. S., Gwinn, W. D., Jolly, W. L., and Thorne, L. R., *J. Amer. Chem. Soc.*, 1978, **100**, 7762.
[40] Brotherton, R. J., Can. Patent 629,812 (1961).
[41] Brotherton, R. J., quoted in ref. 373.
[42] Brotherton, R. J., and Buckman, T., *Inorg. Chem.*, 1963, **2**, 424.
[43] Brotherton, R. J., Manasevit, H. M., and McCloskey, A. L., *Inorg. Chem.*, 1962, **1**, 749.
[44] Brotherton, R. J., McCloskey, A. L., Peterson, L. L., and Steinberg, H., *J. Amer. Chem. Soc.*, 1960, **82**, 6242.
[45] Brotherton, R. J., and Peterson, L. L., Can. Patent 622,990 (1961).
[46] Brotherton, R. J., and Steinberg, H., *J. Org. Chem.*, 1961, **26**, 4632.
[47] Brotherton, R. J., Steinberg, H., and McCloskey, A. L., Fr. Patent 1,241,499 (1960).
[48] Brotherton, R. J., Steinberg, H., and McCloskey, A. L., U.S. Patent 2,974,165 (1961).
[48a] Brown, C., and Cragg, R. H., unpublished work, 1977.
[48b] Brown, C., Cragg, R. H., Miller, T., Smith, D. O'N., and Steltner, A., *J. Organometallic Chem.*, 1978, **149**, C34.
[49] Brown, J. F., *J. Amer. Chem. Soc.*, 1952, **74**, 1219.
[50] Brown. M. P., Dann, A. E., Hunt, D. W., and Silver, H. B., *J. Chem. Soc.*, 1962, 4648.
[51] Brown, M. P., and Heseltine, R. W., *J. Inorg. Nucl. Chem.*, 1967, **29**, 1197.
[52] Brown, M. P., Heseltine, R. W., and Johnstone, D. W., *J. Chem. Soc. (A)*, 1967, 597.
[53] Brown, M. P., Heseltine, R. W., and Sutcliffe, L. H., *J. Chem. Soc. (A)*, 1968, 612,
[54] Brown, H. C., Midland, M. M., and Levy, A. B., *J. Amer. Chem. Soc.*, 1972, **94**, 2114, 3662; 1973, **95**, 2394.
[55] Brown, C. A., and Osthoff, R. C., *J. Amer. Chem. Soc.*, 1952, **74**, 2340.
[56] Buchheit, P., Dissertation, Munich, 1942.
[57] Bullen, G. J., and Clark, N. H., *Chem. Comm.*, 1967, 676; *J. Chem. Soc. (A)*, 1970, 2213.
[58] Bullen, G. J., and Clark, N. H., *J. Chem. Soc. (A)*, 1969, 404.

[59] Bullen, G. J., and Clark, N. H., *J. Chem. Soc. (A)*, 1970, 992.
[60] Bullen, G. J., and Wade, K., *Chem. Comm.*, 1971, 1122.
[61] Burch, J. E., Gerrard, W., and Mooney, E. F., *J. Chem. Soc.*, 1962, 2200.
[62] Burg, A. B., *J. Amer. Chem. Soc.*, 1952, **74**, 1340.
[63] Burg, A. B., *Inorg. Chem.*, 1972, **11**, 2283.
[64] Burg, A. B., and Banus, J., *J. Amer. Chem. Soc.*, 1954, **76**, 3903.
[65] Burg, A. B., and Boone, J. L., *J. Amer. Chem. Soc.*, 1956, **78**, 1521.
[66] Burg, A. B., and Good, C. D., *J. Inorg. Nucl. Chem.*, 1956, **2**, 237.
[67] Burg, A. B., and Kuljian, E. S., *J. Amer. Chem. Soc.*, 1950, **72**, 3103.
[68] Burg, A. B., and Randolph, C. L., *J. Amer. Chem. Soc.*, 1949, **71**, 3451.
[69] Burg, A. B., and Randolph, C. L., *J. Amer. Chem. Soc.*, 1951, **73**, 953.
[70] Burg, A. B., and Sandhu, J. S., *J. Amer. Chem. Soc.*, 1967, **89**, 1626.
[71] Bürger, H., and Höfler, F., *Spectrochim. Acta*, 1970, **26A**, 31.
[71a] Bushnell, G. W., and Rivett, G. A., *Can. J. Chem.*, 1977, **55**, 3294.
[72] Busse, P. J., and Niedenzu, K., *Synth. Inorg. Metal-Org. Chem.*, 1973, **3**, 23.
[73] Campbell, G. W., and Johnson, L., *J. Amer. Chem. Soc.*, 1959, **81**, 3800.
[74] Carmody, D. R., and Zeitz, A., U.S. Patent 3,001,361 (1961).
[75] Caserio, F. F., Cavallo, J. J., and Wagner, R. I., *J. Org. Chem.*, 1961, **26**, 2157.
[76] Caujolle, R., and Quan, D. Q., *Compt. Rend.*, 1970, **271C**, 754.
[77] Çetinkaya, B., King, G. H., Krishnamurthy, S. S., Lappert, M. F., and Pedley, J. B., *Chem. Comm.*, 1971, 1370.
[78] Chakrabarty, M. R., Thompson, C. C., and Brey, W. S., *Inorg. Chem.*, 1967, **6**, 518.
[79] Chang, C. H., Porter, R. F., and Bauer, S. H., *Inorg. Chem.*, 1969, **8**, 1677.
[79a] Christmas, B. K., and Niedenzu, K., *Z. Naturforsch.*, 1977, **32b**, 157.
[80] Clark, A. H., and Anderson, G. A., *Chem. Comm.*, 1969, 1082.
[81] Clippard, F. B., and Bartell, L. S., *Inorg. Chem.*, 1970, **9**, 2439.
[82] Coates, G. E., *J. Chem. Soc.*, 1950, 3481.
[83] Coates, G. E., Green, M. L. H., and Wade, K., 'Organometallic Compounds', Methuen, London, Vol. 1, Ch. III, 1967.
[84] Coates, G. E., and Livingstone, J. G., *J. Chem. Soc.*, 1961, 1000.
[85] Cohen, E. A., and Beaudet, R. A., *Inorg. Chem.*, 1973, **12**, 1570.
[86] Conklin, G. W., and Morris, R. C., U.S. Patent 2,886,575 (1959).
[86a] Cook, W. L., and Niedenzu, K., *Synth. Inorg. Metal-Org. Chem.*, 1972, **2**, 267.
[87] Cook, W. L., and Niedenzu, K., *Synth. Inorg. Metal-Org. Chem.*, 1974, **4**, 53.

[88] Cragg, R. H., *J. Chem. Soc. (A)*, 1968, 2962.
[89] Cragg, R. H., *J. Inorg. Nucl. Chem.*, 1968, **30**, 395.
[90] Cragg, R. H., *J. Inorg. Nucl. Chem.*, 1968, **30**, 711.
[91] Cragg, R. H., *Chem. Comm.*, 1969, 832.
[92] Cragg, R. H., and Lappert, M. F., *Adv. Chem. Ser.*, 1964, **42**, 221.
[93] Cragg, R. H., Lappert, M. F., and Tilley, B. P., *J. Chem. Soc.*, 1964, 2108.
[94] Cragg, R. H., and Lockhart, J. C., *J. Inorg. Nucl. Chem.*, 1969, **31**, 2282.
[94a] Cragg, R. H., and Miller, T. J., *J. Organometallic Chem.*, 1978, **154**, C3.
[95] Cragg, R. H., and Weston, A. F., *Chem. Comm.*, 1972, 79.
[95a] Cragg, R. H., and Weston, A. F., *J.C.S. Dalton*, 1973, 93.
[96] Cragg, R. H., Weston, A. F., Todd, J. F., and Turner, R. B., *Chem. Comm.*, 1972, 206.
[96a] Curry, K. K., and Gilje, J. W., *J. Amer. Chem. Soc.*, 1976, **98**, 8262.
[96b] Curry, K. K., and Gilje, J. W., *J. Amer. Chem. Soc.*, 1978, **100**, 1442.
[97] Dagl, G. H., and Schaeffer, R., *J. Amer. Chem. Soc.*, 1961, **83**, 3032.
[98] Davies, A. G., Hook, S. C. W., and Roberts, B. P., *J. Organometallic Chem.*, 1970, **23**, C11.
[99] Davies, A. G., and Kennedy, J. D., *J. Chem. Soc. (C)*, 1970, 759.
[100] Davies, F. A., Dewar, M. J. S., Jones, R., and Worley, S. D., *J. Amer. Chem. Soc.*, 1969, **91**, 2094.
[101] Davies, F. A., Turchi, I. J., and Greely, D. N., *J. Org. Chem.*, 1971, **36**, 1300.
[102] Dawson, J. W., Fritz, P., and Niedenzu, K., *J. Organometallic Chem.*, 1966, **5**, 211.
[103] Dawson, J. W., Fritz, P., and Niedenzu, K., Abstr. 148th Meeting Amer. Chem. Soc., Chicago, Sept., 1964, 22-0.
[104] Dewar, M. J. S., *Progr. Boron Chem.*, 1964, **1**, 235.
[105] Dewar, M. J. S., Gleicher, G. J., and Robinson, B. P., *J. Amer. Chem. Soc.*, 1964, **86**, 5698.
[106] Dewar, M. J. S., and Kubba, V. P., *Tetrahedron*, 1959, **7**, 213.
[107] Dewar, M. J. S., Kubba, V. P., and Pettit, R., *J. Chem. Soc.*, 1958, 3073.
[108] Dewar, M. J. S., Kubba, V. P., and Pettit, R., *J. Chem. Soc.*, 1958, 3076.
[109] Dewar, M. J. S., and Rona, P., *J. Amer. Chem. Soc.*, 1965, **87**, 5510.
[110] Dewar, M. J. S., and Rona, P., *J. Amer. Chem. Soc.*, 1969, **91**, 2259.
[111] Dobson, J., and Schaeffer, R., *Inorg. Chem.*, 1970, **9**, 2183.
[112] Dornow, A., and Gehrt, H. H., *Z. Anorg. Chem.*, 1958, **294**, 81.
[113] Dorokhov, V. A., and Lappert, M. F., *J. Chem. Soc. (A)*, 1969, 433.
[114] Dorokhov, V. A., Lavrinovich, L. I., Yakovlev, I. P., and Mikhailov, B. M., *Zh. Obshch. Khim.*, 1971, **41**, 2501.
[115] Dorokhov, V. A., and Mikhailov, B. M., *Bull. Acad. Sci. U.S.S.R., Chem. Sect.*, 1966, 364.
[116] Dorokhov, V. A., and Mikhailov, B. M., *Proc. Acad. Sci. U.S.S.R.*, 1969,

187, 1300.
[116a] Dorokhov, V. A., and Mikhailov, B. M., *J. Gen. Chem. U.S.S.R.*, 1973, **21**, 1847.
[117] Elter, G., Glemser, O., and Herzog, W., *J. Organometallic Chem.*, 1972, **36**, 257.
[118] Elter, G., Glemser, O., and Herzog, W., *Inorg. Nucl. Chem. Letters*, 1972, **8**, 191.
[119] Elter, G., Glemser, O., and Herzog, W., *Chem. Ber.*, 1972, **105**, 115.
[119a] Emerick, D. P., Gragg, B. R., and Niedenzu, K., *J. Organometallic Chem.*, 1978, **153**, 9; Gragg, B. R., and Niedenzu, K., *Synth. Inorg. Metal-Org. Chem.*, 1976, **6**, 275.
[119b] Emerick, D. P., Komorowski, L., and Niedenzu, K., *J. Organometallic Chem.*, 1978, **154**, 147.
[120] English, W. D., U.S. Patent 3,052,718 (1962).
[121] English, W. D., McCloskey, A. L., and Steinberg, H., *J. Amer. Chem. Soc.*, 1961, **83**, 2122.
[122] English, W. D., Steinberg, H., and McCloskey, A. L., U.S. Patent 3,068,182 (1962).
[123] Erickson, C. E., and Gunderloy, F. C., *J. Org. Chem.*, 1959, **24**, 1161; Abstr. 136th Meeting, Amer. Chem. Soc., Atlantic City, Sept., 1959, 54-N.
[124] Eubanks, I. D., and Lagowski, J. J., *J. Amer. Chem. Soc.*, 1966, **88**, 2425.
[125] Finch, A., and Gardner, P. J., *Progr. Boron Chem.*, 1970, **3**, 177.
[126] Finch, A., Gardner, P. J., Lockhart, J. C., and Pearn, E. J., *J. Chem. Soc.*, 1962, 1428.
[127] Finch, A., Leach, J. B., and Morris, J. H., *Organometallic Chem. Rev. (A)*, 1969, **4**, 1.
[128] Francois, M., *Compt. Rend.*, 1966, **262C**, 1091.
[129] Fried, J., Bergeim, J., Yale, H. L., and Bernstein, J., U.S. Patent 3,293,252 (1968).
[130] Fritz, P., Niedenzu, K., and Dawson, J. W., *Inorg. Chem.*, 1964, **3**, 1677; 1965, **4**, 886.
[130a] Fuss, W., *Z. Naturforsch.*, 1974, **29b**, 514.
[130b] Fussstetter, H., Huffman, J. C., Nöth, H., and Schaeffer, R., *Z. Naturforsch.*, 1976, **31b**, 1441.
[130c] Fussstetter, H., Nöth, H., and Winterstein, W., *Chem. Ber.*, 1977, **110**, 1931.
[130d] Fussstetter, H., Nöth, H., Wrackmeyer, B., and McFarlane, W., *Chem. Ber.*, 1977, **110**, 3172.
[131] Geisler, I., and Nöth, H., *Chem. Comm.*, 1969, 775.
[132] Geisler, I., and Nöth, H., *Chem. Ber.*, 1970, **103**, 2234.
[132a] Geisler, I., and Nöth, H., *Chem. Ber.*, 1973, **106**, 1943.

[133] George, T. A., and Lappert, M. F., *Chem. Comm.*, 1966, 463.
[134] George, T. A., and Lappert, M. F., *J. Chem. Soc. (A)*, 1969, 992.
[135] Gerrard, W., *The Organic Chemistry of Boron,* Academic Press, New York, 1961.
[136] Gerrard, W., Hudson, H. R., and Mooney, E. F., *J. Chem. Soc.*, 1960, 5168.
[137] Gerrard, W., Hudson, H. R., and Mooney, E. F., *J. Chem. Soc.*, 1962, 113.
[138] Gerrard, W., Lappert, M. F., and Mountfield, B. A., *J. Chem. Soc.*, 1959, 1529.
[139] Gerrard, W., Lappert, M. F., and Pearce, C. A., *J. Chem. Soc.*, 1957, 381.
[140] Gerrard, W., Lappert, M. F., and Pearce, C. A., *Chem. Ind.*, 1958, 292.
[141] Gerrard, W., and Mooney, E. F., *J. Chem. Soc.*, 1960, 4028.
[142] Geymayer, P., and Rochow, E. G., *Monatsh.*, 1966, **97**, 429.
[143] Geymayer, P., Rochow, E. G., and Wannagat, U., *Angew. Chem. Internat. Edn.*, 1964, **3**, 633.
[143a] Gmelin, L., *Handbuch der Anorganischen Chemie,* New Supplement Series to 8th Edition, Vol. 13 (Part 1 of 'Boron Compounds') pp. 86–245, Berlin, 1974.
[143b] Gmelin, L., *Handbuch der Anorganischen Chemie,* New Supplement Series to 8th Edition, Vol. 13 (Part 1 of 'Boron Compounds.) pp. 245–331, Berlin, 1974.
[143c] Gmelin, L., *Handbuch der Anorganischen Chemie,* New Supplement Series to 8th Edition, Vol. 22 (Part 4 of 'Boron Compounds'), Berlin, 1975.
[143d] Gmelin, L., *Handbuch der Anorganischen Chemie,* New Supplement Series to 8th Edition, Vol. 23 (Part 5 of 'Boron Compounds'), Berlin, 1975.
[143e] Gmelin, L., *Handbuch der Anorganischen Chemie,* New Supplement Series to 8th Edition, Vol. 37 (Part 10 of 'Boron Compounds'), Berlin, 1976.
[143f] Gmelin, L., *Handbuch der Anorganischen Chemie,* New Supplement Series to 8th Edition, Vol. 46 (Part 15 of 'Boron Compounds'), Berlin, 1977.
[144] Giesler, E., and Schulze, G., Ger. Patent 1,174,796 (1964).
[145] Giesler, E., and Schulze, G., *Chem. Abstr.*, 1964, **61**, 10592e.
[146] Glogowski, M. E., Grisdal, P. J., Williams, J. L. R., and Regan, T. H., *J. Organometallic Chem.*, 1973, **54**, 51.
[146a] Goetze, R., and Nöth, H., *Z. Naturforsch.*, 1975, **30b**, 875.
[146b] Goetze, R., and Nöth, H., *Chem. Ber.*, 1976, **109**, 3247.
[147] Gal'chenko, G. L., Brykina, E. P., Shchegoleva, N. N., Vasil'ev, L. S., and Mikhailov, B. M., *Izv. Akad. Nauk S.S.S.R. Ser. Khim.*, 1973, 200.
[148] Goldsmith, H., and McCloskey, A. L., Fr. Patent 1,200,953 (1959).

[149] Goubeau, J., *Fiat. Rev. of Ger. Sci.*, 1939–46, Inorganic Chemistry, Part I.

[150] Goubeau, J., *Angew. Chem.*, 1957, **69**, 77.

[151] Goubeau, J., Bessler, E., and Wolff, D., *Z. Anorg. Chem.*, 1967, **352**, 285.

[152] Goubeau, J., and Keller, H., *Z. Anorg. Chem.*, 1951, **267**, 1.

[153] Goubeau, J., Rahtz, M., and Becher, H. J., *Z. Anorg. Chem.*, 1954, **275**, 161.

[154] Goubeau, J., and Schneider, H., *Ann.*, 1964, **675**, 1.

[154a] Gragg, B. R., and Niedenzu, K., *J. Organometallic Chem.*, 1978, **149**, 141, 271.

[154b] Gragg, B. R., Layton, W. J., and Niedenzu, K., *J. Organometallic Chem.*, 1977, **132**, 29.

[154c] Graham, D. M., Bowser, J. R., Moreland, C. G., Neilson, R. H., and Wells, R. L., *J. Amer. Chem. Soc.*, 1978, **100**, 2028.

[155] Greenwood, N. N., and Robinson, B. H., *J. Chem. Soc. (A)*, 1968, 180.

[156] Greenwood, N. N., and Walker, J., *J. Chem. Soc. (A)*, 1967, 959.

[157] Gropen, O., and Seip, H. S., *Chem. Phys. Letters*, 1974, **25**, 206.

[158] Grosse-Ruyken, H., and Kleesaat, R., *Z. Anorg. Chem.*, 1961, **308**, 122.

[159] Gunderloy, F. C., and Erickson, C. E., *Inorg. Chem.*, 1962, **1**, 349.

[159a] Gupta, S. K., *J. Organometallic Chem.*, 1978, **156**, 95.

[160] Haasnoot, J. G., and Groenenveld, W. L., *Z. Naturforsch.*, 1974, **29b**, 52.

[161] Hahn, G. A., and Schaeffer, R., *J. Amer. Chem. Soc.*, 1964, **86**, 1503.

[162] Hall, R. E., and Schram, E. P., *Inorg. Chem.*, 1969, **8**, 270.

[163] Hancock, K. G., and Dickinson, D. A., *J. Amer. Chem. Soc.*, 1972, **94**, 4396.

[164] Hancock, K. G., and Kramer, J. D., *J. Amer. Chem. Soc.*, 1973, **95**, 6463.

[165] Hancock, K. G., and Uriarte, A. K., *J. Amer. Chem. Soc.*, 1970, **92**, 6374.

[166] Hancock, K. G., Uriarte, A. K., and Dickinson, D. A., *J. Amer. Chem. Soc.*, 1973, **95**, 6980.

[167] Harshbarger, W., Lee, G. H., Porter, R. F., and Bauer, S. H., *Inorg. Chem.*, 1969, **8**, 1683.

[168] Haworth, D. T., and Matushek, E. S., *Chem. Ind.*, 1971, 130.

[168a] Haubold, W., and Schaeffer, R., *Chem. Ber.*, 1971, **104**, 513.

[169] Hawthorne, M. F., *J. Amer. Chem. Soc.*, 1959, **81**, 5836.

[170] Hawthorne, M. F., *J. Amer. Chem. Soc.*, 1961, **83**, 2671.

[171] Hengge, E., and Wolfer, D., *Angew. Chem. Internat. Edn.*, 1973, **12**, 315.

[172] Hermannsdörfer, K. H., Matejčikova, E., and Nöth, H., *Chem. Ber.*, 1970, **103**, 516.

[173] Hess, H., *Acta Cryst.*, 1963, **16A**, 74.

[174] Hess, H., *Angew. Chem. Internat. Edn.,* 1967, **6**, 975; *Acta Cryst.,* 1969, **25B**, 2342.
[175] Hess, H., *Acta Cryst.,* 1969, **25B**, 2334.
[176] Hesse, G., and Haag, A., *Tetrahedron Letters,* 1965, 1123.
[177] Hesse, G., and Witte, H., *Angew. Chem. Internat. Edn.,* 1963, **2**, 617.
[178] Hesse, G., Witte, H., and Haussleiter, H., *Angew. Chem. Internat. Edn.,* 1966, **5**, 723.
[179] Hessett, B., Leach, J. B., Morris, J. H., and Perkins, P. G., *J.C.S. Dalton,* 1972, 131.
[180] Heying, T. L., and Smith, H. D., *Adv. Chem. Ser.,* 1964, **42**, 201.
[181] Hohnstedt, L. F., and Pellicciotto, A. M., Abstr. 137th Meeting, Amer. Chem. Soc., Cleveland, April, 1960, 7-O.
[182] Holliday, A. K., Marsden, F. J., and Massey, A. G., *J. Chem. Soc.,* 1961, 3348.
[183] Holliday, A. K., and Massey, A. G., *Chem. Rev.,* 1962, 303.
[184] Holmes, R. R., and Wagner, R. P., *J. Amer. Chem. Soc.,* 1962, **84**, 357.
[185] Hough, W. V., Schaeffer, G. W., Dzurus, M., and Stewart, A. C., *J. Amer. Chem. Soc.,* 1955, **77**, 864.
[185a] Huffman, J. C., Fussstetter, H., and Nöth, H., *Z. Naturforsch.,* 1976, **31b**, 289.
[186] Imbery, D., Jaeschke, A., and Friebolin, H., *Org. Mag. Res.,* 1970, **2**, 271.
[186a] Jefferson, R., and Lappert, M. F., *Intra-Sci. Chem. Rept.,* 1973, **7** (No. 2), 123.
[187] Jefferson, R., Prokai, B., Lappert, M. F., and Tilley, B. P., *J. Chem. Soc. (A),* 1966, 1584.
[188] Jenne, H., and Niedenzu, K., *Inorg. Chem.,* 1964, **3**, 68.
[188a] Jonás, K., Nöth, H., and Storch, W., *Chem. Ber.,* 1977, **110**, 2783.
[189] Jones, R. G., and Kinney, C. R., *J. Amer. Chem. Soc.,* 1939, **61**, 1378.
[190] Kaempfer, K., Nöth, H., Petz, W., and Schmid, G., Internat. Symp. New Aspects Chem. Metal Carbonyl Deriv., Venice, 1968, D7.
[191] Keller, P. C., *Synth. Inorg. Metal-Org. Chem.,* 1973, **3**, 307.
[191a] Keller, P. C., *Inorg. Chem.,* 1971, **10**, 1528.
[192] Keller, P. C., *J. Amer. Chem. Soc.,* 1974, **96**, 3073.
[193] Keller, P. C., *J. Amer. Chem. Soc.,* 1972, **94**, 4020.
[194] Keller, P. C., and Schwarz, L. D., *J. Amer. Chem. Soc.,* 1972, **94**, 3015.
[195] Kelley, H. C., and Edwards, J. O., *J. Amer. Chem. Soc.,* 1960, **82**, 4842.
[196] King, G. H., Krishnamurthy, S. S., Lappert, M. F., and Pedley, J. B., *Discussions Faraday Soc.,* 1972, **54**, 70.
[197] Kinney, C. R., and Kolbezen, M. J., *J. Amer. Chem. Soc.,* 1942, **64**, 1584.
[198] Köster, R., Bellut, H., and Hattori, S., *Ann.,* 1969, **720**, 1.
[199] Köster, R., Bellut, H., and Ziegler, E., *Angew. Chem. Internat. Edn.,*

1964, **3**, 514; 1967, **6**, 255.
[200] Köster, R., Iwasaki, K., Hattori, S., and Morita, Y., *Ann.*, 1968, **720**, 23.
[201] Köster, R., and Iwasaki, K., *Adv. Chem. Ser.*, 1964, **42**, 148.
[202] Kotz, J. C., Zanden, R. J. V., and Cooks, R. G., *Chem. Comm.*, 1970, 923.
[203] Krohmer, P., and Goubeau, J., *Chem. Ber.*, 1971, **104**, 1347.
[203a] Kroner, J., Nölle, D., Nöth, H., and Winterstein, W., *Chem. Ber.*, 1975, **108**, 3807.
[203b] Kroner, J., Nöth, H., and Niedenzu, K., *J. Organometallic Chem.*, 1974, **71**, 165.
[203c] Kroto, H. W., Lappert, M. F., Pedley, J. B., and Rogers, A. J., unpublished work, 1977.
[204] Kuck, M. A., and Urry, G., *J. Amer. Chem. Soc.*, 1966, **88**, 426.
[205] Kwon, C. T., and McGee, H. A., *Inorg. Chem.*, 1973, **12**, 696.
[206] Kyker, G. S., and Schram, E. P., *J. Amer. Chem. Soc.*, 1968, **90**, 3672.
[207] Kyker, G. S., and Schram, E. P., *J. Amer. Chem. Soc.*, 1968, **90**, 3678.
[208] Lang, K., and Schubert, F., U.S. Patent 3,090,809 (1963).
[209] Lappert, M. F., *Chem. Rev.*, 1956, **56**, 959.
[210] Lappert, M. F., *Proc. Chem. Soc.*, 1959, 59.
[211] Lappert, M. F., Ch. 2 in *Developments in Inorganic Polymer Chemistry*, (Eds.) Lappert, M. F., and Leigh, G. J., Elsevier, Amsterdam, 1962.
[212] Lappert, M. F., Ch. 8 in *The Chemistry of Boron and its Compounds*, (Ed.) E. L. Muetterties, Wiley, New York, 1967.
[213] Lappert, M. F., Litzow, M. R., Pedley, J. B., Riley, P. N. K., Spalding, T. R., and Tweedale, A., *J. Chem. Soc (A)*, 1970, 2320.
[214] Lappert, M. F., and Majumdar, M. K., *Proc. Chem. Soc.*, 1961, 425.
[215] Lappert, M. F., and Majumdar, M. K., *Proc. Chem. Soc.*, 1963, 88.
[216] Lappert, M. F., and Majumdar, M. K., *Adv. Chem. Ser.*, 1964, **42**, 208.
[217] Lappert, M. F., and Majumdar, M. K., *J. Organometallic Chem.*, 1966, **6**, 316.
[218] Lappert, M. F., Majumdar, M. K., and Tilley, B. P., *J. Chem. Soc. (A)*, 1966, 1590.
[219] Lappert, M. F., Pedley, J. B., Riley, P. N. K., and Tweedale, A., *Chem. Comm.*, 1966, 788.
[220] Lappert, M. F., and Prokai, B., *J. Chem. Soc.*, 1963, 4223.
[221] Lappert, M. F., and Prokai, B., *Adv. Organometallic Chem.*, 1967, **5**, 225.
[221a] Lappert, M. F., Pyszora, H., and Rieber, M., *J. Chem. Soc.*, 1965, 4256.
[222] Lappert, M. F., and Srivastava, G., *Proc. Chem. Soc.*, 1964, 120.
[223] Lappert, M. F., and Srivastava, G., *Inorg. Nucl. Chem. Letters*, 1965, **1**, 53.
[224] Lappert, M. F., and Srivastava, G., *J. Chem. Soc. (A)*, 1967, 602.
[225] Lappert, M. F., and Srivastava, G., unpublished work, 1966.

[226] Laurent, J. P., *Compt. Rend.*, 1963, **256C**, 3283; 1964, **258C**, 1233.
[227] Letsinger, R. L., and Hamilton, J. B., *J. Amer. Chem. Soc.*, 1958, **80**, 5411.
[228] Lovas, F. J., and Johnson, D. R., *J. Chem. Phys.*, 1973, **59**, 2347.
[229] McCloskey, A. L., Ch. 4 in *Inorganic Polymers,* (Eds.) F. G. A. Stone, and W. A. G. Graham, Academic Press, 1962.
[230] McCloskey, A. L., Collins, G. G., and English, W. D., Fr. Patent 1,3000,002 (1962).
[231] McCloskey, A. L., and Goldsmith, H., Brit. Patent 837,076 (1960).
[232] McCloskey, A. L., Goldsmith, H., and Larrucia, D., quoted in ref. 373.
[233] McCloskey, A. L., Goldsmith, H., and Peterson, L. L., quoted in ref. 373.
[234] McGee, H. A., and Kwon, C. T., *Inorg. Chem.*, 1970, **9**, 2458.
[235] Maitlis, P. M., *Chem. Rev.*, 1962, **62**, 223.
[235a] Maringgele, W., and Meller, A., *Z. Anorg. Chem.*, 1977, **436**, 173.
[236] Mehrotra, R. C., and Srivastava, G., *J. Indian Chem. Soc.*, 1961, **38**, 1.
[237] Meller, A., *Organometallic Chem. Rev.*, 1967, **2**, 1.
[238] Meller, A., *Monatsh.*, 1967, **98**, 2014.
[239] Meller, A., and Gerger, W., *Monatsh.*, 1974, **105**, 684.
[240] Meller, A., Maringgele, W., and Hirninger, F. J., *J. Organometallic Chem.*, 1977, **136**, 289.
[241] Meller, A., Maringgele, W., and Marech, G., *Monatsh.*, 1974, **105**, 637.
[241a] Meller, A., Maringgele, W., and Sicker, U., *J. Organometallic Chem.*, 1977, **141**, 249.
[242] Meller, A., and Ossico, A., *Monatsh.*, 1972, **103**, 577.
[243] Mellon, E. K., and Lagowski, J. J., *Adv. Inorg. Chem. Radiochem.*, 1963, **5**, 259.
[243a] Merriam, J. S., and Niedenzu, K., *J. Organometallic Chem.*, 1973, **51**, C1.
[244] Michaelis, A., and Luxembourg, K., *Ber.*, 1896, **29**, 710.
[245] Mikhailov, B. M., *Uspekhi Khim.*, 1959, **28**, 1450.
[246] Mikhailov, B. M., *Russ. Chem. Rev.*, 1960, 459.
[247] Mikhailov, B. M., and Aronovich, P. M., *Bull. Acad. Sci. U.S.S.R., Div. Chem. Sci.*, 1957, 1146.
[248] Mikhailov, B. M., and Aronovich, P. M., *J. Gen. Chem. U.S.S.R.*, 1959, **29**, 3090.
[249] Mikhailov, B. M., and Bubnov, Yu. N., *Bull. Acad. Sci. U.S.S.R., Div. Chem. Sci.*, 1960, 1872.
[250] Mikhailov, B. M., and Bubnov, Yu. N., *J. Gen. Chem. U.S.S.R.*, 1961, 577.
[251] Mikhailov, B. M., and Bubnov, Yu. N., *J. Gen. Chem. U.S.S.R.*, 1962, 1969.
[252] Mikhailov, B. M., and Dorokhov, V. A., *Proc. Acad. Sci. U.S.S.R., Chem. Sect.*, 1961, **136**, 51.

[253] Mikhailov, B. M., and Dorokhov, V. A., *Bull. Acad. Sci. U.S.S.R., Div. Chem. Sci.*, 1961, 1082.

[254] Mikhailov, B. M., and Dorokhov. V. A., *Bull. Acad. Sci. U.S.S.R., Div. Chem. Sci.*, 1961, 1944.

[255] Mikhailov, B. M., and Dorokhov, V. A., *Bull. Acad. Sci. U.S.S.R., Div. Chem. Sci.*, 1962, 623, 1213.

[256] Mikhailov, B. M., and Dorokhov, V. A., *J. Gen. Chem. U.S.S.R.*, 1961, **31**, 3504.

[257] Mikhailov, B. M., and Dorokhov, V. A., *J. Gen. Chem. U.S.S.R.*, 1962, **32**, 1497.

[258] Mikhailov, B. M., Dorokhov, V. A., and Shchegoleva, T. A., *Bull. Acad. Sci. U.S.S.R., Div. Chem. Sci.*, 1963, 446.

[259] Mikhailov, B. M., and Fedotov, N. S., *Bull. Acad. Sci. U.S.S.R., Div. Chem. Sci.*, 1956, 1511.

[260] Mikhailov, B. M., and Fedotov, N. S., *Bull. Acad. Sci. U.S.S.R., Div. Chem. Sci.*, 1960, 1590.

[261] Mikhailov, B. M., and Fedotov, N. S., *J. Gen. Chem. U.S.S.R.*, 1962, **32**, 93.

[262] Mikhailov, B. M., and Kostroma, T. V., *Bull. Acad. Sci. U.S.S.R., Div. Chem. Sci.*, 1957, 659.

[263] Mikhailov, B. M., and Kostroma, T. V., *J. Gen. Chem. U.S.S.R.*, 1959, **29**, 1477.

[264] Mikhailov, B. M., and Kozminskaya, T. K., *J. Gen. Chem. U.S.S.R.*, 1960, **30**, 3619.

[265] Mikhailov, B. M., and Kozminskaya, T. K., *Bull. Acad. Sci. U.S.S.R., Div. Chem. Sci.*, 1962, 256.

[266] Mikhailov, B. M., and Pozdnev, V. F., *Proc. Acad. Sci. U.S.S.R.*, 1963, 340.

[267] Mikhailov, B. M., Sheludyakov, V. D., and Shchegoleva, T. A., *Bull. Acad. Sci. U.S.S.R., Div. Chem. Sci.*, 1962, 1475.

[268] Mikhailov, B. M., Sheludyakov, V. D., and Shchegoleva, T. A., *Bull. Acad. Sci. U.S.S.R., Div. Chem. Sci.*, 1962, 1559.

[269] Mikhailov, B. M., and Tutorskaya, F. B., *Bull. Acad. Sci. U.S.S.R., Div. Chem. Sci.*, 1961, 1158.

[270] Mikhailov, B. M., and Vasil'ev, L. S., *Bull. Acad. Sci. U.S.S.R., Div. Chem. Sci.*, 1962, 628.

[271] Mikhailov, B. M., Vasil'ev, L. S., and Bezmenov, A. Ya., *Bull. Acad. Sci. U.S.S.R., Div. Chem. Sci.*, 1965, 712.

[272] Mikhailov, B. M., Vasil'ev, L. S., and Dmitrikov, V. P., *Bull. Acad. Sci. U.S.S.R., Div. Chem. Sci.*, 1968, 2059.

[273] Mikhailov, B. M., and Povarov, L. S., *Zh. Obshch. Khim.*, 1971, **41**, 1540.

[274] Molinelli, R., Smith, S. R., and Tanaka, J., *J.C.S. Dalton*, 1972, 1363.

[275] Mooney, E. F., and Henderson, W. G., *Ann. Rev. N.M.R. Spectroscopy*, 1969, **2**, 219.

[276] Mostovoi, N. V., Dorokhov, V. A., and Mikhailov, B. M., *Bull. Acad. Sci. U.S.S.R., Div. Chem. Sci.*, 1966, 70.

[276a] Muckle, G., Nöth, H., and Storch, W., *Chem. Ber.*, 1976, **109**, 2572.

[276b] Müller, K.-D., Komorowski, L., and Niedenzu, K., *Synth. Inorg. Metal-Org. Chem.*, 1978, **8**, 149.

[277] Musgrave, O. C., *J. Chem. Soc.*, 1956, 4305.

[278] Nagasawa, K., Yoshizaki, T., and Watanabe, H., *Inorg. Chem.*, 1965, **4**, 275.

[279] Neilson, R. H., and Wells, R. L., *Inorg. Chem.*, 1977, **16**, 7.

[280] Niedenzu, K., Beyer, H., and Dawson, J. W., *Inorg. Chem.*, 1962, **1**, 738.

[281] Niedenzu, K., Beyer, H., Dawson, J. W., and Jenne, H., *Chem. Ber.*, 1963, **96**, 2653.

[281a] Niedenzu, K., Blick, K. E., and Boenig, I. A., *Z. Anorg. Chem.*, 1972, **387**, 107.

[281b] Niedenzu, K., Boenig, I. A., and Rothgery, E. F., *Chem. Ber.*, 1972, **105**, 2258.

[282] Niedenzu, K., Busse, P. J., and Miller, C. D., *Inorg. Chem.*, 1970, **9**, 977.

[283] Niedenzu, K., and Dawson, J. W., *J. Amer. Chem. Soc.*, 1959, **81**, 5553.

[284] See ref. 283.

[285] Niedenzu, K., and Dawson, J. W., Abstr. 135th Meeting, Amer. Chem. Soc., Boston, April, 1959, 34-M.

[286] Niedenzu, K., and Dawson, J. W., *J. Amer. Chem. Soc.*, 1960, **82**, 4223.

[287] Niedenzu, K., and Dawson, J. W., *Boron-Nitrogen Compounds*, Springer, Berlin, 1964.

[288] Niedenzu, K., Dawson, J. W., Fritz, P., and Jenne, H., *Chem. Ber.*, 1965, **98**, 3050.

[289] Niedenzu, K., Dawson, J. W., Fritz, P., and Weber, W., *Chem. Ber.*, 1967, **100**, 1898.

[290] Niedenzu, K., Dawson, J. W., Reece, G. A., Sawodny, W., Squire, D. R., and Weber, W., *Inorg. Chem.*, 1966, **5**, 2161.

[291] Niedenzu, K., and Fritz, P., *Z. Anorg. Chem.*, 1965, **340**, 329.

[292] Niedenzu, K., and Fritz, P., *Z. Anorg. Chem.*, 1966, **344**, 329.

[293] Niedenzu, K., Fritz, P., and Dawson, J. W., *Inorg. Chem.*, 1964, **3**, 1077.

[294] Niedenzu, K., Harrelson, D. H., George, W., and Dawson, J. W., *J. Org. Chem.*, 1961, **26**, 3037.

[294a] Niedenzu, K., and Miller, C. D., *Fortschr. Chem. Forsch.*, 1970, **15**, 191.

[295] Niedenzu, K., Miller, C. D., and Smith, S. L., *Z. Anorg. Chem.*, 1970, **372**, 337.

[295a] Nölle, D., and Nöth, H., *Angew. Chem. Internat. Edn.*, 1971, **10**, 126.

[296] Nölle, D., and Nöth, H., *Z. Naturforsch.*, 1972, **27b**, 1425.

[296a] Nölle, D., Nöth, H., and Taeger, T., *Chem. Ber.*, 1977, **110**, 1643.
[296b] Nölle, D., Nöth, H., and Winterstein, W., *Z. Anorg. Chem.*, 1974, **406**, 235.
[297] Nöth, H., *Z. Naturforsch.*, 1961, **16b**, 470.
[298] Nöth, H., *Z. Naturforsch.*, 1961, **16b**, 618.
[299] Nöth, H., Abstr. 142nd Meeting, Amer. Chem. Soc., New York, September, 1962, 29-N.
[300] Nöth, H., quoted in ref. 373.
[301] Nöth, H., *Progr. Boron Chem.*, 1970, **3**, 211.
[302] Nöth, H., *Chem. Ber.*, 1971, **104**, 558.
[302a] Nöth, H., *Pure Appl. Chem.*, 1974, **71**, 165.
[303] Nöth, H., and Abeler, G., *Angew. Chem. Internat. Edn.*, 1965, **4**, 522.
[304] Nöth, H., and Abeler, G., *Chem. Ber.*, 1968, **101**, 969.
[305] Nöth, H., Dorokhov, V. A., Fritz, P., and Pfab, F., *Z. Anorg. Chem.*, 1962, **318**, 293.
[306] Nöth, H., and Fritz, P., *Z. Anorg. Chem.*, 1963, **322**, 297.
[307] Nöth, H., and Fritz, P., *Z. Anorg. Chem.*, 1963, **324**, 270.
[308] Nöth, H., and Geisler, I., see ref. [132a].
[309] Nöth, H., and Hermannsdörfer, K. H., *Angew. Chem.*, 1964, **76**, 377.
[310] Nöth, H., and Hollerer, G., *Chem. Ber.*, 1966, **99**, 2197.
[311] Nöth, H., and Lukas, S., *Chem. Ber.*, 1962, **95**, 1505.
[312] Nöth, H., and Meister, W., *Chem. Ber.*, 1961, **94**, 509.
[313] Nöth, H., and Meister, W., *Z. Naturforsch.*, 1962, **17b**, 714.
[314] Nöth, H., and Mikulaschek, G., *Chem. Ber.*, 1961, **94**, 634.
[315] Nöth, H., and Regnet, W., *Z. Anorg. Chem.*, 1967, **352**, 1.
[315a] Nöth, H., Reichenbach, W., and Winterstein, W., *Chem. Ber.*, 1977, **110**, 2158.
[316] Nöth, H., Reiner, D., and Storch, W., *Chem. Ber.*, 1973, **106**, 1508.
[317] Nöth, H., Schick, H., and Meister, W., *J. Organometallic Chem.*, 1964, **1**, 401.
[318] Nöth, H., and Schmid, G., *J. Organometallic Chem.*, 1966, **5**, 109.
[319] Nöth, H., and Schmid, G., *Z. Anorg. Chem.*, 1966, **345**, 69.
[320] Nöth, H., and Schrägle, W., *Angew. Chem. Internat. Edn.*, 1962, **1**, 457.
[321] Nöth, H., and Schrägle, W., *Chem. Ber.*, 1964, **97**, 2218.
[322] Nöth, H., and Schrägle, W., *Chem. Ber.*, 1964, **97**, 2374.
[323] Nöth, H., and Schweizer, P., *Chem. Ber.*, 1964, **97**, 1464.
[324] Nöth, H., and Schweizer, P., *Chem. Ber.*, 1969, **102**, 161.
[325] Nöth, H., and Sprague, M. J., *J. Organometallic Chem.*, 1970, **22**, 11.
[326] Nöth, H., and Storch, W., *Chem. Ber.*, 1974, **107**, 1028.
[327] Nöth, H., and Storch, W., *Chem. Ber.*, 1976, **109**, 884.
[327a] Nöth, H., and Storch, W., *Chem. Ber.*, 1977, **110**, 2607.
[327b] Nöth, H., and Schuchardt, U., *Z. Anorg. Chem.*, 1975, **418**, 97.
[328] Nöth, H., and Tinhof, W., *Chem. Ber.*, 1975, **108**, 3109.

[328a] Nöth, H., Tinhof, W., and Taeger, T., *Chem. Ber.*, 1974, **107**, 3113.
[329] Nöth, H., Tinhof, W., and Wrackmeyer, B., *Chem. Ber.*, 1974, **107**, 518.
[329a] Nöth, H., and Ullmann, R., *Chem. Ber.*, 1975, **108**, 3125.
[330] Nöth, H., Ullmann, R., and Vahrenkamp, H., *Chem. Ber.*, 1973, **106**, 1165.
[331] Nöth, H., and Vahrenkamp, H., *Chem. Ber.*, 1967, **100**, 3353.
[332] Nöth, H., and Vahrenkamp, H., *J. Organometallic Chem.*, 1969, **16**, 357.
[333] Nöth, H., and Vetter, H. J., *Chem. Ber.*, 1963, **96**, 1298.
[333a] Nöth, H., and Wrackmeyer, B., *Chem. Ber.*, 1973, **106**, 1145.
[333b] Nöth, H., and Wrackmeyer, B., *Chem. Ber.*, 1974, **107**, 3070.
[333c] Nöth, H., and Wrackmeyer, B., *Chem. Ber.*, 1974, **107**, 3089.
[333d] Nöth, H., and Wrackmeyer, B., *Nuclear Magnetic Resonance Spectroscopy of Boron Compounds*, Springer, Berlin, 1978.
[334] Nyilas, E., and Soloway, A. H., *J. Amer. Chem. Soc.*, 1959, **81**, 2681.
[335] Osthoff, R. C., and Brown, C. A., *J. Amer. Chem. Soc.*, 1952, **74**, 2378.
[336] Paetzold, P. I., *Fortschr. Chem. Forsch.*, 1967, **8**, 437.
[337] Paetzold, P. I., and Maier, G., *Chem. Ber.*, 1970, **103**, 281.
[338] Paige, H. L., and Wells, R. L., *Inorg. Chem.*, 1971, **10**, 1526.
[339] Pasdeloup, M., Cros, G., Commenges, G., and Laurent, J. P., *Bull. Soc. Chim. Fr.*, 1971, 754.
[340] Pasdeloup, M., Laurent, J. P., and Lefoge, P., *Bull. Soc. Chim. Fr.*, 1970, 102.
[340a] Pedley, J. B., and Rylance, J., *Sussex-N.P.L. computer analysed thermochemical data: organic and organometallic compounds*, University of Sussex, 1977.
[341] Perkins, P. G., and Wall, D. H., *J. Chem. Soc. (A)*, 1966, 1207.
[342] Phillips, W. D., Miller, H. C., and Muetterties, E. L., *J. Amer. Chem. Soc.*, 1959, **81**, 4496.
[343] Price, W. C., Fraser, R. D. B., Robinson, T. S., and Longuet-Higgins, H. C., *Discussions Faraday Soc.*, 1950, **9**, 131.
[344] Proux, Y., and Clement, R., *Compt. Rend.*, 1967, **264C**, 2135.
[345] Purdun, W. R., and Kaiser, E. M., *J. Inorg. Nucl. Chem.*, 1974, **36**, 1465.
[346] Robiette, A. G., Sheldrick, G. M., Sheldrick, W. S., Beagley, B., Cruickshank, D. W. J., Monaghan, J. J., Aylett, B. J., and Ellis, I. A., *Chem. Comm.*, 1968, 909; Robiette, A. G., Sheldrick, G. M., and Sheldrick, W. S., *J. Mol. Struct.*, 1970, **5**, 423.
[347] Rothgery, E. F., McGee, H. A., and Pusatcioglu, S., *Inorg. Chem.*, 1975, **14**, 2236.
[348] Ruff, J. K., *J. Org. Chem.*, 1962, **27**, 1020.
[349] Ruff, J. K., in *Developments in Inorganic Nitrogen Chemistry*, (Ed.) C. E. Colburn, Elsevier, Amsterdam. 1966, 470.
[350] Russ, C. R., and MacDiarmid, A. G., *Angew. Chem. Internat. Edn.*, 1964, **3**, 509.

[351] Ryschkewitsch, G. E., Brey, W. S., and Saji, A., *J. Amer. Chem. Soc.*, 1961, **83**, 1010.
[352] Schaeffer, G. W., and Anderson, E. R., *J. Amer. Chem. Soc.*, 1949, **71**, 2143.
[353] Schaeffer, R., and Gaines, D. F., *J. Amer. Chem. Soc.*, 1963, **85**, 395.
[354] Schlesinger, H. I., *et. al.*, University of Chicago, 'Hydrides and Borohydrides of Light Weight Elements', Final Reports, July 1, 1947–June 30, 1948.
[355] Schlesinger, H. I., *et al.*, University of Chicago, 'Hydrides and Borohydrides of Light Weight Elements', Final Reports, August 1, 1954–July 31, 1955.
[356] Schmid, G., *Chem. Ber.*, 1970, **103**, 528.
[357] Schmid, G., *Angew. Chem. Internat. Edn.*, 1970, **9**, 819.
[358] Schmid, G., Nöth, H., and Deberitz, J., *Angew. Chem. Internat. Edn.*, 1968, **7**, 293.
[358a] Schmid, G., Petz, W., and Nöth, H., *Inorg. Chim. Acta*, 1970, **4**, 423.
[358b] Schmid, G., and Schulze, J., *Angew. Chem. Internat. Edn.*, 1977, **16**, 249.
[359] Schmid, G., and Weber, L., *Inorg. Chim. Acta*, 1970, **4**, 423.
[360] Schram, E. P., *Inorg. Chem.*, 1966, **5**, 1291.
[361] Scott, K. N., and Brey, W. S., *Inorg. Chem.*, 1969, **8**, 1703.
[362] Semlyen, J. A., and Flory, P. J., *J. Chem. Soc. (A)*, 1966, 191.
[363] Shchegoleva. T. A., Shashkova, E. M., and Mikhailov, B. M., *Bull. Acad. Sci. U.S.S.R., Div. Chem. Sci.*, 1961, 848.
[364] Shchukovskaya, L. L., Voronkov, M. G., and Pavlova, O. V., *Proc. Acad. Sci. U.S.S.R., Chem. Sect.*, 1962, **143**, 297.
[365] Shchukovskaya, L. L., Voronkov, M. G., and Pavlova, O. V., *Bull. Acad. Sci. U.S.S.R., Div. Chem. Sci.*, 1962, 341.
[365a] Shaver, A., *J. Organometallic Chem. Library*, 1977, **3**, 157.
[366] Sheldon, J. C., and Smith, B. C., *Quart. Rev.*, 1960, **14**, 200.
[367] Skinner, H. A., and Smith, N. B., *J. Chem. Soc.*, 1953, 4025; 1954, 2324.
[368] Soulier, J., and Cadiot, P., *Bull. Soc. Chim. Fr.*, 1966, 3846.
[369] Soulier, J., and Cadiot, P., *Bull. Soc. Chim. Fr.*, 1966, 3850.
[370] Soulier, J., and Cadiot, P., *Compt. Rend.*, 1966, **262C**, 376.
[371] Spitsyn, V. I., Kolli, I. D., Rodionov, R. A., and Ivakin, Y. D., *Bull. Acad. Sci. U.S.S.R.*, 1967, 473.
[371a] Spitsyn, V. I., Kolli, I. D., and Sevast'yanova, T. G., *Bull. Acad. Sci. U.S.S.R., Div. Chem. Sci.*, 1973, **22**, 1165.
[372] Srivastava, G., *J. Organometallic Chem.*, 1974, **69**, 179.
[373] Steinberg, H., and Brotherton, R. J., *Organoboron Chemistry*, 1966, Vol. 2, Interscience, London.
[374] Steinberg. H., and McCloskey, A. L., U.S. Patent 3,139,453 (1964).
[374a] Storch, W., Jackstiess, W., Nöth, H., and Winter, W., *Angew. Chem.*

Internat. Edn., 1977, **16**, 478.
[375] Storch, W., and Nöth, H., *Angew. Chem. Internat. Edn.*, 1976, **15**, 235.
[375a] Storch, W., and Nöth, H., *Chem. Ber.*, 1977, **110**, 1636.
[375b] Storr, A., Thomas, B. S., and Penland, A. D., *J.C.S. Dalton*, 1972, 326.
[376] Sugihara, J. M., and Bowman, C. M., *J. Amer. Chem. Soc.*, 1958, **80**, 2443.
[377] Sujishi, S., and Witz, S., *J. Amer. Chem. Soc.*, 1957, **79**, 2447.
[378] Summerford, C., and Wade, K., *J. Chem. Soc. (A)*, 1969, 1487: 1970, 2010.
[379] Sujishi, S., and Witz, S., *J. Amer. Chem. Soc.*, 1954, **76**, 4631.
[379a] Suliman, M. R., and Schram, E. P., *Inorg. Chem.*, 1973, **12**, 920, 923.
[380] Thomas, P. C., and Paul, I. C., *Chem. Comm.*, 1968, 1130.
[381] Toporcer, L. H., Dessy, R. E., and Green, S. I. E., *Inorg. Chem.*, 1965, **4**, 1649.
[382] Totani, T., Watanabe, H., Nagasawa, K., Yoshizaki, T., and Malagawa, T., *Adv. Chem. Ser.*, 1964, **42**, 108.
[383] Trefonas, L. M., and Lipscomb, W. N., *J. Amer. Chem. Soc.*, 1959, **81**, 4435.
[384] Trofimenko, S., *J. Amer. Chem. Soc.*, 1967, **88**, 3165, 3903, 4948.
[385] Trofimenko, S., *J. Org. Chem.*, 1971, **36**, 1161.
[385a] Trofimenko, S., *Accounts Chem. Research*, 1971, **4**, 17; *Chem. Rev.*, 1972, **72**, 497.
[386] Turner, H. S., and Warne, R. J., *Proc. Chem. Soc.*, 1962, 69; *J. Chem. Soc.*, 1965, 6421.
[387] Umland, F., *Angew. Chem. Internat. Edn.*, 1967, **6**, 574.
[388] Vasil'ev, L. S., Dmitrikov, V. P., Bogdanov, V. S., and Mikhailov, B. M., *Zh. Obshch. Khim.*, 1972, **42**, 1318.
[389] Voronkov, M. G., and Shchukovskaya, L. L., Russ. Patent 148,047 (1961).
[390] Voronkov, M. G., and Shchukovskaya, L. L., Russ. Patent 148,048 (1961).
[390a] Wagner, R. I., and Bradford, J. L., *Inorg. Chem.*, 1962, **1**, 93.
[391] Wang, T. T., Busse, P. J., and Niedenzu, K., *Inorg. Chem.*, 1970, **9**, 2150.
[392] Wannagat, U., Eisele, G., and Schlingmann, M., *Z. Anorg. Chem.*, 1977, **429**, 83.
[393] Wannagat, U., and Schmidt, P., *Inorg. Nucl. Chem. Letters*, 1968, **4**, 355.
[394] Watanabe, H., Nagasawa, K., Totani, T., Yoshizaki, T., and Nakagawa, T., *Adv. Chem. Ser.*, 1964, **42**, 116.
[394a] Weber, L., and Schmid, G., *Angew. Chem. Internat. Edn.*, 1974, **13**, 467.
[395] Weber, W., Dawson, J. W., and Niedenzu, K., *Inorg. Chem.*, 1966, **5**, 726.
[396] Wells, R. L., and Collins, A. L., *Inorg. Chem.*, 1966, **5**, 1327.

[397] Wells, R. L., and Collins, A. L., *Inorg. Chem.*, 1968, **7**, 419.
[397a] Wells, R. L., and Neilson, R. H., *Synth. Inorg. Metal-Org. Chem.*, 1973, **3**, 137.
[398] Wells, R. L., Paige, H. L., and Moreland, C. G., *Inorg. Nucl. Chem. Letters*, 1971, **1**, 177.
[399] Wiberg, E., *Naturwiss.*, 1948, **35**, 182, 212.
[400] Wiberg, E., and Bolz, A., *Z. Anorg. Chem.*, 1948, **257**, 131.
[401] Wiberg, E., Bolz, A., and Buchheit, P., *Z. Anorg. Chem.*, 1948, **256**, 285.
[402] Wiberg, E., Bolz, A., and Schuster, K., *Angew. Chem.*, 1938, **51**, 835.
[403] Wiberg, E., and Hertwig, K., *Z. Anorg. Chem.*, 1947, **255**, 141.
[404] Wiberg, E., Hertwig, K., and Bolz, A., *Z. Anorg. Chem.*, 1948, **256**, 177.
[405] Wiberg, E., and Horeld, G., *Z. Naturforsch.*, 1961, **16b**, 338.
[406] Wiberg, E., and Schuster, K., *Z. Anorg. Chem.*, 1933, **213**, 77.
[407] Wiberg, E., and Schuster, K., *Z. Anorg. Chem.*, 1933, **213**, 89.
[408] Wiberg, E., and Sturm, W., *Z. Naturforsch.*, 1955, **10b**, 109.
[409] Wiberg, N., and Schwek, G., *Angew. Chem. Internat. Edn.*, 1969, **8**, 755.
[410] Wiedermann, K., and Voitländer, J., *Z. Naturforsch.*, 1969, **24a**, 566.
[411] Wille, H., and Goubeau, J., *Chem. Ber.*, 1972, **105**, 2156.
[412] Witte, H., *Tetrahedron Letters*, 1965, 1127.
[413] Wittig, G., Gousier, L., and Vogel, H., *Ann.*, 1965, **688**, 1.
[413a] Wrackmeyer, B., and Nöth, H., *Chem. Ber.*, 1976, **109**, 1075.
[413b] Wrackmeyer, B., and Nöth, H., *Chem. Ber.*, 1976, **109**, 3480.
[413c] Wrackmeyer, B., and Nöth, H., *Chem. Ber.*, 1977, **110**, 1086.
[414] Zhakharkin, L. I., Kalinin, V. N., Gedymin, V. V., and Dzarasova, G. S., *J. Organometallic Chem.*, 1970, **23**, 303.
[415] Zettler, F., and Hess, H., *Chem. Ber.*, 1975, **108**, 2269.
[416] Zhigach, A. F., Kazakova, Y. B., and Antanov, I. S., *J. Gen. Chem. U.S.S.R.*, 1956, **27**, 1655.
[417] Zhigach, A. F., Sobolev, E. S., Svitsyn, R. A., and Nikitin, V. S., *Zh. Obshch. Khim.*, 1973, **43**, 1966.
[418] Ziegler, M. L., Weidenhammer, K., Autzenreith, K., Friebolin, H., *Z. Naturforsch.*, 1978, **33b**, 200.

2. Aluminium

[1] Abeler, G., Bayrhuber, H., and Nöth, H., *Chem. Ber.*, 1969, **102**, 2249.
[1a] Ahmed, A., Schwarz, W., and Hess, H., *Acta. Cryst.*, 1977, **B33**, 3574.
[1b] Ahmed, A., Schwarz, W., and Hess, H., *Z. Naturforsch.*, 1978, **33b**, 43.
[2] Alford, K. J., Gosling, K., and Smith, J. D., *J.C.S. Dalton*, 1972, 2197, 2203.
[2a] Amero, B. A., and Schram, E. P., *Inorg. Chem.*, 1976, **15**, 2842.
[2b] Amirkhalili, S., Hitchcock, P. B., and Smith, J. D., unpublished work, 1978.

[3] Arduini, A., and Storr, A., *J.C.S. Dalton,* 1974, 503.
[4] Ashby, E. C., and Beach, R. G., *Inorg. Chem.*, 1971, **10**, 1888.
[5] Ashby, E. C., and Kovar, R. A., *J. Organometallic Chem.*, 1970, **22**, C34.
[5a] Ashby, E. C., and Kovar, R. A., *Inorg. Chem.*, 1971, **10**, 893.
[5b] Ashby, E. C., and Lin, J. J., *Tetrahedron Letters,* 1976, 3865.
[6] Atwood, J. L., and Stucky, G. D., *J. Amer. Chem. Soc.*, 1970, **92**, 285.
[6a] Bartke, T. C., Haaland, A., and Novak, D. P., *Acta Chim. Scand.*, 1975, **29A**, 273.
[6b] Basha, A., Lipton, M., and Weinreb, S. M., *Tetrahedron Letters,* 1977, 4171.
[7] Beachley, O. T., Coates, G. E., and Kohnstam, G., *J. Chem. Soc.*, 1965, 3241, 3248.
[8] Beachley, O. T., and Racette, K. C., *Inorg. Chem.*, 1975, **14**, 2110, 2534.
[9] Bowen, R. E., and Gosling, K., *J.C.S. Dalton,* 1974, 964.
[10] Brec, R., Palvadeau, P., and Herpin, P., *Compt. Rend.*, 1972, **266C**, 274.
[11] Brown, H. C., and Davidson, N., *J. Amer. Chem. Soc.*, 1942, **64**, 316.
[12] Buckley, G. D., Lewis, D. A., Small, P. A., and Vichy, E. L., Brit. Patent 924,573 (1963).
[12a] Bürger, H., Cichon, J., Goetze, U., Wannagat, U., and Wismar, H.-J., *J. Organometallic Chem.*, 1971, **33**, 1.
[12b] Busetto, C., Cesari, M., Cucinella, S., and Salvatori, T., *J. Organometallic Chem.*, 1977, **132**, 339; *Inorg. Chim. Acta,* 1978, **26**, L51.
[12c] Cesari, M., Perego, G., Del Piero, G., Cucinella, S., and Cernia, E., *J. Organometallic Chem.*, 1974, **78**, 203.
[12d] Cesari, M., Perego, G., Del Piero, G., Corbellini, M., and Immirzi, A., *J. Organometallic Chem.*, 1975, **87**, 43.
[13] Cesca, S., Santostasi, M. L., Marconi, W., and Palladino, N., *Ann. Chim. (Rome),* 1965, **55**, 704.
[14] Ciobanu, A., Vozniuc, I., Bostan, M., Vioculescu, N., and Popescu, I., *Rev. Roum. Chim.*, 1968, **13**, 1497.
[15] Clemens, D. F., Brey, W. S., and Sisler, H. H., *Inorg. Chem.*, 1963, **2**, 1251.
[16] Coates, G. E., Green, M. L. H., and Wade, K., *Organometallic Compounds,* Methuen, London, Vol. 1, Ch. III, 1967.
[17] Coates, G. E., and Graham, J., *J. Chem. Soc.*, 1963, 233.
[18] Cohen, M., Gilbert, J. K., and Smith, J. D., *J. Chem. Soc.*, 1965, 1092.
[18a] Cucinella, S., Dozzi, G., Bruzzone, M., and Mazzei, A., *Inorg. Chim. Acta,* 1975, **13**, 73.
[18b] Cucinella, S., Dozzi, G., Perego, G., and Mazzei, A., *J. Organometallic Chem.*, 1977, **137**, 257; *ibid.*, 1973, **63**, 17.
[18c] Cucinella, S., Dozzi, G., and Mazzei, A., *J. Organometallic Chem.*, 1975, **84**, C19.
[18d] Cucinella, S., Dozzi, G., Mazzei, A., and Salvatori, T., *J. Organometallic*

Chem., 1975, **90**, 257.
[18e] Cucinella, S., Marco, C., and Salvatori, T., Ger. Patent 2,529,367 (1976); *Chem. Abstr.*, 1976, **85**, 45949.
[18f] Cucinella, S., Mazzei, A., and Marconi, W., *Inorg. Chim. Acta Rev.*, 1970, **4**, 51.
[18g] Cucinella, S., Salvatori, T., Busetto, C., and Mazzei, A., *J. Organometallic Chem.*, 1976, **108**, 13.
[18h] Cucinella, S., Salvatori, T., Busetto, C., and Cesari, M., *J. Organometallic Chem.*, 1976, **121**, 137.
[18i] Cucinella, S., Salvatori, T., Busetto, C., Perego, G., and Mazzei, A., *J. Organometallic Chem.*, 1974, **78**, 185.
[18j] Cucinella, S., Salvatori, T., Dozzi, G., Busetto, C., and Mazzei, A., *J. Organometallic Chem.*, 1976, **113**, 233.
[19] Davidson, N., and Brown. H. C., *J. Amer. Chem. Soc.*, 1942, **64**, 316.
[19a] Del Piero, G., Cesari, M., Dozzi, G., and Mazzei, A., *J. Organometallic Chem.*, 1977, **129**, 281.
[19b] Del Piero, G., Cesari, M., and Mazzei, A., *J. Organometallic Chem.*, 1977, **137**, 265.
[19c] Del Piero, G., Cesari, M., Perego, G., Cucinella, S., and Cernia, E., *J. Organometallic Chem.*, 1977, **129**, 289.
[19d] Del Piero, G., Perego, G., Cucinella, S., Cesari, M., and Mazzei, A., *J. Organometallic Chem.*, 1977, **136**, 13.
[19e] Dozzi, G., Cucinella, S., Mazzei, A., and Salvatori, T., *Inorg. Chim. Acta*, 1975, **15**, 179.
[20] Ehrlich, R. D., *Inorg. Chem.*, 1970, **9**, 146.
[20a] Ehrlich, R. D., Young, A. R., Lichstein, B. M., and Perry, D. D., *Inorg. Chem.*, 1964, **3**, 628.
[21] Eisch, J. J., and Healy, M. E., *J. Amer. Chem. Soc.*, 1964, **86**, 4221.
[21a] Fetter, N. R., Brinckman, F. E., and Moore, D. W., *Can. J. Chem.*, 1962, **40**, 2185.
[22] Fetter, N. R., and Barthocha, B., *Can. J. Chem.*, 1962, **40**, 342.
[23] Fetter, N. R., and Barthocha, B., U.S. Patent 3,320,296 (1967).
[23a] Fischer, P., Gräf, R., Stezowski, J. J., and Weidlein, J., *J. Amer. Chem. Soc.*, 1977, **99**, 6131.
[24] Fishwick, M., Smith, C. A., and Wallbridge, M. G. H., *J. Organometallic Chem.*, 1970, **21**, P9.
[25] Furukawa, J., Tsuruta, T., Yamamoto, N., and Kawabata, N., *Kogyo Kagaku Zasshi*, 1958, **61**, 1528.
[26] Gilbert, J. K., and Smith, J. D., *J. Chem. Soc. (A)*, 1968, 233.
[26a] Giongo, G. M., Di Gregorio, F., Palladino, N., and Marconi, W., *Tetrahedron Letters*, 1973, 3195.
[26b] Glidewell, C., and Rankin, D. W. H., *J. Chem. Soc. (A)*, 1970, 279.
[27] Glore, J. D., Hall, R. E., and Schram, E. P., *Inorg. Chem.*, 1972, **11**, 550,

1532.
[28] Gosling, K., and Bhuiyan, A. L., *Inorg. Nucl. Chem. Letters,* 1972, **8**, 329.
[29] Gosling, K., Bhuiyan, A. L., and Mooney, K. R., *Inorg. Nucl. Chem. Letters,* 1971, **7**, 913.
[29a] Gosling, K., and Bowen, R. E., *J.C.S. Dalton,* 1974, 1961.
[30] Gosling, K., McLaughlin, G. M., Sim, G. A., and Smith, J. D., *Chem. Comm.,* 1970, 1617.
[31] Gosling, K., Smith, J. D., and Wharmby, D. H. W., *J. Chem. Soc. (A),* 1969, 1738.
[31a] Hall, R. E., and Schram, E. P., *Inorg. Chem.,* 1971, **10**, 192.
[32] Hagenmuller, P., Rouxel, J., David, J., and Cohn, A., *Compt. Rend.,* 1961, **253C**, 667.
[33] Hagenmuller, P., Rouxel, J., David, J., Cohn, A., and Le Neidre, B., *Z. Anorg. Chem.,* 1963, **323**, 1.
[34] Hall, R. E., and Schram, E. P., *Inorg. Chem.,* 1969, **8**, 270.
[34a] Hausen, H. D., Gerstner, F., and Schwarz, W., *J. Organometallic Chem.,* 1978, **145**, 277.
[35] Hess, H., Hinderer, A., and Steinhauser, S., *Z. Anorg. Chem.,* 1970, **377**, 1.
[36] Higuchi, T., Concha, J., and Kuramoto, R., *Anal. Chem.,* 1952, **24**, 685.
[37] Hirabayashi, T., Imaeda, H., Itoh, K., Sakai, S., and Ishii, Y., *J. Organometallic Chem.,* 1969, **19**, 299; 1970, **25**, 33.
[38] Hirabayashi, T., Itoh, K., Sakai, S., and Ishii, Y., *J. Organometallic Chem.,* 1970, **21**, 273.
[38a] Hitchcock, P. B., McLaughlin, G. M., Smith, J. D., and Thomas, K. M., *J.C.S. Chem. Comm.,* 1973, 934.
[38b] Hitchcock, P. B., Smith, J. D., and Thomas, K. M., *J.C.S. Dalton,* 1976, 1434.
[39] Hoberg, H., *Ann.,* 1971, **748**, 163.
[40] Hoberg, H., *Ann.,* 1972, **766**, 142.
[41] Hoberg, H., *Ann.,* 1971, **746**, 86.
[41a] Hoberg, H., and Griebsch, U., *Ann.,* 1977, 1516.
[41b] Hoberg, H., and Mur, J. B., *J. Organometallic Chem.,* 1969, **17**, P28.
[42] Horder, J. R., and Lappert, M. F., *J. Chem. Soc. (A),* 1968, 2004.
[43] Jennings, J. R., Lloyd, J. E., and Wade, K., *J. Chem. Soc.,* 1965, 5083.
[44] Jennings, J. R., Wade, K., and Wyatt, B. K., *J. Chem. Soc. (A),* 1968, 2535.
[45] Jones, J. I., and McDonald, W. C., *Proc. Chem. Soc.,* 1962, 366.
[46] Jordan, D. E., *Anal. Chim. Acta,* 1964, **30**, 297.
[47] Kai, Y., Yasuoka, N., Kasai, N., Kakudo, M., Yasuda, H., and Tani, H., *Chem. Comm.,* 1968, 1332.
[48] Kawai, M., Ogawa, T., and Hiroto, H., *Bull. Chem. Soc. Japan,* 1964, **37**,

1302.
[49] Keller, P. C., *Inorg. Chem.*, 1972, **11**, 256.
[50] Keller, P. C., *J. Amer. Chem. Soc.*, 1974, **96**, 3073.
[51] Keller, P. C., *J. Amer. Chem. Soc.*, 1972, **94**, 4020.
[52] Köster, R., and Binger, P., *Adv. Inorg. Chem. Radiochem.*, 1965, **7**, 263.
[52a] Kříž, O., Stuchlik, T., and Časensky, B., *Z. Chem.*, 1977, **17**, 18.
[53] Laboratori Riuniti Studie e Richerche S.p.A., Belg. Patent 617,165 (1962).
[53a] Lang, R. F., *Makromol. Chem.*, 1965, **83**, 274.
[53b] Lappert, M. F., Pedley, J. B., and Rogers, A. J., unpublished work, 1978.
[54] Laubengayer, A. W., *Chem. Soc. Spec. Publ.* No. 15, 1961, p. 78.
[55] Laubengayer, A. W., Smith, J. D., and Ehrlich, G. G., *J. Amer. Chem. Soc.*, 1961, **83**, 542.
[56] Laubengayer, A. W., Wade, K., and Lengnick, G., *Inorg. Chem.*, 1962, **1**, 632.
[57] Leone, P., *Gazz. Chim. Ital.*, 1925, **55**, 306.
[58] Lloyd, J. E., and Wade, K., *J. Chem. Soc.*, 1965, 2662.
[59] Longi, P., Mazzanti, G., and Bernardini, G., *Gazz. Chim. Ital.*, 1960, **90**, 180.
[59a] Maijs, L., Zhinkin, D. Ya., and Sobolevskii, M. V., *Latv. P.S.R. Zinat. Akad. Vestis, Kim. Ser.*, 1966, 453: *Chem. Abstr.*, 1967, **67**, 15066h.
[60] McLaughlin, G. M., Sim, G. A., and Smith, J. D., *J.C.S. Dalton*, 1972, 2197.
[61] Magnuson, V. R., and Stucky, G. D., *J. Amer. Chem. Soc.*, 1968, **90**, 3269; 1969, **91**, 2544.
[62] McDonald, T. R. R., and McDonald, W. S., *Proc. Chem. Soc.*, 1963, 382.
[62a] McDonald, T. R. R., and McDonald, W. S., *Acta Cryst.*, 1972, **B28**, 1619.
[63] McDonald, W. S., *Acta Cryst.*, 1969, **B25**, 1385.
[64] Marconi, W., Cesca, S., and Della Fortuna, G., *Chim. Ind. (Milan)*, 1964, **46**, 1287.
[65] Marconi, W., and Cesca, S., Ger. Patent 2,022,658 (1970).
[66] Marconi, C., and Marconi, R., Fr. Patent 1,533,535 (1969).
[67] Marconi, W., Mazzei, A., Bonati, F., and de Malde, M., *Gazz. Chim. Ital.*, 1962, **92**, 1062.
[68] Mazzei, A., Bonati, F., Marconi, W., Manara, G., and Castelfranchi, G., Ital. Patent 684,730 (1967).
[69] Mazzei, A., Cucinella, S., and Marconi, W., *Inorg. Chim. Acta*, 1968, **2**, 305; *Makromol. Chem.*, 1969, **122**, 168.
[70] Meller, A., and Batka, H., *Monatsh.*, 1970, **101**, 627.
[71] Mironov, V. F., Sobolev, E. S., and Antipin, L. M., *Zh. Obshch. Khim.*, 1967, **37**, 2573.

[71a] Mole, T., and Jeffery, E. A., 'Organoaluminium Compounds', Elsevier, Amsterdam, 1972.
[72] Molinie, P., Brec, R., Rouxel, R., and Herpin, P., *Acta Cryst.*, 1973, **B29**, 925.
[73] Montecatini, Neth. Patent 6,509,160 (1966).
[73a] Müller, J., Marigolis, K., and Dehnicke, K., *J. Organometallic Chem.*, 1972, **46**, 219.
[73b] Muraki, M., and Mukaiyama, T., *Chem. Letters*, 1974, 1447.
[74] Natta, G., Mazzanti, G., Longi, P., and Sempio, C., Ital. Patent 632,822 (1962).
[75] Neumann, W. P., *Ann.*, 1963, **667**, 1.
[76] Neumann, W. P., *Ann.*, 1963, **667**, 12.
[77] Nöth, H., and Konrad, P., *Chem. Ber.*, 1968, **101**, 3423.
[78] Nöth, H., and Wiberg, E., *Fortschr. Chem. Forsch.*, 1967, **8**, 323.
[78a] Nöth, H., and Wolfgardt, P., *Z. Naturforsch.*, 1976, **31b**, 697.
[78b] Nöth, H., and Wolfgardt, P., *Z. Naturforsch.*, 1976, **31b**, 1201.
[78c] Nöth, H., and Wolfgardt, P., *Z. Naturforsch.*, 1976, **31b**, 1447.
[78d] Nyathi, J. Z., B.Sc. Thesis, University of Sussex, 1973.
[78e] Nyathi, J. Z., Ressner, J. M., and Smith, J. D., *J. Organometallic Chem.*, 1974, **70**, 35.
[78f] Perego, G., Cesari, M., Del Piero, G., Balducci, A., and Cernia, E., *J. Organometallic Chem.*, 1975, **87**, 33.
[78g] Perego, G., Del Piero, G., Cesari, M., Zazzetta, A., and Dozzi, G., *J. Organometallic Chem.*, 1975, **87**, 53.
[78h] Perego, G., Del Piero, G., Corbellini, M., and Bruzzone, M., *J. Organometallic Chem.*, 1977, **136**, 301.
[78i] Peters, F. M., and Fetter, N. R., *J. Organometallic Chem.*, 1965, **4**, 181.
[78j] Petz, W., and Schmid, G., *Angew. Chem. Internat. Edn.*, 1972, **11**, 934.
[78k] Petz, W., and Schmid, G., *J. Organometallic Chem.*, 1972, **35**, 321.
[79] Pump, J., Rochow, E. G., and Wannagat, U., *Angew. Chem. Internat. Edn.*, 1963, **2**, 264.
[80] Reinert, K., Bayer, O., and Oertel, G., Ger. Patent 1,167,833 (1964).
[81] Reinheckel, H., and Jahnke, D., Ger. (East) Patent 47,130.
[82] Rie, J. E., and Oliver, J. P., *J. Organometallic Chem.*, 1974, **80**, 218.
[82a] Rie, J. E., and Oliver, J. P., *J. Organometallic Chem.*, 1977, **133**, 147; Oliver, J. P., *Adv. Organometallic Chem.*, 1977, **16**, 111.
[83] Rouxel, J., *Ann. Chim. (Paris)*, 1962, **7**, 49.
[84] Ruff, J. K., *J. Amer. Chem. Soc.*, 1961, **83**, 1798.
[85] Ruff, J. K., *J. Amer. Chem. Soc.*, 1961, **83**, 2835.
[86] Ruff, J. K., *J. Org. Chem.*, 1962, **27**, 1020.
[87] Ruff, J. K., *Inorg. Chem.*, 1962, **1**, 612.
[88] Ruff, J. K., *Inorg. Chem.*, 1963, **2**, 515.
[89] Ruff, J. K., and Hawthorne, M. F., *J. Amer. Chem. Soc.*, 1960, **82**,

2141; 1961, **83**, 535.

[90] Sakakibara, T., Hirabayashi, T., and Ishii, Y., *J. Organometallic Chem.*, 1972, **46**, 231.

[91] Schmall, E. A., and Mirviss, S. B., U.S. Patent 3,531,419 (1970).

[92] Schmidbaur, H., and Kammel, G., *J. Organometallic Chem.*, 1968, **14**, P28.

[93] Schmidbaur, H., and Schmidt, M., *Angew. Chem. Internat. Edn.*, 1962, **1**, 327; 1965, **4**, 152; *J. Amer. Chem. Soc.*, 1962, **84**, 1069.

[94] Schmidbaur, H., Schwirten, K., Pickel, H. H., and Wolfsberger, W., *Chem. Ber.*, 1969, **102**, 556, 564; *J. Organometallic Chem.*, 1969, **16**, 188.

[94a] Schram, E. P., *Inorg. Chem.*, 1966, **5**, 1291.

[95] Schram, E. P., Hall, R. E., and Glore, J. D., *J. Amer. Chem. Soc.*, 1969, **91**, 6643.

[96] Schwering, H. U., Weidlein, J., and Fischer, P., *J. Organometallic Chem.*, 1975, **84**, 17.

[97] Semenenko, K. N., Lobkovskii, E. B., and Dorosinski, E. A., *J. Struct. Chem.*, 1972, **13**, 696.

[97a] Semenenko, K. N., Lobkovskii, E. B., Tarnopol'skii, B. L., and Simonov, M. A., *J. Struct. Chem.*, 1976, **17**, 915.

[97b] Sen, B., and White, G. L., *J. Inorg. Nucl. Chem.*, 1973, **85**, 2207.

[98] Shearer, H. M. M., and Willis, J., cited in ref. 16.

[99] Sheldrick, G. M., and Sheldrick, W. S., *J. Chem. Soc. (A)*, 1969, 2279.

[100] Snaith, R., Summerford, C., Wade, K., and Wyatt, B. K., *J. Chem. Soc. (A)*, 1970, 2635.

[101] Snaith, R., Wade, K., and Wyatt, B. K., *Inorg. Nucl. Chem. Letters*, 1970, **6**, 311.

[102] Snam, S. A., Belg. Patent 654,406 (1965).

[103] Storr, A., and Thomas, B. S., *J. Chem. Soc. (A)*, 1971, 3850; Harrison, W., and Thomas, B. S., *J. Chem. Soc. (A)*, 1971, 3850.

[103a] Storr, A., Thomas, B. S., and Penland, A. D., *J. Chem. Soc. (A)*, 1972, 326.

[104] Tanaka, A., Hozumi, Y., Endo, S., Kudo, T., and Taniguchi, K., *Kobunshi Kagaku*, 1963, **20**, 687.

[105] Tani, H., and Kokuni, N., Jap. Patent 4774 ('67) (1967).

[106] Thewalt, U., and Kawada, I., *Chem. Ber.*, 1970, **103**, 2754.

[106a] Vahrenkamp, H., and Nöth, H., *J. Organometallic Chem.*, 1968, **12**, 281.

[107] Wade, K., *Chem. Brit.*, 1968, **4**, 503

[108] Wade, K., and Wyatt, B. K., *J. Chem. Soc. (A)*, 1967, 1339; 1969, 1121.

[108a] Wakatsuki, K., Takeda, Y., and Tanaka, J., *Inorg. Nucl. Chem. Letters*, 1974, **10**, 383.

[108b] Wakatsuki, K., and Tanaka, J., *Bull. Chem. Soc. Japan*, 1975, **48**, 1475.

[109] Weiss, K., and Marsel, C. J., *U.S. Dept. Com., Office Tech. Ser. A.D.*, 274,499 (1961).
[109a] Wharmby, D. H. W., D.Phil. Thesis, University of Sussex, 1969.
[110] Wiberg, E., and May, A., *Z. Naturforsch.*, 1955, **10b**, 232.
[111] Wiberg, E., and May, A., *Z. Naturforsch.*, 1955, **10b**, 234.
[112] Wiberg, N., Baumeister, W., and Zahn, P., *J. Organometallic Chem.*, 1972, **36**, 267, 277.
[113] Wolfsberger, W., and Schmidbaur, H., *J. Organometallic Chem.*, 1969, **17**, 41.
[114] Yee, D. Y., and Ehrlich, R., *J. Inorg. Nucl. Chem.*, 1965, **27**, 2681.
[115] Yoshikara, Y., Jo, T., Imai, Y., and Kurihara, S., Jap. Patent 71 1081 (1971).
[116] Young, A. R., and Ehrlich, R., *J. Amer. Chem. Soc.*, 1964, **86**, 5359.
[116a] Zhakharkin, L. I., and Khorlina, I. M., *Bull. Acad. Sci. U.S.S.R., Div. Chem. Sci.*, 1959, 523.
[117] Zhakharkin, L. I., and Savina, L. A., *Bull. Acad. Sci. U.S.S.R., Div. Chem. Sci.*, 1959, 420.
[118] Zhakharkin, L. I., and Savina, L. A., *Bull. Acad. Sci. U.S.S.R., Div. Chem. Sci.*, 1962, 768.
[119] Zhakharkin, L. I., and Savina, L. A., *Bull. Acad. Sci. U.S.S.R., Div. Chem. Sci.*, 1964, 1600.
[120] Zhinkin, D. Ya., Korneeva, G. K., Korneev, N. N., and Sobolevskii, M. V., *J. Gen. Chem. U.S.S.R.*, 1966, **36**, 360.
[121] Ziegler, K., Brit. Patent, 799,823 (1955).
[122] Ziegler, K., and Gellert, H.-G., *Ann.*, 1960, **629**, 20.

3. Gallium, indium and thallium

[1] Allmann, R., Henke, W., Krommes, P., and Lorberth, J., *J. Organometallic Chem.*, 1978, **162**, 283.
[1a] Amirkhalili, S., Hitchcock, P. B., and Smith, J. D., unpublished work, 1978.
[1b] Arduini, A., and Storr, A., *J.C.S. Dalton*, 1975, 503.
[2] Beachley, O. T., Coates, G. E., and Kohnstam, G., *J. Chem. Soc.*, 1965, 3241, 3248.
[2a] Bradley, D. C., Lappert, M. F., Pedley, J. B., and Sharp, G. J., *J.C.S. Dalton*, 1976, 1737.
[3] Breakell, K., Patmore, D. J., and Storr, A., *J.C.S. Dalton*, 1975, 749.
[4] Breakell, K., Rendle, D. F., Storr, A., and Trotter, J., *J.C.S. Dalton*, 1975, 1584.
[5] Bürger, H., Cichon, J., Goetze, U., Wannagat, U., and Wismar, H.-J., *J. Organometallic Chem.*, 1971, **31**, 1.
[6] Chung, H. L., and Tuck, D. G., *Can. J. Chem.*, 1975, **53**, 3492.
[7] Coates, G. E., *J. Chem. Soc.*, 1951, 2003.

[8] Coates, G. E., and Graham, J., *J. Chem. Soc.*, 1963, 233.
[9] Coates, G. E., Green, M. L. H., and Wade, K., *Organometallic Compounds*, Methuen, London, Vol. 1, Ch. III, 1967.
[10] Coates, G. E., and Whitcombe, R. A., *J. Chem. Soc.*, 1956, 3351.
[10a] Fischer, P., Gräf, R., Stezowski, J. J., and Weidlein, J., *J. Amer. Chem. Soc.*, 1977, **99**, 6131.
[11] Franklin, E. C., *Proc. Nat. Acad. Sci.*, 1915, **1**, 68; *J. Amer. Chem. Soc.*, 1915, **37**, 2279.
[12] Freudenburg, K., and Uthemann, G., *Ber.*, 1919, **52B**, 1509.
[13] Greenwood, N. N., *Adv. Inorg. Chem. Radiochem.*, 1963, **5**, 91.
[14] Greenwood, N. N., Ross, E. J. F., and Storr, A., *J. Chem. Soc. (A)*, 1966, 706.
[15] Greenwood, N. N., Storr, A., and Wallbridge, M. G. H., *Inorg. Chem.*, 1963, **2**, 1036.
[16] Greenwood, N. N., and Wade, K., *J. Chem. Soc.*, 1958, 1671.
[17] Harrison, W., Storr, A., and Trotter, J., *Chem. Comm.*, 1971, 1101; *J.C.S. Dalton*, 1972, 1554.
[18] Harrison, W., and Thomas, B. S., *J. Chem. Soc. (A)*, 1971, 3850.
[18a] Hausen, H. D., Gerstner, F., and Schwarz, W., *J. Organometallic Chem.*, 1978, **145**, 277.
[18b] Herring, F. G., Patmore, D. J., and Storr, A., *J.C.S. Dalton*, 1975, 711.
[19] Jennings, J. R., Pattison, I., Wade, K., and Wyatt, B. K., *J. Chem. Soc. (A), 1967, 1608.*
[20] Jennings, J. R., and Wade, K., *J. Chem. Soc. (A)*, 1967, 1222.
[21] Jennings, J. R., Wade, K., and Wyatt, B. K., *J. Chem. Soc. (A)*, 1968, 2535.
[22] Kochetkova, A. P., and Tronev, V. G., *Russ. J. Inorg. Chem.*, 1957, **2**, 2043.

[23] Kraus, C. A., and Toonder, F. E., *J. Amer. Chem. Soc.*, 1933, **55**, 3547.
[24] Krommes, P., and Lorberth, J., *J. Organometallic Chem.*, 1977, **131**, 415.
[25] Lile, W. J., and Menzies, R. J., *J. Chem. Soc.*, 1950, 617.
[26] Mertz, K., Schwartz, W., Eberwein, B., Weidlein, J., Hess, H., and Hansen, H. J., *Z. Anorg. Chem.*, 1977, **429**, 99.
[27] Molinie, P., Brec, R., Rouxel, J., and Herpin, P., *Acta Cryst.*, 1973, **B29**, 925.
[28] Molinie, P., Brec, R., and Rouxel, J., *Compt. Rend.*, 1972, **274C**, 1388.
[29] Müller, J., Marigolis, K., and Dehnicke, K., *J. Organometallic Chem.*, 1972, **46**, 219.
[30] Nöth, H., and Konrad, P., *Z. Naturforsch.*, 1975, **30b**, 681.
[31] Patmore, D. J., *J.C.S. Dalton*, 1975, 718.
[32] Rendle, D. F., Storr, A., and Trotter, J., *Can. J. Chem.*, 1975, **53**, 2930,

2944.
[33] Scherbakov, V. I., Zhil'tsov, S. F., and Druzhov, O. N., *Zh. Obshch. Khim.*, 1970, **40**, 1542.
[34] Schmidbaur, H., Schwirten, K., Pickel, H. H., and Wolfsberger, W., *Chem. Ber.*, 1969, **102**, 556, 564; *J. Organometallic Chem.*, 1969, **16**, 188.
[34a] Schwering, H. U., Weidlein, J., and Fischer, P., *J. Organometallic Chem.*, 1975, **84**, 17.
[35] Sen, B., and White, G. L., *J. Inorg. Nucl. Chem.*, 1973, **35**, 2207.
[36] Sheka, L. A., Chause, I. S., and Mityureva, T. T., *The Chemistry of Gallium*, Elsevier, Amsterdam, 1966.
[37] Storr, A., *J. Chem. Soc. (A)*, 1968, 2605.
[38] Storr, A., and Penland, A. D., *J. Chem. Soc. (A)*, 1971, 1237.
[39] Storr, A., and Thomas, B. S., *J. Chem. Soc. (A)*, 1971, 3850.
[40] Storr, A., Thomas, B. S., and Penland, A. D., *J.C.S. Dalton*, 1972, 326.
[41] Sutton, G. J., *Austral. Chem. Inst. J. and Proc.*, 1948, **15**, 356; 1949, **16**, 115; *Chem. Abstr.*, 1949, **43**, 6932.
[41a] Tada, H., and Okawara, R., *J. Organometallic Chem.*, 1970, **35**, 1666.
[42] Tada, H., Yasuda, K., and Okawara, R., *J. Organometallic Chem.*, 1969, **16**, 215.
[43] Wolfsberger, W., and Schmidbaur, H., *J. Organometallic Chem.*, 1969, **17**, 41.

5

Amides of Silicon and the Group 4B Metals: Synthesis, Physical Properties, and Structures

A INTRODUCTION TO SILICON–NITROGEN CHEMISTRY

Stable amides of silicon invariably have the element in oxidation state +4 and a four-co-ordinate environment. They are among the most widely studied amides of all the elements. In his 1964 review [336a], Wannagat drew attention to more than 600 publications on SiN chemistry, of which about 50 dealt with amides. The reader is referred to two surveys which, by and large, cover areas outside our scope, dealing with: (i) cyclic SiN compounds [110a] and (ii) five-membered rings containing Si, N, and a hetero-element such as B, Ge, Sn, P, As, Sb, S, Ti, or Zr [359a]. Many of these ring compounds are mentioned in the text and Tables, if they can properly be regarded as silicon amides, but are not covered comprehensively. To obtain an idea of the range, we draw attention to some representative compounds in (I)–(XIIc), e.g., refs. [340, 341, 346, 351, 356, 357, 359]. Another review very much relevant to the amide theme, deals with metal and metalloid complexes containing one or more $(Me_3Si)_2N^-$ or $Me_3Si(Bu^t)N^-$ ligand(s) [143a]. {Compounds such as $B_2H_5[N(Me)SiMe_3]$ [391], $FB[N(SiMe_3)_2]_2$ [99], $\overline{O(CH_2)_2OB}NHSiMe_3$ [304], $Bu^t_2AsNHSiMe_3$ [283], or $Me_3Si(Me_3Sn)NPh$ [291] are more conveniently treated under the appropriate heteroelement, e.g., B, As, or Sn.} Others deal with SiN compounds [107, 385], cyclosiloxazanes [13], π-bonding [20a, 62b], silicon hydrides [21a, 210a], silicon–halogen–nitrogen compounds [279a], organoelement amines [279b], stereochemistry [21b], SiN polymers [21], and organosilicon chemistry [37a, 67a, 94a, 94b, 94d].

Silicon–nitrogen compounds have considerable potential in organic synthesis (Chapters 10–18), e.g., in preparation of carboxylic acid amides [252, 271] or enamines [86].

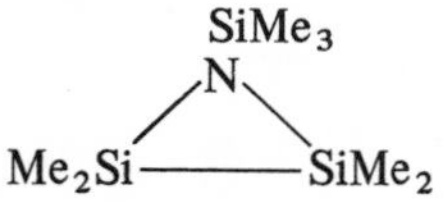

(I) 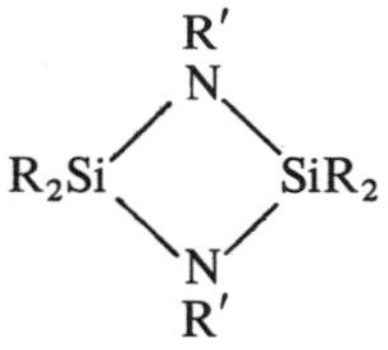 **(II)** $(R_2Si\text{-}NR')_n$ **(III)** (n = 3 or 4)

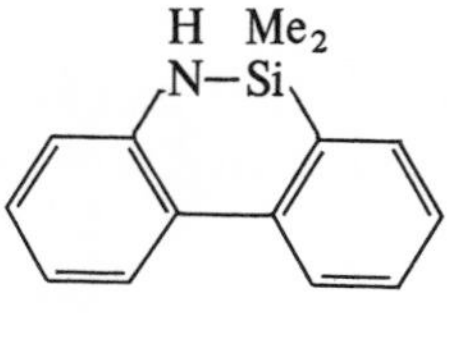

(IV) 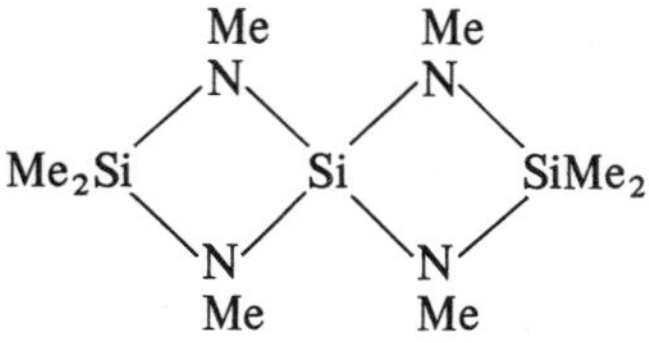 **(V)**

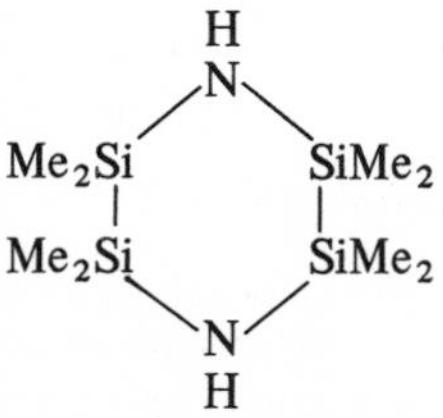

(VI) 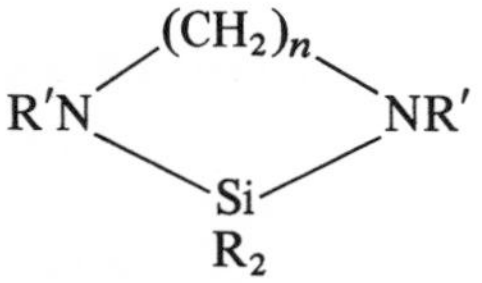 **(VII)** (n = 2, 3, or 4) 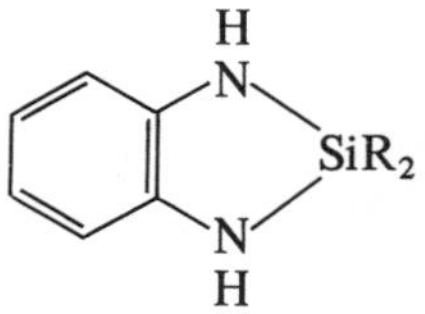 **(VIII)**

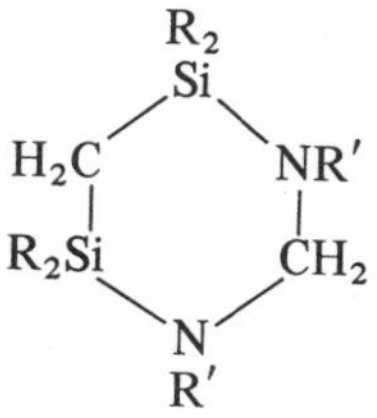

(IX)

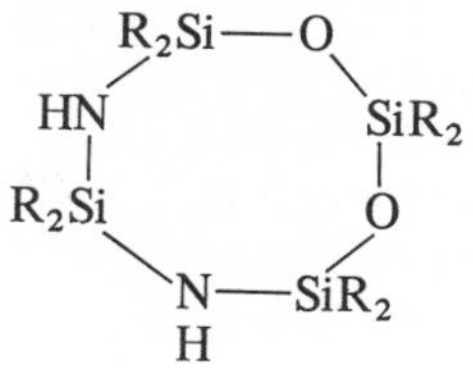

(X)

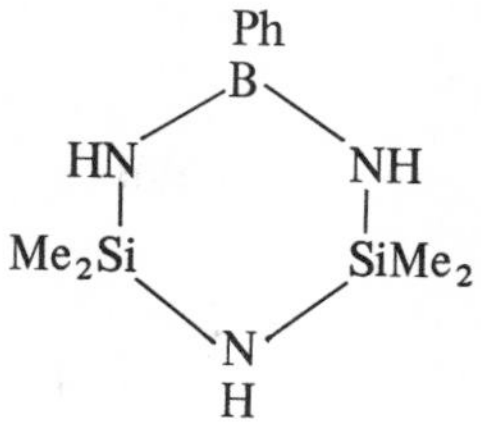

(XI) [242a]

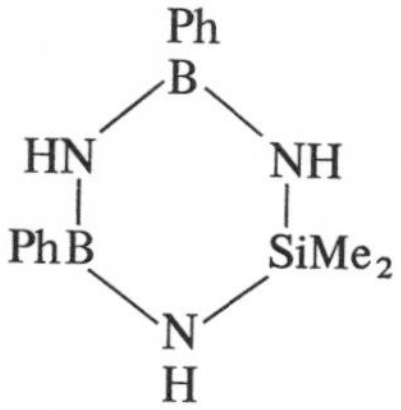

(XII) [242a]

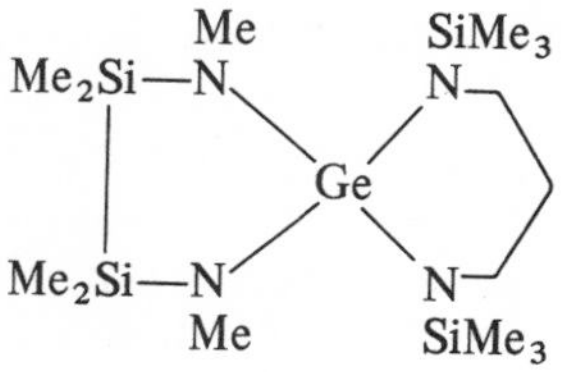

(XIIa) [367a, 367b]
(see Section D)

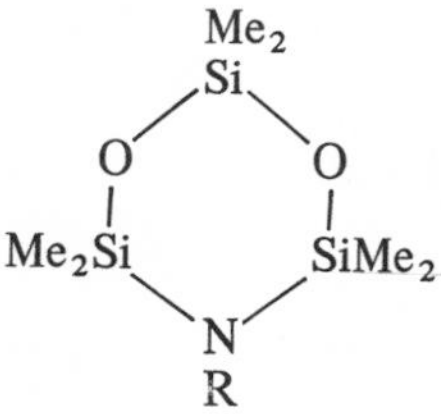

(XIIb) [329]

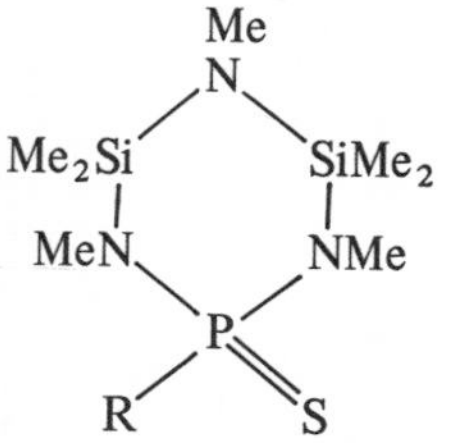

(XIIc) [103]

Mononuclear compounds are generally referred to as aminosilanes, dinuclear as disilazanes or disilylamines, and trinuclear as trisilazanes or trisilylamines. Illustrative examples are dimethylaminotrimethylsilane for Me_3SiNMe_2, heptamethyldisilazane or bis(trimethylsilyl)methylamine for $(Me_3Si)_2NMe$, trisilazane or trisilylamine for $(H_3Si)_3N$; and a cyclodi-, cyclotri-, or cyclotetra-silazane for compounds (II), (III) ($n = 3$), or (III) ($n = 4$), respectively. Steric hindrance allows a stable NH_2-derivative, $Me_2Bu^tSiNH_2$, to be isolated [50a].

B PREPARATION OF SILYLAMINES

1. Reaction of an amine with a monohalogenosilane

The most commonly used synthetic procedure for the title compounds involves the reaction of ammonia, or a primary or secondary amine, with a halogenosilane, Eq. (1). This reaction is strongly influenced by steric factors.

$$R_3SiX + 2HNR'_2 \longrightarrow R_3SiNR'_2 + [R'_2NH_2]X \qquad (1)$$

For example, in the reaction between NH_3 and the chlorosilane R_3SiCl, the product in the case where R = H is the trisilylamine $N(SiH_3)_3$ [65]; for R = Et, it is mostly the monosilylamine Et_3SiNH_2 with a small percentage of $(Et_3Si)_2NH$ [30a]; and for R = Pr or Ph the reaction affords R_3SiNH_2 exclusively and in high yield [178a, 200].

Similarly, in the reaction between a monochlorosilane and a primary amine, Eq. (2), there is a strong tendency to form the disilylamine. Only

$$R_3SiX + 3H_2NR' \longrightarrow (R_3Si)_2NR' + 2[R'NH_3]X \qquad (2)$$

the latter is isolated via Eq. (2) for either R = H with R′ = Me or Et, or R = Me with R′ = Me [100, 101, 349]. However, when R is more sterically demanding, i.e., R ⩾ Et, the monosilylamine is obtained.

The reaction between a secondary amine and a monochlorosilane is normally straightforward and proceeds smoothly, affording a high yield of product. Two recent examples are Eqs. (3) [350] and (4) [291a].

$$R{-}N{<}^{SiMe_2Cl}_{SiMe_2Cl} \xrightarrow{2HNR'R''} R{-}N{<}^{SiMe_2Cl}_{SiMe_2NR'R''} + [R'R''NH_2]Cl \qquad (3)$$

$$Me_3SiCl + 2HNMeBu^t \longrightarrow Me_3SiNMeBu^t + [Bu^t(Me)NH_2]Cl \qquad (4)$$

2. Reaction of an amine with a polyhalogenosilane

The reaction between an amine and a polyhalogenosilane is also controlled by steric influences.

With a secondary amine and a dihalogenosilane, except when R and R′ are very bulky, [Eq. (5)], the bis(amino)silane is formed. Reaction of a dialkyl-

$$R_2SiX_2 + 4HNR'_2 \longrightarrow R_2Si(NR'_2)_2 + 2[R'_2NH_2]X \qquad (5)$$

dichlorosilane with ammonia yields $R_2Si(NH_2)_2$, which is unstable except where R is bulky. Usually condensation takes place and the cyclic oligomer $(R_2SiNH)_n$ is obtained; n is often 2, 3, or 4.

Use of a primary amine, Eq. (6), also illustrates the effect of a bulky alkyl group on the product type; thus, for R = Me or Et = R′ the bis(amino)silane is the major product, unless forcing conditions are used, Eq. (7) [201].

$$R_2SiCl_2 \xrightarrow{H_2NR'} R_2Si(NHR')_2 \text{ (mostly)} \quad (6)$$

$R_2SiNR'SiR_2$ with $R'NH$ and $NR'H$ terminal groups (trace); cyclic $(R_2SiNR')_3$ + cyclic $(R_2SiNR')_4$ (traces)

$$Me_2SiCl_2 \xrightarrow[-[PhNH_3]Cl]{H_2NPh} \begin{cases} \xrightarrow{CCl_4,\ reflux} (Me_2SiNPh)_3 \\ \xrightarrow{C_6H_6,\ 20°C} Me_2Si(NHPh)_2 \end{cases} \quad (7)$$

Conversely, if R and R′ [cf. Eq. (6)] are both fairly bulky, it is possible to isolate either the bis(amino)silane or the amino(halogeno)silane [Eq. (8)] by employing the appropriate stoicheiometry of reagents [59, 62, 317, 319].

$$Et_2SiCl_2 + 2HNEt_2 \longrightarrow Et_2Si(Cl)NEt_2 + [Et_2NH_2]Cl \quad (8)$$

Interaction of a trihalogenosilane and an amine follows a similar pattern to that described above for the dihalogenosilanes. Thus, ammonia affords only a polysilazane. These, the nitrogen analogues of polysiloxanes, have been mentioned as possible water repellents,* anti-foaming agents, or silicone resin modifiers, but in general, their lack of hydrolytic stability has hampered their exploitation. With either a primary or a secondary amine, the reaction proceeds in a stepwise manner, Eqs. (9)-(11). In some cases by employing stoicheiometric

$$RSiCl_3 + 2HNR'_2 \longrightarrow RSiCl_2NR'_2 + [R'_2NH_2]Cl \quad (9)$$

$$RSiCl_2NR'_2 + 2HNR'_2 \longrightarrow RSiCl(NR'_2)_2 + [R'_2NH_2]Cl \quad (10)$$

$$RSiCl(NR'_2)_2 + 2HNR'_2 \longrightarrow RSi(NR'_2)_3 + [R'_2NH_2]Cl \quad (11)$$

amounts of reagents it is possible to obtain the bis- or the mono-halogenosilane [59, 62, 317, 319]. The degree of aminolysis is mainly dependent on steric

*Cf., ref. [139].

factors, as illustrated by Eqs. (12) and (13) [319]; in each case only the mono- or bis-amido compound is isolated. However, use of $HNMe_2$ yields the tris-amide, Eq. (14) [370] and Eq. (15) [371]. Control of the extent of aminolysis may, in

$$HSiCl_3 + \underset{(excess)}{HNEt_2} \longrightarrow HSi(NEt_2)_2Cl + 2[Et_2NH_2]Cl \quad (12)$$

$$HSiCl_3 + \underset{(excess)}{H_2NBu^t} \longrightarrow HSi(NHBu^t)_2Cl + 2[Bu^tNH_3]Cl \quad (13)$$

$$HSiCl_3 + 6HNMe_2 \longrightarrow HSi(NMe_2)_3 + 3[Me_2NH_2]Cl \quad (14)$$

$$MeSiCl_3 + 6HNMe_2 \longrightarrow MeSi(NMe_2)_3 + 3[Me_2NH_2]Cl \quad (15)$$

some cases, be achieved by the use of a bromo- or iodo- rather than a chloro-silane, Eq. (16) [317]; this clearly illustrates that electronic factors are also important [cf. Eqs. (12), (13)]. The product of reaction between $PhSiCl_3$ and $HN(SiMe_3)_2$ depends on stoicheiometry [202a].

$$PrSiX_3 + 6HNEt_2 \longrightarrow PrSi(NEt_2)_3 + 3[Et_2NH_2]X \quad (16)$$
$$(X = Br \text{ or } I)$$

It is possible to form a tetrakisaminosilane by the interaction of a primary or secondary amine and a silicon tetrahalide, Eq. (17) [9, 30]. The scope of this

$$SiCl_4 + 8HNMe_2 \longrightarrow Si(NMe_2)_4 + 4[Me_2NH_2]Cl \quad (17)$$

reaction is limited; with a sterically-demanding amine, only partial substitution is obtained, Eqs. (18) and (19) [59]. Yields can be improved by using bromo- or

$$SiCl_4 + 4HNPr^i_2 \longrightarrow SiCl_2(NPr^i_2)_2 + 2[Pr^i_2NH_2]Cl \quad (18)$$

$$SiCl_4 + 6H_2NBu^t \longrightarrow SiCl(NHBu^t)_3 + 3[Bu^tNH_3]Cl \quad (19)$$

iodo- rather than a chloro-silane [319], but the more readily available and cheaper chloro-silane has obvious counter-advantages.

3. Reaction of an amine with a fluorosilane

Among the silicon halides, fluorosilanes are, being the strongest Lewis acids, unique in that they often form a stable adduct with an amine, which usually undergoes dehydrofluorination only under forcing conditions. For instance, $SiF_4.2HNEt_2$ melts at 66° C and slowly decomposes irreversibly above that temperature [83]. Heating the adduct in a thick-walled glass tube at 110–120° C for 4.5 h quantitatively affords F_3SiNEt_2. Similar procedures give homologues F_3SiNR_2 (R = Me, Pr, or Bu).

In the case of cyclic secondary amines the yields are poor, consistent with the proposition that steric assistance with open-chain compounds is a significant factor. This is also illustrated by the observation that in general a primary amine reacts less readily than a secondary analogue [83], and often affords only the bis-silylamine, e.g., $SiF_4.HN_2Pr$ yields $(F_3Si)_2NPr$. With a bulky primary amine only the monoaminosilane is obtained, as for $F_3SiNHBu^t$ [23].

An alternative procedure of converting a fluorosilane-amine adduct into an aminofluorosilane $F_{4-n}Si(NR_2)_n$ (R = Me, $n = 1$ or 2; R = Et, $n = 1$) is by treating the adduct with $Li[AlH_4]$ (200° C), $Na[BH_4]$ (200° C), or B_2H_6 (190–200° C) [26, 75]. Similarly, heating SiF_4 and a primary amine with $Li[AlH_4]$, or in some cases $Na[BH_4]$, yields the corresponding bis-silyl compound, and in certain instances there is evidence for partly substituted compounds such as F_3SiNHR, or condensed species, such as $F_3SiNR(SiF_2NR)_nSiF_3$ [26, 75].

4. Preparation of silicon amides by transmetallation

Transmetallation is a widely used technique in the preparation of aminosilanes. This route generally involves the reaction of an alkali metal or, in some cases a magnesium, amide with a halogenosilane to afford a metal halide and the aminosilane. The method is especially useful where steric interactions hinder formation of the required product, illustrated by contrasting Eq. (20) [59] with (21) [60]. Similarly, Ph_3SiCl does not react with $HNBu_2$, but the amide is obtained according to Eq. (22) [60].

$$SiCl_4 + H_2NBu^t\,(\text{excess}) \longrightarrow ClSi(NHBu^t)_3 + 3[Bu^tNH_3]Cl \quad (20)$$

$$SiBr_4 + 4BrMgNHBu^t \longrightarrow Si(NHBu^t)_4 + 4MgBr_2 \quad (21)$$

$$Ph_3SiCl + LiNBu_2 \longrightarrow Ph_3SiNBu_2 + LiCl \quad (22)$$

$$SiCl_4 + LiN(SiMe_3)_2\,(\text{excess}) \xrightarrow[\text{reflux}]{Et_2O} Cl_2Si[N(SiMe_3)_2]_2 \quad (23)$$

$$ClSi(NEt_2)_3 + LiNEt_2 \not\longrightarrow \text{no reaction} \quad (24)$$

However, even the use of the more powerful nucleophile, the incipient amide ion rather than amine, by means of transmetallation has inevitable steric constraints, as illustrated by Eqs. (23) [187b] (see also [227, 347a]) and (24) [61].

It is possible that the use of a fluoride substrate may give additional flexibility. This proposal is based on analogy with the synthesis of a tri-substituted halogenosilane, Eq. (25) [92a].

$$SiCl_4 \xrightarrow{\text{excess } LiBu^t} Cl_2SiBu^t_2 \xrightarrow{AsF_3} F_2SiBu^t_2 \xrightarrow{LiBu^t} FSiBu^t_3 \quad (25)$$

Alternatively, the heavier halides are better leaving groups, cf. Eqs. (24) and (26) [49].

$$BrSi(NEt_2)_3 + BrMgNEt_2 \longrightarrow Si(NEt_2)_4 + MgBr_2 \qquad (26)$$

The nature of the solvent may affect the course of transmetallation, as illustrated by Eqs. (27) and (28) [70].

$$Me_3SiCl + LiN(Bu^t)SiMe_3 \xrightarrow{Et_2O} Bu^tN(SiMe_3)_2 + LiCl \qquad (27)$$

$$Me_3SiCl + LiN(Bu^t)SiMe_3 \xrightarrow{T.H.F.} \underset{(30\%)}{Me_3SiCH_2Si(Me)_2NHBu^t} + \underset{(trace)}{Bu^tN(SiMe_3)_2} + LiCl \qquad (28)$$

5. Transamination

The scope of this reaction is limited mainly by the criterion of volatility. Thus, [105, 318] the more volatile amine in the $\equiv SiNR_2/HNR'_2$ system is generally released; however, steric factors also play a significant part in determining the products. Examples of transamination are in Eqs. (29)-(31).

$$(Me_3Si)_2NH + 2HN\langle\text{pyrrolidine}\rangle \longrightarrow 2Me_3SiN\langle\text{pyrrolidine}\rangle + NH_3 \qquad (29)$$

$$Et_3SiNH_2 + H_2NR \longrightarrow Et_3SiNHR + NH_3 \qquad (30)$$

$$(Me_2SiNH)_3 + 6H_2NPh \longrightarrow 3Me_2Si(NHPh)_2 + 3NH_3 \qquad (31)$$

6. Reaction of an alkali-metal-silane with an amine

Few examples of this reaction are known. This is probably due to the relative difficulty in synthesising the alkali-metal salt. Examples are in Eqs. (32) and (33). Silylmetallic species react with azo-compounds [384].

$$Ph_3SiLi + HNR_2 \longrightarrow Ph_3SiNR_2 + LiH \qquad (32)$$

$$Ph_3SiK + Ph_2C{=}NPh \xrightarrow[\text{hydrolysis}]{\text{and subsequent}} Ph_3SiN(Ph)CHPh_2 \qquad (33)$$

7. Reaction of an amine with a silanol or silthiane

Only one example of a reaction between an amine and a silanol, resulting in Si–O cleavage, Eq. (34) [125], has been reported. Such a reaction is thermodynamically implausible because of the far greater bond strength difference between Si–O and Si–N on the one hand, and N–H and O–H on the other. Several examples of the reaction between a silthiane and an amine have been

$$(p\text{-}MeNC_6H_4)_3SiOH + HNBu_2 \longrightarrow (p\text{-}Me_2NC_6H_4)_3SiNBu_2 + H_2O \quad (34)$$

described, but it is rarely employed as a synthetic route; an example is in Eq. (35) [1]; the silthianes are generally made from halogenosilanes, which themselves are convenient precursors to aminosilanes (Section B.1–B.4).

$$Me_3SiSEt + H_2NR \longrightarrow Me_3SiNHR + EtSH \quad (35)$$

8. Preparation of silicon amides by elimination of hydrogen, alkane, or an alkyl halide

Such reactions are rarely used. The hydrogen-elimination is perhaps the best known, as in Eqs. (36), (37) [111].

$$Ph_3SiH + HNR_2 \longrightarrow Ph_3SiNR_2 + H_2 \quad (36)$$

$$R_2R'SiH + (HNR''CH_2)_2 \longrightarrow \text{R''N–Si(R)(R')–NR''} \text{ (ring, } -CH_2CH_2-) + H_2 + RH \quad (37)$$

The other reaction types, all of which involve (as indeed also methods B.1–B.7) nucleophilic attack at Si, are illustrated for the cases of allyl, phenyl, or methyl as leaving groups, Eqs. (38) [324], (39) [42], and (40) [54].

$$PhMe_2SiCH_2CHCH_2 + HN(CH_2)_2 \xrightarrow{NaH} PhMe_2SiN(CH_2)_2 \quad (38)$$

$$Ph_3SiNHEt + H_2NEt \xrightarrow{Li} Ph_2Si(NHEt)_2 + C_6H_6 \quad (39)$$

$$SiBr_4 + PhNMe_2 \longrightarrow Br_3SiNMePh + MeBr \quad (40)$$

9. Preparation of silicon amides by C → N rearrangement

Anionic rearrangements involving silicon (e.g., N → N migration in silylhydrazines) have been known since 1964. Recently it has been shown that transfer of a silyl substituent from carbon to nitrogen may be effected, or alternatively the converse reaction, depending on the nature of the substituents in the substrate [62c, 62d].

The first system to be described was the catalytic rearrangement of an aminomethylsilane to a methylaminosilane, which proceeds by successive deprotonation of the amino group, a 1,2-anionic shift, and a reprotonation [62c], see Scheme. When R′ = Ph, the stabilised nitrogen-centred anion inhibits the C → N transformation, but this effect may be counterbalanced by two

$$Me_3SiCH(R){-}NHR' \xrightarrow[(-H^+)]{LiBu} Me_3SiCH(R){-}\bar{N}R' \rightleftarrows R\bar{C}H{-}N(R')SiMe_3 \xrightarrow{H^+} RCH_2{-}N(R')SiMe_3$$

Scheme The C $\rightleftarrows$ N reversible 1,2-silyl shift

phenyl substituents at carbon, e.g., $Me_3SiCHPh\text{-}NHPh \rightarrow Me_3SiCHPh\text{-}\bar{N}Ph$ (no rearrangement), but $Me_3SiCPh_2\text{-}NHPh \rightarrow Me_3SiCPh_2\text{-}\bar{N}Ph \rightarrow Me_3SiN(Ph)\text{-}CHPh_2$. A strongly polar solvent, such as T.H.F. or T.M.E.D.A., stabilises the carbanion [62d].

Heating $(R_3Si)_2NOSiR_3$ yields $R_3SiN(R)SiR_2OSiR_3$ (R = Me or Et) [376].

C PHYSICAL PROPERTIES AND STRUCTURES OF SILYLAMINES

1. Introduction

Typically, aminosilanes are colourless, volatile liquids or solids. Silicon-nitrogen compounds are thermally stable and Si-N possesses the highest Group 4b element-nitrogen bond strength [34]. This trend (C < Si > Ge > Sn > Pb) is also true for Group 4 element (M) bonds M-O, M-F, or M-Cl (see Fig. 5), but not M-H or M-C bonds, and the C < Si inequality may be due to $M \leftarrow \ddot{X}$ $d \leftarrow p$ π-bonding for Si, but not C.

Thermal stability is indicated by the fact that many aminosilanes may be distilled (without decomposition) at relatively high temperatures (see Tables 10-21). However, Si-N compounds are readily attacked by moisture and a wide variety of protic reagents (Chapter 11). The rate of nucleophilic displacement of amide at Si is largely determined by steric and solubility factors. Important exceptions to the general Si–N lability are the bis- and tris-silylamines, which are inert, for example, to boiling water or aqueous alkali.

Some physical and chemical properties of the tris-silylamines are often cited in support of the existence of $\mathrm{Si}\overset{\curvearrowleft}{-}\ddot{\mathrm{N}}$ $d \leftarrow p$ π-bonding. Such arguments are frequently central for many of the large number of publications on aminosilanes which appear each year. There is, however, considerable current controversy as to whether or not such *d*-orbital participation in Si–N (or Si–O) bonding is as significant as was formerly believed. Definitive evidence remains elusive, despite the large body of data. Nevertheless, in our view there exists substantial evidence (C.2–C.4) in support of the $\mathrm{Si}\overset{\curvearrowleft}{-}\ddot{\mathrm{N}}$ $d \leftarrow p$ π-bonding hypothesis.

2. Structural data and $\mathrm{Si}\overset{\curvearrowleft}{-}\ddot{\mathrm{N}}$ $d \leftarrow p$ π-bonding

One of the most convincing arguments in favour of there being SiN multiple bond character in silicon amides comes from the many electron diffraction (ed) and X-ray studies. Salient features for the majority of cases are that (i) the SiN bond lengths fall in the range 1.65–1.75 Å, and (ii) the nitrogen atom is in a three-co-ordinate, approximately planar, environment. A selection of structural data is summarised in Table 1. The SiN single bond distance (from covalent radii) has been estimated as 1.82–1.87 Å [336a]. A short Si–N bond (see Table 1) may then be accounted for in terms of delocalisation of the $\ddot{\mathrm{N}}$ electrons of π-symmetry into vacant Si *d*-orbitals. Such delocalisation has geometrical onstraints, although this view has been challenged [94c]. For example, in $N(SiH_3)_3$, delocalisation is maximised by having NSi_3 coplanarity as illustrated by the experimental results summarised in Fig. 1. I.r. and Raman data on $CH_3N(SiH_3)_2$ and $CD_3N(SiH_3)_2$ lead to an upper limit of 3.3 and 0.45 kcal mol^{-1} for barriers to rotation about the Si–N and C–N bonds, respectively [94].

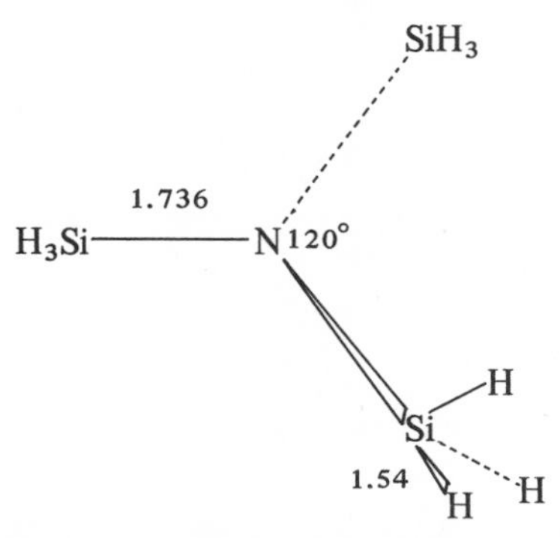

Figure 1 Structure of $N(SiH_3)_3$ showing planar NSi_3 skeleton and short N–Si bond lengths (Å) [37b, 144].

Electron diffraction data on F_3SiNCO [4a] and Me_3SiNCX ($X = O$ or S) [163a] have also been discussed in terms of SiN π-bonding, as has $(H_3Si)_3M$ ($M = P$ or As) for SiM π-bonding [38].

Similarly, structural data for $HN(SiH_3)_2$, $HN(SiMe_3)_2$, and $F_2BN(SiH_3)_2$, shown in Fig. 2 [264], demonstrate that in each case the nitrogen atom is in a planar (rather than pyramidal) environment and the Si–N bond distance is short. The molecule $F_2BN(SiH_3)_2$ even possesses extended coplanarity involving the F_2BNSi_2 skeleton; the B–N bond length, at 1.485 Å, is unremarkable (Chapter 4). The molecule $(Me_3SiNSiMe_2)_2$ (Fig. 3) [378] has a four-membered $(SiN)_2$ planar ring and both the endo- and exo-cyclic Si–N bonds are short.

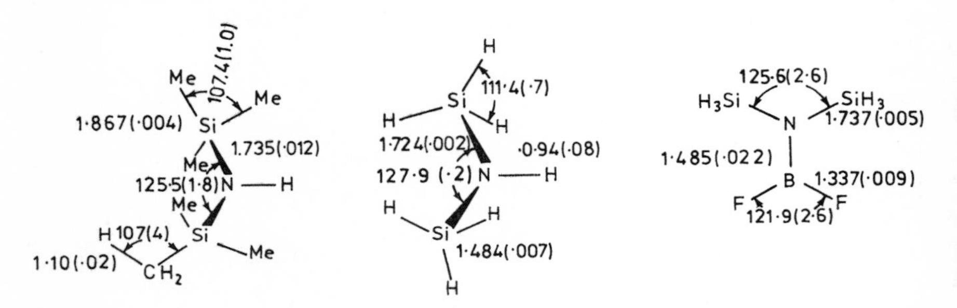

Fig. 2 – The conformation and structures of three disilazanes. *[Reproduced, with permission, from Robiette, A. G., Sheldrick, G. M., Sheldrick, W. S., Beagley, B., Cruickshank, D. W. J., Monaghan, J. J., Aylett, B. J., and Ellis, I. A.,* Chem. Comm., *1968, 909.]*

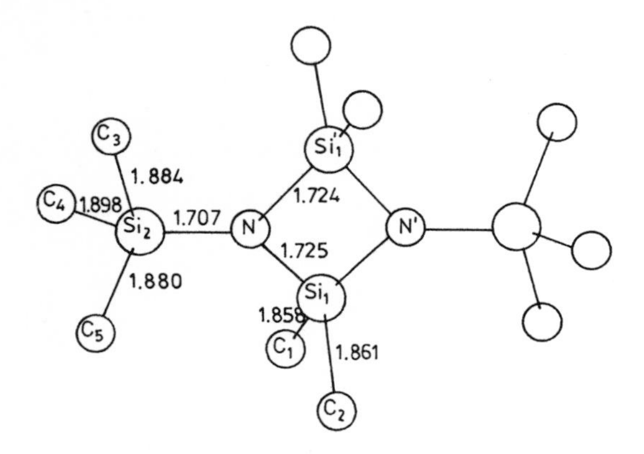

Bond angles

Si_1NSi_1'	91.7°	$C_3Si_2C_4$	106.3°	NSi_1C_1	112.2°
$N'SiN$	88.3	$C_3Si_2C_5$	106.0	NSi_1C_2	116.8
Si_1NSi_2	135.4	$C_4Si_2C_5$	106.4	$N'Si_1C_1$	113.9
$Si_1'NSi_2$	132.8	$C_1Si_1C_2$	108.4	$N'Si_1C_2$	116.4

Fig. 3 – Molecular structure of $(Me_3SiNSiMe_2)_2$. *[Reproduced, with permission, from Wheatley, P. J.,* J. Chem. Soc., *1962, 1721.]*

Table 1 Structural Data on Aminosilanes

Compound	N–Si (Å)	Comments	Method	References
$NH(SiH_3)_2$	1.724	Planar $NHSi_2$ skeleton	ed	[259b,264]
$NH(PF_2)SiH_3$	1.720		ed	[20]
$N(SiH_3)_3$	1.736	Planar NSi_3 skeleton Si–$\widehat{N}$–Si, ca. 120°	ed	[37b,144]
$NH(SiMe_3)_2$	1.735	Planar $NHSi_2$ skeleton	ed	[264]
NMe_2SiH_3	1.715	Non-planar $SiNC_2$ skeleton	ed	[128a]
NMe_2SiF_3	1.654	Planar $SiNC_2$ skeleton	ed	[4b]
NMe_2SiCl_3	1.657	Planar $SiNC_2$ skeleton	ed	[4b]
NMe_2SiMe_3	1.95	Pentaco-ordinate N	X-ray	[266]
$NMe(SiH_3)_2$	1.726	Planar $CNSi_2$ skeleton	ed	[128]
$NBF_2(SiH_3)_2$	1.737	Planar F_2BNSi_2 skeleton	ed	[263a,264]
$Si(Cl)(NMe_2)_3$	1.715	Planar SiNCS system	ed	[330]
$(Me_2Si\text{–}NSiMe_3)_2$	1.74	Planar N_2Si_2 ring	X-ray	[378]
$(Me_2Si\text{–}NSiMe_3)_3$	1.74	$SiNSi_2$ almost planar	X-ray	[3a]
$(Cl_2SiNH)_3$	1.68	Puckered ring	X-ray	[227]
$Me_3SiNHMe$	1.72	Large Si$\widehat{N}$C angle, ca. 130°	ed	[265]
$(Me_2SiNH)_{3\text{ or }4}$	1.78	Puckered rings	X-ray ed	[303a,389] [388a]
SiMe$_3$ SiMe$_3$ N N Me_2Si Si $SiMe_2$ CH N SiMe$_3$ SiMe$_3$	1.74		X-ray	[227]
$N_2(SiH_3)_4$	1.73	Each N atom is planar	ed	[128b]
Me$_2$ Si Ph–C(O)N N Me_2Si O Ph	1.78(av)	Incipiently pentaco-ordinate Si due to Si–O interaction, NSiNC(O) nearly co-planar	X-ray	[49b]
Me$_2$ Si $MePh_2Si$–N $NSiMe_2NHSiMePh_2$ Si Me$_2$	1.74(av)	Planar N_2Si_2 ring	X-ray	[79a]
Ph$_2$ Si—O EtN $SiPh_2$ Si—O Ph$_2$	1.75		X-ray	[119a]

The Si–$\ddot{N}$ bond contraction attributed to Si–N π-orbital overlap should be enhanced by σ-withdrawing groups on silicon, because they would tend to lower the energy of the Si d-orbitals. This hypothesis receives some support from structural data for X_3SiNMe_2 (X = H, F, or Cl) [4b, 128a] summarised in Table 2. We note that an increase in electronegativity at Si influences also the energy of the σ-bonding sp^3-hybrid at Si and thus causes better σ-overlap with nitrogen in the sequence F > Cl > H.

Table 2 Bond Lengths and Angles in X_3SiNMe_2 and $ClSi(NMe_2)_3$

Compound	Bond length (Å)		Angle (°)	
	Si–N	Si–X	$\widehat{CNC}$	$\widehat{CNSi}$
H_3SiNMe_2	1.715	1.485	111.1	120
F_3SiNMe_2	1.654	1.567	120.5	118.7
Cl_3SiNMe_2	1.657	2.023	113.1	123.1
$ClSi(NMe_2)_3$	1.715	2.082	118.5	120.5

There is significant shortening in the SiN bond length between the silylamine (X = H) and the fluoro- or chloro-analogues. It may also be important that the nitrogen atom for X = H shows the greatest deviation from planarity and that the angle between the NC_2 plane and Si–N bond is ca. 28°. However, we note that for these X_3SiNMe_2 molecules, although the Si–N bond distances are rather insensitive to the nature of the halogen (X = F or Cl), the $\widehat{CNC}$ angle for Cl_3SiNMe_2 is close to that in H_3SiNMe_2.

Comparison of data for Cl_3SiNMe_2 and $ClSi(NMe_2)_3$ (Fig. 4) [330] is instructive. The latter, like the former compound, has a nearly planar N environment and trends in the Si–N and Si–Cl bond distances are in opposite directions. However, it is not clear that the differences are best accommodated in terms of a π-model.

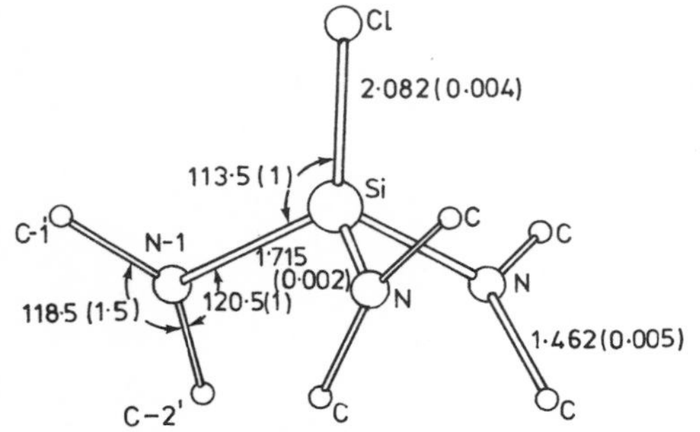

Fig. 4 – The molecular structure of $Si(Cl)(NMe_2)_3$. *[Reproduced, with permission, from Vilkov, L. V., and Tarasenko, N. A.,* Chem. Comm., *1969, 1176.]*

The compound initially formulated [176a] as the Si amide MeC(O)N-$(SiMe_2CH_2Cl)_2$ has been shown by X-ray analysis to be

O Me

$ClMe_2Si$ N $SiMe_2Cl$ [243a].

3. E.s.r. studies on Si–N bonded compounds

We now refer to a number of e.s.r. studies on paramagnetic Si–N bonded compounds [49a, 138a, 372, 374], which have involved the formation and characterisation of silyl-substituted aminyl radicals, $:\dot{N}R_2$, nitroxides, $R\dot{N}(O)R$, or cationic hydrazine radicals, $(RR'NNR''R''')^{+\cdot}$. In these compounds, $a(^{14}N)$ (^{14}N hyperfine coupling to the spin of the unpaired electron) is lower than in the corresponding organic compounds. This is illustrated in Tables 3 and 4. The reduction of $a(^{14}N)$ is particularly noticeable for the nitroxides (Table 3), for which there appears to be an inverse correlation with the electronegativity of the substituent at N. The low $a(^{14}N)$ values for silyl-substituted radicals (Tables 3 and 4) have been ascribed to $\ddot{N}$–Si π-bonding effects. The *g*-values of most organic nitroxides are in the range 2.0060 ± 0.0005; however, there is a significant increase in this parameter in the silylnitroxides to ca. 2.0092 ± 0.0001 and indeed also for the aminyl radicals. For the nitroxides, this has been attributed to the increase in unpaired electron density at O (which has a higher spin-orbit coupling constant), thereby increasing *g*. This explanation is clearly not applicable to the aminyl radicals; it may be that the increased *g*-value is due to the higher value of the spin-orbit coupling constant for Si than C, although ^{29}Si coupling has not yet been observed, and hence there is no clear evidence of significant unpaired electron density at Si.

Table 3 E.s.r. parameters of selected nitroxide radicals [373, 375]

R in $R_2\dot{N}O$	$a(^{14}N)$ (Gauss)	*g*-value
Me	17.1	2.0055
Bu^t	15.2	2.0065
Ph	9.66	2.0056
CF_3	9.3	2.0075
R′MeHSi (R′ = Me or Bu^t)	6.5	–
$R'Me_2Si$ (R′ = Me, Et, Bu^t, or Ph)	6.5	2.0093
Me_3Ge	9.3	2.0075

Table 4 E.s.r. parameters of selected aminyl radicals [138a]

R in $:\dot{N}R_2$	$a(^{14}N)$ (Gauss)	g-value
Me	14.78	2.044
Pr^i	14.30	—
F	16.0	2.0020
H	10.3	2.0048
Ph	8.8	—
Me_3Si	12.71	2.0078
Et_3Si	12.69	2.0071
$(Me_3Si)Bu^t$ $(=R_2)$	14.4	2.0074
Me_3Ge	12.04	2.0079

4. Thermochemistry and He(I) photoelectron spectroscopy

The heats of hydrolysis in 1*M*-hydrochloric acid of several Si–N compounds have been measured [33], e.g., Eq. (41). From these and appropriate heats

$$Me_3SiNMe_2(l) + \tfrac{1}{2}H_2O(l) \longrightarrow \tfrac{1}{2}(Me_3Si)_2O(l) + [Me_2NH_2]Cl(55H_2O) \quad (41)$$

of solution and standard heats of formation [i.e., for Eq. (41), $(Me_3Si)_2O$, and $(Me_2NH_2)Cl$], the standard heat of formation of the SiN compound is calculated, which in turn leads to a mean (Si–N) thermochemical bond energy term $\bar{E}$. Some data are in Table 5.

Table 5 Some thermochemical data for aminosilanes

Compound	$\Delta H_f^\circ(l)$ (kcal mol^{-1})	$\bar{E}$(Si–N) (kcal mol^{-1})	Reference
Me_3Si–NHMe	−62.5 (± 0.8)	73.1	[33]
Me_3Si–NMe_2	−66.6 (± 0.8)	79.1	[33]
$(Me_3Si)_2NH$	−123.9 (± 1.6)	75.8	[33]
$(Me_3Si)_2NMe$	−116.1 (± 1.6)	75.1	[33]
$N(SiMe_3)_3$	−173.3 (± 2.2)	74.1	[33]
$Bu^t(Me_3Si)NH$	−89.9	70.6	[137a]

It was argued [33], making the simplifying assumption that the SiN σ-bond strength is constant for the compounds of Table 5, that for the series $(Me_3Si)_nNMe_{3-n}$ ($n = 3$, 2, or 1) π-delocalisation involves the Si_3N, Si_2N, and SiN framework molecular orbitals, whence the *average* SiN π-contribution to bond strength should increase in the same sequence. Likewise, comparison of $Me_3SiNHMe$ with $(Me_3Si)_2NH$ leads to the expectation that the SiN π-bond contribution to the SiN bond energy term is greater in the former compound. Electron impact data, e.g., on Me_3SiNEt_2, give unrealistically high values for $\bar{E}$(Si–N) [146].

A comparison with nitrogen derivatives of other group 4 elements is interesting [33, 34]. Thus, similar data, ΔH°_f and $\bar{E}$(M–X), are available for the series Me_3M–X (M = C, Si, Ge, Sn, or Pb*; X = F, Cl, OMe, NMe_2, Me, or H). For X = H or Me, there is a monotonic decrease in $\bar{E}$(M–X) with increase in atomic number of M (i.e., M = C > Si > Ge > Sn > Pb*), whereas for the other values of X (each of which has one or more lone pair of electrons and hence at least the possibility of M–X π-bonding) there is a discontinuity at Si (i.e., C < Si > Ge > Sn > Pb*) (Fig. 5).

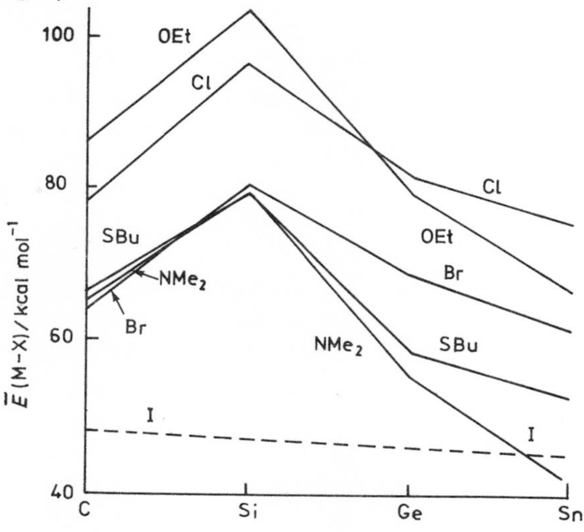

Fig. 5 – Trends in mean bond energy terms $\bar{E}$(M–X) in the series Me_3M–X. *[Reproduced, with permission, from Baldwin, J. C., Lappert, M. F., Pedley, J. B., and Poland, J. S.,* J.C.S. Dalton, *1972, 1943.]*

Comparison of He(I) photoelectron spectra of open-shell transition-metal dimethylamides and disilylamides, e.g., the d^1 [$V(NMe_2)_4$] and [$Ti\{N(SiMe_3)_2\}_3$] [187a] has been made in Chapter 8, where it is demonstrated that the $(Me_3Si)_2N$ group is electron-withdrawing relative to Me_2N, a result which is consistent with, but not necessarily evidence for, Si–N π-bonding. Similarly, the first ionisation potential of $HN(Bu^t)SiMe_3$ is 8.41 eV compared with 8.79 eV for $HN(SiMe_3)_2$ [143b]. The spectra of $Cl_{4-n}Si(NMe_2)_n$ (n = 4, 2, or 1) have been recorded [134].

*Data on Pb are extrapolated; see also p. 271.

5. Other studies on SiN bonded compounds

Molecular orbital calculations [84, 157] lead to an estimate that the π-bond energy contribution to Si–N bond strength is ca. 6 kcal mol^{-1} [41a, 249, and references therein]. I.r. vibrational spectroscopic data provide the basis for assigning the SiN force constant in aminosilanes to be ca. 4.0 mdyne Å^{-1}, although the calculations assuming no π-bonding give somewhat lower values, ca. 3.3 mdyne Å^{-1} [96, 179, 249, 254a, 352]. The magnitude of the SiN force constant is reduced in a Lewis-acid aminosilane adduct. Whereas this may be attributed to the loss of SiN π-bonding in the adduct (four co-ordinate nitrogen) compared with the free silane [382b], it may just as convincingly reflect changes in σ-bond strengths (sp^3 versus sp^2 at N). The activation free energy to Si$\nleftrightarrow$N rotation in RR′R″SiNMe$_2$ has been estimated at ca. 7.2 kcal mol^{-1} [78] and ca. 3 kcal mol^{-1} in *N*-acyl derivatives [176].

Similarly, $N(SiH_3)_3$ is a very weak Lewis base toward BMe_3 [312] and further experimental data [e.g., 95, 245, 246, and 247] on the Lewis-acid-base character of Si–N compounds have been rationalised in terms of π-bonding (Chapter 16) [1a]. Adducts such as $(Me_3Si)_2NH.X$ ($X = AlCl_3$, $SnCl_4$, or $VOCl_3$), $Me_2Si(NMe_2)_2.ZrCl_4$, and $(Me_3SiNMe_2)_2.SnCl_4$ are isolable [189, 382a], but N–Si cleavage occurs upon reaction of BF_3 with $(H_3Si)_nNMe_{3-n}$ ($n = 1, 2,$ or 3) [313].

The molecular electric dipole moments of a wide range of silicon amides have been measured, from which low SiN bond dipole moments have been calculated [174a]. These have been explained on the basis that $Si^{\delta+}$–$N^{\delta-}$ σ-polarity is counterbalanced by reverse π-charge distribution.

The ^{14}N chemical shifts of 39 silylamines Me_3SiNR_2, $(Me_3Si)_2NR$, and $N(SiMe_3)_3$ and silylaminoboranes have been recorded [242b]; the former fall in the range $+340 \pm 40$ p.p.m. (relative to NO_3^-) except for $(Cl_3Si)_2NH$ (+293) and $(Me_3Si)_2NPh$ (+287), and the effect of borylation is to deshield the nitrogen. Trends were interpreted in terms of a SiN π-bonding model. Data are available on $(H_3Si)_3N$, $(H_3Si)_2NH$, and $H_3Si(F_2P)NH$ [11], and $Me_{4-n}Si(NMe_2)_n$ [43].

D GERMANIUM AMIDES

1. Introduction

Although the first germanium-nitrogen compounds were synthesised around 1930 [24a, 25, 26], it was not until the early sixties that interest in metal amides grew and the subject began to develop. This is surprising because silicon-nitrogen chemistry was well-established much earlier, but the cost of the germanium starting materials may have been a factor in the former neglect of GeN compounds. Reviews deal with GeN compounds in the context of related derivatives of Sn and Pb [29] or organic compounds of germanium [28f]. Some quite complex compounds have been made e.g., (XIIa) [58a, 58b].

The chemistry of silicon and germanium shows many similarities. However, this aspect has often, in our view, been overemphasised. Differences may be attributed to the higher electronegativity of Ge and the lower availability of Ge 4*d*-orbitals relative to the Si 3*d*-orbitals for participation in π-bonding. In the context of the metal-nitrogen bond, there is, for instance, a large difference in the thermochemical Ge-N and Si-N bond energies of 55 and 78 kcal mol^{-1}, respectively in $Me_3M\text{-}NMe_2$ [5]. The Ge-N bond is more susceptible than Si-N to cleavage by protic (Chapter 11) nucleophiles; it has been suggested that the greater basicity (see Chapter 16) of amides of Ge rather than Si is a contributory factor [29].

2. Preparation of germanium(IV) amides

As with amides of silicon, the main preparative route involves the treatment of a germanium-halogen compound with an amine (amine hydrochloride elimination). One of the earliest reports [26] describes the conversion of bromotriphenylgermane with ammonia in diethyl ether into triphenylgermylamine, Eq. (42). This compound is one of the few having a relatively stable $GeNH_2$ link.

$$Ph_3GeBr + 2NH_3 \longrightarrow Ph_3GeNH_2 + [NH_4]Br \qquad (42)$$

A more recent paper refers to $Pr^i_3GeNH_2$ [16], Eq. (43). Compounds having one

$$Pr^i_3GeBr + KNH_2 \longrightarrow Pr^i_3GeNH_2 + KBr \qquad (43)$$

or more NH_2 groups attached to Ge are not well-characterised. The series R_3GeNH_2 is thus rather similar to R_3SiNH_2, in that bulky alkyl substituents at Ge promote stability. With less sterically-demanding alkyl substituents, associative mechanisms lead to condensation products containing $\geqslant Ge\text{-}\overset{|}{N}\text{-}Ge \leqslant$ units, i.e., a mixture of bis- and tris-germylamines [45]; e.g., Me_3GeCl and NH_3 in Et_2O at $-60°$ C afford $(Me_3Ge)_2NH$ (50%) and $N(GeMe_3)_3$ (32%) [45].

The reaction between a halogenogermane and a primary amine H_2NR is also often complicated by the instability of the products which may undergo condensation to afford a species such as $(R'_3Ge)_2NR$ or $[Ge(NR)_2]_n$ [from $Ge(NHR)_4$]. An exception is the reaction of H_2NR ($R = C_6H_{11}$ or Ph) with $GeCl_4$, which affords $ClGe(NHR)_3$ [12]. This illustrates the proposition that bulky groups at N likewise favour stabilisation with respect to associative decomposition pathways. Other thermally-stable compounds R'_3GeNHR are often synthesised by transamination.

The majority of the known aminogermanes have a tertiary nitrogen atom, i.e. of formula $\geqslant GeNRR'$. Such compounds are often prepared by the reaction of a germanium halide with a secondary amine [1, 3, 29, 57]. As in the case of the silicon analogues, the extent of Cl/NR_2 exchange is limited by steric factors. The reaction is exemplified by Eqs. (44)-(47) [e.g., refs. 34, 61, 64].

$$GeCl_4 + 8HNMe_2 \longrightarrow Ge(NMe_2)_4 + 4[Me_2NH_2]Cl \quad (44)$$

$$Et_2GeCl_2 + 4HNMe_2 \longrightarrow Et_2Ge(NMe_2)_2 + 2[Me_2NH_2]Cl \quad (45)$$

$$Me_2GeCl_2 + 2HNEt_2 \longrightarrow Me_2Ge(NEt_2)Cl + [Et_2NH_2]Cl \quad (46)$$

$$GeCl_4 + 4RNH(CH_2)_2NHR \longrightarrow Ge[N(R)(CH_2)_2N(R)]_2 + 2[RNH_2(CH_2)_2N \quad (47)$$

The transmetallation technique, utilising in the main the reaction of a lithium or sodium amide with a germanium halide, is widely employed [e.g., refs. 2, 29, 44, 45, 52, 62]. Examples are in Eqs. (48) [37], (49) [32], (50) [24], and (51) [54].

$$Me_3GeCl + LiNHPh \longrightarrow Me_3GeNHPh + LiCl \quad (48)$$

$$Me_3GeCl + LiNHSiMe_3 \longrightarrow Me_3GeN(H)SiMe_3 + LiCl \quad (49)$$

$$(EtO)_2GeCl_2 + 2LiN(SiMe_3)_2 \longrightarrow (EtO)_2Ge[N(SiMe_3)_2]_2 + 2LiCl \quad (50)$$

$$GeCl_4 + 4KNC_4H_4 \longrightarrow Ge(NC_4H_4)_4 + 4KCl \quad (51)$$

Transamination, first reported in 1964 [39, 45], is an attractive route, especially when the displaced amine is volatile, thus simplifying the purification of the aminogermane, as in Eqs. (52) [39], (53) [45], and (54) [39, 45]. In some

$$Et_3GeNMe_2 + H_2NOct \longrightarrow Et_3GeNHOct + HNMe_2 \quad (52)$$

$$Et_3GeNMe_2 + H_2NCO_2Me \longrightarrow Et_3GeNHCO_2Me + HNMe_2 \quad (53)$$

$$(Bu_3Ge)_2NH + 2H_2NOct \longrightarrow 2Bu_3GeNHOct + NH_3 \quad (54)$$

cases, surprisingly forcing conditions are required, for example 160–170° C and $(NH_4)_2SO_4$ as catalyst for the transformation in Eq. (54). The analogous reaction with $(Me_3Ge)_2NH$ proceeds smoothly at lower temperature with no catalyst, presumably due to the less stringent steric requirements [44]. Similarly the trigermylamine undergoes reaction with octylamine, Eq. (55). The imidazolidine

$$N(GeMe_3)_3 + 3H_2NOct \longrightarrow 3Me_3GeNHOct + NH_3 \quad (55)$$

derivatives are also obtained by this procedure, Eq. (56) [62]. Compound (XIII)

$$Me_2Ge(NEt_2)_2 + \underset{H}{MeN}(CH_2)_2\underset{H}{NMe} \longrightarrow Me_2Ge\langle N(Me)CH_2CH_2N(Me)\rangle + 2HNEt_2 \quad (56)$$

(XIII)

undergoes ring-chain polymerisation in the presence of ammonium sulphate to afford the polymer $\{\underset{Me_2}{Ge}\text{-}N(Me)(CH_2)_2N(Me)\}_m$: the monomer is regenerated by thermal cracking of the polymer [63]. An S_N2 mechanism has been proposed.

Several other methods of Ge–N bond formation have been reported; amongst these are the reaction of (i) an amine with a compound containing a Ge–O bond, as in Eq. (57) [29, 38], (ii) an NR_2/OR' exchange, as in Eqs. (58) or (59) [24], and (iii) a scrambling reaction of the type illustrated in Eq. (60) [23]. An interesting insertion reaction is that leading to $\overline{Ph_2GeN(Me)GePh_2N(Me)CH_2CH_2}$ or $Et_3GeN(Me)GePh_2NMe_2$ from $\overline{Ph_2GeN(Me)CH_2CH_2}$ by heating or reaction with Et_3GeNMe_2, respectively [40a]. Oxidative addition (Chapter 18) is available, as in the $Ge[\overline{NCMe_2(CH_2)_3CMe_2}]_2$–PhCOCl system to yield $Ge(NR_2)_2Cl(COPh)$ [28a].

$$(R_3Ge)_2O + 2HN\langle N{=}{=}N\rangle \longrightarrow 2R_3GeN\langle N{=}{=}N\rangle + H_2O \quad (57)$$

$$Me_2Ge(OPh)_2 + \text{excess } LiNEt_2 \longrightarrow Me_2Ge(NEt_2)_2 + Ge(OPh)_4 \quad (58)$$

$$Ge(OEt)_4 + LiN(SiMe_3)_2 \longrightarrow (EtO)_3GeN(SiMe_3)_2 \quad (59)$$

$$nGeX_4 + (4-n)Ge(NMe_2)_4 \longrightarrow 4X_nGe(NMe_2)_{4-n} \quad (60)$$

3. Structures of germanium(IV) amides

Only two structural papers on germanium-nitrogen compounds have been published.* These relate to the electron-diffraction study of the unstable molecule

*For a Ge^{II} compound, see p. 257.

$N(GeH_3)_3$, Fig. 6 [15a] and the X-ray crystallographic study of $(Cl_2GeNMe)_3$, Fig. 7 [66]. In the trigermylamine, the nitrogen atom is in a trigonal and planar NGe_3 environment. The Ge–N bond length is 1.836 ± 0.005 Å, which is slightly shorter than the predicted bond length of 1.876 Å [15a]. In $(Cl_2GeNMe)_3$, the skeletal structure is a hexagon and the ring is almost planar, with the Ge–N bond lengths in the range 1.78–1.811 Å. The planarity and short Ge–N bonds have been taken as evidence for Ge–N $d \leftarrow p$ π-bonding. For p.e. spectra, see Section E.3 [14a].

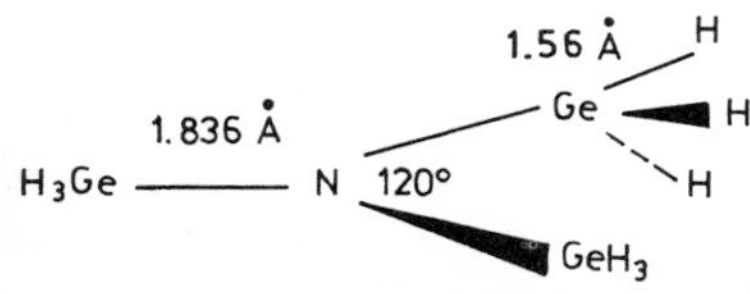

Fig. 6 – Diagrammatic representation of $N(GeH_3)_3$, from reference [15a].

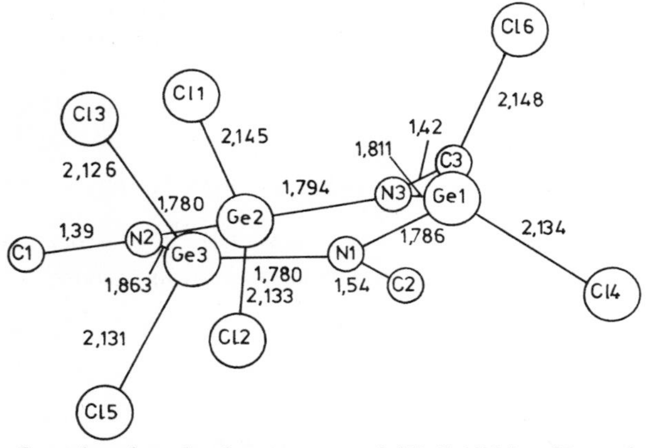

Fig. 7 – Crystal and molecular structure of $(Cl_2GeNMe)_3$. *[Reproduced, with permission, from Ziegler, M. and Weiss, J.,* Z. Naturforsch., *1971,* **26b**, *735.]*

4. Subvalent germanium amides

A significant development in GeN chemistry has been the discovery of some kinetically stabilised Ge^{II} and Ge^{III} amides. The first publication, in 1974, disclosed the coloured, low melting diamagnetic, monomeric (in C_6H_6 or by mass spectrometry) diamides $Ge(NRR')_2$ [R = Bu^t, R′ = Me_3Si (orange); or R = Me_3Si = R′ (yellow)], obtained according to Eq. (61) [19]. Subsequently,

$$GeCl_2.dioxan + 2LiNRR' \longrightarrow Ge(NRR')_2 + 2LiCl \qquad (61)$$

some analogues (R = R′ = Me_3Ge, Et_3Si, Et_3Ge, or Ph_3Ge) were prepared by a similar procedure, and are also yellow and monomeric [in cyclohexane; the bis(triphenylgermyl)amide is insoluble] [17].

These germanium(II) amides, like their Sn^{II} and Pb^{II} analogues and the isoelectronic alkyls $M[CH(SiMe_3)_2]_2$, are thermochromic, becoming colourless at −196° C, and reverting to coloured compounds when the temperature is raised, distilling as reddish vapours. The tendency to redness is most marked for Pb and least for Ge. Some spectroscopic data are summarised in Table 6. From the sharp n.m.r. spectra it is concluded that the compounds are bent singlets (XIV) in the electronic ground state rather than linear triplets (XV). That these bulky substituents at nitrogen can tolerate an angular arrangement at Ge is inferred from

RR'N–Ge(:)–NRR' (XIV)

RR'N —— Ge —— NRR'

(XIV) **(XV)**

the electron diffraction study on $Sn[N(SiMe_3)_2]_2$ (see Section E) and the formation of 'germylene' complexes such as $[Mo(CO)_5(Ge\{N(SiMe_3)_2\}_2)]$ or products of insertion into a metal-halogen bond [28, 34a] (Chapters 16 and 18). An X-ray analysis of bis(2,2,6,6-tetramethylpiperidino)germanium reveals a monomeric structure in the crystal, with Ge–N, 1.88 Å; N–C 1.51 Å; $\widehat{NGeN}$, 111.4°, $\widehat{CNC}$, 118.3°; $\widehat{GeNC}$, 117.6° [28e].

Table 6 Some spectroscopic data for some Ge^{II} amides

Compound	^1H n.m.r. (τ, Me)^a	I.r. (ν_{max}, cm^{-1})^a	1st. I.p. (eV)^b	U.v. [λ_{max}/nm (ϵ/dm^{3}mol^{-1} cm^{-1})] in n-C_6H_{14}^{a}
$Ge[N(SiMe_3)_2]_2$	9.56	428	8.68	364 (2050), ca. 300sh, 228 (6300)
$Ge[N(Bu^t)SiMe_3]_2$	9.51, 8.41	422	8.27	392 (1300), 325 (620), 230 (60,900)
$Ge[N(GeMe_3)_2]_2$	9.59	418		
$Ge(NC_5H_6Me_4)_2$ (2,2,6,6-tetramethylpiperidino, Me_2 / Me_2)^c	9.27, 8.67, 8.21	435	6.8	426 (670), 250 (7600), 217 (16,100)
$Ge(NBu^t_2)_2$^{d}	8.26	410	6.8	445 (420), 227 (7,100)

^{a}From ref. [17]. ^{b}From ref. [20]. ^{c}From ref. [28e]. ^{d}From ref. [28d].

Irradiation of the germanium(II) amide in the cavity of an e.s.r. spectrometer generally produces the Ge^{III} amide, presumably by photochemical disproportionation Eq. (62) [10a, 17], unless steric effects supervene. Thus, the germanium-centred radicals of Table 7 are obtained, but not the analogues with $(Et_3Si)_2N^-$, $(Et_3Ge)_2N^-$, $(Ph_3Ge)_2N^-$, $Bu^t_2N^-$, or $\overline{Me_2C(CH_2)_3CMe_2N}^-$ as ligands. Photolysis of the corresponding Ge^{II} amide gives an e.s.r. signal characteristic of

$$2Ge(NRR')_2 \xrightarrow[n\text{-}C_6H_{14},\ 20°\ C]{h\nu} \dot{Ge}(NRR')_3 + \frac{1}{n}[M(NRR')]_n \qquad (62)$$

the aminyl radical $:\dot{N}(SiEt_3)_2$, $:\dot{N}(GeEt_3)_2$, $:\dot{N}Bu^t_2$, or $\overline{Me_2C(CH_2)_3CMe_2\dot{N}}:$ [17, 28c, 28d]. The Ge^{III}-centred radicals possess a remarkable stability in n-hexane solution at 25° C, the half-life ($t_{1/2}$) for $\dot{Ge}[N(SiMe_3)_2]_3$ being ca. 6 months. The e.s.r. parameters for the Ge-centred radicals in n-hexane at 20° C are shown in Table 7. The salient features of the e.s.r. spectrum (Fig. 8) [17] are a central binomial septet arising from coupling to three equivalent nitrogen atoms (^{14}N, $I = 1$). In addition, ten satellite lines are seen due to coupling with the ^{73}Ge ($I = 9/2$, abundance = 7.8%) nucleus. The large values of $a(^{73}Ge)$ indicate that each radical is pyramidal and that the unpaired electron is in an approximately sp^3 hybrid-orbital, with the $N\text{-}\widehat{Ge}\text{-}N$ angle estimated at ca. 111° [22a].

Table 7 E.s.r. parameters of some kinetically-stabilised germanium(III) compounds [17] in n-C_6H_{14} at 25° C

Compound	g-value	$a(^{14}N)$ (Gauss)	$a(^{73}Ge)$ (Gauss)	$t_{1/2}$(25°C)
$\dot{Ge}[N(SiMe_3)_2]_3$	1.9991	10.6	171	> 5 months
$\dot{Ge}[N(Bu^t)SiMe_3]_3$	1.9998	12.9	173	~ 5 min
$\dot{Ge}[N(GeMe_3)_2]_3$	1.9994	11.0	143	> 22 h

The high value of $a(^{14}N)$ in the radical $\dot{Ge}[N(Bu^t)SiMe_3]_3$ is attributed [20a, 27] to reduction of nitrogen lone-pair delocalisation compared with $\dot{Ge}[N(SiMe_3)_2]_3$. The lower $a(^{73}Ge)$ value for $\dot{Ge}[N(GeMe_3)_2]_3$ may be due to the increased steric effect of the $GeMe_3$ group favouring planarity at the central Ge atom and consequently near sp^2-hybridisation.

E AMIDES OF TIN

1. Introduction

The first tin(IV) amide, $Sn(NEt_2)_4$, was prepared in 1961 [133]. Since then the field has expanded rapidly and the main lines of development had emerged by the early seventies. Many of the observations first made for Sn–N compounds,

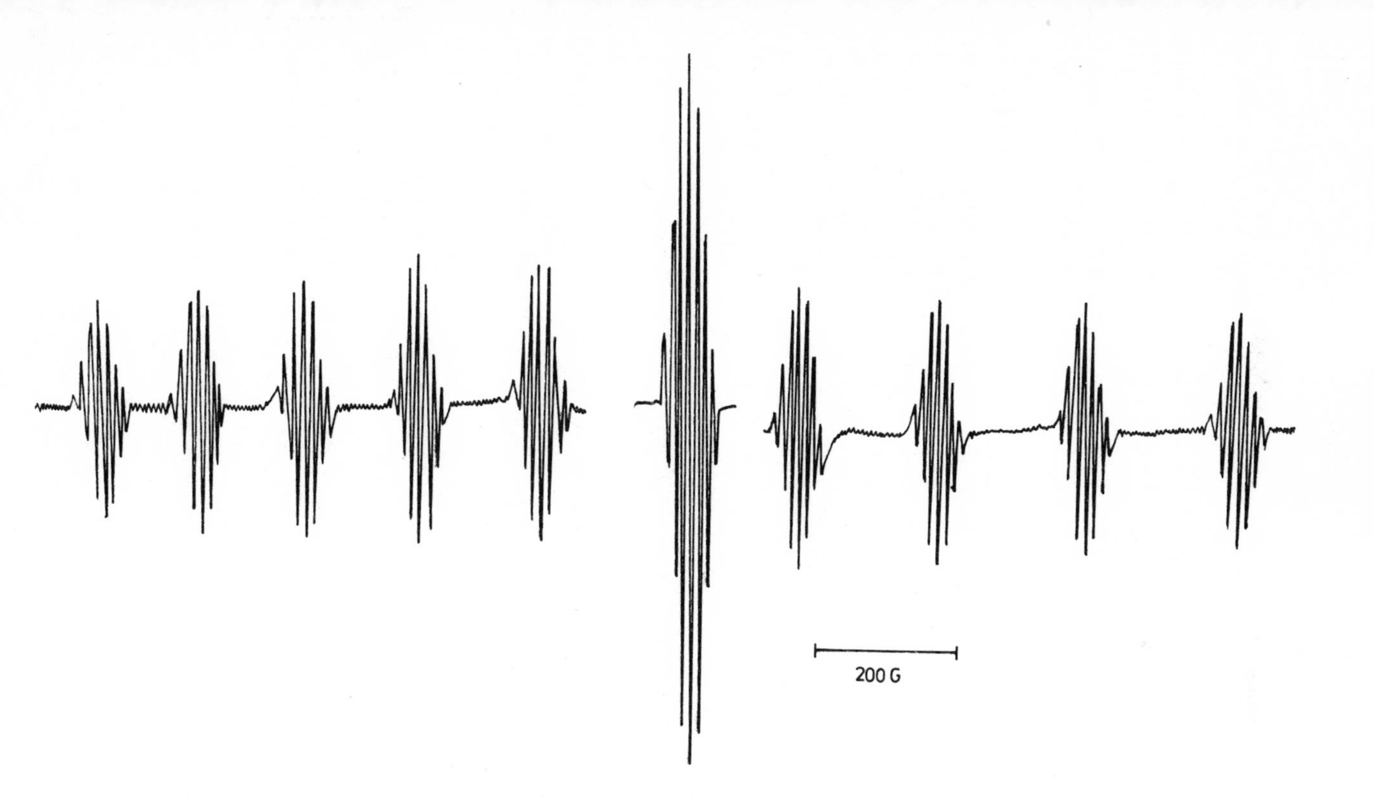

Fig. 8 – E.s.r. spectrum of $\dot{G}e[N(SiMe_3)_2]_3$ at 300K in n-C_6H_{14}. (N.B. One satellite septet is masked by the central septet.) [17]

especially with Me_3SnNMe_2, have served as a model for the chemistry of other metal amides, as predicted at an early stage [20, 21, 33, 37, 38,39, 52, 60–63, 73]. The reactivity of an aminostannane has been compared with that of the Grignard reagent [61, 64] and attributed to a combination of a weak and highly polar $\overset{\delta+}{Sn}–\overset{\delta-}{N}$ bond, which is thus responsive to reaction with a polar substrate. This is the basis for the utility of metal amides in synthesis (see Chapters 10–18).

Compounds such as Me_3SnNMe_2 may be regarded as stannylamines and compared as tertiary bases with tertiary amines and Si^{IV}, Ge^{IV}, or Pb^{IV} analogues. It was predicted [37, 62–64] that (i) metal-nitrogen bond strengths decrease in the sequence $Si > C > Ge > Sn > Pb$, (ii) Sn^{IV} is more of a class 'b' (or softer) acceptor than Si^{IV}, and (iii) basicity decreases in the order $C > Pb > Sn > Ge > Si$ (see Chapter 16). These considerations were based on the assumptions of relative M–N σ-bond strengths (decreasing with increasing atomic number of M) and maximum $M \overset{..}{-} N$ π-bonding for M = Si, followed by Ge, Sn, Pb, and C.

Renewed recent interest in tin amides may in part be attributed to the discovery of stable subvalent tin amides, e.g., $Sn[N(SiMe_3)_2]_{2 \text{ or } 3}$ (see Section E. 4).

The range and number of cyclic Sn–N compounds known at present is significantly smaller than that for Si or Ge. Reviews of tin(IV) amides have appeared [34a, 64, 88].

2. Synthesis of tin(IV) amides

We have seen that the amine hydrohalide elimination reaction provides a convenient procedure for synthesis of Si^{IV} or Ge^{IV} amides, e.g., Eq. (63). By contrast, although a tin(IV) (like Pb^{IV}) halide does react with ammonia or a primary or secondary amine, this leads generally to a stable 1:2 adduct (XVI), e.g., Eq. (64) [4]. On the other hand, SnX_4 is ammonolysed to $XSn(NH_2)_3$ in liquid ammonia [64]. Adducts such as (XVI) have greater M–N bond strengths

$$\equiv\text{M-Cl} + 2\text{HNR}_2 \xrightarrow{\text{M = Ge or Si}} \equiv\text{M-NR}_2 + [\text{R}_2\text{NH}_2]\text{Cl} \qquad (63)$$

$$\equiv\text{M-Cl} + 2\text{HNR}_2 \underset{}{\overset{\text{M = Sn or Pb}}{\rightleftharpoons}} \underset{\textbf{(XVI)}}{\equiv\text{M-Cl.2R}_2\text{NH}} \qquad (64)$$

than Si or Ge analogues (thus, the Lewis acidity of Group 4 metal halides increases with atomic number of M) and the thermolysis usually leads to dissociation into its factors. Reaction according to Eq. (63) has been calculated to be significantly more favourable in terms of ΔH (kcal mol^{-1}) for M = Me_3Si (−22.3) or Me_3Ge (−13.0) than Me_3Sn (−2.4) [6].

The most versatile and effective route to a tin(IV) amide involves transmetallation, whereby the tin(IV) halide is treated with an amide of an alkali-metal or magnesium, as illustrated in Eqs. (65) [133], (66) [129], and (67) [89]. The only limitations appear to be steric. For example, interaction of an excess

$$SnCl_4 + 4LiNEt_2 \longrightarrow Sn(NEt_2)_4 + 4LiCl \quad (65)$$

$$Pr_3SnCl + BrMgNEt_2 \longrightarrow Pr_3SnNEt_2 + MgBrCl \quad (66)$$

$$Me_3SnCl + KNC_4H_4 \longrightarrow Me_3SnNC_4H_4 + KCl \quad (67)$$

of $LiN(SiMe_3)_2.Et_2O$ and $SnCl_4$ in diethyl ether under prolonged reflux affords only the bis-amide $Cl_2Sn[N(SiMe_3)_2]_2$ [75], although three of these bulky amide groups can be introduced by an alternative route, Eq. (68) [74]. The salt elimination technique has also been applied to derivatives of primary amines. However,

$$Sn[N(SiMe_3)_2]_2 + BrN(SiMe_3)_2 \longrightarrow BrSn[N(SiMe_3)_2]_3 \quad (68)$$

the initial product, R_3SnNHR', generally undergoes condensation to $(R_3Sn)_2NR'$, unless steric effects supervene.

Stable compounds having a secondary nitrogen include $Me_3SnNHAr$ (but not Me_3SnNHR, $R = Me$ or Et) [62], $Bu_3SnNHBu^t$ [62], and $Bu^t_2Sn(NHR)_2$ ($R = Me$, CH_2Ph, or Bu^t) [46, 47]. The synthesis of the tristannylamines $(R_3Sn)_3N$ ($R = Me$, Et, or Pr) may be effected by transmetallation, Eq. (69) ($X =$ halide or alkoxide, $M =$ Li or Na) [114, 130].

$$3R_3SnX + 3MNH_2 \longrightarrow (R_3Sn)_3N + 3MX + 2NH_3 \quad (69)$$

Similarly, treatment of lithium nitride with Me_3SnCl affords $(Me_3Sn)_3N$ [79, 134]. A further variant of the transmetallation reaction has been claimed, Eq. (70) [2], but is unsubstantiated and indeed the reverse is known [40].

$$Me_3SiNMe_2 + Me_3SnCl \longrightarrow Me_3SnNMe_2 + Me_3SiCl \quad (70)$$

Considerable use has been made of the transamination reaction for the synthesis of stannylamines, as exemplified by Eq. (71). As for many other

$$R_3SnNR'_2 + HNR''_2 \longrightarrow R_3SnNR''_2 + HNR'_2 \quad (71)$$

amides, the reaction depends on the relative volatilities of the amines HNR'_2 and HNR''_2 [60, 62], as illustrated in Eqs. (72) [67], (73) [85], and (74) [116]. The yields are in most cases quantitative, and one application has been for the synthesis of an isotopically-substituted stannylamine [59], whilst another yields $(Me_3Sn)_3N$ from Me_3SnNMe_2 and ammonia [60, 62].

$$Me_3SnNMe_2 + HN\!\!\vartriangleleft \longrightarrow Me_3SnN\!\!\vartriangleleft + HNMe_2 \tag{72}$$

$$MeSn(NMe_2)_3 + HN(SiMe_3)_2 \longrightarrow MeSn(NMe_2)_2N(SiMe_3)_2 + HNMe_2 \tag{73}$$

$$Me_2Sn(NEt_2)_2 + C_6H_4(NHSiMe_3)_2\text{-}1,2 \longrightarrow C_6H_4(NSiMe_3)_2SnMe_2 + 2HNEt_2 \tag{74}$$

A stannylamine may also react with a primary amine to afford either another aminostannane [Eq. (75)] or a distannylamine [Eq. (76)], depending on steric considerations. For instance, if $R = Me$ and $R' = Me$ or Et, the stannylamine is

$$R_3SnNR'_2 + H_2NR'' \longrightarrow R_3SnNHR'' + HNR'_2 \tag{75}$$

$$2R_3SnNR'_2 + H_2NR'' \longrightarrow (R_3Sn)_2NR'' + HNR'_2 \tag{76}$$

obtained, but if $R = Me$ or Et and $R'' =$ aryl, only the distannylamine is formed [60, 62]. From $Me_2Sn(NMe_2)_2$ and H_2NEt, the cyclic stannazane $(Me_2SnNEt)_3$ is isolated [60, 62], and from $Bu^t_2Sn(NHBu^t)_2$ at 100° C the Sn_2N_2 ring $(Bu^t_2SnNBu^t)_2$ [47].

Insertion of a variety of substrates into the SnH or SnN bond has been utilised in the synthesis of Sn-N compounds (but not tin amides), for example Eqs. (77) [22], (78) [59], (79) [37], (80) [9], (81) [24], or (82) [123].

$$Me_3SnH + PhNCO \longrightarrow Me_3SnN(Ph)C(O)H \tag{77}$$

$$Me_3SnNMe_2 + PhNCO \longrightarrow Me_3SnN(Ph)C(O)NMe_2 \tag{78}$$

$$Me_3SnNMe_2 + PhCN \longrightarrow Me_3SnNC(Ph)NMe_2 \tag{79}$$

$$R_3SnOR' + PhNCO \longrightarrow R_3SnN(Ph)C(O)OR' \tag{80}$$

$$Ph_3SnPPh_2 + 2PhN_3 \longrightarrow Ph_3SnN(Ph)P(NPh)Ph_2 + 2N_2 \tag{81}$$

$$Ph_3SnPPh_2 + PhNCS \longrightarrow Ph_3SnN(Ph)C(S)PPh_2 \tag{82}$$

Certain stannylamines have been prepared from compounds containing Sn-O bonds, Eq. (83), but this procedure is generally thermodynamically

$$(R_3Sn)_2O + 2HN< \longrightarrow 2R_3Sn\text{-}N< + H_2O \qquad (83)$$

unfavourable unless the product has tin in an increased co-ordination number, as (XVII), or is resistant to hydrolysis, as in Eq. (84) [102a]; $(Me_3Sn)_3N$ has been obtained from $(Me_3Sn)_2O$ and $NaNH_2$ [115].

$$\begin{array}{cc} {}^{+}O{=}CX & \\ | \quad\quad | & \textbf{(XVII)} \\ R_3\bar{S}n{-}NR' & \end{array}$$

$$Sn(O_2CEt)_4 + 4HNMe_2 \xrightarrow{-20°\,C} 2H_2O + 2EtCONMe_2 + (EtCO_2)_2Sn(NMe_2)_2 \qquad (84)$$

Cleavage of a SnPh bond (in Bu^t_3SnPh) has also been effected by attack of KNH_2 in liquid NH_3, thus affording a rare example of a compound ($Bu^t_3SnNH_2$) with a Sn–NH_2 bond, Eq. (85) [41]. Another example of Sn–C cleavage is in Eq. (86) [139a].

$$Bu^t_3SnPh + NH_3 \xrightarrow{KNH_2} Bu^t_3SnNH_2 \qquad (85)$$

$$R_4Sn + HN(R')NO_2 \longrightarrow R_3Sn\text{-}N(R')NO_2 + RH \qquad (86)$$

Metathetical exchange reactions (Chapter 13) have been used, e.g., Eqs. (87) [30] and (88) [122a].

$$Bu_2SnCl_2 + Bu_2Sn[NEtC(O)OMe]_2 \longrightarrow 2Bu_2(Cl)Sn[NEtC(O)OMe] \qquad (87)$$

$$Me_2Sn(NEt_2)_2 + Me_3SiX \xrightarrow{X = Cl \text{ or } Br} Me_2(X)SnNEt_2 + Me_3SiNEt_2 \qquad (88)$$

Oxidative additions (Chapter 18) are exemplified in Eqs. (68) [74] and (89) [43, 43a]. The latter reactions have been studied in some detail (see Chapter 18);

$$Sn[N(SiMe_3)_2]_2 + RX \longrightarrow R(X)Sn[N(SiMe_3)_2]_2 \qquad (89)$$

examples of RX include MeBr, MeI, Bu^tBr, and PhI [43], whilst in the presence of a trace of EtBr as catalyst, RX may also be BuCl [43a], and addition of PhBr is likewise catalysed by EtBr. Bivalent tin amides are also convenient sources of acyltin(IV) compounds, as in the $Sn[\overline{NCMe_2(CH_2)_3CMe_2}]_2$-PhCOCl system, leading to $Sn(NR_2)_2Cl(COPh)$ [71b].

Other examples of oxidative or redox reactions are depicted in the Scheme [136].

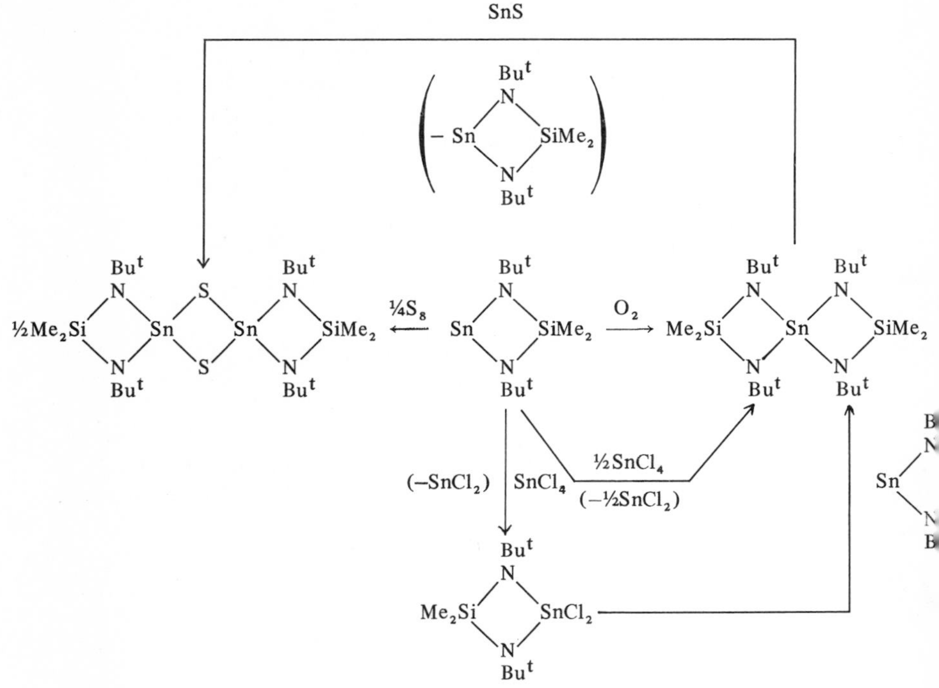

Scheme: Some $Sn^{II} \rightleftharpoons Sn^{IV}$ amide interconversions [136].

3. Physical properties and structures of tin(IV) amides

As tin(IV) is a better acceptor than the lighter group 4 elements, there is in principle the possibility of molecular association for a tin(IV) amide, as for Al analogues (see Chapter 4). For related compounds the tin may be five-co-ordinate, e.g., in Me_3SnCl in the solid state there are bridging chloride ligands between tin atoms. For amides, this is generally not the case, at any rate in the liquid state or in solution, although *N*-tributylstannylimidazole appears to be a co-ordination polymer (XVIII) [58], which is depolymerised by addition of a base such as

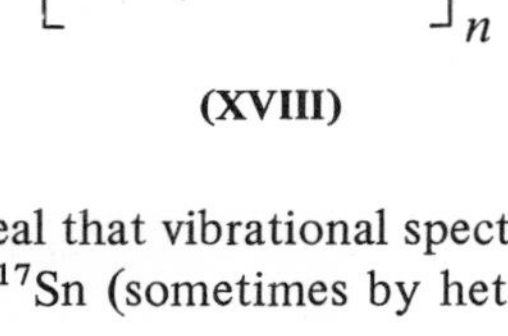

(XVIII)

pyridine. Tables 26–28 reveal that vibrational spectra, as well as ^{119}Sn Mössbauer or ^{1}H-, ^{14}N-, and ^{119}Sn or ^{117}Sn (sometimes by heteronuclear double resonance) spectra have been recorded for many tin(IV) amides. There is still not general agreement relating to the location of SnN_n normal modes, despite ^{15}N studies, or to all the implications of n.m.r. chemical shift or coupling constant data; the reader is referred to an earlier review [64]. Basicity studies are discussed elsewhere (Chapter 16). Dipole moments for R_3SnNMe_2, R_3SnNEt_2, $R_2Sn(NMe_2)_2$, $RSn(NEt_2)_3$, and $Sn(NMe_2)_4$ in benzene are low (1.1 ± 0.3 D) [86].

Thermochemical data, derived from heats of acid hydrolysis, are summarised in Table 8 [6, 42a]. From these we infer that (i) the tin-nitrogen bond strength is low (see Section E.1), (ii) the Sn^{II}-N bond is significantly stronger than the Sn^{IV}-N bond, and (iii) the Sn-N$(SiMe_3)_2$ bond is stronger than Sn-N$(Bu^t)SiMe_3$. Point (iii) is probably a manifestation of a steric, rather than electronic effect, because a similar 'β-silicon effect' is noted for the trend $\bar{E}(M\text{-}CH_2SiMe_3) > \bar{E}(M\text{-}CH_2CMe_3)$ in complexes $[MR_4]$ (M = Ti, Zr, or Hf; and R = Me_3CCH_2 or Me_3SiCH_2) [71c]. The variations in $\bar{E}$(Sn-N) for the series $(Me_3Sn)_nNMe_{3-n}$ (n = 1, 2, or 3) are not readily assessed, particularly in the absence of knowledge regarding the geometry at N or Sn-N lengths.

Table 8 Some thermochemical data for tin(IV) and tin(II) amides

Compound	ΔH° (g) (kcal mol^{-1})	$\bar{E}$(Sn-N) (kcal mol^{-1})	References
$Me_3Sn\text{-}NMe_2$	-4.3	41	[6]
$(Me_3Sn)_2NMe$	-19.5	48	[6]
$(Me_3Sn)_3N$	-14.2	42	[6]
$Me_3Sn\text{-}N(SiMe_3)_2$	-112.9	39	[42a]
$Me_3Sn\text{-}N(Bu^t)SiMe_3$	-64.0	32	[42a]
$Sn[N(SiMe_3)_2]_2$	-191.0	60	[42a]
$Sn[N(Bu^t)SiMe_3]_2$	-103.7	49	[42a]

The He(I) photoelectron spectra of the series $M(NMe_2)_4$ have been examined for M = C, Si, Ge, or Sn (as well as some transition metals, see Chapter 8) [40a]. The first ionisation potentials are 7.19 (C), 8.69 (Si), 8.46 (Ge), and 7.67 eV (Sn). Assuming restriction to M$\nleftarrow$N rotation, the molecular symmetry would be D_{2d}, whence the lone pair combinations would transform as E_1, B_1, and A_1. Experimentally, excepting for $C(NMe_2)_4$, only a very small splitting of the first band is observed. For $C(NMe_2)_4$ there are three distinct and well separated bands in intensity ratio 2:1:1 and hence these probably correspond to the e_1, b_1, and a_1 combinations. The degeneracy shown for the heavier group 4 metal congeners suggests free rotation about the M-N bonds, the difference with C being therefore probably of steric origin, which is also reflected in the low first i.p. for C. It is interesting that the transition metal amides, like carbon, show loss of degeneracy of the N lone pair combinations. On the assumption then of T_d symmetry for $M(NMe_2)_4$ molecules, the MN_4 bonding molecular orbital has symmetry T_2 and is expected to decrease monotonically down the group, cf., the experimental 11.2 (Si), 11.0 (Ge), and 10.8 eV (Sn). The first i.p. for Me_3SnNMe_2, by mass spectrometry, is 7.6 eV [72].

The structures of five tin(IV) compounds have been reported: $Sn(NMe_2)_4$, by electron-diffraction [136a], and by X-ray diffraction, $Me_3SnN(Me)NO_2$ [35], the spiro compound $Sn(\langle N(Bu^t)N(Bu^t) \rangle SiMe_2)_2$ [135a], $Me_3SnN(OSO_2CF_3)S(Me){=}NSO_2CF_3$ [107a], and $BrSn[N(SiMe_3)_2]_3$ [75a]. The structure of $Sn(NMe_2)_4$ is depicted in Fig. 9. $Sn(NMe_2)_4$ possesses a tetrahedral SnN_4 skeleton and the Sn-N distance (2.045 Å) is in good agreement with that calculated (2.04 Å). The $SnNC_2$ unit is almost planar and the N-C and C-H bond lengths are within the normal ranges.

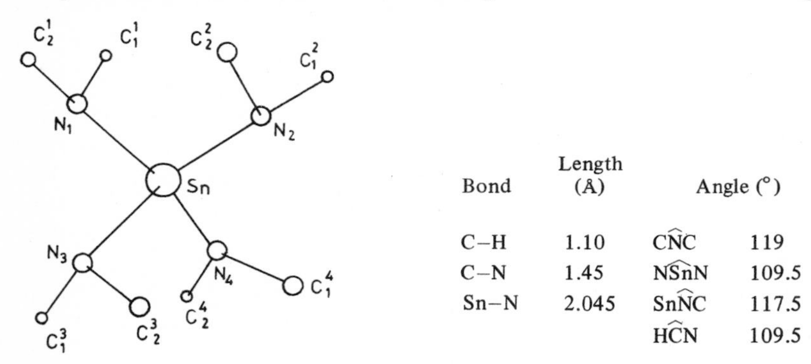

Bond	Length (Å)	Angle (°)	
C–H	1.10	$C\widehat{N}C$	119
C–N	1.45	$N\widehat{Sn}N$	109.5
Sn–N	2.045	$Sn\widehat{N}C$	117.5
		$H\widehat{C}N$	109.5

Fig. 9 – The molecular structure of $Sn(NMe_2)_4$ in the gas phase (electron diffraction). *[Reproduced, with permission, from Vilkov, L. V., Tarasenko, N. A., and Prokof'ev, A. K.,* J. Struct. Chem., *1970,* **11**, *114.]*

Characteristic bond lengths and angles in the molecule (the values given are the average values of chemically equivalent bond lengths, the deviations listed in parentheses are in units of the last decimal place).

Sn–N	2.033 (5) Å	$Si\widehat{N}Sn$	95.2°
Si–N	1.727 (7) Å	$N\widehat{Si}N$	93.4°
C–N	1.469 (7) Å	$N\widehat{Sn}N$	76.3°

Fig. 10 – The crystal and molecular structure of $Sn(\langle N(Bu^t)N(Bu^t) \rangle SiMe_2)_2$.

[Reproduced, with permission, from Veith, M., Angew. Chem. Internat. Edn., *1975,* **14**, *263.]*

The structure of the spirocyclic compound is illustrated in Fig. 10 [135a]. The four-membered ring is near planar and the Sn^{IV}-N bond length of 2.03 Å is in close agreement with that found in $Sn(NMe_2)_4$. The Si-N bond length (1.71 Å) is typical. $BrSn[N(SiMe_3)_2]_3$ is a discrete monomer in the crystal: Sn-Br, 2.52 Å; Sn-N, 2.08 Å; N-Si, 1.77 Å; $Br\widehat{Sn}N$, 102°; $N\widehat{Sn}N$, 116°; and $N\widehat{Si}N$, 120.2° [75a]. The 'nitro' compound is composed of an infinite chain $-SnMe_3-NMe-N(O)-OSnMe_3-$ with Sn-N, 2.33 Å [35].

4. Subvalent tin amides

These are compounds of the general formula $Sn(NR_2)_2$ or $\dot{S}n(NR_2)_3$, and are stable only where NR_2 is a sterically-demanding amido group, generally a silylamide or substituted cyclic amide. Their physical and chemical properties differ significantly from those of the tin(IV) amides.

Under ambient conditions, the tin(II) amides are thermally-stable, orange to red oils {$Sn[N(SiMe_3)_2]_2$ and $Sn[N(Bu^t)SiMe_3]_2$ have the following u.v. λ_{max} in n-C_6H_{14} (ϵ in parentheses): 389 (3100), 287 (4600), and 230 nm (23,000); and 433 (1200), ca. 330(sh), 305 (2800), and 236 nm (15,000), respectively [44, 49]} or low melting solids [e.g., [76] $Sn\{\overline{NC(Me_2)(CH_2)_3C}Me_2\}_2$, m.p. 44–48°, λ_{max} (in n-C_6H_{14}) 475 (720) and 222 (80,000) nm], monomeric (but see ref. [109]) in benzene or cyclohexane solutions, diamagnetic, thermochromic, and hydrocarbon soluble [44, 75]. Like the germanium (Section D.4) or lead (Section F) analogues, they are obtained from the metal(II) chloride and Li salt of the amine, LiNRR'{RR' = $(Me_3Si)_2$ [44, 49, 109], $Me_3Si(Bu^t)$ [44, 49], $(Et_3Si)_2$ [44], $(R''_3Ge)_2$ (R'' = Me, Et, or Ph) [44], or $CMe_2(CH_2)_3CMe_2$ [76]}. Likewise, $LiNBu^t_2$ and $SnCl_2$ yield $Sn(NBu^t_2)_2$ or $[Sn(Cl)(NBu^t_2)]_n$ [77]. The tin(II) amides possess a stereochemically-active lone pair with a low first ionisation potential of ca. 8 eV for $Sn[N(MMe_3)(SiMe_3)]_2$ (M = C or Si) [53] or 6.8 eV for $Sn(NBu^t_2)_2$ [77] or $Sn[\overline{NCMe_2(CH_2)_3C}Me_2]_2$ [76], and undergo a wide range of reactions (Chapters 15, 16, and 18). Further tin(II) amides to have been obtained are cyclic and may be monomeric (cf. Fig. 12) if steric conditions are favourable or dimeric (*N*, *N'*-Pr^i_2-analogue) [135b, 135c]. Other compounds are not monomeric and are colourless, i.e., $[Sn(NMe_2)_2]_2$, formulated as a μ-$(NMe_2)_2$-bridged dimer [36], and $[Sn(Cl)N(SiMe_3)_2]_2$ [44, 49], which is probably μ-Cl_2-bridged [74, 75].

Structural information has been obtained on two of these compounds, by electron diffraction on $Sn[N(SiMe_3)_2]_2$ [76] and X-ray crystallography on

$Sn\langle N(Bu^t) \rangle_2 SiMe_2$ [135c]. The structure of $Sn[N(SiMe_3)_2]_2$ is pictured schematically in Fig. 11 and shows that the compound exists exclusively as a monomer in the vapour phase. It is interesting that even with the bulky amido ligand $N(SiMe_3)_2^-$ the $N\widehat{Sn}N$ angle is low (96°). This V-shaped structure is consistent with a bent singlet formulation expected from the diamagnetism (see Section D.4). The molecule has C_{2v} symmetry, and both $SnNSi_2$ skeletons are planar. The Si–N bond length (1.74 Å) indicates significant π-bonding between the Si (*d*) and N(*p*) orbitals, but the Sn–N bond length (2.09 Å) is in agreement with that expected from the sum of the covalent radii. The isoelectronic $Sn[CH(SiMe_3)_2]_2$ is a diamagnetic Sn–Sn bonded dimer in the crystal [50].

The structure of $Sn\langle N(Bu^t) \rangle_2 SiMe_2$ (Sn and $SiMe_2$ bridged by two NBu^t groups) is interesting in that it exists as monomer and dimer in the solid state, Fig. 12(a) and 12(b) [135c]. The Sn–N bond length in both of these compounds is ca. 2.09 Å, which is slightly longer than the Sn–N bond length in $Sn(NMe_2)_4$ [136a]. Whereas the former has both monomeric and dimeric phases, the bis-isopropyl analogue crystallises exclusively as dimer [135b].

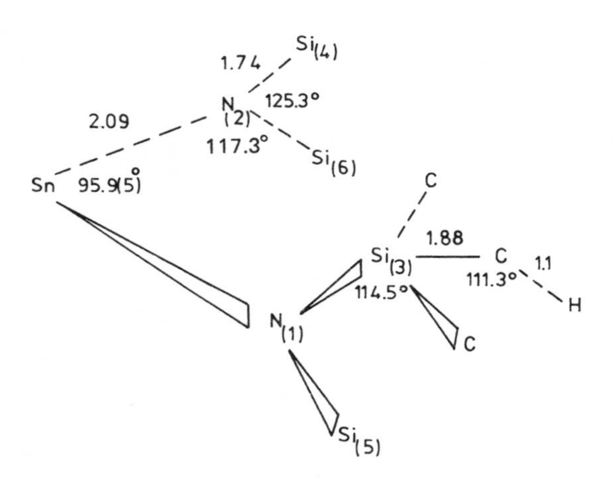

$SnN_{(1)}\,Si_{(4)}\,Si_{(6)}$ are coplanar.
Both NSi_2 planes are parallel.
Angle between Si–N–Si planes and N–Sn–N planes = 90°.
$N_{(1)}$–$N_{(2)}$ = 3.1 Å.
$Si_{(3)}$–$Si_{(5)}$ = 3.09 Å.
$Si_{(3)}$–$Si_{(4)}$=$Si_{(5)}$–$Si_{(6)}$ = 4.29 Å.

Fig. 11 – Preliminary electron diffraction data on $Sn[N(SiMe_3)_2]_2$ [76].

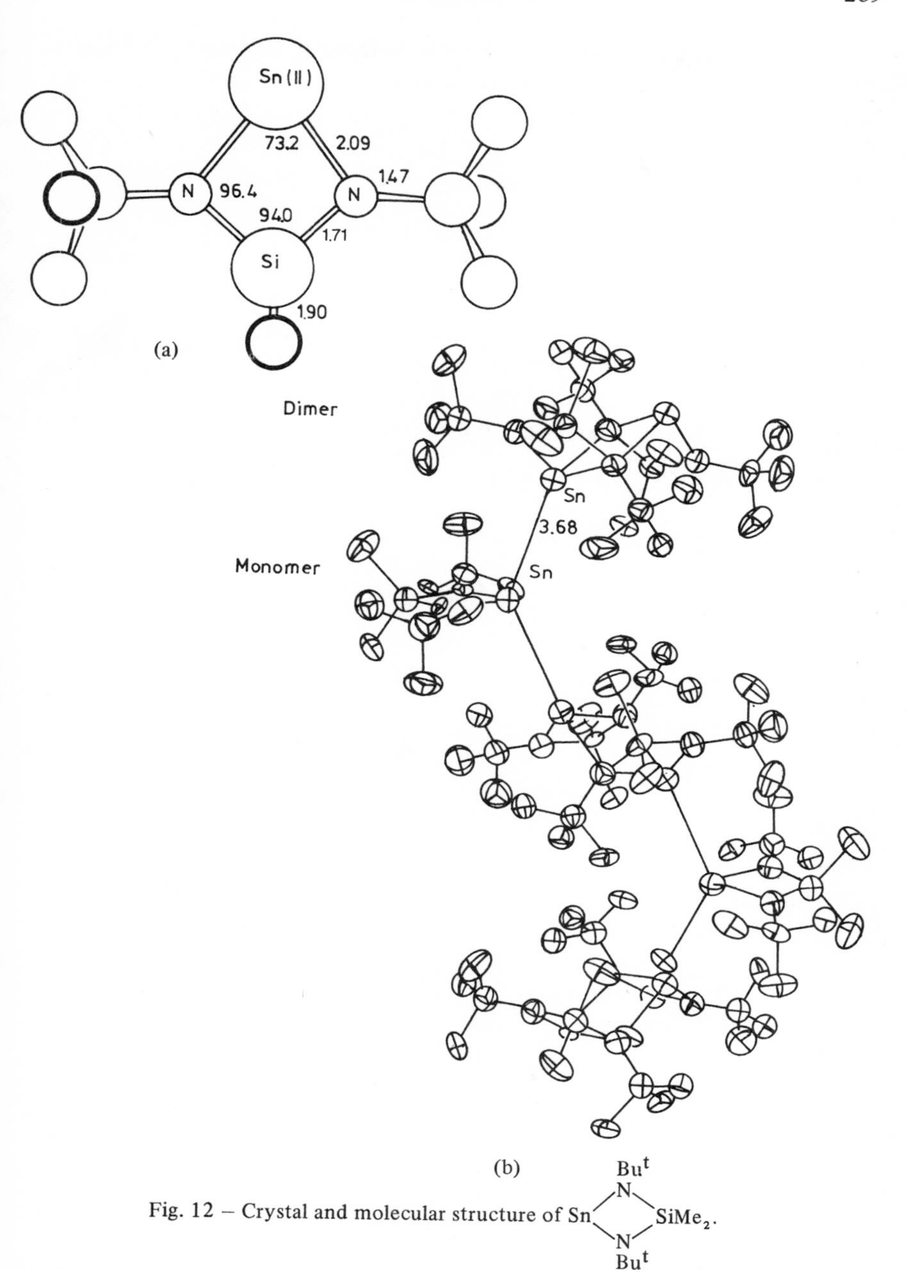

Fig. 12 – Crystal and molecular structure of $Sn(NBu^t)_2SiMe_2$ (cyclic Sn–N(Bu^t)–Si–N(Bu^t)).

(a) The monomer (Sn–N, 2.09 Å), (b) the unit cell, showing monomer and dimer (Sn–N, 2.39 Å). *[Reproduced, with permission from Veith, M.,* Z. Naturforsch., *1978,* **33b,** *7.]*

The photolysis [21d, 44, 51, 56] of a number of the $Sn(NRR')_2$ compounds has led to the formation and e.s.r. characterisation of four tin(III) centred radicals, which on the basis of their hyperfine splitting constants are found (as in the case of Ge, Section D.4) to be pyramidal. The e.s.r. parameters are in Table 9. Photolysis of the compounds $Sn[N(SiEt_3)]_2$, $Sn[N(GePh_3)_2]_2$, or

$$Sn\begin{matrix}\diagup N(SiMe_3) \diagdown \\ \diagdown N(SiMe_3) \diagup\end{matrix}SiMe_2$$

gave no e.s.r. signals [74]. The e.s.r. spectrum of $\dot{S}n[N(SiMe_3)_2]_3$, Fig. 13 [44], consists of a central septet due to coupling with three ^{14}N ($I = 1$) nuclei. The $a(^{14}N)$ value is significantly increased on substituting Bu^t for $SiMe_3$ (Table 9) for reasons explained in Section D.4. The extremely large $a(^{117/119}Sn)$ coupling is due to the pyramidal nature of the radical. Photolysis of $Sn(NBu^t_2)_2$ [77] or $Sn[\overline{NCMe_2(CH_2)_3CMe_2}]_2$ [76] gives the appropriate aminyl radical $:\dot{N}R_2$.

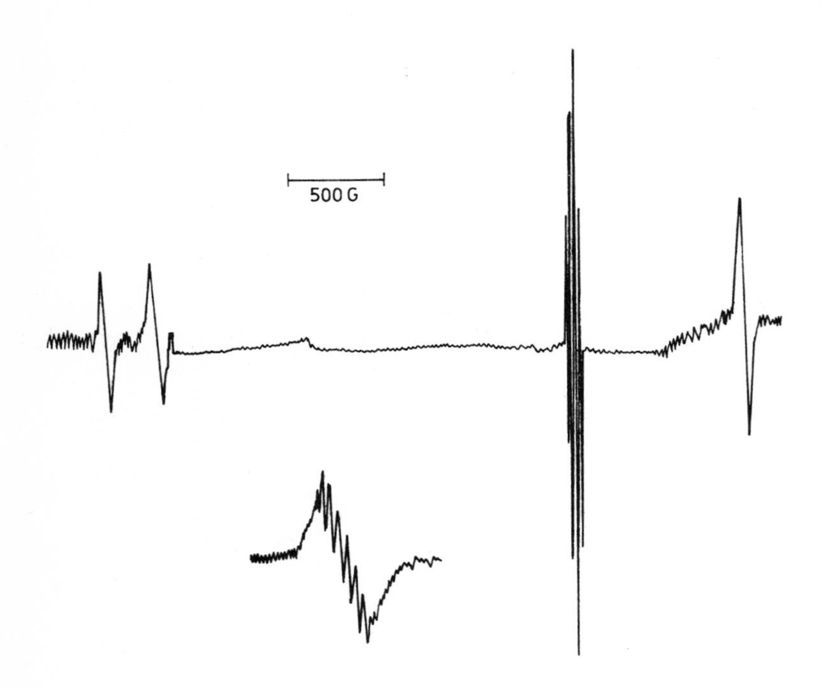

Fig. 13 – E.s.r. spectrum of $\dot{S}n[N(SiMe_3)_2]_3$ at 300K in n-C_6H_{14} [44] showing $^{117/119}Sn$ satellite lines. Inset – × 10 expansion of high field satellite.

Table 9 E.s.r. parameters of some kinetically-stabilised tin(III) compounds [44] in n-C_6H_{14} at 25° C

Compound	*g*-value	$a(^{14}N)$ (Gauss)	$t_{1/2}$
$\dot{Sn}[N(SiMe_3)_2]_3$ [a]	1.9912	10.9	*ca.* 3 months
$\dot{Sn}(NBu^tSiMe_3)_3$	1.9928	12.7	5 min
$\dot{Sn}[N(GeMe_3)_2]_3$	1.9924	10.7	*ca.* 10 h
$\dot{Sn}[N(GeEt_3)_2]_3$	1.9939	11.9	*ca.* 20 h

[a] $a(^{119}Sn) = 3139.5$ G, $a(^{117}Sn) = 3317.5$ G.

F LEAD AMIDES

Lead amides have been the least studied among the nitrogen compounds of the group 4b metals, probably because the lead-nitrogen bond is very weak. Thermochemical measurements have been carried out on some simple Pb^{II} and Pb^{IV} amides and these indicate that $\bar{E}$(Pb–N) is ca. 20-30 kcal mol^{-1} [3a]. Their potential as anti-knocks has been mentioned [19].

The first lead amide was synthesised according to a patent published in 1959 [2]. Simple lead amides have been prepared by the transmetallation technique, e.g. Eq. (90) [8], and (91) [3a, 4].

$$Bu_3PbCl + LiNEt_2 \longrightarrow Bu_3PbNEt_2 + LiCl \qquad (90)$$

$$PbCl_2 + 2LiN(Bu^t)SiMe_3 \longrightarrow Pb[N(Bu^t)SiMe_3]_2 + 2LiCl \qquad (91)$$

The insertion reaction, involving isocyanates or carbodi-imides, has also been utilised to synthesise PbN compounds, Eqs. (92) [11], (93) [1], and (94) [1,11].

$$Bu_3PbNEt_2 + C_6H_{11}NCO \longrightarrow Bu_3PbN(C_6H_{11})CONEt_2 \qquad (92)$$

$$Ph_3PbOMe + 1\text{-}C_{10}H_7NCO \longrightarrow Ph_3PbN(1\text{-}C_{10}H_7)CO_2Me \qquad (93)$$

$$Ph_3PbOMe + (1\text{-}C_{10}H_7N)_2C \longrightarrow Ph_3PbN(1\text{-}C_{10}H_7)\underset{\substack{\| \\ NC_{10}H_7\text{-}1}}{C}OMe \qquad (94)$$

Derivatives of phthalimide have been reported in patents [3, 7], as have those of sulphonamides [14] and imines [9].

The lead(II) amides are related to the Ge^{II} (Section D.4) and Sn^{II} compounds (Section E.4). They are probably less stable thermally than the Sn^{II} analogues, although $Pb[N(SiMe_3)_2]_2$ can be studied in the gas phase, for example, by p.e.

spectroscopy [4a]. Thus, whereas $Sn[\overline{NCMe_2(CH_2)_3C}Me_2]_2$ is indefinitely stable in an inert atmosphere, the lead compound decomposes at below 0° C [6b]. Similarly, photolysis of $Pb[N(SiMe_3)_2]_2$ or $Pb[N(Bu^t)SiMe_3]_2$ leads to deposition of a lead mirror and an aminyl radical, even at low temperature, rather than a Pb^{III} amide (cf. Ge^{III} and Sn^{III} amides, Sections D.4 and E.4). The amide $Pb[N(SiMe_3)_2]_2$ is an effective base [3c] (see Chapter 16), but also inserts into a Pt–Cl bond (Chapter 18). The reaction of $Pb(OAc)_4$ with Ph_3CNH_2 to yield $Ph_2C{=}NPh$ may involve a lead amide as an intermediate [16].

G TABLES OF COMPOUNDS

Table 10† Mono(amino)silanes ($\equiv$Si—N<)

Compound	Method	B.p. (°C/mm.) (m.p., °C)	Comments	References
$H_3SiNHPh$	K		ir, mass	[28]
H_3SiNMe_2	C	(3.3–3.4)		[312]
	C	19.2 (2.2)	d_4^{20} 0.700, ir, mass	[25]
			nmr	[97]
			ed	[128a]
^{14}N, ^{15}N; ^1H, ^2H			ir	[74]
$.BH_3$		d(O)		[211]
$.AlMe_3$				[211]
$.GaMe_3$				[211]
H_3SiNEt_2	C	78 (−149)	d_4^{20} 0.751, ir, mass	[25]
H_3SiNPh_2	C		ir	[27]
H_3SiN(pyrrolidine ring)		81.5 (−46)	d_4^{20} 0.810, ir	[26]
H_3SiN(piperidine ring)		104 (−109)	d_4^{20} 0.775, ir	[26]
$H_3SiSiH_2NMe_2$			nmr	[327]
$H_2MeSiNMe_2$	C	45 (150–60)		[95]
$H_2BuSiNHPh$	C	229	d_4^{20} 0.932,	
$H_2BuSiNHPh$	C	229	d_4^{20} 0.932, n_D^{20} 1.5219	[10]
$H_2PhSiNHPh$	D			[76]
p-$(H_2PhSiNH)_2C_6H_4$	B			[114]
p-$(H_2PhSiNHC_6H_4)_2$	B			[114]
p-$(H_2PhSiNHC_6H_4)_2CH_2$	B			[114]
$HMe_2SiNHEt$	C	74	n_D^{20} 1.3898, ir	[158]
$HMe_2SiNHPr^i$	C	87	n_D^{20} 1.3943, ir	[158]
$HMe_2SiNHBu^t$	C	108	n_D^{20} 1.3940, ir	[158]

Table 10 *(C'td.)*

Compound	Method	B.p. (°C/mm.) (m.p., °C)	Comments	References
$HMe_2SiNHBu$	D	88–91/40	ir, nmr	[91]
$HMe_2SiNHPh$	D	91/12	ir, nmr	[91]
HMe_2SiNMe_2	A, D	68	ir, nmr	[91, 241]
HMe_2SiNEt_2	D	110	ir, nmr	[91]
p-$(PhHMeSiNHC_6H_4)_2$	B			[114]
p-$(PhHMeSiNHC_6H_4)_2CH_2$	B			[114]
HMe(1-Np)SiNHPh	D			[76]
$HEt_2SiNHMe$	C	116/728	n_D^{20} 1.4157	[225]
$HEt_2SiNHPh$	C	113–5/12	n_D^{20} 1.5259	[225]
HEt_2SiNEt_2		72–3.3/36	d_4^{20} 0.7799, n_D^{20} 1.4259	[294]
$HPr_2SiNHPr$	C	186–7	d_4^{20} 0.7906, n_D^{20} 1.4302	[104]
$HPr_2SiNHPr^i$	C	176–8	d_4^{20} 0.7804, n_D^{20} 1.4250	[104]
$HPr_2SiNHPh$	C	166	d_4^{20} 0.9187, n_D^{20} 1.5130	[104]
HPr_2SiNEt_2	C	81.5–2	d_4^{20} 0.7948, n_D^{20} 1.4285	[104]
$HPh_2SiNHMe$	C	95–6/0.02	n_D^{20} 1.5755	[225]
$HPh_2SiNHPh$	C	164–6/10^{-3}	n_D^{20} 1.6264	[225]
p-$(HPh_2SiNH)_2C_6H_4$	B			[74]
p-$(HPh_2SiNHC_6H_4)_2CH_2$	B			[74]
p-$(HPh_2SiNHC_6H_4)_2O$	B			[74]
$Me_3SiNHMe$	C	71	d_4^{20} 0.7395, n_D^{20} 1.3905	[277]
(N) ^1H, ^2H			ir, Raman, ed	[31, 132, 265]
			ΔH_f	[34]
$Me_3SiNHEt$	C	90.1–0.8	n_D^{20} 1.3912	[277]
(N) ^1H, ^2H			ir, Raman, μ	[31, 82a]
$Me_3SiNHPr$			ir, μ	[2, 82a]
$Me_3SiNHPr^i$	I			[120]
			ir, μ	[2, 81a]
$Me_3SiNHBu$	B	134–5	d_4^{20} 0.765, n_D^{20} 1.4094	[185]
	G	135	n_D^{20} 1.4097	[1]
	G	46/29	n_D^{20} 1.4103	[269]
			ir	[2]
			mass	[390]
$Me_3SiNHBu^i$	C	128	d_4^{20} 0.758, n_D^{20} 1.4073	[3]
			ir, μ	[2, 82a]
$Me_3SiNHBu^s$	C	124	d_4^{20} 0.756, n_D^{20} 1.4060	[3]
			ir	[2]
$Me_3SiNHBu^t$	C	118–27	n_D^{25} 1.4060, ΔH_f	[90, 137a]
			ir	[2]
$Me_3SiNHPent$	B	161.4–2.4	d_4^{20} 0.772, n_D^{20} 1.4154	[185]
			mass	[390]
$Me_3SiNHC_7H_{13}$		92–3/26		[223]
$Me_3SiNH(CH_2)_{11}SiHPh_2$		95/0.01	d_4^{25} 0.902, n_D^{25} 1.4952	[307]
$Me_3SiNHCH_2CH_2Ph$	B	107/20		[223]
$Me_3SiNHCH_2CH_2NHPh$	A	74–5/1	n_D^{20} 1.5178	[377]

Table 10 *(C'td.)*

Compound	Method	B.p. (°C/mm.) (m.p., °C)	Comments	References
$Me_3SiNHCH_2C_2H_3$	C	111–5	n_D^{20} 1.4112, ir, nmr	[210]
$(Me_3SiNHCH_2)_2$	B	70/8	n_D^{20} 1.4256	[377]
	B	191	n_D^{20} 1.427	[250]
			nmr	[182]
$Me_3SiNH(CH_2)_3SiMe_2OSiMe_3$		90–4/4	d_4^{20} 0.8366, n_D^{20} 1.4200, ir, nmr	[210]
$Me_3SiNH(CH_2)_3Si(OEt)_3$	B		ir	[326]
$Me_3SiNHC_6H_4SiMe_3$	I		ir, nmr	[381]
$Me_3SiNHC_6H_3(SiMe_3)X$ (X = *m*- or *p*-Me)	I		ir, nmr	[381]
$Me_3SiNH(CH_2)_6NHSiMe_3$	C	110/2	ir, nmr	[333]
$Me_3SiNH(CH_2C_6H_4CH_2)NHSiMe_3$	C	120/1.5	ir, nmr	[333]
$Me_3SiNH(C_6H_4CH_2C_6H_4)NHSiMe_3$	C	186/1.5	ir, nmr	[333]
$Me_3SiNH(C_6H_3Me)NHSiMe_3$	C	110/1	ir, nmr	[333]
$Me_3SiNHCH_2CH_2NMeSiMe_3$	B	76/9	n_D^{20} 1.4289	[377]
$Me_3SiNHCH_2CH_2NHSiMe$ bonded to two $NSiMe_3$ groups (ring: $NSiMe_3$—$NSiMe_3$)			nmr	[182]
$Me_3SiNHCH_2CH_2NPhSiMe_3$	A	104–7/1, 79–80/0.1	n_D^{20} 1.4950	[377]
$Me_3SiNHCH_2CH_2N(SiMe_3)_2$	B	103–4/18	n_D^{20} 1.4420	[377]
	A, C	79–80/3, 237–9	n_D^{20} 1.4437, uv, nmr	[183]
			nmr	[182]
		99/8	n_D^{20} 1.4438	[250]
$Me_3SiNHCH_2CH_2N(SiMe_3)SiMe_2Ph$	E	101/0.3		[226]
$Me_3SiNHCH_2CH_2N(SiMe_3)SiMePh_2$	E	137/0.05		[226]
$Me_3SiNHCH_2N(SiMe_3)_2$	A			[226]
$Me_3SiNHCH{:}C{:}NSiMe_3$		117/30	d_4^{20} 0.822, n_D^{20} 1.4573	[255]
$Me_3SiNH(CH_2)_3OSiMe_3$	B			[206]
$Me_3SiNHCH_2CH_2OC_2H_3$	B	49–53/12		[89]
$Me_3SiNHCMe_2OC_2H_3$	C			[89]
$Me_3SiNH(CH_2)_5CO_2SiMe_3$		114/2.5	d_4^{20} 0.8925, n_D^{20} 1.4331	[267]
$Me_3SiNH(CH_2)_3CO_2SiMe_3$		115/1.1	d_4^{20} 0.9539, n_D^{20} 1.4390	[267]
$Me_3SiNH(CH_2)_2CO_2Et$		96–100/14		[45, 46]
$Me_3SiNHCHMeCH_2CO_2Et$		90–5/20		[45, 46]
$Me_3SiNHCH(CO_2SiMe_3)CHMeOSiMe_3$	B	60–2/0.2		[174]

Table 10 *(C'td.)*

Compound	Method	B.p. (°C/mm.) (m.p., °C)	Comments	References
$Me_3SiNHCH(CO_2SiMe_3)CH_2$–(3-isoquinolyl)	B	140–2/0.2		[174]
$Me_3SiNHCH_2CO_2Et$	C	68–71/13		[45,46]
$Me_3SiNHCH_2CO_2SiMe_3$	A	97/22	d_4^{20} 0.8975, n_D^{20} 1.4229	[267]
	B	97/22	d_4^{20} 0.8975, n_D^{20} 1.4229	[48]
$Me_3SiNHCHMeCO_2SiMe_3$	B	73/10	d_4^{20} 0.8831, n_D^{20} 1.4177	[48]
$Me_3SiNHCHPr^iCO_2SiMe_3$	B	93/10	n_D^{20} 1.4226	[48]
$Me_3SiNHCHBu^iCO_2Et$		107–12/12		[46]
$Me_3SiNHCHBu^iCO_2SiMe_3$	B	105/12	n_D^{20} 1.4236	[39, 48]
$Me_3SiNHCHBu^sCO_2SiMe_3$		64/0.7	d_4^{20} 0.8804, n_D^{20} 1.4267	[267]
$Me_3SiNHCH(CH_2Ph)CO_2SiMe_3$		110/1.1	d_4^{20} 0.9930, n_D^{20} 1.4853	[267]
$Me_3SiNHC_3H_5$	B	110	d_4^{25} 0.7675, n_D^{25} 1.4130	[312]
$Me_3SiNHCH_2Ph$	B	95–6/15	n_D^{25} 1.4918	[223]
$(Me_3SiNHCH_2)_2C_6H_4$	B	109–10/1	ir, nmr	[332]
$Me_3SiNHCMe{:}CHCN$	A	58–60/0.15 (22)	ir, nmr, mass	[188]
$Me_3SiNHC_6H_{11}$	G	47/3	n_D^{20} 1.4453	[269]
Me_3SiNH–(2-furyl)	C	85–6/31	d_4^{20} 0.9246, n_D^{20} 1.4570	[209]
$Me_3SiNHCR{=}CHR'$	D			[81a]
(R R'				
Et Me		42–9/10		
Bu H)				
$Me_3SiNHPh$	C	96–8/24	d_4^{20} 0.940	[8, 244]
	B	96–8/24	d_4^{20} 0.940	[223]
	G	208	n_D^{20} 1.5222	[1]
			uv	[254]
			nmr	[259]
(N) ^1H, ^2H			ir, Raman	[68, 259a]
^{14}N, ^{15}N			ir	[258, 259a]
o-$Me_3SiNHC_6H_4CO_2SiMe_3$		104/0.7 (34)	d_4^{20} 1.0566, n_D^{20} 1.5348	[267]
p-$Me_3SiNHC_6H_4CO_2SiMe_3$		125/0.6	n_D^{20} 1.5416	[267]

Table 10 *(C'td.)*

Compound	Method	B.p. (°C/mm.) (m.p., °C)	Comments	References
Me_3SiNH $NHMe$ Me_3SiNH $NMeSiMe_3$	A	96–8/2.3	ir, nmr	[310]
Me_3SiNH $NHMe$ Cl Me_3SiNH $NMeSiMe_3$ Cl	A	103–5/0.4	n_D^{22} 1.5215, ir, nmr	[310]
Me_3SiNH $NHPh$	A, B	128–30/0.3	ir, nmr	[310]
Me_3SiNH $NHSiMe_3$	B	106–8/2 (32–4.5)	ir, nmr	[310]
Me_3SiNH $NHSiMe_3$ Cl	B	110–5/0.8	n_D^{20} 1.5189, ir, nmr	[310]
Me $NHSiMe_3$ $NHSiMe_3$	B	110–1/1	ir, nmr	[332]
Me_3SiNH $NPhSiMe_3$	A	135–9/1.0	ir, nmr	[310]
Me_3SiNH $NPhSiMe_2Bu^t$	A	157–64/10	ir, nmr	[310]

Table 10 *(C'td.)*

Compound	Method	B.p. (°C/mm.) (m.p., °C)	Comments	References
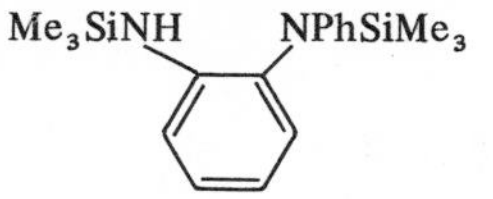 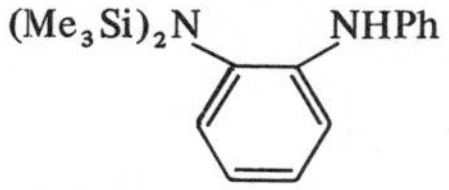 2-(Me_3SiNH)-1-($NPhSiMe_3$)-benzene / 2-($(Me_3Si)_2N$)-1-(NHPh)-benzene	A	134–6/0.5	ir, nmr	[310]
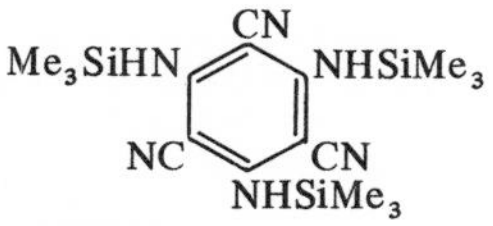 2,4-bis(Me_3SiHN)... Me_3SiHN, CN, $NHSiMe_3$, NC, CN, $NHSiMe_3$ substituted benzene	D	(170–1)		[171]
Me_3SiNMe_2	C	85–6	d^{20}_4 0.7446, n^{20}_D 1.3984	[270]
	C	82		[85]
	A	86/760		[122]
			ir, Raman, X-ray, ΔH_f, pes, mass	[2, 34, 66, 69, 132, 134, 187, 211a, 266]
$Me_3SiNMeBu^t$	A	144/762	ir, nmr	[291a]
$Me_3SiNMePh$	I		ir, nmr	[381]
$Me_3SiNMeCH_2CH_2OC_2H_3$	B	56–8/12		[89]
$(Me_3SiNMeCH_2)_2$		211–2	$n^{21.5}_D$ 1.4350	[250]
			nmr	[182]
$Me_3SiNMeCH_2CH_2N(SiMe_3)_2$	B	114–5/8	n^{20}_D 1.4466	[377]
$Me_3SiNMeCPh{:}NMe$	A	63–4/0.1	ir, nmr	[282]
$Me_3SiNMeCPh{:}NSiMe_3$	A	70–1/0.1	ir, nmr	[282]
$Me_3SiNMeCPh{:}NGeMe_3$	A	79–80/0.1	ir, nmr	[282]
$Me_3SiNMeCPh{:}NSnMe_3$	A	90–1/0.1	ir, nmr	[282]
$Me_3SiNMeCPh{:}NPbMe_3$	A	121/0.1	ir, nmr	[282]
$Me_3SiNMeCPh{:}NAsMe_2$	A	88–92/0.1	ir, nmr	[282]
$Me_3SiNMeCPh{:}NLi$	D			[282]
$Me_3SiNMeCN$	K			[150]
Me_3SiNEt_2	C	126.8–7.1	n^{20}_D 1.4109	[90, 185, 277]
	A	32/24	n^{20}_D 1.4089	[87]
			mass, ΔH_f	[146]
$Me_3SiNEtCPh{:}NSiMe_3$	A	80–2/0.1	ir, nmr	[282]
Me_3SiNPr_2	B	163–4	n^{20}_D 1.4218	[106]
			ir	[2]
$Me_3SiNPr^i_2$	C	157	d^{20}_4 0.786, n^{20}_D 1.4241	[3]
			ir	[2]
Me_3SiNBu_2	G	200	n^{20}_D 1.4291	[1]
	A	93/21	n^{26}_D 1.4262	[87]
	B	201–2	n^{20}_D 1.4762	[106]
			ir	[2]

Table 10 *(C'td.)*

Compound	Method	B.p. (°C/mm.) (m.p., °C)	Comments	References
$Me_3SiNBu^i_2$	B	190–2	n_D^{20} 1.4181	[106]
			ir	[2]
$Me_3SiNBu^s_2$	C	195	d_4^{20} 0.817, n_D^{20} 1.4385	[3]
			ir	[2]
$Me_3SiN(CH_2CH_2OSiMe_3)_2$	B	74/2	d_4^{20} 0.8692, n_D^{20} 1.4290	[206]
Me_3SiN	C, B	95–6	d_4^{20} 0.7926, n_D^{26} 1.4082	[57, 314]
	E	83/760		[324]
Me_3SiN $SiEt_3$	I+ $(-N_2)$	84.5/15	n_D^{20} 1.4248	[102]
Me_3SiN	I+ $(-N_2)$	58–60/15	n_D^{20} 1.4443	[102]
Me_3SiN	B	114–6	ir, nmr	[314]
Me_3SiN	B	139–40	n_D^{20} 1.4297	[106]
Me_3SiN	A, B		uv, μ	[105, 230]
Me Me_3SiN Me			uv, μ	[230]
Me_3SiN	B	158–60	n_D^{20} 1.4403	[106]
Me_3SiN	D	43/70	n_D^{20} 1.4527, ir, nmr	[82]
Me_3SiN	D	49/6.5	n_D^{20} 1.4568, ir, nmr	[82]
Me_3SiN	D	181	n_D^{20} 1.4852, ir, nmr	[82]
Me_3SiN	D	57/6.7	n_D^{20} 1.4839, ir, nmr	[82]
Me_3SiN Me	D	70/6.1	n_D^{20} 1.4884, ir, nmr	[82]

Table 10 *(C'td.)*

Compound	Method	B.p. (°C/mm.) (m.p., °C)	Comments	References
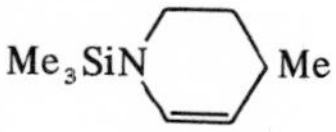 Me_3SiN (ring) Me	D	195	n_D^{20} 1.4639, ir, nmr	[82]
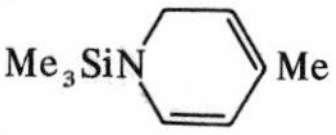 Me_3SiN (ring) Me	D	200	n_D^{20} 1.4883, ir, nmr	[82]
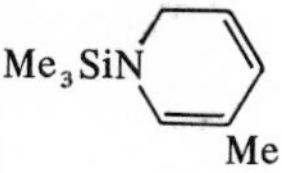 Me_3SiN (ring) Me	D	70/6.1	ir, nmr	[82]
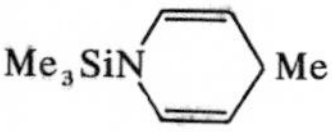 Me_3SiN (ring) Me	D	202	n_D^{20} 1.4825, ir, nmr	[82, 259c]
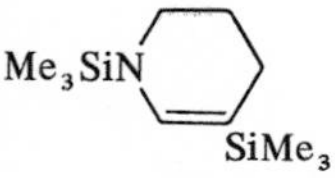 Me_3SiN (ring) $SiMe_3$	D	72/2.6	n_D^{20} 1.4714, ir, nmr	[82]
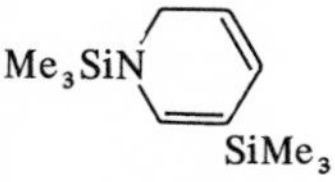 Me_3SiN (ring) $SiMe_3$	D	70/0.7	n_D^{20} 1.4772, ir, nmr	[82]
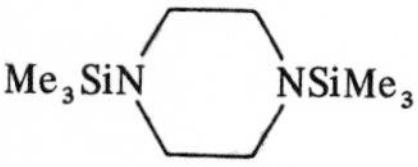 Me_3SiN (ring) $NSiMe_3$		217–8 (23.5–4)	$n_D^{21.5}$ 1.4499	[377]
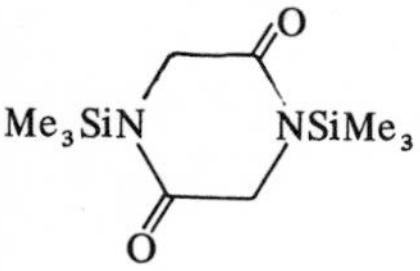 Me_3SiN (ring, O, O) $NSiMe_3$	B			[58]
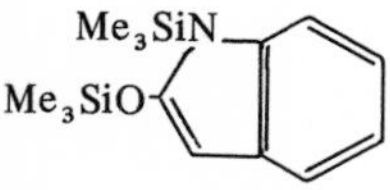 Me_3SiN, Me_3SiO (ring)	B	85–90/0.5		[58]
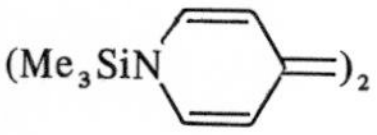 $(Me_3SiN$ (ring) $=)_2$	D	(167)	ir, nmr	[82]
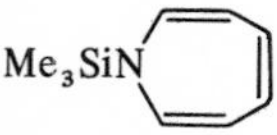 Me_3SiN (ring)	B	74–6/21	d_4^{20} 0.8547, n_D^{20} 1.4525	[207]
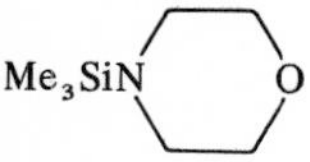 Me_3SiN (ring) O	B	61–2/18	d_4^{20} 0.9014, n_D^{20} 1.4385	[207]
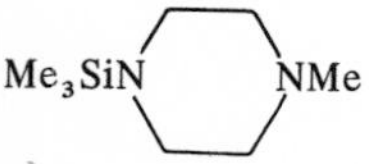 Me_3SiN (ring) NMe	B	65/35	d_4^{20} 0.8590, n_D^{20} 1.4461	[207]

Table 10 *(C'td.)*

Compound	Method	B.p. (°C/mm.) (m.p., °C)	Comments	References
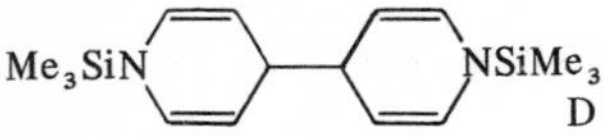 Me_3SiN … $NSiMe_3$	D	(38)	ir, nmr	[40, 41]
	D	(46–8)	ir, nmr	[281]
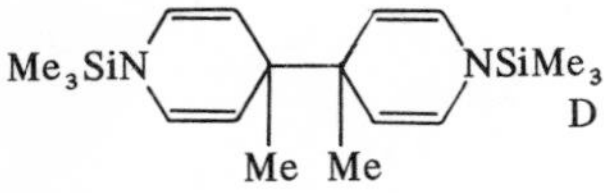 Me_3SiN … $NSiMe_3$ (Me Me)	D	(108)	ir, nmr	[40, 41, 281
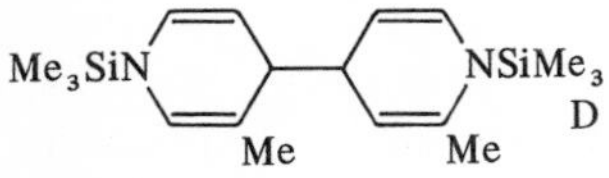 Me_3SiN … $NSiMe_3$ (Me, Me)	D	oil	ir, nmr	[40]
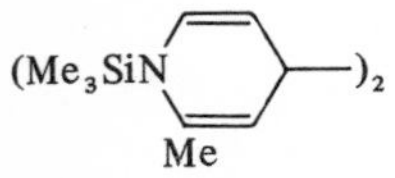 $(Me_3SiN$ … $)_2$ (Me)	D	$143–7/10^{-4}$	ir, nmr	[281]
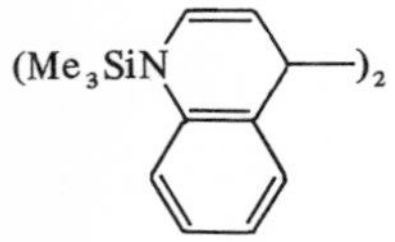 $(Me_3SiN$ … $)_2$	D	$183–7/10^{-4}$	ir, nmr	[281]
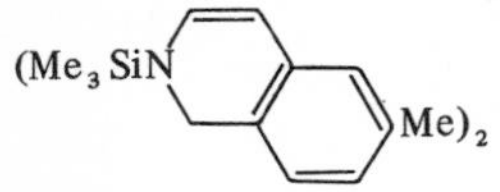 $(Me_3SiN$ … $Me)_2$	D	(162–4)	ir, nmr	[281]
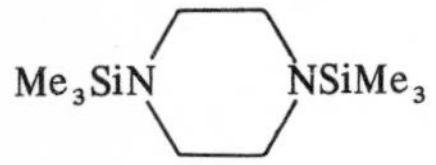 Me_3SiN … $NSiMe_3$	B	(68)	ir, nmr	[281]
Me_3SiNPh_2	A	113–5 (46–8)	ir	[347]
$Me_3SiN(CF_3)_2$	A	−2	ir, nmr, mass	[19, 240]
$Me_3SiCH_2SiMe_2NHBu^t$	A			[72]
$Me_2SiEtNHCH_2CH_2NHPh$	A	89–90/0.1	n_D^{20} 1.5166	[377]
$Me_2SiEtNHCH_2CH_2NPhSiMe_2Et$	A	91–2/0.1	n_D^{20} 1.4977	[377]
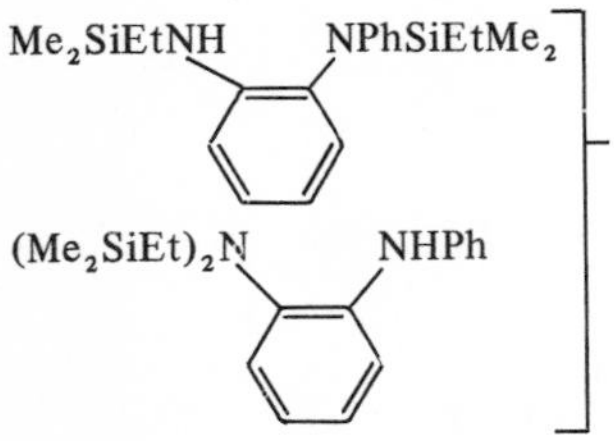 $Me_2SiEtNH$, $NPhSiEtMe_2$ (benzene ring)	A	157–62/10	ir, nmr	[310]
$(Me_2SiEt)_2N$, NHPh (benzene ring)				

Table 10 *(C'td.)*

Compound	Method	B.p. (°C/mm.) (m.p., °C)	Comments	References
$Me_2SiEtN\triangleleft$	E	117/760	n_D^{20} 1.4192	[324]
$(CH_2SiMe_2NEt_2)_2$	C			[131]
$CH_2(CH_2SiMe_2NMe_2)_2$	C			[131]
$Me_3SiN[(CH_2)_2NHR]_2$	B			
(R = H		80/2	ir, nmr	[129]
$SiMe_3$)		130/1	ir, nmr	[129]
$Me_2Si[N(CH_2)_2NHSiMe_3][(CH_2)_2NH_2]$	B	140/20	ir, nmr	[129]
HN–$SiMe_2$ (biphenyl-bridged ring)	A	152/6.5		[121a]
Me_2Si–NH–$SiMe_2$–NMe– (ring)	C			[238]
$PhNHCH_2SiMe_2NHPh$	C	~145/0.4	n_D^{20} 1.5831	[238]
MeN–$SiMe_2$–NMe– (ring)			nmr	[303]
Me_2Si–NMe–$SiMe_2$–NMe– (ring)			nmr	[303]
Me_2Si–NEt–$SiMe_2$–NEt– (ring)		51.2/3.5	$n_D^{21.5}$ 1.4561	[250]
PrN–$SiMe_2$–NPr– (ring)			nmr	[303]
Me_2Si–NPr–$SiMe_2$–NPr– (ring)			nmr	[303]
C_3H_5N–$SiMe_2$–NC_3H_5– (ring)			nmr	[303]

Table 10 *(C'td.)*

Compound	Method	B.p. (°C/mm.) (m.p., °C)	Comments	References
NC_3H_5 / Me_2Si $SiMe_2$ / NC_3H_5 (cyclic)			nmr	[303]
Me_2Si / $MeOCH_2CH_2N$ NCH_2CH_2OMe / $SiMe_2$ (cyclic)		115–7/4	$n_D^{21.5}$ 1.4605	[250]
Me_3SiO / N N / Me_3SiN N N / H H (purine ring)			mass	[379]
Me_3SiNH / N N / N N / $SiMe_3$ (purine ring)			mass	[379]
N N / N N / $SiMe_3$ (purine ring)			mass	[379]
$BrCH_2SiMe_2NHPh$	C	115–2.5	n_D^{20} 1.5572	[361]
$BrCH_2SiMe_2NEt_2$	C	76/10	n_D^{20} 1.4632, nmr	[361]
$XCH_2SiMe_2NMe_2$ [X = MgCl, ZnCl, $AlCl_2$, PbCl, $PtCl(C_2H_3)_2$]				[308]
$Bu\{CH_2CH[Me_2SiNMe_2]\}Li$ (and organic copolymers)				[308]
$C_5H_{11}CH(Li)SiMe_2NMe_2$				[308]
$Me_2Si(C_2H_3)NHMe$	C			[98]
$Me_2PhSiNHMe$	C	83/10	ir, nmr	[52]
$Me_2PhSiNHPh$	F	98–9/0.02	ir, nmr	[52]
$Me_2(n\text{-}Hex)SiNMe_2$	C	103–8/30	ir, nmr	[91]
$Me_2(n\text{-}Hex)SiNEt_2$	C	72/1	ir, nmr	[91]
$Me_2Bu^tSiNH_2$	C	75–9/200	ir, nmr	[50a]
$Me_2Bu^tSiNHMe$	C	87/180	ir, nmr	[50a]

Table 10 *(C'td.)*

Compound	Method	B.p. (°C/mm.) (m.p., °C)	Comments	References
$Me_2Bu^tSiNMe_2$			ion cyclotron	[294a]
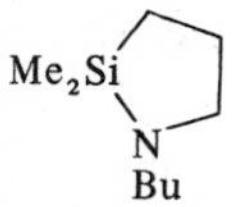 Me_2Si (ring: $Me_2Si(CH_2)_3N$Bu, pyrrolidine-type ring with N-Bu)	A, I	58/10	ir, nmr	[236]
$Me_2Si(C_2H_3)NMe_2$	C	107/748	n_D^{20} 1.4170	[91, 308]
$Me_2SiPhNHBu$		121/11	n_D^{20} 1.4929	[272]
$Me_2SiPhNMe_2$	C	52/2.3		[85]
$Me_2(p\text{-}Me_2SiHC_6H_4)SiNMe_2$	C	100/0.5	ir, nmr	[245]
$Me_2SiPhNEt_2$	C	83.5–8.5/2.5	d_4^{20} 0.9014, n_D^{20} 1.4937	[177]
$Me_2SiPhNBu_2$	F	72–4/0.04	d_4^{20} 0.902, n_D^{20} 1.4933	[206]
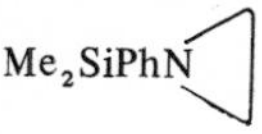 Me_2SiPhN (aziridine ring)	E	104/0.20		[324]
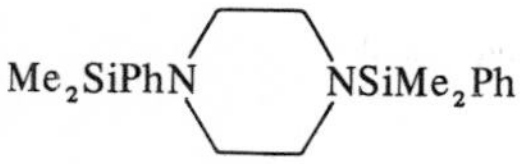 Me_2SiPhN (piperazine ring) $NSiMe_2Ph$	I	133–6/3.5	d_4^{20} 1.0045, n_D^{20} 1.5400	[231]
$Me_2Si(C_6H_4SiMe_2NHPr^i)NHPr^i$ 	C	110/0.2		[131]
$O(C_6H_4SiMe_2NMe_2)_2$ 	C			[131]
MeO_2CSiMe_2NHBu	A, C I, K	47.5–8/1	d_4^{20} 0.906, n_D^{20} 1.4192	[274]
$MeO_2CSiMe_2NHBu^i$	A, K	77/18	d_4^{20} 0.902, n_D^{20} 1.4179	[274]
$MeO_2CSiMe_2NHC_6H_{11}$	A, C I, K	85/3	d_4^{20} 0.970, n_D^{20} 1.4471	[274]
MeO_2CSiMe_2NHPh	A, K	108–10/3 (45–7)		[274]
$MeO_2CSiMe_2NEt_2$	A, K, I	39–41/2–3	d_4^{20} 0.915, n_D^{20} 1.4162	[274]
$MeO_2CSiMe_2NBu^i_2$	A	67/1	d_4^{20} 0.891, n_D^{20} 1.4300	[274]
$PhO_2CSiMe_2NHC_6H_{11}$	K	135–6/0.9	d_4^{20} 1.037, n_D^{20} 1.5062	[274]
$HCC(Me)RSiNR'R''$	C			[175a]
(R R'R''				
Me $(CH_2)_5$		72–3/22		
CH_2CH $(CH_2)_5$		78–9/11		
Me Et_2		64–5/54		
CH_2CH Et_2		80–1/42		
Me H,C_3H_5		70/68		
CH_2CH H,C_3H_5)		76–8/42		

Table 10 *(C'td.)*

Compound	Method	B.p. (°C/mm.) (m.p., °C)	Comments	References
$Me(PhCH_2)_2SiN\langle\rangle NSiMe(CH_2Ph)_2$ (piperazine ring)	I	215–20/115	d_4^{20} 1.383, n_D^{20} 1.5670	[274]
$MeSiPh_2NHBu$		135/0.5	n_D^{20} 1.5439	[274]
$MeSiPh_2NHPh$	F	200–15/0.5 (96)		[274]
$MeSiPh_2NEt_2$	C	145.5–6/2.5	d_4^{20} 0.9862, n_D^{20} 1.5497	[274]
$MeSiPh_2NBu_2$	F	117–8/0.02	d_4^{20} 0.963, n_D^{20} 1.5337	[126]
$MeSiPh_2NPh_2$				[117]
$MeSiPh_2N\langle$ (aziridine ring)	E	94/0.03		[324]
$MeSiPh_2N\langle\rangle NSiMePh_2$ (piperazine ring)	I	220/5 (121)		[231]
(+)1–NpMePhSiNH$_2$				[163]
(—)1–NpMePhSiNHMe			ir, nmr	[162,163]
	C	(63)	$[\alpha]_D^{20}$ −7.1	[272]
(+)	B	(63)	ir, nmr, $[\alpha]_D^{20}$ −3.7	[162]
(—)1–NpMePhSiNHBu	C	176/0.5	n_D^{20} 1.5985, $[\alpha]_D^{20}$ −4.4	[272]
(+)1–NpMePhSiNHBu^i	C		n_D^{20} 1.5970, ir	[305]
			$[\alpha]_D^{20}$ −5.9	[306]
(—)1–NpMePhSiNHC$_6$H$_{11}$	C	195/0.01	n_D^{20} 1.6105, $[\alpha]_D^{20}$ −4.95	[272]
(—)1–NpMePhSiNHCH$_2$Ph	C	215/0.01	n_D^{20} 1.6361, $[\alpha]_D^{20}$ −5.46	[272]
(—)1–NpMePhSiNEt$_2$		(55–7)	$[\alpha]_D^{20}$ −24.5	[163,163c, 254]
(+)1–NpMePhSi$\overline{N(CH_2)_4}$	C		n_D^{20} 1.5206, ir	[305]
$Et_3SiNHMe$	C	151–3	d_4^{20} 0.8011, n_D^{20} 1.4295	[10]
$Et_3SiNHEt$	C	166–7	d_4^{20} 0.7995, n_D^{20} 1.4300	[203]
	A, F	166–7	d_4^{20} 0.7995, n_D^{20} 1.4300	[77,163c]
$Et_3SiNHPr$	C	189.4	d_4^{20} 0.8038, n_D^{20} 1.4332	[93a]
$Et_3SiNHPr^i$	F	179.6	d_4^{20} 0.7962, n_D^{20} 1.4282	[93a]
$Et_3SiNHBu$	B	85–6/12		[196]
$Et_3SiNHBu^t$	I			[120]
	F	193	d_4^{20} 0.8082, n_D^{20} 1.4360	[93a]
$Et_3SiNHCMe_2Et$	F	212–4	d_4^{20} 0.8194, n_D^{20} 1.4400	[93a]
$Et_3SiNHHex$	B	117/15		[196,199]
$Et_3SiNHC_7H_{15}$	B	125/13		[196,199]
$Et_3SiNHCH_2CH_2Ph$	B			[199]
$Et_3SiNHCHMeCH_2Ph$		146/3		[199]
$Et_3SiNHCH_2Ph$	C	148/20		[200]
	B	133/9		[196,199]
	A	124/6	d_4^{20} 0.9130, n_D^{20} 1.4970	[50]
$Et_3SiNHCHMePh$	C	138/13		[200]
	B	142/14		[199]

Table 10 *(C'td.)*

Compound	Method	B.p. (°C/mm.) (m.p., °C)	Comments	References
$Et_3SiNHCHBu^iCO_2SiEt_3$	B	162/7		[48]
$Et_3SiNHC_6H_{11}$	B	112–3/10		[196]
	G	84–5/1.8	n_D^{20} 1.4609	[268]
Et_3SiNH–[ring with O]	C	89/2	d_4^{20} 0.9218, n_D^{20} 1.4656	[209]
Et_3SiNH–[ring with N, S]	B			[196]
$Et_3SiNHPh$	C	134–5/12	d_4^{20} 0.934	[8,200]
	F	134–5/12	d_4^{20} 0.934	[93a]
	B	134–5/12	d_4^{20} 0.934	[196]
$Et_3SiNHC_6H_4Me$-*o*	B	137/9		[196]
$Et_3SiNHC_6H_4Me$-*p*	B	143/12		[196, 200]
Et_3SiNMe_2	C	166–7	d_4^{20} 0.8044, n_D^{20} 1.4325	[160, 251]
Et_3SiNEt_2	C	199.5–201	d_4^{20} 0.8167, n_D^{20} 1.4400	[160, 251]
	F	199.5–201	d_4^{20} 0.8167, n_D^{20} 1.4400	[93a]
			ir, Raman	[276]
$Et_3SiNEtCH_2CH_2CN$	C, D	262–5/740	d_4^{20} 0.8914, n_D^{20} 1.4575	[219]
$Et_3SiNPr^i_2$	F			[93a]
Et_3SiNBu_2	F	253–4	d_4^{20} 0.8280, n_D^{20} 1.4475	[93a]
$Et_3SiNPhCH_2Ph$	F	206/25, 154/1.5	d_4^{20} 1.006, n_D^{20} 1.5630, uv	[119b]
Et_3SiN[three-membered ring]	C	89/60	d_4^{20} 0.8410, n_D^{26} 1.4425	[156]
	E	105/95		[324]
Et_3SiN[six-membered ring]$NSiEt_3$	I	146–8/4	d_4^{20} 0.8905, n_D^{20} 1.4740	[231]
[three-membered ring]$NCH_2CH_2SiEt_2N$[three-membered ring]	A + D	107/9	d_4^{20} 0.9103, n_D^{20} 1.4689	[232]
$Et_2Si(C_2H_3)N$[three-membered ring]	A	82/32	d_4^{20} 0.8519, n_D^{20} 1.4510	[232]
Et_2SiPhN[three-membered ring]	E	130/25	n_D^{20} 1.5095	[324]
Et_2SiPhN[six-membered ring]$NSiPhEt_2$	I	195–200/5	d_4^{20} 0.9999, n_D^{20} 1.5413	[231]

Table 10 *(C'td.)*

Compound	Method	B.p. (°C/mm.) (m.p., °C)	Comments	References
$EtPrPhSiNHC_6H_4Me$-*p*	C	260–5/70		[163b]
$PhCH_2EtPrSiNHC_6H_4Me$-*p*	C			[163c]
$Pr_3SiNHMe$	C	195–6		[197]
$Pr_3SiNHEt$	C	75–6/3		[197]
$Pr_3SiNHPr$	C	95/6		[197]
$Pr_3SiNHPr^i$	C	78–80/9		[197]
	I			[120]
$Pr_3SiNHBu$	A	115–6/11	d_4^{20} 0.8094, n_D^{20} 1.4410	[18]
$Pr_3SiNHCH_2CH_2Ph$	B	154–5/3		[197]
$Pr_3SiNHCH_2Ph$	B	147–8/6		[197]
$Pr_3SiNHC_6H_{11}$	A	143–4/11	d_4^{20} 0.8568, n_D^{20} 1.4610	[50]
	G	103–4/0.9	d_4^{20} 0.857, n_D^{20} 1.4621	[269]
$Pr_3SiNHPh$	B	142–3/5		[197]
	A	125/2	d_4^{20} 0.9094, n_D^{20} 1.5085	[50]
Pr_3SiNMe_2	C	58–60/3		[197]
Pr_3SiNEt_2	C	95–6/7		[197]
Pr_3SiN (aziridine ring)	E	85/5	n_D^{20} 1.4454	[324]
$Pr^i_3SiNHC_6H_4X$	A		nmr	[36]
(X = H				
o-Me		86–8/0.1	n_D^{20} 1.5178	
m-Me		100/0.18	n_D^{20} 1.5190	
p-Me		102/0.2	n_D^{20} 1.5120	
2,3-Me_2		106/0.3	n_D^{20} 1.5141	
2,6-Me_2		98/0.2	n_D^{20} 1.5167	
o-Et		92/0.2	n_D^{20} 1.5182	
p-Et		81/0.1	n_D^{20} 1.5171	
o-F		79/0.3	n_D^{20} 1.5078	
p-F		92/0.3	n_D^{20} 1.5041	
o-Cl		116/0.5	n_D^{20} 1.5195	
m-Cl		105/0.2	n_D^{20} 1.5233	
p-Cl		106/0.2	n_D^{20} 1.5233	
o-OMe		90/0.14	n_D^{20} 1.5235	
p-OMe)		104/0.25 (35–6)		
$Bu_3SiNHMe$	C	117–8/9		[194]
$Bu_3SiNHEt$	C	125/9		[194]
$Bu_3SiNHPr$	C	131–3/8		[194]
$Bu_3SiNHPr^i$	C	126–7/9		[194]
$Bu_3SiNHBu$	C	150–2/16		[194]
$Bu_3SiNHCH_2Ph$	C	148/1		[194]
$Bu_3SiNHPh$	B	190–2/19		[194]
Bu_3SiNEt_2	C			[194]
$Pent_3SiNHMe$	C	128–9/0.5	d_4^{20} 0.8195, n_D^{20} 1.4482	[322]

Table 10 *(C'td.)*

Compound	Method	B.p. (°C/mm.) (m.p., °C)	Comments	References
NCH$_2$CH$_2$SiPh$_2$N (aziridine rings)	A + D	150/1	d_4^{20} 1.0690, n_D^{20} 1.5755	[233]
(C$_2$H$_3$)SiPh$_2$N (aziridine ring)	A	127/1	d_4^{20} 1.0577, n_D^{20} 1.5785	[233]
Ph$_3$SiNHEt	C	163–5/0.5		[42]
Ph$_3$SiNHPr^i	C	(63)		[158]
Ph$_3$SiNHBu	F	138–9/0.02		[126]
Ph$_3$SiNHPh	B			[195]
	A	(134–5)		[347]
Ph$_3$SiNMe$_2$	F	(80–1)		[125]
Ph$_3$SiNEt$_2$	F	(84–5)		[125]
Ph$_3$SiNBu$_2$	C, F	(62)		[125]
	F	(64–5)		[126]
Ph$_3$SiN (piperidine ring)	F	(131–2)		[126]
Ph$_3$SiN NSiPh$_3$ (piperazine ring)	F	(287–8)		[126]
Ph$_3$SiN O (morpholine ring)	F	(179–81)		[126]
Ph$_3$SiNPh$_2$	A	(224–6)		[37, 81]
(C$_6$F$_5$)$_3$SiNMe$_2$	A	120/10^{-2} (124–30)		[186]
(o-MeC$_6$H$_4$)$_3$SiNBu$_2$	A	190/2 (123.5–5)		[127]
(p-Me$_2$NC$_6$H$_4$)$_3$SiNBu$_2$	F	(62–4)		[125]
(1-Np)$_3$SiNHPh		(198–200)		[80]
Me$_3$SiNMe(CH$_2$)$_2$NMeTeF$_5$	A			[121]
Me$_3$SiNMeTeF$_5$	A			[121]
[Me$_3$SiNHMe$_2$]$^+$[CoL$_4$]$^-$ (L = CO or PF$_3$)	D			[148]

† Footnote to Table 10: For explanation of abbreviations, see footnotes to Table 1, p. 57.

Table 11† Bis(amino)silanes $\rangle Si(N\langle)_2$

Compound	Method	B.p. (°C/mm.) (m.p., °C)	d_4^{20}	n_D^{20}	Comments	References
$H_2Si(NMe_2)_2$	C	93/760 (extr.)				
		(−105–3)	0.788		ir	[30]
					ir, mass	[25]
$H_2Si[N(SiH_3)_2]_2$	I				ir, nmr, mass	[278]
$H_2Si(NMeSiH_3)_2$	C				ir, nmr, mass	[7, 278]
$H_2Si(NEt_2)_2$					ir, mass	[25]
$H_2Si(NPh_2)_2$	C				ir	[27]
$(MeNHSiHMe)_2NMe$	C	48.5/5	0.8871	1.4425		[293]
$(MeNHSiHEt)_2NMe$	C	68.5/1.5	0.8929	1.4520		[293]
$HMeSi(NHPh)_2$	I					[115]
$HMeSi(NMe_2)_2$	C	112–3/767			ir, nmr	[371]
$HEtSi(NMe_2)_2$	C	48/30			ir, nmr	[91]
$HMeSi(NEt_2)_2$		73.8–5/18	0.8141	1.4317		[294]
$HEtSi(NEt_2)_2$		78.5/9–10	0.8287	1.4391		[294]
$HPhSi(NHPh)_2$	D					[76]
$H_3SiSiH(NMe_2)_2$					nmr	[328]
$Me_3SiNMeSiMe_2NH_2$	A	45–9/10			nmr	[278a]
$Me_2Si(NHMe)_2$	C	66/165				[201]
	C	108/750.5	0.8219	1.4140		[293]
$(MeHNSiMe_2)_2NH$	A	70/12	0.8833	1.4352		[348]
$Me_3SiNMeSiMe_2NHMe$	A	53–7/10			nmr	[280]
$(MeNHSiMe_2NH)_2SiMe_2$	B					[51]
$(MeNHSiMe_2)_2NMe$	C	34.5/0.5	0.8866	1.4435		[293]
	B, C	77–8/10	0.8907	1.4435		[342]
Et_2NSiMe_2NHMe	C	40–1/13	0.8139	1.4238		[362]
Pr_2NSiMe_2NHMe	C	77/13	0.8207	1.4949		[362]
$(MeNHSiMe_2)_2NBu$	C	67/1.0	0.8800	1.4483		[363]
$(MeNHSiMe_2NSiMe_2)_2$	C					[51]
$Me_2Si(NHEt)_2$	C	143–6	0.8460	1.4310		[18]
$(EtNHSiMe_2)_2NH$	A	46–8/1	0.8534	1.4330		[364]
$(EtNHSiMe_2)_2NMe$	B, C	54–5/0.5	0.8625	1.4415		[342]
$(EtNHSiMe_2)_2NEt$	C	82–3/1	0.8810	1.4417		[18]
$(EtNHSiMe_2)NSiMe_2NEt_2$	A	100–2/0.5	0.8941	1.4494		[366]
$(PrNHSiMe_2)_2NH$	A	74–5/1	0.8554	1.4372		[364]
$(PrNHSiMe_2)_2NMe$	B, C	71–2/0.5	0.8587	1.4443		[342]
$ClMe_2SiNPrSiMe_2NHPr$	C	53–5/0.1	0.9443	1.4509		[363]
$Me_2Si(NHPr^i)_2$	A	47–8/11	0.7983	1.4158		[364]
$(Pr^iNHSiMe_2)_2NMe$	B, C	52–3/0.2	0.8506	1.4400		[342]
$Me_2Si(NHBu)_2$	C	92–3/12	0.818	1.4311		[274]
	A	86–8/11	0.8259	1.4314		[364]
	C	70–1/7			ir, nmr	[91]
$(BuNHSiMe_2)_2NH$	A	91–2/2	0.8587	1.4401		[364]
$(BuNHSiMe_2)_2NMe$	B, C	86–7/0.5	0.8512	1.4469		[342]
$ClMe_2SiNBuSiMe_2NHBu$	C	69–72/0.2	0.9274	1.4515		[363]

Table 11 *(C'td.)*

Compound	Method	B.p. (°C/mm.) (m.p., °C)	d_4^{20}	n_D^{20}	Comments	References
$Me_2Si(NHBu^i)_2$	C	85–6/15	0.810	1.4280		[374]
$(PentNHSiMe_2)_2NMe$	B, C	112–4/0.8	0.8618	1.4494		[342]
$(HexNHSiMe_2)_2NMe$	B, C	129–30/0.5	0.8620	1.4509		[342]
$Me_2Si(NHC_7H_{15})_2$	B	169–70/12	0.8297	1.4425		[201]
$Me_2Si(NHCH_2CH_2OC_2H_3)_2$	B					[89]
Me_2Si ($NHCH_2CH_2O$)($NHCH_2CH_2O$) $SiMe_2$ (ring)	B	54/2	1.0318	1.4540		[206]
$Me_2Si(NHCH_2Ph)_2$	B	174–8/5	1.0090	1.5409		[201]
$(PhCH_2NHSiMe_2)_2NMe$	B, C	168–71/0.9	1.0167	1.5395		[342]
$Me_2Si(NHC_6H_{11})_2$	C	113–4/1.3	0.918	1.4763		[274]
$(C_6H_{11}NHSiMe_2)_2NH$	A	119/1	0.9321	1.4769		[364]
$(C_6H_{11}NHSiMe_2)_2NMe$	B, C	137–8/1.0	0.9298	1.4798		[342]
$Me_2Si(NH$–furan ring (O)$)_2$	C	120–1/2	1.0593	1.4971		[209]
$Me_2Si(NHPh)_2$	C	185–6/9 (45)	$1.07^{30°}$			[201]
	B	185–95/2 (63)				[16]
Me_2Si ring with N($SiMe_3$) and N(R) fused to benzene ring					ir, nmr	[310]
(R = H		130–4/4.2 (119–20)				
$SiMe_3$		(72–6)				
Me		107–11/0.4 (52–3.5)				
Ph)		159–61/0.8				
Me_2Si ring with N($SiMe_3$) and N(R) fused to benzene ring bearing X, Y					ir, nmr	[310]
(X = Cl, Y = H, R = H		(28–34)				
Cl, H, $SiMe_3$		133–4/0.5				
H, Cl, H						
Cl, H, Me)		138–40/1.4				
$Me_3SiNMeSiMe_2NHLi$						[360]

Table 11 *(C'td.)*

Compound	Method	B.p. (°C/mm.) (m.p., °C)	d_4^{20}	n_D^{20}	Comments	References
$Me_3SiNMeSiMe_2NHGeMe_3$						[360]
$Me_2Si(NMe_2)_2$	C	128.4 (−98)	$0.809^{22°}$			[9]
	C	124–7/760			ir, nmr	[371]
	I					[126]
					ir	[69]
$Me_2Si(NMe_2)_2/Me_2SiCl_2$					nmr	[328]
$(Me_2NSiMe_2)_2NMe$	B, C	85/10 44/0.3	0.8704	1.4441		[342]
$Me_2Si(NMeSiMe_2NHR)_2$	C					
(R = Et		82–4/0.5			ir, nmr	[343]
Pr		116–9/2			ir, nmr	[343]
Pr^i		62–72/0.05			ir, nmr	[343]
Bu		117–20/1			ir, nmr	[343]
Hex		138–42/2			ir, nmr	[343]
C_6H_{11})		139–42/1			ir, nmr	[343]
$Me_2Si(NMeSiMe_2NMeSiMe_3)_2$	A	118/0.2			ir, nmr	[123]
$Me_2Si[NMeB(NMe_2)_2]_2$	A	(44–6)			ir, nmr	[242]
$Me_2Si(NMeBRNMe_2)_2$	B					
(R = Me		96/11			ir, nmr	[242]
Bu)		76–9/0.01			ir, nmr	[242]
$Me_2Si(NHC_6F_5)_2$	A				ir, mass	[141]
$Me_2Si(NLiC_6F_5)_2$	E				ir, mass	[141]
$Me_2Si[NH(CH_2)_3Si(OEt)_3]_2$	B				ir, mass	[393]
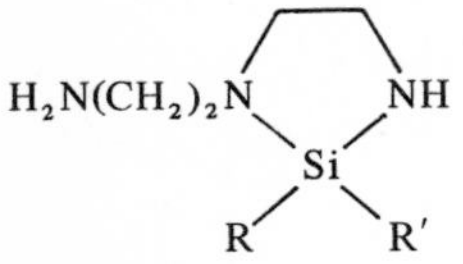 $H_2N(CH_2)_2N$ — NH (five-membered ring with CH_2CH_2), Si, R, R' (R R' Me Me H $CHCH_2$ H Ph)	B				ir, nmr	[129]
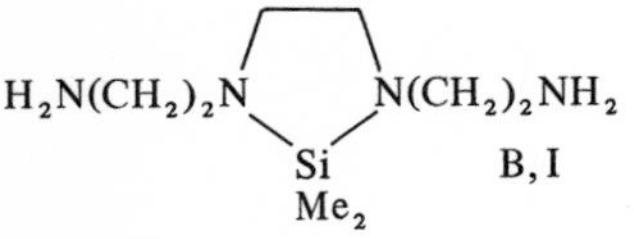 $H_2N(CH_2)_2N$ — $N(CH_2)_2NH_2$ (five-membered ring with CH_2CH_2), Si Me_2	B, I				ir, nmr	[129]

Table 11 *(C'td.)*

Compound	Method	B.p. (°C/mm.) (m.p., °C)	d_4^{20}	n_D^{20}	Comments	References
$SiMe_3N[(CH_2)_2NR]_2SiMeR'$ (cyclic: N(SiMe₃) linked by two $(CH_2)_2$ to RN and NR, both on Si(Me)(R'))	I				ir, nmr	[129]
(R R'						
Me Me		95/5				
Me $CHCH_2$		95/1				
Me Ph						
$SiMe_3$ Me)		100/1				
$Me_2Si(NHR)(N\langle SiMe_2NMe\rangle_2SiMe_2)$ (cyclic)						
(R = H	A	128–130/10			ir, nmr	[354]
Me)		138–140/10			ir, nmr	[354]
$(Me_2NSiMe_2)_3N$	C					[338]
$(Me_2NSiMe_2NMeCH_2)_2$	C	84–6			nmr	[286]
$(Et_2NSiMe_2NMeCH_2)_2$	C	112–4			nmr	[286]
$\text{-(}Me_2SiNMeCPh{:}N\text{)-}_2$	A	(77–9)			ir, nmr	[281]
$Me_2Si(NMeSiMe_3)_2$	A	63–6/1			nmr	[282, 285]
$Me_3SiNMeSiMe_2N(GeMe_3)SnMe_3$						[360]
$Me_3SiNMeSiMe_2NLiGeMe_3$						[360]
$Me_2Si(NMeSiMe_2Cl)_2$		127–8/10	1.0418	1.4643	ir, nmr	[226]
$Me_2Si(NEt_2)_2$	C	57/3–4	0.823	1.4352		[274]
$H_2NSiMe_2NEt_2$	A	38–40/12–3	0.8186	1.4228		[364]
$Me_3SiNHSiMe_2NEt_2$	A	75–8/12		1.4318		[366]
$PhOSiMe_2NHSiMe_2NEt_2$	A	98–9/2				[345]
$NaNHSiMe_2NEt_2$	A					[365]
(piperidino) $NSiMe_2NEt_2$	B	73/2	0.8839	1.4595		[208]
$(Me_3Si)_2NSiMe_2NEt_2$	A	98–100/1	0.8838	1.4600		[366]
$(EtOSiMe_2)_2NSiMe_2NEt_2$	A					[365]
$PhSiMe_2(N\langle SiMe_2, CH_2CH_2\rangle NSiMe_2)_nPh$ (cyclic)						
(n = 1, 2, 3)						[118]

Table 11 *(C'td.)*

Compound	Method	B.p. (°C/mm.) (m.p., °C)	d_4^{20}	n_D^{20}	Comments	Reference
$Me_2Si[N(SiMe_2Ph)]_2$ bridged by —Me— (ring: Me₂Si, two N(SiMe₂Ph) linked)						[118]
$Me_2Si[N(SiMe_2R)]_2$ cyclic (R = Ph, OSiMe₃)						[118]
$(Ph_2MeSiN{-}SiMePh{-}NCH_2)_2$ (cyclic)						[118]
$(Ph_3SiN{-}SiPh_2{-}NCH_2)_2$ (cyclic)						[118]
$Ph_2Si[N(SiMePh_2)]_2$ (cyclic, four-membered link)						[118]
$Ph_2Si[N(SiMePh_2)]_2$ (cyclic)						[118]
$Et_2NSiMe_2(N(SiMe_2)_2NSiMe_2)_nNEt_2$						
(n = 1	C	164/12 (7–8)				[116]
2	C	152/0.01 (−2)				[116]
3	C	224/0.01 (65)				[116]
4)	C	280/0.005 (91)				[116]
$Et_2NSiMe_2(N{-}SiMe_2{-}NSiMe_2)_nNEt_2$ (ring with CH_2CH_2)						
(n = 1	C	110–4/0.02		1.4632		[113]
2)	C	180/0.03		1.4748		[113]
$H_2NSiMe_2NPr_2$	A	66–8/11	0.8236	1.4310		[364]
$(Pr_2NSiMe_2)_2NH$	A	135–40/0.5	0.8659	1.4489		[364]

Table 11 *(C'td.)*

Compound	Method	B.p. (°C/mm.) (m.p., °C)	d_4^{20}	n_D^{20}	Comments	References
$Pr_2NSiMe_2NHSiMe_2NBu_2$	A	120–2/0.1	0.8580	1.4498		[366]
$ClMe_2SiNHSiMe_2NPr_2$	A	77–8/10				[366]
$H_2NSiMe_2NPr^i_2$	A	46–8/11	0.8311	1.4319		[364]
$H_2NSiMe_2NBu_2$	A	56/1.5	0.8199	1.4364		[364]
$Me_3SiNHSiMe_2NBu^i_2$	A	76–6.5/0.5	0.8352	1.4393		[366]
$Me_2Si(N\langle$aziridine ring$\rangle)_2$	C	99–100				[147]
	E	85/100		1.4405		[324]
$Me_2Si(NMe)_2$ (ring: Si–NMe–NMe)	B				nmr	[286]
	C	131/740				[387]
$Me_2Si(NMe)(NSiMe_3)$ (ring: Si–NMe–NSiMe₃)	I	80–4/22		1.4413		[377]
$Me_2Si(NR)_2$ (ring: Si–NR–NR)						
(R = Et	C	162–3/740		1.4408		[387]
Pr	C	81/8		1.4447		[387]
Ph	C	(124–6)				[387]
p-MeC_6H_4	C	(104–6)				[387]
p-$MeOC_6H_4$)	C	(169–72)				[387]
$Me_2Si(N\langle$piperidine ring$\rangle)_2$	B	91–2/3	0.9279	1.4760		[208]
$Me_2Si(N\langle$tetrahydropyridine ring$\rangle)_2$	B	117–20/2	0.9380	1.4860		[207]
$Me_2Si(N\langle$morpholine ring$\rangle O)_2$	B	106–10/4	1.0163	1.4743		[207]
$Me_2Si(N\langle\rangle N)_2SiMe_2$ (bicyclic: two N–N rings bridging Me_2Si and $SiMe_2$)	B				ir, nmr	[388]

Table 11 *(C'td.)*

Compound	Method	B.p. (°C/mm.) (m.p., °C)	d_4^{20}	n_D^{20}	Comments	References
$SiMe_2$ Me_3SiN $NSiMe_3$ (ring)	E	36/0.03				[225]
	I	119–20/23		1.4452		[377]
	A	47/1, 63/4		1.4438		[183]
					nmr	[182]
$SiMe_2$ $BuMe_2Si(N$ $NSiMe_2)_nBu$ (ring)						
(n = 1	C	108–9/0.03		1.4582		[113]
2	C	175–7/0.02		1.4710		[113]
3	C	220/0.03		1.4792		[113]
4)	C	290–3/0.03		1.4830		[113]
$SiMe_2$ $BuMe_2SiN$ $NSiMe_2Cl$ (ring)	C	90/0.5		1.4612		[113]
$SiMe_2$ $PhMe_2Si(N$ $NSiMe_2)_nPh$ (ring)						
(n = 1	E	145–6/0.03				[226]
	C	145–6/0.03				[113]
2	C	217/0.04				[113]
3)	C	265–72/0.06				[113]
$SiMe_2$ $PhMe_2SiN$ $NSiMe_2Ph$ (ring)	B				nmr	[114]
	E	155/0.05				[226]
$SiMe_2$ $Me_2Si(N$ $NSiMe_2Ph)_2$ Me (ring)	E	228–40/0.05				[226]
$SiMe_2$ RMe_2SiN $NSiMe_2R$ (ring)	C	172/0.01		1.5123		[113]
(R = m-$MeOC_6H_4$)						
$SiMe_2$ $ClMe_2Si(N$ $NSiMe_2)_nCl$ (ring)						

Table 11 *(C'td.)*

Compound	Method	B.p. (°C/mm.) (m.p., °C)	d_4^{20}	n_D^{20}	Comments	References
(n = 1	C	87/0.1				[113, 119]
2	C	148–50/0.05		1.4819		[113, 119]
3	C	225–7/0.1		1.4878		[113, 119]
4	C	281/0.05		1.4910		[113, 119]
5	C	304–8/0.06				[113, 119]
9)	C					[113]
$ClPh_2SiN$ (ring: $SiMe_2$, $NSiPh_2Cl$)	C	229/0.01				[113]
$MeSi(C_3H_5)(NHMe)_2$	C	155/752	0.8525	1.4450		[396]
$MeSi(C_3H_5)(NEt_2)_2$	C	79.5/5	0.8435	1.4550		[396]
$MeSi(C_2H_3)(NHMe)_2$	C	131/743.5	0.8423	1.4360		[396]
$[MeSi(C_2H_3)NHMe]_2NMe$	C	79.5/1.5	0.9100	1.4665		[396]
$MeSi(C_2H_3)(NMe_2)_2$	C	147/771			ir, nmr	[371]
	C	83/100		1.4337		[308]
$MeSi(C_2H_3)(NEt_2)_2$	C	76–7/5	0.8310	1.4480		[396]
MePhSi(NR–)(NR–) ring (R = Me	E	110/11				[226]
	C	110/11		1.5155		[387]
Et)	C	140–1/14		1.5075		[387]
$Me(Ph)Si[NH(SiMe_3)]_2$	C	112/4			ir, nmr	[135]
$MePhSi(NR^1R^2)_2$	C					
(R^1 R^2		(108–9)			ir, nmr	[52]
H Me					ir, nmr	[52]
H Ph		(71–4)			ir, nmr	[52]
Me Me)		108–9/11			ir, nmr	[52]
$MePhSi(NEt_2)_2$	C	152/14		1.4935		[387]
(MePhSi(N$SiMePh_2$–)(N–) ring, N–$CH_2)_2$	F	347/0.04				[111]
MePhSi(NPh–)(NPh–) ring	F	110/11		1.5213		[111]
($PhSiMe_2N$ (ring: SiMePh) $NCH_2)_2$	E	347/0.04				[226]
MePhSi(NR–)(NR–) ring						

Table 11 *(C'td.)*

Compound	Method	B.p. (°C/mm.) (m.p., °C)	d_4^{20}	n_D^{20}	Comments	References
(R = $SiMe_3$	E	88/0.03				[226]
					nmr	[111]
$SiMe_2Ph$	E	262/0.05				[226]
$SiMePh_2$)	F	235–6				[111]
MePhSi($NSiMePh_2$–)$_2$ (cyclic, bridged)	B				nmr	[114]
	E					[226]
MePhSi($NSiMePh_2$)$_2$–C_6H_4 (cyclic, o-phenylene bridged)	E	(232–4)				[226]
$Et_2Si(NHMe)_2$	C	156/745	0.8421	1.4330		[293]
$Me(p\text{-}Me_2HSiC_6H_4)Si(NMe_2)_2$	C	106/0.5			ir, nmr	[245]
$Me[CF_3(CH_2)_2]Si[NH(SiMe_3)]_2$	C	222			ir, nmr	[135]
$Et_2Si(NHEt)_2$	C	66–7/10	0.8280	1.4324		[322]
$Et_2Si(NHPr)_2$	C	95/11	0.8329	1.4384		[321]
$Et_2Si(NHPh)_2$	C	172–3/1 (58)				[8]
$Et_2Si(NEt_2)_2$	C	140–2/50	0.853	1.4485		[160]
(aziridinyl)NCH_2CH_2SiEt(N-aziridinyl)$_2$	A + D	112/10	0.9703	1.4784		[233]
$EtSi(C_3H_5)$(N-aziridinyl)$_2$	A	93/32	0.9270	1.4670		[233]
$Bu_2Si(NSiBu_3)_2$ (cyclic, bridged)	E	179/0.03				[226]
$BuSi(C_2H_3)(NHPh)_2$	C	149–50/1.5			ir, X-ray	[316]
$PhSi(C_2H_3)(NHPh)_2$	C	(87–8)			ir, X-ray	[316]
$Ph_2Si(NHMe)_2$	C	155–6/3				[195]
		103/0.009			ir, nmr	[52]
$Ph_2Si(NHEt)_2$		120–2/1				[42]
$Ph_2Si(NHPr)_2$	C	178–82/2				[195]
		123/0.009			ir, nmr	[52]
$Ph_2Si(NHHex)_2$	C	215/2				[195]
$Ph_2Si(NHCH_2Ph)_2$	C					[195]

Table 11 *(C'td.)*

Compound	Method	B.p. (°C/mm.) (m.p., °C)	d_4^{20}	n_D^{20}	Comments	References
$Ph_2Si(NHPh)_2$	C	413d (153)				[8]
	C	(160)				[158]
	C	(161–2)			ir, X-ray	[316]
$Ph_2Si(NMe_2)_2$	C	168/10		1.5568		[387]
	C	138–40/1.9			ir, nmr	[371]
$Ph_2Si(NHSiMe_3)_2$	C	146/3.5			ir, nmr	[135]
$(Ph_2Si$ (ring: $NSiPh_3$, N) $CH_2)_2$	F	(265–7)				[111]
Ph_2Si (ring: NR, NR)						
(R = Me	C	175/10		1.5683		[387]
Ph	E	(233)				[226]
	F	(233)				[111]
$SiMe_3$	E	135–6/0.005				[226]
$SiPh_3$	E	(315–20) (302–24)				[226]
$SiPh_2Cl)$	C	(220)				[113]
	C	(218–20)				[119]
$ClSiPh_2N$ (ring: $SiPh_2$) $NSiPh_2Cl$	B				nmr	[114]
$(PhCH_2)_2Si(N\lhd)_2$	C	168–70				[147]
$(PhCH_2)_2Si(N\lhd)_2$ (Me)	C	173/1				[147]
$1\text{-}Np_2Si(NHPh)_2$	C	(184–6)				[80]

† Footnote to Table 11: For explanation of abbreviations, see footnotes to Table 1, p. 57.

Table 12† Tris(amino)silanes RSi(N⟨)$_3$

Compound	Method	B.p. (°C/mm.) (m.p.,°C)	d_4^{20}	n_D^{20}	Comments	References
$HSi(NHEt)_3$	C	62–3/10				[320]
$HSi(NHBu^t)_3$	C					[319]
$HSi(NHPh)_3$	C	(114)d				[275]
$HSi(NMe_2)_3$	C	145–8/765			ir, nmr	[370]
	C	142/760 (extr.)				
		(−91–89)	0.850		ir, mass	[25, 30]
	C	62/45				[5, 24]
$HSi(NEt_2)_3$	C					[319]
HSi(N⟨pyrrole ring⟩)$_3$	A, F	220/25				[262]
$MeSi(NHMe)_3$	C	61/40	0.8942	1.4339		[320]
$MeSi(NHEt)_3$	C	62–3/10	0.8522	1.4300		[320]
$MeSi(NHPr)_3$	C	110/13	0.8523	1.4381		[321]
$MeSi(NHPr^i)(NHBu)NPr_2$	C	140–200/55				[130]
$MeSi(NHBu^t)_3$	C	96–7/12	0.8413	1.4370		[319]
(⟨ring: CH_2–$NSiMe_3$–$SiMeNHCH_2$–$NSiMe_3$–CH_2⟩)$_2$					nmr	[183]
$MeSi$(–$NSiMe_3CH_2CH_2NSiMe_3$–)(–NH–CH_2–CH_2–NH–)(–$NSiMe_3CH_2CH_2NSiMe_3$–)$SiMe$					nmr	[183]
$MeSi(NHPh)_3$		212/1	$1.14^{30°}$			[8]
$[MeSi(NHPh)_2]_2NPh$	I	(137–8)				[115]
$[MeSi(NHPh)NPh]_2$	I	(229–30)				[115]
$[MeSi(NHPh)(NPh)]_2Si$	I	(186–7)				[115]
$MeSi(NMe_2)_3$	C	161 (−11)	0.850	1.4324		[9]
					ir	[69]
	C	64.5–65/25			ir, nmr	[371]
$MeSi(NMe_2)_3/MeSiCl_3$					nmr	[328]
$MeSi(NEt_2)_3$	C	115–6/12	0.8672	1.4515		[319]
MeSi(N⟨piperidine ring⟩)$_2$NEt$_2$	B + C	103–4/2	0.8715	1.4650		[208]
MeSi(N⟨piperidine ring⟩)$_3$	C	161/5	0.9799	1.4980		[208]
MeSi(N⟨morpholine ring⟩O)$_3$	B	(109–12)				[207]

Table 12 *(C'td.)*

Compound	Method	B.p. (°C/mm.) (m.p., °C)	d_4^{20}	n_D^{20}	Comments	References
$EtSi(NHMe)_3$	C	51–2/10	0.8966	1.4410		[320]
$EtSi(NHEt)_3$	C	78–9/10	0.8581	1.4360		[320]
$EtSi(NHPr)_3$	C	114–5/10	0.8572	1.4429		[321]
$EtSi(NHBu)_3$	C	153/11	0.8541	1.4459		[321]
$EtSi(NHBu^t)_3$	C	115–6/14	0.8459	1.4415		[319]
$EtSi(NHCH_2Ph)_3$	C	250–2/4	1.0614	1.5715		[321]
$EtSi(NHPh)_3$	C	232–4/1				[8,161]
$EtSi(NMe_2)_3$	C	66–7/12			ir, nmr	[245]
$C_2H_3Si(NMe_2)_3$	C	69/8			ir, nmr	[245]
$PhSi(NMe_2)_3$	C	121/8			ir, nmr	[245]
p-$Me_2HSiC_6H_4Si(NMe_2)_3$	C	120/0.3			ir, nmr	[245]
$EtSi(NEt_2)_3$	C	133–4/14	0.8773	1.4595		[319]
$PrSi(NH_2)(NHBu^t)_2$	C	104/15	0.8539	1.4414		[317]
$PrSi(NH_2)(NEt_2)_2$	C	108–9/15	0.8684	1.4489		[317]
$PrSi(NHMe)_3$	C	101/50	0.8884	1.4427		[317]
$PrSi(NHEt)_3$	C	90/12	0.8545	1.4380		[161, 317]
$EtNHSiPr(NHBu^t)_2$	C	109–10/12	0.8471	1.4415		[317]
$PrSi(NHPr)_3$	C	120/11	0.8545	1.4436		[161, 317]
$PrSi(NHPr^i)_3$	C	101–2/12	0.8332	1.4339		[317]
$PrSi(NHPh)_3$	C	178–80/10^{-2}				[161]
$Bu^tNHSiPr(NHPr^i)_2$	C	105–7/11	0.8385	1.4375		[317]
$PhNMeSiPr(NHPr^i)_2$	C	153/10	0.9352	1.5030		[317]
$BuSi(NHPh)_3$	C	192–6/5 × 10^{-3}				[161]
$Et_2NSiPr(NHPr^i)_2$	C	111–2/13	0.8469	1.4412		[317]
$Pr^i_2NSiPr(NHPr^i)_2$	C	131–2/15	0.8564	1.4480		[317]
$(C_6H_{11})_2NSiPr(NHPr^i)_2$	C	188–9/8		1.494		[317]
$(Bu^tNH)_2SiPrNHPr^i$	C	118–9/14	0.8430	1.4404		[317]
$(PhNMe)_2SiPrNHPr^i$	C	204–5/10	1.0155	1.5608		[317]
$(Et_2N)_2SiPrNHPr^i$	C	125–6/15	0.8595	1.4494		[317]
$PrSi(NHBu)_3$	C	117–9/2	0.8527	1.4474		[317]
	B	117–9/2				[318]
$PrSi(NHBu^i)_3$	C	140–1/12	0.8466	1.4424		[317]
$PrSi(NHBu^s)_3$	C	130–1/12	0.8473	1.4436		[317]
$PrSi(NHBu^t)_3$	A, C	119–20/11	0.8452	1.4427		[317, 319]
$PrSi(NHPent)_3$	B	164–5/3	0.8516	1.4503		[318]
$PrSi(NHHex)_3$	B	182–3/2	0.8507	1.4527		[318]
$PrSi(NHCH_2Ph)_3$	B	127–9/2–3				[318]
$PrSi(NHC_6H_{11})_3$	C	187–9/2	0.9478	1.4914		[317]
	B					[318]
$PrSi(NHPh)_3$	C	235/3				[317]
	B					[318]
$PrSi(NMe_2)_3$	C	72–4/13	0.8695	1.4423		[317]
$PrSi(NMePh)_3$	C					[317]
$PrSi(NEt_2)_3$	A, C	136–7/10	0.8752	1.4599		[319]

Table 12 *(C'td.)*

Compound	Method	B.p. (°C/mm.) (m.p., °C)	d_4^{20}	n_D^{20}	Comments	References
$PrSi(N\bigcirc)_3$	C	172–5/2	0.9696	1.5007		[317]
$BuSi(NHCH_2CH_2NH_2)_3$	F	(242–3) d				[256]
$C_{18}H_{37}Si(NEt_2)_3$	C	211–6/0.16				[160]
$C_6H_{11}Si(NHPr)_3$	C	166–7/10	0.9011	1.4672		[321]
$[CH_2Si(NMe_2)_3]_2$	C					[85]
$C_2H_3Si(NHPh)_3$	C	(86–7)			ir, X-ray	[316]
$C_2H_3Si(NMe_2)_3$	C					[85]
	C	82/30		1.4447		[308]
$PhSi(NHEt)_3$	C	140/13				[161]
$PhSi(NHPr)_3$	C	169–70/12	0.9358	1.4942		[161, 321]
$PhSi(NHPent)_3$	C	$148–50/5 \times 10^{-2}$				[161]
$PhSi(NHPh)_3$	I					[115]
	C	(138)			ir, X-ray	[316]
$[PhSi(NHPh)NPh]_2$						
(cis)	I	(285–6)				[115]
(trans)	I	(220–1)				[115]
$PhSi(NMe_2)_3$	C	143–4/32			ir, nmr	[371]
	C	121/8.0				[245]
$PhSi(NHSiMe_3)_3$	C	97/1				[202a]
$1\text{-}NpSi(NHPh)_3$	C	(176–8)				[80]
$N_3Si(NMe_2)_3$	A	76/11		1.4494		[380]
$p\text{-}Me_2SiHC_6H_4Si(NMe_2)_3$	C	120/0.3			ir, nmr	[245]
$Me_2Si(N(SiMe_3))_2Si(N(SiMe_3))(CH(SiMe_3))SiMe_2$ (spiro, two four-membered rings)					X-ray	[227]
$Si_2(NMe_2)_6$	C	subl. $70–80/10^{-4}$				[379a]

† Footnote to Table 12: For explanation of abbreviations, see footnotes to Table 1, p. 57.

Table 13† Tetrakis(amino)silanes Si(N⟨)$_4$

Compound	Method	B.p. (°C/mm.) (m.p., °C)	Comments	References
$(MeNH)_2Si(NMe_2)_2$	A	62–3/11	d_4^{20} 0.9055, n_D^{20} 1.4430	[368]
$Si(NHEt)_4$	C			[203]
$Si(NHBu^t)_4$		127–31/12 (45–6)		[60]
$Si(NHPh)_4$	C	(136)		[260]
$Si(NHC_6H_4Me\text{-}o)_4$	C			[260]
$Si(NHC_6H_4Me\text{-}p)_4$	C	(131–2)		[260]
Si(NHNp-1)$_4$	C			[260]
Si(NHNp-2)$_4$	C			[260]
$Si(NMe_2)_4$	B+C			[57]
	C	74–5/19 (−2)	d_4^{22} 0.972	[9]
	C	196/760 (extr.) (16–8)	d_4^{20} 0.910	[30]
			ir	[69]
			pes	[134]
$/SiCl_4$			nmr	[224, 328a]
			ΔH_f	[186a]
$/SiCl_4/Si(OMe)_4$			nmr	[328a, 332]
$HN[Si(NMe_2)_3]_2$	C	248		[361]
	C	284 (−20)	n_D^{20} 1.4617	[367]
$NaN[Si(NMe_2)_3]_2$		(~160)		[367]
$[HNSi(NMe_2)_2]_3$	C	(94)		[361]
	B	(92–3)		[367]
$Si(NEt_2)_4$	A	156–8/14 (3–4)	d_4^{20} 0.9026, n_D^{20} 1.4670	[58]
$Et_2NSi(N\ \ )_3$	C	100/10		[147]
Si(NSiMe$_3$ / NSiMe$_3$)$_2$			nmr	[183]
Si(N)$_4$	C	181–2/2 (81–2)		[57]
Si(N)$_4$	A, F	(173.4)		[261, 262]
Si(MeN—SiMe$_2$ / MeN—SiMe$_2$)$_2$	A	(107)	ir, nmr	[358]

Table 13 *(C'td.)*

```
        Me        Me
Me2Si—N          N—SiMe2
 |      \       /      \
 |       Si            Y   A
 |      /       \      /
Me2Si—N          N—SiMe2
        Me        Me
```

Compound	Method	B.p. (°C/mm.) (m.p., °C)	Comments	References
(Y = NMe		99/0.3	ir, nmr	[358]
O)		85/0.1	ir, nmr	[358]

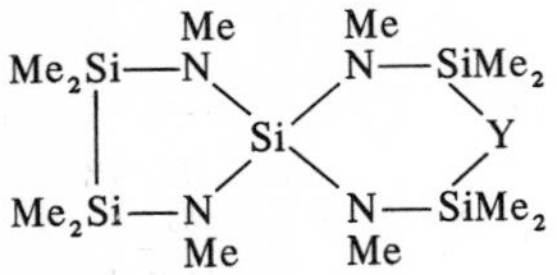

```
        Me        Me        Me        Me
Me2Si—N          N         N         N—SiMe2
 |      \       / \       / \       /     |
 |       Si        Si        Si           |
 |      /       \ /       \ /       \     |
Me2Si—N          N         N         N—SiMe2
        Me        Me        Me        Me
```

Compound	Method	B.p. (°C/mm.) (m.p., °C)	Comments	References
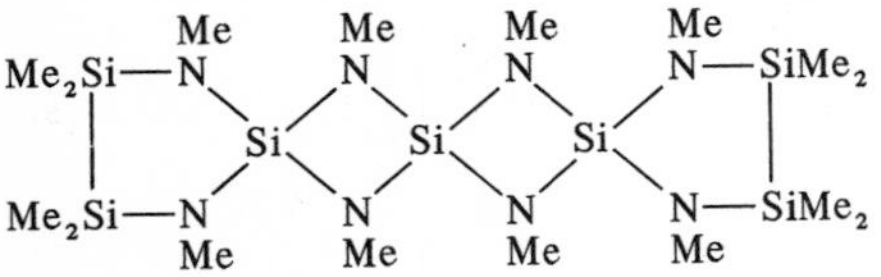 (structure above)		(129)	ir, nmr	[358]

† Footnote to Table 13: For explanation of abbreviations, see footnotes to Table 1, p. 57.

Table 14† Disilylamines (or disilazanes)

Compound	Method	B.p. (°C/mm.) (m.p., °C)	Comments	References
$(H_3Si)_2NH$	C	36 (extrap)	ir, nmr	[29]
			ed	[259b, 264]
$(Me_3Si)_2NH$	C	125–6	ir, Raman, nmr, ed	[124, 132, 244, 264]
			μ	[81a, 174a, 204]
			ΔH_f	[33, 34]
$(Et_3Si)_2NH$	C	100/1		[178]
$[(R)(CH_2)_3Si]_2NH$	C			
(R = Me		91–2/14	ir, nmr, μ	[174a, 236, 237]
Et)			ir, nmr, μ	[174a, 236]
$[Me(CH_2)_4Si]_2NH$	C	111–2/6	ir, nmr	[234]
$(Ph_3Si)_2NH$	C	(175)	ir, nmr	[264]
$[(H)R_2Si]_2NH$	C			
(R = Me			ir, nmr	[180]
Et		102–3/50	ir, nmr	[298]
Ph)			ir, nmr	[216]
$(EtMe_2Si)_2NH$	C	174–5	ir, nmr	[62a, 299]
$(CH_2CHMe_2Si)_2NH$	C		μ	[174a]
$(Et_2MeSi)_2NH$	C	112–4/40	ir, nmr	[299]
$(PrMe_2Si)_2NH$	C	67/5	ir, nmr, μ	[174a, 236
$Me_3Si(ClMe_2Si)NH$	C	70.5–2/32		[302]
$[(MeO)Me_2Si]_2NH$	C	157–8/690		[302]

Table 14 *(C'ntd.)*

Compound	Method	B.p. (°C/mm.) (m.p.,°C)	Comments	References
$[(CH_2CHCH_2)Me_2Si]_2NH$	C		μ	[174a]
$(PhMe_2Si)_2NH$	C	128–9/2	ir, nmr, μ	[174a, 236, 300, 302, 392]
$(Ph_2MeSi)_2NH$	C	(90–1)	ir, nmr	[14]
$(PhEtMeSi)_2NH$	C	169/4	ir, nmr	[14]
$(PhPrMeSi)_2NH$	C	182/6	ir, nmr	[15]
$(PhBuMeSi)_2NH$	C	192/3	ir, nmr	[15]
$(R_3SiCH_2SiR_2)_2NH$	C		ir, nmr	[235]
(R = Me		93/2		
Et		188/2		
Pr		214/2.5		
Bu)		256/3.5		
$(ClCH_2Me_2Si)_2NH$	C	78–9/26	μ	[174a, 302]
$[Et_3Si(CH_2)_3Et_2Si]_2NH$	C	250/2	ir, nmr	[234]
$[Me_3Si(CH_2)_3PhMeSi]_2NH$	C	220/2	ir, nmr	[234]
$(PhMe_2Si)(Me_3Si)NH$	C	79–80/1	ir, nmr	[14, 302, 392]
$(Ph_2MeSi)(Me_3Si)NH$	C	173/3.5	ir, nmr	[14, 392]
$(Et_3Si)(Me_3Si)NH$	C	194–6	ir, nmr	[392]
$(Bu_3Si)(Me_3Si)NH$	C	109–12/5	ir, nmr	[392]
$(Ph_3Si)(Me_3Si)NH$	C	186–7/2	ir, nmr	[392]
$(PhPrMeSi)(Me_3Si)NH$	C	115/2	ir, nmr	[15]
$(PhBuMeSi)(Me_3Si)NH$	C	128/3	ir, nmr	[15]
$[(CF_3CH_2CH_2)_3Si](Me_3Si)NH$	C	41/0.2	ir, nmr	[135]
$[(HC{\equiv}C)R'RSi]_2NH$				
(R R'				
Me Me		74–6/28	ir, nmr	[175]
H Me		73–5/25	ir, nmr	[175]
Me Et)		75–7/9	ir, nmr	[175]
$[MeR(X)Si]_2NH$	C			[53]
(R X				
$C_3H_4F_3$ Cl		72–4/0.3	ir, nmr	[53]
Ph Cl		158–60/0.2	ir, nmr	[53]
C_2H_3 Cl		120–2/18	ir, nmr	[53]
C_2H_5 Cl		88–90/3	ir, nmr	[53]
CH_3 Cl		96–8/48	ir, nmr	[53, 86]
$C_3H_4F_3$ NH_2		82–4/0.7	ir, nmr	[53]
$C_3H_4F_3$ NMe_2		104–7/0.3	ir, nmr	[53]
Ph NMe_2)		168–9/0.4	ir, nmr	[53]
$(Me_3Si)[(Me_2N)_2MeSi]NH$	C	67/5	ir, nmr	[348]
$[(Et_3SiO)_2Si(NH_2)]_2NH$	A	165–6/1		[12]

† Footnote to Table 14: For explanation of abbreviations, see footnotes to Table 1, p. 57.

Table 15† *N*-Substituted disilylamines ($(\equiv Si)_2NR$ (or disilazanes)

Compound	Method	B.p. (°C/mm.) (m.p.,°C)	d_4^{20}	n_D^{20}	Comments	References
$(H_3Si)_2NMe$	C	(−124.6–4.1)				[312]
^{14}N; ^{1}H, ^{2}H. ^{15}N, ^{1}H					ir, Raman	[75, 94]
					ed	[128]
$(H_3Si)_2NMe.BH_3$		d(0)				[211]
$.AlMe_3$						[211]
$.GaMe_3$						[211]
$(H_3Si)_2NEt$	C					[101]
$H_3SiNEtSiH_2Et$					ir, mass	[3]
$(H_3Si)_2NPh$	C	189 (−69)	0.801		ir, mass	[10]
$(MeSiH_2)_2NMe$	C	80 (−115)				[95]
$[Me_2(H)Si]_2N(CH_2)_6N[Si(H)Me_2]_2$	C	129/2			ir, nmr	[333]
$[Me_2(H)Si]_2NCH_2C_6H_4CH_2N[Si(H)Me_2]_2$	C	148/2			ir, nmr	[333]
$[Me_2(H)Si]_2NC_6H_4(CH_2)C_6H_4N(H)Si(H)Me_2$	C	170/1.5			ir, nmr	[333]
$[Me_2(H)Si]_2N[C_6H_3(Me)]N(H)Si(H)Me_2$	C	113/1			ir, nmr	[333]
$(H_3SiSiH_2)_2NMe$					nmr	[327]
$(Me_3Si)_2NMe$	C	148	$0.794^{25°}$	1.4190		[239, 244, 277, 285]
	C	148/760	0.7980	1.4220		[268]
	C	145	0.810	1.4180		[270]
					ir, μ	[2, 81a]
					ir, Raman	[132]
					ΔH_f	[33, 34]
$MeN(SiMe_2Ph)SiMePh_2$	C	143-5/0.02				[52]
$(Me_2PhSi)_2NMe$	C	105-6/0.02			ir, nmr	[52]
$H_2NSiMe_2NMeSiMe_3$	C					[280]
$Me_2NNHSiMe_2NMeSiMe_3$	A	69–72/10			nmr	[280]
$N_3SiMe_2NMeSiMe_3$	A	70–3/10			nmr	[280]
$Me_3SiNMeSiMe_2Cl$	A	52–8/10			nmr	[280]
	I	60–2/25				[301]
$Me_3SiNMeSiCl_3$	I	75–7/27				[137]
$(Me_3Si)_2NEt$	C	164		1.4289		[3]
	A					[31]
	B					[149]
	C	84/50	0.8061	1.4283		[268]
					ir	[2]
$(Me_3SiNEt)_2SiMe_2$						[71]
$(Me_3Si)_2NPr$	B					[149]
	C	69/14	0.8084	1.4310		[268]
$Me_3SiNPrSiMe_2Ph$	F	194/11		1.5359		[111]
$(Me_3Si)_2NPr^i$	B					[149]

Table 15 *(C'ntd.)*

Compound	Method	B.p. (°C/mm.) (m.p.,°C)	d_4^{20}	n_D^{20}	Comments	References
$(Me_3Si)_2NBu$	B					[149]
	C	40/0.6	0.8120	1.4342		[268]
$(Me_3Si)_2NBu^t$	A	75/10				[73]
	B					[149]
$(Me_3Si)_2NCH_2Ph$	C	148/760	0.7980	1.4220		[268]
$(Me_3Si)_2NCH{=}CHR$	D					[81a]
(R = H		48–9/8				
Et)		82–6/20				
$XCH_2CH_2N(SiMe_3)_2$	D				ir, nmr	[253]
(X = Cl		46–7/2.2				
Br		40–1/0.25				
I		57–9/0.3				
Et_2N)		68–9/0.21				
$(EtO)_3SiCH_2CH_2CH_2N(SiMe_3)_2$	G				nmr	[326]
$(Me_3Si)_2NCH_2CH_2NH_2$	K	107–8/24		1.4488		[377]
$(Me_3Si)_2NCH_2CH_2NHMe$	K	113/24		1.4450		[377]
$(Me_3Si)_2NCH_2CH_2NHPh$	I	113–4/0.3		1.5120		[377]
$(Me_3Si)_2NCH_2CH_2NMePh$	A	134–7/1		1.5127		[377]
$[(Me_3Si)_2NCH_2]_2$	B	145–6/12 (49–50)				[149]
	A, C	106–8/1 (51–2)			ir, nmr	[183]
		118–9/8		1.438		[250]
	B				nmr	[114,183]
$[(Me_3Si)_2NCH_2CH_2CH_2]_2$	B	134–6/1			ir, nmr	[332]
$(Me_3Si)_2NCMe{:}CHCN$	A				nmr	[181]
$(Me_3Si)_2NCPh{:}NSiMe_3$	A	101/0.2 (50–50.5)			ir, nmr, mass	[188]
$(Me_3Si)_2NCPh{:}NLi$	D				ir	[188]
$(Me_3Si)_2N(CH_2)_3Me_2Si$–(2,5-thienylene)–$SiMe_2(CH_2)_3N(SiMe_3)_2$	B	187–9/1.5			ir, nmr	[331]
$(Me_3Si)_2NCH_2CH_2OSiMe_3$	B	80/4	0.8551	1.4345		[206]
	B	101/12		1.4360		[149]
$(Me_3Si)_2N(CH_2)_3SiH_2Ph$		184/25	$0.896^{25°}$	$1.4900^{25°}$		[309]
$(Me_3Si)_2NC_3H_5$		179	$0.820^{25°}$	$1.4363^{25°}$		[309]
$(Me_3Si)_2NCH_2CO_2SiMe_3$	B	108–9/12		1.4346		[149]
$(Me_3Si)_2NCHMeCO_2Et$	A	110/111		1.4423		[273]
$(Me_3Si)_2NCHMeCO_2SiMe_3$	A					[273]
$(Me_3Si)_2NCHEtCO_2Et$	A	115–7/12		1.4460		[273]
$(Me_3Si)_2NCHEtCO_2SiMe_3$	A					[273]
$(Me_3Si)_2NCH(CH_2Ph)CO_2Et$	A	179–81/10		1.4952		[273]

Table 15 *(C'ntd.)*

Compound	Method	B.p. (°C/mm.) (m.p.,°C)	d_4^{20}	n_D^{20}	Comments	References
$(Me_3Si)_2NCH(CH_2Ph)CO_2SiMe_3$	A					[273]
$(Me_3Si)_2NCH(CH_2SiHMe_2)CO_2Et$	A	133–4/10		1.4518		[273]
$(Me_3Si)_2NCH(CH_2SiHMe_2)CO_2SiMe_3$	A					[273]
$(Me_3Si)_2NCH{:}C(OEt)OSiMe_3$	A	118/9		1.4430		[273]
$(Me_3Si)_2NCH{:}C(OSiMe_3)_2$	A	122–6/10		1.4397		[273]
$(Me_3Si)_2NCH(CO_2Et)CHPhOSiMe_3$	A	115/0.5		1.4703		[273]
$(Me_3Si)_2NC_6H_{11}$	B					[149]
$(Me_3Si)_2NPh$	C	40–1/0.3 (16)	0.888	1.4846		[3]
	B					[149]
	A	88–9/11 (16–17)	0.8951	1.4855	ir	[347]
					uv	[254]
$(Me_3Si)_2NC_7H_6R$					ir, nmr	[212]
(R = Me		92/2.5				
Pr^i		115–7/2.5				
C_2H_3		74/3				
$SiHMe_2$		82–4/3				
Cl		125/7				
OMe)		113/3				
$Me_3SiNPhSiMe_2Ph$						[118]
$Me_3SiNPhSiCl_3$	A	114/9 (15)	$1.1980^{24°}$	1.5080		[347]
p-$(Me_3Si)_2NC_6H_4CO_2SiMe_3$	I	129–31/1				[63]
$(Me_3Si)_2N$–C_6H_3(MeO)–CO_2SiMe_3	I	136–7/1				[63]
o-$ClC_6H_4N(SiMe_3)_2$		81/1.25		1.4990		[335]
p-$BrC_6H_4N(SiMe_3)_2$	A	86–7/1				[63]
		106/1.2		1.4514		[335]
$(Me_3Si)_2N$–C_6H_3(MeO)–Br	A	99–100/1				[63]

Table 15 *(C'ntd.)*

Compound	Method	B.p. (°C/mm.) (m.p.,°C)	d_4^{20}	n_D^{20}	Comments	References
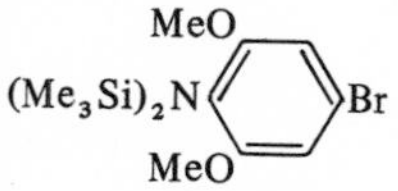 $(Me_3Si)_2N$–C_6H_2(MeO)$_2$–Br (2,6-di-MeO, 4-Br ring)	A	112–3/1 (73)				[63]
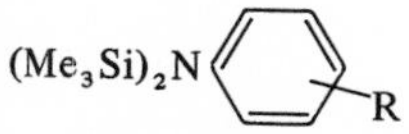 $(Me_3Si)_2N$–C_6H_4–R						
(R = *m*-Me_3Si	A	81/0.4		1.4780		[336]
p-Me_3Si	A	72–4/35		1.4830		[336]
p-Me_3Ge	A	111/0.35		1.4996		[336]
m-Cl	A	81/1.25		1.4990		[336]
p-Br	A	106/1.2		1.5140		[336]
o-NHPh	A, B	157–9/10			ir, nmr	[310]
o-NMePh)	A	(40–2.5)			ir, nmr	[310]
$(Me_3Si)_2NC_6H_4X$-*p*	B				ir, nmr	[383]
[X = Me		98–9/6				
Et		82–5/0.33				
Ph		125–7/0.25				
CN		106–8/0.01				
F		55–7/0.15				
Cl		85–7/1.2				
Br		84/0.3				
I		103–4/0.8				
OMe		79–81/0.25				
OAc		90–5/0.2				
$OSiMe_3$		99–100/0.4				
$N(SiMe_3)_2$]		112–6/0.25				
$(Me_3Si)_2NC_6F_5$	A					[140]
o-$(Me_3Si)_2NC_6H_4SiMe_3$	A					[335]
p-$(Me_3Si)_2NC_6H_4SiMe_3$	A	72–4/0.35		1.4830		[335]
p-$(Me_3Si)_2NC_6H_4GeEt_3$	A					[335]
p-$(Me_3Si)_2NC_6H_4Li$	A					[335]
$(Me_3Si)_2NCN$	A	35–7/10^{-3}			ir, nmr	[379b]
$(Me_2EtSi)_2NCH_2CH_2NHPh$	I	130–3/0.3		1.5151		[377]
$(C_2H_3SiMe_2)_2NMe$	C	116/100		1.4490		[308]
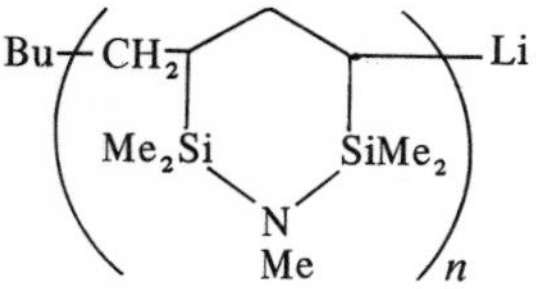 Bu–[CH_2–CH(–Me_2Si–N(Me)–$SiMe_2$–)–CH–]$_n$–Li						[308]
$(BrCH_2Me_2Si)_2NMe$	C	120/3.5		1.5052		[232]

Table 15 *(C'ntd.)*

Compound	Method	B.p. (°C/mm.) (m.p.,°C)	d_4^{20}	n_D^{20}	Comments	References
RN(–$SiMe_2$–NR–$SiMe_2$–) (cyclic)	C					[238]
(R = Me						
Pr		118/13		1.4572		
C_3H_5)		113/15		1.4740		
$[(Me_2SiPh)_2NCH_2]_2CH_2$	B				nmr	[114]
$(Me_2SiPh)_2NPh$						[118]
Cl_2Si(–$NMeSiMe_2$–O–$SiMe_2NMe$–) (cyclic)	A	89/10 (38)			ir, nmr	[355]
Br_2Si(–$NMeSiMe_2$–O–$SiMe_2NMe$–) (cyclic)	A	63/0.2 (45-6)			ir, nmr	[355]
$[(MeSiPh_2)_2N]_2(CH_2)_4$	B				nmr	[114]
$MeSiPh_2NPhSiBu_3$	F	203/0.05		1.5572		[111]
$(MeSiPh_2)_2NPh$						[118]
$(Et_3Si)_2NEt$	I	135-7/5	0.8620	1.4625		[232]
$(Hex_3Si)_2NPh$	F	218-20/0.02		1.4793		[111]
$(Ph_3Si)_2NPh$						[118]
Ph_2Si–$SiPh_2$–NR–$SiPh_2$–$SiPh_2$ (cyclic)	C					[145]
(R = Me		(237-41)				
Et)		(226-8)				

† Footnote to Table 15: For explanation of abbreviations, see footnotes to Table 1, p. 57.

Table 16† Trisilylamines [($\equiv$Si)$_3$N]

Compound	Method	B.p. (°C/mm.)	(m.p.,°C)	Comments	References
$N(SiH_3)_3$				ir, nmr	[311]
				ed	[37b,144]
$N(SiH_3)(SiMe_3)_2$	A	156/734		ir, nmr	[337]
$N(SiMe_3)_3$	A	35/13		ir, nmr	[204, 296, 353]
				μ,	[174a]
				ΔH_f	[33, 34]
$N[\overline{Si(CH_2)_3}(Me)]_3$	C	111-3/2		ir, nmr, μ	[174a,178]
$N(SiMe_2Et)_3$	C			ir, nmr	[296]
$N(SiMe_3)_2[\overline{Si(CH_2)_3}(Me)]$	C	155-7/2		ir, nmr	[237]
$N(SiMe_3)_2(SiMe_2H)$	C	84/26		ir, nmr	[70]
$N(SiMe_3)_2(SiMe_2Ph)$	C	85/4		ir, nmr	[70]
$N(SiMe_3)(SiMe_2H)_2$	C	60/40		ir, nmr	[70]
$N(SiMe_2H)_3$	C	153/760		ir, nmr	[70]
$N(SiMeH_2)_3$	C	(−107)		ir, nmr	[369]
$N(SiMe_3)_2(SiPr_3)$	A	102-3/1		ir, nmr	[353]
$N(SiMe_3)(SiEt_3)(SiPr_3)$	C	130-35/1		ir, nmr	[353]
$N(SiMe_3)_2(SiF_3)$	A	162/760		ir, nmr	[337]
$N(SiMe_3)_2(SiCl_3)$	A	102-3/13		ir, nmr	[204, 337, 353]
$N(SiMe_3)_2(SiBr_3)$	A	110/5	(85.9)	ir, nmr	[337]
$N(SiMe_3)_2(SiI_3)$	A	118/6		ir, nmr	[337]
$N(SiMe_3)_2[Si(NCO)_3]$	A	81/1		ir, nmr	[337]
$N[SiMe(OMe)_2]_2(SiMeCl_2)$	A	78-9/2		ir, nmr	[342a]
$N(SiMe_2OEt)_2(SiMe_2Cl)$	A	108/10		ir, nmr	[342a]
$N[SiMe(OMe)_2]_3$	A	89/3		ir, nmr	[342a]
$N(SiMe_2OEt)_3$	A	111/1		ir, nmr	[342a]
$N[Si(CHCH_2)(OMe)_2]_3$	A	102/2		ir, nmr	[342a]
$N[SiMe(CHCH_2)OPr^i]_3$	A	108/1		ir, nmr	[342a]

† Footnote to Table 16: For explanation of abbreviations, see footnotes to Table 1, p. 57.

Table 17† Halogeno(amino)silanes

Compound	Method	B.p. (°C/mm.) (m.p.,°C)	d_4^{20}	n_D^{20}	Comments	References
$HSi(Me)FN(SiMe_3)_2$	A	61/15			ir, nmr	[165]
$HSi(Ph)FN(SiMe_3)_2$	A	(72)			ir, nmr	[165]
$HSi(Me)ClNMe_2$	C	85–7/767			ir, nmr	[371]
$Me_2Si(F)NMe_2$	C	78.5				[136]
		(−86)			ir, nmr	[229]
$Me_2SiFN(SiMe_3)_2$	A	58/12			ir, nmr	[279]
$RR'SiFN(H)Bu^t$	A					
(R R'						
Me Ph		68/0.01			ir, nmr	[166]
Ph Ph		117/0.01			ir, nmr	[166]
Bu Ph)		62/0.01			ir, nmr	[166]
$Bu^t_2Si(F)NHBu^t$	A				ir, nmr	[167]
$Bu^t_2Si(F)NLiBu^t$	E	(125)d				[167]
$Ph_2Si(F)N(SiMe_3)_2$	A	96/10			ir, nmr	[165]
$RSi[N(SiMe_3)_2](F)OR'$	A				ir, nmr	[165]
(R R'						
H Me		81/11				
H Et		86/17				
H Pr		137/20				
$CHCH_2$ Me		48/0.1				
$CHCH_2$ Et		53/0.1				
$CHCH_2$ Pr		88/0.01				
$CHCH_2$ Pr^i		$44/10^{-3}$				
$CHCH_2$ Ph		$71/10^{-3}$				
Et Me		91/14				
Et Et		149/11				
Ph Me		74/0.01				
Ph Et)		74/0.01				
$Me_2SiClNHPr$	C	39–41/12–13	0.9205	1.4250		[362]
$Me_2SiClNHPr^i$	C	25–6/11	0.9026	1.4198		[362]
$Me_2SiClNHBu$	I	65–6/18		1.4298		[274]
	C	54–5/11	0.9182	1.4290		[362]
$Me_2SiClNHBu^i$	C	57/11–2	0.9092	1.4282		[362]
$Me_2SiClNHBu^t$	A	54/20			ir, nmr	[166]
$Me_2Si[NMe(SiMe_2Cl)]_2$		127–8/10			ir, nmr	[339]
$Me_2Si[NH(SiMe_2F)]_2$		73/15			ir, nmr	[339]
$Me_2SiClNHC_6H_{11}$	C	87/12	0.9815	1.4624		[362]
		(−3)				
$Me_2SiClNHPh$	C	75–6/1	1.0912	1.5402		[362]
$Me_2SiClNMe_2$	A					[241]
	I					[137]
	C	105/765			ir, nmr	[371]
$R'N(SiR_2X)(SiR''_2X)$	C				ir, nmr	[348]
(X R R' R''						
Cl Me Me Me		68–70/12				
Cl Me Et Me		72–4/8				

Table 17 *(C'ntd.)*

Compound	Method	B.p (°C/mm.) (m.p.,°C)	d_4^{20}	n_D^{20}	Comments	References
Cl Me Bu Me		40–2/0.01				
Cl Me Et Ph)		134/0.1				
$RN(SiMe_2Cl)SiMe_2(NR'R'')$						
	C				ir, nmr	[350]
(R R' R''						
Me H Et		75/5				
Me H Pr		60/2				
Me H Pr^i		77/5				
Me H Bu		84/5				
Me H Bu^t		66/2				
Me H C_6H_{11}		82/1				
Me H Hex		77/6.5				
Me Me Me		74/10				
Me Et Et		56/1				
Et H Et		74/4				
Et H Pr		64/2				
Et H Pr^i		57/1.5				
Et H Bu		59/1				
Et H Bu^t		56/1				
Et H C_6H_{11}		75/0.5				
Et H Hex		81/0.5				
Et Me Me		77/10				
Et Et Et)		62/1				
$(Me_2SiClNMeCH_2)_2$	C	66–8/10			nmr	[286]
$(Me_2SiCl)_2NMe$	C	68–70/12	1.0526	1.4511		[363]
	I	30–5/5				[301]
					ir, Raman	[67]
$Me_2SiClNEt_2$	C	52/18	0.912	1.4299		[274]
	C	44.5/13	0.8945	1.4300		[362]
$Me_2SiClNPr_2$	C	40–5/13	0.8945	1.4300		[362]
$Me_2SiClNPr^i_2$	C	61–3/9	0.9143	1.4389		[362]
$Me_2SiClNBu_2$	C	62/1.5	0.9002	1.4410		[362]
$(Me_2SiCl)_2NBu$	C	40–2/0.01	1.0177	1.4540		[363]
$Me_2SiClNBu^i_2$	C	69/2–3		1.4432		[274]
	C	85–7/11	0.8957	1.4411		[362]
$MeSiCl(C_2H_3)NMe_2$	C	128–30/765			ir, nmr	[371]
$Et_2SiClNEt_2$	C	86–8/17				[160]
$Ph_2SiFNEt_2$	C	120–1/3	1.0533	1.5352	MR_D18.90	[79]
$MeBu^tSiFNHPh$	A	68/0.01			ir, nmr, mass	[169a]
$Ph_2SiClNMe_2$	C	137/1.0			ir, nmr	[371]
$HSiF_2N(SiMe_3)_2$	A	103/760			ir, nmr	[168]
$MeSiF_2N(SiMe_3)_2$	A	154/760			ir, nmr	[165]
$EtSiF_2N(SiMe_3)_2$	A	73/11			ir, nmr	[168]
$Pr^iSiF_2N(SiMe_3)_2$	A	80/10			ir, nmr	[168]
$PhSiF_2N(SiMe_3)_2$	A	61/11			ir, nmr	[165]

Table 17 *(C'ntd.)*

Compound	Method	B.p. (°C/mm.) (m.p.,°C)	d_4^{20}	n_D^{20}	Comments	Reference
$CH_2CHSiF_2N(SiMe_3)_2$	A	38/110			ir, nmr	[165]
$PhSiF_2N(H)Bu^t$	A	56/0.1			ir, nmr	[166]
$Bu^tSiF_2NBu^tSiMe_3$	A	54/0.01			ir, nmr	[166]
$PhSiF_2NBu^tSiMe_3$	A	65/0.01			ir, nmr	[166]
$RF_2SiNR'SiMe_3$	A				ir, nmr,	
(R R'					mass	[170]
Ph C_6H_4Me-4		$84/5 \times 10^{-2}$				
Ph $C_6H_2Me_3$-2,4,6		$92/10^{-2}$				
Ph $C_6H_4NMe_2$-4		$125/10^{-2}$				
Bu^i $C_6H_4NMe_2$-4		$85/10^{-2}$				
$HSiCl_2NMe_2$	C	49–50/768			ir, nmr	[370]
$MeSiCl_2NMe_2$	I					[137]
	C	52–3/45			ir, nmr	[371]
$PrSiCl_2NHPr^i$	C					[317]
$PrSiCl_2NHBu^t$	C	73–4/11				[317]
$PrSiCl_2NMePh$	C	125/11				[317]
$PrSiCl_2NEt_2$	C	86–8/17				[317]
$PrSiCl_2NPr^i_2$	C	105/15	1.0192	1.4557		[317]
$PrSiCl_2N(C_6H_{11})_2$	C	180–1/10		1.499		[317]
$PrSiBr_2NPr^i_2$	C	114–5/10				[317]
$PhSiF_2NHPr$	K					[323]
$PhSiCl_2NMe_2$	C	118–21/25			ir, nmr	[371]
$PhSiCl_2N$–$SiMe_2$–$NSiPhCl_2$ (ring)	C	178–81/0.1				[113]
$HSi[N(SiMe_3)_2](F)NRR'$	C				ir, nmr	[164]
(R R'						
H Me		90/20				
Me Me		74/10				
Et Et)		70/11				
$RSi(F)[N(SiMe_3)_2]NMe(SiMe_3)$					ir, nmr	[168]
(R = H	A	45/0.05				
Me		55/0.01				
Et		85/0.05				
Pr		87/0.01				
$CHCH_2$		69/0.05				
Ph)		103/0.01				
$RSi(F)(NR''R''')NR'SiMe_3$	A				ir, nmr, mass	[170]
(R R'	R''		R'''			
Ph C_6H_4Me-4	Bu^t		$SnMe_3$			
		$125/10^{-2}$				
Ph Bu^t	C_6H_4Me-4		$SiMe_3$			
		$115/10^{-2}$				
Ph Bu^t	C_6H_4Me-4		SiF_2Ph			
		$172/5 \times 10^{-2}$				

Table 17 *(C'ntd.)*

Compound	Method	B.p. (°C/mm.) (m.p.,°C)	d_4^{20}	n_D^{20}	Comments	References
Ph $C_6H_2Me_2$-2,4,6 Bu^t SnMe$_3$		155/10^{-2}				
Ph $C_6H_4NMe_2$-4 Bu^t SnMe$_3$		175/10^{-2}				
Bu^i $C_6H_4NMe_2$-4 Bu^t SiF_2Ph)		165/10^{-2}				
$HSiCl(NHBu^t)_2$	C	74–5/12	0.9201	1.4379		[319]
$HSiCl(NMe_2)_2$	C	88–90/768			ir, nmr	[370]
$HSiCl(NEt_2)_2$	C	77–8/12	0.9354	1.4433		[319]
$MeSiCl(NHBu^t)_2$	C	81–2/12	0.9259	1.4419		[319]
$MeSiCl(NMe_2)_2$	I					[137]
	C	56–7/25			ir, nmr	[371]
$MeSiCl(NEt_2)_2$	C	84–5/12	0.9343	1.4453		[319]
$(Cl_3Si)_2NPh$	A	113–4 (29–30)	$1.4715^{23°}$	1.5234	ir	[347]
Br_3SiNMe_2	C	89–90/0.65 (–15)	1.61–3			[287]
$Br_3SiNMePh$		140–5/12				[54]
$Br_3SiNEtPh$		150–4/13				[54]
Br_3SiNPh_2	A	170–2 (38–9)	$1,6645^{21°}$		ir	[347]
$F_2Si(NHMe)_2$	B					[5]
$F_2Si(NMe_2)_2$	C	96 (–69.5)				[22,136]
		(–59)				[4]
$SiF_2N(SiMe_3)_2N(H)Me$	C	81/14			ir, nmr	[164]
$SiF_2N(SiMe_3)_2NMe_2$	C	80/14			ir, nmr	[164]
$SiF_2N(SiMe_3)_2NMe(SiMe_3)$	A	45/0.01			ir, nmr	[168]
$SiF_2N(SiMe_3)_2NRR'$	C					
(R R'						
Bu^i Bu^i		64/0.01			ir, nmr	[164]
H Bu^t		49/0.01			ir, nmr	[164]
Bu^t $SiMe_3$		65/0.01				[164]
$Cl_2Si(NHBu^t)_2$	C	98–101/14 (–9–6)	1.0358	1.4484		[59,62]
$Cl_2Si(NHPh)_2$	C					[142,143]
$Cl_2Si(NHC_6H_4Me\text{-}o)_2$	C					[142,143]
$Cl_2Si(NHC_6H_4Me\text{-}p)_2$	C				ir	[184]
$Cl_2Si(NHNp\text{-}1)_2$	C					[142]
$Cl_2Si(NHNp\text{-}2)_2$	C					[142]
$Cl_2Si(NMe_2)_2$	C	59–61/20				[77]
	I	59–61/20				[137]
					ir	[69]
					pes	[134]
$Cl_2Si(NEt_2)_2$	C	99–104/11	1.0324	1.5414		[57]

Table 17 *(C'ntd.)*

Compound	Method	B.p. (°C/mm.) (m.p.,°C)	d_4^{20}	n_D^{20}	Comments	References
	C	101–4/14		1.4507	ir	[55]
$Cl_2Si(NEt_2)NPr_2$	C	126–8/16		1.4539		[55]
$Et_2NSiCl_2NPrCH{:}CHMe$	C	113–7/11		1.4692		[56]
	C				ir	[55]
$Cl_2Si(NEtPh)_2$	C	185–7/9				[61]
$Cl_2Si(NPr^i_2)_2$	C	125–6/8 (32–40)				[61]
$Cl_2Si(NPr_2)_2$	B, C	191–2/14		1.4560		[55]
$MeClSi<(NSiMe_3)_2$ (cyclic, NSiMe₃ groups linked)	A	53/1 (28–30)				[183]
					nmr	[182]
$EtSiCl(NHBu^t)_2$	C	95–6/12	0.9280	1.4465		[319]
$EtSiCl(NEt_2)_2$	C	100–1/12	0.9395	1.4517		[319]
		90–1/5–6	0.9379	1.4575		[294]
$PrSiCl(NHBu^t)_2$		118–9/15	0.9153	1.4468		[317]
$PrSiCl(NMePh)_2$	C	195–7/11		1.5675		[319]
$PrSiCl(NEt_2)_2$	C	121/17	0.9323	1.4530		[319]
$PhSiCl(NMe_2)_2$	C	127–9/23			ir, nmr	[371]
$F_3SiNHBu^t$	C, I	(−80)			ir, nmr	[4, 24]
F_3SiNMe_2	C	21				[24, 136]
	A	21.7			ir, nmr	[228]
		(−11.7)			ed	[4b]
F_3SiNR_2 (NR_2 = NEt_2, NPr_2, NBu_2, NC_4H_8, NC_5H_{10})	I				ir, nmr	[4, 23]
$(F_3Si)_2NMe$	C	35/760 (−88)				[6, 7, 22]
$(F_3Si)_2NEt$	C	48 (−80)				[6, 7]
$(F_3Si)_2NPr$						[23]
$(F_3Si)_2NBu$						[23]
$(F_3Si)_2NBu^s$	C, I	(−120)				[23, 24]
$Cl_3SiNHEt$	C					[217]
$Cl_3SiNHBu^t$	C	48–51/12 (−51–48)				[59, 62]
Cl_3SiNMe_2	I					[35, 137]
					pes	[4b, 66, 134]
					ir, ed	[69]
Cl_3SiNEt_2	C	55–60/17				[57]
		104/80				
	C	45–6/7			ir	[55]
$Cl_3SiNEtPh$	C	100–5/9				[61]
$Cl_3SiNPr^i_2$	C	73–5/13				[61]
$Cl_3SiNBu^i_2$	C	120–4/30				[217]

Table 17 *(C'ntd.)*

Compound	Method	B.p. (°C/mm.) (m.p.,°C)	d_4^{20}	n_D^{20}	Comments	References
Cl_3SiN (piperidino ring)	C	101–6/17				[57]
Cl_3SiNPh_2	A	125–7 (26–8)			ir	[347]
$Cl_2Si(NSiMe_3)_2$ (ring via $NSiMe_3$–$NSiMe_3$)	A	95/4 (82–3)				[183]
					nmr	[182]
Cl_2Si(NSiMe₃)(N) linked to ClSi(NSiMe₃)(NSiMe₃) (bicyclic)	C					[183]
$Cl_2Si(N$ (piperidino ring)$)_2$	C	150–5/17				[57]
$Br_2Si(NMe_2)_2$	C					[289]
$FSi(NMe_2)_3$	C	144 (−68.5)				[136]
$ClSi(NHBu^t)_3$	C	138–41/34 (17–18)	0.9309	1.4460		[59, 62]
$ClSi(NMe_2)_3$	C	62–3/12	0.9741	1.4423		[57]
					ir, pes	[69, 134]
					ed	[330]
$ClSi(NEt_2)_3$	C	131–6/13	0.9516	1.4568		[57]
$ClSi(N$ (piperidino ring)$)_3$	C	155–8/3				[57]
$BrSi(NMe_2)_3$	C	113–3.5/0.65 (26)	1.23–4			[289]
$XN(SiMe_3)_2$	A				ir, nmr, μ	[382]
(X = Cl		149				
Br		176				
I)		196				
$XN(SiMe_3)Bu^t$	A				ir, nmr, μ	[382]
(X = Cl		157				
Br		180				
$Bu^tN(SiMe_3)SiF_2R$	A				ir, nmr, mass	[169]
(R = PhMeN		90/0.01				
p-$Me_2NC_6H_4$		134/0.01				

† Footnote to Table 17: For explanation of abbreviations, see footnotes to Table 1, p. 57.

Table 18† Alkoxy(amino)silanes

Compound	Method	B.p. (°C/mm.) (m.p., °C)	d_4^{20}	n_D^{20}	Comments	References
$Me_2Si(OC_2H_3)NEt_2$		72–4/18	0.8624	1.4310		[32]
$Me_2Si(OPh)NHMe$	C	95/12	0.9692	1.4896		[344]
$Me_2Si(OPh)NHEt$	C, G	55–6/1	0.9628	1.4848		[290]
$Me_2Si(OPh)NHPr$	C, G	69–71/1	0.9464	1.4825		[290]
$Me_2Si(OPh)NHBu$	C, G	81–3/1	0.9281	1.4829		[290]
$Me_2Si(OPh)NHBu^i$	C, G	74–5/1	0.9359	1.4797		[290]
$Me_2Si(OPh)NHC_6H_{11}$	C, G	96–7/0.4	0.9907	1.5039		[290]
$(Me_2SiOPh)_2NMe$	C	125/1	1.039	1.5202		[344]
$Me_2Si(OPh)NEt_2$	C, G	61–3/1	0.9351	1.4811		[290]
$Me_2Si(OPh)NPr_2$	C, G	95–8/2	0.9420	1.4819		[290]
$Me_2Si(OPh)NPr^i_2$	A	95–7/1				[345]
o-$Et_2NCOC_6H_4CO_2SiMe_2NEt_2$						[243]
$Me_2N(SiMe_2O)_4SiMe_2NMe_2$	C					[85]
$Me_3SiOSiMe_2(N\langle SiMe_2 \rangle NSiMe_2)_nOSiMe_3$ (ring: N–SiMe₂–N bridged)						[113]
($n = 1$	C	80/0.03		1.4295		
2	C	153/0.02		1.4505		
3	C	205–6/0.02		1.4622		
4	C	255/0.01		1.4697		
5)	C	280–300/0.02				
$MePh_2SiOSiMe_2(N\langle SiMe_2 \rangle NSiMe_2)_nOSiMePh_2$ (ring: N–SiMe₂–N bridged)						[113]
($n = 1$	C	226/0.02		1.5420		
2)	C	275/0.02		1.5325		
$MeNH(SiMe_2O)_nSiMe_2NHMe$						
($n = 1$	C	145.5–7/760	0.8817	1.4131		[17]
	C	67–8/20		1.4138		[53a]
2	C	44–5/0.5–1	0.9054	1.4108		[17]
	C	97–8/20		1.4102		[53a]
3	C	71.5–2.5/1–2	0.9205	1.4092		[17]
	C	114–5/9		1.4100		[53a]
4	C	105.5–7.5/2	0.9333	1.4092		[17]
5	C	118–21/0.5–1	0.9425	1.4090		[17]
6	C	143–7/0.5–1	0.9429	1.4088		[17]
7	C	158–60/0.5–1	0.9453	1.4086		[17]
8	C	180.5/0.5–1	0.9488	1.4083		[17]
9)	C	201–3/0.5–1	0.9514	1.4082		[17]
$R_3SiNRSiR_2OSiR_3$	I					
(R = Me, Et)					nmr	[374]
$Me_2EtCOSiH(Me)NEt_2$	K				ir	[138]

able 18 *(C'ntd.)*

Compound	Method	B.p. (°C/mm.) (m.p., °C)	d_4^{20}	n_D^{20}	Comments	References
$I_2C{=}C(R)OSi(H)(Me)NEt_2$ (R = Me, $CH{=}CH_2$)	K				ir	[138]
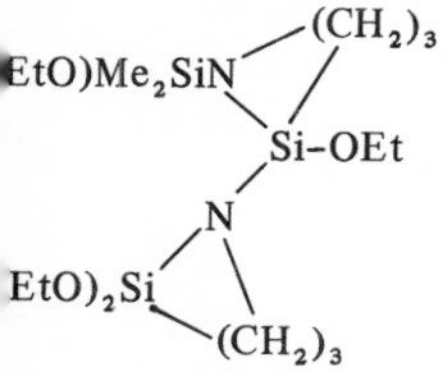 $EtO)Me_2SiN$ / $(CH_2)_3$ / $Si{-}OEt$ / N / $EtO)_2Si$ / $(CH_2)_3$	B+G				nmr, mass	[23]
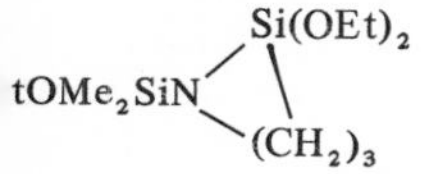 $tOMe_2SiN$ / $Si(OEt)_2$ / $(CH_2)_3$	B+G	70–3/5			nmr, mass	[23]
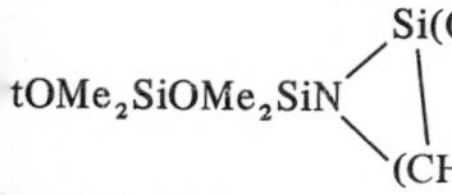 $tOMe_2SiOMe_2SiN$ / $Si(OEt)_2$ / $(CH_2)_3$	B+G				mass	[23]
$:t_3SiO)_2Si(NH_2)_2$	A	116–7/2				[12]
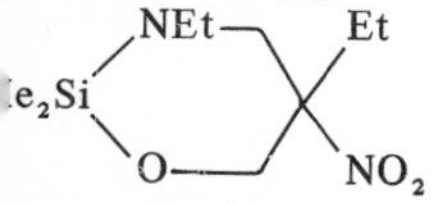 e_2Si / NEt / Et / O / NO_2					nmr, μ	[315]
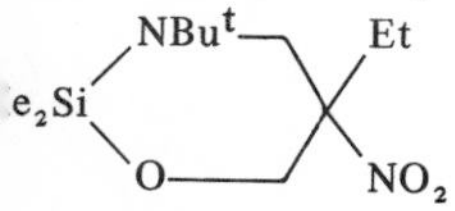 e_2Si / NBu^t / Et / O / NO_2					nmr, μ	[315]
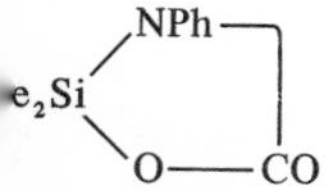 e_2Si / NPh / O / CO	B+K	(84–6)			nmr	[173]
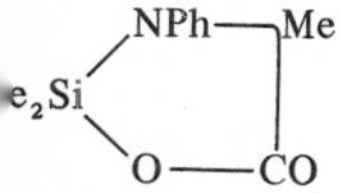 e_2Si / NPh / Me / O / CO	B+K	(78–81)			nmr, $[\alpha]_D^{25}$	[173]
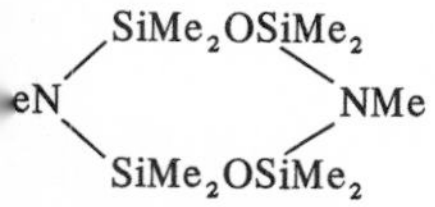 eN / $SiMe_2OSiMe_2$ / NMe / $SiMe_2OSiMe_2$	C	87–8/5 (27–8)		1.4321		[53a]
	C	232.5–4/760	0.9463	1.4314		[17]
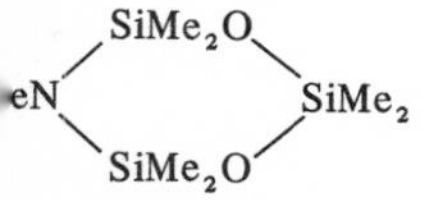 eN / $SiMe_2O$ / $SiMe_2$ / $SiMe_2O$	C	163–4/760	0.9254	1.4100		[17]
	C	60–1/17		1.4100		[53a]
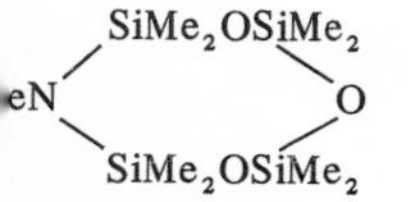 eN / $SiMe_2OSiMe_2$ / O / $SiMe_2OSiMe_2$	C	114–5/9		1.4202		[53a]

Table 18 *(C'ntd.)*

Compound	Method	B.p. (°C/mm.) (m.p., °C)	d_4^{20}	n_D^{20}	Comments	Reference
$Me_2Si(O-SiMe_2)_2NC_6F_5$ (cyclic)	A					[140]
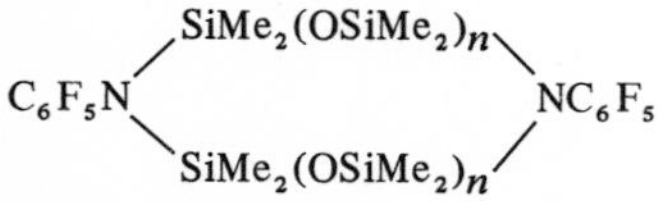 $C_6F_5N[SiMe_2(OSiMe_2)_n]_2NC_6F_5$ (cyclic) ($n = 1, 2$)	A					[140]
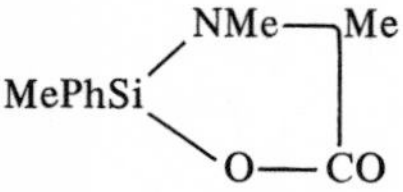 MePhSi(NMe–CHMe–CO–O) (cyclic)	B+K				nmr, $[\alpha]_D^{25}$	[173]
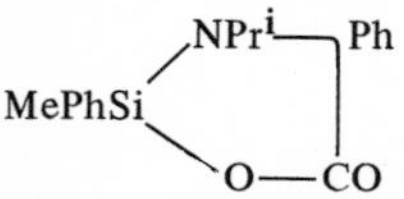 MePhSi(NPr^i–CHPh–CO–O) (cyclic)	B+K (Racemate)	(113–22)			nmr, $[\alpha]_D^{25}$	[173]
MePhSi(NPh–CHMe–CO–O) (cyclic)	(Racemate) B+K	(115–7)				
	(+) B+K	(125–8)			nmr, $[\alpha]_D^{25}$	[173]
	(–) B+K	(129–32)				
MePhSi(NPh–CHPr^i–CO–O) (cyclic)	(Racemate) B+K	(109–12)				
	(+) B+K	(120–5)			nmr, $[\alpha]_D^{25}$	[203]
	(–) B+K	(115–8)				
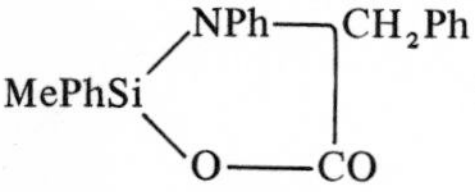 MePhSi(NPh–CH(CH_2Ph)–CO–O) (cyclic)	B+K	(128–30)			nmr, $[\alpha]_D^{25}$	[203]
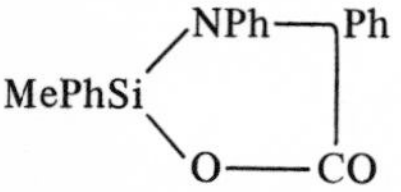 MePhSi(NPh–CHPh–CO–O) (cyclic)	B+K	(171–5)			nmr, $[\alpha]_D^{25}$	[203]
p-$MeNH(SiPh_2OC_6H_4CMe_2C_6H_4O)_3SiPh_2NHMe$		(102)				[153]
p-$PhNH(SiPh_2OC_6H_4CMe_2C_6H_4O)_3SiPh_2NHPh$		(142)				[153]
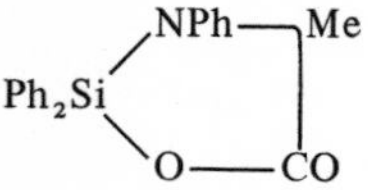 Ph_2Si(NPh–CHMe–CO–O) (cyclic)	B+K	(156–60)			nmr, $[\alpha]_D^{25}$	[203]

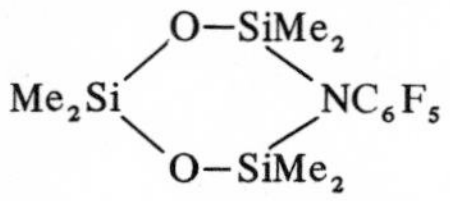

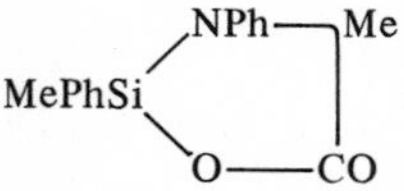

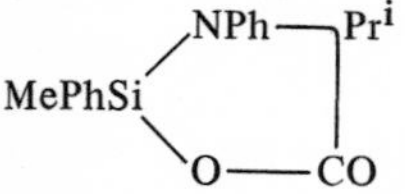

Table 18 *(C'ntd.)*

Compound	Method	B.p. (°C/mm.) (m.p., °C)	d_4^{20}	n_D^{20}	Comments	References
$Ph_2Si(OSiPh_2)_2NMe$ (cyclic: Ph_2Si–$OSiPh_2$–NMe–$OSiPh_2$)						[118]
$AcOSiMe(OMe)NHPr^i$						[93]
$PrCO_2SiHex(OMe)NHBu$						[93]
$MeSi(OPh)_2NHEt$	C, G	135/2	1.0744	1.5321		[290]
$MeSi(OPh)_2NHBu$	C, G	141/1		1.5108		[290]
$MeSi(OPh)_2NEt_2$	C, G	121–2/1	1.0426	1.5238		[290]
$MeSi(OPh)_2NPr^i_2$	C, G	125–6/0.01	1.0327	1.5200		[290]
$MeSi(OPh)_2NBu_2$	C, G	159–61/0.4	1.0522	1.5251		[290]
$MeSi(OPh)_2NPh_2$	C, G	168/1	1.1172	1.5831		[290]
	A	162/2				[345]
$PhSi(OAc)_2NHPr^i$						[93]
$MeSi(OMe)(NHPr^i)_2$		53				[93]
$AcOSiMe(NHPr^i)_2$						[93]
$(MeO)_3SiNPhSiMe_3$	A	103–4/3	$1.0442^{23°}$	1.4821	ir	[347]
$[(MeO)_3Si]_2NPh$	A	120–2/1	1.1310	1.4718	ir	[347]
$AcOSi(OMe)_2NHPr^i$						[93]
$(EtO)_3SiNHPh$	C, G	84/0.88		$1.4797^{25°}$		[248]
$(EtO)_3SiNHC_6H_4Cl$-*m*	C, G	93/0.2		$1.4932^{23.5°}$		[248]
$(EtO)_3SiNHC_6H_4Cl$-*p*	C, G					[248]
$(EtO)_3SiNMe_2$	C	166–7	$0.9125^{15°}$			[263]
$[(EtO)_3Si]_2NPh$	C, G	98/0.15		$1.4501^{25°}$		[248]
$AcOSi(OEt)_2NHPr^i$						[93]
$(BuO)_3SiNHMe$	C	136–7/13				[191, 246]
$(BuO)_3SiNHEt$	C	139–40/15				[190, 191, 246]
$(BuO)_3SiNHPr$	C	149/15				[191, 246]
$(BuO)_3SiNHBu$	C	161–3/15				[191, 246]
$(BuO)_3SiNHCH_2Ph$	C	202/15				[246]
$(BuO)_3SiNHPh$	C	198–201/25	0.9598	1.484		[159]
$(BuO)_3SiNEt_2$	C	145–7/13				[159, 246]
$(BuO)_3SiN$ (aziridine ring)	C	100/1				[147]
$AcOSi(OBu^t)(NHPr^i)_2$						[93]
$AcOSi(OBu^t)(NMe_2)_2$						[93]
$(PhO)_3SiNH(CH_2)_6NH_2$	C					[292]
$(PhO)_3SiNH(CH_2)_6NHSi(OPh)_3$	C					[292]
$\{(CH_2)_3N[Si(OPh)_3]_2\}_2$	C					[292]
$(PhO)_3SiN$ (hexagonal ring)	C	(54)				[292]

Table 18 *(C'ntd.)*

Compound	Method	B.p. (°C/mm.) (m.p., °C)	d_4^{20}	n_D^{20}	Comments	References
$(MeO)_2Si(N\triangleleft)_2$	C	49–50/11				[147]
$(EtO)_2Si(NHC_6H_4Cl\text{-}m)_2$	C, G					[248]
$(EtO)_2Si(NHC_6H_4Cl\text{-}p)_2$	C, G					[248]
$(BuO)_2Si(NHMe)_2$	C	122–3/16				[192]
$(BuO)_2Si(NHEt)_2$	C	129–32/15				[190, 192]
$(BuO)_2Si(NHPr)_2$	C	150–1/15				[192]
$(BuO)_2Si(NHBu)_2$	C	170/15				[192]
$(BuO)_2Si(NHCH_2Ph)_2$	B	196/1				[192]
$(BuO)_2Si(NHPh)_2$	C	183–4/30	0.9550	$1.4646^{25°}$		[159]
$(BuO)_2Si(N\triangleleft)_2$	C	93–4/2				[147]
$(Bu^tO)_2Si(NHEt)_2$	C	108/22	0.8815	1.4168		[202]
$(Bu^tO)_2Si(NHC_7H_{15})_2$	C	212–4	0.8697	1.4348		[202]
$(Bu^tO)_2Si(NHCH_2Ph)_2$	C	217–9/11	1.0003	1.4987		[202]
$(Bu^tO)_2Si(NHPh)_2$	C	(72)				[202]
$H_2N(Bu^tO)_2SiNHPh$						[110]
$NH[(Bu^tO)_2SiNHPh]_2$	I	subl. 80/0.01 (68–9)				[110]
$(Ph_3SiO)_2Si(NHPh)_2$	C	(159–60)			ir, X-ray	[316]
$(Ph_3SiO)Si(NHPh)_3$	C	(201–2)			ir, X-ray	[316]
$BuOSi(NHMe)_3$	C	98–101/15				[193]
$BuOSi(NHEt)_3$	C	115–6/16				[190, 193]
$BuOSi(NHPr)_3$	C	146–7/15				[193]
$BuOSi(NHBu)_3$	C	134–6/1.5				[193]
$BuOSi(NHCH_2Ph)_3$	B	218–33/0.5 (32)				[193]
$BuOSi(NHPh)_3$	C	215–8/0.5				[193]
$O[Si(NEt_2)_3]_2$	C	138–9/1 (−16)				[9]
$Me_3SiN(Me)OSiMe_3$	I	85/97			ir, nmr	[372, 374]
$(Me_3Si)_2NOMe$	I	90/96			ir, nmr	[372, 374]

† Footnote to Table 18: For explanation of abbreviations, see footnotes to Table 1, p. 57.

Table 19† Thio- and phosphino-(amino)silanes‡

Compound	B.p. (°C/mm.) (m.p., °C)	d_4^{20}	n_D^{20}	References
$HSSiMe_2NHMe$				[198]
$MePHSiMe_2NEt_2$	75–6/19	0.8711	1.4702	[32]
$MeP(SiMe_2NEt_2)_2$	106–8/1	0.9184	1.5020	[32]
$H_2PSiMe(NEt_2)_2$	80/5	0.8756	1.4606	[32]
$MePhPSiMe(NEt_2)_2$	82–3/2.5	0.8962	1.4800	[32]
$F_2OPN(SiMe_3)_2$		(ir, nmr)		[88]
H N Me_2Si S		(diamagn. susc.)		[222]

† Footnote to Table 19: For explanation of abbreviations, see footnotes to Table 1, p. 57.
‡ See also Chapter 6.

Table 20† *N*-Alkali-metal derivatives of silylamines‡

Compound	Method	References
$Me_3SiNMeLi$	E	[287]
$Me_3SiNMeNa$	E	[287]
$[(Me_2N)_3Si]_2NNa$	E	[361]
$Me_3SiNEtSiMe_2NEtLi$	K	[71]
Me_3SiNBu^tLi	E	[72]
$(Me_3SiNLiCH_2)_2$	K	[184]
$Me_3SiNHCH_2CH_2NLiSiMe_3$	K	[184]
$Me_3SiNHCH_2NLiSiMe_3$	F	[226]
$(EtNHSiMe_2)_2NNa$	A	[365]
$(PrNHSiMe_2)_2NNa$	A	[365]
$LiNPhSiMe_2NPhLi$	E	[133]
$LiNPhSiPh_2NPhLi$	E	[133]
$RR'SiN(Li)R''$ ($R = Bu^t$, $R' = Ph$, $R'' = C_6H_2Me_3$-2,4,6; $R = Me = R'$, $R'' = C_6H_2Me_3$-2,4,6; $R = Me$, $R' = Ph$, $R'' = C_6H_4Me$-4)	E	[170a]

† Footnote to Table 20: For explanation of abbreviations, see footnotes to Table 1, p. 57.
‡ See also Chapter 2.

Table 21† Silicon derivatives of amides, urea, guanidine, and related compounds

Compound	Method	B.p. (°C/mm.) (m.p., °C)	Comments	References
$Me_3SiNHCHO$		84–5/0.1		[151]
	B	33/0.05	d_4^{20} 0.916, n_D^{20} 1.4322, ir, nmr	[288]
$Me_3SiNHCOMe$	C	185–6 (52–4)		[151]
	B	83–4/11		[25]
	B	80/15	ir, nmr	[288]
	B	124/83.5 (48–9)	ir	[257]
			uv	[254]
$Me_3SiNHCOCH_2Br$				[25]
$Me_3SiNHCOCHBr_2$				[25]
$Me_3SiNHCOEt$	B	132/98–100 (66–7)	ir	[257]
$Me_3SiNHCOPr^i$				[25]
$Me_3SiNHCOCMe_2Br$				[25]
$Me_3SiNHCOCMe{:}CH_2$		(65–8)		[338]
$Me_3SiNHCOC_{17}H_{35}$				[338]
$Me_3SiNHC(O)OSiMe_3$	C	80	ir, nmr, mass	[49]
$(Pr^iO)_3SiNHCO(CH_2)_4CONHSi(OPr^i)_3$				[292]
$(Pr^iO)_3SiNHCO(CH_2)_4CON[Si(OPr^i)_3]_2$				[292]
$\{(CH_2)_2CON[Si(OPr^i)_3]_2\}_2$				[292]
$Me_3SiNHCO$-(2-furyl)	C	(86–8)		[209]
$Me_3SiNHCOPh$	B	134–7/0.25	ir	[25]
	B	(117–21)	ir	[250]
$Me_3SiNHCONH_2$	B	(160–71)		[90]
$Me_3SiNHCONHPh$		subl. 60/0.01 (69)		[109]
$(Me_3SiNH)_2CO$	B	(108–10)	ir	[257]
$(Me_3SiNH)_2CS$	D	(155–8)		[171]
$Me_3SiNHCONMe_2$	D	(125–9)		[171]
$Me_3SiNHCONPhSiMe_3$	D	(69–71)		[171, 172]
	D	(68–70)	ir	[395]
$Me_3SiNHCSNPhSiMe_3$	D	(100–1)		[394]
$Me_3SiNHCO_2Me$	B	66–7/15	d_4^{20} 0.9630, n_D^{20} 1.4267, ir	[257]
$Me_3SiNHCO_2Et$	B	73/12	d_4^{20} 0.9356, n_D^{20} 1.4261, ir	[257]
$Me_3SiNHCO_2Pr^i$	B	80/13	d_4^{20} 0.9100, n_D^{20} 1.4420, ir	[257]
$(Me_3SiNH)_2C{:}NSiMe_3$	B	40/0.5		[171]
$Me_3SiNMeCOMe$		48–9/11	n_D^{24} 1.4379	[151]
	C	154/770	d_4^{20} 0.9009, n_D^{20} 1.4382	[120]
$Me_3SiNMeCOPh$	D	87–9/0.9	ir, nmr	[214]
$Me_3SiNMeCONHMe$		(77–9)		[90]
$Me_3SiNMeCONMePh$	D	(35–8)		[172]

Table 21 *(C'ntd.)*

Compound	Method	B.p. (°C/mm.) (m.p., °C)	Comments	References
$(Me_3SiNMe)_2CO$	D	56/0.1 (22–3)	n_D^{20} 1.4523	[172]
$(Me_3SiNMeCO)_2NMe$	D	95–8/0.1	n_D^{20} 1.4729	[172]
$Me_3SiNMeCOPhSiMe_3$	D		n_D^{20} 1.5005	[172]
	D		ir	[390]
	D	105–8/0.2	ir, nmr	[214]
$Me_3SiNMeCSN(SiMe_3)COPh$	D		ir, nmr	[214]
$Me_3SiNEtCONEt_2$	D	65.5/1	n_D^{20} 1.4478	[109]
$Me_3Si(NEtCO)_2NEt_2$	D		n_D^{20} 1.4630	[112, 171]
$Me_3SiNEtCON$ (aziridine ring)	D	34–6/0.1		[284]
$Me_3SiNBuCOMe$	D	67/0.15		[205]
$Me_3SiNBuCONEt_2$	D	84/1	n_D^{20} 1.4510	[109]
$Me_3Si(NBuCO)_2NEt_2$	D		n_D^{20} 1.4685	[112, 171]
$Me_3SiNBuCON$ (aziridine ring)	D	55–7/0.1		[284]
Me_3SiN–CO–NEt (ring)	I	53–5/0.1		[284]
$Me_3SiN(CO_2Me)(CH_2)_3SiMe_2OMe$	C	97/2	d_4^{20} 0.9617, n_D^{20} 1.4456	[220]
Me_3SiN(CO)(CO)$NSiMe_3$ (ring)	B	(94–5)		[174]
$Me_3SiN(CH_2Ph)CO_2Me$	C	100–1/2	d_4^{20} 1.0223, n_D^{20} 1.5020	[221]
Me_3SiN–CO (ring)		77–81/6		[151]
Me_3SiN, OC (benzo-fused ring)	B	60–1/0.1		[174]
Me_3Si, N, Me_3SiO (benzo-fused ring)	B	98–100/0.2		[174]
$Me_3SiNPhCONHMe$	D	(72–5)		[171, 172]
$Me_3SiNPhCSNHMe$	D	(132–4)		[172]

Table 21 *(C'ntd.)*

Compound	Method	B.p. (°C/mm.) (m.p., °C)	Comments	References
$Me_3SiNPhCONHPh$	D	(77–9)		[171, 172]
	D	(68–9)	ir	[395]
$Me_3SiNPhCONMe_2$	D	(56–9)		[171]
	D	(46–9)	nmr	[172]
$Me_3SiNPhCSNMe_2$	D	(52–3)		[172]
$Me_3SiNPhCONMePh$	D	(56–60)		[171]
$(Me_3SiNPhCO)_2NMe$	D	(104–5)		[172]
	D	(88–9)	ir	[395]
$Me_3SiNPhCONEt_2$	D	95/0.04 (24–6)	n_D^{20} 1.5058	[109]
	D		ir	[395]
$Me_3SiNPhCON$ (azetidine ring)	D	70–2/0.1 (33–5)		[284]
$Me_3SiNPhCON$ (piperidine ring)	D	95–6/3 (42–4)		[172]
$Me_3Si(NPhCO)_2NEt_2$	D	(108–9)		[112, 171]
	D	(97–9)	ir	[395]
$PhC(=O)N(MMe_3)C(NBu^t)N(Me)R$	D		ir, nmr	[215]
(M R				
Si Me		97–100/0.1		
Si $SiMe_3$)		97–8/0.06		
$Me_3M(Me)NC(OSiMe_3)(NMe)$ (M = Ge, Sn)	D		ir, nmr	[154]
$(Me_3SiNPh)_2CO$	B			[171]
	D	(57–9)		[172]
$Me_3SiNPhCOSEt$	D	(69.5–70)	ir, nmr	[155]
	D	94–6/2		[152]
$Me_3SiNPhCOSSiMe_3$	D	(54)	ir, nmr	[155]
	D	(54)	nmr	[152]
$Me_3SiN(C_6H_4Me\text{-}m)COMe$	B	53–5/0.2		[174]
$Me_3SiN(C_6H_4Me\text{-}m)CONHMe$	D	(72–80)		[171, 172]
$Me_3SiN(C_6H_4Me\text{-}m)CONHPh$	D	(58–9)		[171, 172]
$Me_3SiN(C_6H_4Me\text{-}p)COMe$	B	50–3/0.2		[174]
$Me_3SiN(C_6H_4Me\text{-}p)CONHMe$	B	(78–91)		[174]
$Me_3SiN(C_6H_4Me\text{-}p)CONMe_2$	D	80–2/0.5	nmr	[172]

Table 21 *(C'ntd.)*

Compound	Method	B.p. (°C/mm.) (m.p., °C)	Comments	References
$Me_3SiN(C_6H_4Cl\text{-}m)COMe$	B	64–6/0.2		[174]
$Me_3SiN(C_6H_4Cl\text{-}p)COMe$	B	61–3/0.2		[174]
$Me_3SiN(C_6H_4OMe\text{-}m)COMe$	B	69–70/0.3		[174]
$Me_3SiN(C_6H_4OMe\text{-}p)COMe$	B	75–6/0.2		[174]
$Me_3SiN(C_6H_4NO_2\text{-}m)COMe$	B	85–8/0.2		[174]
$Me_3SiN(C_6H_4NO_2\text{-}p)COMe$	B	88–9/0.2		[174]
$Me_3SiN(COPh)CONMe_2$	D		ir, nmr	[213]
$[Me_3SiN(COPh)CO]_2NMe$	D		ir, nmr	[214]
$Me_3SiN(COPh)CONEt_2$	D		ir, nmr	[213]
Me_3SiN CO CO	B	72–3/2	d_4^{20} 1.0635, n_D^{20} 1.4749, ir	[257, 276a]
	B	108–9/0.5	n_D^{20} 1.4742	[25, 44]
Me_3SiN CO CO (benzo)	B	(68–9)		[174, 295a]
Me_3SiN CO CO (benzo) NO_2	C	(214–5)		[297]
$(Me_3Si)_2NCOH$	B	71–3/35	n_D^{20} 1.4395	[171]
	B+A	57/13 154/760	d_4^{20} 0.883, n_D^{20} 1.4388	[288]
$(Me_3Si)_2NCOMe$	D	71–3/35		[171]
			uv	[254]
$RC(O)N(SiMe_3)_2$	B		ir, nmr	[386]
(R = Et		55/10		
Pr^i		52/9		
Bu^t		72–4/15		
CF_3)				
$O{=}C[N(Me)SiMe_3][N(Ph)SnMe_3]$	D	93/0.08	ir, nmr	[154]
$S{=}C[N(Me)SiMe_3][N(Ph)MMe_3]$ (M = Ge, Sn)	D	87–8/0.5	ir, nmr	[154]
$S{=}C(SSnMe_3)[N(Me)SiMe_3]$	D	(67–9)	ir, nmr	[154]
$Me_3Si(Me)NC(NMe)SMMe_3$ (M = Ge	D	59–61/0.04	ir, nmr	[154]
Sn)	D	70/0.12	ir, nmr	[154]

Table 21 *(C'ntd.)*

Compound	Method	B.p. (°C/mm.) (m.p., °C)	Comments	References
$MeC(S)-N(R)SiMe_3$ ⇌ $MeC(SSiMe_3)(NR)$ ($R = Pr^i$, CH_2Bu^i, Me, Et)	B			[334]
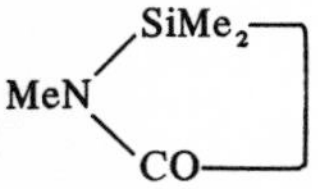 MeN(SiMe₂—CO) (ring: MeN, $SiMe_2$, CO)	C	76–7/2 (45–7)		[218]
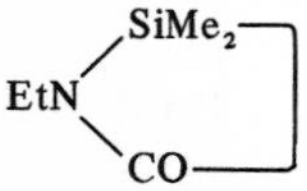 EtN (ring: $SiMe_2$, CO)	C	95–8/10	d_4^{20} 0.9715, n_D^{20} 1.4732	[218]
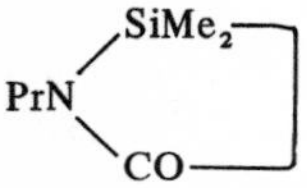 PrN (ring: $SiMe_2$, CO)	C	78–9/1.5	d_4^{20} 0.9491, n_D^{20} 1.4687	[218]
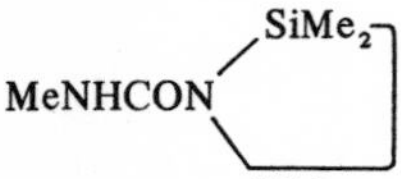 MeNHCON (ring: $SiMe_2$)	C+D	110–3/2.5	nmr	[295]
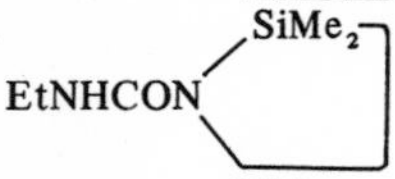 EtNHCON (ring: $SiMe_2$)	C+D	110–5/2 (64–6)	nmr	[295]
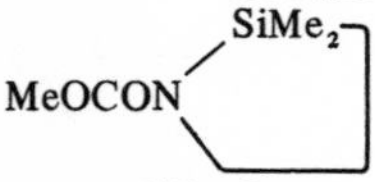 MeOCON (ring: $SiMe_2$)	C	52–3/3	d_4^{20} 1.0096, n_D^{20} 1.4593	[221]
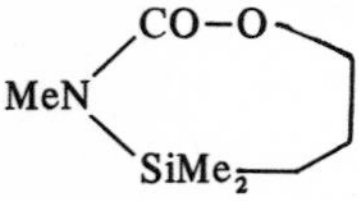 MeN (ring: CO—O, $SiMe_2$)	C			[221]
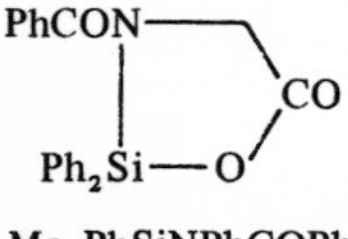 PhCON (ring: CO, O, Ph_2Si)				[64]
$Me_2PhSiNPhCOPh$	I+(—N_2)	142–3/1	n_D^{20} 1.5809	[102]
$Me_2PhSiNPhCSNHBu$	D	(87–8)		[172]
$Me_2PhSiNPhCONHC_6H_{11}$	D	(77)		[171]
$Me_2PhSiNPhCONHPh$	D	(111–3)		[171, 172]
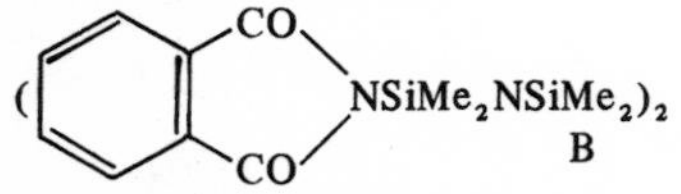 ($C_6H_4(CO)_2NSiMe_2NSiMe_2)_2$	B	(148)		[116]
$EtSSiMe_2NPhCOSEt$	D	109/1		[152]
$MePh_2SiNPhCOPh$	I+(—N_2)	203–5/5		[102]

Table 21 *(C'ntd.)*

Compound	Method	B.p. (°C/mm.) (m.p., °C)	Comments	References
$MePh_2SiN(COPh)_2$	F	278/12	n_D^{20} 1.6072	[111]
(+)-MePh-1-NpSiNMeCONHPh	D		ir, nmr	[162]
(+)-MePh-1-NpSiNPhCONHMe	D		ir, nmr $[a]_D^{20}=6.1$	[162]
(+)-*N*-(MePh-1-NpSi)NPhCSNHMe	D	(126–8)	ir, nmr $[a]_D^{20}-21.3$ (C_6H_6) -27.8 (THF)	[162]
(−)-*N*-(MePh-1-NpSi)NPhCONMe$_2$	D		ir, nmr $[a]_D^{20}-18.5$ (C_6H_6) -24.8 (THF)	[162]
(+)-*N*-(MePh-1-NpSi)NPhCSNMe$_2$	D		ir, nmr $[a]_D^{20}-118$	[162]
(+)-*N*-(MePh-1-NpSi)(NPhCO)$_2$NMe$_2$	D		ir, nmr	[162]
$Et_3SiNPrCOMe$	F	109–10/28	d_4^{20} 0.895, n_D^{20} 1.4480, uv	[119b]
$Et_3SiNPhCOMe$	F	128/4	d_4^{20} 0.961, n_D^{20} 1.5034, uv	[119b]
$Et_3SiNPhCOPh$	F	164/2	d_4^{20} 1.020, n_D^{20} 1.5523, uv	[119b]
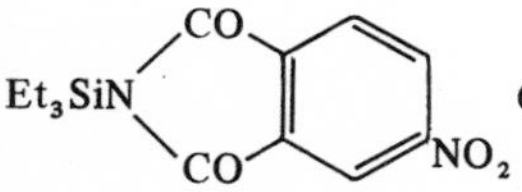 Et_3SiN, CO, CO, NO_2 (structure)	C	(204–5)		[297]
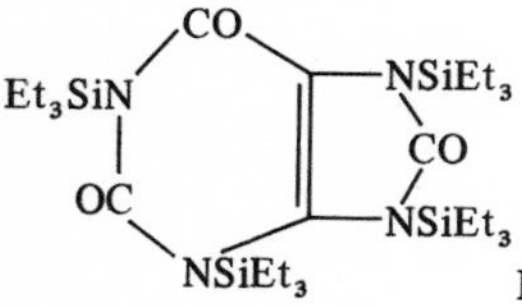 Et_3SiN, CO, OC, $NSiEt_3$, $NSiEt_3$, CO, $NSiEt_3$ (structure)	B	200–5/0.04 (43)		[47]
$Ph_3SiNHCOMe$	C	(159–63)		[338]
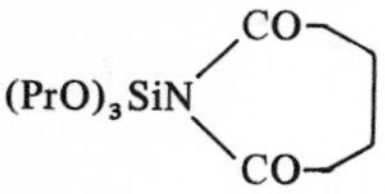 $(PrO)_3SiN$, CO, CO (structure)		(65–7)		[292]
 $Me_2Si(NEtCON$ (aziridine ring) $)_2$	D	(52–4)		[284]
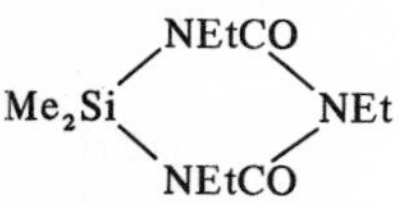 Me_2Si, NEtCO, NEt, NEtCO (structure)	D	(92–3)		[108]

Table 21 *(C'ntd.)*

Compound	Method	B.p. (°C/mm.) (m.p., °C)	Comments	Reference
$(NEtCO)_2$ Me_2Si NH $(NEtCO)_2$	D	103–5/0.04	n_D^{20} 1.4847	[108]
$(NEtCO)_2$ Me_2Si NH $(NEtCO)_3$	D	125–7/0.01		[108]
$NBuCO$ Me_2Si NBu $NBuCO$	D	110–8/0.015		[108]
$Me_2Si[(NBuCO)_6NEt_2]_2$	D			[112]
$NRCO$ Me_2Si NR $NRCO$ $[R = (CH_2)_6NCO]$	D			[108]
O Me_2Si $SiMe_2$ N $COMe$		28–9/6	ir, nmr, tautomerism	[92]
CO $Me_2Si(N$ $NEt)_2$	I	125–30/0.1		[284]
$Me_2Si(NPhCONHEt)_2$	D	(89–91)	ir	[83]
$Me_2Si(NPhCSNHEt)_2$	D	(113–5)	ir	[83]
$NPhCO$ Me_2Si NH $NPhCO$	D			[108]
$Me_2Si(NPhCONEt_2)_2$	D	(50–2)	ir	[83]
$Me_2Si(NPhCSNEt_2)_2$	D	(86–90)	ir	[83]
$NPhCONBu$ Me_2Si $NPhCONBu$	D	(178–80)	ir	[83]
$Me_2Si[(NPhCO)_2NEt_2]_2$	D			[112]
$NPhCO$ Me_2Si NPh $NPhCO$	D	(213–5)		[108]
$Me_2Si[(NPhCO)_3NEt_2]_2$	D			[109]
$Me_2Si(NPhCOSEt)_2$	D	129–31/0.5	nmr	[152]

Table 21 *(C'ntd.)*

Compound	Method	B.p. (°C/mm.) (m.p., °C)	Comments	References
$Me_2Si(NRCO)_2NR'$ (cyclic: Me₂Si bonded to two NRCO groups bridged by NR')	D			[108]
(R = R' = 4-Me-3-OCN-C₆H₃–				
R = R' = p-$ONCC_6H_4CH_2C_6H_4$				
R = p-BrC_6H_4; R' = H)		(330)		

† Footnote to Table 21: For explanation of abbreviations, see footnotes to Table 1, p. 57.

Table 22† Aminogermanes

Compound	Method	B.p. (°C/mm.) (m.p., °C)	d_4^{20}	n_D^{20}	Comments	References
H_3GeNH_2	F					[41,56]
$H_2EtGeNH_2$	F					[15]
$Pr^i_3GeNH_2$	A	35–7/0.2			ir, nmr	[16]
$Pr^i_3GeND_2$	A				ir, nmr	[16]
$Pr^i_3Ge^{15}NH_2$	A				ir, nmr	[16]
Ph_3GeNH_2	A				ir, nmr	[25,26,55]
$Et_2HGeNBu_2$	I				ir, nmr	[40]
$Et_2HGeNMe_2$	I				ir, nmr	[40]
PhH_2GeNBu_2	I				ir, nmr	[40]
$Ph_2HGeNBu_2$	I				ir, nmr	[40]
$Me_3GeN(H)Ph$	B	85–6/4			ir, nmr	[37]
$Me_3GeN(H)SiMe_3$	A	133–4/742			ir, nmr	[33]
$Me_3GeN(H)SiEt_3$	A	81–3/10			ir, nmr	[48]
Me_3GeNMe_2	A	103/760		1.4266	ir, Raman	[30a, 31,44]
					nmr, basicity	[30,46]
					ΔH_f	[5,14a]
$[Me_2Ge(Cl)NMeCH_2]_2$		90/9				[63]
Me_3GeNEt_2	A	138–9	1.01	1.4304		[2]
$Me_3GeN(Me)Bu^t$	A	81/50			ir, nmr	[51]
$Me_3GeN(SiMe_3)_2$	A	(54–6)			ir, nmr	[48]
$Me_3GeN(Bu^t)SiMe_3$	A	47/1			ir, nmr	[53]
$Me_3GeN(Me)GeMe_3$	A	172/735			ir, nmr	[43]
$Me_3GeN(Me)SiMe_3$	A	42/12			ir, nmr	[43,51]
$Me_3GeN(Me)SnMe_3$	A	28/2			ir, nmr	[43,51]
$Me_3GeN(Me)PbMe_3$	A	49/2			ir, nmr	[43,51]

Table 22 *(C'ntd.)*

Compound	Method	B.p. (°C/mm.) (m.p., °C)	d_4^{20}	n_D^{20}	Comments	References
$Me_3GeN(Ph)SnMe_3$	A	86/0.05				[50]
$Me_3GeN(Me)C(NMe)OSiMe_3$	D	48–9/0.08			ir, nmr	[21]
$Me_3GeN(Me)C(NPh)OSiMe_3$ + $Me_3GeN(Ph)C(NMe)OSiMe_3$	D	70–8/0.06			ir, nmr	[21]
$Me_3GeN(Me)C(NPh)OGeMe_3$	D	78/0.05			ir, nmr	[22]
$Me_3GeN(Me)C(NPh)SGeMe_3$	D	92–4/0.01			ir, nmr	[22]
$Me_3GeN(Ph)C(O)OGeMe_3$	D				ir, nmr	[22]
$Me_3GeN(Ph)C(S)N(Me)SiMe_3$ + $Me_3GeN(Ph)C(NMe)SSiMe_3$	D	87–8/0.05			ir, nmr	[21]
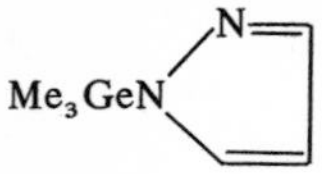 Me_3GeN (ring, N=)	A	106/65			ir, nmr	[14]
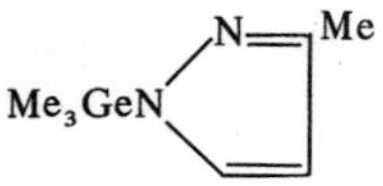 Me_3GeN (ring, N=, Me)	A	95/25			ir, nmr	[14]
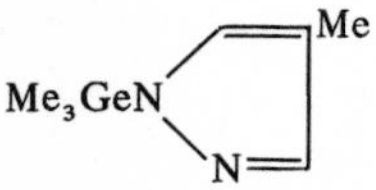 Me_3GeN (ring, Me, N=)	A	109/28			ir, nmr	[14]
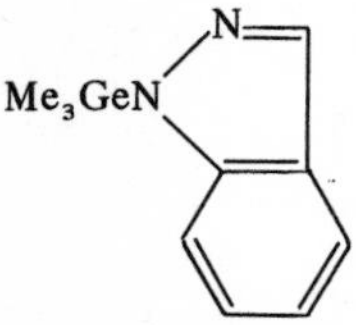 Me_3GeN (fused benzo ring, N=)	A	146/15			ir, nmr	[14]
$Me_3GeN(R')C(O)R$	B, C				ir, nmr	[65]
(R R'						
Me H		93/8				
H H		85/9				
H Ph		70/0.18				
H p-ClC_6H_4		131–2/0.67				
H p-$Me_2NC_6H_4$		155–6/0.54				
Me Ph		129/8				
Me p-ClC_6H_4		106/0.15				
Me p-$MeOC_6H_4$)		176/8				
Ph_3GeNPh_2	A	(153.5–5)				[9]
$(Me_3Ge)_2NOct$	B	147/13		1.4604		[44]
$Et_3GeN(H)Hex$	B	135/20			ir, nmr	[45]
$Et_3GeN(H)Oct$	B	160–2/20			ir, nmr	[7]

Table 22 *(C'ntd.)*

Compound	Method	B.p. (°C/mm.) (m.p., °C)	d_4^{20}	n_D^{20}	Comments	References
Et_3GeN 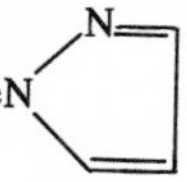 (ring, N=)	B	121/23			ir, nmr	[14]
Et_3GeN 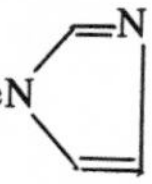 (ring, =N)	C	155-6/13			ir, nmr	[38, 39]
Et_3GeN 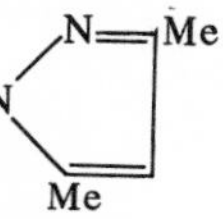 (ring, N=, Me, Me)	B	121/10			ir, nmr, Raman	[14, 31]
$Et_3GeNHC_6H_{11}$	B	135/20	0.9863	1.4526		[45]
$Et_3GeNHPh$	B	141/11	1.1282	1.5373		[45]
Et_3GeNMe_2	A	176/160	1.0234	1.4498	ir, Raman	[31, 45]
	A	65/15		1.4524		[11]
Et_3GeNEt_2	A	86/10	1.0010	1.4551		[45]
	A	90/16		$1.4521^{21°}$		[11]
$Et_3GeN(H)COR$						
(R = H	B	129-31/12			ir, nmr	[7]
Me)	D	80/0.2			ir, nmr	[45]
$Et_3GeN(Me)C(O)Me$	D				nmr	[8]
$Et_3GeN(Ph)C(S)NMe_2$		135/0.5			ir, nmr	[45]
$Et_3GeN(Ph)C(X)NEt_2$	A					
(X = O		119/0.2			ir, nmr	[45]
S)		143/0.4			ir, nmr	[45]
$Et_3GeN(Ph)C(O)N(H)GeEt_3$		134/0.1			ir, nmr	[7]
$Bu_3GeNHOct$	B	130-4/0.06		1.4594		[38]
Bu_3GeNMe_2	A	134-5/15		1.4588		[38]
	A	144-5/19	0.9607	1.4583		[45]
Bu_3GeNEt_2	A	84-8/0.15		1.4604		[38]
	A	150/9	0.9575	1.4609		[45]
Bu_3GeN  (ring)	B	158-60/12		1.4729		[38]
Bu_3GeN  (ring, =)	A	110-4/0.53		1.4882		[38]
Bu_3GeN 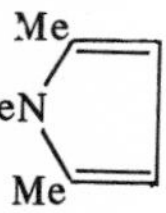 (ring, Me, Me)	B	172-4/10		1.4932		[38]

Table 22 *(C'ntd.)*

Compound	Method	B.p. (°C/mm.) (m.p., °C)	d_4^{20}	n_D^{20}	Comments	References
Bu_3GeN 	A	125–30/0.0015		1.5364		[38]
Bu_3GeN, 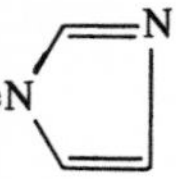 N	C	117/0.07		1.4892		[38]
$Bu_3GeN(H)Oct$	B	130–4/0.06			ir, nmr	[38, 39]
Bu_3GeN,  N	A	105–9/0.53			ir, nmr	[38, 39]
Bu_3GeN, 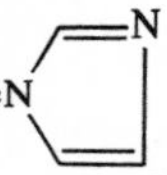 N	A	117/0.07			ir, nmr	[38, 39]
Bu_3GeN, 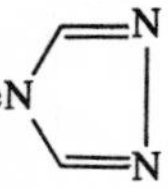 N, N	A	155/13			ir, nmr	[38, 39]
Bu_3GeN, 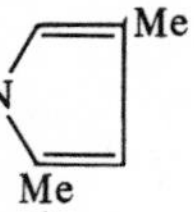 Me, Me	A	172–4/10			ir, nmr	[38, 39]
Bu_3GeN,  O, O	A	177/2			ir, nmr	[38, 39]
Bu_3GeN, 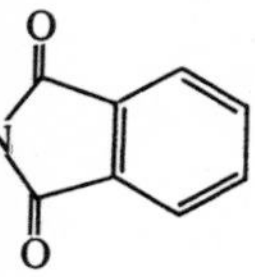 O, O	A	159/0.3			ir, nmr	[38, 39]
Ph_3GeN, 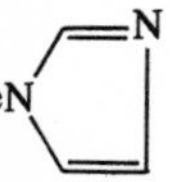 N	A	(139–40)			ir, nmr	[35]
Ph_3GeN, 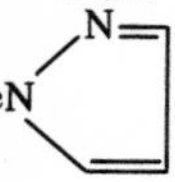 N	A	(147–8)			ir, nmr	[35]

Table 22 *(C'ntd.)*

Compound	Method	B.p. (°C/mm.) (m.p., °C)	d_4^{20}	n_D^{20}	Comments	References
Ph_3GeN (ring: N=, =N)	A	(158–70)			ir, nmr	[35]
Ph_3GeN (ring: O, O)	A	(142–3)			ir, nmr	[35]
Ph_3GeN (ring: O, O, benzo)	A	(169–70)			ir, nmr	[35]
Me_2Ge (ring: NH, X, benzo)	B					
(X = O		115/0.14			ir, nmr	[13, 59, 60]
S						[59, 60]
NH)		144/5				[59, 60]
Et_2Ge (ring: NH, O, benzo)	B	105/0.16			ir, nmr	[13]
Et_2Ge (ring: NH, O)	B	82/0.4			ir, nmr	[13]
Et_2Ge (ring: NH, OC(=O), benzo)	B	148–56/0.7			ir, nmr	[13]
Et_2Ge (ring: NH, S)	B	116/14			ir, nmr	[13]

Table 22 *(C'ntd.)*

Compound	Method	B.p. (°C/mm.) (m.p., °C)	d_4^{20}	n_D^{20}	Comments	References
Et_2Ge ring with N(H) and S fused to a benzene ring	B	114/0.18			ir, nmr	[13]
Pr^i_2Ge ring: N(H)–CH_2CH_2–O	B	135/14			ir, nmr	[13]
$Me_2Ge(Cl)NEt_2$		177/743				[63]
$(PhO)Me_2GeN(SiMe_3)_2$	A	135/1			ir, nmr	[24]
$[-GeMe_2NH(CH_2)_2NH-]_n$	B					[63]
$H_2Ge(NH_2)_2$	C, I					[41]
$Me_2Ge(NHSiEt_3)_2$	A	95–8/0.2			ir, nmr	[48]
$Me_2Ge[N(SiMe_3)_2]_2$	A	135/1			ir, nmr	[47]
$Me_2Ge(\mu\text{-}NSiEt_3)_2SnMe_2$ (four-membered ring)	A	124–8/0.2			ir, nmr	[47]
$Ph_2Ge(\mu\text{-}NC_6F_5)_2GePh_2$ (four-membered ring)	A	(331–2)				[18]
$(Bu_2GeNH)_3$ (six-membered ring)	C	172/4 × 10⁻⁴			ir, nmr	[38]
$(Me_2GeNMe)_3$ (six-membered ring)	C	80/2			ir, nmr	[43]
$[-GeMe_2NH(CH_2)_2NHGeMe_2N(CH_2CH_2)(GeMe_2)N-]_n$ (ring N–CH_2CH_2–N bridged by $GeMe_2$)	B					[63]

Table 22 *(C'ntd.)*

Compound	Method	B.p. (°C/mm.) (m.p., °C)	d_4^{20}	n_D^{20}	Comments	References
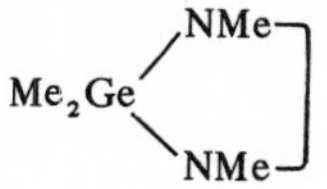 $Me_2Ge(NMe)_2$ (cyclic, NMe–NMe bridged)	A, B	152–3/742				[62]
					nmr	[36]
					ir	[64]
$Me_2Ge(NEt_2)_2$	A	87/13				[62]
					nmr	[36]
					ir	[64]
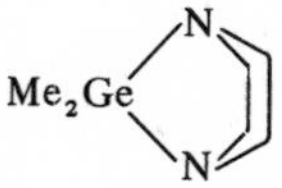 Me_2Ge (cyclic, N=N bridged)	B					[63]
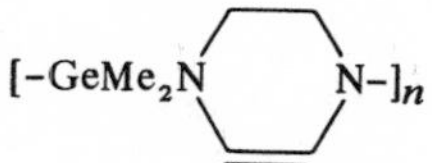 [–$GeMe_2N$ (piperazine ring) N–]$_n$	B					[63]
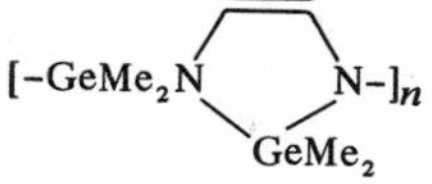 [–$GeMe_2N$ (ring with $GeMe_2$) N–]$_n$	B					[63]
$Et_2Ge(NMe_2)_2$	C	81–2/17			ir, nmr	[34]
$Bu_2Ge(NMe_2)_2$		249/760	1.001	1.4605		[4]
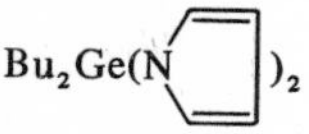 $Bu_2Ge(N$ (pyrrolyl ring) $)_2$	A	156–60/1.1		1.5222		[38]
$Ph_2Ge(NMe_2)_2$		181–2/13				[63]
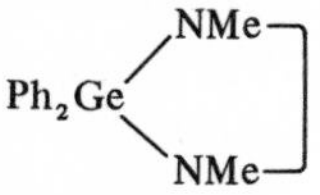 $Ph_2Ge(NMe)_2$ (cyclic, NMe–NMe bridged)		172/13				[63]
					ir	[64]
$EtGe(NH_2)_3$	C				ir, nmr	[15]
$EtGe(NMe_2)_3$	C	191/760 (—46)	1.049			[3]
$EtGe(NEt_2)_3$	C	249/760	1.108			[3]
$Ge(NHC_6H_{11})_4$	C					[12]
$Ge(NMe_2)_4$	C	203/740 (14)	1.069			[3]
					nmr, basicity	[30, 46]
					ΔH_f, pes	[14a, 28b]
$Ge(NMe_2)_4/GeCl_4$						[10]
$Ge(NMe_2)_4/Ge(OMe)_4$						[10]
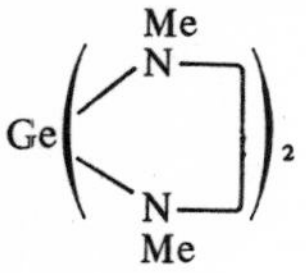 Ge(MeN–NMe cyclic)$_2$	C	106/16		1.4859		[61]
					nmr	[36]
					ir	[64]

Table 22 *(C'ntd.)*

Compound	Method	B.p. (°C/mm.) (m.p., °C)	d_4^{20}	n_D^{20}	Comments	Reference
$Ge\left(\begin{matrix} N(Me)-SiMe_2 \\ \vert \\ N(Me)-SiMe_2 \end{matrix}\right)_2$	A	(96)			ir, nmr	[58]
$Ge(NEt_2)_4$	C	108–10/2				[1]
			1.215	1.4726		[3]
$Ge\left(\begin{matrix} NEt \\ \vert \\ NEt \end{matrix}\right)_2$	C	140–2/9		1.4904		[61]
					ir	[64]
$Ge\left(\begin{matrix} NPh \\ \vert \\ NPh \end{matrix}\right)_2$	C	(305–10)				[61]
$Ge\left(\begin{matrix} N(C_6H_4Me\text{-}p) \\ \vert \\ N(C_6H_4Me\text{-}p) \end{matrix}\right)_2$	C	(317–20)				[61]
$Ge(NC_4H_4)_4$ (pyrrolyl)	A	(202)				[54]
$Ge(NC_5H_{10})_4$ (piperidino)	C					[57]
$Me_2(Cl)GeNMe_2$	A				nmr	[40,46]
$Me_2(Cl)GeNEt_2$	A	177/743			ir, nmr	[63]
$Me_2(Cl)GeN(Me)(CH_2)_2N(Me)Ge(Cl)Me_2$	C	90/9			ir, nmr	[63]
$Ph(Cl)Ge(NMe_2)_2$	A	162/0.25			ir, nmr	[40]
$ClGe(NMe_2)_3$	I				ir, nmr	[10]
$ClGe(NHPh)_3$	C				ir	[12]
$Cl_2Ge(NMe_2)_2$	I				ir, nmr	[10]
$Cl_2Ge(NHPh)_2$	C				ir	[12]
$Cl_2Ge(NHC_6H_{11})_2$	C				ir	[10]
$Me_2Si(NR)-Si(Me_2)(NR)$ ring bonded to Ge, with Ge bonded to $N(Me)-N(Me)$ ring (spiro)	A					
(R = Me		subl. 45/10^{-4}	(32)		ir, nmr	[58a]
Et)		81/0.03	(26)		ir, nmr	[58a]

Table 22 *(C'tnd.)*

Compound	Method	B.p. (°C/mm.) (m.p., °C)	Comments	References
Spiro-Ge: $Me_2Si-N(Me)-Ge-N(Me)-SiMe_2$ ring fused with $N(R)-X-CH_2-X-N(R)$ ring	A			
(R = Me, X = $SiMe_2$;		subl. $70/10^{-4}$ (29)	ir, nmr	[58a]
R = $SiMe_3$, X = CH_2)		122/0.07 (28)	ir, nmr	[58a]
Spiro-Ge: $Me_2Si-N(Me)-Ge-N(Me)-SiMe_2$ ring fused with $N(Me)-SiMe_2-O-SiMe_2-N(Me)$ ring	A	(48)	ir, nmr, mass	[58b]
Spiro-Ge: two $N(Me)-SiMe_2-X-SiMe_2-N(Me)$ rings	A		ir, nmr, mass	[58b]
(X = O,		(69)		
NMe		(38)		
Ph_2Ge azetidine ring with N–Me	A		ir, nmr	[40a]
Ph_2Ge–CH_2CH_2–NMe–$GePh_2$–N(Me) ring	D	145/0.15	ir, nmr	[40a]
$Ph_2Ge(NMe)_2GePh_2$ four-membered ring	A	(178–80)	ir, nmr	[40a]
Me_2NGe ring with N(Ph), $(CH_2)_4$, O	D	$62/4 \times 10^{-2}$	ir, nmr	[39a]

Table 22 *(C'tnd.)*

Compound	Method	B.p. (°C/mm.) (m.p., °C)	d_4^{20}	n_D^{20}	Comments	References
$(Pr^iO)_2Ge$ ring (NH, O)						[32a]
Cl_3GeNMe_2	C				ir, nmr	[12]
$(Cl_2GeNMe)_3$					X-ray	[65]
Cl_2Ge ring (NMe–$SiMe_2$–O–$SiMe_2$–NMe)		92–3/1			ir, nmr	[57a]
$ClGe(NHPh)_3$	C					[12]
$(MeO)_3GeNMe_2$	C				ir, nmr	[24]
$(EtO)_3GeN(SiMe_3)_2$	A	87/1			ir, nmr	[24]

†Footnote to Table 22: for explanation of abbreviations, see footnotes to Table 1, p. 57.

Table 23† Bisgermylamines

Compound	Method	B.p. (°C/mm.) (m.p., °C)	Comments	References
$HN(GeR_3)_2$	A			
(R = Me		47/17	ir, nmr, Raman, ed	[15a, 31, 42, 43]
Et		90–2/0.2	ir, nmr, Raman	[23, 31]
Pr		128–30/1.5	ir, nmr	[23]
Pr^i		117–20/0.2	ir, nmr	[16]
Bu		144–6/0.2	ir, nmr	[23]
Ph)			ir, nmr	[26]
$HN(GeEt_2H)_2$	C	99/2	ir, nmr	[32]
$HN(SiMe_3)(GeMe_3)$*	C	133/752	ir, nmr	[33]
$HN(SiEt_3)(GeMe_3)$*	C	138–41/10	ir, nmr	[47]
$MeN(GeMe_3)_2$	A	172/735	ir, nmr	[43]
$OctN(GeMe_3)_2$	A	147/13	ir, nmr	[7, 49, 52]
$PhN(GeEt_3)_2$	A	120/0.2	ir, nmr, Raman	[7, 31]
$OctN(GeEt_3)_2$	A	152/0.3	ir, nmr	[6]
$Me(O)CN(GeEt_3)_2$		191–2/18	ir, nmr	[6]
$MeN[Ge(Ph_2NMe_2)]GeEt_3$	D	168/0.27	ir, nmr	[40a]

†Footnote to Table 23: for explanation of abbreviations, see footnotes to Table 1, p. 57.
*These compounds are included here for convenience.

Table 24† Trisgermylamines

Compound	Method	B.p. (°C/mm.) (m.p., °C)	Comments	References
$N(GeH_3)_3$	C	unstable		[15a]
$N(GeH_2Et)_3$	C	99/12	ir, nmr	[32]
$N(GeHEt_2)_3$	C	159/15	ir, nmr	[32]
$N(GeR_3)_3$	A			
(R = Me		60/2	ir, nmr	[7, 42]
Et		148–51/0.04	ir, nmr	[23]
Pr		198–200/0.50	ir, nmr	[23]
Bu		222–3/0.05	ir, nmr	[23]
Ph)				[24a]

† Footnote to Table 25: for explanation of abbreviations, see footnotes to Table 1, p. 57.

Table 25† Subvalent germanium amides

Compound	Method	B.p. (°C/mm.) (m.p., °C)	Comments	References
Ge^{III}				
$\dot{G}e[N(SiMe_3)_2]_3$	I		esr	[10a, 11a, 20a, 27]
$\dot{G}e[N(Bu^t)SiMe_3]_3$	I		esr	[10a, 20a]
$\dot{G}e[N(GeMe_3)_2]_3$	I		esr	[17]
Ge^{II}				
$Ge(NRR')_2$	A			
(R R′				
$SiMe_3$ $SiMe_3$		(32–3)	ir, nmr, uv, pes	[10, 17, 19, 20]
Bu^t $SiMe_3$		(21–2)	ir, nmr, uv, pes	[17, 19, 20]
$GeMe_3$ $GeMe_3$		(28–9)	ir, nmr	[17]
$SiEt_3$ $SiEt_3$		$150/10^{-3}$	ir, nmr	[17]
$GeEt_3$ $GeEt_3$		$150/10^{-3}$	ir, nmr	[17]
$GePh_3$ $GePh_3$		d(150)	ir, nmr	[17]
Bu^t Bu^t)		(2)	ir, nmr, pes, mass	[28d]
Ge(N[2,2,6,6-tetramethylpiperidino])$_2$	A	(60–2)	ir, nmr, uv, pes, mass X-ray	[28a, 28c] [28e]

† Footnote to Table 25: for explanation of abbreviations, see footnotes to Table 1, p. 57.

Table 26† Aminostannanes

Compound	Method	B.p. (°C/mm.) (m.p., °C)	Comments	References
$Me_3SnNHEt$	B	153/760	n_D^{20} 1.4639	[26]
$Me_3SnNHPh$	A, B	77/0.05	d_4^{20} 1.4255, n_D^{20} 1.5721	[60,62]
			ir, nmr	[106]
	A		ir	[103]
			Mössbauer	[27]
$Me_3SnNHC_6H_4Cl$-*p*	A			[26]
			Mössbauer	[27]
$Me_3SnNHC_6H_4Me$-*p*	A	106/0.5	d_4^{20} 1.3536, n_D^{20} 1.5622	[62]
$Me_3SnNHC_6F_5$			nmr	[55]
Me_3SnNMe_2	A	126/760	d_4^{20} 1.2173, n_D^{20} 1.4572	[62]
	A	126/760		[5,60]
	A	128/720	nmr	[82,85,139]
			ΔH_f	[6,21a-c,58a, 71a]
			ir, nmr	[90,92,105]
			μ	[86]
			Mössbauer	[7]
.*trans*-$(Bu_3P)_2PtCl_2$				[66]
$Me_3SnNMePh$	A	83/0.1	d_4^{20} 1.3645, n_D^{20} 1.5757	[60,62]
			Mössbauer	[27]
$Me_3SnNMeBu^t$	A	91/30	nmr	[126]
$Me_3SnNMeSiMe_3$	A	59–61/11	ir, nmr	[113,119]
$Me_3SnNMeGeMe_3$	A	28/2	ir, nmr	[122,127]
$Me_3SnNBu^tSiMe_3$	A	50/0.25	ir, nmr	[128]
			ΔH_f	[42a]
$Me_3SnN(SiMe_3)_2$	A	48–50/1	ir, nmr	[83,110,117, 118,121]
		(20–2)	ΔH_f	[42a]
$Me_3SnNPhMMe_3$	A		ir, nmr	[124]
(M = Si		69/0.05		
Ge		86/0.05		
Sn		99/0.05		
Pb)		(130)		
$Me_3SnNMeCPh{:}NMe$	A	70/0.1 (20–2)	ir, nmr	[114]
Me_3SnNEt_2	A	162/760	n_D^{20} 1.4651	[26]
	A, B	43/8	n_D^{20} 1.4618	[62]
	A	140/720	nmr	[69]
	A	156–62		[130]
	A	45/1	nmr	[82,140]
	A		ir, basicity	[1,18]
	A		μ	[80,81,86]
$Me_3SnNPr^i_2$	A	63/8	d_4^{20} 1.1539, n_D^{20} 1.4645	[60,62]
Me_3SnNBu_2	A, B	74/2-5	d_4^{20} 1.1068, n_D^{20} 1.4559	[60,62]
$Me_3SnN(C_6H_{11})_2$	A	96/0.2	d_4^{20} 1.1972, n_D^{20} 1.5055	[62]
Me_3SnNPh_2	A	108/0.1	d_4^{20} 1.3176, n_D^{20} 1.6096	[60,62]

Table 26 *(C'ntd.)*

Compound	Method	B.p. (°C/mm.) (m.p., °C)	Comments	References
	A		Mössbauer	[26, 27]
$Me_3SnNBu^tSiMe_3$	A	50/2.5	ir, nmr	[128]
$Me_3SnN(GeMe_3)SiMe_3$	A	138–41/10	ir, nmr	[112]
$Me_3SnN(GeMe_3)SiMe_2NMeSiMe_3$	A	84–6/0.2	nmr	[111]
Me_3SnN (aziridine ring)	B	53–5/16		[67]
Me_3SnN (pyrrole ring)	A	101.5–2/17	n_D^{20} 1.5302, nmr	[67, 89]
Me_3SnN (imidazole ring)	A	(234–6)	ir	[66, 87]
	A	(235–8)		[89]
Me_3SnN (piperidine ring)	A, B	48/1		[62]
Me_3SnN (benzimidazole ring)	A	(subl. 200)	ir	[66, 87, 89]
$Et_3SnNHPh$	A	100/0.2		[62]
Et_3SnNMe_2	A	76/9	n_D^{20} 1.4783	[60, 62]
	A	55–7/4	nmr	[82]
	A			[21a, 63]
	A		nmr	[105]
			μ	[86]
Et_3SnNEt_2	A	72/2	d_4^{20} 1.1692, n_D^{20} 1.472	[62]
	A	40/0.1	nmr	[84]
	A	114–7/23	ir	[129]
	A		mass	[80, 81]
	A	225/720		[139]
	A	76/0.1	nmr	[82]
	B			[23]
			μ	[86]
Et_3SnNPh_2	B			[23]
$Et_3SnNPhCH_2Ph$	D	149–51/0.002	n_D^{20} 1.5907	[96]
$Et_3SnN(CH_2Ph)C_6H_4Me$-*p*	D	142/0.001	n_D^{20} 1.5843	[94, 96, 97]
Pr_3SnNEt_2	A	118–20/13	ir	[129]
Pr_3SnN (imidazole ring)	A	(152–4)		[66, 87]
	C	(149–50)		[89]

Table 26 *(C'ntd.)*

Compound	Method	B.p. (°C/mm.) (m.p., °C)	Comments	References
$Bu_3SnNHPh$	D	(140)	ir, nmr	[14,124]
$Bu_3SnNHBu^t$	B	124/1	d_4^{20} 1.6577, n_D^{20} 1.4773	[62]
Bu_3SnNMe_2	A	86/0.1	n_D^{20} 1.4737	[60,62]
	A	94/1	n_D^{20} 1.4728	[25]
	A		nmr	[21a,71a,105]
	A	86/0.1	nmr	[82]
			μ	[86]
Bu_3SnNEt_2	A	95/0.1	nmr	[84]
	A	115-20/0.1		[120]
	A	124-34/8	ir	[129]
	A	95-100/0.16		[135]
	A	113/0.6	n_D^{24} 1.4753	[25]
	A		mass	[80,81]
	A	96/0.1	nmr	[82]
			μ	[86]
Bu_3SnNBu_2	A	134/0.1	n_D^{18} 1.4742	[25]
$Bu_3SnN(SiMe_3)_2$	A	140-5/1	ir, nmr	[117]
$Bu_3SnNPhCH_2Ph$	D	160/0.002		[96]
Bu_3SnN (pyrrolyl ring)	A	139-41/0.62		[89]
Bu_3SnN (imidazolyl ring, 2-X) (X = H	C	(65-7)	polymer	[66,87]
	C	(64-4.5)		[89]
Ph)	C			[68]
Bu_3SnN (benzimidazolyl ring)	C	(137.5-9)		[66,87,89]
$Bu^i_3SnNEt_2$	A	75-7/0.2	d_4^{20} 1.08	[98,131,132]
$Bu^t_3SnNH_2$	E	53/0.15	ir, nmr	[41,42]
ND_2				
$(C_6H_{11})_3SnNEt_2$	A	158-61/10^{-3}		[98]
		(79-80)		[132]
$Ph_3SnNHPh$	A	94-6	ir, nmr	[124]
Ph_3SnNMe_2	A	166/0.1		[60,62]
	A	(65-8)	nmr	[82]
	A	(62)		[21a,63]
			Mössbauer	[7,27]
			μ	[86]
Ph_3SnNEt_2	A	165-70/0.1	nmr	[69]
	A	(40)	nmr	[82]
	A	(40)	mass	[80,81]
			μ	[86]

Table 26 *(C'ntd.)*

Compound	Method	B.p. (°C/mm.) (m.p., °C)	Comments	References
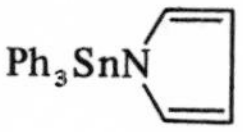 Ph_3SnN (pyrrolyl ring)	A	(205–6)		[89]
	A	(203.2–4)		[89]
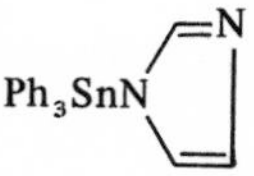 Ph_3SnN (imidazolyl ring, =N)	C	(304–5.5)		[87]
	C	(310–1)		[89]
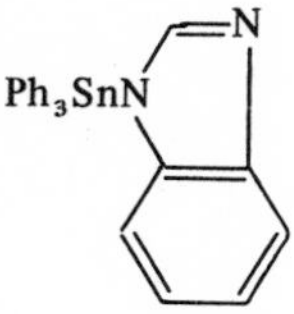 Ph_3SnN (benzimidazolyl ring, =N)	C	(298)d		[87, 89]
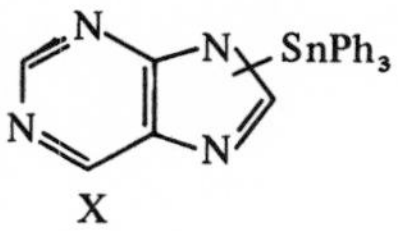 (purine ring: N, N, $N{-}SnPh_3$, N; X)				
(X = H	C	(304–7)d	ir, uv	[71]
Cl)	C	(101)	ir, uv	[71]
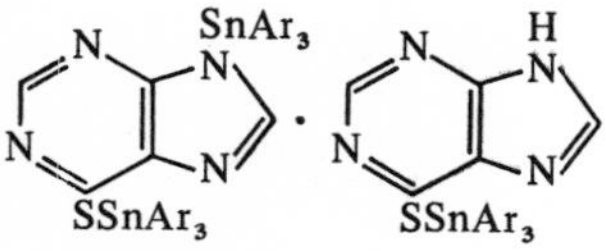 (purine ring: N, N–$SnAr_3$, N, N; $SSnAr_3$) · (purine ring: N, N–H, N, N; $SSnAr_3$)				
(Ar = Ph	C	(216–7)	ir, uv	[71]
p-C_6H_4Cl	C	(210–1)	ir, uv	[71]
p-C_6H_4F	C	(226–7)	ir, uv	[71]
p-C_6H_4Me)	C	(203–4)	ir, uv	[71]
$Me_2Sn(NMe_2)_2$	A	138/760	d_4^{20} 1.1482, n_D^{20} 1.4463	[60, 62]
	A	45/0.1	nmr	[85]
	A	147–50/720	nmr	[82]
	B		nmr	[105]
			μ	[86]
$Me_2Sn(NMeEt)_2$	A		nmr	[105]
$Me_2Sn(NEt_2)_2$	A	78/4		[62]
	A	65/0.1	nmr	[69]
	A	65/1	nmr	[82]
	A		nmr	[104, 105]
	A		ir	[141]
	A		mass	[80, 81]
			Mössbauer	[27]
			μ	[86]
$Me_2Sn(NPr^i_2)_2$	A	66/0.05	d_4^{20} 1.1060, n_D^{20} 1.4685	[60, 62]

Table 26 *(C'ntd.)*

Compound	Method	B.p. (°C/mm.) (m.p., °C)	Comments	References
Me_2Sn ring: two NMe groups joined by a chain (cyclic structure)	B	88–90/18 (1–3)	nmr	[116]
$Me_2Sn(NMeSiMe_3)_2$	A	61–3/0.5	ir, nmr	[113,119]
$Me_2Sn(NBu^tSiMe_3)_2$	A	133/1.5	ir, nmr	[128]
$Me_2Sn(NMe_2)N(SiMe_3)_2$	B	58/0.1	ir, nmr	[89]
$Me_2Sn(Cl)NBu^tSiMe_3$	A	74/0.1	ir, nmr	[128]
$Me_2Sn(Cl)N(SiMe_3)_2$	A	(9–10)	ir, nmr	[128]
$Cl_3SnN(SiMe_3)_2$	A		ir, nmr	[48,136b]
$Me_2SnClNEt_2$	I	(53–5)	ir, nmr	[26, 32, 122a]
$Me_2SnBrNEt_2$	I	(49–51)	ir, nmr	[32]
$Me_2SnINEt_2$	I	(28–34)	ir, nmr	[32]
$(Me_2NCO_2)_2Sn(NMe_2)_2$	K	(160)d		[28]
$RSn(X)[N(SiMe_3)_2]_2$	H			
(X R				
Br Me		(36–7)	ir, nmr	[74]
Br Et		(48–54)	ir, nmr	[74]
Br Pr		(78–84)	ir, nmr	[74]
Br Bu^t		(160–80)	ir, nmr	[74]
Br Ph				[74]
I Me		(78–84)	ir, nmr	[74]
I Et		(94–104)d	ir, nmr	[74]
I Pr^i		(160–70)d	ir, nmr	[74]
I Bu			ir, nmr	[74]
I Ph)			ir, nmr	[74]
$SnCl_2[N(SiMe_3)_2]_2$	A		ir, nmr	[136b]
$(Me_2SnNMe)_3$	B	114/0.2	ir, nmr	[60]
$(Me_2SnNEt)_3$	B	104/0.05	ir, nmr	[60,62]
$Et_2Sn(NMe_2)_2$	A	67–9/5	nmr	[82]
			μ	[86]
$(ClMeSnNEt)_3$	A	(163–8)	ir, nmr	[32]
$(Me_2SnNBu^t)_2$				[132a]
Me_2Sn ring: two $N(SiMe_3)$ groups joined by $(CH_2)_n$	B			
(n = 2		121–3/18	ir, nmr	[116]
3)		70–1/0.1	ir, nmr	[116]
Me_2Sn ring: two $N(SiMe_3)$ groups on a benzene ring	B	114–7/0.1	ir, nmr	[116]

Table 26 *(C'ntd.)*

Compound	Method	B.p. (°C/mm.) (m.p., °C)	Comments	References
$Et_2Sn(NEt_2)_2$	A	77/0.1,		
		108/1	nmr	[84, 86]
	A	77/0.1	nmr	[82]
	A	58–60/0.2		[132]
			μ	[86]
$SiEt_3$ N Me_2Sn $GeMe_2$ N $SiEt_3$	A	124–8/0.2	ir, nmr	[112]
H N Et_2Sn O	B		ir, nmr	[108]
Me N Et_2Sn O	B	(97.5–8.5)	ir, nmr	[108]
$Bu_2Sn(NMe_2)_2$	A	72/0.05	d_4^{20} 1.1247, n_D^{20} 1.4747	[60, 62]
	A	65/0.1	nmr	[82]
			μ	[86]
$Bu_2Sn(NEt_2)_2$	A	98/0.1	nmr	[82, 84]
	A	95–105/0.1		[120]
	A		mass	[80, 81]
			μ	[86]
Me N Bu_2Sn N H	B	150–2/0.3	ir	[108]
Me N Bu_2Sn N Me	B		ir, nmr	[108]
$Bu^i_2Sn(NEt_2)_2$	A	83–6/10^{-3}		[132]
$Bu^t_2Sn(NHR)_2$	A			
(R = Me,			ir, nmr	[47]
CH_2Ph,			ir, nmr	[47]
Bu^t)		86/0.01	ir, nmr	[46]
$(Bu^t_2SnNH)_2$	C	(150)d	ir, nmr	[46]
$(Bu^t_2SnNR)_2$	A		ir, nmr	[47]

Table 26 *(C'ntd.)*

Compound	Method	B.p. (°C/mm.) (m.p., °C)	Comments	References
(R = Me		(155)		
CH_2Ph		(220)		
Bu^t)		(24)	ir, nmr	[46]
$Bu^t_2Sn(NH_2)(NHSiMe_3)$	A			[45]
$Bu^t_2Sn(Cl)NHSiMe_3$	I	81/0.3	ir, nmr	[26]
$Bu^t_2Sn(NHBu^t)(NBu^tSnPh_3)$		(180–5)d	ir, nmr	[46]
$Ph_2Sn(NMe_2)_2$	A	128/0.2		[60,62]
	A	128/0.1	nmr	[82]
			Mössbauer	[27]
			μ	[86]
$Ph_2Sn(NEt_2)_2$	A	145–50/0.1	nmr	[69]
	A	150–60/0.1	nmr	[82]
			mass	[80,81]
			μ	[86]
$MeSn(NMe_2)_3$	A	50/0.1	nmr	[85]
	A	45–50/1	nmr	[82]
	A	45–50/0.1		[70]
			μ	[86]
$MeSn(NMeSiMe_3)_3$	A	83–5/0.5	nmr	[113]
$MeSn(NMe_2)_2N(SiMe_3)_2$	B	78/0.1	nmr	[89]
$MeSn(NEt_2)_3$	A	92/0.1	nmr	[69]
	A	96/0.1	nmr	[82]
	A		mass	[80]
			μ	[86]
$EtSn(NMe_2)_3$	A	51–3/1	nmr	[82]
			μ	[86]
$EtSn(NEt_2)_3$	A	76/0.1	nmr	[82,84]
	A		mass	[80]
			μ	[86]
$BuSn(NMe_2)_3$	A	67/0.1		[60,62]
	A	60–5/0.1	nmr	[82]
			μ	[86]
$BuSn(NEt_2)_3$	A	96/0.1	nmr	[82,84]
	A	115/1.0		[120]
	A		mass	[80]
			μ	[86]
$PhSn(NMe_2)_3$	A	80/0.1		[62,139]
	A	95–100/0.1	nmr	[82]
			Mössbauer	[27]
			μ	[86]
$PhSn(NEt_2)_3$	A	130/0.1	nmr	[69]
	A	130–5/0.1	nmr	[82]
	A		mass	[80]
			μ	[86]

Table 26 *(C'ntd.)*

Compound	Method	B.p. (°C/mm.) (m.p., °C)	Comments	References
$Cl_2Sn[N(Me)SiMe_2]_2O$ (ring: Me–N—$SiMe_2$–O–$SiMe_2$—N–Me)	A	52–3	ir, nmr	[137]
$SnBr[N(SiMe_3)_2]_3$	H	(226–8)	ir, nmr	[74]
			X-ray	[75a]
$Sn(NMe_2)_4$	A	51/0.15	d_4^{20} 1.687, n_D^{20} 1.4774	[60, 62]
	A	53–5/0.1	nmr	[82]
	A		ir	[19, 54]
	A		Mössbauer	[26, 27]
			nmr	[90]
			μ	[86]
			ed, pes	[40a, 136a]
$.VOCl_3$	I			[78]
$Sn(NEt_2)_4$	A	90/0.1		[133]
	A	90/0.05	d_4^{20} 1.1042, n_D^{20} 1.4800	[62]
	A	116/1	nmr	[69]
	A	116/0.1		[139]
	A	110/0.1	nmr	[82]
	A		Mössbauer	[26, 27]
$Cl_2Sn(NBu^t)_2SiMe_2$ (ring)				[136]
$Sn((NBu^t)_2SiMe_2)_2$ (rings)	A	(197)	ir, nmr, X-ray	[136]
$Sn((NBu^t)_2SiMe_2)((NPr^i)_2SiMe_2)$ (rings)	A, H	(57–9)	ir, nmr	[136]
$Sn(N(Me)SiMe_2SiMe_2N(Me))_2$ (rings)	A	(86)	ir, nmr	[138]
$Sn(NMeSiMe_3)_4$	A	(90–2)	ir, nmr	[113]

Table 26 *(C'ntd.)*

Compound	Method	B.p. (°C/mm.) (m.p., °C)	Comments	References
(S–Sn(–N(Bu^t)–)$_2$SiMe$_2$)$_2$ [cyclic: Sn bridged to SiMe$_2$ by two NBu^t groups]	H	(275–280)	ir, nmr	[136]
$ClSn[\overline{NCMe_2(CH_2)_3CMe_2}]_2COPh$		(115–8)	ir, nmr, mass	[71b]

† Footnote to Table 26: for explanation of abbreviations, see footnotes to Table 1, p. 57.

Table 27† Bis- and tris-stannylamines

Compound	Method	B.p. (°C/mm.) (m.p., °C)	Comments	References
$(R_3Sn)_2NH$	E		ir, nmr	[42]
(R = Pr^i		95–8/0.2		
Bu^i				
Neopent		(205–8)		
Neophyl)		(128–33)		
$(Me_3Sn)_2NMe$	A, B	64/4	d_4^{20} 1.4794, n_D^{20} 1.4901	[60, 62]
	B		Mössbauer, ΔH_f	[26, 27]
$(Me_3Sn)_2NEt$	B	93/15	d_4^{20} 1.4805, n_D^{20} 1.4968	[60, 62]
$(Me_3Sn)_2NPr^i$	B	95–7/11		[122]
$(Me_3Sn)_2NPh$	D	99–100/1	ir	[124, 125]
$(Bu_3Sn)_2NEt$	B		ir, Raman	[93]
$(Bu_3Sn)_2NPh$	I	170/0.1	ir, nmr	[31]
$(Bu_3Sn)_2NC_6H_4Me$-*p*	I	160/0.05	ir, nmr	[31]
$(Me_3Sn)_2NR$	A		ir, nmr	[107b]
[R = $SiMe_3$		52/0.4		
$SiMe_2Cl$		65/0.1		
$SiMeCl_2$		74/0.1		
$SiCl_3$		84/0.4		
$SiMe_2NCO$		66/0.1		
$SiMe(NCO)_2$		80/0.1		
$SiMe_2NCS$		80/0.1		
$SiMe(NCS)_2$]		102/0.1		
$(ClMe_2Sn)_2NEt$	A	(158–62)	ir, nmr	[32]
$(ClMe_2Sn)(ClPh_2Sn)NEt$	A	(72–9)	ir, nmr	[32]
$(ClMe_2Sn)(Cl_2MeSn)NEt$	A	(91–8)	ir, nmr	[32]
$(ClBu^t_2Sn)_2NMe$	C		ir, nmr	[93]
$(R_3Sn)_3N$	A			
(R = Me		70/0.2 (26–8)	ir, nmr, ΔH_f	[6, 60, 62, 115, 130]
Et		86–8/9	ir, nmr	[130]
Pr)		122–3/10	ir, nmr	[130]

† Footnote to Table 27: for explanation of abbreviations, see footnotes to Table 1, p. 57.

Table 28† Stannyl derivatives of amides, ureas, carbamates, and related compounds

Compound	Method	B.p. (°C/mm.) (m.p., °C)	Comments	References
Me_3SnN (cyclic, CO–…–CO ring)	J	(148)	ir, nmr	[91]
$Me_3SnN(Me)C(NMe)OSiMe_3$	D		ir, nmr	[57]
$Me_3SnN(Ph)C(O)NMeSiMe_3$	D		ir, nmr	[57]
$Me_3SnN(Ph)C(S)NMeSiMe_3$	D		ir, nmr	[57]
$Me_3SnN(Ph)C(NMe)SSiMe_3$	D		ir, nmr	[57]
$Bu_3SnN(Et)COMe$	B		ir, nmr	[107]
$Me_nSn[NMeS(Me)(NMe_2)_2]_{4-n}$				
($n = 1$	B	98/100	ir, nmr	[47a]
2		112/115	ir, nmr	[47a]
3		75/0.02	ir, nmr	[47a]
0)		(152)d	ir, nmr	[47a]
$Ph_3SnN(Me)S(Me)(NMe_2)_2$	B	(137–8)	ir, nmr	[47a]
$Et_3SnNBuCHO$	B			[23]
Et_3SnNBu^tCHO	B			[23]
$Et_3SnN(p\text{-}C_6H_4Cl)CHO$	D	120–2/0.1 (77–9)	ir	[23,102]
$Et_3SnN(p\text{-}C_6H_4NO_2)CHO$	D	(116–9)	ir	[23,102]
$Et_3SnNPhCHO$	D	171–2/12 (50–3)		[101]
	C	97/0.1 (49.5–52.5)		[34]
	D	110/0.3 (56–8)		[100]
	D	171–2/13 (50–3)	ir	[102] [99,102]
$Et_3SnNHexCHO$	C	105–9/0.3	n_D^{20} 1.4918	[100]
	D	105–6/0.1 (28)	n_D^{20} 1.4910, ir	[101,102]
$Bu_3SnNMeCHO$	C	130–40/0.4 (37–7.5)		[34]
$Bu_3SnNHexCHO$	C	167–70/4 (35–7)		[34]
$Bu_3SnNPhCHO$	D	170/0.1 (64)	ir	[9,102]
		(64–7)		[34]
$Et_2Sn(NPhCHO)_2$	D		ir	[102]
Et_3SnN (phthalimido: CO–C_6H_4–CO ring)	C	(71–3)		[65, 79]

Table 28 *(C'ntd.)*

Compound	Method	B.p. (°C/mm.) (m.p., °C)	Comments	References
$Et_3SnNPhCSH$	D	115/0.2	n_D^{20} 1.5910, ir	[101,102]
$Et_3SnNPhCOMe$	D	111-2/0.15 (46-8)		[100]
Bu_3SnN(CO)(CO) (cyclic imide, N bonded to two CO groups joined in a ring)	C	144/0.2 (21)	d_4^{20} 1.23, n_D^{20} 1.5086	[24]
$Bu_3SnNHCOMe$	C	142-4/0.1		[34]
$Bu_3SnNPhCOMe$	C	104-6/0.03		[34]
$Ph_3SnNPhCOPPh_2$	D			[123]
$Ph_3SnNPhCSPPh_2$	D			[123]
$Et_3SnNBuCO_2Et$	D	57.5/0.01		[9,14]
$Et_3SnNPhCO_2Me$	C	105-8/0.4	n_D^{20} 1.5388	[100]
$Et_3SnNPhCO_2Bu^i$	D	(45-7)		[95]
$Et_3SnNPhCO_2SnEt_3$	D	(58-60)	ir	[12,15]
$Et_3SnNHexCO_2Me$	G	92-3/0.2		[100]
$Et_3SnN(Np\text{-}1)CO_2Et$	D	104-8/0.05		[9,14]
$Pr_3SnNPhCO_2SnPr_3$	D			[10]
$Bu_3SnNMeCO_2Me$	D	101.5/0.05		[14]
$Bu_3SnNMeCO_2SnBu_3$	D		ir	[13,15]
$Bu_3SnNEtCO_2Me$	D	89-90/0.05		[14]
	C	90-95/0.2		[34]
$Bu_3SnNEtCO_2SnBu_3$	D	oil, d.		[12]
$Bu_3SnN(C_3H_5)CO_2SnBu_3$	D			[17]
$Bu_3SnNPhCO_2Me$	D	99-100/0.1		[11]
	C	118-20/0.1		[34]
$Bu_3SnNPhCO_2Ph$	D	solid, d.		[14]
$Bu_3SnNPhCO_2SnBu_3$	D	150/0.02	ir	[12,17,31]
$Bu_3SnN(Np\text{-}1)CO_2SnBu_3$	D	120/0.01		[9,14]
	D	oil, d.	ir	[12,15]
$Et_2SnClNPhCO_2Me$	I			[30]
$Bu_2SnXNEtCO_2Me$ ($X = Cl, Br, I, OAc, O_2CC_{11}H_{23}, NCS$)	I			[30]
$Bu_2SnClNPhCO_2Me$	I			[30]
$Bu_2Sn(NPhCO_2Me)_2$	D	154/0.2		[30]
$Bu_2Sn(NEtCO_2Me)_2$	D	d.		[30]
$Bu_3SnNPhCSOMe$	D	68/0.02		[9,10]
$Ph_3SnNMeCO.OOBu^t$	D	(88-90)d	ir, nmr	[8]
$Ph_3SnNEtCO.OOBu^t$	D	(90-100)d	ir, nmr	[8]
$Ph_3SnNBuCO.OOBu^t$	D	(60-8)d	ir, nmr	[8]
$Me_3SnNPhCONMe_2$	D	103/0.5	ir	[37,60]
			Mössbauer	[27]
$Me_3SnNPhCSNMe_2$	D		ir	[37,61]
$(Bu_3SnNMe)_2CO$	D	195/0.05	ir, nmr	[16]
$Bu_3SnNMeCONBuSnBu_3$	D	150/0.1	ir, nmr	[17a]
$Bu_3SnNMeCONPhSnBu_3$	D	148-50/0.05	nmr	[16]

Table 28 *(C'ntd.)*

Compound	Method	B.p. (°C/mm.) (m.p., °C)	Comments	References
$Bu_3SnNEtCONEt_2$	D	149/0.2	ir, nmr	[17a]
$Bu_3SnNEtCONPhSnBu_3$	D	167/0.05	nmr	[16]
			ir	[31]
$(Bu_3SnNPr^i)_2CO$	D	150/0.05		[17]
$Bu_3SnNBuCONMeCH(CCl_3)OSnBu_3$	D	105/0.1	ir, nmr	[17a]
$Bu_3SnNBuCONMeCHO$	D		ir, nmr	[17a]
$(Bu_3SnNBu)_2CO$	D	170/0.1	ir, nmr	[17a]
$(Bu_3SnNPh)_2CO$	D, I	130/0.05		[12, 31]
	D	200/0.1		[15]
	C	130/0.05		[34]
$(Bu_3SnNC_6H_{11})_2CO$	D	155/0.1		[17]
$Bu_3SnNPhCON(Np\text{-}1)SnBu_3$	D	155/0.1		[17]
$(Bu_3SnNNp\text{-}1)_2CO$	D	220/0.05		[12]
	I			[15]
	D		ir, nmr	[17a]
	C			[34]
$(Bu_3SnNPhCO)_2NEt$	D	oil	ir	[13,16]
$Bu_3SnNPhCONMeCON(Np\text{-}1)SnBu_3$	D	oil		[13,16]
$Bu_3SnNPhCONEtCON(Np\text{-}1)SnBu_3$	D	oil		[13,16]
CO Bu_3SnN $NSnBu_3$ OC CO N $SnBu_3$	C	300/0.04		[34]
$Me_3SnN(p\text{-}C_6H_4Me)C(NEt_2){:}NC_6H_4Me\text{-}p$	D	168/0.1	ir	[37, 61]
$Et_3SnN(C_6H_{11})CH{:}NC_6H_{11}$	D	126/0.0001	n_D^{20} 1.5233	[96]
$Bu_3SnN(Np\text{-}1)C(OMe){:}NNp\text{-}1$	D	140/0.02		[10,17]
$Bu_3SnN(p\text{-}C_6H_4Me)C(OMe){:}NC_6H_4Me\text{-}p$	D			[17]
$Me_2SnClNPhCSNEt_2$	I	(18–26)	ir, nmr	[26]

† Footnote to Table 28: for explanation of abbreviations, see footnotes to Table 1, p. 57.

Table 29† Subvalent tin amides

Compound		Method	B.p. (°C/mm.) (m.p., °C)	Comments	References
Sn^{III}					
$\dot{S}n[N(SiMe_3)_2]_3$		I		esr	[21d, 44, 50, 51, 56, 75]
$\dot{S}n[N(Bu^t)SiMe_3]_3$		I		esr	[44, 51, 56, 75]
$\dot{S}n[N(GeMe_3)_2]_3$		I		esr	[44, 75]
$\dot{S}n[N(GeEt_3)_2]_3$		I		esr	[44, 75]
Sn^{II}					
$Sn(NMe_2)_2$	dimer	A	subl. $70/10^{-4}$	ir, dnmr Mössbauer	[36]
Sn(N[ring])$_2$	dimer				[142]
$Sn(NPr^i_2)_2$				ir, nmr	[75,142]
$Sn(NBu^t_2)_2$	monomer	A	(44–7)	ir, nmr, uv, pes	[77]
$Sn(NPhSiMe_3)_2$					[142]
$Sn(NPh_2)_2$				ir, nmr	[75,142]
$Sn[N(MR_3)_2]_2$		A			
(M R					
Si Me	monomer		(37–8)	ir, nmr, uv, mass, μ	[29, 44, 49, 64a, 75, 109]
				ed, ΔH_f, pes	[42a, 43a, 76]
Si Et	monomer		$150/10^{-3}$	ir, nmr, uv	[44, 49, 75]
Ge Me	monomer		(35–7)	ir, nmr, uv	[44, 49, 75]
Ge Et	monomer		$150/10^{-3}$	ir, nmr, uv	[44, 49, 75]
Ge Ph)	monomer			ir, nmr, uv	[44, 49, 75]
$Sn[N(Bu^t)SiMe_3]_2$	monomer	A	(31–2)	ir, nmr, uv, mass; ΔH_f, pes	[44, 49, 75]; [42a, 44]
$Sn\langle N(Bu^t) \rangle_2 X$ (cyclic: Sn–N(Bu^t)–X–N(Bu^t)–Sn)	monomer	A			
(X = $SiMe_2$			55–7/0.1	ir, nmr, X-ray, (monomer and dimer)	[135a, 135b, 135c]
$SnBu^t_2$)			(5–8)	ir, nmr	[45]
$Sn\langle N(SiMe_3) \rangle_2 X$ (cyclic: Sn–N(SiMe$_3$)–X–N(SiMe$_3$)–Sn) [X = $(CH_2)_n$]		A		ir, nmr	[109]
(n = 2,			(134–5)		
3,			(108–9)		
4)			(83–4)		
(X = 1,2-phenyl.THF			(88)	ir, nmr, Mössbauer	[109]
1,8-naphthyl.THF)			(82.5)	ir, nmr, Mössbauer	[109]

Table 29 *(C'ntd.)*

Compound	Method	(M.p., °C)	Comments	References
Sn(N(2,2,6,6-tetramethylpiperidino))$_2$ monomer	A	(48–50)	ir, nmr, uv, pes	[71b,76,142]
$Sn(Cl)N(MMe_3)_2$ dimer	A			
(M = Si		(>150)d	ir, nmr	[49]
Ge)		(>150)d	ir, nmr	[74]

† Footnote to Table 29: for explanation of abbreviations, see footnotes to Table 1, p. 57.

Table 30† Compounds with lead-nitrogen bonds

Compound	Method	M.p.(°C)	Comments	References
Pb^{IV}				
$Et_3PbNHBu^s$	A			[2]
Et_3PbNEt_2	A			[10]
Pr_3PbNEt_2	A			[10]
Bu_3PbNH_2	A	−2	ir, nmr	[2a]
Bu_3PbNEt_2	A			[10]
$Bu^i_3PbNEt_2$	A			[10]
$(C_6H_{11})_3PbNEt_2$	A			[10]
Ph_3PbNEt_2	A			[10]
$Ph_2Pb(NEt_2)_2$	A			[10]
$Bu_3PbN(C_6H_{11})CONEt_2$	D			[11]
$Bu_3PbNPhCONEt_2$	D			[11]
$Bu_3PbN(Np\text{-}1)CO_2Me$		72–5		[1]
$Ph_3PbN(Np\text{-}1)CO_2Me$		135–41d		[1]
$Ph_3PbN(Np\text{-}1)C(OMe){:}NNp\text{-}1$	D	120–40d		[1]
$R_3PbNEtCSNEt_2$	D			[11]
$R_3PbN(C_6H_{11})C(NEt_2){:}NC_6H_{11}$	D			[11]
(R = Et, Bu, C_6H_{11}, Ph claimed)				
$Me_3PbN(CO)_2C_2H_4$ (succinimido)	E	177–8		[12]
$Me_3PbN(CO)_2C_6H_4$ (phthalimido)				[3]

Table 30 *(C'ntd.)*

Compound	Method	M.p.(°C)	Comments	References
$MeEtPrPbN(CO)_2C_6H_4$ (phthalimide)				[3]
$Et_3PbN(CO)_2C_6H_4$ (phthalimide)				[3]
$Pr_3PbN(CO)_2C_6H_4$ (phthalimide)				[3]
$Pent_3PbN(CO)_2C_6H_4$ (phthalimide)				[3]
6-Z-9-($PbPh_3$)purine (Z = H	C	275–7d	ir, uv	[7]
Cl	C	217–9d	ir, uv	[7]
SMe)	C	131–2	ir, uv	[7]
$Ph_3PbN(CO)_2C_2H_4$ (succinimide)	E	195–7		[12]
	C	195–7		[5]
$Ph_3PbN(CO)_2C_2H_2$ (maleimide)	C	172–4		[5]
$Ph_3PbN(CO)_2C_6H_4$ (phthalimide)	J	174–6		[5]
6-(Ph_3PbS)-9-($PbPh_3$)purine; 6-(Ph_3PbS)-9H-purine	C	197–8	ir, uv	[6]

Table 30 *(C'ntd.)*

Compound	Method	B.p. (°C,mm.) (m.p., °C)	Comments	References
Et_3PbN (CO)₂ cyclic imide with tetrachlorobenzene ring (tetrachlorophthalimide)				[7]
Me_3PbNMe_2	A			[6a]
$Me_3PbNMeBu^t$	A	87/10	ir, nmr	[18]
$Me_3PbN(Me)SiMe_3$	A	60–5/25 (10–15)	ir, nmr, ΔH_f	[3a,18]
$Me_3PbN(Me)GeMe_3$	A	49/2	ir, nmr	[18]
$Me_3PbN(SiMe_3)_2$	A	85–7/3	ir, nmr, ΔH_f	[3a,15]
$Me_3PbN(Ph)SnMe_3$	A	subl. 30		[17]
$R_3Pb[N(H)C(=NH)N(H)CN]$ (R = Me, Et, Bu, Bu^i)	G			[13]
R_3PbN (five-membered ring: N–X=C(R')–N=C) (R = Me, Et, Bu, Bu^i; R' = H, Me, Ph, NH_2; X = N, CR″; R″ = H, Ph)	G			[13]
R_3PbN (benzo-fused ring: N–X=N, fused benzene) (R = Me, Et, Bu, Bu^i; X = N, CR″; R″ = H, Ph)	G			[13]
Pb^{II}				
$Pb[N(SiMe_3)_2]_2$ monomer	A	50/0.04 (37–8)	ir, nmr, uv, mass ΔH_f, pes, ed	[3b, 3c, 4] [3a, 4a]
$Pb[N(Bu^t)SiMe_3]_2$ monomer	A	50/0.04 (21–2)	ir, nmr, uv, mass ΔH_f	[3a, 3b, 4, 4a]
$Pb(N$ 2,2,6,6-tetramethylpiperidino$)_2$		unstable	ir, nmr	[6b, 6c]

† Footnote to Table 30: for explanation of abbreviations, see footnotes to Table 1, p. 57.

H. REFERENCES TO CHAPTER 5

1. Silicon

[1] Abel, E. W., *J. Chem. Soc.,* 1961, 4933.

[1a] Abel, E. W., Armitage, D. A., and Tyfield, S. P., *J. Chem. Soc. (A),* 1967, 554.

[2] Abel, E. W., Armitage, D. A., and Willey, G. R., *Trans. Faraday Soc.,* 1964, **60**, 1257.

[3] Abel, E. W., and Willey, G. R., *J. Chem. Soc.,* 1964, 1528.

[3a] Adamson, G. W., and Daly, J. J., *J. Chem. Soc. (A),* 1970, 2724.

[4] Airey, W., Aylett, B. J., Ellis, I. A., and Sheldrick, G. M., *Spectrochim. Acta (A),* 1971, **27**, 1505.

[4a] Airey, W., Glidewell, C., Robiette, A. G., and Sheldrick, G. M., *J. Mol. Struct.,* 1971, **8**, 435.

[4b] Airey, W., Glidewell, C., Robiette, A. G., Sheldrick, G. M., and Freeman, J. M., *J. Mol. Struct.,* 1971, **8**, 423.

[5] Allan, M., Aylett, B. J., and Ellis, I. A., *Chem. Ind. (London),* 1966, 1417.

[6] Allan, M., Aylett, B. J., and Ellis, I. A., *J.C.S. Dalton,* 1973, 2675.

[7] Allan, M., Aylett, B. J., and Ellis, I. A., *Inorg. Nucl. Chem. Letters,* 1966, **2**, 261.

[8] Anderson, H. H., *J. Amer. Chem. Soc.,* 1951, **73**, 5802.

[9] Anderson, H. H., *J. Amer. Chem. Soc.,* 1952, **74**, 1421.

[10] Anderson, H. H., *J. Amer. Chem. Soc.,* 1960, **82**, 1323.

[11] Anderson, D. W. W., Bentham, J. E., and Rankin, D. W. H., *J.C.S. Dalton,* 1973, 1215.

[12] Andrianov, K. A., and Delazar, N. V., *Izvest. Akad. Nauk S.S.S.R. Ser. Khim.,* 1972, 2748.

[13] Andrianov, K. A., Emel'yanov, V. N., and Rudman, E. V., *Dokl., Akad. Nauk S.S.S.R.,* 1972, **204**, 855.

[14] Andrianov, K. A., Kononov, A. M., and Makarova, N. N., *Zh. Obshch. Khim.,* 1966, **36**, 895.

[15] Andrianov, K. A., Kononov, A. M., Klimov, A. K., and Tikhonova, G. I., *Izvest. Akad. Nauk S.S.S.R. Ser. Khim.,* 1969, 617.

[16] Andrianov, K. A., and Rumba, G. Ya., *J. Gen. Chem. U.S.S.R.,* 1962, **32**, 1993.

[17] Andrianov, K. A., Talanov, V. N., Khananashvili, L. M., Minakova, T. V., and Gashnikova, N. P., *Bull. Acad. Sci. U.S.S.R., Div. Chem. Sci.,* 1968, 640.

[18] Andrianov, K. A., Vasil'eva, T. V., and Minaeva, A. A., *Bull. Acad. Sci. U.S.S.R., Div. Chem. Sci.,* 1963, 2227.

[19] Ang, H. G., *J. Chem. Soc. (A),* 1968, 2734.

[20] Arnold, D. E. J., Ebsworth, E. A. V., Jessep, H. F., and Rankin, D. W. H., *J.C.S. Dalton,* 1972, 1681.

[20a] Attridge, C. J., *Organometallic Chem. Rev. (A)*, 1970, **5**, 323.
[21] Aylett, B. J., *Organometallic Chem. Rev. (A)*, 1968, **3**, 151.
[21a] Aylett, B. J., *Adv. Inorg. Chem. Radiochem.*, 1968, **11**, 249.
[21b] Aylett, B. J., *Progr. Stereochem.*, 1969, **4**, 213.
[22] Aylett, B. J., and Burnett, G. M., U.S. Patent 3,234,148 (1966).
[23] Aylett, B. J., Ellis, I. A., and Porritt, C. J., *Chem. Ind. (London)*, 1970, 499.
[24] Aylett, B. J., Ellis, I. A., and Porritt, C. J., *J.C.S. Dalton*, 1973, 83.
[25] Aylett, B. J., and Emsley, J., *J. Chem. Soc. (A)*, 1967, 652.
[26] Aylett, B. J., and Emsley, J., *J. Chem. Soc. (A)*, 1967, 1918.
[27] Aylett, B. J., and Hakim, M. J., *J. Chem. Soc. (A)*, 1969, 636.
[28] Aylett, B. J., and Hakim, M. J., *J. Chem. Soc. (A)*, 1969, 800.
[29] Aylett, B. J., and Hakim, M. J., *Inorg. Chem.*, 1966, **5**, 167.
[30] Aylett, B. J., and Peterson, L. K., *J. Chem. Soc. (A)*, 1964, 3429.
[30a] Bailey, D. L., Sommer, L. H., and Whitmore, F. C., *J. Amer. Chem. Soc.*, 1948, **70**, 435.
[31] Bailey, R. E., and West, R., *J. Organometallic Chem.*, 1965, **4**, 430.
[32] Balashova, L. D., Broker, A. B., and Soborovskii, L. Z., *J. Gen. Chem. U.S.S.R.*, 1966, **36**, 73.
[33] Baldwin, J. C., Lappert, M. F., Pedley, J. B., and Poland, J. S., *J.C.S. Dalton*, 1972, 1943.
[34] Baldwin, J. C., Lappert, M. F., Pedley, J. B., and Treverton, J. A., *J. Chem. Soc. (A)*, 1967, 1980.
[35] Banford, T. A., and MacDiarmid, A. G., *Inorg. Nucl. Chem. Letters*, 1972, **8**, 733.
[36] Bassindale, A. R., Eaborn, C., and Walton, D. R. M., *J. Organometallic Chem.*, 1970, **25**, 57.
[37] Baum, G., Lehn, W. L., and Tamborski, C., *J. Org. Chem.*, 1964, **29**, 1264.
[37a] Bažant, V., Chvalovský, V., and Rathouský, J., *Organosilicon Compounds*, Czech. Academy of Sciences, Prague, 1965.
[37b] Beagley, B., and Conrad, A. R., *Trans. Faraday Soc.*, 1970, **66**, 2740.
[38] Beagley, B., Sheldrick, G. M., and Robiette, A. G., *Chem. Comm.*, 1967, 601.
[39] Becke-Göhring, M., and Wunsch, G., *Ann.*, 1958, **618**, 43.
[40] Becker, H. P., and Neumann, W. P., *J. Organometallic Chem.*, 1969, **20**, 3.
[41] Becker, H. P., and Neumann, W. P., *J. Organometallic Chem.*, 1972, **37**, 57.
[41a] Bekiarogluo, P., *Z. Anorg. Chem.*, 1966, **345**, 290.
[42] Benkeser, R. A., Robinson, R. E., and Landesman, H., *J. Amer. Chem. Soc.*, 1952, **74**, 5699.
[43] Berghe, E. V., and van der Kelen, J. P., *J. Organometallic Chem.*, 1976, **122**, 329.

[44] Birkofer, L., and Dickopp, H., *Chem. Ber.*, 1968, **101**, 2585.
[45] Birkofer, L., and Ritter, A., *Angew. Chem.*, 1956, **68**, 461.
[46] Birkofer, L., and Ritter, A., *Ann.*, 1958, **612**, 22.
[47] Birkofer, L., and Ritter, A., *Angew. Chem.*, 1959, **71**, 372.
[48] Birkofer, L., and Ritter, A., *Chem. Ber.*, 1960, **93**, 424.
[49] Birkofer, L., and Sommer, P., *J. Organometallic Chem.*, 1972, **35**, C15.
[49a] Bock, H., Kaim, W., and Connolly, J. W., *Angew. Chem. Internat. Edn.*, 1976, **15**, 708.
[49b] Boer, F. P., and van Remoortere, F. P., *J. Amer. Chem. Soc.*, 1969, **91**, 2377.
[50] Bolotov, B. A., Kharutonov, N. P., Batyaev, E. A., and Rumyantseva, E. G., *J. Gen. Chem. U.S.S.R.*, 1967, **37**, 1602.
[50a] Bowser, J. R., Neilson, R. H., and Wells, R. L., *Inorg. Chem.*, 1978, **100**, 1882.
[51] Breed, L. W., Budde, W. L., and Elliott, R. L., *J. Organometallic Chem.*, 1966, **6**, 676.
[52] Breed, L. W., Elliott, R. L., and Wiley, J. C., *J. Organometallic Chem.*, 1971, **31**, 179.
[53] Breed, L. W., and Wiley, J. C., *Inorg. Chem.*, 1972, **11**, 1634.
[53a] Breed, L. W., Whitehead, M. E., and Elliott, R. L., *Inorg. Chem.*, 1967, **6**, 1254.
[54] Breederveld, H., *Rec. Trav. Chim.*, 1959, **78**, 589.
[55] Breederveld, H., *Rec. Trav. Chim.*, 1960, **79**, 1197.
[56] Breederveld, H., Ger. Patent 1,129,946 (1962).
[57] Breederveld, H., and Waterman, H. I., *Research*, 1952, **5**, 537.
[58] Breederveld, H., and Waterman, H. I., *Research*, 1953, **6(1)**, 1S.
[59] Breederveld, H., and Waterman, H. I., *Research*, 1953, **6(7)**, 43S.
[60] Breederveld, H., and Waterman, H. I., *Research*, 1953, **6(7)**, 48S.
[61] Breederveld, H., and Waterman, H. I., *Research*, 1954, **7**, 5S.
[62] Breederveld, H., and Waterman, H. I., U.S. Patent 2,807, 635 (1957).
[62a] Brewer, S. D., and Haber, C. P., *J. Amer. Chem. Soc.*, 1948, **70**, 3888.
[62b] Brill, T. B., *J. Chem. Ed.*, 1973, **50**, 392.
[62c] Brook, A. G., and Duff, J. M., *J. Amer. Chem. Soc.*, 1974, **96**, 4692.
[62d] Results cited in West, R., *Adv. Organometallic Chem.*, 1977, **16**, 1.
[63] Broser, W., and Harrer, W., *Angew. Chem. Internat. Edn.*, 1965, **4**, 1081.
[64] Buchwald, H., Seidel, C., Kunzek, H., Ludwig, P., Frölich, B., and Rühlmann, K., *J. Organometallic Chem.*, 1972, **37**, C1.
[65] Burg, A. B., and Kuljian, E. S., *J. Amer. Chem. Soc.*, 1950, **72**, 3103.
[66] Bürger, H., *Inorg. Nucl. Chem. Letters*, 1965, **1**, 11.
[67] Bürger, H., *Monatsh.*, 1966, **97**, 869.
[67a] Bürger, H., *Angew. Chem. Internat. Edn.*, 1973, **12**, 474.
[68] Bürger, H., and Goetze, U., *Monatsh.*, 1968, **99**, 155.
[69] Bürger, H., and Sawodny, W., *Inorg. Nucl. Chem. Letters*, 1966, **2**, 209.
[70] Bush, R. P., Lloyd, N. C., and Pearce, C. A., *Chem. Comm.*, 1967, 1240.

[71] Bush, R. P., Lloyd, N. C., and Pearce, C. A., *Chem. Comm.*, 1967, 1269.
[72] Bush, R. P., Lloyd, N. C., and Pearce, C. A., *Chem. Comm.*, 1967, 1270.
[73] Bush, R. P., Lloyd, N. C., and Pearce, C. A., *J. Chem. Soc. (A)*, 1969, 253.
[74] Buttler, M. J., and McKean, D. C., *Spectrochim. Acta*, 1965, **21**, 485.
[75] Buttler, M. J., McKean, D. C., Taylor, R., and Woodward, L. A., *Spectrochim. Acta*, 1965, **21**, 379.
[76] Carey, F. A., and Hsu, C.-L. W., *Tetrahedron Letters*, 1970, **44**, 3885.
[77] Cass, R., and Coates, G. E., *J. Chem. Soc.*, 1952, 2347.
[78] Chan, S., DiStefano, S., Fong, F., Goldwhite, H., Gysegem, P., and Mazzola, E., *Inorg. Chem.*, 1973, **12**, 51.
[79] Chernyshev, E. A., and Dolgaya, M. E., *Sinetz. i Svoista Monomer ov. Akad. Nauk U.S.S.R., Inst. Neftekhim. Sinteza, Sb. Rabot 12-oi [Dvenadtsatoi] Konf. po. Vysokomolekul. Soedin.*, 1962, 123; *Chem. Abstr.*, 1965, **62**, 6503e.
[79a] Chioccola, G., and Daly, J. J., *J. Chem. Soc. (A)*, 1968, 1658.
[80] Chugunov, V. S., *J. Gen. Chem. U.S.S.R.*, 1953, **23**, 777.
[81] Chugunov, V. S., *J. Gen. Chem. U.S.S.R.*, 1956, **26**, 2474.
[81a] Churakov, A. M., Ioffe, S. L., Khasapov, B. N., and Tartakovskii, V. A., *Bull. Acad. Sci. U.S.S.R.*, 1978, **27**, 113.
[82] Cook, N. C., and Lyons, J. E., *J. Amer. Chem. Soc.*, 1966, **88**, 3396.
[82a] Cook, R. L., and Mills, A. P., *J. Phys. Chem.*, 1961, **65**, 252.
[83] Cragg, R. H., and Lappert, M. F., *J. Chem. Soc. (A)*, 1966, 82.
[84] Craig, D. P., Maccoll, A., Nyholm, R. S., Orgel, L. E., and Sutton, L. E., *J. Chem. Soc.*, 1954, 332.
[85] Creamer, C. E., Ger. Patent 1,800,968 (1969).
[86] Comi, R., Franck, R. W., Reitano. M., and Weinreb, S. M., *Tetrahedron Letters*, 1973, 3107.
[87] Cuvigny, T., and Normant, H., *Compt. Rend.*, 1969, **268C**, 834.
[88] Czieslik, G., Flaskerud, G., Höfer, R., and Glemser, O., *Chem. Ber.*, 1973, **106**, 399.
[89] De Benneville, P. L., and Hurwitz, M. J., U.S. Patent 2,847,409 (1958).
[90] De Benneville, P. L., and Hurwitz, M. J., U.S. Patent 2,906,756 (1959).
[91] Dennis, W. E., and Speier, J. L., *J. Org. Chem.*, 1970, **35**, 3879.
[92] Dejak, B., and Lasocki, Z., *J. Organometallic Chem.*, 1972, **44**, C39.
[92a] Dexheimer, E. M., Spialter, L., and Smithson, L. D., *J. Organometallic Chem.*, 1975, **102**, 21.
[93] Di Paolo, J. F., Fr. Patent 1,541,542 (1968).
[93a] Dolgov, B. N., Kharutonov, N. P., and Voronkov, M. G., *J. Gen. Chem. U.S.S.R.*, 1954, **24**, 678.
[94] Durig, J. R., and Cooper, P. J., *J. Mol. Struct.*, 1977, **41**, 183.
[94a] Eaborn, C., *Organosilicon Compounds*, Butterworths, 1960, p. 339.
[94b] Ebsworth, E. A. V., *Volatile Silicon Compounds*, Pergamon Press, 1963.

[94c] Ebsworth, E. A. V., *Chem. Comm.*, 1966, 530.
[94d] Ebsworth, E. A. V., Ch. 1 in *Organometallic Compounds of the Group IV Elements,* (Ed.) A. G. MacDiarmid, Dekker, Vol. 1, Part I, 1968.
[95] Ebsworth, E. A. V., and Emeléus, H. J., *J. Chem. Soc.*, 1958, 2150.
[96] Ebsworth, E. A. V., Hall, J. R., MacKillop, M. J., McKean, D. C., Sheppard, N., and Woodward, L. A., *Spectrochim. Acta,* 1958, **13**, 202.
[97] Ebsworth, E. A. V., and Sheppard, N., *J. Inorg. Nucl. Chem.*, 1959, **9**, 95.
[98] Elliott, R. L., and Breed, L. W., *J. Chem. Eng. Data,* 1966, **11**, 604.
[99] Elter, G., Glemser, O., and Herzog, W., *Chem. Ber.*, 1972, **105**, 115.
[100] Emeléus, H. J., and Miller, N., *Nature,* 1938, **142**, 996.
[101] Emeléus, H. J., and Miller, N., *J. Chem. Soc.*, 1939, 819.
[102] Ettenhuber, E., and Rühlmann, K., *Chem. Ber.*, 1968, **101**, 743.
[103] Falius, H., Giesen, K. P., and Wannagat, U., *Z. Anorg. Chem.*, 1973, **402**, 139.
[104] Federova, G. T., and Kharutonov, N. P., *Khim. Prakt. Primen. Kremniiorg. Soedin., Tr. Sovesch.*, 1968, **1966**, 32.
[105] Fessenden, R., and Crowe, D. F., *J. Org. Chem.*, 1960, **25**, 598.
[106] Fessenden, R., and Crowe, D. F., *J. Org. Chem.*, 1961, **26**, 4638.
[107] Fessenden, R., and Fessenden, J. S., *Chem. Rev.*, 1961, **61**, 361.
[108] Fink, W., *Chem. Ber.*, 1964, **97**, 1424.
[109] Fink, W., *Chem. Ber.*, 1964, **97**, 1433.
[110] Fink, W., *Helv. Chim. Acta,* 1964, **47**, 498.
[110a] Fink. W., *Angew. Chem. Internat. Edn.*, 1966, **5**, 760.
[111] Fink, W., *Helv. Chim. Acta,* 1966, **49**, 1408.
[112] Fink, W., Ger. Patent 1,238,027 (1967).
[113] Fink, W., *Helv. Chim. Acta,* 1967, **50**, 1131.
[114] Fink, W., *Helv. Chim. Acta,* 1968, **51**, 954.
[115] Fink, W., *Helv. Chim. Acta,* 1968, **51**, 974.
[116] Fink, W., *Helv. Chim. Acta,* 1968, **51**, 1011.
[117] Fink, W., *Helv. Chim. Acta,* 1969, **52**, 1833.
[118] Fink, W., *Helv. Chim. Acta,* 1969, **52**, 1841.
[119] Fink, W., U.S. Patent 3,509,194 (1970).
[119a] Fink, W., and Wheatley, P. J., *J. Chem. Soc. (A),* 1967, 1517.
[119b] Frainnet, E., Basouin, A., and Calas, R., *Compt. Rend.*, 1963, **257C**, 1304.
[120] Frainnet, E., and Duboudin, F., *Compt. Rend.*, 1966, **262C**, 1693.
[121] Fraser, G. W., Peacock, R. D., and Watkins, P. M., *J. Chem. Soc. (A),* 1971, 1125.
[121a] Gaidis, J. M., and West, R., *J. Amer. Chem. Soc.*, 1964, **86**, 5699.
[122] George, T. A., and Lappert, M. F., *J. Chem. Soc. (A),* 1969, 992.
[123] Gerschler, L., and Wannagat, U., *J. Organometallic Chem.*, 1971, **29**, 217.

[124] Gierut, J. A., Sowa, F. J., and Nieuwland, J. A., *J. Amer. Chem. Soc.*, 1936, **58**, 786.
[125] Gilman, H., Hoffert, B., Melvin, H. W., and Dunn, G. E., *J. Amer. Chem. Soc.*, 1950, **72**, 5767.
[126] Gilman, H., and Lichtenwalter, G. D., *J. Amer. Chem. Soc.*, 1960, **82**, 3319.
[127] Gilman, H., and Smart, R., *J. Org. Chem.*, 1951, **16**, 424.
[128] Glidewell, C., Rankin, D. W. H., Robiette, A. G., and Sheldrick, G. M., *J. Mol. Struct.*, 1969, **4**, 215.
[128a] Glidewell, C., Rankin, D. W. H., Robiette, A. G., and Sheldrick, G. M., *J. Mol. Struct.*, 1970, **6**, 231.
[128b] Glidewell, C., Rankin, D. W. H., Robiette, A. G., and Sheldrick, G. M., *J. Chem. Soc. (A)*, 1970, 318.
[129] Gol'din, G. S., Baturina, L. S., and Gavrilova, T. N., *J. Gen. Chem. U.S.S.R.*, 1975, **10**, 2152.
[130] Goossens, J. C., Fr. Patent 1,426,346 (1966).
[131] Goossens, J. C., Fr. Patent 1,548,896 (1969).
[132] Goubeau, J., and Jiminez-Barbera, J., *Z. Anorg. Chem.*, 1960, **303**, 217.
[133] Green, M. C., Lappert, M. F., and Lynch, J., unpublished results, 1969.
[134] Green, M. C., Lappert, M. F., Pedley, J. B., Schmidt, W., and Wilkins, W. T., *J. Organometallic Chem.*, 1971, **31**, C55.
[135] Gorislavskaya, Zh. V., Mal'nova, G. N., Zhinkin, D. Ya., Andronov, V. F., and Ainshtein, A. N., *J. Gen. Chem. U.S.S.R.*, 1972, **42**, 2200.
[136] Grosse-Ruyken, H., and Kleesaat, R., *Z. Anorg. Chem.*, 1961, **308**, 122.
[137] Grosse-Ruyken, H., and Schaarschmidt, K., *Chem. Tech. (Berlin)*, 1959, **11**, 451.
[137a] Gümrükçü, I., Lappert, M. F., and Pedley, J. B., unpublished work, 1978.
[138] Gverdsiteli, I. M., Baramidze, L. V., and Dzheliya, M. I., *Zh. Obshch. Khim.*, 1972, **42**, 2019.
[138a] Gynane, M. J. S., Lappert, M. F., and Power, P. P., unpublished results, 1976.
[139] Haber, C. P., U.S. Patent 2,553,314 (1951).
[140] Haiduc, I., and Gilman, H., *J. Organometallic Chem.*, 1969, **18**, P5.
[141] Haiduc, I., and Gilman, H., *Synth. Inorg. Metal-Org. Chem.*, 1971, **1**, 69 and 75.
[142] Harden, A., *J. Chem. Soc.*, 1887, **51**, 40.
[143] Harold, J. F. X., *J. Amer. Chem. Soc.*, 1898, **20**, 13.
[143a] Harris, D. H., and Lappert, M. F., *J. Organometallic Chem. Library*, 1976, **2**, 13.
[143b] Harris, D. H., Lappert, M. F., Pedley, J. B., and Sharp, G. J., *J.C.S. Dalton*, 1976, 945.
[144] Hedberg, K., *J. Amer. Chem. Soc.*, 1955, **77**, 6491.
[145] Hengge, E., and Wolfer, D., *J. Organometallic Chem.*, 1974, **66**, 413.

[146] Hess, G. G., Lampe, L. W., and Sommer, L. H., *J. Amer. Chem. Soc.*, 1965, **87**, 5327.
[147] Heyna, J., and Bauer, A., Ger. Patent 834,990 (1952).
[148] Highsmith, R. E., Bergerund, J. R., and MacDiarmid, A. G., *Chem. Comm.*, 1971, 48.
[149] Hils, J., Hagen, V., Ludwig, H., and Rühlmann, K., *Chem. Ber.*, 1966, **99**, 776.
[150] Hundeck, J., and Volkamer, K., *Internat. Symp. Organosilicon Chem., Sci. Comm. (Prague)*, 1965, 320.
[151] Hurwitz, M. J., and De Benneville, P. L., U.S. Patent 2,876,234 (1959).
[152] Ishii, Y., and Itoh, K., *Asaki Garasu Kogyo Gijutsu Shoreikai Kenkyu Hokoku*, 1968, **14**, 39.
[153] Ismail, R. M., and Kötzsch, H. J., *Z. Anorg. Chem.*, 1968, **363**, 305.
[154] Itoh, K., Matsuzaki, K., and Ishii, Y., *J. Chem. Soc. (C)*, 1968, 2709.
[155] Itoh, K., Katsuura, T., Matsuda, I., and Ishii, Y., *J. Organometallic Chem.*, 1972, **34**, 63.
[156] Ivin, S. Z., Provnonenko, V. K., and Konopatova, G. V., *J. Gen. Chem. U.S.S.R.*, 1967, **37**, 1600.
[157] Jaffé, H. H., *J. Chem. Phys.*, 1954, **58**, 185.
[158] Jarvie, S. W., and Lewis, D., *J. Chem. Soc.*, 1963, 4758.
[159] Joffe, I., and Post, H. W., *J. Org. Chem.*, 1949, **14**, 421.
[160] Johannson, O. K., U.S. Patent 2,429,883 (1947).
[161] Kannengiesser, G., and Damm, F., *Bull. Soc. Chim. Fr.*, 1969, 891.
[162] Kaufmann, K. D., Bormann, H., Rühlmann, K., Engelhardt, G., and Kriegsmann, H., *J. Organometallic Chem.*, 1970, **23**, 385.
[163] Kaufmann, K. D., Mann, U., and Rühlmann, K., *Z. Chem.*, 1965, **5**, 188; *Chem. Abstr.*, 1965, **63**, 14896c.
[163a] Kimura, K., Katada, K., and Bauer, S. H., *J. Amer. Chem. Soc.*, 1966, **88**, 416.
[163b] Kipping, F. S., *J. Chem. Soc.*, 1907, **91**, 209.
[163c] Kipping, F. S., *J. Chem. Soc.*, 1907, **91**, 717.
[164] Klingebiel, U., Fischer, O., and Meller, A., *Monatsh.*, 1975, **106**, 459.
[165] Klingebiel, U., and Meller, A., *Chem. Ber.*, 1975, **108**, 155.
[166] Klingebiel, U., and Meller, A., *Chem. Ber.*, 1976, **109**, 2430.
[167] Klingebiel, U., and Meller, A., *Angew. Chem. Internat. Edn.*, 1976, **15**, 312.
[168] Klingebiel, U., and Meller, A., *J. Organometallic Chem.*, 1975, **88**, 149.
[169] Klingebiel, U., and Meller, A., *Z. Naturforsch.*, 1977, **32b**, 537.
[170] Klingebiel, U., *Z. Naturforsch.*, 1978, **33b**, 521.
[170a] Klingebiel, U., *Chem. Ber.*, 1978, **111**, 2735.
[171] Klebe, J. F., Fr. Patent 1,442,585 (1966).
[172] Klebe, J. F., Bush, J. B., and Lyons, J. E., *J. Amer. Chem. Soc.*, 1964, **86**, 4400.

[173] Klebe, J. F., and Finkbeiner, H., *J. Amer. Chem. Soc.*, 1968, **90**, 7255.
[174] Klebe, J. F., Finkbeiner, H., and White, D. M., *J. Amer. Chem. Soc.*, 1966, **88**, 3390.
[174a] Kokoreva, I. Y., Syrkin, Ya. K., Babich, E. D., and Vdovin, V. M., *J. Struct. Chem.*, 1967, **8**, 978.
[175] Komarov, N. V., and Loi, L. I., *J. Gen. Chem. U.S.S.R.*, 1976, **46**, 1169.
[175a] Komarov, N. V., and Ol'khovskaya, L. I., *J. Gen. Chem. U.S.S.R.*, 1977, **47**, 1099.
[176] Komoriya, A., and Yoder, C. H., *J. Amer. Chem. Soc.*, 1972, **94**, 5285.
[176a] Kowalski, J., and Lasocki, Z., *J. Organometallic Chem.*, 1976, **116**, 75.
[177] Kozlova, L. V., Zhinkin, D. Ya., and Pozhidaev, E. D., *Zh. Obshch. Khim.*, 1969, 2511.
[178] Kraus, C. A., and Nelson, W. K., *J. Amer. Chem. Soc.*, 1934, **56**, 195.
[178a] Kraus, C. A., and Rosen, R., *J. Amer. Chem. Soc.*, 1925, **47**, 2739.
[179] Kriegsmann, H., *Z. Elektrochem.*, 1957, **61**, 1088.
[180] Kriegsmann, H., and Engelhardt, G., *Z. Anorg. Chem.*, 1961, **310**, 100.
[181] Kruger, C., *J. Organometallic Chem.*, 1967, **9**, 125.
[182] Kummer, D., and Baldeschwieler, J. D., *J. Phys. Chem.*, 1963, **67**, 98.
[183] Kummer, D., and Rochow, E. G., *Z. Anorg. Chem.*, 1963, **321**, 21.
[184] Lamming, F. C., and Emmanuel, V. K., *J. Chem. Eng. Data*, 1962, **7**, 558.
[185] Langer, S. H., Connell, S., and Wender, I., *J. Org. Chem.*, 1958, **23**, 50.
[186] Lappert, M. F., and Lynch, J., *Chem. Comm.*, 1968, 750.
[186a] Lappert, M. F., Patil, D. S., and Pedley, J. B., *J.C.S. Chem. Comm.*, 1975, 830.
[187] Lappert, M. F., Pedley, J. B., and Riley, P. N. K., unpublished work, 1967.
[187a] Lappert, M. F., Pedley, J. B., Sharp, G. J., and Bradley, D. C., *J.C.S. Dalton*, 1976, 1737.
[187b] Lappert, M. F., and Power, P. P., unpublished observations, 1977.
[188] Lappert. M. F., and Sanger, A. R., unpublished results (see also ref. [276b]).
[189] Lappert, M. F., and Srivastava, G., *Inorg. Nucl. Chem. Letters*, 1965, **1**, 53.
[190] Larsson, E., *Acta Chem. Scand.*, 1954, **8**, 1084.
[191] Larsson, E., *Kgl. Fysiogr. Sällsk. Lund Forh.*, 1954, **24**, 139.
[192] Larsson, E., *Kgl. Fysiogr. Sällsk. Lund Forh.*, 1954, **24**, 145.
[193] Larsson, E., *Kgl. Fysiogr. Sällsk. Lund Forh.*, 1954, **24**, 149.
[194] Larsson, E., *Kgl. Fysiogr. Sällsk. Lund Forh.*, 1958, **28**, 1.
[195] Larsson, E., and Bjellerup, L., *J. Amer. Chem. Soc.*, 1953, **75**, 995.
[196] Larsson, E., and Carlson, C., *Acta Chim. Scand.*, 1950, **4**, 45.
[197] Larsson, E., and Marin, R. E. I., *Acta Chim. Scand.*, 1951, **5**, 1173.
[198] Larsson, E., and Marin, R. E. I., Swed. Patent 138,357 (1952).
[199] Larsson, E., and Mjörne, O., *Svensk. Kem. Tidskr.*, 1949, **61**, 59.
[200] Larsson, E., and Mjörne, O., *Trans. Chalmers Univ., Gothenburg*, 1949, **87**, 29; *Chem. Abstr.*, 1950, **44**, 1402.

[201] Larsson, E., and Smith, B., *Acta Chim. Scand.*, 1949, **3**, 48.
[202] Larsson, E., and Smith, B., *Svensk. Kem. Tidskr.*, 1950, **62**, 141.
[202a] Lebedev, E. P., and Valimukhametova, R. G., *J. Gen. Chem. U.S.S.R.*, 1977, **47**, 978.
[203] Lengfield, F., *Amer. Chem. J.*, 1899, **21**, 531.
[204] Levy, H., *J. Inorg. Nucl. Chem.*, 1967, **29**, 1859.
[205] Limburg, W. W., and Post, H. W., *Rec. Trav. Chim.*, 1962, **81**, 430.
[206] Lukevics, E., Liberts, L., and Voronkov, M. G., *J. Gen. Chem. U.S.S.R.*, 1968, 1838.
[207] Lukevics, E., Pestunovich, A. E., Gaile, R., Pestunovich, V. A., and Voronkov, M. G., *Zh. Obshch. Khim.*, 1970, **40**, 620.
[208] Lukevics, E., Pestunovich, A. E., and Voronkov, M. G., *Khim. Geterosikl. Soedin. Akad. Nauk Latv. S.S.S.R.*, 1969, 647.
[209] Lukevics, E., and Voronkov, M. G., *Khim. Geterosikl. Soedin., Akad. Nauk Latv. S.S.S.R.*, 1965, 36: *Chem. Abstr.*, 1965, **63**, 2994.
[210] Lyashenko, I. N., Nametkin, N. S., and Chernysheva, T. I., *Bull. Acad. Sci. U.S.S.R., Div. Chem. Sci.*, 1969, 1132.
[210a] MacDiarmid, A. G., *Adv. Inorg. Chem. Radiochem.*, 1961, **3**, 207.
[211] Manasevit, H. M., *U.S. Dept. Com., Office Tech. Ser., P.B. Rept.*, 1959, 143,572,1.
[211a] Marchand, A., Forel, M.-T., Rivière-Baudet, M., and Soulard, M.-H., *J. Organometallic Chem.*, 1978, **156**, 341.
[212] Martel, B., and Aly, E., *J. Organometallic Chem.*, 1971, **29**, 61.
[213] Matsuda, I., Itoh, K., and Ishii, Y., *J. Chem. Soc. (C)*, 1969, 701.
[214] Matsuda, I., Itoh, K., and Ishii, Y., *J. Organometallic Chem.*, 1969, **19**, 339.
[215] Matsuda, I., Itoh, K., and Ishii, Y., *J. Organometallic Chem.*, 1974, **69**, 353.
[216] Metras, F., and Valade, J., *Bull. Soc. Chim. Fr.*, 1965, 1423.
[217] Michaelis, A., and Luxembourg, K., *Ber.*, 1896, **29**, 710.
[218] Mironov, V. F., Fedetova, N. S., and Rybalka, I. G., *Khim. Geterosikl. Soedin.*, 1968, 1124.
[219] Mironov, V. F., Petrov, A. D., and Pogonkina, N. A., *Bull. Acad. Sci. U.S.S.R., Div. Chem. Sci.*, 1955, 768.
[220] Mironov, V. F., Sheludyakov, V. D., and Kozyukov, V. P., *J. Gen. Chem. U.S.S.R.*, 1969, **39**, 220.
[221] Mironov, V. F., Sheludyakov, V. D., Kozyukov, V. P., and Khatuntsev, G. D., *Proc. Acad. Sci. U.S.S.R.*, 1968, **181**, 115.
[222] Mital, R. L., Goyal, R. D., and Gupta, R. R., *Inorg. Chem.*, 1972, **11**, 1924.
[223] Mjörne, O., *Svensk. Kem. Tidskr.*, 1950, **62**, 120.
[224] Moedritzer, K., and Van Wazer, J. R., *Inorg. Chem.*, 1964, **3**, 268.
[225] Monsanto Co., Neth. Patent 6,507,996 (1965).

[226] Monsanto Co., Brit. Patent 1,135,248 (1969).
[227] Mootz, D., Fayos, J., and Zinnius, A., *Angew. Chem. Internat. Edn.*, 1972, **11**, 58.
[228] Moscony, J. J., and MacDiarmid, A. G., *Chem. Comm.*, 1965, 307.
[229] Müller, M., and Van Wazer, J. R., *J. Organometallic Chem.*, 1970, **23**, 395.
[230] Nagy, J., Hencsei, P., and Gergö, E., *Z. Anorg. Chem.*, 1969, **367**, 293.
[231] Nametkin, N. S., Grushevenko, I. A., Perchenko, V. N., and Kamneva, G. L., *Proc. Acad. Sci. U.S.S.R.*, 1972, **207**, 920.
[232] Nametkin, N. S., Perchenko, V. N., and Batalova, L. G., *Proc. Acad. Sci. U.S.S.R.*, 1965, **160**, 1087.
[233] Nametkin, N. S., Perchenko, V. N., and Kuzovkina, M. E., *Proc. Acad. Sci. U.S.S.R.*, 1968, **182**, 842.
[234] Nametkin, N. S., Topchiev, A. V., Chernysheva, T. I., and Kirtasheva, L. I., *Izv. Akad. Nauk S.S.S.R., Otdel. Khim. Nauk,* 1963, 654.
[235] Nametkin, N. S., Topchiev, A. V., Povarov, L. S., and Garneshevskaya, G. V., *Dokl. Akad. Nauk S.S.S.R.*, 1956, **109**, 787.
[236] Nametkin, N. S., Vdovin, V. M., and Babich, E. D., *Khim. Geterosikl. Soedin. Akad. Nauk Latv. S.S.S.R.*, 1967, 148; *Chem. Abstr.*, 1967, **67**, 32732q.
[237] Nametkin, N. S., Vdovin, V. M., Babich, E. D., and Oppengeim, W. D., *Khim. Geterosikl. Soedin, Akad. Nauk Latv. S.S.S.R.*, 1965, **3**, 455; *Chem. Abstr.*, 1965, **63**, 13308b.
[238] Niederprüm, H., and Simmler, W., *Ber.*, 1963, **96**, 965.
[239] Noll, J. E., Speier, J. L., and Daubert, B. F., *J. Amer. Chem. Soc.*, 1951, **73**, 3867.
[240] Noskov, V. G., Kirpichnikov, A. H., and Sokalskii, M. A., *J. Gen. Chem. U.S.S.R.*, 1973, **43**, 2079.
[241] Nöth, H., *Z. Naturforsch.*, 1961, **16b**, 618.
[242] Nöth, H., and Tinhof, W., *Chem. Ber.*, 1975, **108**, 3109.
[242a] Nöth, H., Tinhof, W., and Taeger, T., *Chem. Ber.*, 1974, **107**, 3113.
[242b] Nöth, H., Tinhof, W., and Wrackmeyer, B., *Chem. Ber.*, 1974, **107**, 518.
[243] Oertel, G., Holtschmidt, H., and Maltz, H., Ger. Patent 1,157,226 (1963).
[243a] Onan, K. D., McPhail, A. T., Yoder, C. H., and Hillyard, R. W., *J.C.S. Chem. Comm.*, 1978, 209.
[244] Osthoff, R. C., and Kantor, S. W., *Inorg. Synth.*, 1957, **5**, 55.
[245] Pacl, Z., Jakonblova, M., Papouskova, Z., and Chvalovský, V., *Coll. Czech. Chem. Comm.*, 1971, **36**, 1588.
[246] Paul, R. C., Dhindsa, K. S., Ahluwalia, S. C., and Narula, S. P., *J. Inorg. Nucl. Chem.*, 1972, **34**, 1813.
[247] Paul, R. C., Kapula, B., Aggarwal, U. K., and Narula, S. P., *J. Inorg. Nucl. Chem.*, 1972, **34**, 2968.
[248] Pepe, E. H., and Kanner, B., U.S. Patent 3,054,818.
[249] Perkins, P. G., *Chem. Comm.*, 1967, 268.

[250] Pfleger, H., *U.S. Dept., Com. Office Tech. Ser.*, AD263,736 (1961).
[251] Pike, R. M., *J. Org. Chem.*, 1961, **26**, 232.
[252] Pike, R. M., *Rec. Trav. Chim.*, 1961, **80**, 819.
[253] Piper, F., and Rühlmann, K., *J. Organometallic Chem.*, 1976, **121**, 149.
[254] Pitt, C. G., and Fowler, M. S., *J. Amer. Chem. Soc.*, 1967, **89**, 6792.
[254a] Plazank, J., Metras, F., Marchand, A., and Valade, J., *Bull. Soc. Chim. Fr.* 1967. 1920.
[255] Prober, M., *J. Amer. Chem. Soc.*, 1956, **78**, 2274.
[256] Prokhorova, V. A., and Reikhsf'eld, V. O., *J. Gen. Chem. U.S.S.R.*, 1963, **33**, 2617.
[257] Pump, J., and Wannagat, U., *Monatsh.*, 1962, **93**, 352.
[258] Randall, E. W., *Inorg. Nucl. Chem. Letters*, 1965, **1**, 109.
[259] Randall, E. W., Ellner, J. J., and Zuckerman, J. J., *J. Amer. Chem. Soc.*, 1966, **88**, 622.
[259a] Randall, E. W., and Zuckerman, J. J., *J. Amer. Chem. Soc.*, 1968, **90**, 3167.
[259b] Rankin, D. W. H., Robiette, A. G., Sheldrick, G. M., Sheldrick, W. S., Aylett, B. J., Ellis, I. A., and Monaghan, J. J., *J. Chem. Soc. (A)*, 1969, 1224.
[259c] Reuter, K., and Neumann, W. P., *Tetrahedron Letters*, 1978, 5235.
[260] Reynolds, J. E., *J. Chem. Soc.*, 1889, **55**, 475.
[261] Reynolds, J. E., *J. Chem. Soc.*, 1909, **95**, 505.
[262] Reynolds, J. E., *J. Chem. Soc.*, 1909, **95**, 508.
[263] Reynolds, H. H., Bigelow, L. A., and Kraus, C. A., *J. Amer. Chem. Soc.*, 1929, **51**, 3067.
[263a] Robiette, A. G., Sheldrick, G. M., and Sheldrick, W. S., *J. Mol. Struct.*, 1970, **5**, 423.
[264] Robiette, A. G., Sheldrick, G. M., Sheldrick, W. S., Beagley, B., Cruickshank, D. W. J., Monaghan, J. J., Aylett, B. J., and Ellis, I. A., *Chem. Comm.*, 1968, 909.
[265] Roper, W. R., and Wilkins, C. J., *Trans. Faraday Soc.*, 1962, **58**, 1686.
[266] Rudman, R., Hamilton, W. C., Novick, S., and Goldfarb, T. D., *J. Amer. Chem. Soc.*, 1967, **89**, 5157.
[267] Rühlmann, K., *J. Prakt. Chem.*, 1959, **9**, 86.
[268] Rühlmann, K., *Chem. Ber.*, 1961, **94**, 2311.
[269] Rühlmann, K., *J. Prakt. Chem.*, 1962, **16**, 172.
[270] Rühlmann, K., Ger. Patent 1,163,826 (1964).
[271] Rühlmann, K., *Chem. Ber.*, 1965, **98**, 1814.
[272] Rühlmann, K., Kauffmann, K. D., and Mann, U., *Z. Chem.*, 1965, **5**, 107; *Chem. Abstr.*, 1965, **63**, 14895h.
[273] Rühlmann, K., and Kuhrt, G., *Angew. Chem. Internat. Edn.*, 1968, **7**, 809.
[274] Rühlmann, K., and Mansfield, M., *J. Prakt. Chem.*, 1964, **24**, 226.

[275] Ruff, O., *Ber.*, 1908, **41**, 3738.
[276] Ryskin, Ya. I., and Voronkov, M. G., *Coll. Czech. Chem. Comm.*, 1959, **24**, 3816.
[276a] Sakurai, H., Hosomi, A., Nakajima, J., and Kumada, M., *Bull. Chem. Soc. Jap.*, 1966, **39**, 2263.
[277] Sauer, R. O., and Hasek, R. H., *J. Amer. Chem. Soc.*, 1946, **68**, 241.
[278] Scantlin, W., and Norman, A. D., *Chem. Comm.*, 1971, 1246.
[278a] Schaarschmidt, K., *Z. Anorg. Chem.*, 1961, **310**, 69.
[279] Schaeffer, C. D., Griffith, D. R., and Yoder, C. H., *J. Inorg. Nucl. Chem.*, 1970, **32**, 3689.
[279a] Scherer, O. J., *Organometallic Chem. Rev.*, 1968, **3**, 281.
[279b] Scherer, O. J., *Angew. Chem. Internat. Edn.*, 1969, **8**, 861.
[280] Scherer, O. J., Biller, D., and Schmidt, M., *Inorg. Nucl. Chem. Letters*, 1966, **2**, 103.
[281] Scherer, O. J., and Hornig, P., *Angew. Chem. Internat. Edn.*, 1967, **6**, 89.
[282] Scherer, O. J., and Hornig, P., *Chem. Ber.*, 1968, **101**, 2533.
[283] Scherer, O. J., and Janssen, W., *J. Organometallic Chem.*, 1969, **16**, P69.
[284] Scherer, O. J., and Schmidt, M., *Chem. Ber.*, 1965, **98**, 2243.
[285] Scherer, O. J., and Schmidt, M., *J. Organometallic Chem.*, 1965, **3**, 156.
[286] Scherer, O. J., and Schmidt, M., *Inorg. Nucl. Chem. Letters*, 1966, **2**, 101.
[287] Scherer, O. J., and Schmidt, M., *Internat. Symp. Organosilicon Chem. Sci. Comm., (Prague)*, 1965, 315.
[288] Schirawski, G., and Wannagat, U., *Monatsh.*, 1969, **100**, 1901.
[289] Schott, G., and Gastmeier, G., *Z. Chem.*, 1961, **1**, 123.
[290] Schreiner, G., Pohl, J., and Wannagat, U., *Monatsh.*, 1965, **96**, 1909.
[291] Schumann, H., and Ronecker, S., *J. Organometallic Chem.*, 1970, **23**, 451.
[291a] Schumann, H., Schumann-Ruidisch, I., and Ronecker, S., *Z. Naturforsch.*, 1970, **25b**, 565.
[292] Schwarz, R., and Weigel, F., *Z. Anorg. Chem.*, 1951, **268**, 291.
[293] Semenova, E. A., Zhinkin, D. Ya., and Andrianov, K. A., *Bull. Acad. Sci. U.S.S.R., Div. Chem. Sci.*, 1962, 2036.
[294] Sergeeva, Z. I., and Hsieh, C.-L., *J. Gen. Chem. U.S.S.R.*, 1962, **32**, 1987.
[294a] Shea, K. J., Gobeille, R., Bramblett, J., and Thompson, E., *J. Amer. Chem. Soc.*, 1978, **100**, 1611.
[295] Sheludyakov, V. D., Kozyukov, V. P., and Mironov, V. F., *Zh. Obshch. Khim.*, 1970, **40**, 1170.
[295a] Sheludyakov, V. D., Tkachev, A. S., Sheludyakova, S. V., and Mironov, V. F., *J. Gen. Chem. U.S.S.R.*, 1977, **47**, 974.
[296] Shiina, K., *J. Amer. Chem. Soc.*, 1972, **94**, 9266.
[297] Shikhiev, I. A., and Shiraliev, V. M., *Zh. Obshch. Khim.*, 1970, **40**, 1075.

[298] Shostakovskii, M. F., Kochkin, D. A., and Rogov, U. M., *Izvest. Akad. Nauk S.S.S.R., Otdel. Khim. Nauk,* 1956, 1062; *Chem. Abstr.*, 1957, **51,** 4983.
[299] Shostakovskii, M. F., Kochkin, D. A., Shikieva, I. A., and Vlasov, U. M., *Zh. Obshch. Khim.*, 1955, **25,** 622; *Chem. Abstr.*, 1956, **50,** 3270.
[300] Shostakovskii, M. F., and Kondrat'ev, K. I., *Izvest. Akad. Nauk S.S.S.R., Otdel. Khim. Nauk,* 1956, 81; *Chem. Abstr.*, 1957, **51,** 3486.
[301] Silbiger, J., and Fuchs, J., Israeli Patent 22,431.
[302] Silbiger, J., and Fuchs, J., *Inorg. Chem.*, 1965, **4,** 1371.
[303] Simmler, W., Walz, H., and Niederprüm, H., *Chem. Ber.*, 1963, **96,** 1495.
[303a] Smith, G. S., and Alexander, L., *Acta Cryst.*, 1963, **16,** 1015.
[304] Srivastava, G., *J. Organometallic Chem.*, 1974, **69,** 179.
[305] Sommer, L. H., and Citron, J. D., *J. Amer. Chem. Soc.*, 1967, **89,** 5797.
[306] Sommer, L. H., and Frye, C. L., U.S. Patent 3,024,262 (1962).
[307] Speier, J. L., U.S. Patent 2,762,823 (1956).
[308] Speier, J. L., U.S. Patent 3,485,857 (1969).
[309] Speier, J. L., Zimmerman, R., and Webster, J., *J. Amer. Chem. Soc.*, 1956, **78,** 2278.
[310] Steward, H. F., Koepsell, D. G., and West, R., *J. Amer. Chem. Soc.*, 1970, **92,** 846.
[311] Stock, A., and Somieski, K., *Chem. Ber.*, 1921, **54,** 740.
[312] Sujishi, S., and Witz, S., *J. Amer. Chem. Soc.*, 1954, **76,** 4631.
[313] Sujishi, S., and Witz, S., *J. Amer. Chem. Soc.*, 1957, **79,** 2447.
[314] Sytov, G. A., Ledina, L. E., Sul, V., Krip, A. H., Perchenko, V. N., and Nametkin, N. S., *Doklady. Chem.*, 1973, **212,** 718.
[315] Szretter-Szmid, M., and Urbanski, T., *Tetrahedron Letters,* 1967, 2131.
[316] Takiguchi, T., and Susuki, M., *Bull. Chem. Soc. Japan,* 1969, **42,** 2708.
[317] Tansjo, L., *Acta Chim. Scand.*, 1957, **11,** 1613.
[318] Tansjo, L., *Acta Chim. Scand.*, 1959, **13,** 29.
[319] Tansjo, L., *Acta Chim. Scand.*, 1959, **13,** 35.
[320] Tansjo, L., *Acta Chim. Scand.*, 1960, **14,** 2097.
[321] Tansjo, L., *Acta Chim. Scand.*, 1961, **15,** 1583.
[322] Tansjo, L., *Acta Chim. Scand.*, 1964, **18,** 456.
[323] Tansjo, L., *Acta Chim. Scand.*, 1964, **18,** 465.
[324] Thames, S. F., McClesky, J. E., and Kelly, P. L., *J. Heterocyclic Chem.*, 1968, **5,** 749.
[325] See ref. [119b].
[326] Tsai, T. T., and Marshall, C. J., *J. Org. Chem.*, 1969, **34,** 3676.
[327] Van Dyke, C. H., and MacDiarmid, A. G., *Inorg. Chem.*, 1964, **3,** 1071.
[328] Van Wazer, J. R., and Moedritzer, K., *J. Inorg. Nucl. Chem.*, 1964, **26,** 737.
[328a] Van Wazer, J. R., and Norval, S., *Inorg. Chem.*, 1965, **4,** 1294.
[329] Varezhkin, Yu. M., Zhinkin, D. Ya., Morgunova, M. M., and Bochkarev,

V. N., *J. Gen. Chem. U.S.S.R.*, 1975, **45**, 2410.
[330] Vilkov, L. V., and Tarasenko, N. A., *Chem. Comm.*, 1969, 1176.
[331] Vostokov, I. A., Dergunov, Yu. I., Kozyukov, V. P., Sheludyakov, V. D., Mironov, V. F., and Anisimova, V. Z., *J. Gen. Chem. U.S.S.R.*, 1973, **43**, 621.
[332] Vostokov, I. A., Dergunov, Yu. I., Mironov, V. F., Sheludyakov, V. D., and Kozyukov, V. P., *J. Gen. Chem. U.S.S.R.*, 1974, **44**, 2116.
[333] Vostokov, I. A., Dergunov, Yu. I., Mironov, V. F., Sheludyakov, V. D., and Kozyukov, V. P., *J. Gen. Chem. U.S.S.R.*, 1975, **45**, 1989.
[334] Walter, W., and Lüke, H., *Angew. Chem. Internat. Edn.*, 1975, **14**, 428.
[335] Walton, D. R. M., *J. Chem. Soc. (C)*, 1966, 1706.
[336] Walton, D. R. M., Brit. Patent 1,153,132 (1969).
[336a] Wannagat, U., *Adv. Inorg. Chem. Radiochem.*, 1964, **6**, 225.
[337] Wannagat, U., Behmel, K., Bürger, H., *Chem. Ber.*, 1964, **97**, 2029.
[338] Wannagat, U., and Bogusch, E., *Inorg. Nucl. Chem. Letters*, 1966, **2**, 97.
[339] Wannagat, U., Bogusch, E., and Hofler, F., *J. Organometallic Chem.*, 1967, **7**, 203.
[340] Wannagat, U., Bogusch, E., and Geymayer, P., *Monatsh.*, 1971, **102**, 1825.
[341] Wannagat, U., Bogusch, E., Geymayer, P., and Rabet, F., *Monatsh.*, 1971, **102**, 1844.
[342] Wannagat, U., and Braun, R., *Monatsh.*, 1969, **100**, 1910.
[342a] Wannagat, U., and Bürger, H., *Angew. Chem.*, 1964, **74**, 497.
[343] Wannagat, U., and Gerschler, L., *Ann.*, 1971, **744**, 111.
[344] Wannagat, U., Geymayer, P., and Bogusch, E., *Monatsh.*, 1965, **96**, 585.
[345] Wannagat, U., Geymayer, P., and Schreiner, G., *Angew. Chem.*, 1964, **76**, 99.
[346] Wannagat, U., Giesen, K. P., and Falius, H., *Monatsh.*, 1973, **104**, 1444.
[347] Wannagat, U., Krüger, C., and Niederprüm, H., *Z. Anorg. Chem.*, 1962, **314**, 80.
[347a] Wannagat, U., Herzig, J., and Bürger, H., *J. Organometallic Chem.*, 1970, **23**, 373.
[348] Wannagat, U., Herzig, J., Schmidt, P., and Schulze, M., *Monatsh.*, 1971, **102**, 1306, 1817.
[349] Wannagat, U., and Labuhn, D., *Z. Anorg. Chem.*, 1973, **402**, 147.
[350] Wannagat, U., and Labuhn, D., *Monatsh.*, 1973, **104**, 1457.
[351] Wannagat, U., and Meier, S., *Z. Anorg. Chem.*, 1972, **392**, 179.
[352] Wannagat, U., Moretto, H. H., and Schmidt, P., *Z. Anorg. Chem.*, 1972, **394**, 125.
[353] Wannagat, U., and Niederprüm, H., *Z. Anorg. Chem.*, 1961, **308**, 387.
[354] Wannagat, U., and Paul, U., *Monatsh.*, 1974, **105**, 1240.
[355] Wannagat, U., and Rabet, F., *Inorg. Nucl. Chem. Letters*, 1969, **5**, 789.
[356] Wannagat, U., and Schlingmann, M., *Z. Anorg. Chem.*, 1974, **406**, 7.

[357] Wannagat, U., and Schlingmann, M., *Z. Anorg. Chem.*, 1976, **419**, 108.
[358] Wannagat, U., and Schlingmann, M., *Z. Anorg. Chem.*, 1976, **419**, 1115.
[359] Wannagat, U., and Schlingmann, M., *Z. Anorg. Chem.*, 1976, **419**, 48.
[359a] Wannagat, U., Schlingmann, M., and Autzen, H., *Z. Naturforsch.*, 1976, **31b**, 621.
[360] Wannagat, U., Schmidt, P., and Schulze, M., *Angew. Chem. Internat. Edn.*, 1967, **6**, 446.
[361] Wannagat, U., Schmidt, P., and Schulze, M., *Angew. Chem. Internat. Edn.*, 1967, **6**, 447.
[362] Wannagat, U., and Schreiner, G., *Monatsh.*, 1965, **96**, 1889.
[363] Wannagat, U., and Schreiner, G., *Monatsh.*, 1965, **96**, 1895.
[364] Wannagat, U., and Schreiner, G., *Monatsh.*, 1965, **96**, 1916.
[365] Wannagat, U., and Schreiner, G., *Monatsh.*, 1968, **99**, 1372.
[366] Wannagat, U., and Schreiner, G., *Monatsh.*, 1968, **99**, 1376.
[367] Wannagat, U., and Schulze, M., *Inorg. Nucl. Chem. Letters,* 1969, **5**, 789.
[367a] Wannagat, U., and Seifert, R., *Monatsh.*, 1978, **109**, 209.
[367b] Wannagat, U., Seifert, R., and Schlingmann, M., *Z. Naturforsch.*, 1977, **32b**, 869.
[368] Wannagat, U., Smrekar, O., and Braun, R., *Monatsh.*, 1969, **100**, 1916.
[369] Wannagat, U., and Wismar, H.-J., *Monatsh.*, 1973, **104**, 1465.
[370] Washburne, S. S., and Peterson, W. R., *Inorg. Nucl. Chem. Letters,* 1969, **5**, 17.
[371] Washburne, S. S., and Peterson, W. R., *J. Organometallic Chem.*, 1970, **21**, 59.
[372] West, R., and Boujouk, P., *J. Amer. Chem. Soc.*, 1973, **95**, 3987.
[373] West, R., and Boujouk, P., *J. Amer. Chem. Soc.*, 1973, **95**, 3983.
[374] West, R., and Boujouk, P., *J. Amer. Chem. Soc.*, 1971, **93**, 5901.
[375] West, R., and Boujouk, P., *J. Amer. Chem. Soc.*, 1971, **93**, 5901.
[376] West, R., and Gornowicz, G. A., *J. Amer. Chem. Soc.*, 1971, **93**, 1714.
[377] West, R., Ishikawa, M., and Murai, S., *J. Amer. Chem. Soc.*, 1968, **90**, 727.
[378] Wheatley, P. J., *J. Chem. Soc.*, 1962, 1721.
[379] White, E., Krueger, P. M., and McCloskey, J. A., *J. Org. Chem.*, 1972, **37**, 430.
[379a] Wiberg, E., Stecher, O., and Neumaier, A., *Inorg. Nucl. Chem. Letters,* 1965, **1**, 33.
[379b] Wiberg, N., and Hübler, G. Z., *Z. Naturforsch.*, 1976, **31b**, 1317.
[380] Wiberg, N., and Neruda, B., *Chem. Ber.*, 1966, **99**, 740.
[381] Wiberg, N., and Pracht, H. J., *J. Organometallic Chem.*, 1972, **40**, 289.
[382] Wiberg, N., and Raschig, F., *J. Organometallic Chem.*, 1967, **10**, 15.
[382a] Wiberg, N., and Schmid, K. H., *Z. Anorg. Chem.*, 1966, **345**, 93.
[382b] Wiberg, N., and Uhlenbroch, W., *Angew. Chem. Internat. Edn.*, 1970, **9**, 70.

[383] Witke, K., Reich, P., and Kriegsmann, H., *Z. Anorg. Chem.*, 1971, **380**, 164; **381**, 280.
[384] Wittenberg, D., George, M. V., Wu, T. C., Miles, D., and Gilman, H., *J. Amer. Chem. Soc.*, 1958, **80**, 4532.
[385] Wittenberg, D., and Gilman, H., *Quart. Rev.*, 1959, **13**, 116.
[386] Yoder, C. H., Copenhafer, W. C., and Du Beshten, B., *J. Amer. Chem. Soc.*, 1974, **96**, 4283.
[387] Yoder, C. H., and Zuckerman, J. J., *Inorg. Chem.*, 1965, **4**, 116.
[388] Yoder, C. H., and Zuckerman, J. J., *J. Amer. Chem. Soc.*, 1966, **88**, 4831.
[388a] Yokoi, M., *Bull. Chem. Soc. Japan*, 1957, **30**, 100.
[389] Yokoi, M., and Yamasaki, K., *J. Amer. Chem. Soc.*, 1953, **75**, 4139.
[390] Zahn, C., Sharkey, A. G., and Wender, I., *U.S. Bur. Mines, Rept. Invest. No. 5976 (1962); Chem. Abstr.*, 1962, **57**, 4038.
[391] Zhigach, A. F., Sobolev, E. S., Svitsyn, R. A., and Nikitin, V. S., *J. Gen. Chem. U.S.S.R.*, 1973, **43**, 1949.
[392] Zhinkin, D. Ya., Mal'nova, G. N., and Gorislavspryn, Zh. U., *Zh. Obshch. Khim.*, 1968, **38**, 2800.
[393] Zhinkin, D. Ya., Mal'nova, G. N., and Solov'eva, T. P., *Zh. Obshch. Khim.*, 1971, **41**, 870.
[394] Zhinkin, D. Ya., Morgunova, M. M., and Andrianov, K. A., *Proc. Acad. Sci. U.S.S.R.*, 1965, **165**, 114.
[395] Zhinkin, D. Ya., Morgunova, M. M., Popkov, K. K., and Andrianov, K. A., *Proc. Acad. Sci. U.S.S.R.*, 1964, **158**, 641.
[396] Zhinkin, D. Ya., Semenova, E. A., and Markova, N. V., *J. Gen. Chem. U.S.S.R.*, 1963, **33**, 3736.

2. Germanium

[1] Abel, E. W., *J. Chem. Soc.*, 1958, 3746.
[2] Abel, E. W., Armitage, D. A., and Brady, D. B., *J. Organometallic Chem.*, 1966, **5**, 130.
[3] Anderson, H. H., *J. Amer. Chem. Soc.*, 1952, **74**, 1421.
[4] Anderson, H. H., *J. Amer. Chem. Soc.*, 1961, **83**, 547.
[5] Baldwin, J. C., Lappert, M. F., Pedley, J. B., and Poland, J. S., *J.C.S. Dalton*, 1972, 1943.
[6] Baudet, M., Thesis, Toulouse, France, 1966.
[7] Baudet, M., and Satgé, J., *Compt. Rend.*, 1966, **263C**, 435.
[8] Baukov, Yu. I., Burlachenko, G. S., Kostyuk, A. S., and Lutsenko, I. F., *Zh. Obshch. Khim.*, 1970, **40**, 707.
[9] Baum, G., Lehn, W. L., and Tamborski, C., *J. Org. Chem.*, 1964, **29**, 1264.
[10] Burch, G. M., and Van Wazer, J. R., *J. Chem. Soc. (A)*, 1966, 586.
[10a] Cotton, J. D., Cundy, C. S., Harris, D. H., Hudson, A., Lappert, M. F., and Lednor, P. W., *J.C.S. Chem. Comm.*, 1974, 651.

[11] Cuvigny, T., and Normant, H., *Compt. Rend.*, 1969, **268C**, 834.
[11a] Davidson, P. J., Harris, D. H., and Lappert, M. F., *J.C.S. Dalton*, 1976, 2286.
[12] Davidsohn, W. E., Dissertation, Braunschweig, 1961.
[13] Dousse, G., Satgé, J., and Rivière-Baudet, M., *Synth. Inorg. Metal-Org. Chem.*, 1973, **3**, 11.
[14] Elguero, J. M., Rivière-Baudet, M., and Satgé, J., *Compt. Rend.*, 1968, **266C**, 44.
[14a] Gibbins, S. G., Lappert, M. F., Pedley, J. B., and Sharp, G. J., *J.C.S. Dalton*, 1975, 72.
[15] Glarum, S. N., and Kraus, C. A., *J. Amer. Chem. Soc.*, 1950, **72**, 5398.
[15a] Glidewell, C., Rankin, D. W. H., and Robiette, A. G., *J. Chem. Soc. (A)*, 1970, 2935.
[16] Götze, H.-J., *Chem. Ber.*, 1975, **108**, 938.
[17] Gynane, M. J. S., Harris, D. H., Lappert, M. F., Power, P. P., Rivière, P., and Rivière-Baudet, M., *J.C.S. Dalton*, 1977, 2004.
[18] Haiduc, I., and Gilman, H., *J. Organometallic Chem.*, 1969, **18**, P5.
[19] Harris, D. H., and Lappert, M. F., *J.C.S. Chem. Comm.*, 1974, 895.
[20] Harris, D. H., Lappert, M. F., Pedley, J. B., and Sharp, G. J., *J.C.S. Dalton*, 1976, 945.
[20a] Hudson, A., Lappert, M. F., and Lednor, P. W., *J.C.S. Dalton*, 1976, 2369.
[21] Itoh, K., Katsuura, T., Matsuda, I., and Ishii, Y., *J. Organometallic Chem.*, 1972, **34**, 63.
[22] Itoh, K., Katsuura, T., Matsuda, I., and Ishii, Y., *J. Organometallic Chem.*, 1972, **34**, 73.
[22a] Kagan, M., B.Sc. Thesis, University of Sussex, 1976.
[23] Koester-Pflugmacher, A., and Hirsch, A., *J. Organometallic Chem.*, 1968, **12**, 349.
[24] Koester-Pflugmacher, A., and Termin, E., *Naturwiss.*, 1964, **51**, 552.
[24a] Kraus, C. A., and Foster, L. S., *J. Amer. Chem. Soc.*, 1927, **49**, 457.
[25] Kraus, C. A., and Wooster, C. B., *J. Amer. Chem. Soc.*, 1930, **52**, 372.
[26] Kraus, C. A., and Nutting, H. S., *J. Amer. Chem. Soc.*, 1932, **54**, 1622.
[27] Lappert, M. F., and Lednor, P. W., *Adv. Organometallic Chem.*, 1976, **14**, 345.
[28] Lappert, M. F., Miles, S. J., Power, P. P., Carty, A. J., and Taylor, N. J., *J.C.S. Chem. Comm.*, 1977, 458.
[28a] Lappert, M. F., Onyszchuk, M., and Slade, M. J., unpublished results, 1978.
[28b] Lappert, M. F., Patil, D. S., and Pedley, J. B., *J.C.S. Chem. Comm.*, 1975, 830.
[28c] Lappert, M. F., Power, P. P., Slade, M. J., Hedberg, L., Hedberg, K., and Schomaker, V., *J.C.S. Chem. Comm.*, 1979, 369.
[28d] Lappert, M. F., and Slade, M. J., unpublished results, 1978.

[28e] Lappert, M. F., Slade, M. J., Atwood, J. L., and Zaworotko, M. J., unpublished results, 1978.

[28f] Lesbre, M., Mazerolles, P., and Satgé, J., *The Organic Compounds of Germanium,* Interscience-Wiley, New York, 1971.

[29] Luijten, J. G. A., Rijkens, F., and van der Kerk, G. J. M., *Adv. Organometallic Chem.*, 1965, **3**, 397.

[30] Mack, J., and Yoder, C. H., *Inorg. Chem.*, 1969, **8**, 278.

[30a] Marchand, A., Forel, M.-T., Rivière-Baudet, M., and Soulard, M.-H., *J. Organometallic Chem.*, 1978, **156**, 341.

[31] Marchand, A., Rivière-Baudet, M., and Satgé, J., *J. Organometallic Chem.*, 1976, **107**, 33.

[32] Massol, M., and Satgé, J., *Bull. Soc. Chim. Fr.*, 1966, 2737.

[32a] Mehrotra, R. C., and Chandra, G., *Indian J. Chem.*, 1965, **3**, 497.

[33] Mironov, V. F., Sobolev, E. S., and Antipin, L. M., *Zh. Obshch. Khim.*, 1967, **37**, 1707.

[34] Pacl, Z., Jakovbuka, M., Rericha, R., and Chvalovský, V., *Coll. Czech. Chem. Comm.*, 1971, **36**, 2181.

[34a] Power, P. P., D.Phil. Thesis, University of Sussex, 1977.

[35] Quane, D., and Roberts, S. D., *J. Organometallic Chem.*, 1976, **108**, 27.

[36] Randall, E. W., Yoder, C. H., and Zuckerman, J. J., *Inorg. Chem.*, 1967, **6**, 744.

[37] Randall, E. W., and Zuckerman, J. J., *J. Amer. Chem. Soc.*, 1968, **90**, 3167.

[38] Rijkens, F., Janssen, M. J., and van der Kerk, G. J. M., *Rec. Trav. Chim.*, 1965, **84**, 1597.

[39] Rijkens, F., and van der Kerk, G. J. M., *Investigations in the Field of Organogermanium Chemistry*, T.N.O., Utrecht, 1964.

[39a] Rivière, P., Cazes, A., Castel, A., Rivière-Baudet, M., and Satgé, J., *J. Organometallic Chem.*, 1978, **155**, C58.

[40] Rivière, P., Rivière-Baudet, M., Couret, C., and Satgé, J., *Synth. Inorg. Metal-Org. Chem.*, 1974, **4**, 295.

[40a] Rivière-Baudet, M., Rivière, P., and Satgé, J., *J. Organometallic Chem.*, 1978, **154**, C23.

[41] Royden, P., Rocktäschel, C., and Mosch, W., *Angew. Chem.*, 1964, **76**, 860.

[42] Ruidish, I., and Schmidt, M., *Angew. Chem. Internat. Edn.*, 1964, **3**, 231.

[43] Ruidisch, I., and Schmidt, M., see ref. [49].

[44] Satgé, J., and Baudet, M., *Compt. Rend.*, 1966, **263C**, 435.

[45] Satgé, J., Lesbre, M., and Baudet, M., *Compt. Rend.*, 1964, **259C**, 4733.

[46] Schaeffer, C. D., and Griffith, D. R., *J. Inorg. Nucl. Chem.*, 1970, **32**, 3689.

[47] Scherer, O. J., and Biller, D., *Z. Naturforsch.*, 1967, **22b**, 1079.

[48] Scherer, O. J., and Schmidt, M., *Angew. Chem. Internat. Edn.*, 1963, **2**, 478.
[49] Schmidt, M., and Ruidisch, I., *Angew. Chem. Internat. Edn.*, 1964, **3**, 637.
[50] Schumann, H., and Ronecker, S., *J. Organometallic Chem.*, 1970, **23**, 451.
[51] Schumann, H., Schumann-Ruidisch, I., and Ronecker, S., *Z. Naturforsch.*, 1970, **25b**, 565.
[52] Schumann-Ruidisch, I., and Jutzi-Mebert, B., *J. Organometallic Chem.*, 1968, **11**, 77.
[53] Schumann-Ruidisch, I., Kalk, W., and Brüning, R., *Z. Naturforsch.*, 1968, **23b**, 307.
[54] Schwarz, R., and Reinhardt, W., *Ber.*, 1932, **65**, 1743.
[55] Smith F. B., and Kraus, C. A., *J. Amer. Chem. Soc.*, 1952, **74**, 1418.
[56] Teal, G. K., and Kraus, C. A., *J. Amer. Chem. Soc.*, 1950, **72**, 4706.
[57] Thomas, J. S., and Southwood, W. W., *J. Chem. Soc.*, 1931, 2083.
[57a] Wannagat, U., and Rabet, F., *Inorg. Nucl. Chem. Letters*, 1970, **6**, 155.
[58] Wannagat, U., and Schlingmann, M., *Z. Anorg. Chem.*, 1976, **419**, 115.
[58a] Wannagat, U., and Seifert, R., *Monatsh.*, 1978, **1**, 209.
[58b] Wannagat, U., Seifert, R., and Schlingmann, M., *Z. Naturforsch.*, 1977, **32b**, 869.
[59] Wieber, M., and Schmidt, M., *Z. Naturforsch.*, 1963, **18b**, 849.
[60] Wieber, M., and Schmidt, M., *Angew. Chem.*, 1963, **75**, 1116.
[61] Yoder, C. H., and Zuckerman, J. J., *Inorg. Chem.*, 1964, **3**, 1329.
[62] Yoder, C. H., and Zuckerman, J. J., *J. Amer. Chem. Soc.*, 1966, **88**, 2170.
[63] Yoder, C. H., and Zuckerman, J. J., *J. Amer. Chem. Soc.*, 1966, **88**, 4831.
[64] Yoder, C. H., and Zuckerman, J. J., *Inorg. Chem.*, 1966, **5**, 2055.
[65] Yoder, C. H., Moore, W. S., Copenhafer, W. C., and Sigel, J., *J. Organometallic Chem.*, 1974, **82**, 353.
[66] Ziegler, M., and Weiss, J., *Z. Naturforsch.*, 1971, **26b**, 735.

3. Tin

[1] Abel, E. W., Armitage, D. A., and Brady, D. B., *Trans. Faraday Soc.*, 1966, **62**, 3459.
[2] Abel, E. W., Brady, D. B., and Lerwill, B. R., *Chem. Ind. (London)*, 1962, 1333.
[3] Abel, E. W., and Dunster, M. O., *J. Organometallic Chem.*, 1973, **49**, 435.
[4] Anderson, H. H., *J. Amer. Chem. Soc.*, 1952, **74**, 1421.
[5] Anderson, J. W., and Drake, J. E., *J. Inorg. Nucl. Chem.*, 1973, **35**, 1032.
[6] Baldwin, J. C., Lappert, M. F., Pedley, J. B., and Poland, J. S., *J.C.S. Dalton*, 1972, 1943.
[7] Bird, S. R. A., Donaldson, J. D., Keppie, S. A., and Lappert, M. F., *J. Chem. Soc. (A)*, 1971, 1311.

[8] Bloodworth, A. J., *J. Chem. Soc. (C)*, 1968, 2380.
[9] Bloodworth, A. J., and Davies, A. G., *Proc. Chem. Soc.*, 1963, 264.
[10] Bloodworth, A. J., and Davies, A. G., *Proc. Chem. Soc.*, 1963, 315.
[11] Bloodworth, A. J., and Davies, A. G., *Chem. Comm.*, 1965, 24.
[12] Bloodworth, A. J., and Davies, A. G., *Chem. Ind. (London)*, 1965, 900.
[13] Bloodworth, A. J., and Davies, A. G., *Chem. Ind. (London)*, 1965, 1868.
[14] Bloodworth, A. J., and Davies, A. G., *J. Chem. Soc.*, 1965, 5238.
[15] Bloodworth, A. J., and Davies, A. G., *J. Chem. Soc.*, 1965, 6245.
[16] Bloodworth, A. J., and Davies, A. G., *J. Chem. Soc. (C)*, 1966, 299.
[17] Bloodworth, A. J., Davies, A. G., and Vasishtha, S. C., *J. Chem. Soc. (C)*, 1967, 1309.
[17a] Bloodworth, A. J., Davies, A. G., and Vasishtha, S. C., *J. Chem. Soc. (C)*, 1968, 2640.
[18] Bulten, E. J., and Buding, H. A., *J. Organometallic Chem.*, 1978, **157**, C3.
[19] Bürger, H., and Sawodny, W., *Inorg. Nucl. Chem. Letters*, 1966, **2**, 209.
[20] Cardin, D. J., and Lappert, M. F., *Chem. Comm.*, 1966, 506.
[21] Cardin, D. J., and Lappert, M. F., *Chem. Comm.*, 1967, 1034.
[21a] Cardin, D. J., Keppie, S. A., and Lappert, M. F., *J. Chem. Soc. (A)*, 1970, 2594.
[21b] Chandra, G., George, T. A., and Lappert, M. F., *J. Chem. Soc. (C)*, 1969, 2565.
[21c] Chandra, G., Jenkins, A. D., Lappert, M. F., and Srivastava, R. C., *J. Chem. Soc. (A)*, 1970, 2550.
[21d] Cotton, J. D., Cundy, C. S., Harris, D. H., Hudson, A., Lappert, M. F., and Lednor, P. W., *J.C.S. Chem. Comm.*, 1974, 651.
[22] Creemers, H. M. J. C., Noltes, J. G., and van der Kerk, G. J. M., *Rec. Trav. Chim.*, 1964, **83**, 1284.
[23] Creemers, H. M. J. C., Verbeek, F., and Noltes, J. G., *J. Organometallic Chem.*, 1967, **8**, 469.
[24] Cummins, R. A., and Dunn, P., *Australian J. Chem.*, 1964, **17**, 411.
[25] Cuvigny, T., and Normant, H., *Compt. Rend.*, 1969, **268C**, 834.
[26] Dalton, R. F., and Jones, K., unpublished results, 1970.
[27] Dalton, R. F., and Jones, K., *Inorg. Nucl. Chem. Letters*, 1969, **5**, 785.
[28] Dalton, R. F., and Jones, K., *J. Chem. Soc. (A)*, 1970, 520.
[29] Davidson, P. J., Harris, D. H., and Lappert, M. F., *J.C.S. Dalton*, 1976, 2268.
[30] Davies, A. G., and Harrison, P. G., *J. Chem. Soc. (C)*, 1967, 1313.
[31] Davies, A. G., and Kennedy, J. D., *J. Chem. Soc. (C)*, 1968, 2630.
[32] Davies, A. G., and Kennedy, J. D., *J. Chem. Soc. (C)*, 1970, 759.
[33] Davies, A. G., and Mitchell, T. N., *J. Organometallic Chem.*, 1966, **6**, 568.
[34] Davies, A. G., Mitchell, T. N., and Symes, W. R., *J. Chem. Soc. (C)*, 1966, 1311.
[34a] Dergunov, Yu. I., Gerega, V. F., and D'yachkovskaya, O. S., *Russ. Chem.*

Rev., 1977, **46**, 1132.
[35] Domingos, A. M., and Sheldrick, G. M., *J. Organometallic Chem.*, 1974, **69**, 207.
[36] Foley, P., and Zeldin, M., *Inorg. Chem.*, 1975, **14**, 2264.
[37] George, T. A., Jones, K., and Lappert, M. F., *J. Chem. Soc.*, 1965, 2157.
[38] George, T. A., and Lappert, M. F., *Chem. Comm.*, 1966, 463.
[39] George, T. A., and Lappert, M. F., *J. Organometallic Chem.*, 1968, **14**, 327.
[40] George, T. A., and Lappert, M. F., *J. Chem. Soc. (A)*, 1969, 992.
[40a] Gibbins, S. G., Lappert, M. F., Pedley, J. B., and Sharp, G. J., *J.C.S. Dalton*, 1975, 72.
[41] Götze, H.-J., *Angew. Chem. Internat. Edn.*, 1974, **13**, 88.
[42] Götze, H.-J., *J. Organometallic Chem.*, 1973, **47**, C25.
[42a] Gümrükçü, I., Lappert, M. F., Pedley, J. B., Power, P. P., and Rai, A. K., unpublished results, 1977.
[43] Gynane, M. J. S., Lappert, M. F., Miles, S. J., and Power, P. P., *J.C.S. Chem. Comm.*, 1976, 256.
[43a] Gynane, M. J. S., Lappert, M. F., Miles, S. J., and Power, P. P., *J.C.S. Chem. Comm.*, 1978, 192.
[44] Gynane, M. J. S., Harris, D. H., Lappert, M. F., Power, P. P., Rivière, P., and Rivière-Baudet, M., *J.C.S. Dalton*, 1977, 2004.
[45] Hänssgen, D., Kuna, J., and Ross, B., *Chem. Ber.*, 1976, **109**, 1797.
[46] Hänssgen, D., Kuna, J., and Ross, B., *J. Organometallic Chem.*, 1975, **92**, C49.
[47] Hänssgen, D., and Pohl, I., *Angew. Chem. Internat. Edn.*, 1974, **13**, 607.
[47a] Hänssgen, D., and Roelle, M. W., *J. Organometallic Chem.*, 1974, **71**, 231.
[48] Harris, D. H., D.Phil. Thesis, University of Sussex, 1975.
[49] Harris, D. H., and Lappert, M. F., *J.C.S. Chem. Comm.*, 1974, 895.
[50] Harris, D. H., Golberg, D. E., Lappert, M. F., and Thomas, K. M., *J.C.S. Chem. Comm.*, 1976, 261.
[51] Harris, D. H., and Lappert, M. F., *J. Organometallic Chem. Library*, 1976, **2**, 13.
[52] Harris, D. H., Keppie, S. A., and Lappert, M. F., *J.C.S. Dalton*, 1973, 1653.
[53] Harris, D. H., Lappert, M. F., Pedley, J. B., and Sharp, G. J., *J.C.S. Dalton*, 1976, 945.
[54] Hester, R. E., and Jones, K., *Chem. Comm.*, 1966, 317.
[55] Hogben, M. G., Oliver, A. J., and Graham. W. A. G., *Chem. Comm.*, 1967, 1183.
[56] Hudson, A., Lappert, M. F., and Lednor, P. W., *J.C.S. Dalton*, 1976, 2369.
[57] Itoh, K., Katsuura, T., Matsuda, I., and Ishii, Y., *J. Organometallic Chem.*, 1972, **34**, 63.
[58] Janssen, M. J., Luijten, J. G. A., and van der Kerk, G. J. M., *J. Organo-*

metallic Chem., 1964, **1**, 286.
[58a] Jenkins, A. D., Lappert, M. F., and Srivastava, R. C., *J. Organometallic Chem.*, 1970, **23**, 165.
[59] Jones, K., unpublished results, 1967.
[60] Jones, K., and Lappert, M. F., *Proc. Chem. Soc.*, 1962, 358.
[61] Jones, K., and Lappert, M. F., *Proc. Chem. Soc.*, 1964, 22.
[62] Jones, K., and Lappert, M. F., *J. Chem. Soc.*, 1965, 1944.
[63] Jones, K., and Lappert, M. F., *J. Organometallic Chem.*, 1965, **3**, 295.
[64] Jones, K., and Lappert, M. F., *Organometallic Chem. Rev.*, 1966, **1**, 67; Jones, K., and Lappert, M. F., in *Organotin Compounds*, Ed. A. K. Sawyer, Marcel Dekker, Vol. 2, 1971.
[64a] Kagan, M., B.Sc. Thesis, University of Sussex, 1976.
[65] van der Kerk, G. J. M., and Luijten, J. G. A., *J. Appl. Chem. (London)*, 1956, **6**, 49.
[66] van der Kerk, G. J. M., Luijten, J. G. A., and Janssen, M. J., *Chimia*, 1962, **16**, 10.
[67] Kostyanovskii, R. G., and Prokof'ev, A. K., *Izv. Akad. Nauk S.S.S.R., Ser. Khim.*, 1967, 473.
[68] Kozima, S., Itano, T., Mihara, M., Sisido, K., and Isida, T., *J. Organometallic Chem.*, 1972, **44**, 117.
[69] Kula, M.-R., Kreiter, C. G., and Lorberth, J., *Chem. Ber.*, 1964, **97**, 1294.
[70] Kula, M.-R., Lorberth, J., and Amberger, E., *Chem. Ber.*, 1964, **97**, 2087.
[71] Kupchik, E. J., and McInerney, E. F., *J. Organometallic Chem.*, 1968, **11**, 291.
[71a] Lappert, M. F., Lorberth, J., and Poland, J. S., *J. Chem. Soc. (A)*, 1970, 2954.
[71b] Lappert, M. F., Onyszchuk, M., and Slade, M. J., unpublished results, 1978.
[71c] Lappert, M. F., Patil, D. S., and Pedley, J. B., *J.C.S. Chem. Comm.*, 1975, 830.
[72] Lappert, M. F., Pedley, J. B., and Riley, P. N. K., unpublished results, 1966.
[73] Lappert, M. F., and Poland, J. S., *Chem. Comm.*, 1969, 156.
[74] Lappert, M. F., and Power, P. P., unpublished results, 1975.
[75] Lappert, M. F., and Power, P. P., *Adv. Chem. Ser., (Amer. Chem. Soc.)*, 1976, **157**, 70.
[75a] Lappert, M. F., Power, P. P., Atwood, J. L., and Rogers, R. D., unpublished results, 1978.
[76] Lappert, M. F., Power, P. P., Slade, M. J., Hedberg, L., Hedberg, K., and Schomaker, K., *J.C.S. Chem. Comm.*, 1979, 369.
[77] Lappert, M. F., and Slade, M. J., unpublished results, 1978.
[78] Lappert, M. F., and Srivastava, G., *Inorg. Nucl. Chem. Letters*, 1965, **1**, 53.

[79] Lehn, W. L., *J. Amer. Chem. Soc.*, 1964, **86**, 305.
[80] Lorberth, J., Dissertation, Munich, 1965.
[81] Lorberth, J., *Chem. Ber.*, 1965, **98**, 1201.
[82] Lorberth, J., *J. Organometallic Chem.*, 1969, **16**, 235.
[83] Lorberth, J., *J. Organometallic Chem.*, 1969, **19**, 435.
[84] Lorberth, J., and Kula, M.-R., *Chem. Ber.*, 1964, **97**, 3444.
[85] Lorberth, J., and Kula, M.-R., *Chem. Ber.*, 1965, **98**, 520.
[86] Lorberth, J., and Nöth, H., *J. Organometallic Chem.*, 1969, **19**, 203.
[87] Luijten, J. G. A., Janssen, M. J., and van der Kerk, G. J. M., *Rec. Trav. Chim.*, 1962, **81**, 202.
[88] Luijten, J. G. A., Rijkens, F., and van der Kerk, G. J. M., *Adv. Organometallic Chem.*, 1965, **3**, 397.
[89] Luijten, J. G. A., and van der Kerk, G. J. M., *Rec. Trav. Chim.*, 1963, **82**, 1181.
[90] Mack, J., and Yoder, C. H., *Inorg. Chem.*, 1969, **8**, 278.
[91] Maire, J. C., Prosperini, R., and Van Rietschoten, J., *J. Organometallic Chem.*, 1970, **21**, P41.
[92] Marchand, A., Forel, M.-T., Rivière-Baudet, M., and Soulard, M.-H., *J. Organometallic Chem.*, 1978, **156**, 341.
[93] Marchand, A., Lemerle, C., and Forel, M.-T., *J. Organometallic Chem.*, 1972, **42**, 353.
[94] Neumann, W. P., and Heymann, E., *Angew. Chem. Internat. Edn.*, 1963, **2**, 100.
[95] Neumann, W. P., and Heymann, E., *Ann.*, 1965, **683**, 11.
[96] Neumann, W. P., and Heymann, E., *Ann.*, 1965, **683**, 24.
[97] Neumann, W. P., Niermann, H., and Sommer, R., *Angew. Chem.*, 1961, **73**, 768.
[98] Neumann, W. P., Schneider, B., and Sommer, R., *Ann.*, 1966, **692**, 1.
[99] Noltes, J. G., *Rec. Trav. Chim.*, 1964, **83**, 515.
[100] Noltes, J. G., *Rec. Trav. Chim.*, 1965, **84**, 799.
[101] Noltes, J. G., and Janssen, M. J., *Rec. Trav. Chim.*, 1963, **82**, 1055.
[102] Noltes, J. G., and Janssen, M. J., *J. Organometallic Chem.*, 1964, **1**, 346.
[102a] Okawara, R., and Ohara, M., in *Organotin Compounds*, Ed. A. K. Sawyer, Marcel Dekker, New York, Vol. 2, p. 253.
[103] Randall, E. W., Ellner, J. J., and Zuckerman, J. J., *Inorg. Nucl. Chem. Letters*, 1966, **1**, 109.
[104] Randall, E. W., Yoder, C. H., and Zuckerman, J. J., *Inorg. Chem.*, 1967, **6**, 744.
[105] Randall, E. W., Yoder, C. H., and Zuckerman, J. J., *J. Amer. Chem. Soc.*, 1967, **89**, 3438.
[106] Randall, E. W., and Zuckerman, J. J., *J. Amer. Chem. Soc.*, 1968, **90**, 3167.
[107] Roubineau, A., and Pommier, J. C., *J. Organometallic Chem.*, 1976, **107**,

63.
[107a] Roesky, H. W., Diehl, M., Fuess, H., and Bats, J.W., *Angew. Chem. Internat. Edn.*, 1978, **17**, 58.
[107b] Roesky, H. W., and Wiezer, H., *Chem. Ber.*, 1974, **107**, 3186.
[108] Sakai, S., Fujimura, Y., and Ishii, Y., *J. Organometallic Chem.*, 1973, **50**, 113.
[109] Schaeffer, C. D., and Zuckerman, J. J., *J. Amer. Chem. Soc.*, 1974, **96**, 7160.
[110] Scherer, O. J., *Angew. Chem. Internat. Edn.*, 1969, **8**, 861.
[111] Scherer, O. J., and Biller, D., *Angew. Chem. Internat. Edn.*, 1967, **6**, 446.
[112] Scherer, O. J., and Biller, D., *Z. Naturforsch.*, 1967, **22b**, 1079.
[113] Scherer, O. J., and Hornig, P., *J. Organometallic Chem.*, 1967, **8**, 465.
[114] Scherer, O. J., and Hornig, P., *Chem. Ber.*, 1968, **101**, 2533.
[115] Scherer, O. J., Schmidt, J. F., and Schmidt, M., *Z. Naturforsch.*, 1964, **19b**, 447.
[116] Scherer, O. J., Schmidt, J. F., Wokulat, J., and Schmidt, M., *Z. Naturforsch.*, 1965, **20b**, 183.
[117] Scherer, O. J., and Schmidt, M., *Angew. Chem. Internat. Edn.*, 1963, **2**, 478.
[118] Scherer, O. J., and Schmidt, M., *J. Organometallic Chem.*, 1964, **1**, 490.
[119] Scherer, O. J., and Schmidt, M., *J. Organometallic Chem.*, 1965, **3**, 156.
[120] Schmid, G., Dissertation, Munich, 1963.
[121] Schmidbaur, H., *J. Amer. Chem. Soc.*, 1963, **85**, 2336.
[122] Schmidt, M., and Ruidisch, I., *Angew. Chem. Internat. Edn.*, 1964, **3**, 637.
[122a] Schumann, H., du Mont, W. W., and Kroth, H. J., *Chem. Ber.*, 1976, **109**, 23.
[123] Schumann, H., and Jutzi, P., *Chem. Ber.*, 1968, **101**, 24.
[124] Schumann, H., and Ronecker, S., *J. Organometallic Chem.*, 1970, **25**, 451.
[125] Schumann, H., and Roth, A., *J. Organometallic Chem.*, 1968, **11**, 125.
[126] Schumann, H., Schumann-Ruidisch, I., and Ronecker, S., *Z. Naturforsch.*, 1970, **25b**, 565.
[127] Schumann-Ruidisch, I., and Jutzi-Mebert, B., *J. Organometallic Chem.*, 1968, **11**, 77.
[128] Schumann-Ruidisch, I., Kalk, W., and Brüning, R., *Z. Naturforsch.*, 1968, **23b**, 307.
[129] Sisido, K., and Kozima, S., *J. Org. Chem.*, 1962, **27**, 4051.
[130] Sisido, K., and Kozima, S., *J. Org. Chem.*, 1964, **29**, 907.
[131] Sommer, R., Neumann, W. P., and Schneider, B., *Tetrahedron Letters*, 1964, 3875.
[132] Sommer, R., Schneider, B., and Neumann, W. P., *Ann.*, 1966, **692**, 12.
[132a] Storch, W., Jackstiess, W., Nöth, H., and Winter, G., *Angew. Chem.*

Internat. Edn., 1977, **16**, 478.
[133] Thomas, I. M., *Can. J. Chem.*, 1961, **39**, 1386.
[134] Tsai, T. T., Sicree, A. J., and Lehn, W. L., *Tech. Rept. AFML-TR-66-108, Air Force Materials Laboratory*, Wright-Patterson A. F. B., Ohio, 1966.
[135] Tzschach, A., and Reiss, E., *J. Organometallic Chem.*, 1967, **8**, 255.
[135a] Veith, M., *Angew. Chem. Internat. Edn.*, 1975, **14**, 263.
[135b] Veith, M., *Z. Naturforsch.*, 1978, **33b**, 1.
[135c] Veith, M., *Z. Naturforsch.*, 1978, **33b**, 7.
[136] Veith, M., Recktenwald, O., and Humpfer, E., *Z. Naturforsch.*, 1978, **33b**, 14.
[136a] Vilkov, L. V., Tarasenko, N. A., and Prokof'ev, A. K., *J. Struct. Chem.*, 1970, **11**, 114.
[136b] Wannagat, U., *Adv. Inorg. Chem. Radiochem.*, 1964, **6**, 225.
[137] Wannagat, U., and Rabet, F., *Inorg. Nucl. Chem. Letters*, 1970, **6**, 155.
[138] Wannagat, U., and Schlingmann, M., *Z. Anorg. Chem.*, 1976, **419**, 115.
[139] Wiberg, E., and Rieger, R., Ger. Patent 1,121,650 (1960).
[139a] Winters, L. J., and Hill, D. T., *Inorg. Chem.*, 1965, **4**, 1433.
[140] Wright, C. M., and Muetterties, E. L., *Inorg. Synth.*, 1967, **10**, 137.
[141] Yoder, C. H., and Zuckerman, J. J., *Inorg. Chem.*, 1966, **5**, 2055.
[142] Zuckerman, J. J., Internat. Conf. on the Chemistry of Ge, Sn and Pb Nottingham, July 1977.

4. Lead

[1] Davies. A. G., and Puddephatt, R. J., *J. Organometallic Chem.*, 1966, **5**, 590.
[2] DePree, D. O., U.S. Patent 2,893,857 (1959).
[2a] Götze, H.-J., and Garbe, W., *Chem. Ber.*, 1978, **111**, 2051.
[3] Gorsich, R. D., U.S. Patent 3,261,806 (1966).
[3a] Gümrükçü, I., Lappert, M. F., Pedley, J. B., Power, P. P., and Rai, A. K., unpublished results, 1977.
[3b] Gynane, M. J. S., Harris, D. H., Lappert, M. F., Power, P. P., Rivière, P., and Rivière-Baudet, M., *J.C.S. Dalton*, 1977, 2004.
[3c] Gynane, M. J. S., Lappert, M. F., Miles, S. J., Power, P. P., Carty, A. J., and Taylor, N. J., *J.C.S. Chem. Comm.*, 1977, 458.
[4] Harris, D. H., and Lappert, M. F., *J.C.S. Chem. Comm.*, 1974, 895.
[4a] Harris, D. H., Lappert, M. F., Pedley, J. B., and Sharp, G. J., *J.C.S. Dalton*, 1976, 945.
[5] Kümmel, R., and Meissner, I., *Z. Chem.*, 1970, **10**, 71.
[6] Kupchik, E. J., and McInerney, E. F., *J. Organometallic Chem.*, 1968, **11** 291.
[6a] Lappert, M. F., Lorberth, J., and Poland, J. S., *J. Chem. Soc. (A)*, 1970, 2954.

[6b] Lappert, M. F., Power, P. P., Slade, M. J., Hedberg, L., Hedberg, K., and Schomaker, K., *J.C.S. Chem. Comm.*, 1979, 369.
[6c] Lappert, M. F., and Slade, M. J., unpublished results, 1978.
[7] Liggett, W. B., Closson, R. D., and Wolf, C. N., U.S. Patents 2,595,798 (1952) and 2,640,006 (1953).
[8] Mandal, H., *Ber.*, 1921, **54B**, 703.
[9] Neumann, W. P., and Kühlein, K., *Tetrahedron Letters*, 1966, 3415.
[10] Neumann, W. P., and Kühlein, K., *Tetrahedron Letters*, 1966, 3419.
[11] Neumann, W. P., and Kühlein, K., *Tetrahedron Letters*, 1966, 3423.
[12] Pant, B. C., and Davidsohn, W. E., *J. Organometallic Chem.*, 1969, **19**, P3.
[13] Pevilstein, W. L., and Beatty, M. A., Ger. Patent 2,008,067 (1971).
[14] Saunders, B. C., *J. Chem. Soc.*, 1950, 684.
[15] Scherer, O. J., and Schmidt, M., *J. Organometallic Chem.*, 1964, **1**, 490.
[16] Sisti, A. J., *Chem. Comm.*, 1968, 1272.
[17] Schumann, H., and Ronecker, S., *J. Organometallic Chem.*, 1970, **23**, 451.
[18] Schumann, H., Schumann-Ruidisch, I., and Ronecker, S., *Z. Naturforsch.*, 1970, **25b**, 565.
[19] Van Peski, A. J., U.S. Patent 2,196,447 (1940).

6

Amides of Phosphorus: Synthesis, Physical Properties, and Structures

A INTRODUCTION, CLASSIFICATION, AND REVIEWS

The variety and number of phosphorus–nitrogen bonded compounds i great. In 1950 a review [156] of organophosphorus compounds containing P–N bond or a P=N bond encompassed some seven hundred compounds an one hundred and forty-two pertinent references. In recent years the output o information on these systems is measured in hundreds of papers and patents Because the subject is so vast, and phosphorus may only marginally be describe as a metalloid element, this chapter will attempt to describe only the salien aspects of the classification, preparation, physical properties, and structures of P–NRR′ bonded compounds. No attempt is made at comprehensive coverage, nor do we provide here tables of compounds as elsewhere (Chapters 2-5 and 7-9). The reactions of phosphorus amides are referred to in Chapters 10–18, but again are not listed in tables. Reference will frequently be made in this chapter to publications in which a number of papers concerning particular aspects of the chemistry of these systems are gathered.

Phosphorus-nitrogen compounds are conveniently classified into the following types (i)–(iv):

(i) Derivatives of di-, tri-, tetra-, or penta-co-ordinate phosphorus in which there is at least one P–NR_2 linkage. Stable compounds of this type are known for the +2, +3, and +5 oxidation states of phosphorus.

(ii) The monophosphazenes, e.g., $L_3P{=}NR$, with a single PN double bond; and the diphosphazenes, e.g., $[N(PL_3)_2]^+$ (see Chapter 15) or $R_2NP(:NR')_2$, containing two PN multiple bonds.

(iii) The cyclodiphosphazanes $(L_3PNR)_2$, i.e., the dimers of the monophosphazenes, which are four-membered ring compounds, the diazaphosphetidines, $(XPNR)_2$.

(iv) The cyclopolyphosphazenes or phosphonitrilic compounds $(NPL_2)_n$, $n = 3$, 4, 5, etc.

This chapter focuses on compounds of type (i), and those of types (ii)–(iii) will be discussed only if they also contain one or more P–NR_2 linkage. Compounds of type (iv), which are becoming of some significance as thermally-stable materials (see some of the reviews cited below), have been known for more than a century [83, 165, 215]. They may undergo ring-opening to form versatile polymers having –P–N–P–N– backbones [6]. Some may have pendant NR_2 group(s) at P, e.g., ref. [45]. They will not be discussed further.

There are many reviews of aspects of P–N chemistry, and some major works are listed below:

G. M. Kosolapoff, *Organophosphorus Compounds,* Wiley, (1950).

Organic Phosphorus Compounds, ed. G. M. Kosolapoff and L. Maier, Wiley, (1973).

Topics in Phosphorus Chemistry, ed. M. Grayson and E. J. Griffith, Wiley, [Vols. 4 (1967) and 6 (1969)].

Organophosphorus Chemistry, Sen. Reporter S. Trippett, Specialist Periodical Reports, The Chemical Society, London: appears annually.

M. T. P. Internat. Rev. of Science, Ser. One, Vol. 2, ed. C. C. Addison and D. B. Sowerby, Butterworth, (1972–5) – includes reviews concerning phosphonitrilics and halogeno(amino)phosphines.

D. E. C. Corbridge, *The Structural Chemistry of Phosphorus,* Elsevier, (1974).

R. F. Hudson, *Structure and Mechanism in Organo-Phosphorus Chemistry,* Academic Press, (1965).

K. Sasse, *Methoden der Organischen Chemie,* 12/I, G. Thieme, Stuttgart (1963) – a review of reactions of organophosphorus compounds, including dimethylaminolysis of compounds of P^{III} and P^{V}.
Ye. L. Gefter, *Organophosphorus Monomers and Polymers,* trans. J. Burton, Pergamon Press, (1962) – a review of unsaturated organophosphorus acids and their amides, as well as polymers derived therefrom.
M. Bermann, 'The Phosphazotrihalides', *Adv. Inorg. Chem. Radiochem.* 1972, **14,** 1.
J. F. Nixon, *Adv. Inorg. Chem. Radiochem.,* 1970, **13,** 363 – a review of fluorophosphines, including P^{III} amides.
H. R. Allcock, *Phosphorus-Nitrogen Compounds,* Academic Press, (1972) – subtitled 'Cyclic, Linear, and High Polymeric Systems', concerning phosphonitrilics, etc. An article which briefly reviews the literature concerning metal complexes of phosphonitrilics is by N. L. Paddock, T. M. Ranganathan, and J. N. Wingfield, *J. Chem. Soc. Dalton,* 1972, 1578.
S. S. Krishnamurthy, A. C. Sau, and M. Woods, *Adv. Inorg. Chem. Radiochem.* 1978, **21,** 41 – 'Cyclophosphazenes'.
M. M. Crutchfield, C. H. Dungan, J. H. Letcher, V. Mark, and J. R. Van Wazer *Topics in Phosphorus Chemistry.* Vol. 5. ^{31}P Nuclear Magnetic Resonance Wiley, (1967).
L. C. Thomas, *Interpretation of the Infrared Spectra of Organophosphorus Compounds,* Heyden, (1974).
J. Emsley and D. Hall, *The Chemistry of Phosphorus,* Harper and Row (1976).

B SYNTHESIS

1. Reaction of a phosphorus(III) halide with a primary or secondary amine

Reaction of a phosphorus chloride (or bromide, or iodide) with an excess of a secondary amine is normally straightforward and gives the fully substituted aminophosphorus compound [96a, 156, 157, 218]. This reaction is exemplified by Eq. (1). The reaction is carried out in an inert hydrocarbon solvent, and the PN

$$PCl_3 + 6HNEt_2 \longrightarrow P(NEt_2)_3 + 3[Et_2NH_2]Cl \qquad (1)$$

product is usually isolated by distillation after removal of the ammonium salt by filtration. The scope of this method has been extended to include the preparation of biologically important systems [95, 164, 223]. However, substitution of less than all halogen atoms is not readily achieved by this method unless the amine is bulky [Eq. (2)].

$$PCl_3 \begin{cases} \xrightarrow{4HNR_2} \text{mixture of } Cl_nP(NR_2)_{3-n} \\ \xrightarrow{2HNBu^t_2} Cl_2PNBu^t_2 \text{ only} \end{cases} \qquad (2)$$

To obtain, for example, the compound Cl_2PNR_2 (R = Me, Et, Pr, or Bu), it is necessary to use the corresponding ammonium halide under reflux conditions in the absence of free amine [Eq. (3)] [156, 174]; $(Cl_2P)_2NR$ (R = Me or Et) may similarly be obtained and, by treatment with PF_3, $(F_2P)_2NR$ [190d].

$$PCl_3 + [R_2NH_2]Cl \longrightarrow Cl_2PNR_2 + 2HCl. \qquad (3)$$

The reaction of a halogenophosphorus compound with a primary amine [Eq. (4)] is frequently very complicated. Compounds containing the –NH–PX–

$$PCl_3 + 6H_2NR \longrightarrow P(NHR)_3 + 3[RNH_3]Cl \qquad (4)$$

group react further, when warmed, distilled, or in the presence of base (amine), to afford condensation products [143a, 156, 235], usually with the 1,3-diamino-2,4,1,3-diazadiphosphetidines, (I), predominating [Eq. (5)]. Methylamine is (to date)

$$2PCl_3 + 10H_2NR \longrightarrow \begin{array}{c} RHNP\text{—}NR \\ |\quad\;\; | \\ RN\text{—}PNHR \end{array} + 6[RNH_3]Cl \qquad (5)$$

(I)

unique in that the product is the cage-like compound $P_4(NMe)_6$, (II), [Eq. (6)] [130, 253]. Similarly, reaction of $PCl(NRR'_2)_2$ with $ArSO_2NH_2$ yields $[(ArSO_2)NP(NRR')]_2$ [38]. However, rarely and only under carefully controlled conditions, compounds $P(NHR)_3$ may also be obtained [48].

$$4PCl_3 + 18H_2NMe \longrightarrow 12[MeNH_3]Cl + P_4(NMe)_6 \qquad (6)$$

P, MeN, NMe, NMe, P, P, MeN, MeN, NMe, P

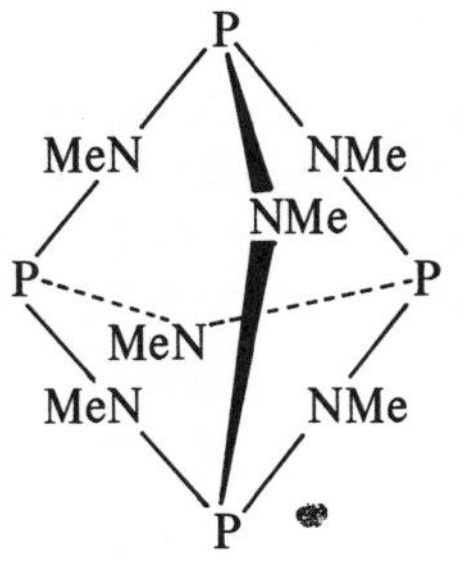

(II)

Reaction of a primary aromatic amine with an excess of PCl_3 was reported [176] to give $[ArNPCl]_2$, and this was accepted for many years [156]. Recently such systems have been reinvestigated [82, 140]. At ambient temperature, the initial product is either Cl_2PNHAr ($Ar = p\text{-}NO_2C_6H_4$) or $(Cl_2P)_2NAr$ [82]. Upon thermolysis (80–150° C) the diazadiphosphetidine is formed by elimination of PCl_3 [Eq. (7)].

$$H_2NAr \xrightarrow[-HCl]{PCl_3} Cl_2PNHAr \xrightarrow[-HCl]{PCl_3} (Cl_2P)_2NAr \xrightarrow[80\text{–}150°\,C]{-PCl_3} \tfrac{1}{2}[ArNPCl]_2. \quad (7)$$

In contrast to this tendency to complexity, some substituted halogeno-phosphorus compounds react in a straightforward manner with primary amines [Eqs. (8), (9)], e.g., [58, 75, 94]. The lack of subsequent reaction is usually due

$$Ph_2PNHR + PhPCl_2 \longrightarrow Ph_2PN(R)PPhCl + HCl \quad (8)$$

$$RPCl_2 + [(MeHN)SiMe_2]_2NMe \longrightarrow RP\begin{matrix} N(Me)Si(Me_2) \\ \\ N(Me)Si(Me_2) \end{matrix}NMe + 2HCl \quad (9)$$

to a constraining factor, such as operation at low temperature or the presence of a bulky substituent, in preventing condensation. It is noteworthy that amido(halogeno)phosphorus compounds react with amines preferentially to eliminate HCl rather than undergo transamination, e.g., [30, 35]. Hydrazines often react similarly to amines, e.g., [112, 142]. Compounds $R(H)PNR'_2$ (=L) are unknown, but metal complexes, e.g., $[Mn(\eta\text{-}C_5H_5)(CO)_2L]$ are stable [137].

2. Reaction of a primary or secondary amine with a phosphoryl or thiophosphoryl halide

A secondary amine reacts smoothly with a phosphoryl halide to form the corresponding amide [Eq. (10)], under similar conditions to those used for the

$$P(O)Cl_3 + 6HNR_2 \longrightarrow P(O)(NR_2)_3 + 3[R_2NH_2]Cl \quad (10)$$

phosphorus(III) halide, except that the mixture is usually heated. This is in order to effect complete reaction, because of the lower reactivity of the phosphoryl halide to nucleophilic attack. The general applicability of such reactions for the per-amination of various phosphoryl halides is illustrated by Eqs. (11) [90] and (12) [209]. It is possible in some instances to prepare phosphoryl derivatives of a

$$Me_2\overline{CCHMeCMe_2P}(O)Cl + 2HNR_2 \longrightarrow Me_2\overline{CCHMeCMe_2P}(O)NR_2 + [R_2NH_2]Cl \quad (11)$$

$$[Cl_2P(O)]_2NMe + 8HNR_2 \longrightarrow [(R_2N)_2P(O)]_2NMe + 4[R_2NH_2]Cl \quad (12)$$

primary amine [e.g., Eqs. (13), (14), (15)] [90, 156, 207, 220]. Reaction (15)

$$Me_2\overline{CCHMeCMe_2P}(O)Cl + 2H_2NR \longrightarrow Me_2\overline{CCHMeCMe_2P}(O)NHR + [RNH_3]Cl \quad (13)$$

$$RP(O)Cl_2 + 4H_2NR' \longrightarrow RP(O)(NHR')_2 + 2[R'NH_3]Cl \quad (14)$$

$$P(O)Cl_3 + 2H_2NAr \longrightarrow P(O)Cl_2(NHAr) + [ArNH_3]Cl \quad (15)$$

gave satisfactory results only with slow mixing of the cooled reagents [156].

In the preparation of a mono- or bis- primary aminophosphoryl compound it is more usual to heat the phosphoryl chloride and a stoicheiometric quantity of a primary ammonium salt [e.g., Eqs. (16) and (17)].

$$P(O)Cl_3 + [RNH_3]Cl \longrightarrow P(O)Cl_2(NHR) + 2HCl \quad (16)$$

$$P(O)Cl_3 + 2[RNH_3]Cl \longrightarrow P(O)Cl(NHR)_2 + 4HCl \quad (17)$$

Thiophosphoryl chlorides react in a similar manner to the oxygen analogues. Heating is required in many cases due to the low reactivity of the sulphur compounds. In some cases, e.g., $P(S)Cl_2(NR_2)$, the product is not amenable to nucleophilic attack (e.g., by alcohols or thiols) and can be steam-distilled. The tris(amino)thiophosphoryl derivatives of secondary amines are formed by refluxing $P(S)Cl_3$ with an excess (> six-fold) of HNR_2 [Eq. (18)]. It is sometimes

$$P(S)Cl_3 + {>}6HNR_2 \longrightarrow P(S)(NR_2)_3 + 3[R_2NH_2]Cl \quad (18)$$

possible to achieve higher yields, especially for mono- (secondary) amides, by treatment of $P(S)Cl_3$ with the appropriate amine hydrochloride. Further examples are in Eqs. (19) [49, 220], (20) [156], and (21) [157]. Attack of $HNMe_2$ on $Cl_2P(O)NMeP(S)Cl_2$ is at the phosphoryl group in diethyl ether, but on the thiophosphoryl group in a non-polar solvent [49].

$$[Cl_2P(S)]_2NR \xrightarrow{HNMe_2} Me_2NP(S)(Cl)NRP(S)Cl_2 \xrightarrow{HNMe_2} [Me_2NP(S)Cl]_2NR$$

$$\xrightarrow{2HNMe_2} [(Me_2N)_2P(S)]_2NR \quad (19)$$

$$P(S)Cl_3 + 6HNMe_2 \longrightarrow P(S)(NMe_2)_3 + 3[Me_2NH_2]Cl \quad (20)$$

$$P(S)Cl_3 + 2H_2NR \longrightarrow P(S)(NHR)Cl_2 + [Me_2NH_2]Cl \quad (21)$$

Although these reactions may be used to introduce amino groups with substituents in the side-chain [80, 81, 219], the use of bulky substituents may be restricted [228].

3. Interaction of a primary or secondary amine and a tri- or tetra- co-ordinate phosphorus–fluorine compound

We have seen that chloride (or Br^- or I^-) is readily displaced from P by the -NHR or -NR_2 group; however, this is not the case for fluoride. The relatively inert P–F bond has been utilised in many instances to prepare partially aminated halogenophosphorus compounds, derived from secondary, [e.g., Eqs. (22)–(25)] [92, 97, 99] or primary amines [e.g., Eqs. (26), (27)] [72, 127, 213].

$$PF_3 + HNMe_2 \text{ (excess)} \longrightarrow FP(NMe_2)_2 \quad (22)$$

$$PF_3 + MeNH(CH_2)_2NHMe \text{ (excess)} \longrightarrow FP\overset{\text{N(Me)}}{\underset{\text{N(Me)}}{\langle\,\,\rangle}} \quad (23)$$

$$F_2P(CH_2)_2PF_2 + 2HNMe_2 \longrightarrow F_2P(CH_2)_2P(F)NMe_2 + [Me_2NH_2]F \quad (24)$$

$$F_2P(O)Cl + 2H\overline{N(CH_2)_n} \longrightarrow F_2P(O)\overline{N(CH_2)_n} + [H_2\overline{N(CH_2)_n}]Cl \quad (25)$$

$$F_2P(O)Cl + 2H_2NR \longrightarrow F_2P(O)NHR + [RNH_3]Cl \quad (26)$$

$$FP(O)Cl_2 + 4H_2NR \longrightarrow FP(O)(NHR)_2 + 2[RNH_3]Cl \quad (27)$$

The remaining fluorine ligand(s) may subsequently be replaced, and thus PF bonds are used as protecting groups.

4. Reaction of a primary or secondary amine with a phosphorus(V) halide

The reaction of a primary amine with PCl_5 generally leads to a mono-phosphazene, Cl_3PNR, [Eq. (28)] [153], or the diphosphazane $(Cl_3PNR)_2$.

$$PCl_5 + H_2NAr \longrightarrow Cl_3PNAr + 2HCl \quad (28)$$

Similarly the reaction of a secondary amine with PCl_5 affords a phosphonium-type compound Cl_4PNR_2 or its adduct with PCl_5. Reaction of a dialkylamine with a partially substituted chlorophosphorane sequentially replaces chloride ligands [Eqs. (29), (30)] [57, 203, 243]. A fluorophosphorane, like the fluoro-phosphine analogue, is even more resistant to attack by a secondary [64, 243] or primary [72] amine. In many cases, a stable 1:1 adduct may be isolated which subsequently affords the aminophosphorane on heating, [Eq. (31)] [64]. With the exception of $HNMe_2$, a secondary amine does not displace more than one F ligand from PF_5, although further substitution is possible by transmetallation (see below). A primary aliphatic amine reacts with R_3PCl_2 to afford the product

$$(CF_3)_2PCl_3 + 2HNMe_2 \longrightarrow (CF_3)_2PCl_2(NMe_2) + [Me_2NH_2]Cl \quad (29)$$

$$(CF_3)_3PCl_2 + HNMe_2 \longrightarrow (CF_3)_3PCl(NMe_2) + [Me_2NH_2]Cl \xrightarrow{2HNMe_2} (CF_3)_3P(NMe_2)_2 + [Me_2NH_2]Cl \quad (30)$$

$$PF_5.Pr^i{}_2NH \longrightarrow F_4PNPr^i{}_2 + HF \quad (31)$$

$R_3P{=}NR'$; this reaction [Eq. (32)] is known as the Kirsanov reaction. Although

$$R_3PCl_2 + H_2NR' \longrightarrow R_3P{=}NR' + 2HCl \quad (32)$$
$$(\text{e.g., } R = Ph, OPh, F)$$

hydrogen halide is usually eliminated, this is not always the case, Eq. (32a) [163].

$$Ph_3PBr_2 + H_2NR + NEt_3 \longrightarrow [Ph_3PNHR]Br + [Et_3NH]Br \quad (32a)$$

5. Synthesis of a tri-, tetra-, or penta-co-ordinate phosphorus amide by transmetallation

Transmetallation reactions are frequently of use in obtaining materials which would otherwise not be available or present problems of purification if prepared by an alternative procedure. A good example is the high yield preparation of a bis-amidophosphorus compound by reacting the appropriate (P^{III} or P^V) trihalide with two equivalents of the lithium or sodium derivative of the amine, Eq. (33). If the nitrogen substituent is very bulky, e.g., a silyl group,

$$P(X)Cl_3 + 2LiNR_2 \longrightarrow P(X)(NR_2)_2Cl + 2LiCl \quad (33)$$
$$(X = \text{lone pair, O, or S})$$

complications may ensue, e.g., Eqs. (34), (35) [12, 187–189, 204a]. $LiN(SiMe_3)_2$ and $ClP(O)X_2$ ($X = F$, Cl, or Ph) give $Me_3SiOP(NSiMe_3)X_2$ [e.g., Eqs. (44) (60)] rather than the isomeric P^V amide [183a] (see Chapter 18).

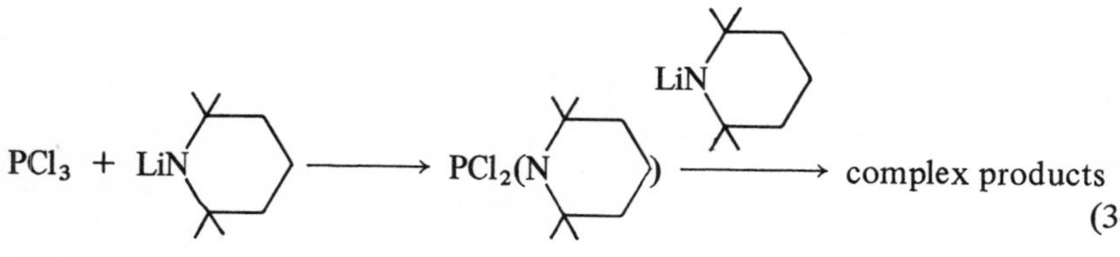

PCl_3 + LiN(2,2,6,6-tetramethylpiperidino) ⟶ PCl_2(N(2,2,6,6-tetramethylpiperidino)) $\xrightarrow{\text{LiN(2,2,6,6-tetramethylpiperidino)}}$ complex products (34)

$$PCl_3 + 2LiN(SiMe_3)_2 \longrightarrow ClP[N(SiMe_3)_2]_2 \xrightarrow[25^\circ C]{slow} P(NSiMe_3)N(SiMe_3)_2 + Me_3SiCl \quad (35)$$

The use of silylamines as transmetallating reagents has assumed increasing importance, especially with regard to the synthesis of P^V amino-compounds. It can be seen [Eq. (36)] that a phosphorus(III) chloride reacts with $(Me_3Si)_2NH$ preferentially to eliminate Me_3SiCl [23, 34, 191]. The related reaction with

$$X_2PCl + (Me_3Si)_2NH \longrightarrow X_2PNHSiMe_3 + Me_3SiCl \quad (36)$$
$$(X = Cl\ [23],\ OR\ [34],\ or\ Ph\ [191])$$

silyl-, disilyl-, or trisilyl-amine may be designed stoicheiometrically to give, with a high degree of specificity and in good yield, a wide variety of aminophosphorus compounds. The following equations exemplify the range of materials so prepared. Depending on the precise conditions, the reaction of PCl_3 with $(Me_3Si)_2NMe$ has variously been reported to yield a diphosphinoamine [Eq. (37)] [143] or a diazadiphosphetidine, Eq. (38) [2]. Me_3SiCl is more readily eliminated than Me_3SiF [Eqs. (39), (40)] [99], but alkoxysilanes are not readily lost [Eq. (41)] [34].

$$2PCl_3 + (Me_3Si)_2NMe \longrightarrow (Cl_2P)_2NMe + 2Me_3SiCl \quad (37)$$

$$2PCl_3 + 2(Me_3Si)_2NMe \longrightarrow (ClPNMe)_2 + 4Me_3SiCl \quad (38)$$

$$P(O)F_3 + Me_3Si\overline{N(CH_2)_n} \longrightarrow Me_3SiF + F_2P(O)\overline{N(CH_2)_n} \quad (39)$$

$$P(O)ClF_2 + Me_3SiNPh_2 \longrightarrow Me_3SiCl + F_2P(O)NPh_2 \quad (40)$$

$$(RO)_2PCl + (Me_3Si)_2NH \longrightarrow Me_3SiCl + (RO)_2PNHSiMe_3 \quad (41)$$

Further variants are in Eqs. (42) [87], (43), (44), [79], (45), (46), [57, 243], (47) [232], (48) [230], (49) [139, 148], (50) [148], (51) [150, 151], (52) [82], and (53) [100]; related to Eq. (45) is the $Me(CF_3)_3PCl$–Me_3SiNMe_2 system [243a].

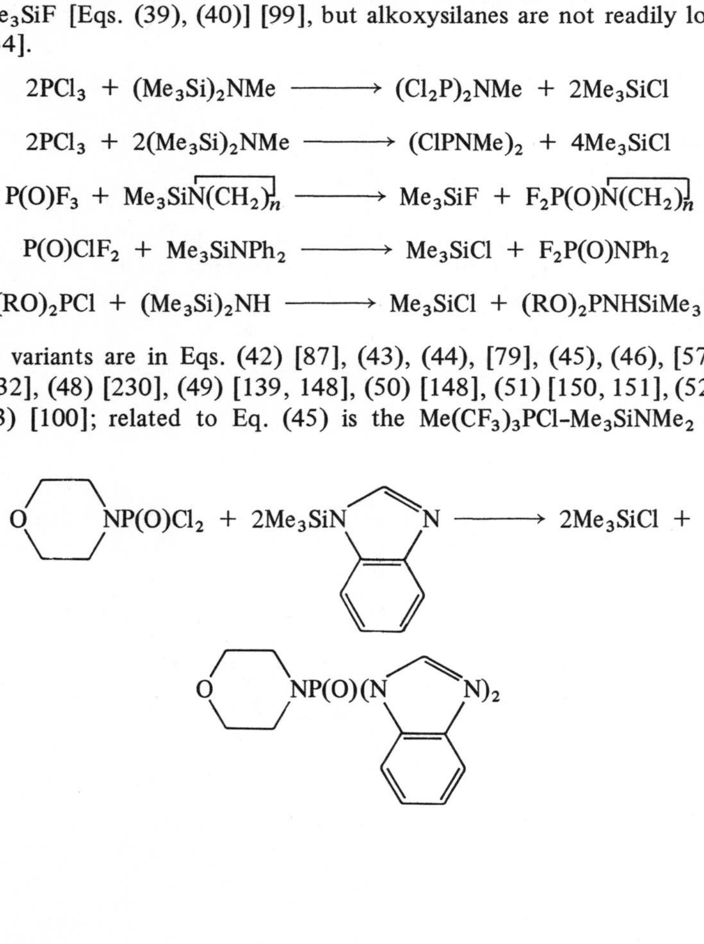

$$O\overline{\quad}NP(O)Cl_2 + 2Me_3SiN\overline{\quad}N \longrightarrow 2Me_3SiCl + O\overline{\quad}NP(O)(N\overline{\quad}N)_2 \quad (42)$$

$$(Me_3Si)_3N\ [(or\ LiN(SiMe_3)_2] \xrightarrow[-Me_3SiF\ (or\ -LiF)]{P(O)F_3} (Me_3Si)_2NP(O)F_2 \xrightarrow{P_2O_3F_4} Me_3SiN[P(O)F_2]_2$$

$$(Me_3Si)_3N\ [(or\ LiN(SiMe_3)_2] \xrightarrow[-(Me_3Si)_2O]{2P_2O_3F_4} Me_3SiN[P(O)F_2]_2 \qquad (43)$$

$$(Me_3Si)_3N \text{ or } LiN(SiMe_3)_2 + P(O)Cl_3 \longrightarrow (Me_3Si)_2NP(O)Cl_2 + Me_3SiOP(NSiMe_3)Cl_2 + Me_3SiCl \text{ or } LiCl \qquad (44)$$

$$(CF_3)_3PCl_2 \xrightarrow{Me_3SiNMe_2} (CF_3)_3P(Cl)NMe_2 \xrightarrow{Me_3SiNMe_2} (CF_3)_3P(NMe_2)_2 \qquad (45)$$

$$(CF_3)_3PF_2 + Me_3SiNMe_2 \longrightarrow (CF_3)_3P(F)NMe_2 + [(CF_3)_3PNMe_2][(CF_3)_3PF_3] \qquad (46)$$

$$PF_5 \xrightarrow[(-Me_3SiF)]{Me_3SiNR_2} F_4PNR_2 \xrightarrow[(-Me_3SiF)]{Me_3SiNR'_2} F_3P(NR_2)(NR'_2) \qquad (47)$$

$$2MePF_4 + (Me_3Si)_2NMe \longrightarrow 4Me_3SiF + F_2(Me)P(\mu\text{-}NMe)_2PF_2Me \qquad (48)$$

$$Cl_2P(O)NMe_2 + (Me_3Si)_2NMe \longrightarrow Me_3SiCl + Me_3SiNMeP(O)Cl(NMe_2) \xrightarrow{P(O)Cl_3} Cl_2P(O)NMeP(O)Cl(NMe_2) \qquad (49)$$

$$P(X)Cl_3 \xrightarrow{(Me_3Si)_2NMe} Cl_2P(X)NMeSiMe_3 \xrightarrow{P(O)Cl_3} Cl_2P(X)NMeP(O)Cl_2$$

(X = lone pair, O, or S) (50)

$$Me_3SiNMeP(O)Cl(NMe_2) + Cl_2PNMe_2 \longrightarrow Me_2N(Cl)PNMeP(O)Cl(NMe_2) \longrightarrow Cl_2PNMeP(O)(NMe_2)_2 \qquad (51)$$

$$2PhN(PCl_2)_2 + 2Me_3SiNMe_2 \longrightarrow (PhNPNMe_2)_2 + 2PCl_3 + 2Me_3SiCl \qquad (52)$$

(c.f., $+ H_2NMe \longrightarrow P_4(NMe)_6$ [82])

$$[F_2P(O)]_2O + (Me_3Si)_2NMe \longrightarrow F_2P(O)NMeSiMe_3 + Me_3SiOP(O)F_2$$

$$\downarrow [F_2P(O)]_2O$$

$$[F_2P(O)]_2NMe + Me_3SiOP(O)F_2 \qquad (53)$$

It is surprising that, in some circumstances, elimination of Me_3SiF is favoured even over elimination of LiF [Eq. (54)] [100]. However, reaction of H_3SiBr with F_2PNH_2 does not give $H_3SiPF(NH_2)$, but instead yields $H_3SiNHPF_2$ [16].

$$2P(O)F_3 + LiN(SiMe_3)_2 \longrightarrow LiN[P(O)F_2]_2 + 2Me_3SiF$$

$$\downarrow CF_3SO_3H$$

$$[F_2P(O)]_2NH \qquad (54)$$

If the amine from which the silylamine is derived possesses a potentially reactive hydrogen atom, the final product may not include a P–N bond [e.g., Eq. (55)] [128].

$$PF_5 + Me_3SiN\text{(2-methylpyrrole ring, Me)} \longrightarrow Me_3SiF + HN\text{(pyrrole ring bearing } F_4P\text{, Me)} \qquad (55)$$

As already mentioned, in cases where the aminophosphorus product contains both P–Cl (or F) and N–Si bonds Me_3SiCl elimination occurs (see Chapter 18). This process has been utilised to form the first examples of a two-co-ordinate P=N bonded system [Eqs. (56)-(58)] [187, 225-227], which are yellowish-green, distillable liquids.

$$F_2PN(SiMe_3)_2 + LiN(SiMe_3)_2 \longrightarrow LiF + Me_3SiF +$$

$$(Me_3Si)_2NP{=}NSiMe_3 \qquad (56)$$

$$Cl_2PN(SiMe_3)_2 + LiN(SiMe_3)R \longrightarrow LiCl + (Me_3Si)_2N(Cl)PN(SiMe_3)R$$

$$\xrightarrow[-Me_3SiCl]{} (Me_3Si)_2NP{=}NR \qquad (57)$$

$$(R = Bu^t \text{ or } SiMe_3)$$

$$PBr_3 + 2LiNBu^t(SiMe_3) \longrightarrow Me_3SiBr + 2LiBr + Me_3Si(Bu^t)NP{=}NBu^t \qquad (58)$$

Phosphorus(V) compounds may behave in a similar manner, e.g., Eq. (59) [214], but may also undergo rearrangement, Eq. (60) [79], see Chapter 18.

$$RPF_4 + (Me_3Si)_2NPF_2 \xrightarrow{} 2Me_3SiF + RF_2P{=}NPF_2 \quad (R = F \text{ or } Ph) \tag{59}$$

$$P(O)Cl_3 + LiN(SiMe_3)_2 \longrightarrow (Me_3SiO)Cl_2P{=}NSiMe_3 \tag{60}$$

The more normal course of events, when the *N*- substituents are less bulky, is formation of diazadiphosphetidine or cyclodiphosphazane rings, e.g., Eqs. (61)-(63).

$$2PCl_3 + 2(Me_3Si)_2NMe \longrightarrow [MeNPCl]_2 + 4Me_3SiCl \tag{61}$$

$$2PhPF_4 + 2(Me_3Si)_2NMe \longrightarrow [PhPF_2NMe]_2 + 4Me_3SiF \tag{62}$$

$$2Me_3SiNRP(Cl)Bu^t(:NR) \xrightarrow{-2Me_3SiCl} [RN{=}P(Bu^t)NR]_2 \tag{63}$$

As well as the above systems, which employ mainly silylated or lithiated amines, examples of the use of derivatives of other metals are known, but are less familiar, e.g., Eq. (64) [208].

$$2F_2PBr + Ag_2(CN_2) \longrightarrow (F_2PN)_2C + 2AgBr \tag{64}$$

6. Preparation of phosphorus amides by transamination

Transamination [Eq. (65)] has been utilised to synthesise homoleptic as well as heteroleptic aminophosphorus compounds.

$$L_nPNR_2 + HNR'_2 \longrightarrow L_nPNR'_2 + HNR_2 \tag{65}$$

The *rate* of transamination depends directly on the relative nucleophilicities of the attacking and leaving amines [161]. Thus, in contrast to some other amino-metalloids (e.g., B, Chapter 4) for which volatility is an important factor, pyrrole does not displace dimethylamine [161]. We have previously noted (Section B.1) that transamination reactions are rarely able to compete with dehydro-halogenations; however, an example to the contrary is the synthesis of a cyclo-diphosphazane containing dissimilar phosphorus substituents [Eq. (66)] [230].

$$(MePF_2NMe)_2 + (PhPF_2NMe)_2 \longrightarrow 2Me\overline{PF_2NMePF_2(Ph)N}Me \tag{66}$$

Reaction of $P(NMe_2)_3$ with a urea in refluxing dioxan gives the transamination product [Eq. (67)], but with acetamide at 115-120° C a mixture is obtained [Eq. (68)] [85].

$$P(NMe_2)_3 + (RHN)_2CO \longrightarrow OC\begin{matrix} NR \\ \\ NR \end{matrix}PNMe_2 + 2HNMe_2 \quad (67)$$

$$P(NMe_2)_3 + MeCONH_2 \longrightarrow (Me_2N)_2P(O)H + HNMe_2 + MeCN + [MeCONP(NMe_2)]_2 \quad (68)$$

An interesting exchange of a chelating amido- ligand has been achieved without cleaving the $P{-}NMe_2$ bond [Eq. (69)] [52].

$$[(C_6H_4(O)(NMe))_2PH] + (\text{pinacolato})PNMe_2 \longrightarrow C_6H_4(O)(NMe)P{-}NMe_2 \quad (69)$$

Ammonia displaces both HCl and a secondary amine from $ClP(NR_2)_2$ to give a cyclotriphosphazene containing a P–H bond, Eq. (70) [231].

$$ClP(NR_2)_2 + NH_3 \longrightarrow [R_2NP(H){=}N]_3 + [R_2NH_2]Cl \quad (70)$$

The cyclic PN compounds (III) undergo ring-opening with a variety of reagents, e.g., Eqs. (71)-(74) [156, 175], without transamination occurring. An unusual product from this type of reaction is $F_2P(O)\overline{NSN{=}S{=}NS}$, from $(Me_3SiN)_2S$ and $[F_2P(O)]_2O$ [212].

$$[RP(O)NR']_2 \ \textbf{(III)} \xrightarrow{H_2NR''} 2RP(O)(NHR'')NHR' \quad (71)$$

$$[RP(O)NR']_2 \xrightarrow{HNR''_2} 2RP(O)(NR''_2)NHR' \quad (72)$$

$$[RP(O)NR']_2 \xrightarrow{R''OH} 2RP(O)(OR'')NHR' \quad (73)$$

$$[RP(O)NR']_2 \xrightarrow{NaOH} Na[RP(O)(NHR')O] \quad (74)$$

7. **Preparation of phosphorus amides** via **scrambling reactions or reactions with hydrogen halides**

Preparation of (amido)halogenophosphorus compounds is generally effected by one of three methods. We have already seen that use of silylamines in appropriate amounts under moderate conditions permits partial halide substitution in polyhalogenophosphorus compounds by one or more amino- groups (Section B.5). Two further procedures involve either a 'scrambling' (or metathesis) reaction, e.g., Eq. (75) [192], or treatment of a poly(amido)phosphorus compound with a limited amount of a hydrogen halide or pseudohalide, e.g., Eq. (76) [17]. Such reactions may even be performed with a co-ordinated phosphine [129]; or alternatively the amide group is used merely as a protecting group to prepare otherwise elusive products, e.g., Eqs. (77), (78) [4, 50, 86, 89, 178]. A complication is that heating a heteroleptic compound, such as (IV), may cause disproportionation to give the most volatile homoleptic product, e.g., Eq. (79) [156]. Another type of ligand exchange involves different elements, as in the $B(NEt_2)_3$-PCl_3 or -$P(X)Cl_3$ (X=O or S) systems [63a]. Hydrogen chloride causes the condensation of $OP(NH_2)_3$ to $[OP(NH_2)_2]_2NH$, etc. [210].

$$P(NMeNMe)_3P + PCl_3 \longrightarrow ClP(NMeNMe)_2PCl \xrightarrow{PCl_3} Cl_2PNMeNMePCl_2 \quad (75)$$

$$(F_2P)_3N + HX \longrightarrow F_2PX + HN(PF_2)_2 \quad (76)$$

$$CF_3P(NR_2)_2 + 4HX \longrightarrow CF_3PX_2 + 2[R_2NH_2]X \quad (77)$$

$$Me_2PNMe_2 + 2HX \longrightarrow Me_2PX + [Me_2NH_2]X \quad (78)$$

$$\underset{\textbf{(IV)}}{3(ArO)_2P(O)NR_2} \xrightarrow{\text{heat}} P(O)(NR_2)_3 + 2P(O)(OAr)_3 \quad (79)$$

8. **Preparation of phosphorus-nitrogen compounds** via **insertion reactions**

Insertion reactions have occasionally been used to synthesise P—N bonded compounds. Such reactions may involve insertion of an unsaturated N=Y substrate into a P–X bond, e.g., Eqs. (80), (81), [133, 167] or addition of an amine to a phosphorus isocyanate, e.g., Eqs. (82)–(84) [246, 247].

$$(RO)_2PNR'R'' + R'''NCO \longrightarrow (RO)_2PNR'''C(O)NR'R'' \quad (80)$$

$$PhP(X)RH + R'N{=}NR' \longrightarrow RP(X)PhNR'NR'H \quad (81)$$

$$(X = O \text{ or } S;\ R' = CO_2Et)$$

$$PhP(O)(NCO)_2 + H_2NR \xrightarrow{-50° C} PhP(O)(NCO)NHC(O)NHR \xrightarrow{110° C} Ph(O)P\begin{matrix} NHCO \\ \quad \\ NHCO \end{matrix}NR \qquad (82)$$

$$(Me_2N)_2P(O)NCO \xrightarrow{R'OH} (Me_2N)_2P(O)NHC(O)OR' \qquad (83)$$

$$(Me_2N)_2P(O)NCO \xrightarrow{H_2NR'} (Me_2N)_2P(O)NHC(O)NHR' \qquad (84)$$

It is noteworthy [Eq. (83)] that insertion may occur preferentially over amine displacement. Insertion of either the dione or lactone form of dimethylketene dimer takes place into a terminal [Eq. (85)] but not an endocyclic [Eq. (86)] P–N bond [26, 27]. Aldehydes or ketones condense with systems $PhP(H)(CH_2)_4NHR$ to give cyclic or polymeric P–C–N systems [141].

$$P(NMe_2)_3 + \begin{matrix} Me_2C - CO \\ | \quad\quad | \\ OC - CMe_2 \end{matrix} \longrightarrow \begin{matrix} (Me_2N)_2P - C(NMe_2)CMe_2 \\ \| \quad\quad | \quad\quad\quad | \\ O \quad\quad CMe_2 - CO \end{matrix} \qquad (85)$$

$$Me_2NP\begin{matrix} O \\ \quad \\ N \\ R \end{matrix} + \begin{matrix} Me_2C - C{=}O \\ | \quad\quad | \\ Me_2C{=}C - O \end{matrix} \longrightarrow \begin{matrix} NR \\ P{-}O \\ O \end{matrix} \begin{matrix} C{=}CMe_2 \\ CMe_2 \\ O{=}C \\ NMe_2 \end{matrix} \qquad (86)$$

Boron halides add across the P=N bond of $(Me_3Si)_2NP{=}NSiMe_3$ to give the cyclic compounds (V), as well as $(Me_3Si)_2NPX_2$, and polymers, Eq. (87) [186]. A similar reaction of the ion $[Cl_3PNPCl_3]^+$ in the presence of methylammonium

$$(Me_3Si)_2NP{=}NSiMe_3 + BX_3 \longrightarrow Me_3SiN\begin{matrix} PX \\ \quad \\ BX \end{matrix}NSiMe_3 + Me_3SiX \qquad (87)$$

(X = Cl or Br) (V)

chloride gives a remarkable zwitterionic heterocycle (VI), Eq. (88) [24], which can be fully fluorinated by reaction with SbF_3 or AsF_3 [33].

$$[Cl_3PNPCl_3]^+ + 2[MeNH_3]Cl + BCl_3 \longrightarrow Cl_2\overset{-}{B}\begin{matrix} \diagup NMe-PCl_2 \diagdown\!\!\diagdown \\ \diagdown NMe-\overset{+}{P}Cl_2 \diagup \end{matrix}N \quad \textbf{(VI)} \quad + H^+ + 5HCl \tag{88}$$

The reaction of $(Me_3Si)_2NP(:NSiMe_3)_2$ with metal halides, MCl_n, e.g., $SnCl_4$, $AlCl_3$, $FeCl_3$, $TiCl_4$, or $NbCl_5$, also affords zwitterionic complexes [Eq. (88a)] [190a].

$$RN{=}P\begin{matrix} \diagup NR_2 \\ \diagdown\!\!\diagdown NR \end{matrix} \xrightarrow{MCl_n} \begin{matrix} & NR_2 & \\ & | & \\ Cl-&\overset{}{P^+}&-NR \\ & | & | \\ & RN & -\,{}^{-}MCl_{n-1} \end{matrix} \tag{88a}$$

9. Synthesis of phosphorus amides involving cleavage of P–O, P–S, or P–C bonds

Phosphorus-nitrogen bonds may be formed by cleavage of P–O or P–S bonds [156]. Indeed, amines react directly with P_4O_{10} [42] or P_4S_{10} [41] as well as with P–OH or P–SH compounds, Eqs. (89)–(91a) [19], or Eq. (91b) [156].

$$P_4O_{10} \xrightarrow{H_2NPh} (PhNH)_2P(O)OH + \text{'}HPO_3\text{'} \tag{89}$$

$$P_4S_{10} \xrightarrow[R = Ph]{H_2NR/heat} (RHN)_3P(S) + H_2S; \quad P_4S_{10} \searrow (RHN)_2P(S)SH \nearrow (RHN)_3P(S) \tag{90}$$

$$[(RO)_2P(O)]_2O + NH_3 \longrightarrow (RO)_2P(O)NH_2 + (RO)_2P(O)OH \tag{91a}$$

$$(RHN)_2P(S)SH + H_2NR \longrightarrow (RHN)_3P(S) + H_2S \tag{91b}$$

Trimethylsilyl azide reacts similarly with a sulphur-bridged heterocycle, Eq. (92) [211]. Reaction of ammonia with phosphorus in the presence of hydrogen peroxide gives the salts (VII), Eq. (93) [93].

$$[RP(S)S]_2 + Me_3SiN_3 \xrightarrow{-S,\,-N_2} R(S)P\begin{matrix} \diagup S \diagdown \\ \diagdown N \diagup \end{matrix}P(S)R \quad (N\text{–}SiMe_3) \tag{92}$$

$$2P + 4NH_3 + 4H_2O_2 \longrightarrow \underset{\textbf{(VII)}}{[NH_4]_2[H_2NPO_2PO_2NH_2]} + 4H_2O \quad (93)$$

Cleavage of a P–C bond by reaction with an amine has been reported, Eq. (94) [56], but this method is inapplicable to most RP< compounds. The CF_3 group, in this as in many other contexts, behaves as a pseudohalide.

$$(CF_3)_2POSiMe_3 + HNMe_2 \longrightarrow CF_3P(NMe_2)OSiMe_3 + CF_3H \quad (94)$$

Reaction of *o*-aminophenol with hexachlorotriphosphazene causes both displacement of HCl and ring-opening to give the spirophosphorane (VIII), Eq. (95) [7].

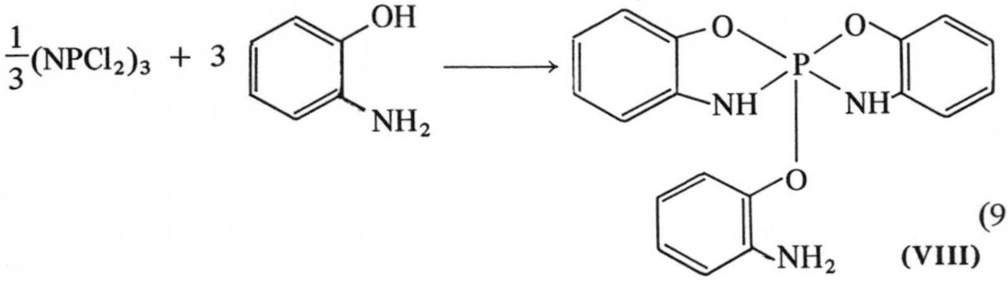

$$\frac{1}{3}(NPCl_2)_3 + 3\ o\text{-}C_6H_4(OH)NH_2 \longrightarrow (o\text{-}C_6H_4(O)NH)_2P\text{–}O\text{-}C_6H_4\text{-}o\text{-}NH_2 \quad (95)$$

(VIII)

Reaction of the same amine with a triarylphosphine gives the phosphazene (IX) probably via the phosphorane (X), Eq. (96) [241].

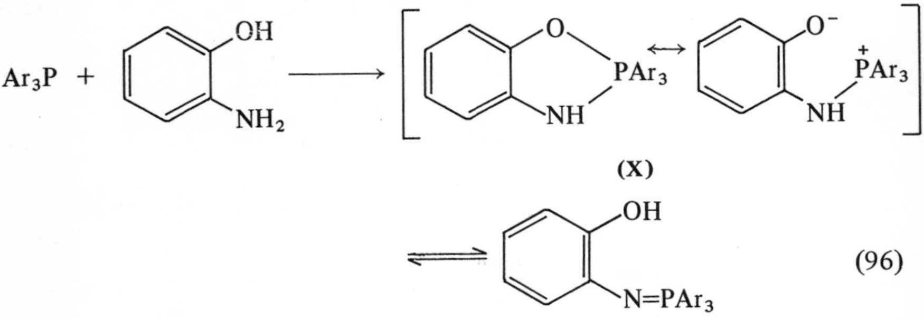

$$Ar_3P + o\text{-}C_6H_4(OH)NH_2 \longrightarrow \left[o\text{-}C_6H_4(O)(NH)PAr_3 \longleftrightarrow o\text{-}C_6H_4(O^-)(NH)\overset{+}{P}Ar_3 \right] \rightleftharpoons o\text{-}C_6H_4(OH)N{=}PAr_3 \quad (96)$$

(X) **(IX)**

10. Use of CCl_4 in phosphorus-nitrogen bond formation

Displacement of hydrogen from a diester of phosphorous acid by means of an amine occurs in the presence of carbon tetrachloride, Eq. (97) [18].

$$(RO)_2P(O)H + 2H_2NR' + CCl_4 \longrightarrow (RO)_2P(O)NHR' + CHCl_3 + [R'NH_3]C \quad (97)$$

The use of CCl_4 as a reagent for chlorination or dehydrogenation of phosphorus compounds and for P–N bond formation has been reviewed [8]. The variety of such uses is illustrated in Eqs. (98)-(107) [8-10, 13, 14].

$$R_3P + CCl_4 + HNRR' \longrightarrow [R_3PNRR']Cl + CHCl_3 \quad (98)$$

$$Ph_3P + CCl_4 + RSO_2NH_2 \longrightarrow Ph_3P{=}NSO_2R \quad (99)$$

$$\underset{(X = O \text{ or } S)}{Ph_3P + CCl_4 + R_2P(X)NH_2} \longrightarrow Ph_3P{=}NP(X)R_2 \quad (100)$$

$$Ph_3P + CCl_4 + R_2P(O)OH + 2HNR'_2 \longrightarrow [Ph_3POP(O)R_2]^+$$

$$\longrightarrow R_2P(O)NR'_2 + CHCl_3 + [R'_2NH_2]Cl + Ph_3PO \quad (101)$$

$$\underset{(X = O \text{ or } S)}{R_2P(X)NH_2 + PPh_3 + CCl_4} \longrightarrow [R_2P(X)NHPPh_3]Cl + CHCl_3 \quad (102)$$

$$(EtO)_2P(O)NH_2 + PPh_3 + CCl_4 \longrightarrow \frac{1}{n}[(EtO)_2PN]_n + Ph_3PO + CHCl_3 + HCl \quad (103)$$

$$(Ph_2P)_2CH_2 + NH_3 + CCl_4 \longrightarrow [Ph_2P(NH_2)N{=}PMePh_2]\,Cl \quad (104)$$

$$Ph_2PCH_2CH_2PPh_2 + NH_3 + CCl_4 \longrightarrow \left[\begin{array}{c} \text{cyclic } CH_2CH_2(PPh_2)_2N \end{array}\right] Cl^- \quad (105)$$

$$(Ph_2P)_2 + HNR_2 + CCl_4 \longrightarrow [Ph_2P(NR_2)CH_2Cl]Cl \quad (106)$$

$$\underset{(X = O \text{ or } S)}{R_2PP(X)R_2 + HNR'_2 + CCl_4} \longrightarrow [R_2P(NR'_2)_2]Cl + R_2P(X)NR'_2 \quad (107)$$

11. Synthesis of P–N compounds not involving P–N bond formation

Groups other than amido bonded to phosphorus may be exchanged under a variety of conditions without destroying the P–N linkage, e.g., Eqs. (108)-(112) [59, 146, 156, 159, 206].

$$Et_2NPCl_2 + 2M(NCO) \longrightarrow Et_2NP(NCO)_2 \xrightarrow[(NO_2)]{O_2} Et_2NP(O)(NCO)_2 \quad (108)$$

$$(M = Na, Ag, \text{ or } Me_3Si)$$

$$(R_2N)_2PCl + BrMgC{\equiv}CR' \longrightarrow (R_2N)_2PC{\equiv}CR' + MgBrCl \quad (109)$$

$$2R_2NP(R')Cl + BrMgC{\equiv}CMgBr \longrightarrow R_2NP(R')C{\equiv}CP(R')NR_2 + 2MgBrCl \quad (110)$$

$$(Me_2N)_{3-n}PCl_n + n\,MeOH \xrightarrow{Et_3N} (Me_2N)_{3-n}P(OMe)_n \quad (111)$$

$$R_2NP(O)Cl_2 + 2NaF \longrightarrow R_2NP(O)F_2 + 2NaCl \quad (112)$$

Perfluoroacetone oxidises phosphorus(III) compounds to the corresponding perfluoropinacol derivatives, Eqs. (113)-(115) [106, 107, 250]. If the P–N bonded compound contains an N-$SiMe_3$ linkage, condensation can occur, Eq. (114) [106, 107].

$$F_2PN(SiMe_3)_2 + 2(CF_3)_2CO \longrightarrow (Me_3Si)_2N{-}PF_2[OC(CF_3)_2C(CF_3)_2O] \quad (113)$$

$$F_2PN(SiMe_3)_2 + (CF_3)_2CO \xrightarrow{150^\circ C} [(CF_3)_2C(CF_3)_2\text{-}O_2]PF(NSiMe_3)_2PF[O_2\text{-}(CF_3)_2C(CF_3)_2] \quad (114)$$

$$(PhO)_2PNR_2 + 2(CF_3)_2CO \longrightarrow R_2N(PhO)(OPh)P[OC(CF_3)_2C(CF_3)_2O] \quad (115)$$

A primary amido-phosphorus compound may be *N*-chlorinated with Bu^tOCl [207], Eq. (116). The product may internally condense or, on reaction with a nucleophile, may rearrange, Eq. (117) [207].

$$R'P(O)(NHR)_2 + Bu^tOCl \xrightarrow{Eq.\ (116)} R'P(O)NHR(NClR) \xrightarrow{-HCl} R'\overline{P(O)NRNR}$$

$$R'P(O)(NHR)_2 + Bu^tOCl \xrightarrow{NaOMe} R'P(O)(OMe)NRNHR; \quad R'\overline{P(O)NRNR} \xrightarrow{MeOH} R'P(O)(OMe)NRNHR \quad (117)$$

The reaction of an alcohol with a cyclic tris(amino)phosphine causes preferential displacement of an acyclic ligand, as in Eq. (118), unless the ring system is strained or otherwise labilised, e.g., Eq. (119) [51].

$$Me_2NP\langle NMe\text{-}CH_2CH_2\text{-}NMe\rangle + MeOH \xrightarrow{80^\circ C} MeOP\langle NMe\text{-}CH_2CH_2\text{-}NMe\rangle + HNMe_2 \quad (118)$$

$$Me_2NP\langle NMe\rangle_2C{=}O + MeOH \longrightarrow (MeO)_2PNMe_2 + (MeO)_2PNMeCONHMe + MeOP(NMe_2)NMeCONHMe \quad (119)$$

An alkyl-lithium reagent behaves as a base (Chapter 16) towards $P(O)(NMe_2)_3$, rather than as an alkylating agent; the lithiated compound (XI) then rearranges, Eq. (120) [1, 222].

$$P(O)(NMe_2)_3 + LiR \longrightarrow RH + \underset{\textbf{(XI)}}{LiCH_2NMeP(O)(NMe_2)_2} \longrightarrow LiOP(NMe_2)_2 + CH_2{=}NMe \quad (120)$$

A derivative of a primary amine can be *N*-lithiated, and a further nitrogen substituent may subsequently be introduced, Eqs. (121) and (122) [221, 229]. For derivatives of diamines, this may be complicated by ring closure, Eq. (122) [221]. $[Me_3PN(SiMe_3)_2]I$ and LiBu gives $Me_2(Me_3SiCH_2)P{=}NSiMe_3$ [253b].

$$(EtO)_2PNHR + LiBu \longrightarrow BuH + (EtO)_2PN(Li)R \xrightarrow{R'Cl} (EtO)_2PNRR' + LiCl \quad (121)$$

$$(EtO)_2PNHCH_2CH_2NHR + 2LiBu \longrightarrow (EtO)_2PN(Li)CH_2CH_2N(Li)R$$

$$\xrightarrow{R'Cl} (EtO)_2PNR'CH_2CH_2NRR' + EtOP\langle NR\text{-}CH_2CH_2\text{-}NR\rangle \quad (122)$$

$$Bu^t_2PNHSiMe_3 + LiBu \longrightarrow Bu^t_2PN(Li)SiMe_3 \xrightarrow{Me_3SiCl} \underset{\textbf{(XII)}}{Bu^t_2PN(SiMe_3)_2} \quad (123)$$

Compound (XII) [184] was originally formulated as $Bu^t_2P(NSiMe_3)SiMe_3$ [229].

Oxidation of a P^{III} compound to the corresponding $P^V(O)$ compound is normally accomplished by reaction with O_2, HgO, Me_2SO, H_2O_2, or NO_2, [e.g., 59, 143a, 146, 159]. Formation of a $P^V(S)$ or $P^V(Se)$ analogue is effected by heating the P^{III} substrate with sulphur or selenium, e.g., [110, 143a, 149, 159], or alternatively another P^V sulphide, as in Eq. (124) [36, 110].

$$[(R_2N)_2PNCO]_2 + 2Cl_3P(S) \longrightarrow 2(R_2N)_2P(S)NCO + 2PCl_3 \quad (124)$$

The formal analogy between the Wittig reaction and addition of a ketone to a P^V=N compound includes the formation of an intermediate phosphorane, as in Eq. (124a) (e.g., X = Me = R′ = R) [230a].

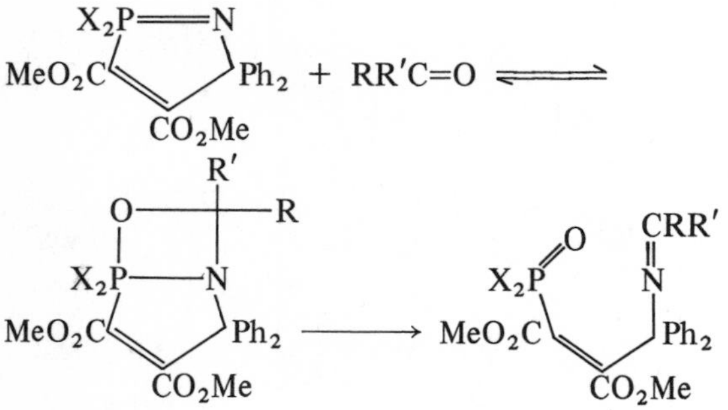

$X_2P{=}N$ ring with MeO_2C, Ph_2, CO_2Me + $RR'C{=}O$ ⇌ [O–$C(R)(R')$–N / X_2P–N four-membered ring fused with MeO_2C, Ph_2, CO_2Me] ⟶ $X_2P({=}O)$, MeO_2C, CO_2Me, Ph_2, $N{=}CRR'$ (124a)

An unusual reaction whereby a co-ordinated amide ligand rearranges and a new P–C bond is made is illustrated in Eq. (124b) [126].

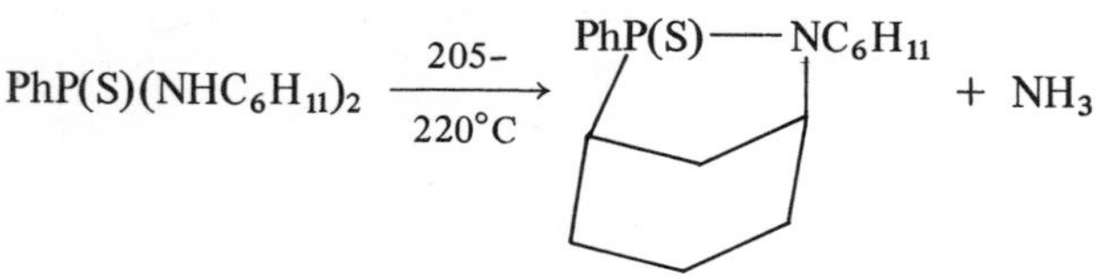

$$PhP(S)(NHC_6H_{11})_2 \xrightarrow[220°C]{205-} PhP(S)\text{—}NC_6H_{11}\ (\text{cyclohexane-fused ring}) + NH_3 \quad (124b)$$

An aminophosphine has ligand properties, e.g., ref. [198]. Further discussion and references are found in Chapter 16. Amino(fluoro)phosphine complexes, such as $[Ni(PF_2NR_2)_4]$ were first described in 1964 [231a], and early reviews are available [147, 190f]. Complexes include $H_2BH_2BP(F)_2N(Me)PF_2$ [199a], $F_nP(NMe_2)_{3-n}.BX_3$ [98], $F_2PFe(\eta\text{-}C_5H_5)N(Me)P(F)_2Fe(\eta\text{-}C_5H_5)P(F)_2N(Me)PF_2$ [185a], $[Co(L)\{PF_2N(Me)PF_2\}_3Co(L)]$ ($L = PF_2NMe_2$ [57b] or Br_2 [185b]) (X-ray data), $[Fe(CO)_4\{FP(NMe_2)_2\}]$ [87a]; an analogue, $[Fe(CO)_4\text{-}\{FPN(Me)(CH_2)_2NMe\}]$, of the last complex, on treatment with PF_5, yields $[Fe(CO)_4\{PN(Me)(CH_2)_2NMe\}]^+$ [178a], also obtained from $[Fe(CO)_5]$ and

$[\overline{PN(Me)(CH_2)_2NMe}][PF_6]$. Similar compounds $[Mo(\eta\text{-}C_5H_5)(CO)_3\text{-}\{\overline{PN(Me)(CH_2)_2NMe}\}]$ and $[Fe(CO)_4\{\overline{PN(Me)(CH_2)_2NMe}\}_2]$ are obtained by salt elimination using the metallate anion and a $P^{III}F$ compound as starting materials [165a]. Complexes $[M(CO)_5\{P(NH_2)_3\}]$ (M = Cr, Mo, or W) are of interest because the free ligand $P(NH_2)_3$ is unstable [191a]. An unusual cluster compound $\{MeN[PPh_2AgBr]_2\}_2$ has an $Ag_4Br_2^{2+}$ unit clamped by $MePN(PPh_2)_2$ [241a].

12. Quaternisation and related reactions of P–N compounds

Phosphonium salts containing P–N bonds are readily prepared (see also Chapters 16 and 18). Quaternisation of the phosphorus atom and not the nitrogen atom inevitably occurs for stable systems, e.g., Eq. (125) [50], (126) [91], (127) [65], (128) [94], (129) [135]. A chloroamine reacts readily with a P^{III} compound to give the phosphonium salt, Eq. (130) [61, 103]. Quaternisation can occur by oxidation of the P^{III} compound with $SbCl_5$, Eqs. (131), (132) [205, 206]. Similarly $P(NR_2)_3$ and I_2 gives $I_nP(NR_2)_3$ (n = 2 or 4) [111a].

$$Me_2PNMe_2 + MeI \longrightarrow [Me_3PNMe_2]I \tag{125}$$

$$PhP(NEt_2)_2 + MeI \longrightarrow [PhPMe(NEt_2)_2]I \tag{126}$$

$$(C_6F_5)_2PNHPr^i + MeI \xrightarrow{BF_4^-} [(C_6F_5)_2PMeNHPr^i]BF_4 + I^- \tag{127}$$

$$RP\left\langle\begin{matrix} NMeSiMe_2 \\ NMeSiMe_2 \end{matrix}\right\rangle NMe + MeI \longrightarrow \left[RMeP\left\langle\begin{matrix} NMeSiMe_2 \\ NMeSiMe_2 \end{matrix}\right\rangle NMe\right] I \tag{128}$$

$$P_4(NMe)_6 + MeI \longrightarrow [MeP_4(NMe)_6]I \tag{129}$$

$$R_3P + ClNR'_2 \longrightarrow [R_3PNR'_2]Cl \tag{130}$$

$$Cl_2PNMe_2 + 2SbCl_5 \longrightarrow SbCl_3 + [Cl_3PNMe_2][SbCl_6] \tag{131}$$

$$(MeO)_2PNMe_2 + 2SbCl_5 \longrightarrow SbCl_3 + [(MeO)_2PCl(NMe_2)][SbCl_6] \tag{132}$$

Acyl halides react with P–N bonded compounds to give P–Hal bonded products. The mechanism of the reaction has been postulated to involve formation of an aminophosphonium salt or a phosphinoammonium salt as intermediate, depending on the nature of the phosphorus substituents [40].

Secondary N-halogenoamides react with trialkyl phosphites to eliminate alkyl halides, Eq. (133) [84], but primary analogues give P–Cl bonded products, Eq. (134) [84]. However, the reaction between a secondary N-chlorosilylamine,

$$P(OR)_3 + \overline{\underline{|}}\begin{matrix}CO \\ CO\end{matrix}\!\!>NCl \longrightarrow RCl + \overline{\underline{|}}\begin{matrix}CO \\ CO\end{matrix}\!\!>NP(O)(OR)_2 \quad (133)$$

$$P(OEt)_3 + R'CONHHal \longrightarrow (EtO)_3PO + R'CN + HHal$$

$$(EtO)_2P(O)Hal \xleftarrow{R'CONHHal} (EtO)_2P(O)H \quad (134)$$

Eq. (135), and an appropriate phosphorus(III) compound affords a monophosphazene (XIII, R = Ph, Bu, OEt, OPr, or OBu) with elimination of trimethylchlorosilane [201a].

$$R_3P + ClN(SiMe_3)_2 \longrightarrow Me_3SiCl + \underset{\textbf{(XIII)}}{Me_3SiN{=}PR_3} \quad (135)$$

The Staudinger reaction, Eq. (136) [156], is a versatile and common method of formation of P=N bonded systems e.g., Eqs. (137) [111], (138) [158], or (139) [31, 32]. The reaction is influenced most strongly by the basicity of R_3P [158]. o-$HOC_6H_4N_3$ with Ph_2POMe and $ClNPr^i_2$ yields o-$\overline{OC_6H_4N(H)P}Ph_2OMe$, and with $P(NMe_2)_3$ affords o-$\overline{OC_6H_4N{=}P}(NMe_2)_2$ [53a]. The compound $R_2NP{=}NR'$ ($R = Pr^i$, $R' = Bu^t$; or $R = Me_3Si$, $R' = Bu^t$) with $R'N_3$ gives $R'\overline{NN{=}NN(R')P}NR_2$ [190b, 202c].

$$R_3P + R'N_3 \longrightarrow R_3P{=}NR' + N_2 \quad (136)$$

$$P(OEt)_3 + EtN_3 \longrightarrow EtN{=}P(OEt)_3 + N_2 \quad (137)$$

$$P(NR_2)_3 + R'N_3 \longrightarrow R'NNNP(NR_2)_3 \longrightarrow N_2 + R'N{=}P(NR_2)_3 \quad (138)$$

$$P_4(NMe)_6 + nPhN_3 \longrightarrow (PhNP)_nP_{4-n}(NMe)_6 + nN_2 \quad (139)$$

In a series of similar reactions, azido-phosphorus or -silicon compounds eliminate dinitrogen to give P=N bonded products, e.g., Eqs. (140) [32, 239], (141) [15], (142) [15], (143) [188, 189].

$$R_3P + Ph_2P(O)N_3 \longrightarrow N_2 + Ph_2P(O)N{=}PR_3 \quad (140)$$

$$Me_2Si(N_3)_2 + (Ph_2P)_2(CH_2) \longrightarrow \left\langle \begin{matrix} PPh_2{=}N \\ PPh_2{=}N \end{matrix} \right\rangle SiMe_2 + 2N_2 \quad (141)$$

$$2R_3SiN_3 + (Ph_2P)_2(CH_2)_n \xrightarrow[(n = 1\text{-}3)]{} (R_3SiN{=}PPh_2)_2(CH_2)_n + 2N_2 \quad (142)$$

$$(Me_3Si)_2NP{=}NSiMe_3 + Me_3SiN_3 \longrightarrow (Me_3Si)_2NP(:NSiMe_3)_2 + N_2 \quad (143)$$

The product of reaction (141) cannot be obtained from $(R_3SiN{=}PPh_2)_2CH_2$ and Me_2SiCl_2 [15].

Co-ordinated PPh_2Pr reacts with the azide of a sulphinic acid to give a co-ordinated P=N bonded moiety, Eq. (144) [234].

$$[MoCl_4(PPh_2Pr)_2] + p\text{-}MeC_6H_4SO_2N_3 \longrightarrow N_2 + [MoCl_4(OPPh_2Pr)(NPPh_2Pr)] + p\text{-}MeC_6H_4SSO_2C_6H_4Me\text{-}p \quad (144)$$

The *N*-chloro-*N*-sodio- derivative of a sulphonamide reacts with a phosphine to form a P–N [Eq. (145)] or a P=N [Eq. (146)] linkage [168].

$$Ar_3P + p\text{-}MeC_6H_4SO_2N(Cl)Na \xrightarrow[98\%]{EtOH} Ar_3P(OH)NHSO_2C_6H_4Me\text{-}p \quad (145)$$

$$Ar_3P + p\text{-}MeC_6H_4SO_2N(Cl)Na \xrightarrow{\text{absolute EtOH}} Ar_3P{=}NSO_2C_6H_4Me\text{-}p \quad (146)$$

Reaction of a phosphazene [Eq. (147)] [216, 217] or pentaphenylphosphorane [Eq. (148)] [216] with potassamide gives ionic products.

$$2Et_3P{=}NH + 2KNH_2 \longrightarrow K_2[Et_2P(:N)NHP(:N)Et_2] + NH_3 + C_2H_6 \quad (147)$$

$$PPh_5 + KNH_2 \longrightarrow K_4[PhP(:NH)(:N)NHPPh(:NH)(:N)] \quad (148)$$

Reactions related to that of Eq. (143) are of some generality as illustrated by Eqs. (149)-(152) ($R = Me_3Si$) [11, 12, 188-189a, 227]. Evidence for the product type of Eq. (149) has been obtained by using $Me(R')CN_2$ [190c]; the methylenephosphorane reacts with excess of the diazoalkane to yield successively $R_2NP(:NR)$-$[N{=}\overset{+}{N}{=}C(R')Me][C(R')Me]$ and $R_2N(:NR)P[NHN{=}C(R')Me][C(R')Me]$.

$$R_2NP{=}NR + CH_2N_2 \longrightarrow R_2NP(:NR){=}CH_2 \quad (149)$$

$$R_2NP{=}NR + (CF_3)_2CN_2 \xrightarrow{(1:1)} R_2NP(:NR){=}NN{=}C(CF_3)_2 \xrightarrow{(2:1)}$$

$$R_2NP\langle NR, N(NC(CF_3)_2)\rangle P(NR_2){=}NR \qquad (150)$$

$$R_2NP{=}NR + S \longrightarrow R_2NP(S){=}NR \xrightarrow{R_2NPNR}$$

$$\begin{array}{cc} RN{-}P(S)NR_2 \\ | \quad\; | \\ R_2NP{-}NR \end{array} \xrightarrow{S} \begin{array}{cc} RN{-}P(S)NR_2 \\ | \quad\; | \\ R_2N(S)P{-}NR \end{array} \qquad (151)$$

$$R_2NP{=}NR + Me_3SiN_3 \xrightarrow{-N_2} R_2NP(:NR)_2 \xrightarrow{CH_2N_2} R_2NP{=}NR \text{ (with } CH_2{-}NR \text{ ring)}$$

$$R_2NP(:NR)_2 \xrightarrow{R_2NP=NR} R_2NP\langle NR, NR\rangle P(=NR)NR_2$$

$$R_2NP(:NR)_2 \xrightarrow{MeOH} R_2NP(OMe)(NHR){=}NR \qquad (152)$$

It is noted that $[(Me_3Si)_2N]_2PCl$ reacts with Me_3SiN_3 to give $[(Me_3Si)_2N]_2PN_3$ [227]. An interesting method of PN formation is illustrated in Eq. (153) ($Ar = C_6H_3Me_2$-2,6) [53, 54].

$$PhP\langle O{-}CH_2CH_2{-}O\rangle + C_6H_4(NO_2)(SAr) \longrightarrow$$

$$C_6H_4\langle S, N(Ar)\rangle P(Ph)\langle O{-}CH_2CH_2{-}O\rangle + SO_2 \qquad (153)$$

13. Synthesis of stable phosphorus(II) amides

Although phosphorus(II) compounds in the form of phosphinyl radicals have been known for many years, they are all unstable species, being studied, in the main, at low temperatures via matrix isolation. These radicals have been synthesised by two main routes involving either thermolysis or photolysis of a suitable precursor, although recently radical displacement reactions have become more widely used.

With regard to amido- substituted P^{II} compounds, $\dot{P}(NMe_2)_2$ was the first to be well characterised in solution [116a] and was synthesised as in Eq. (154).

$$(OBu^t)_2 \xrightarrow{h\nu} 2\dot{O}Bu^t + (Me_2N)_4P_2 \longrightarrow Bu^tO(Me_2N)_2\dot{P}P(NMe_2)_2$$

$$\longrightarrow Bu^tOP(NMe_2)_2 + \dot{P}(NMe_2)_2 \qquad (154)$$

This radical is very reactive and has a very short half-life which is attributed to rapid dimerisation affording $P_2(NMe_2)_4$.

It is possible to prevent the dimerisation reaction by employing large substituents at the nitrogen. Using very sterically-demanding ligands, it has been possible to prepare a series of stable amido- substituted phosphinyl radicals via reaction of the phosphorus(III) halogen- substituted radical precursor with a reducing agent [an electron-rich olefin (e.r.o.), e.g., (tetraaminoethylene structure: two six-membered rings each with N(Et) atoms joined by a C=C bond)] in toluene. Photolysis of this reaction mixture affords a persistent e.s.r. signal attributable to a stable phosphinyl species [Eq. (155)] [117].

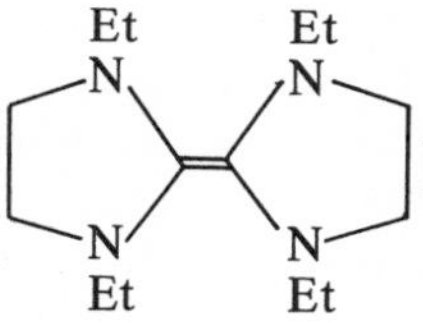

$$ClP(NR_2)_2 \xrightarrow[\text{PhMe, 25° C}]{\text{e.r.o.}} \dot{P}(NR_2)_2 \qquad (155)$$

It can be seen (half-lives) from the following table, that these compounds are extremely long-lived, due to kinetic stabilisation, which is attributed to the sterically-demanding substituents preventing dimerisation. The e.s.r. parameters of these radicals are discussed in Section C.6.

Table 1 Kinetically-stabilised bivalent phosphorus amides [117]

Compound	Half-life at 25° C in PhMe[a]
$:\dot{P}[CH(SiMe_3)_2]NMe_2$	ca. 5 min
$:\dot{P}[CH(SiMe_3)_2][N(SiMe_3)_2]$	10 d
$:\dot{P}[CH(SiMe_3)_2](NPr^i_2)$	1 d
$:\dot{P}[N(SiMe_3)_2]_2$	5 d
$:\dot{P}(NPr^i_2)N(SiMe_3)_2$	1 d
$:\dot{P}(NBu^tSiMe_3)_2$	5 d
$:\dot{P}(NPr^i_2)N(Bu^tSiMe_3)$	3 d

[a] d = day

C STRUCTURE AND BONDING IN AMIDOPHOSPHORUS COMPOUNDS

This is described in detail in reviews [63, 132, 170, 253a] dealing with the structure and bonding of a large variety of such compounds, the spectral techniques used to study them, and structures of relevant reaction intermediates. In general, amides of P^{III} and P^{V} show trigonal hybridisation at N (bond angles ca. 120°).

1. Five-co-ordinate phosphorus(V) amides, the phosphoranes

The structures of phosphoranes are normally close to trigonal bipyramidal (Fig. 1). Much has been made of the relative tendencies of groups preferentially to adopt apical positions (apicophilicity) [e.g., 197, 249, 250]; electronic factors favour the preferential placing of the most electronegative ligands in the apical positions. The structures of these five-co-ordinate compounds are, however, often determined by the requirements of other groups for the equatorial positions. A single secondary amino-group has a marked tendency to adopt an equatorial location. Further, such an amino-group is normally close to planar, and close to co-planar with the axial groups. In this manner interactions between the lone-pair on nitrogen (virtually 2*p*) and the electron density of all other ligands are minimised (Fig. 2).

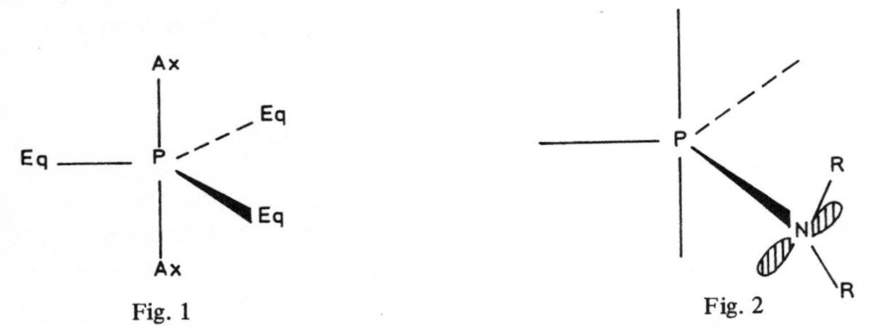

Fig. 1

Fig. 2

The barrier to rotation about the P–N bond in F_4PNH_2 (>15 kcal mol^{-1}) is much greater than in F_4PNMe_2 (9 kcal mol^{-1}) [72]. This effect has been attributed to an interaction between the protons of the NH_2 groups and the axial fluorines [72]. The barrier to rotation for $F_3P(NH_2)_2$ is 12.3 kcal mol^{-1}, which was regarded as appropriate for P=N double bonding [181]. The mass, i.r., and n.m.r. spectra of a number of similar fluoro-(amino)- or -(dialkylamino)-phosphoranes have been measured, and in each case the amino-group is equatorial [e.g., 57, 64, 73, 196, 204, 232, 233]. The n.m.r. spectra of $F_4PNPr^i_2$ show that not only is rotation about the P–N bond restricted, but that rotations about N–C bonds are restricted also, probably due to interactions with axial fluorines (Fig. 3) [64]. N.m.r. spectra are generally used for assignment of configuration, e.g., ref. [199c]. Dynamic ^{13}C and ^{31}P n.m.r. studies show that $\Delta G^{\ddagger}_{298}$ for CF_3 site exchange in $(CF_3)_3MePNMe_2$ is 16.5 kcal mol^{-1} [55].

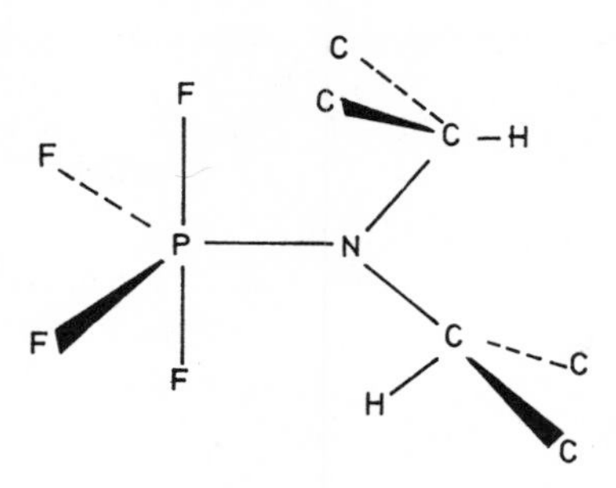

Fig. 3

The molecular structure of $F_2P(NMe_2)_3$ has been determined by electron diffraction (Fig. 4) [196]. The amino-groups are, as expected, equatorial and close to planar. However, the mutual interactions of these groups, and the lone pairs, cause the NMe_2 groups to be at an angle (~20°) to the NPF_2 plane.

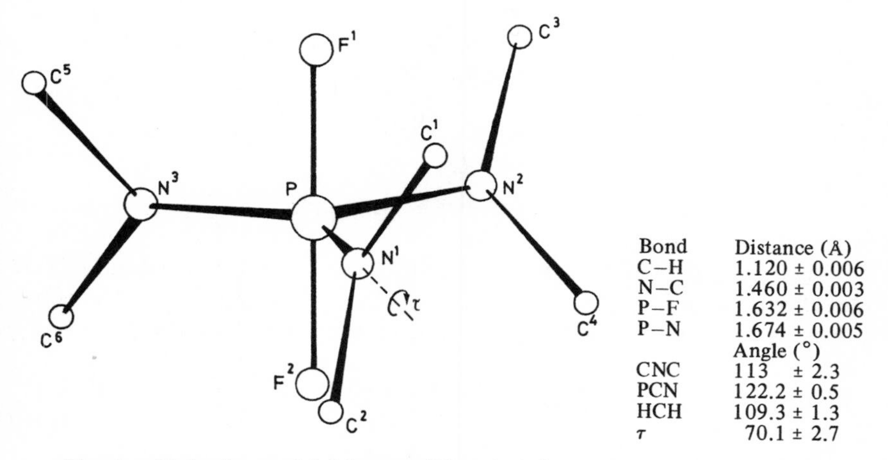

Bond	Distance (Å)
C–H	1.120 ± 0.006
N–C	1.460 ± 0.003
P–F	1.632 ± 0.006
P–N	1.674 ± 0.005
	Angle (°)
CNC	113 ± 2.3
PCN	122.2 ± 0.5
HCH	109.3 ± 1.3
τ	70.1 ± 2.7

Fig. 4 – Molecular model (electron diffraction) for $F_2P(NMe_2)_3$. *[Reproduced, with permission, from Oberhammer, H., and Schmutzler, R.,* J.C.S. Dalton, *1976, 1454.]*

An interesting feature of the structure is the relatively long P–N bond length of 1.674(5) Å, which is considerably longer than the value of 1.595 Å found in the cyclodiphosphazene $[(Me)NPF_3]_2$ [7a]. The bond is similar in length to the values found for three-co-ordinate phosphorus (ca. 1.65–1.70 Å), but shorter than the value quoted for a P–N single bond (1.77 Å) [75a]. The photoelectron spectra of this compound, $F_3P(NMe_2)_2$, and F_4PNMe_2 have been studied, and dependence of the spectra on lone-pair interactions has been established [67].

The trifluoromethyl-phosphorus bonded analogues of many of the above systems have been studied [57, 57a, 204, 243]. Again, the R_2N group(s) are inevitably equatorial, and F, Cl, or CF_3 ligands occupy the remaining axial and equatorial positions.

When ring-strain is introduced into the system, the structures of the phosphoranes can be much distorted from the trigonal bipyramid, almost to square pyramidal. The relative axial or equatorial preferences of many groups in such systems have been reviewed [249]. The 'apicophilicity' of NMe_2 has been placed in the series H>OPh>NMe_2>Me>Pr^i>CH=CMe_2>Ph [197, 250]. The difference between OPh and NMe_2 was considered large [197]. A group such as the perfluorobut-1,2-diolato bidentate ligand is stereochemically required to bond to one axial and one equatorial position. In such a case, when an OR group is forced to occupy an axial position, a fluorine is displaced to an equatorial position (e.g., Fig. 5) [106, 107].

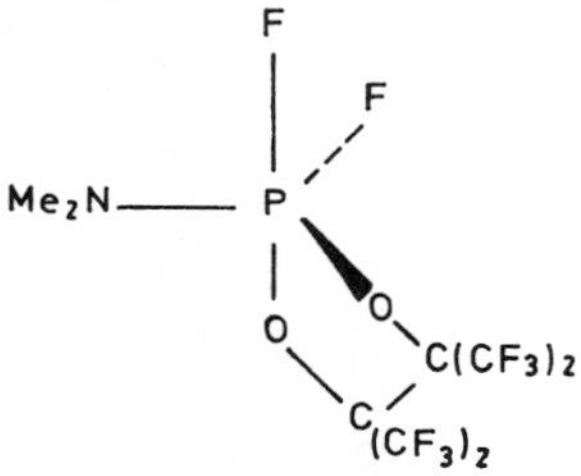

Fig. 5

A similar series of compounds in which PhO replace P–F groups has been described [250].

Hydroxy(amino)alkanes, such as $RNHCH_2CH_2OH$ or (-)-ephedrine, chelate phosphorus atoms [e.g., 52, 62, 185]. The synthesis of these compounds involves a final insertion reaction, forming a P–H bond, Eq. 155 [52, 62, 185].

$$2R(H)NCH_2CH_2OH + P(NMe_2)_3 \longrightarrow (\overline{RNCH_2CH_2O)_2P}H + 3HNMe_2 \quad (155)$$

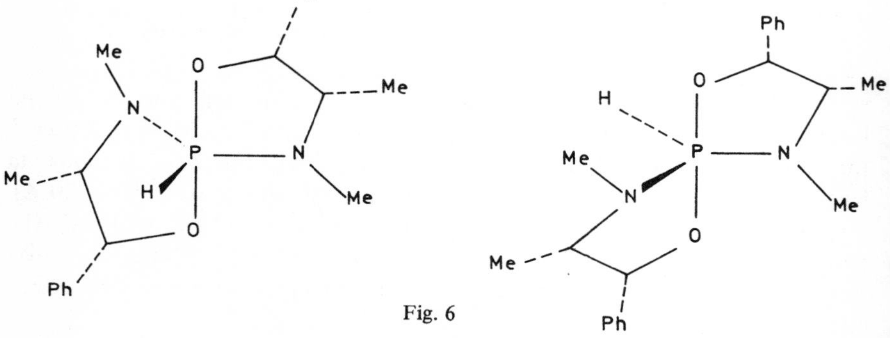

Fig. 6

The five-membered rings require that co-ordination of each bidentate group to phosphorus must occur at one axial and one equatorial position. In each case, the hydrogen and the amino-groups are equatorial. The crystal structure of the (-)-ephedrine derivative showed the presence of two isomers, due to the optical activity of the ligand (Fig. 6) [185].

An amide in which geometric constraint enforces a strict geometry upon the system is that of $(PN)_2$ cyclic compounds. (The cyclodiphosphazanes have been reviewed [113].) The $(PN)_2$ ring must be close to planar, or else the $\widehat{PNP}$ or $\widehat{NPN}$ angles must drop significantly below 90°. The crystal structures of $(RPF_2NMe)_2$ (R = Me, Et, or Ph) [74, 96] have shown that these and similar amides [e.g., 123, 232] adopt a *trans*-configuration (Fig. 7) in the solid state, although the *cis*-isomer may exist in solution.

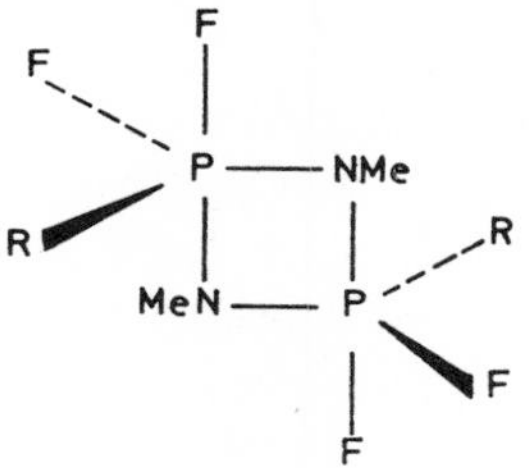

Fig. 7

The constraints imposed by the $(PN)_2$ ring geometry oblige each NMe group to be equatorial to one phosphorus, but axial to the other, displacing fluorine to an equatorial position. When the phosphorus atoms are chelated as in Fig. 8, it is the axial fluorine that is replaced [106, 107]. A similar system derived from diacetyl has been described (Fig. 9) [255].

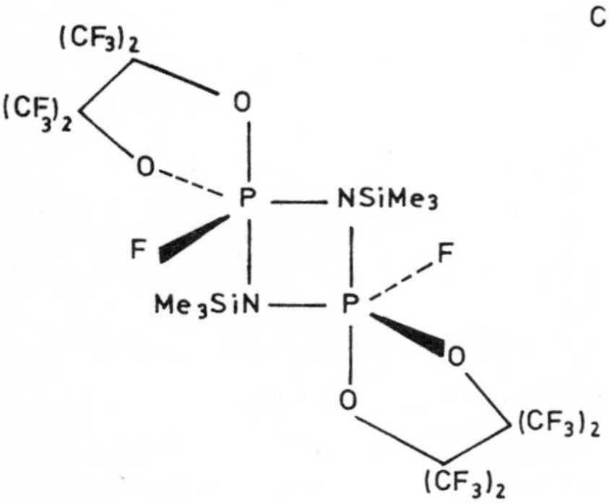

Fig. 8

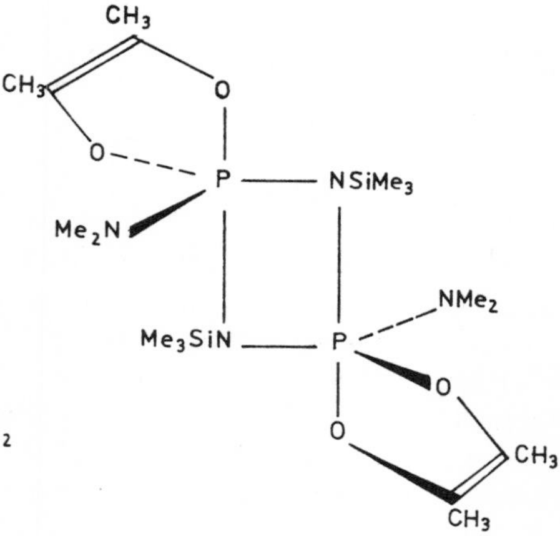

Fig. 9

The compound shown in Fig. 10 is obtained from $P(NMe_2)_3$ and 1,4,7,10-tetraazacyclododecane [208a], the macrocycle effect stabilising the P^V tautomer of Fig. 10. Related compounds have been made, based on a series of four nitrogen 12- to 16-membered macrocycles, to study the effect of ring size on the P^{III}-P^V tautomerism [20a].

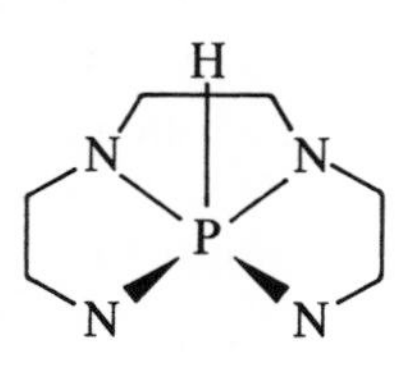

Fig. 10

2. Four-co-ordinate phosphorus(V) amides

The amides $[RP(X)NR']_2$ (X = O [e.g., 152], or S [e.g., 47, 138, 149, 200]) may in principle exist in *cis*- and *trans*- forms. The $(PN)_2$ ring is planar. The compound $[ClP(O)NBu^t]_2$ has been shown by n.m.r. to exist only as the *trans*-isomer (Fig. 11a) [152]. In the thiophosphorus systems repulsions between diametrically opposed atoms are important [47, 200]. Nevertheless, the compound $[PhP(S)NR]_2$ exists in the solid state as both *trans*- and *cis*-isomers (Figs. 11a, 11b) [138]. {$(ClPNBu^t)_2$ has the *cis*-structure (Fig. 11b, X = a lone pair of electrons) [182]}.

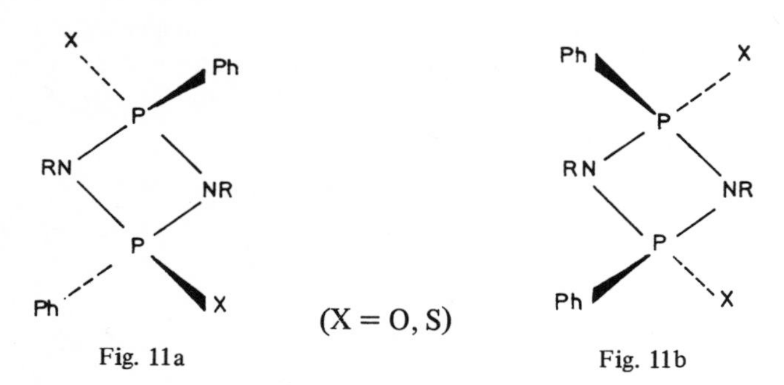

Fig. 11a Fig. 11b

Similar compounds, with inequivalent NR groups [224] or with X = NR [202, 224], have been described. The amide $[(Me_3Si)_2NP(S)NBu^t]_2$ contains the $(PS)_2$ ring and exocylic :NBu^t groups [202].

Phosphoryl amides are among the best-known of phosphorus compounds. Recent advances in the chemistry of organophosphorus P–N bonded compounds, especially of compounds of biological interest, have been reviewed [173]. Carbohydrate derivatives may be used to study phosphorus stereochemistry, e.g., Fig. 12 [124].

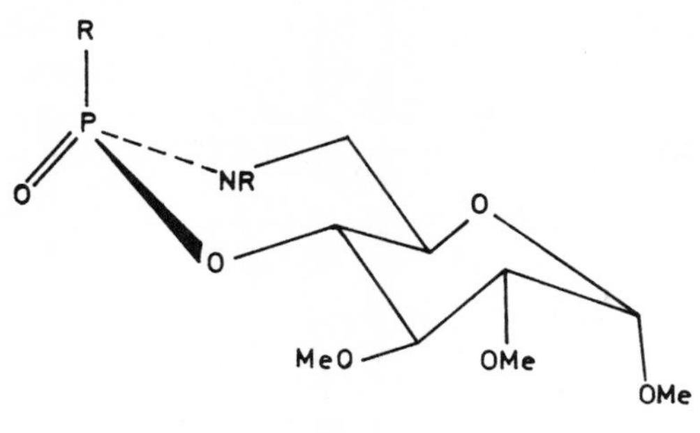

Fig. 12

The crystal structure of $PhP(O)(NH_2)_2$ has shown that one nitrogen atom is planar and coplanar with the P=O bond, but that the other is pyramidal [43]. In contrast, the carcinostat of Fig. 13 ($R = ClCH_2CH_2$) contains exocyclic and ring nitrogen atoms, each of which is close to planar [60].

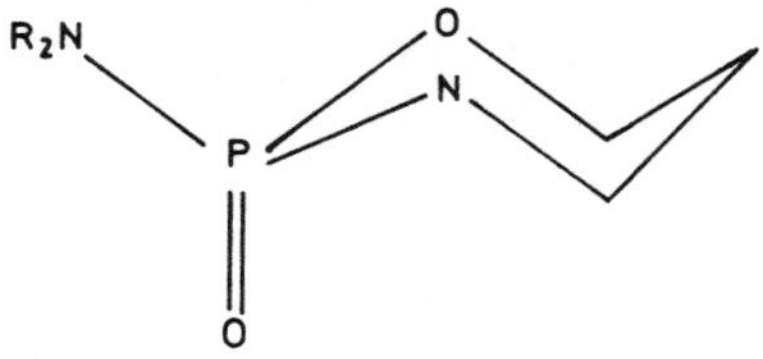

Fig. 13

Dipole moment and n.m.r. studies have shown that exocyclic secondary amino-groups bonded to phosphorus(V) prefer the equatorial (Fig. 14) conformation of the chair form of a six-membered ring, although only by 2:1 (axial) in the example shown [180]. In contrast, the anilino-derivative is in the axial form, Fig. 15 [240].

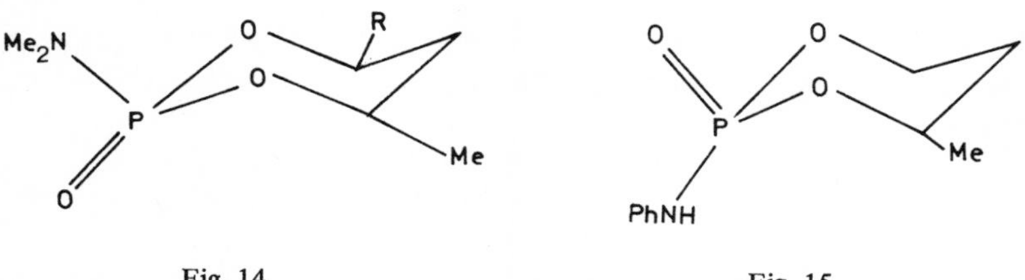

Fig. 14

Fig. 15

The nitrogen atom in $Ph_2P(O)NMe_2$ is reported to be non-planar [120], one of the few P–N compounds in which N deviates markedly from planarity (Fig. 16).

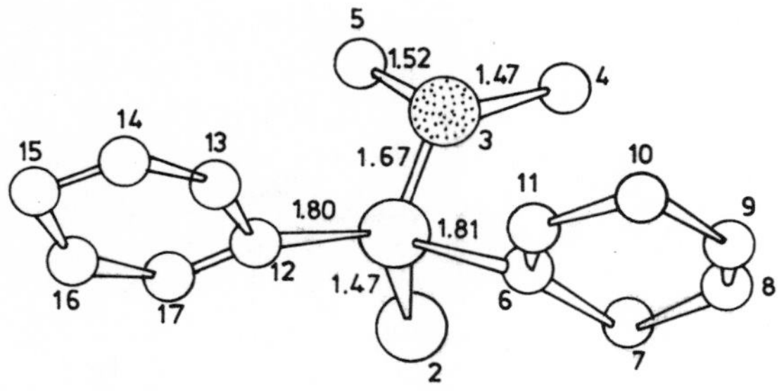

Fig. 16 – X-ray crystal structure of $Ph_2P(O)NMe_2$. *[Reproduced, with permission, from Haque, M., and Caughlaw, C. N.,* Chem. Comm., *1969, 921.]*

The primary amino-phosphorus compound $Cl_2P(O)NHMe$ is (i.r.) dimeric, due to hydrogen-bonding [236]. The i.r. spectra of XYP(O)NHR (X, Y = RO, ArO, Halogen, NHR) show only one N–H stretching mode [193]. The data indicate that the favoured conformation has the N–H bond *cis* to P=O [193].

Rotation about the P–N bond has been shown (n.m.r.) to be hindered in similar systems ($XY = CMe_2CHMeCMe_2$); the cyclic molecule shown in Fig. 17 shows *two* doublets for NMe_2 in the ^1H n.m.r. spectrum [90]. The barriers to rotation about the P–N bond in $PhP(O)ClNPr^i_2$ (9.4 kcal mol^{-1}) and $PhP(S)ClNPr^i_2$ (9.9 kcal mol^{-1}) are each considerable [144].

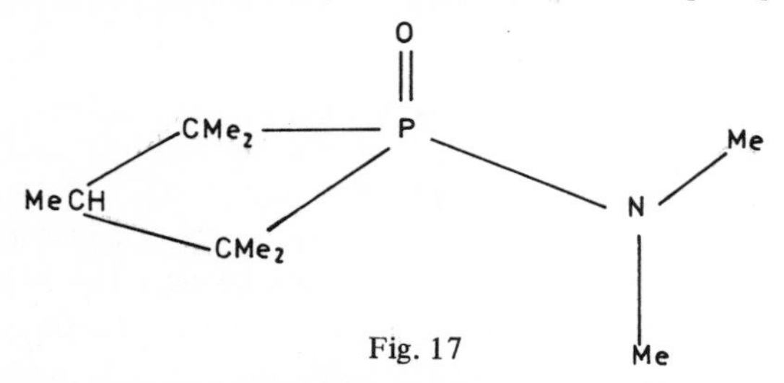

Fig. 17

The i.r. spectra of the analogous thiophosphoryl systems XYP(S)NHR show *two* N–H stretching modes which are assigned to the conformers in which the N–H bond is *cis* (Fig. 18) or *trans* (Fig. 19) to the P=S bond [172]. Some of the N–H (or N–D) bending modes are not purely N–H, but also involve nearby atoms [194, 195].

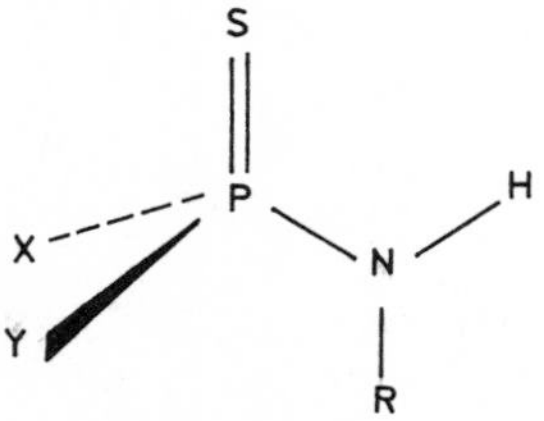

Fig. 18

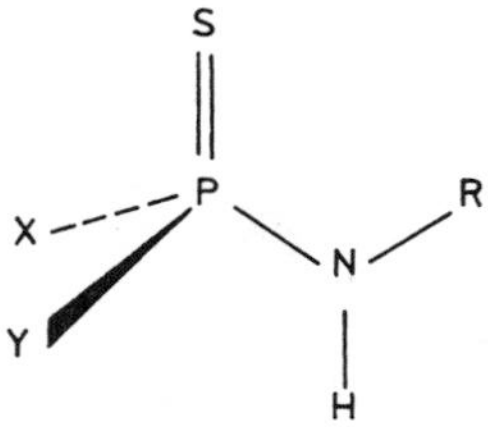

Fig. 19

The orbital energies of the compounds $P(O)X_3$ and $P(S)X_3$, including $X = NR_2$, have been investigated by photoelectron spectroscopy [101].

The series shown in Fig. 20 (R^1-R^4 = Cl, NMe_2; R^5 = Me, Ph; X, Y = O, S) has been studied by i.r. and n.m.r. spectroscopy [118, 139, 148, 150, 151, and references therein]. X-ray data are available on related compounds (Figs. 21 and 22) [105] and on $Ph(Cl)(O)PNPr^i_2$ [199b].

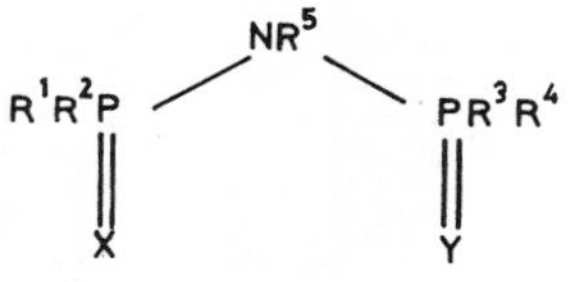

Fig. 20

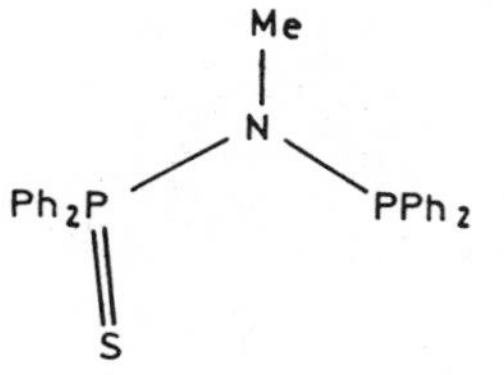

Fig. 21

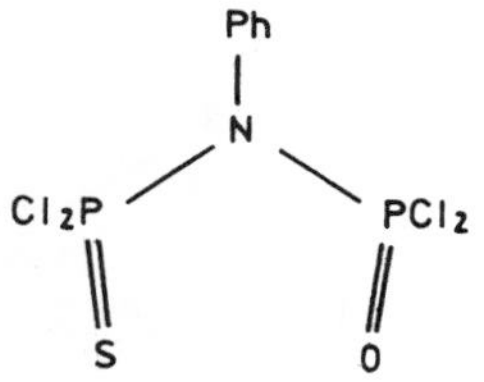

Fig. 22

Dimethylaminolysis of $[Cl_2P(O)]_2NMe$ gives the mono-, *PP'*-di-, and tetra- substituted products [139]. The geminally di-substituted compound is obtained by the reaction of $(Me_2N)_2P(O)NHMe$ with $P(O)Cl_3$ [139]. For similar amides in which the phosphorus atoms are in different oxidation states, the phosphorus(V) atom is preferentially aminated [148, 150, 151]. Indeed, $Me_2NP(Cl)NMeP(O)Cl(NMe_2)$ rearranges to form $(Me_2N)_2P(O)NMePCl_2$ [150]. Such an exchange reaction also occurs between $(Me_2N)_2PCl$ and $Me_2NP(O)Cl_2$ to give the products $ClP(O)(NMe_2)_2$ and Cl_2PNMe_2 [150]. It should be noted that the symmetrically substituted product $[Me_2NP(O)Cl]_2NMe$ has two chiral centres [139], as does $[RP(O)(NMe_2)]_2O$ [145]. The stereoisomers are distinguishable by n.m.r. spectroscopy [139, 145]. The compounds $[X_2P(O)]_2Y$ ($X = NMe_2$, OPr^i; Y = O, NMe, CH_2) have been used as bidentate ligands, and crystal structures have been determined [177].

The P—N bond length and the geometry about nitrogen in $OP(NMeNMe)_3PO$ have been discussed in terms of P—N (d_π-p_π) bonding [109].

Substituent effects on the two-co-ordinate nitrogen system (the mono-phosphazenes) $Ar_3CN{=}PAr_3$ and $Ar_3SiN{=}PPh_3$ have been studied by n.m.r. [254]. A derived system $[RuCl_3(PEt_2Ph)_2(NPEt_2Ph)]$ contains a virtually linear Ru-N-P link [201]. The P—N bond length (1.571 Å) is typical of a P=N double bond [201].

A four-co-ordinate P^V system of an unusual nature is the zwitterio $Cl_2\bar{B}[N(Me)P(Cl)_2]_2\overset{+}{N}$ [24], related to $[Cl_3PNPCl_3]^+$ [29]. Further examples o zwitterionic P–N rings have been synthesised (Eq. 88a) via reaction of a met halide $[MCl_n]$ with $R_2NP(NR)_2$ (R = $SiMe_3$), affording a four-membered MNPI metallocycle. The X-ray crystal structure for one such compound is show in Fig. 23.

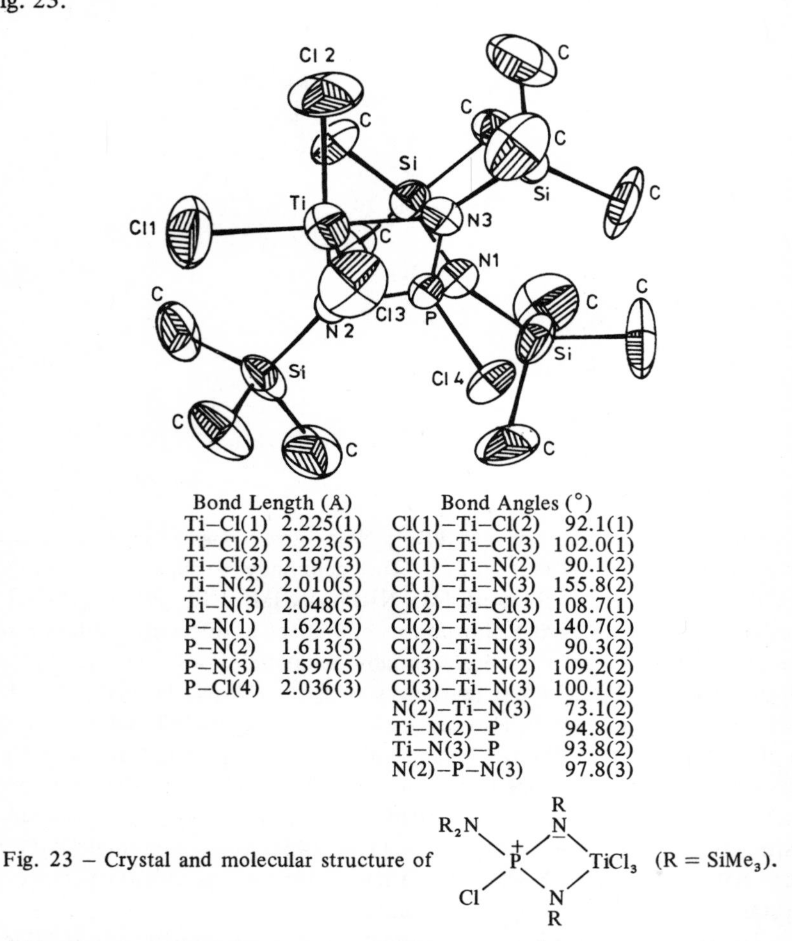

Bond Length (Å)	
Ti–Cl(1)	2.225(1)
Ti–Cl(2)	2.223(5)
Ti–Cl(3)	2.197(3)
Ti–N(2)	2.010(5)
Ti–N(3)	2.048(5)
P–N(1)	1.622(5)
P–N(2)	1.613(5)
P–N(3)	1.597(5)
P–Cl(4)	2.036(3)

Bond Angles (°)	
Cl(1)–Ti–Cl(2)	92.1(1)
Cl(1)–Ti–Cl(3)	102.0(1)
Cl(1)–Ti–N(2)	90.1(2)
Cl(1)–Ti–N(3)	155.8(2)
Cl(2)–Ti–Cl(3)	108.7(1)
Cl(2)–Ti–N(2)	140.7(2)
Cl(2)–Ti–N(3)	90.3(2)
Cl(3)–Ti–N(2)	109.2(2)
Cl(3)–Ti–N(3)	100.1(2)
N(2)–Ti–N(3)	73.1(2)
Ti–N(2)–P	94.8(2)
Ti–N(3)–P	93.8(2)
N(2)–P–N(3)	97.8(3)

Fig. 23 – Crystal and molecular structure of $R_2N(Cl)\overset{+}{P}(\overset{-}{N}R)(NR)TiCl_3$ (R = $SiMe_3$).

[Reproduced, with permission, from Niecke, E., Kröher, R., and Pohl, S., Angew. Chem. Internat. Edn., *1977,* **16**, *864.]*

First discovered in 1897, the structure of the tricyclic compound shown i Fig. 24 has only recently been determined [125]. The three bridgehead chlorin atoms lie on the same side of the molecule, which has a three-fold axis c symmetry [125].

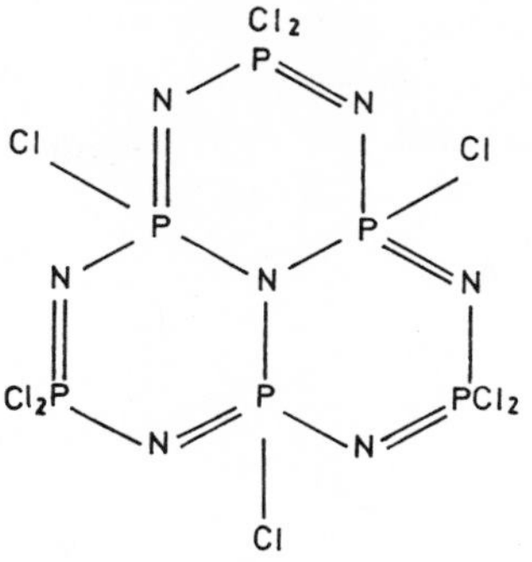

Fig. 24

Unusual P^V compounds having a $P{=}NSiMe_3$ unit as an integral part are illustrated in Figs. 25–27 [12, 188, 189, 189a]. A bridgehead compound has the structure $EtC(CH_2O)_3P{=}NC(CF_3)_2NC(CF_3)_2$ [108].

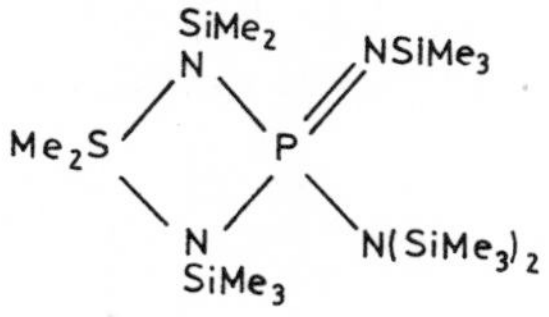

Fig. 25

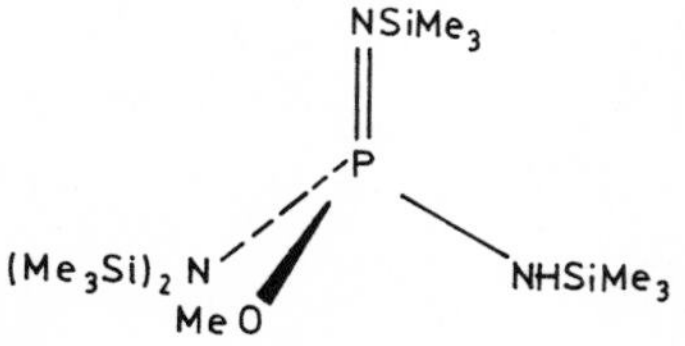

Fig. 26

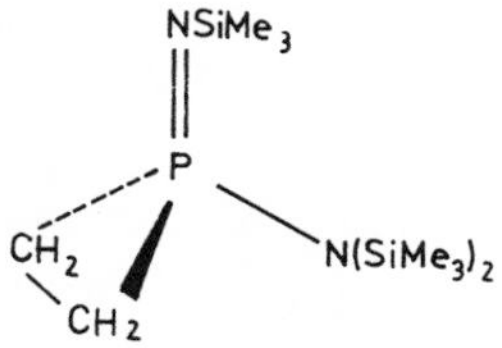

Fig. 27

The photoelectron spectrum of $[(Ph_3P)_2N]^+$(abbreviated as PPN^+) indicates a bent PNP arrangement (Fig. 28) [242], as confirmed by X-ray analysis of $[PPN][Cr_2(CO)_{10}I]$ [119a] (see Chapter 15). The related cyclic ion (Fig. 29) adopts a similar geometry [13]. Further X-ray data are in ref. [253c].

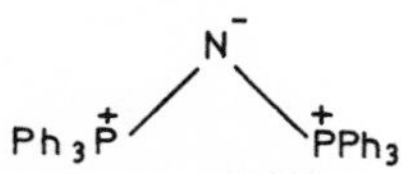

Fig. 28

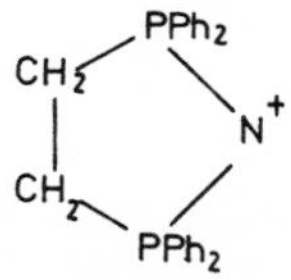

Fig. 29

3. Three-co-ordinate phosphorus(V) amides

This is a most unusual co-ordination number for phosphorus(V) compounds, and is almost entirely restricted to *N*-silylamido derivatives, e.g., Fig. 30 with $X=NSiMe_3$ [12, 188, 189, 190c, 227] or S [189a]; a t-butylimide, Fig. 31 [225] is also known.

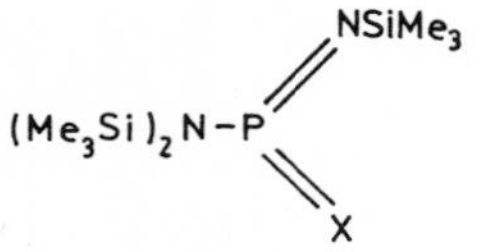

Fig. 30

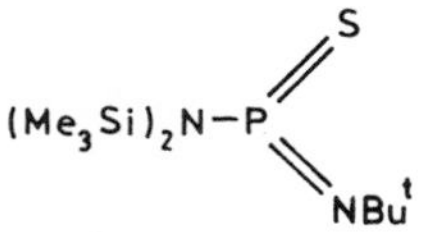

Fig. 31

The structure of $(Me_3Si)_2NP(NSiMe_3)_2$ has been elucidated by X-ray analysis (Fig. 32) [202b]. The molecule has C_2 symmetry. The PN_3 system is planar and the P–N imino distance (1.503 Å) indicates extensive P–N multiple bonding. The P–N amino distance is relatively short, indicating d_π-p_π P–N̈ interaction in this bond and this may account for the slight lengthening in the corresponding Si–N bond due to the reduction in Si–N̈ d_π-p_π interaction. The strengthening of the Si–N imino π-bond is consistent with a short (1.697 Å) bond and the wide Si–N̂–P angle of 148.5°.

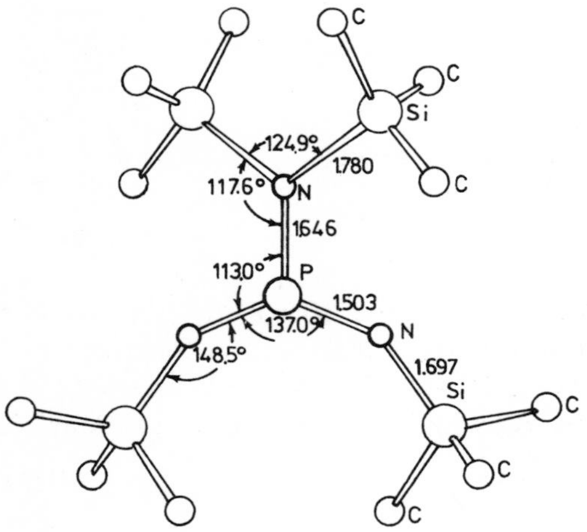

Fig. 32 – The molecular structure of $(Me_3Si)_2NP(NSiMe_3)_2$. *[Reproduced, with permission, from Pohl, S., Niecke, E., and Krebs, B.,* Angew. Chem. Internat. Edn., *1975,* **14,** *261.]*

4. Three-co-ordinate phosphorus(III) amides: the aminophosphines

This co-ordination number is by far the most common for amino-phosphorus(III) compounds. Phosphorus is always pyramidal, but the geometry at nitrogen is generally close to planar [e.g., 39, 77, 78, 88, 102, 160, 179]. ^{14}N n.m.r. data have recently been summarized [22a].

Photoelectron spectroscopic studies [66, 67, 162] on aminophosphines indicate that torsional barriers arise predominantly from steric and lone-pair electronic interactions. The symmetry of $P(NMe_2)_3$ is *not* C_{3v} [67]. The first ionisation potential for $F_nP(NMe_2)_{3-n}$ ($n = 0$–3) has been correlated with the coupling constant (J_{P-B}) for the borane adducts [162].

CNDO calculations have indicated that the P–N bond order in F_2PNMe_2 is greater than 1 [22], a result apparently supported by n.m.r. [71, 73, 160, 233] and magneto-optical and dipole moment studies [160]. A π-component in the P–N bond is further indicated by X-ray crystallographic data [179], which show the P–N bond length to be 1.628 Å and the environment at nitrogen to be planar (Fig. 32). P–N bond lengths have often been compared with that (1.77 Å) found in $P(O)\overset{+}{N}H_3(\bar{O})_2$ [75a], which is thought to have no π-character. Because most P–N bond lengths are less than this value, it is generally held that very few P–N links can be thought of as merely σ-bonds. However, compounds X_2PNMe_2 are often good nitrogen-centred bases (see Chapter 16) [166].

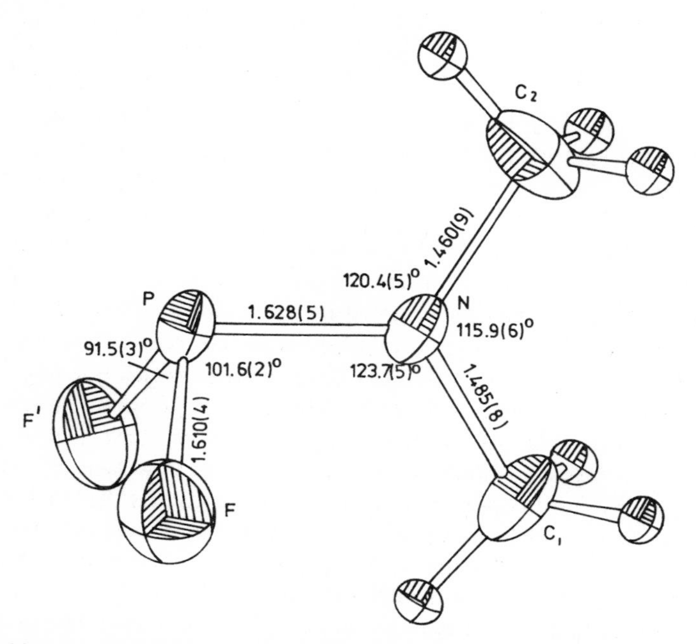

Fig. 33 – X-ray crystal and molecular structure of F_2PNMe_2. *[Reproduced, with permission, from Morris, E. D., and Nordman, C. E.,* Inorg. Chem., *1969,* **8**, *1673.]*

Ab initio molecular orbital calculations indicate that in H_2PNH_2 the $\widehat{HPN}$ angle is 99.4°, and that nitrogen is planar [77, 78]; the barrier to inversion at phosphorus (42.6 kcal mol^{-1}) is higher than that for PH_3 (41.4 kcal mol^{-1}) due to electron-withdrawal by the amino-group.

The structures of F_2PNH_2, determined by microwave spectroscopy, Fig. 34 (X = H) [179], and of F_2PNMe_2, Fig. 34 (X = Me), by both microwave spectroscopy [102] and crystallography [179], show that in each case nitrogen is planar (see also ref. [131]).

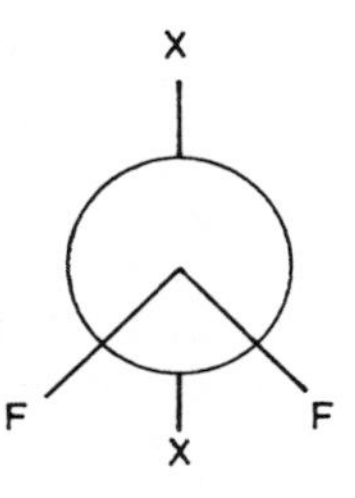

Fig. 34

In both molecules the P–N bond is short (1.628 Å) [179]. The n.m.r. [69] and vibrational spectra [88] of Cl_2PNMe_2 likewise show that this molecule also has a gauche conformation with a planar nitrogen atom, analogous to Fig. 34. The coupling constants $^3J_{P-H}$ and $^2J_{P-^{13}C}$ are greater for the methyl group *cis* to the lone-pair than for the *trans* methyl group [69], and of opposite sign [238]. X-ray data are available for $Bu^tNN{=}NN(Bu^t)PN(SiMe_3)_2$ [202c] and *cis*-$(C_5H_{10}NPNBu^t)_2$ [151a]; n.m.r. data on the latter and related compounds are available [152a]. The temperature dependance of $J(PP')$ in $(F_2P)_2NEt$ has been described [190, 190e].

As we have seen, the barriers to rotation about a P^V–N bond in five-co-ordinate [64, 72, 181] or four-co-ordinate P amides [90, 144, 172, 193–195] have been measured by variable temperature n.m.r. spectroscopy. The barriers to rotation about P^{III}–N bonds are of similar magnitude. N.m.r. data indicate that in F_2PNMe_2 [160], $(MeO)_2PNMe_2$ [237], or related compounds [e.g., 70, 73] there is a substantial barrier to rotation about the P–N bond. However other n.m.r. studies on P–N, As–N, S–N, or Si–N amides show that it may be unnecessary to suggest π-bonding to explain the origin of these barriers (6–12 kcal mol^{-1}) [73]. Further, the basicity of X_2PNMe_2 indicates that such a π-bond is not of predominant significance [166]. At low temperatures, the methyl groups of $PhP(Cl)NMe_2$ become magnetically inequivalent and the barrier to rotation about the P–N bond is estimated to be ca. 12 kcal mol^{-1} [68]. It has also been shown that the topomerisation of H_2PNH_2 is a hybrid process consisting of P–N bond rotation and inversion at nitrogen [73a]. The complexity of P–N rotation is supported by several studies. Increasing the bulk of the nitrogen substituents, the barrier to P–N bond rotation [70, 86a] is raised, but upon increasing the steric crowding at phosphorus this barrier is lowered [86a]. This has been explained [70, 116] by assuming that steric effects at P cause deformation at N toward pyramidicity, effectively lowering the P–N barrier. When the substituent at P is very bulky, steric influences are of over-riding importance and the configuration at nitrogen has little influence on the size of the P–N rotational barrier [116, 204a].

For compounds X_2PNHMe two N–H stretching modes are observed in the i.r. spectra, indicating the presence of two isomers, e.g., those of Fig. 35 or Fig. 36 [21, 171]. However, for X_2PNHBu^t only one structure (Fig. 36) is observed [171].

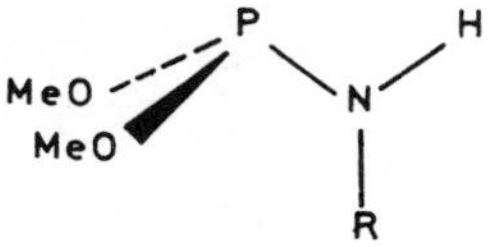

Fig. 35

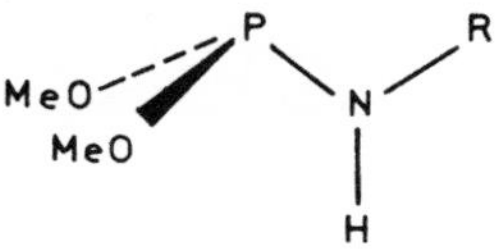

Fig. 36

The reaction of $Bu^t_2PNHSiMe_3$ sequentially with LiBu and then Me_3SiCl gives a material which shows two distinct Me_3Si signals in the n.m.r. spectrum [229]; it was thus assigned the formula $Bu^t_2P(SiMe_3)$=$NSiMe_3$ [229]. Subsequent studies of the systems $(CF_3)_2PN(SiMe_3)_2$ and $(CF_3)_2PN(Bu^t)SiMe_3$ indicate that restricted rotation about the P–N bond of the isomeric P^{III} amide is the more probable origin of the distinct Me_3Si signals [184], with a pyramidal structure at P (Figs. 37, 38), but a planar nitrogen environment (Figs. 39, 40). In the case of the $N(Bu^t)SiMe_3$ compound, each conformer was observed (n.m.r.) at ambient temperature [184].

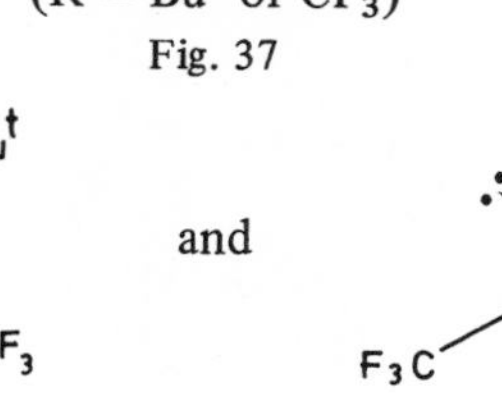

(R = Bu^t or CF_3)

Fig. 37

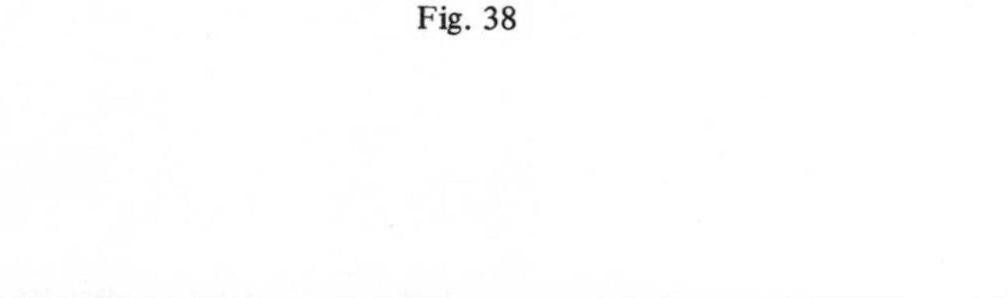

and

Fig. 38

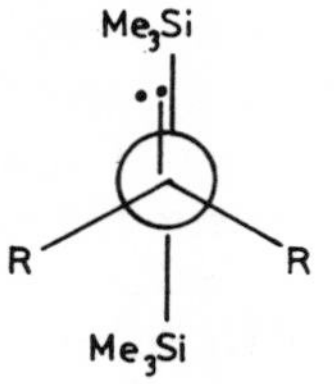

$(R = Bu^t \text{ or } CF_3)$

Fig. 39

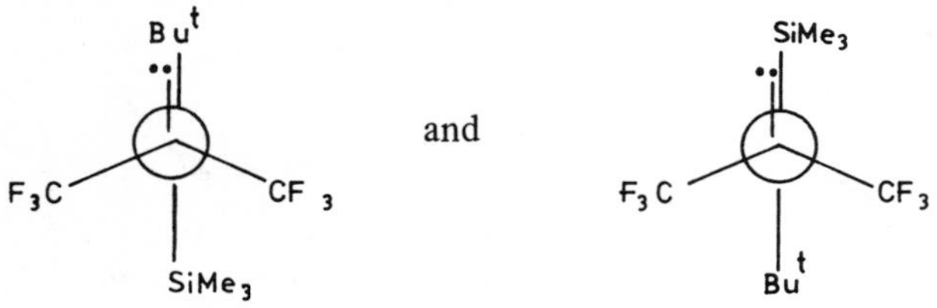

and

Fig. 40

Conformational effects on *PNCH, PNC,* and *PNSi* spin coupling constants in various P^{III} amides have been noted [48a].

Hindered rotation about the P–N bond has also been determined for the more exotic systems $F_nP(NMeOMe)_{3-n}$ ($n = 1$ or 2) and the chloro-analogues [134].

The NR_2 group bonded to a heterocyclic phosphorus(III) atom preferentially occupies an equatorial position (Fig. 41) [28] in a similar manner to the P^V analogues.

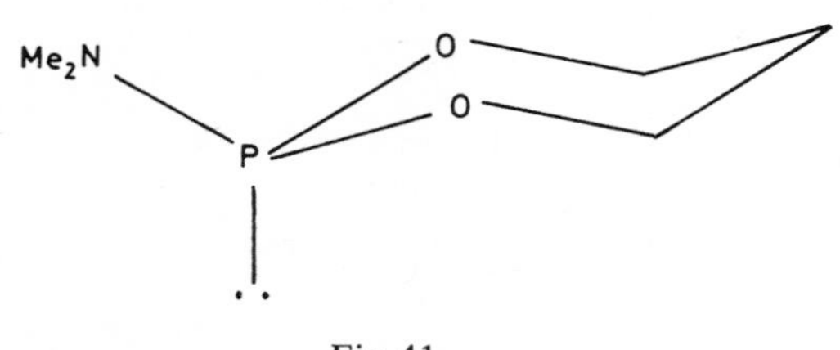

Fig. 41

The conformations of a number of exocyclic substituted heterocyclic compounds with P–N bonds [Fig. 42] have been determined using n.m.r. spectroscopy [136].

The structure of $P(NMeNMe)_3P$ has been examined by electron diffraction [252], as has that of the carbodi-imide $(F_2PN)_2C$ [208].

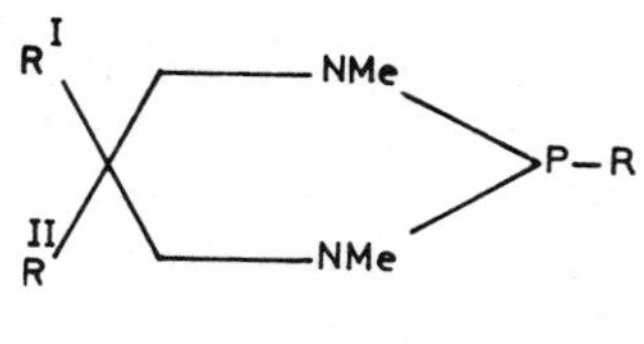

Fig. 42

^{13}C n.m.r. spectra on a series of 2-dialkylamino-1,3-dimethyl-1,3,2-phospholanes indicate that the low temperature conformer is that shown in Fig. 43a rather than Fig. 43b [120a], contrary to a suggestion based on M.O. calculations [113a].

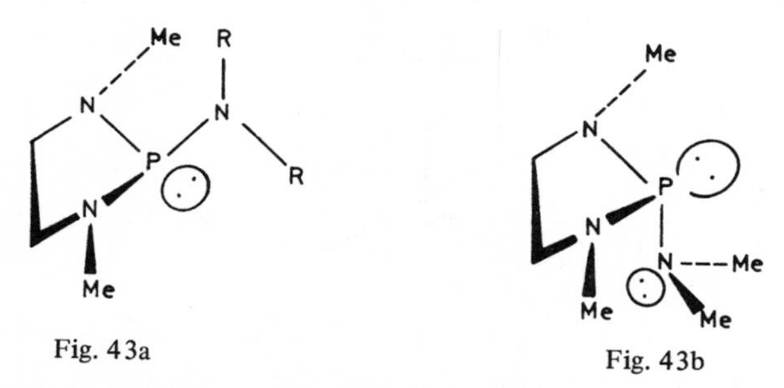

Fig. 43a

Fig. 43b

Polycyclic cage P^{III} amides have been characterised, Fig. 44 ($n=3=m$; $n=2$, $m=3$; and $n=3$, $m=2$) [20].

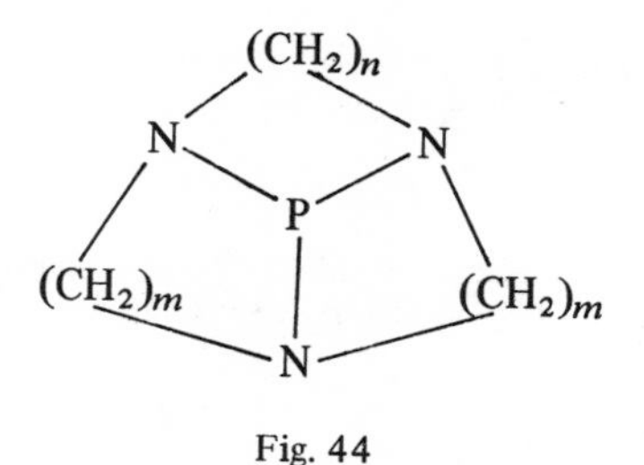

Fig. 44

5. Two-co-ordinate phosphorus(III) amides

We have already noted that the substructure $Me_3SiN{=}P$ is the basis of a number of unusual compounds of phosphorus(V). The first two-co-ordinate phosphorus(III) compound to be synthesised was $(Me_3Si)_2NP{=}NSiMe_3$ [12, 187–189], the parent of this family. The only other related compounds are the t-butyl analogues $(Me_3Si)_2NP{=}NBu^t$ [225] and $Me_3Si(Bu^t)NP{=}NBu^t$ [226]. Two-fold co-ordination at phosphorus in $Me_3Si(Bu^t)NP{=}NBu^t$ has been verified by X-ray crystallography (Fig. 45) [202a]. On the basis of these data, resonance contributions to the structure have been ruled out. A significant feature is the size of the $N\widehat{P}N$ angle (104.9°) due to extensive participation of *p*-orbitals in the phosphorus-nitrogen bonds. The ions $[P(NMe_2)_2]^+$ [245, 245a] and the cyclic analogues (XIV) [97] and (XV) [169] are further examples of cationic two-co-ordinate phosphorus(III) compounds, as is $[P(NPr^i_2)_2]^+$ for which X-ray data are available: $l(P–N)_{av}$, 1.613(4) Å; $N\widehat{P}N$, 114.8(2)° [65a]. Dynamic ^1H n.m.r. measurements on $[P(NRR')_2]^+$ {R=R'=Me [245a] or Pr^i [65a]; or R=Me, R'=Pr^i [65a]} suggest restricted rotation about the P–N bond rather than inversion at P.

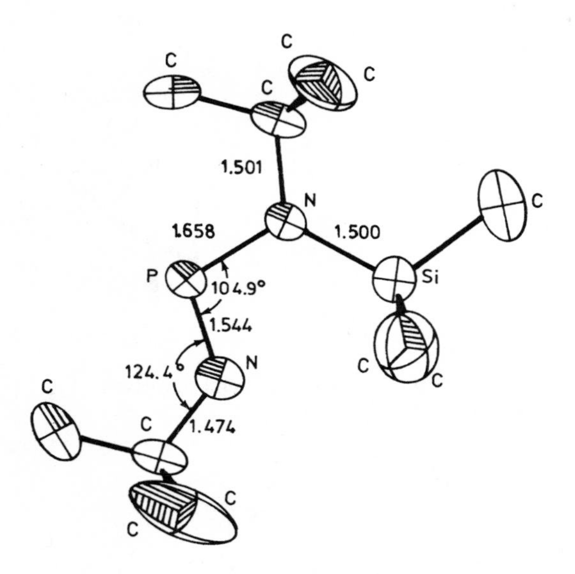

Fig. 45 – Molecular structure of $Me_3Si(Bu^t)NP{=}NBu^t$. *[Reproduced, with permission, from Pohl, S.,* Angew. Chem. Internat. Edn., *1976,* **15,** *687.]*

Me
N
P⁺
N
Me

(XIV)

NMe
P⁺
NMe

(XV)

6. Two-co-ordinate phosphorus(II) amides

A further recent example of two-co-ordination is for a series of stable amido-substituted phosphinyl radicals [117, 204a], the structure of which is established from e.s.r. spectra (see Table 2 and Figs. 46 and 47).

Table 2 E.s.r. parameters for some novel phosphinyl radicals [117, 204a]

Radical	$a(^{31}P)$ (G)	$a(^{14}N)$ (G)
$:\dot{P}[N(SiMe_3)_2]_2$	91.8	–
$:\dot{P}[CH(SiMe_3)_2]N(SiMe_3)_2$	93	–
$:\dot{P}[CH(SiMe_3)_2]NPr^i_2$	63	3.75
$:\dot{P}(NPr^i_2)N(SiMe_3)_2$	77	5.25
$:\dot{P}(NPr^i_2)[N(Bu^t)SiMe_3]$	74	5.12
$:\dot{P}[N(Bu^t)SiMe_3]_2$	101.5	
$:\dot{P}[CH(SiMe_3)_2]NMe_2$	65	

The solution e.s.r. spectra at ca. 300K consist essentially of a large doublet due to splitting by the ^{31}P nucleus ($I = ½$). The $a(^{31}P)$ is similar in magnitude to previously reported values for matrix isolated transient species in the solid state. Thus, these are bent π-radicals, i.e., the unpaired electron, is in a phosphorus $2p$-orbital, the angle at phosphorus being ca. 96°.

A feature of interest in the e.s.r. parameters is that there is no unpaired electron-nitrogen (^{14}N, $I = 1$) -coupling when there is a silyl substituent at N (Figs. 46 and 47). A possible explanation is that $Si\overset{\frown}{-}\ddot{N}$ $d \leftarrow p$ π-bonding lowers the interaction between the unpaired electron and the nitrogen; this has received some support from parallel studies on the alkyl- and silyl-aminyl radicals $:\dot{N}X_2$ where silyl substituents have been shown to produce a significant lowering in the $a(^{14}N)$ value [204a].

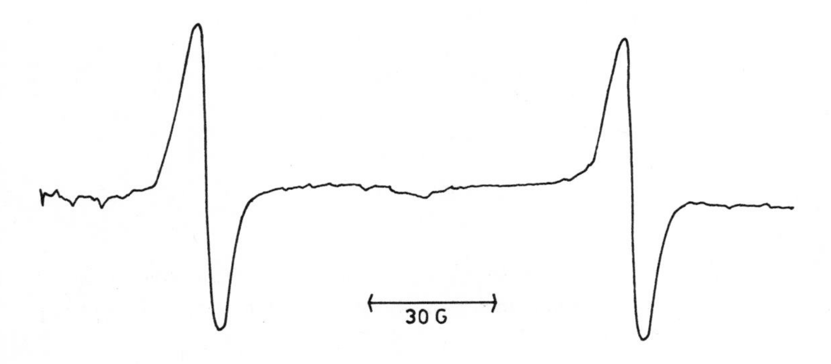

Fig. 46 – E.s.r. spectrum of $\dot{P}[N(SiMe_3)_2]_2$ at 300K in PhMe [117].

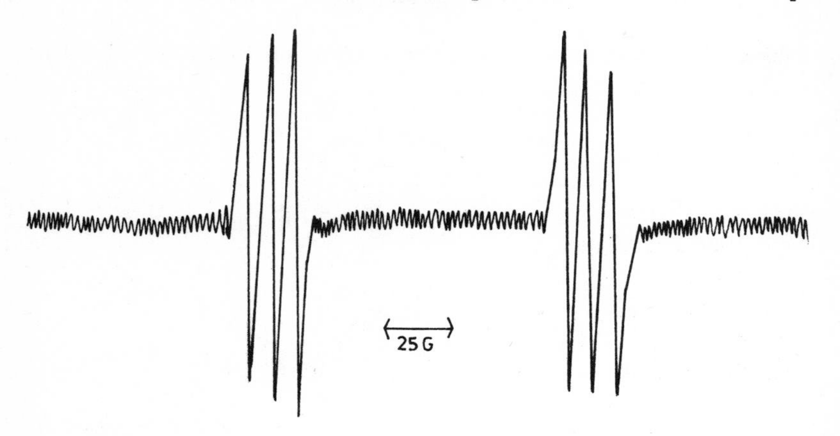

Fig. 47 – E.s.r. spectrum of $\dot{P}(NPr^i_2)[N(SiMe_3)_2]$ at 300K in PhMe [117].

7. Miscellaneous PN compounds

The crystal structures of numerous phosphonitrilic compounds (polyphosphazenes), $[X(X')P{=}N]_n$, have been elucidated and other structural information derived from n.m.r. spectroscopy. Structures involving X,X′ = NHR [e.g., 46, 183] and NR_2 [e.g., 25, 44, 46, 183] groups have also been determined.

The reactions of P–N bonded compounds have been widely studied, and the structures of many intermediates have been proposed. Of biological interest are the mechanisms of certain hydrolyses. Such processes are exceedingly complex [e.g., 37, 115, 119, 121, 154, 155, 251]. For the alkaline hydrolysis of cyclic phosphoramidates, a square pyramidal (rather than trigonal bipyramidal) intermediate is suggested (Fig. 48) [37].

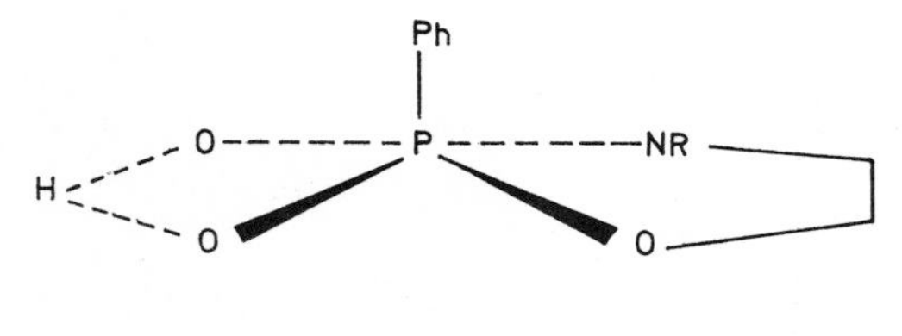

Fig. 48

The reaction of β-keto-sulphides with $P(NMe_2)_3$ to give a ketone or keto-ester apparently proceeds via a phosphonium moiety ($R_3PSR'^+$) [122].

D. REFERENCES

[1] Abatjoglou, A. G., and Eliel, E. L., *J. Org. Chem.*, 1974, **39**, 3042.
[2] Abel, E. W., Armitage, D. A., and Willey, G. R., *J. Chem. Soc. (A)*, 1969, 57; Abel, E. W., and Willey, G. R., *Proc. Chem. Soc.*, 1962, 308.
[3] Addison, C. C., and Sowerby, D. B. (editors), *M.T.P. Internat. Rev. Sci., Ser. One, Vol. 2,* Butterworths, 1972; *Ser. Two,* 1975.
[4] Adler, O., and Kober, F., *J. Organometallic Chem.*, 1974, **72**, 351.
[5] Allcock, H. R., *Phosphorus-Nitrogen Compounds,* Academic, 1972.
[6] Allcock, H. R., *Science,* 1976, **193**, 1214.
[7] Allcock, H. R., Kugel, R. L., and Moore, G. Y., *Inorg. Chem.*, 1975, **14**, 2831.
[7a] Almenningen, A., Andersen, B., and Astrup, E. E., *Acta Chim. Scand.*, 1969, **23**, 2179.
[8] Appel, R., *Angew. Chem. Internat. Edn.*, 1975, **14**, 801.
[9] Appel, R., and Einig, H., *Chem. Ber.*, 1975, **108**, 914.
[10] Appel, R., and Einig, H., *Z. Anorg. Chem.*, 1975, **414**, 241.
[11] Appel, R., and Halstenberg, M., *J. Organometallic Chem.*, 1975, **99**, C25.
[12] Appel, R., and Halstenberg, M., *Angew. Chem. Internat. Edn.*, 1975, **14**, 768.
[13] Appel, R., Kleinstück, R., and Ziehn, K. D., *Chem. Ber.*, 1972, **105**, 2476.
[14] Appel, R., and Milker, R., *Chem. Ber.*, 1975, **108**, 2349.
[15] Appel, R., and Ruppert, I., *Z. Anorg. Chem.*, 1974, **406**, 131.
[16] Arnold, D. E. J., Ebsworth, E. A. V., Jessep, H. F., and Rankin, D. W. H., *J.C.S. Dalton,* 1972, 1681.
[17] Arnold, D. E. J., and Rankin, D. W. H., *J.C.S. Dalton,* 1975, 889.
[18] Atherton, F. R., Openshaw, H. T., and Todd, A. R., *J. Chem. Soc.*, 1945, 660.
[19] Atherton, F. R., and Todd, A. R., *J. Chem. Soc.*, 1947, 674.
[20] Atkins, T. J., *Tetrahedron Letters,* 1978, 4331.
[20a] Atkins, T. J., and Richman, J. E., *Tetrahedron Letters,* 1978, 5149.
[21] Ayed, N., Mathis, R., Burgada, R., and Mathis, F., *Compt. Rend.*, 1974, **278C**, 1085.
[22] Bach, M. C., Brian, C., Crasnier, F., Labarre, J. F., Leibovici, C., and Dargelos, A., *J. Mol. Struct.*, 1973, **17**, 23.
[22a] Barlos, K., Nöth, H., and Wrackmeyer, B., *Z. Naturforsch.*, 1978, **33b**, 515.
[23] Becke-Goehring, M., and Krill, H., *Chem. Ber.*, 1961, **94**, 1059.
[24] Becke-Goehring, M., and Müller, H. J., *Z. Anorg. Chem.*, 1968, **362**, 51.
[25] Begley, M. J., Millington, D., King, T. J., and Sowerby, D. B., *J.C.S. Dalton,* 1974, 1162; Millington, D., King, T. J., and Sowerby, D. B., *ibid,* 1973, 396.

[26] Bentrude, W. G., Johnson, W. D., and Khan, W. A., *J. Amer. Chem. Soc.*, 1972, **94**, 923.
[27] Bentrude, W. G., Johnson, W. D., and Khan, W. A., *J. Org. Chem.*, 1972, **37**, 642.
[28] Bentrude, W. G., and Tan, H. W., *J. Amer. Chem. Soc.*, 1972, **94**, 8222.
[29] Bermann, M., *Adv. Inorg. Chem. Radiochem.*, 1972, **14**, 1.
[30] Bermann, M., and Van Wazer, J. R., *J.C.S. Dalton*, 1973, 813.
[31] Bermann, M., and Van Wazer, J. R., *Inorg. Chem.*, 1973, **12**, 2186.
[32] Bermann, M., and Van Wazer, J. R., *Inorg. Chem.*, 1974, **13**, 737.
[33] Binder, H., *Phosphorus*, 1972, **1**, 287.
[34] Binder, H., and Fischer, R., *Chem. Ber.*, 1974, **107**, 205.
[35] Boden, G., Geissler, J., Grosskreutz, W., Kessler, G., and Scheler, H., *Z. Chem.*, 1972, **12**, 298.
[36] Boden, G., Grosskreutz, W., Kessler, G., and Scheler, H., *Z. Chem.*, 1972, **12**, 299.
[37] Boudreau, J. A., Brown, C., and Hudson, R. F., *J.C.S. Chem. Comm.*, 1975, 679.
[38] Bowden, F. L., Dronsfield, A. T., Haszeldine, R. N., and Taylor, D. R., *J.C.S. Perkin (I)*, 1973, 516.
[39] Brittain, A. H., Smith, J. E., Lee, P. L., Cohn, K., and Schwenderman, R. H., *J. Amer. Chem. Soc.*, 1971, **93**, 6772.
[40] Brown, C., Hudson, R. F., and Searle, R. J. G., *Phosphorus*, 1973, **2**, 287.
[41] Buck, A. C., Bartleson, J. D., and Lankelma, H. P., *J. Amer. Chem. Soc.*, 1948, **70**, 744.
[42] Buck, A. C., and Lankelma, H. P., *J. Amer. Chem. Soc.*, 1948, **70**, 2398.
[43] Bullen, G. J., and Dann, P. E., *Acta Cryst.*, 1973, **B29**, 331.
[44] Bullen, G. J., and Dann, P. E., *J.C.S. Dalton*, 1973, 1453; 1974, 705.
[45] Bullen, G. J., and Tucker, P. A., *J.C.S. Dalton*, 1972, 2437.
[46] Bullen, G. J., and Mallinson, P. R., *J.C.S. Dalton*, 1972, 1412.
[47] Bullen, G. J., Rutherford, J. S., and Taylor, P. A., *Acta Cryst.*, 1973, **B29**, 1439.
[48] Bulloch, G., and Keat, R., *J.C.S. Dalton*, 1974, 2010.
[48a] Bulloch, G., Keat, R., and Rycroft, D. S., *J.C.S. Dalton*, 1978, 764.
[49] Bulloch, G., Keat, R., and Tennent, N. H., *J.C.S. Dalton*, 1974, 2329.
[50] Burg, A. B., and Slota, P. J., *J. Amer. Chem. Soc.*, 1958, **80**, 1107; Burg, A. B., and Wagner, R. I., U.S. patent 2,934,564 (1957).
[51] Burgada, R., *Bull. Soc. Chim. France*, 1971, 136.
[52] Burgada, R., and Laurenço, C., *J. Organometallic Chem.*, 1974, **66**, 255.
[53] Cadogan, J. I. G., Gould, R. O., and Tweddle, N. J., *J.C.S. Chem. Comm.*, 1975, 773.
[53a] Cadogan, J. I. G., Stewart, N. J., and Tweddle, N. J., *J.C.S. Chem. Comm.*, 1978, 182.
[54] Cadogan. J. I. G., Tait, B. S., and Tweddle, N. J., *J.C.S. Chem. Comm.*,

1975, 847.
[55] Cavell, R. G., Gibson, J. A., and The, K. I., *Inorg. Chem.*, 1978, **17**, 2880.
[56] Cavell, R. G., Leary, R. D., Sanger, A. R., and Tomlinson, A. J., *Inorg. Chem.*, 1973, **12**, 1374.
[57] Cavell, R. G., Poulin, D. D., The, K. I., and Tomlinson, A. J., *J.C.S. Chem. Comm.*, 1974, 19.
[57a] Cavell, R. G., and The, K. I., *Inorg. Chem.*, 1978, **17**, 355.
[57b] Chang, M., Newton, M. G., King, R. B., and Lotz, T. J., *Inorg. Chim. Acta*, 1978, **28**, L153.
[58] Charlton, T. L., and Cavell, R. G., *Inorg. Chem.*, 1970, **9**, 379.
[59] Charrier, C., and Simonnin, M. P., *Compt. Rend.*, 1967, **264C**, 995.
[60] Clardy, J. C., Mosbo, J. A., and Verkade, J. G., *J.C.S. Chem. Comm.*, 1972, 1163.
[61] Clemens, D. F., and Perkinson, W. E., *Inorg. Chem.*, 1974, **13**, 333.
[62] Contreras, R., Brazier, J. F., Klaebe, A., and Wolf, R., *Phosphorus*, 1972, **2**, 67.
[63] Corbridge, D. E. C., *The Structural Chemistry of Phosphorus*, Elsevier, 1974.
[63a] Costes, J.-P., Cros, G., and Laurent, J.-P., *J. Inorg. Nucl. Chem.*, 1978, **40**, 829.
[64] Cowley, A. H., Braun, R. W., and Gilje, J. W., *J. Amer. Chem. Soc.*, 1975, **97**, 434.
[65] Cowley, A. H., Cushner, M. C., Fild, M., and Gibson, J. A., *Inorg. Chem.*, 1975, **14**, 1851.
[65a] Cowley, A. H., Cushner, M. C., and Szobota, J. S., *J. Amer. Chem. Soc.*, 1978, **100**, 7784.
[66] Cowley, A. H., Dewar, M. J. S., Gilje, J. W., Goodman, D. W., and Schweiger, J. R., *J.C.S. Chem. Comm.*, 1974, 340.
[67] Cowley, A. H., Dewar, M. J. S., Goodman, D. W., and Schweiger, J. R., *J. Amer. Chem. Soc.*, 1973, **95**, 6506.
[68] Cowley, A. H., Dewar, M. J. S., and Jackson, W. R., *J. Amer. Chem. Soc.*, 1968, **90**, 4185.
[69] Cowley, A. H., Dewar, M. J. S., Jackson, W. R., and Jennings, W. B., *J. Amer. Chem. Soc.*, 1970, **92**, 1085.
[70] Cowley, A. H., Dewar, M. J. S., Jackson, W. R., and Jennings, W. B., *J. Amer. Chem. Soc.*, 1970, **92**, 5206.
[71] Cowley, A. H., and Schweiger, J. R., *J.C.S. Chem. Comm.*, 1970, 1492.
[72] Cowley, A. H., and Schweiger, J. R., *J.C.S. Chem. Comm.*, 1972, 560.
[73] Cowley, A. H., and Schweiger, J. R., *J. Amer. Chem. Soc.*, 1973, **95**, 4179.
[73a] Cowley, A. H., Taylor, M.W., Whangbo, M. H., and Wolfe, S., *J.C.S. Chem. Comm.*, 1976, 838.
[74] Cox, J. W., and Corey, E. R., *Chem. Comm.*, 1967, 123.

[75] Cross, R. J., Green, T. H., and Keat, R., *J.C.S. Dalton,* 1976, 1424.
[75a] Cruickshank, D. W. J., *Acta Cryst.*, 1964, **17**, 671.
[76] Crutchfield, M. M., Dungan, C. H., Letcher, J. H., Mark, V., and Van Wazer, J.R., *Topics in Phosphorus Chemistry, Vol. 5.* ^{31}*P Nuclear Magnetic Resonance,* Wiley, 1967.
[77] Csizmadia, I. G., Cowley, A. H., Taylor, M. W., and Wolfe, S., *J.C.S. Chem. Comm.*, 1974, 432.
[78] Csizmadia, I. G., Cowley, A. H., Taylor, M. W., Tel, L. M., and Wolfe, S., *J.C.S. Chem. Comm.*, 1973, 1147.
[79] Czieslik, G., Flaskerud, G., Höfer, R., and Glemser, O., *Chem. Ber.*, 1973, **106**, 399.
[80] Das, R. N., Shaw, R. A., Smith, B. C., and Woods, M., *J.C.S. Dalton,* 1973, 709, 1883.
[81] Das, S., Shaw, R. A., and Smith, B. C., *J.C.S. Dalton,* 1974, 1610.
[82] Davies, A. R., Dronsfield, A. T., Haszeldine, R. N., and Taylor, D. R., *J.C.S. Perkin (I),* 1973, 379.
[83] Davy, H., *Phil. Trans. Royal Soc. London,* 1810, **100**, 231.
[84] Desmarchelier, J. M., and Fukuto, T. R., *J. Org. Chem.*, 1972, **37**, 4218.
[85] Devillers, J., Willson, M., and Burgada, R., *Bull. Soc. Chim. France,* 1968, 4670.
[86] Dietz, E. A., Jr., and Martin, D. R., *Inorg. Chem.*, 1973, **12**, 241.
[86a] DiStefano, S., Goldwhite, H., and Mazzola, E., *Org. Mag. Res.*, 1974, **6**, 1.
[87] Dorn, H., Graubaum, H., Zeigan, D., and Radeglia, R., *Z. Chem.*, 1975, **15**, 486.
[87a] Douglas, W. M., and Ruff, J. K., *J. Chem. Soc. (A),* 1971, 3558.
[88] Durig, J. R., and Casper, J. M., *J. Phys. Chem.*, 1971, **75**, 3837.
[89] Ellis, K., Smith, D. J. H., and Trippett, S., *J.C.S. Perkin (I),* 1972, 1184.
[90] Emsley, J., and Williams, J. K., *J.C.S. Dalton,* 1973, 1576.
[91] Ewart, G., Payne, D. S., Porte, A. L., and Lane, A. P., *J. Chem. Soc.*, 1962, 3984.
[92] Falardeau, E. R., Morse, K. W., and Morse, J. G., *Inorg. Chem.*, 1975, **14**, 132.
[93] Falius, H., *Z. Anorg. Chem.*, 1973, **396**, 245.
[94] Falius, H., Giesen, K. P., and Wannagat, U., *Z. Anorg. Chem.*, 1973, **402**, 139.
[95] Feldmann, V., and Thilo, E., *Z. Anorg. Chem.*, 1964, **327**, 159.
[96] Fild, M., Sheldrick, W. S., and Stankiewicz, T., *Z. Anorg. Chem.*, 1975, **415**, 43.
[96a] Fischer, S., Peterson, L. K., and Nixon, J. F., *Can. J. Chem.*, 1974, **52**, 3981.
[97] Fleming, S., Lupton, M. K., and Jekot, K., *Inorg. Chem.*, 1972, **11**, 2534.
[98] Fleming, S., and Parry, R. W., *Inorg. Chem.*, 1972, **11**, 1.
[99] Fluck, E., and Benerle, E., *Z. Anorg. Chem.*, 1975, **411**, 125.

[100] Fluck, E., and Benerle, E., *Z. Anorg. Chem.*, 1975, **412**, 65.
[101] Fluck, E., and Weber, D., *Z. Anorg. Chem.*, 1975, **412**, 47.
[102] Forti, P., Damiani, D., and Favero, P. G., *J. Amer. Chem. Soc.*, 1973, **95**, 756.
[103] Frazier, S. E., and Sisler, H. H., *Inorg. Chem.*, 1972, **11**, 1431.
[104] Gefter, Ye. L., *Organophosphorus Monomers and Polymers*, trans. J. Burton, Pergamon, 1962.
[105] Ghouse, K. M., Keat, R., Mills, H. H., Roberts, J. M., Cameron, T. S., Howlett, K. D., and Prout, C. K., *Phosphorus*, 1972, **2**, 47.
[106] Gibson, J. A., and Röschenthaler, G.V., *J.C.S. Chem. Comm.*, 1974, 694.
[107] Gibson, J. A., and Röschenthaler, G. V., *J.C.S. Dalton*, 1976, 1440.
[108] Gieren, A., Narayanan, P., Burger, K., and Thenn, W., *Angew. Chem. Internat. Edn.*, 1974, **13**, 343.
[109] Gilje, J. W., and Seff, K., *Inorg. Chem.*, 1972, **11**, 1643.
[110] Gloe, K., Kessler, G., and Scheler, H., *Z. Chem.*, 1972, **12**, 337.
[111] Goldwhite, H., Gysegem, P., Schow, S., and Swyke, C., *J.C.S. Dalton*, 1975, 12.
[111a] Gorbatenko, I., and Feshenko, N. G., *J. Gen. Chem. U.S.S.R.*, 1977, **47**, 1752.
[112] Goya, A. E., Rosario, M. D., and Gilje, J.W., *Inorg. Chem.*, 1969, **8**, 725.
[113] Grapov, A. F., Mel'nikov, N. N., and Razvodovskaya, L. V., *Russ. Chem. Rev.*, 1970, **39**, 20.
[113a] Gray, G. A., and Albright, T. A., *J. Amer. Chem. Soc.*, 1977, **99**, 3243.
[114] Grayson, M., and Griffith, E. J., editors, *Topics in Phosphorus Chemistry*, Vol. 6, Wiley, 1969.
[115] Greenhalgh, R., and Hudson, R. F., *Phosphorus*, 1972, **2**, 1.
[116] Goldwhite, H., and Power, P. P., *Org. Mag. Res.*, 1978, **11**, 499.
[116a] Griller, D., Roberts, B. P., Davies, A. G., and Ingold, K. U., *J. Amer. Chem. Soc.*, 1974, **96**, 554.
[117] Gynane, M. J. S., Hudson, A., Lappert, M. F., Power, P. P., and Goldwhite, H., *J.C.S. Chem. Comm.*, 1976, 623; *J.C.S. Dalton*, 1980, in the press.
[118] Hägele, G., Harris, R. K., Wazeer, M. I. M., and Keat, R., *J.C.S. Dalton*, 1974, 1985.
[119] Hamer, N. K., and Tack, R. D., *J.C.S. Perkin (II)*, 1974, 1184; and refs. therein.
[119a] Handy, L. B., Ruff, J. K., and Dahl, L. F., *J. Amer. Chem. Soc.*, 1970, **92**, 7327.
[120] Haque, M., and Gaughlan, C. N., *Chem. Comm.*, 1966, 921.
[120a] Hargis, J. H., Worley, S. D., Jennings, W. B., and Tolley, M. S., *J. Amer. Chem. Soc.*, 1977, **99**, 8090.
[121] Harger, M. J. P., *J.C.S. Chem. Comm.*, 1973, 774.
[122] Harpp, D. N., and Vines, S. M., *J. Org. Chem.*, 1974, **39**, 647.
[123] Harris, R. K., Lewellyn, M., Wazeer, M. I. M., Woplin, J. R., Dunmur, R. E.,

Hewson, M. J. C., and Schmutzler, R., *J.C.S. Dalton,* 1975, 61.
[124] Harrison, J. M., Inch, T. D., and Lewis, G. J., *J.C.S. Perkin (I),* 1975, 1892.
[125] Harrison, W., and Trotter, J., *J.C.S. Dalton,* 1972, 623.
[126] Healy, J. D., Ibrahim, E. H. M., Shaw, R. A., Cameron, T. S., Howlett, K. D., and Prout, C. K., *Phosphorus,* 1971, **1**, 157.
[127] Heap, R., and Saunders, B. C., *J. Chem. Soc.,* 1948, 1313.
[128] Hewson, M. J. C., Schmutzler, R., and Sheldrick, W. S., *J.C.S. Chem. Comm.,* 1973, 190.
[129] Höfler, M., and Schitzler, M., *Chem. Ber.,* 1972, **105**, 1133.
[130] Holmes, R. R., and Forstner, J. A., *Inorg. Synth.,* 1966, **8**, 63.
[131] Holywell, G. C., Rankin, D.W. H., Beagley, B., and Freeman, J. M., *J. Chem. Soc. (A),* 1971, 785.
[132] Hudson, R. F., *Structure and Mechanism in Organo-Phosphorus Chemistry,* Academic, 1965.
[133] Hudson, R. F., and Mancuso, A., *Phosphorus,* 1972, **1**, 265, 271.
[134] Hung, A., and Gilje, J. W., *J.C.S. Chem. Comm.,* 1972, 662.
[135] Hunt, G. W., and Cordes, A. W., *Z. Chem.,* 1974, **13**, 1688.
[136] Hutchins, R. O., Maryanoff, B. E., Albrand, J. P., Cogne, A., Gagnaire, D., and Robert, J. P., *J. Amer. Chem. Soc.,* 1972, **94**, 9151.
[137] Huttner, G., and Müller, H.-D., *Angew. Chem. Internat. Edn.,* 1975, **14**, 571.
[138] Ibrahim, E. H. M., Shaw, R. A., Smith, B. C., Thakur, C. P., Woods, M., Bullen, G. J., Rutherford, J. S., Tucker, P. A., Cameron, T. S., Howlett, K. D., and Prout, C. K., *Phosphorus,* 1971, **1**, 153.
[139] Irvine, I., and Keat, R., *J.C.S. Dalton,* 1972, 17.
[140] Issleib, K., and Handke, R., *Z. Anorg. Chem.,* 1975, **413**, 109.
[141] Issleib, K., Oehme, H., and Mohr, K., *Z. Chem.,* 1973, **13**, 139.
[142] Johns, H. J., and Thielemann, L., *Z. Anorg. Chem.,* 1973, **397**, 47.
[143] Jefferson, R., Nixon, J. F., and Painter, T. M., *J.C.S. Chem. Comm.,* 1969, 622.
[143a] Jefferson, R., Nixon, J. F., Painter, T. M., Keat, R., and Stobbs, L., *J.C.S. Dalton,* 1973, 1414.
[144] Jennings, W. B., *J.C.S. Chem. Comm.,* 1971, 867.
[145] Joesten, M. D., and Chen, Y. T., *Inorg. Chem.,* 1972, **11**, 429.
[146] Jurgen, W., Kessler, G., and Scheler, H., *Z. Chem.,* 1972, **12**, 337.
[147] Karayannis, N. M., Mikulski, C. M., and Pytlewski, L. L., *Inorg. Chim. Acta Rev.,* 1971, **5**, 69.
[148] Keat, R., *J. Chem. Soc. (A),* 1970, 2732.
[149] Keat, R., *J.C.S. Dalton,* 1972, 2189.
[150] Keat, R., *Phosphorus,* 1972, **1**, 253.
[151] Keat, R., *J.C.S. Dalton,* 1974, 876.
[151a] Keat, R., Keith, A. N., Macphee, A., Muir, K. W., and Thompson, D. G., *J.C.S. Chem. Comm.,* 1978, 372.

[152] Keat, R., Manojlović-Muir, L., and Muir, K. W., *Angew. Chem. Internat. Edn.*, 1973, **12**, 311.
[152a] Keat, R., and Thompson, D. G., *J.C.S. Dalton*, 1978, 634.
[153] Klein, H. A., and Latscha, H. P., *Z. Anorg. Chem.*, 1974, **406**, 214.
[154] Koizumi, T., and Haake, P., *J. Amer. Chem. Soc.*, 1973, **95**, 8073.
[155] Koizumi, T., Kobayashi, Y., and Yoshii, E., *J.C.S. Chem. Comm.*, 1974, 678.
[156] Kosolapoff, G. M., *Organophosphorus Compounds*, Wiley, 1950.
[157] Kosolapoff, G. M., and Maier, L., editors, *Organic Phosphorus Compounds*, Wiley, 1973.
[158] Kroshefsky, R. D., and Verkade, J. G., *Inorg. Chem.*, 1975, **14**, 3090.
[159] Kuchen, W., and Koch, K., *Z. Anorg. Chem.*, 1972, **394**, 74.
[160] Labarre, M. C., and Coustures, Y., *J. Chim. Phys.*, 1973, **70**, 534.
[161] Lafaille, L., Burgada, R., and Mathis, F., *Compt. Rend.*, 1970, **270C**, 1138.
[162] Lappert, M. F., Pedley, J. B., Wilkins, B. T., Stelzer, O., and Unger, E., *J.C.S. Dalton*, 1975, 1207.
[163] Lee, K. W., and Singer, L. A., *J. Org. Chem.*, 1974, **39**, 3780.
[164] Letsinger, R. L., Wilkes, J. S., and Dumas, L. B., *J. Amer. Chem. Soc.*, 1972, **94**, 292.
[165] Liebig, J., and Wohler, F., *Ann. Pharm.*, 1834, **11**, 139.
[165a] Light, R. W., and Paine, R. T., *J. Amer. Chem. Soc.*, 1978, **100**, 2230.
[166] Lines, E. L., and Centofanti, L. F., *Inorg. Chem.*, 1974, **13**, 2796.
[167] Linke, K. H., and Brandt, W., *Angew. Chem. Internat. Edn.*, 1975, **14**, 643.
[168] Mann, F. G., and Chaplin, E. J., *J. Chem. Soc.*, 1937, 527.
[169] Maryanoff, B. E., and Hutchins, R. O., *J. Org. Chem.*, 1972, **37**, 3475.
[170] Mathis, F., *Phosphorus and Sulfur*, 1976, **1**, 109.
[171] Mathis, R., Ayed, N., Charbonnel, Y., and Burgada, R., *Compt. Rend.*, 1973, **277C**, 493.
[172] Mathis, R., Khemdoubi, J., and Bouisson, T., unpublished results, quoted by Mathis, F., ref. 170.
[173] Mel'nikov, N. N., *Z. Chem.*, 1972, **12**, 201.
[174] Michaelis, A., *Ann.*, 1903, **326**, 129.
[175] Michaelis, A., *Ann.*, 1915, **407**, 290.
[176] Michaelis, A., and Schroeter, G., *Chem. Ber.*, 1894, **27**, 490.
[177] Miller, P. T., Lenhert, P. G., and Joesten, M. D., *Inorg. Chem.*, 1972, **11**, 2221.
[178] Montemayor, R. G., and Parry, R. W., *Inorg. Chem.*, 1973, **12**, 2482.
[178a] Montemayor, R. G., Sauer, D. T., Fleming, S., Bennett, D. W., Thomas, M. G., and Parry, R. W., *J. Amer. Chem. Soc.*, 1978, **100**, 2231.
[179] Morris, E. D., and Nordman, C. E., *Inorg. Chem.*, 1969, **8**, 1673.
[180] Mosbo, J. A., and Verkade, J. G., *J. Amer. Chem. Soc.*, 1972, **94**, 8224.
[181] Muetterties, E. L., Meakin, P., and Hoffmann, R., *J. Amer. Chem. Soc.*,

1972, **94**, 5674.
[182] Muir, K. W., *J.C.S. Dalton,* 1975, 259.
[183] Nabi, S. N., and Shaw, R. A., *J.C.S. Dalton,* 1974, 1618.
[183a] Neilson, R. H., Jacobson, R. D., Scheirman, R. W., and Wilburn, J. C., *Inorg. Chem.*, 1978, **17**, 1880.
[184] Neilson, R. H., Lee, R. C. Y., and Cowley, A. H., *J. Amer. Chem. Soc.*, 1975, **97**, 5302.
[185] Newton, M. G., Collier, J. E., and Wolf, R., *J. Amer. Chem. Soc.*, 1974, **96**, 6888.
[185a] Newton, M. G., King, R. B., Chang, M., and Gimeno, J., *J. Amer. Chem. Soc.*, 1978, **100**, 1635.
[185b] Newton, M. G., Pantaleo, N. S., King, R. B., and Lotz, T. J., *J.C.S. Chem. Comm.*, 1978, 514.
[186] Niecke, E., and Bitter, W., *Angew. Chem. Internat. Edn.*, 1975, **14**, 56.
[187] Niecke, E., and Flick, W., *Angew. Chem. Internat. Edn.*, 1973, **12**, 585.
[188] Niecke, E., and Flick, W., *Angew. Chem. Internat. Edn.*, 1974, **13**, 134.
[189] Niecke, E., and Flick, W., *Angew. Chem. Internat. Edn.*, 1975, **14**, 363.
[189a] Niecke, E., and Flick, W., *J. Organometallic Chem.*, 1976, **104**, C23.
[190] Niecke, E., and Nixon, J. F., *Z. Naturforsch.*, 1972, **27b**, 467.
[190a] Niecke, E., Kröher, R., and Pohl, S., *Angew. Chem. Internat. Edn.*, 1977, **16**, 864.
[190b] Niecke, E., and Schäfer, H.-G., *Angew. Chem. Internat. Edn.*, 1977, **16**, 783.
[190c] Niecke, E., and Wildbredt, D. A., *Angew. Chem. Internat. Edn.*, 1978, **17**, 199.
[190d] Nixon, J. F., *J. Chem. Soc. (A),* 1968, 2689.
[190e] Nixon, J. F., *J. Chem. Soc. (A),* 1969, 1087.
[190f] Nixon, J. F., *Adv. Inorg. Chem. Radiochem.*, 1970, **13**, 363.
[191] Nöth, H., and Meinel, L., *Z. Anorg. Chem.*, 1967, **349**, 225.
[191a] Nöth, H., Reith, H., and Thorn, V., *J. Organometallic Chem.*, 1978, **159**, 165.
[192] Nöth, H., and Ullmann, R., *Chem. Ber.*, 1974, **107**, 1019.
[193] Nyquist, R. A., *Spectrochim. Acta,* 1963, **19**, 713.
[194] Nyquist, R. A., Blair, E. H., and Osbourne, D. W., *Spectrochim. Acta,* 1967, **23A**, 2505.
[195] Nyquist, R. A., Wass, M. N., and Mulder, W. W., *Spectrochim. Acta,* 1970, **26A**, 611.
[196] Oberhammer, H., and Schmutzler, R., *J.C.S. Dalton,* 1976, 1454.
[197] Oram, R. K., and Trippett, S., *J.C.S. Perkin (I),* 1973, 1300.
[198] Ozari, Y., and Jagur-Grodzinski, J., *J.C.S. Chem. Comm.*, 1974, 295.
[199] Paddock, N. L., Ranganathan, T. N., and Wingfield, J. N., *J.C.S. Dalton,* 1972, 1578; and refs. therein.
[199a] Paine, R. T., *J. Amer. Chem. Soc.*, 1977, **99**, 3884.

[199b] Paxton, K., and Hamor, T. A., *J.C.S. Dalton,* 1978, 647.
[199c] Peake, S. C., Hewson, M. J. C., Schlak, O., Schmutzler, R., Harris, R. K., and Wazeer, M. I. M., *Phosphorus and Sulfur,* 1978, **4**, 67.
[200] Peterson, M. B., and Wagner, A. J., *J.C.S. Dalton,* 1973, 106.
[201] Phillips, F. L., and Skapski, A. C., *J.C.S. Chem. Comm.,* 1973, 49.
[201a] Pinchuk, A. M., Filonenko, L. P., and Suleimanova, M. G., *Zh. Obshch. Khim.,* 1972, **42**, 2115.
[202] Pohl, S., *Chem. Ber.,* 1976, **109**, 3122.
[202a] Pohl, S., *Angew. Chem. Internat. Edn.,* 1976, **15**, 687.
[202b] Pohl, S., Niecke, E., and Krebs, B., *Angew. Chem. Internat. Edn.,* 1975, **14**, 261.
[202c] Pohl, S., Niecke, E., and Schäfer, H.-G., *Angew. Chem. Internat. Edn.,* 1978, **17**, 136.
[203] Poulin, D. D., and Cavell, R. G., *Inorg. Chem.,* 1974, **13**, 2324; corrigenda: *ibid,* 1975, **14**, 2022.
[204] Poulin, D. D., and Cavell, R. G., *Inorg. Chem.,* 1974, **13**, 3012.
[204a] Power, P. P., D.Phil. Thesis, University of Sussex, 1977.
[205] Pressl, K., and Schmidt, A., *Chem. Ber.,* 1972, **105**, 3518.
[206] Pressl, K., and Schmidt, A., *Chem. Ber.,* 1973, **106**, 2217.
[207] Quast, H., Heuschmann, M., and Abdel-Rahman, M. O., *Angew. Chem. Internat. Edn.,* 1975, **14**, 486.
[208] Rankin, D. W. H., *J.C.S. Dalton,* 1972, 869.
[208a] Richman, J. E., and Atkins, T. J., *Tetrahedron Letters,* 1978, 4333.
[209] Riesel, L., Pätzmann, H. H., and Bartich, H. P., *Z. Anorg. Chem.,* 1974, **404**, 219.
[210] Riesel, L., and Somieski, R., *Z. Anorg. Chem.,* 1975, **415**, 1.
[211] Roesky, H. W., and Dietl, M., *Angew. Chem. Internat. Edn.,* 1973, **12**, 425.
[212] Roesky, H. W., and Grimm, L. F., *Angew. Chem. Internat. Edn.,* 1972, **11**, 642.
[213] Roesky, H. W., and Niecke, E., *Z. Naturforsch.,* 1969, **24b**, 1101.
[214] Röschenthaler, G. V., and Schmutzler, R., *Z. Anorg. Chem.,* 1975, **416**, 289.
[215] Rose, H., *Ann. Phys. Chem.,* 1832, **24**, 295.
[216] Ross, B., and Dyroff, W., *Z. Anorg. Chem.,* 1973, **401**, 57.
[217] Ross, B., and Reetz, K. P., *Chem. Ber.,* 1974, **107**, 2720.
[218] Sasse, K., *Methoden der Organischen Chemie,* 12/I, G. Thieme, Stuttgart, 1963.
[219] Savignac, P., Chenault, J., and Dreux, M., *J. Organometallic Chem.,* 1974, **66**, 63.
[220] Savignac, P., Dreux, M., and Plé, G., *J. Organometallic Chem.,* 1973, **60**, 103.
[221] Savignac, P., Lavielle, G., and Dreux, M., *J. Organometallic Chem.,* 1974,

72, 361.
[222] Savignac, P., and Leroux, Y., *J. Organometallic Chem.*, 1973, **57**, C47.
[223] Schattka, K., and Jastorff, B., *Chem. Ber.*, 1974, **107**, 3043.
[224] Scherer, O. J., Klusmann, P., and Kuhn, N., *Chem. Ber.*, 1974, **107**, 552.
[225] Scherer, O. J., and Kuhn, N., *J. Organometallic Chem.*, 1974, **82**, C3.
[226] Scherer, O. J., and Kuhn, N., *Angew. Chem. Internat. Edn.*, 1974, **13**, 811.
[227] Scherer, O. J., and Kuhn, N., *Chem. Ber.*, 1974, **107**, 2123.
[228] Scherer, O. J., and Kuhn, N., *Chem. Ber.*, 1975, **108**, 2478.
[229] Scherer, O. J., and Schneider, G., *Chem. Ber.*, 1968, **101**, 4148.
[230] Schlak, O., Schmutzler, R., Harris, R. K., and Murray, M., *J.C.S. Chem. Comm.*, 1973, 23.
[230a] Schmidpeter, A., and von Crieghorn, T., *Angew. Chem. Internat. Edn.*, 1978, **17**, 55.
[231] Schmidpeter, A., and Rossknecht, H., *Chem. Ber.*, 1974, **107**, 3146.
[231a] Schmutzler, R., *Inorg. Chem.*, 1964, **3**, 415.
[232] Schmutzler, R., *J.C.S. Dalton*, 1973, 2687.
[233] Schweiger, J. R., Cowley, A. H., Cohen, E. A., Kroon, P. A., and Manatt, S. L., *J. Amer. Chem. Soc.*, 1974, **96**, 7122.
[234] Scott, D., and Wedd, A. G., *J.C.S. Chem. Comm.*, 1974, 527.
[235] Shaw, R. A., and Ibrahim, E. H. M., *Angew. Chem.*, 1967, **79**, 575.
[236] Shihada, A. F., *Z. Anorg. Chem.*, 1975, **411**, 135.
[237] Simonnin, M. P., Charrier, C., and Burgada, R., *Org. Mag. Res.*, 1972, **4**, 113.
[238] Simonnin, M. P., Lequan, R. M., and Wehrli, F. W., *J.C.S. Chem. Comm.*, 1972, 1204.
[239] Singh, G., and Zimmer, H., *Organometallic Chem. Rev. (A)*, 1967, **2**, 279.
[240] Stec, W. J., and Okruszek, A., *J.C.S. Perkin (I)*, 1975, 1828.
[241] Stegmann, H. B., Stöker, F., and Bauer, G., *Ann.*, 1972, **755**, 17.
[241a] Schmidbaur, H., Aly, A. A. M., and Schubert, U., *Angew. Chem. Internat. Edn.*, 1978, **17**, 846.
[242] Schwarz, W. E., Jr., Ruff, J. K., and Hercules, D. M., *J. Amer. Chem. Soc.*, 1972, **94**, 5227.
[243] The, K. I., and Cavell, R. G., *Inorg. Chem.*, 1976, **15**, 2518.
[243a] The, K. I., and Cavell, R. G., *Inorg. Chem.*, 1977, **16**, 2887; Cavell, R. G., Gibson, J. A., and The, K. I., *J. Amer. Chem. Soc.*, 1977, **99**, 7841.
[244] Thomas, L. C., *Interpretation of the Infra-Red Spectra of Organophosphorus Compounds*, Heyden, 1974.
[245] Thomas, M. G., Kopp, R. W., Schultz, C. W., and Parry, R. W., *J. Amer. Chem. Soc.*, 1974, **96**, 2646.
[245a] Thomas, M. G., Schultz, C. W., and Parry, R. W., *Inorg. Chem.*, 1977, **16**, 994.

[246] Tomaschewski, G., and Kühn, G., *Z. Chem.*, 1972, **12**, 332.
[247] Tomaschewski, G., and Kühn, G., *Z. Chem.*, 1972, **12**, 418.
[248] Trippett, S., senior reporter, *Organophosphorus Chemistry*, Specialist Periodical Reports, The Chemical Society, London (annual).
[249] Trippett, S., *Phosphorus and Sulfur*, 1976, **1**, 89.
[250] Trippett, S., and Whittle, P. J., *J.C.S. Perkin (I)*, 1973, 2302.
[251] Tyssee, D. A., Bausher, L. P., and Haake, P., *J. Amer. Chem. Soc.*, 1973, **95**, 8066.
[252] VanDoorne, W., Hunt, G. W., Perry, R. W., and Cordes, A. W., *Inorg. Chem.*, 1971, **10**, 2591.
[253] Vetter, H.-J., and Nöth, H., *Chem. Ber.*, 1963, **96**, 1308.
[253a] Vilkov, L. V., and Khaikin, L. S., *Topics Current Chem.*, 1975, **53**, 25.
[253b] Wilburn, J. C., and Neilson, R. H., *J.C.S. Chem. Comm.*, 1977, 308.
[253c] Wilson, R. D., and Bau, R., *J. Amer. Chem. Soc.*, 1974, **96**, 7601.
[254] Yolles, S., and Woodland, J. H. R., *J. Organometallic Chem.*, 1975, **93**, 297.
[255] Zeiss, W., *Angew. Chem. Internat. Edn.*, 1976, **15**, 555.

7

Amides of Arsenic, Antimony, and Bismuth: Synthesis, Physical Properties, and Structures

A ARSENIC

1. Introduction

The subject has been reviewed [43], and a collection of papers has been assembled together with brief editorial comments [44a]. An earlier article summarised data on 'elemento-organic amines' of formula $Me_3MN<$ (M = Si, Ge, or Sn), which included such amides of As and Sb [77]; another dealt with organoarsenic compounds [12], and $N(CF_3)_2$ derivatives of As, Sb, and Bi (as well as of other elements) are in ref. [6b].

The vast majority of arsenic amides which have been characterised (see Tables 1-5) have three-co-ordinate arsenic in the oxidation state +3. These include mono-, bis-, and tris-amides, [e.g., X_2AsNMe_2, $XAs(NMe_2)_2$, and $As(NMe_2)_3$ (X = R, halogen, or OR)], as well as diarsenyl(III)amines, such as $[(CF_3)_2As]_2NH$. A simple compound in which N is surrounded by three arsenic atoms has not yet been reported, although $As_4(NMe)_6$ [64, 99], of adamantane structure (Fig. 1) is known. Compounds related to this, but of As^V, have empirical formulae $(MeO)_8As_4(NPh)_6$ and $(PhNH)_8As_4(NPh)_6$ [27]. Cycloarsazanes are also known, as are compounds having alternating As and N atoms in 4-, 6-, and 8-membered rings. Examples include $(MeAsNPh)_2$ [45] and $(ClAsNMe)_3$ (see Fig. 2). An arsenic(V) ring compound is $(Ph_2AsN)_4$, obtained from thermolysis of Ph_2AsN_3 [74]; R_2AsN_3 derivatives give $(R_2As)_2$ [75].

Amides in which arsenic has the oxidation state +5 may involve the As atom in 3-, 4-, or 5-co-ordination, as in $OAs(NHPh)_3$ and $F(MeO)_3AsNHPr$, and the unusual nitride $(MeO)_2As{\equiv}N$, obtained from $(MeO)_3AsO$ and treatment successively with NH_3 and heating *in vacuo* [26].

Simple As-N compounds beyond the scope of this book include H_2AsNH_2, $HAs(NH_2)_2$, and $As(NH_2)_3$ (obtained by acid hydrolysis of a mixture of Mg_3As_2

and Mg_3N_2 [76]) and $K[As(NH_2)_2]$, from $AsPh_3$ and KNH_2 in liquid ammonia [85]. It is interesting that while $SbPh_3$ behaves similarly to afford $K[Sb(NH_2)_2]$, $BiPh_3$ yields an explosive bismuth nitride. Other compounds not discussed further are the hydrazides, such as $(CF_3)_2AsNRNR'R''$ (R, R', R'' = H or Me) [51, 70], arsine-imines, such as $Ph_3As{:}NCOR$ (R = Me or Ph) [18], and imido derivatives $>As{-}N{=}C<$ [91]

The As–N bond is easily cleaved by electrophiles, such as HCl, PhCOCl, $PhCH_2Cl$, or Ph_2BCl, to afford the arsenic halide [97]. This has been used preparatively [e.g., Eq. (1)] and is stoicheiometric in all proportions.

$$As(NMe_2)_3 + 2PhCH_2Cl \longrightarrow ClAs(NMe_2)_2 + [(PhCH_2)_2NMe_2]Cl \quad (1)$$

The As–N bond is also cleaved by methyl-lithium [Eq. (2)] [82].

$$(Me_2As)_2NMe + LiMe \longrightarrow Me_2AsNMeLi + AsMe_3 \quad (2)$$

Transamination reactions afford arsenic derivatives of hindered primary amines [e.g., Eq. (15)]; otherwise cycloarsazanes are obtained [e.g., Eq. (16)] (see Section A.2). Arsenic derivatives of primary amines may condense when heated to yield a cycloarsazane [e.g., Eq. (4)] or an arsaza-adamantane [95, 96]. The reactions of a primary amine with an arsenic(III) halide also usually afford such a compound [Eq. (5)], although aniline behaves differently [Eq. (6)] [23].

$$2As(NMe_2)_3 + 2H_2NPh \longrightarrow (Me_2NAsNPh)_2 + 4HNMe_2 \quad (3)$$

$$6As(NMe_2)_3 + 6H_2NBu \longrightarrow 3[Me_2NAsNBu]_2 \xrightarrow{\text{heat}} As_4(NBu)_6 + 2As(NMe_2)_3 \quad (4)$$

$$4AsI_3 + 18H_2NR \longrightarrow As_4(NR)_6 + 12[RNH_3]I \quad (5)$$

$$AsI_3 + 2H_2NPh \longrightarrow IAs(NHPh)_2 + 2HI \quad (6)$$

The chemistry of As-N compounds has been explored quite extensively (see Chapters 10-18). The As–N bond is cleaved preferentially to the P–N bond by HCl in compounds such as $(CF_3)_2AsN(Me)P(CF_3)_2$ [87].

2. Synthesis

The first arsenic amide was isolated in 1896 [Eq. (7)] [53]. Amine hydrochloride elimination has since proved versatile and general for the preparation

$$AsCl_3 + 2HNBu^i_2 \longrightarrow Cl_2AsNBu^i_2 + [Bu^i_2NH_2]Cl \quad (7)$$

of many aminoarsines, e.g., Eq. (8) [54] or (9) [37, 38]. However, AsI_3 and HNMePh afforded $[PhMeNH_2]_3[As_2I_9]$ [23].

$$MeAsCl_2 + 4HNMe_2 \longrightarrow MeAs(NMe_2)_2 + 2[Me_2NH_2]Cl \quad (8)$$

$$CF_3AsI_2 + 4HNR_2 \longrightarrow CF_3As(NR_2)_2 + 2[R_2NH_2]I \quad (9)$$

(e.g., R = Me, Et, Pr or Bu)

Salt elimination or metathetical exchange reactions afford compounds with As–N bonds [1, 8, 20, 42, 61]. The technique has not been very widely used. However, a complete series of methyl(trimethylsilyl)aminoarsines was prepared by this method, Eqs. (10) [78], (11) [63]. Variants are shown in Eqs. (12) [83] and (13) (M, M′ = Si, Ge, or Sn) [81] and (14) [101a].

$$Me_{3-n}AsCl_n + nLiNMeSiMe_3 \longrightarrow Me_{3-n}As(NMeSiMe_3)_n + nLiCl \quad (10)$$

$$AsF_3 + (3\text{-}n)LiNRSiMe_3 \longrightarrow AsF_n(NRSiMe_3)_{3-n} + (3\text{-}n)LiF \quad (11)$$

$$Me_2As(S)SAsMe_2 + NaN(SiMe_3)_2 \longrightarrow Me_2AsN(SiMe_3)_2 + Me_2As(S)SNa \quad (12)$$

$$Bu^t{}_2AsNH_2 \xrightarrow[\text{(ii) } Me_3MCl]{\text{(i) BuLi}} Bu^t{}_2AsNHMMe_3 \xrightarrow[\text{(ii) } Me_3M'Cl]{\text{(i) BuLi}} Bu^t{}_2AsN(MMe_3)(M'Me_3) \quad (13)$$

$$AsCl_3 + MeN(SiMe_2NMeLi)_2 \longrightarrow ClAs\begin{matrix}N(Me)\text{—}SiMe_2\\ \\ N(Me)\text{—}SiMe_2\end{matrix}NMe + 2LiCl \quad (14)$$

Transamination reactions have been applied to the synthesis of a variety of As–N compounds {e.g., Eqs. (15) [100], (16) [98], (17) [8], or (18) [39]}, (see also ultimate paragraph of Section A.1); NMe_2 exchange in the systems $Me_2AsNMe_2/HNMe_2$ and $MeAs(NMe_2)_2/HNMe_2$ was observed by ^1H n.m.r. using $H^{15}NMe_2$ [4a].

$$As(NMe_2)_3 + 2H_2NBu^t \longrightarrow (Bu^tHN)_2AsNMe_2 + 2HNMe_2 \quad (15)$$

$$2As(NMe_2)_3 + 4H_2NBu \longrightarrow (Me_2NAsNBu)_2 + 2HNMe_2 \quad (16)$$

$$As(NMe_2)_3 + 3PhNHCN \longrightarrow As(NPhCN)_3 + 3HNMe_2 \quad (17)$$

$$Me_2AsNMe_2 + HNR_2 \longrightarrow Me_2AsNR_2 + HNMe_2 \quad (18)$$

$$[R_2 = Et_2, (CH_2)_{4,5}, \text{ or } (CH)_4]$$

The reaction of $PhCH_2NH_2$ or an arylamine with a trialkyl arsenate(V) yields the arsenic(V) amide [Eq. (19)] [25]. Similarly $As(OMe)_5$ and an amine gives an unstable 4-membered $(AsN)_2$ ring compound [27].

$$(EtO)_3AsO + 3H_2NPh \longrightarrow OAs(NHPh)_3 + 3EtOH \quad (19)$$

Tris(trifluoromethyl)arsine eliminates 1 or 2 moles of F_3CCl upon reaction with bis(trifluoromethyl)chloramine [Eq. (20)] [7]. $(F_3C)_2AsN(CF_3)_2$ is also obtained via Eq. (21) [6a].

$$As(CF_3)_3 + (F_3C)_2NCl \longrightarrow (F_3C)_2AsN(CF_3)_2 + F_3CCl \quad (20)$$

$$As(CF_3)_3 + O[N(CF_3)_2]_2 \longrightarrow (F_3C)_2AsN(CF_3)_2 + (F_3C)_2NOCF_3 \quad (21)$$

Redistribution reactions afford alkyl(amino)arsines, halogeno(amino)-arsines, or alkyl(halogeno)(amino)arsines [e.g., Eq. (22)] [97].

$$As(NMe_2)_3 + 2AsCl_3 \longrightarrow 3Cl_2AsNMe_2 \quad (22)$$

The distribution of products and thermodynamic parameters of redistribution reactions have been studied for: $As(NMe_2)_3/AsX_3$ (X = F, Cl, Br, or OMe) [97], $As(NMe_2)_3$ with As_2O_3 or As_2S_3 [98], and $As_4(NMe)_6$ with AsF_3 [73].

The Lewis basicity of $As(NMe_2)_3$ is demonstrated (see Chapter 16) by its reaction with $[Ni(CO)_4]$ [Eq. (23)] [97], and of F_2AsNMe_2 by its reaction with BH_3 [59].

$$As(NMe_2)_3 + [Ni(CO)_4] \longrightarrow [Ni(CO)_3\{As(NMe_2)_3\}] + CO \quad (23)$$

Although $As(NMe_2)_3$ inserts 3 moles of CS_2 to afford $As(S_2CNMe_2)_3$, only an addition product (2:3) is isolated from the reaction with BuNCO [96]. However, reactions with several isocyanates afford ureidoarsines [e.g., Eqs. (24), (25)] [66, 67]; with $C_4F_9SO_2NCO$, $Me_2NAs[N(SO_2C_4F_9)(CONMe_2)]_2$ is obtained [75a], and use of Me_3SiNCO or $SO_2(NCO)_2$ leads to the four- or six-membered ring compounds $[(Me_2N)AsN(CONMe_2)]_n$ ($n = 2$ or 3) and $[(Me_2N)AsN(CONMe_2)S(O_2)N(CONMe_2)]_2$.

$$As(NMe_2)_3 + 3MeNCS \longrightarrow As(NMeCSNMe_2)_3 \quad (24)$$

$$As(NMe_2)_3 + 3PhNCO \longrightarrow As(NPhCONMe_2)_3 \quad (25)$$

A recent development is the synthesis of the first examples of stable two-co-ordinate As^{II} compounds. As in the corresponding phosphorus compounds (Chapter 6), synthesis was achieved by reduction of an M^{III} precursor, as in Eq. (26) [21]. The radicals possess a half-life of >15 days at 25° C.

$$ClAs[N(SiMe_3)_2]_2 \xrightarrow[\text{in PhMe, 25°C}]{=[\overline{CNEt(CH_2)_2NEt}]_2,\ h\nu} :\dot{A}s[N(SiMe_3)_2]_2 \quad (26)$$

Attempts to synthesise a two-co-ordinate arsenic(III) compound similar to $(Me_3Si)_2N{-}P{=}NSiMe_3$ have resulted in the formation of the 4-membered ring compound of Eq. (27) [63]. An analogue, $(Bu^tNAsCl)_2$, is made by thermolysis of $Cl_3As{=}NBu^t$ (with loss of Cl_2) [70a].

$$2FAs[N(SiMe_3)_2]_2 \xrightarrow{150°\ C} [(Me_3Si)_2NAsNSiMe_3]_2 + 2Me_3SiF \quad (27)$$

The Staudinger reaction is not as well-established in As^V as in P^V chemistry. However, $Ph_3SnAsPh_2$ with PhN_3 gives the amide $Ph_2As(NPh)[N(Ph)SnPh_3]$ and dinitrogen; likewise $(Ph_3Sn)_2AsPh$ yields $PhN{=}AsPh[N(Ph)SnPh_3]_2$ [85a].

3. Physical properties and structures

The structural problems are perhaps somewhat less complex for this group of elements, since the problem of association to form aggregates (cf. Chapters 2-4, or 8) does not arise. Apart from routine spectroscopic examination of compounds, and of physical studies of redistribution reactions, other results relate to activation parameters for rotation about As–N bonds and to crystallographic data. As for the latter, the only three compounds examined are cyclic derivatives, as illustrated in Figs. 1, 2, and 3. It is interesting that the 6-membered ring (Fig. 2) is non-planar. K-Edge X-ray absorption experiments on $As(NMe_2)_3$ and $As_4(NMe)_6$ suggest As–N bond lengths of 1.90 and 1.69 Å and As 1*s* binding energies of 92 and 87 eV, respectively [49].

The structure of $[Ph_2AsN]_3$ [4] is shown in Fig. 3. It has a 6-membered $(AsN)_3$ framework. A notable feature is the short As–N bond length of ca. 1.76 Å, which is about 0.1 Å shorter than the predicted bond length of 1.87 Å [confirmed by the X-ray structure of $As_4(NMe)_6$, Fig. 1]. The short As–N bond length, coupled with the near planarity of the $(As{-}N)_3$ ring has been rationalised in terms of delocalised π-bonding.

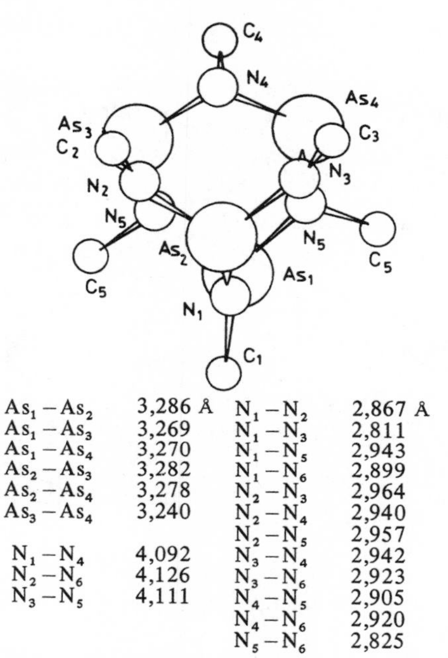

As_1-As_2	3,286 Å	N_1-N_2	2,867 Å
As_1-As_3	3,269	N_1-N_3	2,811
As_1-As_4	3,270	N_1-N_5	2,943
As_2-As_3	3,282	N_1-N_6	2,899
As_2-As_4	3,278	N_2-N_3	2,964
As_3-As_4	3,240	N_2-N_4	2,940
		N_2-N_5	2,957
N_1-N_4	4,092	N_3-N_4	2,942
N_2-N_6	4,126	N_3-N_6	2,923
N_3-N_5	4,111	N_4-N_5	2,905
		N_4-N_6	2,920
		N_5-N_6	2,825

Fig. 1 – Molecular structure of $As_4(NMe)_6$. H atoms omitted for clarity. *[Reproduced, with permission, from Weiss, J., and Eisenhuth, W.,* Z. Anorg. Chem., *1967,* **350,** *9.]*

As_1-N_1	1,79 Å	As_1-Cl_1	2,28 Å
As_1-N_3	1,85	As_2-Cl_2	2,24
As_2-N_1	1,83	As_3-Cl_3	2,28
As_2-N_2	1,88	N_1-C_1	1,49
As_3-N_2	1,76	N_2-C_2	1,51
As_3-N_3	1,86	N_3-C_3	1,35

$As_1-N_1-As_2$	131,5°	$N_1-As_1-N_3$	102,5°
$As_2-N_2-As_3$	128,4	$N_2-As_2-N_1$	99,7
$As_3-N_3-As_1$	121,9	$N_3-As_3-N_2$	105,5
$As_1-N_1-C_1$	110,4	$N_1-As_1-Cl_1$	101,8
$As_2-N_1-C_1$	117,2	$N_3-As_1-Cl_1$	99,6
$As_2-N_2-C_2$	109,7	$N_2-As_2-Cl_2$	95,9
$As_3-N_2-C_2$	121,5	$N_1-As_2-Cl_2$	99,6
$As_1-N_3-C_3$	123,0	$N_2-As_3-Cl_3$	100,0
$As_3-N_3-C_3$	113,4	$N_3-As_3-Cl_3$	95,6

Fig. 2 – Molecular structure of $(ClAsNMe)_3$. H atoms omitted for clarity. *[Reproduced, with permission, form Weiss, J., and Eisenhuth, W.,* Z. Naturforsch., *1967,* **22b,** *454.]*

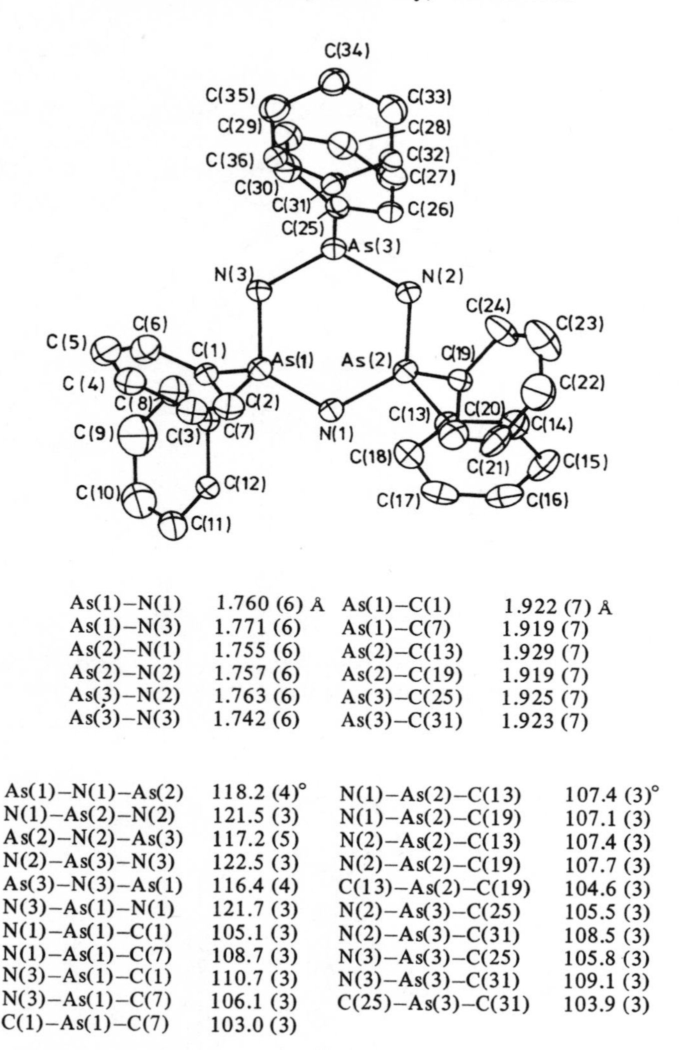

As(1)–N(1)	1.760 (6) Å	As(1)–C(1)	1.922 (7) Å
As(1)–N(3)	1.771 (6)	As(1)–C(7)	1.919 (7)
As(2)–N(1)	1.755 (6)	As(2)–C(13)	1.929 (7)
As(2)–N(2)	1.757 (6)	As(2)–C(19)	1.919 (7)
As(3)–N(2)	1.763 (6)	As(3)–C(25)	1.925 (7)
As(3)–N(3)	1.742 (6)	As(3)–C(31)	1.923 (7)

As(1)–N(1)–As(2)	118.2 (4)°	N(1)–As(2)–C(13)	107.4 (3)°
N(1)–As(2)–N(2)	121.5 (3)	N(1)–As(2)–C(19)	107.1 (3)
As(2)–N(2)–As(3)	117.2 (5)	N(2)–As(2)–C(13)	107.4 (3)
N(2)–As(3)–N(3)	122.5 (3)	N(2)–As(2)–C(19)	107.7 (3)
As(3)–N(3)–As(1)	116.4 (4)	C(13)–As(2)–C(19)	104.6 (3)
N(3)–As(1)–N(1)	121.7 (3)	N(2)–As(3)–C(25)	105.5 (3)
N(1)–As(1)–C(1)	105.1 (3)	N(2)–As(3)–C(31)	108.5 (3)
N(1)–As(1)–C(7)	108.7 (3)	N(3)–As(3)–C(25)	105.8 (3)
N(3)–As(1)–C(1)	110.7 (3)	N(3)–As(3)–C(31)	109.1 (3)
N(3)–As(1)–C(7)	106.1 (3)	C(25)–As(3)–C(31)	103.9 (3)
C(1)–As(1)–C(7)	103.0 (3)		

Fig. 3 – The crystal and molecular structure of $[Ph_2AsN]_3$. *[Reproduced, with permission, from Krannich, L. K., Thewalt, U., Cook, W. J., Jain, S. R., and Sisler, H. H.,* Inorg. Chem., *1973,* **12,** *2304.]*

^1H n.m.r. spectra at various temperatures, and the magnetic non-equivalence of Me protons in $PhAs(Cl)NMe_2$ [10], $Bu^t_2AsN(SiMe_3)_2$ [81], and Bu^t_2As-$N(SiMe_3)SbBu^t_2$ [81] at low temperatures, leads to the following values for the rotational activation energies in the three compounds, 8.2, 24.2, and 22.8 kcal mol^{-1}, respectively. For the phenyl compound, the barrier is ca. 2.5 kcal mol^{-1} lower than in the P–N analogue [10] and substantially lower than in the corresponding BN compound (see Chapter 4); it is regarded as too high to represent a *N*-inversion barrier. The AsNSi and AsNSb compounds, as well as $Bu^t{}_2SbN(SiMe_3)_2$ ($\Delta G^{\ddagger} = 19.6$ kcal mol^{-1}) [81] are regarded as of the *gauche*- rather than the *anti*- conformation, as shown in Newman projections (circle = As) for sp^2 N in (I) and (II) respectively. For $Bu^t{}_2AsN(SiMe_3)(GeMe_3)$, $Bu^t{}_2AsN(SiMe_3)(SnMe_3)$, $Bu^t{}_2AsN(GeMe_3)(SnMe_3)$, and $Bu^t{}_2AsN(SiMe_3)(SbBu^t{}_2)$, there is evidence for the presence of two rotamers at 35° C for all but the last-named compound.

(I) (II)

The e.s.r. spectrum of $\dot{A}s[N(SiMe_3)_2]_2$ consists of a 1:1:1:1 quartet due to ^{75}As ($I = 3/2$) coupling to the unpaired electron [21], as illustrated in Fig. 4 [70b]. Coupling to nitrogen (^{14}N, $I = 1$) was not observed, as in the phosphorus analogue (Chapter 6). The $a(^{75}As)$ value differs little from that in matrix-isolated $\dot{A}sMe_2$ and hence the $N\widehat{As}N$ bond angle is probably ca. 95°, and it is a π-radical with the unpaired electron in a *p*-orbital orthogonal to the AsN_2 plane.

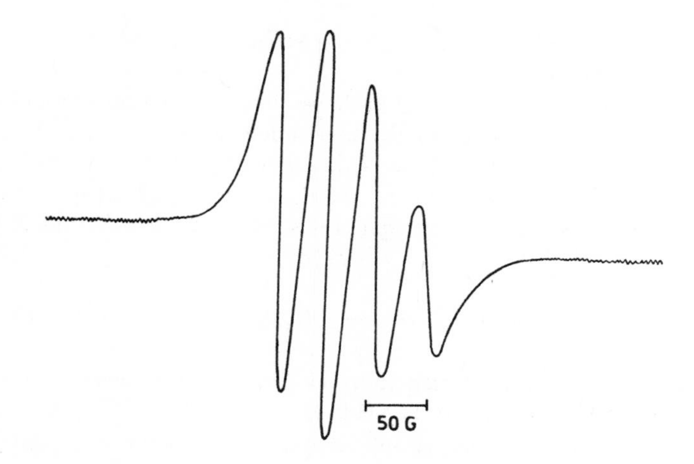

Fig. 4 – E.s.r. spectrum of $\dot{A}s[N(SiMe_3)_2]_n$ at 300K in PhMe [70b].

B ANTIMONY AND BISMUTH

Amides of antimony and bismuth have not been much studied and the number of such Bi compounds, in particular, is quite small (see Tables 1-3 and 5). Organobismuth compounds have been reviewed [22].

Antimony nitride was identified by its absorption spectrum when SbI_3 vapour was photolysed in presence of small amounts of N_2 [9].

Among Sb-N compounds, representatives of Sb oxidation states of +3 and +5, and corresponding co-ordination numbers of three and five respectively have been reported. Examples are R'_2SbNR_2, $R'Sb(NR_2)_2$, $Sb(NR_2)_3$, $F_2Sb(OEt)_2NHPr$, and $(R_3SbCl)_2NH$. Cyclostibazanes are ill-defined, cf. $(FSbNR)_n$ [29]. For bismuth, oxidation state +3 with three-co-ordination, as in Cl_2BiNEt_2 and $Bi[N(Me)SiMe_3]_3$, are the rule (however, see Table 7 and ref. [14a]). Nevertheless (presumably) cyclic derivatives [e.g., Eq. (28)] are known [24].

$$n\mathrm{IBi(NHPr)_2} \xrightarrow{\text{heat}} \mathrm{(IBiNPr)}_n + n\mathrm{H_2NPr} \quad (28)$$

Reaction of a halogen derivative with an amine is a common synthetic route {e.g., Eqs. (29) [19], (30) [25]}; antimony(V) compounds are also available through this procedure [30].

$$\mathrm{C_6H_4O_2SbCl} + 3\mathrm{HNC_5H_{10}} \longrightarrow \mathrm{C_6H_4O_2Sb(L)NC_5H_{10}}$$

$$(\mathrm{L = C_5H_{10}NH}) + [\mathrm{C_5H_{10}NH_2}]\mathrm{Cl} \quad (29)$$

$$\mathrm{I_3Bi} + 4\mathrm{H_2NBu} \longrightarrow \mathrm{IBi(NHBu)_2} + 2[\mathrm{BuNH_3}]\mathrm{I} \quad (30)$$

Salt elimination or metathetical exchange has provided Sb amides [20, 55]. A series of Sb and Bi methyl(trimethylsilyl)amides has been obtained, as shown for the As analogues in Eq. (10), [78]. Likewise, $Bu^t_2AsN(SiMe_3)SbBu^t_2$ is prepared by a variant of Eq. (12) [81], and $R_nSb(NEt_2)_{3-n}$ [52] and $Bu^t_2SbNR_2$ (R = Me or $SiMe_3$) [81] species by Eq. (31) (X = Cl or Br, R′ = Et, and R = alkyl or Ph [52]; or R = Bu^t, X = Cl, and n = 2) [81].

$$\mathrm{R}_n\mathrm{SbX}_{3-n} + (3\text{-}n)\mathrm{LiNR'_2} \longrightarrow \mathrm{R}_n\mathrm{Sb(NR'_2)}_{3-n} + (3\text{-}n)\mathrm{LiX} \quad (31)$$

Transamination is not at present known in Sb-N or Bi-N chemistry. Oxidation of R_3Sb with $ClNH_2$ gives $(R_3SbCl)_2NH$ [50].

Apart from routine spectroscopic characterisation, the only physical or structural studies relate to rotational barriers in some Sb-N compounds [81] (see Section A.2).

C TABLES OF COMPOUNDS

Table 1† Amino-arsines and -stibines

Compound	Method	B.p. (°C/mm.) (m.p., °C)	Comments	References
Me_2AsNMe_2	C	112/760	ir	[4a, 14]
	C	108 (<-110)	ir	[54]
	I	105-7		[97]
Me_2AsNR_2	B			
[NR_2 = $\overline{N(CH_2)_4}$			ir, nmr	[39]
$\overline{N(CH)_4}$			ir, nmr	[39]
$N(CH_2)_4O$,			ir, nmr	[39]
$\overline{N(CH_2)_2CHMe(CH_2)_2}$		70/0.1	ir, nmr	[39]
$N(CH_2CHCH_2)_2$		53/0.1	ir, nmr	[39]
NPr_2			ir, nmr	[39]
NBu_2			ir, nmr	[39]
NBu^s_2		65/0.1	ir, nmr	[39]
NBu^i_2		63/0.1	ir, nmr	[39]
$NHBu^t$]		50/0.1	ir, nmr	[41]
$(Me_2AsNMeCH_2)_2$	C	94-6/1		[82]
$[(CH_2)As(Me)]_2NMe$	C	74-6/12		[88]
Me_2AsN(piperidino ring)	B, C	75/8	ir, nmr	[15, 39]
$H_2C[As(Me)NR_2]_2$	C			
[NR_2 = NMe_2		76/0.4	ir, nmr	[38]
NEt_2		86/1.8	ir, nmr	[38]
NPr_2		98/0.6	ir, nmr	[38]
$\overline{N(CH_2)_4}$		105/0.5	ir, nmr	[38]
$\overline{N(CH_2)_5}$		125/0.6	ir, nmr	[38]
$N(CH_2)_4O$]		140/0.5	ir, nmr	[38]
p-$MeC_6H_4MeAsNMe_2$		(<8)		[62]
Me_2AsNEt_2	A, B	61/35	ir, nmr	[39, 79]
Et_2AsNEt_2	C	(45-7)		[92]
		81/35		[79]
Et_2AsN(aziridine ring)	C	47-8/18	d^{20}_4 1.1399, n^{20}_D 1.4820, ir	[33]
Et_2AsN(piperidino ring)	C	95-7/5		[92]
Et_2AsNPr_2	C			[92]
Pr_2AsNEt_2		66/1		[79]

Table 1 *(C'ntd.)*

Compound	Method	B.p. (°C/mm.) (m.p., °C)	Comments	References
Pr_2AsN (aziridine ring)	C	48/2	d_4^{20} 1.1023, n_D^{20} 1.4927, ir	[33]
Pr_2AsN (piperidine ring)	C	95–105/1		[92]
Bu_2AsNEt_2	A, C	110–2/6		[92]
		66/0.025		[79]
Bu_2AsN (aziridine ring)	C	74/1	d_4^{20} 1.0803, n_D^{20} 1.4970, ir	[33]
Bu_2AsN (piperidine ring)	C	122–6/3–4		[92]
$Bu^t_2AsNMe_2$	A	44–5/0.04	m.p. −38 to −36, nmr	[81]
$(C_5H_{11})_2AsNEt_2$	A	67/0.03	ir, nmr	[79]
Ph_2AsNMe_2	I	108–10/vac.		[97]
$(F_3C)_2AsNHMe$	C	84	ir	[13]
$(F_3C)_2AsNH_2$			nmr	[11]
$[(F_3C)_2As]_2NH$	C	126.5	nmr	[11, 13, 87]
$(F_3C)_2AsNHEt$	C	98.5	ir	[13]
$(F_3C)_2AsNMe_2$	C	89	ir	[13,43]
$(F_3C)_2AsNR_2$ (R = Me, Et, Pr, Pr^i, Bu, Bu^i, Bu^s, C_3H_5, C_6H_{11})	A, B		ir, nmr	[3, 42, 43]
$(F_3C)_2AsN(CF_3)_2$	I, J	70		[6a, 7]
Et_2SbNEt_2	A	77–80/12	ir, nmr	[52]
Pr_2SbNEt_2	A	53/0.1	ir, nmr	[52]
Bu_2SbNEt_2	A	100–5/1.4	ir, nmr	[52]
$Bu^t_2SbNMe_2$	A	25/0.3	m.p. <−78°, nmr	[81]
Ph_2SbNEt_2	A	130–3/0.3	ir, nmr	[52]
$EtSb(NEt_2)_2$	A	70–2/1.3	ir, nmr	[52]
$PrSb(NEt_2)_2$	A	58–60/0.1	ir, nmr	[52]
$BuSb(NEt_2)_2$	A	67–70/0.1	ir, nmr	[52]
$PhSb(NEt_2)_2$	A	94–5/0.1	ir, nmr	[52]
$MeAs(NHBu^t)_2$	B	(40–1)		[96]
$MeAs(NMe_2)_2$	A, C	148 (−62)	ir, nmr	[4a, 37, 41, 54]
	I	145–9		[97]
$MeAs(NR_2)_2$ [NR_2 =	A			
NEt_2		110/0.01	ir, nmr	[37, 41]
NPr_2		160/0.01	ir, nmr	[37, 41]
NBu_2		165/0.01	ir, nmr	[37, 41]
$\overline{N(CH)_4}$		110/0.01	ir, nmr	[37]

Table 1 *(C'ntd.)*

Compound	Method	B.p. (°C/mm.)	Comments	References
$\overline{N(CH_2)_4}$		120/0.01	ir, nmr	[37]
$\overline{N(CH_2)_5}$		110/0.01	ir, nmr	[37]
$N(CH_2)_4O]$		130/0.01	ir, nmr	[37]
$RAs(NMe_2)_2$	C		ir, nmr	[41]
(R = Me		50/0.1		
Et		52/0.1		
Pr		55/0.1		
Pr^i		70/0.1		
Bu		60/0.1		
Bu^i		65/0.1		
C_6H_{11}		55/0.01		
Ph)		75/0.01		
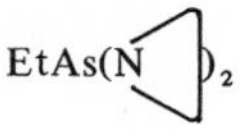 $EtAs(N\langle\rangle)_2$ (azetidinyl)	C	54–5/8	d_4^{20} 1.2149, n_D^{20} 1.5051, ir	[33]
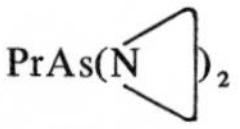 $PrAs(N\langle\rangle)_2$ (azetidinyl)	C	48–9/2	d_4^{20} 1.1992, n_D^{20} 1.5080, ir	[33]
$PhAs(NMe_2)_2$	I	71–3/2		[97]
	C	70/0.2	n_D^{20} 1.5587	[17,41]
$PhAs(NEt_2)_2$	C	108–10/0.2	n_D^{24} 1.5339	[17]
$PhAs(NPr_2)_2$	C	126–8/0.2	n_D^{20} 1.5179	[17]
$PhAs(NBu_2)_2$	C	148–50/0.2		[17]
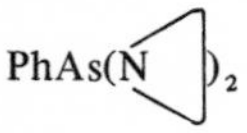 $PhAs(N\langle\rangle)_2$ (azetidinyl)	C	100–1/1	d_4^{20} 1.6010, n_D^{20} 1.3276, ir	[33]
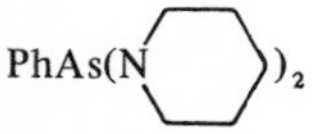 $PhAs(N\langle\rangle)_2$ (piperidino)	C	141–4/0.2	n_D^{24} 1.5718	[17]
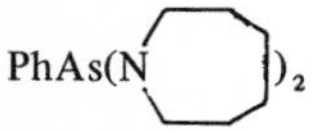 $PhAs(N\langle\rangle)_2$ (hexamethyleneimino)	C	168–70/0.2	n_D^{20} 1.5691	[17]
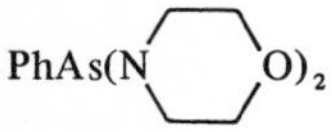 $PhAs(N\langle\rangle O)_2$ (morpholino)	C			[17]
$CF_3As(NR_2)_2$	C		ir, nmr	[4]
$[NR_2 = NMe_2$		48/20		
NEt_2		92/20		
NPr_2		75/1		
NBu_2		122/1		
$\overline{N(CH_2)_4}$		60/0.5		
$\overline{N(CH_2)_5}$		80/3		
$N(CH_2)_4O]$		90/0.5		

Table 1 *(C'ntd.)*

Compound	Method	B.p. (°C/mm.) (m.p., °C)	Comments	References
$CH_2[As(NR_2)_2]_2$	C		ir, nmr	[40, 43]
[NR_2 = NMe_2		105/0.2		
NEt_2		120/0.2		
NPr_2		120/0.3		
$\overline{N(CH_2)_4}$		130/0.9		
$\overline{N(CH_2)_5}$		140/0.9		
NBu_2		130/0.6		
NBu^s_2]		100/0.2		
$CF_3As[N(CF_3)_2]_2$	J	109		[7]
$(Bu^tNH)_2AsNMe_2$	B, C	67/0.01	n_D^{20} 1.4717	[100]
$(Bu^tNHAsNBu^t)_2$	C	125/0.01 (80–5)		[99, 100]
$As(NMe_2)_3$	A			[6, 20, 103, 104]
	C	42–3/3.5	d_4^{20} 1.1248, n_D^{20} 1.4848	[31]
	C	55–7/10; 170/760 (extr.) (−53)	ir	[54]
			nmr	[94]
			X-ray absorption	[49]
$.Ni(CO)_3$				[97]
$/AsF_3$			nmr	[56]
$/AsCl_3$			nmr	[56]
$/AsBr_3$			nmr	[56]
$/As(OMe)_3$			nmr	[56]
$/As_2O_3$			nmr	[57]
$/As_2S_3$			nmr	[57]
$(Me_2N)_2AsN(CO)_2C_6H_4$ (phthalimido)	B			[98]
$[Me_2NAsNCO\text{-}C_6H_4\text{-}CONMe_2]_n$	B			[98]
$Me_2NAs[N(SO_2C_4F_9)(CONMe_2)]_2$	D, I	(185–7)	ir, nmr	[75a]
$[(Me_2N)AsN(CONMe_2)S(O)_2N(CONMe_2)]_2$	D, I	(142–4)	ir, nmr	[75a]
$[(Me_2N)AsN(CONMe_2)]_2$	D, I	(120–2)	ir, nmr	[75a]
$(Me_2NAsNBu)_2$	B	111–3/0.5	n_D^{20} 1.5148	[98]
$(Me_2NAsNBu^t)_2$	B, C	111–3/0.5	n_D^{20} 1.5148	[100]
$(Me_2NAsNCH_2Ph)_2$	B	subl. 165/10^{-3} (213–5)		[95, 98]

Table 1 *(C'ntd.)*

Compound	Method	B.p. (°C/mm.) (m.p., °C)	Comments	References
$(Me_2NAsNPh)_2$	B	subl. $110/10^{-3}$ (163–5)		[95, 98]
$(Me_2NAsNCOPh)_n$	D			[96]
$Me_2NAsNCOPhCS_2$	D	(164–8)		[96]
$(Me_2NAsNNMe_2)_2$	B	106/2		[98]
$As(NMeCH_2)_3CMe$ (As—NMeCH$_2$ ×3 —CMe, cage)	B	91–4/10		[48]
$As(NEt_2)_3$	C	91–3/2	d_4^{20} 1.2418, n_D^{20} 1.4825	[31]
	B	73/0.5	n_D^{20} 1.4839	[98, 100]
$As(NBu_2)_3$	B	159–61/2	n_D^{20} 1.5789	[98]
	B	159–61/0.01	n_D^{20} 1.4789	[100]
$As(NPhCN)_3$	A	(250)		[8]
As(N◁)$_3$ (aziridino)	B			[60]
As(N⬡)$_3$ (piperidino)	C	173/3		[19]
	B	139–42/0.5 (49–51)		[98]
	B	139–42/0.01 (49–51)		[100]
As(N□=O)$_3$ (2-oxopyrrolidino)				[5]
As(N⬡O)$_3$ (morpholino)	C	(113–4)		[19]
$Sb(NMe_2)_3$	A	30/0.5		[20]
	A	32–4/0.45		[55]
Sb(N□=O)$_3$ (2-oxopyrrolidino)				[5]
$(Me_2As)_2NMe$	C	70–2/14 (−34)		[82]
$(PhNH)_3AsO$	G			[25]
$(p\text{-}MeC_6H_4NH)_3AsO$				[101]
$(PhCH_2NH)_3AsO$	G			[25]
$As_4(NMe)_6$	C	(122–5)		[33]
			X-ray absorption	[49]
			X-ray	[106]
$/As_2O_3$			nmr	[73]

Table 1 *(C'ntd.)*

Compound	Method	B.p. (°C/mm.) (m.p., °C)	Comments	References
$As_4(NPr^i)_6$	C	150/high vac. (47–51)		[99]
$As_4(NBu)_6$	C	197–201/1		[99]
$As_4F_2(NMe)_5$	C	142–5/0.3 (60–3)		[86]
$As_4(NR)_4(CH_2)_2$	C			
(R = Me		130/0.1	ir, nmr	[40, 43, 88a]
Et		140/0.1	ir, nmr	[40, 43, 88a]
Pr		180/0.6	ir, nmr	[40, 43, 88a]
Pr^i		130/0.1	ir, nmr	[40, 43, 88a]
Bu		184/0.1	ir, nmr	[40, 43, 88a]
Bu^i)		185/0.8	ir, nmr	[40, 43, 88a]
$(NAsPh_2)_3$	C		ir, nmr, X-ray	[45]

† Footnote to Table 1; for explanation of abbreviations, see footnotes to Table 1, p. 57.

Table 2† Halogeno(amino)-arsines, -stibines, and -bismuthines

Compound	Method	B.p. (°C/mm.) (m.p., °C)	n_D^{20}	Comments	References
$Me_2AsNHBu^t$	C	50/0.1			[41]
$MeAsFNMe_2$	I	112/760		ir, nmr	[35]
$MeAsClNMe_2$	I	112/760			[35, 97]
EtAsClN(cyclohexane ring)	C	108/8			[15]
$PhAsClNMe_2$	C	49–50/0.05		nmr	[10]
$MeAsBrNMe_2$	I	140/220			[35]
$MeAsINMe_2$	I	108/4			[35]
EtAsINHPh	C	110/10			[15]
m-$O_2NC_6H_4AsClNHMe$		(18)			[62]
$F_2AsNHC_6H_{11}$	C	(65–6)			[69]
F_2AsNMe_2	C	43–4/40 (2–4)	1.4461,	ir, nmr	[59, 69, 86]
$F_2AsNMe_2.BF_3$				ir, nmr	[59]
$F_2AsNMe_2.BH_3$				ir, nmr	[59]
F_2AsNEt_2	C	39–40/20	1.4857,	ir, nmr	[44, 69, 86]

Table 2 *(C'ntd.)*

Compound	Method	B.p. (°C/mm.) (m.p., °C)	d_4^{20}	n_D^{20}	Comments	References
F_2AsNR_2	C				ir, nmr	[44]
[R_2 = Pr_2		$60/10^{-2}$				
Pr^i_2		85/50				
Bu_2		$70/10^{-2}$				
Bu^i_2		60/5				
Bu^s_2		80/4				
$(C_3H_5)_2$		50/27				
$(C_6H_{11})_2$		$90/10^{-2}$				
$(CH_2)_4$		$90/10^{-2}$				
$(CH_2)_4O$		60/60				
$(CH_2)_4NMe$]		125/3				
F_2AsN(piperidino ring)	C	44–5/10		1.4920,	ir, nmr	[44, 69]
$F_2AsN(SiMe_3)_2$	A				ir, nmr	[63]
$F_2AsN(SiMe_3)Bu^t$	A				ir, nmr	[63]
$Cl_2AsNHC_6H_{11}$	C					[68]
$Cl_2AsNHPh$	C	(89)				[15]
Cl_2AsNMe_2	C	73–5/25	1.6560	1.5532		[31]
	C	50–1/10		1.5564		[68]
	I, X	80–1/30		1.5556		[97]
					nmr	[94]
					ir, Raman	[16]
$Cl_2AsNMePh$	C	116/3				[15]
Cl_2AsNEt_2	A	36–7/0.1		1.5341		[1]
	C	107/33				[15]
	C	83–4/13	1.4727	1.5335		[31, 32]
	C	70–1/9		1.5345		[68]
	C	(70–2)				[92]
$Cl_2AsNBu^i_2$	C	125/15				[53]
Cl_2AsN(piperidino ring)	C	91/10		1.5675		[68]
	C	95–7/5				[94]
Cl_2AsN(morpholino ring)O	C	125–7/13				[68]
Cl_2BiNEt_2	C	(204)				[47]
Br_2AsNMe_2	I, X	70/3		1.6428	nmr	[97]
Br_2BiNMe_2	K					[58]
$FAs[N(SiMe_3)_2]_2$	A	83–4/0.1			ir, nmr	[63]
$FAs[N(Bu^t)SiMe_3]_2$	A	88–9/0.1			ir, nmr	[63]
$ClAs(NMe_2)_2$	C	42–3/13	1.3460	1.5255		[31]
	K	69–70/10		1.5477		[97]
					nmr	[94]

Table 2 *(C'ntd.)*

Compound	Method	B.p. (°C/mm.) (m.p., °C)	d_4^{20}	n_D^{20}	Comments	References
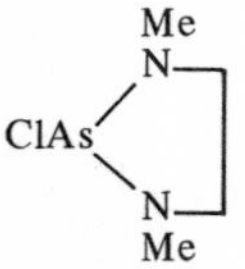	C	103–5/14 (19–21)				[82]
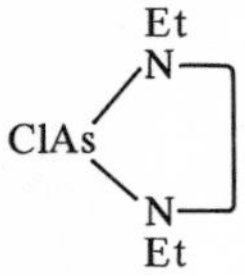	A	86/0.7				[1a]
$ClAs(NEt_2)_2$	A	45–6/0.08		1.5094		[1]
	A	48–9/0.1		1.5096		[1]
	C	118–20/17	1.2225	1.5098		[31, 32]
$ClAs[N(SiMe_3)_2]_2$	A	(70–2)			ir, nmr	[21, 63]
$FSb(NHR)_2$ ($R = Me, Bu, PhCH_2$)						[29]
$ClSb[N(SiMe_3)_2]_2$	A	142–4/1.5			ir, nmr	[21]
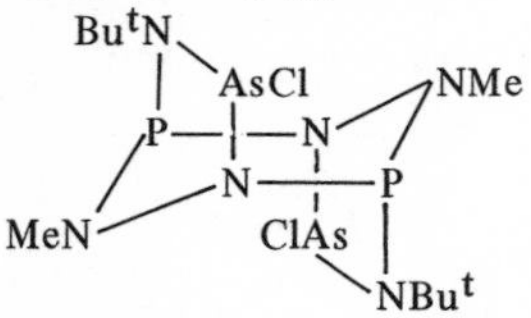	A				ir, nmr, X-ray	[84]
$BrAs(NMe_2)_2$	C					[94]
	I, K	87–8/10				
		70–2/3		1.5660		[97]
$(ClAsNMe)_3$		(103–6)			X-ray	[99, 105]
$(FAsNR)_n$	B					[44]
$(ClAsNPh)_3$						[45a]
$IAs(NHPh)_2$	C					[23]
$IBi(NHPr)_2$	C					[24]
$IBi(NHBu)_2$	C					[24]
$F_2Sb(OEt)_2NHPr$	G					[30]
$F_2Sb(OEt)_2NHPh$	G					[30]
$(EtO)_2Sb(NCH_2Ph)NHCH_2Ph$						[30]
$(EtO)_3Sb(NHPr)_2$	C, G					[30]
$(EtO)_3Sb(NHCH_2Ph)_2$	C, G					[30]
$(Me_3SbCl)_2NH$		(>200d)			ir, nmr	[50]
$(Et_3SbCl)_2NH$		(128–9.5)			ir, nmr	[50]
$(Pr_3SbCl)_2NH$		(83–5)			ir, nmr	[50]
$(Bu_3SbCl)_2NH$		(49–50)			ir, nmr	[50]
$(Ph_3SbCl)_2NH$		(219–20.5)			ir, nmr	[50]
$As_4(NMe)_5Cl_2$		(104–6)				[99]
$[As(OMe)_3NPh]_2$						[27]

Table 2 *(C'ntd.)*

Compound	Method	References
$FAs(OMe)_3NHPr^i$	C	[28]
$[FAs(OMe)_2NPr]_2$	C	[28]
$[PhCH_2NHAs(OMe)_2NCH_2Ph]_2$	C	[27]
$[PrNHAs(OMe)_2NPr]_2$	C	[27]
$[BuNHAs(OMe)_2NBu]_2$	C	[27]
$[(PhCH_2NH)_2As(OMe)NCH_2Ph]_2$	C	[27]

† Footnote to Table 2; for explanation of abbreviations, see footnotes to Table 1, p. 57.

Table 3† Alkoxy(amino)-arsines and -stibines

Compound	Method	B.p. (°C/mm.) (m.p., °C)	d_4^{20}	n_D^{20}	References
$(EtO)_2AsNHCH_2CONHCH(CO_2Et)CH_2Ph$	C				[93]
$(EtO)_2AsNHCHPr^iCO_2Et$	C				[93]
$(EtO)_2AsNHCH(CO_2Et)CH_2Ph$	C				[93]
$(EtO)_2AsNHCH(CO_2Et)CH_2C_6H_4OH$-*p*	C				[93]
$(EtO)_2AsNHPh$	C				[93]
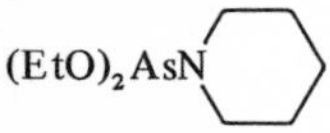 $(EtO)_2AsN$ (piperidino ring)	C				[93]
$(C_6H_{11}O)_2AsNEt_2$	C	156–7/4	1.1530	1.5000	[34]
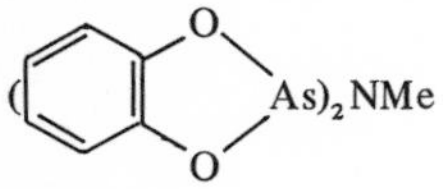 ($C_6H_4O_2$As)$_2$NMe (o-phenylenedioxy ring)	C	(154–5)			[19]
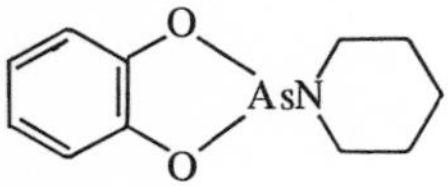 $C_6H_4O_2$AsN (o-phenylenedioxy; piperidino ring)	C	142/3			[19]
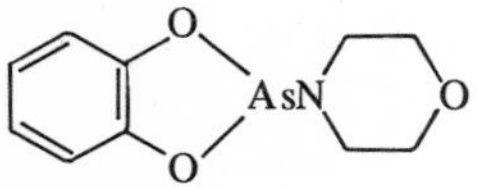 $C_6H_4O_2$AsN (o-phenylenedioxy; morpholino ring)	C				[19]
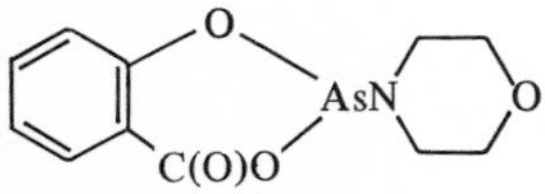 C_6H_4(O)(C(O)O)AsN (morpholino ring)	C	d(76)			[19]
$(MeO)_2PO_2AsNMe_2$	A	104–6/0.06			[61]

Table 3 *(C'ntd.)*

Compound	Method	B.p. (°C/mm.) (m.p., °C)	d_4^{20}	n_D^{20}	Comments	References
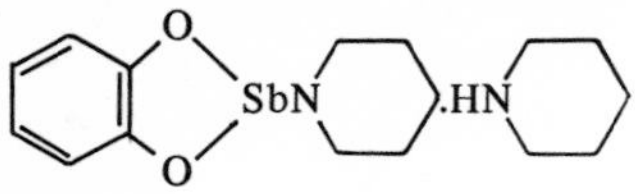 (o-phenylenedioxy)SbN(piperidine).HN(piperidine) [structure]	C					[19]
$C_6H_{11}OAs(NEt_2)_2$	C	163/12	1.0870	1.4873		[34]
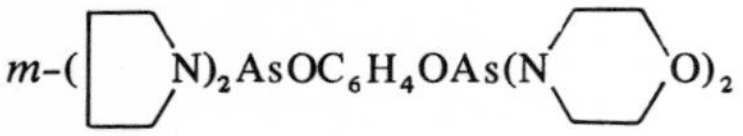 m-(morpholino-N)$_2$AsOC$_6$H$_4$OAs(N-morpholino)$_2$ [structure]	C	(105–7)				[19]
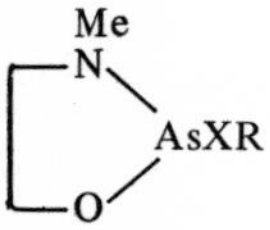 3-methyl-1,3,2-oxazarsolidine AsXR [structure: Me–N, O ring]	G or I					
(X R						
O Me		57/9			ir, nmr	[2]
O Et		40/2			ir, nmr	[2]
O Pr		62/3			ir, nmr	[2]
O Pr^i		55/5			ir, nmr	[2]
O Bu		58/0.5			ir, nmr	[2]
O Bu^s		50/1			ir, nmr	[2]
O Bu^i		62/1			ir, nmr	[2]
O Bu^t		49/2			ir, nmr	[2]
O C_6H_{11}		104/3			ir, nmr	[2]
O Ph		86/0.01			ir, nmr	[2]
S Me		54/0.01			ir, nmr	[2]
S Et		55/0.01			ir, nmr	[2]
S Pr		74/0.01			ir, nmr	[2]
S Ph)		105/0.01			ir, nmr	[2]
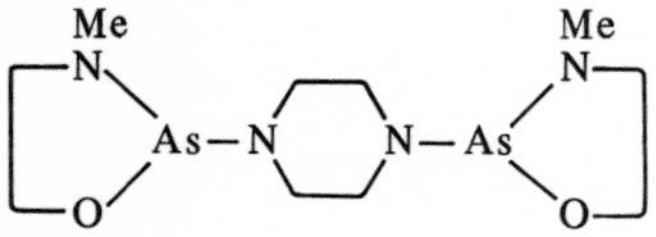 (3-methyl-oxazarsolidinyl)As–N(piperazine)N–As(3-methyl-oxazarsolidinyl) [structure]	B	89/0.01			ir, nmr	[2]
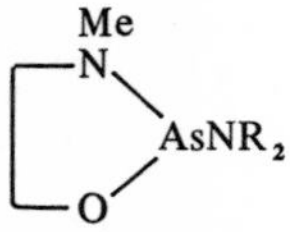 3-methyl-1,3,2-oxazarsolidine AsNR$_2$ [structure]	B					
[NR$_2$ = NEt$_2$		(58)				[2]
NPr$_2$		(57)				[2]
NBu$^i_2$		69/0.01				[2]
NBu$_2$		77/0.01				[2]
$\overline{N(CH_2)_5}$		60/0.01				[2]

Table 3 *(C'ntd.)*

Compound	Method	B.p. (°C/mm.) (m.p., °C)	d_4^{20}	n_D^{20}	Comments	References
$\overline{N(CH_2)_4}$		51/0.01				[2]
$N(C_6H_{11})_2$		125/0.01				[2]
N(ring)Me		75/0.01				[2]
N(ring)		73/0.01				[2]
NPh_2]		145/0.01				[2]
$AsNMe_2(ONCRR')_2$						
[RR' = Me, Me		93/1			ir, nmr	[36]
$-(CH_2)_5-$]		74/0.2			ir, nmr	[36]
$As_4(NR)_4(CH_2)_2$	C					
(R R'						
H Me		90/10			ir, nmr	[36]
$-(CH_2)_5-$		105/1			ir, nmr	[36]
Me Et		105/16			ir, nmr	[36]
Et Et)		105/8			ir, nmr	[36]
$Me_2NOC(O)As(NMe_2)_2$	K	85-7/0.4				[65, 66]
$(Me_2N)_{3-n}As(ONCRR')_n$					ir, nmr	[36]
[R = 1, 2; R = R' = Me, Et;						
RR' = $(CH_2)_4$; R = Me, R' = H, Et]						
$(Me_2N)_2AsC(N_2)CO_2Et$					ir, nmr	[46]
$MeAs[C(N_2)CO_2Et]NMe_2$		68/0.01			ir, nmr	[46]
$(MeO)_3As$–O–$As(OMe)_3$ / N(Pr) ring	G, I				nmr, mass	[21, 71, 72]
$(MeO)_3As$–N(Bu)–$As(OMe)_3$ / O ring	G, I				nmr, mass	[72]
$(MeO)_2As$=N–$As(OMe)_2$=N–$As(OMe)_2$=N ring	G				ir, Raman	[71]
$[NAs(OMe)_2]_n$	G, I				ir, mass	[72]
$[NAs(OEt)_2]_n$	G, I				ir, mass	[72]

† Footnote to Table 3; for explanation of abbreviations, see footnotes to Table 1, p. 57.

Table 4† **Arsenic or antimony derivatives of ureas**

Compound	Method	M.p. (°C)	Comments	References
$PhAs(NPhCONR_2)_2$	D			[17]
$As(NHCONH_2)_3$	B	144–6	ir, nmr	[89]
$As(NMeCONHMe)_3$	B	198	ir, nmr	[89]
$As(NMeCSNMe_2)_3$	D	115		[67]
$As(NEtCONMe_2)_3$	D			[67]
$As[N(CONMe_2)(CH_2)_6Cl]_3$	D			[66,67]
$As(NPhCONMe_2)_3$	D	143		[66,67]
$As(NPhCSNMe_2)_3$	D	140d		[66,67]
$As[N(p\text{-}C_6H_4Me)CONMe_2]_3$	D	160		[67]
$As[N(p\text{-}C_6H_4NO_2)CONMe_2]_3$	D	195d		[66,67]
$As[N(m\text{-}C_6H_4Cl)CONMe_2]_3$	D	158–60		[66,67]
$As[N(p\text{-}C_6H_4Cl)CONMe_2]_3$	D	166d		[67]
$As[N(3,4\text{-}C_6H_3Cl_2)CSNMe_2]_3$	D			[67]
$As[N(p\text{-}C_6H_4OEt)CSNMe_2]_3$	D			[67]
$As_2(NMeCONMe)_3$	D	185–90		[86]
$As(NRSO_2NR)_3As$	B			
(R = H		172–4	ir, nmr	[89]
Me)		251–3	ir, nmr	[89]
$SbPh_3\overline{NCO(CH_2)_2CO}$	A	112	ir	[90]
$SbAr_3\overline{NCOC_6H_4CO}$	A		ir	[90]
(Ar = Ph		167–8		
$C_6H_4Me\text{-}p$)		214		
$MR_3(X)(\overline{NCOYCO})$				[14a]
(R = Me, Ph; M = As, Sb, Bi; X = Cl, Br; Y = $(CH_2)_2$, $o\text{-}C_6H_4$)				

†Footnote to Table 4; for explanation of abbreviations, see footnotes to Table 1, p. 57.

Table 5† ***N*-metalloamino-arsines, -stibines, and -bismuthines**

Compound	Method	B.p. (°C/mm) (m.p., °C)	Comments	References
$Me_2AsNMeLi$	K	45–6/1	ir, nmr	[82]
$Me_2AsN(SiMe_3)_2$	A	(—46)	nmr	[83]
$Me_2AsNMeSiMe_3$	A	44–6/11 (—50)	nmr	[78]
	A	66–8/14 (—50)		[82]
$MeAs(NMeSiMe_3)_2$	A	55–9/0.1	nmr	[78]
$As(NMeSiMe_3)_3$	A	67–70/0.1 (11–3)	nmr	[78]
$[CH_2N(Me)]_2PN(Me)AsMe_2$	A	76–8/1 (—38)	ir, nmr	[83]
$[CH_2N(Me)]_2PN(Me)SbMe_2$	A	88–90/0.1	ir, nmr	[83]

Table 5 *(C'ntd.)*

Compound	Method	B.p. (°C/mm) (m.p., °C)	Comments	References
$[CH_2N(Me)]_2PN(Me)As[N(Me)CH_2]_2$	A	87/0.01 (9–10)		[83]
$Me_2SbNMeSiMe_3$	A	44–6/1	nmr	[78]
$MeSb(NMeSiMe_3)_2$	A	59–61/0.1	nmr	[78]
$Sb(NMeSiMe_3)_3$	A	78–9/0.1 (9–11)	nmr	[78]
$Me_2BiNMeSiMe_3$	A	31–2/0.1	nmr	[78]
$MeBi(NMeSiMe_3)_2$	A	70–1/0.1	nmr	[78]
$Bi(NMeSiMe_3)_3$	A	90–2/0.1	nmr	[78]
$Bi[N(SiMe_3)_2]_3$	A	(90)d	nmr	[48a]
$[MeBu^t_2AsNMeSiMe_3]^+I^-$				[80]
$(CF_3)_2AsN(R)P(CF_3)_2$	A, C			[87]
$Bu^t_2AsNHSiMe_3$	A	36–8/0.02	(−6 to −5), nmr	[81]
$Bu^t_2AsNHGeMe_3$	A	40/0.06	(−15 to −13), nmr	[81]
$Bu^t_2AsNHSnMe_3$	A	54–6/0.01	(8–10), nmr	[81]
$Bu^t_2AsN(SiMe_3)SbBu^t_2$	A		(105–6), nmr	[81]
$Bu^t_2AsN(SiMe_3)_2$	A	95–105/0.02	(87–9), nmr	[81]
$Bu^t_2AsN(SiMe_3)GeMe_3$	A	72–80/0.01	(54–6), nmr	[81]
$Bu^t_2AsN(SiMe_3)SnMe_3$	A	75–85/0.01	(65–7), nmr	[81]
$Bu^t_2AsN(GeMe_3)_2$	A	82–5/0.01	(55–8), nmr	[81]
$Bu^t_2AsN(GeMe_3)SnMe_3$	A	70–5/0.01	(57–60), nmr	[81]
$Bu^t_2SbN(SiMe_3)_2$	A	73–5/0.05		[81]
$(CF_3)_2AsNHP(CF_3)_2$		110 (extr.)	(−44), ir	[87]
$(CF_3)_2AsNMeP(CF_3)_2$		127 (extr.)	(−30), ir	[87]
$(CF_3)_2AsN[P(CF_3)_2]_2$		161 (extr.)	(26), ir	[87]
Cyclic $[(Me_3Si)_2NAs-NSiMe_3]_2$: As(N(SiMe₃)₂) and As(N(SiMe₃)₂) bridged by two $NSiMe_3$ groups (Me_3SiN, $NSiMe_3$)		(208)	ir, nmr	[63]
Ring Me_2Si—$SiMe_2$, each Si bonded to N(R), both N bonded to M(R′): (M R R′				
As Me OMe			ir, nmr	[102a]
As Et NMe_2		94/1.5	ir, nmr	[102a]
As Pr Me		71/0.1	ir, nmr	[102a]
As Me Cl		83/0.1	ir, nmr	[102a]
As Et Cl		95/0.3	ir, nmr	[102a]
As Pr Cl		97/0.1	ir, nmr	[102a]
Sb Me Cl			ir, nmr	[102a]
Sb Et Cl		86/0.2	ir, nmr	[102a]
Sb Pr Cl)			ir, nmr	[102a]

Table 5 *(C'ntd.)*

Compound	Method	B.p. (°C/mm) (m.p., °C)	Comments	References
$Me_2Si(X)SiMe_2(NMe)M(Cl)(NMe)$ — six-membered ring: X bridging the two $SiMe_2$ groups, each Si bonded to MeN, both N bonded to M–Cl	A			
(M = As, X = NMe		78–80/0.2	ir, nmr	[101a]
As, X = O		69/0.30	ir, nmr	[102]
Sb, X = NMe)		111–13/1	ir, nmr	[101a]
$[ClAsNBu^t]_2$ — four-membered ring: two Cl–As bridged by two Bu^tN	I	(62–4)		[70a]
$Ph_2As(NPh)[N(Ph)SnPh_3]$		(160–2)	ir	[85a]
$PhAs(NPh)[N(Ph)SnPh_3]$		(152–5)d	ir	[85a]

†Footnote to Table 5: for explanation of abbreviations, see footnotes to Table 1, p. 57.

D REFERENCES

[1] Abel, E. W., Armitage, D. A., and Willey, G. R., *J. Chem. Soc.,* 1965, 57.
[1a] Abel, E. W., and Bush, R. P., *J. Organometallic Chem.,* 1965, **3**, 245.
[2] Adler, O., and Kober, F., *Z. Naturforsch.,* 1976, **31b**, 246.
[3] Adler, O., and Kober, F., *J. Fluorine Chem.,* 1975, **5**, 231.
[4] Adler, O., and Kober, F., *J. Organometallic Chem.,* 1974, **72**, 351.
[4a] Adler, O., and Kober, F., *Chem. Z.,* 1976, **100**, 235.
[5] Akiyama, Y., Jap. Patent 547 (1965).
[6] Anderson, R. H., and Cragg, R. H., *Inorg. Nucl. Chem. Letters,* 1971, **7**, 583, and *Chem. Comm.,* 1971, 1414.
[6a] Ang, H. G., and Lien, W. S., *J. Fluorine Chem.,* 1974, **4**, 447.
[6b] Ang, H. G., and Syn, Y. C., *Adv. Inorg. Chem. Radiochem.,* 1974, **16**, 1.
[7] Ang, H. G., and Emeléus, H. J., *Chem. Comm.,* 1966, 460.
[8] Biechler, J., *Compt. Rend.,* 1936, **202C**, 666.
[9] Ciach, S., and Thistlethwaite, P. J., *J. Chem. Phys.,* 1970, **53**, 3381.
[10] Cowley, A. H., Dewar, M. J. S., Jackson, W. R., and Jennings, W. B., *J. Amer. Chem. Soc.,* 1970, **92**, 5206.
[11] Cowley, A, H., and Schweiger, J. R., *Chem. Comm.,* 1970, 1492.
[12] Cullen, W. R., *Adv. Organometallic Chem.,* 1966, **4**, 145.
[13] Cullen, W. R., and Emeléus, H. J., *J. Chem. Soc.,* 1959, 372.
[14] Cullen, W. R., and Walker, L. G., *Can. J. Chem.,* 1960, **38**, 472.

[14a] Dahlmann, J., and Winsel, K., Ger. Patent, 83, 136 (1971); *Chem. Abstr.*, 1973, **78**, 43715q.
[15] Doak, G. O., *J. Amer. Pharm. Assoc.*, 1935, **24**, 453; *Chem. Abstr.*, 1935, **29**, 6563.
[16] Durig, J. R., and Casper, J. M., *J. Mol. Struct.*, 1971, **10**, 427.
[17] Fluck, E., and Jakobson, G., *Z. Anorg. Chem.*, 1969, **369**, 178.
[18] Frøyen, P., *Acta Chim. Scand.*, 1969, **23**, 2935.
[19] Funk, H., and Kohler, H., *J. Prakt. Chem.*, 1961, **14**, 226.
[20] George, T. A., and Lappert, M. F., *J. Chem. Soc. (A)*, 1969, 992.
[21] Gynane, M. J. S., Goldwhite, H., Hudson, A., Lappert, M. F., and Power, P. P., *J.C.S. Chem. Comm.*, 1976, 623; *J.C.S. Dalton*, 1980, in the press.
[22] Harrison, P. G., *Organometallic Chem. Rev. (A)*, 1970, **5**, 183.
[23] Hass, D., *Z. Anorg. Chem.*, 1963, **325**, 139.
[24] Hass, D., *Z. Chem.*, 1964, **4**, 185.
[25] Hass, D., *Z. Chem.*, 1964, **4**, 353.
[26] Hass, D., *Z. Chem.*, 1967, **7**, 395.
[27] Hass, D., *Z. Anorg. Chem.*, 1966, **347**, 123.
[28] Hass, D., and Cech, D., *Z. Chem.*, 1969, **9**, 432.
[29] Hass, D., and Cech, D., *Z. Chem.*, 1970, **10**, 33.
[30] Hass, D., and Cech, D., *Z. Chem.*, 1970, **10**, 75.
[31] Kamai, G., and Khisamova, Z. L., *Proc. Acad. Sci. U.S.S.R.*, 1955, **105**, 489.
[32] Kamai, G., and Khisamova, Z. L., *J. Gen. Chem., U.S.S.R.*, 1956, **26**, 125.
[33] Kamai, G., and Khisamova, Z. L., *Proc. Acad. Sci. U.S.S.R.*, 1964, **156**, 365.
[34] Kamai, G., and Miftakhova, R. G., *J. Gen. Chem. U.S.S.R.*, 1962, **32**, 2839.
[35] Kaufmann, J., and Kober, F., *J. Organometallic Chem.*, 1975, **96**, 243.
[36] Kaufmann, J., and Kober, F., *Z. Anorg. Chem.*, 1975, **416**, 152.
[37] Kober, F., *Z. Anorg. Chem.*, 1973, **397**, 97.
[38] Kober, F., *Z. Anorg. Chem.*, 1973, **412**, 202.
[39] Kober, F., *Z. Anorg. Chem.*, 1973, **400**, 285.
[40] Kober, F., *J. Organometallic Chem.*, 1975, **94**, 393.
[41] Kober, F., *Z. Anorg. Chem.*, 1973, **401**, 243.
[42] Kober, F., *Chem. Z.*, 1976, **100**, 197.
[43] Kober, F., *Chem. Z.*, 1976, **100**, 313.
[44] Kober, F., and Adler, O., *J. Fluorine Chem.*, 1974, **4**, 73.
[44a] Krannich, L. K., (ed.), *Compounds Containing As–N Bonds*, Dowden, Hutchinson, and Ross, Inc., Stroudsburg, Pa., 1976.
[45] Krannich, L. K., Thewalt, U., Cook, W. J., Jain, S. R., and Sisler, H. H., *Inorg. Chem.*, 1973, **12**, 2304.
[45a] Krieg, V., and Weidlein, J., *Angew. Chem. Internat. Edn.*, 1971, **10**, 516.

[46] Krommes, P., and Lorberth, J., *J. Organometallic Chem.,* 1975, **93**, 339.
[47] Lal, K., Bhatia, I. P., and Kaushik, R. L., *Current Sci. (India),* 1960, **29**, 272; *Chem. Abstr.,* 1961, **55**, 3625.
[48] Lambe, B. L., Bertrand, R. D., Casedy, G. A., Crompton, R. D., and Verkade, J. G., *Inorg. Chem.,* 1967, **6**, 173.
[48a] Lappert, M. F., and Power, P. P., unpublished results, 1977.
[49] Levy, R. M., Van Wazer, J. R., and Simpson, J., *Inorg. Chem.,* 1966, **5**, 332.
[50] McKenney, R. L., and Sisler, H. H., *Inorg. Chem.,* 1967, **6**, 1178.
[51] Märkl, G., Baier, H., and Martin, C., *Tetrahedron Letters,* 1974, 1977.
[52] Meinema, H. A., and Noltes, J. G., *Inorg. Nucl. Chem. Letters,* 1970, **6**, 241.
[53] Michaelis, A., and Luxembourg, K., *Ber.,* 1896, **29**, 710.
[54] Moedritzer, K., *Chem. Ber.,* 1959, **92**, 2637.
[55] Moedritzer, K., *Inorg. Chem.,* 1964, **3**, 609.
[56] Moedritzer, K., and Van Wazer, J. R., *Inorg. Chem.,* 1964, **3**, 139.
[57] Moedritzer, K., and Van Wazer, J. R., *Inorg. Chem.,* 1965, **4**, 893.
[58] Moedritzer, K., Van Wazer, J. R., and Weingarten, H., U.S. Patent 3,504,005 (1970).
[59] Morse, J. G., and Morse, K. W., *Inorg. Chem.,* 1973, **12**, 2119.
[60] Murakami, T., Morisaki, S., and Hiro, K., Brit. Patent 1,030,394 (1966).
[61] Nagasawa, M., and Imamiya, Y., Jap. Patent 8116 (1963).
[62] Nagasawa, M., and Totsuka, T., Jap. Patent 2299 (1959).
[63] Niecke, E., and Bitter, W., *Synth. Inorg. Metal-Org. Chem.,* 1975, **5**, 231.
[64] Nöth, H., and Vetter, H. J., *Z. Naturforsch.,* 1961, **16b**, 553.
[65] Oertel, G., Holtschmidt, H., and Malz, H., Ger. Patent 1,170,393 (1963).
[66] Oertel, G., Malz, H., and Holtschmidt, H., *Chem. Ber.,* 1964, **97**, 891.
[67] Oertel, G., Malz, H., Holtschmidt, H., and Degener, E., Ger. Patent 1,155,433 (1963).
[68] Olah, G., and Oswald, A., *Can. J. Chem.,* 1960, **38**, 1428.
[69] Olah, G., and Oswald, A., U.S. Patent 3,024,238 (1962).
[70] Peterson, L. K., and Thé, K. I., *Chem. Comm.,* 1967, 1056; *Can. J. Chem.,* 1969, **47**, 339.
[70a] Pinchuk, A. M., Khranovskii, V. A., Kuplennik, Z. I., and Filonenko, L. P., *J. Gen. Chem. U.S.S.R.,* 1977, **47**, 49.
[70b] Power, P. P., D.Phil. Thesis, University of Sussex, 1977.
[71] Priess, H., and Hass, D., *Z. Anorg. Chem.,* 1974, **404**, 190.
[72] Priess, H., and Hass, D., *Z. Anorg. Chem.,* 1974, **404**, 199.
[73] Rausch, M. D., Van Wazer, J. R., and Moedritzer, K., *J. Amer. Chem. Soc.,* 1964, **86**, 814.
[74] Reichle, W. T., *Tetrahedron Letters,* 1962, 61.
[75] Revitt, D. M., and Sowerby, D. B., *Inorg. Nucl. Chem. Letters,* 1969, **5**, 459.

[75a] Roesky, H. W., and Sidiropoulos, G., *Z. Naturforsch.*, 1977, **32b**, 628.
[76] Royen, P., Rocktäschel, C., and Mosch, W., *Angew. Chem. Internat. Edn.*, 1964, **3**, 703.
[77] Scherer, O. J., *Angew. Chem. Internat. Edn.*, 1969, **8**, 861.
[78] Scherer, O. J., Hornig, P., and Schmidt, M., *J. Organometallic Chem.*, 1966, **6**, 259.
[79] Sagan, L. S., Zingaro, R. A., and Irgolic, K. J., *J. Organometallic Chem.*, 1972, **39**, 301.
[80] Scherer, O. J., and Janssen, W., *J. Organometallic Chem.*, 1969, **16**, P69.
[81] Scherer, O. J., and Janssen, W., *Chem. Ber.*, 1970, **103**, 2784.
[82] Scherer, O. J., and Schmidt, M., *Angew. Chem. Internat. Edn.*, 1964, **3**, 702.
[83] Scherer, O. J., and Wokulat, J., *Z. Anorg. Chem.*, 1968, **357**, 92; **361**, 296.
[84] Scherer, O. J., Glässel, W., and Huttner, G., *Angew. Chem. Internat. Edn.*, 1976, **14**, 702.
[85] Schmitz-Du Mont, O., and Ross, B., *Z. Anorg. Chem.*, 1967, **349**, 328.
[85a] Schumann, H., and Roth, A., *Chem. Ber.*, 1969, **102**, 3731.
[86] Singer, R. J., Eisenhut, M., and Schmutzler, R., *J. Fluorine Chem.*, 1971/2, **1**, 193.
[87] Singh, J., and Burg, A. B., *J. Amer. Chem. Soc.*, 1966, **88**, 718.
[88] Sommer, K., *Z. Anorg. Chem.*, 1970, **376**, 150.
[88a] Sommer, K., *Z. Anorg. Chem.*, 1970, **377**, 120.
[89] Sommer, K., and Lauer, W., *Z. Anorg. Chem.*, 1970, **378**, 310.
[90] Srivastava, R. C., and Bajpai, K., *Synth. Inorg. Metal-Org. Chem.*, 1979, in press.
[91] Swindell, R. F., Ouelette, T. J., Babb, D. P., and Shreeve, J. M., *Inorg. Nucl. Chem. Letters*, 1971, **7**, 239.
[92] Tzschach, A., and Lange, W., *Z. Anorg. Chem.*, 1964, **326**, 280.
[93] Vaughan, J. R., U.S. Patent 2,631,158 (1953).
[94] Vetter, H.-J., *Naturwiss.*, 1964, **51**, 240.
[95] Vetter, H.-J., and Nöth, H., *Angew. Chem.*, 1962, **74**, 943.
[96] Vetter, H.-J., and Nöth, H., *Z. Naturforsch.*, 1964, **19b**, 166.
[97] Vetter, H.-J., and Nöth, H., *Z. Anorg. Chem.*, 1964, **330**, 233.
[98] Vetter, H.-J., Nöth, H., and Hayduk, U., *Z. Anorg. Chem.*, 1964, **331**, 35.
[99] Vetter, H.-J., Nöth, H., and Jahn, W., *Z. Anorg. Chem.*, 1964, **328**, 144.
[100] Vetter, H.-J., Strametz, H., and Nöth, H., *Angew. Chem. Internat. Edn.*, 1963, **2**, 218.
[101] Voitenko, G. A., *Farmakol. i Toksikol.*, 1964, **27**, 156; *Chem. Abstr.*, 1966, **64**, 11790.
[101a] Wannagat, U., Bogusch, E., and Braun, R., *J. Organometallic Chem.*, 1969, **19**, 367.
[102] Wannagat, U., and Rabet, F., *Inorg. Nucl. Chem. Letters*, 1970, **6**, 155.

[102a] Wannagat, U., and Schlingmann, M., *Z. Anorg. Chem.*, 1976, **424**, 87.
[103] Weingarten, H., *Chem. Comm.*, 1966, 293.
[104] Weingarten, H., and White, W. A., *J. Org. Chem.*, 1966, **31**, 2874, 4041; *J. Amer. Chem. Soc.*, 1966, **88**, 850.
[105] Weiss, J., and Eisenhuth, W., *Z. Naturforsch.*, 1967, **22b**, 454.
[106] Weiss, J., and Eisenhuth, W., *Z. Anorg. Chem.*, 1967, **350**, 9.

8
Amides of Transition Metals and f-Block Metals: Synthesis, Physical Properties, and Structures

A INTRODUCTION

This chapter deals with the preparation, physical properties, and structures of compounds with transition metal to nitrogen σ-bonds of the type $[M(NR_2)_n]$ or $[M(L)_m(NR_2)_n]$ where R is an alkyl or trimethylsilyl group and L may be any of a large variety of donor ligands. The wide range of transition metal amides is illustrated by reference to the known homoleptic neutral complexes, in Table 1. {Homoleptic complexes have been defined as those in which all the ligands attached to the central metal are the same, whereas heteroleptic analogues are mixed-ligand species: e.g., $[Ti(NMe_2)_4]$ belongs to the former class and $[TiCl(NMe_2)_3]$ to the latter.} A significant feature is that the number of transition metal amides becomes very small toward the right-hand side of the Table; this phenomenon is discussed in Section A.3. Excellent general reviews of homoleptic transition metal amides have appeared [35, 39], while other accounts have dealt with specific aspects of the subject: steric factors [36, 43, 150b], titanium amides [89], reactions [204] or disilylamides [171].

Other nitrogen derivatives of the transition metals which are of interest include N_2 complexes, e.g., of Ti, Re, Fe, Ru, or Os [5, 20, 116, 118, 162]; oxyamides, e.g., of Fe [170]; nitrides and arylimides, e.g., of Re [117, 162];

Table 1 Stable neutral homoleptic transition metal and *f*-block metal amides,[a]

Group 3	Group 4	Group 5	Group 6	Group 7
$[Sc(NR_2)_3]$ Me_3Si	$[\{Ti(NR_2)\}_x]$ Me	$[\{V(NR_2)_2\}_x]$ Me, Et	$[\{Cr(NR_2)_2\}_x]$ E.g., Et	$[\{Mn(NR_2)_2\}_2]$ Me_3Si
	$[\{Ti(NR_2)_2\}_x]$ Me, Et	$[\{V(NR_2)_3\}_x]$ Me, Et, or ($x = 1$) Me_3Si		
	$[\{Ti(NR_2)_3\}_x]$ E.g., Me ($x = 2$), Me_3Si ($x = 1$)	$[V(NR_2)_4]$ E.g., Me, Ph	$[\{Cr(NR_2)_3\}_x]$ Me, or ($x = 1$) Me_3Si or Pr^i	
	$[Ti(NR_2)_4]$ E.g., Me, Pr^i, or Ph		$[Cr(NR_2)_4]$ Et, Pr, Bu	
$[Y(NR_2)_3]$ Me_3Si, Pr^i	$[Zr(NR_2)_4]$ E.g., Me, Pr^i; or NR_2 = $\overline{NCHMe(CH_2)_3CHMe}$	$[Nb(NR_2)_4]$ E.g., Et	$[\{Mo(NR_2)_3\}_2]$ E.g., Me	
		$[Nb(NR_2)_5]$ E.g., Me	$[Mo(NR_2)_4]$ Me, Et	
Lanthanides $[M(NR_2)_3]$ Me_3Si; M = La, Ce, Pr, Nd, Sm, Eu, Gd, Ho, Yb, Lu	$[Hf(NR_2)_4]$ E.g., Me	$[Ta(NR_2)_4]$ (MeBu)	$[\{W(NR_2)_3\}_2]$ Me, Et	
		$[Ta(NR_2)_5]$ E.g., Me	$[W(NR_2)_6]$ Me	
Actinides $[Th(NR_2)_4]$ Me, Et, Bu^i				
$[\{U(NR_2)_4\}_x]$ E.g., Et ($x = 2$), Ph				

[a]Named groups, such as Me, Et, or Me_3Si, refer to R in the cited formulae immediately above.

ammines, e.g., of Ti [75]; carboxylic acid amide complexes, e.g., of Zr [234]; biurets [199]; salen derivatives, e.g., of Ni^{II} [152, 231]; Schiff-base complexes, e.g., $[Ti\{OC_6H_4(CH{=}NEt)\text{-}o\}_4]$ [57], and other chelates, e.g., having the unit $M{-}N(R)C{=}C(R)(CH)_3C(R)C{=}NR$ [29, 50, 70, 143–148, 210, 239].

$[\{M(NR_2)_n\}_x]^a$

Group 8a	Group 8b	Group 8c	Group 1
$[\{Fe(NR_2)_2\}_x]$ Et (?)	$[Co(NR_2)_2]$ Me_3Si	$[Ni(NR_2)_2]$ Me_3Si	$[\{Cu(NR_2)\}_4]$ Me_3Si
	$[Co(NR_2)_2]$ Me_3Si		$[Cu(NR_2)]$ 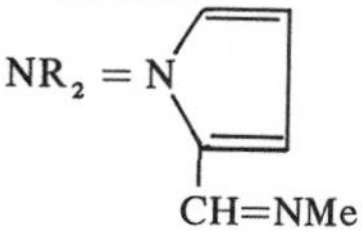 $NR_2 = N$ (pyrrole ring) with $CH{=}NMe$
$[Fe(NR_2)_3]$ Me_3Si	$[\{Co(NR_2)_3\}_x]$ 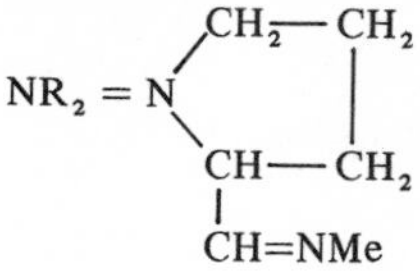 $NR_2 = N$ (ring: $CH_2{-}CH_2$, $CH{-}CH_2$) with $CH{=}NMe$		
			$[\{Ag(NR_2)\}_x]$ Me_3Si
			$[\{Au(NR_2)\}_3]$ 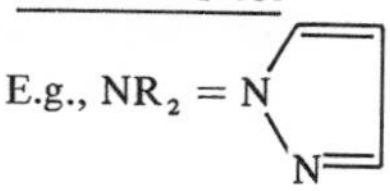 E.g., $NR_2 = N$ (pyrazole ring, $N{=}$)

Transition metal amides are of wide interest for several reasons, as briefly outlined in Sections A.1-A.4.

1. Bonding

For a σ-bonded alkylamido ligand there are three bonding possibilities. Structure (I) shows a metal-nitrogen σ-bond with a pyramidal nitrogen and approximately sp^3 hybridisation. Structure (II) illustrates the possibility of π-bonding involving overlap of the p-orbital orthogonal to the trigonal planar nitrogen atom and appropriate d-orbitals of the transition metal. Many spectroscopic and structural investigations have indicated that such π-bonding does take place.

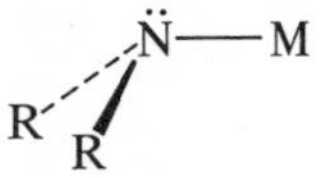

(I) 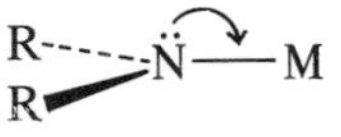 (II) 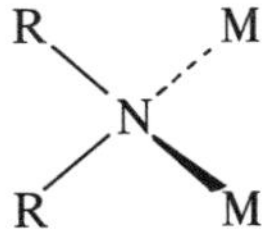 (III)

Examples of the amido ligand behaving as a bridging moiety, structure (III), are also found. It is generally possible to preclude such bonding by employing bulky substituents on nitrogen, e.g., with the ligands $NPr^i_2{}^-$ or $N(SiMe_3)_2^-$; cf., $[\{Ti(NMe_2)_3\}_2]$ and $[Ti\{N(SiMe_3)_2\}_3]$.

2. Co-ordination number

The use of extremely bulky amido ligands has allowed the isolation of complexes having unusually low metal co-ordination numbers, e.g., $[Ti\{N(SiMe_3)_2\}_3]$, $[Cr(NPr^i_2)_3]$, and $[Co\{N(SiMe_3)_2\}_2]$. Detailed studies of their structures, electronic absorption spectra, and the magnetic properties are of interest with regard to ligand field theory.

Conversely, using the less bulky amido ligands (e.g., NMe_2^-) it is possible to isolate complexes of higher co-ordination numbers. For example, $[Nb(NMe_2)_5]$ or $[W(NMe_2)_6]$ represent rare examples of complexes of Nb^V and W^{VI} which are discrete monomers under ambient conditions and which may be studied as vapours (e.g., for photoelectron spectroscopy) or in solution in inert solvents such as hydrocarbons.

3. Amides of the later transition metals

An interesting generalisation about transition metal amides is the dearth of well-characterised complexes of the later metals. There does not appear to be any simple reason for this phenomenon. Some authors have pointed out that the possibility of $M\overset{\frown}{-}\ddot{N}$ π-bonding is less than in their earlier congeners, because the d-orbitals, say in a d^8 Pt^{II} complex, are less likely to behave as electron-acceptors. An important exception to the generalisation is the silylamido complexes. An electronic explanation may be that the nitrogen *p*-electrons are delocalised onto silicon, which thus allows the $N(SiMe_3)_2^-$ ligand to behave as a weak π-acceptor. However, more significant factors, applicable also for rationalisation of the existence of low co-ordination number bis(trimethylsilyl)-amides (and related amides) of the early transition metals, are kinetic in origin. Thus, such amides are stable because normally accessible decomposition path-ways are energetically unattractive. Especially significant is the fact that a bulky ligand causes the metal environment to be sterically hindered, e.g., a bulky Ni^{II} amide cannot associate. Another factor is the absence of β-hydrogen (see Section F and Chapter 18).

4. Comparison between metal amides and alkyls

Metal alkyls and amides are isoelectronic, because CH and $\ddot{N}$ have this relationship. A comparison of the two classes of compounds (and indeed alkoxides or fluorides, which complete the isoelectronic series of first-row ligands) is therefore in order.

Whereas the β-decomposition pathway for decomposition of a transition metal alkyl involves migration of a β-hydrogen or alkyl group to the metal [137a] in a concerted manner and is very common, it has now also been reported in the amides, although studies on appropriate compounds, e.g., diethylamides, are not extensive. Further discussion is in Section F and Chapter 18.

Although homoleptic metal alkyls are not well represented among the later transition metals, there are numerous examples of heteroleptic alkyl complexes. This is in marked contrast to the very few reports of corresponding amido compounds. We anticipate that a substantial chemistry will be developed involving, for example, d^4, d^6, d^8, or d^{10} metal heteroleptic amides, in which associated ligands include tertiary phosphines or $C_5H_5^-$; many of these will be 16- or 18-electron species.

The similarities between transition metal alkyls and amides are more notable than their differences. There are many structural parallels, e.g., $[\{Cu(CH_2SiMe_3)\}_4]$ and $[\{CuN(SiMe_3)_2\}_4]$, $[WMe_6]$ and $[W(NMe_2)_6]$, and $[Cr\{CH(SiMe_3)_2\}_3]$ and $[Cr\{N(SiMe_3)_2\}_3]$. Furthermore, unusual imides, such as $[Ta(NR_2)_3NR]$ (R = alkyl, but not Me) find their counterparts in the recently synthesised alkylidenes, such as $[Ta(CH_2R)_3(CHR)]$. Finally, alkyls and amides undergo many similar types of reactions: e.g., involving insertion, metathesis, or reaction with protic reagents, and products are similar. Perhaps this is due to polar mechanisms, metal alkyls and amides behaving as incipient carbanions and as NR_2^-, respectively.

B SYNTHESIS

In earlier chapters, we have classified preparative methods for metal amides into several distinct types. Using the same abbreviations (see summarising Tables in Chapters 2-5 and 7), the most important methods for transition metal amides involve transmetallation (C), amine hydrohalide elimination (A), transamination (B), or disproportionation (I).

1. Transmetallation

The method, e.g., Eq. (1), is the most common route to transition metal

$$MCl_n + nLiNR_2 \longrightarrow [M(NR_2)_n] + nLiCl \qquad (1)$$

amides and it is an almost exclusive process for preparing the homoleptic compounds. The reaction of $[TiCl_4]$ with $Na[NPh_2]$ was an early example [138], but others did not emerge until the early sixties, when Bradley and Thomas published the first [64] of a series of papers on complexes of some early transition metals.

This synthetic procedure has been employed to prepare amides of Ti^{III} [42, 205, 206], Ti^{IV} [e.g., 34, 51, 178, 207], Zr^{IV} [2, 21, 51, 65, 102], Hf^{IV} [51], V^{III} [11, 42], V^{IV} [11], Nb^{IV} [38, 51, 66], Nb^{V} [51, 203], Ta^{IV} [67], Ta^{V} [51, 73], Cr^{II} [23, 105], Cr^{III} [6, 40, 63, 69], Cr^{IV} [22, 23], Mo^{III} [123, 124, 130], Mo^{IV} [39a], W^{III} [41, 122, 124, 126, 127, 129b], W^{VI} [41, 41a, 127], Mn^{II} [91], Fe^{III} [99], Co^{II} [60, 80], Ni^{I} [60], Ni^{II} [97], Cu^{I} [97], Ag^{I} [93], and Au^{I} [247a]. The wide applicability of transmetallation reactions is illustrated in Scheme 1.

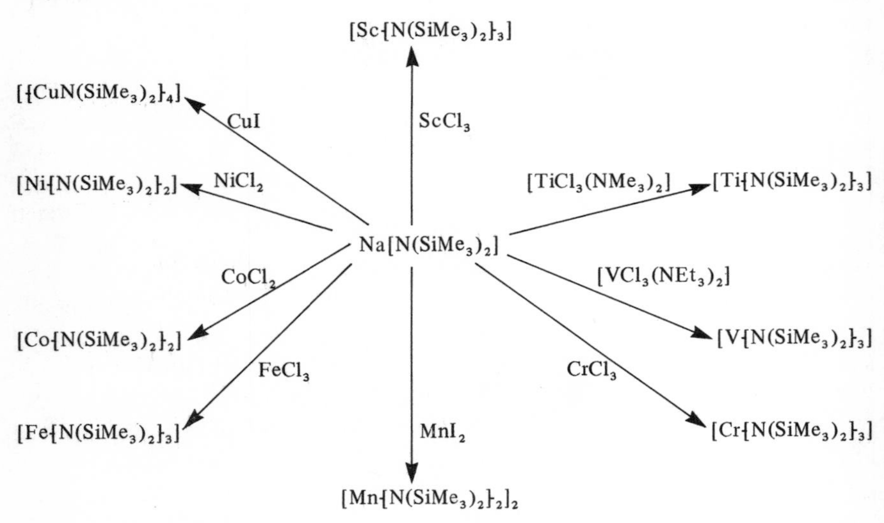

Scheme 1 – Synthesis of first-row transition metal bis(trimethylsilyl)amides.

It is not inevitable that transmetallation provides (i) the amide of the same metal oxidation state as the starting transition metal chloride, or (ii) a homoleptic metal amide. However, departures from the norm, i.e., Eq. (1), are probably due either to steric effects or to disproportionation of the first formed [Eq. (1)] homoleptic amide. Examples are in Eqs. (2) [39a] and (3) [124, 127].

$$MoCl_5 + 5LiNMe_2 \longrightarrow [Mo(NMe_2)_4] + LiCl + \text{other products} \quad (2)$$

$$WCl_4 + 5LiNMe_2 \longrightarrow [W(NMe_2)_6] + [W_2(NMe_2)_6] \quad (3)$$

The reaction of $TaCl_5$ with $LiNR_2$ affords $[Ta(NMe_2)_5]$ or $[Ta(NR_2)_3(NR)]$ (e.g., R = Et [51]). Reaction of a Co^{II}, Ni^{II}, or Cu^{II} halide with $LiNEt_2$ [33] leads to unusual di-iminato complexes of the type

$$[M(N(Et)=C(Me)\cdots N(Et))_2]$$

2. Elimination of amine hydrohalide

The reactions of a transition metal halide with ammonia or an amine to form a metal ammine or amido complex have been known for many years. These are closely related to the amine hydrohalide elimination route to transition metal amides, which has wide application for the higher oxidation state halides of the early transition metals (which are covalent and have polar $\overset{\delta+}{M}-\overset{\delta-}{X}$ bonds): Ti^{IV} [100, 133, 139, 156], V^{III} [157], V^{IV} [14, 158], Nb^{V} [159, 163], Ta^{V} [108, 163], Mo^{V} [149, 156], W^{IV} [71, 72], W^{V} [72], and W^{VI} [72, 212, 213], e.g., Eq. (4). In general, the degree of solvolysis decreases in the sequence

$$[TiCl_4] + 8HNMe_2 \longrightarrow [Ti(NMe_2)_4] + 4[Me_2NH_2]Cl \quad (4)$$

$NH_3 > RNH_2 > R_2NH$ [149]. Secondary, rather than primary, amines are more likely to give amides with the central metal atom in a lower oxidation state [16, 71, 72, 149, 156].

The main limitation in the method is that solvolysis rarely goes to completion, and exceptions usually involve primary amines [34, 178]. A further drawback is the possible formation of donor complexes, e.g., $[MoCl_3(NHMe_2)(NMe_2)_2]$ [149, 156].

3. Transamination

Transamination reactions are limited in their application by steric factors and amine volatility. The more volatile amine is usually displaced. The influence of steric factors on such a reaction is illustrated in Eq. (5) [65].

$$[Ti(NMe_2)_4] \begin{cases} \xrightarrow{\text{excess } HNPr^i_2} [Ti(NMe_2)_3NPr^i_2] \\ \xrightarrow{3HNPr_2} [TiNMe_2(NPr_2)_3] \end{cases} \quad (5)$$

In rare cases reduction may also take place, as in Eq. (6) [66]. Where

$$[Nb(NMe_2)_5] \begin{cases} \xrightarrow{\text{excess } HNEt_2} [Nb(NEt_2)_3NMe_2] \\ \xrightarrow{2HNEt_2} [Nb(NEt_2)_2(NMe_2)_3] \end{cases} \quad (6)$$

applicable, transamination proceeds smoothly and in high yield; if the eliminated amine is $HNMe_2$, which is a vapour under ambient conditions, the procedure is particularly attractive as the complex is the sole non-gaseous product.

A special case of transamination yields $[Mo_2Cl_2(NMe_2)_4]$ from $[Mo_2(NMe_2)_6]$ and $2Me_3SiCl$ [3].

C AMIDES OF THE TITANIUM GROUP

1. Titanium

Group 4a metal amides are more numerous than those of any other transition metal group. Homoleptic and heteroleptic titanium amides have been studied in greatest detail. The reader is referred to the review by Bürger and Neese for an account of titanium amides up to 1969 [89].

Amides of titanium are known for all four oxidation states of that element, but Ti^{I} and Ti^{II} complexes are rare and ill-characterised [11, 197]. These low valent compounds are obtained by disproportionation, which occurs above the boiling point of the corresponding Ti^{III} or Ti^{IV} compound. Thus, $[Ti(NMe_2)_4]$ (b.p. 50°/0.04 mmHg) was displaced at 60°/0.05 mmHg upon heating $[\{Ti(NEt_2)(NMe_2)_2\}_2]$ or $[\{Ti(NMe_2)_3\}_2]$, leaving a residue of $[\{Ti(NEt_2)_2\}_m]$ or $[\{Ti(NMe_2)_2\}_m]$, respectively.

Titanium halides show similar properties at higher temperatures, e.g., Eqs. (7), (8).

$$2TiCl_3 \xrightarrow{440° C} [TiCl_4] + TiCl_2 \quad (7)$$

$$2TiCl_2 \xrightarrow{600°/0.02 \text{ mmHg}} [TiCl_4] + Ti \quad (8)$$

The only Ti^{I} compound which has been reported is $TiNMe_2$ [197]. The amides of Ti^{I} and Ti^{II} are apparently polymeric, probably with bridging NR_2 groups.

Many homoleptic titanium(III) amides have been synthesised by the transmetallation method. Compounds of type $Ti(NR_2)_3$ are known for $NR_2 = NMe_2$ [11, 205], NEt_2 [205], NPh_2 [230], $N(SiMe_2)_2$ [42], and N◁ [221]. Heteroleptic amides $Ti(NR_2)_x(NR'_2)_{3-x}$ (synthesised by transamination) and $TiL_x(NR_2)_{3-x}$ (where $L = \eta\text{-}C_5H_5$ or alkyl and R = alkyl) are also well known. All except the most bulky, e.g., the monomeric $[Ti\{N(SiMe_3)_2\}_3]$, are dimeric in the condensed phase and essentially dimeric in solution. This has been shown in some cases by cryoscopy, but more widely by n.m.r. spectroscopy. For instance, the ^{1}H n.m.r. spectrum of $[\{Ti(NMe_2)_3\}_2]$ shows a sharp singlet at room temperature, but at $-80°$ C two distinct peaks 2.2 Hz apart are seen in approximately 2:1 ratio. Thus, the structure is probably that shown in (IV) rather than $[(Me_2N)_3Ti-Ti(NMe_2)_3]$. Similarly, the complexes $[\{Ti(\eta\text{-}C_5H_5)_n(NMe_2)_{3-n}\}_2]$ ($n = 1$ or 2) have a pair of bridging NMe_2 groups, with cyclopentadienyl ligands occupying terminal sites [11, 205, 206].

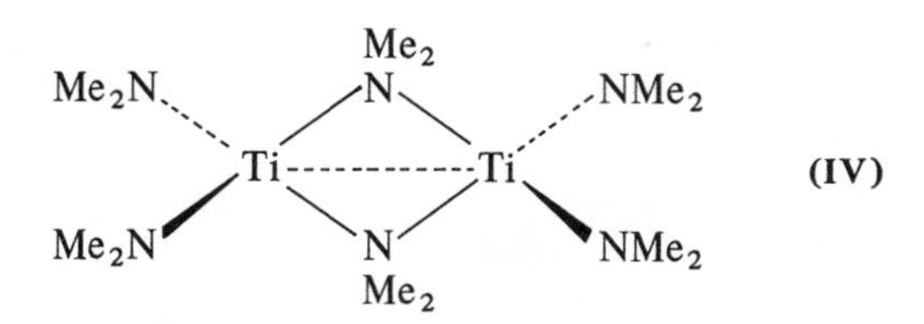

(IV)

Further evidence for the dimeric nature of the liquid Ti^{III} complexes comes from their e.s.r. spectra; the pure complexes are virtually diamagnetic but solutions give broad e.s.r. signals. The e.s.r. spectrum of $[\{Ti(NMe_2)_3\}_2]$ shows only one broad band at room temperature or at $-196°$ C. This supports structure (IV), but not the dimer without amide bridging. A small concentration of the triplet form of (IV), with each unpaired electron associated with an approximately tetrahedral titanium atom is consistent with the spectrum, the absence of fine structure being due to the triplet-broadening. In heteroleptic titanium(III) amides the loss of symmetry is reflected by the presence of more than one e.s.r. signal.

The monomeric $[Ti\{N(SiMe_3)_2\}_3]$ [2, 42, 172] is prevented from dimerising by steric hindrance. The e.s.r. spectrum shows a signal at room temperature indicative of a $^2A_1'$ ground state and frozen solutions (135K) exhibited g anistropy confirming the axial symmetry of the trigonal structure. The e.s.r. spectral patterns of both $[Ti(NPh_2)_3]$ [120, 230] and $[Ti(NPr^i_2)_3]$ [42, 120] have also been reported, but the pure compounds are not well characterised [120]. $[TiCl_4]$ and $(Me_3Si)_2NH$ were said (however, see ref. [96]) to yield the imide $HN{=}TiCl_2$ with elimination of Me_3SiCl [13]. Pyrolysis of $TiCl_4.NH_3$ gives titanium nitride, possibly via amide intermediates [245, 246].

Amido derivatives of Ti^{IV} are the most numerous of the titanium amides. The main synthetic route is transmetallation, especially for homoleptic amides. Homoleptic Ti^{IV} amides to have been synthesised via transmetallation are: $NR_2 = NMe_2$, NEt_2, N, NPr_2, NPr^i_2, NBu_2, NBu^i_2, NHBu, NMeBu, N, N, NHPh, NPh_2, NHOct, and $NHC_{18}H_{37}$.
Me

An interesting claim for a transamination reaction of a Ti^{IV} amide with a primary amine is the formation of an imido-bridged complex [Eqs. (9), (10)] [68].

$$[\mathrm{Ti(NEt_2)_4}] \xrightarrow{\mathrm{H_2NBu}} [(\mathrm{Et_2N})_2\mathrm{Ti}\left(\begin{matrix} & \mathrm{Bu} & \\ & \mathrm{N} & \\ \diagup & & \diagdown \\ & & \mathrm{Ti} \\ \diagdown & & \diagup \\ & \mathrm{N} & \\ & \mathrm{Bu} & \end{matrix}\right)_n \left(\begin{matrix} & \mathrm{Bu} & \\ & \mathrm{N} & \\ \diagup & & \diagdown \\ & & \mathrm{Ti(NEt_2)_2} \\ \diagdown & & \diagup \\ & \mathrm{N} & \\ & \mathrm{Bu} & \end{matrix}\right)] \quad (9)$$

$$(n = 4, 12, \text{ or } 14)$$

$$[\mathrm{Ti(NMe_2)_4}] \xrightarrow{\mathrm{H_2NBu^t}} [\mathrm{Ti_2(NBu^t)_2(NHBu^t)_2(NMe_2)_2}] + [\{\mathrm{Ti(NBu^t)(NMe_2)_2}\}_2] \quad (10)$$

The course of reaction (10) may be via (V) followed by intermolecular elimination of butylamine, or via (VI) followed by intermolecular elimination of dimethylamine. There is support for a structure such as that proposed in Eq. (9), from the X-ray data for a related compound $[\{\mathrm{TiCl_2(NSiMe_3)}\}_n]$, (VIII), which

$[\mathrm{Ti(NHBu^t)_3NMe_2}]$ **(V)** $[\mathrm{Ti(NHBu^t)_2(NMe_2)_2}]$ **(VI)**

has (a) rigid planar $(\mathrm{Ti\text{-}N})_2$ rings (see Fig. 1) and (b) trigonal bipyramidal Ti and planar nitrogen geometry [4]. A five-membered ring is found in compound (XI) (see Fig. 2 and Table 2) [99].

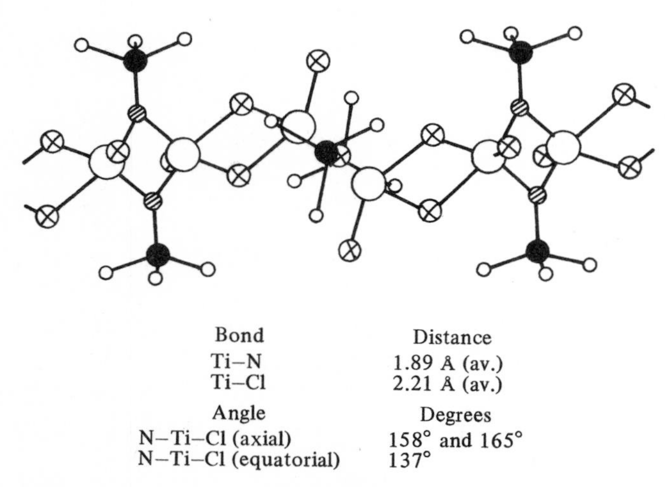

Bond	Distance
Ti–N	1.89 Å (av.)
Ti–Cl	2.21 Å (av.)
Angle	**Degrees**
N–Ti–Cl (axial)	158° and 165°
N–Ti–Cl (equatorial)	137°

Fig. 1 – X-ray crystal structure of $[\{\mathrm{TiCl_2(NSiMe_3)}\}_n]$, (VIII). The circles are in decreasing order of size Ti, Cl, Si, N, and C. *[Reproduced, with permission, from Alcock, N. W., Pierce-Butler, M., and Willey, G. R.*, J.C.S. Chem. Comm., *1974, 627.]*

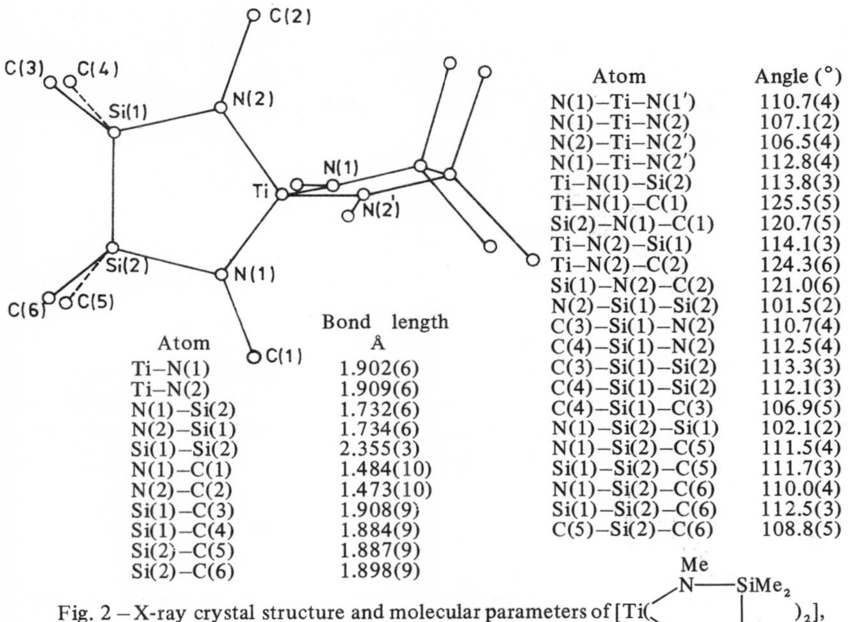

Atom	Bond length Å
Ti–N(1)	1.902(6)
Ti–N(2)	1.909(6)
N(1)–Si(2)	1.732(6)
N(2)–Si(1)	1.734(6)
Si(1)–Si(2)	2.355(3)
N(1)–C(1)	1.484(10)
N(2)–C(2)	1.473(10)
Si(1)–C(3)	1.908(9)
Si(1)–C(4)	1.884(9)
Si(2)–C(5)	1.887(9)
Si(2)–C(6)	1.898(9)

Atom	Angle (°)
N(1)–Ti–N(1')	110.7(4)
N(1)–Ti–N(2)	107.1(2)
N(2)–Ti–N(2')	106.5(4)
N(1)–Ti–N(2')	112.8(4)
Ti–N(1)–Si(2)	113.8(3)
Ti–N(1)–C(1)	125.5(5)
Si(2)–N(1)–C(1)	120.7(5)
Ti–N(2)–Si(1)	114.1(3)
Ti–N(2)–C(2)	124.3(6)
Si(1)–N(2)–C(2)	121.0(6)
N(2)–Si(1)–Si(2)	101.5(2)
C(3)–Si(1)–N(2)	110.7(4)
C(4)–Si(1)–N(2)	112.5(4)
C(3)–Si(1)–Si(2)	113.3(3)
C(4)–Si(1)–Si(2)	112.1(3)
C(4)–Si(1)–C(3)	106.9(5)
N(1)–Si(2)–Si(1)	102.1(2)
N(1)–Si(2)–C(5)	111.5(4)
Si(1)–Si(2)–C(5)	111.7(3)
N(1)–Si(2)–C(6)	110.0(4)
Si(1)–Si(2)–C(6)	112.5(3)
C(5)–Si(2)–C(6)	108.8(5)

Fig. 2 – X-ray crystal structure and molecular parameters of [Ti(N(Me)–$SiMe_2$–$SiMe_2$–N(Me))$_2$], (XI). *[Reproduced, with permission, from Bürger, H., Wiegel, K., Thewalt, U., and Schomberg, D. J.,* J. Organometallic Chem., *1975,* 8, *301.]*

Table 2 Ti–N bond lengths in some titanium amides

Compounds	Ti–N (Å)	References
[Ti{N($SiMe_3$)$_2$}$_3$], (VII)	1.929	[172]
[{$TiCl_2$($NSiMe_3$)}$_n$], (VIII)	1.89	[4]
[$TiCl_3$(NEt_2)], (IX)	1.852	[154a]
[{TiF_2(NMe_2)$_2$}$_4$], (X)	2.14 bridging	[247d]
	2.19 bridging	
	1.99 terminal	
[Ti(N(Me)–$SiMe_2$–$SiMe_2$–N(Me))$_2$], (XI)	1.902-1.909	[99]
[Fe{η-C_5H_4Ti(NEt_2)$_3$}$_2$],	1.89	[254a]
[Ti(CH_2Ph)$_2${N($SiMe_2NMe$)$_2SiMe_2$}$_2$]	1.92	[70a]

Longer Ti-N bonds are observed in $[\{TiF_2(NMe_2)_2\}_4]$, (X), [247d] (see Fig. 3); the titanium atoms achieve six-co-ordination through both Ti–F–Ti and Ti-NMe_2-Ti bridges. The Ti-N bond length is 1.99(3) Å, even though the sum of the bond angles at terminal N's is 360° which indicates that the nitrogen lone pair displays *p*-character and is available for $(p{\rightarrow}d)_\pi$ bonding.

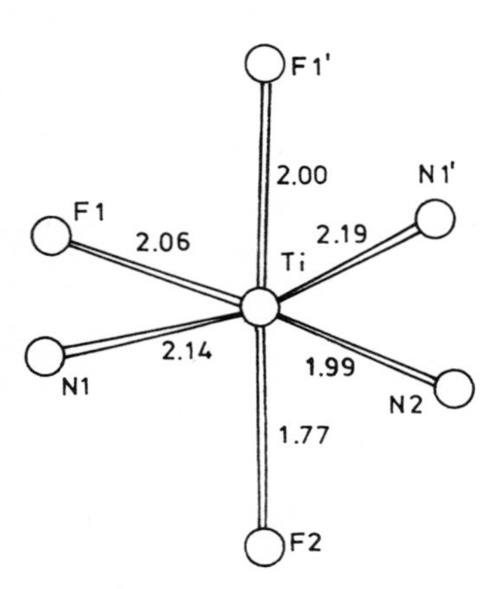

Interatomic distances (Å) and bond angles (degrees)

Ti–F(1)	2.06(2)	Ti–F(1′)	2.00(2)
Ti–F(2)	1.77(2)	Ti–N(2)	1.99(3)
Ti–N(1)	2.14(3)	Ti–N(1′)	2.19(3)
N(1)–C(11)	1.55(3)	N(1)–C(12)	1.45(4)
N(2)–C(21)	1.45(4)	N(2)–C(22)	1.51(4)
F(1)–Ti–N(1)	75.7(8)	F(1)–Ti–F(1′)	93.1(5)
F(1)–Ti–N(1′)	84.2(5)	F(1)–Ti–F(2)	85.5(8)
F(1′)–Ti–N(1′)	75.7(8)	F(1′)–Ti–N(2)	89.8(8)
F(1′)–Ti–N(1)	83.7(7)	N(1)–Ti–N(2)	98.7(10)
N(1)–Ti–F(2)	102.0(10)	N(1′)–Ti–N(2)	102.3(7)
N(1′)–Ti–F(2)	98.0(8)	F(2)–Ti–N(2)	92.2(9)
Ti′–F(1)–Ti	107.7(8)	Ti′–N(1)–Ti	98.1(11)
C(11)–N(1)–Ti	108(2)	C(12)–N(1)–Ti	120(2)
C(11)–N(1)–Ti′	107(2)	C(12)–N(1)–Ti′	116(2)
C(11)–N(1)–C(12)	106(3)	C(21)–N(2)–Ti	120(2)
C(22)–N(2)–Ti	120(2)	C(21)–N(2)–C(22)	120(3)

Fig. 3 – X-ray crystal structure and molecular parameters of $[\{TiF_2NMe_2\}_4]$, (X). *[Reproduced, with permission, from Sheldrick, W. S.,* J. Fluorine Chem., *1974,* **4**, *415.]*

In general, Ti^{IV} amides are stable compounds and show no tendency to polymerise. Insertion reactions are prominent (Chapter 10), e.g., ref. [132], as are catalytic polymerisations of activated olefins such as acrylonitrile (Chapter 17), e.g., ref. [180].

Heteroleptic Ti^{IV} amides are more numerous than the homoleptic analogues. A large variety of compounds has been obtained; most of these are monomeric, but the halogeno derivatives in many cases are oligomers, e.g., $[\{TiF_2(NMe_2)_2\}_4]$ [100, 247d], with bridging fluorine atoms (Fig. 3).

X-ray structural studies have been carried out on seven titanium amides and six of these have been reported in detail [70a, 99, 154a, 172, 247d, 254a] (Table 2).

A Ti-N σ-bond length has been estimated to be 1.981 Å [150b]. It can be seen from Table 2 that in all cases, except $[\{TiF_2(NMe_2)_2\}_4]$, the Ti-N bond length is less than this value. Furthermore, the nitrogen atoms are found to be in a planar or nearly planar environment in each complex. This has led to a proposal of $Ti \overset{\frown}{-} \ddot{N}$ $d_\pi \leftarrow p_\pi$ interactions in titanium amides.

2. Zirconium and hafnium

Homo- and hetero-leptic amides of Zr^{IV} and Hf^{IV} are prepared by similar methods to those used for Ti^{IV}. However, amides of the lower oxidation states of Zr or Hf have not been reported. It is a common feature of the chemistry of the Group 4a metals that lower oxidation states, especially M^{III}, are much more readily accessible for Ti than the heavier metals.

The tetrakis(dialkylamides) of Zr^{IV} or Hf^{IV} may not contain the four-co-ordinate metal in the solid state, and the possibility of some polymerisation of $[Zr(NMe_2)_4]$ in the liquid state has been proposed on the basis of i.r. and n.m.r. data [51]. Nevertheless, its boiling point and mass spectrum favour a monomer, at any rate for the gaseous molecule. Homoleptic bis(trimethylsilyl)-amides have not been synthesised, but as with titanium, the spirocyclic derivatives, e.g., $[Zr\{(N(Me)SiMe_2)_2O\}_2]$ [82, 102] are obtained, from $ZrCl_4$ and the dilithium reagent $[LiN(Me)SiMe_2]_2O$.

D AMIDES OF THE VANADIUM GROUP

1. Vanadium

Homoleptic amides of vanadium are known for the oxidation states two, three, and four, but not V^V, for which only heteroleptic complexes are known. Compounds such as $[\{V(NR_2)_2\}_n]$ (R = Me or Et) [11, 168] have been synthesised, as in the case of titanium, by disproportionation. These compounds are probably polymeric and not well-characterised.

The dimethyl- and diethyl-amides [11] of V^{III} appear to be dimeric in solution. They are synthesised by salt elimination from the reaction of VCl_3 with the appropriate lithium dialkylamide. They are unstable and readily disproportionate to give the more stable quadrivalent amides $[V(NR_2)_4]$ [11, 51, 168]. The compound $[V\{N(SiMe_3)_2\}_3]$ [8, 42, 172] has been prepared from $[VCl_3(NEt_3)_2]$ and three equivalents of $LiN(SiMe_3)_2$. Magnetic susceptibility data and results from i.r. or electronic spectra [8, 9, 36, 42, 44, 203], and X-ray diffraction [172] are consistent with a trigonal molecule of skeletal D_{3h} symmetry. The p.e. spectrum [203] is discussed in Section H.

Amides of V^{IV} are the most numerous and well-studied. Transmetallation combined with disproportionation [11] or amine hydrochloride elimination [e.g., 141, 157, 158] are employed in their synthesis. The latter technique is, as for Ti^{IV}, useful for partial substitution, but donor complexes are frequently formed if excess base is present [e.g., 141, 157, 158].

The e.s.r. spectrum of $[V(NEt_2)_4]$ has been studied in several different solvents over the temperature range +20° to −150° C [175]. At 20° C, the pure compound exhibits an asymmetric signal, $g = 1.97$ due to hyperfine interaction, while a dilute sample gave an 8-line spectrum due to ^{51}V ($I = 7/2$) hyperfine coupling; $g_{\parallel}$ and $g_{\perp}$ values were evaluated from the spectra at 20° and −150° C.

An i.r. and e.s.r. study [7, 51, 62] of $[V(NMe_2)_4]$ and $[V(NEt_2)_4]$ implies a splitting of the ground state ^2E term due to distortion of the expected T_d structure to D_{2d} by $\overset{\frown}{N{-}V}$ bonding. This is confirmed by the electronic spectra, where complex *d–d* transitions indicate splitting of ^2E and $^2T_2$ levels [165].

Reactions between $[VCl_{3-n}(OPr^i)_n]$ ($n = 1$ or 2) and $NaN(SiMe_3)_2$ are complex [94]: simple salt elimination to give Cl/NR_2 exchange results, but distillation causes decomposition to give polymeric products. The 1:1 reaction between $[VOCl_3]$ and $NaN(SiMe_3)_2$ gives the thermally unstable compound $[V(O)Cl_2N(SiMe_3)_2]$ [94]; the imido derivative $[V(NSiMe_3)\{N(SiMe_3)_2\}_2(OSiMe_3)]$ is obtained from the reaction between three equivalents of $NaN(SiMe_3)_2$ and $[VOCl_3]$ [94, 258]. Compounds of type $[VO(NR_2)_3]$ (R = Me or Et) are obtained [196, 197] via the reaction of $[VOCl_3]$ with the appropriate lithium dialkylamide. Several heteroleptic derivatives of type $[VCl_n(NR_2)_{4-n}]$ have been prepared by amine hydrochloride elimination [e.g., 41, 157, 158].

2. Niobium and tantalum

Amides are known for Nb^{IV}, Nb^{V}, and Ta^{V}, although one tetrakisamide of tantalum has been reported [67]. The dearth of Ta^{IV} derivatives may be due to the relative instability of the +4 oxidation state for Ta. Thus, sterically demanding Ta^{V} amides do not disproportionate, whereas Nb^{V} analogues generally yield Nb^{IV} complexes. For example, $[Ta(NMe_2)_5]$ is readily prepared, but attempts to obtain homologues or Nb analogues afford the Ta^{V} imides $[Ta(NR)(NR_2)_3]$ (R = Et, Pr, or Bu) [51] or Nb^{IV} amides $[Nb(NR_2)_4]$ [38, 51, 66], respectively.

X-ray crystallographic studies have been carried out on $[Nb(NMe_2)_5]$ [173] and [Nb(N⬡)$_5$] [41, 166a, 173]; they have very similar structures (Fig. 4). The basic shapes are distorted square pyramidal, tending towards trigonal bipyramidal. The basal and axial bond lengths in both compounds differ by about 0.07 Å. It has been suggested that there is some $d \leftarrow p$ π-interaction in the M–N axial bond, because a 'normal' Nb–N bond length has been estimated as 2.04–2.08 Å [173]. The X-ray analysis of [Ta(NMe$_2$)$_2$(O$_2$CNMe$_2$)$_3$] reveals seven co-ordinate Ta in a pentagonal bipyramid [121c], whilst that of [Ta(NMe$_2$)$_3$(NBu^t)] has a three-fold axis [230a]. X-ray data are available on [NbCl$_3${N(Me)C(S)Me}$_2$] [140].

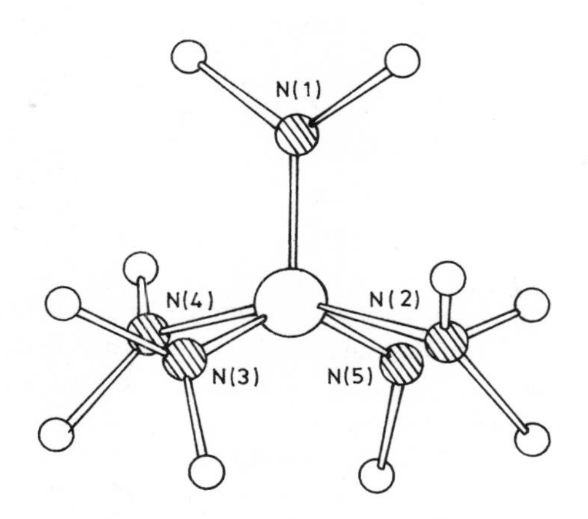

Molecular parameters for the NbN$_5$ *units*

Bond lengths (Å)

	Nb(NMe$_2$)$_5$	Nb(pip)$_5$[1]	Nb(pip)$_5$[2]
Nb–N(1) (axial)	1.977(17)[a]	1.991(12)	1.981(13)
Nb–N(2) (basal)	2.044(14)	2.056(13)	2.043(15)
Nb–N(3) ,,	2.040(15)	2.051(14)	2.032(13)
Nb–N(4) ,,	2.044(14)	2.046(14)	2.048(16)
Nb–N(5) ,,	2.040(15)	2.055(14)	2.046(14)
Mean M–N (basal)	2.042	2.052	2.042

Bond angles (°)

	Nb(NMe$_2$)$_5$	Nb(pip)$_5$[1]	Nb(pip)$_5$[2]
N(1)–Nb–N(2)	101.5(4)	100.5(5)	97.3(6)
N(1)–Nb–N(3)	109.1(4)	106.6(5)	109.0(5)
N(1)–Nb–N(4)	101.5(4)	100.2(5)	99.5(6)
N(1)–Nb–N(5)	109.1(4)	110.1(5)	112.0(5)
N(2)–Nb–N(3)	87.3(6)	87.9(6)	84.4(6)
N(2)–Nb–N(5)	86.1(6)	85.4(6)	88.8(6)
N(3)–Nb–N(4)	86.1(6)	86.0(6)	89.9(6)
N(4)–Nb–N(5)	87.3(6)	87.7(6)	85.0(6)

[a]E.s.d.'s in parentheses. [1][41]. [2][166a].

Fig. 4 – X-ray crystallographic data for [Nb(NR$_2$)$_5$][NR$_2$ = NMe$_2$ or N(CH$_2$)$_5$]. *[Reproduced, with permission, from Heath, C., and Hursthouse, M. B.,* Chem. Comm., *1971, 145.]*

E AMIDES OF THE CHROMIUM GROUP

1. Chromium

Amides are known for Cr^{II}, Cr^{III}, and Cr^{IV}, but many of the two former species are not well-characterised. Exceptions are the compounds $[Cr\{N(SiMe_3)_2\}_3(NO)]$ and $[Cr\{N(SiMe_3)_2\}_2(T.H.F.)_2]$ for which X-ray structures have been reported [56] (see Figs. 6 and 7). The Cr^{III} amides are synthesised by transmetallation [Eq. (11)]. They appear, in most cases, to be dimeric in solution with bridging dialkylamido groups. The Cr^{III} compounds readily disproportionate, Eq. (12), affording, except where the amide is bulky, the Cr^{IV} complex.

$$CrCl_3 + 3LiNR_2 \xrightarrow{T.H.F.} \frac{1}{m}[\{Cr(NR_2)_3\}_m] + 3LiCl \qquad (11)$$

$$\frac{2}{m}[\{Cr(NR_2)_3\}_m] \xrightarrow[3h/10^{-3}\ mmHg]{40-60°\ C} [Cr(NR_2)_4] + \frac{1}{n}[\{Cr(NR_2)_2\}_n] \qquad (12)$$

$$(NR_2 = NEt_2, NPr_2, NMeBu, \text{ or } N\langle\text{piperidino}\rangle)$$

Curiously $[Cr(NMe_2)_4]$ has not been made by this method, perhaps because in this instance the Cr^{III} intermediate does not disproportionate. On the other hand, compounds such as $[Cr(NPr^i_2)_3]$ [6] or $[Cr\{N(SiMe_3)_2\}_3]$ [59, 97] are stable and are rare examples of three-co-ordinate Cr^{III} complexes.

The structure of $[Cr(NPr^i_2)_3]$ is represented in Fig. 5 and confirms the three-co-ordinate monomeric nature of the compound [55]. The $CrNC_2$ system is planar and the dihedral angles between the CrN_3 and the three amide planes are very similar at 68°, 72°, and 73°. The Cr–N bonds, at ca. 1.87 Å, are somewhat shorter (ca. 0.1 Å) than might have been expected. This, together with the planarity at each N, suggests ligand to metal π-bonding.

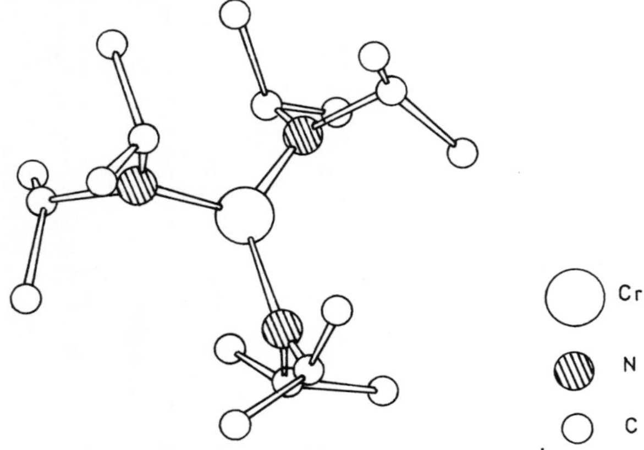

Fig. 5 – The crystal and molecular structure of $[Cr(NPr^i_2)_3]$. *[Reproduced, with permission, from Bradley, D. C., Hursthouse, M. B., and Newing, C. W.,* Chem. Comm., *1971, 411.]*

Although sterically hindered, the trisamides undergo reaction with small rod-like molecules, specifically O_2 or NO [63, 121], to afford the adducts [Cr{N(SiMe$_3$)$_2$}$_3$(NO)] or [Cr(NPr$^i_2$)$_3$(O$_2$)] (see Fig. 6). Extensive spectroscopic [44, 63, 121] and X-ray crystallographic data [56] support both M—N amide π-bonding as well as the π-acceptor nature of NO (see Chapter 16). The molecule has a skeletal tetrahedral structure. The shortness of the Cr–NO bond coupled with the increase in N–O bond length upon co-ordination, suggest M—NO π-bonding. The $CrNSi_2$ group is planar and the Cr–N(SiMe$_3$)$_2$ bond length (1.79 Å) [56, 172] is shorter than that found in [Cr{N(SiMe$_3$)$_2$}$_3$], suggesting increased Cr—N(SiMe$_3$)$_2$ π-bonding. The chromium may be regarded as being in the +2 oxidation state, considering the nitrosyl as NO$^+$.

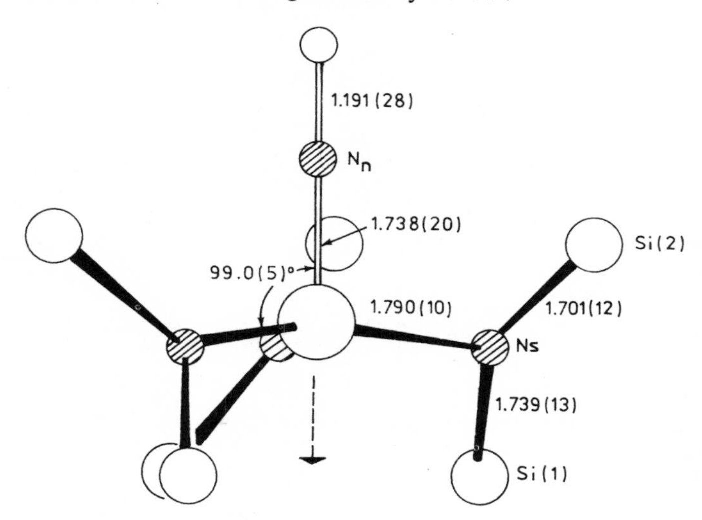

Fig. 6 – Crystal and molecular structure of [Cr{N(SiMe$_3$)$_2$}$_3$(NO)]. *[Reproduced, with permission, from Bradley, D. C., Hursthouse, M. B., Newing, C. W., and Welch, A. J.,* J.C.S. Chem. Comm., *1972, 567.]*

The molecule [Cr{N(SiMe$_3$)$_2$}$_2$(T.H.F.)$_2$] is interesting as a unique example of a square planar Cr^{II} d^4 complex, Fig. 7. Each NSi_2 group is planar and is inclined at a dihedral angle of 73° with respect to the CrN_2O_2 square plane. Short N–Si bonds suggest π-bonding, but at present there are no comparative data.

The Cr^{IV} amides are paramagnetic, containing two unpaired d-electrons. Electronic spectra [23] suggest distortion from tetrahedral to D_{2d} symmetry and the possibility of Cr—N π-bonding. This distortion is further discussed in Section E.2; it is shown that for D_{2d} symmetry, the metal orbitals, except $d_{x^2-y^2}$ (b_1), may be involved in covalent bonding. The greatest stabilisation occurs in the diamagnetic $[Mo(NMe_2)_4]$. It is presumed that in $[Cr(NEt_2)_4]$ the pairing energy is greater than the energy difference between the $d_{x^2-y^2}$ and the d_{z^2} orbitals, thus allowing two unpaired electrons in the triplet ground state.

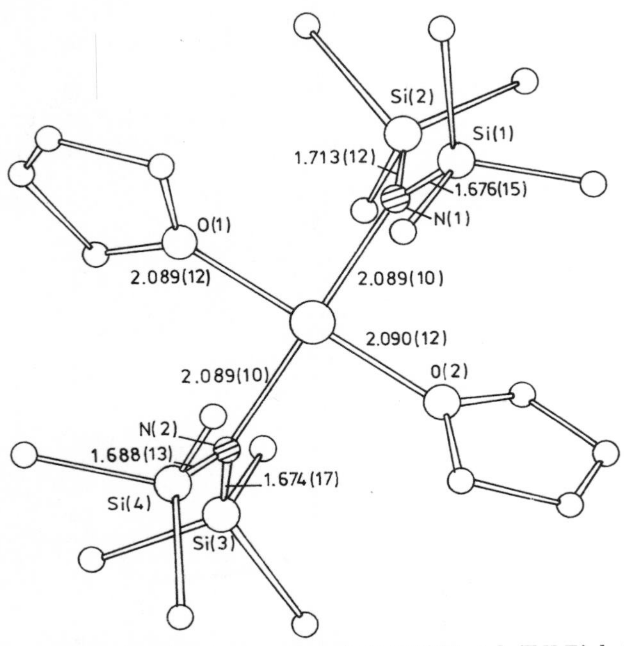

Fig. 7 – Crystal and molecular structure of $[Cr\{N(SiMe_3)_2\}_2(T.H.F.)_2]$. *[Reproduced, with permission, from Bradley, D. C., Hursthouse, M. B., Newing, C. W., and Welch, A. J.,* J.C.S. Chem. Comm., *1972, 567.]*

2. Molybdenum and tungsten

A wide variety of stable oxidation states (i.e., Mo^{III}, Mo^{IV}, Mo^{V}, W^{III}, W^{IV}, W^{V}, W^{VI}) is known for Mo and W amides, with the greatest number of compounds in the +3 oxidation state, e.g., $[M_2(NR_2)_6]$ (M = Mo or W, R = Me or Et).

The Mo^{III} compounds are synthesised by reaction of the metal(III) chloride with three equivalents of the lithium amide, or from the metal(V) chloride with five equivalents of lithium amide [124, 130]. The amide $[Mo_2(NR_2)_6]$ may be sublimed and is thermally stable. Cryoscopic and n.m.r. data, and the Raman spectrum indicate that the structure is dimeric with a Mo≡Mo triple bond. Metal-metal bonding is confirmed by the X-ray crystal structure of $[Mo_2(NMe_2)_6]$ (Fig. 8) [123]. The salient features of the structure are (a) the short Mo–Mo distance, (b) the absence of any bridging amido groups, (c) the staggered conformation, and (d) the planar $MoNC_2$ system having somewhat shorter (ca. 0.03–0.04 Å) Mo–N bonds than expected. Reviews on metal–metal bonding in these and related compounds are in refs. [121a, 131a].

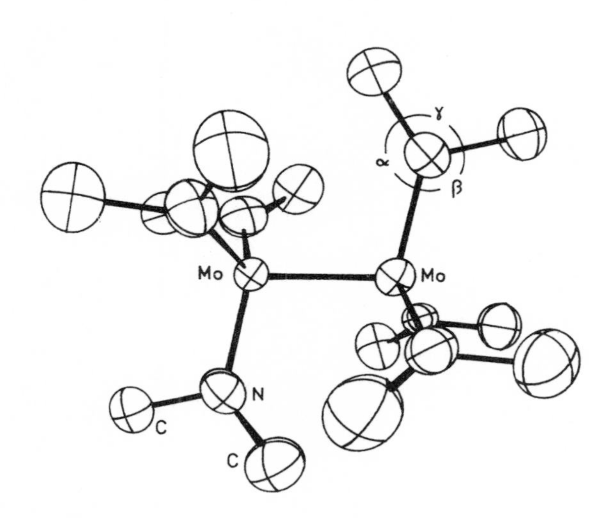

Fig. 8 – The molecular structure of $[Mo_2(NMe_2)_6]$. The important dimensions are: Mo–Mo, 2.214(2); Mo–N, 1.98(1); N–C (both sets together), 1.48(2) Å; <Mo–Mo–N, 103.7(3)°; α, 133(1)°; β, 116(1)°; γ, 110(1)°. *[Reproduced, with permission, from Chisholm, M. H., Cotton, F. A., Frenz, B. A., and Shive, L.,* J.C.S. Chem. Comm., *1974, 480.]*

Variable temperature n.m.r. studies have been performed: the ^1H n.m.r. spectrum shows a sharp singlet in the methyl region at 20° C, which below −20° C separates into a sharp doublet with a substantial (ca. 2 p.p.m.) chemical shift difference. This has been attributed to distinct signals from proximal and distal methyl protons (the former being directed toward, and the latter being directed away from, the Mo≡Mo bond) with the large shift difference arising because of the magnetic anisotropy of the Mo≡Mo triple bond. The molecule has S_6 symmetry; in an attempt to resolve the S_6 and D_3 conformers, solutions were further cooled to −90° C, but there was no further spectral change [130].

The corresponding tungsten compound $[W_2(NMe_2)_6]$ [127] has a similar structure and chemical properties to the molybdenum analogue. It was prepared either by reaction of four equivalents of $LiNMe_2$ with $[WCl_4(OEt_2)_2]$ to afford a mixture of $[W_2(NMe_2)_6]$ and $[W(NMe_2)_6]$, or alternatively from WCl_6 and five equivalents of $LiNMe_2$. These preparative methods are discussed fully in ref. [127]. The molecular parameters for $[M_2(NMe_2)_6]$ (M = M or W) are shown in Table 3.

Table 3 Comparison table of selected bond lengths and angles in group 6 metal complexes $[M_2(NMe_2)_6]$

Distance (Å) or angle (°)	$[Mo_2(NMe_2)_6]$[1]	$[W_2(NMe_2)_6]$[2]	$[W_2(NMe_2)_6]$[1]
M–M	2.214(2)	2.294	2.292
M–N	1.97	1.982	1.97
N–C	1.48	1.466	1.48
M–$\hat{M}$–N	103.7	103.8	103.3
N–$\hat{M}$–N	114.6	114.5	114.8
M–$\hat{N}$–C (proximal)	133	132.6	134
M–$\hat{N}$–C (distal)	116	116.5	116

[1] Values for pure crystalline molecules.
[2] Values for $[W_2(NMe_2)_6]$ in the co-crystallised molecule $[2W_2(NMe_2)_6.W(NMe_2)_6]$.

In addition to these homoleptic derivatives, other amides have also been synthesised. For example, the mixed chloro(dialkylamido)tungsten compound $[W_2Cl_2(NEt_2)_4]$ is prepared from the reaction of WCl_4 with a deficiency of $LiNEt_2$. X-ray diffraction [125] shows that the structure has an *anti* rotamer conformation, Fig. 9. The ^1H n.m.r. spectrum varies considerably with temperature, due to proximal-distal interconversion. Additionally, the methylene groups are diastereotopic. At −20° C the methyl groups are observed as two triplets, while the methylene groups are multiplets; this is attributed to exclusive presence of the *anti* rotamer. At higher temperature, proximal-distal exchange results in signal averages for the two methyl protons. Other tungsten analogues, $[W_2X_2(NEt_2)_4]$ (X = Me [125a], Br, I, or CH_2SiMe_3 [126]) have been prepared by transmetallation or halogen exchange, and several of their structures have been elucidated, as illustrated by Fig. 9; important parameters are summarised in Table 4, which also includes data for the molybdenum analogues.

The MNC_2 units in each of the above compounds are planar and the short M-N bond distances are indicative of $M \leftarrow \ddot{N}$ π-bonding. All the compounds have the staggered conformation. Substitution of a non-π-donating ligand for an amide produces a significant decrease in M-M distance: i.e., compare *l*(M-M) in $[M_2(NMe_2)_6]$ and $[M_2R_6]$ (Table 4). Similarly, substitution of two NMe_2 groups by two halide ligands results in a small increase in M-M distance and a decrease in the remaining M-N bond lengths.

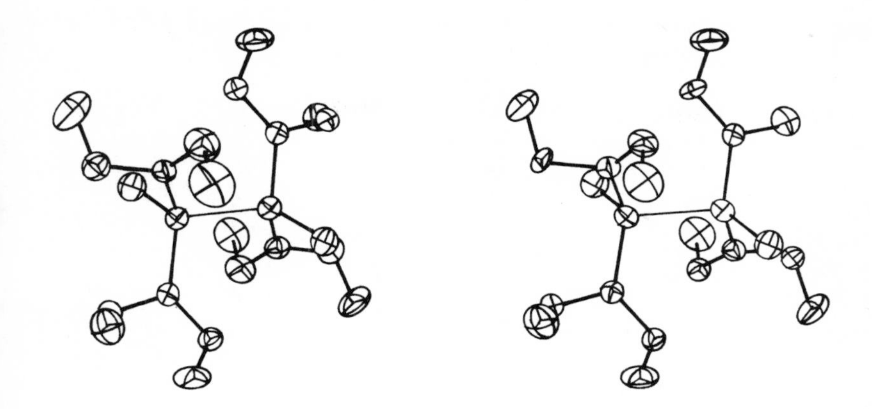

Bond Distances and Angles for $[W_2Cl_2(NEt_2)_4]$

	Distances, Å	Averages
W–W′	2.3012 (8)	
W–Cl	2.332 (8)	
W–N1	1.935 (8)	1.936
W–N2	1.937 (9)	
N1–C(1)	1.48 (1)	1.49
–C3	1.50 (1)	
N2–C5	1.50 (1)	
–C7	1.46 (1)	
C(1)–C(2)	1.54 (2)	1.53
C3–C4	1.52 (2)	
C5–C6	1.52 (2)	
C7–C8	1.53 (2)	
	Bond Angles, degrees	
W′–W–Cl	107.93 (9)	
W′–W–N1	103.5 (2)	102.8
W′–W–N2	102.1 (3)	
Cl–W–N1	115.3 (3)	114.4
Cl–W–N2	113.4 (3)	
N1–W–N2	113.0 (4)	
C(1)–N1–C3	112.8 (9)	113.3
C5–N2–C7	113.7 (9)	
N1–C(1)–C2	111.9 (10)	111.5
N1–C3–C4	111.5 (10)	
N2–C5–C6	111.5 (11)	
N2–C7–C8	110.9 (11)	
W–N1–C1	114.3 (4)	
W–N1–C3	132.1 (3)	
W–N2–C5	112.9 (4)	
W–N2–C7	133.2 (3)	

Fig. 9 – X-ray crystal and molecular structure of $[W_2Cl_2(NEt_2)_4]$. *[Reproduced, with permission, from Chisholm, M. H., Cotton, F. A., Extine, M., Millar, M., and Stults, B. R.,* J. Amer. Chem. Soc., *1976,* **98,** *4486.]*

Table 4 Some molecular parameters in $[M_2L_6]$ and $[M_2L_4X_2]$ compounds (X = halogen, L = amide)

Compound	Mo–Mo (Å)	Mo–N (Å)	W–W (Å)	W–N (Å)	References
$[M_2(NMe_2)_6]$	2.214	1.98	2.294	1.98	[123,127]
$[M_2Cl_2(NMe_2)_4]^a$	2.201	1.93	2.285	1.94	[3]
$[M_2Cl_2(NEt_2)_4]^a$			2.301	1.936	[125]
$[M_2Br_2(NEt_2)_4]^a$			2.301	1.90	[126]
$[M_2I_2(NEt_2)_4]^a$			2.300	1.92	[126]
$[M_2Me_2(NMe_2)_4]^a$	2.201	1.95			[121d]
$[M_2Me_2(NEt_2)_4]^a$			2.291	1.97	[125a]
$[M_2(CH_2SiMe_3)_6]$	2.167		2.225		[3]

^{a}These are symmetrical dimers $[\{MX(NR_2)_2\}_2]$.

Stable amides are also known for Mo^{IV}. These compounds are prepared by the addition of five equivalents of the lithium amide to a suspension of $MoCl_5$ in T.H.F.; heating the residue affords the purple $[Mo(NMe_2)_4]$ or $[Mo(NEt_2)_4]$. They are monomeric and their ^1H n.m.r. spectra over a wide temperature range (−90° to +90° C) give only a singlet. The former compound belongs to the D_{2d} space group (X-ray) with Mo-N, 1.926(6) Å; N-C, 1.466(15) Å; N-$\hat{Mo}$-N, 109.5(19)°; Mo-$\hat{N}$-C, 124(1)°; and C-$\hat{N}$-C, 110(1)° [121b]; calculations indicate a singlet ground state.

The compounds are diamagnetic, in contrast to the Cr^{IV} homologues. This is unusual for a d^2 system in a tetrahedral field, but can be rationalised [39a] by considering the symmetry of the molecule. If π-bonding is assumed and each $MoNC_2$ unit is planar, the highest symmetry is D_{2d}. In this point group, the metal orbitals transform as a_1 (s, p_z, d_{z^2}), b_1 $(d_{x^2-y^2})$, b_2 (d_{xy}), e (p_x, p_y), and $e(d_{xz}, d_{yz})$. Thus, molybdenum can form four σ-bonds of A_1, B_2, and E symmetry and four π-bonds of A_1, B_2, and E symmetry, leaving the $b_1(d_{x^2-y^2})$ non-bonding orbital containing both electrons. Greater stability results because both electrons are paired in a non-interacting b_1-orbital. In $[Cr(NEt_2)_4]$ the bonding picture is almost identical except that in this case there is an energy gap between the b_1 and the lowest antibonding a_1-orbital.

I.r. and electronic spectra of $[Mo(NR_2)_4]$ support the π-bonding hypothesis. The spectrum resembles that of $[Ti(NMe_2)_4]$ but shows significant differences: e.g., (i) the TiN_4 asymmetric stretching mode is a strong symmetric band, whereas the ν_{as} (MoN_4) has in addition a shoulder, and (ii) the NC_2 stretching vibration at 1100–1150 cm^{-1} has an extra band. Thus the i.r. spectrum shows a closer resemblance to that of $[V(NMe_2)_4]$, which is thought to have D_{2d} symmetry on the basis of i.r. and e.s.r. evidence (see Section D.1).

The compound $[W(NMe_2)_6]$ is prepared [41, 127] from $LiNMe_2$ and $[WCl_4(OEt_2)_2]$. The structure of this compound, Fig. 10 [41a], consists of an octahedral WN_6 skeleton with each WNC_2 unit co-planar and arranged so as to give three mutually perpendicular planes involving planar *trans*-C_2N-W-NC_2 units. The symmetry is thus O_h, Fig. 10 [41a]. The tungsten obtains its 18-electron valence shell via π-bonding; there are six W–N σ-bonds and three $\ddot{N}$–W π-bonds which are delocalised over the WN_6 framework giving an average W–N bond order of 1.5. The bond lengths, 2.03 Å, are somewhat short, supporting the π-bonding hypothesis. Higher homologues have not been obtained, possibly due to steric constraints. However, $[W(NMe_2)_3(O_2CNMe_2)_3]$ [129b] has been characterised by X-ray crystallography (Fig. 11). The main feature of interest in this compound is the short W–N bond length of 1.922 Å, cf., *l*(W–N) in $[W(NMe_2)_6]$, 2.032 Å. This may be due to replacement of three NMe_2 ligands by the weakly π-donating moiety from the monodentate carbamate [129b] Several heteroleptic derivatives of W^{III} are known and these were synthesised by the amine hydrohalide elimination technique. Tungsten-(IV) and -(V) derivatives have also been obtained by similar methods.

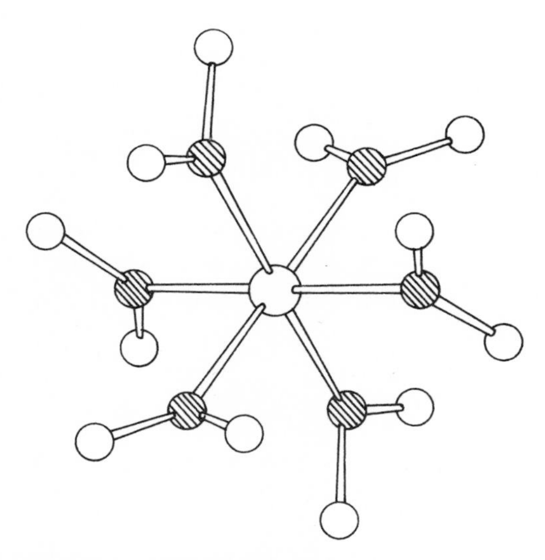

Fig. 10 – Molecular structure of $[W(NMe_2)_6]$ deduced from single-crystal X-ray studies. W–N = 2.017(6) Å, 2.032(25) Å, N–C = 1.515(15) Å, α = 104(1)°. The molecule has O_h symmetry. *[Reproduced, with permission, from Bradley, D. C., Chisholm, M. H., and Extine, M. W.,* Inorg. Chem., *1977,* **16,** *1791.]*

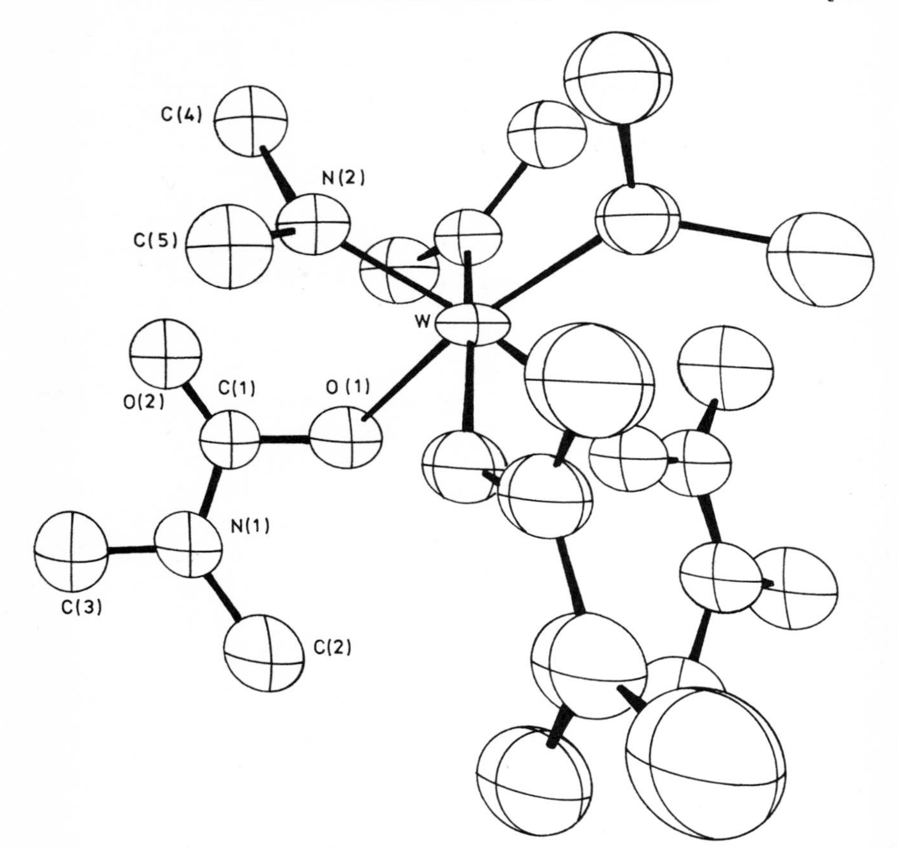

Fig. 11 – Molecular structure of $[W(NMe_2)_3(O_2CNMe_2)_3]$ showing the *fac*-WN_3O_3 moiety; the molecule has C_3 symmetry. Bond lengths: W–N(2) 1.922(7) Å and W–O(1) 2.041(6) Å; angles: N–$\widehat{W}$–N 94.8°, O–$\widehat{W}$–O 82.1°, N–$\widehat{W}$–O 170.3°, N–$\widehat{W}$–O 93.2°, and N–$\widehat{W}$–O 89.3°. *[Reproduced, with permission, from Chisholm, M. H., Extine, M. W., and Reichert, W. W.,* Adv. Chem. Ser., *1976,* **150,** *273.]*

F AMIDES OF THE LATER TRANSITION METALS

1. Introduction

Apart from a few compounds involving bis(trimethylsilyl)amido ligands, simple dialkylamides of the later transition metals are unknown. This is indicative of the reluctance of the later transition metals to form stable compounds having M-NR_2 bonds, which contrasts with the numerous metal alkyls, e.g., *cis*-$[PtMe_2(PEt_3)_2]$. Attempts at the synthesis of some homoleptic compounds using the transmetallation method have sometimes led to the isolation of di-iminato complexes [33] or to co-ordination polymers of indefinite structure.

These observations have been accounted for in terms of the π-donor properties of the NR_2^- ligand; the latter may thus act not only as a two-electron σ-donor, but also as a two-electron π-donor (Fig. 12). This π-donation to metal clearly requires suitable vacant metal *d*-orbitals which generally are unavailable for the complexes of the later transition metals.

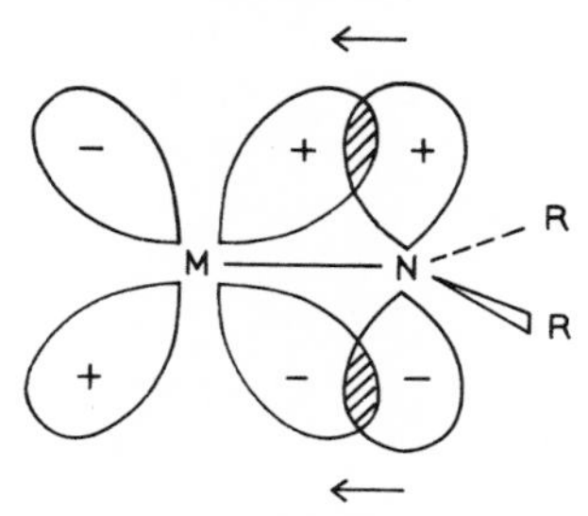

Fig. 12 – Schematic representation of π-bonding in a metal amide.

Although homoleptic alkylamides are unknown, several heteroleptic compounds have been prepared in which the amido ligand is either electronegative, e.g., NCl_2^- or $N(CN)_2^-$ (see Table 6), or co-ordination saturation is achieved, as in the Pt^{IV} compound (XII), for which there is X-ray as well as ^1H n.m.r. and ESCA evidence [246a]. Compound (XII) was obtained under mild conditions [of oxidation (possibly by air) and deprotonation] from *cis,cis*-tri(amino)cyclohexane and $[Pt^{II}(bipy)Cl_2]$.

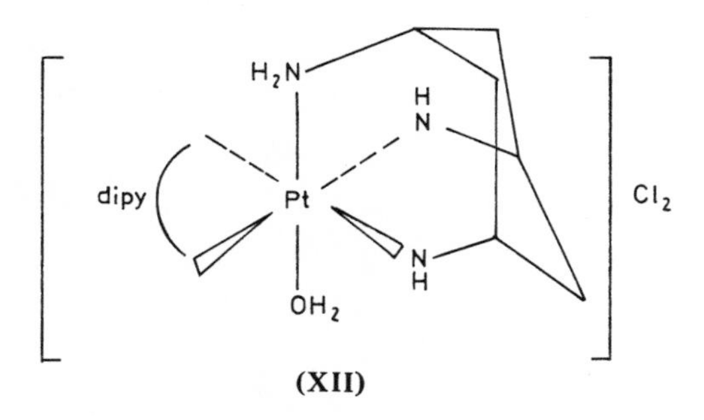

(XII)

Dichloroamides of platinum(IV), $[PtCl_2(NCl_2)_2py_2]$ and $[PtCl_2(NH_3)(NCl_2)py_2]^+$, were obtained by addition of Cl_2 to *cis*- or *trans*-$[PtCl_2(NH_3)_2py_2]^{2+}$, respectively [198a]. The X-ray structure of a related compound $[PtCl(NH_3)_3(NCl_2)_2]Cl$ has been determined [264].

The effect of electron-withdrawing substituents at N may have two consequences: loss of basicity at N and hence little tendency for oligomerisation, and low $\ddot{N}\rightarrow M$ π-donation which may alternatively allow for amide compatibility with soft co-ligands such as CO or PR_3.

2. Bis(trimethylsilyl)amides

Bis(trimethylsilyl)amides of some of the later transition metals are noteworthy examples of stable amides of such elements. Compounds incorporating this ligand are known for Cr^{III}, Mn^{II}, Fe^{III}, Co^{I}, Co^{II}, Ni^{I}, Ni^{II}, Cu^{I}, Ag^{I}, Rh^{I}, Ru^{II}, and Au^{I}. For a review, see ref. [171]. It is interesting that stable, but highly reactive, low co-ordination number disilylamides of Rh^{I} and Ru^{II} have been made, $[Rh\{N(SiMe_3)_2\}(PPh_3)_2]$ and $[RuH\{N(SiMe_3)_2\}(PPh_3)_2]$ [109]; corresponding preparative routes to similar dimethylamides of Rh^{I} or Ru^{II} from $[RhCl(PPh_3)_3]$ or $[RuCl(H)(PPh_3)_3]$ yield the hydrido-metal products of β-elimination [138a] (see Chapter 18).

Only one derivative of a group 7 transition metal has been reported, the buff, air-sensitive, volatile, crystalline bis(bistrimethylsilylamido)manganese(II), which is obtained as in Eq. (13). It forms an adduct with T.H.F. to give

$$2NaN(SiMe_3)_2 + MnI_2 \xrightarrow{\text{T.H.F.}} [Mn\{N(SiMe_3)_2\}_2]_2 + 2NaI \qquad (13)$$

$[Mn\{N(SiMe_3)_2\}_2(\text{T.H.F.})]$ for which X-ray crystallographic data are available [35, 36]. It is not very stable, the T.H.F. being removable by pumping *in vacuo*,* whereas the Cr^{II} analogue is the stable $[Cr\{N(SiMe_3)_2\}_2(\text{T.H.F.})_2]$ [56]. The difference may be due to electronic, rather than steric, effects. Thus the d^5 Mn^{II} complex gains no C.F.S.E. by formation of a T.H.F. adduct in contrast to the square planar d^4 Cr^{II} complex [36].

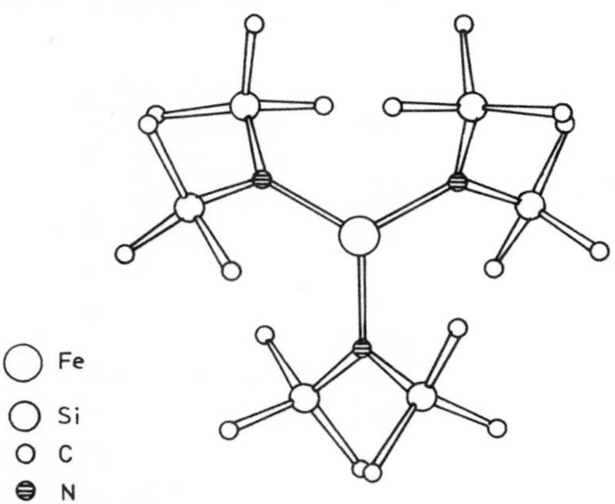

Fig. 13 – Crystal and molecular structure of $[Fe\{N(SiMe_3)_2\}_3]$. Bond distances are Fe–N, 1.918 Å; Si–N, 1.731 Å; angles Fe–$\hat{N}$–Si, 119.4; Si–$\hat{N}$–Si, 121.24°. *[Reproduced, with permission, from Bradley, D. C., Hursthouse, M. B., and Rodesiler, P. F.,* Chem. Comm., *1969, 14.]*

*The solvent-free complex is a dimer (X-ray), Eq. (13), with (t = terminal, b = bridging) Mn–N_t, 1.994 Å; Mn–N_b = 2.184 Å; Si–N_t, 1.723 Å, Si–N_b, 1.774 Å; Mn$\hat{N}_b$Mn, 81.1°; and $N_b\widehat{Mn}N_b$, 98.9° [54a].

[Fe{N(SiMe$_3$)$_2$}$_3$] has been prepared [177] by salt elimination, with the stoicheiometry adjusted to allow for the formation of Na$_3$[FeCl$_6$]. Spectroscopic data [6, 8] indicate a trigonal high spin d^5 structure with D_{3h} symmetry. This is confirmed by an X-ray diffraction study [58], which shows that the FeN$_3$ and FeNSi$_2$ groups are planar (as with the chromium analogue), the dihedral angle between the two planes being 49°, Fig. 13.

Similarly, the green, air-sensitive solid, [Co{N(SiMe$_3$)$_2$}$_2$], is obtained via salt elimination. Spectroscopic studies [46, 98] support a linear two-co-ordinate structure. It forms adducts with pyridine, 4-methylpyridine, or 2,4-dimethylpyridine [155].

The amide [Co{N(SiMe$_3$)$_2$}$_2$(PPh$_3$)] is a d^7 high spin three-co-ordinate metal complex, as confirmed by X-ray crystallography [60] (Fig. 14).

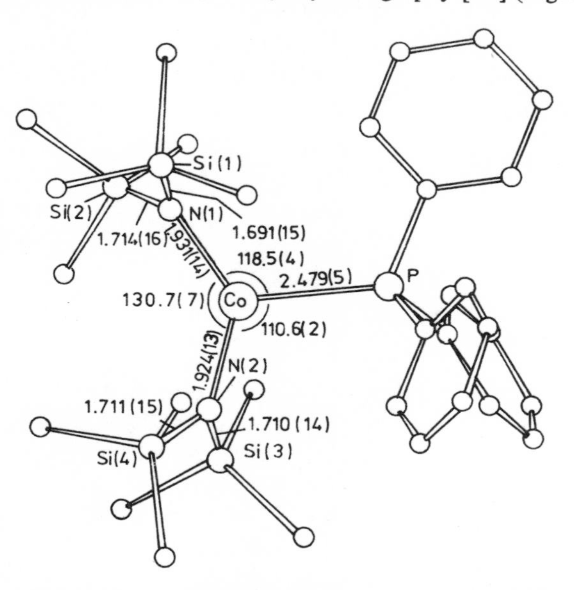

Fig. 14 – View of the [Co{N(SiMe$_3$)$_2$}$_2$(PPh$_3$)] molecule in a direction perpendicular to the CoN$_2$P mean plane. Angles (°) at the N atoms (omitted from the Fig. for clarity) are: Co–$\widehat{N}$(1)–Si(1) 119(1), Co–$\widehat{N}$(1)–Si(2) 116(1), Si(1)–$\widehat{N}$(1)–Si(2) 125(1), Co–$\widehat{N}$(2)–Si(3) 115(1), Co–$\widehat{N}$(2)–Si(4) 120(1), Si(3)–$\widehat{N}$(2)–Si(4) 124(1)°. *[Reproduced, with permission, from Bradley, D. C., Hursthouse, M. B., Smallwood, R. J., and Welch, A. J.,* J.C.S. Chem. Comm., *1972, 872.]*

Two nickel derivatives containing the bis(trimethylsilyl)amido ligand, both prepared by salt elimination, Eq. (14), are known. The reaction of NiI$_2$ with NaN(SiMe$_3$)$_2$ in T.H.F. gave a red, presumably monomeric, oil, possibly with a linear structure [Ni{N(SiMe$_3$)$_2$}$_2$] [94]. It is unstable and decomposes at 20° C;

a stable T.H.F. adduct has been isolated, but dissociates into its factors at ca. 80° C/0.2 mmHg. The stable Ni^{I} species $[NiN(SiMe_3)_2(PPh_3)_2]$ [60] is surprisingly obtained from a Ni^{II} precursor, Eq. (14).

$$x\text{LiN(SiMe}_3)_2 + [\text{MCl}_2(\text{PPh}_3)_2] \longrightarrow [\text{M}\{\text{N(SiMe}_3)_2\}_x(\text{PPh}_3)_{3-x}] + x\text{LiCl} \quad (14)$$

$$(\text{M} = \text{Co}, x = 2; \text{M} = \text{Ni}, x = 1)$$

X-ray crystallographic data show that the metal is three-co-ordinate in these compounds (Fig. 15). The N-Si bond lengths are short, possibly due to enhanced $\ddot{N}$–Si and consequently low $\ddot{N}$–M π-bonding. Nevertheless, the Ni^{I}-N bond length is rather short. This may be due to the amide behaving as a π-acceptor [36] or due to the π-acceptor character of the phosphine ligands. The latter view is supported by comparison of X-ray data on $[Co\{N(SiMe_3)_2\}_2PPh_3]$, (Fig. 14), and $[Co\{N(SiMe_3)_2\}(PPh_3)_2]$ (Table 5), with the Co^{I} compound having a very short Co-N bond.

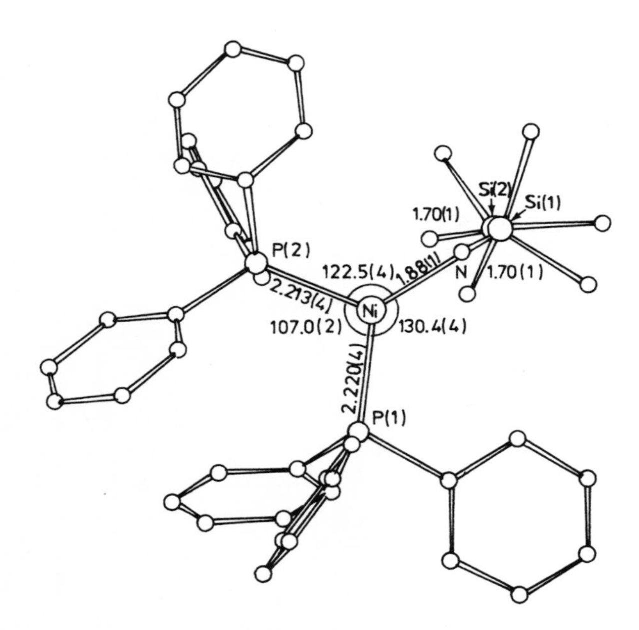

Fig. 15 – View of the $[Ni\{N(SiMe_3)_2\}(PPh_3)_2]$ molecule in a direction perpendicular to the $NiNP_2$ mean plane. Angles (°) at the nitrogen atom are: Ni–$\widehat{N}$–Si(1) 117(1), Ni–$\widehat{N}$–Si(2) 117(1), Si(1)–$\widehat{N}$–Si(2) 120(1)°. *[Reproduced, with permission, from Bradley, D. C., Hursthouse, M. B., Smallwood, R. J., and Welch, A. J.,* J.C.S. Chem. Comm., *1972, 872.]*

Table 5 Some molecular geometry parameters for mixed silylamidophosphine complexes of Co and Ni

Compound	$[Co\{N(SiMe_3)_2\}_2PPh_3]$	$[Co\{N(SiMe_3)_2\}(PPh_3)_2]$	$[Ni\{N(SiMe_3)_2\}(PPh_3)_2]$
M–N (Å)	1.934(8) 1.917(8)	1.924(8)	1.870(8)
M–P (Å)	2.472(3)	2.257(3) 2.251(3)	2.222(3) 2.209(3)
$N-\widehat{M}-N$ (°)	131.0(3)		
$N-\widehat{M}-P$ (°)	117.7(2) 111.2(2)	128.9(2) 124.1(2)	130.3(2) 122.8(2)
$P-\widehat{M}-P$ (°)		106.5(2)	106.8(2)
N–Si (Å)	1.71(1), 1.73(1) 1.69(1), 1.69(1)	1.66(1) 1.72(1)	1.71(1) 1.71(1)
$Si-\widehat{N}-Si$ (°)	125.5(5)	128.0(5)	125.4(5)

Reaction of CuI or $CuCl_2$ with $NaN(SiMe_3)_2$ produces the volatile colourless $[\{CuN(SiMe_3)_2\}_4]$ (XIII), which is a tetramer (X-ray) in the crystalline state [176].

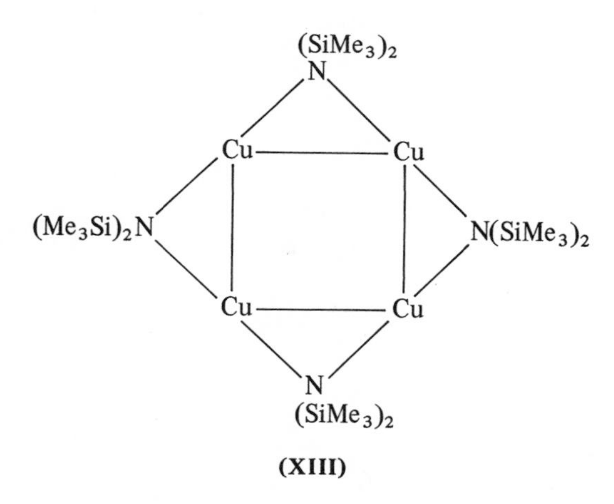

(XIII)

It forms an adduct with T.H.F. which dissociates at 100° C *in vacuo* [8, 9, 36a]. A similar silver complex $[AgN(SiMe_3)_2]$ [93] has been obtained from $KN(SiMe_3)_2$ and $[Ag(NH_3)_2]Cl$ in liquid ammonia. Gold(I) analogues may be isolated as phosphine complexes, [Eq. (15)] [247a].

$$NaN(SiMe_3)_2 + [Au(Cl)PR_3] \longrightarrow [Au\{N(SiMe_3)_2\}PR_3] + NaCl \quad (15)$$

(R = Me or Ph)

G AMIDES OF Sc, Y, THE LANTHANIDES, AND ACTINIDES

Few homoleptic dialkylamides of lanthanides or actinides are known [18, 216, 259]. The compound $[Th(NEt_2)_4]$ is formed from $ThCl_4$ and the lithium dialkylamide [18, 259]. The 2,5-dimethylpyrrolidinato derivative of uranium(IV), (XIV), is formulated, on the basis of n.m.r. data, as having three mono- and one penta-hapto liganded amide [216]. The heteroleptic thorium compound $[ThCl\{N(SiMe_3)_2\}_3]$ has been prepared [49], but due to steric hindrance (cf. Ti) it is not possible to substitute the remaining chloride by a further bis(trimethylsilyl)amido ligand.

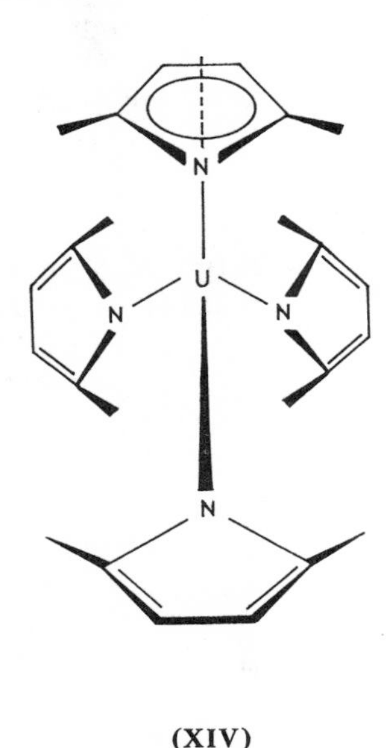

(XIV)

Recently there have been several publications on uranium amides [179a, 216, 241a–e], which describe interesting and unusual structural features. For example, $[\{U(NEt_2)_4\}_2]$ (Fig. 16) [241e] is dinuclear in the solid state, with two bridging diethylamido groups and an unusual five-co-ordination for uranium(IV). The five N atoms are at the corners of a distorted trigonal bipyramid with N(1) and N(4′) in axial positions, and N(2), N(3), and N(4) meridional. The U–N distances are longer for bridging than for terminal nitrogen atoms. The bridging nitrogen atoms are approximately tetrahedral and the terminal nitrogens are almost planar with the $C\text{–}\widehat{N}\text{–}C$ angle (112–116°) intermediate between sp^3 and sp^2 hybridisation.

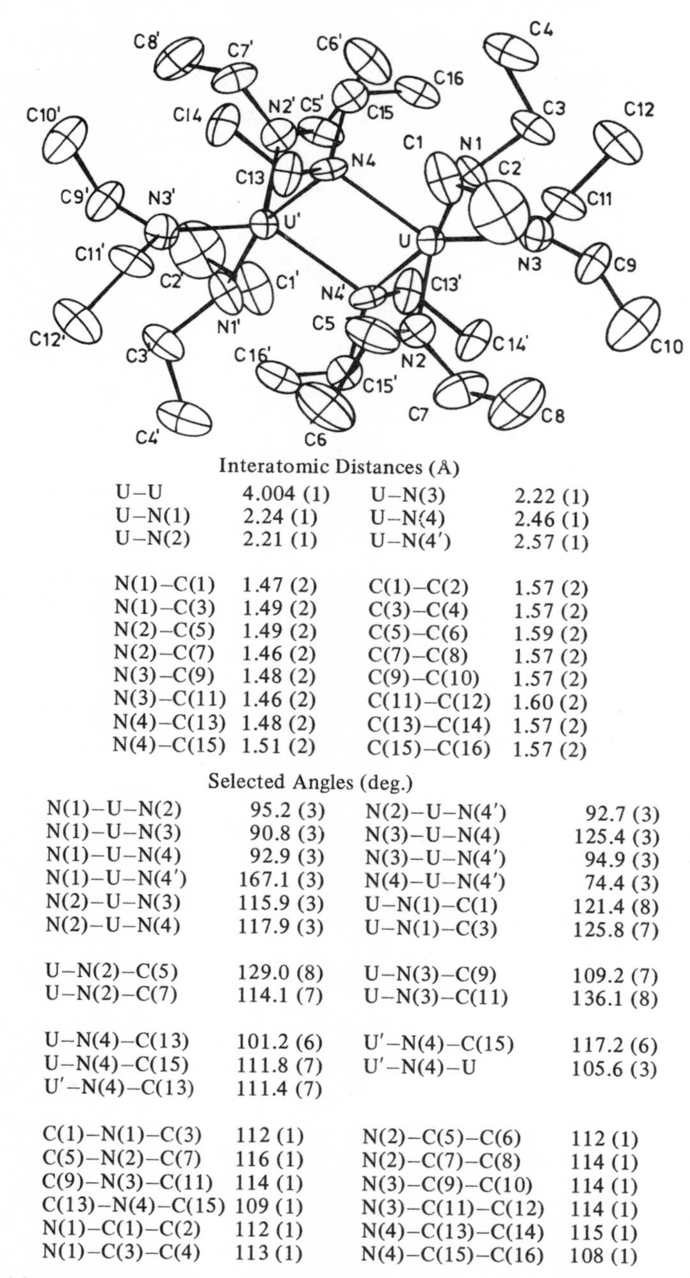

Interatomic Distances (Å)

U–U	4.004 (1)	U–N(3)	2.22 (1)
U–N(1)	2.24 (1)	U–N(4)	2.46 (1)
U–N(2)	2.21 (1)	U–N(4′)	2.57 (1)
N(1)–C(1)	1.47 (2)	C(1)–C(2)	1.57 (2)
N(1)–C(3)	1.49 (2)	C(3)–C(4)	1.57 (2)
N(2)–C(5)	1.49 (2)	C(5)–C(6)	1.59 (2)
N(2)–C(7)	1.46 (2)	C(7)–C(8)	1.57 (2)
N(3)–C(9)	1.48 (2)	C(9)–C(10)	1.57 (2)
N(3)–C(11)	1.46 (2)	C(11)–C(12)	1.60 (2)
N(4)–C(13)	1.48 (2)	C(13)–C(14)	1.57 (2)
N(4)–C(15)	1.51 (2)	C(15)–C(16)	1.57 (2)

Selected Angles (deg.)

N(1)–U–N(2)	95.2 (3)	N(2)–U–N(4′)	92.7 (3)
N(1)–U–N(3)	90.8 (3)	N(3)–U–N(4)	125.4 (3)
N(1)–U–N(4)	92.9 (3)	N(3)–U–N(4′)	94.9 (3)
N(1)–U–N(4′)	167.1 (3)	N(4)–U–N(4′)	74.4 (3)
N(2)–U–N(3)	115.9 (3)	U–N(1)–C(1)	121.4 (8)
N(2)–U–N(4)	117.9 (3)	U–N(1)–C(3)	125.8 (7)
U–N(2)–C(5)	129.0 (8)	U–N(3)–C(9)	109.2 (7)
U–N(2)–C(7)	114.1 (7)	U–N(3)–C(11)	136.1 (8)
U–N(4)–C(13)	101.2 (6)	U′–N(4)–C(15)	117.2 (6)
U–N(4)–C(15)	111.8 (7)	U′–N(4)–U	105.6 (3)
U′–N(4)–C(13)	111.4 (7)		
C(1)–N(1)–C(3)	112 (1)	N(2)–C(5)–C(6)	112 (1)
C(5)–N(2)–C(7)	116 (1)	N(2)–C(7)–C(8)	114 (1)
C(9)–N(3)–C(11)	114 (1)	N(3)–C(9)–C(10)	114 (1)
C(13)–N(4)–C(15)	109 (1)	N(3)–C(11)–C(12)	114 (1)
N(1)–C(1)–C(2)	112 (1)	N(4)–C(13)–C(14)	115 (1)
N(1)–C(3)–C(4)	113 (1)	N(4)–C(15)–C(16)	108 (1)

Fig. 16 – The crystal and molecular structure of [{$U(NEt_2)_4$}$_2$]. *[Reproduced, with permission, from Reynolds, J. G., Zalkin, A., Templeton, D. H., Edelstein, N. M., and Templeton, L. K.,* Inorg. Chem., *1976,* **15**, *2498.]*

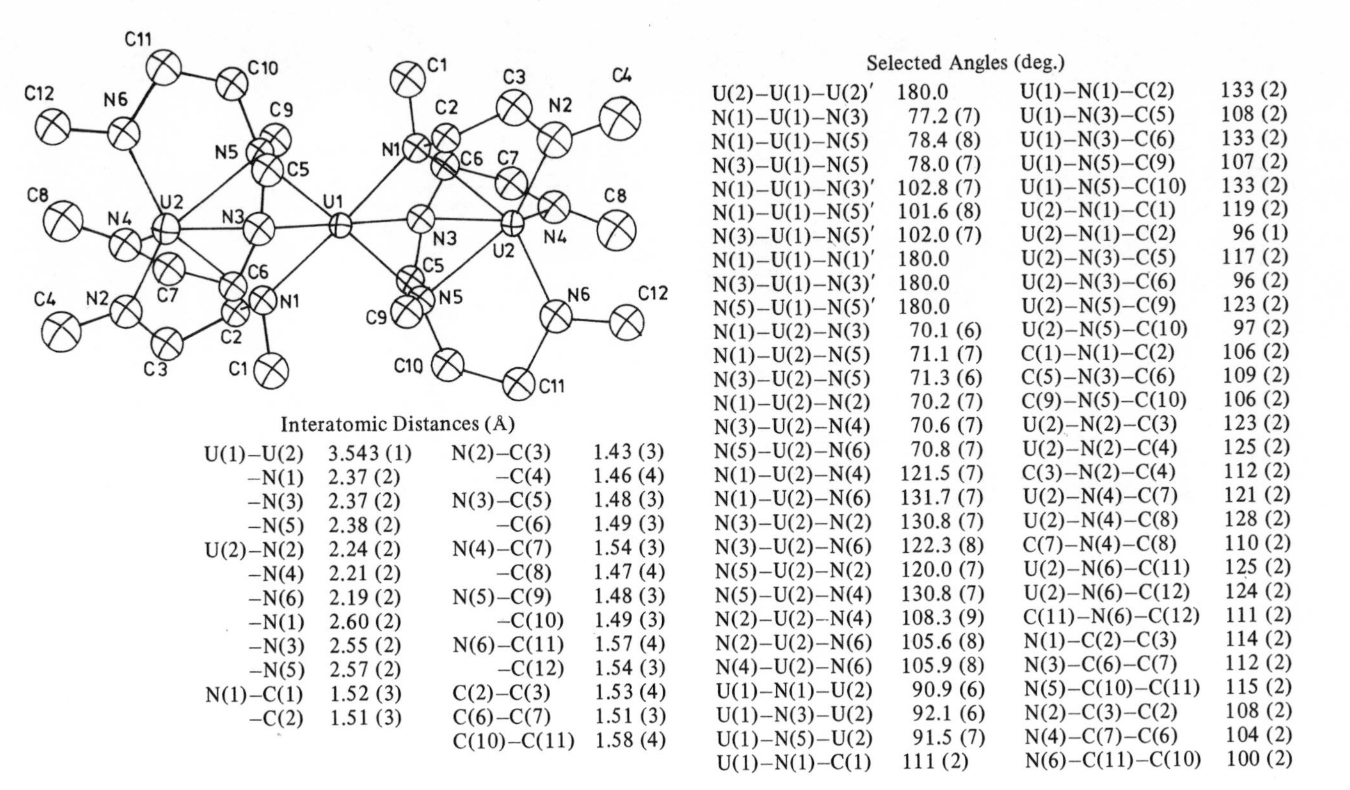

Interatomic Distances (Å)

U(1)–U(2)	3.543 (1)	N(2)–C(3)	1.43 (3)
–N(1)	2.37 (2)	–C(4)	1.46 (4)
–N(3)	2.37 (2)	N(3)–C(5)	1.48 (3)
–N(5)	2.38 (2)	–C(6)	1.49 (3)
U(2)–N(2)	2.24 (2)	N(4)–C(7)	1.54 (3)
–N(4)	2.21 (2)	–C(8)	1.47 (4)
–N(6)	2.19 (2)	N(5)–C(9)	1.48 (3)
–N(1)	2.60 (2)	–C(10)	1.49 (3)
–N(3)	2.55 (2)	N(6)–C(11)	1.57 (4)
–N(5)	2.57 (2)	–C(12)	1.54 (3)
N(1)–C(1)	1.52 (3)	C(2)–C(3)	1.53 (4)
–C(2)	1.51 (3)	C(6)–C(7)	1.51 (3)
		C(10)–C(11)	1.58 (4)

Selected Angles (deg.)

U(2)–U(1)–U(2)′	180.0	U(1)–N(1)–C(2)	133 (2)
N(1)–U(1)–N(3)	77.2 (7)	U(1)–N(3)–C(5)	108 (2)
N(1)–U(1)–N(5)	78.4 (8)	U(1)–N(3)–C(6)	133 (2)
N(3)–U(1)–N(5)	78.0 (7)	U(1)–N(5)–C(9)	107 (2)
N(1)–U(1)–N(3)′	102.8 (7)	U(1)–N(5)–C(10)	133 (2)
N(1)–U(1)–N(5)′	101.6 (8)	U(2)–N(1)–C(1)	119 (2)
N(3)–U(1)–N(5)′	102.0 (7)	U(2)–N(1)–C(2)	96 (1)
N(1)–U(1)–N(1)′	180.0	U(2)–N(3)–C(5)	117 (2)
N(3)–U(1)–N(3)′	180.0	U(2)–N(3)–C(6)	96 (2)
N(5)–U(1)–N(5)′	180.0	U(2)–N(5)–C(9)	123 (2)
N(1)–U(2)–N(3)	70.1 (6)	U(2)–N(5)–C(10)	97 (2)
N(1)–U(2)–N(5)	71.1 (7)	C(1)–N(1)–C(2)	106 (2)
N(3)–U(2)–N(5)	71.3 (6)	C(5)–N(3)–C(6)	109 (2)
N(1)–U(2)–N(2)	70.2 (7)	C(9)–N(5)–C(10)	106 (2)
N(3)–U(2)–N(4)	70.6 (7)	U(2)–N(2)–C(3)	123 (2)
N(5)–U(2)–N(6)	70.8 (7)	U(2)–N(2)–C(4)	125 (2)
N(1)–U(2)–N(4)	121.5 (7)	C(3)–N(2)–C(4)	112 (2)
N(1)–U(2)–N(6)	131.7 (7)	U(2)–N(4)–C(7)	121 (2)
N(3)–U(2)–N(2)	130.8 (7)	U(2)–N(4)–C(8)	128 (2)
N(3)–U(2)–N(6)	122.3 (8)	C(7)–N(4)–C(8)	110 (2)
N(5)–U(2)–N(2)	120.0 (7)	U(2)–N(6)–C(11)	125 (2)
N(5)–U(2)–N(4)	130.8 (7)	U(2)–N(6)–C(12)	124 (2)
N(2)–U(2)–N(4)	108.3 (9)	C(11)–N(6)–C(12)	111 (2)
N(2)–U(2)–N(6)	105.6 (8)	N(1)–C(2)–C(3)	114 (2)
N(4)–U(2)–N(6)	105.9 (8)	N(3)–C(6)–C(7)	112 (2)
U(1)–N(1)–U(2)	90.9 (6)	N(5)–C(10)–C(11)	115 (2)
U(1)–N(3)–U(2)	92.1 (6)	N(2)–C(3)–C(2)	108 (2)
U(1)–N(5)–U(2)	91.5 (7)	N(4)–C(7)–C(6)	104 (2)
U(1)–N(1)–C(1)	111 (2)	N(6)–C(11)–C(10)	100 (2)

Fig. 17 – The crystal and molecular structure of $[\{U(MeN(CH_2)_2NMe)_2\}_3]$. *[Reproduced, with permission, from Reynolds, J. G., Zalkin, A., Templeton, D. H., and Edelstein, N. M.,* Inorg. Chem., *1977,* **16,** *599.]*

The structures of $[U(NPh_2)_4]$ [241c] (monomeric, distorted tetragonal) and the cluster compounds $[U_3\{MeN(CH_2)_2NMe\}_6]$ [241a] and $[U_4\{MeN(CH_2)_2NMe\}_8]$ [241b] have been reported. The trimer $[\{U(MeN(CH_2)_2NMe)_2\}_3]$ has the structure illustrated in Fig. 17 [241a]. It has three colinear uranium(IV) atoms bridged by nitrogen atoms. The central U atom is on a centre of symmetry and is bridged by three equivalent nitrogen atoms (which form the corners of an elongated trigonal antiprism) to the terminal metal atoms. For each ligand, one nitrogen is bridging and the other terminal. For the bridging nitrogens, angles at N are tetrahedral, but are planar for the terminal N atoms.

For the lanthanides only a small number of dialkylamides has been reported [46a], but several tris(bis-trimethylsilylamido)metal(III) species $[M\{N(SiMe_3)_2\}_3]$, (M = Sc, Y, La, Ce, Pr, Nd, Sm, Eu, Gd, Ho, Yb, or Lu) have been characterised. All these complexes are monomeric, with a three-co-ordinate metal environment. The scandium, yttrium, lanthanum, and lutetium species are diamagnetic and show sharp singlets in their ^1H n.m.r. spectra. The cerium, praseodymium, samarium, and europium analogues display considerable paramagnetic shifts due to the pseudo-contact mechanism, although there may be some contact contribution. Spectroscopic studies support three-co-ordination with a high C.F.S.E. due to the large trigonal field.

X-ray diffraction studies [165] show the scandium Eu, and Nd [3a] derivatives to be isostructural, and confirm the metal three-co-ordination; however, the metal atom resides just above or below the MN_3 plane (Fig. 18). The deviation from planarity is attributed to rather ionic M–N bonds, making the complex stereochemically flexible. The distortion was thought to occur upon crystallisation, since dipole moment studies on solutions yielded zero values. Support for this hypothesis comes from the wide SiN_2 angle (ca. 130°) in the Eu complex and the short Si–N bond length, similar to those found in $K[N(SiMe_3)_2]$.dioxan (see Chapter 2). However, the Sc–N bond lengths in the Sc compound are not indicative of its ionic character.

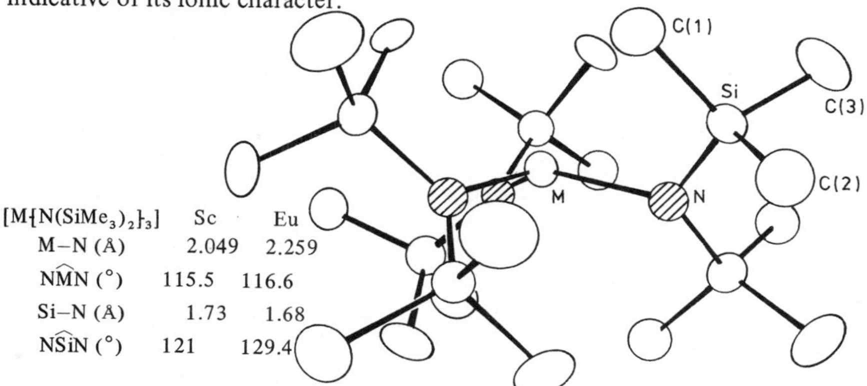

$[M\{N(SiMe_3)_2\}_3]$	Sc	Eu
M–N (Å)	2.049	2.259
$N\widehat{M}N$ (°)	115.5	116.6
Si–N (Å)	1.73	1.68
$N\widehat{Si}N$ (°)	121	129.4

Fig. 18 – Molecular structure and parameters for $[M\{N(SiMe_3)_2\}_3]$ (M = Sc or Eu). *[Reproduced, with permission, from Ghotra, J. S., Hursthouse, M. B., and Welch, A. J.,* J.C.S. Chem. Comm., *1973, 669.]*

These lanthanide (Ln) compounds undergo interesting reactions with $(O)PPh_3$ to give the unusual complexes (Figs. 19 and 20) shown in Eq. 16 [48, 48a] (see Chapter 15).

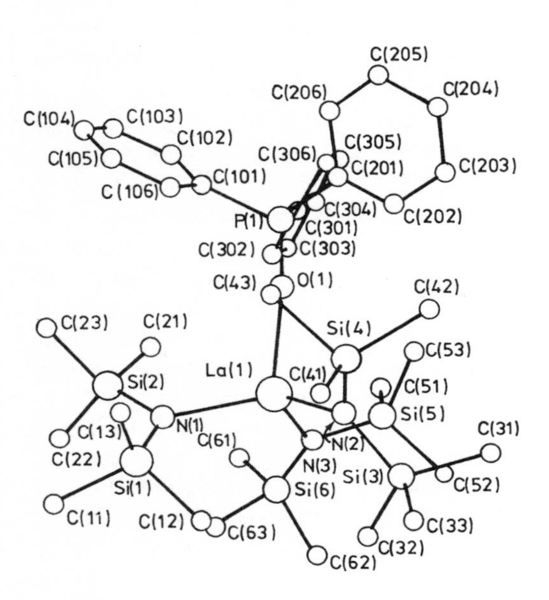

Some important bond lengths (Å) and angles (°) for $[La\{N(SiMe_3)_2\}_3(OPPh_3)]$

(*a*) Lengths

La(1)–N(1)	2.41(2)	La(1)–N(2)	2.40(3)
La(1)–N(3)	2.38(2)	La(1)–O(1)	2.40(2)
N(1)–Si(1)	1.73(3)	N(2)–Si(3)	1.76(3)
N(1)–Si(2)	1.74(3)	N(2)–Si(4)	1.65(2)
N(3)–Si(5)	1.78(3)	N(3)–Si(6)	1.65(2)
O(1)–P(1)	1.52(2)		

(*b*) Angles

O(1)–La(1)–N(1)	107.8(7)	La(1)–N(2)–Si(3)	119.3(10)
O(1)–La(1)–N(2)	105.2(7)	La(1)–N(2)–Si(4)	119.7(14)
O(1)–La(1)–N(3)	104.8(8)	La(1)–N(3)–Si(5)	116.1(8)
N(1)–La(1)–N(2)	109.2(9)	La(1)–N(3)–Si(6)	124.1(14)
N(1)–La(1)–N(3)	116.4(6)	Si(1)–N(1)–Si(2)	119.6(13)
N(2)–La(1)–N(3)	112.7(8)	Si(3)–N(2)–Si(4)	120.9(16)
La(1)–N(1)–Si(1)	123.7(12)	Si(5)–N(3)–Si(6)	119.8(15)
La(1)–N(1)–Si(2)	116.6(14)	La(1)–O(1)–P(1)	174.6(9)

Fig. 19 – Crystal and molecular structure of $[La\{N(SiMe_3)_2\}_3(OPPh_3)]$. *[Reproduced, with permission, from Bradley, D. C., Ghotra, J. S., Hart, F. A., Hursthouse, M. B., and Raithby, P. R.,* J.C.S. Dalton, *1977, 1166.]*

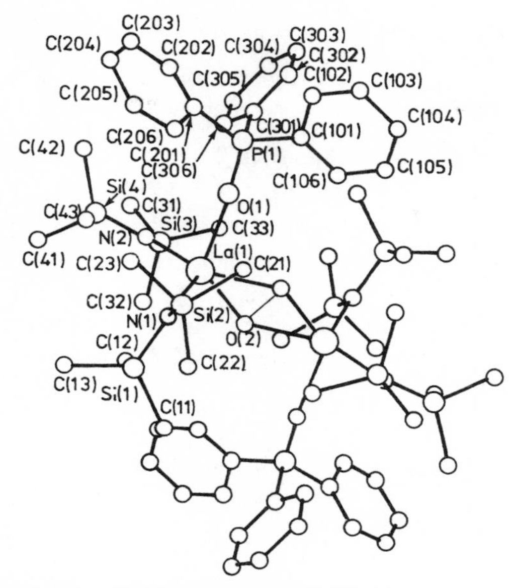

Some important bond lengths (Å) and angles (°) for $[La\{N(SiMe_3)_2\}_4(O_2)(OPPh_3)_2]$

(*a*) Lengths

La(1)–N(1)	2.37(2)	La(1)–N(2)	2.49(2)
La(1)–O(1)	2.42(2)	O(1)–P(1)	1.51(2)
La(1)–O(2)	2.33(3)	La(1)–O(2′)[a]	2.34(2)
N(1)–Si(1)	1.73(3)	N(1)–Si(2)	1.72(2)
N(2)–Si(3)	1.69(3)	N(2)–Si(4)	1.66(3)
La(1)···La(1′)	4.36(1)		

(*b*) Angles

N(1)–La(1)–N(2)	119.8(11)	N(1)–La(1)–M	109.4(7)
N(1)–La(1)–O(1)	92.1(8)	N(2)–La(1)–M	115.4(8)
N(1)–La(1)–O(2)	124.4(7)	O(1)–La(1)–M	107.0(7)
N(1)–La(1)–O(2′)	103.8(8)	La(1)–N(1)–Si(1)	114.4(9)
N(2)–La(1)–O(1)	111.2(7)	La(1)–N(1)–Si(2)	122.9(17)
N(2)–La(1)–O(2)	111.9(10)	La(1)–N(2)–Si(3)	118.1(15)
N(2)–La(1)–O(2′)	104.4(8)	La(1)–N(2)–Si(4)	115.3(8)
O(1)–La(1)–O(2)	87.4(9)	Si(1)–N(1)–Si(2)	122.5(14)
O(1)–La(1)–O(2′)	126.3(10)	Si(3)–N(2)–Si(4)	126.0(18)
O(2)–La(1)–O(2′)	41.5(10)	La(1)–O(1)–P(1)	172.6(14)
		La(1)–O(2)–La(1′)	138.5(14)

Fig. 20 – Crystal and molecular structure of $[La_2\{N(SiMe_3)_2\}_4(O_2)(OPPh_3)_2]$. *[Reproduced, with permission, from Bradley, D. C., Ghotra, J. S., Hart, F. A., Hursthouse, M. B., and Raithby, P. R.,* J.C.S. Dalton, *1977, 1166.]*

$$[Ln\{N(SiMe_3)_2\}_3] + (O)PPh_3 \xrightarrow{C_6H_6} [Ln\{N(SiMe_3)_2\}_3(OPPh_3)] \tag{16}$$

$$[Ln\{N(SiMe_3)_2\}_3] \xrightarrow{2(O)PPh_3,\ H_2O_2} [Ln_2\{N(SiMe_3)_2\}_4(O_2)(OPPh_3)_2]$$

Significant features of Fig. 19 are (a) the variations in the O-$\widehat{La}$-N and the N-$\widehat{La}$-N angles from respectively 105 to 108° and 109 to 116°, (b) the small deviation from planarity in the LaN_3 framework, which is not as marked as in the parent amides $[M\{N(SiMe_3)_2\}_3]$ (M = Sc or Eu), and (c) the La-$\widehat{O}$-P angle, which is almost 180° (actually 174°), and the orientation of the phenyl rings which is unsymmetrical about the La-O-P axis. Finally, (d), the Si-N distance (ca. 1.73 Å) and Si-$\widehat{N}$-Si bond angles (ca. 120°) indicate smaller ionic components than in $[Eu\{N(SiMe_3)_2\}_3]$. The complex shown in Fig. 20 is a unique double-bidentate-peroxo-bridging dimetal complex, and the O-O distance (1.65 Å) is longer than expected for O_2^{2-}, attributed to crystal decomposition [48a].

H METAL-NITROGEN (M$\overset{\frown}{-}$N) $d\leftarrow p$ π-BONDING

Throughout this chapter various physical properties of transition metal amides have been rationalised in terms of M$\overset{\frown}{-}$N ($d\leftarrow p$)-π-bonding. These properties include i.r. data, relating to unexpected distortions from stereo-regularity [51], or to trends in either $\nu(MN_n)$ asymmetric stretching modes of $[M(NR_2)_n]$ complexes or ν(NO) in the series $[Cr(NR_2)_3(NO)]$ [$R_2 = Pr^i_2$, $(SiMe_3)_2$, or $(CH_2)_5$] and $[Cr(NPr^i_2)(NO)(OBu^t)_2]$ [63]. Electronic or e.s.r. spectra, see refs. [36, 44], and results cited earlier especially on $[M(NR_2)_3]$ (M = V, Cr, or Mo), are consistent with such π-bonding. Finally, analysis of many of the finer points of X-ray structures have relied on a similar hypothesis. Data using two other techniques, photoelectron spectroscopy and calorimetry, warrant further discussion.

The He(I) photoelectron (p.e.) spectra of the compounds $[M(NR_2)_4]$ (M = C, Si, Ge, Sn, Ti, Zr, Hf, or V, with R = Me; or M = Ti, Zr, or Hf, with R = Et), $[W(NMe_2)_6]$ [166] and $[M\{N(SiMe_3)_2\}_3]$ (M = Sc, Ti, Cr, Fe, Ga, or In) [203] have been examined. For the dialkylamides, the only open-shell compound, $[V(NMe_2)_4]$, has the single unpaired electron in a *d*-orbital corresponding to an ionisation potential (i.p.) of 6.19 eV [166]. By contrast, the open-shell di(silyl)-amides do not show a metal 3*d*-orbital at i.p. < ca. 9 eV [203]. This suggests a strong $-I$ effect of the $(Me_3Si)_2N^-$ ligand, in contrast to the situation obtaining for Me_2N^- in $[V(NMe_2)_4]$. Thus, for the former ligand N$\overset{\frown}{-}$M π-bonding is weak, possibly because of the competing claims of N$\overset{\frown}{-}$Si π-interactions. Ionisation in the d^0-d^3 (Sc-Cr) di(silyl)amides at ca. 8 and ca. 8.7 eV are assigned to the N-lone pair region, corresponding to orbitals of A_2 and E symmetry in the D_3 point group; the band at 9.45 eV, observed only in the Sc compound, may reflect its higher ionicity, Sc being the most electronegative of the metals under discussion; the orbital is essentially nitrogen-centred rather than a Sc-N *e* bonding orbital. For the dialkylamides $[M(NR_2)_4]$ and $[W(NMe_2)_6]$, the nitrogen lone pair orbitals are found at lower ionisation energies: e.g., 7.13, 7.36, 7.75, and 8.00 eV in $[Ti(NMe_2)_4]$ (corresponding to E, B_1, and A_1 symmetry combinations in the D_{2d} point group and hence restricted M–N rotation) and 6.73 and 7.92 eV in the W^{VI} compound [166], assigned to an essentially non-bonding (T_u) and a slightly bonding (T_g) lone pair combination, respectively for tetrahedral symmetry assuming M–N restricted rotation.

The heats of alcoholysis in isopropyl alcohol of the d^0 $[M(NR_2)_4]$ complexes (M = Ti or Zr, and R = Me or Et; or M = Hf, R = Et) were determined [202a]. From these, standard heats of formation (ΔH_f°) of the gaseous molecules were derived. Similar experiments led to ΔH_f° $[MR_4]$ (g) (R = Me_3CCH_2, Me_3SiCH_2, or $PhCH_2$, with M = Ti or Zr; or R = Me_3CCH_2, M = Hf) and ΔH_f° $[M(OPr^i)_4]$ (g) (M = Ti, Zr, or Hf), using available literature data on corresponding tetrachlorides. Thus, these results lead to standard heats of formation for the gaseous tetrahedral molecules $[MX_4]$ (M = Ti, Zr, or Hf; X = R, NR'_2, OR″, F, or Cl), and hence to thermochemical bond energy terms $\bar{E}$(M–X). For the amides, bond

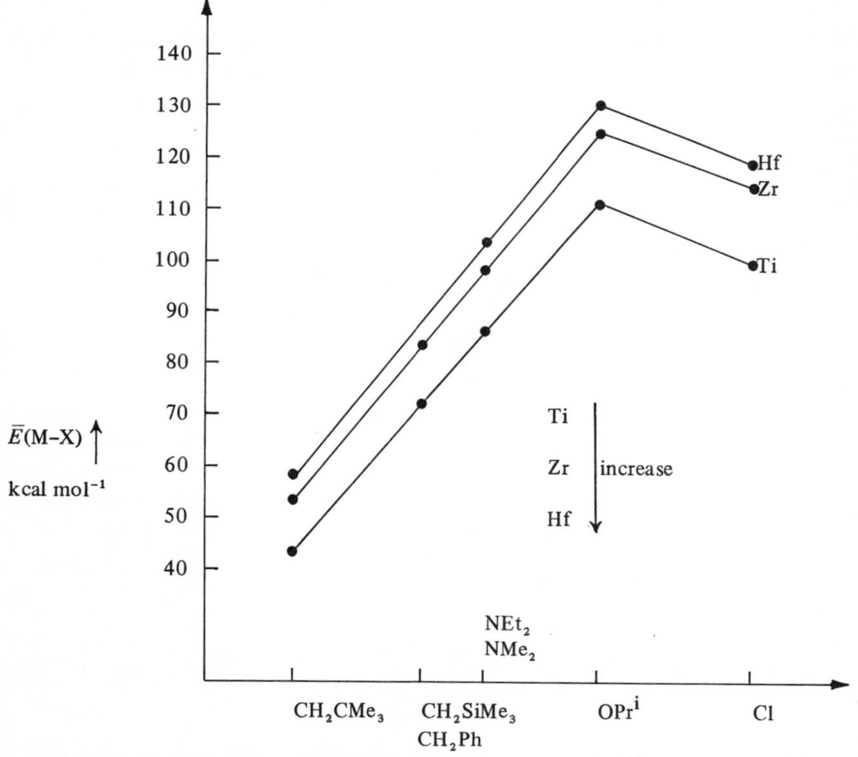

Fig. 21 – Trends in mean thermochemical bond energy terms $\bar{E}$(M–X) in the series $[MX_4]$ (M = Ti, Zr, or Hf; X = CH_2SiMe_3, NMe_2, NEt_2, OPr^i, Cl) (from ref. [202a]).

strengths, expressed as $\bar{E}$ are (kcal mol^{-1}) Ti-NMe$_2$, 81; Ti-NEt$_2$, 81; Zr-NMe$_2$, 91; Zr-NEt$_2$, 89; and Hf-NEt$_2$, 95. Trends in $\bar{E}$ are (i) Ti$\ll$Zr$<$Hf, (ii) R$<$NR$'_2$$<OR''$$<$F, and (iii) NR$'_2$$<Cl<OR''$; furthermore, bonds to Zr are ca. 12% stronger than to Ti and ca. 6% weaker than to Hf, irrespective of the ligand X. Further illustration is by reference to Fig. 21. These data suggest that (a) bond strengths are dominated by the σ-bond [N.B. $\bar{E}$(M–X) increases with increasing electronegativity of X] and (b) the π-bond effects are small in terms of bond strengths, unless the improbable assumption is made that the π contribution to the strength of each M–X bond is invariant for the widely differing ligands X and metals M. The standard heat of formation of liquid [Ti(NEt$_2$)$_4$] has been obtained from heat of alcoholysis [54]. The standard heats of formation of [Mo(NMe$_2$)$_4$], [W(NMe$_2$)$_6$], and [M$_2$(NMe$_2$)$_6$] (M = Mo or W) have been obtained and mean M–N [Mo^{IV}–N, 61; and W^{VI}–N, 53 kcal mol^{-1}] and M^{III}≡M^{III} (Mo, 48; W, 82 kcal mol^{-1}) bond dissociation energies ($\bar{D}$) were calculated [131].

In summary, whereas M–N̈ π-bonding appears to have a profound effect on molecular geometry, there is little evidence that it contributes noticeably to bond strengths. With reference to the latter in terms of bond lengths, arguments cannot be decisive because they rely upon assumptions of metal and nitrogen radii which are inherently incapable of experimental verification from non-existent 'ideal' compounds taken as models.

I TABLE OF COMPOUNDS

Table 6† Transition metal amides

Compound	Method	B.p. (°C/mm.) (m.p., °C)	Comments	References
Titanium: Ti(I)				
$TiNMe_2$	I			[197]
Ti(II)				
$Ti[N(CN)_2]_2$			ir	[189]
$Ti(NMe_2)_2$	I			[11, 205]
$Ti(NEt_2)_2$	I			[11, 205]
$ClTiNMe_2$	I			[11]
Ti(III)				
Cp_2TiNMe_2	A		ir, nmr, esr, dimer, mag	[11, 205, 206]
$CpTi(NMe_2)_2$	K		ir, nmr, dimer	[11, 205, 206]
$Ti(NMe_2)_3$	A	d50/0.01	ir, nmr, dnmr, esr, dimer, mag	[11, 205, 206]
$.OEt_2$	A	d50/0.01	ir, nmr, esr, dimer, mag	[11, 205, 206]
$(Me_2N)_2TiNEt_2$	B		ir, nmr	[205]
$(Me_2N)_2TiNPr^i_2$	B		ir, nmr	[205]
$(Me_2N)_2TiN(SiMe_3)_2$	B		ir, nmr	[205]
$Ti_2(MeNCH_2CH_2NMe)_3$	B		ir, nmr	[205]
$Ti(NEt_2)_3$	A	d100/0.01	ir, nmr, dimer	[11, 205, 206]
$Ti(NPr^i_2)_3$			esr	[42, 120]
Ti(N◻)$_3$				[205, 221]
$Ti(NPh_2)_3$	A		esr	[120, 230]
$Ti[N(SiMe_3)_2]_3$	A		ir, nmr, esr, X-ray, mass, monomer, mag, uv, pes	[8, 9, 36, 42, 44, 172, 203]
$ClTi(NMe_2)_2$				[264]
$ClTi(NEt_2)_2$				[264]
$ClTi(NPr^i_2)_2$	A		ir, nmr	[11, 205]
$ClTi(NBu^s_2)_2$	A			[11]
OTiNHEt				[196]
$Cp_2TiNPhCONMe_2$	D		ir, mag	[206]
$CpTi(NPhCONMe_2)_2$	D		ir	[206]
$Ti(NPhCONMe_2)_3$	D		ir, mag	[206]
$Ti(NPhCSNMe_2)_3$	D		ir, mag	[206]
Ti(IV)				
$(PhCH_2)_3TiNEt_2$	A	(5-10)d	ir, nmr	[79]

Table 6 *C'ntd.*

Compound	Method	B.p. (°C/mm.) (m.p., °C)	Comments	References
$(PhCH_2)_3TiN$⬡	A	(11–12)d	ir, nmr	[79]
Cp_2Ti ring with CH_2, $SiMe_2$ and N–$SiMe_3$	A, E	subl. 110/0.001		[28]
$Cp_2Ti[N(MMe_3)SiMe_3]$ (M = C, Si)	A		ir, nmr, esr, mass	[170a]
$Me_2Ti(NMe_2)_2$	A	(−35)d	ir, nmr	[79]
$(PhCH_2)_2Ti(NMe_2)_2$	A	(−20)d	ir, nmr	[79]
$(Me_3SiCH_2)_2Ti(NMe_2)_2$	A	(<0)	ir, nmr	[79]
$Me_2Ti(NEt_2)_2$	A	(−30)	ir, nmr	[79]
$(PhCH_2)_2Ti(NEt_2)_2$	A	(−10)d	ir, nmr	[79]
$(Me_3SiCH_2)_2Ti(NEt_2)_2$	A	(<0)	ir, nmr	[79]
$Me_2Ti(N$⬡$)_2$	A	(−30)	ir, nmr	[79]
$(PhCH_2)_2Ti(N$⬡$)_2$	A	(20)	ir, nmr	[79]
$(HOCH_2CH_2)_2Ti(NEt_2)_2$				[201]
$PhC{:}CTi(C_5H_4Me)(NMe_2)_2$	K		ir, nmr	[181]
$Cp_2Ti(NMe_2)_2$	A	120–8/0.2		[114, 115]
$Cp_2Ti(NPh_2)_2$	A	(248)	ir, nmr	[25, 230]
$Cp_2Ti(N$⬠$)_2$	A	(175–7)	ir, mag	[177]
$Cp_2Ti(N$⬠$)_2$ (benzo-fused)	A	(195)	ir	[177]
$Cp_2Ti(N$⬠$)_2$ (dibenzo-fused)	A	(205–8)	ir	[177]
$Cp_2Ti[N(H)(C_6H_4\text{-}p\text{-}NO_2)]_2$	A	(272)	ir	[25]
$Cp_2Ti[N(H)Ph]_2$	A		ir	[25]

Table 6 *C'ntd.*

Compound	Method	B.p. (°C/mm.) (m.p., °C)	Comments	References
$Me_2Ti\langle N(Me)-SiMe_2-SiMe_2-N(Me)\rangle$ (cyclic)	A		ir, nmr	[103]
			ir	[92]
$[CH_2C(Me)CH_2]_2Ti(NEt_2)_2$	A		ir, nmr	[88]
$MeTi(NMe_2)_3$	A	subl. $20/10^{-3}$	ir, Raman, nmr	[84]
		(8–9)	uv	[85]
$MeTi(NEt_2)_3$	A	(4–5)	ir, Raman, nmr,	[84]
			uv, pes	[85, 247b]
$CD_3Ti(NMe_2)_3$	A	subl. $20/10^{-3}$		[84]
		(9–10)		
$CH_2C(Me)CH_2Ti(NMe_2)_3$	A			[88]
$PhCH_2Ti(NMe_2)_3$	A	(<0)	ir, nmr	[79]
$Me_3SiCH_2Ti(NMe_2)_3$	A	(<0)	ir, nmr	[79]
$(RC_5H_4)Ti(NMe_2)_3$	K		ir, nmr	[76]
(R = Et		$78/10^{-3}$		
Pr^i		$64/10^{-3}$		
$CHEt_2$		$85/10^{-3}$		
$CHPh_2$		–		
Bu^t		(25)		
$SiMe_3$		(30)		
$SiMe_2Ph$		$85/10^{-3}$		
$GeMe_3$)		(30)		
$Ph_3GeTi(NMe_2)_3$	A			[87]
$R R'C_5H_3Ti(NMe_2)_3$	K		ir, nmr	[76, 78]
(R R'				
Me Bu^t		$70/10^{-3}$		
Me $SiMe_3$		75/0.01		
Bu^t $GeMe_3$)		90/0.01		
$CD_3Ti(NEt_2)_3$	A	subl. $60\text{–}70/10^{-3}$		[84]
		(5)		
$EtTi(NMe_2)_3$	A	95/0.01		
		subl. $20/10^{-3}$	ir, Raman, nmr	[84]
		(~ –5)		
$EtTi(NEt_2)_3$	A		ir, Raman, nmr	[84]
			uv	[85]
$PrTi(NEt_2)_3$	A	(13–14)	ir, Raman, nmr	[84]
$Pr^iTi(NEt_2)_3$	A		ir, Raman, nmr	[84]
$BuTi(NEt_2)_3$	A	(–4)	ir, Raman, nmr	[84]
$Bu^tTi(NEt_2)_3$	A		ir, Raman, nmr	[84]
$PhCH_2Ti(NEt_2)_3$	A	(13)	ir, nmr	[79]
$Me_3SiCH_2Ti(NEt_2)_3$	A	(<0)	ir, nmr	[79]
			pes	[247b]
$CH_2C(Me)CH_2Ti(NEt_2)_3$	A			[88]
$CH_2CHCH_2Ti(NEt_2)_3$	A	$70/10^{-4}$		[88]

Table 6 *C'ntd.*

Compound	Method	B.p. (°C/mm.) (m.p., °C)	Comments	References
$MeCHCHCH_2Ti(NEt_2)_3$	A			[88]
$CH_2CHC(H)MeTi(NEt_2)_3$	A			[88]
$R'H_4C_5Ti(NC_5H_{10})_3$ (piperidino)	K		ir, nmr	[76]
(R' = H		(135)		
Me		$110/10^{-3}$		
Pr^i		$110/10^{-3}$		
Bu^t		$153/10^{-3}$		
$SiMe_3$		$115/10^{-3}$		
$GeMe_3$)		$130/10^{-3}$		
$PhC{:}CTi(NMe_2)_3$	A	subl. $80/10^{-3}$ (56–8)	ir, Raman, nmr	[84]
$PhTi(NEt_2)_3$	A		ir, Raman, nmr	[84]
$CpTi(NMe_2)_3$	K	95/0.05		[114, 115]
$H_4BTi(NMe_2)_3$				[197]
$CpTi(NEt_2)_3$	K	146/0.2		[113, 114]
$(MeC_5H_4)Ti(NEt_2)_3$	B	84/0.02		[113]
$Ti(NHBu)_4$	C			[34]
$Ti_2(NHBu^t)_2(NMe_2)_2(NBu^t)_2$	B	subl. 140–60/0.1		[68]
$Ti(NHOct)_4$	C			[178, 179]
$Ti(NHC_{18}H_{37})_4$	C			[178, 179]
$Ti(NHPh)_4$	C			[34]
$Ti(NMe_2)_4$	A		ir, nmr, pes	[51, 82, 166]
	A	50/0.05		[64, 65, 95]
			ΔH_f	[202a]
$/TiCl_4$				[260]
$/Ti(OMe)_4$				[260]
$.2Fe(CO)_5$			ir, mag	[37]
$/SnCl_4$				[208]
$.2B_{10}H_{14}$				[115]
$(Me_2N)_3TiNPr^i_2$	B	80/0.05		[65]
$(Me_2N)_3TiN{:}CMeNMe_2$	D		ir	[45]
$(Me_2N)_3TiN$ (2-methylpiperidino, Me)	B	160/0.1		[64]
$(Me_2N)_3TiN$ (2,6-dimethylpiperidino, Me, Me)	B	120–5/0.1		[64, 65]
$(Me_2N)_3TiN(SiMe_3)_2$	B	subl. 110–20/0.02		[114]
	A	110–20/0.2	nmr	[84, 86]
$Ti_2(NMe_2)_4(NBu^t)_2$	B			[68]
$(Me_2N)_2Ti(N{:}CMeNMe_2)_2$	D			[45]
$(Me_2N)_2Ti(NPhCONMe_2)_2$	D	(123)d	ir, nmr	[113]

Table 6 *C'ntd.*

Compound	Method	B.p. (°C/mm.) (m.p., °C)	Comments	References
$[p\text{-}MeC_6H_4N{:}C(NMe_2)N(p\text{-}MeC_6H_4)]_2Ti(NMe_2)_2$	D	(>150)d	ir	[113]
$[C_6H_{11}N{:}C(NMe_2)NC_6H_{11}]_2Ti(NMe_2)_2$	D	(148)d	ir	[113]
$Me_2NTi(NPr_2)_3$	B	95/0.05		[65]
$Me_2NTi(NBu^i_2)_3$	B	170/0.1		[65]
$Me_2NTi(N\bigcirc)_3$ (ring with Me)	B	160/0.1		[65]
$Ti(NMeBu)_4$	A		ir, nmr	[51]
$Ti(NEt_2)_4$	A		ir, nmr, pes	[50, 51, 52, 82, 166]
	A	108/0.1		[137]
	A	110/0.1		[64]
			ΔH_f	[54, 202a]
$(Me_3Si)_2NTi(NEt_2)_3$	A	subl. $100/10^{-4}$		[86]
$Ti_2(NEt_2)_6(NBu)$	B			[68]
$(Et_2N)_2Ti(N\bigcirc)_2$ (pyrrolyl)	B	135/0.1	ir	[40]
$(Et_2N)_2Ti(N\bigcirc)_2$ (2,5-Me pyrrolyl)	B			[40]
$(Et_2N)_2Ti[(NBu)_2Ti]_n(NBu)_2Ti(NEt_2)_2$ ($n = 4, 12, 14, 98$)	B			[68]
$Ti(NPr_2)_4$	A		ir, nmr	[51]
	A	150/0.1		[65]
$Ti[N(CH_2)_4]_4.HCl$				[214]
$Ti(NPr^i_2)_4$	A	(82–5)	ir, nmr	[2]
$(Pr_2N)_2Ti(N\bigcirc)_2$ (pyrrolyl)	B	155/0.25	ir	[40]
$Ti(NBu_2)_4$	C			[34]
	A		ir, nmr	[51]
$(Bu_2N)_2Ti(N\bigcirc)_2$ (pyrrolyl)	B	210/0.1	ir	[40]
$Ti(NBu^i_2)_4$	A		ir, nmr	[51]
	A	170/0.1		[65]
$Ti(N\square)_4$	B	(>110)	nmr	[207]
	B	d>150/0.005	nmr	[181]

Table 6 *C'ntd.*

Compound	Method	B.p. (°C/mm.) (m.p., °C)	Comments	References
$(Me_2N)_2Ti(NC_4H_4)_2$	B	120/0.1	ir	[40]
$(Me_2N)_2Ti(NC_4H_2Me_2)_2$ (2,5-Me)	B			[40]
$Z_2Ti(N(SiMe_2NMe)_2SiMe_2)_2$ (ring: N–$SiMe_2$–NMe–$SiMe_2$–NMe–$SiMe_2$)	A		ir, nmr	[70a]
(Z = Cl		(121)		
Br		(86)		
CH_2Ph)		(131)	X-ray	
$Z_2Ti[N(Me)SiMe_2]_2Y$ (ring: Ti–NMe–$SiMe_2$–Y–$SiMe_2$–NMe)	A			
(Z Y		(50)	ir, nmr	[103]
NMe_2 NMe		$35/10^{-4}$	ir, nmr	[103]
NMe_2 O		$80/10^{-4}$	ir, nmr	[103]
NEt_2 NMe		$70/10^{-4}$	ir, nmr	[103]
NEt_2 O			ir, nmr	[103]
CH_2Ph O			ir, nmr	[70a]
CH_2Ph NMe			ir, nmr	[70a]
CH_2Ph CH_2)			ir, nmr	[70a]
$X_2Ti[N(R)CH_2CH_2CH_2N(R)]$ (ring: RN–$(CH_2)_3$–NR chelating Ti)	A		ir, nmr	[77]
(R X				
Me NEt_2		$60–70/10^{-3}$		
Pr^i NEt_2		$50/10^{-4}$		
$SiMe_3$ NEt_2)		$60/10^{-4}$		
$SiMe_3$ NC_5H_{10})		$75/10^{-3}$		
$Ti(NC_5H_{10})_4$	A		ir, nmr	[51]
	A	180/0.1 (60)		[64]
	A, B	180/0.1 (100)		[65]

Table 6 *C'ntd.*

Compound	Method	B.p. (°C/mm.) (m.p., °C)	Comments	References
$Ti(N(CH_2)_4CHMe)_4$ (ring structure: 2-methylpiperidino)	A		ir, nmr	[51]
$Ti[N(Me)SiMe_2SiMe_2N(Me)]_2$ (ring structure)	A	(105)	X-ray	[99]
$Ti[N(Me)SiMe_2\text{–}Y\text{–}SiMe_2N(Me)]_2$ (ring structure)	A		ir, nmr	[101]
(Y				
NMe		(36)		
O		(78)		
CH_2)		(59)		
$Ti[N(R)(CH_2)_3N(R)]_2$ (spiro ring structure)	A		ir, nmr	[77]
(R				
Me		$60/10^{-3}$		
Pr^i		(~40)		
Bu^t		(<0)		
$SiMe_3$)		$85/10^{-4}$		
$Ti(NPh_2)_4$	A			[138, 167, 230]
$CpTiCl_2(NHMe).4MeNH_2$	C			[12]
F_3TiNMe_2 polymer	I, K	(>140)d	ir, nmr	[100]
F_3TiNEt_2 polymer	I, K	(>130)d	ir, nmr	[100]
F_3TiNR_2 ($R = Pr, Pr^i$)	C			[110, 111]
$Cl_3TiNH(CH_2)_6NH_2.n(H_2N)_2(CH_2)_6$ (n = 1, 2, 3) ?	C			[254]
Cl_3TiNMe_2	I			[30]
	A		ir, Raman, nmr	[82a]
	A		ir	[202]
$Cl_3TiNMePh$	A		ir, Raman, nmr	[82a]
Cl_3TiNEt_2	A	(85–6)	ir, Raman, nmr	[82a]
	A	(76–8)	ir, nmr	[256]
			X-ray	[154a]
Cl_3TiNPr_2	A	subl. $50/10^{-4}$	ir, Raman, nmr	[82a, 89]

Table 6 *C'ntd.*

Compound	Method	B.p. (°C/mm.) (m.p., °C)	Comments	References
Cl_3TiNPh_2	I			[30]
	A		ir, Raman, nmr	[82a]
$Cl_3Ti[N(SiMe_3)_2]$	A	(75–7)	ir, nmr	[74, 96]
$Cp_2TiNMe_2(Cl)$	C		ir	[24]
$Cp_2TiN(H)Et(Cl)$	C		ir	[24]
$Cp_2TiN(C_5H_{10})(Cl)$ (piperidino)	C		ir	[24]
$Cp_2TiN(H)Ph(Cl)$	A		ir	[25]
$Cp_2TiNPh_2(Cl)$	A	(253)d	ir	[25]
$Cp_2TiN(H)(C_6H_4\text{-}p\text{-}NO_2)(Cl)$	A	(288)d		[25]
$[Me_3SiNTiCl_2]_n$	I	(142)d	ir, nmr	[238]
			X-ray	[4]
Br_3TiNMe_2	I	subl. $60/10^{-4}$ (109–10)	ir, nmr	[83]
Br_3TiNEt_2	I	$98\text{–}9/10^{-3}$ (1)	ir, nmr	[83]
$Br_3TiN(C_5H_{10})$ (piperidino)	A	(93–5)	ir, nmr	[79]
$F_2Ti(NMe_2)_2$	I, K	(>130)d	ir, nmr, X-ray	[100, 247d]
$F_2Ti(NEt_2)_2$	I, K	(68)	ir, nmr	[100]
$F_2Ti[N(Me)SiMe_2]_2$ (cyclic: F_2Ti bonded to two N(Me), linked by $SiMe_2$–$SiMe_2$)	A	(140)d	ir, nmr	[103]
$Cl_2Ti(NMe_2)O_2SNMe_2$	D			[195]
$Cl_2Ti(NMe_2)S_2CNMe_2$	D			[194]
$Cl_2Ti(NHMe)_2$	C			[133]
$.2MeNH_2$	C			[133]
$.4MeNH_2$	C			[134]
$Cl_2Ti(NHEt)_2$	C			[133]
$.EtNH_2$	C			[134]
$.3EtNH_2$	C			[134]
$Cl_2Ti(NHPr)_2$	C		ir	[134]
$.4PrNH_2$	C			[134]
$Cl_2Ti(NHBu)_2$	C		ir	[134]
$.4BuNH_2$	C			[134]
$Cl_2Ti(NMe_2)_2$	I	subl. 70/vac		[30]
	A		ir, Raman, nmr	[82a]
$Cl_2Ti(NEt_2)_2$	A	90–100/0.1	ir, Raman, nmr	[82a]
	I	96–8/0.57 (67–9)		[256]

Table 6 *C'ntd.*

Compound	Method	B.p. (°C/mm.) (m.p., °C)	Comments	References
$Cl_2Ti[N(Me)SiMe_2]_2NMe$ (cyclic: Cl_2Ti bonded to two N(Me)–$SiMe_2$ groups bridged by NMe)	A	(68)	ir, nmr	[103]
$Cl_2Ti[N(Me)SiMe_2]_2$ (cyclic: Cl_2Ti bonded to two N(Me) groups, $SiMe_2$–$SiMe_2$ linked)	A		ir, nmr	[68,103]
$Br_2Ti(NHMe)_2$	C			[134]
$.4MeNH_2$	C			[134]
$Br_2Ti(NHEt)_2.4EtNH_2$	C			[134]
$Br_2Ti(NHPr)_2$	C			[134]
$.4PrNH_2$	C			[134]
$Br_2Ti(NHBu)_2.4BuNH_2$	C			[134]
$Br_2Ti(NMe_2)_2$	C			[156]
	I	$113\text{–}4/10^{-3}$ (43–4)	ir, nmr	[83,90]
$Br_2Ti(NEt_2)_2$	I	$108\text{–}9/10^{-3}$	ir, nmr	[83,90]
$Br_2Ti(NC_5H_{10})_2$ (piperidino)	I	(86–7)	ir, nmr	[79]
$Br_2Ti[N(Me)SiMe_2]_2X$ (cyclic: Br_2Ti bonded to two N(Me)–$SiMe_2$ groups bridged by X)	A		ir, nmr	[103]
(X = NMe		(72)		
CH_2)		(65)		
$Br_2Ti[N(Me)SiMe_2]_2$ (cyclic: Br_2Ti bonded to two N(Me) groups, $SiMe_2$–$SiMe_2$ linked)	A	(85)	ir, nmr	[103]
			ir	[92]
$I_2Ti(NMe_2)_2$	I	(62–4)	ir, nmr	[82]
$I_2Ti(NEt_2)_2$	I	subl. 75/0.0002	ir, nmr	[82]
$I_2Ti(NHMe)_2.4MeNH_2$	C			[134]
$I_2Ti(NHEt)_2.4EtNH_2$	C			[134]
$FTi(NEt_2)_3$	I, K	(10)	ir, nmr, pes	[100, 247b]
ClTi(NHMe)(NMe)	I			[133]
$ClTi(NMe_2)_3$	I	subl. 40–50/0.05 (60–2)		[30]
	A	$82\text{–}4/10^{-3}$		[30]

Table 6 *C'ntd.*

Compound	Method	B.p. (°C/mm.) (m.p., °C)	Comments	References
$ClTi(NMe_2)(O_2SNMe_2)_2$	D			[195]
$ClTi(NEt_2)_3$	I	95/0.02		[30]
			pes	[247b]
$ClTi(NPh_2)_3$	I			[30]
			pes	[83]
$ClTi[N(SiMe_3)_2]_3$	A	(202–5)	ir, nmr	[2]
$BrTi(NMe_2)_3$	I	subl. $80/10^{-4}$ (99–100)	ir, nmr	[82]
$BrTi(NEt_2)_3$	I	$123–5/10^{-3}$	ir, nmr	[82, 83, 88]
		(51)	pes	[247b]
$BrTi(N\langle piperidino\rangle)_3$ 	I	(105)	ir, nmr	[79]
$ITi(NMe_2)_3$	I	(130)	ir, nmr	[82]
$ITi(NEt_2)_3$		(49–50)	ir, nmr, pes	[82, 247b]
$Ti(NMe_2)_3[MCp(CO)_3]$ (M = Cr, Mo, W)			ir, nmr	[61]
$MeO[MeOCOC(NMe_2){:}C(COMe)]Ti(NMe_2)_2$	D		ir	[113]
$(Pr^iO)_3TiNMe_2$	K	48–50/0.02		[115]
$(RO)_3TiNEt_2$	K		ir, nmr	
(R = Pr^i				[130c, 130d]
Bu^t				[130c]
Am^t)				[130c]
$(Bu^iO)_3TiNMe_2$	K	185–210		[154]
$(Pr^iO)_2Ti(NMe_2)_2$	I	60/0.02		[30]
$(BuO)_2Ti(NHOct)_2$	C			[178]
$(Bu^iO)_2Ti(NMe_2)_2$	K			[153, 154]
$(Bu^tO)_2Ti(NEt_2)_2$	D	76–9/0.03		[114]
$(PhO)_2Ti(NEt_2)_2$	A			[230]
$(PhO)_2Ti(NPh_2)_2$	A			[230]
$Pr^iOTi(NMe_2)_3$	I	87/0.1		[30]
$EtSTi(NMe_2)_3$	K		ir, nmr	[53]
$(EtS)_2Ti(NMe_2)_2$	K		ir, nmr	[53]
$(Pr^iS)_2Ti(NMe_2)_2$	K		ir, nmr	[53]
$(Me_2NCS_2)_2Ti(NMe_2)_2$				[194]
$Et_2PTi(NMe_2)_3$		$51–2/10^{-3}$		[83, 86]
$(Me_3Si)_2PTi(NMe_2)_3$		$70–2/10^{-2}$		[83, 86]
$Et_2AsTi(NMe_2)_3$		$54–5/10^{-3}$		[83, 86]
		(−25)		[86]
$Et_2PTi(NEt_2)_3$		subl. $90/10^{-4}$		[86]
$Cp_2Ti(N\langle phthalimido: CO, CO, C_6H_4\rangle)_2$ 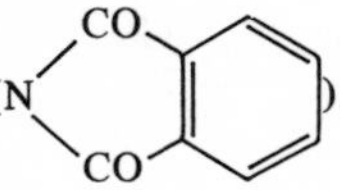	A	(175)	ir	[84]

Table 6 *C'ntd.*

Compound	Method	B.p. (°C/mm.) (m.p., °C)	Comments	References
$Ti(NPhCONMe_2)_4$	D			[114]
$.C_6H_6$	D	(149)	ir, nmr	[113]
$Ti(NPhCSNMe_2)_4$	D			[114]
$(EtO)_3TiNPhCO_2Et$	D	(92–5)	ir	[73]
$(EtO)_3TiN(Np\text{-}1)CO_2Et$	D	(71–3)	ir	[73]
$(PrO)_3TiNPhCO_2Pr^i$	D	(93–4)d	ir	[73]
$Ti_2(\mu\text{-}PMe_2)_2(NMe_2)_4$	E			[247]
(R-substituted O,O-chelate ring) $Ti[N(SiMe_3)_2]_2$	A			
(R = H			ir, nmr, esr, mag, uv	[215]
Bu^t		(134–6)	ir, nmr, esr, mag, uv	[215a]
Ph)		(134–8)	ir, nmr, esr, mag, uv	[215a]
$CpFe[C_5H_4Ti(NR_2)_3]$	A			
(R = Me		(18–20)	ir, nmr	[81]
Et)		(31–2)		[81]
$M[C_5H_4Ti(NR_2)_3]_2$	A			
(M R				
Fe Et		(40–50)d		[81]
Ru Me)		(90)d		[80]
$(CpFeC_5H_4)_2Ti(NEt_2)_2$	A	(72)d		[81]
$M(CO)_xC(NMe_2)OTi(NMe_2)_3$	K		ir, nmr	[237]
(M *x*				
Mo 5				
W 5				
Fe 4				
Ni 3)				
Zirconium: Zr(IV)				
$Cp_2Zr(NMe_2)_2$	K	subl. 120/0.1		[112]
	A, K	subl. 110–20/0.1		[115]
$Cp_2Zr(N\langle\text{aziridinyl}\rangle)_2$	B	subl. $110/10^{-3}$	nmr	[207]
	B	subl. $100/5 \times 10^{-3}$	nmr	[181]
$(MeC_5H_4)_2Zr(NMe_2)_2$	K	subl. 125–35/0.1		[115]
$Cp_2Zr(NEt_2)_2$	K	subl. 120–30/0.03		[115]
$Cp_2Zr(N\langle\text{pyrrolyl}\rangle)_2$	A	(170–1)	ir, mag	[177]

Table 6 *C'ntd.*

Compound	Method	B.p. (°C/mm.) (m.p., °C)	Comments	References
$Cp_2Zr(N\ \ )_2$	A	(59–62)	ir	[177]
$Cp_2Zr(NPh_2)_2$	A		ir, mag	[177]
$Cp_2Zr(N\ \ )_2$	A	(216–9)	ir	[177]
$CpZr(NMe_2)_3$	K	94–6/0.05		[115]
$IndZr(NMe_2)_3$	K	144/0.15		[115]
$Zr(NHMe)_4$	B			[21]
$(MeNH)_2Zr(NMe)$	B	(>360)		[21]
$(EtNH)_2Zr(NEt)$	B	(203)		[21]
$(PrNH)_2Zr(NPr)$	B	(208)		[21]
$(BuNH)_2Zr(NBu)$	B	(229)		[21]
$Zr(NHPh)_4$	B			[21]
$(PhNH)_2Zr(NPh)$	B	(>360)		[21]
$Zr(NMe_2)_4$			ir, nmr, pes	[51,166]
	A	120/0.1		[64]
	A	80/0.05 (70)		[65]
			ΔH_f	[202a]
$.2Ni(CO)_4$				[37]
$.2Fe(CO)_5$				[37]
$.2Mo(CO)_6$				[37]
$(Me_2N)_2Zr(NPr^i_2)_2$	B	subl. 100/0.05		[65]
$(Me_2N)_2Zr[N(SiMe_3)_2]_2$	B	subl. 120–40/0.04		[115]
$[C_6H_{11}N{:}C(NMe_2)NC_6H_{11}]_2Zr(NMe_2)_2$	D	(227)d	ir	[113]
$Zr(NMeBu)_4$			ir, nmr	[51]
$Zr(NEt_2)_4$			ir, nmr, pes	[51,166]
	A	120/0.1		[64]
	A, B	120/0.1		[65]
			ΔH_f	[202a]
$(Et_2N)_3ZrNPr^i_2$	B	112/0.05 (40)		[65]
$Zr(NPr_2)_4$			ir, nmr	[51]
	B	165/0.1		[65]
$Zr(NPr^i_2)_4$	A	subl. 120/10^{-3}	ir, nmr	[2]
$Zr(NBu_2)_4$			ir, nmr	[51]
$Zr(NBu^i_2)_4$			ir, nmr	[51]
	B	subl. 180/0.1		[65]

Table 6 *C'ntd.*

Compound	Method	B.p. (°C/mm.) (m.p., °C)	Comments	References
$Zr(N\bigcirc)_4$			ir, nmr	[51]
	A	190/0.2 (60)		[64]
	B	190/0.2 (80)		[65]
$Zr(N\bigcirc)_4$ (ring with Me)			ir, nmr	[51]
	B	190/0.1		[65]
$Zr(N\bigcirc)_4$ (ring with Me, Me)	B	200/0.05		[65]
$Zr(N(Me)-SiMe_2-Y-SiMe_2-N(Me))_2$	A		ir, nmr	[102]
(Y = NMe		(86)		
O		(95)		
CH_2)		(106)		
$Cl_2Zr(N(Me)-SiMe_2-Y-SiMe_2-N(Me))$	A		ir, nmr	[102]
(Y = NMe		(140)		
O		(140–1)d		
CH_2)		(150)d		
$R_2Zr(N(Me)-SiMe_2-SiMe_2-N(Me))$ (R = Cl, Br, Me)	A		ir	[92]
$Cl_3Zr(NHEt).EtNH_2$	C			[139]
$Cl_3Zr(NHPr).PrNH_2$	C			[139]
$Cl_3Zr(NHBu).BuNH_2$	C			[139]
$Cl_2Zr(NHMe)_2.MeNH_2$	C			[139]
$Zr(NPhCONMe_2)_4$	D	(158–60)	ir, nmr	[113]
$Cp_2Zr(N(CO)_2C_6H_4)_2$	A	(200)	ir	[177]

Table 6 *C'ntd.*

Compound	Method	B.p. (°C/mm.) (m.p., °C)	Comments	References
Hafnium: Hf(IV)				
$Cp_2Hf(NMe_2)_2$	K	subl. 120–5/0.02		[114]
$Cp_2Hf(NEt_2)_2$	K	subl. 120–30/0.03		[114]
$Hf(NMe_2)_4$			ir, nmr, pes	[51, 166]
	A			[107]
$.2Ni(CO)_4$				[37]
$.2Fe(CO)_5$				[37]
$.2Mo(CO)_6$				[37]
$Hf(NMeBu)_4$			ir, nmr	[51]
$Hf(NEt_2)_4$			ir, nmr, pes	[51, 166]
			ΔH_f	[202a]
$Hf(NPr_2)_4$			ir, nmr	[51]
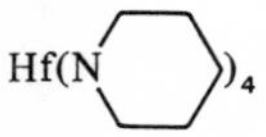 Hf(N)$_4$			ir, nmr	[51]
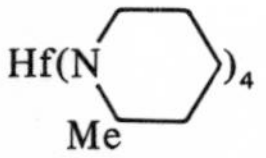 Hf(N)$_4$ Me			ir, nmr	[51]
$[C_6H_{11}N{:}C(NMe_2)NC_6H_{11}]_2Hf(NMe_2)_2$	D		ir	[113]
$Hf(NPhCONMe_2)_4$	D	(177)	ir	[113]
Vanadium: V(II)				
$V(NMe_2)_2$	I			[11]
$V(NEt_2)_2$	I			[11]
V(III)				
$V[N(CN)_2]_3$				[82]
$V(NMe_2)_3$	A			[11]
$V(NEt_2)_3$	A			[11]
$V[N(SiMe_3)_2]_3$	A		esr, X-ray, mass	[8, 9, 36, 42, 44, 172, 203]
$ClV(NPr^i_2)_2$	A		dimer	[11]
$OVNEt_2$	A+K			[197]
$KV(NPh_2)_4$	I		mag	[162a]
V(IV)				
$V(NMe_2)_4$			ir, nmr	[11]
			ir, nmr, pes	[51, 168]
			uv, esr	[7, 62]
	A			[255]
			pes	[168]

Table 6 *C'ntd.*

Compound	Method	B.p. (°C/mm.) (m.p., °C)	Comments	References
$V(NEt_2)_4$	I			[11]
			ir, nmr	[51]
			uv, esr, mag	[7]
			uv, esr, pes	[168, 175]
$V(NPh_2)_4$	A	(215)	ir	[162a]
$V(N(Ph)C_6H_4N(Ph))_2$ (cyclic, N-Ph)	A	(230)d	ir	[162a]
$V[(NMe-SiMe_2)_2Y]_2$	A		ir, nmr, esr, mag	[260b]
(Y = O,				
CH_2		(44)		
NMe)		(52)		
$Cl_2V(NHMe)_2$	C		ir, uv	[157, 158]
$.2MeNH_2$	C			[141]
$.4MeNH_2$	C		mag	[158]
$Cl_2V(NHEt)_2$	C		ir, uv	[157]
$.4EtNH_2$	C			[141]
$Cl_2V(NHPr)_2$	C		ir, uv	[157]
	C			[141]
$Cl_2V(NHBu)_2.4BuNH_2$	C			[141]
$Cl_2V(NMe_2)_2$	C		polymer, with $HNMe_2$	[141, 158]
			ir, uv	[157]
$Cl_2V(NEt_2)_2$	C		polymer, with $HNEt_2$	[136, 141]
			ir, uv	[157]
$Cl_2V(NPr_2)_2$	C		polymer, with $HNPr_2$	[141]
$Cl_3VNMe_2.Me_2NH$	C			[158]
$OV(NEt_2)_2$	A+K			[196, 197]
$(RC(O)CHC(O)R)V[N(SiMe_3)_2]_2$	A		ir, nmr, esr, mag, uv	[215a]
(R = Bu^t		subl. $125/10^{-4}$		
Ph)		(146–7)		
V(V)				
$OV(NMe_2)_3$	A	(40)		[196]
	A	subl. 10^{-3} mm. (40)		[197]
$OV(NEt_2)_3$	A	100/0.3		[130e, 196, 197]
$OVCl(NEt_2)_2$	A			[196, 197]

Table 6 *C'ntd.*

Compound	Method	B.p. (°C/mm.) (m.p., °C)	Comments	References
$V(OSiMe_3)(NSiMe_3)[N(SiMe_3)_2]_2$	A	(68)	ir, nmr	[94, 258]
$OVCl_2N(SiMe_3)_2$	A		ir, nmr	[94]
$OV(OPr^i)_2N(SiMe_3)_2$	A	81/0.5	ir, nmr	[94]
$OVOPr^i[N(SiMe_3)_2]_2$	A	106/0.6	ir, nmr	[94]
$OV(OPr^i)_2N[Si(OPr^i)_3]_2$	A	125/0.3	ir, nmr	[94]
$OV(OPr^i)[N\{Si(OPr^i)_3\}_2]_2$	A		ir, nmr	[94]
$OVOPr^i(NEt_2)_2$		$74/10^{-2}$	ir, nmr	[130e]
$OV(OPr^i)_2NEt_2$		$60\text{–}1/10^{-2}$	ir, nmr	[130e]
$Li[OV(OPr^i)_3NEt_2]$			ir, nmr	[130e]
Niobium: Nb(IV)				
$CpNb(NMe_2)_3$	K	d 150/0.005	nmr	[207]
	K	d>150/0.005	nmr	[181]
$Me_2NNb(NEt_2)_3$	B	120/0.1		[66]
$Nb(NMeBu)_4$	A	150/0.1		[66]
$Nb(NEt_2)_4$			ir, nmr, esr, mag, uv	[38, 51]
	A	120/0.1d	esr, mag, uv	[38, 66, 67]
$(Et_2N)_3NbN(CH_2)_5$ (piperidino)	B	170/0.1		[66]
$(Et_2N)_2Nb(N(CH_2)_5)_2$ (piperidino)	B	170/0.1d		[66]
$Nb(NPr_2)_4$	A	155/0.1		[66]
			esr, mag, uv	[38]
$Nb(NBu_2)_4$	A	175/0.1		[66]
$Nb(N(CH_2)_5)_4$ (piperidino)	B	~170/0.1d		[66]
Nb(V)				
$Nb(NMe_2)_5$			ir, nmr, X-ray	[51, 173]
	A	subl. 100/0.1		[66]
$(Me_2N)_3Nb(NEt_2)_2$	B	110/0.1d		[66]
$Nb(NMeBu)_5$	A	150/0.1d		[66]
$(BuMeN)_3Nb{:}NBu$	B			[66]
$(Me_2N)_3Nb{:}NBu^t$	A	(58–60)		[230a]
$Nb(N(CH_2)_5)_5$ (piperidino)	A, B	~170/0.1d	X-ray, ir, nmr	[66, 173]
F_4NbNEt_2	A	(150)	ir	[163]
$F_3Nb(NEt_2)_2$	A	(150)	ir	[163]
$F_3Nb(NEt_2)_2.C_5H_5N$	A	(56)	ir	[163]
$Cl_4NbN(Me)C(S)Me$	D		ir, nmr	[261]
$Cl_3Nb[N(Me)C(S)Me]_2$	D		ir, nmr, X-ray	[140, 261]

Table 6 *C'ntd.*

Compound	Method	B.p. (°C/mm.) (m.p., °C)	Comments	References
$Cl_2Nb[N(Me)C(S)Me]_3$	D		ir, nmr	[261]
$Cl_3Nb[N(R)C(O)Me]_2$	D		ir, nmr	[262]
(R = Me				
Ph)				
$Cl_3Nb(NHPr)_2.PrNH_2$	C			[108]
$Cl_3Nb(NHBu)_2.BuNH_2$	C			[108]
$Cl_3Nb(NMe_2)_2$	C			[159]
$.Me_2NH$	C			[108,159]
$Br_3Nb(NMe_2)_2.Me_2NH$	C			[108]
$Cl_2Nb(NHMe)_3$	C			[159]
$Cl_2(Me)Nb[NRC(O)Me]_2$	D			
(R = Me, Ph)			ir, nmr	[262]
$(O)Br_2NbN(H)R$	C			
(R = Et, Bu, Oct)			ir	[104]
$(O)Br_2NbNR_2.NHR_2$	C			
(R = Et, Bu, Oct)			ir	[104]
$(RO)_{5-n}Nb[N(Ph)C(O)OR]_n$	D		ir	[218]
(R = Me, Et, or Pr^i, n = 1, 2, 3, 4 or 5)				
(R = Me, n = 1)		130/0.1		
(R = Et, n = 1)		145/0.3		
(R = Pr^i, n = 1)		100/0.2		
(R = Bu^t, n = 1)		130–40/0.1		
Tantalum: Ta(IV)				
$Ta(NMeBu)_4$	A	155–8/0.1		[67]
Ta(V)				
$Ta(NMe_2)_5$			ir, nmr	[51]
	A	subl. 110/0.1		[64,67]
$(Me_2N)_4TaN{:}CMeNMe_2$	D		ir	[45]
$Ta(NMeBu)_5$	A			[67]
$(BuMeN)_3Ta{:}NBu$			ir, nmr	[51]
	I	150–5/0.1		[67]
$Ta(NEt_2)_5$	A			[73]
$(Et_2N)_3Ta{:}NEt$			ir, nmr	[51]
	I	130/0.1		[64,67]
$Ta(NPr_2)_5$	A			[67]
$(Pr_2N)_3Ta{:}NPr$			ir, nmr	[51]
	I	150/0.1		[67]
	A			[255]
$Ta(NBu_2)_5$	A			[67]
$(Me_2N)_3Ta{:}NBu^t$	A, B	(68–9)	nmr, X-ray	[230a]
$(Bu_2N)_3Ta{:}NBu$			ir, nmr	[51]
	I	175/0.1		[67]

Table 6 *C'ntd.*

Compound	Method	B.p. (°C/mm.) (m.p., °C)	Comments	References
Ta(N(CH2)5)5 [piperidino ring]			ir, nmr	[51]
	A, B			[67]
F_4TaNEt_2	A	(123)	ir, nmr	[163]
$F_4TaNEt_2.C_5H_5N$	A	(115)	ir, nmr	[163]
$F_4TaNEt_2.4\text{-}C_6H_7N$	A	(122)	ir, nmr	[163]
$F_3Ta(NEt_2)_2$	A	(203)	ir, nmr	[163]
$F_3Ta(NEt_2)_2.C_5H_5N$	A	(51)	ir, nmr	[163]
$F_3Ta(NEt_2)_2.4\text{-}C_6H_7N$	A	liquid	ir, nmr	[163]
$Cl_3Ta(NEt_2)_2.Et_2NH$	C			[108]
$Br_3Ta(NMe_2)_2.Me_2NH$	C			[108]
$Cl_4TaN(Me)C(S)Me$	D		ir, nmr	[261]
$Cl_4TaN(Ph)C(S)Me$	D		ir, nmr	[261]
$Cl_3Ta[N(Me)C(O)Me]_2$	D		ir, nmr	[261]
$Cl_3Ta[N(Ph)C(O)Me]_2$	D		ir, nmr	[261]
$Cl_2(Me)Ta[N(Me)C(O)Me]_2$	D		ir, nmr	[261]
$Cl_2Ta[N(Me)C(O)Me]_3$	D		ir, nmr	[261]
$Cl_2(Me)Ta[N(Ph)C(O)Me]_2$	D		ir, nmr	[261]
$Cl_2(Me)Ta[N(Me)C(S)Me]_2$	D		ir, nmr	[261]
$Cl_2(Me)Ta[N(Ph)C(S)Me]_2$	D		ir, nmr	[261]
$Ta(NMe_2)_2\{OC(O)NMe_2\}_3$			X-ray, dnmr	[121c, 129a]
Chromium: Cr(O)				
$[Ph_4As][(OC)_5CrN(CO)_2(CH_2)_2]$ [succinimido ring]	D			[27]
$[Ph_4As][(OC)_5CrN(CO)_2C_6H_4]$ [phthalimido ring]	D			[27]
Cr(II)				
$Cr(NR_2)_2$ $[R_2 = Et_2, Pr_2, Bu_2, (CH_2)_5]$	I			[23]
$CpCr(NO)NMe_2$	A		mass, dimer	[1, 241]
$Cr(NC_4H_3CH_2NH_2)_2$ [2-(aminomethyl)pyrrolyl, $H_2N \rightarrow Cr$ chelate]			ir, nmr	[174]
$[Cr\{N(H)C(CH)_3CMeN\}_2]_2$	A		X-ray	[131b]

Table 6 *C'ntd.*

Compound	Method	B.p. (°C/mm.) (m.p., °C)	Comments	References
$[CpCr(NO)NMe_2]_2$				
(cis)			X-ray	[105]
(trans)			X-ray	[105]
$Cr_2(O_2CNEt_2)_4(HNEt_2)_2$	D		X-ray	[121e]
$Cr[N(SiMe_3)_2]_2(THF)_2$	A		X-ray, mag	[56, 98]
Cr(III)				
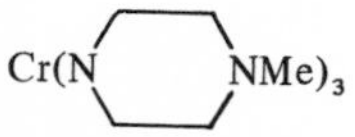 $Cr(N\ \ \ NMe)_3$ (piperazine ring)	C	(325)d		[69]
$Cr(NR_2)_3$ $[R_2 = Me_2, Et_2, Pr_2, Bu_2, (CH_2)_5]$	A			[23]
$Cr[N(SiMe_3)_2]_3$	A		esr, uv, pes X-ray	[8, 9, 44, 203] [36, 172]
$Cr(NPr^i_2)_3$	A		X-ray, esr, monomer mag, uv	[6, 55, 120]
$Cr(NPr^i_2)_3(O_2)$			ir, esr, mass, mag, uv	[63, 121]
$Cr(NPr^i_2)_3(NO)$		subl. $75/10^{-4}$	ir	[63]
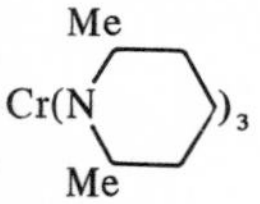 $Cr(N\ \ \ )_3$ (2,6-Me ring) .NO		subl. $90/10^{-4}$	ir	[63]
$Cr(OBu^t)_2NPr^i_2(NO)$		subl. $52/10^{-4}$ (61–2)	ir	[63]
$Cr[N(SiMe_3)_2]_3(NO)$	H	subl. $110/10^{-4}$	ir, nmr, X-ray, mass, uv	[56, 63]
$Cr[N(COMe)_2]_3$	A	(180–0.5)	ir, uv	[198]
$Cr[N(COPh)_2]_3$	A	(273–4.5)	ir, uv	[198]
$Cr_2(\mu\text{-}NEt_2)_2(O_2CNEt_2)_4$	D		X-ray	[121e]
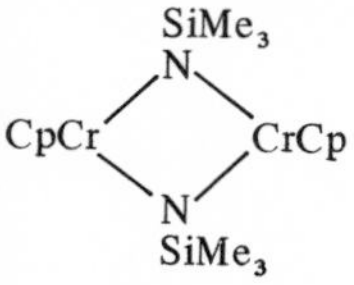 $CpCr(\mu\text{-}NSiMe_3)_2CrCp$	A		X-ray	[260a]
Cr(IV)				
$Cr(NEt_2)_4$	I		mass, mag, uv	[22, 23]
$Cr(NR_2)_4$ (R = Pr, Bu)	I		mag, uv	[23]
$OCr(NPr^i_2)_2$	K		esr	[121]
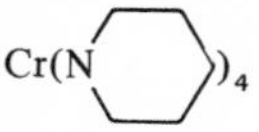 $Cr(N\ \ \ )_4$ (piperidine ring)	I	(~60)	mag, uv	[23]

Table 6 *C'ntd.*

Compound	Method	B.p. (°C/mm.) (m.p., °C)	Comments	References
Molybdenum: Mo(O)				
$[Ph_4As][(OC)_5Mo\overline{NCOCH_2CH_2CO}]$	D			[27]
$[Ph_4As][(OC)_5Mo\overline{NCOC_6H_4CO}]$	D			[27]
Mo(II)				
$[Mo\{N(H)\overline{C(CH)_3CMeN}\}_2]_2$	A		X-ray	[131b]
Mo(III)				
$Mo_2(NMe_2)_6$	A	subl. $100/10^{-6}$	ir, nmr, X-ray	[123, 130]
			ΔH_f	[131]
$Mo_2(NMeEt)_6$	A	$100–130/10^{-4}$		[124]
$Mo_2(NEt_2)_6$	A	$140–170/10^{-4}$		[124]
$Mo_2Cl_2(NMe_2)_4$	A		ir, nmr, X-ray	[3]
$Mo_2(O_2CNMe_2)_4(NMe_2)_2$	D		ir, nmr, mass	[130a]
$Mo_2Me_2(NMe_2)_4$	A		ir, nmr, X-ray	[121d]
$Mo_2Et_2(NMe_2)_4$	A	subl. $60–70/10^{-4}$	ir, dnmr, mass	[129c]
Mo(IV)				
$Mo(NMe_2)_4$	A	50	ir, nmr, uv,	
			X-ray	[39a, 121b, 130b]
			ΔH_f	[131]
$Mo(NEt_2)_4$	A	liquid	ir, nmr, uv	[39a]
Mo(V)				
$Cl_3Mo(NHMe)_2.MeNH_2$	C			[156]
$Cl_3Mo(NHEt)_2$	C		dimer	[156]
$Cl_3Mo(NHPr)_2.PrNH_2$	C			[156]
$Cl_3Mo(NHBu)_2.BuNH_2$	C	(128)d		[149, 156]
$Cl_3Mo(NMe_2)_2.Me_2NH$	C	(162)d		[149, 156]
$Cl_3Mo(NEt_2)_2$	C	(147)	dimer	[149, 156]
$Cl_3Mo(NPr_2)_2$	C	(206)	dimer	[149, 156]
$Cl_2Mo(NHMe)_3.MeNH_2$	C			[156]
$Cl_2Mo(NHPr)_3$	C	(74)d	dimer	[149, 156]
$.PrNH_2$	C			[149]
$Mo(O)Cl_2(NRR')$ (R = Me, R' = Et; R = R' = Me)	C			[201a]
$Mo(O)Cl(NEt_2)_2$	C			[201a]

Table 6 *C'ntd.*

Compound	Method	B.p. (°C/mm.)	Comments	References
Tungsten: W(O)				
$[Ph_4As][(OC)_5WN(CO)_2C_2H_4]$ (cyclic imide: N bonded to two CO groups bridged by CH_2CH_2)	D			[27]
$[Ph_4As][(OC)_5WN(CO)_2C_6H_4]$ (phthalimide: N bonded to two CO groups bridged by a benzene ring)	D			[27]
W(III)				
$W(NMe_2)_3$	A+I		polymer	[41]
$W_2(NMe_2)_6$	A	subl. 100–120/10^{-4}	nmr, dnmr, mass, X-ray, monomer	[122, 124, 127, 129]
			ΔH_f	[131]
$W_2(NMeEt)_6$	A	subl. 100–130/10^{-4}	dnmr, mass	[124, 127, 129]
$W_2(NEt_2)_6$	A	subl. 140–170/10^{-4}	dnmr, mass	[124, 127, 129]
$W_2(NMe_2)_2(NEt_2)_4$	A		ir, nmr, dnmr	[125]
$W_2Cl_2(NMe_2)_4$	A		ir, nmr, mass, X-ray	[3]
$W_2Me_2(NEt_2)_4$	A		X-ray	[125, 125a]
$W_2Cl_2(NEt_2)_4$	A		ir, nmr, dnmr, mass, X-ray	[3, 125, 126] 127]
$W_2Cl(NMe_2)_5$	I		ir, nmr	[3]
$W_2Br_2(NEt_2)_4$	A		ir, nmr, dnmr, mass, X-ray	[126]
$W_2I_2(NEt_2)_4$	A		ir, nmr, dnmr, mass, X-ray	[126]
$W_2(CH_2SiMe_3)_2(NEt_2)_4$	A		ir, dnmr, mass	[126]
$W_2(O_2CNMe_2)_3(NMe_2)_3$	A		X-ray	[129b]
W(IV)				
$Cl_3WNMe_2.2Me_2NH$	C		mag, uv	[72]
$Cl_3WNEt_2.2Et_2NH$	C		ir, mag, uv	[71, 72]
$Br_3WNEt_2.2Et_2NH$	C		ir, mag	[71]
$Cl_2W(NMe_2)_2.Me_2NH$	C		ir, mag	[71]
$Br_2W(NMe_2)_2.Me_2NH$	C		ir, mag	[71]
W(V)				
$Cl_2W(NHPr)_3$	C		ir, mag	[71]
$Cl_2W(NHBu)_3.BuNH_2$	C		ir, mag	[71]
$Br_2W(NHBu)_3.1.5BuNH_2$	C		ir, mag	[71]

Table 6 *C'ntd.*

Compound	Method	B.p. (°C/mm.) (m.p., °C)	Comments	References
W(VI)				
$W(NMe_2)_6$	A		ir, nmr, Raman, X-ray, mass, mag, uv	[41,41a,127]
			ΔH_f	[131]
			pes	[166]
$W[N(CD_3)_2]_6$	B, C		ir, nmr	[41a]
$W(NMe_2)_3(OMe)_3$	I		ir, nmr	[41a]
$W(NEt)_2(NEt_2)_2$	A, I		mass	[41a]
$W(NMe_2)_3(O_2CNMe_2)_3$	D		X-ray	[128,129a]
$Cl_2W(NHPr)_4$	C	(90)		[72]
$Cl_2W(NHBu)_4$	C	(133–6)		[72]
F_5WNMe_2	A		ir, nmr	[211]
F_5WNEt_2	A		ir, nmr	[211,212]
$WF_4(NEt_2)_2$	A		ir, nmr	[213]
$WF_2(NEt_2)_4$	A		ir, nmr	[213]
$WF_4\,OMeNEt_2$	A		ir, nmr	[212]
$WF_4OPhNEt_2$	A		ir, nmr	[212]
Manganese: Mn(I)				
$(OC)_5MnNHMe.MeNH_2$	C			[14]
$(OC)_5MnNMe_2$	A			[164]
Mn(II)				
$Mn[N(CN)_2]_2$	A		ir, mag	[186]
.2py	A			[186]
.2DMSO	A	(143)	ir	[187]
Mn(N(CO)₂C₆H₄ [phthalimido])$_2$.4RNH$_2$ (R = H, Me, Et, Pr^i, Bu, Bu^i, Pent, Pent^i)			ir, uv	[227]
.4R$_2$NH			ir, uv	[227]
$Mn[N(SiMe_3)_2]_2$	A	100/0.2	ir, nmr	[98]
			dimer (X-ray)	[54a]
$Mn[N(SiMe_3)_2]_2$.T.H.F.	A		ir, nmr, X-ray	[35, 150b]
Rhenium				
$Re_3Cl_6(NH_2)_3(NH_3)_3$	C		ir	[150]
$[Re(O)(NH_2)_4]_n$	A		ir	[150a]
Iron: Fe(I)				
$(OC)_4FeN$(cyclic) (?)	E			[106]

Table 6 *C'ntd.*

Compound	Method	B.p. (°C/mm.) (m.p., °C)	Comments	References
Fe(II)				
$Fe(NEt_2)_2$ (?)	A			[182]
.3bipy	A			[135]
$Fe[N(CN)_2]_2.2py$	A		ir, mag	[186]
$CpFe(CO)_2N$	A	subl. 60–70/0.1 (91)	ir, nmr	[235]
$CpFe(CO)_2N$ OCOMe	A	subl. 80–90/0.01	ir, nmr	[235]
$CpFe(CO)_2N$	A	(114–5)	ir, nmr	[235]
$CpFe(CO)_2N$	A	(150–5)d		[235]
$CpFe(CO)_2N$	A			[235]
$CpFeN(Me)PF_2Fe(PF_2)PF_2N(Me)PF_2$		(162–4)	X-ray	[229a]
Fe(III)				
$Fe[N(SiMe_3)_2]_3$	C	(60)d	ir, nmr, X-ray, mag, uv, esr, pes, Mössbauer	[8, 9, 10, 44, 97, 172, 203]
FeONHMe	C	d(60)		[240]
$Fe(N(CO)_2C_6H_4)_3$				[185]
$.3RNH_2$			ir	[223, 224]

Table 6 *C'ntd.*

Compound	Method	B.p. (°C/mm.) (m.p., °C)	Comments	References
(R = H, Me, Et, Pr, Pr^i, Bu, Bu^i)				
$.3R_2NH$			ir	[223, 224]
(R = Me, Et)				
ClFe(N-3-methylindolyl)$_2$	A			[232]
Ruthenium: Ru(II)				
$Ru(H)[N(SiMe_3)_2](PPh_3)_2$	A	(123–5)	ir, nmr, diamag, monomer	[109]
Cobalt: Co(I)				
$Co[N(SiMe_3)_2](PPh_3)_2$			X-ray	[150b]
Co(II)				
$Co[N(SiMe_3)_2]_2$	A	(73)	ir, nmr, uv	[46, 98]
$.NC_5H_5$	A	(105)	ir, nmr	[155]
$.NC_6H_7$	A	(106)	ir, nmr	[155]
$.NC_7H_9$	A	(61)	ir, nmr	[155]
$Co[N(SiMe_3)_2]_2(PPh_3)$	A	(>90d)	X-ray	[60]
$Co(NEtCMeCHCHNEt)_2$	A, I	120/0.1	ir, nmr, uv	[32]
$Co[N(CN)_2]_2$	A		ir, mag	[186]
.2py	A		ir, mag, uv	[186]
.2DMSO	A	(198–200)	ir, mag	[187]
.2(2-pic)	A		mag	[188]
.2b	A		ir, mag, uv	[192]
(b = quin, isoquin, 2,3-lut, 2,4-lut, 3-pic, 4-pic)				
.4b	A		ir, mag, uv	[192]
(b = quin, isoquin)				
Co(N(CO)$_2$C$_6$H$_4$)$_2$.b (phthalimido)			ir	[222, 228]
(b = $2NH_3$, en, bipy, *o*-phenanthroline)				
$[(NH_3)_5CoNHCOH]^{2+}$			ir, nmr, uv	[19]
Co(III)				
Co(N-pyrrolyl-2-CH:NMe)$_3$	C	(163–5)		[151]
$Cp_2Co_2(CO)(NBu^t)_2$	I		X-ray	[217, 233]

Table 6 *C'ntd.*

Compound	Method	B.p. (°C/mm.) (m.p., °C)	Comments	References
Rhodium: Rh(I)				
$Rh[N(SiMe_3)_2](PPh_3)_2$	A	(128–30)	ir, nmr, monomer	[109]
Nickel: Ni(I)				
$NiN(SiMe_3)_2(L)_2$				
($L = PPh_3$	A	(>80)d	X-ray	[60]
PMe_2Ph, $PMePh_2$, PEt_2Ph, $PEtPh_2$, PEt_3)			esr	[60]
Ni(N(2,5-dimethylpyrrolyl))(PPh_3)			ir, esr	[43]
Ni(II)				
$Ni[N(SiMe_3)_2]_2$		80/0.2		[98]
$(Et_3P)_2Ni(N(pyrrolyl))_2$	A		uv	[209]
$(Bu_3P)_2Ni(N(pyrrolyl))_2$	A		uv	[209]
$[(C_6H_{11})_3P]_2Ni(N(pyrrolyl))_2$	A		uv	[209]
H_2N—CH_2 → Ni—(N(pyrrolyl))$_2$			ir, nmr	[174]
$Ni[N(CN)_2]_2$	A		ir, mag	[186]
.2py	A		ir, mag	[186]
.2DMSO	A	(234)	ir, mag	[187]
.2(2–pic)	A		mag	[190]
.2b	A		ir, mag, uv	[188]
(b = quin, isoquin, 2,3–lut, 2,4–lut, 3–pic, 4–pic)				
.4b	A		ir, mag, uv	[188]
(b = quin, isoquin)				
$(CpNi)_2NBu^t$	I			[233]
$Ni[N(CONHC_6H_{11})C(NH_2){:}NH]_2$	C			[142]
Ni(N(CO)(CO))$_2(RNH_2)_2$ (R = H, Me, Pr, Pr^i, Bu, Bu^i, Bu^t, CH_2CH_2Ph, CH_2CH_2OH)			ir, uv, conductivity	[251]

Table 6 *C'ntd.*

Compound	Method	B.p. (°C/mm.) (m.p., °C)	Comments	References
Ni(N(CO)₂C₆H₄)₂(RNH₂)₂ (R = H, Me, Pr, Prⁱ, Bu, Buⁱ, Buᵗ, CH₂CH₂Ph, CH₂CH₂OH)				[251]
Ni(NC₄H₂Me₂)₂ (2,5-dimethylpyrrolyl)			ir, nmr, mass	[43]
Ni(N(CO)₂C₆H₄)₂.2b (b = NH₃, PrNH₂, PrⁱNH₂, en, pn, bipy, *o*-phenanthroline)				[248]
Palladium: Pd(II)				
Pd[N(CONHC₆H₁₁)C(NH₂):NH]₂	C			[142]
cis-Pd(NN=NN=C(Me))₂(PMe₂Ph)₂ (5-methyltetrazolyl)	A		X-ray	[15]
Platinum: Pt(II)				
trans-(Ph₃P)₂Pt(H)N(CO)₂C₂H₄ (succinimido)	H	(220–3)d	ir, nmr	[243, 244]
trans-(Ph₃P)₂Pt(H)N(CO)₂C₆H₄ (phthalimido)	H	(194–215)d	ir, nmr	[243, 244]
(Ph₃P)₂Pt(H)N(CO—NH)(CO—CO)	D	(257–61)	ir, nmr	[244]
(Et₃P)₂Pt(Cl)NHR	D		ir	[26]
[R = Ph		(105–6)		
p-O₂NC₆H₄		(210)		
PhCO		(104–5)		
CO₂Et		(69)		
CO₂Buᵗ		(109)		
C₃N₃(N₃)₂]		(168–9)		

Table 6 *C'ntd.*

Compound	Method	B.p. (°C/mm.) (m.p., °C)	Comments	References
$Pt(NHCOMe)_2$	A		ir	[200]
Pt(IV)				
$[Pt(Cl)(NCl_2)_2(NH_3)_3]Cl$	H		X-ray	[198a, 264]
$[PtCl_2(NCl_2)(NH_3)py_2]Cl$	H		ir, nmr	[198a]
$[PtCl_2(NCl_2)_2py_2]$	H			[198a]
$Pt(en)(NHMe)(NO_2)_2Cl$			$[M]_D^0$ 121.67	[119]
$Pt(en)(NHEt)(NO_2)_2Cl$			$[M]_D^0$ 60.10	[119]
$Pt(en)(NHPr)(NO_2)_2Cl$			$[M]_D^0$ 42.00	[119]
Copper: Cu(I)				
$[Cu\{N(SiMe_3)_2\}]_4$	A	180/0.2	X-ray	[36a, 98,176]
$CuNHPh.2NH_3$	B			[161]
p-$CuNHC_6H_4Me.2NH_3$	B			[161]
CuN	A			[184]
CuN O	A			[184]
$CuNPhCONPh_2.NH_3$	A		ir	[229]
Cu(II)				
$Cu(N\ )_2$ CH:NMe	C	(163–5)		[151]
$Cu(N\ )_2$ CHO	C			[151]
$Cu(N\ )_2$ CH:NBu		(160)	mag	[220]
$Cu(N\ )_2$ $CH:NC_6H_4Me$-*p*		(198)	mag	[220]
$Cu(N\ )_2$ CH:N– Me Me Me		(230)	mag	[220]

Table 6 *C'ntd.*

Compound	Method	B.p. (°C/mm.) (m.p., °C)	Comments	References
H_2N—CH_2 ↓ Cu—(N pyrrolyl ring $)_2$			ir, nmr	[174]
$Cu(NPhCONPh_2)_2.4NH_3$	A		mag	[229]
$Cu(N(CO)_2)_2.H_2O.2b$ (b = 2,4-lut, 4-pic)			ir	[219]
$.H_2O.b$ (b = py, 4-pic)			ir	[219]
$.2b$ (b = H_2O; py, 2,4-lut, 3,5-lut, 2-pic, 3-pic, 4-pic)			ir	[219]
$.H_2O, MeOH.2$-pic			ir	[219]
$.2H_2O.2$(3-pic)			ir	[219]
$.2RNH_2$			uv	[226]
			mag	[225]
(R = H, Me, Et, Pr, Pr^i, Bu, Bu^i, Pent)	F		ir	[263]
$.2MN(CO)_2.nH_2O$ (M,*n* = Li,*1*; Na,*4*; K,*6*; Rb,*2*; Cs,*2*)	C			[257]
$K_2[Cu(N(CO)_2)_4].6H_2O$			ir	[244, 257]
$Cu(N(CO)_2C_2H_4)_2.2RNH_2$ (R = H, Me, Pr, Pr^i, Bu, Bu^i, Bu^t, CH_2CH_2Ph, CH_2CH_2OH)			ir, uv, conductivity	[251, 252]
$Cu(N(CO)_2C_6H_4)_2.2RNH_2$ (R = H, Me, Pr, Pr^i, Bu, Bu^i, Bu^t, CH_2CH_2Ph, CH_2CH_2OH)			ir, uv, conductivity	[251, 252]

Table 6 *C'ntd.*

Compound	Method	B.p. (°C/mm.) (m.p., °C)	Comments	References
$Cu(N(CO)_2C_6H_4)_2.2RNH_2$			ir	[249,250]
(R = H, Me, Et, Pr, Bu, Pent)				
$.2R_2NH$ (R = Et, Pr, Pr^i, Bu)				[240]
$.RNH_2$ (R = Me, Et, Pr, Pr^i, Bu, Pent)			mag, uv	[240,253]
$.R_2NH$ (R = Et, Pr)			mag, uv	[253]
$Cu[N(CONHC_6H_{11})C(NH_2):NH]_2$	C		mag	[142]
$Cu[N(CN)CMe:NCN]_2$	A			[247c]
$Cu[N(CN)_2]_2$	A		ir, mag	[186]
.2py	A		ir, mag	[186]
.2DMSO	A	(99–100)		[187]
.2b				[191]
.4b				[191]
$ClCuNHCSNH_2$				[242]
$ClCuNEtCSNH_2$				[242]
Silver: Ag(I)				
AgNHCMe:NH	A		ir	[200]
AgNHCOMe	B			[161]
$.2NH_3$	B			[160,161]
AgNHCOPh				[17]
AgNPhCOMe	B			[161]
$.NH_3$	B			[161]
$AgNPhCONPh_2.NH_3$	A			[229]
AgNPhCN	A			[236]
$AgN(CO)_2C_2(CN)_2$	C			[236]
$AgN(CO)_2C_6H_4$	A			[236]
$(AgNCO)_3$	C			[169]
$(AgNCO)_2(HNCO)$	C	subl. $250/10^{-3}$		[169]
$(AgNCO)(HNCO)_2$	C			[169]
AgNPhCOMe	B			[160]
$.NH_3$	B			[160]
$[AgN(SiMe_3)_2]_n$	A			[93]

Table 6 *C'ntd.*

Compound	Method	B.p. (°C/mm.) (m.p., °C)	Comments	References
Gold: Au(I)				
$Au(L)[N(SiMe_3)_2]$	A		ir, nmr	[247a]
(L = PMe_3		(35–37)		
PPh_3		(120–123)		
$AsPh_3$)		(41–43)		
Au(III)				
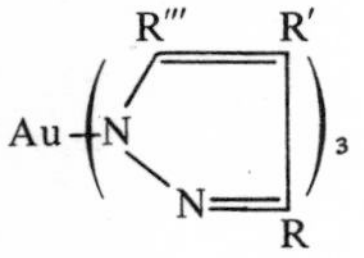 R''' R' Au–N N= R (ring)$_3$	C		ir, nmr, mass, trimer	[31]
(R R' R''				
H H H				
Me H Me				
Me H Ph				
Me Et Me)				
Lanthanides and Actinides				
$Y(NPr^i_2)_3$	A			[46a]
$Yb(NPr^i_2)_3$	A			[46a]
$Nd(NPr^i_2)_3$	A			[46a]
$Sc[N(SiMe_3)_2]_3$	A	(172–4)	ir, nmr, mass, uv, esr, X-ray	[8, 9, 44, 165]
$M[N(SiMe_3)_2]_3$	A			
(M = La or Ce		(145–9)	ir, nmr, mag, mass	[47, 47a]
Pr		(155–8)	ir, nmr	[47, 47a]
Nd		(161–4)	ir, nmr, X-ray	[3a, 47, 47a]
Sm		(155–8)	ir, nmr	[47, 47a]
Eu		(160–2)	ir, nmr, X-ray	[47, 47a, 165]
Gd		(160–3)	ir, nmr	[47, 47a]
Ho		(160–4)	ir, nmr	[47, 47a]
Y		(180–4)	ir, nmr	[47, 47a]
Yb		(162–5)	ir, nmr, X-ray	[47, 47a, 150b]
Lu)		(167–70)	ir, nmr	[47, 47a]
$M[N(SiMe_3)_2]_3(OPPh_3)$				
(M = La		(185–7)	ir, nmr, X-ray	[48, 48a]
Eu		(141–4)	ir, nmr	[48, 48a]
Lu)		(165–6)	ir, nmr	[48, 48a]
$M_2[N(SiMe_3)_2]_4O_2(OPPh_3)_2$				
(M = La			ir, nmr, X-ray	[48, 48a]
Pr			ir, nmr	[48, 48a]
Sm			ir, nmr	[48, 48a]
Eu			ir, nmr	[48, 48a]
Lu)			ir, nmr	[48, 48a]

Table 6 *C'ntd.*

Compound	Method	B.p. (°C/mm.) (m.p., °C)	Comments	References
Uranium: U(IV)				
$U(\eta\text{-}C_5H_5)_2(NEt_2)_2$	K		ir, nmr, mass	[179a]
$U(NR_2)_4$	A			
(R = Me, Et, Bu^i)			ir	[18]
$U(NEt_2)_4$	A	115–25/0.06	ir, nmr, X-ray	[183, 241c]
$U(NPr_2)_4$	A	40–50/10^{-4}	ir, nmr, mag	[241a]
$U(NBu_2)_4$	A	99–100/10^{-4}	ir, nmr, mag	[241a]
U(N(2,5-dimethylpyrrolyl))$_4$ [structure: pyrrole ring with Me, Me]	A	(98–102)	ir, nmr	[216]
$U(NPh_2)_4$	A		ir, nmr, X-ray	[241c]
$Li.UO(NPh_2)_3.Et_2O$	I		ir, nmr, X-ray	[241c]
$U_3[MeN(CH_2)_2NMe]_6$	B		ir, nmr, X-ray	[241b]
$U_4[MeN(CH_2)_2NMe]_6$	B		ir, nmr, X-ray	[241d]
Thorium: Th(IV)				
$Th(NEt_2)_4$	A	110/0.05	ir, nmr	[51, 259]
$Th(Cl)[N(SiMe_3)_2]_3$	A	(178–181)	ir, nmr	[49]
$Th(NR_2)_4$ (R = Me, Et, Pr, Bu, or Bu^i)	A		ir, nmr, mag	[18, 241a]

†Footnotes to Table 6: for explanation of abbreviations, see footnotes to Table 1, p. 57.

J REFERENCES

[1] Ahmed, M., Bruce, R., and Knox, G., *Z. Naturforsch.*, 1966, **21b**, 289.

[2] Airoldi, C., and Bradley, D. C., *Inorg. Nucl. Chem. Letters*, 1975, **11**, 155.

[3] Akiyama, M., Chisholm, M. H., Cotton, F. A., Extine, M. W., and Murillo, C. A., *Inorg. Chem.*, 1977, **16**, 2407.

[3a] Andersen, R. A., Templeton, D. H., and Zalkin, A., *Inorg. Chem.*, 1978, **17**, 2317.

[4] Alcock, N. W., Pierce-Butler, M., and Willey, G. R., *J.C.S. Chem. Comm.*, 1974, 627.

[5] Allen, A. D., and Bottomley, F., *Accounts Chem. Res.*, 1968, **1**, 360; Murray, R., and Smith, D. C., *Co-ord. Chem. Rev.*, 1968, **3**, 429.

[6] Alyea, E. C., Basi, J. S., Bradley, D. C., and Chisholm, M. H., *Chem. Comm.*, 1968, 495; *J. Chem. Soc. (A)*, 1971, 772.

[7] Alyea, E. C., and Bradley, D. C., *J. Chem. Soc. (A)*, 1969, 2330.

[8] Alyea, E. C., Bradley, D. C., and Copperthwaite, R. G., *J.C.S. Dalton*, 1972, 1580.

[9] Alyea, E. C., Bradley, D. C., Copperthwaite, R. G., and Sales, K. D., *J.C.S. Dalton,* 1973, 185.
[10] Alyea, E. C., Bradley, D. C., Copperthwaite, R. G., Sales, K. D., Fitzsimmons, B., and Johnson, C. E., *Chem. Comm.*, 1970, 1715.
[11] Alyea, E. C., Bradley, D. C., Lappert, M. F., and Sanger, A. R., *Chem. Comm.*, 1969, 1064.
[12] Anagnostopoulos, A., and Nicholls, D., *J. Inorg. Nucl. Chem.*, 1966, **28**, 3045.
[13] Andrianov, K. A., Astakhin, V. V., Kochkin, D. A., and Sukhanova, I. V., *J. Gen. Chem. U.S.S.R.*, 1961, **31**, 3178.
[14] Angelici, R. J., *Chem. Comm.*, 1965, 486.
[15] Ansell, G. B., *J.C.S. Dalton,* 1973, 371.
[16] Antler, M., and Laubengayer, A. W., *J. Amer. Chem. Soc.*, 1955, **77**, 5250.
[17] Arbuzov, A. E., and Shapshinskaya, O. M., *Tr. Kazank. Khim.-Technol. Inst.*, 1962, **30**, 22.
[18] Bagnall, K. W., and Yanir, E., *J. Inorg. Nucl. Chem.*, 1974, **36**, 777.
[19] Balahura, R. J., and Jordan, R. B., *J. Amer. Chem. Soc.*, 1970, **92**, 1533.
[20] Bancroft, G. M., Mays, M. J., and Prater, B. E., *Chem. Comm.*, 1969, 585.
[21] Bartlett, R. K., *J. Inorg. Nucl. Chem.*, 1966, **28**, 2448.
[22] Basi, J. S., and Bradley, D. C., *Proc. Chem. Soc.*, 1963, 305.
[23] Basi, J. S., Bradley, D. C., and Chisholm, M. H., *J. Chem. Soc. (A)*, 1971, 1433.
[24] Baye, L. J., *Synth. Inorg. Metal-org. Chem.*, 1972, **2**, 47.
[25] Baye, L. J., *Synth. Inorg. Metal-org. Chem.*, 1975, **5**, 95.
[26] Beck, W., and Bauder, M., *Chem. Ber.*, 1970, **103**, 583.
[27] Beck, W., and Shier, E., *Z. Naturforsch.*, 1970, **25b**, 211.
[28] Bennett, C. R., and Bradley, D. C., *J.C.S. Chem. Comm.*, 1974, 29.
[29] Benson, R. E., Eaton, D. R., Josey, A. D., and Phillips, W. D., *J. Amer. Chem. Soc.*, 1961, **83**, 3714.
[30] Benzing, E. P., and Kornicker, W. A., *Chem. Ber.*, 1961, **94**, 2263.
[31] Bonati, F., Minghetti, G., and Banditelli, G., *J.C.S. Chem. Comm.*, 1974, 88.
[32] Bonnett, R., Bradley, D. C., and Fisher, K. J., *Chem. Comm.*, 1968, 886.
[33] Bonnett, R., Bradley, D. C., Fisher, K. J., and Rendall, I. F., *J. Chem. Soc. (A)*, 1971, 1622.
[34] Boyd, T., U.S. Patent 2,579,413 (1951).
[35] Bradley, D. C., *Adv. Inorg. Chem. Radiochem.*, 1972, **15**, 259.
[36] Bradley, D. C., *Chem. in Brit.*, 1975, **11**, 393.
[36a] Bradley, D. C., unpublished results cited in ten Hoedt, R. W. M., Noltes, J. G., van Koten, G., and Spek, A. L., *J.C.S. Dalton,* 1978, 1800.
[37] Bradley, D. C., Charalambous, J., and Jain, S., *Chem. Ind. (London)*, 1965, 1730.
[38] Bradley, D. C., and Chisholm, M. H., *J. Chem. Soc. (A)*, 1971, 1511.

[39] Bradley, D. C., and Chisholm, M. H., *Accounts Chem. Res.*, 1976, **9**, 273.
[39a] Bradley, D. C., and Chisholm, M. H., *J. Chem. Soc. (A)*, 1971, 2741.
[40] Bradley, D. C., and Chivers, K. J., *J. Chem. Soc. (A)*, 1968, 1967.
[41] Bradley, D. C., Chisholm, M. H., Heath, C. E., and Hursthouse, M. B., *Chem. Comm.*, 1969, 1261.
[41a] Bradley, D. C., Chisholm, M. H., and Extine, M. W., *Inorg. Chem.*, 1977, **16**, 1791, 1794.
[42] Bradley, D. C., and Copperthwaite, R. G., *Chem. Comm.*, 1971, 764.
[43] Bradley, D. C., *Adv. Chem. Ser.*, 1976, **150**, 266.
[44] Bradley, D. C., Copperthwaite, R. G., Cotton, S. A., Gibson, J. F., and Sales, K. D., *J.C.S. Dalton*, 1973, 191.
[45] Bradley, D. C., and Fisher, K. J., *J. Amer. Chem. Soc.*, 1971, **93**, 2058.
[46] Bradley, D. C., and Ganorkar, M. C., *Chem. Ind. (London)*, 1968, 1521.
[46a] Bradley, D. C., Ghotra, J. S., and Hart, F. A., unpublished results.
[47] Bradley, D. C., Ghotra, J. S., and Hart, F. A., *J.C.S. Dalton*, 1973, 1021.
[47a] Bradley, D. C., Ghotra, J. S., and Hart, F. A., *J.C.S. Chem. Comm.*, 1972, 349.
[48] Bradley, D. C., Ghotra, J. S., Hart, F. A., Hursthouse, M. B., and Raithby, P. R., *J.C.S. Chem. Comm.*, 1974, 40.
[48a] Bradley, D. C., Ghotra, J. S., Hart, F. A., Hursthouse, M. B., and Raithby, P. R., *J.C.S. Dalton*, 1977, 1166.
[49] Bradley, D. C., Ghotra, J. S., and Hart, F. A., *Inorg. Nucl. Chem. Letters*, 1974, **10**, 209.
[50] Bradley, D. C., and Gitlitz, M. H., *Chem. Comm.*, 1965, 289.
[51] Bradley, D. C., and Gitlitz, M. H., *J, Chem. Soc. (A)*, 1969, 980.
[52] Bradley, D. C., and Gitlitz, M. H., *J. Chem. Soc. (A)*, 1969, 1152.
[53] Bradley, D. C., and Hammersley, P. A., *J. Chem. Soc. (A)*, 1967, 1894.
[54] Bradley, D. C., and Hillyer, M. J., *Trans. Faraday Soc.*, 1966, **62**, 2374.
[54a] Bradley, D. C., Hursthouse, M. B., Malik, K. M. A., and Möseler, R., *Transition Met. Chem.*, 1978, **3**, 253.
[55] Bradley, D. C., Hursthouse, M. B., and Newing, C. W., *Chem. Comm.*, 1971, 411.
[56] Bradley, D. C., Hursthouse, M. B., Newing, C. W., and Welch, A. J., *J.C.S. Chem. Comm.*, 1972, 567.
[57] Bradley, D. C., Hursthouse, M. B., and Rendall, I. F., *Chem. Comm.*, 1969, 672.
[58] Bradley, D. C., Hursthouse, M. B., and Rodesiler, P. F., *Chem. Comm.*, 1969, 14.
[59] Bradley, D. C., Hursthouse, M. B., and Rodesiler, P. F., *J.C.S. Dalton*, 1972, 2100.
[60] Bradley, D. C., Hursthouse, M. B., Smallwood, R. J., and Welch, A. J., *J.C.S. Chem. Comm.*, 1972, 872.
[61] Bradley, D. C., and Kasenally, A. S., *Chem. Comm.*, 1968, 1430.

[62] Bradley, D. C., Moss, R. H., and Sales, K. D., *Chem. Comm.*, 1969, 1255.
[63] Bradley, D. C., and Newing, K., *Chem. Comm.*, 1970, 219.
[64] Bradley, D. C., and Thomas, I. M., *Proc. Chem. Soc.*, 1959, 225.
[65] Bradley, D. C., and Thomas, I. M., *J. Chem. Soc.*, 1960, 3857.
[66] Bradley, D. C., and Thomas, I. M., *Can. J. Chem.*, 1962, **40**, 449.
[67] Bradley, D. C., and Thomas, I. M., *Can. J. Chem.*, 1962, **40**, 1355.
[68] Bradley, D. C., and Torrible, E. G., *Can. J. Chem.*, 1963, **41**, 134.
[69] Brasen, W. R., Holmquist, H. E., and Benson, R. E., *J. Amer. Chem. Soc.*, 1961, **83**, 3125.
[70] Brasen, W. R., and Howard, E. G., U.S. Patent 3,052,705 (1962).
[70a] Brauer, D. J., Bürger, H., and Wiegel, K., *J. Organometallic Chem.*, 1978, **150**, 215.
[71] Brisdon, B. J., and Fowles, G. W. A., *J. Less-Common Metals*, 1964, **7**, 102.
[72] Brisdon, B. J., Fowles, G. W. A., and Osborne, B. P., *J. Chem. Soc.*, 1962, 1330.
[73] Bürger, H., *Monatsh.*, 1964, **95**, 671.
[74] Bürger, H., *Angew. Chem. Internat. Edn.*, 1964, **3**, 141.
[75] Bürger, H., *Monatsh.*, 1963, **94**, 574.
[76] Bürger, H., and Dämmgen, U., *J. Organometallic Chem.*, 1975, **101**, 295.
[77] Bürger, H., and Dämmgen, U., *Z. Anorg. Chem.*, 1977, **429**, 173.
[78] Bürger, H., and Dämmgen, U., *J. Organometallic Chem.*, 1975, **101**, 307.
[79] Bürger, H., and Kluess, C., *J. Organometallic Chem.*, 1976, **108**, 69.
[80] Bürger, H., and Kluess, C., *Z. Anorg. Chem.*, 1976, **423**, 112.
[81] Bürger, H., and Kluess, C., *J. Organometallic Chem.*, 1973, **56**, 269.
[82] Bürger, H., Kluess, C., and Neese, H.-J., *Z. Anorg. Chem.*, 1971, **381**, 198.
[82a] Bürger, H., and Neese. H.-J., *Z. Anorg. Chem.*, 1969, **365**, 243.
[83] Bürger, H., and Neese, H.-J., *Z. Anorg. Chem.*, 1969, **370**, 275.
[84] Bürger, H., and Neese, H.-J., *J. Organometallic Chem.*, 1969, **20**, 129.
[85] Bürger, H., and Neese, H.-J., *J. Organometallic Chem.*, 1970, **21**, 383.
[86] Bürger, H., and Neese, H.-J., *Inorg. Nucl. Chem. Letters*, 1970, **6**, 299.
[87] Bürger, H., and Neese, H.-J., *J. Organometallic Chem.*, 1971, **32**, 223.
[88] Bürger, H., and Neese, H.-J., *J. Organometallic Chem.*, 1971, **31**, 213.
[89] Bürger, H., and Neese, H.-J., *Chimia*, 1970, **24**, 209.
[90] Bürger, H., and Neese, H.-J., *J. Organometallic Chem.*, 1972, **36**, 101.
[91] Bürger, H., Sawodny, W., and Wannagat, U., *J. Organometallic Chem.*, 1965, **3**, 113.
[92] Bürger, H., Schlingmann, M., and Pawelke, V., *Z. Anorg. Chem.*, 1976, **419**, 116, 121.
[93] Bürger, H., and Seyferth, H., *Angew. Chem. Internat. Edn.*, 1964, **3**, 646.
[94] Bürger, H., Smrekar, O., and Wannagat, U., *Monatsh.*, 1964, **95**, 292.
[95] Bürger, H., Stammreich, H., and Sans, Th.F., *Monatsh.*, 1966, **97**, 1276.
[96] Bürger, H., and Wannagat, U., *Monatsh.*, 1963, **94**, 761.

[97] Bürger, H., and Wannagat, U., *Monatsh.*, 1963, **94**, 1007.
[98] Bürger, H., and Wannagat, U., *Monatsh.*, 1964, **95**, 1099.
[99] Bürger, H., Wiegel, K., Thewalt, U., and Schomburg, D., *J. Organometallic Chem.*, 1975, **87**, 301.
[100] Bürger, H., and Wiegel, K., *Z. Anorg. Chem.*, 1973, **398**, 257.
[101] Bürger, H., and Wiegel, K., *Z. Anorg. Chem.*, 1976, **419**, 157.
[102] Bürger, H., and Wiegel, K., *Z. Anorg. Chem.*, 1976, **426**, 301.
[103] Bürger, H., and Wiegel, K., *J. Organometallic Chem.*, 1977, **124**, 279.
[104] Busla'ev, Yu. A., Simbsyna, S. M., Sinyagen, V. I., and Polikerpova, M. A., *Zh. Neorg. Khim.*, 1970, **15**, 2324.
[105] Bush, M. A., and Sim, G. A., *J. Chem. Soc. (A)*, 1970, 611.
[106] Bulkin, B. J., and Lynch, J. A., *Inorg. Chem.*, 1968, **7**, 2654.
[107] Cardin, D. J., Chandra, G., and Lappert, M. F., *U.S. Govt. Res. Develop. Rep.*, 1968, **68**, 62.
[108] Carnell, P.J.H., and Fowles, G.W.A., *J. Less-Common Metals*, 1962, **4**, 40.
[109] Cetinkaya, B., Lappert, M. F., and Torroni, S., *J.C.S. Chem. Comm.*, 1979, in the press.
[110] Chandler, J. A., and Drago, R. S., *Inorg. Chem.*, 1962, **1**, 356.
[111] Chandler, J. A., Wuller, J. E., and Drago, R. S., *Inorg. Chem.*, 1962, **1**, 65.
[112] Chandra, G., George, T. A., and Lappert, M. F., *Chem. Comm.*, 1967, 116.
[113] Chandra, G., Jenkins, A. D., Lappert, M. F., and Srivastava, R. C., *J. Chem. Soc. (A)*, 1970, 2550.
[114] Chandra, G., and Lappert, M. F., *Inorg. Nucl. Chem., Letters*, 1965, **1**, 83.
[115] Chandra, G., and Lappert, M. F., *J. Chem. Soc. (A)*, 1968, 1940.
[116] Chatt, J., Dilworth, J. R., and Leigh, G. J., *Chem. Comm.*, 1969, 687.
[117] Chatt, J., Garforth, J. D., Johnson, N. P., and Rowe, G. A., *J. Chem. Soc.*, 1964, 1012.
[118] Chatt, J., Leigh, G. J., and Richards, R. L., *Chem. Comm.*, 1969, 515.
[119] Chernyaev, I., Andrianova, O. N., and Leites, N. S. L., *Russ. J. Inorg. Chem.*, 1962, **7**, 749.
[120] Chien, J. C. W., and Kruse, W., *Inorg. Chem.*, 1970, **9**, 2615.
[121] Chien, J. C. W., Kruse, W., Bradley, D. C., and Newing, C. W., *Chem. Comm.*, 1970, 1177.
[121a] Chisholm, M. H., and Cotton, F. A., *Accounts Chem. Res.*, 1978, **11**, 356.
[121b] Chisholm, M. H., Cotton, F. A., and Extine, M. W., *Inorg. Chem.*, 1978, **17**, 1329.
[121c] Chisholm, M. H., Cotton, F. A., and Extine, M. W., *Inorg. Chem.*, 1978, **17**, 2000.
[121d] Chisholm, M. H., Cotton, F. A., Extine, M. W., and Murillo, C. A., *Inorg. Chem.*, 1978, **17**, 2338.
[121e] Chisholm, M. H., Cotton. F. A., Extine, M. W., and Rideout, C. D., *Inorg. Chem.*, 1978, **17**, 3536.
[122] Chisholm, M. H., Cotton, F. A., Extine, M. W., Stults, B. R., and Troup,

J. M., *J. Amer. Chem. Soc.*, 1975, **97**, 1242.
[123] Chisholm, M. H., Cotton, F. A., Frenz, B. A., Reichert, W. W., and Shive, L., *J.C.S. Chem. Comm.*, 1974, 480.
[124] Chisholm, M. H., Cotton, F. A., Frenz, B. A., Reichert, W. W., Shive, L., and Stults, B. R., *J. Amer. Chem. Soc.*, 1976, **98**, 4469.
[125] Chisholm, M. H., Cotton, F. A., Extine, M. W., Miller, M., and Stults, B. R., *J. Amer. Chem. Soc.*, 1976, **98**, 4486.
[125a] Chisholm, M. H., Cotton, F. A., Extine, M. W., Miller, M., and Stults, B. R., *Inorg. Chem.*, 1976, **15**, 2244.
[126] Chisholm, M. H., Cotton, F. A., Miller, M., and Stults, B. R., *Inorg. Chem.*, 1977, **16**, 320.
[127] Chisholm, M. H., Extine, M. W., Cotton, F. A., and Stults, B. R., *J. Amer. Chem. Soc.*, 1976, **98**, 4477.
[128] Chisholm, M. H., and Extine, M. W., *J. Amer. Chem. Soc.*, 1974, **96**, 6214.
[129] Chisholm, M. H., and Extine, M. W., *J. Amer. Chem. Soc.*, 1975, **97**, 5625.
[129a] Chisholm, M. H., and Extine, M. W., *J. Amer. Chem. Soc.*, 1977, **99**, 792.
[129b] Chisholm, M. H., Extine, M. W., and Reichert, W. W., *Adv. Chem. Ser.*, 1976, **150**, 273.
[129c] Chisholm, M. H., Haitko, D. A., and Murillo, C. A., *J. Amer. Chem. Soc.*, 1978, **100**, 6262.
[130] Chisholm, M. H., and Reichert, W. W., *J. Amer. Chem. Soc.*, 1974, **96**, 1249.
[130a] Chisholm, M. H., and Reichert, W. W., *Inorg. Chem.*, 1978, **17**, 767.
[130b] Chisholm, M. H., Reichert, W. W., and Thornton, P., *J. Amer. Chem. Soc.*, 1978, **100**, 2744.
[130c] Choukroun, R., and Gervais, D., *Compt. Rend.*, 1974, **278C**, 1409.
[130d] Choukroun, R., and Gervais, D., *Synth. Inorg. Metal-org. Chem.*, 1978, **8**, 137.
[130e] Choukroun, R., and Gervais, D., *Inorg. Chim. Acta*, 1978, **27**, 163.
[131] Connor, J. A., Pilcher, G., Skinner, H. A., Chisholm, M. H., and Cotton, F. A., *J. Amer. Chem. Soc.*, 1978, **100**, 7738.
[131a] Cotton, F. A., *Accounts Chem. Res.*, 1978, **11**, 225.
[131b] Cotton, F. A., Niswander, R. H., and Sekutowski, J. C., *Inorg. Chem.*, 1978, **17**, 3541.
[132] Coutts, R. S. P., Wailes. P. C., and Kingston, J. V., *Chem. Comm.*, 1968, 1170.
[133] Cowdell, R. T., and Fowles, G. W. A., *J. Chem. Soc.*, 1960, 2522.
[134] Cowdell, R. T., Fowles, G. W. A., and Walton, R. A., *J. Less-Common Metals*, 1963, **5**, 386.
[135] Creemers, H. M. J. C., *Hydrostannolysis*, Organisch Chemisch Instituut T.N.O., Utrecht (1967).

[136] Cucinella, S., and Mazzei, A., Ger. Patent 2,047,060 (1971).
[137] Cuvigny, T., and Normant, H., *Compt. Rend.*, 1969, **268C**, 834.
[137a] Davidson, P. J., Lappert, M. F., and Pearce, R., *Chem. Rev.*, 1976, **76**, 219.
[138] Dermer, O. C., and Fernelius, W. C., *Z. Anorg. Chem.*, 1935, **221**, 83.
[138a] Diamond, S. E., and Mares, F., *J. Organometallic Chem.*, 1977, **142**, C55.
[139] Drake, J. E., and Fowles, G. W. A., *J. Chem. Soc.*, 1960, 1498.
[140] Drew, M. G. B., and Wilkins, J. D., *J.C.S. Dalton*, 1974, 198.
[141] Duckworth, M. W., and Fowles, G. W. A., *J. Less-Common Metals*, 1962, **4**, 338.
[142] Dutta, R. L., and Lahiry, S., *J. Indian Chem. Soc.*, 1960, **37**, 789.
[143] Eaton, D. R., Josey, A. D., Phillips, W. D., and Benson, R. E., *Mol. Phys.*, 1962, **5**, 407.
[144] Eaton, D. R., Josey, A. D., Phillips, W. D., and Benson, R. E., *J. Chem. Phys.*, 1962, **37**, 347.
[145] Eaton, D. R., Josey, A. D., Phillips, W. D., and Benson, R. E., *Discuss. Faraday Soc.*, 1962, **34**, 77.
[146] Eaton, D. R., Josey, A. D., and Sheppard, W. A., *J. Amer. Chem. Soc.*, 1963, **85**, 2689.
[147] Eaton, D. R., and Phillips, W. D., *J. Chem. Phys.*, 1965, **43**, 392.
[148] Eaton, D. R., Phillips, W. D., and Caldwell, D. J., *J. Amer. Chem. Soc.*, 1963, **85**, 397.
[149] Edwards, D. A., and Fowles, G. W. A., *J. Chem. Soc.*, 1961, 24.
[150] Edwards, D. A., and Ward, R. T., *J. Inorg. Nucl. Chem.*, 1973, **35**, 1043.
[150a] Edwards, D. A., and Ward, R. T., *J.C.S. Dalton*, 1972, 89.
[150b] Eller, P. G., Bradley, D. C., Hursthouse, M. B., and Meek, D. W., *Co-ord. Chem. Rev.*, 1977, **24**, 1.
[151] Emmert, B., Diehl, K., and Gollwitzer, F., *Ber.*, 1929, **62B**, 1733.
[152] Ewald, A. H., and Sinn, E., *Inorg. Chem.*, 1967, **6**, 40.
[153] Farbwerke Hoechst, A.-G., Belg. Patent 652,506 (1965).
[154] Farbwerke Hoechst, A.-G., Belg. Patent 661,389 (1965).
[154a] Fayos, J., and Mootz, D., *Z. Anorg. Chem.*, 1971, **380**, 196.
[155] Fisher, K. J., *Inorg. Nucl. Chem. Letters*, 1973, **9**, 921.
[156] Fowles, G. W. A., and Cartell, E., *U.S. Dept. Com. Office Tech. Serv. P.B. Rept.*, 149,423 (1962).
[157] Fowles, G. W. A., and Lanigan, P. G., *J. Less-Common Metals*, 1964, **6**, 396.
[158] Fowles, G. W. A., and Pleass, C. M., *J. Chem. Soc.*, 1957, 1674.
[159] Fowles, G. W. A., and Pleass, C. M., *J. Chem. Soc.*, 1957, 2078.
[160] Franklin, E. C., *J. Amer. Chem. Soc.*, 1915, **37**, 2279.
[161] Franklin, E. C., *Proc. Nat. Acad. Sci.*, 1915, **1**, 68.
[162] Friederich, E., *Z. Physik*, 1925, **31**, 813; Friederich, E., and Sittig, L., *Z. Anorg. Chem.*, 1925, **143**, 293; Klemm, W., and Schuth, W., *ibid.*, 1931, **201**, 24.

[162a] Frölich, H. O., and Märkisch, V., *Z. Chem.*, 1975, **15**, 276.
[163] Fuggle, J. C., Sharp, D. W. A., and Winfield, J. M., *J.C.S. Dalton*, 1972, 1766.
[164] George, T. A., and Lappert, M. F., *Chem. Comm.*, 1966, 463.
[165] Ghotra, J. S., Hursthouse, M. B., and Welch, A. J., *J.C.S. Chem. Comm.*, 1973, 669.
[166] Gibbins, S. G., Lappert, M. F., Pedley, J. B., and Sharp, G. J., *J.C.S. Dalton*, 1975, 72.
[166a] Gillespie, R. J., *J. Chem. Soc.*, 1963, 4672.
[167] Giannini, U., Longi, P., Deluca, D., and Pivatto, B., Ger. Patent 2,030,753 (1971).
[168] Gottardi, W., *Monatsh.*, 1967, **98**, 1613.
[169] Green, M. L. H., in *Organometallic Compounds*, ed. Coates, G. E., Green, M. L. H., and Wade, K., Methuen, London, Vol. 2, 1968.
[170] Hagenmüller, P., Rouxel, J., and Portier, J., *Compt. Rend.*, 1962, **254C**, 2000.
[170a] Harris, D. H., D.Phil. Thesis, University of Sussex, 1975.
[171] Harris, D. H., and Lappert, M. F., *J. Organometallic Chem. Library*, 1976, **2**, 13.
[172] Heath, C. E., and Hursthouse, M. B., unpublished data.
[173] Heath, C. E., and Hursthouse, M. B., *Chem. Comm.*, 1971, 143.
[174] Hennig, H., and Daute, R., *Z. Chem.*, 1969, **9**, 275.
[175] Holloway, C. E., Mabbs, F. E., and Smail, W. R., *J. Chem. Soc. (A)*, 1968, 2980.
[176] Hursthouse, M. B., personal communication.
[177] Issleib, K., and Batz, G., *Z. Anorg. Chem.*, 1969, **369**, 83.
[178] Issleib, K., and Haeckert, H., *Z. Naturforsch.*, 1966, **21b**, 519.
[179] Jacini, G., *Olii minerali, grassi e saponi, colori e vernici*, 1953, **30**, 193; *Chem. Abstr.*, 1955, **49**, 3796.
[179a] Jamerson, J. D., and Takats, J., *J. Organometallic Chem.*, 1974, **78**, 623.
[180] Jenkins, A. D., Lappert, M. F., and Srivastava, R. C., *Polymer Letters*, 1968, **6**, 865; *European Polymer Journal*, 1971, **7**, 289.
[181] Jenkins, A. D., Lappert, M. F., and Srivastava, R. C., *J. Organometallic Chem.*, 1970, **23**, 165.
[182] Jones, K., and Lappert, M. F., *J. Organometallic Chem.*, 1965, **1**, 295.
[183] Jones, R. G., Karmas, G., Martin, G. A., and Gilman, H., *J. Amer. Chem. Soc.*, 1956, **78**, 4285.
[184] Kauffman, T., Albrecht, J., Berger, D., and Legler, J., *Angew. Chem. Internat. Edn.*, 1967, **6**, 633.
[185] Kaufmann, H. P., and Fleiter, L., *Fette, Seifen, Anstrichmittel*, 1964, **66**, 477; *Chem. Abstr.*, 1964, **61**, 14895.
[186] Köhler, H., *Z. Anorg. Chem.*, 1964, **331**, 237.
[187] Köhler, H., *Z. Anorg. Chem.*, 1965, **336**, 245.

[188] Köhler, H., Hartung, H., and Seifert, B., *Z. Anorg. Chem.*, 1966, **347**, 30.
[189] Köhler, H., Laub, P., and Frischkov, A., *J. Less-Common Metals*, 1971, **23**, 171.
[190] Köhler, H., and Seifert, B., *Z. Chem.*, 1965, **5**, 142.
[191] Köhler, H., and Seifert, B., *Z. Naturforsch.*, 1967, **22b**, 238.
[192] Köhler, H., and Seifert, B., *Z. Anorg. Chem.*, 1967, **352**, 265.
[193] Koepf, H., and Block, B., *Z. Naturforsch.*, 1968, **23b**, 1534.
[194] Kornicker, W. A., U.S. Patent 3,297,733 (1967).
[195] Kornicker, W. A., U.S. Patent 3,318,932 (1967).
[196] Kornicker, W. A., Benzing, E. P., and Perry, E., U.S. Patent 3,370,041 (1968).
[197] Kornicker, W. A., Benzing, E. P., and Perry, E., U.S. Patent 3,394,156 (1968).
[198] Kraihanzel, C. S., and Stehly, D. N., *Inorg. Chem.*, 1967, **6**, 277.
[198a] Kukushkin, Y. N., *Russ. J. Inorg. Chem.*, 1957, **2**, 2371; 1959, **4**, 1131; 1960, **5**, 1943.
[199] Kurzer, F., *Chem. Rev.*, 1956, **56**, 95.
[200] Kutzelnigg, W., and Mecke, R., *Spectrochim. Acta*, 1962, **18**, 549.
[201] Kuznetsov, E. V., Ignat'eva, E. K., and Kostromina, S. Ya., U.S.S.R. Patent 182,722 (1966).
[201a] Kuznetsova, A. A., Goryachova, L. F., and Busla'ev, Yu. A., *Izv. Akad. Nauk. S.S.S.R. Ser. Khim.*, 1970, 509.
[202] Kyker, G. S., and Schram, E. P., *J. Amer. Chem. Soc.*, 1968, **90**, 3672.
[202a] Lappert, M. F., Patil, D. S., and Pedley, J. B., *J.C.S. Chem. Comm.*, 1975, 830.
[203] Lappert, M. F., Pedley, J. B., Sharp, G. J., and Bradley, D. C., *J.C.S. Dalton*, 1976, 1737.
[204] Lappert, M. F., and Prokai, B., *Adv. Organometallic Chem.*, 1967, **5**, 225.
[205] Lappert, M. F., and Sanger, A. R., *J. Chem. Soc. (A)*, 1971, 874.
[206] Lappert, M. F., and Sanger, A. R., *J. Chem. Soc. (A)*, 1971, 1314.
[207] Lappert, M. F., and Srivastava, R. C., unpublished results, 1969.
[208] Lappert, M.F., and Srivastava, G., *Inorg. Nucl. Chem., Letters*, 1965, **1**, 53.
[209] Martin, R. L., and Winter, G., *J. Chem. Soc.*, 1965, 4709.
[210] McClellan, W. R., and Benson, R. E., *J. Amer. Chem. Soc.*, 1966, **88**, 5165.
[211] Majid, A., McLean, R. R., Ouellette, T. J., Sharp, D. W. A., and Winfield, J. M., *Inorg. Nucl. Chem., Letters*, 1971, **7**, 53.
[212] Majid, A., Sharp, D. W. A., Winfield, J. M., and Hanby, I., *J.C.S. Dalton*, 1973, 1876.
[213] Majid, A., McLean, R. R., Sharp, D. W. A., and Winfield, J. M., *Z. Anorg. Chem.*, 1971, **385**, 65.
[214] Manoussakis, G. E., and Tossidis, J. A., *J. Inorg. Nucl. Chem.*, 1972, **34**, 2449.

[215] Manzer, L. E., *J. Amer. Chem. Soc.*, 1977, **99**, 276.
[215a] Manzer, L. E., *Inorg. Chem.*, 1978, **17**, 1552.
[216] Marks, T. J., and Kolb, J. R., *J. Organometallic Chem.*, 1974, **82**, C35.
[217] Matsu-ura, Y., Yasuoka, N., Ueki, T., Kasai, N., Kakudo, M., Yoshida, T., and Otsuka, S., *Chem. Comm.*, 1967, 1122.
[218] Mehrotra, R. C., and Rai, A. K., *J. Inorg. Nucl. Chem.*, 1974, **36**, 1887.
[219] Miki, S., and Yamada, S., *Bull. Chem. Soc. Japan*, 1964, **37**, 1044.
[220] Minkin, V. I., Osipov, O. A., and Verkovodova, D. Sh., *Russ. J. Inorg. Chem.*, 1966, **11**, 2829.
[221] Murakami, T., Morisaki, S., and Hiro, K., Brit. Patent 1,030,394 (1966).
[222] Narain, G., *Bull. Chem. Soc. Japan*, 1965, **38**, 2073.
[223] Narain, G., *Z. Anorg. Chem.*, 1966, **343**, 109.
[224] Narain, G., *Can. J. Chem.*, 1966, **44**, 975.
[225] Narain, G., *J. Inorg. Nucl. Chem.*, 1966, **28**, 2441.
[226] Narain, G., and Shukla, P., *Z. Anorg. Chem.*, 1966, **342**, 221.
[227] Narain, G., and Shukla, P., *Austral. J. Chem.*, 1967, **20**, 227.
[228] Narain, G., Shukla, P., and Srivastava, L.N., *J. Prakt. Chem.*, 1966, **31**, 123.
[229] Nast, R., and Danneker, W., *Ann.*, 1966, **693**, 1.
[229a] Newton, M. G., King, R. B., Chang, M., and Gimeno, J., *J. Amer. Chem. Soc.*, 1978, **100**, 1635.
[230] Nicco, A., and Aboulafia, J., Fr. Patent 1,358,503 (1964).
[230a] Nugent, W. A., and Harlow, R. L., *J.C.S. Chem. Comm.*, 1978, 579.
[231] Orioli, P. L., *Co-ord. Chem. Rev.*, 1971, **6**, 285.
[232] Oddo, B., *Gazz. Chim. Ital.*, 1914, **44**, 268.
[233] Otsuka, S., Nakamura, A., and Hoshida, T., *Ann.*, 1969, **719**, 54.
[234] Paul, R. C., Chadha, S. L., and Vasisht, S. K., *J. Less-Common Metals*, 1968, 288.
[235] Pauson, P. L., and Qazi, A. R., *J. Organometallic Chem.*, 1967, **7**, 321.
[236] Permachem Corp., Brit. Patent 847,256 (1960).
[237] Petz, W., *J. Organometallic Chem.*, 1974, **72**, 369.
[238] Pierce-Butler, M., and Willey, G. R., *J. Organometallic Chem.*, 1973, **54**, C19.
[239] Phillips, W. D., and Benson, R. E., *J. Chem. Phys.*, 1960, **33**, 607.
[240] Portier, J., *Rev. Chim. Miner.*, 1966, **3**, 483; *Chem. Abstr.*, 1967, **66**, 101251.
[241] Preston, F. J., and Reed, R. L., *Chem. Comm.*, 1966, 51.
[241a] Reynolds, J. G., and Edelstein, N. M., *Inorg. Chem.*, 1977, **16**, 2822.
[241b] Reynolds, J. G., Zalkin, A., Templeton, D. H., and Edelstein, N. M., *Inorg. Chem.*, 1977, **16**, 599.
[241c] Reynolds, J. G., Zalkin, A., Templeton, D. H., and Edelstein, N. M., *Inorg. Chem.*, 1977, **16**, 1090.
[241d] Reynolds, J. G., Zalkin, A., Templeton, D. H., and Edelstein, N. M., *Inorg. Chem.*, 1977, **16**, 1858.

[241e] Reynolds, J. G., Zalkin, A., Templeton, D. H., Edelstein, N. M., and Templeton, L. K., *Inorg. Chem.*, 1976, **15**, 2498.
[242] Rogers, M. C., and Borth, P. F., U.S. Patent 3,136,670 (1964).
[243] Roundhill, D. M., *Chem. Comm.*, 1969, 567.
[244] Roundhill, D. M., *Inorg. Chem.*, 1970, **9**, 254.
[245] Ruff, O., and Eisner, F., *Ber.*, 1908, **41**, 2250; Ruff, O., *Ber.*, 1909, **42**, 900.
[246] Ruff, O., and Treidel, O., *Ber.*, 1913, **45**, 1346.
[246a] Sarneski, J. E., McPhail, A. T., Onan, K. D., Erickson, L. E., and Reilley, C. N., *J. Amer. Chem. Soc.*, 1977, **99**, 7376.
[247] Schmidbaur, H., Scharf., W., and Füller, H.-J., *Z. Naturforsch.*, 1977, **32b**, 858.
[247a] Schmidbaur, H., and Shiotani, A., *J. Amer. Chem. Soc.*, 1970, **92**, 7003.
[247b] Sharp, G. J., D.Phil. Thesis, University of Sussex, 1975.
[247c] Shaw, J. T., U.S. Patent 3,198,829 (1965).
[247d] Sheldrick, W. S., *J. Fluorine Chem.*, 1974, **4**, 415.
[248] Shukla, P., Khare, M. P., and Srivastava, L. N., *Z. Anorg. Chem.*, 1964, **333**, 165.
[249] Shukla, P., Khare, M. P., and Srivastava, L. N., *J. Prakt. Chem.*, 1965, **28**, 21.
[250] Shukla, P., Khare, M. P., and Srivastava, L. N., *J. Prakt. Chem.*, 1965, **29**, 38.
[251] Slabbert, N. P., and Thornton, D. A., *J. Inorg. Nucl. Chem.*, 1972, **34**, 2449.
[252] Slabbert, N. P., and Thornton, D. A., *Spectrosc. Letters*, 1970, **3**, 83.
[253] Srivastava, L. N., Shukla, P., and Khare, M. P., *Z. Phys. Chem., (Leipzig)*, 1967, **234**, 153, 157.
[254] Sumarokova, T. N., and Sakenova, D. S., *J. Gen. Chem. U.S.S.R.*, 1964, **34**, 2696.
[254a] Thewalt, U., and Schomburg, D., *Z. Naturforsch.*, 1975, **30b**, 636.
[255] Thomas, I. M., *Can. J. Chem.*, 1961, **39**, 1386.
[256] Toyo Rayon Co. Ltd., Fr. Patent 1,386,994 (1965).
[257] Tschugaeff, L. A., *J. Russ. Phys. Chem. Soc.*, 1906, **7**, 1083; *Chem. Abstr.*, 1907, **1**, 1394.
[258] Wannagat, U., *Angew. Chem. Internat. Edn.*, 1964, **3**, 318.
[259] Watt, G. M., and Gadd, K. F., *Inorg. Nucl. Chem. Letters*, 1973, **9**, 203.
[260] Weingarten, H., and Van Wazer, J. R., *J. Amer. Chem. Soc.*, 1965, **87**, 724.
[260a] Wiberg, N., Häring, H.-W., and Schubert, U., *Z. Naturforsch.*, 1978, **33b**, 1365.
[260b] Wiegel, K., and Bürger, H., *J. Organometallic Chem.*, 1977, **129**, 309.
[261] Wilkins, J. D., *J. Organometallic Chem.*, 1974, **65**, 383.
[262] Wilkins, J. D., *J. Organometallic Chem.*, 1974, **67**, 269.
[263] Yamada, S., and Miki, S., *Bull. Chem. Soc. Japan*, 1963, **36**, 680.

[264] Zipprich, M., and Pritzkow, H., *Angew. Chem. Internat. Edn.*, 1976, **15**, 225.

9

Amides of Zinc, Cadmium, and Mercury: Synthesis, Physical Properties, and Structures

A ZINC

1. Introduction

There are no reviews dealing with amides of zinc, but the topic is discussed in a wider context of organozinc chemistry [8b] or of complexes of $\bar{N}(SiMe_3)_2$ or related ligands [16d]. The chemistry of zinc amides has not been extensively examined. As well as amidozinc compounds, a dizincylamine $Bu^tN(ZnEt)_2$, has been reported, prepared from Et_2Zn and H_2NBu^t [34a]; the intermediate $EtZnNHBu^t$ was isolated.

Apart from their insertion reactions [e.g., Eq. (2)], hydrolysis, or reactions with other protic compounds (e.g., diazoalkanes [24a]), and reactions with bases such as pyridine (see below), little has been recorded relating to their chemistry. The compounds $(RZnNR'_2)_2$ are efficient catalysts for the trimerisation of isocyanates [31, 32] (as are other amides, e.g., of Sn^{IV}, see Chapter 10). The mechanism involves initial formation of a 1:1 insertion product (the effective catalyst, acting as a template) which then forms a co-ordination polymer with isocyanate from which the trimer can be displaced by some suitable base [32].

Data on thermochemistry [18a] or photoelectron spectroscopy [16e] are conveniently discussed in connection with similar studies on mercury amides (Section C.3).

2. Synthesis

Zinc derivatives of primary or secondary amines or amides were prepared by alkane elimination as long ago as 1856 [Eq. (1)] [15]. The object of that early

$$Et_2Zn + 2HNR_2 \longrightarrow Zn(NR_2)_2 + 2EtH \quad (1)$$

paper was to show that the same relationships exist between metallo-derivatives of amines and amides as those between amines and amides. Considering the facilities available at that time, this was a remarkable piece of work. Not until 1965 was further systematic investigation of these compounds made: the reaction of an alcohol, thiol, or secondary amine with diethylzinc yields mono- or di-substituted products, $(RZnNR'_2)_n$ and $[Zn(NR'_2)_2]_n$ [9]. The same method was used to prepare various alkyl(toluido)zinc compounds and the cyclohexylamido analogues [2], as well as (from Et_2Zn) $Zn[N(SiCl_3)_2]_2$ [42b], $EtZnNR_2$ [17c], $EtZnNHBu^t$ [34a], and $Bu^tN(ZnEt)_2$.

The reaction of Et_2Zn with $HNPh_2$ is accelerated by equimolar quantities of bipyridine or certain other bases, affording $EtZnNPh_2$.bipy or $Zn(NPh_2)_2$.bipy [17b]. Similarly, zinc derivatives of secondary amines [27, 28, 32], amides [7, 27, 32], and thioamides [7, 27, 32] have been prepared. Whereas $Zn[N(SiMe_3)_2]_2$ (or its Cd or Hg analogue) was originally claimed not to form a complex with pyridine or tetrahydrofuran [8], a crystalline 1:1 adduct with pyridine or γ-picoline (or, in solution only, NMe_3) was later obtained [14].

Insertion of phenyl isocyanate into zinc-carbon, -oxygen, or -nitrogen bonds has yielded *N*-zinc substituted amides, carbamates, or ureas [Eq. (2)] [31].

$$EtZnNPh_2 + PhNCO \longrightarrow EtZnNPhCONPh_2 \quad (2)$$

Reaction (3) affords a route to the interesting $Zn[N(SiMe_3)_2]_2$ [8].

$$ZnCl_2 + 2NaN(SiMe_3)_2 \longrightarrow 2NaCl + Zn[N(SiMe_3)_2]_2 \quad (3)$$

Metathetical exchange of a zinc alkyl with tris(dimethylamino)borane, $B(NMe_2)_3$, or dimethyl(dimethylamino)alane, Me_2AlNMe_2, yields methyl-(dimethylamido)zinc [1]. Bis(dicyanoamido)zinc [22, 23] and bis(triphenyl-ureido)zinc [27] are prepared similarly.

The bischelate complex (I) is obtained by addition of the free amine to an aqueous ammoniacal solution of a zinc salt [11], in an analogous manner to the preparation of the corresponding nickel(II) complex.

The bischelate complex (II) is isolated from the cobalt(II) analogue by successive demetallation (to liberate the free base EtN=CH−CH=CMeNHEt) and reaction with ethanolic zinc chloride [7a]; compound (II) is a di-iminato complex rather than an amide.

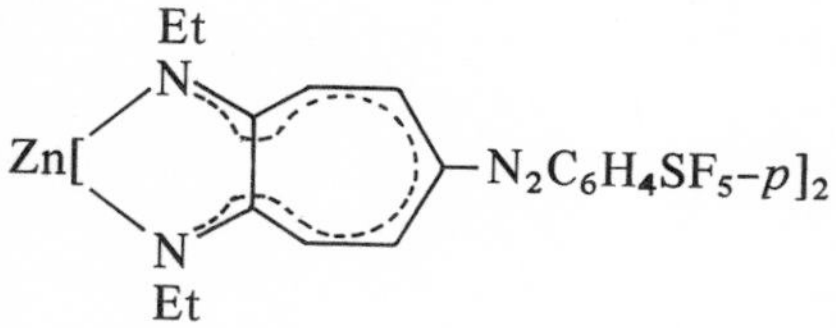

(I)

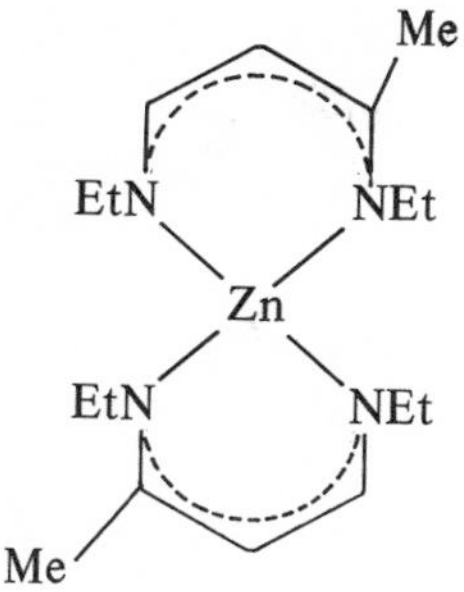

(II)

3. Physical properties and structures

The co-ordination number of zinc in amides appears to be two {in the monomeric D_{2d} or D_{2h} $Zn[N(SiMe_3)_2]_2$ [8]}, three, or four (corresponding to a tetrahedral arrangement of valencies about Zn). Zinc bisdialkylamides are generally polymeric [7, 9, 28, 31-33], e.g., see (III). Alkyl(dialkylamido)zinc compounds are usually dimeric, with bridging amido-groups, as in (IV) [9, 28, 31, 33]. Addition of pyridine to $(MeZnNPh_2)_2$ gives the tetrahedral monomeric complexes $Zn(NPh_2)_2.2py$ (+ Me_2Zn) or $MeZnNPh_2.2py$ [9]. Bis(diphenylamido)zinc is also dimeric, probably due to the bulk of the NPh_2 groups [33], while $Zn[N(SiMe_3)_2]_2$ is monomeric [8], perhaps also for steric reasons.

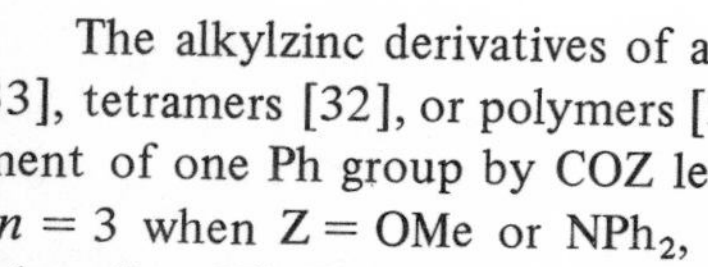

(III)

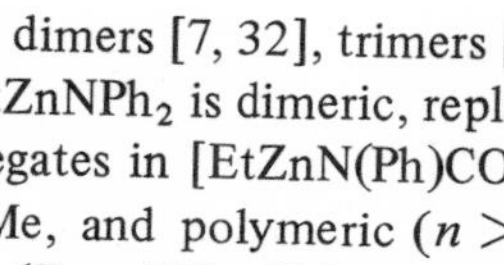

(IV)

The alkylzinc derivatives of amides or ureas are dimers [7, 32], trimers [32, 33], tetramers [32], or polymers [33]. Thus, while $EtZnNPh_2$ is dimeric, replacement of one Ph group by COZ leads to higher aggregates in $[EtZnN(Ph)COZ]_n$ [$n = 3$ when Z = OMe or NPh_2, $n = 4$ when Z = Me, and polymeric ($n > 6$) when Z = H]. These compounds form mononuclear (Z = OMe, SMe, or NPh_2) [EtZnN(Ph)COZ.2py] or binuclear (Z = H or Me) [{EtZnN(Ph)COZ}$_2$.2py] complexes with pyridine [33].

The chelate complexes (I) [11], and (II) are probably tetrahedral [7a]. Structure (V) has been proposed for a 2-aminoethyl derivative [9]. There is analogy with a magnesium compound, the structure of which is firmly based on an X-ray study, see p. 53, and on the data of Fig. 1 [43a].

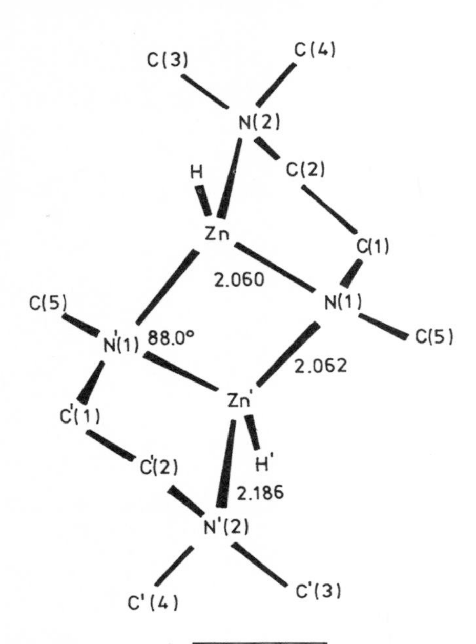

Fig. 1 – Molecular structure of $[HZnN(Me)C_2H_4NMe_2]_2$. *[Reproduced, with permission, from Bradley, D. C.,* Inorg. Macromol. Rev., *1970,* **1**, *141.]* For clarity, H atoms have been omitted.

Although there are spectroscopic data to back up some of these structures, there are as yet few crystallographic or electron diffraction results.

Compounds $EtZnNPhC(=S)X$ [X = H (insoluble in C_6H_6), Me, OMe, SMe, or NPh_2] are dimeric (X = H, believed to be a polymer) in benzene; an unusual three-co-ordinate arrangement about sulphur has been proposed for one derivative, (VI) [7]. With pyridine, monomeric adducts are formed.

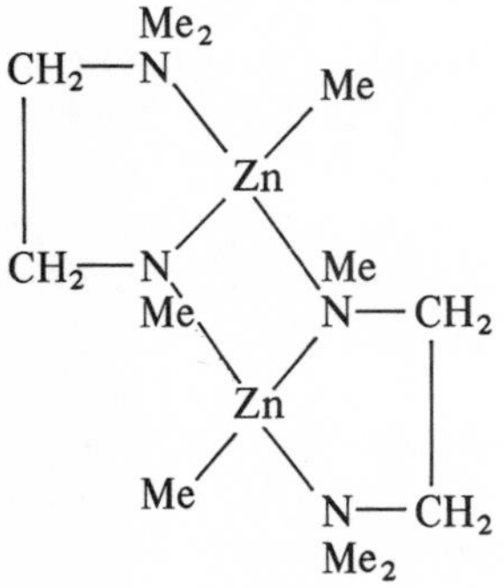

(V)

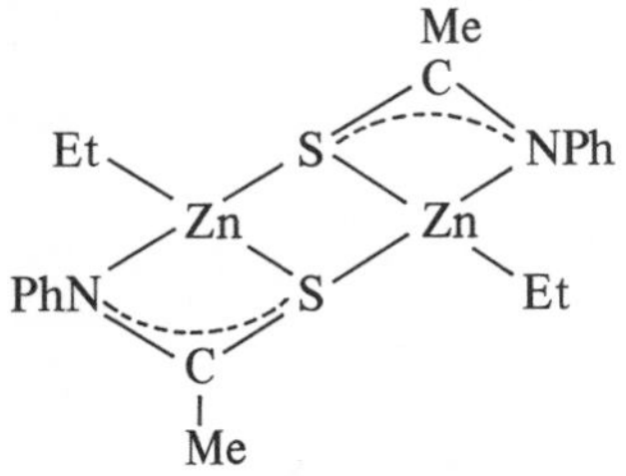

(VI)

Crystal structure data are available for $[MeZnNPh_2]_2$ and $[H\overline{ZnN(Me)C_2H_4N}Me_2]_2$ (see Fig. 1) [43a]. The former compound has a ZnNZnN four-membered ring with bridging NPh_2 groups and Zn–N, 2.07 Å; Zn–C, 1.95 Å; $N\widehat{Zn}N$, 90.6°; and $Zn\widehat{N}Zn$, 89.4°. The latter, obtained from zinc hydride and HN(Me) $C_2H_4NMe_2$, has Zn–H 1.6 Å (from neutron diffraction).

The He(I) photoelectron spectrum of gaseous $Zn[N(SiMe_3)_2]_2$ is consistent with a linear monomeric structure (Fig. 2) [16e] (see Section C.3, also for ΔH_f° [18a]); electron diffraction data are awaited [23c].

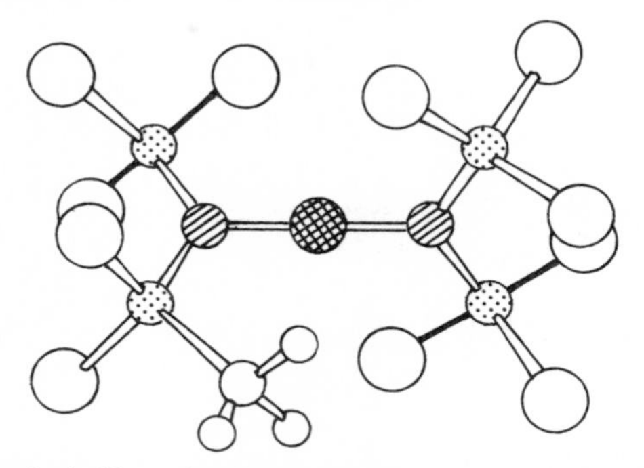

Fig. 2 – Probable molecular structure of gaseous $Zn[N(SiMe_3)_2]_2$.

B CADMIUM

Both bis(dicyanoamido)cadmium, $Cd[N(CN)_2]_2$ [22, 23], and bis(triphenylureido)cadmium, $Cd(NPhCONPh_2)_2$ [27], are prepared by salt elimination. $Cd(NEt_2)_2$ (obtained from Et_2Cd–$HNEt_2$) [9a] and di[bis(trimethylsilyl)-amido]cadmium, $Cd[N(SiMe_3)_2]_2$ [8], are the only reported amides of cadmium (see Table 3). The compound $Cd[N(SiMe_3)_2]_2$ [from $CdI_2 + NaN(SiMe_3)_2$] is monomeric, probably of D_{2d} or D_{2h} symmetry, well-characterised [8], and has been used as a source of diazoalkane derivatives of Cd^{II} [24a]. Like its Zn and Hg analogues it was stated not to form adducts with pyridine (see, however [14], Section A.2); for thermochemical data, see Section C.3.

C MERCURY

1. Introduction

It is well-known that mercury has a well-developed metal(I) chemistry, a property it shares to a lesser extent with cadmium of the Group 2b elements. In their metal–nitrogen chemistry, however, univalent compounds are known only for mercury.

A series of mercury(I) amides [e.g., (VII)] ($R = SO_2F$, CO_2Et, CO_2Me, or $CONEt_2$) is characterised [Eq. (4)] [7c]. Thus the system >N—Hg—Hg—N< is more stable than had been supposed, and the stability of these Hg^I compounds may be attributed to the electron-withdrawing substituents at nitrogen. Related to Eq. (4), carboxylic acid amides react likewise, as do *N,N*-disubstituted

$$Hg_2CO_3 + 2RNHSO_2F \longrightarrow (R)(FO_2S)N{-}Hg{-}Hg{-}N(R)(SO_2F) + H_2O + CO_2 \quad (4)$$

(VII)

hydrazides [7c]. The compounds $[Hg_2NAcNAc]_n$, $[Hg_2N(COCF_3)N(COCF_3)]_n$, and $[HgN(COCCl_3)N(COCCl_3)]_n$ are thought (from X-ray data) to have structures based on chains, as does the ureido derivative (VIII), obtained from *N,N*-bisfluorosulphurylurea and Hg_2CO_3. An attempt to make a Hg^I bis(trimethylsilyl)amide via salt elimination from Hg_2Cl_2 and $LiN(SiMe_3)_2$ gave only the disproportionation products $Hg[N(SiMe_3)_2]_2$ and Hg, as well as LiCl [8].

$$\left[{-}Hg{-}Hg{-}N(SO_2F){-}C(=O){-}N(SO_2F){-}Hg{-}Hg{-}N(SO_2F){-}C(=O){-}N(SO_2F){-} \right]_n$$

(VIII)

Mercury(II)-nitrogen compounds have been known for a long time, as evident from reviews dealing exclusively with that topic [7b, 23e], or in the context of organomercury chemistry [8b], or of complexes of $\bar{N}(SiMe_3)_2$ or related ligands [16d]; the chemistry of $Hg[N(CF_3)_2]_2$ is discussed extensively in a survey of bis(trifluoromethyl)amino compounds for which it is a key intermediate [2a]. Many of the compounds have four-co-ordinate nitrogen, but here we are concerned mainly with those compounds in which nitrogen has the co-ordination number of three. Mercury in amides is generally two-co-ordinate; however, in $(C_5H_5)Hg[N(SiMe_3)_2]$ [41] this may not be the case. The compounds range from low molecular weight to high polymers. The simplest compounds of the former type are $Hg[N(CF_3)_2]_2$ [2a, 11a, 11b, 11c, 17, 47a, 48, 48a], $Hg[N(SiMe_3)_2]_2$ [8, 10b, 16b, 16c, 16e, 24a, 41], $MeHg[N(CF_3)_2]$ [48a], and $RHg[N(SiMe_3)_2]$ [16b, 18b, 24a, 24b, 42a].

The silylamides are useful intermediates, especially as sources, for example, of diazoalkane- [9b, 24a, 24b], porphyrin- [8a], or stannyl- [10b] mercury(II) compounds by reaction with an appropriate protic compound (see Chapters 11 and 12). It has been suggested that to stabilise a monomeric mercury(II) amide it is necessary to have highly σ-electron- (e.g., CF_3) or π-electron- (e.g., $SiMe_3$) withdrawing groups at N [7b]. Presumably such provisions make the usually 'hard' amide ligand become polarisable and hence compatible with the 'soft' Hg^{2+}. However, steric effects may also have a definite role; compound (IX) is a colourless, thermally-labile compound [23d].

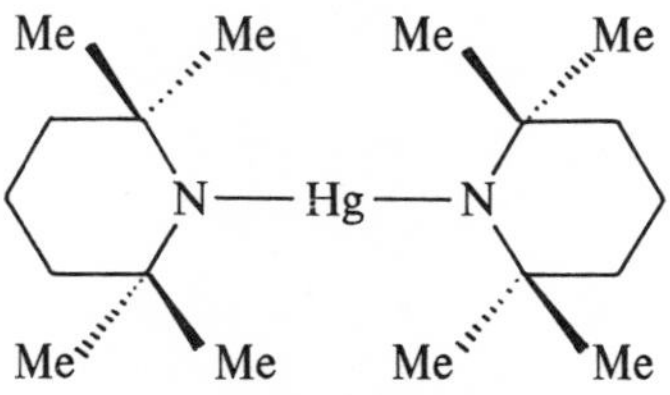

(IX)

Closely related to the amides is the hydrazide $Hg[N(CF_3)N(CF_3)_2]_2$ [10a] and thioamide $Hg[N(CF_3)SF_5]_2$ [45a]. Amides of this type are hydrolysed readily, in contrast to the carboxamides. A typical member of this group is mercury(II) bis(acetamide), $Hg(NHCOMe)_2$ [23a]. Compounds which fall outside our scope (but see ref. [7b]) include the sulphur imides $Hg(NS_7)_2$ [25b], $Hg(NS)_2$ [25a], $Hg_5(NS)_8$ [25a], and $Hg(N_2S)$ [3a], $Hg(NSF_2)_2$ [16a, 23b], and the amido-sulphates, such as $[Hg(NHSO_3)_2]^{2-}$. Although all these compounds are mono-nuclear, a solitary tri(mercuryl)amine $N(HgMe)_3$ has been reported [9b], see Eq. (10). Compounds of formula $HgNR.HNO_3$ have been described [36].

Until a review and reassessment of the literature on mercury–nitrogen chemistry in 1907 [16], various formulae for the many known compounds were postulated. Relationships between the various substances were then rationalised. Literature earlier than 1907 will be referred to in revised (post 1907) form where appropriate. The 'fusible white precipitate', previously formulated as $Hg_2NCl.3NH_4Cl$ in a widely accepted theory, was revised as $HgCl_2.2NH_3$. Similarly, the 'infusible precipitate', $Hg_2NCl.NH_4Cl$, was later taken as $ClHgNH_2$ and the chloride of Millon's base, $Hg_2NCl.H_2O$, was formulated as a mixed derivative, $ClHgOHgNH_2$. X-ray data (see ref. [7b]) put these postulates on a firm basis about twenty years ago. The amido derivatives may be assigned analogous formulae (a similar suggestion of the 1840's had previously been ignored). The structures (X) and (XI) are thus assigned to the succinimido and *N*-formyl-*p*-toluido derivatives, respectively, and the concept of two-co-ordination for Hg^{II} may be taken as firmly established also for amides.

$$\mathrm{ClHgN}\begin{matrix}\diagup \mathrm{CO} \diagdown \\ \diagdown \mathrm{CO} \diagup\end{matrix}$$

(X)

$$\mathrm{MeC(O)OHgN}\begin{matrix}\diagup \mathrm{CHO} \\ \diagdown \mathrm{C_6H_4Me}\text{-}p\end{matrix}$$

(XI)

The mercury(II) nitrogen compounds appear to be of interest as fungicides or herbicides (e.g., see refs. [3] and [5]), which perhaps accounts for the large variety of compounds (see Table 4) which has been prepared.

2. Synthesis

Perfluorobis(dimethylamido)mercury(II) was made from HgF_2 at 100° C [see Eq. (5)] [48], by the fluorination of cyanogen or cyanogen chloride with HgF_2 [11a], or from Hg and Br-$N(CF_3)_2$ [25]. Related derivatives are obtained according to Eq. (6) (R = Me or SCF_3, X = Cl or Br). *N*-Bromo-succinimide or -phthalimide reacts with divinylmercury, eliminating vinyl bromide [Eq. (7)] [45]. A similar reaction is employed for piperidinato- derivatives [12].

$$\mathrm{HgF_2 + 2F_3C{-}N{=}CF_2 \longrightarrow Hg[N(CF_3)_2]_2} \quad (5)$$

$$\mathrm{R_2Hg + XN(CF_3)_2 \longrightarrow RHgN(CF_3)_2 + RX} \quad (6)$$

$$\mathrm{(C_2H_3)_2Hg + BrN(CO)_2C_6H_4 \longrightarrow C_2H_3HgN(CO)_2C_6H_4 + C_2H_3Br} \quad (7)$$

Unlike zinc, mercury derivatives are most commonly prepared by addition of the amine, or more usually the alkali metal amide, to the metal(II) halide or oxide [16, 18, 19, 24, 24a, 24b, 29, 30, 40, 42, 42a, 45-47]. This is illustrated in Eq. (8) [8], or by the synthesis of $MeHg[N(SiMe_3)_2]$ from MeHgCl and $LiN(SiMe_3)_2$ [24b].

$$\mathrm{HgBr_2 + 2NaN(SiMe_3)_2 \longrightarrow 2NaBr + Hg[N(SiMe_3)_2]_2} \quad (8)$$

Other methods include alkane elimination, which has been employed to form mercury-nitrogen bonds [34, 37, 38], but not to the same extent as in zinc chemistry. Transmetallation has similarly been little used [4, 27], although the metathetical exchange reaction of Eq. (9) may be taken as a special case [41].

$$\mathrm{(C_5H_5)_2Hg + Hg[N(SiMe_3)_2]_2 \longrightarrow 2(C_5H_5)Hg[N(SiMe_3)_2]} \quad (9)$$

A single example, Eq. (10), of transamination has been reported [9b].

$$3MeHg[N(SiMe_3)_2] + NH_3 \longrightarrow N(HgMe)_3 + 3HN(SiMe_3)_2 \quad (10)$$

The mercuration reaction yields mercury derivatives of aromatic amines, but these generally do not have Hg-N bonds. The intermediate (XII) in reaction (11) is probably a mercurio-ammonium salt, which rearranges to give the *o*- or *p*-mercury substituted compound (XIII) [Eq. (11)] [21].

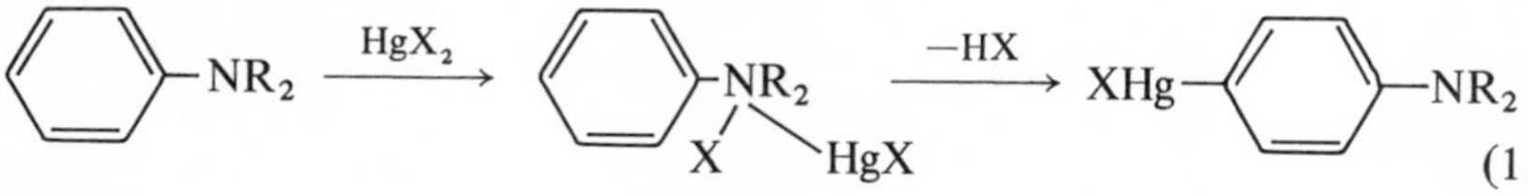

$$C_6H_5NR_2 \xrightarrow{HgX_2} C_6H_5N^{+}R_2(HgX)\,X^{-} \xrightarrow{-HX} XHg{-}C_6H_4{-}NR_2 \quad (11)$$

(XII) (XIII)

(R = alkyl or H, X = Cl or OAc)

An exception to this sequence is the case [Eq. (12)] when excess of mercury(II) acetate is used, and the product is subsequently hydrolysed [43]. Compound (XIV) has been similarly prepared [43].

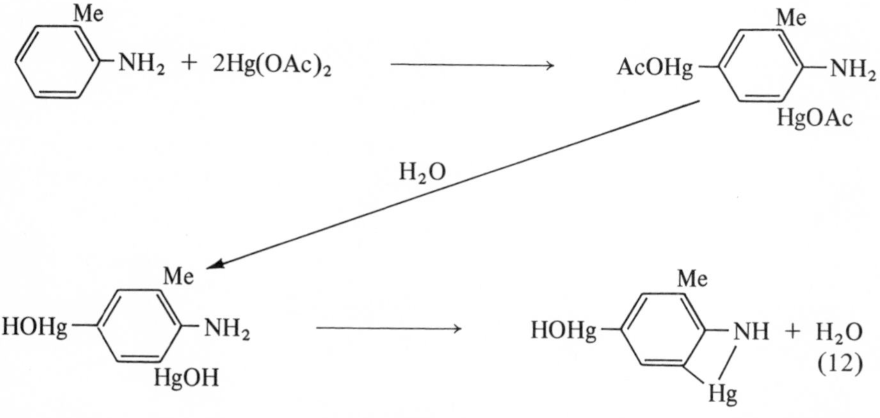

$$2\text{-}MeC_6H_4NH_2 + 2Hg(OAc)_2 \longrightarrow AcOHg{-}C_6H_2(Me)(HgOAc){-}NH_2 \xrightarrow{H_2O} HOHg{-}C_6H_2(Me)(HgOH){-}NH_2 \longrightarrow HOHg{-}C_6H_2(Me)(NH{-}Hg) + H_2O \quad (12)$$

The reaction of the purine (XV) with mercury(II) acetate probably mercurates a methyl group [39].

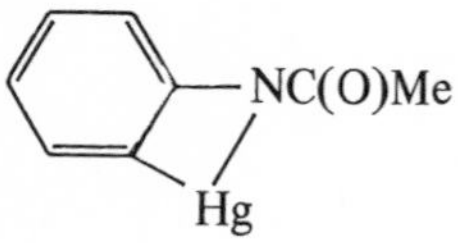

(XIV): $C_6H_4(Hg)NC(O)Me$ (cyclic)

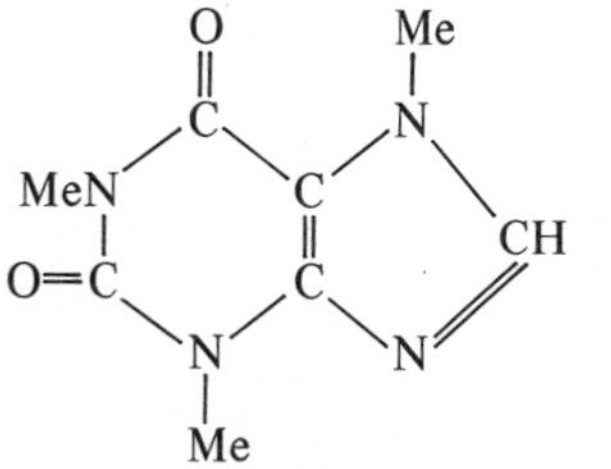

(XV): 1,3,7-trimethylpurine-2,6-dione (MeN, C=O, N–Me, CH, N)

(XIV) (XV)

Reaction of mercury(II) acetate with a carboxylic acid amide affords the mercury carboxamide, e.g., $Hg(NHCOMe)_2$ [23f].

3. Physical properties and structures

The compounds $Hg[N(CF_3)_2]_2$ [11a] and $Hg[N(SiMe_3)_2]_2$ [8] are monomeric and the latter compound, on the basis of vibrational spectral data, was assigned to the D_{2h} or D_{2d} point group.

The bisacetamide $Hg(NHCOMe)_2$ is polymeric in the solid state, as shown by X-ray analysis [19], according to which planar $Hg(NHCOMe)_2$ molecules (point group C_{2h}) with colinear NHgN bonds (Hg-N, 2.06 Å) are linked by O...H...N bonds into ribbons. N.m.r. data confirm the $Hg(NHCOMe)_2$ arrangement rather than the tautomeric $Hg[N{:}C(OH)Me]_2$ [7e]. The diazohydrazides $HgN_2(COMe)_2$ [7d] and $HgN_2(COCF_3)_2$ [47a] are assumed to have a similar structure [7d]. The monomeric $Hg(NSF_2)_2$ has (X-ray) a linear NHgN arrangement (Hg-N, 2.05 Å) [23b].

Thermochemical data are available for $Hg[N(SiMe_3)_2]_2$ and the zinc and cadmium analogues (Table 1) [18a]. They were obtained from heats of hydrolysis in $1M$ aqueous H_2SO_4 and appropriate thermochemical data on $HN(SiMe_3)_2$ and heats of vaporisation. The following trends emerge (Fig. 3): (i) M-N bond strengths decrease in the sequence Zn > Cd > Hg and (ii) the $M{-}\ddot{N}{<}$ bond is stronger than the isoelectronic $M{-}C{\leqslant}$. With regard to (i) it is clear that the trend of decreasing M-ligand bond strength with increasing atomic number is characteristic of *s*- and *p*- block groups of elements, whereas the reverse is the case for *d*- or *f*- block groups (see Chapter 8).

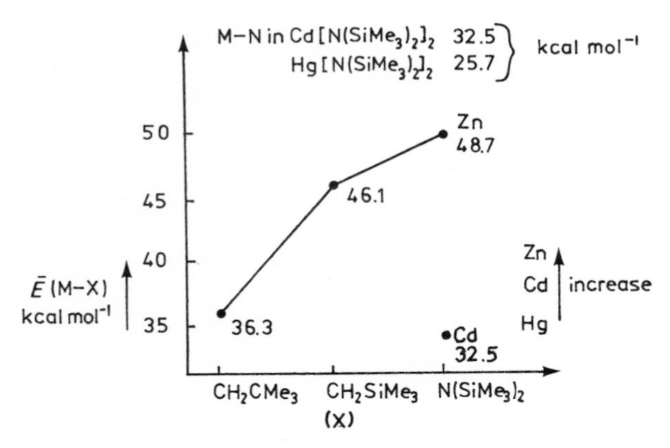

Fig. 3 – Trends in mean thermochemical energy terms $\bar{E}$(M-X) in the series $[MX_2]$ [M = Zn, Cd, or Hg; X = CH_2CMe_3, CH_2SiMe_3, or $N(SiMe_3)_2$] (from ref. [18a]).

Table 1 Thermochemical data for $M[N(SiMe_3)_2]_2$ (M = Zn, Cd, or Hg) [18a]

M in $M[N(SiMe_3)_2]_2$	$\Delta H_f^\circ(l)$ (kcal mol^{-1})	$\Delta H_f^\circ(g)$ (kcal mol^{-1})	$\bar{E}$(M-N) (kcal mol^{-1})
Zn	−227.7 ± 2.7	−210.8 ± 2.8	48.7
Cd	−200.3 ± 2.7	−183.5 ± 2.9	32.5
Hg	−199.1 ± 2.7	−182.0 ± 2.9	25.7

The He(I) photoelectron spectrum of gaseous $Hg[N(SiMe_3)_2]_2$ and its Zn analogue have been recorded [16e]. There is a single band at ionisation energy below 9.3 eV (8.33 for the Hg^{II} and 8.50 eV for the Zn^{II} amide) which shows that the nitrogen lone-pair orbitals are degenerate, consistent with free rotation about the M–N bonds in a linear N–M–N assembly. The next band (9.38 for the Hg^{II} and 9.55 eV for the Zn^{II} amide) is assigned to the antisymmetric N–Hg–N σ-bond combination.

D TABLES OF COMPOUNDS

Table 2† Zinc amides and related compounds

Compound	Method	M.p. (°C) (B.p., °C/mm.)	Comments	References
$MeZnNHC_6H_{11}$	E			[2]
p-$MeZnNHC_6H_4Me$	E			[2]
$MeZnNMe_2$	A			[1]
	E		ir	[9]
$MeZn_2(NMe_2)_3$	A			[1]
$HZnNMe(CH_2)_2NMe_2$	F		X-ray	[43a]
$MeZnNMe(CH_2)_2NMe_2$	E	110–1	dimer, ir	[9]
$MeZnNPh_2$	E		dimer, ir	[9]
	E	106	dimer	[33]
.2py	E	110	ir	[9]
	E	111		[33]
$.(Me_2NCH_2)_2$	E	134 d		[33]
.bipy	E	103		[33]
$EtZnNHBu^t$	E	96		[34a]
$EtZnNHC_6H_{11}$	E			[2]
$EtZnNHC_6H_4Me$-p	E			[2]
$EtZnNEt_2$	E		liquid	[28, 31]
$EtZnNPh_2$	E	99–100	dimer, ir	[9]
	E	105–6	dimer	[28]
	E	104–6	dimer	[31]
	E	108–10	ir	[17b, 17c]
$EtZn[N(SiCl_3)_2]$	E	(72/10⁻²) −7	n_D^{20} 1.5088, ir d_4^{20} 1.6742	[42b]

Table 2 *C'ntd.*

Compound	Method	M.p. (°C) (B.p., °C/mm.)	Comments	References
$EtZnNR_2$	E			
(R = Pr)				[17c]
(NR_2 = —N				[17c]
—N				[17c]
—N)				[17c]
$PrZnNHC_6H_{11}$	E			[2]
$PrZnNHC_6H_4Me$-*p*	E			[2]
$Pr^iZnNHC_6H_{11}$	E			[2]
$Pr^iZnNHC_6H_4Me$-*p*	E			[2]
Pr^iZnNPh_2	E	106–7		[31]
$BuZnNHC_6H_{11}$	E			[2]
$BuZnNHC_6H_4Me$-*p*	E			[2]
$BuZnNPh_2$	E	83–5		[31]
$Bu^tN(ZnEt)_2$		67–9		[34a]
$Me_2CHCHMeZnNPh_2$	E		nmr	[17b]
.dimethylaminoquinoline			ir, nmr	[17b]
$PhZnNEt_2$	E			[28]
$PhZnNPh_2$	E	192–5		[28]
$Zn(NHPr)_2$				[35]
$Zn(NHC_6H_{11})_2$	E			[2]
$Zn(NHPh)_2$	E			[15]
$Zn(NHC_6H_4Me$-*p*$)_2$	E			[2]
$Zn(NMe_2)_2$	E	270–90 d	polymer, ir	[9]
$Zn(NEt_2)_2$	E			[15]
	A			[9a]
$Zn[N(SiCl_3)_2]_2$	E	($84.5/10^{-2}$)	ir	[42b]
		49		
$Zn[N(SiMe_3)_2]_2$	A	12.5	ir, Raman, nmr	[8,14]
		(82°/0.5)	n_D^{20} 1.4506, d_4^{20} 0.952	
			ΔH_f	[18a]
			ed	[23c]
			pes	[16e]

Table 2 *C'ntd.*

Compound	Method	M.p. (°C) (B.p., °C/mm.)	Comments	References
$Zn[N(SiMe_3)_2]_2.L$				
($L = NC_5H_5$,		71	ir, nmr	[14]
NC_5H_4Me)		57	ir, nmr	[14]
$Zn(NPh_2)_2$	E	282	dimer	[33]
			X-ray	[43a]
.2py	I	210–1	ir	[9]
.bipy	E	194		[33]
$.(Me_2NCH_2)_2$	E	>260		[33]
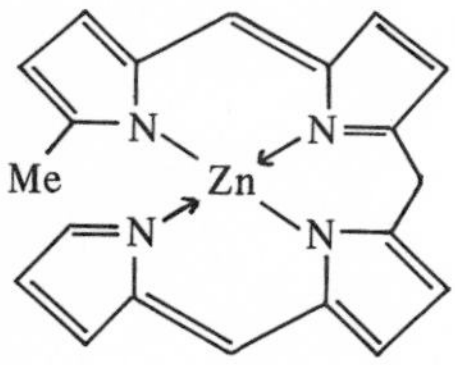 *			ir, nmr	[17a]
$ClZnNHCH_2CH_2C_6H_3(OH\text{-}4)OMe\text{-}3$				[20]
$IZnNHCH_2Ph$	E			[6]
$EtZnNMeCSNPh_2$	D	172–3		[31]
EtZnNPhCOH	D		>pentamer	[32]
	E	132	~hexamer, ir, nmr	[33]
.py	E	66–72	dimer, ir, nmr	[33]
(,, $)_2.(Me_2NCH_2)_2$	E	70–5	ir, nmr	[33]
EtZnNPhCSH	E	101	polymer, nmr	[7]
.2py	E		glass, nmr	[7]
$.(Me_2NCH_2)_2$	E		glass, nmr	[7]
EtZnNPhCOMe	D		tetramer	[32]
	E		ir, nmr	[33]
.py	E	125	dimer, ir, nmr	[33]
(,, $)_2.(Me_2NCH_2)_2$	E	55–60	ir, nmr	[33]
EtZnNPhCSMe	E	105	dimer, nmr	[7]
	D		dimer	[32]
.py	E	70	nmr	[7]
$.(Me_2NCH_2)_2$	E	60	nmr	[7]
$EtZnNPhCONEt_2$	D	117–8		[31]
$EtZnNPhCSNEt_2$	D	124–5		[31]
$EtZnNPhCONPh_2$	D	214–7		[31]
	D		trimer	[32]
	E	239	trimer, ir, nmr	[33]
.2py	E	99	ir, nmr	[33]
$.(Me_2NCH_2)_2$	E	88	ir, nmr	[33]

*Not really within our scope.

Table 2 *C'ntd.*

Compound	Method	M.p. (°C) (B.p., °C/mm.)	Comments	References
$EtZnNPhCSNPh_2$	E	226	dimer, nmr	[7]
	D		dimer	[32]
	D	225–7		[31]
.2py	E	67	nmr	[7]
$.(Me_2NCH_2)_2$	E	130	nmr	[7]
$EtZnNPhCO_2Me$	E	158	trimer, ir, nmr	[33]
.2py	E	48	ir, nmr	[33]
$.(Me_2NCH_2)_2$	E	85–9	ir, nmr	[33]
EtZnNPhCSOMe	E	91	dimer, nmr	[7]
	D		dimer	[32]
.2py	E		oil, nmr	[7]
$.(Me_2NCH_2)_2$	E	55	nmr	[7]
EtZnNPhCOSMe	E	d>5	polymer, ir, nmr	[33]
.py	E		ir, nmr	[33]
$.(Me_2NCH_2)_2$	E		ir, nmr	[33]
$EtZnNPhCS_2Me$	E		dimer, nmr	[7]
.py	E		nmr	[7]
$.(Me_2NCH_2)_2$	E		nmr	[7]
$EtZnNPhCOPPh_2$	D			[28]
$EtZnNPhCSPPh_2$	D	139–42		[28]
$PhZnNPhCONEt_2$	D	160–5		[28]
$PhZnNPhCONPh_2$	D	135–40		[28]
$PhZnNPhCSNPh_2$	D	243–5		[28, 31]
$Zn[N(CN)_2]_2$	A		ir	[22]
.py	A		ir	[22]
.0.5DMSO	A			[23]
$Zn(NR\ \ NR)_2$ (cycloheptatriene ring with X; $R = Et$, $X = m\text{-}F_5SC_6H_4N_2$)	C	219–21	nmr	[11]
$Zn(NHCOMe)_2$	E			[15]
Zn(NHCO–CONH) (cyclic)	E			[15]
$Zn(N(CO)_2C_6H_4)_2$ (phthalimide)				[13]
$Zn(NPhCONPh_2)_2$	A			[27]
$Zn(NEt\text{—}CHMe\text{—}CH_2\text{—}CH_2\text{—}NEt)_2$ (cyclic)	C	(*ca.* $60°/10^{-4}$ mm)	ir, nmr, uv, mass	[7a]

Table 2 *C'ntd.*

Compound	Method	M.p. (°C) (B.p., °C/mm.)	Comments	References
$(MeO)C(=O)-CH-C(=O)(OMe)$ chelate ZnNEt$_2$ [HC(C(OMe)O)$_2$ZnNEt$_2$]	E		ir, nmr	[20a]

†Footnote to Table 2: for explanation of abbreviations, see footnotes to Table 1, p. 57.

Table 3† Cadmium amides and related compounds

Compound	Method	M.p. (°C)	References
$Cd(NEt_2)_2$	A		[9a]
$Cd[N(SiMe_3)_2]_2$	A	8 (b.p. 93°/0.5, n_D^{20} 1.4660, d_4^{20} 1.062)	[8]
		ΔH_f	[18a]
$Cd[N(CN)_2]_2$			
.2py	A		[22]
.2DMSO	A	280d	[23]
$Cd(NPhCONPh_2)_2.0.5NH_3$	A		[27]

†Footnote to Table 3: for explanation of abbreviations, see footnotes to Table 1, p. 57.

Table 4† Mercury amides and related compounds

Compound	Method	M.p. (°C) (B.p., °C/mm.)	Comments	References
$MeHgN(CF_3)_2$	J		ir	[10,12]
MeHgNHC(NH)NHCN	C			[42]
MeHgN(CO)$_2$ (cyclic imide, CO–CO bridged)	E			[37]
$MeHgN(SiMe_3)_2$	A	(40/0.1)	ir, nmr	[9b, 24a, 24b, 42a]
$EtHgN(SiMe_3)_2$	A	(60-3/1)	ir, nmr	[24a, 24b]
$C_5H_5HgN(SiMe_3)_2$	I, K	120d	ir, nmr, mass	[41]

Table 4 *C'ntd.*

Compound	Method	M.p. (°C) (B.p., °C/mm.)	Comments	References
$EtHgN(CO)_2C_6H_4$	A	150–6		[4]
$H_2NCONHCH_2CH(OMe)CH_2HgN(CO)_2(CH_2)_2$	C	159–60		[40]
$H_2NCONHCH(OMe)CHCH_2HgN(CO)_2C_6H_4$	C	188–8.5		[40]
$C_2H_3HgN(CO)_2(CH_2)_2$	J	113–4		[45]
$C_2H_3HgN(CO)_2C_6H_4$	C, J	133–5		[45]
$(NO_2)_3CHgNHPh$	E			[34]
$(NO_2)_3CHgNMe_2$	J			[12]
$(NO_2)_3CHgNC_5H_{10}$	J	153–5d		[12]
$PhHgNHCH_2CH_2NH_2$				[5]
$Hg(NHCOMe)_2$	C	196–7		[16, 47]
$Hg(NHCOEt)_2$	C	201		[47]
$Hg(NHCOPr)_2$	C	222–4		[47]
$Hg(NHCOC_2H_3)_2$ (and polymers)				[26]
$Hg(NHCOPh)_2$	C	222		[16, 47]
$Hg(NHCOC_6H_4Cl\text{-}m)_2$	C	245		[47]
$Hg(NHCOC_6H_4Cl\text{-}p)_2$	C	258		[47]
$Hg(NHCOC_6H_4Br\text{-}o)_2$	C	242		[47]
$Hg(NHCOC_6H_4Br\text{-}m)_2$	C	235		[47]
$Hg(NHCOC_6H_4Br\text{-}p)_2$	C	266		[47]
$Hg(NHCOC_6H_4OH\text{-}o)_2$	C	190		[47]
$Hg(NHCOC_6H_4OMe\text{-}o)_2$	C	241		[47]
$Hg(NHCOC_6H_4OMe\text{-}p)_2$	C	222		[47]
$Hg(NPhCONPh_2)_2$	A			[27]
$MeCOHgNPhCHO$	C			[16]
$MeCOHgN(CHO)C_6H_4Me\text{-}p$	C			[16]
$ClHgNHEt$	C			[16]
$ClHgNHCH_2Ph$	C			[46]

Table 4 *C'ntd.*

Compound	Method	M.p. (°C) (B.p., °C/mm.)	Comments	References
ClHgNHPh	C			[16]
$ClHgNHC_6H_4Me$-*o*	C	d 165		[30]
$ClHgNHC_6H_4Me$-*m*	C	d 180		[30]
$ClHgNHC_6H_4Me$-*p*	C	d 130		[30]
$ClHgNHC_6H_3Me_2$-3,5	C	d 115		[30]
$ClHgNHC_6H_4NO_2$-*o*	C	d 240		[30]
$ClHgNHC_6H_4NO_2$-*m*	C	d 225		[30]
$ClHgNHC_6H_4NO_2$-*p*	C	d 140		[30]
ClHgNHNp-1	C	d 125		[30]
ClHgNHNp-2	C	d 170		[30]
ClHgNMePh	C			[29]
XHgNHR ($X = Cl, NO_3$; R = Me, Et, C_5H_{11})				[36]
$PhHgNHCONH_2$				[3]
$HgC_6H_2Me(HgOH)NH$ (Hg–NH bridged)	I			[43]
$HgC_6H_3MeNCOMe$		229		[43]
m-$MeC_6H_4HgN(CO)_2$ (cyclic)		168		[44]
m-$ClC_6H_4HgN(CO)_2$ (cyclic)		202		[44]
p-$ClC_6H_4HgN(CO)_2$ (cyclic)		165		[44]
$MeCO_2Hg$–C_6H_4–$N(C_2H_3)_2NH(HCl)$–C_6H_4–HgNH–$C_6H_3(NO_2)_2$	E	244		[38]
$N(HgMe)_3$	B		mass, ir, Raman	[9b]
$Hg(NHPh)_2$	C			[24]
$Hg(NHC_6H_4NO_2$-*o*$)_2$	C			[18]
$Hg(NHC_6H_4NO_2$-*m*$)_2$	C			[18]
$Hg(NHC_6H_4NO_2$-*p*$)_2$	C			[18]
Hg(N-pyrrolidinyl)$_2$	C			[24]
Hg(N-piperidinyl)$_2$	C			[24]
	J			[12]
$Hg[N(CF_3)_2]_2$	D			[17, 25]
	D + I	17.5 (127°)		[10, 10a, 11a, 11b, 11c, 48, 48a]

Table 4 *C'ntd.*

Compound	Method	M.p. (°C) (B.p., °C/mm.)	Comments	References
$Hg[N(SiMe_3)_2]_2$	A	11 (78°/0.15)	ir, Raman, nmr	[8,10b,16b,16c, 18b, 24a, 41]
			n_D^{20} 1.4717, d_4^{20} 1.288	
			ΔH_f	[18a]
			pes	[16e]
$Hg(N\langle$ ring with Me_2, $Me_2\rangle)_2$	A		ir, nmr	[23d]
$Hg(NHCOMe)_2$	C		X-ray	[19]
			nmr	[7e, 23f]
ClHgNEtHgOH	C			[16, 29]
ClHgNPhHgOH	C			[16]
$(ClHgNH)_2CO$	C			[16]
ClHgN(CO)(CO) (cyclic)	C			[16]
IHgNHCOPh	C			[16]
$HOHgNHCH_2Ph$	C			[16]
$HOHgNHCONH_2$	C			[16]
(HOHgNH)CO	C			[16]
$MeCO_2HgNHCH_2Ph$	C			[16]
$(MeCO_2HgNH)_2CO$	C			[16]
$NO_3HgNHCONH_2$	C			[16]
$(NO_3HgNH)_2CO$	C			[16]
$SO_4(HgNHCH_2Ph)_2$	C			[16]
$SO_4(HgNH)_2CO$	C			[16]
$F_3CSHgN(CF_3)_2$	J		ir	[10]
Hg(I)				
$Hg_2[N(COR)]_2$ polymer ($R = Me$, CF_3, CCl_3)	C		Raman, ir, X-ray, chain	[7a]
$[HgN(R)SO_2F]_2$ ($R = SO_2F$, CO_2Et, CO_2Me, $CONEt_2$, CO)	A			[7c]
$[Hg_2NAcNAc]_n$	A		X-ray	[7c]
$[Hg_2N(COCF_3)N(COCF_3)]_n$	A		X-ray	[7c]
$[Hg_2N(COCCl_3)N(COCCl_3)]_n$	A		X-ray	[7c]

†Footnote to Table 4: for explanation of abbreviations, see footnotes to Table 1, p. 57.

E REFERENCES

[1] Abeler, G., Bayrhuber, H., and Nöth, H., *Chem. Ber.*, 1969, **102**, 2249.
[2] Abraham, M. H., and Hill, J. A., *J. Organometallic Chem.*, 1967, **7**, 23.
[2a] Ang, H. G., and Syn, Y. C., *Adv. Inorg. Chem. Radiochem.*, 1974, **16**, 1.
[3] Barss, H. P., Clayton, E. C., Goldsworthy, M. C., Haskell, R. J., Heuberger, J. W., Leukel, R. W., McClellan, W. D., and Miller, P. R., *Plant Disease Rept.*, 1948, *Suppl. 176*, 95; *Chem. Abstr.*, 1948, **42**, 8397.
[3a] Berg, W., Goehring, M., and Malz, H., *Z. Anorg. Chem.*, 1956, **283**, 13.
[4] Bialas, J., Eckstein, Z., Ejmocki, Z., Hetnarski, B., Sobotka, W., and Szymaszkierwicz, J., *Prezemysl Chem.*, 1961, **40**, 567; *Chem. Abstr.*, 1962, **57**, 12524.
[5] Bing, A., *Proc. Northeast Weed Control Conf., 8th Meeting*, 1954, 153; *Chem. Abstr.*, 1954, **48**, 5424.
[6] Blazević, K., Houghton, R. P., and Williams, C. S., *J. Chem. Soc. (C)*, 1968, 1704.
[7] Boersma, J., and Noltes, J. G., *J. Organometallic Chem.*, 1967, **17**, 1.
[7a] Bonnett, R., Bradley, D. C., Fisher, K. J., and Rendall, I. F., *J. Chem. Soc. (A)*, 1971, 1622.
[7b] Breitinger, D., and Broderson, K., *Angew. Chem. Internat. Edn.*, 1970, **9**, 357.
[7c] Breitinger, D., Broderson, K., and Jürgen, L., *Chem. Ber.*, 1970, **103**, 2388.
[7d] Broderson, K., and Kunkel, L., *Z. Anorg. Chem.*, 1959, **298**, 34.
[7e] Brown, D. B., and Robin, M. B., *Inorg. Chim. Acta*, 1969, **3**, 644.
[8] Bürger, H., Sawodny, W., and Wannagat, U., *J. Organometallic Chem.*, 1965, **3**, 11.
[8a] Clare, P., and Glockling, F., *Inorg. Chim. Acta*, 1976, **17**, 229.
[8b] Coates, G. E., Green, M. L. H., and Wade, K., *Organometallic Compounds*, Methuen, London, Vol. 1, Ch. II, 1967.
[9] Coates, G. E., and Ridley, D., *J. Chem. Soc.*, 1965, 1870.
[9a] Cuvigny, T., and Normant, H., *Compt. Rend.*, 1969, **268C**, 834.
[9b] Dehnicke, K., Lorberth, J., Thiel, W., and Weller, F., *Z. Anorg. Chem.*, 1971, **381**, 57.
[10] Dobbie, R. C., and Emeléus, H. J., *J. Chem. Soc. (A)*, 1966, 367.
[10a] Dobbie, R. C., and Emeléus, H. J., *J. Chem. Soc. (A)*, 1966, 933.
[10b] Eaborn, C., Thompson, A. R., and Walton, D. R. M., *Chem. Comm.*, 1968, 1051.
[11] Eaton, D. R., Josey, A. D., and Sheppard, W. A., *J. Amer. Chem. Soc.*, 1963, **85**, 2689.
[11a] Emeléus, H. J., and Hurst, G. L., *J. Chem. Soc.*, 1964, 396.
[11b] Emeléus, H. J., and Tattershall, B. W., *J. Chem. Soc.*, 1964, 5892.
[11c] Emeléus, H. J., and Tattershall, B. W., *J. Inorg. Nucl. Chem.*, 1966, **28**, 1823.

[12] Erashko, V. I., Shevelov, S. A., and Fainzil'berg, A. A., *Bull. Acad. Sci. U.S.S.R., Chem. Sect.*, 1967, 2579.
[13] Farbwerke Hoechst A.-G., Fr. Patent 1,355,980 (1964).
[14] Fisher, K. J., *Inorg. Nucl. Chem. Letters,* 1973, **9,** 921.
[15] Frankland, E., *Proc. Roy. Soc.*, 1856-7, **8,** 502.
[16] Franklin, E. C., *J. Amer. Chem. Soc.*, 1907, **29,** 35.
[16a] Glemser, O., Mews, R., and Roesky, H. W., *Chem. Ber.*, 1969, **102,** 1523.
[16b] Glockling, F., and Mahale, V. B., *Inorg. Chim. Acta,* 1977, **25,** L117.
[16c] Glockling, F., and Sweeney, J. J., *J. Chem. Res.*, 1977, (S)35, (M)615.
[16d] Harris, D. H., and Lappert, M. F., *J. Organometallic Chem. Library,* 1976, **2,** 13.
[16e] Harris, D. H., Lappert, M. F., Pedley, J. B., and Sharp, G. J., *J.C.S. Dalton,* 1976, 945.
[17] Haszeldine, R. N., and Tipping, A. E., *J. Chem. Soc. (C),* 1967, 1241.
[17a] Hennig, H., and Daute, R., *Z. Chem.*, 1969, **9,** 275.
[17b] Inoue, S., and Yamada, T., *J. Organometallic Chem.*, 1970, **25,** 1.
[17c] Inoue, S., and Imanaka, Y., *J. Organometallic Chem.*, 1972, **35,** 1.
[18] Jackson, C. L., and Peakes, R. W., *Amer. Chem. J.*, 1908, **39,** 567.
[18a] Jeffery, J., Lappert, M. F., Pedley, J. B., and Rai, A. K., unpublished results, 1977.
[18b] Kalinina, G. S., Petrov, B. I., Kruglaya, O. A., and Vyazankin, N. S., *Zh. Obshch. Khim.*, 1972, **42,** 148.
[19] Kamenar, B., and Grdenić, D., *Inorg. Chim. Acta,* 1969, **3,** 25.
[20] Kametani, T., Takano, S., and Karibe, E., *Yakugaku Zasshi,* 1963, **83,** 1035.
[20a] Kawakami, Y., and Tsuruta, T., *Bull. Soc. Chem. Japan,* 1971, **44,** 247.
[21] Kharasch, M. S., and Jacobsohn, I. M., *J. Amer. Chem. Soc.*, 1921, **43,** 1894.
[22] Köhler, H., *Z. Anorg. Chem.*, 1964, **331,** 237.
[23] Köhler, H., *Z. Anorg. Chem.*, 1965, **336,** 245.
[23a] Kravcov, D. N., *Dokl. Akad. Nauk S.S.S.R.*, 1965, **162,** 581.
[23b] Krebs, B., Meyer-Hussein, E., Glemser, O., and Mews, R., *Chem. Comm.*, 1968, 1578.
[23c] Lappert, M. F., Power, P. P., Slade, M. J., Hedberg, K., and Hedberg, L., unpublished results, 1978.
[23d] Lappert, M. F., and Slade, M. J., unpublished work, 1978.
[23e] Levason, W., and McAuliffe, C. A., *The Chemistry of Mercury,* ed. McAuliffe, C. A., Macmillan, London, 1977, p. 85.
[23f] Ley, H., and Kissel, H., *Ber.*, 1899, **32,** 1358; Ley, K., and Schaefer, K., *ibid.*, 1902, **35,** 1309.
[24] Lattes, A., and Périé, J. J., *Tetrahedron Letters,* 1967, **51,** 5165; *Compt. Rend.*, 1966, **262C,** 1591.
[24a] Lorberth, J., *J. Organometallic Chem.*, 1971, **27,** 303; 1974, **71,** 159.

[24b] Lorberth, J., and Weller, F., *J. Organometallic Chem.*, 1971, **32**, 145.
[25] Makarov, S. P., Shpanskii, V. A., Ginsberg, V. A., Shchekotikhin, A. I., Filatov, A. S., Martynova, L. L., Pavlovskaya, I. V., Golovaneva, A. F., and Yakubovitch, A. Ya., *Proc. Acad. Sci. U.S.S.R.*, 1962, **142**, 596.
[25a] Meuwsen, A., and Lösel, M., *Z. Anorg. Chem.*, 1953, **271**, 217.
[25b] Meuwsen, A., and Schlossnagel, F., *Z. Anorg. Chem.*, 1953, **271**, 226.
[26] Micheli, R. A., U.S. Patent 3,134,757 (1964).
[27] Nast, R., and Danneker, W., *Ann.*, 1966, **693**, 1.
[28] Nederlandse Centrale Organisatie voor Toegepaste-Naturwetenschappelijk Ondersoek, Neth. Patent 6,500,454 (1966).
[29] Neogi, P., and Chatterjee, M. P., *J. Indian Chem. Soc.*, 1928, **5**, 221.
[30] Neogi, P., and Mukherjee, G. K., *J. Indian Chem. Soc.*, 1935, **12**, 211.
[31] Noltes, J. G., *Rec. Trav. Chim.*, 1965, **84**, 126.
[32] Noltes, J. G., and Boersma, J., *J. Organometallic Chem.*, 1967, **7**, P6.
[33] Noltes, J. G., and Boersma, J., *J. Organometallic Chem.*, 1969, **16**, 345.
[34] Novikov, S. I., Godovikova, T. I., and Tartakovskii, V. A., *J. Acad. Sci. U.S.S.R., Div. Chem. Sci.*, 1960, 505.
[34a] Oguni, N., and Tani, H., *J. Polymer Sci. Polymer Chem. Ed.*, 1973, **11**, 573.
[35] Okado, H., Kamio, K., and Kojima, S., Jap. Patent 20,499 (1963).
[36] Raffo, M., and Scarella, A., *Gazz. Chim. Ital.*, 1915, **45**, 123.
[37] Razuvaev, G. A., and Bugaeva, Z. J., *Uchenye Zapiski Gor'kov Univ.*, 1953, 143; *Chem. Abstr.*, 1955, **49**, 8230.
[38] Reitzenstein, F., and Bonitsch, G., *J. Prakt. Chem.*, 1913, **86**, 774.
[39] Rosenthaler, L., *Arch. Pharm.*, 1929, **226**, 694.
[40] Rowland, R. L., Perry, W. L., Foreman, E. L., and Friedman, H. F., *J. Amer. Chem. Soc.*, 1950, **72**, 3595.
[41] Sarraje, I., and Lorberth, J., *J. Organometallic Chem.*, 1978, **146**, 113.
[42] Sawicki, K., Gorska-Poczopko, J., Gorski, A., Stec, M., and Miernik, J., *Przem. Chem.*, 1967, **46**, 221; *Chem. Abstr.*, 1967, **67**, 43900.
[42a] Schmidbaur, H., and Räthlein, K.-H., *Chem. Ber.*, 1974, **107**, 102.
[42b] Schmidt, P., and Wannagat, U., *Inorg. Nucl. Chem. Letters*, 1968, **4**, 331.
[43] Schrauth, W., and Schoeller, W., *Ber.*, 1913, **45**, 2808.
[43a] Shearer, H. M. M., unpublished work cited by Bradley, D. C., *Inorg. Macromol. Rev.*, 1970, **1**, 141.
[44] Spinelli, D., and Saleremini, A., *Ann. Chim. (Rome)*, 1961, **17**, 1296.
[45] Tobler, E., and Foster, D. J., *Z. Naturforsch.*, 1962, **17b**, 135, 136.
[45a] Tullock, C. W., Coffman, D. D., and Muetterties, E. L., *J. Amer. Chem. Soc.*, 1964, **86**, 357.
[46] Voynnet, R., *J. Pharm. Chim.*, 1933, **16**, 344.
[47] Williams, J. W., Rainey, W. T., and Leopold, R. S., *J. Amer. Chem. Soc.*, 1942, **64**, 1738.

[47a] Young, J. A., Durrell, W. S., and Dresdner, R. D., *J. Amer. Chem. Soc.*, 1962, **84**, 2105.
[48] Young, J. A., Tsoukalas, S. N., and Dresdner, R. D., *J. Amer. Chem. Soc.*, 1958, **80**, 3604.
[48a] Young, J. A., Tsoukalas, S. N., and Dresdner, R. D., *J. Amer. Chem. Soc.*, 1960, **82**, 396.

10

Insertion Reactions

A INTRODUCTION

Insertions, as illustrated by Eq. (1), are one of the most fundamental of inorganic or organometallic reactions, and may be classified according to the

$$LM{-}X{-} + Z \longrightarrow LM{-}Z{-}X{-} \tag{1}$$

nature of the migrating group X and the inserting molecule (or moiety) Z [430]. This chapter is concerned with the case of X being an amino group, such as NMe_2, Eq. (2). The generality of the process, especially with Z being a 1,2-dipole A=B,

$$LM{-}NRR' + Z \longrightarrow LM{-}Z{-}NRR' \tag{2}$$

{Z = O=CO, O=SO, $O=SO_2$, S=CS, O=CS, S=CO, RN=CO, S=CNR, RN=CS, RN=CNR′, CH_2=CO, $(CH_2=CO)_2$, CH_2=CNR, RR′C=O, RR′C=S, —N=SNS—, RCH=CR′R″, RC≡CR′, RC≡N, RN=SO, RN=O, RN(O)=O, RNC, CO, $M(CO)_n$, $RCH{-}CH_2$ (episulfide, bridged by S), $MeCH{-}CH_2$ (epoxide, bridged by O),

$$\text{cyclo-}C_6H_{10}O, \ \overline{(CH_2)_3O}, \ \overline{(CH_2)_2OCO}, \ \overline{(CH_2)_3OCO},$$

$$\overline{(CH_2)_2C(O)OC(O)}, \ o\text{-}C_6H_4[C(O)]_2O, \text{ or } \overline{CHC(O)OCOCH} \}$$

was first recognised in 1962 [351]. It was suggested that the essential characteristics of the reagent AB are that it be susceptible to attack by a nucleophile (i.e., $LM\ddot{N}RR'$ acts as a nitrogen-centred donor) but not an electrophile, and that the negative end of the $\overset{\delta+}{M}—\overset{\delta-}{N}$ dipole is well displaced towards the nitrogen. The relative migratory aptitude of an amino group ($X = NRR'$) is high compared with X = H, R, OR, or Cl, in part due to a somewhat low M-N bond strength (see Chapters 1-9) and a high $\overset{\delta+}{M}—\overset{\delta-}{N}$ bond polarity. This is represented by the transition state shown in (I), although the concerted mechanism, depicted in (IIa) or (IIb), is often considered.

LM—NRR′ / B=A **(I)** LM—NRR′ / A=B **(IIa)** RR′N / LM B / A **(IIb)**

Such transformations may be described as aminometallations of the substrate Z, a generic term derived from 'aminoboration' [180], and have been extensively reviewed, especially for tin(IV) amides, such as Me_3SnNMe_2 [205a, 355, 356], and are now summarised in the Table (Section E). Related processes are the phosphinometallations, e.g., ref. [682].

The classification based on the nature of the inserting species Z depends in the first instance on the number of atoms directly separating the NRR′ migrating group from the metal or metalloid atom M. This may be a single atom (Section C), two atoms (Section B), or more than two atoms (Section D); the first type may thus be described as exemplifying geminal or 1,1- addition to Z, whereas the others give vicinal (or 1,2-, 1,3-, 1,4-, or 1,5-) products.

The most prolific reactions are those (Section B) in which the reagent Z is a 1,2-dipole, or an incipient 1,2-dipole, such as diketene or S_4N_4. The heterocumulenes X=Y=X′ have a particularly important role as reagents, especially CO_2, CS_2, RNCO, and RNCS.

Many of these reactions appear to be thermodynamically controlled [255, 351] and proceed under mild conditions. For a heterocumulene insertion, the product often enjoys additional conjugative stabilisation compared with the reagents, and this provides for a favourable free energy change for the reaction, e.g., that of Eq. (3). A further driving force often derives from the generally greater polarity of the insertion adduct in relation to its factors, and the former thus frequently precipitates from solution in a non-polar solvent, cf., (III), in Eq. (3).

$$Me_2B{-}NMe_2 + PhNCO \longrightarrow Me_2\bar{B}{-}N(Ph){-}C(NMe_2){=}\overset{+}{O}{-}\text{(ring to B)} \quad \text{(III)} \tag{3}$$

In addition reactions of an unsymmetrical cumulene X=Y=X′ there is ambiguity about the structure of the 1,1-adduct, as illustrated by Eq. (4).

$$LM{-}NRR' + X{=}Y{=}X' \begin{cases} \xrightarrow{\text{either}} LM{-}X{-}Y({=}X')NRR' \\ \xrightarrow{\text{or}} LM{-}X'{-}Y({=}X)NRR' \end{cases} \tag{4}$$

This arises because the reagent X=Y=X′ is an ambident ligand. In the Table we shall generally write the formula of the addition product of an isocyanate, isothiocyanate, ketene, or *N*-thionylaniline (with COS, spectroscopic data are unequivocal) as involving the N=C, N=C, C=C, or N=S bond, respectively, but this may not invariably be correct. X-ray results are sometimes available, whence it is clear that particularly with RNCS or COS either mode is found under different circumstances and that structure (IV) (X = NR or O) is favoured over (V) when M is a soft centre, i.e., is of pronounced class (*b*) character, whereas the converse is true when M is a hard centre, i.e., belongs to class (*a*).

$$LM{-}S{-}C({=}X)NRR' \quad \text{(IV)} \qquad\qquad LM{-}X{-}C({=}S)NRR' \quad \text{(V)}$$

A further feature with a heterocumulene insertion product is that it is potentially bidentate. In the simplest instance this may lead to a chelate, as shown for a dithiocarbamate in (VI) or (VII); alternatively, the ligand may

$$L\bar{M}\langle S^{+}{=}C(NRR')\text{—}S\rangle \longleftrightarrow L\bar{M}\langle S\text{—}C(NRR'){=}S^{+}\rangle$$

(VI)

$$LM\text{—}S\text{—}C(NRR'){=}S^{+}\text{—}\bar{M}L \longleftrightarrow L\bar{M}\text{—}S^{+}{=}C(NRR')\text{—}S\text{—}ML$$

(VII)

occupy a terminal site, as shown in (IV) or (V) (X = S). The tendency for the new ligand formed as a result of aminometallation to behave in a bidentate fashion, as in (III), (VI) or (VII), depends on the metal or metalloid having adequate Lewis acidity and being otherwise co-ordinatively unsaturated. Consequently among main group elements M, chelation or bridging is most frequently found among the *s*-block, Group III, or electropositive (e.g., Sn^{IV}, rather than Si^{IV}) elements and is commonly encountered in transition metal complexes.

Although insertion reactions usually lead to the 1,1-adduct, because this precipitates from solution, there are cases of multiple insertions, illustrated by Eqs. (5) and (6) [344].

$$o\text{-}C_6H_4O_2B\text{-}NEt_2 + 2p\text{-}MeC_6H_4N{=}C{=}NC_6H_4Me\text{-}p$$

$$\longrightarrow o\text{-}C_6H_4O_2B\text{—}N(C_6H_4Me\text{-}p)\text{—}C({=}NC_6H_4Me\text{-}p)\text{—}N(C_6H_4Me\text{-}p)\text{—}C(NEt_2){=}NC_6H_4Me\text{-}p \quad (5)$$

$$o\text{-}C_6H_4O_2B\text{-}NEt_2 + p\text{-}MeC_6H_4N{=}C{=}NC_6H_4Me\text{-}p + PhNCO$$

$$\longrightarrow o\text{-}C_6H_4O_2B\text{—}N(Ph)\text{—}C({=}O)\text{—}N(C_6H_4Me\text{-}p)\text{—}C(NEt_2){=}NC_6H_4Me\text{-}p \quad (6)$$

Many insertion reactions provide elegant and facile syntheses for a substantial variety of metal or metalloid complexes, some of which are either not accessible by alternative routes or are less conveniently so obtained, e.g., dithiocarbamates, especially of metalloids. Another example is the synthesis of large ring heterocycles, e.g., of the bisureidoborane (VIII) from 2-phenyl-1,3,2-diazaborolidine, (IX), and PhNCO [180], or compound (X) from $As(NMe_2)_3$ and $SO_2(NCO)_2$ [630a].

Ph Ph B—N PhN CO OC NH N CH$_2$ H CH$_2$

(VIII)

H N—CH$_2$ PhB N—CH$_2$ H

(IX)

NMe$_2$ CONMe$_2$ O As—N Me$_2$NCN SO$_2$ O$_2$S NCONMe$_2$ N—As Me$_2$NCO NMe$_2$

(X)

The thermodynamic control of many insertion reactions is further demonstrated by (i) their reversibility and (ii) the occurrence of various displacement reactions which are directed to the most stable products. Point (ii) is neatly indicated by the contrasting behaviour of the hard Si^{IV} or the soft Sn^{IV} reactions of Eq. (7) (M = Si or Sn) which lead to $Me_3Si\text{-}OCONMe_2$ or $Me_3SnSCSNMe_2$, respectively [255]. From thermochemical data (see Chapter 5, Sections C4 and

$$Me_3MO{-}C(=O)NMe_2 + CS_2 \rightleftharpoons Me_3MS{-}C(=S)NMe_2 + CO_2 \qquad (7)$$

E3) [54], it is possible to estimate that for the $Sn^{IV}\text{-}O \rightarrow Sn^{IV}\text{-}S$ transformation of Eq. (7) ΔH is significantly negative as is the $Si^{IV}\text{-}S \rightarrow Si^{IV}\text{-}O$ conversion. Another recent example of a displacement reaction relates to the treatment of $[Ta(NBu^t)\{OC(Ph)_2NMe_2\}_2NMe_2]$ with an excess of CO_2 to yield $[Ta(NBu^t)\{OC(O)NMe_2\}_3]$ and Ph_2CO [556a].

The reverse of aminometallation has only rarely been observed, but is favoured if the retrogression is carried out under non-equilibrium conditions. For example, heating a carbamate *in vacuo* or in an open system directs the equilibrium of Eq. (8) to the elimination of the volatile carbon dioxide.

$$LMO{-}C(=O)NMe_2 \rightleftharpoons LMNMe_2 + CO_2 \qquad (8)$$

Unsaturated compounds need not necessarily behave in the sense of Eq. (2). Thus, (a) if Z also has protic character (e.g., C_5H_6), an amine elimination reaction may occur (see Chapter 11), (b) if Z has a good leaving group (e.g., F, Cl, or OR), a metathetical exchange reaction (with NRR′) is a possibility (see Chapter 13), (c) if Z has a combination of characteristics (a) and (b), an elimination reaction [e.g., dehydrochlorination of $CH{\equiv}CCH_2Cl \rightarrow CH{\equiv}C(CH_2)_2C{\equiv}CCH_2Cl$] may be favoured, while (d) if the product of Eq. (2) is capable of reacting with further molecules of Z, then telomerisation, e.g., Eqs. (5) or (6), or polymerisation (see Chapter 17) of Z becomes feasible. Insertion reactions which are oxidative are considered in Chapter 18. The M–N bond is not usually implicated, but the low valent metal amide behaves as a carbene analogue.

In some cases, adducts may be presumed to be formed initially but decompose irreversibly. Examples may be found in the $NaN(SiMe_3)_2$ reactions [731] with (a) RR′CO ($\rightarrow RR'C{=}NSiR_3 + NaOSiR_3$) [143, 411, 412], (b) $PhCH_2Cl$ [$\rightarrow Me_3Si(Ph)C{=}C{=}NSiMe_3$] [411], (c) CH_3CN [$\rightarrow 3Na^+(C_2N)^{3-}$] [411], (d) CO_2, CS_2, Me_3SiNCO, $Si(NCO)_4$, or Me_3SiNCS [$\rightarrow Me_3SiN{=}C{=}NSiMe_3$] [734, 735], (e) Ph_2CO [$\rightarrow Ph_2C{=}CH_2$] [411], and (f) RCOOR′ [$\rightarrow RC(OSiMe_3){=}NSiMe_3$] [412].

B REACTIONS OF AN AMIDE WITH A 1,2-DIPOLE (VICINAL, OR 1,2 ADDITION)

1. Insertion reactions of an alkene or alkyne

A disubstituted acetylene [352, 355] or olefin, including cyclohexene, norbornadiene, styrene, and α-methylstyrene [255], does not react with a metal or metalloid amide. However, the presence of powerfully electron-withdrawing groups, when conjugated with the acetylenic or ethylenic bond, encourages addition to occur, Eq. (9) (M = Si, Ge, or Sn) [145, 146, 257]. Transition metal carbon σ-bonds are notoriously labile. However, stable σ-organo-titanium or -zirconium compounds have been isolated from insertion reactions, e.g., Eqs. (10) or (11) [146]. The products are believed to contain the metal in a five co-ordinate environment, with a chelating amide carbonyl group ($C\ddot{O}NMe_2$) for the latter. Reactions (10) and (11) were realised at −78° C; at ambient temperature, and in contrast to Group IVB derivatives Me_3MNMe_2, even a trace of amide caused the polymerisation or polycondensation of the acetylenedicarboxylic ester. Insertion may be accompanied by another reaction, as illustrated by Eq. (12) [214].

$$EtOOCC{\equiv}CCOOEt + Me_3MNMe_2 \longrightarrow Me_3M[C(COOEt){=}C(COOEt)NMe_2]$$

$$[Ti(NMe_2)_4] + MeOOCC{\equiv}CCOOMe \longrightarrow [Ti\{C(CONMe_2){=}C(COOMe)NMe_2\}(NMe_2)_2OMe] \quad (10)$$

$$[Zr(NMe_2)_4] + 2MeOOCC{\equiv}CCOOMe \longrightarrow [Zr\{C(CONMe_2){=}C(COOMe)NMe_2\}_2(OMe)_2] \quad (11)$$

$$Pr_2BNH\text{-}(2\text{-pyridyl}) + EtOC{\equiv}CH \longrightarrow \text{(cyclic adduct: } Pr_2\overset{-}{B}\text{–}\overset{+}{N}\text{(pyridine)–C–NH–C(OEt)=CH–B)} \quad (12)$$

From the reaction of Me_3GeNMe_2 with $MeOOCC{\equiv}CCOOMe$ both the two geometrical isomers of $Me_3Ge(MeOOC)C{=}C(COOMe)NMe_2$ were isolated [622], and the formation of the *trans*-isomer points against a concerted mechanism (II), but favours the polar transition state (I), unless (II) is followed by an equilibration.

α,β-Unsaturated carbonyl compounds or nitriles, such as $CH_2{=}CHCOOMe$, $CH_2{=}CHCN$, and $CH_2{=}C(Cl)CN$, also react with amides of Sn^{IV} [146, 257], Ge [146, 621], or Ti [146, 346]. The rapid polymerisation (see Chapter 17) of acrylonitrile, even at $-78°$ C, by a trace amount of the Group IVA metal amide $[M(NMe_2)_4]$ (M = Ti, Zr, or Hf) probably involves a multistep insertion process, which is encouraged by the lability of the metal-carbon bond [346, 348]. The difference between the Group IVA and IVB metal amide does not lie in the different chosen functionalities, since $Sn(NMe_2)_4$ affords only a 1:4-adduct with acrylonitrile [146]. The 1:1-adduct of Me_3SnNMe_2 and $CH_2{=}CHCN$ does not react with further $CH_2{=}CHCN$. An aminoborane is inert towards $CH_2{=}CHX$ (X = CN or COOMe), $PhCH{=}CHCN$, or $CH_2{=}CClY$ (Y = Cl or CN) [146]. Similarly, Me_3SiNMe_2 does not react with $CH_2{=}CHCOOMe$ or $CH_2{=}CHCN$ [146]. These results are attributable to the greater B-N and Si-N bond strengths and to the low basicities of the B and Si amines.

Dimethylamine adds to a methylvinylsilane in the presence of lithium amide, Eq. (13) [281].

$$LiNMe_2 + Me{-}\overset{|}{\underset{|}{Si}}{-}CH{=}CH_2 \longrightarrow Me{-}\overset{|}{\underset{|}{Si}}{-}\underset{\underset{Li}{|}}{CH}{-}CH_2{-}NMe_2$$

$$\xrightarrow{HNMe_2} Me{-}\overset{|}{\underset{|}{Si}}{-}CH_2{-}CH_2{-}NMe_2 + LiNMe_2 \quad (13)$$

2. Insertion reactions of an isocyanate, isothiocyanate, carbodi-imide, ketene, diketene, or ketenimine

Aminometallation of the cumulene RNCO, RNCS, RNCNR, or CH_2CO is quite general [351], and has so far been realised for M = Li, Mg, B, Al, Ga, In, Tl, Si, Ge^{IV}, Sn^{II}, Sn^{IV}, Pb^{IV}, P, As, Sb, Bi, Ti^{III}, Ti^{IV}, Zr^{IV}, Hf^{IV}, Nb^{IV}, Nb^{V}, Ta^{V}, Cr^{III}, and Zn. The product (usually of the form LM—N= for RNCO, RNCS, or RNCNR, but may be LM—S— for RNCS) is frequently stabilised by conjugation (see Section A). Usually these reactions are carried out under mild conditions, being noticeably exothermic at ambient temperature. Yields are essentially quantitative, and it is rare for a mixture of products to be obtained, e.g., $B(NMe_2)_3$ forms 1:1, 1:2, or 1:3 adducts [180]; and $[Ti(NMe_2)_4]$ forms 1:2 or 1:4 adducts [146] with PhNCO, depending on the proportions in which the reagents are mixed. However, the order of mixing of the reagents may have a significant effect on the product formed. A 1:1 adduct is obtained when an equimolar amount of an isocyanate is added to R_3PbNEt_2 while the reverse addition causes the trimerisation of the isocyanate [514]. The latter may result from a pathway involving consecutive insertion processes [Eq. (14)]; with di- or tri-isocyanates, insoluble polymers are obtained. Equation (14) illustrates the use of amides in catalysis. In commercial practice, organotin alkoxides or carboxylates are preferred, being used for the catalytic trimerisation of isocyanates to cyanurates and addition of diols to di-isocyanates to yield polyurethanes [81, 199].

$$R_3PbNEt_2 \xrightarrow{R'NCO} R_3PbN(R')CONEt_2 \xrightarrow{R'NCO} R_3PbN(R')CON(R')CONEt_2$$

$$R_3PbN(R')CON(R')CONEt_2 \xrightarrow{R'NCO} R_3Pb(NR'CO)_3NEt_2 \longrightarrow R_3PbNEt_2 + \text{(R'NCO)}_3 \text{ (cyclic: } R'N\text{–CO–NR'–CO–NR'–CO)} \quad (14)$$

A primary aminophosphine Ph_2PNHR reacts with PhNCO to yield $Ph_2PN(R)C(O)NHPh$ by proton migration, but the same product is obtained from aminophosphonation of RNCO with Ph_2PNHPh [323a]. A related P^{III} amide $(EtO)_2PNHR$ with PhNCO gives two isomers which are in equilibrium: $(EtO)_2PN(R)C(O)NHPh \rightleftharpoons (EtO)_2PN(Ph)C(O)NHR$ [323]. Kinetics show (i) R_2PNHR' is ca. 10^4–10^5 more reactive than $R_2PNR'_2$, attributed to proton mobility; (ii) $(EtO)_2PNHR$ is much more reactive than Ph_2PNHR; (iii) R_2PNHR' is more reactive than R_2PNHPh, ascribed to the greater nitrogen basicity of the former compound; and (iv) very small temperature coefficients: ca. 1 kcal mol^{-1}

for $(EtO)_2PNHR/PhNCO$ and ca. 3–4 kcal mol^{-1} for $Ph_2PNHR/PhNCO$, suggesting a pre-equilibrium and hence possibilities *A* or *B* in the Scheme. The latter was preferred [323] because $\overline{OCH_2CH_2OP}NRR'$ is more reactive and $\overline{OCH_2CH_2N(Me)P}Me$ less reactive than acyclic analogues, consistent with π-stabilisation in the intermediate phosphineimine.

$$(EtO)_2PNHR + PhNCO \overset{A}{\rightleftharpoons} (EtO)_2P\text{—}\overset{R}{\overset{|+}{N}}\text{—}H \ (O{=}C\text{—}\bar{N}Ph) \xrightarrow{A} (EtO)_2PNR(CONHPh)$$

$$(EtO)_2PNHR \overset{B,\ +RNCO}{\rightleftharpoons} (EtO)_2\overset{+}{P}\text{—}NHR\ (RN{=}C\text{—}\bar{O}) \overset{B}{\rightleftharpoons} (EtO)_2\overset{+}{P}\text{—}\bar{N}R\ (RN{=}C\text{—}OH) \xrightarrow[(R = Ph)]{B} (EtO)_2PNR(CONHPh)$$

Scheme: Mechanisms for $(EtO)_2PNHR/PhNCO$ insertion.

The products of reaction between $[Ti(NMe_2)_4]$ and PhNCO are structurally interesting [146]. The 1:4 (X = O) and 1:2 adducts probably involve eight- and six-co-ordination, respectively, at the central metal atom. The Zr and Hf analogues only give the 1:4-adducts. Migration of the Me_3M (M = Si or Ge)

$$Ti\left[\begin{matrix} NPh \\ X \end{matrix} \gt CNMe_2\right]_4 \qquad (Me_2N)_2Ti\left[\begin{matrix} NPh \\ X \end{matrix} \gt CNMe_2\right]_2$$

group between an oxygen and nitrogen atom has been suggested for the adduct of a silyl- or germyl-amine with thiobenzoyl isocyanate, Eq. (15) [332]. In the

$$Me_3MNMe_2 + PhC({=}S)\text{—}N{=}C{=}O \longrightarrow PhC({=}S)\text{—}N(MMe_3)\text{—}C({=}O)\text{—}NMe_2 \rightleftharpoons PhC({=}S)\text{—}N{=}C(NMe_2)\text{—}O\text{—}MMe_3 \qquad (15)$$

case of the analogous reaction of Me_3SnNMe_2 with PhC(S)NCO [338b], PhC(O)NCX (X = O or S) [338, 338a, 472a, 475], or PhC(O)NCNBu^t a six-membered ring, as in (XI), is formed; the contrast with the Si or Ge systems is attributable to the greater Lewis acidity of the heavier metal.

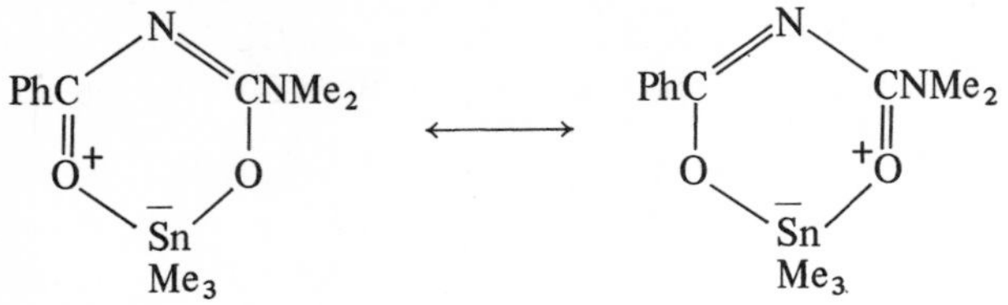

$$\mathrm{PhC(=\overset{+}{O})\text{–}N{=}C(NMe_2)\text{–}O\text{–}\overset{-}{Sn}Me_3} \longleftrightarrow \mathrm{PhC(\text{–}O)=N\text{–}C(NMe_2)=\overset{+}{O}\text{–}\overset{-}{Sn}Me_3}$$

(XI)

As indicated in Section A, the isothiocyanate RNCS may, in principle, insert into the LM-N bond to give either $LMN(R)CSNMe_2$ or $LMSC(NR)NMe_2$. In practice, compounds of the latter type are formed for LM = R_3Sn [255], whilst for the related silicon compound, $R_3SiN(Ph)CSNMe_2$, bonding to Si is through nitrogen [377]. Amides of Ti^{III} [432], Ti^{IV} [146], and Zr^{IV} behave like those of Sn^{IV}; the Ti^{IV} adducts are thus not of the type (V) (X = NR′), but are $\geqslant TiSC(NR_2)NR'$. Whereas fluorosulphonyl isocyanate gives the insertion product with $As(NMe_2)_3$, the betaine $FSO_2\overset{-}{N}C(O)\overset{+}{P}(NMe_2)_3$ is formed with $P(NMe_2)_3$ [631]. Reference has already been made to two-step insertions, Eqs. (5) and (6).

An electron-withdrawing group causes enhancement of reactivity of substrate: e.g., benzoyl or a sulphonyl isocyanate is more reactive than ArNCO and an alkyl isocyanate, RNCO, is least reactive [205a]; however, there is a greater chance of competing metathetical exchange: e.g., Me_3SnNMe_2 and PhCONCS give Me_3SnNCS and $PhCONMe_2$ [338a].

As for the relative migratory aptitude of an amino group compared to some other univalent groups, trends such as $NMe_2 > Ph > NCO$ or NCS are derived from potential competition experiments of the type shown in Eq. (16) [343b]

$$\mathrm{SCNB(NMe_2)_2 + PhNCO \longrightarrow SCNB[N(Ph)CONMe_2]NMe_2} \quad (16)$$

Similar experiments using the carbodi-imide $p\text{-}MeC_6H_4NCNC_6H_4Me\text{-}p$ and a boron reagent lead to the order $NHR > NR_2 > SR > Ph > Cl$ [344].

Reactions of diketene are generalised in terms of Eq. (17) and differ from those of the monomer, Eq. (18) [320]. This may be because of the unsymmetrical structure of the dimer. In contrast, symmetrical phenyl isocyanate dimer behaves similarly to its monomer in its reaction with $B(NMe_2)_3$, Eq. (19).

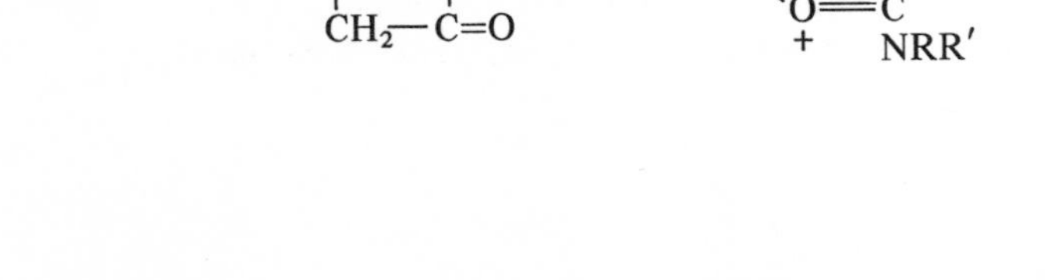

$$\mathrm{LMNRR' + CH_2{=}C\text{–}O\text{–}C(=O)\text{–}CH_2\ (\text{ring}) \longrightarrow \overset{-}{L}M\text{–}O\text{–}C(Me){=}CH\text{–}C(NRR'){=}\overset{+}{O}\text{–}(\text{ring to }LM)} \quad (17)$$

$$LMNRR' + CH_2{=}C{=}O \longrightarrow LMCH_2CONRR' \quad (18)$$

$$B(NMe_2)_3 + (PhNCO)_2 \longrightarrow Me_2NB[N(Ph)CONMe_2]_2 \quad (19)$$

Reaction of a dialkylaminosilane with ketene produces an amide of either β-siloxyvinylacetic acid, (XII), or silylated acetic acid, (XIII) [Eq. (20)]. In each case an intermediate, (XIV), can be isolated [396]. In contrast to the aminosilane reaction, see also ref. [653], a Ge or Sn amide does not form the O–M derivative, analogous to (XIV) [395]. A solitary case of addition to a ketenimine, into Et_2AlNMe_2, leads to compound (XV) (Ar = Ph or p-MeC_6H_4) [721], effectively by a double insertion process, Eq. (21).

$$R_3SiNR'_2 \xrightarrow[-15°\,C]{CH_2=C=O} CH_2{=}C(OSiR_3)(NR'_2)\ \textbf{(XIV)} \longrightarrow R_3SiCH_2CONR'_2\ \textbf{(XIII)}$$

$$CH_2{=}C(OSiR_3)(NR'_2)\ \textbf{(XIV)} \xrightarrow{CH_2=C=O} CH_2{=}C(OSiR_3)(CH_2CONR'_2)\ \textbf{(XII)} \quad (20)$$

$$R_3SiNR'_2 \xrightarrow[50°\,C]{2CH_2=C=O} CH_2{=}C(OSiR_3)(CH_2CONR'_2)\ \textbf{(XII)}$$

$$Et_2AlNMe_2 + 2Me_2C{=}C{=}NAr \longrightarrow \textbf{(XV)} \quad (21)$$

(XV): six-membered ring Me$_2$C=C–N(Ar)–$\overset{-}{Al}$Et$_2$–$\overset{+}{N}$Me$_2$–C(=CMe$_2$)–NAr–, i.e. ring atoms C(=CMe$_2$), N(Ar), AlEt$_2$ (−), NMe$_2$ (+), C(=CMe$_2$), NAr.

(XV)

3. Insertion reactions of carbon dioxide, carbon disulphide, COS, CSe$_2$,* sulphur dioxide, SeO$_2$,† sulphur trioxide, PhNSO, S$_4$N$_4$, or a P=N bonded compound

Insertion of carbon dioxide into an appropriate LM–NRR′ bond (M = B, Al, Tl, Si, Ge, Sn, P, As, Bi, Ti^{IV}, Zr^{IV}, V^{IV}, Nb^{V}, Ta^{V}, Mo^{III}, W^{III}, W^{VI}, U^{IV}, and Zn) gives the corresponding carbamate, LMOCONRR′. Analogous insertions of CS_2 (also with M = Ga, In, Sb, Ti^{III}, and Nb^{IV}) are also established. Carbon dioxide

†An example of SeO_2 insertion into an Si–N bond is known [55].

*Insertions of CSe_2 into Th–N or U–N bonds are recorded [53b].

reacts more readily than carbon disulphide with an aminoborane [179] or aminosilane [177]. With Sn and Ti amides, however, the reverse is true, as is evident from the displacement reactions of Eq. (7) (M = Sn) and Eq. (22) [146, 255] and a competition reaction, Eq. (23) [195]; this may again be seen as a reflection of the class (*b*) character of Sn^{IV} and Ti^{IV}. The COS reactions are less widely studied than those of CO_2 or CS_2. However, addition across the C=S bond is established not only for amides of Sn but also of Ge, P, or As, whereas insertion of COS into an aminosilane yields the SiO thiocarbamate [445, 494, 769]. With $Me(CF_3)_3PNMe_2$, both isomers are formed [714a].

$$[Ti(OCONMe_2)_4] + 3CS_2 \longrightarrow [Ti(OCONMe_2)(SCSNMe_2)_3] + 3CO_2 \quad (22)$$

$$Me_3SnNMe_2 + COS \longrightarrow Me_3Sn{-}SCONMe_2 \quad (23)$$

Insertions of CX_2 into Me_3SiNMe_2 have been suggested to require an ionic chain mechanism, Eq. (24) for CO_2 [100, 101].

$$\left.\begin{array}{l} CO_2 + HNMe_2 \rightleftharpoons Me_2NC(=O)OH \\ Me_3SiNMe_2 + Me_2NCOOH \longrightarrow Me_3SiOC(=O)NMe_2 + HNMe_2 \end{array}\right\} \quad (24)$$

The facts that such reactions (i) only take place in the presence of trace quantities of free amine, and (ii) are inhibited by addition of halogenosilanes (which are known to react with amine) provide evidence for such a postulate. This is further supported by the observation that the reactions of chloroamino-silanes with CS_2 are sluggish and ill-defined [740]. Similar catalysis is found in some related reactions involving B-N or Si-N cleavage (see Table, Section E), but this is not the case with amides of class (*b*) metals such as Sn^{IV} where insertion proceeds even in the presence of Me_3SnCl which reacts instantly with an amine [255]. Insertion reactions of CO_2 into Ti-NR_2 or Mo-NR_2 bonds have also been shown susceptible to catalysis by HNR_2, a trace of which may be formed as a result of adventitious exposure of the amide to moisture; $^{13}CO_2$-labelling experiments demonstrate exchange between the carbamate and $^{13}CO_2$, consistent with a pathway similar to that of Eq. (24) [154b]. The aminoalation of carbon dioxide, using Et_2AlNEt_2, is accelerated by the presence of a complexing agent, such as *N*-methylimidazole or pyridine, perhaps because the $\overset{\delta+}{Al}-\overset{\delta-}{N}$ polarity is thereby increased [327].

The stoicheiometry of the aminometallation of carbon dioxide can be varied, but when an excess of the gas is used the percarboxylated product is not invariably obtained. Interesting illustration is provided in transition metal chemistry, cf., Eqs. (25) (M = Ti or Zr, with $n = 4$; or M = Nb or Ta, with $n = 5$ [154a]) and (26) [90a, 152, 154a] for M = Mo or W. Whereas $[W_2(NMe_2)_6]$ forms $[W_2(OCONMe_2)_6]$ [151a], with the Mo analogue only $[Mo_2(OCONMe_2)_4(NMe_2)_2]$ is obtained [155c]. With $[Mo_2Et_2(NMe_2)_4]$, reductive elimination as well as insertion is observed to yield $[Mo_2(OCONMe_2)_4]$ and presumably C_2H_4 and C_2H_6 [155e]; by contrast $[MoMe_2(NMe_2)_4]$ yields $[Mo_2Me_2(OCONMe_2)_4]$. $[Cr(NEt_2)_4]$ yields Cr^{III} and Cr^{II} products in $[Cr_2(OCONEt_2)_4(\mu\text{-}NEt_2)_2]$ and $[Cr_2(OCONEt_2)_4(HNEt_2)_2]$, respectively, as authenticated by X-ray analysis [151d] (see Chapter 18).

$$[M(NMe_2)_n] + nCO_2 \longrightarrow [M(OCONMe_2)_n] \qquad (25)$$

$$[M(NMe_2)_6] + 6CO_2 \longrightarrow [M(OCONMe_2)_3(NMe_2)_3] \qquad (26)$$

There are structural data for various metal carbamates, including X-ray for compounds (XVI) [714a], (XVII) [154b], $[Ta(OCONMe_2)_3(NMe_2)_2]$ (seven co-ordinate Ta) [151c], $[W_2(OCONMe_2)_6]$ [151a], $[W(OCONMe_2)_3(NMe_2)_3]$ [155a], and $[U(OCONMe_2)_4]$ [127c]. I.r. and/or n.m.r. results indicate a *fac*-configuration for $[M(OCONMe_2)_3(NMe_2)_3]$ (M = Mo or W) [90a, 151a, 152, 154a,

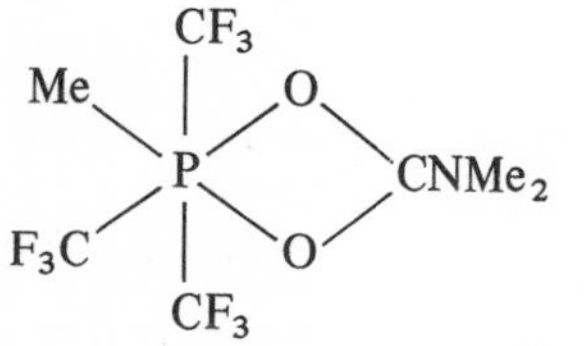

(XVI)

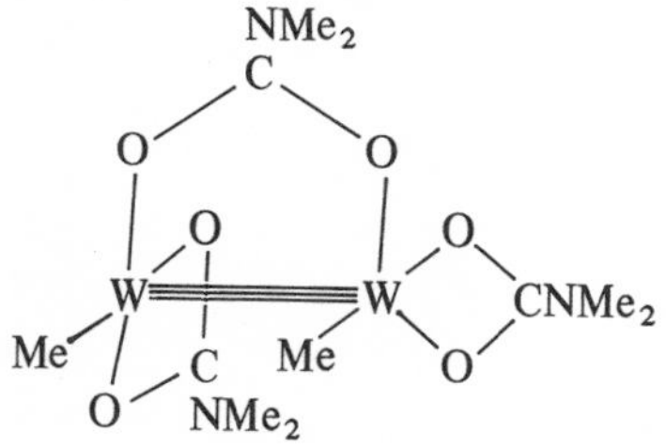

(XVII)

155c], and eight-fold co-ordination about the metal (8O's) for $[Nb(OCONMe_2)_5]$ [154, 154a]. For $[Mo_2(OCONMe_2)_4(NMe_2)_2]$, structure (XVIII) is proposed [155c], in which proximal [*a* in (XIX)] and distal [*b* in (XIX)] methyl groups are inequivalent.

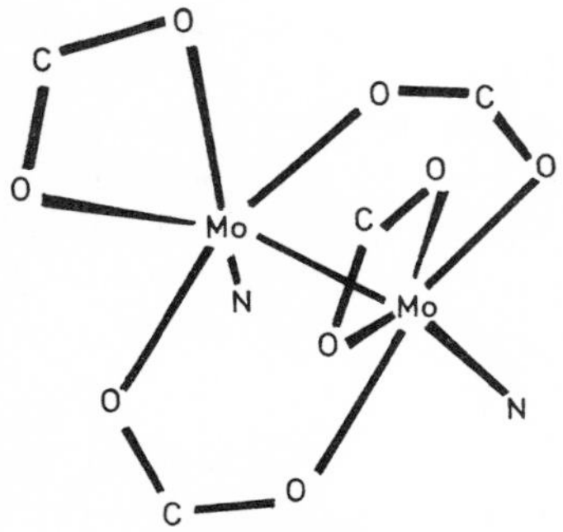

(XVIII)

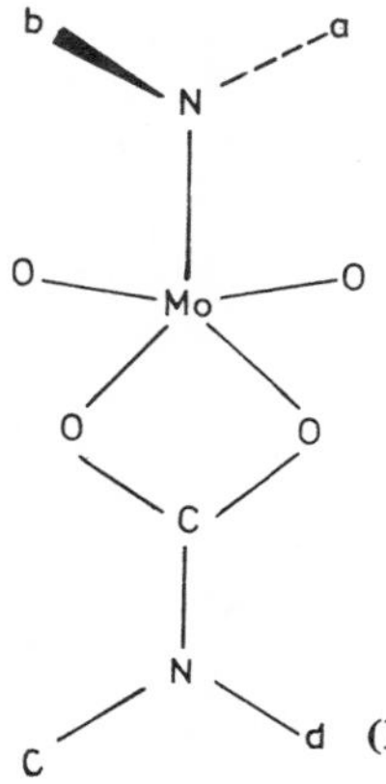

(XIX)

In the reaction of $(Me_3Sn)_2NMe$ or $(Me_3Sn)_3N$ with compounds containing a C=S bond, the addition products are usually not isolated due to the ready formation of $(Me_3Sn)_2S$, Eq. (27) [195, 334]. A similar high SnS affinity is evident from the reaction of $\overline{Bu^t_2SnN(Me)Sn(Bu^t)_2NMe}$ with X=C=S (X = S or PhN) to give $\overline{Bu^t_2SnSSn(Bu^t)_2S}$ and X=C=NMe [705a].

$$(Me_3Sn)_2NR \xrightarrow{>C=S} Me_3Sn{-}S{-}C(<)(NR{-}SnMe_3) \longrightarrow (Me_3Sn)_2S + >C=NR \qquad (27)$$

The CS_2 insertions into dimethylamides of Ti^{IV}, Zr^{IV}, V^{IV}, V^{V}, or Ta^{V} give the expected dithiocarbamatometallanes. With $[\{Ti(NMe_2)_3\}_2]$, $[Nb(NMe_2)_5]$, $[Cr(NEt_2)_4]$, or $[W(NMeR)_6]$, however, the isolated products were $[Ti(SCSNMe_2)_4]$ [25, 432], $[Nb(SCSNMe_2)_4]$ [93], $[Cr(SCSNEt_2)_3]$ [56a], and $[W(SCSNMeR)_4]$ (R = Me or Et) [90a], respectively; redistribution processes (see Chapter 8) may play a role in these reactions.

A type of metathesis which does not involve redox changes is illustrated by the $[V(O)(OPr^i)_2(NEt_2)] + CS_2$ reaction, which furnishes $[V(O)(SCSNEt_2)_3]$ and $[V(O)(OPr^i)_3]$ [157a].

The dithiocarbamates are a well-studied group of compounds. The Ga complex $Ga(SCSNMe_2)_3$ is believed to be a monomer with exclusive metal-to-sulphur binding [538], in contrast to the aluminium analogue (^{27}Al n.m.r. evidence). X-ray data for a related indium compound $In[SCS\overline{N(CH_2)_5}]_3$ shows the co-ordination geometry about the metal to be closer to trigonal prismatic than octahedral [299a].

Insertion products of sulphur dioxide or trioxide and a metal amide are not, in the majority of cases, isolated due to loss of a thionyl or sulphuryl amide with concomitant formation of a compound having an M=O or M—O—M bond. This is illustrated for amides of boron [Eq. (28)] or titanium [Eq. (29)] [144, 546]. It is noteworthy that the metal-oxygen products have exceptional thermodynamic stability, possibly in part as a result of $\ddot{O}{\to}M$ π-bonding. For amides of thallium (Me_2TlNMe_2/SO_2 or SO_3) [729], silicon [Me_3SiNRR'/SO_3 or $(Me_3Si)_3N/SO_3$], or tin (Me_3SnNMe_2/SO_2), adducts have been isolated [45, 255, 355, 398a, 401, 671].

$$3RB(NMe_2)_2 + 3SO_2 \longrightarrow (RBO)_3 \text{ (six-membered B}_3\text{O}_3\text{ ring)} + 3OS(NMe_2)_2 \qquad (28)$$

$$[Ti(NMe_2)_4] + SO_2 \longrightarrow [Ti(O)(OSONMe_2)_2] + OS(NMe_2)_2 \quad (29)$$

There appears to be only a solitary case of aminometallation of *N*-thionylaniline, Eq. (30) [255], but a few more (Si, Sn) with tetrasulphur tetranitride. They may be straightforward, as in Eq. (31) [626], or complex, as in Eqs. (32) [289, 642] or (33) [631a]. The aminobis(imino)phosphorane $P[N(SiMe_3)_2](NSiMe_3)_2$ undergoes cycloaddition with EX_n (=$AlCl_3$, $SnCl_4$, $TiCl_4$, $NbCl_5$, or $FeCl_3$) to yield $X[(Me_3Si)_2N]\overline{PN(SiMe_3)E(X_{n-1})N}SiMe_3$ [520a].

$$Me_3SnNMe_2 + PhNSO \longrightarrow Me_3SnN(Ph)SONMe_2 \quad (30)$$

$$Me_3SiNMe_2 + ½S_4N_4 \longrightarrow Me_3SiN{=}S{=}NSNMe_2 \quad (31)$$

$$S_4N_4 \xrightarrow{Me_3SiNMe_2} Me_3SiNSNSNSNSiMe_3 + S(NMe_2)_2$$
$$S_4N_4 \xrightarrow{Me_3SnNMe_2} Me_3SnN{=}S{=}NSnMe_3 + S(NMe_2)_2 \quad (32)$$

$$(Me_3Sn)_3N + S_4N_4 \longrightarrow Me_2Sn\langle S{-}N{=}S{=}N\rangle \text{ (ring: } Me_2Sn{-}S{-}N{=}S{=}N{-}Sn) + Me_4Sn + Me_3SnN{=}S{=}NSnMe_3 \quad (33)$$

4. Insertion reactions of an aldehyde or ketone, or of a related compound having an isolated >C=O, >C=S, \N=O, or \N(O)=O bond

Aminometallation of a carbonyl compound having adjacent electron-withdrawing groups is well-established [14, 253]. The addition product (XX) (R = Me) of chloral and dimethylaminotrimethylsilane has been isolated, Eq. (34); however, the analogous tin derivative undergoes β-elimination, Eq. (35) [253]. The steric environment of the nitrogen atom in the dialkylaminotrimethylsilane appears to influence the stability of the insertion product, e.g., 23% (XX) and 69.9% (XXI) are obtained when R = Pr [253].

$$Me_3SiNR_2 + Cl_3CCHO \longrightarrow Me_3SiOCH(NR_2)CCl_3 \quad (34)$$

(XX)

$$Me_3SnOCH(NR_2)CCl_3 \longrightarrow Me_3SnCCl_3 + HC(=O)NR_2 \quad (35)$$

(XXI)

Hexafluoroacetone undergoes aminometallation, Eq. (36) (R = Me, M = Si or Sn; or R = Bu, M = Ge) [14].

$$R_3MNMe_2 + (CF_3)_2CO \longrightarrow R_3MOC(CF_3)_2NMe_2 \qquad (36)$$

Reaction with a diazasilacyclopentane results in telomerisation, Eq. (37) [14].

$$3EtN(CH_2CH_2)NEt\,SiMe_2 \text{ (ring)} + 6\,(F_3C)_2C{=}O \longrightarrow \tfrac{1}{3}\left[\begin{array}{l} EtN{-}CH_2{-}CH_2{-}NEt \\ (F_3C)_2C \qquad\qquad C(CF_3)_2 \\ O{-}SiMe_2 \qquad\quad O{-} \end{array}\right]_3 \qquad (37)$$

Carbonyl compounds with α-hydrogen atoms can undergo Claisen-Schmidt-type condensations in the presence of a metal complex having pronounced Lewis base character (e.g., the Me_3SnNMe_2/Me_2CO system) (see Chapters 11, 13, and 14); such carbonyl compounds may react solely as protic species (see Chapter 11). Aromatic aldehydes or ketones usually give straightforward 1,1-adducts. Whereas (2-pyridylamino)diphenylborane with CH_3CHO, PhCHO, Me_2CO or $\overline{(CH_2)_5CO}$ gives a normal insertion product (XXII), [276] acetophenone apparently yields $Ph_2BOB(Ph)NH(C_5H_4N\text{-}2)$ [277]. Et_3GeNMe_2 and $CH_2(CO_2Et)X$ (X = CO_2Et or CN) give both the C=O insertion product and $Et_3GeCH(CO_2Et)X$ [651a] (see Chapter 11).

$$\text{(XXII): six-membered ring } \overset{+}{N}(\text{pyridyl}){-}\overset{-}{B}Ph_2{-}O{-}CRR'{-}NH{-}C(\text{pyridyl})$$

A silylamine may react with an organic carbonyl compound in the presence of a metal catalyst to give an imine, e.g., Eq. (38) [220]. This type of exchange reaction, reminiscent of the Wittig reaction (P and Si share a high affinity for oxygen), is a feature of metal amide chemistry (see Chapter 13).

$$Me_3SiNHMe + PhCHO \xrightarrow{Zn/Cd} PhCH{=}NMe \qquad (38)$$

Examples of aminophosphination include those of Eqs. (39) [522] and (40) (X = Y = NMe_2 or NEt_2 [523]; X = NMe_2, Y = OH, R = Me [522]; or X = NMe_2, Y = NHCOMe, R = Me [601]).

$$HOP(NMe_2)_2 + Me_2CO \longrightarrow HOP(O)[CMe_2NMe_2] \quad (39)$$

$$X(Y)PNR_2 + PhCHO \longrightarrow X(Y)P(O)[CH(Ph)NR_2] \quad (40)$$

Fragmentary reports have described the aminometallation of an ester (see also Chapter 13), thio-ketone or -amide (see Table), and for the most part are unexceptional.

Reaction of nitrosobenzene with dimethylaminotrimethylstannane [Eq. (41)] represents the sole example to date of an insertion of a nitroso compound [257].

$$Me_3SnNMe_2 + PhNO \longrightarrow Me_3SnON(Ph)NMe_2 \quad (41)$$

Nitromethane adducts have so far only been obtained with aminoboranes [178, 179], and undergo amine elimination upon removal of solvent, or on gentle heating, Eq. (42); the crystalline products have tentatively been assigned structure (XXIII) (see also Chapter 11 for other protic reactions of nitroalkanes).

$$>BNR_2 + MeNO_2 \xrightarrow[20°\,C]{Et_2O} >BO{-}\overset{\overset{O^-}{|+}}{\underset{\underset{Me}{|}}{N}}NMe_2 \xrightarrow{-HNMe_2} \underset{\textbf{(XXIII)}}{>BCH_2NO_2} \quad (42)$$

Expected insertion reactions involving aldimides or ketimides, RCH=NR′ or RR″C=NR′, related to those of the isoelectronic aldehydes or ketones, do not appear to have been reported. There has been much recent interest in compounds having $(p\text{-}p)\pi$-bonds between C, N, O, or S on the one hand and B, Si, Ge, or Sn on the other. To date, a pure compound of this type stable at ambient temperature has not been isolated. However, there is substantial evidence for their existence from spectroscopy, kinetics, or trapping experiments, such as those of Eq. (43) [621a], examples of insertion reactions of the transient molecule (XXIV). Related results are in ref. [616a].

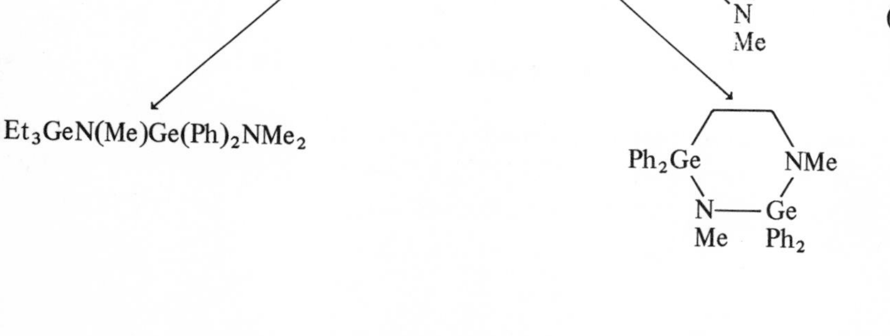

$$Ph_2Ge\langle CH_2CH_2N(Me)\rangle \xrightarrow{-C_2H_4} \underset{\textbf{(XXIV)}}{[Ph_2Ge{=}NMe]} \xleftarrow{-LiCl} Ph_2Ge(CH_2CH_2Cl){-}N^-(Me)\,Li^+$$

(XXIV) + $Et_3Ge{-}NMe_2$ ⟶ $Et_3GeN(Me)Ge(Ph)_2NMe_2$

(XXIV) + $Ph_2Ge\langle CH_2CH_2N(Me)\rangle$ ⟶ six-membered ring: $Ph_2Ge{-}CH_2CH_2{-}NMe{-}GePh_2{-}N(Me){-}$ (43)

5. Insertion reactions of a nitrile

With a nitrile, four types of situations have been realised [Eq. (44)]; pathway (44a) represents insertion.

$$LM{-}NR'R'' \xrightarrow{RC\equiv N} \begin{cases} \xrightarrow{(a)} RC(NR'R'')=NML \\ \xrightarrow{(b)} HNR'R'' + \text{product(s)} \\ \xrightarrow{(c)} (RCN)_3 \\ \xrightarrow{(d)} LMF + R'R''NC_6F_4CN \quad (R = C_6F_5) \end{cases} \quad (44)$$

Such reactions have so far been observed (Table) with amides of lithium, aluminium, tin, or of transition metals. Presence of electron-withdrawing groups, as in CCl_3CN [146, 200], encourages course (44a); Bu^tCN does not react with Me_3SnNMe_2 or $[Ti(NMe_2)_4]$ [146]. Treatment of MeCN with an amide of Ti^{IV}, Zr^{IV}, or Ta^{V} takes route (44a), but the products are, presumably, polymeric [91]. Benzonitrile or *p*-toluonitrile gives a 1:4-adduct with a Group IVA amide $[M(NMe_2)_4]$ (M = Ti, Zr, or Hf) [146], and $[Ti\{N(CPh)NMe_2\}_3]$ is obtained from the Ti^{III} amide [432]. Reaction (44b) represents the pattern for the Me_3SnNMe_2/MeCN system (see Chapter 11); aminoboranes do not react [178, 179]. In petroleum as solvent, $LiNMe_2$ causes the trimerisation [route (44c)] of PhCN [647]; this does not occur in diethyl ether except at high concentrations. The mechanism probably involves three successive PhCN insertions into the Li-N bond and subsequent $LiNMe_2$ elimination. This was confirmed by treating $Li{-}N{=}C(Ph)NMe_2$ with 2 equivalents of p-MeC_6H_4CN to afford the mixed 1,3,5-triazine $(p\text{-}MeC_6H_4)_2PhC_3N_3$ [433, 647]. Pathway (44d) is exemplified by Eq. (45) (see Chapter 13); the product of this reaction then adds to Me_3SnNMe_2, Eq. (46) [146].

$$Me_3SnNMe_2 + C_6F_5CN \longrightarrow Me_3SnF + p\text{-}Me_2NC_6F_4CN \quad (45)$$

$$Me_3SnNMe_2 + p\text{-}Me_2NC_6F_4CN \longrightarrow Me_3SnN{=}C(NMe_2)C_6F_4NMe_2\text{-}p \quad (46)$$

C GEMINAL (OR 1,1-) ADDITION OF AN AMIDE

1. Insertion reactions of an isonitrile

Isonitrile insertion into Sn-N [257] or Pb-N [514] bonds [Eqs. (47), (48)] are rare examples of 1:1 insertion of a stable compound into a main group metal

M–X bond. (Others, not involving X = an amino group, include the $Bu_3B/PhNC$ [304] and $Me_3SiH/C_6H_{11}NC$ [644] systems).

$$Me_3SnNMe_2 + p\text{-}MeC_6H_4NC \longrightarrow p\text{-}MeC_6H_4N{=}C(SnMe_3)(NMe_2) \qquad (47)$$

$$Bu_3PbNEt_2 + PhNC \longrightarrow PhN{=}C(PbBu_3)(NEt_2) \qquad (48)$$

2. Insertion reactions of carbon monoxide or a metal carbonyl (the latter are 1,2-additions)

Although insertion of carbon monoxide into a metal–Me bond is well-known in transition metal chemistry, among metal amides it has only been unambiguously observed for alkali metal derivatives. In the $NaN(SiMe_3)_2/CO$ system [736], the expected product $Na[CON(SiMe_3)_2]$, if formed, decomposes, Eq. (49), whereas $NaNH_2$ and CO afford Na[HCONH] [507]. On the other hand, $[Cr(CO)_6]$ gives [Eq. (50)] the 1,1-adduct, characterised as the carbene complex (XXV) [241, 242]. A similar reaction was postulated to occur between $[Fe(CO)_5]$ and $NaN(SiMe_3)_2$ [731], the adduct decomposing to yield $Na[Fe(CO)_4CN]$ and $(Me_3Si)_2O$. More recent results show a further subtlety, as illustrated in the low temperature reaction of Eq. (51) (M = Li or Na, at −10 to +25° C) [80, 108, 374, 500, 687]; with $[M(\eta\text{-}C_5H_5)(CO)_2NO]$ (M = Cr, Mo, or W], the CO group is attacked preferentially to yield $Na[M(\eta\text{-}C_5H_5)(CO)(NO)CN]$ [410].

$$NaN(SiMe_3)_2 + CO \longrightarrow NaCN + (Me_3Si)_2O \qquad (49)$$

$$LiNEt_2 + [Cr(CO)_6] \xrightarrow{Et_2O} [Cr(CO)_5\{C(OLi)NEt_2\}].OEt_2 \xrightarrow{[Et_3O][BF_4]} [Cr(CO)_5\{C(OEt)NEt_2\}] \qquad (50)$$

(XXV)

$$MN(SiMe_3)_2 + [Fe(CO)_5] \longrightarrow [Fe(CO)_4(CNSiMe_3)] + MOSiMe_3 \qquad (51)$$

The homoleptic dimethylamides of Ti^{IV}, Zr^{IV}, or Hf^{IV} were originally believed to form 1,2-adducts with various metal carbonyls [94]. However, in view of the results with amides of Li [Eq. (50)], Al [578, 581], Sn^{IV} [580], or Ti^{IV} [579], which lead to carbenemetal complexes, this now seems unlikely; even $C(NMe_2)_4$

reacts with $[Ni(CO)_4]$ or $[Fe(CO)_5]$ to give similar complexes [579a]. Examples are in Eqs. (52)–(54), whence it will be clear that these aminometallations are actually examples of vicinal (1,2-) rather than geminal (1,1-) addition.

$$[Ti(NMe_2)_4] + [Mo(CO)_6] \longrightarrow [Mo(CO)_5\{C(NMe_2)OTi(NMe_2)_3\}] \quad (52)$$

$$Sn(NMe_2)_4 + 2[Fe(CO)_5] \longrightarrow (OC)_3Fe\begin{smallmatrix} C(NMe_2)\text{—}O \\ C(NMe_2)\text{—}O \end{smallmatrix}Sn\begin{smallmatrix} O\text{—}C(NMe_2) \\ O\text{—}C(NMe_2) \end{smallmatrix}Fe(CO)_3 \quad (53)$$

$$[Al(NMe_2)_3]_2 + 2[Ni(CO)_4] \longrightarrow (Me_2N)((OC)_3Ni(Me_2N)CO)\overset{-}{Al}(\overset{+}{N}Me_2)_2\overset{-}{Al}(OC(NMe_2)Ni(CO)_3)(NMe_2) \quad (54)$$

D REACTIONS OF AMIDES WITH 1,3-, 1,4-, OR 1,5-DIPOLES BY RING-OPENING

Insertion in a formal 1,3- manner has been observed with epoxides; a trace of $LiNEt_2$ may catalyse the process [718]. These reactions, e.g., Eq. (55), probably proceed via (XXVI) as a propagating species. An aminosilane [341] (without catalyst) or -germane (in presence of $ZnCl_2$) [623] behaves similarly, in the latter case also with the four-membered ring ether. However, reactions of the Sn–N compounds [Eq. (56)] with ethylene or propylene sulphide proceed without any catalyst [146, 346], showing once more the class (*b*) or soft character of Sn^{IV}. A related process leading to the betaine (XXVII) is shown in Eq. (57) [343a].

$$\underset{\text{(epoxide, O bridge)}}{RCH\text{—}CH_2} + LiNEt_2 \longrightarrow \underset{\substack{| \\ OLi}}{RCH}\text{—}CH_2NEt_2 \xrightarrow{Bu_3SnNEt_2} \underset{\substack{| \\ OSnBu_3}}{RCH}\text{—}CH_2NEt_2 + LiNEt_2 \quad (55)$$

(XXVI)

$$R_3SnNMe_2 + \underset{\text{(S bridge)}}{R'CH\text{—}CH_2} \longrightarrow R_3SnSC(R')HCH_2NMe_2 \quad (56)$$

$$P(NMe_2)_3 + \underset{\text{(}SO_2\text{ bridge)}}{PhC{=}CPh} \longrightarrow \underset{\substack{| \\ {}^{-}SO_2}}{PhC}{=}C(Ph)\overset{+}{P}(NMe_2)_3 \quad (57)$$

(XXVII)

The cleavage of β-propiolactone [Eq. (58)] is an example of a 1,4-insertion reaction. With the Sn-N [339], Ti-N, and Hf-N compounds, acyl-oxygen fission of the lactone is preferred while analogous Si-N and Ge-N compounds induce alkyl-oxygen bond-breaking [146, 339].

$$Me_3MNEt_2 + \underbrace{CH_2\text{—}CH_2\text{—}C(=O)\text{—}O}_{\text{ring}} \begin{cases} \xrightarrow{(M = Sn)} Me_3SnOCH_2CH_2CONEt_2 \\ \xrightarrow[(M = Si \text{ or } Ge)]{} Me_3MOCOCH_2CH_2NEt_2 \end{cases} \quad (58)$$

Another 1,4-addition is depicted in Eq. (59) [514], and for the Me_3SnNMe_2/PhC(X)N=C=O (X = O or S) system, see (XI) (or the S analogue) Section B.2 [332, 475].

$$Bu_3PbNEt_2 + PhCH{=}C(CN)_2 \longrightarrow PhCH(NEt_2){-}C(CN){=}C{=}NPbBu_3 \quad (59)$$

1,5-Additions are found in the aminogermanation ring-opening of γ-butyrolactone [624] or the anhydride of succinnic [623, 624], maleic [624], or *o*-phthalic [624] acid, as in Eq. (60).

$$Et_3GeNMe_2 + \begin{matrix} CH_2CO \\ | \quad\; \rangle O \\ CH_2CO \end{matrix} \longrightarrow Et_3GeOCOCH_2CH_2CONMe_2 \quad (60)$$

E TABLE OF REACTIONS

†Insertion Reactions

Reactants		Product	References
1. Reactions with Alkynes or Alkenes			
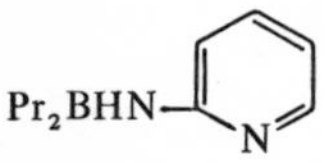	EtOC≡CH	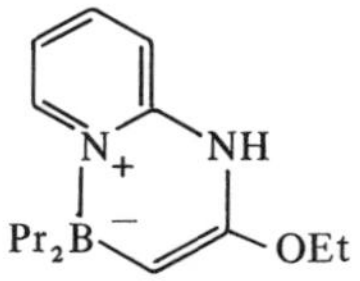	[214]
$B(NMe_2)_3$ (R = Pr, Bu^t, Me_3Si)	EtOOCC≡CCOOEt	$B[C(COOEt){=}C(COOEt)NMe_2]_3$	[145, 257]
Me_3MNMe_2 (M = Si, Ge, Sn)	EtOOCC≡CCOOEt	$Me_3MC(COOEt){=}C(COOEt)NMe_2$	[145, 257]
Me_3SiNHR	MeOOCC≡CCCOOMe	$Me_3SiC(COOMe){=}C(COOMe)NHR$	[701b]
Me_3SiNMe_2	MeOOCC≡CCOOMe	$Me_3SiC(COOMe){=}C(COOMe)NMe_2$	[146]
R_3GeNMe_2 (R = Bu)	MeOOCC≡CCOOMe	$R_3GeC(COOMe){=}C(COOMe)NMe_2$	[146]
(R = Me, Et)		Two isomers	[622]
$Ti(NMe_2)_4$	MeOOCC≡CCOOMe	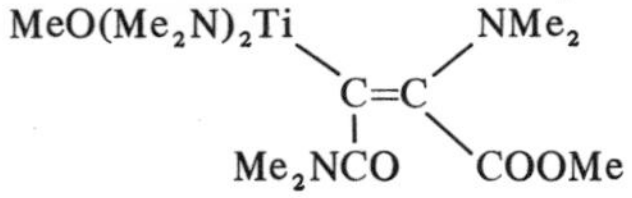	[146]
$Zr(NMe_2)_4$	MeOOCC≡CCOOMe	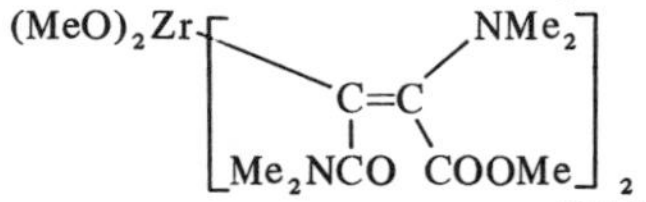	[146]
Me_3SnNMe_2	PhC≡CCl	$Me_3SnC(Ph){=}C(NMe_2)Cl$	[145, 257]
Et_3SnNMe_2	$CH_2{=}CHCOOMe$	$Et_3SnCH(COOMe)CH_2NMe_2$	[145, 257]
Et_3SnNMe_2 (R = H,[1] Me, Ph)	RCH=CHCHO	$Et_3SnCH(CHO)CH(R)NMe_2$	[145, 257]
R_3GeNMe_2 (R = Bu, Et)	$CH_2{=}CHCN$	$R_3GeCH(CN)CH_2NMe_2$	[146, 621]
R_3SnNMe_2 (R = Me and R′ = H or Me; R = Et and R′ = H)	$CH_2{=}C(R')CN$	$R_3SnC(CN)R'CH_2NMe_2$	[145, 146, 257]
$Sn(NMe_2)_4$	$CH_2{=}CHCN$	$Sn[CH(CN)CH_2NMe_2]_4$	[149]
Me_3SnNMe_2	$CH_2{=}C(CN)Cl$	$Me_3SnC(CN)ClCH_2NMe_2$	[146]
$Me_3SnN{=}CPh_2$	$CH_2{=}CHCN$	$Me_3SnCH(CN)CH_2N{=}CPh_2$	[424a]
$M(NMe_2)_4$ (M = Ti, Zr)	$CH_2{=}C(CN)Cl$	$M[C(CN)ClCH_2NMe_2]_4$	[146]

Insertion Reactions *(C'ntd.)*

Reactants		Product	References
2. Reactions with RNCO (R = Me, Et, Bu, C_6H_{11})			
$PhB(NMe_2)_2$	EtNCO	$PhB[N(Et)CONMe_2]_2$	[70]
$(MeBNH)_3$	BuNCO	Me B HN NBu OC CO N Bu	[89]
Me_3SiNEt_2	MeNCO	$Me_3SiN(Me)CONEt_2$	[561]
$Me_3SiNHPh$	MeNCO	$Me_3SiN(Me)CONHPh$	[377]
$(Me_3Si)_2NMe$	MeNCO	$Me_3SiN(Me)CON(Me)SiMe_3$	[377]
$Me_3SiN(Me)CON(Me)SiMe_3$	MeNCO	$Me_3SiN(Me)CON(Me)CON(Me)SiMe_3$	[377]
Et_2AlNMe_2 (R = Me, Et, C_6H_{11})	RNCO	$Et_2AlN(R)CONMe_2 + (RNCO)_3$	[312]
$Me_3SiNEtR'$ (R' = Et, $CONEt_2$; R = Et, Bu)	RNCO	$Me_3SiN(Et)CONEtR'$	[239]
$Me_{4-n}Si[N(CH_2)_2]_n$ (R = Et, Bu, n = 1; R = Et, n = 2)	RNCO	$Me_{4-n}Si[N(R)CON(CH_2)_2]_n$	[662]
$(Me_2SiNH)_3$ (R = Et, Bu)	RNCO	$SiMe_2$ RN NR OC CO N R	[238, 662]
$Bu_3SnN{=}CPh_2$	MeNCO	$Bu_3SnN(Me)CON{=}CPh_2$	[296]
$As(NMe_2)_3$	BuNCO	$Me_2NAs[N(Bu)CONMe_2]_2$	[722]
3. Reactions with PhNCO			
Me–(CH–S, CH₂–S ring)$BNEt_2$		Me–(CH–S, CH₂–S ring)$BN(Ph)CONEt_2$	[175]
$B(NMe_2)_3$		$B[N(Ph)CONMe_2]_3$	[307]
$B_2(NMe_2)_4$		$B_2[N(Ph)CONMe_2]_4$	[307]
(–O, –O ring)$BNEt_2$		(–O, –O ring)$BN(Ph)CONEt_2$	[175]
$(CH_2O)_2BNEt_2$		$(CH_2O)_2BN(Ph)CONEt_2$	[174]
$B(NHBu^t)_3$		$B[N(Ph)CONHBu^t]_3$	[176]

Insertion Reactions *(C'ntd.)*

Reactants	Product	References
$Bu^tHNB(NPr^i_2)_2$	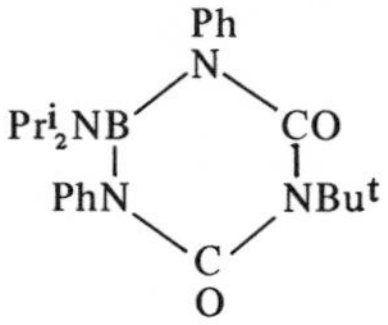 Ph; N; Pr^i_2NB; CO; PhN; NBu^t; C; O	[176]
$Bu^tN(BNHBu^t)_2$	$Bu^tN\{B[N(Ph)CONHBu^t]_2\}_2$	[178]
$Ph_{3-n}B(NHR)_n$ ($R = Bu^t$, $n = 1,2$; $R = Et$, $n = 2$)	$Ph_{3-n}B[N(Ph)CONHR]_n$	[180]
$PhB(X)NMe_2$ ($X = Cl$, $NHBu^t$)	$PhB(X)[N(Ph)CONMe_2]$	[178]
$PhB(NHNHPh)_2$	$PhB[N(Ph)CONHNHPh]_2$	[178]
$PhB(NMe_2)_2$	$PhB[N(Ph)CONMe_2]_2$	[70]
	$PhB(NMe_2)[N(Ph)CONMe_2]$	[307]
$(SCN)B(NMe_2)_2$	$(SCN)B(NMe_2)[N(Ph)CONMe_2]$	[343b]
$PhB(NRCH_2)_2$ ($R = H$, Bu)	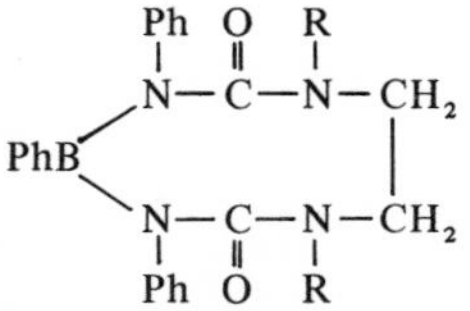 Ph O R; N—C—N—CH_2; PhB; N—C—N—CH_2; Ph O R	[180]
o-$C_6H_4O_2BNRR'$	o-$C_6H_4O_2BN(Ph)CONRR'$	
(R, $R' = H$, Bu^t)		[180]
($R = R' = Et$)		[437]
$(MeBNH)_3$	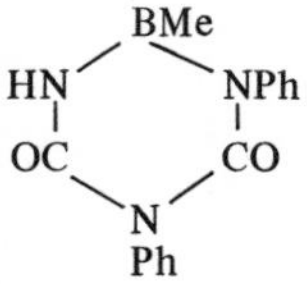 BMe; HN; NPh; OC; CO; N; Ph	[71,89]
$(Et_2NBNR)_3$	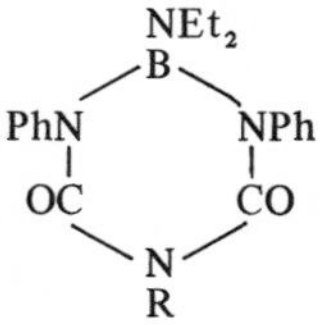 NEt_2; B; PhN; NPh; OC; CO; N; R	
($R = H$)		[70]
($R = Et$)		[176]
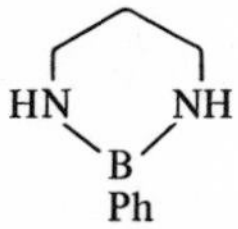 HN; NH; B; Ph	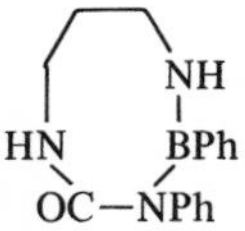 NH; HN; BPh; OC—NPh	[260]
$PhB\overline{N(CH_2)_4CH(CH_2)_2O}$	$PhBN(Ph)CO\overline{N(CH_2)_4CH(CH_2)_2O}$	[181a]

Insertion Reactions *(C'ntd.)*

Reactants	Product	References
Et_2AlNMe_2	$Et_2AlN(Ph)CONMe_2$	[312]
Me_2TlNMe_2 (n = 2–4)	$Me_2Tl[OCNPh]_nNMe_2$	[729]
$Et_2InNPhCOEt$	$Et_2In(NPhCO)_3Et$	[713a]
Me_3SiNRR'	$Me_3SiN(Ph)CONRR'$	[377]
(RR′ = HMe, Me_2)		[239, 377]
(RR′ = Et_2)		[561]
(RR′ = C_5H_{10})		[239]
(RR′ = $(CH_2)_2$)		[662]
$Me_2PhSiNHPh$	$Me_2PhSiN(Ph)CONHPh$	[377, 561]
$Me_2Si(NEtR)_2$	$Me_2Si[N(Ph)CONEtR]_2$	
(R = H, Et)		[177]
(R = Et)		[561]
$Me_2Si(NRCH_2)_2$ (R = H, Et, Bu)	$Me_2Si[N(Ph)CONRCH_2]_2$	[177]
$(Me_3Si)_2NR$	$Me_3SiN(Ph)CON(R)SiMe_3$	[491]
(R = H)		[239]
(R = Me, H)		[377, 496]
(R = Et)		[496]
(R = Ph)		[377, 496]
$(Me_2SiNH)_n$	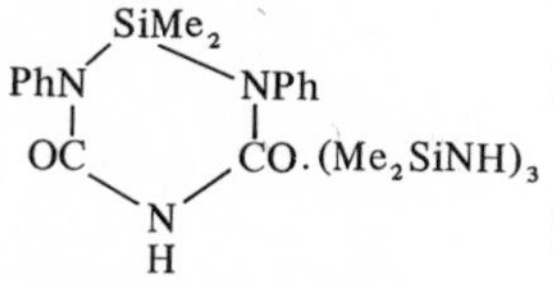 $SiMe_2$ PhN NPh OC CO N H	$[662]^2$
(n = 3,4)	CO.$(Me_2SiNH)_3$	[2]
$Me_3SiN{=}CPh_2$	$Me_3SiN(Ph)CON{=}CPh_2$ and/or $Me_3SiOC({=}NPh)N{=}CPh_2$	[474]
$Me_3SiN(Ph)CONEt_2$	$Me_3SiN(Ph)CON(Ph)CONEt_2$	[239]
$Me_3SiN(Ph)CON(Me)SiMe_3$	$Me_3SiN(Ph)CON(Me)CON(Ph)SiMe_3$	[377]
Et_3GeNEt_2	$Et_3GeN(Ph)CONEt_2$	[446]
$Sn[N(SiMe_3)_2]_2$	$Sn[N(Ph)CON(SiMe_3)_2]_2$	[295]
Me_3SnNMe_2	$Me_3SnN(Ph)CONMe_2$	[255, 351]
$Bu_3SnN(Ph)CON(Et)SnBu_3$	$Bu_3SnN(Ph)CON(Et)CON(Ph)SnBu_3$	[81]
$R_3SnN{=}CR'_2$	$R_3SnN(Ph)CON{=}CR'_2$	
(R = Me, R′ = Ph, CF_3)		[424a]
(R = Bu, R′ = Ph)		[296]
Bu_3PbNEt_2	$Bu_3PbN(Ph)CONEt_2$, $(PhNCO)_3$	[514]
$PhAs(NRR')_2$ (R = R′ = Me, Et, Pr, Bu; RR′ = C_5H_{10})	$PhAs[N(Ph)CONRR']_2$	[245]
$As(NMe_2)_3$	$As[N(Ph)CONMe_2]_3$	[560]
$Bi(NR_2)_3$	$Bi[N(Ph)CONR_2]_3$	[36]

Insertion Reactions *(C'ntd.)*

Reactants	Product	References
$Cp_{3-n}Ti(NMe_2)_n$ (n = 1,2,3)	$Cp_{3-n}Ti[N(Ph)CONMe_2]_n$	[432]
$M(NMe_2)_4$ (M = Ti, Zr, Hf)	$M[N(Ph)CONMe_2]_4$	[146, 148]
	or $(Me_2N)_2Ti[N(Ph)CONMe_2]_2$	[146]
$Ti(NMe_2)_3$	$Ti[N(Ph)CONMe_2]_3$	[432, 433]
$[(PhCH_2O)_2CrNH_2]_n$	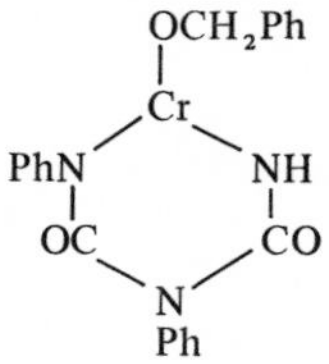 OCH_2Ph / Cr / PhN NH / OC CO / N / Ph	[675]
$[ROCr(NH_2)_2]_n$	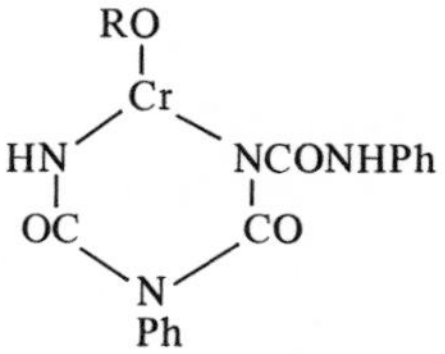 RO / Cr / HN NCONHPh / OC CO / N / Ph	[674]
$EtZnNR_2$ (R = Et, Ph)	$EtZnN(Ph)CONR_2$	[526]

4. *Reactions with* OCN–C_6H_3(NCO)–Me

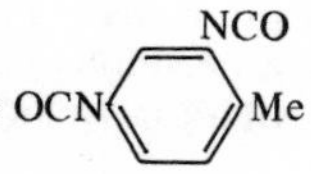

Reactants	Product	References
$(MeBNH)_3$	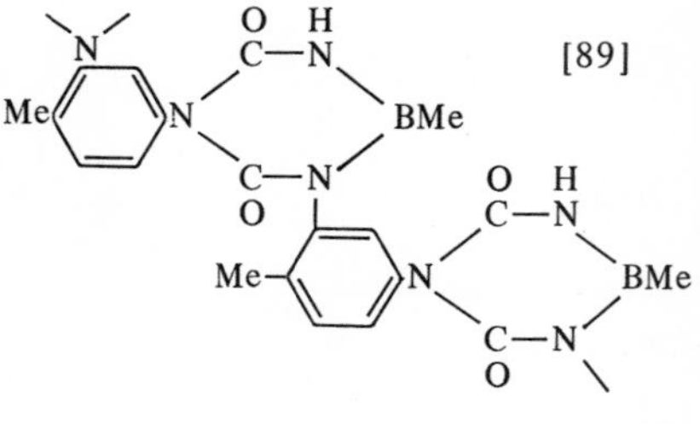 O H / C—N / Me–C_6H_3–N BMe / C—N / O / Me– C_6H_3 –N / O H / C—N / BMe / C—N / O	[89]
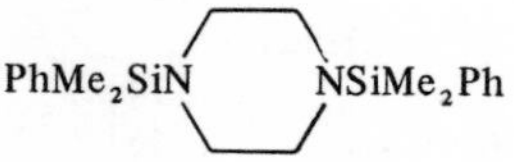 $PhMe_2SiN$(piperazine)$NSiMe_2Ph$	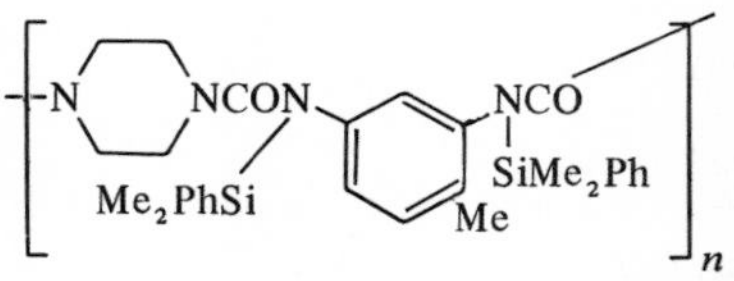 $[$–N(piperazine)NCON(Me_2PhSi)–C_6H_3(Me)–N($SiMe_2Ph$)CO–$]_n$	[376][4]

Insertion Reactions *(C'ntd.)*

Reactants		Product	References
5. Reactions with $CH_2{=}CHNCO$ *or* $o\text{-}C_6H_4O_2BNCO$			
$(Me_3Si)_2NH$	$CH_2{=}CHNCO$	$CH_2{=}CHN(SiMe_3)CONHCH{=}CH_2$	[566]
$o\text{-}C_6H_4O_2BNHBu^t$	$o\text{-}C_6H_4O_2BNCO$	$[o\text{-}C_6H_4O_2BN(CONHBu^t)BO_2C_6H_4\text{-}o]$	[180]
$Me_2Si(NMe_2)_2$		$Me_2NSi(Me_2)N$–[cyclic: N–CO–N, ring]–$NCONMe_2$ + $[Me_2NCON$–[cyclic: N–C(=O)–N, ring]–$N]_2\text{-}SiMe_2$	[661]
6. Reactions of $As(NMe_2)_3$ *with* XNCO			
$As(NMe_2)_3$	$C_4F_9SO_2NCO$	$Me_2NAs[N(SO_2C_4F_9)(CONMe_2)]_2$	[630a]
$2As(NMe_2)_3$	$3FSO_2NCO$	$As[N(SO_2NMe_2)(CONMe_2)]_3 + AsF_3$	
$2As(NMe_2)_3$	$2SO_2(NCO)_2$	eight-membered ring: Me_2NAs–N($CONMe_2$)–S(O_2)–$NCONMe_2$–$AsNMe_2$–N($CONMe_2$)–S(O_2)–N–C(=O)NMe_2	
$2As(NMe_2)_3$	$2Me_3SiNCO$	four-membered ring: Me_2NAs—$NCONMe_2$ / Me_2NCON—$AsNMe_2$	
7. Reactions with $1\text{-}C_{10}H_7NCO$			
$Bu_3SnN(Me)CON(Ph)SnBu_3$		$Bu_3SnN(1\text{-}C_{10}H_7)CON(Ph)SnBu_3$	[81]
8. Reactions with PhCONCO			
Me_3MNR_2 (M = Si, Ge, Sn)		$Me_3MN(PhCO)CONR_2$	[332, 338]
$Me_3SiN(PhCO)CONR_2$		six-membered ring: O–C(Ph)=N–C(=O)–N=C(NR_2)–O + $PhCOOSiMe_3$	[332]
$(Me_3Si)_2NMe$		$Me_3SiN(PhCO)CON(Me)SiMe_3$	[476]
$Me_3SiN(PhCO)CON(Me)(SiMe_3)$		$[Me_3Si\text{-}N(PhCO)CO]_2NMe$	[476]

Insertion Reactions *(C'ntd.)*

Reactants		Product	References
9. Reactions with $YSO_2—N=C=O$			
$PhBN(Me)(CH_2)_2O$ (cyclic)	$Y = p\text{-}MeC_6H_4$	$PhBN(Y)CON(Me)(CH_2)_2O$ (cyclic)	[181a]
$PhBN(CH_2)_4CH(CH_2)_2O$ (cyclic)	$Y = p\text{-}MeC_6H_4$	$PhBN(Y)CON(CH_2)_4CH(CH_2)_2O$ (cyclic)	[181a]
$PhB(NMe_2)_2$	$Y = NCO$	$PhB[N(CONMe_2)]_2SO_2$ (cyclic: PhB and SO_2 bridged by two N–$CONMe_2$)	[628a]
Me_3SiNRR'	$Y = p\text{-}XC_6H_4$	$p\text{-}X\text{-}C_6H_4\text{-}SO_2\text{-}N(SiMe_3)\text{-}CONRR'$	
($X = H$; $R = R' = Me$, Et; and $X = H$, Me, Cl; $R = Me$; $R' = Me_3Si$)			[335]
$As(NMe_2)_3$	$Y = F$	$As[N(CONMe_2)SO_2F]_3$	[631]
10. Reactions with MeNCS			
Et_2AlNMe_2		$Et_2AlN(Me)CSNMe_2$	[312]
Me_2TlNMe_2		$Me_2TlSC(NMe)NMe_2$	[729]
Me_3SiNEt_2		$Me_3SiN(Me)CSNEt_2$	[561]
$Me_2Si(NEt_2)_2$		$Me_2Si(NEt_2)N(Me)CSNEt_2$	[561]
$M(NMe_2)_4$ ($M = Ti$[5], Zr)		$M[SC(NMe_2)NMe]_4$	[146]
$As(NMe_2)_3$		$As[N(Me)CSNMe_2]_3$	[560]
$EtZnNPh_2$		$EtZnN(Me)CSNPh_2$	[526]
11. Reactions with EtNCS			
Me_3SnNMe_2		$Me_3SnSC(NMe_2)NEt$	[257]
Bu_3PbNEt_2		$Bu_3PbN(Et)CSNEt_2$	[514]
12. Reactions with PhNCS			
$PhB(NMe_2)_2$		$PhB[N(Ph)CSNMe_2]_2$, $Ph(NMe_2)BN(Ph)CSNMe_2$	[70]
$B(NMe_2)_3$		$B[N(Ph)CSNMe_2]_3$	[307]
$B_2(NMe_2)_4$		$B_2[N(Ph)CSNMe_2]_4$	[307]
$o\text{-}C_6H_4O_2BNHBu^t$		$o\text{-}C_6H_4O_2BN(Ph)CSNHBu^t$	[437]
Et_2AlNMe_2		$Et_2AlN(Ph)CSNMe_2$	[312]
Me_2TlNMe_2		$Me_2TlSC(NMe_2)NPh$	[729]
$Me_3SiNRMe$ ($R = H$, Me)		$Me_3SiN(Ph)CSNRMe$	[377]
Me_3SiNEt_2		$Me_3SiN(Ph)CSNEt_2$	[560]
$Me_2PhSiNHBu^t$		$Me_2PhSiN(Ph)CSNHBu^t$	[377]
$Me_2Si(NREt)_2$ ($R = H$, Et)		$Me_2Si[N(Ph)CSNREt]_2$	[177]
$(Me_3Si)_2NMe$		$Me_3SiN(Ph)CSN(Me)SiMe_3$	[377]
		$Me_3SiN(Ph)CSN(Me)SiMe_3$ + $Me_3SiN(Me)C(=NPh)SSiMe_3$	[377][6,7]

Insertion Reactions *(C'ntd.)*

Reactants		Product	References
$Me_3SiN{=}CPh_2$		$Me_3SiN(Ph)CSN{=}CPh_2$ and/or $Me_3SiSC({=}NPh)N{=}CPh_2$	[474]
Et_3GeNMe_2		$Et_3GeN(Ph)CSNMe_2$	[446]
Me_3SnNMe_2		$Me_3SnSC(NMe_2)NPh$	[257]
$(Bu^t_2SnNMe)_2$		$(Bu^t_2SnS)_2$ + PhNCNMe	[705a]
$Bi(NR_2)_3$ (R = Me, Et, Pr)		$Bi[SC(NR_2)NPh]_3$	[36]
$RZnNR'_2$ (R = R′ = Ph; R = Et, R′ = Et, Ph)		$RZnN(Ph)CSNR'_2$	[526]
$Ti(NMe_2)_3$		$Ti[SC(NMe_2)NPh]_3$	[432]
$Ti(NMe_2)_4$		$Ti[SC(NMe_2)NPh]_4$	[146, 148]
13. Reactions with PhCONCS			
Me_3SiNR_2		$Me_3SiN(PhCO)CSNR_2$ ⟶ Me_3SiNCS + $PhCONR_2$	[472a]
$(Me_3Si)_2NMe$		$Me_3SiN(PhCO)CSN(Me)SiMe_3$	[476]
Me_3SnNMe_2		Me_3SnNCS + $PhCONMe_2$	[338a]
$(Me_3Sn)_2NMe$		$(Me_3Sn)_2S$ + PhCON=C=NMe	[475]
14. Reactions with ArN=C=NAr			
$PhB(SBu)NEt_2$		$Ph(BuS)BN(Ar)C({=}NAr)NEt_2$	[344][8]
$ClB(NEt_2)_2$		$Cl(Et_2N)BN(Ar)C({=}NAr)NEt_2$	[344][8]
$(XNC)B(NMe_2)_2$ (X = O or S)		$(XNC)B(NMe_2)N(Ar)C({=}NAr)NMe_2$	[343b][8]
Ph_2BNEt_2		$Ph_2BN(Ar)C({=}NAr)NEt_2$	[344][8]
$(Pr^i_2N)_2BNHBu^t$		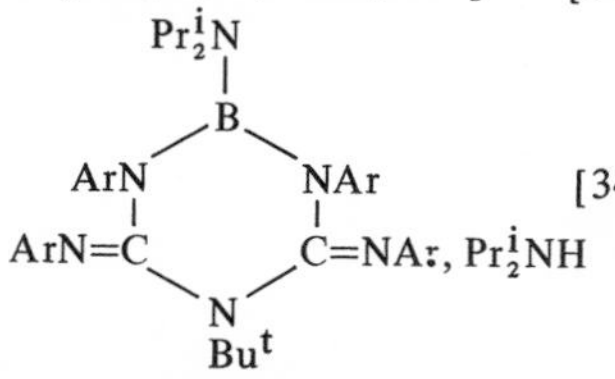	[344][8]
$o\text{-}C_6H_4O_2BNEt_2$		$o\text{-}C_6H_4O_2BN(Ar)C({=}NAr)NEt_2$ + $o\text{-}C_6H_4O_2BN(Ar)C({=}NAr)\text{-}N(Ar)C({=}NAr)NEt_2$	[344][8]
$Me_3MN(Me)R$ (M = Si, R = Me, Me_3Si; M = Ge, R = Me; M = Sn, R = Me, Me_3Si)	$PhCON{=}C{=}NBu^t$	$Me_3MN(COPh)C({=}NBu^t)N(Me)R$	[475]
Me_3MNMe_2 (M = Si, Ge, Sn)	$(PhCON)_2C$	$Me_3MN(COPh)C({=}NCOPh)NMe_2$	[473]
Me_3SnNMe_2		$Me_3SnN(Ar)C({=}NAr)NMe_2$	[255, 355][9]
Bu_3PbNEt_2		$Bu_3PbN(Ar)C({=}NAr)NMe_2$	[514][10]
$Ti(NMe_2)_4$		$Ti[N(Ar)C({=}NAr)NMe_2]_4$	[146][8]

Insertion Reactions *(C'ntd.)*

Reactants	Product	References
$M(NMe_2)_4$ (M = Ti, Zr, Hf)	$M[N(Ar)C(=NAr)NMe_2]_4$	[146][10]
15. Reactions with $H_2C=C=O$		
$o\text{-}C_6H_4O_2BNR_2$ ($R = Me, Et, Pr^i$)	$o\text{-}C_6H_4O_2BCH_2CONR_2$	[313, 320, 438]
Me_3SiNMe_2	$CH_2=C(OSiMe_3)CH_2CONMe_2$ + $Me_3SiCH_2CONMe_2$	[59]
$(MeO)_3SiN(Me)COMe$	$CH_2=C(OSi(OMe)_3)(N(Me)COMe)$	[396]
$Me_3SiNHMe$	$CH_2=C(OSiMe_3)(N(Me)COMe)$	[396]
$Me_3SiNHBu$	$Me_3SiCH_2CONHBu$	[451]
Et_3SiNMe_2	$CH_2=C(OSiEt_3)NMe_2$ + $CH_2=C(OSiEt_3)CH_2CONMe_2$	[395]
$R_3SiNR'_2$ (R = R' = Me, Et; R = Me, R' = Et; R = Et, R' = Me)	$CH_2=C(OSiMe_3)CH_2CONMe_2$	[395][10]
$R_3SiNMeX$ (R = Me, Et; X = CHO, COPr)	$R_3SiOC(NMeX)=CH_2$	[653]
$R_3MN(COMe)Me$ (R = Me, Et; M = Ge, Sn)	$R_3MCH_2CON(COMe)Me$	[395]
$(CH_2)_n$–CO–$NSiMe_3$ (ring) (n = 3 to 5); $CH_2CO + H_2O$	$(CH_2)_n$–CO–$NCOCH_2COMe$ (ring)	[763]
$Me_3SiN(Ac)Me$; $CH_2CO + H_2O$	$Me(Ac)NCOCH_2COMe$	[763]
$R_3MN(COMe)Me$	$R_3MCH_2CON(COMe)Me$	[395]
Me_3SnNMe_2	$Me_3SnCH_2CONMe_2$	[255]
16. Reactions with $CH_2=C$–O–$C(=O)$–CH_2 (diketene)		
$B(NRR')_3$ ($RR' = Me_2, HBu^t$)	$B(NRR')_3 . 3(CH_2CO)$	[320][11]
$o\text{-}C_6H_4O_2BNEt_2$	$o\text{-}C_6H_4O_2\bar{B}$[O–C(Me)=CH–C(NEt_2)=$\overset{+}{O}$] (chelate ring)	[320]

Insertion Reactions *(C'ntd.)*

Reactants		Product	References
Me_3SiNMe_2		$Me_3SiOCMe{=}CHCONMe_2$ + $CH_2{=}C(OSiMe_3)CH_2CONMe_2$	[577, 688]
Me_3SiNEt_2		$Me_3\bar{Si}$ ring: O—C(Me)=CH—C(NEt$_2$)=$\overset{+}{O}$	[396]
Me_3SnNEt_2		$Me_3\bar{Sn}$ ring: O—C(Me)=CH—C(NEt$_2$)=$\overset{+}{O}$	[320]

17. Reactions with CO_2

Reactants		Product	References
$(Bu^tHN)_3B$	H_2NBu^t	Bu^tHNB ring: O—CO—NBu^t—C(=O)—O	[176]
$PhB(NRR')_2$ ($R = R' = Et$; $R = H$, $R' = Et$, Bu^t)		$PhB(OCONRR')_2$	[179]
o-$C_6H_4O_2BNRR'$ ($RR' = Et_2$, HBu^t)	$HNEt_2$	o-$C_6H_4O_2BOCONRR'$	[179]
$XB(NMe_2)_2$ ($X = Me$, Cl, $PhC{\equiv}C$, Ph)		$XB(OCONMe_2)_2$	[179]
$B(NMe_2)_3$		$B(OCONMe_2)_3$	[179]
$B(NHNMe_2)_3$		$B(OCONHNMe_2)_3$	[476c]
$R_2B_2(NMe_2)_2$ ($R = Me$, Bu, Ph)		$R_2B_2(OCONMe_2)_2$	[18a]
$B_2(NMe_2)_4$		$B_2(NMe_2)_2(OCONMe_2)_2$	[18a]
Et_2AlNEt_2		$Et_2AlOCONEt_2$	[327]
$[(CO)_3Br_2Fe(AlNMe_2)]_2$		$(CO)_3Fe(Br_2)\bar{Al}$ ring: O—C(NMe$_2$)=O^+—$\bar{Al}Fe(Br_2)(CO)_3$—$\overset{+}{N}Me_2$	[582]
Me_2TlNMe_2		$Me_2TlOCONMe_2$	[729]
Me_3SiNEt_2	$HNEt_2$	$Me_3SiOCONEt_2$	[100, 560]

Insertion Reactions *(C'ntd.)*

Reactants		Product	References
Me_3SiNMe_2		$Me_3SiOCONMe_2$	[769]
H_3SiNMe_2		$H_3SiOCONMe_2$	[224][12]
$(Me_3Si)_2NH$	HNRR′	$Me_3SiOCONRR'$	[689]
(R = R′ = Et; R = H, R′ = Bu, Ph, C_3H_5)			
Me_3SiNHR		$Me_3SiOCONHR$	[485]
(R = Me, Bu, C_3H_5)			
H_3SiNR_2		$H_3SiOCONR_2$	[269]
(R = Me, Et; R_2 = C_4H_8, C_5H_{10})			
$Me_3SiNR'CH_2SiR_2OMe$		O, O, NR′, R_2Si, SiR_2, R′N, O, O	[690]
$Me_2Si(NHEt)_2$		$Me_2Si(OCONHEt)_2$	[177]
$Me_2Si(NEt_2)_2$	$HNEt_2$	$Me_2Si(OCONEt_2)_2$	[100,177, 559]
Me_3GeNEt_2		$Me_3GeOCONEt_2$	[494]
Et_3GeNEt_2		$Et_3GeOCONEt_2$	[446]
Me_3SnNMe_2		$Me_3SnOCONMe_2$	[255, 351]
Me_3SnNRR'		$Me_3SnOCONRR'$	[195]
(R = R′ = Me, Et; R = H, R′ = Ph; R = Me, R′ = Ph)			
(When R = R′ = Ph, and RR′ = C_4H_4, no reaction was observed)			
$Me_2Sn(NR_2)_2$		$Me_2Sn(OCONR_2)_2$	[195]
(R = Me, Et)			
$Ph_{4-n}Sn(NMe_2)_n$		$Ph_{4-n}Sn(OCONMe_2)_n$	[195]
(n = 1,2)			
$Sn(NMe_2)_4$		$(Me_2N)_2Sn(OCONMe_2)_2$	[195]
$(Me_3Sn)_2NMe$		$Me_3SnOCON(Me)SnMe_3$	[195]
$As(NMe_2)_3$		$As(OCONMe_2)_3$	[560, 562]
$Bi(NR_2)_3$		$Bi(OCONR_2)_3$	[36]
(R = Me, Et, Pr)			
$Ti(NMe_2)_4$		$Ti(OCONMe_2)_4$ or $Ti(NMe_2)_2(OCONMe_2)_2$	[146,148, 154b]
$Ti(OPr^i)_3NEt_2$		$Ti(OPr^i)_3(OCONEt_2)$ + $Ti(OPr^i)_4$	[157b]
$M(NMe_2)_n$		$M(OCONMe_2)_n$	[154a]
(M = Ti, Zr, V, and n = 4; M = Nb, Ta, and n = 5)			
$OV(NEt_2)(OPr^i)_2$		unclear	[157a]
$Nb(NMe_2)_5$		$Nb(OCONMe_2)_5$	[154,154a]
$Bu^tN{=}Ta(OCPh_2NMe_2)_2NMe_2$		$Bu^tN{=}Ta(OCONMe_2)_3$	[556a]
$Cr(NEt_2)_4$		$Cr_2(OCONEt_2)_4(\mu\text{-}NEt_2)_2$ + $Cr_2(OCONEt_2)_4(HNEt_2)_2$	[151d]
$Mo_2(NMe_2)_6$	excess CO_2	$Mo_2(NMe_2)_2(OCONMe_2)_4$	[151a, 155c]
$Mo_2Me_2(NMe_2)_4$		$Mo_2Me_2(OCONMe_2)_4$	[155e]
$Mo_2Et_2(NMe_2)_4$		$Mo(OCONMe_2)_4 + C_2H_4 + C_2H_6$	[155e]

Insertion Reactions *(C'ntd.)*

Reactants	Product	References
$W(NMe_2)_6$	$W(OCONMe_2)_3(NMe_2)_3$	[90a,152, 154a]
$W_2(NMe_2)_6$	$W_2(OCONMe_2)_6$	[151a,151b, 154b]
$W_2(NMeR)_6$ (R = Me, Et)	$W_2(OCONMeR)_6$	[153,155]
$W_2Me_2(NEt_2)_4$	$W_2(OCONEt_2)_4Me_2$	[151b,155]
$UCl_4 + HNR_2$ (R = Me, Et)	$U(OCONR_2)_4$ + $U_4O_2(OCONR_2)_{12}$	[127b,127c]
$+O_2$	$UO_2(OCONR_2)_4$	[127b]
$EtZnNR_2$ (R = Et, Ph)	$EtZnOCONR_2$	[526]
$M(NR_2)_4$ (M = Th and R = Me, Et, Bu^i; M = U and R = Me, Et)	$M(OCONR_2)_4$	[53b]

18. Reactions with COS

Reactants	Product	References
$B(NMe_2)_3$	$Me_2NB(OCSNMe_2)_2$	[476c]
$ClB(NMe_2)_2$	$ClB(NMe_2)OCSNMe_2$	[476c]
Me_3SiNR_2 (R = Me, Et)	$Me_3SiOCSNR_2$	[769]
H_3SiNMe_2	$H_3SiOCSNMe_2$	[224, 268][12]
H_3SiNR_2 (R = Et; $NR_2 = NC_4H_8$, NC_5H_{10})	$H_3SiOCSNR_2$	[269]
R_3GeNEt_2 (R = Me) (R = Cl, no reaction)	$R_3GeSCONEt_2$	[494]
Me_3SnNMe_2	$Me_3SnSCONMe_2$	[195]
$Me_3SnNHPh$	$(Me_3Sn)_2S$ + PhNHCONHPh	[195]
$(Me_3Sn)_2NMe$	$(Me_3Sn)_2S$ + MeNCO	[195, 334]
$Me_3SnOCONMe_2$	$Me_3SnSCONMe_2 + CO_2$	[195]
R_2MNMe_2 (R = Me, Ph, M = P; R = Me, M = As) [$Me_2P(O)NEt_2$, Cl_2PNMe_2, or Cl_2AsNMe_2 – no reaction]	$R_2MSCONMe_2$	[494]
$W_2(NMeR)_6$ (R = Me, Et)	$W_2(SCONMeR)_6$	[153]
$M(NR_2)_4$ (M = Th and R = Me, Et; M = U and R = Et)	$M(OCSNR_2)_4$	[53b]

19. Reactions with CS_2 *or* CSe_2

Reactants	Product	References
$RB(NR'_2)_2$ (R = Me, Ph, R' = Me; R = Ph, R' = Et; R = Ph, NR'_2 = NHEt)	$RB(SCSNR'_2)_2$	[179]
$B_2(NMe_2)_4$	$B_2(SCSNMe_2)_4$	[18a]
$ClB_2(NMe_2)_3$	$ClB_2(SCSNMe_2)_3$	[18a]

Insertion Reactions *(C'ntd.)*

Reactants		Product	References
$R_2B_2(NMe_2)_2$ (R = Me, Bu)		$R_2B_2(SCSNMe_2)_2$	[18a]
$Ga(NMe_2)_3$		$Ga(SCSNMe_2)_3$	[538]
$InCl_3 + Na\overline{N(CH_2)_5} 2H_2O$		$In[SCS\overline{N(CH_2)_5}]_3$	[299a]
Me_2TlNMe_2		$Me_2TlSCSNMe_2$	[729]
Me_3SiNR_2 (R = Me, Et)	$HNEt_2$	$Me_3SiSCSNR_2$	[100,101]
H_3SiNMe_2		$H_3SiSCSNMe_2$	[224][12]
H_3SiNR_2 (R = Me, Et; $R_2 = C_4H_8, C_5H_{10}$)		$H_3SiSCSNR_2$	[224,269]
$Me_2Si(NREt)_2$ (R = H or Et)	$HNEt_2$	$Me_2(REtN)SiSCSNEtR$	[177]
$Ph_2Si(NMe_2)_2$		$Ph_2Si(SCSNMe_2)_2$	[740]
$CH_2{=}CHSi(NMe_2)_2Me$		$Me_2N(CH_2{=}CH)Si(Me)SCSNMe_2$ $Me(CH_2{=}CH)Si(SCSNMe_2)_2$	[740]
$(Me_3Si)_2NMe$	168 h at 150°	$Me_3SiN(Me)C(S)SSiMe_3$	[337][13]
Me_3GeNRR' ($R = H, Me, GeMe_3$; $R' = NMe_2$) (When $R = Me_3Si$, no reaction, R = R' = Et)		$Me_3GeSCSNRR'$	[573] [494]
Et_3GeNR_2 (R = Me, Et)		$Et_3GeSCSNR_2$	[446]
Me_3SnNMe_2		$Me_3SnSCSNMe_2$	[255, 351]
Me_3SnNRR' (R = R' = Me, Et; R = Me, R' = Ph; $RR' = C_4H_8$) (No reaction was observed when R = R' = Ph; and $RR' = C_4H_8$)		$Me_3SnSCSNRR'$	[195]
$Me_3SnNHPh$		$(Me_3Sn)_2S + PhNHCSNHPh$	[195]
$Me_2Sn(NR_2)_2$ (R = Me, Et)		$Me_2Sn(SCSNR_2)_2$	[195]
$Ph_{4-n}Sn(NMe_2)_n$		$Ph_{4-n}Sn(SCSNMe_2)_n$	[195]
$Sn(NMe_2)_4$		$Sn(SCSNMe_2)_4$	[195]
$(Me_3Sn)_2NR$ ($R = Me, Me_3Sn$)		$(Me_3Sn)_2S + RNCS$	[195]
$Sn(NEt_2)_4$		$Sn(SCSNEt_2)_4$	[93]
$(Bu^t_2SnNMe)_2$		$(Bu^t_2SnS)_2 + SCNMe$ (via insertion)	[705a]
$MeAs(NMe_2)_2$		$MeAs(SCSNMe_2)_2$	[724]
$PhAs(NRR')_2$ (R = R' = Me, Et, Pr, Bu; $RR' = C_5H_{10}$)		$PhAs(SCSNRR')_2$	[245]
$As(NR_2)_3$ (R = Me, Bu; $R_2 = C_5H_{10}$)		$As(SCSNR_2)_3$	[558,723, 724]

Insertion Reactions *(C'ntd.)*

Reactants		Product	References
$(Me_2NAsNCOPh)_n$		$SCSNMe_2$ / As / PhCON COPh NCOPh / N / As As / N / COPh	[722]
$Sb(NMe_2)_3$		$Sb(SCSNMe_2)_3$	[558]
$Bi(NR_2)_3$ (R = Me, Et, Pr)		$Bi(SCSNR_2)_3$	[36]
$Ti(NMe_2)_3$		$Ti(SCSNMe_2)_4$	[25, 432]
Cp_2TiNMe_2		$Cp_2TiSCSNMe_2.CS_2$	[432]
$M(NR_2)_4$ (M = Ti, Zr; R = Me, Et, Pr; and M = V, R = Me, Et)		$M(SCSNR_2)_4$	[93]
$OV(NEt_2)_3$		$OV(SCSNEt_2)_3$	[157a]
$OV(OPr^i)_2NEt_2$		1/3 $OV(SCSNEt_2)_3$ + 2/3 $OV(OPr^i)_3$	[157a]
$Nb(NR_2)_n$ (R = Me, $n = 5$; R = Et, $n = 4$)		$Nb(SCSNR_2)_n$	[93]
$Nb(NMe_2)_5$		$Nb(SCSNMe_2)_4$	[93]
$Me_3Ta(NMe_2)_2$		$Me_3Ta(SCSNMe_2)_2$	[648]
$Ta(NMe_2)_5$		$Ta(SCSNMe_2)_5$	[93]
$Cr(NEt_2)_4$		$Cr(SCSNEt_2)_3$	[56a]
$W_2(NMeR)_6$ (R = Me, Et)		$W_2(SCSNMeR)_6$	[153]
$W(NMe_2)_6$		$W(SCSNMe_2)_4$	[90a]
$UCl_4 + HNR_2$ (R = Me, Et)		$U(SCSNR_2)_4$	[127a]
$M(NR_2)_4$ (M = Th, R = Me, Et; M = U, R = Et)		$M(SCSNR_2)_4$	[53b]
	$+O_2$	$UO_2(SCSNR_2)_4$	[127a]
$M(NEt_2)_4$ (M = Th, U)	CSe_2	$M(SeCSeNEt_2)_4$	[53b]
$EtZnNR_2$ (R = Et, Ph)		$EtZnSCSNR_2$	[526]

20. Reactions with $Ph(X)C{=}S$ (X = Ph, NMe_2) *or a* Ge=N *Compound.*

Reactants		Product	References
$(Me_3Sn)_2NMe$	PhC(:S)X	$(Me_3Sn)_2S$ + $Ph(X)C{=}NMe$	[334]
Et_3GeNMe_2	$Ph_2Ge{=}NMe$	$Et_3GeN(Me)GePh_2NMe_2$	[621a]

Insertion Reactions *(C'ntd.)*

Reactants		Product	References
$PhGe\overline{N(Me)(CH_2)_2}$	$Ph_2Ge{=}NMe$	$Ph_2Ge(CH_2CH_2)NMe\text{–}GePh_2\text{–}NMe$ (ring: Ph_2Ge–CH_2–CH_2–NMe–$GePh_2$–N(Me)–)	[621a]

21. Reactions with a Ketenimine.

Reactants		Product	References
Et_2AlNMe_2 (Ar = Ph, *p*-tolyl)	$Me_2C{=}C{=}NAr$	ring: $Me_2C{=}C$–N(Ar)–$AlEt_2$–NMe_2–$C({=}CMe_2)$–NAr–	[721]

22. Reactions with Aldehydes, Ketones, Esters, Amides, or Carbonyls

Reactants		Product	References
$R_2BHNC_5H_4N$ (2-pyridyl)	R'R"CO	$R_2BOC(R'R'')HNC_5H_4N$	[276]
(R = Ph; R' = H, R" = Me or Ph; R' = R" = Me; R'R" = C_5H_{10})			[276]
(R' = H, R" = CCl_3)			[167]
(When R=R'=Ph and R" = Me, the product was $Ph_2BOB(Ph)HNC_5H_4N$)			
$o\text{-}C_6H_4O_2BNEt_2$	MeCH=CHCHO	$o\text{-}C_6H_4O_2BOCH(NEt_2)CH{=}CHMe$	[146]
$3Et_2B\overline{N(CH)_4}$ (R = Me, R' = Et; R = R' = Et, Pr)	RR'CO	$Et_2\overset{-}{B}$(pyrrolyl)$_2$CRR' (cyclic, $\overset{+}{N}$) + $(Et_2B)_2O$ + $H\overline{NC_4H_4}$	[66]
$Me_2AlNMeSiMe_3$	RR'CO	$[Me_2AlOCRR'NMeSiMe_3]$	
		$(Me_2AlOMe)_n + (Me_3Si)_2NMe + RR'CNMe$	[646]
$Me_2AlNMeSiMe_3$	RCOOMe	$Me_2AlOSiMe_3 + RR'CNMe$	[646]
$Me_2AlNMeSiMe_3$	$MeCONMe_2$	$Me_2AlOCMe(NMe_2)NMeSiMe_3$	[646]
$Me_3SiNMeCOR$ (R = H, R' = Pr, Ph; R = Me, R' = Et, Ph, *p*-MeC_6H_4)	R'CHO	$Me_3SiOCH(NMeCOR)R'$	[77]
$Me_3SiNMeP(OEt)_2$ ($n = 2, m = 0$, X = O; $n = 1, m = 1$, X = S)	nPhCHO + $m$$CS_2$	$Me_3SiOCHPhP(X)(OEt)_2$	[603]
$Me_3Si\overline{NCO(CH_2)_3}$	PhCHO	$Me_3SiOCH[\overline{NCO(CH_2)_3}]Ph$	[77]

Insertion Reactions *(C'ntd.)*

Reactants		Product	References
$Me_2Si(N\langle CH_2, CH_2 \rangle)_2$ (aziridinyl)	MeCHO	$Me_2Si[OCH(Me)N\langle CH_2, CH_2 \rangle]_2$ (aziridinyl)	[503]
Me_3SiNRR' (R = R' = Me, Et, Pr) (R = Ph, R' = P(O)XY (X = Y = OMe, Me_3SiO; X = Me, Y = EtO, Me_3SiO)	Cl_3CCHO	$Me_3SiOCH(NRR')CCl_3$	[253] [509]
Me_3SiNRR' [R = Me, Et, R' = C_6H_{11}; RR' = $(CH_2)_4$, $(CH_2)_5$]	Cl_3CCHO	$Me_3SiOCH(NRR')CCl_3$	[253]
Me_3SiNRR' (R = R' = Me; R = H, R' = Ph)	$(CF_3)_2CO$	$Me_3SiOC(CF_3)_2NRR'$	[14]
Me_3SiNMe_2	$CF_3(ClCF_2)CO$	$Me_3SiOC(CF_3)(NMe_2)CF_2Cl$	[14]
Me_3SiNMe_2	$(CF_2Cl)_2CO$	$Me_3SiOC(CF_2Cl)_2NMe_2$	[14]
$Me_2\overline{SiNEtCH_2CH_2N}Et$ (cyclic)	$(CF_3)_2CO$	$Me_2SiOC(CF_3)_2NEtCH_2CH_2NEt$–$OC(CF_3)_2$ (ring, Si–OC(CF₃)₂–N)	[14]
$(Me_3Si)_2NH$ (X = F, NO_2)	CF_3COCF_2X	$Me_3SiOC(CF_3)(CF_2X)NHSiMe_3$	[693]
Et_3GeNMe_2 (X = CO_2Et, CN)	$CH_2(CO_2Et)X$	$Et_3GeOC(NMe_2)(OEt)CH_2X$ + $Et_3GeCH(CO_2Et)X$	[651a]
$R_3SnNR'_2$	Me_2CO	$R_3SnOCMe_2NR'_2$ ↓ $R_3SnOH + CH_2 = CMeNR'_2$	[593]
$Bu_3SnN{=}CPh_2$ (R' = R'' = CCl_3; R' = CCl_2F, R'' = $CClF_2$)	R'R''CO	$Bu_3SnOC(R'R'')N{=}CPh_2$	[296]
Bu_3PbNEt_2 (R = Pr, p-MeC_6H_4)	RCHO	$Bu_3PbOCH(NEt_2)R$	[514]
$Bu^tN{=}Ta(NMe_2)_3$	Ph_2CO	$Bu^tN{=}Ta(OCPh_2NMe_2)_2NMe_2$	[556a]
$EtZnNPh_2$	RCHO	$EtZnOC(R)HNPh_2$	[86]
$[(OC)_3(Me_2NAl)FeBr_2]_2$	Bu_3P	$[(OC)_2(Bu_3P)(Me_2NAl)FeBr_2]_2$ + 2CO → $[(OC)_2(Bu_3P)(Me_2NCOAl)FeBr_2]_2$	[582]
$LiNMe_2$	$Cr(CO)_6$	$(CO)_5Cr[C(NMe_2)OLi]$	[242]
$LiNEt_2$	$Cr(CO)_6$	$(OC)_5CrC(OLi.OEt_2)(NEt_2)$	[241][16]
$[Al(NMe_2)_3]_2$ (M = Fe, n = 5) (M = Ni, n = 4)	$2M(CO)_n$	$[(OC)_{n-1}M[C(NMe_2)OAl(NMe_2)_2]_2$	[578, 581] [578]
Me_3SnNMe_2	$Fe(CO)_5$	$(OC)_4Fe{-}C(NMe_2)(OSnMe_3)$	[580]

Insertion Reactions *(C'ntd.)*

Reactants		Product	References
$Me_nSn(NMe_2)_{4-n}$ ($n = 0, m = 2; n = 2, m = 1$)	$m Fe(CO)_5$	$[(OC)_3Fe[C(NMe_2)-O]_2]_m SnMe_n$ (ring: $(OC)_3Fe$ bridged by two $C(NMe_2)$–O units to $SnMe_n$)	[580]
$Sn(NMe_2)_2$	$Fe(CO)_5$	$(OC)_3Fe[C(NMe_2)-O]_2Sn$ (ring: $(OC)_3Fe$ bridged by two $C(NMe_2)$–O units to Sn)	[580]
$Ti(NMe_2)_4$ ($M = Mo, W, n = 6; M = Fe, n = 5; M = Ni, n = 4$)	$M(CO)_n$	$(OC)_{n-1}M[C(NMe_2)OTi(NMe_2)_3]$	[579]
23. Reaction with Nitrosobenzene			
Me_3SnNMe_2		$Me_3SnON(Ph)NMe_2$	[257]
24. Reactions with Nitromethane			
$B(NHBu^t)_3$		1:3 adduct	[179] [14]
$o\text{-}C_6H_4O_2BNEt_2$		1:1 adduct	[179] [14]
$o\text{-}C_6H_4O_2BNHBu^t$		1:1 adduct	[178] [14]
25. Reactions with Nitriles			
$LiN(SiMe_3)_2$	PhCN	$LiN{=}C(Ph)N(SiMe_3)_2$	[647]
$LiNR_2$ ($R = Me, Me_3Si$)	ArCN	$LiN{=}C(Ar)NR_2 + (ArCN)_3$	[433, 647]
LiNHMe	PhCN	$LiN{=}C(Ph)NHMe$	[659]
Et_2AlNMe_2 [$R = p\text{-}XC_6H_4$ ($X = NO_2$, Me, OMe, NMe_2), Ph, Me, $PhCH_2$]	RCN	$Et_2AlN{=}C(R)NMe_2$	[313]
$Cl_2AlN(Me)Ph$	PhCN	$Cl_2AlN{=}C(Ph)N(Me)Ph$	[316]
$(Et)_{2-n}Cl_nAlNEt_2$	PhCN	$(Et)_{2-n}Cl_nAlN{=}C(Ph)NEt_2$	[316]
Me_3SnNMe_2	PhCN	$Me_3SnN{=}C(Ph)NMe_2$	[255, 355]
Me_3SnNMe_2 ($R = CCl_3, p\text{-}Me_2NC_6F_4$)	RCN	$Me_3SnN{=}C(R)NMe_2$	[146]
$[Bu^t_2Sn{-}NMe]_2$ (four-membered ring: Bu^t_2Sn–NMe–$SnBu^t_2$–MeN)	PhCN	eight-membered ring: Bu^t_2Sn–N=C(Ph)–NMe–$SnBu^t_2$–N=C(Ph)–MeN–	[288, 705a]
Me_3SnNMe_2	$(CN)_2C{=}C(CN)_2$	$Me_3SnN{=}C(NMe_2)C(CN){=}C(CN)_2$	[257]
$Ti(NMe_2)_3$	3PhCN	$Ti[N{=}C(NMe_2)Ph]_3$	[432] [15]

Insertion Reactions *(C'ntd.)*

Reactants		Product	References
$Ti(NMe_2)_4$ (R = Ph, p-MeC_6H_4)	4RCN	$Ti[N{=}C(NMe_2)R]_4$	[146][5]
$M(NMe_2)_4$ (M = Zr, Hf; R = Ph, p-MeC_6H_4)	4RCN	$M[N{=}C(NMe_2)R]_4$	[146]
$M(NMe_2)_4$ (M = Ti, n = 1 or 2; M = Zr, n = 4)	nMeCN	$(Me_2N)_{4-n}M[N{=}C(NMe_2)Me]_n$	[91]
$OV(OPr^i)_2NEt_2$	MeCN	unclear	[157a]
$Ta(NMe_2)_5$	MeCN	$(Me_2N)_4Ta[N{=}C(NMe_2)Me]$	[91]

26. Reactions with Isonitriles

Reactants		Product	References
Me_3SnNMe_2	p-MeC_6H_4NC	p-$MeC_6H_4N{=}C(NMe_2)(SnMe_3)$	[257]
Bu_3PbNEt_2	PhNC	$PhN{=}C(NEt_2)(PbBu_3)$	[514]

27. Reactions with $RCH{-}CH_2$ (three-membered ring with X)

Reactants		Product	References
$Me_3SiNR'_2$ (R' = Me, Et; R = CCl_3; $R'_2 = C_5H_{10}$; R = CCl_3, $ClCH_2$)	$RCH{-}CH_2$ (epoxide, O)	$Me_3SiOC(R)HCH_2NR'_2$	[341]
Et_3GeNR_2 (R = Me, n = 2,3,4; R = C_6H_{11}, n = 2)	$(CH_2)_n$–O ring + $ZnCl_2$	$Et_3GeO(CH_2)_nNR_2$	[623]
Et_3GeNMe_2	CH_3CHCH_2 (epoxide, O)	$Et_3GeOC(Me)HCH_2NMe_2$	[623]
R_3SnNMe_2 (R' = H or Me)	$R'CH{-}CH_2$ (episulfide, S)	$R_3SnSC(R')HCH_2NMe_2$	[146]
Bu_3SnNEt_2 (R = Et, Ph)	$RCH{-}CH_2$[15] (epoxide, O)	$Bu_3SnOCH(R)CH_2NEt_2$	[718]
Bu_3SnNEt_2	cyclohexene oxide	2-(NEt_2)cyclohexyl-$OSnBu_3$	[718]

28. 1,4 additions

Reactants		Product	References
Me_3SiNR_2	$CH_2{-}C{=}O$ / $CH_2{-}O$ (β-propiolactone)	$Me_3SiOCOCH_2CH_2NEt_2$	[339, 340]

(R = Me, Et, Pr; R_2 = N(pyrrolidine), N(piperidine), N(morpholine)O)

Insertion Reactions *(C'ntd.)*

Reactants		Product	References
Me_3GeNEt_2	CH_2—C=O / CH_2—O (β-propiolactone)	$Me_3GeOCOCH_2CH_2NEt_2$	[339]
Me_3SnNR_2 (R = Me, Et)	CH_2—C=O / CH_2—O	$Me_3SnOCH_2CH_2CONR_2$	[146, 339]
Bu_3SnNEt_2	PhCH=CHCOPh	$Bu_3SnOC(Ph)$=CHCH(NEt_2)Ph	[104]
$M(NR_2)_4$ (M = Ti, R = Me; M = Hf, R = Et)	4CH_2—C=O / CH_2—O	$M(OCH_2CH_2CONR_2)_4$	[146]
29. 1,5 additions			
Et_3GeNMe_2	CH_2—C=O / CH_2 / CH_2—O	$Et_3GeO(CH_2)_3CONMe_2$	[624]
Et_3GeNMe_2	CH_2CO / O / CH_2CO (succinic anhydride)	$Et_3GeOOCCH_2CH_2CONMe_2$	[623, 624]
Et_3GeNMe_2	C_6H_4(CO)$_2$O (phthalic anhydride)	Et_3GeOOC–C_6H_4–$CONMe_2$ (ortho)	[624]
Et_3GeNMe_2	CH–CO / O / CH–CO (maleic anhydride)	$Et_3GeOOCCH$=CHCONMe$_2$	[624]
30. Oxides and Nitrides of Sulphur and SeO_2*			
Me_3SnNMe_2	PhNSO	$Me_3SnN(Ph)SONMe_2$	[255]
$2Me_3SiNR_2$ (R = Me, Et)	S_4N_4	Me_3SiN=S=NSNR$_2$	[626]
$2Me_3SiNMe_2$	S_4N_4	Me_3SiN=S=N–S–N=S=NSiMe$_3$ + $(Me_2N)_2S$	[642]
Me_3SnNMe_2	S_4N_4	Me_3SnN=S=NSnMe$_3$ + $(Me_2N)_2S$	[289]
$2Me_2Si(NMe_2)_2$	S_4N_4	$2Me_2Si(NMe_2)N$=S=NSNMe$_2$	[630]
$(Me_3Sn)_3N$	S_4N_4	Me_2Sn ring (Sn–S–N=S–N=, five-membered: Me_2Sn, S, N, S, N) + Me_4Sn + $(Me_3SnN)_2S$	[631a]
Me_2TlNMe_2	SO_2	$Me_2TlOSONMe_2$	[729]
Me_2TlNMe_2	SO_3	$Me_2TlOSO_2NMe_2$	[729]
Me_3SiNEt_2	SO_3	$Me_3SiOSO_2NEt_2$	[671]

Insertion Reactions *(C'ntd.)*

Reactants		Product	References
Me_3SiNRR' (R = H, R' = Me_3Si; R = Me, R' = Ph)	SO_3	Me_3SiOSO_2NRR'	[401]
$(Me_3Si)_3N$	SO_3	$Me_3SiOSO_2N(SiMe_3)_2$	[45]
$(Me_3SnN{=})_2SMe_2$	SO_2	$Me_3SnOSON{=}S({=}NSnMe_3)Me_2$	[287]
Me_3SnNMe_2	SO_2	$Me_3SnOSONMe_2$	[255]
$Ti(NMe_2)_4$	SO_2	$OTi(OSONMe_2)_2 + SO(NMe_2)_2$	[144]

*For insertion of SeO_2 into Si–N bonds, see ref. [55].

†*Footnotes to Table:*

1. $Et_3Sn[CH(CHO){-}CH_2{-}]_nNMe_2$ (n = 6–12).
2. When $n = 4$.
3. Also forms 1:2 adduct.
4. Polymers of this nature were obtained by using other similar reagents and substrates.
5. Products were identified by characterising the hydrolysis products.
6. Product decomposed to give MeNCS (46%) and $(Me_3Si)_2NPh$ (30%).
7. Product decomposed to give $(Me_3Si)_2S$ (47%) + MeN=C=NPh (→ Polymer).
8. Ar = p-MeC_6H_4.
9. Ar = $C_{10}H_7$.
10. Ar = C_6H_{11}.
11. Structure not yet elucidated.
12. Unstable at room temperature.
13. Product decomposed to give MeNCS (80%) and $(Me_3Si)_2S$ (80%).
14. These probably have structure >B–O–N(Me)(O)–N< [178, 179].
15. $LiNEt_2$ was used as catalyst.
16. Product of reaction + $Et_3O^+BF_4^-$ ⟶ $(OC)_5CrC(OEt)(NEt_2)$ [241].

11

Reactions of Metal or Metalloid Amides with Protic Compounds

A INTRODUCTION

The reactions discussed in this chapter usually take the form shown in Eq. (1) and data are summarised in the Table (Section G). The generality of the process was first explicitly recognised for amides of tin [352], reviewed in refs. [205a, 355, 356], such as Me_3SnNMe_2, the acid HA being oxygen-, nitrogen-, phosphorus-, arsenic-, or carbon-centred, or a hydrogen halide. From the results shown in the Table, it is evident that acidic hydrides of sulphur and selenium are also reactive. There may be synthetic advantages in reactions of this type for preparation of compounds LM–A of Eq. (1), not least because if R=Me=R′ the by-product, dimethylamine, is conveniently volatile.

$$LM{-}NRR' + HA \longrightarrow LM{-}A + HNRR' \qquad (1)$$

Further applications in synthesis depend on the use of the amino function either as a protecting group or as a disguised functional group [see Eq. (1)]. The former situation is exemplified by consideration of routes to perfluorophenylsilanes such as $(C_6F_5)_3SiA$ [423]: $SiCl_4$ with LiC_6F_5 leads exclusively to $(C_6F_5)_4Si$

irrespective of stoicheiometry, whereas Eq. (2) shows a high yield preparation of $(C_6F_5)_3SiNMe_2$, whence use of Eq. (1) leads to a wide range of silanes $(C_6F_5)_3SiA$. Similar examples of the use of NMe_2 as a protecting group are available for heteroleptic phosphorus(III) compounds [231a], and an interesting case is the preparation of F(Cl)PBr from $F(Cl)PNMe_2$ and HBr [492].

$$SiCl_4 \xrightarrow{LiNMe_2} Cl_3SiNMe_2 \xrightarrow{C_6F_5Li} (C_6F_5)_3SiNMe_2 \qquad (2)$$

In order to avoid duplication, we exclude from this chapter consideration of (i) transaminations (i.e., HA is ammonia or a primary or secondary amine), but see Chapters 1-9, (ii) reactions with metal hydrides (see Chapter 12), and (iii) cases where the amide simply acts as a Brønsted base (see Chapters 14 and 15). Hydrazine derivatives are excluded, but see e.g., ref. [574].

The relative reactivities of metal amides are controlled both by electronic and steric effects. Thus, for an isoleptic series of amides, the most reactive are those of the least electronegative elements, e.g., reactivities increase in the sequences Si$<$Ge$<$Sn or P$\ll$As$<$Sb$<$Bi for Me_3M-NMe_2 or Me_2M-NMe_2, respectively. For the metalloid elements, the order is P$\approx$Si$\ll$B, and Si-N cleavage is preferred over SN in $O_2S[N(SiMe_3)_2]_2$ [44]. It appears that the acid HA behaves as a nucleophile, and relevant electronic factors include the energetic availability of an unfilled orbital on the metal or metalloid and the polarity of the $\overset{\delta+}{M}-\overset{\delta-}{N}<$ bond. Further support for this analysis is (i) the high reactivity of transition metal amides towards protic compounds, (ii) a rough correlation of reactivity with the pK_A of the acid HA, and (iii) the observation in some cases of acid catalysis: e.g., aminosilanes, although slow to hydrolyse in water, react readily with aqueous mineral acid especially in a homogeneous medium such as acetone. Steric effects are particularly noted if the central metal or metalloid is small, or the metal is in a high co-ordination state. For instance, there are striking examples in the chemistry of aminoboranes to illustrate the former point, while $[W(NMe_2)_6]$ undergoes alcoholysis with decreasing facility in the sequence MeOH$>$EtOH$>$PrOH$\gg$Bu^{t}OH, Me_3CCH_2OH, or Et_3SiOH (which do not react at ambient temperature) [90a].

B REACTIONS OF AMIDES WITH HYDROXY COMPOUNDS

1. Hydrolysis

Metal or metalloid amides are usually very susceptible to hydrolysis; corresponding hydroxides or oxides are formed. Reactions of this type are so common that they are omitted from the Table. Compounds containing larger substituents on metal or nitrogen are somewhat more resistant to attack by moisture; $B(NMePh)_3$ is not attacked by water or alkali [50] and in the series

$[Me_3SiN(Me)]_nSnMe_{4-n}$, the water-sensitivity and degree of steric hindrance are related [659a]. Amides of the more electronegative elements (Si, P, and S) are somewhat less readily hydrolysed than others. P^V amides are less reactive than P^{III} analogues [231a]; for example, $Cl_2(S)PNMe_2$ may be steam-distilled without decomposition. A unique nitrogen compound of a rare gas $FXeN(SO_2F)_2$ hydrolyses to yield Xe, XeF_2, and $[N(SO_2F)]_2$ [444]. Some authors have interpreted resistance to hydrolysis as implying some degree of π-delocalisation involving the $\ddot{N}\!-\!M$ system, but such arguments are unconvincing. A case in point is that of compound (I), which may be precipitated from solution in acetone by

(I) [pyrrole ring, N–$GePh_3$]

addition of water [615]; however, compound (I) is more reactive than the $Bu_3GeNC_4H_4$ analogue towards various nucleophiles [608].

Aminostannanes, when exposed to the atmosphere, afford the carbonate [Eq. (3)] [354, 513, 697, 698] probably in a step-wise manner [698] [Eq. (4)].

$$2R_3SnNR'_2 + CO_2 + H_2O \longrightarrow (R_3Sn)_2CO_3 + 2HNR'_2 \qquad (3)$$

$$2R_3SnNR'_2 + H_2O \longrightarrow 2HNR'_2 + (R_3Sn)_2O \xrightarrow{CO_2} (R_3Sn)_2CO_3 \qquad (4)$$

2. Reactions with alcohols, silanols, phenols, carboxylic acids, diols, or miscellaneous hydroxy compounds

With an alcohol, an aminometallane readily gives the corresponding alkoxide [Eq. (5)]. The relative rates of condensation of various straight-chain 1-hydroxy-alkanes with diethylaminotrimethylsilane vary inversely with the chain length; branched alcohols react at a slower rate [587]. Trimethylstannyl methoxide is easily prepared by this method [27]. Similarly, reactions with silanols [Eq. (6)] provide routes to metallosiloxanes [583, 715].

For aminosilanes, Si-N cleavage by reaction with water or an alcohol may be catalysed by acid or base, but the former is more effective. Mechanisms of Si-N solvolysis have been critically assessed [84] and two-step pathways were ruled out, but preference is given to S_N2-Si and S_Ni-Si schemes.

Co-condensation of a bifunctional aminosilane and a silanediol offers an approach to the synthesis of silicon polymers or co-polymers, often with controlled structural features [40, 302, 567, 583, 589, 611, 686].

Formation of pentavalent derivatives from the reaction [Eq. (7)] of a niobium(IV) dialkylamide with triethylsilanol or an alcohol is surprising.

Similarly, treatment of a Ti^{III} amide, $Ti(NMe_2)_3$, with phenol gives [Eq. (8)] a Ti^{IV} product, although various alcohols afford Ti^{III} alkoxides [432]; and several V^{IV} amides [715], or $[Mo(NMe_2)_4]$ [155d], with an alcohol or a trialkylsilanol give tetra- derivatives, interestingly as the *trans*-octahedral bis-amine adduct for $[Mo(OSiR_3)_4(HNMe_2)_2]$ (R = Me or Et) [155d]. Reactions leading to products having the metal in a changed oxidation state may involve disproportionation (see Chapter 8), e.g., the synthesis of a Nb^{IV} amide from $NbCl_5$ or disproportionation of $Ti(NMe_2)_3$ into Ti^{IV} and Ti^{II} complexes.

$$\text{LM—NRR}' + \text{R}''\text{OH} \longrightarrow \text{LM—OR}'' + \text{HNRR}' \quad (5)$$

$$\text{LM—NRR}' + \text{R}''_3\text{SiOH} \longrightarrow \text{LM—O—SiR}''_3 + \text{HNRR}' \quad (6)$$

$$[\text{Nb(NR}_2)_4] + 5\text{R}'\text{OH} \longrightarrow [\text{Nb(OR}')_5] + 4\text{HNR}_2 \quad (7)$$

$$\tfrac{1}{2}[\{\text{Ti(NMe}_2)_3\}_2] + 4\text{PhOH} \longrightarrow [\text{Ti(OPh)}_4(\text{HNMe}_2)] + 2\text{HNMe}_2 \quad (8)$$

Metal bis(trimethylsilyl)amides are useful precursors for some unusual aryloxymetallanes $Sn(OAr)_2$ [141a], $Pb(OAr)_2$, $[Ni(OAr)(PPh_3)_2]$ [141c], and $[Rh(\eta^5\text{-}OAr)(PPh_3)_2]$ [141d] ($Ar = 2,6\text{-}Bu^t_2\text{-}4\text{-}MeC_6H_2$; η^5-OAr = 2,6-di-t-butyl-4-methyl-1-ketocyclohexadienyl).

Organo-germanium or -arsenic heterocycles are prepared from the reaction of the appropriate aminometallane, e.g., Eqs. (9) and (10), [34, 217]. Similar reactions, or minor variants, are used for synthesis of related ring compounds of phosphorus [68, 166, 221, 465, 466, 606, 673] and other elements: e.g., compounds (II), from $PhB(NEt_2)_2$ using $HOCH_2CH_2NH_2$ [178, 182], or (III), using $Sn(NMe_2)_2$ and $MeN(CH_2OH)_2$ [246a].

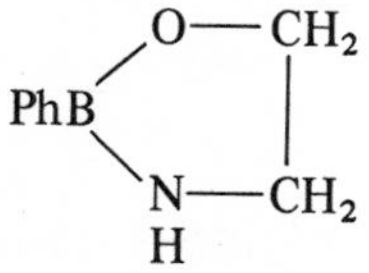

(II): PhB ring with O—CH₂ and N(H)—CH₂ (five-membered ring: PhB–O–CH₂–CH₂–NH–)

$$\text{PhB}\begin{smallmatrix}\text{O—CH}_2\\ \text{N—CH}_2\\ \text{H}\end{smallmatrix}$$

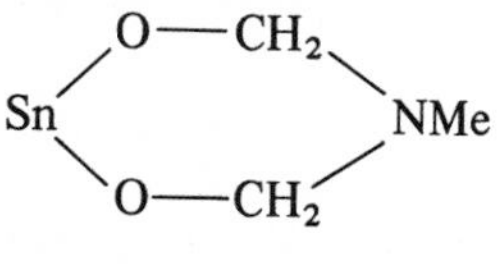

(III)

$$\text{Sn}\begin{smallmatrix}\text{O—CH}_2\\ \text{O—CH}_2\end{smallmatrix}\text{NMe}$$

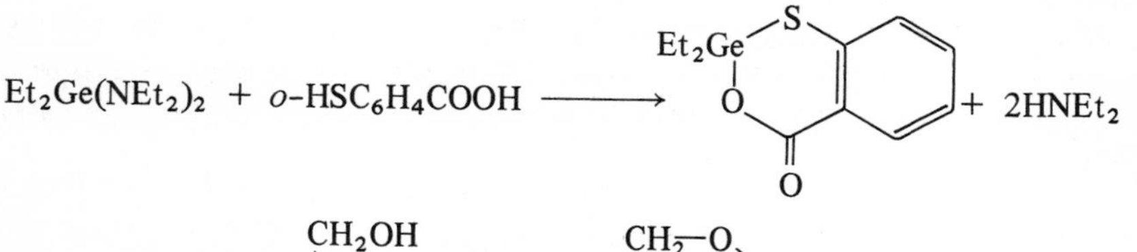

$$\text{Et}_2\text{Ge(NEt}_2)_2 + o\text{-HSC}_6\text{H}_4\text{COOH} \longrightarrow \text{Et}_2\text{Ge}\begin{smallmatrix}\text{S—C}_6\text{H}_4\\ \text{O—C(=O)}\end{smallmatrix} + 2\text{HNEt}_2 \quad (9)$$

$$\text{As(NEt}_2)_3 + \begin{smallmatrix}\text{CH}_2\text{OH}\\ |\\ \text{MeCHOH}\end{smallmatrix} \longrightarrow \begin{smallmatrix}\text{CH}_2\text{—O}\\ |\\ \text{MeCH—O}\end{smallmatrix}\text{AsNEt}_2 + 2\text{HNEt}_2 \quad (10)$$

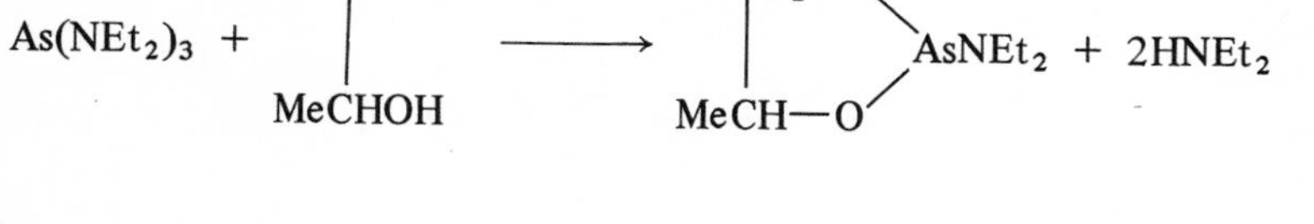

The reaction of an aminometallane with a mono- or bi-functional oxime provides a one step route to an organometallic oxime, e.g., of Si [249, 250, 695, 696, 719, 720, 728], Ge [728], Sn [297], or As [359, 360, 362], as in Eqs. (11) (M = Si or Ge) and (12).

$$(R_3M)_2NH + \begin{array}{l} R'C{=}NOH \\ | \\ R'C{=}NOH \end{array} \longrightarrow \begin{array}{l} R'C{=}NOMR_3 \\ | \\ R'C{=}NOMR_3 \end{array} + NH_3 \quad (11)$$

$$Me_2AsNMe_2 + RR'C{=}NOH \longrightarrow Me_2AsONCRR' + HNMe_2 \quad (12)$$

Hydroxylamines also behave as oxygen-centred acids, e.g., for B [532] or Si [251, 758], as in Eq. (13) ($n = 1, 2,$ or 3); and peroxides Et_3MOOBu^t (M = Si or Sn) are obtained from Et_3MNEt_2 and Bu^tOOH [613].

$$R_{3-n}B(NMe_2)_n + nEt_2NOH \longrightarrow R_{3-n}B(ONEt_2)_n + nHNMe_2 \quad (13)$$

Organotin acetates [144] are conveniently prepared from the appropriate tin(IV) amide and acetic acid, and the procedure is available also for other elements, e.g., Si [78]. Reactions between a carboxylic acid and an amide of B, Al, Ti, or As are discussed further in Chapter 13. Oxyacids of phosphorus react similarly [226, 725]: e.g., the $(Me_3Si)_2NH/H_2(O)POH$ mixture yields $(Me_3SiO)_2PH$ [725]. Ammonium salts of oxyacids may be silylated using $(Me_3Si)_2NH$, as for $[NH_4]_n[H_{3-n}PO_4]$ (n = 1, 2, or 3) to give $P(O)(OSiMe_3)_3$ [727], $[NH_4][OOCH]$ to yield $HC(O)OSiMe_3$, or $[NH_4][VO_3]$ to afford $[V({=}NSiMe_3)(OSiMe_3)_3]$ [65]; some salts $[NH_4]X$ [$X^- = Cl^-, Br^-, I^-$, or $\frac{1}{2}(COO^-)_2$] do not react [727]. *p*-Benzoquinone behaves as the dienol, to yield *p*-$(Me_3SnO)_2C_6H_4$ and the imine $RN{=}NMMe_3$, with $Me_3Sn(R)NN(MMe_3)SnMe_3$ (M = Si or Ge) [756].

C REACTIONS WITH HYDROGEN HALIDES OR PSEUDOHALIDES

In general, the M–N bond in aminometallanes is readily cleaved [Eq. (14)] by hydrogen halide (but see Chapter 15), as exemplified in Eq. (15) (R = Ph or *m*-MeC_6H_4) [768a]. Organotin cyanides [400, 460] or azides [400, 455] are readily prepared, using hydrogen cyanide or hydrazoic acid. Aminophosphines react similarly with a hydrogen halide or cyanide, e.g., refs. [208, 235, 364, 600, 602].

$$LM{-}NRR' + 2HX \longrightarrow LM{-}X + [RR'NH_2]X \quad (14)$$

$$R(Et)AsNH_2 + HCl \longrightarrow R(Et)AsCl + NH_3 \quad (15)$$

D REACTIONS WITH HYDRIDES OF NITROGEN, PHOSPHORUS, ARSENIC, OR ANTIMONY

Interaction of a dimethylaminotrialkylstannane and a hydride of a Group V element [Eqs. (16) (M = P or As) and (17) (M = P, As, or Sb)] offers a simple route to 4-co-ordinate tin-phosphorus, tin-arsenic [199, 306, 352, 354], or tin-antimony [306] compounds. The procedure has been extended to tin(II) phosphides or arsenides, using $Sn[N(SiMe_3)_2]_2$ and Ph_2PH, Bu^t_2PH, or Ph_2AsH [594]. Mixed germanio-stannio-arsines, e.g., $[(Me_3Ge)_2As]_2SnMe_2$ from $Me_2Sn(NMe_2)_2$ and $(Me_3Ge)_2AsH$, are also accessible [35].

$$R_3SnNMe_2 + Ph_2MH \longrightarrow R_3SnMPh_2 + HNMe_2 \qquad (16)$$

$$3R_3SnNMe_2 + MH_3 \longrightarrow (R_3Sn)_3M + 3HNMe_2 \qquad (17)$$

These syntheses are clearly related to transamination (see Chapters 1-9) (e.g., refs. [391a, 551a]) or to the reaction between an amide and a metal hydride (Chapter 12).

With reference to Eqs. (16) and (17), the following displacement orders are noteworthy: (i) phosphine > amine, and (ii) arsine > amine. Evidently, tin(IV) is a class '*b*' acceptor (or soft acid) with preference for Sn–P and Sn–As rather than Sn–N bonds [354]; this is confirmed by chemical observations on Sn–O and Sn–S compounds [255] (see Chapter 10) and by thermochemical data [54]. By contrast, Si, Ti, and Zr appear to be class '*a*' acceptors (or hard acids), as amides of Si [354], Ti [147], and Zr do not react with phosphines. More complex PH or AsH compounds often behave in the sense of Eq. (1): e.g., Me_3SiNEt_2 reacts with $H(O)P(NEt_2)_2$ to give $Me_3SiP(O)(NEt_2)_2$ [58].

Related to transaminations are the reactions with a ketimine [64, 480], hydrazine [423a, 540a, 540b, 755a], or phosphineimine [463]: thus, metal amides prove to be convenient reagents for preparing metal imides LM—N=C<, hydrazino-metallanes, or e.g., $MeHgNP(NMe_2)_3$. Transamination may compete with other reactions, as in the $(C_6F_5)_3SiNMe_2/H_2NC_6F_5$ system: C–Si cleavage gives rise to C_6F_5H [423]. The same amide with water also yields C_6F_5H [280]. Acidic hydrogen may be provided internally, as in Eq. (18) [423]; further examples are found in Chapters 1-9.

$$2(C_6F_5)_2Si(NHPh)Me \xrightarrow{\text{heat}} C_6F_5(Me)Si\begin{matrix} \diagup N(Ph) \diagdown \\ \diagdown N(Ph) \diagup \end{matrix}Si(Me)C_6F_5 + 2C_6F_5H \qquad (18)$$

The protic hydrogen atoms of aminoacids may be wholly [Eq. (19), R = Bu or Ph, $n = 1, 2, 3, 4$, or 5], or partially, silylated or stannylated [78, 217].

$$3R_3SnNEt_2 + H_2N(CH_2)_nCOOH \longrightarrow (R_3Sn)_2N(CH_2)_nCOOSnR_3 + 3HNEt_2 \qquad (19)$$

E REACTIONS WITH THIOLS, HYDROGEN SULPHIDE, OR HYDRIDES OF SELENIUM

Bis(diethylamino)borane reacts with BuSH at 100–110° C to yield diethylaminobutylthioborane [479]. Reaction of an aminosilane with a thiol proceeds, in general, only when the thiol has a higher boiling point than the amine produced, e.g., Eq. (20); ethanethiol does not react with diethylaminotrimethylsilane [2]. With aminoboranes there is a delicate balance, but using either a potentially chelating thiol or in presence of HCl the thiaborane (IV) (Q = O or S) [175] or (V) [467] is obtained.

$$Me_3SiNEt_2 + BuSH \longrightarrow Me_3SiSBu + HNEt_2 \qquad (20)$$

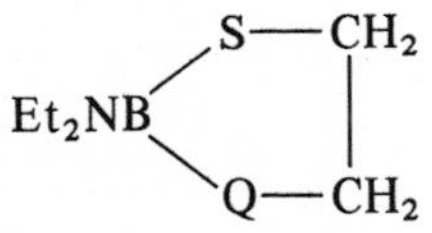

Et_2NB (ring: S—CH_2 / Q—CH_2)

(IV)

S, S ring with B(NMe_2)—B(NMe_2), B(NMe_2)—B(NMe_2)

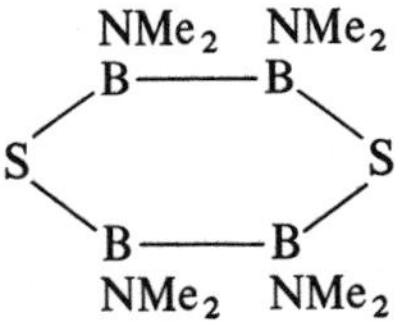

(V)

Whereas the interconversion of Si–N and Si–S bonds appears to be reversible and depends upon ease of removal of one of the products [3], diethylaminotrimethylgermane or diethylaminotrimethylstannane reacts exothermically with a thiol [Eq. (21)] (M = Ge or Sn) and it is unlikely that these reactions may be reversed [7]. This again demonstrates the class '*b*' character of Sn relative to the class '*a*' behaviour of Si. Likewise the Sn^{II} amide $Sn[N(SiMe_3)_2]_2$ readily yi[illegible]s $[Sn(SPh)_2]_n$ upon treatment with PhSH [428]. $(Me_3Si)_2S$ or compound (V[illegible] obtained from H_2S and $(Me_3Si)_2NH$ (with Me_3SiCl) or $(Me_2SiNH)_3$ (w[illegible] Me_2SiCl_2) [441, 442]. The PN bond in an aminophosphine is also readily cleav[illegible] by treatment with an alcohol, thiol, or H_2S [150, 209, 225, 413].

$$Me_3MNEt_2 + BuSH \longrightarrow Me_3MSBu + HNEt_2 \qquad (21)$$

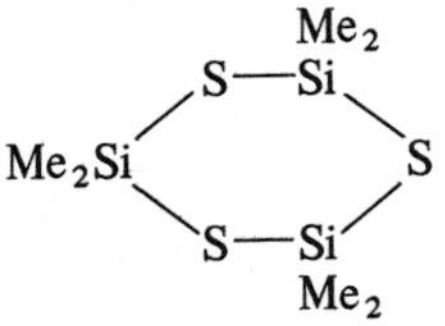

Me_2Si ring: S—$SiMe_2$—S—$SiMe_2$—S

(VI)

A titanium(IV) dialkylamide $[Ti(NR_2)_4]$ (R = Me or Et) with an excess of a thiol R'SH (R' = Et or Pr^i) gives an insoluble non-volatile red titanium tetra-alkylthiolate containing strongly addended thiol and amine, Eq. (22) [94, 147]; these compounds are polymeric and probably have $\overline{S}R$ bridges with octa-co-ordinated titanium. Compounds of the type $[Ti(NMe_2)_n(SR)_{4-n}]$ may be prepared by mixing the reagents in stoicheiometric ratios and are polymeric, except for $[Ti(NMe_2)_3SEt]$, which is a monomer [94]. However, monomeric bis(cyclopentadienyl)bis(t-butylthio)zirconium(IV) and the Hf analogue are known, Eq. (23) (M = Zr or Hf) [147]. Reaction of PhSH with $[\{Ti(\eta\text{-}C_5H_5)_2NMe_2\}_2]$ yields $[\{Ti(\eta\text{-}C_5H_5)_2SPh\}_2]$ [432], but $[V(O)(OPr^i)_2(NEt_2)_2]$ yields the disproportionation products $[V(O)(OPr^i)_3]$ and $[V(O)(SR)_3]$ when treated with RSH (R = Pr^i or CF_3CH_2) [157a]. A reductive reaction is noted in the conversion of $[W(NMe_2)_6]$ into $[\{W(SR)_3\}_n]$ (R = Me or Ph) [90a].

$$[Ti(NR_2)_4] + R'SH \longrightarrow Ti(SR')_4.(R'SH)_x(R_2NH)_y \qquad (22)$$

$$[M(\eta\text{-}C_5H_5)_2(NMe_2)_2] + 2Bu^tSH \longrightarrow [M(\eta\text{-}C_5H_5)_2(SBu^t)_2] + 2HNMe_2 \qquad (23)$$

Various U^{IV} thiolates are obtained from $[U(\eta\text{-}C_5H_5)_2(NEt_2)_2]$, e.g., by treatment with EtSH, $(CH_2SH)_2$, o-HSC_6H_4OH, or 4-Me, 2-HSC_6H_3SH [342a].

Trimethyl-silicon, -tin, or -germanium methylselenide is obtained from the appropriate amide and MeSeH [32], Eq. (24) (M = Si, Ge, or Sn), and $H_3SiSeH.HNMe_2$ is prepared by treatment of H_3SiNMe_2 with H_2Se; H_2S gives the analogous alkylthiosilane [173].

$$Me_3MNMe_2 + MeSeH \longrightarrow Me_3MSeMe + HNMe_2 \qquad (24)$$

F REACTIONS WITH CARBON ACIDS AND A NOTE ON BORANES AND SILANES

1. Introduction

Lithium acetylides are normally obtained from the acetylene and a conveniently available alkyl such as LiBu. However, almost forty years ago it was shown that a lithium amide serves as well [128]. The first pointer to the use of metal amides LM–NRR' for obtaining corresponding organometallic compounds LM–A, Eq. (1) where $\bar{A}$ is a conjugatively stabilised carbanion, came in 1964 [352]. Some of the products, e.g., $Me_3Sn\text{-}C_5H_5$, $Me_3SnCF{=}CF_2$, or $Me_3SnC{\equiv}CR$, are themselves valuable reagents for transfer of the group A (in these cases A = C_5H_5, $CF_2{=}CF$, or $PhC{\equiv}C$) to another metal (such as Pt^{II}) site [15a, 128a]. For other materials, such as the organometallic diazoalkanes (of electropositive metals), the metal amide/HA (in this case $RHCN_2$) reaction is the method of choice [421, 421a].

2. Acetylenes

Acetylenes and $LiNMe_2$ behave in the sense of Eq. (1) [128, 576, 702]. Amino-boranes and -silanes do not react with phenylacetylene, and amino-germanes react only under forcing conditions [Eq. (25)] [236, 446]. By contrast, aminostannanes readily liberate amine when treated with acetylene (reactions proceed directly to the second stage) [Eq. (26)] [352, 354] or a monosubstituted acetylene [Eq. (27)] [347, 352, 354]. These reactions are quantitative and provide a useful synthetic route to alkynylstannanes, see also refs. [185, 236, 261]; by employing amides of other electropositive elements, further alkynylmetallanes are accessible.

$$Bu_3GeNEt_2 + PhC{\equiv}CH \xrightarrow{180^\circ C} Bu_3GeC{\equiv}CPh + HNEt_2 \qquad (25)$$

$$2R_3SnNR'_2 + HC{\equiv}CH \longrightarrow R_3SnC{\equiv}CSnR_3 + 2HNR'_2 \qquad (26)$$

$$R_{4-n}Sn(NR'_2)_n + nR''C{\equiv}CH \longrightarrow R_{4-n}Sn(C{\equiv}CR'')_n + nHNR'_2 \qquad (27)$$

Disubstituted organometallic acetylenes containing two different Group IVB metal atoms are obtained either (i) by treating the product of the reaction between a metal amide and sodium acetylide (which can be isolated in some cases) with an organometallic halide, e.g., Eq. (28); or (ii) from the reaction of a metal amide with an organometallic acetylide, Eq. (29) [236].

$$R_3MNEt_2 + NaC{\equiv}CH + R'_3M'Cl \longrightarrow R'_3M'C{\equiv}CMR_3 + NaCl + HNEt_2 \qquad (28)$$

$$R_3SnNEt_2 + R'_3MC{\equiv}CH \longrightarrow R_3SnC{\equiv}CMR'_3 + HNEt_2 \qquad (29)$$

Dimethylamine is liberated upon mixing phenylacetylene and tetrakis-(dimethylamido)titanium, but the product (having a Ti-C bond) was not isolated [147]. However, the compounds $[M(\eta\text{-}C_5H_5)_2(C{\equiv}CPh)_2]$ (M = Zr or Hf) are obtained by interaction of $[M(\eta\text{-}C_5H_5)_2(NMe_2)_2]$ and phenylacetylene, and $[Ti(\eta\text{-}MeC_5H_4)(C{\equiv}CPh)(NMe_2)_2]$ is prepared from the same acetylene and $[Ti(\eta\text{-}MeC_5H_4)(NMe_2)_3]$ [347].

3. Cyclopentadienes or some boron acids

With cyclopentadiene (CpH) and indene (IndH) (which are acidic hydrocarbons, forming stable aromatic anions), organometallic derivatives of Sn^{IV} [Eq. (30)] [295, 352, 354], Sn^{II} [295], Sb^{III} [404], Bi^{III}, Ti^{III} [25, 431], Ti^{IV} [Eq. (31)] [147], Zr^{IV} [Eqs. (32) and (33)], Hf^{IV} [Eq. (32), M = Zr or Hf], Nb^{IV} [Eq. (34)] [347], and U^{IV} [Eq. (35)] [342a] are obtained. It is interesting to note that even with excess cyclopentadiene, only one NR_2 group is directly replaced in the case of titanium or niobium, whereas two NR_2 groups are replaced for Zr,

Hf, or U. The Ti^{III} dimethylamide also has only one group replaceable in yielding $[\{Ti(\eta\text{-}C_5H_5)(NMe_2)_2\}_2]$ [25, 431]. A similar trend has been observed in the reaction between Ti^{IV} or Zr^{IV} amides and hexamethyldisilazane [Eq. (36), $M = Ti$, $n = 1$; or $M = Zr$ when $n = 2$] [147]. The differences between Ti^{IV} on the one hand and Zr^{IV} or Hf^{IV} on the other are undoubtedly attributable to steric effects which clearly are more restricting on the smaller (Ti^{IV}) reacting centre. Synthesis of ferrocene in high yield by a related reaction [Eq. (37) (HA = CpH or IndH)] is noteworthy [354]; similar reductions are also found for Nb^{V} [347] [Eq. (34)] and Cr^{IV} [56a] (Chapter 18).

$$R_3SnNR'_2 + HA \longrightarrow R_3SnA + HNR'_2 \tag{30}$$

$$[Ti(NR_2)_4] + 3C_5H_6 \longrightarrow [Ti(\eta\text{-}C_5H_5)(NR_2)_3] + HNR_2 \tag{31}$$

$$[M(NR_2)_4] + 3C_5H_6 \longrightarrow [M(\eta\text{-}C_5H_5)_2(NR_2)_2] + 2HNR_2 \tag{32}$$

$$[Zr(NMe_2)_4] + 2IndH \longrightarrow [Zr(\eta\text{-}Ind)(NMe_2)_3] + HNMe_2 \tag{33}$$

$$[Nb(NMe_2)_5] + 4C_5H_6 \longrightarrow [Nb(\eta\text{-}C_5H_5)(NMe_2)_3] + HNMe_2 \tag{34}$$

$$[U(NEt_2)_4] + 2C_5H_6 \longrightarrow [U(\eta\text{-}C_5H_5)_2(NEt_2)_2] + 2HNEt_2 \tag{35}$$

$$[M(NMe_2)_4] + n(Me_3Si)_2NH \xrightarrow{-nHNMe_2} [M\{N(SiMe_3)_2\}_n(NMe_2)_{4-n}] \tag{36}$$

$$FeCl_3 \xrightarrow{LiNEt_2} \underset{\text{(not isolated)}}{Fe(NEt_2)_n} \xrightarrow{C_5H_6} [Fe(\eta\text{-}C_5H_5)_2] + nHNEt_2 \tag{37}$$

Ferrocene (FcH) can be silylated under mild conditions by treatment with an aminosilane or an aminogermane in the presence of aluminium chloride, e.g., Eqs. (38) (39) [572, 699], and $Sn(C_5H_5)_2$ reacts with Me_3SnNEt_2 to yield $Sn(C_5H_4SnMe_3)_2$ [110a].

$$Ph_3SiNMe_2 + FcH + AlCl_3 \xrightarrow{H_2O} FcSiPh_3 \tag{38}$$

$$Ge(NMe_2)_4 \xrightarrow[\text{(ii) } H_2O]{\text{(i) } 4FcH,\, 4AlCl_3} (Fc_3Ge)_2O \tag{39}$$

Several mono- or di-organosilyl-, -germyl-, or -stannyl-substituted cyclopentadienes react with amides of Sn^{IV} [298, 598] or Ti^{IV} [116] to yield polymetallated cyclopentadienes, Eqs. (40) ($M = M' = Sn$; or $M = Si$, Ge, or Sn; $M' = Si$) and (41) ($R = R' = Me_3Si$ or Me_3Ge; or $R = Me_3Si$, $R' = Me_3Ge$); these show fluxional behaviour [598]. Surprisingly $C_5H_3(SnMe_3)_3$ reacts with further Me_3SnNEt_2 to yield $C_5H_2(SnMe_3)_4$.

$$C_5H_4(MMe_3)(M'Me_3) + Me_3SnNEt_2 \longrightarrow (Me_3Sn)_2C_5H_2(MMe_3)(M'Me_3) + HNEt_2 \quad (40)$$

$$[Ti(NR_2)_4] + C_5H_4R'R'' \longrightarrow [Ti(\eta\text{-}C_5H_3R'R'')(NR_2)_3] + HNR_2 \quad (41)$$

Organostannylcarboranes $R'CB_{10}H_{10}CSnR_3$ are obtained from R_3SnNEt_2 and *o*–, *m*–, or *p*–$R'CB_{10}H_{10}CH$ [486]; by contrast $B_{10}H_{14}$ reacts as a source of $B_{10}H_{13}^-$ which probably behaves as a non-co-ordinating anion [147] (see Chapter 16).

4. Polyhalogenohydrocarbons

An appropriate halogenated hydrocarbon with an aminostannane $R_3SnNR'_2$ gives R_3SnCCl_3 [157, 202], R_3SnCBr_3 [202] [Eq. (42), X = Cl or Br], or $Me_3SnCCl{=}CCl_2$ [156, 157] [Eq. (43)]. Formation of $Me_3SnC_6F_5$ and $Et_3SnC_6F_5$, in similar fashion from C_6F_5H is surprising [347]; but Group IVA or silicon amides are unreactive as are (i) Me_3SiNMe_2 with an acetylene, or (ii) Me_3SnNMe_2 with C_6Cl_5H. The pentafluorobenzene/tin amide results were interpreted as showing that the *kinetic* rather than *thermodynamic* acidity of the protic species is important [347].

$$R_3SnNR'_2 + HCX_3 \longrightarrow R_3SnCX_3 + HNR'_2 \quad (42)$$

$$Me_3SnNMe_2 + HCCl{=}CCl_2 \longrightarrow Me_3SnCCl{=}CCl_2 + HNMe_2 \quad (43)$$

5. Diazoalkanes or Me_3PCH_2

Diazomethane and Me_3SnNMe_2 give the volatile distannyldiazomethane $(Me_3Sn)_2CN_2$ [422, 425], Eq. (44), initially misformulated as Me_3SnCHN_2 [421].

$$Me_3SnNMe_2 + CH_2N_2 \xrightarrow{-HNMe_2} Me_3SnCHN_2 \xrightarrow{-HNMe_2} (Me_3Sn)_2CN_2 \quad (44)$$

A number of related reactions [Eq. (45)] (M = Ge, Sn, or Pb, $n = 3$) have been realised (for a review of α-heterodiazoalkane chemistry, see [427]) [421, 426, 427, 456] for Ge^{IV}, Sn^{IV}, and Pb^{IV} analogues.

$$R_nMNR'_2 + HC(X)N_2 \longrightarrow R_nMC(X)N_2 + HNR'_2 \quad (45)$$

Mechanistically, reactions (44) and (45) may be of the type shown in Eq. (1) [422]. Diazomethane certainly shows protic character in its behaviour towards RLi. Moreover, an electron-releasing substituent X [see Eq. (45)] decreases the reactivity of the diazoalkane, whereas an electron-withdrawing group has the converse effect. A limitation on procedure (45) is that while Ge, Sn, Pb, Tl [407], As [403, 405], Sb, and Bi amides react with CH_2N_2, amides of B or Si generally do not even react with diazoacetic ester [422, 427]. However, Zn [421a], Cd, and Hg^{II} amides $M[N(SiMe_3)_2]_2$ or $RHg[N(SiMe_3)_2]$ (R = Me or Et) afford diazoalkanes with $CH(CO_2Et)N_2$ and $[Zr(\eta\text{-}C_5H_5)_2\{C(N_2)SiMe_3\}_2]$ is obtained [426] from $[Zr(\eta\text{-}C_5H_5)_2(NMe_2)_2]$ and Me_3SiCHN_2.

An alternative model for reaction (45) is that of a double two-step process, with initial 1,3-insertion to afford $R_3SnCHX{-}N{=}N{-}NR'_2$ and subsequent HNR'_2 elimination. Whereas Me_3SnNMe_2 reacts with Me_3SiCHN_2 to give the mixed ligand diazoalkane $Me_3Sn(Me_3Si)CN_2$ [422], tris-dimethylaminophosphine and Me_2AsCRN_2 react according to Eq. (46) [402]; and this suggests that the initial step also in the other reactions is the formation of a donor-acceptor adduct, such as (VI).

$$P(NMe_2)_3 + Me_2AsCRN_2 \longrightarrow \underset{\textbf{(VI)}}{Me_2AsCR{=}N{-}N{=}P(NMe_2)_3} \quad (46)$$

$$(R = H \text{ or } CO_2Et)$$

The Ti^{IV} amide $[TiCl_2(NMe_2)_2]$ reacts with Me_3PCH_2 to yield $(Me_2N)_2\overline{TiC(PMe_3)Ti(NMe_2)_2C}(PMe_3)$ [669a].

6. Cyano-, nitro-, keto-, or carbethoxy-alkanes having α-hydrogen

Acetonitrile liberates hexamethyldisilazane from $NaN(SiMe_3)_2$ according to Eq. (47). The α-alkylation of nitriles containing α-H has also been achieved by successive use of $LiNR_2$ and R′Hal [135]. Reaction (47) was established by demonstrating that addition of alkyl chloride (RCl) affords RCH_2CN, while use of concentrated solution and an excess of MeCN followed by addition of Me_3SiCl furnishes $Me_3SiNHC(Me){=}CHCN$ [410]. A similar product is also obtained from $LiN(SiMe_3)_2$ and successively CH_3CN and Me_3SiCl [433]; the same reactions but using $LiNMe_2$ yield a product believed to be (VII).

$$NaN(SiMe_3)_2 + MeCN \longrightarrow NaCH_2CN + HN(SiMe_3)_2 \quad (47)$$

The Thorpe reaction [e.g., Eq. (48)] may be catalysed by sodamide [135] or various lithium amides [292, 300] (for related use of alkali metal amides, see Chapters 2 and 14).

$$2RR'CHCN \longrightarrow RR'C(CN)C(=NH)CHRR' \quad \textbf{(VII)} \tag{48}$$

Nitriles containing an α-hydrogen atom appear to behave as protic compounds when treated with an aminostannane, Eq. (49); the amine is invariably isolated in high yield, but the residue gives a mixture of nitrile-containing products, suggesting thermal redistribution [354]. Reaction of an aminogermane with MeCN similarly furnishes a mixture of mono- and di-substituted α-germylated nitriles, Eq. (50) [621].

$$R_3SnNR'_2 + R''R'''CHCN \longrightarrow R_3SnC(R'')(R''')-CN + HNR'_2 \tag{49}$$

$$R_3GeNMe_2 + MeCN \longrightarrow R_3GeCH_2CN + (R_3Ge)_2CHCN + HNMe_2 \tag{50}$$

β-Nitropropane and an aminostannane afford the appropriate β-nitropropylstannane, Eq. (51) [455, 462]; however, with nitromethane, although a high yield of amine was obtained, the tin product was not characterised [354]. There is some doubt whether the nitroalkylmetallane [e.g., (VIII)] may not in fact be the isomeric nitronate, as has been proposed for a lead analogue $Me_3PbON(O)CMe_2$ [462]. Aminosilanes, as usual, are less reactive than their heavier Group IVB congeners. However, the readiness to effect Si-N cleavage is enhanced by *N*-acylation or related *N*-protection by an electronegative substituent; thus $Me_3SiN(Ph)CONHPh$ and $HC(NO_2)_3$ yield $Me_3SiC(NO_2)_3$ [327a]. $As(NMe_2)_3$ reacts with $MeNO_2$ in a more complex fashion to give As_2O_3, $HNMe_2$, and Me_2NCN [740a].

$$R_3SnNR'_2 + Me_2CHNO_2 \longrightarrow \underset{\textbf{(VIII)}}{R_3SnCMe_2NO_2} + HNR'_2 \tag{51}$$

Ketones [300, 773] or carboxylic esters [292] having α-H atom(s), give aldol condensation products with $LiNR_2$. Reaction of acetone with an aminotrialkylstannane, Eq. (52) [354], or tetrakis(dimethylamido)titanium, Eq. (53) [354], affords amine and mesityl oxide, presumably by a base-catalysed reaction. With isopropenyl acetate and $R_3SnNR'_2$, *N,N*-dialkylacetamide, mesityl oxide, and bis(trialkylstannyl) oxide, Eq. (54), are obtained [354].

$$2R_3SnNR'_2 + 2Me_2CO \longrightarrow (R_3Sn)_2O + 2HNR'_2 + Me_2C{=}CHCOMe \tag{52}$$

$$2[Ti(NMe_2)_4] + 2Me_2CO \longrightarrow [Ti(OH)_{4-n}(NMe_2)_n] + HNMe_2 + Me_2C{=}CHCOMe \quad (53)$$

$$R_3SnNR'_2 + CH_2{=}CMeOCOMe \longrightarrow R'_2NCOMe + (R_3Sn)_2O + Me_2C{=}CHCOMe \quad (54)$$

It will be evident that organic species which owe their acidity to an α-CO, CN, NO_2, or related unsaturated group may act not as protic species but by virtue of their functionality, as discussed in Chapters 10 and 13.

In contrast to the above reactions, a ketone, β-diketone, or carboxylic acid amide [unless this has *N*-hydrogen: MeCONHR′ (R′ = Me, Et, or Pr) with Bu_3SnNR_2 (R = Me or Et) gives $Bu_3SnN(R')COMe$] may behave in the sense of Eq. (1) when treated with a metal amide. This is illustrated in Eqs. (55) [591], (56) (R = Me or Bu^t) [22], and (57) (R = Me or Et; R′ = Me, Et, or Pr) [637].

$$Bu_3SnNEt_2 + Bu^t(Me)CO \longrightarrow Bu_3SnCH_2COBu^t + Bu_3SnOC(:CH_2)Bu^t + HNEt_2 \quad (55)$$

$$Hg[N(SiMe_3)_2]_2 + 2CH_2(COR)_2 \longrightarrow Hg[CH(COR)_2]_2 + 2HN(SiMe_3)_2 \quad (56)$$

$$Bu_3SnNR_2 + Me(R'_2N)CO \longrightarrow Bu_3SnCH_2CONR'_2 + HNR_2 \quad (57)$$

In some instances there is competition between protonolysis and another process, e.g., insertion. This is illustrated by Eq. (58) ($X = CO_2Et$ or CN) [651a].

$$Et_3GeNMe_2 \xrightarrow{CH_2(CO_2Et)X} \begin{cases} \xrightarrow[-HNMe_2]{\text{Protonolysis}} Et_3GeCH(CO_2Et)X \\ \xrightarrow{\text{Insertion}} Et_3GeOC(NMe_2)(OEt)CH_2X \end{cases} \quad (58)$$

7. Reactions with silanes

In general Si-H compounds are insufficiently acidic to cause cleavage of an M-N bond, except under severe conditions (see also Chapter 12), as has been demonstrated for $M[N(SiMe_3)_2]_2$ with $(C_6F_5)_3SiH$, Eq. (59) (M = Cd or Hg) [357a]. Trichlorosilane Cl_3SiH is also significantly acidic, but the presence of the Si-Cl functionality causes a metathetical exchange reaction, Eq. (60), to take place with $MeHg[N(SiMe_3)_2]$ [269b] (for further comments, see Chapter 13).

$$M[N(SiMe_3)_2]_2 + 2(C_6F_5)_3SiH \longrightarrow M[Si(C_6F_5)_3]_2 + 2HN(SiMe_3)_2 \quad (59)$$

$$MeHg[N(SiMe_3)_2] + Cl_3SiH \longrightarrow MeHgCl + Cl_2(H)SiN(SiMe_3)_2 \quad (60)$$

G TABLE OF REACTIONS

Reactions of Protic Compounds with Metal Amides[1]

Metal Amide	Reagent	Product[2]	References
Group 1 Metal Amides			
$LiNMe_2$	RCCH	LiCCR	[128, 576, 702]
$LiNR_2$	$R'CH_2C(O)R''$	aldol condensation products	[292, 300, 773]
$LiNR_2$	$R'CH_2CN$	LiCH(R')CN	[135, 433]
$LiNMe_2$	MeCN	$LiNC(Me)CH_2CN$	[433, 647]
$MN(SiMe_3)_2$ (M = Na or K; X^1 = H, X^2 = X^3 = Cl or Br; X^1 = H, X^2 = Cl, X^3 = Br or I)	$HCX^1X^2X^3$	$M(CX^1X^2X^3)$	[471, 472]
$NaN(SiMe_3)_2$ (R = R′ = H; R = H, R′ = Ph)	RR′CHCN	NaCRR′CN	[406]
$MNMe_2.2BH_3$ (M = Na, K)	MeOH, HCl	$B(OMe)_3 + [Me_2NH_2]Cl + 6H_2 + MCl$	[54a]
Group 3 Metal Amides			
R_2BNEt_2 (R = Pr, Bu, Ph; R′ = R″ = Ph, *p*-Cl–, *p*-Br–, or *p*-MeC_6H_4; R′R″C = fluorenylidene)	R′R″CNH.HCl	$R_2BNCR'R''$	[480, 481, 481a]
Me_2BNMe_2	Bu^tOH	Me_2BOBu^t	[408]
Et_2BNMe_2	HN(imidazole ring)=N	Et_2BN(imidazole ring)=N	[499a]
Ph_2BNEt_2 (R = Ph, Me)	$RCONH_2$	$Ph_2BNHCOR$	[540a, 540b]
Ph_2BNEt_2	$Ph_2C{=}NH$	$Ph_2BN{=}CPh_2$	[540b]
$2Ph_2BNEt_2$ (R = H, R′ = H, Me; R = R′ = Me)	RHNNHR′	$Ph_2BNRNR'BPh_2$	[540b]
$(BuO)_2BNEt_2$	BuOH	$B(OBu)_3$	[259]
$(BuO)_2BNEt_2$	2HCl	$(BuO)_2BCl$	[259]

Table of Reactions *(C'ntd.)*

Metal Amide	Reagent	Product2	References
o-$C_6H_4O_2BNEt_2$ (R = R′ = Me; R = Ph, R′ = H)	RR′NNH_2	o-$C_6H_4O_2$BNHNRR′	[423a]
$RB(NMe_2)_2$ (R = H, Me, Ph; R′ = H, Me)	MeHNNHR′	Me R′ N—N RB BR N—N Me R′	[540a]
$PhB(NEt_2)_2$	$PhHNNH_2$	$PhB(NHNHPh)_2$	[423a]
$S(CH_2)_3SBNEt_2$	$PhHNNH_2$	$S(CH_2)_3$SBNHNHPh	[181b]
$PhB(NEt_2)_2$	$HOCH_2CH_2NH_2$	NH—CH_2 PhB | O—CH_2	[178, 182]
$2BuOB(NEt_2)_2$	7HCl	$(BuO)_2BCl + 2Et_2NBCl_2.HCl$	[258]
$BuOB(NEt_2)_2$	2BuOH	$B(OBu)_3$	[258]
$R_3SiB(NMe_2)_2$ (R = Me, Ph)	3HCl	$R_3SiBCl_2.HNMe_2$	[536, 537]
$R_3SiB(NMe_2)_2$ (R = Me, Ph)	2HCl	$Cl(Me_2N)BSiR_3$	[536, 537]
$Mn(CO)_4(PPh_3)B(NMe_2)_2$	3HCl	$Mn(CO)_4(PPh_3)BCl_2.HNMe_2$	[543]
$Mn(CO)_5B(NMe_2)_2$	3HCl	$BCl_3 + Mn(CO)_5H$	[543]
$Mn(CO)_5B(NMe_2)_2$	4HCl	$Mn(CO)_5BCl_2$	[543]
$MeB(NMe_2)_2$	$2Bu^tOH$	$MeB(OBu^t)_2$	[408]
$RB(NMe_2)_2$ (R = Me, Bu)	$Me_2Si(NHMe)_2$	Me_2 Si MeN NMe RB BR N Me	[551a]
$PhB(NEt_2)_2$	$Ph_2CNH.HCl$	$PhB(NCPh_2)_2$	[481, 481a]
$HB(NEt_2)_2$ (R = Pr, Bu)	RSH	$H(RS)BNEt_2$	[479]
$B(NMe_2)_3$	$3Bu^tOH$	$B(OBu^t)_3$	[408]
$B(NMe_2)_3$	2HCl	$ClB(NMe_2)_2$	[531, 540]
$B(NMe_2)_3$	3MeSH + 3HCl	$B(SMe)_3$	[105]
$B(NEt_2)_3$	$3Bu^tOH$	$B(OBu^t)_3$	[408]
$B(NEt_2)_3$	$3Et_3SiOH$	$B(OSiEt_3)_3$	[16]
$B(NEt_2)_3$ (Q = Q′ = O, S; Q = O, Q′ = S)	$HQCH_2CH_2Q'H$	Q—CH_2 Et_2NB | Q′—CH_2	[175]

Table of Reactions *(C'ntd.)*

Metal Amide	Reagent	Product[2]	References
$B(NEt_2)_3$	BuOH	$BuOB(NEt_2)_2$	[258]
$B(NEt_2)_3$	4HCl	Cl_2BNEt_2	[531, 540]
$B(NEt_2)_3$	$PhN{=}\overline{C(CH_2)_4CH_2}$	$(Et_2N)_2BN(Ph)\overline{CH(CH_2)_3CH{=}CH}$	[213]
$R_nB(NMe_2)_{3-n}$	$3-n HONEt_2$	$R_nB(ONEt_2)_{3-n}$	[532]
$B_2(NMe_2)_4$	10HF	$B_2F_4.2HNMe_2$	[467]
$B_2(NMe_2)_4$ (R = Me, Et, Pr^i, Ph)	4ROH + 4HCl	$B_2(OR)_4$	[106]
$B_2(NMe_2)_4$	2HCl	$B_2(NMe_2)_4.2HCl$	[467]
$B_2(NMe_2)_4$ (X = Cl, Br)	6HX	$B_2X_4.2HNMe_2$	[467]
$B_2(NMe_2)_4$	4HCN	$B_2(CN)_4.2HNMe_2$	[467]
$B_2(NMe_2)_4$	$2H_2S$ + 2HCl	Me_2N NMe_2 B—B S S B—B Me_2N NMe_2	[467]
$B_2(NHSiEt_3)_4$	HCl	Et_3SiCl	[23]
Me_2InNMe_2	C_5H_6	$Me_2InC_5H_5$	[404]
$(Me_2TlNMe_2)_2$	CH_2N_2	$(Me_2Tl)_2CN_2$	[407]
Silicon Amides			
H_3SiNMe_2 (X = S, Se)	H_2X	$H_3SiXH.HNMe_2$	[173]
HMe_2SiNEt_2 (R' = Me, CH_2CHCH_2)	CHCC(Me)R'OH	$CHCC(Me)R'OSiMe_2H$	[285]
$2HMe_2SiNEt_2$	$HO(CH_2)_2OH$	Me_2Si $\langle O(CH_2)_2O$ / $O(CH_2)_2O \rangle$ $SiMe_2$	[39]
$2HMe_2SiNEt_2$	$HOCH_2CH(Me)OH$	Me_2Si $\langle OCH_2CH(Me)O$ / $O(Me)CHCH_2O \rangle$ $SiMe_2$	[39]
$Me_3SiNHMe$ (R = Et, Ph)	ROH	Me_3SiOR	[234]
$Me_3SiNHEt$	$[Ph(OH)\overline{SiO]_4}$	$[Ph(OSiMe_3)\overline{SiO]_4}$	[42]
$Me_3SiNHPh$ (R = Me, Et)	ROH	Me_3SiOR	[234]
Me_3SiNMe_2	2HCl	Me_3SiCl	[223]
Me_3SiNMe_2	$H_2NCONHCH_2OH$	$H_2NCONHCH_2OSiMe_3$	[609]
$8Me_3SiNMe_2$	$H_2NCONHCH_2OH$	$Me_3SiNHCON(SiMe_3)CH_2OSiMe_3$	[609]
$2Me_3SiNMe_2$	$(HOCH_2NH)_2CO$	$(Me_3SiOCH_2NH)_2CO$	[609]
$5Me_3SiNMe_2$	$(HOCH_2NH)_2CO$	$Me_3SiOCH_2NHCON(SiMe_3)CH_2OSiMe_3$	[609]

Table of Reactions *(C'ntd.)*

Metal Amide	Reagent	Product[2]	References
$2Me_3SiNMe_2$	PhC=NOH / PhC=NOH (vicinal)	$Me_3SiON{=}CPh$ / $Me_3SiON{=}CPh$ (vicinal)	[361]
$2Me_3SiNMe_2$	1,2-cyclo-$C_6H_8(NOH)_2$	1,2-cyclo-$C_6H_8(NOSiMe_3)_2$	[361]
Me_3SiNMe_2	MeAs($ONCMe_2$)($O(CH_2)_4OH$)	MeAs($ONCMe_2$)($O(CH_2)_4OSiMe_3$)	[359]
Me_3SiNMe_2	MeSeH	Me_3SiSeH	[32]
Me_3SiNMe_2	$CH_2(MeCNOH)_2$	$(Me_3SiONCH_2)_2CH_2$	[361]
Me_3SiNEt_2 (R = R′ = Et; R = Me, R′ = Et, Pr, Pr^i)	RR′CNOH	$Me_3SiONCRR'$	[696]
Me_3SiNEt_2 (R = H, Me)	2-HO-6-R-pyridine	2-Me_3SiO-6-R-pyridine	[109]
Me_3SiNEt_2 (R = H, Me)	2-HO-4,6-R_2-pyrimidine.HCl	2-$OSiMe_3$-4,6-R_2-pyrimidine	[109]
Me_3SiNEt_2 (M = C, Si, Ge)	Ph_3MOH	$Me_3SiOMPh_3$	[586]
Me_3SiNEt_2 (R = Et, Ph)	$R_2Si(OH)_2$	$(Me_3SiO)_2SiR_2$	[584]
Me_3SiNEt_2 (R = Me, n = 3; R = Ph, n = 2)	$HO(SiR_2O)_nH$	$Me_3SiO(SiR_2O)_nSiMe_3$	[584]
Me_3SiNEt_2	$(Et_2N)_2P(O)H$	$Me_3SiP(O)(NEt_2)_2$	[58]
Me_3SiNEt_2 (n = 1, R = H; n = 7, R = Me_3Si)	$nHN(CH_2CH_2NR_2)_2$	$Me_3SiN(CH_2CH_2NR_2)_2$	[273]
Me_3SiNEt_2 (R = Ph, X = O; R = Pr^i, X = NCH_2Pr^i)	cyclic CH_2CH_2–NR–P(O)H–X	$Me_3Si(O)P$ cyclic (NR–CH_2CH_2–X)	[605]
Me_3SiNEt_2	PrOH	Me_3SiOPr	[583]
Me_3SiNEt_2	$(HONH_3)_2SO_4$	$Me_3SiON(SiMe_3)_2$	[758]
$3Me_3SiNEt_2$	$MeCH(OH)CH(NH_2)CO_2H$	$MeCH(OSiMe_3)CHNH(SiMe_3)CO_2SiMe_3$	[311]
$2Me_3SiNEt_2$	$H_2S + Me_3SiCl$	$(Me_3Si)_2S$	[441]
Me_3SiNRR' [R = R′ = Et; Z = $(CH_2)_3$, $CH(Me)CH_2$, CH_2CMe_2, CH_2CMeH; R″ = H, Me, Et)	$HO(Z)NR''_2$	$Me_3SiO(Z)NR''_2$	[464]

Table of Reactions *(C'ntd.)*

Metal Amide	Reagent	Product[2]	References
Me_3SiNRR' (R = R' = Et, R'' = Et, Ph; R = H, R' = Bu, R'' = Et, Ph; R = H, R' = Bu^t, R'' = Ph)	R''_3SiOH	$Me_3SiOSiR''_3$	[583]
Me_3SiNRR' (R = R' = Et; R = H, R' = Bu^t; R'' = Et, Pr)	$(R''O)_2P(O)H$	$Me_3Si(O)P(OR'')_2$	[604]
Me_3SiNEt_2	$2HCN + Me_3SiCl$	$2Me_3SiCN$	[325]
Me_3SiNEt_2	BuSH	Me_3SiSBu	[2]
$Me_3SiNHPr$	NH_4VO_3	$(Me_3SiO)_3V{=}NPr$	[65]
$Me_3SiN(R)CH_2Ph$ [R = C(Me)=NCH($SiMe_3$)Ph]	MeOH	Me_3SiOMe	[663]
$Me_3SiN(Ph)CONHPh$	$HC(NO_2)_3$	$(O_2N)_2CN(O)OSiMe_3$	[327a]
2Me₃SiN(pyrazolyl ring: N=, CH=CH)	H_2S	$(Me_3Si)_2S$	[47]
$Me_2(C_6Cl_5)SiNMe_2$	$Me_2(C_6Cl_5)SiOH$	$[Me_2(C_6Cl_5)Si]_2O$	[265]
Et_3SiNEt_2	Bu^tOOH	$Et_3SiOOBu^t + Bu^tOH$	[613]
RMe_2SiNMe_2 (R = o-$CH_2CHCH_2C_6H_4$)	$HOSiMe_2$–C_6H_4–$SiMe_2OH$	$RMe_2SiOSiMe_2$–C_6H_4–$SiMe_2OSiMe_2R$	[567]
$R_3Si\overline{N(CH_2)_k}$ (k = 2,3; R = C_mH_{2m+1}, C_mH_{2m-1}, C_mH_{m-1}; R' = H, C_nH_{2n+1}, C_nH_{2n-1})	R'OH	R_3SiOR'	[504]
$R_3SiNR'Et$ (R = Et, R' = H, Et; R = Bu, R' = Et)	Ph_3SiOH	$R_3SiOSiPh_3$	[583]
$2R_3SiNEt_2$ (R = Me, Et)	$P(O)(OH)H_2$	$(R_3SiO)_2PH$	[725]
R_3SiNHR' (R = Me, Et; R' = H, $SiMe_3$; R'' = e.g., Me)	$R''C(NH_2)HCOOH$	$R_3SiOOCCH(NH_2)R''$	[78, 610]
R_3SiNEt_2 (R = Me, R' = R'' = Et; R = Me, R' = R'' = Et, Pr, Pr^i; R = Et, R' = R'' = Et; R' = Me, R'' = Et, Pr)	R'R''CNOH	$R_3SiONCR'R''$	[249, 297]
$(C_6F_5)_2MeSiNHPh$	HCl	$(C_6F_5)_2MeSiCl$	[280]
$(C_6F_5)_3SiNR_2$ (R = Me, Et)	HCl	$(C_6F_5)_3SiCl$	[280]
$(C_6F_5)_3SiNMe_2$	$C_6F_5NH_2$	C_6F_5H	[423]
$(EtO)_3SiNMe_2$	HCl	$(EtO)_3SiCl$	[636]
$Me_2NSiMe_2(C_6H_4)SiMe_2NMe_2$	$HOSiMe_2C_6H_4SiMe_2OH$	polymer	[590]

Table of Reactions *(C'ntd.)*

Metal Amide	Reagent	Product[2]	References
$Me_2NMe_2Si(OSiMe_2)_nNMe_2$ (n = 0,1,2; R = p-CH_2=$CHCH_2C_6H_4$)	$RSiMe_2OH$	$R(SiMe_2O)_{n+1}SiMe_2R$	[567]
$F_{4-n}Si(NMe_2)_n$ (n = 1,2)	nHI	I_nSiF_{4-n}	[53]
$R_{4-n}Si(NHPh)_n$ (X = Cl, Br, I)	HX	$R_{4-n}SiX_n$	[234]
$R_{4-n}Si(NHR')_n$	H_2S	$R_{4-n}Si(SH)_n$	[439]
$HMeSi(NEt_2)_2$	$2Et_3SiOH$	$HMeSi(OSiEt_3)_2$	[38]
$HMeSi(NEt_2)_2$ (R = Me, Et)	HC≡CC(Me)(R)OH	HC≡CC(Me)ROSi(H)-$MeNEt_2$	[285]
$Me_2Si(NHPh)_2$ (R = Me, Et)	2ROH	$Me_2Si(OR)_2$	[234]
$Me_2Si(NMe_2)_2$	m-$[Me_2(HO)Si]_2C_2B_{10}H_{10}$	polymer	[302]
$Me_2Si(NEt_2)_2$	$Ph_2Si(OH)_2$	polymer	[584]
$Me_2Si(NEt_2)_2$	$HO(SiMe_2O)_3H$	polymer	[584]
$Me_2Si(NEt_2)_2$	$HO(CH_2)_2O(CH_2)_2OH$	Me_2Si ring (O, O, O)	[39]
$MePhSi(NEt_2)_2$ (R = Me, H; R' = Me, Et, Ph; R'' = Ph, CH_2Ph)	RR'C(OH)CHC(SR'')-C(OH)RR'	Ph(Me)Si ring (O, O; R R'; SR''; R' R)	[524]
$MeRSi(NEt_2)_2$* (R = Me, Ph; R', R'' = Me, Et, Ph, OEt)	H_2S + $R'R''SiCl_2$	$(SSiR'R'')_3$	[442]
R_2Si(NSiR''_3)$_2SiR_2$ (cyclic) (e.g., R = Me = R' = R'')	R'OH	$R''_3SiN[Si(OR')R_2]_2$ + $R_2Si(OR')_2$	[99]
$R_2Si(NMe_2)_2$	Fe(η-$C_5H_4CH_2OH$)$_2$	Fe(η-$C_5H_4CH_2OH$)-(η-$C_5H_4CH_2OSiR_2NMe_2$)	[589]
p-$[R_2(Me_2N)Si]_2C_6H_4$	Fe(η-$C_5H_4CH_2OH$)$_2$	polymer ($\overline{MW}$ 26,900)	[589]
$R_2Si(NHEt)_2$ (R = Me, Ph)	$[HO(Ph)SiO]_4$	R_2Si and SiR_2 bridged by O–Si(Ph)–O–Si(Ph)–O chains, with O linking the Si(Ph) atoms	[42]

*Really a metathetical exchange reaction.

Table of Reactions *(C'ntd.)*

Metal Amide	Reagent	Product2	References
$RR'Si(NMe_2)_2$ ($RR' = Me_2, MePh, Ph_2$)	$HOSiMe_2C_6H_4SiMe_2OH$	polymer	[590]
$MeSi(NMe_2)_3$ ($n = 0, 1, 2$)	$RSiMe_2OH$	$RSiMe_2OSiMe(SiMeR_2)_n$-$OSiMe_2R$	[567]
$(H_3Si)_2NMe$ ($X = S, Se$)	$2H_2X$	$(H_3Si)_2X + 2H_3SiXH.HNMe_2$	[173]
$(H_3Si)_2NMe$ ($X = Cl, Br$)	HX	H_3SiX	[230, 707]
$(H_2MeSi)_2NR$	$4HCl$	$3MeSiH_2Cl$	[223]
$(Me_3Si)_2NH$	$2ROH$	$2Me_3SiOR$	[234]
$3(Me_3Si)_2NH$ (X = …)	$2X(OH)_3$	$2X(OSiMe_3)_3$	[206, 207]
$(Me_3Si)_2NH$	o-$HOC_6H_4CONH_2$	o-$Me_3SiOC_6H_4CONH_2$	[232a]
$(Me_3Si)_2NH$	$H(C{\equiv}C)_2CH_2OH$	$H(C{\equiv}C)_2CH_2OSiMe_3$	[163]
$(Me_3Si)_2NH$ ($R = Me, Et, Bu$)	$(RO)_2POH$	$Me_3SiOP(OR)_2$	[151]
$(Me_3Si)_2NH$ ($R = Me, Et, Bu^t$; $Q = O, S$)	$R_2P(X)QH$	$Me_3SiQP(X)R_2$	[704]
$(Me_3Si)_2NH$	NH_4VO_3	$(Me_3SiO)_3V{=}NSiMe_3$	[65]
$(Me_3Si)_2NH$	$HONCR(CH_2)_nRCNOH$	$(CH_2)_n[RCNOSiMe_3]_2$	[720]
$(Me_3Si)_2NH$ (Y = …)	$HON(Y)NOH$	$Me_3SiON(Y)NOSiMe_3$	[720]
$(Me_2Si)_2NH$	$RR'CNOH + Me_3SiCl$	$Me_3SiONCRR'$	[719]
$(Me_3Si)_2NH$ ($R = Me, Et$)	$2R_2CNOH$	$2Me_3SiONCR_2$	[249]
$(Me_3Si)_2NH$	$H_2S + Me_3SiCl$	$(Me_3Si)_2S$	[441, 442]
$(Me_3Si)_2NCOMe$ ($n = 2, 3$)	HO, $(CH_2)_n$, OH	Me_3SiO, $(CH_2)_n$, $OSiMe_3$	[440]

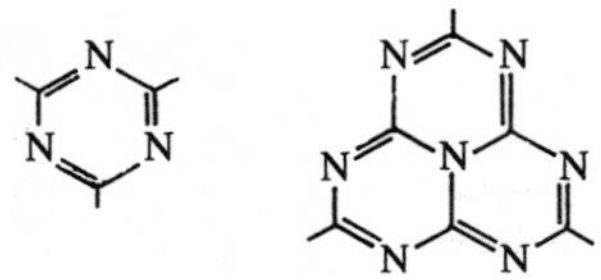

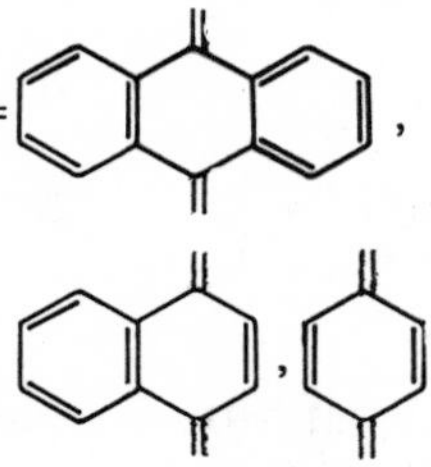

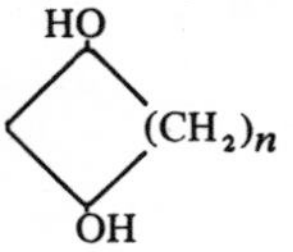

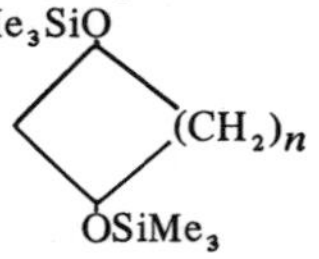

Table of Reactions *(C'ntd.)*

Metal Amide	Reagent	Product2	References
$(Me_3Si)_2NC_6H_4X$ ($X = p$-Me_3Si, p-Et_3Ge, m-Me_3Si)	alk-MeOH	$2Me_3SiOMe$	[730]
$(Me_3Si)_2NH$	o-$(HOOC)_2C_6H_4$	o-$(Me_3SiOOC)_2C_6H_4$	[452]
$(R_3Si)_2NH$	2HCl	R_3SiCl	[234]
$(R_3Si)_2NH$ (R = Me, Et, Ph; oxH = 8-OH-quinoline)	2Hox	R_3Siox	[728]
$(H_3Si)_3N$ (X = S, Se)	H_2X	$(H_3Si)_2X$	[173]
$(H_2MeSi)_3N$	HCl	$MeSiH_2Cl$	[223]
$(Me_3SiNH)_3$	$H_2S + Me_2SiCl_2$	Me_2 S—Si Me_2Si S S—Si Me_2	[442]
$(Me_3SiNH)_3$ (oxH = 8-OH-quinoline)	Hox	Me_3Siox	[728]
$SiMe_3$ $SiMe_3$ N N Me_2Si Si $SiMe_2$ CH N $SiMe_3$ $SiMe_3$	12HCl	$3NH_4Cl + 3Me_3SiCl + Me_2SiCl_2$ $+ C_6H_{16}Cl_4Si_3$	[732c]
Cl_2 Si Cl_3SiN $NSiCl_3$* Si Cl_2	10ROH + $10NEt_3$ (R = Me or Et)	$(OR)_2$ Si $(RO)_3SiN$ $NSi(OR)_3$ Si $(OR)_2$	[734a]
Cl_2 Si Cl_3SiN $NSiCl_3$* Si Cl_2	HNR_2 (R = Me or Et)	Cl NMe_2 Si $Cl(R_2N)_2SiN$ $NSi(NR_2)_2Cl$ Si Me_2N Cl	[734a]
$Me(H)N(SiMe_2)_2NHMe$	H_2O	Me_2 Me_2 Si—Si O O Si—Si Me_2 Me_2	[732b]

*Actually an MCl_n reaction.

Table of Reactions *(C'ntd.)*

Metal Amide	Reagent	Product[2]	References
Germanium Amides			
Me_3GeNMe_2	MeSeH	$Me_3GeSeMe$	[32]
Me_3GeNMe_2 (R = H, $GeMe_3$)	$HC(R)N_2$ + Me_3SnCl	$Me_3GeC(R)N_2$	[407]
Me_3GeNMe_2	$HC(N_2)COOEt$	$M_3GeC(N_2)COOEt$	[422]
Me_3GeNEt_2	BuSH	Me_3GeSBu	[7]
Et_3GeNMe_2	RCOOH	$Et_3GeOOCR$	[446]
Et_3GeNMe_2	Ph_2PH	Et_3GePPh_2	[446]
Et_3GeNMe_2 (X = CO_2Et, CN)	$CH_2(CO_2Et)X$	$Et_3GeCH(CO_2Et)X$ + $Et_3GeOC(NMe_2)CH_2X$	[651a]
$R_3GeNR'_2$ (R = Et, Bu; R' = Me, Et; R'' = Pr^i, Bu^t, Oct)	R''SH	R_3GeSR''	[446, 651]
Bu_3GeNEt_2	PhCCH	$Bu_3GeCCPh$	[446]
Ph_3GeNEt_2	NaCCH + Ph_3SiCl	$Ph_3GeCCSiPh_3$	[236]
Ph_2Ge (four-membered ring with CH_2CH_2 and N–Me)	MeOH	$Ph_2Ge(OMe)CH_2CH_2NHMe$	[621a]
R_3GeNR_2 (e.g., R = Et)	ROH	R_3GeOR	[446]
R_3GeNMe_2	CH_2N_2	$(R_3Ge)_2CN_2$	[422]
R_3GeNMe_2 (R = Et, Ph)	MeCN	R_3GeCH_2CN + $(R_3Ge)_2CHCN$	[621]
$Me_2Ge(NMe_2)_2$	$2(Me_3Ge)_2AsH$	$Me_2Ge[As(GeMe_3)_2]_2$	[35]
$Et_2Ge(NEt_2)_2$	$HO(CH_2)_2NH_2$	Et_2Ge (ring: O–CH_2CH_2–NH)	[650]
$Et_2Ge(NEt_2)_2$ (Q = NH, S)	o-HQC_6H_4COOH	Et_2Ge (ring: Q–C_6H_4–C(=O)–O)	[217]
$Et_2Ge(NEt_2)_2$ (Y = O, S; n = 1, 2, 3)	$HY(CH_2)_nSH$	Et_2Ge (ring: Y–$(CH_2)_n$–S)	[217]
$R_2Ge(NEt_2)_2$ (R = Me, Et, Pr; X = O, S)	$HXCH_2CH_2NH_2$	R_2Ge (ring: X–CH_2CH_2–NH)	[217]
$Et_2Ge(NEt_2)_2$ (n = 2, 3, 4)	$HO(CH_2)_nOH$	Et_2Ge (ring: O–$(CH_2)_n$–O)	[650]

Table of Reactions *(C'ntd.)*

Metal Amide	Reagent	Product2	References
$(R_3Ge)_2NH$	2Hox	$2R_3Geox$	[728]
(R = Me, Et, Ph; oxH = 8-OH-quinoline)			
$(R_2GeNH)_n$	2Hox	$R_2Ge(ox)_2$	[728]
(R = Me, Et, or Ph; oxH = 8-OH-quinoline)			
$EtGe(NEt_2)_3$	6HI	$EtGeI_3$	[30]
Tin(IV) Amides			
Me_3SnNEt_2	RR'CNOH	$Me_3SnONCRR'$	[297]
$2Me_3SnNMe_2$	CH_2N_2	$(Me_3Sn)_2CN_2$	[421, 422, 425]
Me_3SnNMe_2	Me_3SiCHN_2	$(Me_3Sn)(Me_3Si)CN_2$	[422, 426]
Me_3SnNR_2 (R = Et, $SiMe_3$)	$CH(N_2)CO_2Et$	$Me_3SnC(N_2)CO_2Et$	[680]
Me_3SnNMe_2	RCCH	Me_3SnCCR	[347, 352, 354]
[R = Pr, Bu, Ph, $Me(CH_2{=})C$]			
Me_3SnNMe_2	$HC{\equiv}C(CH_2)_3C{\equiv}CH$	$Me_3SnCC(CH_2)_3CCSnMe_3$	[347]
Me_3SnNMe_2	HX (X = Cl, CH_2NO_2, Me_2CNO_2)	Me_3SnX	[352, 354, 455, 462]
Me_3SnNMe_2	MeSeH	$Me_3SnSeMe$	[32]
Me_3SnNEt_2	MeOH	Me_3SnOMe	[27]
Me_3SnNEt_2	$HO(CH_2)_2COOMe$	$Me_3SnO(CH_2)_2COOMe$	[336]
Me_3SnNMe_2	Ph_2AsH	$Me_3SnAsPh_2$	[352, 354]
Me_3SnNMe_2 (R = Me, Ph, $GeMe_3$)	$Me_3GeAs(R)H$	$Me_3SnAs(R)GeMe_3$	[35]
Me_3SnNMe_2	MH_3 (M = P, As, Sb)	$(Me_3Sn)_3M$	[350b]
Me_3SnNMe_2	$RAsH_2$	$(Me_3Sn)_2AsR$	[35]
Me_3SnNEt_2	Me_3SiCCH	$Me_3SnCCSiMe_3$	[236]
Me_3SnNEt_2	$MeO(CH_2)_2COOH$	$Me_3SnOOC(CH_2)_2OMe$	[336]
Me_3SnNEt_2	BuSH	Me_3SnSBu	[7]
Me_3SnNEt_2	$3C_5H_6$	$C_5H_5SnMe_3$	[298, 598]
Me_3SnNEt_2	$Sn(C_5H_5)_2$	$Sn(C_5H_4SnMe_3)_2$	[110a]
$2Me_3SnNEt_2$	C_5H_6	$C_5H_4(SnMe_3)_2$	[298, 598]
Me_3SnNMe_2	CCl_3H	Me_3SnCCl_3	[156, 157]
Me_3SnNMe_2	$ClCHCCl_2$	$Me_3SnCClCCl_2$	[156, 157]
Et_3SnNEt_2	Bu^tOOH	$Et_3SnOOBu^t$	[613]
R_3SnNMe_2 (R = Me, Et)	Ph_2PH	R_3SnPPh_2	[199, 352, 354]
$Et_2Sn(GePh_3)NEt_2$	PhOH	$Et_2Sn(GePh_3)OPh$	[185]
Bu_3SnNR_2 (R = Me, Et; R' = Me, Et, Pr)	MeCONHR'	$Bu_3SnN(R')COMe$	[637]
Bu_3SnNR_2 (R = Me, Et; R' = Me, Et, Pr)	$MeCONR'_2$	$Bu_3SnCH_2CONR'_2$	[637]
Bu_3SnNMe_2	$Et_3Si(C{\equiv}C)_2H$	$Et_3Si(C{\equiv}C)_2SnBu_3$	[261]
Bu_3SnNMe_2	$HC{\equiv}CH$	$Bu_3SnCCSnBu_3$	[354]

Table of Reactions *(C'ntd.)*

Metal Amide	Reagent	Product[2]	References
Bu_3SnNEt_2	PhC(Me)CHCOPh	$PhC(CH_2)CHCPhOSnBu_3$	[104]
Bu_3SnNMe_2	MeCOOH	$Bu_3SnOOCMe$	[144]
Bu_3SnNEt_2	$Me(Bu^t)CO$	$Bu_3SnCH_2COBu^t$ + $Bu_3SnOC(=CH_2)Bu^t$	[591]
R_3SnNMe_2 (R = Me, Bu)	$2Me_2CO$	$(R_3Sn)_2O + Me_2C{=}CHCOMe$	[354]
R_3SnNEt_2 (R = Me, Pr, Bu)	HCl	R_3SnCl	[697]
R_3SnNEt_2 (R = Me, Ph)	NaCCH + R_3GeCl	$R_3SnCCGeR_3$	[236]
Ph_3SnNEt_2	Ph_3GeCCH	$Ph_3SnCCGePh_3$	[236]
Ph_3SnNEt_2	NaCCH + Ph_3SiCl	$Ph_3SnCCSiPh_3$	[236]
R_3SnNMe_2 (R = Et, Ph)	PhCCH	$R_3SnCCPh$	[352, 354, 458]
Ph_3SnNMe_2	IndH	$IndSnPh_3$	[352, 354]
$2R_3SnNMe_2$ (R = Et, Bu, Ph)	CH_2N_2	$(R_3Sn)_2CN_2$	[422]
$R_3SnNR'_2$ (R = Me, Et, Bu, Ph; R' = Me, Et)	$CH(N_2)COR'$	$R_3SnC(N_2)COR'$	[397, 456]
$R_3SnNR'_2$ (R = R' = Et, R'' = R''' = H; R = Bu, R' = Me, R'' = R''' = H; R = R' = Me, R'' = H, R''' = Ph; R = R' = Me, Et, and R'' = R''' = Ph)	R''R'''CHCN	$R_3SnCR''R'''CN$	[354]
R_3SnNMe_2 (R = Et, Bu, Ph)	MeOH	R_3SnOMe	[26, 347, 352, 354]
R_3SnNMe_2 (R = Me, Et)	C_6F_5H	$R_3SnC_6F_5$	[347]
R_3SnNMe_2 (R = Me, Bu, Ph)	C_5H_6	$C_5H_5SnR_3$	[352, 354]
R_3SnNEt_2 [R = Me, Et, Bu; R' = H, Bu_3Sn, CH_2CH, $CH_2C(Me)_n$]	*o*-, *m*-, or *p*-$HCB_{10}H_{10}CR'$	*o*-, *m*-, or *p*-$R'CB_{10}H_{10}CSnR_3$	[486]
R_3SnNMe_2	Bu^tOOH	$R_3SnOOBu^t$	[613]
R_3SnNEt_2 (R = Me, Bu; X = Cl, Br)	CX_3H	R_3SnCX_3	[202]
$Me_2Sn(NMe_2)_2$	$2(Me_3Ge)_2AsH$	$Me_2Sn[As(GeMe_3)_2]_2$	[35]
$Me_2Sn(NMe_2)_2$	$(Me_3Ge)_2AsH$	$Me_2Sn[As(GeMe_3)_2]_2$	[35]
$Bu_2Sn(NMe_2)_2$	HC≡CH	polymer	[354]
$Et_2Sn(NEt_2)_2$	*o*-$(CH_2{=}CR)CB_{10}H_{10}CH$	$Et_2Sn[CB_{10}H_{10}CRCH_2\text{-}o]_2$	[486]
$Et_2Sn(NEt_2)_2$	*o*-$HCB_{10}H_{10}CH$	$[Et_2SnCB_{10}H_{10}C]_n$, polymer	[486]
$Me_nSn(NR_2)_{4-n}$	$Me_2S(NH)_2$	$Me_2S[NSnMe_n(NR_2)_{3-n}]_2$	[278]

Table of Reactions *(C'ntd.)*

Metal Amide	Reagent	Product[2]	References
$R_{4-n}Sn(NR'_2)_n$ (R = Me, Et, Bu; R' = Me, Et, n = 1, 2)	nHC(N_2)COOEt	$R_{4-n}Sn[C(N_2)COOEt]_n$	[422, 456]
$R_{4-n}Sn(NEt_2)_n$ (R = Me, Et, Bu, Ph; n = 2, 3)	2nHCN 2nHN_3 2nPhCCH	$R_{4-n}Sn(CN)_n$ $R_{4-n}Sn(N_3)_n$ $R_{4-n}Sn(CCPh)_n$	[460] [460] [458]
$Et_{3-n}Sn(GePh_3)(NEt_2)_n$ (n = 1, 2)	nPhCCH	$Et_{3-n}Sn(GePh_3)(CCPh)_n$	[185]
$Sn(NEt_2)_4$	4Pr^iOH	$Sn(OPr^i)_4$	[715]
Tin(II) Amides			
$Sn[N(SiMe_3)_2]_2$ (A = OAc, OEt, Cl, C_5H_5)	2HA	$\frac{1}{x}[SnA_2]_x$	[295]
$Sn[N(SiMe_3)_2]_2$ (R = Bu^t, Ph, M = P; R = Ph, M = As)	R_2MH	$\frac{1}{x}[Sn(MR_2)_2]_x$	[594]
$Sn[N(SiMe_3)_2]_2$	PhSH	$\frac{1}{x}[Sn(SPh)_2]_x$	[428]
$Sn[N(SiMe_3)_2]_2$ (Ar = 2, 6-Bu^t_2-4-MeC_6H_2)	2ArOH	$Sn(OAr)_2$	[141a]
$Sn(NMe_2)_2$	EtOH	$\frac{1}{x}[Sn(OEt)_2]_x$	[246a]
$Sn(NMe_2)_2$	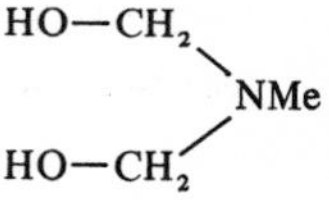 HO—CH_2 NMe HO—CH_2	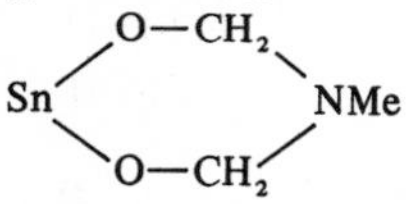 O—CH_2 Sn NMe O—CH_2	[246a]
Lead Amides			
Me_3PbNEt_2 (M = Si, Ge, Sn)	NaCCH + Ph_3MCl	$Me_3PbCCMPh_3$	[236]
Me_3PbNEt_2	Me_2CHNO_2	$Me_3PbON(O)CMe_2$	[462]
Me_3PbNEt_2	$CH(N_2)CO_2Et$	$Me_3PbC(N_2)CO_2Et$	[680]
$Me_3PbN(SiMe_3)_2$ (R = CO_2Et, Ac, Bz, Me_3Pb)	$CH(N_2)R$	$Me_3PbC(N_2)R$	[284, 421a]
$Me_3PbN(SiMe_3)_2$	CH_2N_2	$(Me_3Pb)_2CN_2$	[422]
$Pb[N(SiMe_3)_2]_2$ (Ar = 2, 6-Bu^t_2-4-MeC_6H_2)	2ArOH	$Pb(OAr)_2$	[141a]
Arsenic, Antimony, and Bismuth Amides			
Me_2AsNMe_2 (X = X' = O, S; X = O, X' = S; R = R' = H, Me; R = H, R' = CH_2Cl; $-CR_2CR'_2$ 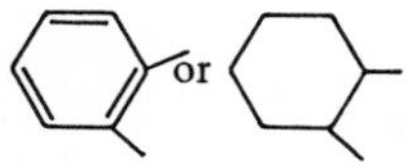 = or)	$HXCR_2CR'_2X'H$	$Me_2AsXCR_2CR'_2X'AsMe_2$	[307]
Me_2AsNMe_2 (n = 2, 3)	MeAs($ONCMe_2$)O-(CH_2)$_n$OH	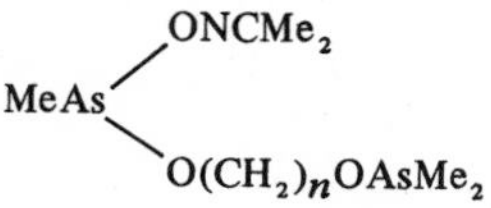 $ONCMe_2$ MeAs O(CH_2)$_n$O$AsMe_2$	[359]

Table of Reactions *(C'ntd.)*

Metal Amide	Reagent	Product[2]	References
$2Me_2AsNMe_2$	$MeC(CH_2OH)_3$	$Me(HOCH_2)C(CH_2OAsMe_2)_2$	[639]
$3Me_2AsNMe_2$	$N(CH_2CHROH)_3$	$N(CH_2CHROAsMe_2)_3$	[386]
Me_2AsNMe_2	$MeAs(SCH_2CH_2OH)_2$	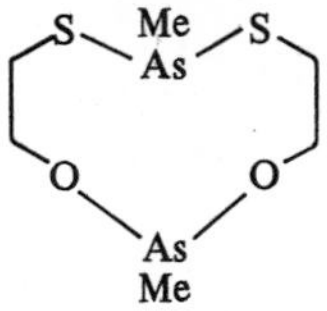 ring: S–As(Me)–S, O–As(Me)–O	[386]
$2Me_2AsNMe_2$	$MeAs(SCH_2CH_2OH)_2$	$MeAs(SCH_2CH_2OAsMe_2)_2$	[386]
Me_2AsNMe_2 (X = O, R = Me, Et, Bu, Ph; X = S, R = Et, Pr, Ph)	RXH	Me_2AsXR	[384]
Me_2AsNMe_2	$CH_2N_2 + Me_3SnCl$	$Me_2AsC(H)N_2$	[402]
Me_2AsNMe_2	$(CH_2SH)_2$	$(Me_2AsSCH_2)_2$	[384]
$2Me_2AsNMe_2$	R·CNOH | R·CNOH	$RCNOAsMe_2$ | $RCNOAsMe_2$	[362]
$2Me_2AsNMe_2$	R_2COH | RCNOH	$R_2COAsMe_2$ | $RCNOAsMe_2$	[362]
Me_2AsNMe_2 (R = R' = Me, Et, Pr; RR' = C_6H_{10}, C_5H_8, C_3H_8, $C_{12}H_{22}$; RR' = MeH, BuMe, PhMe, PhH; R = Me, R' = MeCO, PhCO)	RR'CNOH	$RR'CNOAsMe_2$	[360] [362]
$Me_2AsNMeR$	$HC(N_2)R'$	$Me_2AsC(N_2)R'$	[405]
R_2AsNEt_2 (R = Me, Et, Pr; X = O, S, Se; R' = Me, Et, Ph, CH_2Ph)	HXR'	R_2AsXR'	[645]
Me_2AsNMe_2*	$H^{15}NMe_2$	$Me_2As^{15}NMe_2$	[20]
$(CF_3)_2AsNMe_2$	2HCl	$(CF_3)_2AsCl$	[191]
$MeAs(NMe_2)_2$*	$H^{15}NMe_2$	$MeAs(^{15}NMe_2)_2$	[20]
$EtRAsNH_2$ (R = Ph, *m*-Tol)	HCl	EtRAsCl	[768a]
$RAs(NR'_2)_2$	$(CH_2SH)_2$	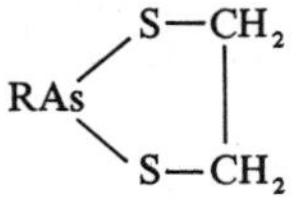 RAs(S–CH_2)(S–CH_2) ring	[34]
$MeAs(X)NMe_2$ (X = NMe_2, Y = F, Cl, Br, I; X = F, Cl, Br, I; Y = Cl, Br)	HY	MeAs(X)Y	[363]
$R_2AsNR'_2$ (R = R' = Me; X = F, Cl, Br, I; R = F; R' = Me, X = Cl; R = R' = Et; X = Cl)	HX	R_2AsX	[384] [718a]

*These are transaminations

Table of Reactions *(C'ntd.)*

Metal Amide	Reagent	Product2	References
$(MeAsNR_2)_2CH_2$ (X = OR', SR'; R' = Me, Et, Bu^t, Ph)	HX	$(MeAsX)_2CH_2$	[383]
$(Me_2As)_2NR$	2R'OH	$2Me_2AsOR'$	[379]
$(Me_2As)_2NR$ (X = F, Cl, Br, I)	2HX	$2Me_2AsX$	[379]
$MeAs(NMe_2)_2$	$HC(N_2)CO_2Et$	$MeAs[C(N_2)CO_2Et]_2$	[403]
$MeAs(NMe_2)_2$	$2HS(CH_2CH_2)OH$	$MeAs(SCH_2CH_2OH)_2$	[386]
$CF_3As(NMe_2)_2$ (Y = OMe, OPh, SPh)	2HY	CF_3AsY_2	[20]
$CF_3As(NMe_2)_2$	HCl	CF_3AsCl_2	[20]
$MeAs(NR_2)_2$ (R = Me, Et)	CH_2SH–CH_2SH	MeAs(–S–CH_2–CH_2–S–) (ring)	[380]
$MeAs(NR_2)_2$ (R = Me, Et; R' = Et, Pr, Ph)	R'SH	$MeAs(SR')_2$	[380]
$RAs(NR'_2)_2$ (R = R' = Me; R = Ph, R' = Et)	CH_2OH–CH_2OH	RAs(–O–CH_2–CH_2–O–) (ring)	[34]
$MeAs(NR_2)_2$ (R = Me, Et; R' = Me, Et, Ph)	R'OH	$MeAs(OR')_2$	[380]
$RAs(NMe_2)_2$ (R = Me, NMe_2; R' = H, Me, $CH_2CHR''OH$; R'' = H, Me)	$R'N(CH_2CHR''OH)_2$	R'N($CH_2CHR''O$)$_2$AsR (ring)	[386]
Me–N(ring, –O–)AsNMe$_2$	Me_2CNOH	Me–N(ring, –O–)AsONCMe$_2$	[359]
$[(R_2N)_2As]_2CH_2$ [R = alkyl; R_2N = $\overline{N(CH_2)_4}$, $\overline{N(CH_2)_5}$]	HCl	$[Cl_2As]_2CH_2$	[381]
$[(R_2N)_2As]_2CH_2$ (R = alkyl, R' = Me, Et; R_2N = $\overline{N(CH_2)_4}$, $\overline{N(CH_2)_5}$)	4R'OH	$[(R'O)_2As]_2CH_2$	[381]
$Me_nAs(NMe_2)_{3-n}$ (n = 1, 2)	Fe(η-C_5H_5)-(η-$C_5H_4CH_2OH$)	{Fe(η-C_5H_5)(η-$C_5H_4CH_2O$)}$_{3-n}$ – $AsMe_n$	[385]
$As(NMe_2)_3$	$HC(N_2)CO_2Et$	$(Me_2N)_2AsC(N_2)CO_2Et$	[403]
$As(NMe_2)_3$	$CH_2(CH_2OH)_2$	Me_2NAs(–O–CH_2–CH_2–CH_2–O–) (ring)	[34]

Table of Reactions *(C'ntd.)*

Metal Amide	Reagent	Product[2]	References
$As(NMe_2)_3$ (R = Me, CH_2OH)	$RC(CH_2OH)_3$	$As(-O-CH_2-)_3CR$ (cyclic: As bonded to three O–CH_2 groups joined at CR)	[639]
$As(NMe_2)_3$ (n = 1, 2, or 3)	$n Me_2AsOCH_2CH_2OH$	e.g., $Me_2As-O(CH_2)_2-OAs<(O-CH_2)(O-CH_2)$ (cyclic)	[638]
$As(NEt_2)_3$	CH_2OH–$C(Me)HOH$	$Et_2NAs<(O-CH_2)(O-CHMe)$ (cyclic)	[34]
Me_2MNMeR (M = As, Sb; R = H, Me; R′ = H, CO_2Et; R″ = Me_2M, CO_2Et; M = Bi, R = Me_3Si, R′ = H, CO_2Et, R″ = Me_2Bi, CO_2Et)	$HC(N_2)R'$	$Me_2MC(N_2)R'$	[405]
Me_2MNMeR (M = Sb, R = Me; M = Bi, R = Me_3Si)	C_5H_6	$Me_2MC_5H_5$	[404]
Transition Metal Amides			
$Ti(NMe_2)_3$ (R = Me, Et)	ROH	$Ti(OR)_3$	[25, 432]
$Ti(NMe_2)_3$	$MeCOCH_2COMe$ [i.e., (acacH)]	$Ti(acac)_3$	[25]
$Ti(NMe_2)_3$	PhOH	$Ti(OPh)_4.HNMe_2$	[432]
$Ti(NMe_2)_3$	o-$C_6H_4(OH)_2$	$Ti_2(C_6H_4O_2)_3.2HNMe_2$	[432]
$Ti(NMe_2)_3$	C_5H_6	$Ti(\eta-C_5H_5)(NMe_2)_2$	[25, 431]
$Ti(\eta-C_5H_5)_{3-n}(NMe_2)_n$ (n = 1, 2)	nEtOH	$Ti(\eta-C_5H_5)_{3-n}(OEt)_n$	[25, 432]
$Ti(\eta-C_5H_5)_2NMe_2$	PhXH (X = O, S)	$Ti(\eta-C_5H_5)_2(XPh)$	[432]
$Ti(NMe_2)_4$ (R = Et, Pr^i; n = 1, 2)	nRSH	$(RS)_nTi(NMe_2)_{4-n}$	[94]
$Ti(NMe_2)_4$ (R = Me, n = 0.55; R = Et, n = 0.67)	4RSH	$Ti(SR)_4.(HNMe_2)_n$	[94]
$Ti(NMe_2)_4$	Me_2CO	$Ti(OH)_{4-n}(NMe_2)_n$ + $Me_2C{=}CHCOMe$	[144]
$Ti(NMe_2)_4$	$4Ph_2C{=}NH$	$Ti(NCPh_2)_4$	[164]
$Ti(NEt_2)_4$	4EtSH	$Ti(SEt)_4$	[147]
$Ti(NEt_2)_4$	$4Et_3SiOH$	$Ti(OSiEt_3)_4$	[715]
$Ti(NPr_2)_4$	4BuOH	$Ti(OBu)_4$	[715]
$Ti(NR'_2)_4$ (R = Et, Bu, Am; R′ = Me, Et)	4ROH	$Ti(OR)_4$	[95]

Table of Reactions *(C'ntd.)*

Metal Amide	Reagent	Product[2]	References
$Ti(NR_2)_4$ (R = Me, Et)	C_5H_6	$Ti(\eta\text{-}C_5H_5)(NR_2)_3$	[147]
$Ti(NMe_2)_4$	C_5H_5Me	$Ti(\eta\text{-}C_5H_4Me)(NMe_2)_3$	[147]
$Ti(NR_2)_4$ [NR_2 = NMe_2, NEt_2; R′, R″ = H, alkyl, Me_3Si, Me_3Ge]	$C_5H_4R'R''$	$Ti(\eta\text{-}C_5H_3R'R'')(NR_2)_3$	[116,147]
$Ti(\eta\text{-}C_5H_5)(NMe_2)_3$	$3Pr^iOH$	$Ti(\eta\text{-}C_5H_5)(OPr^i)_3$	[147]
$Ti(\eta\text{-}C_5H_5)(NMe_2)_3$	$4Pr^iOH$	$Ti(OPr^i)_4$	[147]
$Ti(OPr^i)_3NMe_2$	C_5H_6	$Ti(\eta\text{-}C_5H_5)(OPr^i)_3$	[147]
$Ti(OPr^i)_3(NEt_2)$ (A = Pr^iO, CF_3CH_2O, EtS, C_5H_5)	HA	$Ti(OPr^i)_3A$	[157b]
$TiCl_2(NMe_2)_2$	Me_3PCH_2	$(Me_2N)_2\overline{TiC(PMe_3)Ti(NMe_2)_2C}(PMe_3)$	[669a]
$Zr(NMe_2)_4$	C_5H_6	$Zr(\eta\text{-}C_5H_5)(NMe_2)_3$	[147]
$Zr(NMe_2)_4$	excess C_5H_5Me	$Zr(\eta\text{-}C_5H_4Me)_2(NMe_2)_2$	[147]
$Zr(NMe_2)_4$	2IndH	$Zr(Ind)(NMe_2)_3$	[147]
$Zr(NEt_2)_4$	$4Et_3SiOH$	$Zr(OSiEt_3)_4$	[715]
$Zr(NR_2)_4$ (R = Me, Et)	excess C_5H_6	$Zr(\eta\text{-}C_5H_5)_2(NR_2)_2$	[147]
$Zr(\eta\text{-}C_5H_5)_2(NMe_2)_2$	$2Me_3SiCHN_2$	$Zr(\eta\text{-}C_5H_5)_2[C(N_2)SiMe_3]_2$	[426]
$Zr(\eta\text{-}C_5H_5)_2(NMe_2)_2$	$2Bu^tOH$	$Zr(\eta\text{-}C_5H_5)_2(OBu^t)_2$	[147]
$M(\eta\text{-}C_5H_5)_2(NMe_2)_2$ (M = Zr, Hf)	$Ph_2C{=}NH$	$M(\eta\text{-}C_5H_5)_2(NCPh_2)_2$	[164]
$M(\eta\text{-}C_5H_5)_2(NMe_2)_2$ (M = Zr, Hf)	Bu^tSH	$M(\eta\text{-}C_5H_5)_2(SBu^t)_2$	[147]
$M(\eta\text{-}C_5H_4R)_2(NMe_2)_2$ (M = Zr, Hf, and R = H; M = Zr and R = Me)	PhCCH	$M(\eta\text{-}C_5H_4R)_2(CCPh)_2$	[347]
$Hf(NR_2)_4$ (R = Me or Et)	C_5H_6	$Hf(\eta\text{-}C_5H_5)_2(NR_2)_2$	[147]
$Hf(\eta\text{-}C_5H_5)_2(NMe_2)_2$	Bu^tOH	$Hf(\eta\text{-}C_5H_5)_2(OBu^t)_2$	[147]
$V(NMe_2)_4$ (R = Pr^i, Bu^t)	4ROH	$V(OR)_4$	[25,715]
$V(NEt_2)_4$	$4Et_3SiOH$	$V(OSiEt_3)_4$	[715]
$V(\eta\text{-}C_5H_5)_2N(SiMe_3)_2$	HCl	$V(\eta\text{-}C_5H_5)_2Cl$	[614]
$V(\eta\text{-}C_5H_5)_2N(SiMe_3)_2$	MeCOOH	$V(\eta\text{-}C_5H_5)(OOCMe)_2$	[614]
$V(\eta\text{-}C_5H_5)_2N(SiMe_3)_2$	MeOH	$C_5H_6 + HN(SiMe_3)_2$ + unidentified products	[614]
$V(O)(OPr^i)_2(NEt_2)_2$ (R = Pr^i or CF_3CH_2)	RSH	$V(O)(SR)_3 + V(O)(OPr^i)_3$	[157a]
$Nb(NMe_2)_5$	C_5H_6	$Nb(\eta\text{-}C_5H_5)(NMe_2)_3$	[347]
$Nb(NEt_2)_4$ (R = Pr^i, Bu^t)	4ROH	$Nb(OR)_4$	[715]
$Nb(NPr_2)_4$	$4Et_3SiOH$	$Nb(OSiEt_3)_4$	[715]
$Ta(NEt_2)_3NEt$	Et_3SiOH	$Ta(OSiEt_3)_5$	[715]

Table of Reactions *(C'ntd.)*

Metal Amide	Reagent	Product[2]	References
$Ta(NPr_2)_3NPr$	ROH	$Ta(OR)_5$	[715]
(R = Et, Bu)			
$Cr(NEt_2)_4$	$4R_3SiOH$	$Cr(OSiR_3)_4$	[56]
$Mo_2(NMe_2)_6$	$6Bu^tOH$	$Mo_2(OBu^t)_6$	[155b]
$Mo(NMe_2)_4$	ROH	$Mo(OR)_4$	[155d]
(R = Me, Et, Pr^i, Bu^t, or Bu^tCH_2)			
$Mo(NMe_2)_4$	R_3SiOH (R = Me or Et)	$Mo(OSiR_3)_4(HNMe_2)_2$	[155d]
$MoEt_2(NMe_2)_4$	$6Bu^tOH$	$MoEt(OBu^t)_5$	[155e]
$W_2(NRR')_6$	R″OH	$[W(OR'')_3]_x$	[153]
(R″ = small alkyl)			
$W_2(NRR')_6$	R″OH	$W_2(OR'')_6$	[153]
(R = R′ = Et; R = Me, R′ = Et; R″ = Bu^t, $SiMe_3$)			
$W(NMe_2)_6$	ROH, 20° C	$W(OR)_6$	[90a]
(R = Me, Et, Pr)			
$W(NMe_2)_6$	MeOH, 10° C	$W(NMe_2)_3(OMe)_3$	[90a]
$W(NMe_2)_6$	Bu^tOH, C_6H_6, reflux	$OW(OBu^t)_4$	[90a]
$W(NMe_2)_6$	RSH	$W(SR)_3$	[90a]
(R = Me, Ph)			
$M[N(SiMe_3)_2](PPh_3)_2$	ArOH	$M(OAr)(PPh_3)_2$	
(M = Ni, Ar = 2, 6-Bu^t_2-4-MeC_6H_2;			[141c]
M = Rh, Ar = 2, 6-Bu^t_2-4-MeC_6H_2)		$Rh(\eta^5\text{-}OAr)(PPh_3)_2$	[141d]
$U(\eta\text{-}C_5H_5)_2(NEt_2)_2$	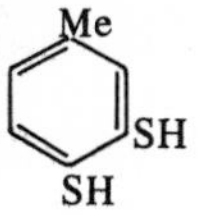 Me, SH, SH	$(\eta\text{-}C_5H_5)_2U$ 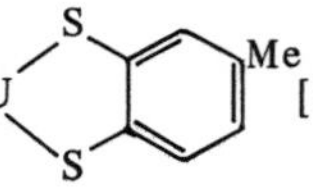 S, S, Me	[342a]
$U(\eta\text{-}C_5H_5)_2(NEt_2)_2$	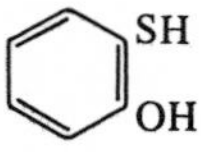 SH, OH	$(\eta\text{-}C_5H_5)_2U$ 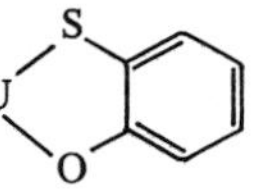 S, O	[342a]
$U(\eta\text{-}C_5H_5)_2(NEt_2)_2$	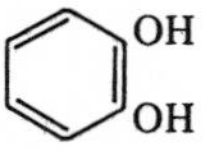 OH, OH	$(\eta\text{-}C_5H_5)_2U$ 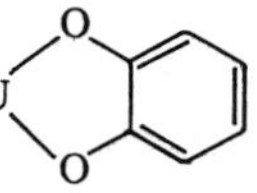 O, O	[342a]
$U(\eta\text{-}C_5H_5)_2(NEt_2)_2$	$HS(CH_2)_2SH$	$(\eta\text{-}C_5H_5)_2U$ 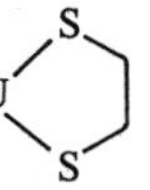 S, S	[342a]
$U(\eta\text{-}C_5H_5)_2(NEt_2)_2$	EtSH	$(\eta\text{-}C_5H_5)_2U(SEt)_2$ + $(\eta\text{-}C_5H_5)_3USEt$	[342a]
$U(\eta\text{-}C_5H_5)_2(NEt_2)_2$	Bu^tSH	$(\eta\text{-}C_5H_5)_2U(SBu^t)_2$	[342a]
$U(NEt_2)_4$	$2C_5H_6$	$U(\eta\text{-}C_5H_5)_2(NEt_2)_2$	[342a]
$U(NEt_2)_4$	4ROH	$U(OR)_4$	[357]
(R = Me, Et)			

Table of Reactions *(C'ntd.)*

Metal Amide	Reagent	Product[2]	References
Group 2 Metal Amides			
$M[N(SiMe_3)_2]_2$ (M = Zn, Cd, Hg; R = H, CO_2Et)	$CH(R)N_2$	$M[C(N_2)R]_2$	[421a]
$MeHgN(SiMe_3)_2$	ArOH	MeHgOAr	[269a]
$MeHgN(SiMe_3)_2$	NH_3	$N(HgMe)_3$	[204a]
$MeHgN(SiMe_3)_2$	Cl_3SiH	MeHgCl + $Cl_2(H)SiN(SiMe_3)_2$	[269b]
$MeHgN(SiMe_3)_2$ (X = H, F)	$CX_3COCH_2COCH_3$	$MeHgCH(COMe)COCX_3$	[157c]
$MeHgN(SiMe_3)_2$	$HNP(NMe_2)_3$	$MeHgNP(NMe_2)_3$	[461]
$RHgN(SiMe_3)_2$ (R = Me, Et; Y = C_5H_5, N_3, HNCN)	HY	RHgY	[462,463]
$RHgN(SiMe_3)_2$ (R = Me, Et; R' = H, CO_2Et, RHg)	$CH(R')N_2$	$RHgC(N_2)R'$	[421a]
$Hg[N(SiMe_3)_2]_2$ (R = Me, Bu^t)	$2CH_2(COR)_2$	$Hg[CH(COR)_2]_2$	[22]
$M[N(SiMe_3)_2]_2$ (M = Cd, Hg)	$(C_6F_5)_3SiH$	$M[Si(C_6F_5)_3]_2$	[269b]

Footnotes to Table:

1. Reactions with water are not included.
2. The liberated amine is not listed.

12

Reactions with Metal Hydrides

A DISCUSSION

The reactions of metal amides discussed in this chapter developed naturally from those described in Chapter 11. Having found that protic compounds react with amides in the sense of Eq. (1) [352], it was logical to examine the case [Eq. (2)] of the protic compound being a metal hydride and then to extend the study to neutral and even basic hydrides [132]. The most detailed investigations have been of amides of tin and the earliest publications were in the mid 1960s [132, 183, 516, 700]; data are available also for amides of Si^{IV}, Ge^{IV}, Sn^{II}, Pb^{IV}, Ti^{III}, Ti^{IV}, Zr^{IV}, and Hg^{II} (see Table). The reactions are important in providing a rather general route to compounds having metal-metal bonds and to polymetal (mainly polystannanes) compounds of specific chain-branching. The reactions are frequently essentially quantitative [131], and relatively mild reaction conditions are required; if a dimethylamide is used, evolution of the gaseous $HNMe_2$ is an advantage for the subsequent purification of the metal-metal bonded compound. Interaction of amido- and hydrido-metal compounds need not lead to metal-metal bonded products but may instead give rise to ligand exchange [as with Me_3SnNMe_2 and $Me_2Si(H)Cl$] (see Chapters 13 and 14) or dehydrochlorination {as with Me_3SnNMe_2 and $[IrHCl_2(PPh_3)_3]$} [133] (see Chapter 14). As for the latter, the reaction between a Pt^{II} hydride and a tin amide (but not the analogous less reactive Si or Ge compound) was originally formulated [Eq. (3)] as a deamination [131, 132] but is now believed to be [221a] a dehydrochlorination followed by oxidative addition [see paragraph after Eq. (14)].

$$LM{-}N{=} + HA \longrightarrow LM{-}A + HN{=} \qquad (1)$$

$$LM{-}N{=} + HM'L' \longrightarrow LM{-}M'L' + HN{=} \qquad (2)$$

$$\left[\begin{matrix} Ph_3P & & H \\ & Pt & \\ Cl & & PPh_3 \end{matrix}\right] + Me_3SnNMe_2 \xrightarrow{-HNMe_2} \begin{cases} \not\longrightarrow [Pt(Cl)(PPh_3)_2(SnMe_3)] \\ \longrightarrow cis\text{-}[Pt(Me)(PPh_3)_2(SnMe_2Cl)] \end{cases} \qquad (3)$$

The reactivity of hydrides R_3Sn–H, with respect to reactions of the type shown in Eq. (2) with tin amides, increases in the order $R = Bu^s < Bu^i < Bu < Ph$ [517] and increases in the series $R_3Sn_2R_2H < R_3SnH < R_2SnH_2$ [701]. Amide $R_3SnNR'R''$ reactivity increases with increasing basicity of nitrogen: i.e. $R_3SnNHR' < R_3SnNR'_2$ [517]; $(Me_3Sn)_2NPr^i$ is also highly reactive. A polar mechanism, with polarisation in the sense $\equiv \overset{\delta-}{Sn}-\overset{\delta+}{H}$ is therefore indicated [184, 188]. In polar solvents an S_E2 mechanism, and in non-polar an S_F2 (four-centre) mechanism, was proposed [188], and possible transition states are shown in (I) and (II).

$$\begin{array}{c} \overset{\delta+}{R_3Sn}\text{-----}\overset{\delta-}{N}< \\ \vdots\ \delta+ \\ H \\ \vdots\ \delta- \\ R_3Sn \end{array} \qquad \begin{array}{c} \overset{\delta+}{R_3Sn}\text{-----}\overset{\delta-}{N}< \\ \vdots\ \delta- \qquad \vdots\ \delta+ \\ R_3Sn\text{-----}H \end{array}$$

(I) **(II)**

Exchange reactions of the type shown in Eq. (4a) may sometimes be competing processes for the Sn–Sn forming reactions [Eq. (4b)], but are not [188] (cf. [517, 701]) implicated in the reaction sequence leading to Sn–Sn bond formation.

$$R_3Sn{-}H + Et_3Sn{-}N< \begin{cases} \xrightarrow{(a)} Et_3Sn{-}H + R_3Sn{-}N< \\ \xrightarrow{(b)} Et_3Sn{-}SnR_3 + HN< \end{cases} \tag{4}$$

For the synthesis of assemblies of more than two tin atoms, the use of dihydrides R_2SnH_2 rather than diamides $R'_2Sn(NR''_2)_2$ is preferred, as in Eqs. (5), (6) and (7) [701]. This is partly due to a question of relative reactivities, but also due to a competing exchange reaction in the $R_3SnH/R'_2Sn(NEt_2)_2$ system, leading to $R'_2Sn(H)NEt_2 + R_3SnNEt_2$.

$$R_2SnH_2 + 2R'_3SnNEt_2 \longrightarrow R'_3Sn{-}\underset{\underset{R}{|}}{\overset{\overset{R}{|}}{Sn}}{-}SnR'_3 + 2HNEt_2 \tag{5}$$

$$HSnR_2SnR_2H + 2R'_3SnNEt_2 \longrightarrow R'_3Sn{-}\left[\underset{\underset{R}{|}}{\overset{\overset{R}{|}}{Sn}}\right]_2{-}SnR'_3 + 2HNEt_2 \tag{6}$$

The tin amide $(R_3SnNR'_2)$–metal hydride reaction has been extended to Ph_3GeH [Eqs. (8) and (9)]; R_3GeH and Ar_3SiH proved unreactive [517]. Phenylgermanes containing electron-withdrawing groups $PhGeX_2H$ (X = Cl or Br) react with $PhGe(NMe_2)_3$ to yield the appropriate digermane [618].

Compounds with more than two metal (e.g., Sn) [185, 187, 512] atoms, may be prepared by the reactions of dihydrides and compounds of the type $R_{4-n}M(NR'_2)_n$ [Eq. (7)]. By varying the ratio of the reagents it is possible to achieve hydrogenolysis of all the Sn-N bonds or to prepare compounds which contain not only the stable Ge-Sn or Sn-Sn bond, but also the labile Sn-N or Sn-H bonds which then easily undergo their normal chemical reactions [e.g., Eqs. (8) and (9)] [185]. Similar reaction sequences have led to the formation of a tin-zirconium compound with nine inter-joined metal atoms [Eq. (10)] and a polymer with alternating zirconium and tin atoms in the chain [Eq. (11)] [187]. Instead of a metal amide, other metal-nitrogen compounds, such as that shown in Eq. (10) or a ketimide, e.g., $Me_3SnN{=}C(CF_3)_2$ [424a], may be used as substrates for metal-metal bond formation in the sense of Eq. (2); reactivity decreases in the sequence $Me_3SnNMe_2 \approx Me_3SnN{=}CBu^t_2 > Me_3SnN{=}CAr_2 > Me_3SnN{=}C(CF_3)_2$ [424a]. Side-reactions including ketimide/H exchange [141b] and dehydrochlorination [424a] may take place: in the $Me_3SnN{=}C(CF_3)_2$/*trans*-$[PtCl(H)(PPh_3)_2]$ system leading to $[Pt\{\eta\text{-}HN{=}C(CF_3)_2\}(PPh_3)_2]$ and *trans*-$[Pt\{N{=}C(CF_3)_2\}(PPh_3)_2(SnMe_3)]$, respectively.

$$2R_3SnNEt_2 + R'_2SnH_2 \longrightarrow R_3Sn{-}SnR'_2{-}SnR_3 + 2HNEt_2 \qquad (7)$$

$$Et_2Sn(NEt_2)_2 + Ph_3GeH \longrightarrow \underset{\textbf{(III)}}{Ph_3Ge{-}Sn(Et)_2NEt_2} + HNEt_2 \qquad (8)$$

$$\textbf{(III)} \longrightarrow \begin{cases} \xrightarrow{Ph_2GeH_2} (Ph_3Ge{-}SnEt_2)_2GePh_2 + 2HNEt_2 \\ \xrightarrow{PhOH} Ph_3Ge{-}Sn(Et)_2OPh + HNEt_2 \\ \xrightarrow{HC\equiv CPh} Ph_3Ge{-}Sn(Et)_2C{\equiv}CPh + HNEt_2 \end{cases} \qquad (9)$$

$$\left.\begin{array}{l} [Zr(NEt_2)_4] + 3Ph_3SnH \longrightarrow [ZrNEt_2(SnPh_3)_3] + 3HNEt_2 \\ [ZrNEt_2(SnPh_3)_3] + HN(Ph)COH \longrightarrow [Zr\{N(Ph)COH\}(SnPh_3)_3] \\ \qquad\qquad + HNEt_2 \\ 2[Zr\{N(Ph)COH\}(SnPh_3)_3] + Ph_2SnH_2 \longrightarrow SnPh_2[Zr(SnPh_3)_3]_2 \end{array}\right\} \qquad (10)$$

$$\left.\begin{array}{l}[Zr(NEt_2)_4] + 2Ph_3SnH \longrightarrow [Zr(NEt_2)_2(SnPh_3)_2] + 2HNEt_2 \\ [Zr(NEt_2)_2(SnPh_3)_2] + 2HN(Ph)COH \longrightarrow [Zr\{N(Ph)COH\}_2(SnPh_3)_2] + 2HNEt_2 \\ n[Zr\{N(Ph)COH\}_2(SnPh_3)_2] + nPh_2SnH_2 \\ \qquad \longrightarrow \left[-Zr(SnPh_3)_2-Sn(Ph)_2- \right]_n + 2HN(Ph)C(H){=}O \end{array}\right\} \quad (11)$$

Reaction (2) has been used to prepare compounds containing Si, Ge, or Sn bonded to Cr, Mo, or W [131]; Sn bonded to Hg [222], Ti [187], Zr [187], Pt, and Mn [131, 132, 221a]; and Ti^{IV} bonded to Mo [131, 132]. These are exemplified by Eqs. (3), (12), (13), and (14). Other examples relate to Mo bonded to Sn^{II} [594a] or Ti^{III} [432].

$$[M(\eta{-}C_5H_5)(CO)_3H] + R_3M'{-}NMe_2 \xrightarrow[M' = \text{Si, Ge, or Sn})]{(M = \text{Cr, Mo, or W};} [M(\eta{-}C_5H_5)(CO)_3{-}M'R_3] + HNMe_2 \quad (12)$$

$$2R_3SnH + Hg[N(SiMe_3)_2]_2 \longrightarrow Hg(SnR_3)_2 + 2HN(SiMe_3)_2 \quad (13)$$

$$[Mo(\eta{-}C_5H_5)(CO)_3H] + [Ti(NMe_2)(OPr^i)_3] \longrightarrow [Mo(\eta{-}C_5H_5)(CO)_3{-}Ti(OPr^i)_3] + HNMe_2 \quad (14)$$

The transition metal hydride need not necessarily be protic, but the reaction conditions are less severe if this is the case [131, 132]. Amides of tin are more reactive than those of silicon or germanium [131, 370]; consistent with this, reactivity orders for the system $[M(\eta{-}C_5H_5)(CO)_3H]/Me_3M'{-}NMe_2$ are M=Cr>Mo>W and M′=Sn>Ge>Si. In the reaction of an amide with a metal hydride containing halogen, further elimination of metal halide is possible. Thus, $[RhH(PF_3)_4]$ and Me_3SnNMe_2 initially probably afford $[Rh(PF_3)_4{-}SnMe_3]$ but subsequently Me_3SnF [132]. This type of reaction is closely related to those described in Chapter 13. The reaction with $[CoH(PF_3)_4]$ and Me_3SiNMe_2 gives the salt $[Me_3SiHNMe_2]^+[Co(PF_3)_4]^-$ [69]. Some reactions have been carried out with lower valent amides of Ti using Ph_3SnH or $[M(\eta{-}C_5H_5)(CO)_3H]$ (M = Mo or W) [432]. These appear to give the appropriate d^1 or d^2 Ti-metal bonded complexes; dehydrochlorination was, however, observed with *trans*-$[PtCl(H)(PPh_3)_2]$ [432]. As noted in Chapter 14, this was also believed to be the reaction mode in the Me_3SnNMe_2/*trans*-$[PtCl(H)(PPh_3)_2]$ system [221a]. Thus, while an earlier report had claimed that the reagents, when heated in boiling xylene, gave $[Pt(Cl)(PPh_3)_2(SnMe_3)]$ [131, 132], the product is now reformulated [221a] as the isomeric *cis*-$[Pt(Me)(PPh_3)_2(SnMe_2Cl)]$, probably resulting from initial dehydrochlorination [132] and subsequent oxidative addition of Me_3SnCl to $[Pt(PPh_3)_2]$ [Eq. (3)].

An interesting development is the extension of reaction (2) to the transition-metal dihydrides $[M(\eta\text{-}C_5H_5)_2H_2]$ (M = Mo or W), which can behave as bases, and the neutral trihydride $[Ta(\eta\text{-}C_5H_5)_2H_3]$ [294a], as illustrated by Eqs. (15) and (16); the reactions were carried out in tetrahydrofuran under reflux. It was not possible to replace more than one hydrogen atom from these hydrido-complexes. Relative reactivities decreased in the order Mo>W>Ta. The hydrogen replaced in the tantalum trihydride [Eq. (16)] is the central unique ligand (i.e., ^1H n.m.r., shows an A_2B pattern for the starting trihydride and an A_2 pattern for the product).

$$Me_3SnNMe_2 + [M(\eta\text{-}C_5H_5)_2H_2] \xrightarrow{(M = Mo \text{ or } W)} [M(\eta\text{-}C_5H_5)H(SnMe_3)] \quad (15)$$

$$Me_3SnNMe_2 + [Ta(\eta\text{-}C_5H_5)_2H_3] \longrightarrow [Ta(\eta\text{-}C_5H_5)_2H_2(SnMe_3)] \quad (16)$$

B TABLE OF REACTIONS

Reactions with Metal Hydrides

Reagents		M—M′ Product	References
Reactions with Organo-germanium or -tin Hydrides (with elimination of amine or related compound)			
$PhGe(NMe_2)_3$	$RGeX_2H$ (R = Ph, Cl; X = Cl, Br)	$RX_2Ge\text{-}GePh(NMe_2)_2$	[618]
Bu_3GeNEt_2	Ph_3SnH	$Bu_3Ge\text{-}SnPh_3$	[185]
Bu_3GeNMe_2	Bu_2SnH_2	$Bu_3Ge\text{-}SnBu_2H$	[185]
R_3SnNEt_2	Ph_3GeH (R = Me, Bu, Et, Ph)	$R_3Sn\text{-}GePh_3$	[185, 517, 700]
R_3SnNEt_2	R'_3SnH (R = Bu^i, R′ = Bu^i, Ph; R = Bu, R′ = Et; R = Et, R′ = Bu^i)	$R_3Sn\text{-}SnR'_3$	[700]
R_3SnNEt_2	R'_3SnH (R = R′ = Me, Et, Bu, Bu^i, C_6H_{11}; R = Et, R′ = Me, Bu^i; R = C_6H_{11}, R′ = Ph; R = Bu, R′ = Et; R = Me, Bu^i; R′ = C_6H_{11}, Ph)	$R_3Sn\text{-}SnR'_3$	[517]
Et_3SnX	Ph_3SnH [X = N(p-ClC_6H_4)CHO, N(Ph)CHO, N(Ph)N(Ph)H, N(C_6H_{13})CHO, NEt_2]	$Et_3Sn\text{-}SnPh_3$	[184]

Table of Reactions *(C'ntd.)*

Reagents		M—M′ Product	References
$R_3SnN(Ph)CHO$	Ph_3SnH (R = Me, Et, Bu, Oct)	$R_3Sn-SnPh_3$	[186]
$R_3Sn-N(Ph)-NHPh$	R_3SnH (R = Et, Ph)	$R_3Sn-SnR_3$	[525]
$[Et_3Sn-N(COOEt)]_2$	Ph_3SnH	$2Et_3Sn-SnPh_3$	[525]
R_3SnNEt_2	$BuGe(Cl)H_2$ (R = Et, Ph)	$BuGe(SnR_3)_2Cl$	[185]
R_3SnNEt_2	Ph_2GeH_2 (R = Et, Ph)	$Ph_2Ge(SnR_3)_2$	[185]
R_3SnNEt_2	R'_2SnH_2 (R = R′ = Me, Et, Bu, Bu^i; R = Et, R′ = Bu; R = Et, Ph, C_6H_{11}, R′ = Bu^i; R = Et, R′ = Et, Bu, Bu^i; R = R′ = Bu^i)	$R'_2Sn(SnR_3)_2$	[516, 700, 701]
R_3SnNEt_2	$R'_2(H)Sn-Sn(H)R'_2$ (R = Me, Et, C_6H_{11}, Ph, R′ = Bu^i; R = Me, R′ = Bu)	$R_3Sn(SnR'_2)_2SnR_3$	[701]
$R_3SnN(Ph)CHO$	Ph_2SnH_2 (R = Me, Et, Bu, Oct)	$Ph_2Sn(SnR_3)_2$	[186]
$Bu_3SnN(Ph)CHO$	$PhSnH_3$	$(Bu_3Sn)_2Sn(H)Ph$	[183]
$R_3SnN(Ph)CHO$	$PhSnH_3$ (R = Et, Oct)	$(R_3Sn)_3SnPh$	[186]
$Ph_3GeSnEt_2NEt_2$	Ph_3MH (M = Ge, Sn)	$Ph_3GeSnEt_2MPh_3$	[185]
$Ph_3GeSnEt_2NEt_2$	Ph_2GeH_2	$Ph_2Ge[Sn(GePh_3)Et_2]_2$	[185]
$Ph_3GeSnEt_2N(Ph)CHO$	Ph_2SnH_2	$Ph_2Sn[Sn(GePh_3)Et_2]_2$	[185]
$Ph_3GeSnEt_2$-N(Ph)CHO	R_2SnH_2 (R = Et, Ph)	$R_2Sn[Sn(GePh_3)Et_2]_2$	[185]
$EtSn(GePh_3)_2$-N(Ph)CHO	Ph_2SnH_2	$Et(Ph_3Ge)_2Sn(SnPh_2)_2Sn(GePh_3)_2Et$	[185]
$EtSn(GePh_3)_2$-N(Ph)CHO	R_2SnH_2 (R = Bu, Ph)	$R_2Sn[Sn(GePh_3)_2Et]_2$	[185]
$R_3SnSnR_2N(Ph)CHO$	R'_3SnH (R = R′ = Et, Bu; R = Bu, R′ = Ph)	$R_3SnSnR_2SnR'_3$	[183]
$Et_3SnSnEt_2N(Ph)CHO$	Et_2SnH_2	$Et_2Sn[Sn(SnEt_3)Et_2]_2$	[183]
$R_2Sn(NEt_2)_2$	Ph_3GeH (R = Et, Ph)	$Ph_3Ge-SnR_2NEt_2$	[185]
$R_2Sn(NEt_2)_2$	Ph_2GeH_2	$[-SnPh_2-GePh_2-]_n$	[185]
$R_2Sn(NEt_2)_2$	R'_3SnH (R = Bu, R′ = Et, Bu; R = Me, R′ = Et)	$R_2Sn(SnR'_3)_2$	[700, 701]
$R_2Sn(NEt_2)_2$	R_2SnH_2 (R = Et, Bu, Bu^i)	$(R_2Sn)_6$	[700]
$R_2Sn[N(Ph)CHO]_2$	Ph_3SnH (R = Et, Bu, Ph)	$R_2Sn(SnPh_3)_2$	[186]
$Et_2Sn[N(Ph)CHO]_2$	Ph_2SnH_2	$[-Et_2Sn-SnPh_2-]_n$	[186]
$Et(Ph_3Ge)Sn$-$[N(Ph)CHO]_2$	Ph_3SnH	$EtSn(SnPh_3)_2GePh_3$	[185]
$EtSn(NEt_2)_3$	Ph_3GeH (n = 1, 2, 3)	$Et(Et_2N)_{3-n}Sn(GePh_3)_n$	[185]

Table of Reactions *(C'ntd.)*

Reagents		M—M′ Product	References
R_3PbNEt_2	Ph_3GeH (R = Pr, Bu^i, C_6H_{11}, Ph)	$R_3PbGePh_3$	[515]
Ph_3PbNEt_2	Ph_2GeH_2	$(Ph_3Pb)_2GePh_2$	[515]
$(\eta\text{-}C_5H_5)_2TiNMe_2$	Ph_3SnH	$(\eta\text{-}C_5H_5)_2Ti\text{-}SnPh_3.THF$	[432]
$Ti(NMe_2)_3$	Ph_3SnH	$Ti(SnPh_3)_3$	[432]
$Ti(NMe_2)_4$	Ph_3SnH	$Ti(SnPh_3)_4$	[187]
$Ti(NEt_2)_4$	R_3SnH (R = Et, Ph)	$(R_3Sn)_2Ti(NEt_2)_2$	[187]
$Zr(NEt_2)_4$	Ph_3SnH	$Zr(SnPh_3)_4$	[187]
Reactions with Transition Metal Hydrides			
Me_3SiNMe_2	$M(\eta\text{-}C_5H_5)(CO)_3H$ (M = Mo, W)	$M(\eta\text{-}C_5H_5)(CO)_3\text{-}SiMe_3$	[129,131]
Me_3SiNMe_2	$(PF_3)_4CoH$	$[Me_3SiNHMe_2]^+[Co(PF_3)_4]^-$	[69]
Me_3GeNMe_2	$M(\eta\text{-}C_5H_5)(CO)_3H$ (M = Cr, Mo, W)	$M(\eta\text{-}C_5H_5)(CO)_3\text{-}GeMe_3$	[130,131]
Me_3SnNMe_2	$M(\eta\text{-}C_5H_5)(CO)_3H$ (M = Cr, Mo, W)	$M(\eta\text{-}C_5H_5)(CO)_3\text{-}SnMe_3$	[130,131, 132]
$Sn[N(SiMe_3)_2]_2$	$Mo(\eta\text{-}C_5H_5)(CO)_3H$	$Sn[Mo(\eta\text{-}C_5H_5)(CO)_3]_2$	[429, 594a]
Me_3SnNMe_2	$Ta(\eta\text{-}C_5H_5)_2H_3$	$Ta(\eta\text{-}C_5H_5)_2H_2SnMe_3$	[294a]
Me_3SnNMe_2	$M(\eta\text{-}C_5H_5)_2H_2$ (M = Mo, W)	$M(\eta\text{-}C_5H_5)_2HSnMe_3$	[294a]
Me_3SnNMe_2	*trans*-$[PtCl(H)(PPh_3)_2]$	*cis*-$[Pt(Me)(PPh_3)_2SnMe_2Cl]$	[221a, see also 131,132]
$Ti(\eta\text{-}C_5H_5)_2NMe_2$	$M(\eta\text{-}C_5H_5)(CO)_3H$ (M = Mo, W)	$Ti(\eta\text{-}C_5H_5)_2[M(\eta\text{-}C_5H_5)(CO)_3]$	[432]
$Ti(OPr^i)_3NMe_2$	$Mo(\eta\text{-}C_5H_5)(CO)_3H$	$Mo(\eta\text{-}C_5H_5)(CO)_3Ti(OPr^i)_3$	[131,132]
$Hg[N(SiMe_3)_2]_2$	$(C_6F_5)_3GeH$	$Hg[Ge(C_6F_5)_3]_2$	[82]
$Hg[N(SiMe_3)_2]_2$	Ph_3MH (M = Ge, Sn)	$Hg(MPh_3)_2$	[83, 222]

13

Metathetical Exchange Reactions: Metal or Metalloid Amides as Aminating Reagents

A INTRODUCTION AND GENERAL CONSIDERATIONS

There has been an increasing awareness of the prevalence of metathetical exchange reactions in inorganic and organometallic chemistry. Indeed in 1970 a book appeared which was devoted entirely to this topic [453a], under the title 'Redistribution Reactions'. Other terms which are essentially synonymous include 'scrambling reactions', 'disproportionation', or the old-fashioned 'double decomposition'. Many of the systems are readily amenable to kinetic and, especially, thermodynamic studies, and are followed by ^{1}H, ^{11}B, or ^{31}P n.m.r. spectroscopy [170, 490a].

We shall define a metathetical exchange reaction in terms of Eq. (1) [but see also Eq. (6)], in which L represents the sum of all ligands other than a group Y

$$LM{-}Y + X{-}A \longrightarrow LM{-}X + Y{-}A \qquad (1)$$

attached to the metal or metalloid M and X-A is a molecule in which there is significant polarity in the single σ-bond joining the moieties X and A. In this chapter our concern is with the particular case of Y being an amino group NRR′, Eq. (2), and hence an alternative way of describing such a metathetical exchange reaction is as an 'amination' [see also Eq. (6)]. This places emphasis on the synthetic utility often found for reaction (2), whereby the function of the metal or metalloid residue LM is to provide a vehicle for the transfer of an amino group from an organic primary or secondary amine via LM to an electrophilic site, A in Eq. (2).

$$\text{LM—NRR}' + \text{X—A} \longrightarrow \text{LM—X} + \text{A—NRR}' \tag{2}$$

The further classification adopted hereafter stresses the exchange nature of the reactions. In Section B we consider the example of XA taking the form L′M′-X, which may further be subdivided into the various NRR′/X exchanges: X = H, Hal, R (= an alkyl or other carbon-centred ligand, i.e., hydrocarbyl), OR, SR, ½-O-, ½-S-, or NR″R‴. Some of the reactions, notably those with (i) XA = L′M′-H, or (ii) XA = L′M′-NR″R‴, either lead to the opposite exchange process, [NRR′/L′M′ for (i), see Chapters 11 or 12], or are discussed elsewhere, [as for transamination, which represents case (ii), see Chapters 1-9]. In Section C.1 we deal with organic substrates XA in which the bond polarisation makes A an electrophilic site, $\overset{\delta-}{\text{X}}\text{—}\overset{\delta+}{\text{A}}$, because of an adjacent C=O or S=O bond: for example A = MeC(O) with XA = MeC(O)OR″, and NRR′/OR″ exchange; thus X is required to be an effective leaving group.

It is often convenient to subdivide the reactions of Section B into homo- or hetero-metallic types, i.e., XA = L′M-X or L′M′-X, respectively, depending on whether redistribution of ligands takes place on the same metal or metalloid M or between different elements M and M′. The former are particularly useful to prepare mixed ligand complexes, as in Eq. (3) [258]. This equation serves to illustrate a general feature of the above metathetical exchanges: namely that

$$2\text{B(NEt}_2)_3 + \text{BCl}_3 \longrightarrow 3\text{ClB(NEt}_2)_2 \tag{3}$$

they represent reactions in which bonds change in relative position but not in type. A corollary might be held to be that such processes should be close to thermoneutral [e.g., in Eq. (3), both the starting materials and the product have six B-NEt$_2$ and three B-Cl bonds]. However, this is not usually the case; mixed ligand or heteroleptic complexes generally have higher negative standard heats of formation (i.e., are more stable) than would be expected from consideration of bond energy additivity based on homoleptic analogues, as for $\text{Si(NMe}_2)_n\text{Cl}_{4-n}$ (n = 1-3) [721a]. At times this has been attributed to π-bonding effects; e.g., in the case of Eq. (3), the argument starts with the premise of constancy of σ-bond energy contributions in each of $\text{Cl}_n\text{B(NEt}_2)_{3-n}$, then assumes $\ddot{\text{N}}\text{—B} > \ddot{\text{Cl}}\text{—B}$ π-bonding, whence in the heteroleptic ClB(NEt$_2$)$_2$ the mean BN bond energy term is greater than in B(NEt$_2$)$_3$, in which three rather than two nitrogen lone pair orbitals are competing for the single vacant orbital on boron (see Chapter 4, Section C.2). In reactions which are close to thermoneutral, as when the amido and its exchanging ligand are of similar electronegativity, the free energy change will depend on the entropy term. Thermodynamic and, to some extent, kinetic implications of metathetical exchange reactions are discussed in ref. [490a].

It is often the case that the more effectively the central atom M or M′ can confer Lewis acidity on the compound LM-NRR′, L′M-X, or L′M′-X, the more rapid are the redistribution reactions, whereas among the exchanging groups donor power is a measure of kinetic lability. Mechanisms are frequently discussed in terms of a concerted process involving a four-centre transition state, as in (I). However, the vast majority of reactions are polar in character (unless NRR′ and X are of comparable electronegativity), and many are more likely to involve a rate-determining bimolecular attack by the amide, functioning as a nucleophile, on the electrophilic end A of the AX dipole, (II).

RR′ N LM A X

(I)

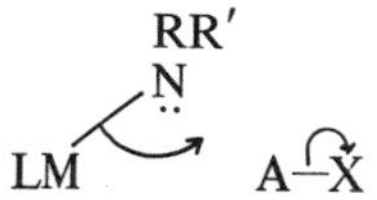

RR′ N LM A–X

(II)

There have been few detailed mechanistic studies. However, an exception relates to Ti^{IV} compounds. Dimethylamido and Bu^tO groups are scrambled on Ti^{IV} in a second-order reaction in toluene, being first-order in each of $[Ti(NMe_2)_4]$ and $[Ti(OBu^t)_4]$. Rates were determined by following the appearance of $[Ti(NMe_2)_3(OBu^t)]$, the first scrambling product, for which $k_2 = 4.2 \pm 0.4 \times 10^{-5}$ l mol^{-1} s^{-1} at 25° C and $\Delta G^{\ddagger} = 40.4$ kJ mol^{-1} and $\Delta S^{\ddagger} = -197$ JK^{-1}. The extraordinarily high activation entropy points to a four-centre process (I) which would be constrained for the bulky ligands in this reaction. The rate is five orders of magnitude greater if the less bulky Pr^iO group is used in place of Bu^tO [745]. The equilibrium situation was also explored [744]: $K_1 = 0.24$, $K_2 = 0.40$, and $K_3 = 0.11$, compared with statistical values of $K_1 = K_3 = 0.375$ and $K_2 = 0.44$, where $K_1 = [TiX_4][TiX_2Y_2]/[TiX_3Y]^2$, $K_2 = [TiX_3Y][TiXY_3]/[TiX_2Y_2]^2$, and $K_3 = [TiX_2Y_2][TiY_4]/[TiXY_3]^2$ ($X = OBu^t$, $Y = NMe_2$). The corresponding $[Ti(NMe_2)_4]/[Ti(OPr^i)_4]$ exchange is virtually thermoneutral.

An approach has been made to a linear free energy relationship, using the reaction of Eq. (4) in CCl_4 to provide the data [54b]. The equilibrium constant K is obtained by n.m.r. spectroscopy, and a plot of log K versus the sum of the Taft substituents $\Sigma\sigma^{\star}$ for the R groups is linear, obeying the relationship log $K = 0.89\ \Sigma\sigma^{\star}$.

$$Me_3SiNMe_2 + RR'R''SiCl \longrightarrow RR'R''SiNMe_2 + Me_3SiCl \qquad (4)$$

For purposes of synthesis, amination of the type shown in Eq. (2) is perhaps most effective when LMNRR′ is an amide of tin(IV), such as Me_3SnNMe_2 (for reviews, see refs. [205a, 356]). This was first recognised in 1966 [256]. Attention was drawn to (i) the high donor strength of the tin(IV) amide, (ii) the weak and polar $\gtrdot\overset{\delta+}{Sn}-\overset{\delta-}{N}RR'$ bond, (iii) the large heat of formation often a characteristic of the co-product, which may also be insoluble, and (iv) the high yields and mild reaction conditions. These features are illustrated by Eq. (5). A mixture of *cis*- and *trans*-enamines (III) (79%) is obtained at 0° C; the co-product Me_3SnF (76%) not only has a strong Sn-F bond but is a polymer and insoluble in the reaction medium, and hence is readily separated. Organo-silicon compounds,

$$Me_3SnNMe_2 + CF(Cl){=}CF_2 \longrightarrow Me_3SnF + \underset{\textbf{(III)}}{CF(Cl){=}C(F)NMe_2} \quad (5)$$

such as Me_3SiNMe_2, are also widely employed, cf. ref. [5], but are less reactive, e.g., there is no exchange between Me_3SiNMe_2 and severally $CF(Cl){=}CF_2$, Me_3SnCl, or $BF_3.OEt_2$ [256]. Me_3SiNMe_2 has, however, the advantage of the ease of separation of the volatile by-product, such as Me_3SiCl. The compound $Hg[N(CF_3)_2]_2$ is particularly useful as a bis(trifluoromethyl)amino group transfer reagent [210, 230a, 298a, 771, 772, 772a].

An interesting mechanistic complication is demonstrated in the contrasting behaviour between Me_3SnNMe_2 and either the enol ester $CH_2{=}CHOCOMe$ or PhOAc. The former yields initially the enamine (NMe_2/OAc exchange: alkenyl-oxygen fission) $CH_2{=}CHNMe_2$, whereas the latter furnishes acetanilide (NMe_2/OPh exchange: acyl-oxygen fission) [145].

Finally, in Section C.2 we consider amination reactions which do not conform to the pattern of Eq. (2). Instead of an NRR′ exchange with a univalent ligand X, we now have two amino groups displacing the bivalent O^{2-} of an organic carbonyl or of SO_2. The conversions: $>C{=}O$ into $>C(NMe_2)_2$ using $[Ti(NMe_2)_4]$ [748] (or $[TiCl_4] + HNMe_2$ [748a]), and $-C(O)OR$ into $-C(O)NMe_2$ employing Me_3SnNMe_2 [256], were first described in 1966. Amides of B, P^{III}, or As^{III} are also effective; each of B, Sn^{IV}, P^{III}, or As^{III} forms strong M—O or M=O bonds and the reactivity order appears to be Ti>Sn>As>B>P [144]. These transformations probably proceed via an initial insertion step to afford the intermediate (IV). The carbonyl reactions may thus be represented by Eq. (6). If the amide is at least bifunctional, i.e., $L'M(NRR')_2$, a molecular rearrangement

$$LM{-}NRR' + >C{=}O \longrightarrow \underset{\textbf{(IV)}}{>C<^{OML}_{NRR'}} \longrightarrow \text{products} \quad (6)$$

of (IV) leads to $>C(NRR')_2 + L'MO$. When an effective leaving group is available on the organic substrate, e.g., OR″, compound (IV) fragments to yield $>C(O)NRR'$

+ LMOR″. The reaction of Ph_2PNR_2 with ethylene carbonate (A) obeys a second-order rate law, being first-order in each of $[Ph_2PNR_2]$ and [A] [692].

The immediate product of the reaction between a carboxylic acid, anhydride, or ester and a metal or metalloid amide is the carboxylic amide and this conforms with Eq. (2) (see Section C.1). The latter gives the *gem*-triamine or enamine, whereas an aldehyde or ketone yields a *gem*-diamine or enamine, depending on the absence or presence of a β-hydrogen atom [144, 747]. We note that in some cases the organic carbonyl compound behaves as a protic compound (see Chapter 11), and in others as a substrate for formation of a stable 1,1-insertion adduct (see Chapter 10). Steric effects have a role, as is neatly illustrated by Eq. (7), in which the insertion [145, 257] requires the minimum of steric hindrance, otherwise the metathetical exchange process [145] supervenes.

$$Me_3SnNMe_2 \xrightarrow[\text{insertion}]{CH_2{=}CHCO_2Me} Me_2NCH_2{-}CH(CO_2Me)(SnMe_3)$$

$$Me_3SnNMe_2 \xrightarrow[\text{metathetical exchange}]{CH_2{=}C(Me)CO_2Me} Me_3SnOMe + CH_2{=}C(Me)CONMe_2 \quad (7)$$

A preparative advantage of amination reactions may derive from the difficulty, in some cases, of achieving alternative syntheses, and in others it is the co-product which is the object of the synthesis. Examples are in Eqs. (8) (M = Zr or Hf) [218] and (9) [197]. Thus $[M(\eta\text{-}C_5H_5)_2F_2]$ (M = Zr or Hf) cannot be obtained from $[M(\eta\text{-}C_5H_5)_2Cl_2]$ and NaF in acetone and attempted preparation of $Ge[CH(SiMe_3)_2]_2$ from $Li[CH(SiMe_3)_2]$ and $GeCl_2$.dioxan or GeI_2 is unsuccessful, see also Eq. (48). Procedures of the type illustrated by Eq. (1) other than Y = NRR′ are exemplified by Eqs. (10) [219], (11) [172], (12) [8], (13) [31], and (14) [256]; $Me_3Sn[N{=}C(CF_3)_2]$ behaves similarly to Me_3SnNMe_2 [141b].

$$[M(\eta\text{-}C_5H_5)_2(NMe_2)_2] + 2C_6F_5CN \longrightarrow [M(\eta\text{-}C_5H_5)_2F_2] + 2p\text{-}C_6F_4(NMe_2)CN \quad (8)$$

$$Ge[N(SiMe_3)_2]_2 + 2Li[CH(SiMe_3)_2] \longrightarrow Ge[CH(SiMe_3)_2]_2 + 2LiN(SiMe_3)_2 \quad (9)$$

$$4BI_3 + 3TiCl_4 \longrightarrow 4BCl_3 + 3TiI_4 \quad (10)$$

$$AlBr_3 + (Me_3Si)_2O \longrightarrow Me_3SiBr + Me_3SiOAlBr_2 \quad (11)$$

$$3Me_3SiSEt + POCl_3 \longrightarrow 3Me_3SiCl + PO(SEt)_3 \quad (12)$$

$$2Et_3GeNCS + PhPCl_2 \longrightarrow 2Et_3GeCl + PhP(NCS)_2 \quad (13)$$

$$3Me_3SnPPh_2 + AsF_3 \longrightarrow 3Me_3SnF + As(PPh_2)_3 \quad (14)$$

Reaction according to Eq. (15) probably proceeds via initial $SiMe_3/SnEt_3$ exchange and subsequent decomposition of $Hg(SnEt_3)_2$ [490].

$$2Et_3SnNEt_2 + (Me_3Si)_2Hg \longrightarrow Hg + 2Me_3SiNEt_2 + (Et_3Sn)_2 \quad (15)$$

B REACTIONS OF AMIDES WITH METAL OR METALLOID COMPOUNDS: THE LM–NRR′/L′M′–X SYSTEM

1. The NRR′/H exchange reactions

Few reactions involving NR_2/H exchange have been reported; however, they are probably implicated in the catalytic NEt_2/Et exchange $[Et_3B/B(NEt_2)_3]$ in the presence of $C_4H_8O.BH_3$ [283a]. An incomplete hydride exchange is observed between an aluminium hydride and $Be(NMe_2)_2$ [571]. An aminoborane and diborane furnish dimethylaminodiborane in good yield [113]. Reduction by an aluminium hydride of $B_2(NMe_2)_4$ leads to a compound with empirical formula $Al_3B_3(NMe_2)_7H_5$ which is said to contain 3B–Al bonds on the basis of i.r., ^1H and ^{11}B n.m.r. spectra, and analytical and molecular weight data, Eq. (16) [291].

$$B_2(NMe_2)_4 + AlH_3.2NMe_3 \longrightarrow Al_3B_3(NMe_2)_7H_5 + H_2 + NMe_3 + HB(NMe_2)_2 + (H_2BNMe_2)_2 + BH_3.NMe_3 \quad (16)$$

An amino-alane or -stannane reacts with a boron hydride to give an amino-borane, Eqs. (17), (18) [256, 340, 367], or a complex mixture, Eq. (19) [368]. Tris(diethylamino)borane is reduced to bis(diethylamino)borane in the presence of $Li[BH_4]$ and HCl [Eq. (20)] or diethylamineborane [534]. The hydrides LiH, $Na[BH_4]$, or $Li[AlH_4]$ do not react with an aminostannane; dibutylalane, Bu_2AlH, however, gives the aminoalane, Eq. (21) ($n = 1$, 2, or 3) [414]. An amino(chloro)-silane is reduced to silane in the presence of $Li[AlH_4]$, Eq. (22) ($n = 1$ or 2) [136], but with Et_3SnH at high temperature under aerobic conditions a complex mixture is obtained [726].

$$Al(NMe_2)_3 + BuBH_2.NMe_3 \longrightarrow BuB(NMe_2)_2 + H_2AlNMe_2 + NMe_3 \quad (17)$$

$$3Me_3SnNMe_2 + BH_3.NEt_3 \longrightarrow 3Me_3SnH + B(NMe_2)_3 + NEt_3 \quad (18)$$

$$Al(NMe_2)_3 + B_2H_6 \xrightarrow{Et_2O} H_2B(NMe_2)_2Al[BH_4]_2 + Al[BH_4]_2NMe_2 + Al[BH_4]_3.OEt_2 + \mu\text{-}(Me_2N)_2B_3H_7 + (H_2BNMe_2)_2 + \mu\text{-}Me_2NB_2H_5 \quad (19)$$

$$2B(NEt_2)_3 + Li[BH_4] + HCl \longrightarrow 3HB(NEt_2)_2 + LiCl + H_2 \quad (20)$$

$$R_{4-n}Sn(NEt_2)_n + nBu_2AlH \longrightarrow R_{4-n}SnH_n + nBu_2AlNEt_2 \quad (21)$$

$$Cl_{4-n}Si(NMe_2)_n + Li[AlH_4] \longrightarrow SiH_4 + \text{trace of } H_2 \quad (22)$$

2. The NRR′/Hal exchange reactions

With metal halides, two types of reaction [Eq. (23)] have been observed, [657]. Pathway (23a) represents donor-acceptor bond formation as described in Chapter 16. In some cases, these complexes (e.g., with BX_3 [112, 708]) are stable only at low temperature, whereafter disproportionation occurs according to pathway (23c). Reaction (23b) represents ligand-exchange, and has been realised for amides of alkali metals [121, 731, 737], Mg [358], B, Al, Si, Ge, Sn^{II},

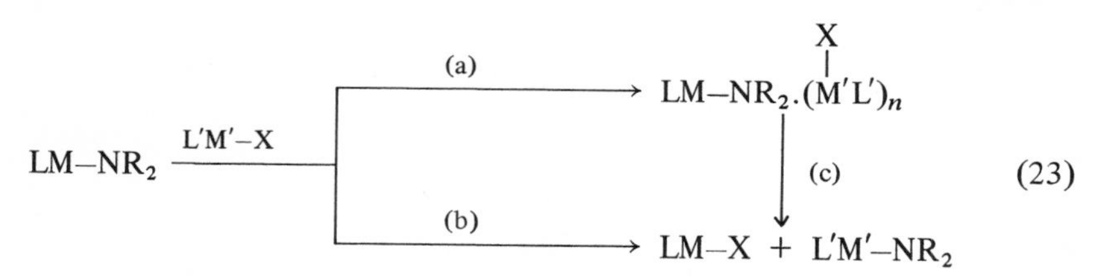

Sn^{IV}, P, As, Ti, Zr, Hf, Cu, and Hg (see Table). Some quantitative data in the Ti^{IV} system are discussed in Section A.

The NR_2/Hal exchange reaction has been widely used for synthesis of complicated boron compounds. An early example is in Eq. (24) [436]. A related process is shown in Eq. (25) [55a], and an extension illustrates stepwise displacements, Eq. (26) [55d]; the tin amide is, as elsewhere, a more powerful reagent, Eq. (27) [706a].

$$(Me_3Si)_2NH + 2o\text{-}C_6H_4O_2BCl \longrightarrow (o\text{-}C_6H_4O_2B)_2NH + 2Me_3SiCl \quad (24)$$

$$(Me_3Si)_3N + 2MeBBr_2 \longrightarrow [Me(Br)B]_2NSiMe_3 + 2Me_3SiBr \quad (25)$$

$$(Me_3Si)_3N \begin{cases} \xrightarrow[0°C]{3BBr_3} [Me_2(Br)Si]_3N + 3MeBBr_2 \\ \xrightarrow[23°C]{BBr_3} Me_2(Br)SiN(SiMe_3)_2 + MeBBr_2 \\ \xrightarrow[50\text{–}70°C]{BBr_3} [Me_2(Br)Si](Me_3Si)NB(Br)Me + Me_3SiBr \end{cases} \quad (26)$$

$$(Me_3Sn)_3N + 3Me_2BBr \longrightarrow (Me_2B)_3N + 3Me_3SnBr \quad (27)$$

Further syntheses relate to (V) from (VI) and SbF_5 or BF_3 [254], (VII) from (VIII) and Me_2BBr [55c], (IX) from (X) and BCl_3 [705a], (XI) from $Me_3SiN[B(Br)Me]_2$ and $Me_3SiOOSiMe_3$ [55b], the mixed borazine (XII) from (XIII) and $MeBBr_2$ [547a], and $Me_2BN(Me)(CH_2)_2CN$ from $Me_3SiN(Me)(CH_2)_2CN$ and Me_2BBr [476b].

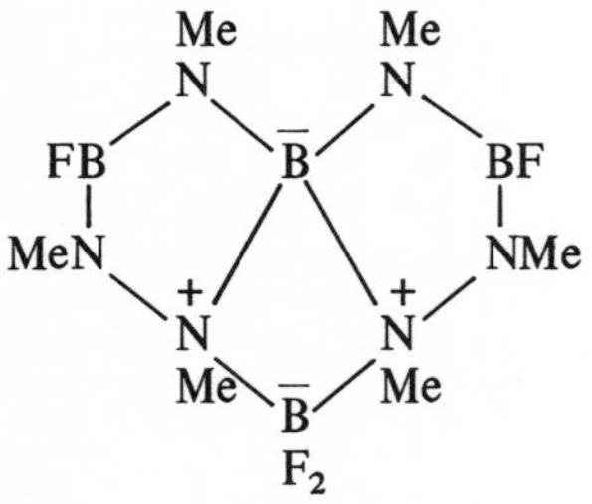

Me Me
N N
FB $\bar{B}$ BF
MeN NMe
$\overset{+}{N}$ $\overset{+}{N}$
Me $\bar{B}$ Me
F_2

(V)

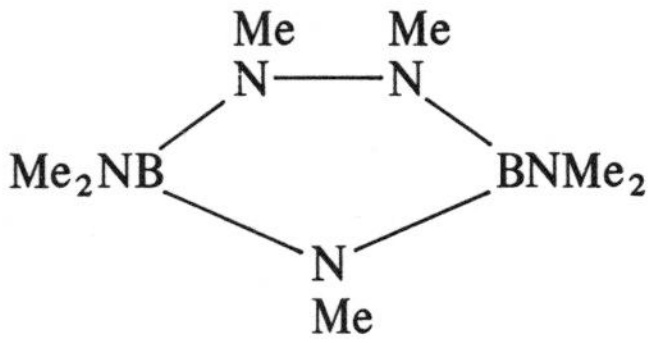

Me Me
N—N
Me_2NB $BNMe_2$
N
Me

(VI)

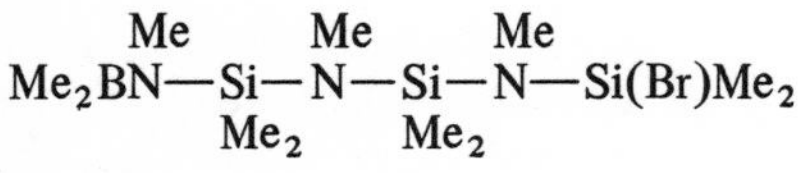

Me Me Me
$Me_2BN—Si—N—Si—N—Si(Br)Me_2$
Me_2 Me_2

(VII)

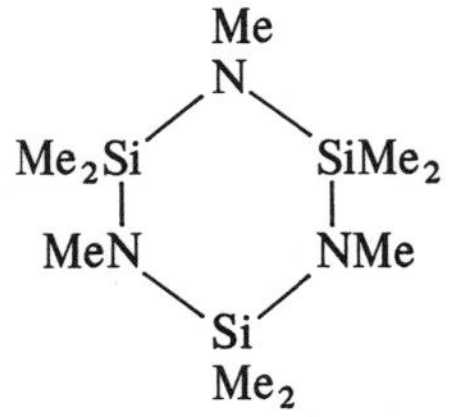

Me
N
Me_2Si $SiMe_2$
MeN NMe
Si
Me_2

(VIII)

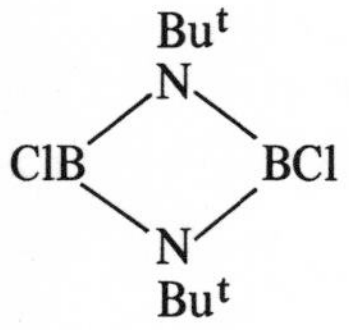

Bu^t
N
ClB BCl
N
Bu^t

(IX)

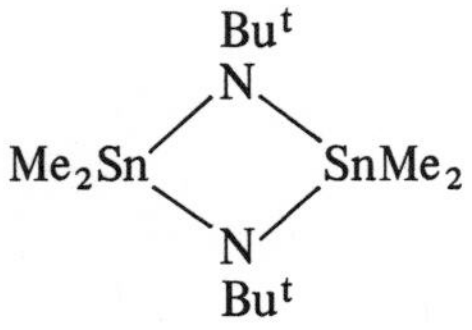

Bu^t
N
Me_2Sn $SnMe_2$
N
Bu^t

(X)

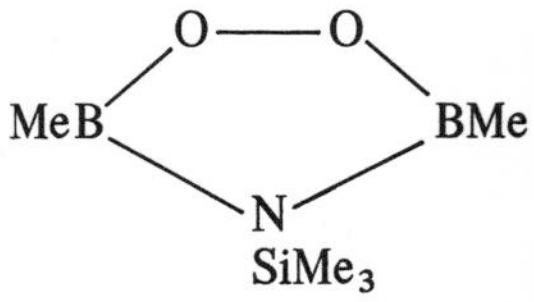

O—O
MeB BMe
N
$SiMe_3$

(XI)

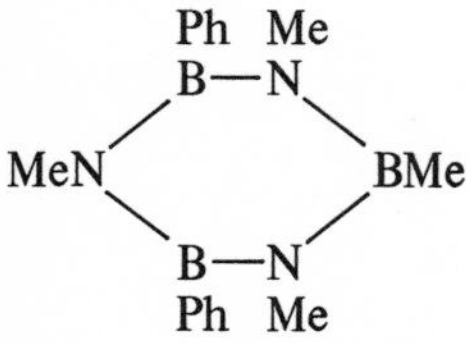

Ph Me
B—N
MeN BMe
B—N
Ph Me

(XII)

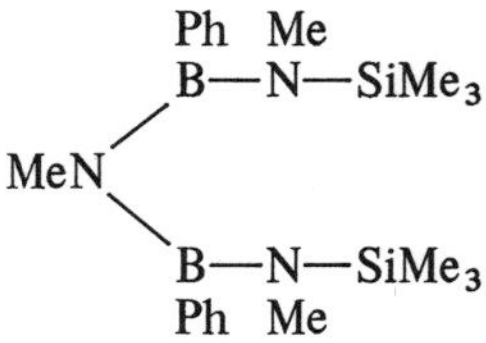

Ph Me
$B—N—SiMe_3$
MeN
$B—N—SiMe_3$
Ph Me

(XIII)

An interesting mild reaction is that whereby $[(Me_2N)_2B]_2$ is converted into $Me_2N(Cl)BB(NMe_2)_2$ by treatment with CCl_4 under irradiation [286d].

The ^{1}H n.m.r. spectrum of a mixture of $(Et_2AlX)_2$ and $(Et_2AlNR_2)_2$ in 1:1 molar ratio shows the presence of a mixed bridged compound; trimethylamine cleaves the bridge, e.g., Eq. (28) (X = Cl, Br, I, OEt, or SMe; R = Me; or X = Br, R = Et) [243].

$$Et_2Al(\mu\text{-}NR_2)(\mu\text{-}X)AlEt_2 + NMe_3 \longrightarrow Et_2AlX.NMe_3 + \tfrac{1}{2}(Et_2AlNR_2)_2 \quad (28)$$

Reaction of an aminosilane with a phosphorus fluoride provides a route to P-N compounds involving P^{III} or P^{V} [205, 678, 679]. Titanium tetrahalides lead to corresponding dialkylamidotitanium compounds $[TiNR_2(X)_3]$ (X = Cl or Br) [117]. Silylaminoboranes $Me_3Si{-}NR{-}BR'_2$ may be prepared by the reaction of $(Me_3Si)_2NR$ and the corresponding monohalogenoborane, e.g., Eq. (29) [547, 549].

$$\text{Me-}C_4H_7BN(SiMe_3)_2 + Me_2BBr \longrightarrow Me_3SiBr + \text{Me-}C_4H_7BN(SiMe_3)BMe_2 \quad (29)$$

An aminosilane readily undergoes NR_2/F exchange with XeF_2. The xenon(II) amide is probably obtained at low temperature, but disproportionates, Eq. (30) [264]; the formation of amine and polymer is attributed to hydrogen abstraction from acetonitrile solvent by the dimethylaminyl radical $\dot{N}Me_2$ which may alternatively decompose by loss of hydrogen (e.g., $\dot{N}Me_2 \rightarrow MeN{=}CH_2 \rightarrow$ polymer).

$$Me_3SiNR_2 + XeF_2 \longrightarrow Me_3SiF + [Xe(NR_2)_2] \longrightarrow Xe + HNR_2 + \text{polymer} \quad (30)$$

Acylaminosilanes may also usefully be employed in scrambling reactions [470a, 691a]. An example is shown in Eq. (31) and leads to compound (XIV), as substantiated by X-ray analysis [564b], and not to the isomer (XV), as originally proposed [398].

$$MeC(O)N(SiMe_3)_2 + ClCH_2Si(Cl)Me_2 \longrightarrow \text{(XIV)} \quad (31)$$

(XIV): $Me_2(Cl)\bar{Si}$ chelated by $\overset{+}{O}{=}C(Me)$ and $CH_2{-}NCH_2Si(Cl)Me_2$

$$MeC(O)N(SiMe_3)_2 + ClCH_2Si(Cl)Me_2 \not\longrightarrow MeC(O)N[SiMe_2(CH_2Cl)]_2 \ \text{(XV)}$$

With an alkyl halide, an aminostannane yields the corresponding halide, Eq. (32) ($n = 1$, 2, or 3; X = Br or I) [457]. The versatility of $(Me_3Sn)_3N$ is illustrated not only by Eq. (27), but also Eq. (33) [633].

$$Bu_{4-n}Sn(NR_2)_n + 2nMeX \longrightarrow Bu_{4-n}SnX_n + n[Me_2NR_2]X \quad (32)$$

$$(Me_3Sn)_3N + Me_3SiCl \longrightarrow (Me_3Sn)_2NSiMe_3 + Me_3SnCl \quad (33)$$

Likewise, dimethylaminotrimethylstannane readily aminates chlorotrimethylsilane, Eq. (34) [256]. A related reverse reaction has been claimed (Me_3SnBr/$Me_3SiNHEt$); it was necessary to heat the initially formed 1:1 complex [12]. A transformation similar to that of Eq. (34), also using Me_3SiCl, is $[Mo_2(NMe_2)_6] \rightarrow [Mo_2Cl_2(NMe_2)_4]$ [20a].

$$Me_3SnNMe_2 + Me_3SiCl \longrightarrow Me_3SnCl + Me_3SiNMe_2 \quad (34)$$

Chloramine, NH_2Cl, reacts in a complicated fashion (probably via a free radical pathway) with a Group 4 metal amide R_3MNMe_2, $(R_3M)_2NH$, or $(R_3M)_3N$ [310]; the trialkyltin chloride is isolated for M = Sn, but N_2, NH_3, and $HNMe_2$ are among the other products. The tin and germanium amides are more reactive than their Si analogues, but Me_2NCl cleaves only the Sn-N bond.

Trifluoro-phosphine, -arsine, or -stibine, or TiF_4 is easily converted into the corresponding amide by means of Me_3SnNMe_2; however, PbF_4, BiF_3, CoF_3, or ZrF_4 do not react [256]. The Sn-N bond is also cleaved by dichlorine, Eq. (35) [256].

$$Et_3SnNMe_2 + Cl_2 \longrightarrow Et_3SnCl + Me_2NCl \quad (35)$$

Redistribution reactions involving amides of phosphorus are much used. Examples are found in refs. [47a, 48, 48a, 158, 244, 301, 483, 556, 607, 662a], e.g., Eqs. (36) [158, 483] and (37) [743].

$$[(Me_2N)_2PN]_4 \text{ (cyclic: } (NMe_2)_2P{=}N{-}P(NMe_2)_2{=}N{-}P(NMe_2)_2{=}N{-}P(NMe_2)_2{=}N) \xrightarrow{SbF_3} P_4N_4F_n(NMe_2)_{8-n} \text{ (including isomers)} \quad (36)$$

$$Cl_2PNMe_2 + P(O)Cl_3 \longrightarrow PCl_3 + Cl_2P(O)NMe_2 \quad (37)$$

The exchange process may be accompanied by further features, such as insertion [Eq. (38)] [630a] (for convenience the NMe_2/NCO system is included in this section), reduction [Eq. (39)] [375e], or oxidation [Eq. (40)] [294] (see also Chapter 18).

$$2As(NMe_2)_3 + 2Me_3SiNCO \longrightarrow 2Me_3SiNMe_2 + \begin{array}{c} O \\ \| \\ Me_2NAs—NCNMe_2 \\ |\quad\quad | \\ Me_2NCN—AsNMe_2 \\ \| \\ O \end{array} \tag{38}$$

$$(CF_3)SF_4Cl \xrightarrow{2Me_3SiNMe_2} (CF_3)S(Cl)(NMe_2)_2 \tag{39}$$

$$Sn[N(SiMe_3)_2]_2 + 2SnBr_4 \longrightarrow SnBr_2 + 2Br_3Sn[N(SiMe_3)_2] \tag{40}$$

3. The NRR′/R″ exchange reactions

Only certain metal or metalloid amides react readily with an alkylating agent such as a Grignard reagent or an organolithium compound. Aminoboranes $B(NR_2)_3$ are suitable substrates for producing organoboranes $R'B(NR_2)_2$, R'_2BNR_2, and R'_3B, but R'/NR_2 exchange is less facile than either R'/Cl or R'/OR [424]. Alkylating reactivities with respect to $B(NMe_2)_3$ decrease in the order $R_3Al > R_3B > R_2Zn > R_2Cd > R_4Sn$ [19]; clearly the less polar the organometallic reagent the lower its reactivity. Various related scrambling reactions $>B—NRR'/>B—X$ have been studied [420, 554]. The $Et_3B/B(NEt_2)_3$ system is catalysed by $C_4H_8O.BH_3$ [283a]. The reagent Li_2CH_2 reacts with $ClB(NMe_2)_2$ not only to provide for Cl_2/CH_2 exchange, yielding $CH_2[B(NMe_2)_2]_2$, but also $Cl, NMe_2/CH_2$ exchange to give (XVI) [401a].

$$\begin{array}{ccc} & NMe_2 & \\ & B & \\ CH_2 & & CH_2 \\ | & & | \\ Me_2NB & & BNMe_2 \\ & CH_2 & \end{array} \qquad \textbf{(XVI)}$$

The reaction of Me_3Al with $B_2(NMe_2)_4$ or H_2BNMe_2 is said to furnish various catenated aluminium species depending on stoicheiometry, Eqs. (40)–(43) [269, 270, 681].

$$3B_2(NMe_2)_4 + 8Me_3Al \longrightarrow Al_4B(NMe_2)_3Me_6 + 5Me_2BNMe_2 + 2(Me_2AlNMe_2)_2 \tag{41}$$

$$B_2(NMe_2)_4 + Me_3Al \longrightarrow Me_3B + Me_2BNMe_2 + (Me_2AlNMe_2)_2 + Al_3Me_3(NMe_2)_2 \tag{42}$$

$$2(H_2BNMe_2)_2 + 12Me_3Al \longrightarrow (Me_2AlNMe_2)_2 + 2(Me_2AlH)_3 + 4Me_3B + Al_4Me_8(NMe_2)_2H_2 \quad (43)$$

Tris(dimethylamino)alane replaces only one alkyl group of R_3B, Eq. (44) (R = Pr or Bu), and does not react with Bu_2BNMe_2 [640]. The $(AlMe_3)_2$-$(Me_2AlNPh_2)_2$ system, in equilibrium with $Me_2Al(NPh_2)(Me)AlMe_2$, has been studied [614b] from the standpoint of mechanism.

The aminosilanes $Me_nSi(NMe_2)_{4-n}$ react only to a small extent [1% (n = 1 or 2) or 7% (n = 3)] with butyl-lithium in hexane at room temperature [705]. However, cleavage of an Si-N bond in $(Me_2SiNEt)_2$ by methyl-lithium [Eq. (45)] has been reported [125]. The tin(IV) amide Me_3SnNMe_2 readily aminates an alkylborane, Eq. (46), or triethylalane, Eq. (47) [256].

$$Al(NMe_2)_3 + 2R_3B \longrightarrow 2R_2BNMe_2 + R_2AlNMe_2 \quad (44)$$

$$EtN(SiMe_2)_2NEt \xrightarrow[\text{(ii) } Me_3SiCl]{\text{(i) MeLi}} Me_3SiNEtSiMe_2N(Et)SiMe_3 + (Me_3SiNEt)_2SiMe_2 \quad (45)$$

$$Me_3SnNMe_2 + Bu_3B \longrightarrow Bu_2BNMe_2 + Me_3SnBu \quad (46)$$

$$2Me_3SnNMe_2 + 2Et_3Al \longrightarrow (Et_2AlNMe_2)_2 + 2Me_3SnEt \quad (47)$$

The germanium(II) amide (XVII) is a useful starting material for preparing the alkyl (XVIII) [see also Eq. (9)] [197, 272]; the amido function thus behaves as a protecting group, Eq. (48).

$$GeCl_2 \xrightarrow{Li[N(SiMe_3)_2]} \underset{\textbf{(XVII)}}{Ge[N(SiMe_3)_2]_2} \xrightarrow{Li[CH(SiMe_3)_2]} \underset{\textbf{(XVIII)}}{Ge[CH(SiMe_3)_2]_2} \quad (48)$$

Cyclopentadienyl groups are readily exchanged, as in the $Sn[N(SiMe_3)_2]_2$/$Sn(C_5H_5)_2$ [295] or $Hg[N(SiMe_3)_2]_2$/$Hg(C_5H_5)_2$ [648a] systems.

4. The NRR′/OR″ exchange reactions

A lithium amide reacts with either $CH_2{=}CHC{\equiv}CCMe_2OMe$ [702] or [192] a compound of the type C-O-M, M-O-M, or M-O-M′ (M and M′ = B, Ti, Si, Ge, Sn, P, or S) to give the corresponding organic, metal, or non-metal amide in high yield. The following examples serve as illustrations. An amino-alane [Eq. (49)] or -stannane [Eq. (50)] undergoes an exchange with an alkoxyborane [256,

640]. The amide $Al(NMe_2)_3$ reacts with a boroxine to give an aminoborane, Eq. (51) [640], and the scrambling on silicon [Eq. (52)] is used to prepare a phenylthioamine [46].

$$Al(NMe_2)_3 + PhB(OEt)_2 \longrightarrow PhB(NMe_2)_2 + (EtO)_2AlNMe_2 \quad (49)$$

$$3Me_3SnNMe_2 + 3B(OMe)_3 \longrightarrow 3Me_3SnOMe + B(NMe_2)_3 + 2B(OMe)_3 \quad (50)$$

$$3Al(NMe_2)_3 + (RBO)_3 \longrightarrow 3RB(NMe_2)_2 + \frac{3}{n}(Me_2NAlO)_n \quad (51)$$

$$Me_3SiNRR' + PhSOMe \longrightarrow Me_3SiOMe + PhSNRR' \quad (52)$$

Redistribution in the system $[Ti(OR)_4]/[Ti(NMe_2)_4]$ (R = Et, Pr^i, or Bu^t) has been investigated [745] (see Section A): the reactions are only mildly exothermic in toluene in contrast to the highly exothermic $[Ti(OR)_4]/[TiCl_4]$ exchanges.

5. The NRR′/SR″ exchange reactions

The borthiin (XIX) is prepared as shown in Eq. (53) (M = B or Al) [749]. Redistribution equilibria in the $B(NMe_2)_3/B(SMe)_3$ system are conveniently followed by ^1H n.m.r. spectroscopy and are essentially statistical with equilibrium attained in ca. 96 h at 120° C to give $(MeS)_nB(NMe_2)_{3-n}$ (n = 1 or 2) [170].

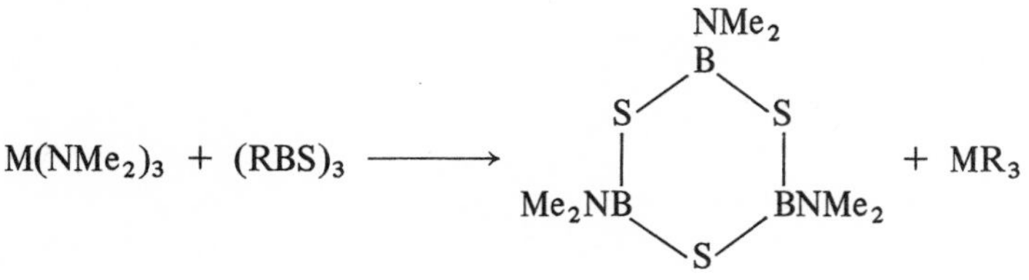

$$M(NMe_2)_3 + (RBS)_3 \longrightarrow \text{(XIX)} + MR_3 \quad (53)$$

(XIX): six-membered ring of alternating B and S atoms, each B bearing NMe_2.

(XIX)

6. The NR_2/NR′R″ exchange reactions

The principal type of amido group exchange reaction is the transamination (see Chapters 1-9), when the reagent is a primary or secondary amine, and these may be classified as reactions of a metal or metalloid amide with a protic reagent (Chapter 11). A rare example of a metathetical exchange reaction is illustrated in Eq. (54) [499a].

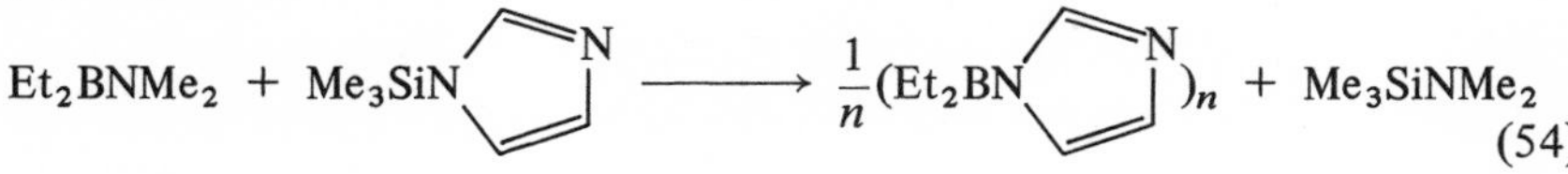

$$Et_2BNMe_2 + Me_3SiN\text{(imidazolyl)} \longrightarrow \frac{1}{n}(Et_2BN\text{(imidazolyl)})_n + Me_3SiNMe_2 \quad (54)$$

C REACTIONS OF AMIDES WITH ORGANIC CARBONYL, SULPHONYL, OR RELATED COMPOUNDS

1. Exchanges of univalent groups NRR′ and X

Amides of Na, B [447, 508], Al [48b, 55f], Si [79, 317, 377a, 504, 646], Ge [446, 624], Sn [144, 256, 591, 593, 634], P [57, 127a, 390, 692], As [34a, 245, 314, 468, 747], Sb [247, 248, 403], Bi [36], and Ti [144, 746, 747] aminate carbonyl compounds, and among some common amides the reactivity order appears to be Ti>Sn>As>B. With a carboxylic acid (but see Chapter 11), anhydride, or ester, a *gem*-triamine or enamine is formed [144, 256, 508, 746]. Aldehydes and ketones yield *gem*-diamines or enamines depending on the absence or presence of a β-hydrogen atom [508, 747]. Highly hindered enamines may be prepared by this method [747]. Compounds $B(NHR)_3$ and ketones (e.g., PhCOMe) afford Schiff's bases, while carboxylic acids yield secondary carboxylic amides [447]. The reactions of an acyloxyborane with an amine have been investigated [568, 569].

The initial step in each of the amide/carbonyl reactions may be a 1,2-dipolar insertion, Eq. (6) [144, 746]. The difference in behaviour of R_3SnNMe_2 and $[Ti(NMe_2)_4]$ towards an α,β-unsaturated carbonyl compound or *N,N*-dimethylformamide probably lies in the monofunctionality of the former. For example, cinnamaldehyde affords the insertion product with the tin amide [145, 257], but with $[Ti(NMe_2)_4]$ the *gem*-diamine is obtained [144]; its formation may result from a rearrangement such as that shown in Eq. (55).

$$(Me_2N)_2Ti(NMe_2)\text{–O–CH(NMe}_2\text{)CH=CHPh} \longrightarrow PhCH{=}CHCH(NMe_2)_2 + Ti(O)(NMe_2)_2 \qquad (55)$$

The initial dipolar adduct of a metal amide and a carboxylic ester has available alternative decomposition paths, Eq. (56). Acyl-oxygen fission (56a) is the more common, but alkyl-oxygen fission (56b) is observed with an enol-ester ($R = CH_2{=}CH$). The low pK_A values for $CH_2(CO_2Et)_2$ and $AcCH_2CO_2Et$ (see C.2) might have led to the expectation that they would behave like acetone, which with $[Ti(NMe_2)_4]$ [144] or Me_3SnNMe_2 [353] affords dimethylamine and mesityl oxide (but see Table for the reaction of Bu_3SnNR_2 and acetone when in different molar ratios [592]), presumably by a base-catalysed reaction. β-Keto-esters, however, give β-enamino-amides with $B(NMe_2)_3$ or $P(NMe_2)_3$; the β-keto-amides are clearly precursors in these reactions [115, 508].

$$\mathrm{-C(=O)OR} \xrightarrow{\mathrm{LMNMe_2}} \begin{cases} \mathrm{-C(O{-}ML)(NMe_2)(OR)} \xrightarrow{(a)} \mathrm{-C(=O)NMe_2} + \mathrm{LMOR} \\ \mathrm{-C(O{-}ML)(NMe_2)(OR)} \xrightarrow{(b)} \mathrm{-C(=O)OML} + \mathrm{RNMe_2} \end{cases} \qquad (56)$$

Aminoalanes smoothly convert carboxylic esters into acid amides in CH_2Cl_2 at 25–41° C, Eq. (57) [R, R′ = H, H; H, Bu^s; H, Bu^t; H, CH_2Ph; NRR′ = $\overline{N(CH_2)_4}$ or $\overline{N(CH_2)_5}$] [55f], and $HAl(NR_2)_2$ (R = Pr^i or Bu) with HBI_2 is a highly regiospecific reagent for conjugate reduction of an enone [48b].

$$\mathrm{R''COOR'''} \xrightarrow{\mathrm{Me_2AlNRR'}} \mathrm{R''CONRR'} \qquad (57)$$

Interesting examples of aminophosphines indulging in NR_2/OR' exchange are in Eqs. (58) [390], (59) [612], and (60) [127a].

$$\mathrm{Ph_2PNMe_2} + \mathrm{PhCNO} \longrightarrow \mathrm{Ph_2POPh} + \mathrm{Me_2NCN} \qquad (58)$$

$$\mathrm{C_3H_5P(NR''_2)_2} + \mathrm{R(R'O)P(O)CH_2CHO} \longrightarrow \mathrm{R(R'O)P(O)CH{=}CHNR''_2} \qquad (59)$$

$$\mathrm{P(NMe_2)_3} + o\text{-}\mathrm{C_6H_4(OCOPh)(N_3)} \longrightarrow \mathrm{C_6H_4(O)(N){=}P(NMe_2)_2} + \mathrm{PhC(O)NMe_2} \qquad (60)$$

2. Exchanges of amide groups for O^{2-} or S^{2-}

N,N-Dimethylformamide does not react with R_3SnNMe_2, but with $Sn(NMe_2)_4$, Eq. (61) [459], or $[Ti(NMe_2)_4]$, a *gem*-triamine is obtained.

$$\mathrm{Sn(NMe_2)_4} + 2\mathrm{Me_2NCH(=O)} \longrightarrow \mathrm{SnO_2} + 2\mathrm{HC(NMe_2)_3} \qquad (61)$$

A β-dicarbonyl compound $RCOCH_2COR'$ reacts with tetrakis(dimethylamido)titanium to give complex (XX), an enamine-amide, enamine-ester, keto-enamine, dienediamine (XXI), or dienetriamine depending on the nature of the groups R and R′ and the ratio of $[Ti(NMe_2)_4]$ to carbonyl compound, Eqs. (62)-(64) [741, 742, 743]. It is interesting that using an alternative aminating reagent leads to the enamine; and thus $[Ti(NMe_2)_4]$ is the compound of highest reactivity. Diethyl malonate or $CH_2(CONMe_2)_2$ behaves merely as a protic compound towards $[Ti(NMe_2)_4]$ [743] (see Chapter 11).

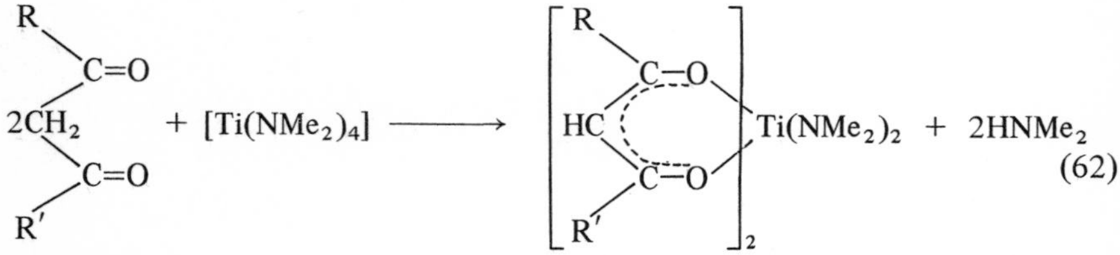

$$2\,RC(=O)CH_2C(=O)R' + [Ti(NMe_2)_4] \longrightarrow [(RC\cdots O)(HC)(R'C\cdots O)Ti(NMe_2)_2]_2 + 2HNMe_2 \qquad (62)$$

(XX)

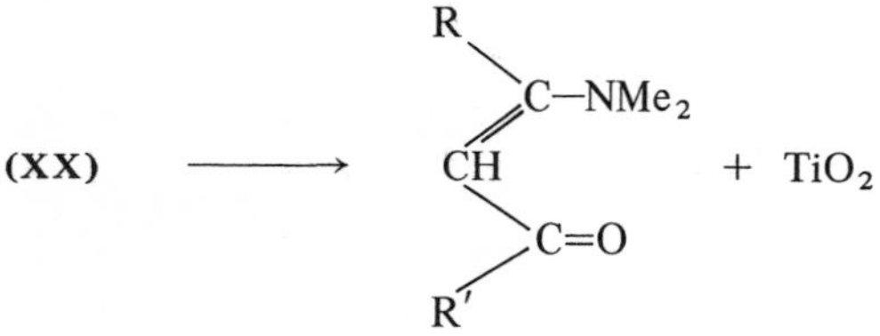

$$\textbf{(XX)} \longrightarrow RC(NMe_2){=}CH{-}C(=O)R' + TiO_2 \qquad (63)$$

(XXI)

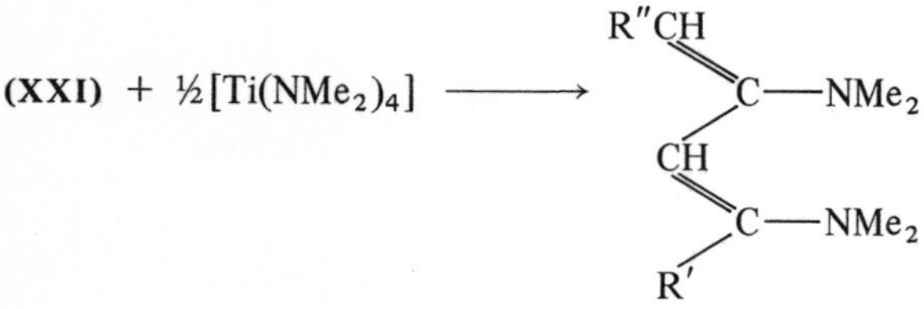

$$\textbf{(XXI)} + \tfrac{1}{2}[Ti(NMe_2)_4] \longrightarrow R''CH{=}C(NMe_2){-}CH{=}C(NMe_2)R' \qquad (64)$$

In the $[Ti(NMe_2)_4]/SO_2$ system an initial insertion step to afford (XXII) is similarly postulated, Eq. (65) [144], and subsequent pyrolysis of the titanyl sulphonamide (XXII) yields further $SO(NMe_2)_2$ [144]. An aminoborane behaves similarly [546].

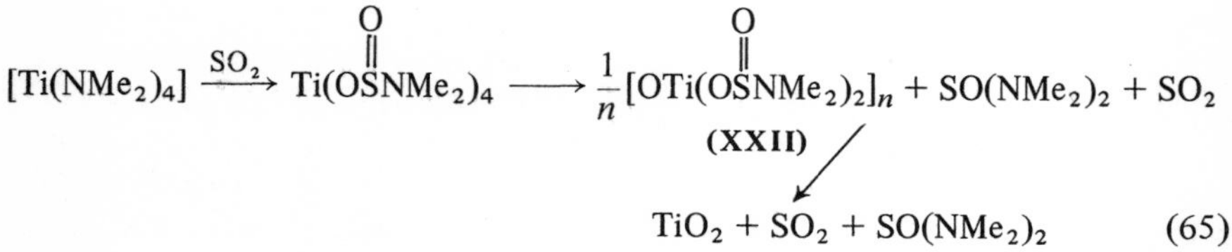

$$[Ti(NMe_2)_4] \xrightarrow{SO_2} Ti(OS(=O)NMe_2)_4 \longrightarrow \frac{1}{n}[OTi(OS(=O)NMe_2)_2]_n + SO(NMe_2)_2 + SO_2$$

(XXII)

$$\textbf{(XXII)} \longrightarrow TiO_2 + SO_2 + SO(NMe_2)_2 \qquad (65)$$

Diethylaminotrimethylstannane is inert towards dimethyl sulphoxide at 150° C; $[Ti(NMe_2)_4]$, however, reacts violently in absence of solvent with a methyl sulphoxide, Eq. (66) (R = Me or C_6H_{11}) [144]. The reaction may be assisted by the capacity of the sulphoxide to behave as a ligand [434]; the fragmentation of an intermediate 1:1-complex may be concerted, (XXIII) and (XXIV), the acidity of Me_2SO [169] being enhanced by co-ordination to an acceptor site.

$$[Ti(NMe_2)_4] + MeSOR \longrightarrow Me_2NCH_2SR + HNMe_2 + \frac{1}{n}[OTi(NMe_2)_2]_n \quad (66)$$

(XXIII)

(XXIV)

The mercury(II) amide $Hg[N(CF_3)_2]_2$ undergoes $(NR_2)/S^{2-}$ exchange with S_8 to yield HgS and *inter alia* $S[N(CF_3)_2]_2$ and $CF_3N{=}CH_2$ [772a].

D TABLE OF REACTIONS

Metathetical Exchange Reactions

Reagents		Products	References
NRR'/H System			
LiNRR'	B_2H_6	$LiBH_4 + B(NRR')_3$	[683]
$Be(NMe_2)_2$ (n = 1, 2, 3)	$Me_{3-n}AlH_n.NMe_3$	$Me_{3-n}AlH_{n-1}NMe_2.HBeNMe_2$ + Al-rich residue + NMe_3	[571]
$2B(NEt_2)_3$	$LiBH_4 + HCl$	$3HB(NEt_2)_2 + LiCl + H_2$	[534]
$XB(NMe_2)_2$ (X = H, NMe_2)	B_2H_6	$Me_2NB_2H_5 + H_2BNMe_2$ (trace)	[113, 369]
$B(NR_2)_3$	B_2H_6	μ-$R_2NB_2H_5 + HB(NR_2)_2$	[683]
$HB(NR_2)_2$	B_2H_6	μ-$R_2NB_2H_5 + H_2BNR_2$	[683]
$2H_2BNMe_2$	B_2H_6	2μ-$Me_2NB_2H_5$	[114]
$Cl_2(Me_2N)B_3N_3H_3$	B_2H_6	$Cl_2(H)B_3N_3H_3$	[60]
$B(NMe_2)_3$	B_2H_6 (excess)	$H_2BNMe_2 + \mu$-$Me_2NB_2H_5$ + cyclic $H_2B(NMe_2\text{-}BH_2)_2H$	[369]

Table of Reactions *(C'ntd.)*

Reagents		Products	References
$B(NEt_2)_3$	$BH_3.HNEt_2$	$2HB(NEt_2)_2 + H_2$	[534]
$Al(NMe_2)_3$	$BuBH_2.NMe_3$	$BuB(NMe_2)_2 + NMe_3 + H_2AlNMe_2$	[640]
$Al(NMe_2)_3$ ($R = Bu^s, Bu^t$)	$2RBH_2.NMe_2$	$2RB(NMe_2)H + NMe_3 + H_2AlNMe_2$	[640]
H_2AlNMe_2	B_2H_6 (excess)	μ-$Me_2NB_2H_5 + Al(BH_4)_3$	[367]
$(BH_4)_2AlNMe_2$	$BH_3.THF$	μ-$Me_2NB_2H_5 + HAl(BH_4)_2.2THF$	[367]
$Al(NMe_2)_3$	$2AlH_3.NMe_3$	$3H_2AlNMe_2 + 2NMe_3$	[640]
$Cl_{4-n}Si(NMe_2)_n$ ($n = 1, 2$)	$LiAlH_4$	SiH_4 + trace of H_2	[136]
$(H_3Si)_3N$	$LiAl(PH_4)_4$	SiH_4 + solid polymer	[269]
$Me_3SiN(Me)COR$	([ring: Me, BH])$_2$	Me_3SiH + [ring: Me, BN(Me)COR]	[470a]
$3Me_3SnNMe_2$	$BH_3.NEt_3$	$3Me_3SnH + NEt_3 + B(NMe_2)_3$	[256]
$2R_{4-n}Sn(NR'_2)_n$ ($R' = Et, n = 4$; $R = R' = Me, n = 1$; $R = Me, R' = Et, n = 2$; $R = Bu, R' = Et, n = 1$; $R = Ph, R' = Et, n = 2$)	$n(BH_3)_2$	$2R_{4-n}SnH_n + 2nH_2BNR'_2$	[414]
$R_{4-n}Sn(NEt_2)_n$	$(Bu_2AlH)_n$	$R_{4-n}SnH_n + nBu_2AlNEt_2$	[414]
NRR'/Hal System			
$LiN(SiMe_3)_2$	$(ClCH_2)_3N$	$LiCl + CH_2[N(SiMe_3)_2]_2$ + polymer	[237]
$LiN(SiCl_3)_2$	SiF_4 or $SiCl_4$	$(Cl_3Si)_2N(SiF_3)$ or cyclo-$[Cl_3SiN{-}SiCl_2{-}NSiCl_3{-}SiCl_2]$	[737]
$(LiNBu^t)_2SnBu^t_2$	$SnCl_2$	$2LiCl + Bu^t_2Sn(\mu\text{-}NBu^t)_2Sn$	[287a]
$LiNRSiMe_3$ ($R = Me, Et, Bu, Ph$)	$Ni(PF_3)_4$	$LiF + (PF_3)_3Ni(PF_2NRSiMe_3)$	[409]
$2NaN(SiMe_3)_2$	$Me_3SiNSOF_2$	$2NaF + (Me_3Si)_2O + S(=NSiMe_3)_3$	[450]
$LiN(SiMe_3)_2$	$PF_2N(SiMe_3)_2$	$LiF + Me_3SiF + (Me_3Si)_2NP = NSiMe_3$	[520]
$nMN(SiMe_3)_2$ [$M = Li, Na; n = 2, X = (=NSiMe_3); n = 1, X = F_2$]	SOF_4 ($-70°$)	$nLiF + nMe_3SiF + SOX(=NSiMe_3)$	[233]

Table of Reactions *(C'ntd.)*

Reagents		Products	References
[When M = Na and n = 3, the products were $3NaF + 3Me_3SiF + (Me_3Si)_2O + S(=NSiMe_3)_3$]			[450]
$2LiN(SiMe_3)_2$	NSF_3	$LiF + Me_3SiF + S(=NSiMe_3)_3 + SF_2(NSiMe_3)_2$	[267, 268]
$MN(SiMe_3)_2$ (M = Li, Na)	$ArN_2^+Cl^-$(<−20°)	$MCl + ArN{=}N{-}N(SiMe_3)_2$	[752, 753]
$MN(SiMe_3)_2$ (M = Li, Na)	NOCl	$(Me_3Si)_2NNO$	[754]
$NaN(SiMe_3)_2$	$2VOCl_3$ + 4L (L = 2,2'-pyridylquinoline [structure])	[structure: dinuclear V complex, (O)(Cl)(L)₂V bridged by N and Cl to V(L)₂(Cl)(O)] $+ 2Me_3SiCl + NaCl$	[119]
$NaN(SiMe_3)_2$	$SiCl_4$	[structure: spiro Si bonded in one ring to N(SiMe₃) and C(H)(SiMe₃) bridged by SiMe₂, and in the other ring to two N(SiMe₃) bridged by SiMe₂]	[732a]
$(BrMg)_2NC_6H_4Br$-p	Me_3SiCl	$BrMgCl + (Me_3Si)_2NC_6H_4Br$-$p$	[768]
$(CH_2BNMe_2)_3$	BCl_3	$Cl_2BNMe_2 + (CH_2BCl)_3$	[401a]
$B(NEt_2)_3$	$2BF_3$	$3F_2BNEt_2$	[258]
$FB(NR_2)_2$ (R = Me, Me_3Si)	BF_3	$2F_2BNR_2$	[111]
$2B(NMe_2)_3$ (X = Cl	BX_3	$3XB(NMe_2)_2$	[107, 534]
Br)			[107]
$2B(NEt_2)_3$	BCl_3	$3ClB(NEt_2)_2$	[258]
$B[N(Me)Ph]_3$	$2BCl_3$	$3Cl_2BN(Me)Ph$	[51]
$B(NR_2)_3$ (R = Me, Et; X = F, Cl, Br, I)	$2BX_3$	$3X_2BNR_2$	[703]
$PhB(NR_2)_2$ (R_2 = H, Bu^t; H, Pr^i; Me_2; Et_2)	$PhBCl_2$	$PhB(Cl)NR_2$	[435a]
$PhB(NEt_2)_2$	BCl_3	$PhBCl_2 + 2Cl_2BNEt_2$	[252]
$B(NMe_2)_3$	$2o$-$C_6H_4O_2BNCS$	$(SCN)_2BNMe_2 + 2o$-$C_6H_4O_2BNMe_2$	[307]
Cl_2BNMe_2	Me_2BNMe_2	$2MeB(Cl)NMe_2$	[283]
$3B(NMe_2)_3$	$2(BrBS)_3$	$3Br_2BNMe_2 + 2(Me_2NBS)_3$	[751]
Me_2BNMe_2 (X = F, Cl)	BX_3	$Me_2BX + X_2BNMe_2$	[61, 111]
$B_2(NMe_2)_4$	$3BF_3$	$B_2(NMe_2)_2F_2 + 2F_2BNMe_2$	[542]
$nB_2(NMe_2)_4$ (n = 1	BCl_3	$nCl_{2-n}(Me_2N)_nB{-}B(NMe_2)Cl + ClB(NMe_2)_2$	[542]
2)			[535, 542]

Table of Reactions *(C'ntd.)*

Reagents		Products	References
$B_2(NMe_2)_4$	$4BCl_3$	$Me_2NB_3Cl_6^* + 3Cl_2BNMe_2$	[542]
$B_2(NMe_2)_4$	$B_2Cl_2(NMe_2)_2$	$2(Me_2N)_2B\text{–}B(NMe_2)Cl$	[542]
$B_2(NMe_2)_4$ (R = Et, Bu, Ph)	$2R_2BCl$	$B_2Cl_2(NMe_2)_2 + R_2BNMe_2$	[542]
$B_2(NMe_2)_4$ (R = Br, Me)	$RBBr_2$	$B_2(NMe_2)_2Br_2 + RB(NMe_2)_2$	[542]
$2B_4(NMe_2)_6$	BCl_3	$2B_4(NMe_2)_5Cl + ClB(NMe_2)_2$	[303a]
$B_2(NMe_2)_4$ (X = H, Cl)	$CXCl_3$	$B_2Cl(NMe_2)_3$	[286d]
$B_2(NMe_2)_4$	CH_2Cl_2	$B(NMe_2)_3 + ClB(NMe_2)_2$ (major)	[286d]
$B(NMeSiMe_3)_3$ ($n = 1, 2, 3$)	Me_2BBr	$Me_3SiBr + (Me_2BNMe)_nB(NMeSiMe_3)_{3-n}$	[548]
$Me_2NB<^{(NMe)_2}_{NMe}>BNMe_2$ (M = Sb, $n = 5$; M = B, $n = 3$)	MF_n	$FB<^{(NMe)_2}_{NMe}>BF$	[254]
$(RN\text{–}BNHR)_3$	BF_3	$(RNBF)_3 + B(NHR)_3$	[482]
$B(NMe_2)_3$	$TiCl_4$	$Cl_2BNMe_2 + ClB(NMe_2)_2 + Cl_3TiNMe_2 + Cl_2Ti(NMe_2)_2$	[417]
$MeB(NMe_2)_2$	$TiCl_4$	$MeB(Cl)NMe_2 + Cl_3TiNMe_2$	[417]
$B_2(NMe_2)_4$	$TiBr_4$	$B_2Br_2(NMe_2)_2.TiBr_3$	[712]
Cl_2BNMe_2	VCl_4	$VCl_3 + BCl_3.MeN{=}CH_2 + BCl_3.HNMe_2$	[371]
$B(NEt_2)_3$ ($n = 1, 2$; X = O, S)	PCl_3 or $P(X)Cl_3$	$Cl_nB(NEt_2)_{3-n}$	[170]
F_2BNHPF_2	F_2PNH_2	$HN(PF_2)_2 + BF_3.NH_3$	[47a]
$B[NMeSiMe_3]_3$ ($n = 1\text{–}3$)	Me_2BBr	$(Me_2BNMe)_nB[NMeSiMe_3]_{3-n}$	[548]
$(CH_2NMe)_2BNMeSiMe_3$ [X = Br, $R_2 = Me_2$; X = Cl, $R_2 = (MeNCH_2)_2, (SCH_2)_2, CH_2CHMe(CH_2)_2$]	R_2BX	$(CH_2NMe)_2BNMeBR_2 + Me_3SiX$	[549]
$Al(NMe_2)_3$ ($n = 2, 3$)	$nBCl_3$	$nCl_{3-n}B(NMe_2)_n + Cl_nAl(NMe_2)_{3-n}$	[641]
$ClAl(NMe_2)_2$	BCl_3	$Cl_2AlNMe_2 + Cl_2BNMe_2$	[641]
$3LiAl(NMe_2)_4$	$AlCl_3$	$3LiCl + 4Al(NMe_2)_3$	[641]
$HAl(NR_2)_2$ (R = Pr^i, Bu)	BHI_2	aminoboranes	[48b]
$2Al(NMe_2)_3$	$3AlCl_3$	$3ClAl(NMe_2)_2$	[641]
$Al(NMe_2)_3$	$2AlCl_3$	$3Cl_2AlNMe_2$	[641]
Et_2AlNR_2 (R = Me, X = Cl, Br, I, OEt, SMe; R = Et, X = Br)	Et_2AlX	$Et_2\overset{-}{Al}<^{\overset{+}{N}R_2}_{\overset{+}{X}}>\overset{-}{Al}Et_2$	[243]

*This may be the complex $B_2Cl_4.Cl_2BNMe_2$

Table of Reactions *(C'ntd.)*

Reagents		Products	References
$2R_2AlNR_2$	AlX_3	$2RXAlNR_2 + R_2AlX$	[315]
R_2AlNR_2	$2AlX_3$	$X_2AlNR_2 + 2RAlX_2$	[315]
Me_3SiNMe_2	C_6F_5CN	$Me_3SiF + p\text{-}Me_2NC_6F_4CN$	[347]
H_3SiNMe_2	BF_3	$H_3SiF + (MeNBF)_3$	[708]
Me_3SiNMe_2	BF_3	$Me_3SiF + F_2BNMe_2$	[223]
Me_3SiNR_2	$F_2BN(SiMe_3)_2$	$(Me_3Si)_2NB(F)NR_2$	[227]
(R = Me, Et, Pr, Bu)			
$(Me_3Si)_2NR$	$F_2BN(SiMe_3)_2$	$Me_3SiNRBFN(SiMe_3)_2$	[227]
(R = Me, Et)			
$(Me_3Si)_2NH$	BF_3 (200°)	$2Me_3SiF + (FBNH)_n$	[529]
$(Me_3Si)_2NH$	BF_3 (0–20°)	$Me_3SiF + Me_3SiNHBF_2$	[529]
$(MeH_2Si)_2NMe$	BF_3	$MeH_2SiF + F_2BNMe_2$	[529]
$(Me_3Si)_2NH$	$F_2BN(SiMe_3)_2$	$Me_3SiF + (Me_3Si)_2NB(F)NHSiMe_3$	[229]
Me_3SiNRR'	$F_2BN(SiMe_3)_2$	$Me_3SiF + (Me_3Si)_2NB(F)NRR'$	[227]
(R = R' = Me, Et, Pr, Bu; R = Me_3Si, R' = H, Me, Et)			
$(H_3Si)_2NMe$	BF_3	$H_3SiF + (MeNBF)_3$	[708]
$(R_3Si)_3N$	BF_3	$R_3SiF + F_2BN(SiR_3)_2$	
(R = Me			[62, 643]
H)			[708]
$(MeH_2Si)_3N$	BF_3	$MeH_2SiF + F_2BN(SiH_2Me)_2$	[223]
$(H_3SiSiH_2)_3N$	BF_3	$H_3SiSiH_2F + F_2BN(SiH_2SiH_3)_2$	[1]
nMe_3SiNMe_2	BX_3	$nMe_3SiX + X_{3-n}B(NMe_2)_n$	[529]
(n = 1, 2, 3; X = Cl, Br)			
$Me_3SiNClR$	$R'PCl_2$	$Me_3SiCl + RN{=}PCl_2R'$	[587a]
(R = Cl, Me, Ph, OPh, Et_2N, $Me(Cl_2P)N$; R' = Bu^t; or R=Cl, Et_2N; R' = Bu^t)			
Me_3SiNR_2	BX_3	$Me_3SiX + X_2BNR_2$	[9]
(R = Et, X = Cl, Br; R = Bu, X = I)			
$(Me_3Si)_2NH$	BCl_3 (0°)	$Me_3SiCl + NH_4Cl + H_2NB[NHSiMe_3]N(SiMe_3)_2$	[165]
$nMe_3SiNREt$	BCl_3	$nMe_3SiCl + Cl_{3-n}B(NREt)_n$	[9]
(R = Et, n = 2, 3; R = H, n = 1)			
$(H_3Si)_2NMe$	BCl_3	$2H_3SiCl + {}^1/_3(MeNBCl)_3$	[112]
$(Me_3Si)_2NH$	BX_3	$2Me_3SiCl + {}^1/_3(HNBX)_3$	[529]
(X = Cl, Br, I)			
$(Me_3Si)_2NH$	BCl_3	$Me_3SiCl + Cl_2BNHSiMe_3$	[37]
$3(Me_3Si)_2NMe$	$3BF_3$	$(MeNBF)_3 + 6Me_3SiF$	[547]
$(Me_3Si)_2NMe$	BX_3	$X_2BNMeSiMe_3 + Me_3SiX$	[547]
(X = Cl, Br)			
$2(Me_3Si)_2NMe$	BF_3	$FB(NMeSiMe_3)_2 + 2Me_3SiF$	[547]
$3(Me_3Si)_2NMe$	BBr_3	$B(NMeSiMe_3)_3 + 3Me_3SiBr$	[547]
$(R_3Si)_3N$	BCl_3	$R_3SiCl + Cl_2BN(SiR_3)_2$	
(R = Me			[62]
H)			[112]
$(Me_3Si)_3N$	BX_3	$Me_3SiX + X_2BN(SiMe_3)_2$	[299]
(X = F, Cl, Br, I)			

Table of Reactions *(C'ntd.)*

Reagents		Products	References
$3Me_3SiNEt_2$	$(ClBNH)_3$	$3Me_3SiCl + (Et_2NBNH)_3$	[9]
$2Me_3SiNEt_2$	$PhBCl_2$	$2Me_3SiCl + PhB(NEt_2)_2$	[9]
$(Me_3Si)_2NH$	$RBCl_2$	$2Me_3SiCl + (RBNH)_3$	
(R = Ph)			[349]
(R = Me, Pr, Bu, Bu^t)			[529]
$(Me_3Si)_2NMe$	$Cl_2BNMeSiMe_3$	$ClB(NMeSiMe_3)_2 + Me_3SiCl$	[547]
$3(Me_3Si)_2NMe$	$3MeBBr_2$	$(MeBNMe)_3 + 6Me_3SiBr$	[547]
$(Me_3Si)_2NMe$	$2MeBBr_2$	$MeN(BMeNMeSiMe_3)_2$ $(MeBNMe)_3 + Me_3SiBr$	[547]
$2(Me_3Si)_2NMe$	$PhBCl_2$	$PhB(NMeSiMe_3)_2 + 2Me_3SiCl$	[547]
$3(Me_3Si)_2NMe$	$2PhBCl_2$	$(Me_3SiNMeBPh)_2NMe + 4Me_3SiCl$	[547]
$3(Me_3Si)_2NMe$ (R = Cl, Ph)	$3RBCl_2$	$6Me_3SiCl + (RBNMe)_3$	[9]
$(Me_2SiNMe)_3$ (R = Cl, Ph)	$3RBCl_2$	$3Me_2SiCl_2 + (RBNMe)_3$	[9]
$(NMe)_2$ MeB $SiMe_2$ NMe (X = Cl, Br, SMe)	BX_3	$(NMe)_2$ MeB BX NMe	[540c]
CH_2—NEt $SiMe_2$ CH_2—NEt	$PhBCl_2$	Me_2SiCl_2 + CH_2—NEt BPh CH_2—NEt	[13]
N $NSiMe_3$ N (R = X = F, Cl; R = Ph, X = Cl)	R_2BX	N $NBR_2 + Me_3SiX$ N	[85]
Me_3SiNMe_2	R_2BCl	$Me_3SiCl + R_2BNMe_2$	[529]
$Me_3SiNR(CH_2)_nCN$ (n = 1, R = Me, Bu, CH_2CN; n = 2, R = Me)	Me_2BBr	$Me_2BNR(CH_2)_nCN + Me_3SiBr$	[476b]
$(Me_3SiNMeBR)_2NMe$ (R = Me, R′ = Ph, X = Cl; R = Me, R′ = Me, X = Br; R = Ph, R′ = Me, X = Br)	$R'BX_2$	$R'B(NMeBR)_2NMe$ $+ Me_3SiX$	[547a]
$(Me_3SiNMeBPh)_2NMe$	Ph_2BCl	$Ph_3B + (PhBNMe)_3 + Me_3SiCl$	[547a]
$Me_3SiNHBPh_2$	Me_2BBr	$Me_2BNHBPh_2$	[350a]
$Me_3SiNMeP(S)Me_2$ (X = F, Cl, Br)	BX_3	$Me_2P(S)NMeBX_2$	[498a]
		$[Me_2P(S)NMe]_2BX$	[498a]
		$[Me_2P(S)NMe]_3B$	[498a]
$Me_3SiNMeP(S)Me_2$	$MeBX_2$	$[Me_2P(S)NMe]_2BMe$	[498a]
$Me_3SiNMePPh_2$ (X = Cl, Br)	BX_3	$Ph_2PNMeBX_2 + Me_3SiX$	[549a]
		$\rightarrow (XBNMe)_3 + Ph_2PX$	[549a]

Table of Reactions *(C'ntd.)*

Reagents		Products	References
$[Me_3SiN(Me)]_2PPh$	Me_2BBr	$PhP[N(Me)BMe_2]_2 + Me_3SiBr$	[549a]
$(Me_3Si)_2NMe$ (R = Me, Ph, NMe_2; X = Cl, Br)	R_2BX	$R_2BNMeSiMe_3 + Me_3SiX$	[547]
$3(Me_3Si)_2NMe$	$6Ph_2BCl$	$(PhBNMe)_3 + 3BPh_3 + 6Me_3SiCl$	[547]
$(Me_3Si)_2NH$	$PhB(Cl)NMe_2$	$Me_3SiCl + Ph(Me_3SiNH)BNMe_2$	[349]
$(Me_3Si)_2NH$	$2PhB(Cl)NMe_2$	$2Me_3SiCl + (Me_2NBPh)_2NH$	[349]
$(H_3Si)_2NMe$	B_2H_5Br	$H_3SiBr + SiH_4 + B_2H_6 + (MeNBH)_3 + (MeNSiH_3)B_2H_5$	[112]
$(Me_3Si)_2NH$ (R = Pr, n = 1, 2; R = Bu, n = 2)	nR_2BCl	$nMe_3SiCl + (R_2B)_nNH(SiMe_3)_{2-n}$	[529]
$Me_3SiNHBPr_2$ (R = Pr, Bu)	R_2BCl	$Me_3SiCl + R_2BN(BPr_2)H$	[529]
$(Me_3Si)_2NR$ (R = H, Me)	$2Ph_2BBr$	$(Ph_2B)_2NR + 2Me_3SiBr$	[350a]
$(Me_3Si)_2NR$ (R = H, Me)	Me-substituted five-membered ring BCl	Me-substituted five-membered ring $BNRSiMe_3 + Me_3SiCl$	[549]
		→ (Me-substituted five-membered ring $B)_2NR$	[549]
$(Me_3Si)_2NH$	$2o\text{-}C_6H_4O_2BCl$	$2Me_3SiCl + (o\text{-}C_6H_4O_2B)_2NH$	[436]
$(Me_3Si)_2NH$	BCl_3	$Me_3SiCl + Me_3SiNHBCl_2$	[62]
$(Me_3Si)_2NMe$ (R = Me, X = Br; R = Ph, X = Cl)	RBX_2	$Me_3SiNMeBRX$	[55c]
$(ClSiMe_2)_2NMe$ (R = Ph, Cl)	$RBCl_2$	$ClSiMe_2NMeBRCl$	[55c]
$(BrSiMe_2)_2NMe$ (R = Br, X = Me; R = Me, Br, Ph, X = Br)	RBX_2	$BrSiMe_2NMeBRX$	[55c]
$Cl_3SiNMeSiMe_3$	BCl_3	$Cl_2BNMeSiCl_3 + Cl_2BNMeSiMe_3$	[55c]
$(H_3Si)_3N$	B_2H_5Br	$\frac{1}{2}[(H_3Si)_2NBH_2]_2 + H_3SiBr + B_2H_6$	[112]
$Me_3SiNR'P(S)R_2$ (R = Me, Ph, Me_2N; R' = Me, Et, Ph)	R_2BX	$Me_3SiX + R_2BNR'P(S)R_2$	[541]
$(Me_3Si)_2NPNSiMe_3$ (X = Cl, Br)	BX_3	Me_3SiX + XB(N($SiMe_3$))$_2$PX (four-membered B–N–P–N ring)	[518a]
$(Me_3Si)_2NH$	ring: –NMe–(CH$_2$)$_2$–NMe–BCl	Me_3SiCl + ring: –NMe–(CH$_2$)$_2$–NMe–BNHR	[549]

Table of Reactions *(C'ntd.)*

Reagents		Products	References
($R = Me_3Si$, $\left[\begin{matrix}NMe\\NMe\end{matrix}\right]B$)			
$(Me_2SiNH)_3$	$n PhBCl_2$	cyclic $PhB(NH)_2X X'NH$ [six-membered ring: B(Ph), NH, X', NH, X, NH] + Me_2SiCl_2	[552]
($n = 1, 3$; $X = X' = Me_2Si$, PhB, respectively)			
$(Me_3Si)_2NH$	$ClH_2B.OEt_2$	$Me_3SiCl + (H_2BNHSiMe_3)$ $H_2 + (BHNSiMe_3)_n$	[774]
$(Me_3Si)_2NMe$	$ClH_2B.OEt_2$	$Me_3SiCl + H_2BN(Me)SiMe_3$	[774]
$(Me_3Si)_2NR$	Me-benzo-1,3,2-dithiaborole-BCl	Me_3SiCl + (Me-benzo-1,3,2-dithiaborol-2-yl)$_2$N(R)	[701a]
($R = H, Me_3Si$)			
$(Me_2SiNH)_3$	Pr_2BCl	$Me_2SiCl_2 + (Pr_2BNHSiMe_2)_2NH$	[529]
$(Me_2SiNMe)_3$	$RBBr_2$	$BrMe_2SiNMeBBrR$	[55c]
($R = Me, Ph$)			
	Ph_2BBr	$(BrMe_2Si)_2NMe + BrMe_2SiNMeBPh_2 + (Ph_2B)_2NMe$	[55c]
	Me_2BBr	$BrMe_2SiNMeBMe_2 + Me_2B(NMeSiMe_2)_3Br$	[55c]
	BCl_3	$ClMe_2SiNMeBCl_2$	[55c]
	BBr_3	$(BrBNMe)_3$	[55c]
$(Me_3Si)_2NB$(Me-cyclo-C$_4$)	R_2BX	Me_3SiX + (Me-cyclo-C$_4$)$BN(SiMe_3)BR_2$	[549]
($R_2B = Me_2B$, $X = Br$; R_2B = (Me-cyclo-C$_4$)B, $X = Cl$)			
$Me_3SiN(BMeBr)_2$	$(Me_3SiO)_2$	cyclic MeB–O–O–BMe–N(SiMe$_3$)	[55b]
$(Me_3Si)_3N$	R_2BBr	$(Me_3Si)_2NBR_2$	[706a]
		$\rightarrow Me_3SiN(BR_2)_2$	[706a]
		$\rightarrow (R_2B)_3N$	[706a]
[$R_2 = Me_2$, $CH_2CHMe(CH_2)_2$]			
$(Me_3Si)_3N$	$MeBX_2$	$(MeBX)_2NSiMe_3 + Me_3SiX$	[55a]
[X_2 = MeO, Br; $(MeO)_2$; Cl, NMe_2; Br, NMe_2; OMe, NMe_2; $(NMe_2)_2$; SMe, NMe_2; $(SMe)_2$; Br, SMe]			
$(Me_3Si)_3N$	BBr_3	$BrSiMe_2(Me_3Si)NBMeBr + Me_3SiBr$	[55d]

Table of Reactions *(C'ntd.)*

Reagents		Products	References
$(Me_3Si)_2NH$ (X = Cl, Br, I)	AlX_3	$2Me_3SiX + (HNAlX)$	[37]
$(Me_3Si)_2NH$ (X = Cl, Br)	AlX_3	$Me_3SiX + Me_3SiNHAlX_2$	[37]
$2(Me_3Si)_2NH$	$2AlCl_3$	$2Me_3SiCl + (Me_3SiNHAlCl_2)_2$	[670]
$(Me_3Si)_3N$ (X = Cl, Br)	AlX_3	$Me_3SiX + (Me_3Si)_2NAlX_2$	[62]
$(Me_3Si)_2NCl$	CF_3COCl	$Me_3SiCl + ClN = C(CF_3)OSiMe_3$	[448]
$Me_3SiN(R)CH_2SiR'R''Me$ (R, R', R'' = Pr^i, Me, MeO; Pr^i, MeO, MeO; Ph, Et, OEt)	ClCN	$Me_3SiCl + MeR'R''SiCH_2N(R)CN$	[399]
$R'''_3SiNR'R''$ (R' = H, Me, Et, C_3H_5, $EtMe_2SiCH_2$; R'' = Me, Et, Ph, Me_3Si, Me_2HSi; R = Me, Et, C_3H_5; R''' = Me, Et)	ROOCCl	$R'''_3SiCl + ROOCNR'R''$	[656]
$ClSiMe_2(CH_2)_3N(R)SiMe_3$	MeOOCCl	Me_3SiCl + (ring: Me_2Si, NCO_2Me)	[656]
Me_3SiNMe_2	Si_2F_6	$F_3SiNMe_2 + (SiF_2)_n$	[102]
$4Me_3SiNHPh$	$SiBr_4$	$4Me_3SiBr + Si(NHPh)_4$	[29]
$3Et_3SiNHPh$	$AmSiI_3$	$3Et_3SiI + AmSi(NHPh)_3$	[29]
$2Me_3SiNMe_2$	$SiCl_4$	$2Me_3SiCl + Cl_2Si(NMe_2)_2$	[282]
$Me_3SiNMeBMe_2$	$MeSiCl_3$	$Me_3SiNMeSiMeCl_2$ + $Me_2BNMeSiMeCl_2$	[55c]
		$\rightarrow (MeSiCl_2)_2NMe$	[55c]
$(Me_3Si)_2NMe$	$SiCl_4$	$Me_3SiCl + Me_3SiNMeSiCl_3$	[282]
$(Me_3Si)_2NH$ (R = R' = Me, Ph; R = Me, R' = Ph)	$RR'SiCl_2$	$Me_3SiCl + (\text{—}HNSiRR'\text{—})_3$	[443]
$MeCON(SiMe_3)_2$	$ClCH_2SiMe_2Cl$	(structure: $\overset{+}{O}$=Me, $Me_2\overset{-}{Si}$, Cl, N, $SiMe_2Cl$)	[398, 564b]
$(Me_3Si)_2NPR(=NSiMe_3)X$ (X = Cl, Br, I; R = CCl_3, Pr^i, Et; X' = Cl, Br)	SiX'_4	Me_3SiX' + XP—$NSiMe_3$, Me_3SiN—SiX'_2 (ring)	[518]
Me_3SiNEt_2	Et_3SiCl	$Me_3SiCl + Et_3SiNEt_2$	[18]
$2R_{4-n}Si(NHPh)_n$ (R = Me, n = 3; R = Et, n = 1)	nPh_2SiCl_2	$2R_{4-n}SiCl_n + nPh_2Si(NHPh)_2$	[29]
(ring: BMe, Me_3SiN, $NSiMe_3$)	Me_2PCl	(ring: BMe, Me_2PN, $NSiMe_3$)	[549a]
$(Me_3SiNMe)_2S$	$(Me_2ClSi)_2$ (excess)	$Me_3SiCl + (Me_2SiClNMe)_2S$	[664]

Table of Reactions *(C'ntd.)*

Reagents		Products	References
$(Me_3SiNMe)_2S$	$(Me_2ClSi)_2$	Me_3SiCl + S(NMe—$SiMe_2$—$SiMe_2$—NMe—S—NMe) (ring: S bonded to NMe—$SiMe_2$ and NMe—S—NMe, closed through $SiMe_2$)	[664]
$Me_3SiN(Me)PPh_2$	$MeSiCl_3$	$Me_3SiCl + Ph_2PN(Me)SiMeCl_2$	[366]
$(RR'_2SiNH)_n$	$RSiCl_3$	$RR'_2SiCl + RR'_2SiNHSiRCl_2$	[733]
(R = Me, R′ = Cl; R = Cl, R′ = Me; n = 3, 4)			
$(R_3Si)_2NH$	$MeSiCl_3$	$R_3SiCl + R_3SiNHSi(Me)Cl_2$	[733]
(R = Me, Cl)			
$3Me_2Si(NHPh)_2$	$2PhSi(NCS)_3$	$3Me_2Si(NCS)_2 + 2PhSi(NHPh)_3$	[31]
$Me_2Si(NHPh)_2$	$Ph_2Si(NCO)_2$	$Me_2Si(NCO)_2 + Ph_2Si(NHPh)_2$	[31]
Me_3SiNRR'	Me_3SiO_3SCl	$Me_3SiCl + Me_3SiO_3SNRR'$	[126]
(R = R′ = Me, Et, Me_3Si; R = Me_3Si, R′ = Me, Et)			
Me_3SiNR_2	TiX_4	$R_2NTiX_3 + Me_3SiX$	
(R = Me, Et; X = F;			[123, 124]
R = Me, Et, Pr^i; X = Cl, Br)			[117]
$2(Me_3Si)_2NH$	$2FeCl_3$	$2Me_3SiCl$ + $Me_3Si\overset{+}{N}H$—$\overset{-}{Fe}Cl_2$ / $Cl_2\overset{-}{Fe}$—$\overset{+}{N}HSiMe_3$ (four-membered ring)	[378]
$(Me_3Si)_2NH$	WF_6	$Me_3SiF + [MeNH_3][A]$	[142]
[A] = $[WF_5NR]$, $[F_4(MeN)WFW(NMe)F_4]$			
Me_3SiNRR'	$WF_6 + MeCN$	$Me_3SiF + WF_4(NMe)(MeCN)$	[293]
(R = Me_3Si, R′ = H; R = Me, R′ = PF_2)			
$2(Me_3Si)_3N$	$2TiCl_4$	$(Me_3SiN—TiCl_2)_4 + 4Me_3SiCl$	[20b, 21]
$Me_3SiNHPh$	$GeBr_4$	$4Me_3SiBr + Ge(NHPh)_4$	[29]
$Me_3SiNHEt$	Me_3SnBr	$Me_3SiBr + Me_3SnNHEt$	[12]
$(Me_3Si)_2NCOMe$	$2ClSiMe_2CH_2X$	$2Me_3SiCl + MeCON(SiMe_2CH_2X)_2$	[398]
(X = Cl, Br)			
$(Me_3Si)_2NSO_2R'$	R_2SnCl_2	Me_3SiCl + R_2Sn—NSO_2R' / $R'SO_2N$—SnR_2 (four-membered ring)	[493]
(R′ = Me_2N, morpholino; R = Me, Ph)			
Me_3SiNR_2	PF_5	$Me_3SiF + F_4PNR_2$	
(R = Me			[205]
Et			[63]
Me, Et, Ph)		also $F_3P(NR_2)_2$	[676]
Me_3SiNR_2	PF_4X	$Me_3SiF + F_3XPNR_2$	[678]
(R = Et, X = NEt_2; R = Me, X = Ph)			
Me_3SiNHR	$Ni(PF_3)_4$	$Me_3SiF + (PF_3)_nNi(PF_2NHR)_{4-n}$	[409]
(R = Me, Bu; n = 0, 1, 3)			
$(Me_3SiN)_2S$	PF_5	bicyclic $S_3N_4PF_3$ (ring: S=N—PF_3, N, S=N, N=S)	[629]

Table of Reactions *(C'ntd.)*

Reagents		Products	References
(pyrrolyl)$NSiMe_3$	$PhPF_4$	Me_3SiF + (pyrrolyl)NPF_3Ph	[678]
Me_3SiNR_2 (R = Me, Et; R_2N = pyrrolyl; R′ = Et, Ph)	$R'PF_4$	$Me_3SiF + R'PF_3NR_2$	[676]
$n(Me_3Si)_2NMe$ (n = 1, 2)	nPF_5	$2nMe_3SiF + (MeNPF_3)_2$	[205] (n = 1), [679] (n = 2)
$Me_3SiNMeNMeSiMe_3$	$2PF_5$	$2Me_3SiF + F_4PNMeNMePF_4$	[274]
$(Me_3Si)_2NMe$	$2RPF_4$	$4Me_3SiF + (RF_2PNMe)_2$	[679]
Si[N(Me)CH$_2$CH$_2$N(Me)]$_2$ (spiro) (R = Et, Ph)	$4RPF_4$	$2\,[$(MeNCH$_2$CH$_2$NMe)P(F)(R)$]^+$ $[RPF_5]^-$ + SiF_4	[679]
$(Me_3Si)_2NR$ (R = Me, n = 1; R = H, n = n)	Ph_2PF_3	$2Me_3SiF + (Ph_2PFNR)_n$	[676, 679] (R = Me), [677] (R = H)
$(Me_3Si)_2NR$ (R = Me, Et, Ph; R′ = F, Me, Et, Ph, $ClCH_2$, 2, 5 − $Me_2C_6H_3$, m-$CF_3C_6H_4$)	$R'PF_4$	Me_3SiF + cyclo-[R′F$_2$P—NR—PF$_2$R′—NR]	[676]
$(Me_3Si)_2NLi$	$(Me_3Si)_2NPF_2$	$Me_3SiF + LiF + (Me_3Si)_2NP{=}NSiMe_3$	[520]
$(Me_3Si)_2NH$	$PhPF_4$	$2Me_3SiF + (PhPF_2NH)_n$	[677]
$(Me_3Si)_3N$	OPF_3	$Me_3SiF + (Me_3Si)_2NPOF_2$	[198]
$(Me_3Si)_2NR$	$nP_2O_3F_4$	$Me_3SiF + Me_3SiN[P(O)F_2]_2$	[193]
Me_3SiNMe_2	$(CF_3)_3PCl_2$	$Me_3SiCl + (CF_3)_3PClNMe_2$	[714]
nMe_3SiNR_2 (R = Me, Et; n = 1, 2; X = Cl, NR_2)	$Cl_2(O)PNMeP(S)Cl_2$	$Me_3SiCl + XCl(O)PNMeP(S)ClNR_2$	[110]
Me_3SiNR_2 [e.g., R may be Me and R′ = $ArSO_2$, $(R''O)_2P(O)$, or EtO_2C]	Cl_2NR'	$Me_3SiNClR' + ClNR_2$	[588]
$Me_3SiNClR$ (e.g., R = Cl or Me, R′ = Bu^{t}or Me_3Si) (really an α-elimination)	$R'PCl_2$	$RN{=}PCl_2R'$	[587a]
$(Me_3Si)_2NCl$ (R = Bu, Ph, OEt, OPr, OBu)	R_3P	$Me_3SiN = PR_3$ (really an α-elimination)	[588a]
$(Me_3Si)_2NH$ (n = 1, 2, 3)	$Cl_2P(O)(N{=}PCl_2)_nCl$	$Me_3SiCl + Cl_2P(O)(N{=}PCl_2)_nNHSiMe_3$	[616]
$Me_3SiNPr^i_2$	PBr_3	$Me_3SiBr + Pr^i_2NPBr_2$	[11]
$2Me_3SiNEt_2$	$PhPCl_2$	$2Me_3SiCl + PhP(NEt_2)_2$	[11]
Me_3SiNEt_2	Ph_2PCl	$Me_3SiCl + Ph_2PNEt_2$	[11]
$Me_3SiN(Me)PPh_2$	Ph_2PCl	$Me_3SiCl + Ph_2PN(Me)PPh_2$	[366]
Me_3SiNHR (R = Me, Me_3Si; R′ = Me, Et)	$(R'O)_2PCl$	$Me_3SiCl + (R'O)_2PNHR$	[76, 605]
$(Me_3Si)_2NMe$	$3MePCl_2$	$Me_3SiCl + MeN(PClMe)_2$	[732]
$Me_3SiNR'BR_2$ [R = Me, Ph, Me_2N, $CH_2(Me)N$; R′ = Me, Et, Ph]	$R_2P(S)Cl$	$Me_3SiCl + R_2P(S)NR'BR_2$	[541]

Table of Reactions *(C'ntd.)*

Reagents		Products	References
$(Me_3SiNMeBPh)_2NMe$	$Cl_2P(O)Me$	Me_3SiCl 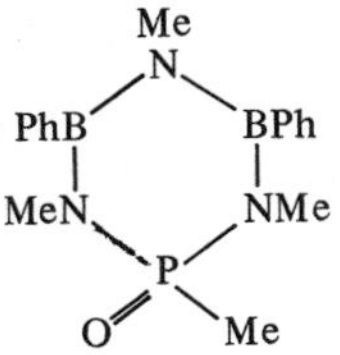 + PhB(NMe)BPh(NMe)P(O)Me(NMe) ring [structure: six-membered ring N(Me), BPh, NMe, P(=O)Me, NMe, BPh]	[551]
$(Me_3Si)_2NMe$	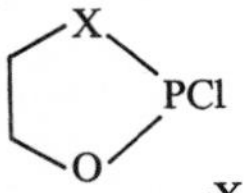 [ring: X, PCl, O]	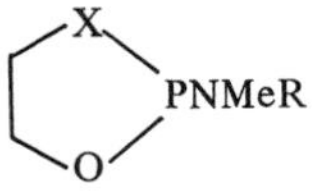 Me_3SiCl + [ring: X, PNMeR, O]	[605]
(X = O, NPh; R = Me_3Si, 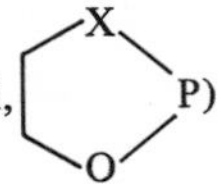 [ring: X, P, O])			
$(Me_3Si)_2NMe$	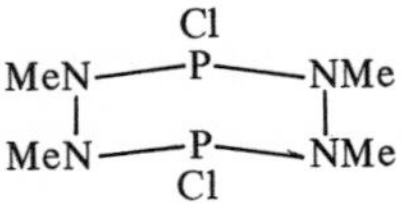 [MeN—P(Cl)—NMe / MeN—P(Cl)—NMe]	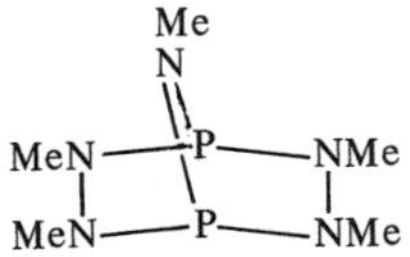 Me_3SiCl + [MeN—P(=NMe)—NMe / MeN—P—NMe]	[553]
Me_3SiNRR'	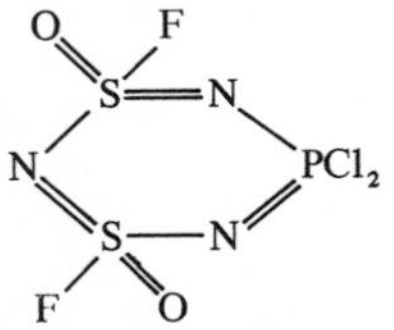 [ring: S(O)(F)=N, PCl₂, N=S(F)(O), N]	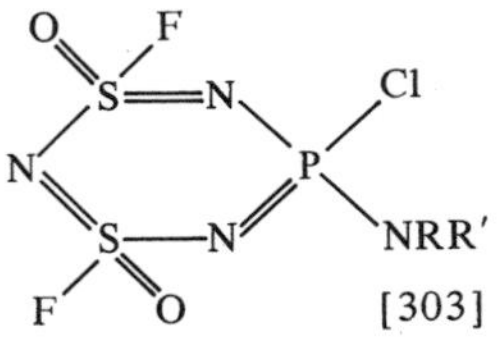 Me_3SiCl + [ring: S(O)(F)=N, P(Cl)(NRR'), N=S(F)(O), N]	[303]
(R = H, Me, Et; R' = Me, Et, Me_3Si)			
$(Me_3Si)_2NMe$	$n(RO)_2PCl$	$Me_3SiCl + (RO)_2PN(R')Me$	[76]
[n = 1, R' = Me_3Si; n = 2, R' = $(RO)_2P$]			
$Me_3SiN(Me)R$	R'Cl	Me_3SiCl + R'N(Me)R	[550]
[R = BMe_2; R' = $R''_2P(S)$ {R'' = Me_2N, Et_2N; R''_2 = [ring: NMe, NMe]}; R = $R''_2P(S)$; R' = BMe_2] 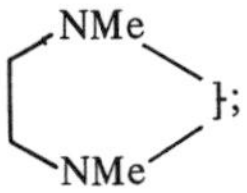			
$(Me_3Si)_2NH$	$R_2P(S)Br$	$Me_3SiBr + R_2P(S)NHSiMe_3$	[704]
$(Me_3Si)_2NMe$	$(Cl_2P)_2NMe$	$Me_3SiCl + Cl_2PNMeSiMe_3 + P_4(NMe_2)_6$	[345]
$(Me_3Si)_2NMe$	Cl_2PR	$Me_3SiCl + ClRPNMeSiMe_3 + (ClRP)_2NMe$	[345]
(R = Ph, p-tolyl)			
$(Me_3Si)_2NH$	YCl	$Me_3SiCl + YNHSiMe_3$	[628]
(Y = OPFXN=PCl_2; X = Cl, F)			
$(Me_3SiN)PMe_3$	$RPCl_2$	$Me_3SiCl + RP(N{=}PMe_3)_2$	[760]
(R = Cl, R' = N=PMe_3; R = Me, R' = Me)			
$2Me_3SiNMe_2$	$[Cl_2P(O)]_2NMe$	$2Me_3SiCl + [Cl(Me_2N)P(O)]_2NMe$	[329]

Table of Reactions *(C'ntd.)*

Reagents		Products	References
$(Me_3Si)_2NMe$	$Cl_2P(O)NMe_2$	$Me_3SiCl + Cl(Me_2N)P(O)NMeSiMe_3$	[329]
$(Me_3Si)_2NX$ ($X = Li, n = 1; X = Me_3Si, n = 2$)	PCl_5	$nMe_3SiCl + Cl_3P{=}NSiMe_3 + 2{-}nXCl$	[519]
$(Me_3Si)_2NR$ ($n = 1, 2$)	$S{=}PCl_2(N{=}PCl_2)_nN{=}PCl_3$	$Me_3SiCl + S{=}PCl_2(N{=}PCl_2)_{n+1}NRSiMe_3$	[625]
$(Me_3Si)_2NPh$ [$X = Cl, Ph(Cl_2P)N$]	$XPCl_2$	Me_3SiCl + PhN—PX / XP—NPh (four-membered ring)	[201]
Me_3SiNMe_2	PhN—PCl / ClP—NPh (four-membered ring)	Me_3SiCl + PhN—$PNMe_2$ / Me_2NP—NPh (four-membered ring)	[201]
$4Me_3SiNMe_2$	$(Cl_2P)_2NPh$	$Me_3SiCl + P(NMe_2)_3$ + PhN—$PNMe_2$ / Me_2NP—NPh (four-membered ring)	[201]
Me_3SiNMe_2	$Cl_2PNMeP(O)Cl_2$	$Me_3SiCl + Cl(Me_2N)PNMeP(O)Cl_2$ ↓ Rearranges	
$Me_3SiNMeP(O)ClNMe_2$	PCl_3	$Me_3SiCl + Cl_2PNMeP(O)ClNMe_2$	[365]
$Me_3SiNMeP(O)ClNMe_2$	Cl_2PNMe_2	$Me_3SiCl + Cl(Me_2N)PNMeP(O)ClNMe_2$ ↓ Rearranges	
Me_3SiNMe_2	$Cl_2PNMeP(O)ClNMe_2$	$Me_3SiCl + Cl_2PNMeP(O)(NMe_2)_2$	[365]
$2Me_3SiNMe_2$	$(Cl_2P)_2NMe$	$2Me_3SiCl + [Cl(Me_2N)P]_2NMe$ (unstable)	[365]
$(Me_3Si)_2NMe$	$Cl_2P(O)OMe$	$Me_3SiCl + Me_3SiNMeP(O)ClOMe$	[365]
nMe_3SiNR_2 ($R = Me, n = 1; R = Et, n = 2$)	$POCl_3$	$nMe_3SiCl + Cl_{3-n}P(O)(NR_2)_n$	[11]
CH_2—NEt / CH_2—NEt ring with $SiMe_2$	$PhPCl_2$	Me_2SiCl_2 + CH_2—NEt / CH_2—NEt ring with PPh	[13]
Me_2—NMe / Me_2—NMe ring with $SiMe_2$	$PhP(O)Cl_2$	Me_2SiCl_2 + Me_2—NMe / Me_2—NMe ring with P(O)Ph	[205]
$3Me_2Si(NHPh)_2$	$2PBr_3$	$3Me_2SiBr_2 + 2P(NHPh)_3$	[353]
$2(Me_3Si)_2NMe$	$2PCl_3$	$4Me_3SiCl + (MeNPCl)_2$	[11]
$(Me_3Si)_2NEt$	PCl_3	$2Me_3SiCl + (EtNPCl)_n$	[11,17]
$(Me_3Si)_2NMe$	$POCl_3$	$2Me_3SiCl + (MeNPOCl)_n$	[11]
$(Me_3Si)_2NH$	$POCl_3$	$Me_3SiCl + OPCl_2NHSiMe_3$	[64]
$(Me_3Si)_2NH$	$ClPO(OPh)_2$	$Me_3SiCl + Me_3SiNHPO(OPh)_2$	[63]
$(Me_3Si)_3N$	PCl_3	$Me_3SiCl + (Me_3Si)_2NPCl_2$	[62]

Table of Reactions *(C'ntd.)*

Reagents		Products	References
Me_3SiNR_2 (M = P, As; R = Me, Et, Pr)	$(CF_3)_2MF$	$Me_3SiF + (CF_3)_2MNR_2$	[382]
Me_3SiNRR' (R = R' = Me, Me_3Si; R = H, R' = Me_3Si)	AsF_3	$Me_3SiF + F_2AsNRR'$	[694]
nMe_3SiNEt_2 (n = 1, 2, 3)	$AsCl_3$	$nMe_3SiCl + Cl_{3-n}As(NEt_2)_n$	[11]
$3Me_2Si(NHPh)_2$ (M = As, Sb)	$2MCl_3$	$3Me_2SiCl_2 + 2M(NHPh)_3$	[31]
CH_2—NMe, CH_2—NMe ring closed by $SiMe_2$	$AsCl_3$	Me_2SiCl_2 + (CH_2—NMe, CH_2—NMe ring closed by AsCl)	[13]
$(Me_3Si)_3N$ (M = As, Sb, Bi)	MCl_3	$Me_3SiCl + Cl_2MN(SiMe_3)_2$	[62]
$RN(SiMe_3)_2$ (R = Me, Ph)	$AsCl_3$	$Me_3SiCl + (RNAsCl)_n$	[5]
Me_3SiNR_2 (R = Me, Me, Et)	SF_4	$Me_3SiF + F_3SNR_2$	[205] [478]
$Me_3SiNHEt$	SF_4	$Me_3SiF + EtNSF_2 + [EtNH_3]F$	[205]
$(Me_3Si)_2NMe$	SF_4	$2Me_3SiF + MeNSF_2$	[205]
Me_3SiNR_2	NSF_3	$Me_3SiF + NSF_2NR_2$	[268]
$(R_3Si)_2NR$	Me_2NSF_3	$2Me_3SiF + RN{=}S(NMe_2)F$	[290]
nMe_3SiNR_2 (n = 1, 2; R = Me, Et; R' = CF_2ClCF_2, $CFCl_2CF_2$)	$R'N{=}SF_2$	$nMe_3SiF + R'N{=}S(NR_2)_nF_{2-n}$	[477]
$(Me_3Si)_2NMe$	$CF_2ClCF_2N = SF_2$	$2Me_3SiF + CF_2ClCF_2N{=}S{=}NMe$	[477]
Me_3SiNMe_2 (X = Cl, Br)	SF_5X	$Me_3SiF + XSF_4NMe_2$	[375e]
Me_3SiNMe_2	CF_3SF_4Cl	$Me_3SiF + CF_3S(Cl)(NMe_2)_2$	[375e]
Me_3SiNMe_2	$[Me_{3-n}Cl_nS][SbCl_6]$	$Me_3SiCl + [Me_{3-n}(Me_2N)_nS][SbCl_6]$	[738]
Me_3SiNMe_2	$CF_2ClCF_2N{=}SCl_2$	$Me_3SiCl + CF_2ClCF_2N{=}S(Cl)NMe_2$	[477]
Me_3SiNEt_2 (R = Me, Ph)	RSO_2Cl	$Me_3SiCl + RSO_2NEt_2$	[4]
$Me_3SiN(Me)PPh_2$	p-$TolSO_2Cl$	Me_3SiCl + p-$TolSO_2N(Me)PPh_2$	[366]
$2Me_3SiNEt_2$ (n = 1, 2)	S_nCl_2	$2Me_3SiCl + S_n(NEt_2)_2$	[4]
$2Me_3SiNEt_2$ (n = 1, 2)	SO_nCl_2	$2Me_3SiCl + O_nS(NEt_2)_2$	[4]
CH_2—NEt, CH_2—NEt ring closed by $SiMe_2$	$SOCl_2$	(CH_2—NEt, CH_2—NEt ring closed by SO) + Me_2SiCl_2	[13]

Table of Reactions *(C'ntd.)*

Reagents		Products	References
$(Me_3Si)_2NMe$ ($n = 1, 2$)	S_nCl_2	$2Me_3SiCl + \frac{1}{x}(MeNS_n)_x$	[4]
$(Me_3Si)_2NH$	SCl_2	$2Me_3SiCl + S_4N_4 + NH_4Cl$	[4]
$(Me_3Si)_2NR$	$SOCl_2$	$2Me_3SiCl + (RNSO)_n$	[4, 449]
(R = Me, Bu, Ph; $n = 1$; R = H, $n = \infty$; R = $OSiMe_3$, $n = 1$)			
$(Me_3SiN)_2S$ (Y = $MeSO_2$, Cl_3CS)	YCl	$Me_3SiCl + YN{=}S{=}NSiMe_3$	[219]
$2(Me_3Si)_2NH$	SO_2Cl_2	$2Me_3SiCl + O_2S(NHSiMe_3)_2$	[63]
$[(Me_3Si)_2N]_2SO_2$	$SOCl_2$	$Me_3SiCl + O_2S(N{=}SO)_2$	[402]
$(Me_3Si)_3N$	$SOCl_2 + AlCl_3$	$Me_3SiNSO + Me_3SiCl$	[658]
$(Me_3Si)_2NH$	TeF_6	$Me_3SiF + F_5TeNHSiMe_3$	[685]
$2Me_3SiNR_2$ (R = Me, Et, Pr)	XeF_2	$2Me_3SiF + Xe + HNR_2$	[264]
$(C_6F_5)_3SiNMe_2$	MeCOCl	$(C_6F_5)_3SiCl + MeCONMe_2$	[423]
$(Me_3Si)_2NCl$	CF_3COCl	$Me_3SiCl + ClN = C(CF_3)OSiMe_3$	[448]
H_3SiNMe_2 (X = NO, CN)	ClX	$H_3SiCl + XNMe_2$	[127, 269]
$Me_3SiN(R)CH_2SiR'R''Me$	ClCN	$Me_3SiCl + MeR'R''SiCH_2N(R)CN$	[399]
(R, R′, R″ = Pr^i, Me, MeO; Pr^i, MeO, MeO; Ph, Et, EtO)			
Me_2Si and $SiMe_2$ joined in a ring by two N(R) groups (R–N, N–R)	XCl	$2ClMe_2SiCH_2N(R)X$	
(R = Bu, Bu^t, Ph, C_3H_5; X = COCl)			
(When R = Me_3Si; X = COCl the product is $ClSiMe_2CH_2NCO + Me_3SiCl$			[487]
R = $CH_2CH{=}CH_2$; X = Ac, MeO_2C)			[691]
$(H_3Si)_2NMe$	NOCl	$(H_3Si)_2O + N_2 + MeCl$	[269]
[$(H_3Si)_3N$ does not react with NOCl or ClCN]			
$Me_{4-n}Si(NMe_2)_n$	PhCOCl	$Me_{4-n}SiCl_n + PhCONMe_2$	[30]
$M(NMe_2)_4$ (M = Si, Ge)	PhCOCl	$MCl_4 + PhCONMe_2$	[30]
$Me_3SiNMeCOR$ (R = H, Me, Pr, OMe, OEt)	Et_3SiCH_2COCl	$Me_3SiCl + Et_3SiCH_2CONMeCOR$	[653]
$3R_3MNMe_2$ (M = Si, Ge, Sn)	$3NH_2Cl$	$3R_3MCl + 3HNMe_2 + NH_3 + N_2$	[310]
$3(R_3M)_2NH$ (M = Ge, Sn)	$6NH_2Cl$	$6R_3MCl + 5NH_3 + 2N_2$	[310]
$2Me_3GeN(SiMe_3)NMe_2$ (R = Me_3Si, Me_3Ge)	$2BF_3$	$Me_3GeF + Me_3SiF + Me_2NN(BF_2)R$	[573]
$Me_3SiN{=}CR_2$	$C_5H_5Mo(CO)_3Cl$	$Me_3SiCl + C_5H_5Mo(CO)_3N{=}CR_2$	[326]
Et_3GeNBu_2	$Ph_2Ge(H)Cl$	$Et_3GeCl + Ph_2Ge(H)NBu_2$	[620]
Et_3GeNR_2	$R'R''GePhCl$	$Et_3GeCl + R'R''GePhNR_2$	[617]
(R = Me, Bu; R′ = Ph, Cl, H; R″ = Ph, NMe_2, H)			

Table of Reactions *(C'ntd.)*

Reagents		Products	References
R_3GeNMe_2	PhGeCl	R_3GeCl + $PhGeNMe_2$	[619]
Me_2 NMe Me_2 NMe MMe_2 (ring) (M = Si, Ge)	$PhPOCl_2$	Me_2 NMe Me_2 NMe P(O)Ph (ring) + Me_2MCl_2	[770]
$(Me_3Sn)_2NSiMe_nR_{3-n}$ (R = Me, n = 0; R = F, n = 2; R = Cl or F, n = 1) (Y = F, N, N, N, F ring)	YF	Me_3SnF + $YN(SnMe_3)SiMe_nR_{3-n}$	[635]
$(Me_3Sn)_3N$	CF_3COCl	$CF_3CON{=}C(CF_3)OSnMe_3$ + Me_3SnF	[448]
Me_3SnNMe_2	C_6F_6	Me_3SnF + $C_6F_5NMe_2$	[256]
Me_3SnNMe_2 (X = Br, CN)	C_6F_5X	Me_3SnF + p-$Me_2NC_6F_4X$	[347]
Me_3SnNMe_2	$HN(CF_3)_2$	Me_3SnF + $(Me_2NCF_2)NH(CF_3)$	[256]
Me_3SnNMe_2	$CF_2{=}CFCl$	Me_3SnF + $CFNMe_2 = CFCl$	[256]
$3Me_3SnNMe_2$	$BF_3.OEt_2$	$3Me_3SnF$ + $B(NMe_2)_3$ + Et_2O	[256]
$2Bu_3SnNEt_2$	$2BF_3.OEt_2$	$2Bu_3SnBF_4$ + $2Et_2O$ + $(F_2BNEt_2)_2$	[457]
Bu_3SnNEt_2	$2[Et_3O][BF_4]$	Bu_3SnBF_4 + $2Et_2O$ + $[Et_4N][BF_4]$	[457]
$3Me_3SnNMe_2$ (M = P, As, Sb)	MF_3	$3Me_3SnF$ + $M(NMe_2)_3$	[256]
$4Me_3SnNMe_2$	TiF_4	$4Me_3SnF$ + $Ti(NMe_2)_4$	[256]
$(Me_3Sn)_2NR$	$(F_2P{-}N)_3$	Me_3SnF + $Me_3SnNRPF$ (ring: N, PF_2, N, N, P, F_2)	[632]
$(Me_3Sn)_2NMe$	$F_2S(NSiMe_3)_2$	Me_3SnF + $(Me_3SiN)_2S(NMeSnMe_3)_2$	[632]
$Me_3SnN(Me)PSCl_2$	$(F_2PO)_2O$	Me_3SnF + $F_2P(O)NMeP(S)Cl_2$	[317]
Me_3SnNMe_2	BCl_3	Me_3SnCl + Cl_2BNMe_2	[256]
$(Me_3Sn)_3N$ (R = Ph, NMe_2)	$R(Cl)BNMe_2$	Me_3SnCl + $(Me_3Sn)_2NBRNMe_2$	[748]
$(Me_3Sn)_3N$ (n = 1, 2, 3)	nBrB (ring) Me	$n$$Me_3SnBr$ + ((ring) Me B$)_nN(SnMe_3)_{3-n}$	[706, 706a]
$(Me_2SnNBu^t)_2$ (R = Me, Cl)	RBX_2	$(RBNBu^t)_2$ + Me_2SnX_2	[705a]
Me_3SnNMe_2	Me_3SiCl	Me_3SnCl + Me_3SiNMe_2	[256]
Me_3SnNMe_2	$Ph_2Si(H)Cl$	Me_3SnCl + $Ph_2Si(H)NMe_2$	[133]

Table of Reactions *(C'ntd.)*

Reagents		Products	References
$(Me_3Sn)_3N$ ($n = 0$ to 3)	Me_nSiCl_{4-n}	$Me_3SnCl + (Me_3Sn)_2NSiMe_nCl_{3-n}$	[633]
$Me_2Sn(NEt_2)_2$ (X = Cl, Br)	Me_3SiX	$Me_2Sn(NEt_2)X + Me_3SiNEt_2$	[682a]
$(Me_3SnN)_2S$	Me_3SiCl	$Me_3SnCl + (Me_3SiN)_2S$	[289]
$Et_3SnN(Me)COMe$	Me_3GeBr	$Et_3SnBr + Me_3GeN(Me)COMe$	[395]
Et_3SnNMe_2	Cl_2	$Et_3SnCl + ClNMe_2$	[256]
$Sn[N(SiMe_3)_2]_2$ (X = Cl, Br	SnX_2	$\frac{2}{n}[Sn(X)N(SiMe_3)_2]_n$	[294, 295]
X = Cp)			[295, 655]
$Sn[N(SiMe_3)_2]_2$	Me_2SnCl_2	$ClSnN(SiMe_3)_2 + Me_2Sn(Cl)N(SiMe_3)_2$	[294]
$Sn[N(SiMe_3)_2]_2$ [$n = 2$, R = Br; $n = 1$, R = $N(SiMe_3)_2$]	$n SnBr_4$	$RSnBr + n SnBr_3N(SiMe_3)_2$	[294]
$Me_2Si(NR)_2Sn$ (cyclic) (R = Bu^t)	$SnCl_4$ (2:1)	$(Me_2Si(NR)_2)_2Sn$ (cyclic)	[721b]
	(1:1)	$Me_2Si(NR)_2SnCl_2$ (cyclic)	[721b]
$(Me_3Sn)_2NR$ [R = Me; Y = $P(S)Cl_2$, $CF_3(CF_2)_3SO$; R = Me_3Sn; Y = $CF_3(CF_2)_3SO$, $FSO_2N{=}PCl_2$, $O_2S(N{=}PCl_2)_2$]	YCl	$Me_3SnCl + Me_3SnNRY$	[632]
$(Me_3Sn)_2NP_3N_3F_5$	$S_3N_2Cl_2$	$2Me_3SnCl + F_5N_3P_3N{=}S$ (ring: S–N≡S=N–S)	[627]
$Me_3SnN{=}C(CF_3)_2$	Cp_2TiCl_2	$Me_3SnCl + Cp_2Ti(Cl)N{=}C(CF_3)_2$	[424a]
Me_3SnNMe_2	$BrMn(CO)_5$	$Me_3SnBr + (OC)_5MnNMe_2$*	[256]
Me_3SnNMe_2	*trans*-$(Ph_3P)_2PtClH$	$Me_3SnCl + [(Ph_3P)_2Pt]_n$*†	[131, 132, 133, 221a]
Me_3SnNMe_2	$(F_3P)_4RhH$	Me_3SnF + ?	[131, 132]
Me_2AsNMe_2	MeI	$Me_2AsI + [Me_4N]I$	[384]
$MeAs(NR_2)_2$ (R = Me, Et)	2MeI	$MeAsI_2 + 2MeNR_2$	[380]
$(Me_2As)_2NR$	AsF_3	$2Me_2AsF + (F_2As)_2NR$	[379]
$As(NR_2)_3$ (R = Me, Et, Pr, Bu; $R_2 = C_5H_{10}$)	BF_3	$FAs(NR_2)_2 + F_2BNR_2$	[530]
$MeAs(NMe_2)_2$ (X = F, Cl, Br, I)	$MeAsX_2$	$2MeAs(X)NMe_2$	[363]
$(C_5H_5)_2M(NMe_2)_2$ (M = Zr, Hf)	$2BF_3.OEt_2$	$(C_5H_5)_2MF_2 + 2F_2BNMe_2$	[692]

*Not isolated.

†See Chapters 12 and 14.

Table of Reactions *(C'ntd.)*

Reagents		Products	References
$Ti(NR_2)_4$	TiF_4	$2F_2Ti(NR_2)_2$	[124]
$Ti(NMe_2)_4$	PF_5	$F_3TiNMe_2 + F_2P(NMe_2)_3^*$	[124]
$Ti(NEt_2)_4$	$PhPF_4$	$FTi(NEt_2)_3 + PhPF_3NEt_2^*$	[124]
$nTi(NR_2)_4$	$(4-n)TiCl_4$	$4Cl_{4-n}Ti(NR_2)_n$	[124]
($R = Me, n = 1-3$; $R = Ph, n = 1, 3$			[67]
$R = Et, n = 1-3$)			[717]
$Mo_2(NMe_2)_6$	$2Me_3SiCl$	$Mo_2Cl_2(NMe_2)_4 + 2Me_3SiNMe_2$	[20a]
$CuN(SiMe_3)_2$	ArI	$Cu_2I_2 + [ArN(SiMe_3)_2]$ $\xrightarrow{MeOH}$ $ArNH_2$	[373]
(Ar = Ph, p-$MeOC_6H_4$, p-tolyl, o-ClC_6H_4, p-$NO_2C_6H_4$, 2-thienyl)			
$CuN(SiMe_3)N(SiMe_3)_2$	RC_6H_4I	$Cu_2I_2 + [(Me_3Si)_2NN(SiMe_3)C_6H_4R]$ $\xrightarrow{MeOH}$ $H_2NNHC_6H_4R$	[372]
$Hg[N(CF_3)_2]_2$	$2Me_3SiCl$	$HgCl_2 + 2Me_3SiF + 2CF_3N{=}CF_2$	[43]
$Hg[N(CF_3)_2]_2$	RCOCl	$HgCl_2 + (CF_3)_2NCOR$	[771]
($R = Me, Ph, CF_3$)			
$Hg[N(CF_3)_2]_2$	RCl	$(CF_3)_2NR$	[230a, 231]
($R = CF_3S$, MeS, ClS)			
$MeHgN(CF_3)_2$	NOCl	$MeHgCl + (CF_3)_2NNO$	[210]
$CF_3SHgN(CF_3)_2$	NOCl	$HgCl_2 + CF_3SSCF_3 + (CF_3)_2NNO$	[210]
$Hg[N(CF_3)_2]_2$	2XY	$HgY_2 + (CF_3)_2NX$	
($X = Y = Cl, Br, I$)			[210, 772]
($X = I, Y = SiMe_3$)			[528]
NRR'/R'' System			
$Be(NMe_2)_2$	$2Me_3Al$	$Me_2Be + (Me_2AlNMe_2)_2$	[49]
$(CH_2)_4BNR'_2$	Me, B, R	$(CH_2)_4BR$ + Me, B, NR'_2	[394]
NH, B, Bu	BuY; H_2O	$\overset{+}{N}H_2$, $\overset{-}{B}$, Bu_2	[212]
(Y = Li, MgCl)			
$B(NMe_2)_3$	Ph_3B	$Ph_2BNMe_2 + PhB(NMe_2)_2$	[319]
$B(NEt_2)_3$	Et_3B (+$C_4H_8O.BH_3$)	Et_2BNEt_2	[283a]
$B(NR_2)_3$	$3R'_3B$	$3R'_2BNR_2$	[521]
$B(NMe_2)_3$	R'_nM	$BR_3 + M(NMe_2)_n$	[19, 424]
($MR_n = SnR'_4, CdR'_2, ZnR'_2, BR'_3, AlR'_3$; $R' = Me, Et$)			
H_2BNMe_2	Me_3Al	$Me_8Al_4(NMe_2)_2H_2 + Me_3B + Me_2AlNMe_2 + Me_2AlH$	[270]

*Not isolated.

Table of Reactions *(C'ntd.)*

Reagents		Products	References
$3B_2(NMe_2)_4$	$4Me_3Al$	$Me_6Al_4B(NMe_2)_3 + 5Me_2BNMe_2 + 2(Me_2AlNMe_2)_2$	[681]
$B_2(NMe_2)_4$	Me_3Al	$Me_3Al_3(NMe_2)_2 + Me_3B + Me_2BNMe_2 + (Me_2AlNMe_2)_2$	[681]
$Al(NMe_2)_3$ (R = Pr, Bu, Ph)	$2R_3B$	$2R_2BNMe_2 + R_2AlNMe_2$	[640]
$M[N(SiMe_3)_2]_2$ [M = Ge, Sn; R = $(Me_3Si)_2CH$]	2RLi (0–20°)	$R_2M + LiN(SiMe_3)_2$	[197, 272]
Me_3SnNMe_2 (R = Bu, Ph)	R_3B	$R_2BNMe_2 + Me_3SnR$	[256]
$2Me_3SnNMe_2$	$(AlEt_3)_2$	$2Me_3SnEt + (Et_2AlNMe_2)_2$	[256]
Bu_3SnNEt_2	MeMgBr	$Bu_3SnMe + Et_2NMgBr$	[457]
$Sn[N(SiMe_3)_2]_2$	$Sn(C_5H_5)_2$	$Sn(C_5H_5)[N(SiMe_3)_2]$	[295]
$Hg[N(SiMe_3)_2]_2$	$Hg(C_5H_5)_2$	$2Hg(C_5H_5)[N(SiMe_3)_2]$	[648a]
NRR'/OR" System			
$LiNR_2$ (R = R' = Me, Et; M = Si, Ge)	$R_3MC{\equiv}COR'$	$LiOR' + R_3MC{\equiv}CNR_2$	[594]
$LiNMe_2$	$CH_2{=}CHC{\equiv}CCMe_2OMe$	$LiOMe + Me_2NCH_2CH{=}C{=}CMe_2$	[702]
$B(NMe_2)_3$	$(BuO)_3B + BCl_3$	$3BuOB(Cl)NMe_2$	[51]
$PhB(NEt_2)_2$	2PhSOMe	$2PhSNEt_2 + PhB(OMe)_2$	[175]
$B(NHMe)_3$	3RCOOH	$B(OH)_3 + 3RCONHR$	[479]
$B(NMe_2)_3$	B_2O_3	six-membered B–O ring: Me_2NB, O, $BNMe_2$ / O, O / B–NMe_2	[275]
$Al(NMe_2)_3$ (R = Me, Et, Bu)	$B(OR)_3$	$Al(OR)_3 + B(NMe_2)_3$	[640]
$Al(NMe_2)_3$ ($n = 1, 2$)	$nPh_nB(OEt)_{3-n}$	$nPh_nB(NMe_2)_{3-n} + (EtO)_2AlNMe_2$	[640]
$3Al(NMe_2)_3$ (R = Bu, Bu^s, Bu^t)	$2(RBO)_3$	$RB(NMe_2)_2 + \frac{3}{n}(Me_2NAlO)_n$	[640]
Me_2AlNRR' (R = H, R' = H, Bu^s, Bu^t, CH_2Ph; RR' = $(CH_2)_4$, $(CH_2)_5$; R", R‴ = alkyl, aryl)	R‴COOR"	R‴CONRR'	[55f]
$R_2Si[C_6H_4N(SiMe_3)_2]_2$ (R = Me, Ph; R_2 = MePh)	2 $C_6H_4(CO)_2O$	$2(Me_3Si)_2O + R_2Si[C_6H_4N(CO)_2C_6H_4]_2$	[596]
Me_3SiNR_2 ($R_2 = Et_2$, MePh; R' = H, Me, CF_3, Ph)	$Me_3SiOCOR'$	$(Me_3Si)_2O + R'CONR_2$	[585]
$Me_3SiN(Me)COMe$	Et_3SnOMe	$Me_3SiOMe + Et_3SnNMeCOMe$	[395]
$(Me_3Si)_2NH$	$ROSiMe_2CH_2NHR'$	$Me_3SiOR + (Me_2SiCH_2NR')_2 + NH_3$	[488]

Table of Reactions *(C'ntd.)*

Reagents		Products	References
$(Me_3SiN)_3S$	$n(CF_3CO)_2O$	$Me_3SiO_2CCF_3 + (Me_3SiN)_nS(=NCOCF_3)_{3-n}$	[317]
$R_3Si{-}N\overline{\ \ \ }(CH_2)_m$ (ring)	Ac_2O	$R_3SiOAc + Ac{-}N\overline{\ \ \ }(CH_2)_m$ (ring)	[504]
Me_3SiNRR' ($R = R' = Me, Et$; $R = Me, R' = Me_3Si$; $R = H, R' = Et, Pr$)	$PhSOMe$	$Me_3SiOMe + PhSNRR'$	[46]
$(Me_3Si)_2NMe$	$(CF_3CO)_2O$	$Me_3SiOOCCF_3 + F_3CCONMeSiMe_3$	[79]
Et_3GeNMe_2 ($R = H, Me$)	$CH_2{=}CROAc$	$Et_3GeOCR{=}CH_2 + AcNMe_2$ (also Et_3GeCH_2COMe when $R = Me$)	[624]
Bu_3GeNEt_2	Ac_2O	$Bu_3GeOAc + AcNMe_2$	[446]
$3Me_3SnNMe_2$	$3B(OMe)_3$	$3Me_3SnOMe + B(NMe_2)_3 + 2B(OMe)_3$	[256]
Me_3SnNMe_2	Ph_2BOMe	$Me_3SnOMe + Ph_2BNMe_2$	[256]
Bu_3SnNEt_2	$EtCH{=}CHOSnBu_3$	$(Bu_3Sn)_2O + EtCH{=}CHNEt_2$	[104]
Me_3SnNEt_2 ($R = Me, Et$)	$Me_3SnC{\equiv}COR$	$Me_3SnOR + Me_3SnC{\equiv}CNEt_2$	[594]
$(Bu_3Sn)_2NEt$ ($R = Et, Pr, Bu, C_5H_{11}, C_7H_{15}$)	$RCH{=}CHOSnBu_3$	$(Bu_3Sn)_2O + Bu_3SnCHRCH{=}NEt +$ $RCH{=}CHNEtSnBu_3$	[103]
Me_3SnNMe_2 ($X = Ac, CH_2 = CH$)	$AcOX$	$Me_3SnOAc + Me_2NX$	[144, 256]
R_3SnNMe_2 [$R = R' = Me; Y = Ac$; $R = Me, R' = Ph, Y = Ac$; $R = Bu, R' = Et, Y = Ac, AcCH_2CO$; $R = Me, R' = Me, Y = CH_2{=}C(Me)CO$; $R = Me, R' = Et, Y = (EtO_2C)C{=}C(CO_2Et)CO$]	$R'OY$	$R_3SnOR' + Me_2NY$	[144, 256]
$2Me_3SnNMe_2$	$CH_2(CO_2Et)_2$	$2Me_3SnOEt + CH_2(CONMe_2)_2$	[144, 256]
$4Bu_3SnNMe_2$	$(EtO_2C)_2C{=}C(CO_2Et)_2$	$4Bu_3SnOEt + (Me_2NOC)_2C{=}C(CONMe_2)_2$	[144, 256]
Me_3SnNEt_2	$2MeO{-}C_6H_2(NO_2)_3$ (2,4,6-NO_2)	$Me_3SnO{-}C_6H_2(NO_2)_3 +$ $\overset{+}{[Me_2NEt_2]}\,[O{-}C_6H_2(NO_2)_3]^-$	[457]
$(Me_3Sn)_2NP_3N_3F_5$	$(CF_3CO)_2CO$	$Me_3SnOOCCF_3 + Me_3SnN(COCF_3)P_3N_3F_5$	[634]
$PhAs(NRR')_2$ ($R = R' = Me, Et, Pr, Bu$; $RR' = C_5H_{10}, C_4H_8O$)	$2Ac_2O$	$PhAs(OAc)_2 + 2AcNRR'$	[245]

Table of Reactions *(C'ntd.)*

Reagents		Products	References
$Bu^t_2SbNMe_2$	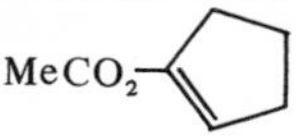 $MeCO_2$–	$MeCONMe_2$ + Bu^t_2SbO– + Bu^t_2Sb– (O) 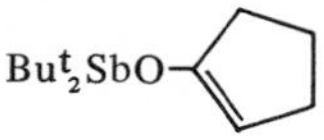 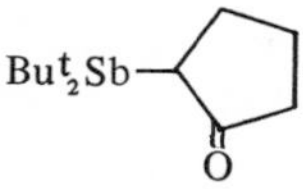	[247]
$M(NR_2)_3$ (M = Sb, R = Me, R′ = Et; M = Bi, R = Me, Et, Pr, R′ = Ac)	AcOR′	$M(OR')_3$ + $AcNR_2$	[403] [36]
$n\text{Ti}(NR_2)_4$ (e.g., R = Me; R′ = Pr^i)	$(4-n)\text{Ti}(OR')_4$	$4\text{Ti}(OR')_{4-n}(NR_2)_n$	[67,744,745]
NRR′/SR″ System			
$B(NEt_2)_3$ (X = Cl, Br)	$B(SBu)_3 + BX_3$	$3XB(SBu)NEt_2$	[181]
$M(NMe_2)_3$ (M = B, Al; X = Br, SH)	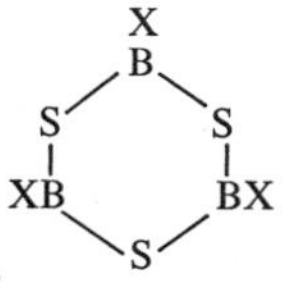 X, B, S, S, XB, BX, S	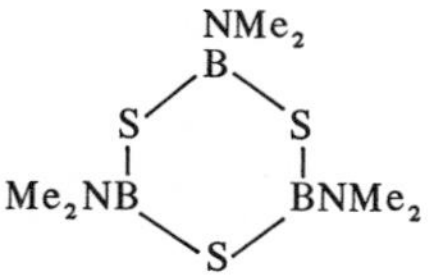 NMe_2, B, S, S, Me_2NB, $BNMe_2$, S + MX_3	[749, 750]
NRR′/S System			
$Hg[N(CF_3)_2]_2$	S_8	$[(CF_3)_2N]_2S$; $CF_3N{=}CF_2$; etc.	[772a]
$(R_3Sn)_2N(CH_2)_nCO_2SnR_3$	CS_2	$(R_3Sn)_2S + SCN(CH_2)_nCO_2SnR_3$	[331]
*NR_2/NRR''_2 System**			
Et_2BNMe_2	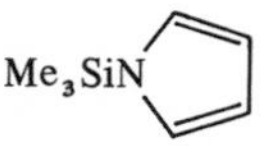 Me_3SiN	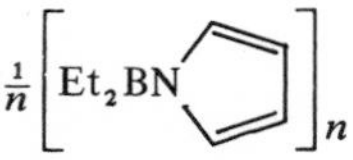 $\frac{1}{n}[Et_2BN]_n$ + Me_3SiNMe_2	[499a]
$RB(NMe_2)_2$	$Me_2Si(NHMe)_2$	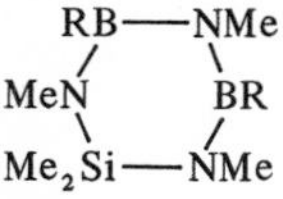 RB—NMe, MeN, BR, Me_2Si—NMe	[551a]
$B(NMe_2)_3$	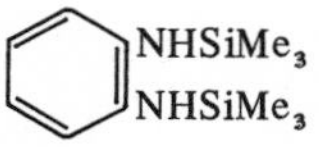 NHSiMe$_3$, NHSiMe$_3$	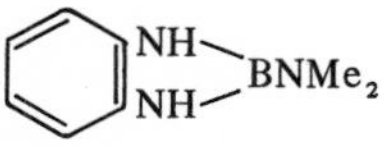 NH, NH, $BNMe_2$ + $2Me_3SiNMe_2$	[271]
Me_3SiNR_2 [R′ = $ArSO_2$ (Ar = Ph, p-Tolyl), $(R''O)_2PO$ (R″ = Et, Pr), EtO_2C]	$R'NCl_2$	$Me_3SiNClR' + R_2NCl$	[589]
Exchange Reactions with Compounds Containing C=O or S=O Bonds (see also Chapter 10)			
$3XB(NMe_2)_2$ (X = NMe$_2$, Ph)	$3SO_2$	$(XBO)_3 + 3OS(NMe_2)_2$	[546]
$2Bu_2BNRR'$	SO_2	$(Bu_2B)_2O + OS(NRR')_2$	[546]

*See also Chapters 1–9 (transamination).

Table of Reactions *(C'ntd.)*

Reagents		Products	References
$Me_2AlNMeSiMe_3$	Pr^i_2CO	$Pr^i_2C{=}NMe + Pr^iCOCMe_2AlMe_2 + (Me_3Si)_2NMe$	[646]
$Me_2AlNMeSiMe_3$	$MeC(Bu^t)O$	$Bu^tMeC{=}NMe + (Me_3Si)_2NMe + Bu^tCOCH{=}CMeBu^t$	[646]
$Me_2AlNMeSiMe_3$	Ph_2CO	$Ph_2C{=}NMe + Me_2AlOSiMe_3 + Ph_2CHOAlMe_2 + (Me_3Si)_2NMe$	[646]
$Me_2AlNMeSiMe_3$	$MeC(Ph)O$	$MePhC{=}NMe + Me_2AlOSiMe_3 + PhMeC(OAlMe_2)NMeSiMe_3 + (Me_3Si)_2NMe$	[646]
$Me_2AlNMeSiMe_3$	RNCO	$RN{=}C{=}NR + R'_2AlOSiMe_3 + (Me_3Si)_2NMe$	[646]
(R = R′ = Me; R′ = Me, Et, R = Ph)			
$Me_2AlNMeSiMe_3$	$MeCONMe_2$	$MeN{=}CMeNMe_2 + Me_3SiNMe_2 +$ Me_2N, Me, $C{=}\overset{+}{N}$, HC, $\overset{-}{Al}Me_2$, C—O, Me	[646]
$Me_2AlNMeSiMe_3$	RCOOMe	$(Me_3Si)_2NMe + Me_3SiOMe +$ $Me_2\overset{-}{Al}$—O—$\overset{-}{Al}Me_2$, $Me\overset{+}{N}$, $\overset{+}{O}Me$, CR	[646]
(R = Me, Bu^t, Ph)			
$RR'(F)SiN(Li)R''$	PhCHO (and H_2O)	$PhCONHR''$	[377a]
[$RR'R''$ = $Me_2(C_6H_2Me_3$-2,4,6), $MeBu^t(C_6H_2Me_3$-2,4,6), $MePh(C_6H_4Me$-*p*)]			
$2Bu_3SnNR_2$	$2Me_2C{=}O$	$Me_2C{=}CHCOMe + (Bu_3Sn)_2O + 2HNR_2$	[593]
$4Bu_3SnNR_2$	$2Me_2C{=}O$	$Me_2C{=}CHC(NR_2){=}CH_2 + 3HNR_2 + 2(Bu_3Sn)_2O$	[593]
$2Bu_3SnNR_2$	$Me_2C{=}CHCOMe$	$Me_2C{=}CHC(NR_2){=}CH_2 + (Bu_3Sn)_2O + HNMe_2$	[593]
$M(NR_2)_4.4HCl$	Et_2CO	$MeCH{=}C(NR_2)Et$ + Metal Oxide	[469, 470]
(M = Sn, Ti, R = Et; M = Si, Ge, Sn, Ti, NR_2 = 1-pyrrolidino) [$SiCl(NEt_2)_3$ and $Ge(NEt_2)_4$ do not react]			
$As(NMe_2)_3$	$(HCHO)_n$	$CH_2(NMe_2)_2$	[314]
$As(NMe_2)_3$	BuCHO	$CH(NMe_2){=}CHCH_2Me$	[314]
$As(NMe_2)_3$	PhCHO	$PhCH(NMe_2)_2$	[314]
$As(NMe_2)_3$	$RCH_2(R')CO$	$R'C(NMe_2){=}CHR$	[34a, 468]

Table of Reactions *(C'ntd.)*

Reagents		Products	References
$As(NMe_2)_3$	MeN⟨ring⟩=O	MeN⟨ring⟩–NMe_2	[314]
$As(NR_2)_3$ (R = Me; NR_2 = NC_4H_8, NC_5H_{10})	⟨ring⟩=O	⟨ring⟩–NR_2	[314]
$As(NR_2)_3$ (R = Me, NR_2 = NC_4H_8, NC_5H_{10})	MeC(Ph)=O	CH_2=C(NR_2)Ph	[314]
$As(NMe_2)_3$	MeC(NC_5H_4)O	CH_2=C(NMe_2)NC_5H_4	[314]
$As(NMe_2)_3$	CH_2–CH–CH_2 / NMe C=O / CH_2–CH–CH_2	CH_2–CH–CH_2 / NMe $CNMe_2$ / CH_2–CH=CH	[314]
$2As(NMe_2)_3$	$2Me_3SiNCO$	Me_2NAs–$NCONMe_2$ / Me_2NOCN–$AsNMe_2$ + $2Me_3SiNMe_2$	[630a]
$R_2SbNR'_2$	$MeCO_2CMe$=CMe_2	$MeCONR'_2$ + $R_2SbOCMe$=CMe_2 + R_2SbCMe_2COMe	[248]
$2R_2SbNR'_2$	$MeCO_2CH$=CR''_2	R''_2C=$CHNR'_2$ + $(R_2Sb)_2O$	[248]
$Ti(NMe_2)_4$	Me_2NCHO	$HC(NMe_2)_3$	[746]
$Ti(NMe_2)_4$	$Me_2NC(Me)O$	CH_2=$C(NMe_2)_2$	[746]
$Ti(NMe_2)_4$ (R = Me, Ph)	$Me_2NC(RCH_2)O$	RCH=$C(NMe_2)_2$	[746]
$Ti(NMe_2)_4$ (X = Me, Cl)	$Me_2NC(X_2CH)O$	X_2C=$C(NMe_2)_2$	[746]
$Ti(NMe_2)_4$	CH_2CO–O–$COCH_2$ (cyclic)	[–CH=$C(NMe_2)_2]_2$	[746]
$Ti(NMe_2)_4$ (R = Me, Ph)	RCHO	$RCH(NMe_2)_2$	[747]
$Ti(NMe_2)_4$	⟨ring⟩=O	⟨ring⟩–NMe_2	[747]
$Ti(NMe_2)_4$	$Pr^iC(Me)O$	CH_2=$C(NMe_2)Pr^i$	[747]
$Ti(NMe_2)_4$	$Bu^iC(Ph)O$	Pr^iCH=$C(NMe_2)Ph$	[747]
$Ti(NMe_2)_4$	$(Pr^i)_2C$=O	Me_2C=$C(NMe_2)Pr^i$	[747]
$Ti(NMe_2)_4$	$Bu^tC(Me)O$	CH_2=$C(NMe_2)Bu^t$	[747]
$Ti(NMe_2)_4$ (R = NMe_2, OMe)	$2RCOCH_2COR$	(R–C=O / HC / R–C–O)$_2$$Ti(NMe_2)_2$ + $2HNMe_2$	[741]
$Ti(NMe_2)_4$ [R = OMe, NMe_2 (excess)]	$MeCOCH_2COR$	$MeC(NMe_2)$=CHCOR	[741]

Table of Reactions *(C'ntd.)*

Reagents		Products	References
$Ti(NMe_2)_4$	$MeCOCH_2CONMe_2$	$CH_2{=}C(NMe_2)CH{=}C(NMe_2)_2$	[741]
$Ti(NMe_2)_4$ (R = OMe, NMe_2)	$MeC(NMe_2){=}CHCOR$	$CH_2{=}C(NMe_2)CH{=}C(NMe_2)_2$	[741]
$Ti(NMe_2)_4$ (R = Bu^t, CF_3)	$RCOCH_2COR$	$RC(NMe_2){=}CHCOR$	[741]
$Ti(NMe_2)_4$ (R = Me, Ph)	$MeCOCH_2COR$	$CH_2{=}C(NMe_2)CH{=}C(NMe_2)R$	[741]
$Ti(NMe_2)_4$	$[Fe(O{-}C(Me){=}CH{-}C(Me){=}O)]_3$	$CH_2{=}C(NMe_2)CH{=}C(NMe_2)Me$	[741]
$Ti(NMe_2)_4$	SO_2	$\frac{1}{n}[OTi(OS(O)NMe_2)_2]_n$ + $SO(NMe_2)_2$	[144]
$Ti(NMe_2)_4$ (R = Me, C_6H_{11})	MeSOR	$Me_2NCH_2SR + HNMe_2 + \frac{1}{n}[OTi(NMe_2)_2]_n$	[144]

Reactions with Carbon Monoxide and Metal Carbonyls (see also Chapter 10)

Reagents		Products	References
$C_2H_4S_2BNMe_2$	$Fe_2(CO)_9$	$Fe_2(CO)_6S_2C_2H_4$	[545a]
$(SBNMe_2)_3$	$Fe_2(CO)_9$	$Me_2NBS_2[Fe(CO)_3]_2$	[545a]
$NaN(SiMe_3)_2$	CO	$NaCN + (Me_3Si)_2O$	[687]
$NaN(SiMe_3)_2$ (M = Ni; $n = 4$ / M = Cr, Mo, W; $n = 6$)	$M(CO)_n$	$Na[M(CO)_{n-1}CN] + (Me_3Si)_2O$	[687] [374]
$NaN(SiMe_3)_2$ (M = Cr, Mo, W)	$CpM(CO)_2NO$	$Na[CpM(CO)(NO)CN] + (Me_3Si)_2O$	[108]
$MN(SiMe_3)_2$ (M = Li, Na)	$Fe(CO)_5$	$(CO)_4FeCNSiMe_3 + MOSiMe_3$	[80]

Reactions with p-Benzoquinone (see Chapter 11)

Reagents		Products	References
$Me_3SnNRN(SnMe_3)MMe_3$ (R = Me, Bu^t, Ph; M = Si, Ge)	$O{=}C_6H_4{=}O$	$RN{=}NMMe_3$ + $Me_3SnOC_6H_4OSnMe_3$	[756]

(When M = Sn and R = Ph the products are $Me_3SnOC_6H_4OSnMe_3 + PhH + N_2$)

Reagents		Products	References
$Me_3SnNPhN(GeMe_3)_2$	$O{=}C_6H_4{=}O$	$PhN{=}NGeMe_3$ + $Me_3SnOC_6H_4OGeMe_3$	[756]

Table of Reactions *(C'ntd.)*

Reagents		Products	References
Miscellaneous			
$B_n(NMe_2)_{n+2}$ (n = 3,4)	O_2	$B(NMe_2)_3$ + $(OBNMe_2)_3$	[303a]
$Et_2AlN{=}CR'R''$	M	$[Et_2AlN(M)CR'R''-]_2$	[314b]
(R' = Ph; R'' = H, Ph; M = alkali metal)			
	+ H_2O	$(Et_2AlNHCR'R''-)_2$	[314b]
$Bu^i_2AlN = CRR'$	M	$[Bu^i_2AlN(M)CRR'-]_2$	[314b]
(R = C_6H_{11}, Me; R' = H; M = alkali metal)			
	+ H_2O	$(Bu^i_2AlNHCRR'-)_2$	[314b]
$2R_3SnNEt_2$	$(Me_3Si)_2Hg$	Me_3SiNEt_2 + R_6Sn_2 + Hg	[489]
$(Me_3Sn)_3N$	$Mn_2(CO)_{10}$	$Me_3SnMn(CO)_5$	[15]

14

Dehydrochlorination and Related Reactions

A INTRODUCTION

We have seen that many metal amides react not only with protic (Chapters 11 and 12) but also with halogen (Chapter 13) compounds. It is therefore not altogether surprising that compounds with both replaceable hydrogen and halogen often undergo dehydrochlorination reactions in presence of amides, as they do with organic bases [557]. The aim of this chapter is to review such reactions, but examples of deprotonation or dehydrochlorination other than those covered in Chapters 11, 12, or 13 (NR_2/H or NR_2/Cl exchange) are also considered.

B DEHYDROCHLORINATION

With rare exceptions [e.g., Eq. (1)] only tin amides have so far been shown to act as powerful dehydrochlorinating agents [Eq. (2)] [133] and clearly this area is open for further research. *trans*-$[PtCl(H)(PPh_3)_2]$ and $[\{Ti(\eta\text{-}C_5H_5)_2NMe_2\}_2]$ react according to Eq. (1) [433]. A similar observation has been made for the $[PtCl(H)(PPh_3)_2]/Me_3SnNMe_2$ system [221a]; the isolated product is *cis*-$[PtMe(PPh_3)_2(SnMe_2Cl)]$ {contrary to an earlier report [131] that it is the isomeric $[PtCl(PPh_3)_2(SnMe_3)]$}, possibly via initial dehydrochlorination (see Chapter 12).

$$trans\text{-}[PtCl(H)(PPh_3)_2] + \tfrac{1}{2}[\{Ti(\eta\text{-}C_5H_5)_2NMe_2\}_2] \longrightarrow \frac{1}{n}[Pt(PPh_3)_2]_n + \tfrac{1}{2}[\{Ti(\eta\text{-}C_5H_5)_2Cl\}_2] + HNMe_2 \quad (1)$$

The efficacy of Me_3SnNMe_2 as a dehydrochlorinating reagent was attributed [133] to a combination of factors which include (i) the weak and highly polar Sn–N bond, (ii) the high basicity of Me_3SnNMe_2, (iii) the high value for the heat of formation of crystalline $Me_3SnCl.HNMe_2$, (iv) the capacity of tin amides to exercise a synergic effect, since they are able to behave both as proton [355, 356] and halide ion [256] abstractors, and (v) for inorganic compounds, the intermediate formation of compounds having metal–tin bonds (see Chapter 12).

$$Me_3SnNMe_2 + Cl{-}(B){-}H \longrightarrow Me_3SnCl.HNMe_2 + B \quad (2)$$

An example is shown in Eq. (3) [133]. The iridium(I) product was not isolated, but evidence for its existence was furnished by characterisation of reaction products with other reagents, e.g., CO afforded *trans*-$[Ir(CO)Cl(PPh_3)_2]$ [134]. The reaction of Me_3SnNMe_2 and $[IrCl_2(H)(PPh_3)_3]$ has been investigated under milder conditions (ambient temperature, 30 min), when $[IrCl_2(PPh_3)_2SnMe_3]$ could be isolated [419]. This may represent the first stage in reaction (3), although neither elimination of Me_3SnCl nor isolation of $[IrCl(PPh_3)_3]$ has been established.

$$[IrCl_2(H)(PPh_3)_3] + Me_3SnNMe_2 \xrightarrow[2\text{ h}]{\text{xylene }(80\text{–}100°\text{ C})} [IrCl(PPh_3)_3] + Me_3SnCl.HNMe_2 \quad (3)$$

As mentioned in the introductory paragraph, alternative modes of reaction to dehydrochlorination between a hydrido-chloro-metal complex and a metal amide are possible; a pertinent example is the NMe_2/Cl exchange (see Chapter 13) of Eq. (4).

$$Ph_2SiH(Cl) + Me_3SnNMe_2 \longrightarrow Ph_2SiH(NMe_2) + Me_3SnCl \quad (4)$$

Dehydrochlorination of organic halides [e.g., Eq. (5)] has been observed in presence of Me_3SnNMe_2 [133]; it is interesting that there is preponderance of the thermodynamically less stable olefins.

$$n\text{-}C_4H_9Cl + Me_3SnNMe_2 \xrightarrow[4\text{ h}]{40°\text{ C}} \textit{cis}\text{-}(5.2\%)\text{ and } \textit{trans}\text{-}(88.5\%)\ MeCH{=}CHMe + (6.3\%)n\text{-}C_4H_8 + Me_3SnCl.HNMe_2 \quad (5)$$

In the systems Me_3SnNMe_2/C_2Cl_5H and $Me_3SnNMe_2/CH{\equiv}CCH_2Cl$, dehydrochlorination was presumably achieved since the dimethylamine adduct of trimethyltin chloride, $Me_3SnCl.HNMe_2$, was isolated from both reactions [157, 347].

C OTHER DEPROTONATION REACTIONS

Organometallic or metal amides are powerful bases (Chapter 16), especially when the compounds are essentially ionic, as with the derivatives of the heavier alkali metals. Then they are indeed among the strongest bases of chemistry. A further feature of salts such as $K[NPr^i_2]$ or $K[N(SiMe_3)_2]$ is that while the amide ion is a very powerful base towards the proton, it is not, being sterically-hindered, an effective nucleophile with respect to carbon. This has considerable use in organic synthesis [234a]; other amides to have been employed include $LiNEt_2$, $MNPr^i_2$ (M = Li or Na), $LiN(C_6H_{11})_2$, $LiN(Pr^i)C_6H_{11}$, $Li\overline{N(CH_2)_4}$, $MN(SiMe_3)_2$ (M = Li or Na), and MN(Me)Ph (M = Li or Na). Hydrogen bonded to atoms other than carbon is replaceable: e.g., $Bu_3SnH + LiNPr^i_2 \rightarrow Bu_3Sn^-$ [704a].

Examples of proton abstraction from acidic hydrocarbons (and consequential reactions) are shown in Eqs. (6) [412]-(9) (reference should also be made to Chapter 11). The metallations are often carried out at low temperature to avoid side-reactions involving a functional group.

$$RCH_2COR' \xrightarrow[\text{(ii) } R''X]{\text{(i) } NaN(SiMe_3)_2} RR''CHCOR' + RCH{=}C(OR'')R' \quad (6)$$

$$CH_3CN \xrightarrow[\text{(ii) } C_6H_{10}]{\text{(i) } NaN(SiMe_3)_2} \text{bicyclo[4.1.0]heptane} + NaCN \quad (7)$$

$$CHX_3 \xrightarrow[\text{(ii) } Me_3SiX]{\text{(i) } MN(SiMe_3)_2} Me_3SiCX_3 + MX \quad (8)$$

Reactions (7) [410], (8) [472], and (9) [471] are carbene insertions.

$$CH_2Cl_2 \xrightarrow[\text{(ii) } C_6H_6\text{–}MN(SiMe_3)_2]{\text{(i) } MN(SiMe_3)_2} \text{7-H-7-}(N(SiMe_3)_2)\text{cycloheptatriene} + (Me_3Si)_2N\text{-cycloheptatriene} \quad (9)$$

The base is useful for promoting sterically-hindered condensations [411]: e.g., ethyl isobutyrate → ethyl isobutyrylisobutyrate, cyclisation of suberodinitrile [Eq. (10)], or in the Wittig reaction (e.g., $Ph_2CO \rightarrow Ph_2C{:}CPh_2$).

$$NC\text{–}CH_2\text{–}(CH_2)_5\text{–}C{\equiv}N \longrightarrow \text{2-cyano-cycloheptanimine } (=NH,\ CN) \quad (10)$$

The well-known Birch reduction (Li or Na in liquid NH_3 and ROH, as in $C_6H_6 \rightarrow$ hexa-1,3-diene), and its many modifications (using amines or diamines in place of ammonia and alternative H-sources to ROH), proceeds by a single electron transfer from the base and subsequent hydrogen atom abstraction by the arene radical anion from the alcohol [234a]; it probably does not involve the metal amide.

Other examples of the reactions of $M[N(SiMe_3)_2]$ (M = Li, Na, or K) and $LiN(CMe_3)SiMe_3$ are found elsewhere [295a]. Interesting 'inorganic' cases of proton abstractions are in Eq. (11) which led [53a] to the first isolation of di-imine possibly via nitrene, and Eq. (12) [66a], in which compound (I) may be formed via $[Ti(\eta\text{-}C_5H_5)_2Cl\{N(SiMe_3)_2\}]$.

$$\text{TosNHNH}_2 \xrightarrow[\text{(ii)} <10^{-4}\text{ mm}]{\text{(i) MN(SiMe}_3)_2} \text{HN=NH} + \text{MTos} \qquad (11)$$

(M = Li, Na, or K)

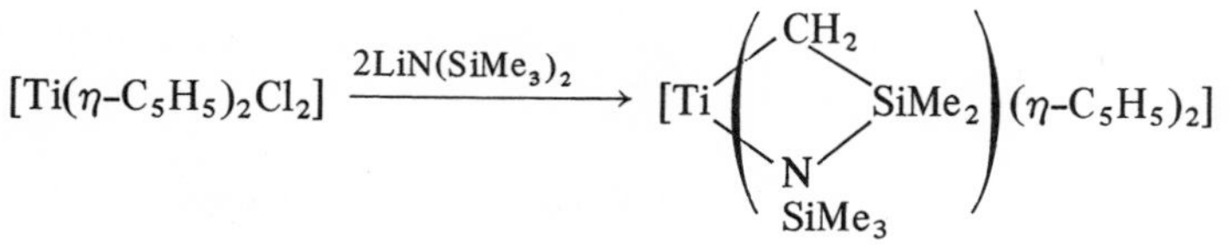

$$[Ti(\eta\text{-}C_5H_5)_2Cl_2] \xrightarrow{2LiN(SiMe_3)_2} [Ti(CH_2SiMe_2N(SiMe_3))(\eta\text{-}C_5H_5)_2] \qquad (12)$$

(I)

The Brønsted basicity of the alkali metal amides is so high that they have been used in studies of acidity of unactivated hydrocarbons, e.g., using $MNHC_6H_{11}$ [242a]. Kinetic acidity of hydrocarbons has been monitored in the $K[ND_2]/RH$ system by the rate of H/D exchange [687a], while equilibrium pK_a's of furan and related heterocycles were obtained with the aid of $Cs[NHC_6H_{11}]$ by spectrophotometry [706b].

The highly hindered lithium amide (II) (LiTMP) has been referred to as a 'H^+arpoon' base and is proton-selective, being particularly useful for α-elimination, as in Eq. (13) [564a]; *trans*-but-2-ene gives only isomer (III).

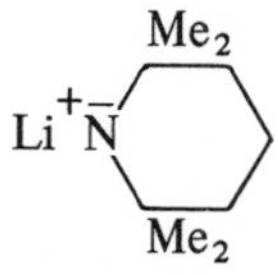

Li^+ N^- ring with Me_2, Me_2

(II)

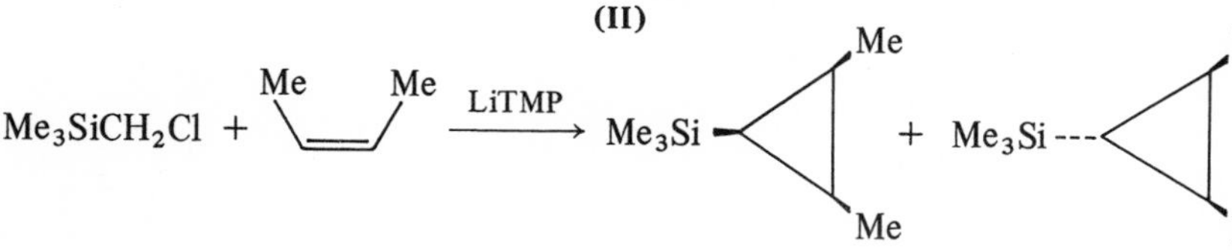

$$Me_3SiCH_2Cl + \text{cis-MeCH=CHMe} \xrightarrow{LiTMP} Me_3Si\text{-cyclopropane(Me)(Me)} + Me_3Si\text{---cyclopropane(Me)(Me)} \qquad (13)$$

(but no Me_3Si–cyclopropane(Me)(Me))

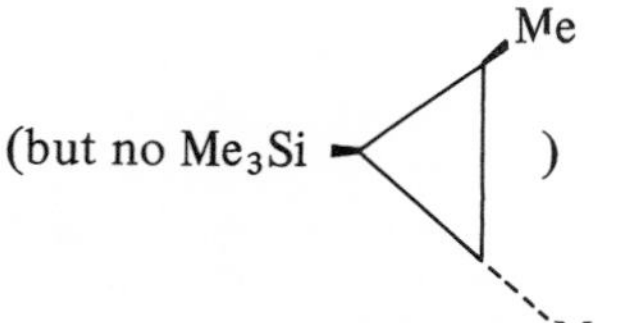

(III)

D TABLE OF REACTIONS

Dehydrochlorination Reactions

Reagents	Experimental conditions	Products	
Reactions of Me_3SnNMe_2 *with* C_4H_9Cl *or* Ir^{III} [133, 134, 419] *or* Pt^{II} [221a] *complexes*[a]			
C_4H_9Cl	30–40°	$EtCH{=}CH_2$	(6.3%)
		cis-$MeCH{=}CHCH_3$	(5.2%)
		trans-$MeCH{=}CHCH_3$	(88.5%)
		+ $Me_3SnCl.HNMe_2$	
$(PPh_3)_3IrHCl_2$	Xylene; 130–140°	$(PPh_3)_3IrCl$? + $Me_3SnCl.HNMe_2$	
$(PPh_3)_3IrHCl_2$	Heat; H_2	$(PPh_3)_3IrH_2Cl + Me_3SnCl.HNMe_2$	
$(PPh_3)_3IrHCl_2$	110°; CO	*trans*-$(PPh_3)_2Ir(CO)Cl + Me_3SnCl.HNMe_2 + PPh_3$	
$(PPh_3)_3IrHCl_2$	Reflux in C_6H_6 or C_6D_6	$(PPh_3)_3IrCl$ (isomer) + $Me_3SnCl.HNMe_2$	
$(PPh_3)_3IrHCl_2$	25°; CO	$(PPh_3)_2Ir(CO)Cl + Me_3SnCl.HNMe_2$	
$(PPh_3)_3IrH_2Cl$	Reflux in toluene	$IrH_3(PPh_3)_2 + PPh_3 + Me_3SnCl.HNMe_2$	
$(PPh_3)_3IrH_2Cl$	80°–100°; CO	$(PPh_3)_3Ir(CO)H + Me_3SnCl.HNMe_2$	
$(PPh_3)_3IrH_2Cl$	Excess amide	$(o\text{-}C_6H_4PPh_2)Ir(SnMe_3)(PPh_3)_2 + 2HNMe_2 + Ph_3P + Me_3SnCl.HNMe_2$	
$(PPh_3)_2PtHCl$	Xylene, heat	*cis*-$(Ph_3P)_2Pt(Me)(SnMe_2Cl)$	
Reactions of $[(\eta\text{-}C_5H_5)_2TiNMe_2]_2$ [433]			
$(R_3P)_2PtHCl$ ($R_3 = Ph_3$ or Me_2Ph, but not $= Et_3$)	C_6H_6; 20°; 7 days	$[(R_3P)_2Pt]_m + [(\eta\text{-}C_5H_5)_2TiCl]_2 + HNMe_2$	

[a]Consistent with our practice in earlier chapters, transition metal complexes are formulated in the Table so as to show the reacting groups adjacent to the metal.

15

Amides as Lewis Acids

A DISCUSSION

In principle, metal amides possess both acceptor (the metal atom) and donor (the nitrogen atom) sites, and are therefore capable of behaving as Lewis acids [Eq. (1)] or bases [Eq. (2)] (see Chapter 16) (where D and A are donor and acceptor molecules, respectively). The integrity of the M–N bond is maintained in these reactions. However, relatively few such reactions have been observed, probably because of (i) the usually ready cleavage of the M–N bond and (ii) the possibility of nitrogen–metal (or metalloid) $p_\pi \rightarrow d_\pi$ (or $p_\pi \rightarrow p_\pi$) bonding [this may explain why, for example, tris(dialkylamino)boranes do not react with pyridine] [437]. Additionally, the metal atom may provide the donor site {Chapter 16; see also Chapter 18 for oxidative addition, as for $M(NMe_2)_3$ (M = P or As) or $M[N(SiMe_3)_2]_2$ (M = Ge, Sn, or Pb)}.

$$LM{-}NR_2 + D \longrightarrow L\overset{-}{M}(NR_2){-}\overset{+}{D} \qquad (1)$$

$$LM{-}NR_2 + A \longrightarrow LM{-}\overset{+}{N}R_2{-}\overset{-}{A} \qquad (2)$$

The beryllium atom in $(RBeNR'_2)_n$ is co-ordinatively unsaturated and therefore these compounds are expected to react with bases, although $MeBeNMe_2$ does not react with Me_3N [160]. With pyridine, two types of complexes (I) and (II) [Eq. (3)] are obtained; in some cases, however, disproportionation of complex (I) occurs [Eq. (4)] [159]. Similar complexes with Zn [161, 240], Cd [240], and Hg amides and pyridine have been reported [161, 240]. Replacement of a Ph group in $EtZnNPh_2$ (dimer) [162] by the group COX increases the co-ordination number to four by further co-ordination involving the COX group. In solution, compounds of the type RZnN(Ph)COX, (III), are trimers (X = OMe or NPh_2), tetramers (X = Me [527], Et [162], or Ph), or higher aggregates

(X = H). Compounds (III) (X = OMe or NPh_2) form monomeric complexes with pyridine or $(CH_2NMe_2)_2$ (i.e., TMEDA) [162, 527]. The thiocarbonyl analogues of (III) (where X = OMe, Me, or SMe) also form 1:1 monomeric complexes with pyridine or TMEDA, except $EtZnN(Ph)CSNPh_2$ which forms a 1:2 complex with pyridine [87].

$$(RBeNR'_2)_n + NC_5H_5 \longrightarrow \text{(I)} \xrightarrow{NC_5H_5} \text{(II)} \quad (3)$$

(I): $R(C_5H_5N)Be(\mu\text{-}NR'_2)_2Be(NC_5H_5)R$

(II): $R(R'_2N)Be(NC_5H_5)_2$

$$(MeBeNPr_2)_2 + 2NC_5H_5 \longrightarrow (I{:}R = Me,\ R' = Pr) \rightleftharpoons Me_2Be.2NC_5H_5 + Be(NPr_2)_2 \quad (4)$$

Several types of addition complexes of aminoboranes with donor molecules (D) have been reported. These are the neutral molecular addition compounds, $\equiv\overset{-}{B}-\overset{+}{D}$, 1:1 electrolytes or boronium salts $[=BD_2]^+Cl^-$, and 1:3 addition complexes containing $[-BD_3]^{2+}$ [533]. Dimethylaminodichloroborane [258] forms a 1:1 complex with Me_3N whereas the complex with pyridine is probably a boronium salt of the type (IV) [437]. The 1:3 addition complex (V) between Et_2NBCl_2 and pyridine is probably the first example of a compound with a doubly-charged boron cation [437]. Borazines of the type $(RBNR')_3$ (where R = Ph, R′ = H; R = H, R′ = Ph; or R = R′ = Me or Pr^i) do not react with pyridine. However, upon adding a borazine having strongly electronegative groups attached to the boron atoms [e.g., $(ClBNH)_3$] to an excess of pyridine, a solid complex (VII) having four moles of pyridine is formed. When the mode of addition is reversed, a 1:2 complex (VI) is produced even in the presence of a large excess of pyridine. This demonstrates that besides steric and electronic factors, solubility criteria are also important in complex formation reactions of B–N compounds [437].

$$[(Me_2N)(Cl)B(NC_5H_5)_2]^+Cl^- \quad \text{(IV)}$$

$$[(Et_2N)(C_5H_5N)B(NC_5H_5)_2]^{++}\ 2Cl^- \quad \text{(V)}$$

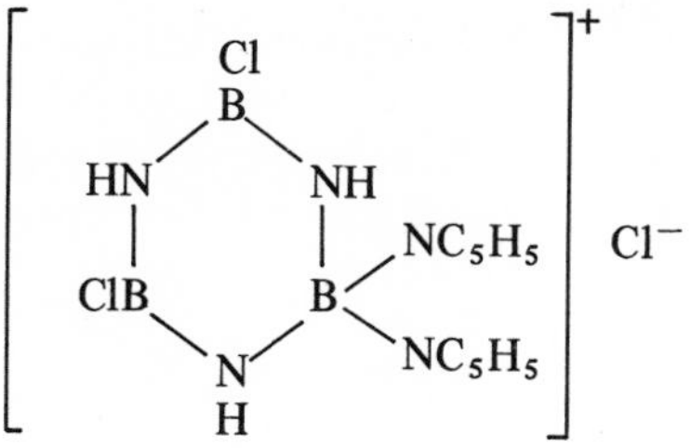

(VI) 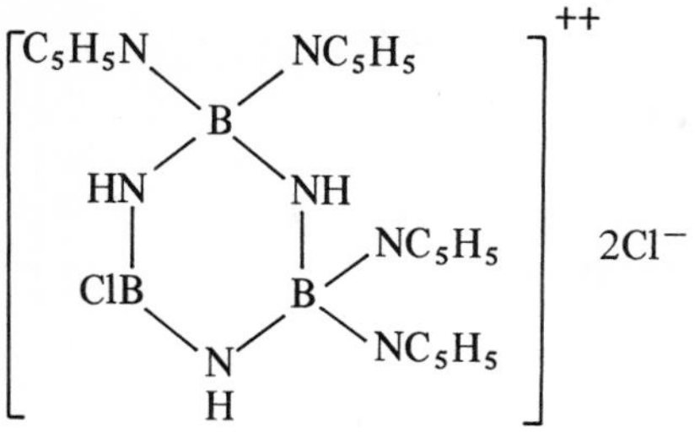 (VII)

The 1:1 adduct (VIII) [Eq. (5)] is quite stable; thermolysis yields (IX), which undergoes only partial hydrolysis in hot T.H.F. to give (X) [277].

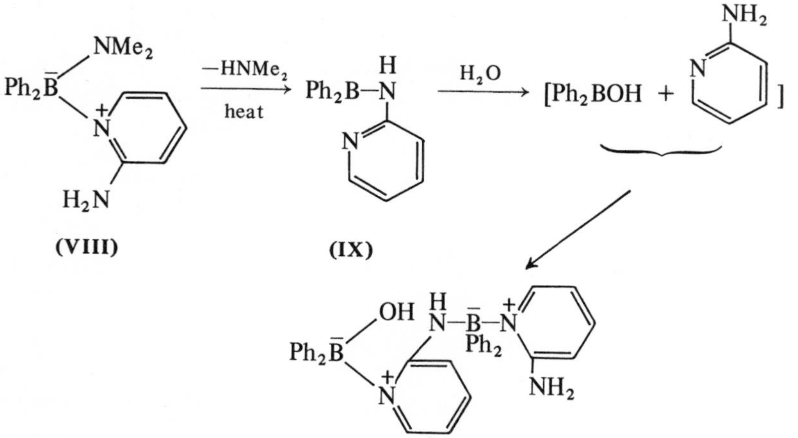

(VIII) (IX) (X) (5)

Dialkylaminodichloroboranes, Cl_2BNR_2, react with hydrogen chloride to give 1:1 complexes, $R_2NBCl_2.HCl$ (R = Me or Et) [258, 565, 682c]. Both these complexes were initially considered to be boronium salts but electrical conductivity measurements on the ethyl homologue showed that it has the structure $Cl_3\overset{-}{B}—\overset{+}{N}HEt_2$ [258]. Further examples of boronium salts include (XI) [540], from HCl and $B(NMe_2)_3$ or $ClB(NMe_2)_2$ and (XII) [54c], from $ClB(NMe_2)_2$, 2,2′-bipyridine, and sodium tetraphenylborate. It is arguable whether reactions of aminoboranes or other amidometal complexes (see Table) with hydrogen halides are appropriately classified as exemplifying Lewis acid rather than base behaviour of the substrate; however, the alternative seems to us artificial: to place the formation of, for example, boronium salts in the Lewis acid category when these salts are derived from an aminoborane and a base, and in the Lewis base when the reagent is HCl.

$$[(Me_2HN)_2BCl_2]^+ \, Cl^-$$

(XI)

$$[(Me_2N)_2B(\text{bipy})]^+ \, [Ph_4B]^-$$

(XII)

Aminoalanes form donor-acceptor complexes with azomethines and some aromatic *N*-heterocycles (see Table) [510]. These yellow-to-green complexes show a broad band at ca. 22,200 cm^{-1}, which seems to be characteristic for the complexes and is considered to arise from a charge transfer mechanism. With other amine donors, this band is found in the ultraviolet. Introduction of an alkyl group into the donor shifts the charge transfer band to lower wave number [511].

The aminosilane $Cl_2Si(NMe_2)_2$ and hydrogen chloride form addition complexes of unknown structure and variable stoicheiometry: $(Me_2N)_2SiCl_2.(HCl)_{3\cdot 2}$ at $-78°$ C and $(Me_2N)_2SiCl_2.(HCl)_2$ at $+20°$ C [136]. However, in the case of $(MeSiH_2)_3N/HCl$, ligand-exchange was observed [Eq. (6)] [223].

$$(MeSiH_2)_3N + 4HCl \longrightarrow 3MeSi(Cl)H_2 + NH_4Cl \qquad (6)$$

Bis(trichlorosilyl)amine reacts with pyridine, or another tertiary amine, to yield the corresponding salt $[pyH]^+[(Cl_3Si)_2N]^-$ or $[R_3NH]^+[(Cl_3Si)_2N]^-$

$$Cl_3\overset{-}{Si}{=}\overset{+}{N}{=}\overset{-}{Si}Cl_3$$

(XIII)

(R = Me or Et), which shows exceptionally strong Si-N bonds attributed to significant contributions from the canonical structure (XIII) [495].

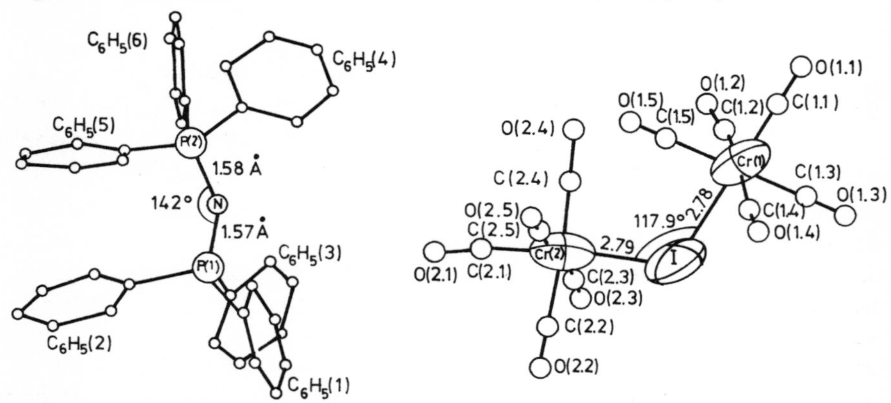

Fig. 1 – View of the cation and anion as found in the structure of $\{[(C_6H_5)_3P]_2N\}$-$[Cr_2(CO)_{10}I]$. *[Reproduced, with permission, from Handy, L. B., Ruff, J. K., and Dahl, L. F.,* J. Amer. Chem. Soc., *1970,* **92,** *7327.]*

Isoelectronic with $[(X_3Si)_2N]^-$ is $[(X_3P)_2N]^+$. The latter, often abbreviated as PPN^+, has gained importance as a cation for stabilising otherwise labile anions especially of carbonylmetallates: examples include $[M_2(CO)_{10}]^{2-}$ (M = Cr, Mo, or W), $[Cr_2(CO)_{10}I]^-$, $[M(CO)_5X]^-$ (M = Cr, Mo, or W; X = O_2CR or SR′), $[M_2Ni_3(CO)_{16}]^{2-}$ (M = Mo or W), and $[(OC)_5M(SMe)M'(CO)_5]$ (M, M′ = Cr, Mo, or W) [665a]. Whereas structure (XIII) would require linearity at N, $[PPN]^+$ may be bent, the lone-pair on the nitrogen clearly being stereochemically active as shown in Fig. 1, or linear as in (X-ray) the $[V(CO)_6]^-$ salt [756b].

N-Tributylstannylimidazole, which exists as a co-ordination polymer, forms monomeric complexes with basic ligands in which the tin is five-co-ordinate [Eq. (7)] [343].

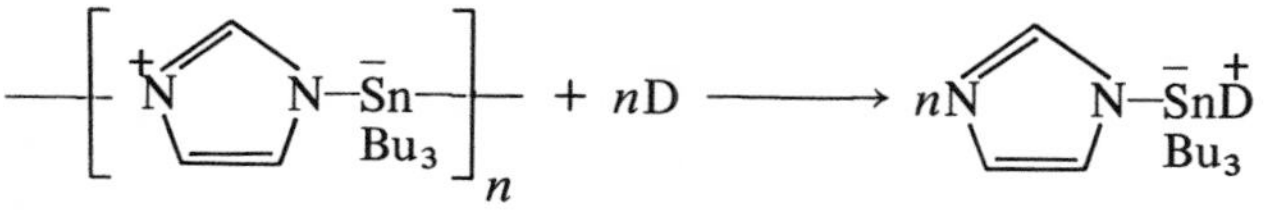

$$\left[-\overset{+}{N}\langle\text{imidazole}\rangle N-\overset{-}{Sn}(Bu_3)-\right]_n + nD \longrightarrow n\,N\langle\text{imidazole}\rangle N-\overset{-}{Sn}(Bu_3)\overset{+}{D} \qquad (7)$$

$$R_3\overset{-}{Sn}\langle O, N(R)\rangle\overset{+}{N}O$$

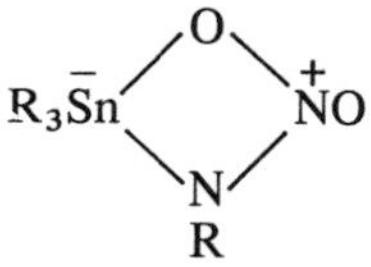

(XIV)

Complexes formed between organotin nitramines $R_3SnN(NO_2)R$ are believed to have the chelate structure (XIV), on the basis of electrical conductivity in nitrobenzene, molecular weight, and infrared and Mössbauer spectra [757].

The dimeric tin(II) amide $[Sn(NMe_2)_2]_2$ yields the adduct $Sn(NMe_2)_2.NC_5H_5$ with pyridine [246a]. The more bulky monomeric tin(II) amide $Sn[N(SiMe_3)_2]_2$ shows only feebly acidic properties forming a 1:1 adduct with 1,10-phenanthroline, but not with pyridine, 4-picoline, or 2,2′-bipyridine [594a]; neither the Ge^{II} nor the Pb^{II} analogue gives complexes even with 1,10-phenanthroline, whereas the isoelectronic $Sn[CH(SiMe_3)_2]_2$ forms a weak adduct with pyridine [428].

Tris(disilylamido)lanthanides afford unusual five-co-ordinate peroxy-complexes with Ph_3PO or $(Ph_3PO)_2(H_2O_2)$ [Eq. (8)] [92]. The lanthanum complex (XV, M = La) has been shown (X-ray) to contain the first reported case of a doubly-bidentate bridging peroxo ligand [92].

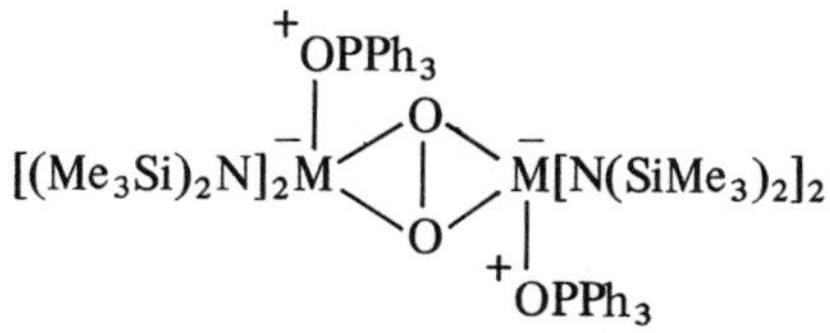

$$[(Me_3Si)_2N]_2\overset{-}{M}(\overset{+}{O}PPh_3)\langle O\text{—}O\rangle\overset{-}{M}(\overset{+}{O}PPh_3)[N(SiMe_3)_2]_2$$

(XV)

$$[M\{N(SiMe_3)_2\}_3] + Ph_3PO \xrightarrow{C_6H_6} [M\{N(SiMe_3)_2\}_3(OPPh_3)] \xrightarrow{H_2O_2} (XV) \quad (M = La, Pr, Sm, \text{ or } Eu) \tag{8}$$

$$[M\{N(SiMe_3)_2\}_3] \xrightarrow{Ph_3PO,\ H_2O_2} (XV)$$

The compound $[Ti\{N(SiMe_3)_2\}Cl_3]$ reacts with pyridine with elimination of Me_3SiCl and formation of $Me_3SiNTiCl_2.nNC_5H_5$ (see also Chapter 18 for other Me_3SiCl eliminations) [120]. Tris(dimethylamido)titanium forms an etherate $[Ti(NMe_2)_3(OEt_2)]$ [25, 431].

Pyridine effects ring-contraction of $(Cl_2Ti\text{–}NSiMe_3)_4$, forming compound (XVI) ($L = NC_5H_5$), in which the titanium atoms have apparently increased their co-ordination number to six [21].

L SiMe$_3$L
N
Cl$_2$Ti TiCl$_2$
N
L SiMe$_3$L

(XVI)

Contrary to an early report, $[Zn\{N(SiMe_3)_2\}_2]$ forms a monomeric 1:1 adduct with pyridine or a substituted pyridine [118]; the Co^{II} analogue behaves similarly. Trimethylamine is a weaker base with respect to these metal amides, because similar complexes were not isolated as solids although there was evidence for their formation in solution [240].

Amides of transition metals in a low metal oxidation state may often be characterised, although labile, as their adducts with a Lewis base. Examples are found for Cr^{II}, Mn^{II}, Co^{II}, Ni^{I}, and Au^{I}: $[Cr\{N(SiMe_3)_2\}_2(T.H.F.)_2]$ [96], $[Mn\{N(SiMe_3)_2\}_2(T.H.F.)]$ [89a], $[Co\{N(SiMe_3)_2\}_2(PPh_3)]$ [96a], $[NiL_2\{N(SiMe_3)_2\}]$ ($L = PPh_3$ [96a] or T.H.F. [122]), and $[Au(MR_3)\{N(SiMe_3)_2\}]$ ($R = Me$ or Ph, $M = P$; $R = Ph$, $M = As$) [671a]. Reference may also be made to Tables in Chapter 8 and ref. [89a].

B TABLE OF REACTIONS

Reactions as Lewis Acids

Reactants		Products	References
$(RBeNR'_2)_n$ (R=R′=Me; R=Me, R′=Ph)	$2NC_5H_5$	R'_2 R N NC_5H_5 Be Be H_5C_5N N R R'_2	[159]
$(RBeNR'_2)_n$ (R=R′=Me; R=Me, Et, R′=Ph)	excess NC_5H_5	R NC_5H_5 Be R'_2N NC_5H_5	[159]
$(MeBeNPr_2)_2$	$2NC_5H_5$	$Me_2Be.2NC_5H_5 + (Pr_2N)_2Be$	[159]
$(Me_2N)_2Be$	bipy	$(Me_2N)_2Be.bipy$	[159]
$(MeBeNMe_2)_3$	bipy	$Me_2Be.bipy + (Me_2N)_2Be.bipy$	[159]
$(RBeNPh_2)_2$	bipy	$R(Ph_2N)Be.bipy$	[159]
$Be(NMe_2)_2$	Bu^tLi	$Li[Be(NMe_2)_2Bu^t]$	[33]
$(Me_3Si)_2NMgBr$	$Me_2C(COOMe)_2$	$Me_2CHC(OMe)=OMg(Br)N(SiMe_3)_2$	[453]
$[(Me_3Si)_2N]_2Mg$	$Me_2CHCOOMe$	OMe O=C $[(Me_3Si)_2N]_2Mg$ CMe_2 O=C OMe	[453]
Me_2NBCl_2 (n = 1, 3)	nNC_5H_5	$[Me_2NB(Cl)(NC_5H_5)_2]^+Cl^-$	[437]
Et_2NBCl_2	$3NC_5H_5$	$[Et_2NB(NC_5H_5)_3]^{++}2Cl^-$	[437]
$Pr^i_2NBCl_2$	$3NC_5H_5$	$Pr^i_2N(Cl_2)B.NC_5H_5$	[437]
Me_2NBCl_2	$2NC_5H_4.Me(4)$	$[Me_2NB(Cl)(NC_5H_4.Me)_2]^+Cl^-$	[437]
$Pr^i_2NBCl_2$	$2NC_5H_4.Me(4)$	$Pr^i_2N(Cl_2)B.NC_5H_4.Me$	[437]
$(HNBCl)_3$	$6.5NC_5H_5$	[C_5H_5N NC_5H_5 B HN NH ClB B NC_5H_5 NC_5H_5 N H]$^{++}$ $2Cl^-$	[437]
$(HNBCl)_3$ (n = 1, 2, 3)	nNC_5H_5	[Cl B HN NH ClB B NC_5H_5 NC_5H_5 N H]$^+$ Cl^-	[437]

Table of Reactions *(C'ntd.)*

Reactants		Products	References
$(Me_2N)_2BCl$	bipy, $Na[BPh_4]$	$[B(NMe_2)_2bipy][BPh_4]$	[54c]
Me_2NBPh_2	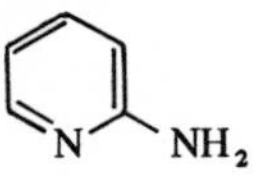 N NH_2	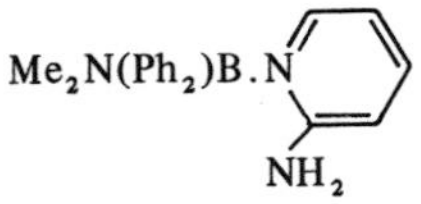 $Me_2N(Ph_2)B.N$ NH_2	[277]
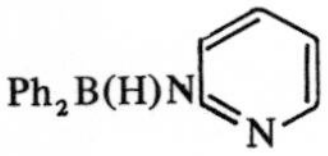 $Ph_2B(H)N$ N	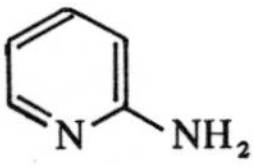 N NH_2 + Ph_2BOH	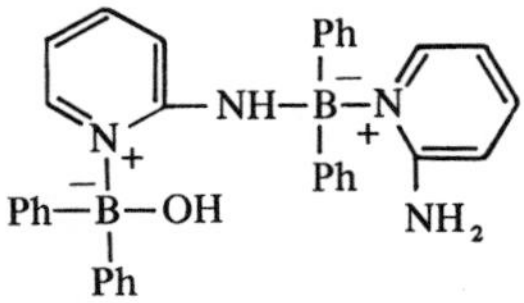 Ph; NH—B—N; +; Ph; NH_2; N^+; Ph—B—OH; Ph	[277]
Me_2NBCl_2 [a] (X=Cl, Me)	Me_2NBXMe	$Me_2NBCl_2.Me_2NBXMe$	[283]
R_2NBCl_2 [a] (R=Me, Et)	HCl	$R_2NBCl_2.HCl$	[258,565]
$XB(NMe_2)_2$ [a] (X=Cl, NMe_2)	HCl	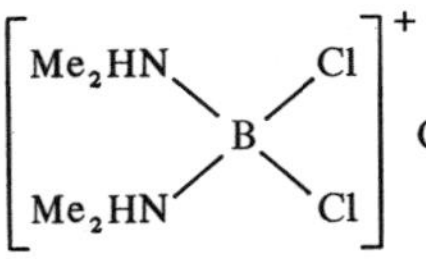 $[(Me_2HN)_2BCl_2]^+$ Cl^-	[540]
$(Me_2N)_3B$ [a]	3HCl	$(Me_2N)_3B.3HCl$	[682b]
$(Et_2N)_3B$ [a]	5HCl	$Et_2NBCl_2.HCl + 2Et_2NH.HCl$	[258]
$2BuOB(NEt_2)_2$ [a]	7HCl	$(BuO)_2BCl + Et_2NBCl_2.HCl + 3Et_2NH.HCl$	[258]
$(Me_2N)_3B$ [a]	2HI	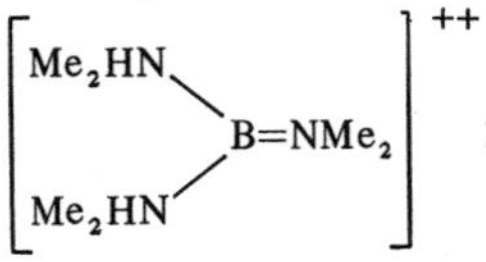 $[(Me_2HN)_2B{=}NMe_2]^{++}$ $2I^-$	[535]
$(Me_2N)_2BMe$ [a]	HI	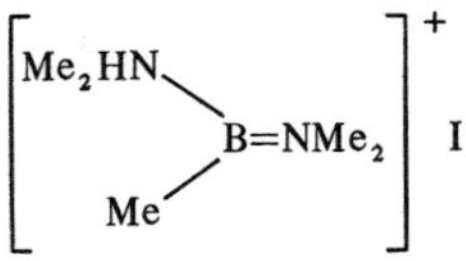 $[(Me_2HN)(Me)B{=}NMe_2]^+$ I^-	[535]
$(Me_2AlNMe_2)_2$ [a]	8HCl	$2(Me_2NH_2)^+AlCl_4^- + 4MeH$	[196]
Bu^i_2AlR [R=N(CH_2Ph)Ph, n=1; R=H, n=2]	nPhN=CHPh	$Bu^i_2[Ph(PhCH_2)N]Al.N(=CHPh)Ph$	[510]
$Bu^i_2AlN(CH_2Ph)Ph$	$HNPhCH_2Ph$	$Bu^i_2[Ph(PhCH_2)N]Al.NHPhCH_2Ph$	[510]
Bu^i_2AlH (R=p-MeC_6H_4, $PhCH_2$, 2-$C_{10}H_7$; 2L=phenanthridine, acridine)	2RN=CHPh(=2L)	$Bu^i_2[R(PhCH_2)N]Al.N(=CHPh)R$	[510]
$Bu^i_2AlN(CH_2R)R$ (R=1-$C_{10}H_7$)	RN=CHR	$Bu^i_2[R(CH_2R)N]Al.N(=CHR)R$	[510]
$Ph_2AlN(R)PPh_2$ (L=o-phenanthroline, acridine; R=p-tolyl, 2-pyridyl)	L	$Ph_2AlN(R)PPh_2.L$	[266]

Table of Reactions *(C'ntd.)*

Reactants		Products	References
$(Me_2N)_nSiCl_{4-n}$[a] $(n=1,2)$	nHCl	$(Me_2N)_nSiCl_{4-n}.nHCl$	[136]
$[\overset{+}{N}C_3H_3N\text{—}\overset{-}{Sn}Bu_3\text{—}]_n$ (imidazolyl ring)	nD	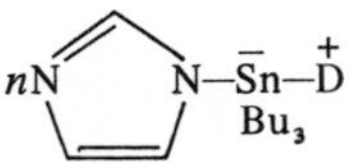 n $NC_3H_3N\text{—}\overset{-}{Sn}Bu_3\text{—}\overset{+}{D}$ (imidazolyl ring)	[343]
($D=C_4H_8O$, C_4H_8S, Bu_3N, Bu_3P, BuCl, BuBr, NC_5H_5, $MeCONMe_2$, $MeCSNMe_2$)			
$[Sn(NMe_2)_2]_2$	NC_5H_5	$Sn(NMe_2)_2.NC_5H_5$	[246a]
$Sn[N(SiMe_3)_2]_2$	*o*-phenanthroline	1:1 adduct	[428, 594a]
$M[N(SiMe_3)_2]_3$ (M=La, Pr, Sm, Eu)	L [$L=2Ph_3PO$ or $(Ph_3PO)_2H_2O_2$]	$M_2O_2(OPPh_3)_2[N(SiMe_3)_2]_4$ or $[M\{N(SiMe_3)_2\}_3(OPPh_3)]$	[92]
$(Me_3Si)_2NTiCl_3$	nNC_5H_5 $(n=2, 2.5)$	$Me_3SiNTiCl_2.nNC_5H_5$	[120]
$(Cl_2Ti\text{–}NSiMe_3)_4$	NC_5H_5	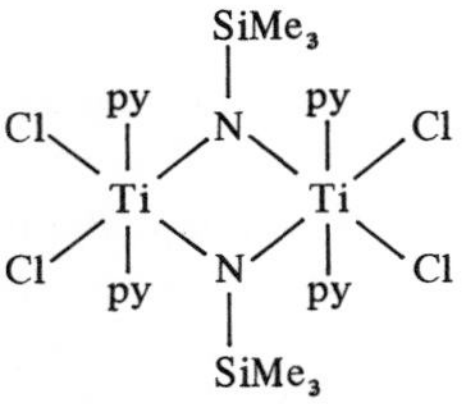 $[Cl_2(py)Ti(\mu\text{-}NSiMe_3)]_2$ (dimer: two Ti each bearing 2 Cl and 2 py, bridged by two $NSiMe_3$)	[21]
$(MeZnNPh_2)_2$	$2NC_5H_5$	$ZnMe_2 + Zn(NPh_2)_2.2NC_5H_5$	[161]
$(MeZnNPh_2)_2$	excess NC_5H_5	$Me(Ph_2N)Zn.2NC_5H_5$	[161]
$M[N(SiMe_3)_2]_2$ (M=Zn, $L=NC_5H_5$, NC_5H_4Me-4; M=Co, $L=NC_5H_5$, NC_5H_4Me-4, $NC_5H_3Me_2$-2,4)	L	$M[N(SiMe_3)_2]_2.L$	[240]
$RZnNPh_2$[a] [R=Et, NPh_2, and ZnX_2=bis(2,2-dimethyl-3,5-hexanedionato)zinc]	ZnX_2	$Zn(NPh_2)_2.ZnX_2$	[88]

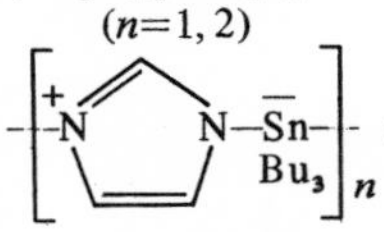

[a]These reactions could equally be designated as illustrating Lewis base properties of the amide.

16

Amides as Lewis Bases

A INTRODUCTION

Most reactions of metal amides illustrate their properties as nitrogen-centred nucleophiles. This is certainly the case for the majority of insertions (Chapter 10), reactions with protic compounds (Chapter 11) including those with many metal hydrides (Chapter 12), metathetical exchanges (Chapter 13), and dehydrochlorination or related reactions (Chapter 14); some of the reactions with hydrogen halides, placed for convenience in Chapter 15, belong to the same category. In this chapter, however, we are solely concerned with those systems in which the amide behaves as base in forming a donor-acceptor adduct, although this may be due either to one or more (I) nitrogen atoms, or the metal, acting as the donor site [Eq. (1)].

$$\mathrm{LM}\left\langle {}^{\overset{+}{N}R_2}_{\underset{+}{N}R_2} \right\rangle \mathrm{A}^{2-} \quad \mathbf{(I)}$$

$$\mathrm{LM{-}NR_2} + \mathrm{A} \begin{cases} \longrightarrow \mathrm{LM{-}\overset{+}{N}(R_2){-}\overset{-}{A}} \\ \longrightarrow \mathrm{L\overset{+}{M}(NR_2){-}\overset{-}{A}} \end{cases} \qquad (1)$$

Photoelectron spectroscopy shows that metal amides have relatively low first ionisation potentials (see Chapter 1), the first I.P. decreasing with increasing number of amido groups. Metals, being electropositive, cause the electron density at the amide nitrogen to be increased compared with the parent amine. Non-bonding electron-pair repulsions become particularly significant if there is more than one amido ligand attached to a small central metal atom or ion, and also cause lowering of the first I.P. These features are illustrated by the following data (first I.P. in eV): $HNMe_2$, 8.85; $C(NMe_2)_4$, 7.19; $Si(NMe_2)_4$, 8.69; $Ge(NMe_2)_4$, 8.48; $Sn(NMe_2)_4$, 7.67; $V(NMe_2)_4$, 7.08; $Ti(NMe_2)_4$, 7.13; $Zr(NMe_2)_4$, 7.23; $Hf(NMe_2)_4$, 7.50; $W(NMe_2)_6$, 6.73 [263a]; $B(NMe_2)_3$, 7.58; $B(NMe_2)_2Me$, 7.63; $B(NMe_2)Me_2$, 8.92 [83a]. For the recently prepared Ge^{II}, Sn^{II}, and Pb^{II} bis(silyl)-amides, although the central metal lone pair (8.68, 8.38, and 8.16 eV for Ge, Sn, Pb compounds $M[N(SiMe_3)_2]_2$) is not of lower orbital energy than the $b_2\ddot{N}_2$-combination (7.71, 7.75, and 7.92 eV for Ge, Sn, and Pb) [295b], the metal is believed to be the donor site in derived complexes such as $[Mo(CO)_5(Sn\{N(SiMe_3)_2\}_2)]$ [428] (see Table 2); for $M[\overline{NCMe_2(CH_2)_3CMe_2}]_2$ the central metal lone pair is energetically more accessible (6.8 eV for M = Ge and Sn) [429a]. For many complexes formed from amides of P^{III} or As^{III}, where a central metal lone pair is again available, P or As is usually the ligating site. The formation of a phosphonium salt, as in Eq. (2) [595, 597], (3) [137], and (4) [333], belongs to a related system, whereas the reaction shown in Eq. (5) [333] is not particularly characteristic of an amide. Cations said to have P–P bonds, $[(Me_2N)_3P{-}P(NMe_2)X]^+$ (X = Cl or NMe_2), were reported to be formed from dimethylamidophosphorus(III) chlorides and $AlCl_3$ [392]. In $[MCl_4(hmpa)_2]$ [M = Ce or U, hmpa = $P(NMe_2)_3O$], the oxygen is the ligating atom in hmpa (X-ray) [220a]. In connection with metal lone-pair basicity reference is also made to oxidative addition reactions (Chapter 18), as in $Sn^{II} \rightarrow Sn^{IV}$ or $P^{III} \rightarrow P^{V}$.

$$X_{3-n}P(NMe_2)_n + SbCl_5 + Cl_2 \longrightarrow [(Me_2N)_nPX_{3-n}Cl]^+[SbCl_6]^- \quad (2)$$

$$(X = Cl \text{ or } OMe,\ n = 1, 2, \text{ or } 3)$$

$$P(NMe_2)_3 + K[PF_6] + Y \longrightarrow [(Me_2N)_3PX]^+[PF_6]^- \quad (3)$$

$$(Y = Br_2 \text{ and } X = Br, \text{ or } Y = CCl_4 \text{ and } X = Cl)$$

$$Me_{3-n}P(NMe_2)_n + ClNHR \longrightarrow [(Me_2N)_nPMe_{3-n}(NHR)]^+Cl^- \quad (4)$$

$$MeN{=}P(Me)(NMe_2)_2 + MeI \longrightarrow [(Me_2N)_3PMe]^+I^- \quad (5)$$

Another interesting system is that in which the nitrene $R\ddot{N}$ may be assumed to be the Lewis acid, yielding ylides as in Eq. (6) (R = Me [286c], Ph [408a], or Me_3Si [665]). A related case is that in which $Me_2(Me_2N)P{=}P(CF_3)$ is formed from Me_2PNMe_2 and $(CF_3P)_4$ [171]. The phosphine-imines, themselves, may behave as Lewis bases, as in $R_3P{=}NSiMe_3.\frac{1}{2}Me_2GeI_2$ [761] or $(Me_3PN)_2SiMe_2.MR_2$ (M = Zn or Cd, R = Me or Et) [672].

$$P(NMe_2)_3 + RN_3 \longrightarrow (Me_2N)_3P{=}NR + N_2 \qquad (6)$$

Rather obscure examples, in which phosphorus nucleophilicity is certainly implicated, are in the reactions of $P(NMe_2)_3$ with $PhC{=}CPh$ (bridged by SO_2, i.e. $\overset{\diagdown\;\diagup}{\underset{O_2}{S}}$) [343a] or 1,3-dibromocyclopent-4-ene [770a] to give $PhC(\overset{-}{S}O_2){=}C(Ph)\overset{+}{P}(NMe_2)_3$ or $[\{P(NMe_2)_3\}_2C_5H_3]Br$, respectively.

Hexamethylphosphoric triamide (HMPT) is the term organic chemists use to describe $OP(NMe_2)_3$ (or alternatively HMPA, hexamethylphosphoramide), which has gained considerable synthetic utility as a powerfully-solvating, high-dielectric constant (30 at 25° C) aprotic solvent [234a]. This is clearly due to its base behaviour, there being four possible donor sites. Amongst its uses are for 'difficult' Grignard or organolithium reactions (involving unreactive halides); displacement of halogen (as X^-) from unreactive halides (e.g., neopentyl, with $LiC{\vdots}CH$; aryl, with $NaNH_2$ *via* a benzyne); dehydrohalogenation of α-bromoketones to α,β-unsaturated ketones; reduction (with Li or K) of the latter to saturated ketones; α-*C*-alkylation (with RMgX) of $R'R''CHCOR'''$; oxidation (with CrO_3) of a steroidal allyl alcohol; in peptide synthesis, with tosic anhydride; and in the dehydration of hindered alcohols, e.g., $Bu^t_2C(OH)CH_2Pr^i \rightarrow Bu^t_2C{=}C(H)Pr^i$. In terms of the 'donor number' scale ($-\Delta H$ for reaction with $SbCl_5$), HMPT at 38.8 is the highest; cf. DMSO, 29.8 [284a].

B AMIDES AS LEWIS BASES, WITH N AS LIGATING ATOM(S)

Formation of stable molecular addition compounds with strong Lewis acids is a general feature of metal amides. However, compounds having powerfully electronegative substituents [e.g., Cl_2BNMe_2 or $(HNBCl)_3$] [405] do not behave in this fashion. Spectroscopic studies (i.r., Raman, u.v., n.m.r.) have been used to determine the base strengths of metal amides, and results have often been interpreted in terms of the significance to $\ddot{N}\curvearrowright M$ π-bonding. For example, the observed shifts in the C–D stretching mode in mixtures of base and $CDCl_3$ have been used to compare the donor properties of various organometallic amides Me_3MNEt_2 and $(Me_3M)_3N$ (M = Si, Ge, or Sn). In aminosilanes, the relative base strengths decrease in the sequence $Me_3SiNR_2 > Me_3SiNHR > (Me_3Si)_2NR > (Me_3Si)_3N$; in aminogermanes, $Me_3GeNEt_2 > (Me_3Ge)_3N$; in aminostannanes, $(Me_3Sn)_3N > (Me_3Sn)_2NR > Me_3SnNR_2$ [6, 10], and in aminoboranes, $B(NHPr^i)_3 > RB(NHPr^i)_2 > R_2BNHR'$ (R = Me or Ph) [533a]. Ion cyclotron resonance measurements have yielded proton affinity data for $Me_3Si(CH_2)_nNMe_2$ (227.6 kcal mol^{-1} for $n = 1$, 2, or 3), $Bu^tMe_2SiNMe_2$ (225.6 kcal mol^{-1}), $Me_3C(CH_2)_nNMe_2$ (226.3 or 225.8 kcal mol^{-1} for $n = 2$ or 1), and $Bu^tMe_2CNMe_2$ (230.7 kcal mol^{-1}) [687b].

Competition and/or displacement reactions [e.g., Eqs. (7)-(9)] have been used to show that, with respect to $SnCl_4$ as acceptor, donor strengths decrease in the series (i) $C_5H_5N > Me_2Si(NMe_2)_2 > B(NMe_2)_3 > ClB(NMe_2)_2 \gg Cl_2BNMe_2$ (ii) $Me_2Si(NMe_2)_2 > B(NMe_2)_3 > Me_3SiNMe_2 > (Me_3Si)_2NH$, and (iii) $C_5H_5N > Ti(NMe_2)_4$. With $Me_2Si(NMe_2)_2$ as base, the Lewis acid character decreases in the series $SnCl_4 > TiCl_4$ or $VOCl_3$ [435].

$$B(NMe_2)_3.SnCl_4 + 2NC_5H_5 \longrightarrow SnCl_4.2NC_5H_5 + B(NMe_2)_3 \qquad (7)$$

$$Me_2Si(NMe_2)_2 + B(NMe_2)_3 + SnCl_4 \longrightarrow Me_2Si(NMe_2)_2.SnCl_4 + B(NMe_2)_3 \qquad (8)$$

$$Me_2Si(NMe_2)_2 + SnCl_4 + VOCl_3 \longrightarrow VOCl_3 + Me_2Si(NMe_2)_2.SnCl_4 \qquad (9)$$

Amides of B, Si, Sn, and Ti give stable molecular addition compounds with strong Lewis acids such as $SnCl_4$, $TiCl_4$, $ZrCl_4$, $VOCl_3$, and $HgCl_2$ [417, 435, 544]. Structures of most of these adducts have not yet been fully examined, partly due to their insolubility in common organic solvents. It has been suggested, on the basis of infrared studies, that the ligand $XB(NMe_2)_2$ (X = Br or Cl) in the complex $2XB(NMe_2)_2.3TiX_4$, (I), utilises both nitrogen atoms as ligands to bridge neighbouring tetrahalogenotitanium moieties, [417, 711]. In $ClB(NMe_2)_2.SnCl_4$, (II), on the other hand, the $ClB(NMe_2)_2$ ligand is assumed to be unidentate [435].

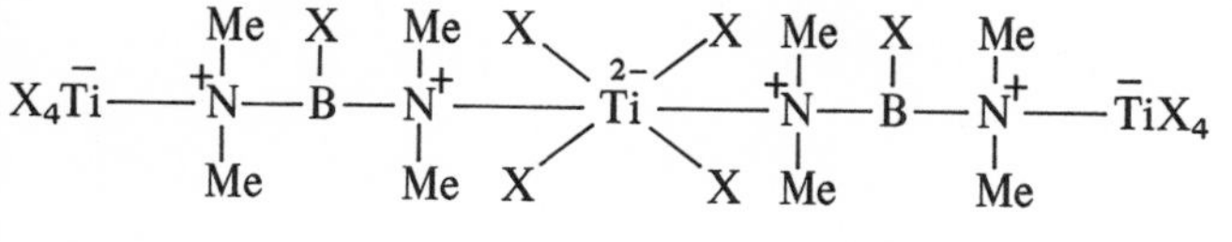

$$\bar{X_4Ti}—\overset{+}{N}Me_2—BX—\overset{+}{N}Me_2—\overset{2-}{Ti}X_4—\overset{+}{N}Me_2—BX—\overset{+}{N}Me_2—\bar{TiX_4}$$

(I) (X = Cl or Br)

$$Cl_4\bar{Sn}—\overset{+}{N}Me_2—B(Cl)\overset{\frown}{—}NMe_2$$

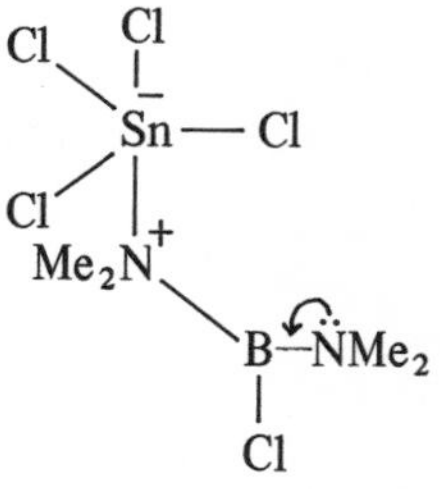

(II)

Tetrakis(dimethylamino)diboron is able to react with trimethylaluminium to form complex $B_6(NMe_2)_{12}.Al_6Me_{12}$ [28]. When the reaction of $B_2(NMe_2)_4$ with excess of trimethylalane is carried out at −15° C, a white crystalline solid, believed to be $AlMe_3.Me_2AlNMe_2$, is formed but disproportionates at room temperature to give $(AlMe_3)_2$ and $(Me_2AlNMe_2)_2$.

$[W(CO)_3\{B(NMe_2)_3\}]$ [545] and $[Ni(CO)_3\{As(NMe_2)_3\}]$ [723] are rare examples of a metal carbonyl having a main group metal amide as ligand. In the complexes (III) and (IV) [Eqs. (10) and (11)], bonding between the B-N π-system and the metal atom has been suggested on the basis of i.r. and ^{11}B n.m.r. spectra

[667] (see Section B). Borazine was believed (cf., Section C) to behave as a π-donor in $[M(CO)_3(R_3B_3N_3R'_3)]$ (M = Cr or Mo) [204, 599, 684]; kinetic studies for the borazine displacement reaction between the chromium complex (R = Me, R′ = Et) and a tertiary phosphite show that the donor–Cr bond dissociation energy (25 ± 3 kcal mol^{-1}) is significantly lower than $D(Cr\text{–}C_6H_6)$ [684a] (for bonding in such compounds, see Section B).

$$XB(NMe_2)Y + [Fe_2(CO)_9] \longrightarrow \underset{\textbf{(III)}}{[Fe(CO)_3(Me_2NBXY)]} \quad (10)$$

$$(X = Me, Y = NMe_2;\ X = CH{=}CH_2, Y = Br)$$

$$Me_2BNMe_2 + [Ni(\eta\text{-}C_3H_5)_2] \longrightarrow \underset{\textbf{(IV)}}{[Ni(\eta\text{-}C_3H_5)(Me_2BNMe_2)]} \quad (11)$$

A tetra-azaboroline, Me_2N_4BMe (L), acts as a unidentate ligand with respect to main group metal Lewis acids: $BCl_3.L$, *trans*-$SnCl_4.2L$, and $SbCl_5.L$ have been isolated [305] and characterised by their i.r. and Raman spectra. Titanium(IV) chloride reacts with L (or its *B*-phenyl or *B*-chloro- analogue) to give complexes of stoicheiometry $4TiCl_4.3L$, formulated as (V). The latter are capable of retaining aromatic solvents in the crystal lattice [497].

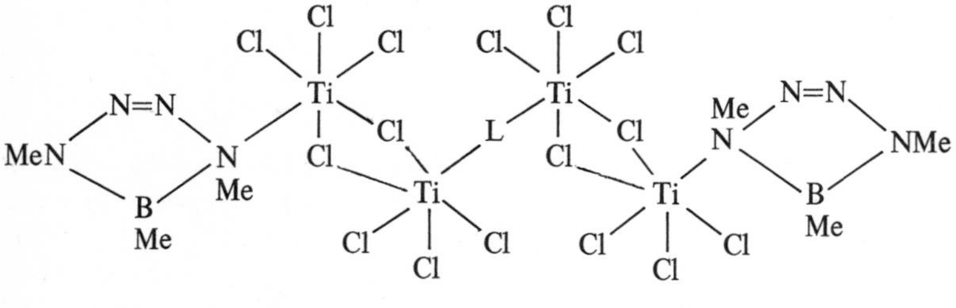

(V) ($L = Me_3N_4B$)

The aminoborane $XB(NMe_2)_2$ (X = Me or Cl) displaces the co-ordinated benzonitrile of $[PdCl_2(NCPh)_2]$, or (X = Br or I) causes the formation of an insertion product [Eq. (12)]; in both cases an η-boraza-allyl ligation is proposed [668] (see Section C).

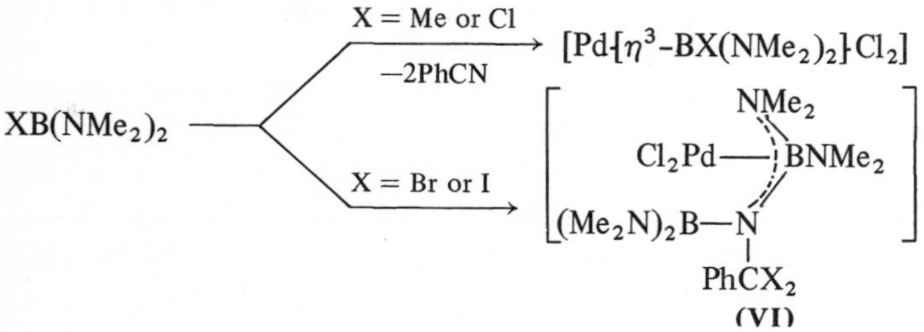

$$XB(NMe_2)_2 \xrightarrow[-2PhCN]{X = Me\ or\ Cl} [Pd\{\eta^3\text{-}BX(NMe_2)_2\}Cl_2] \quad (12)$$

$$XB(NMe_2)_2 \xrightarrow{X = Br\ or\ I} [Cl_2Pd\text{—}\{NMe_2 \cdots BNMe_2 \cdots N(CPhX_2)\text{—}B(NMe_2)_2\}]\ \textbf{(VI)}$$

Aminoboranes behave as amine donors towards iodine [Eq. (13)]: equilibrium constants and heats of formation of the 1:1 complexes have been obtained from near-ultraviolet spectrophotometry [232].

$$R_2BNR'_2 + I_2 \rightleftharpoons R_2BNR'_2.I_2 \qquad (13)$$

Aminosilane-adducts with BH_3, BMe_3, or H_3SiI are invariably unstable [52]. Possible reference acids appear to be limited to these rather weak acceptors, because others such as HHal, $BHal_3$, or $AlHal_3$ have effected Si-N cleavage [228, 229, 710]. Adducts of $AlMe_3$ with $(R_3Si)_2NX$ (X = H or Cl), however, decompose at high temperatures without cleaving the Si-N bond [751]. Phase-diagrams for the systems $(H_3Si)_nNMe_{3-n}$ ($n = 1$, 2, or 3) with B_2H_6, BMe_3, $(AlMe_3)_2$, or $GaMe_3$ have established that the di- and mono-silylamines form adducts at low temperatures [709]. Complexes of $SnCl_4$ with disilylamines (1:2) and with acyclic- and cyclic-diaminosilanes (1:1) have been reported [391]. Donor-acceptor complexes have been suggested as (at least transient) intermediates [435] in chemical transformations in SiN chemistry, as illustrated for example in Eqs. (14) [41] and (15) [654].

$$(R_2SiNR')_3 \xrightarrow[SnCl_4]{AlCl_3,\ TiCl_4,\ or} (R_2SiNR')_4 \qquad (14)$$

$$H_3Si\overset{|}{\underset{|}{N}} + \text{borane} \rightleftharpoons H_3Si\overset{|}{\underset{|}{N}}.\text{borane} \xrightarrow{H_3Si\overset{|}{\underset{|}{N}}} \overset{|}{\underset{|}{N}}SiH_2\overset{|}{\underset{|}{N}} + SiH_4 + \text{borane} \qquad (15)$$

$$(\text{borane} = B_2H_6,\ B_5H_9,\ 1\text{-}BrB_5H_8,\ \text{or}\ 2\text{-}BrB_5H_8)$$

Aminophosphines are ambident nucleophiles and the site utilised for complex formation is dependent upon the nature of the Lewis acid used. In cases of acceptors based on B, Al, or P [140], or transition metals [215, 216, 375] the co-ordination is usually through the P atom of the aminophosphines. There is spectroscopic evidence, however, that the BX_3 (X = F, Cl, or Me) or SbF_3 adducts, which are generally stable at low temperature and undergo P-N cleavage at or above room temperature, involve donation through the N atom of the aminophosphines [244, 279, 555, 575]. The complex $2F_3B.(Me_2N)_2PF$ is said to involve co-ordination through both the N atoms [555], but $X_3B.Me_3SiN(Me)PPh_2$ (X = Cl or Br) has a PB bond from $^3J(^{31}PNC^1H)$ [549a]. In the complex $SnCl_4.2Me_2NPR_2$, however, donation is from the N atom when R = Cl and from the P atom when R = Ph [416].

Dimethylaminodifluoroarsine gives 1:1 adducts with boron trifluoride or BH_3. The BF_3 adduct involves co-ordination through the N atom and is stable only at low temperature. The reduced basicity of the N atom as a result of donation of some electron density to the *d*-orbitals of As in the planar F_2AsNMe_2 presumably permits only weak interaction with the Lewis acid. In the adduct of B_2H_6, however, donation from the As atom has been suggested [498].

Formation of $Me_3SnNMe_2.B_{10}H_{14}$ or $Ti(NMe_2)_4.2B_{10}H_{14}$ from decaborane and dimethylaminotrimethylstannane or tetrakis(dimethylamido)titanium is interesting, as decaborane possesses acidic bridge-hydrogen atoms. Conductivity measurements of the titanium compound suggest an ionic structure of the type $[Ti(NMe_2)_2]^{2+}[B_{10}H_{13}.Me_2NH]_2^-$ or $[Ti(NMe_2)_2(Me_2NH)_2]^{2+}[B_{10}H_{13}]_2^-$. The tin complex may be formulated similarly as $[Me_3Sn]^+[B_{10}H_{13}Me_2NH]^-$ or $[Me_3Sn(Me_2NH)]^+[B_{10}H_{13}]^-$; the former was preferred on the basis of infrared studies [147].

Transition metal(IV) amides give stable adducts with metal carbonyls $[Cr(CO)_6]$, $[Mo(CO)_6]$, $[W(CO)_6]$, $[Fe(CO)_5]$, or $[Ni(CO)_4]$, tentatively formulated [Eqs. (16), (17)] as (VII) or (VIII) [90, 97]. The paramagnetic complex (VII; $M = Ti$, $M' = Ni$, and $n = 4$) may have a polymeric structure involving carbonyl and dimethylamido bridging groups [90]. Steric effects play a role: $[Ti(NEt_2)_4]$ does not react with $[Ni(CO)_4]$ [90]. On the basis of molecular weight determinations it was supposed that the molecular structure of the complexes (VIII) involves three bridging dimethylamido groups joining the $Ti(\eta\text{-}C_5H_5)$ and Group VI metal$(CO)_3$ fragments: however, ^1H n.m.r. data were not consistent with the proposed structures [97], and more recent studies favour their formulation as carbenemetal complexes [579] (see also Chapter 10). The molecular structure of $[\{Fe(CO)_3(NH_2)\}_2]$ has been determined [194]; similar NH_2 or NH bridges are found in $[\{Pt(PPh_3)_2X\}_2][BF_4]_2$ ($X = N{=}NH$ or NH_2) [211].

$$[M(NMe_2)_4] + [M'(CO)_n] \longrightarrow \underset{\textbf{(VII)}}{M(NMe_2)_4.2M'(CO)_n} \qquad (M = Ti, Zr, Hf) \tag{16}$$

$$[Ti(\eta\text{-}C_5H_5)(NMe_2)_3] + [M'(CO)_6] \longrightarrow \underset{\textbf{(VIII)}}{Ti(\eta\text{-}C_5H_5)(NMe_2)_3.2M'(CO)_3} \tag{17}$$

C AMIDES AS LEWIS BASES, WITH THE METAL AS LIGATING ATOM

Transient molecules such as B^INR_2 or Al^INMe_2 have been claimed to be stabilised as complexes (IX) or (X), in which a lone pair on the metal is assumed to be the donor [582, 666, 667a]; details are not available for (IX), which was obtained from $[Fe_2(CO)_9]$ and Br_2BNEt_2 [667a]. For example, (X) was obtained from $[Fe_3(CO)_{12}]$ and $[Br_2AlNMe_2]_2$; reaction with T.H.F. was said to yield $[FeBr_2(CO)_3\{Al(NMe_2)(T.H.F.)\}]$ [582].

$[Fe(CO)_4(BNR_2)]$

(IX)

$(OC)_3Br_2Fe{-}Al(\mu\text{-}NMe_2)_2Al{-}FeBr_2(CO)_3$

(X)

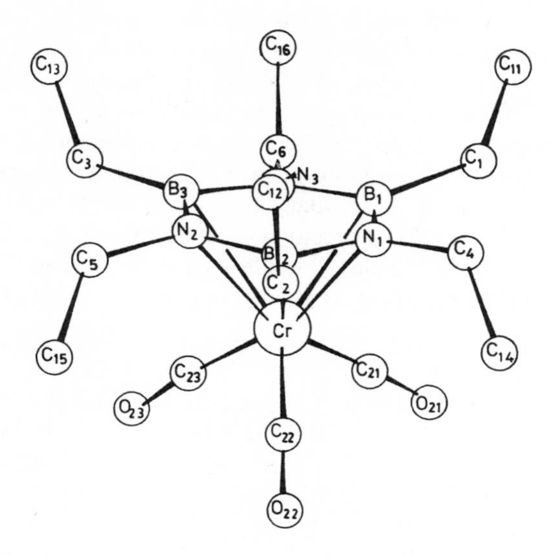

Bond lengths (Å)			
Cr–B(1)	2.30 ± 0.02	Cr–N(1)	2.23 ± 0.01
Cr–B(2)	2.33 ± 0.02	Cr–N(2)	2.24 ± 0.01
Cr–B(3)	2.31 ± 0.02	Cr–N(3)	2.18 ± 0.01
B(1)–N(1)	1.46 ± 0.02	N(1)–B(2)	1.36 ± 0.02
B(2)–N(2)	1.45 ± 0.02	N(2)–B(3)	1.44 ± 0.02
B(3)–N(3)	1.47 ± 0.02	N(3)–B(1)	1.46 ± 0.02

Bond angles (degrees)			
N(1)–B(1)–N(3)	113 ± 1	B(1)–N(1)–B(2)	127 ± 2
N(2)–B(2)–N(1)	116 ± 1	B(2)–N(2)–B(3)	123 ± 2
N(3)–B(3)–N(2)	115 ± 1	B(3)–N(3)–B(1)	123 ± 1

Torsion angles (degrees)			
B(1)–N(1)–B(2)–N(2)	6	N(1)–B(2)–N(2)–B(3)	−8
B(2)–N(2)–B(3)–N(3)	12	N(2)–B(3)–N(3)–B(1)	−14
B(3)–N(3)–B(1)–N(1)	12	N(3)–B(1)–N(1)–B(2)	−8
B(1)–N(1)–B(2)–C(2)	−169	N(1)–B(2)–N(2)–C(5)	179
B(2)–N(2)–B(3)–C(3)	−169	N(2)–B(3)–N(3)–C(6)	168
B(3)–N(3)–B(1)–C(1)	−161	N(3)–B(1)–N(1)–C(4)	177

Fig. 1 – Molecular structure of tricarbonyl(hexaethylborazine)chromium(0). *[Reproduced, with permission, from Huttner, G., and Krieg, B.,* Angew. Chem. Internat. Edn., *1971,* **10,** *512.]*

It has generally been assumed that BN compounds such as borazines or aminoboranes behave as π-ligands towards low oxidation state transition metal acceptor centres. The analogy is clearly with isoelectronic η-arene, ethylene, buta-1,3-diene, or η-allyl complexes. At present we regard the evidence for such assumptions as inadequate. In the case of borazine complexes, even X-ray data do not definitively distinguish an N,N',N''- from a η^6-borazine complex; however, the former is the more likely because in $[Cr(CO)_3\{(EtBNEt)_3\}]$ the ring is puckered, Fig. 1 [325a]. For a number of complexes of benzodiazaboroles or phenylboranes, e.g., $PhB(NMe_2)_2$, n.m.r. data show that the $Cr(CO)_3$ moiety is exclusively fixed to the benzo or phenyl ring, respectively [271a]. The diazaboroline $\overline{MeBNBu^tCH{=}CHNBu^t}$ (L) displaces acetonitrile from $[Cr(CO)_3(NCMe)_3]$ to yield $[Cr(CO)_3L]$ [667b].

The tin(II) amide $Sn[N(SiMe_3)_2]_2$ displaces CO from $[W(CO)_6]$ [Eq. (18)] [428] to give the complex (XI). Other transition metal compounds behave as substrates for insertion of the stannylene into the M–X (Pd–Cl or Pt–Cl) bond [428] (see Chapter 18).

$$Sn[N(SiMe_3)_2]_2 + [W(CO)_6] \xrightarrow{-CO} \underset{\textbf{(XI)}}{[W(CO)_5(Sn\{N(SiMe_3)_2\}_2)]} \quad (18)$$

Further examples of a bivalent Group 4 metal amide as donor are shown in the Scheme [424b] and in ref. 594a (see Table 2); compounds (XII)–(XIV) are donor-acceptor adducts, whereas (XV) is the product of oxidative addition to give a Pb^{IV} complex. In the former, the assumption of the metal(II) as donor is backed up by spectroscopy (mainly i.r.), but also by analogy with the isoelectronic $M[CH(SiMe_3)_2]_2$ (M = Ge or Sn) for which there is X-ray evidence, as in $[Cr(CO)_5(Ge\{CH(SiMe_3)_2\}_2)]$ [424b].

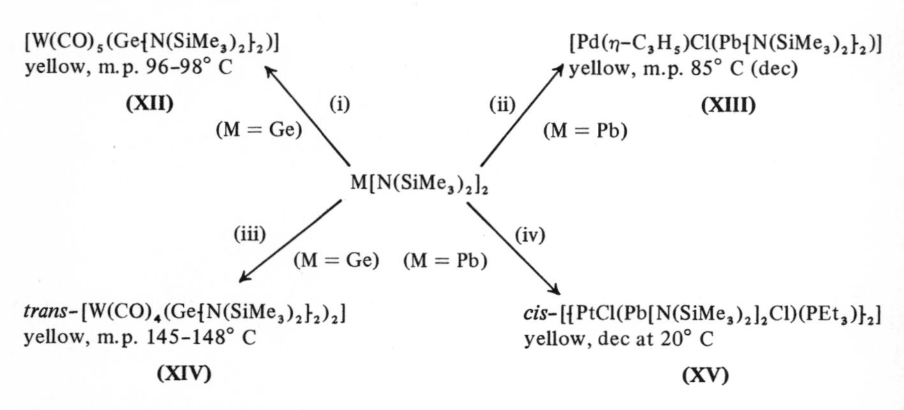

Scheme: (i) $[W(CO)_6]$, n-C_6H_{14}, $h\nu$, 20° C; (ii) $[\{Pd(\eta\text{-}C_3H_5)Cl\}_2]$, n-C_6H_{14}, 20° C; (iii) $[W(CO)_4(\text{norbornadiene})]$, n-C_6H_{14}, reflux; (iv) $[\{PtCl_2(PEt_3)\}_2]$, n-C_6H_{14}, 20° C.

Some cases of amides of P^{III} or As^{III} as donors have been discussed in the introduction and in Section A. However, X-ray data on two complexes having $Bu^tN{=}PN(SiMe_3)_2$ or $Me_3SiN{=}P(NHBu^t)$ as ligand, with $[Cr(CO)_5]$ as acceptor, show unequivocal *P*-ligation [590a], as illustrated for the former compound in Fig. 2. X-ray data are also available for the zwitterionic complex (XVI), obtained from $(Me_3SiN{=})_2PN(SiMe_3)_2$ and $TiCl_4$ [520a]. Likewise $H_8B_4.F_2PNMe_2$ has a PB bond (X-ray) [594b]. For complex (XVII), phosphorus is clearly the ligating atom [566a], and n.m.r. data point to the same conclusion for $Ph_2PN(Me)SiMe_3.BX_3$ (X = Cl or Br) [549a]. The compounds $[M(CO)_5\{P(NH_2)_3\}]$ (M = Cr, Mo, or W), made from $[M(CO)_5(PCl_3)]$ and ammonia, are interesting because the parent ligand $P(NH_2)_3$ is unknown [541a]. Similarly, whereas aminophosphines which contain both a P–H and a P–NR_2 or P–NHR bond are unknown, probably because of α-elimination, the compounds $[Mn(\eta\text{-}C_5H_5)(CO)_2\{PPh(H)X\}]$ ($X = NBu_2$ or NHC_6H_{11}) have been isolated [325b].

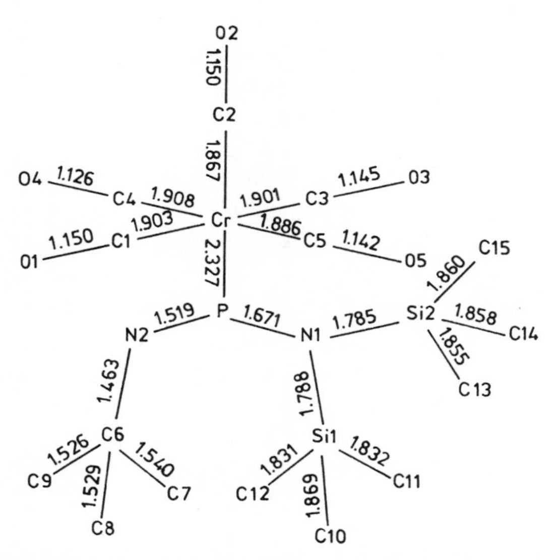

Fig. 2 – Molecular structure (bond lengths in Å) of $[Cr(CO)_5\{Bu^tN{=}PN(SiMe_3)_2\}]$. *[Reproduced, with permission, from Pohl, S.,* J. Organometallic Chem., *1977,* **142,** *185.]*

$N(SiMe_3)_2$
ClP^{+} — $NSiMe_3$
Me_3SiN — $^{-}TiCl_3$

(XVI)

Me
N
F_2P PF_2
B—B
H_2 H_2

(XVII)

The cation (XVIII) was first found as the $[PF_6]^-$ salt [243a]; the latter reacts with $[Fe(CO)_5]$ to yield $[Fe\{\overline{PNMe(CH_2)_2NMe}\}(CO)_4][PF_6]$, which is alternatively formed from (XIX) and PF_5 [492a]. A compound related to (XIX) is $[Fe\{PF(NMe_2)_2\}(CO)_4]$ [215], which similarly yields $[Fe\{P(NMe_2)_2\}(CO)_4]^+$ [492a]. Cations similar to (XVIII) are $[P(NMe_2)_2]^+$ [716] and $[PCl(NMe_2)]^+$ [716a].

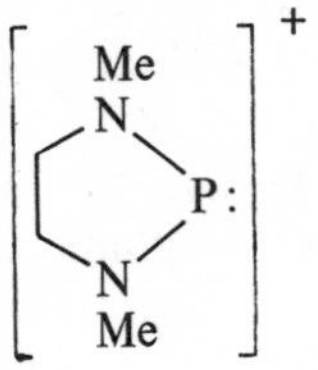

(XVIII) 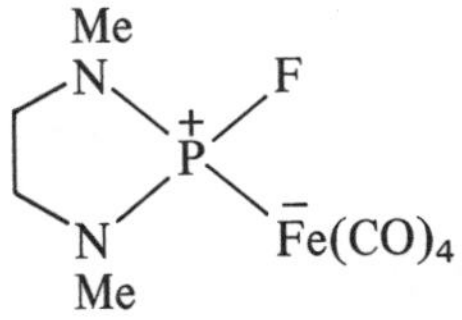 **(XIX)**

The first amino(fluoro)phosphine complex $[Ni(F_2PNMe_2)_4]$ dates from 1964 [676a], and such species are discussed in a review [523b]. Recently there has been a resurgence of such ligands, especially using the bidentate $MeN(PF_2)_2$ [149a, 149b, 149c, 375a, 375b, 375c, 375d, 375e, 517a, 517b, 517c, 517d] and transition metal carbonyls. Examples of interesting complexes are (XX) $\{L = F_2PNMe_2$ (acting as a phosphorus donor), obtained by co-condensation of cobalt vapour into a mixture of $MeN(PF_2)_2$ and F_2PNMe_2 [149c], or from $[Co_2\{MeN(PF_2)_2\}_3(CO)_2]$ and Br_2 [517d]$\}$ and (XXI), obtained from $[\{Fe(\eta\text{-}C_5H_5)(CO)_2\}_2]$ and $MeN(PF_2)_2$ [517b], all of which are structurally

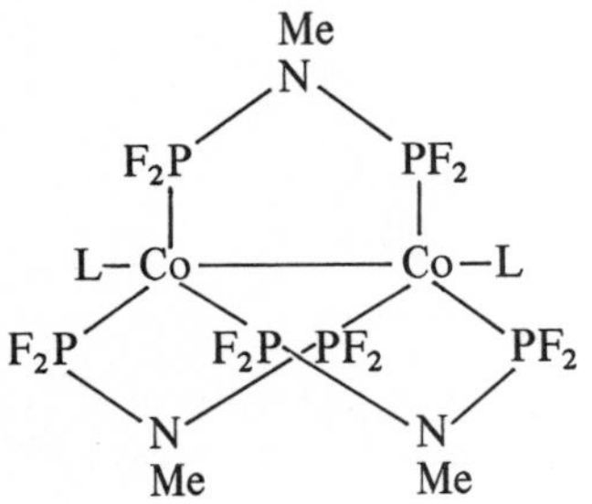

(XX) 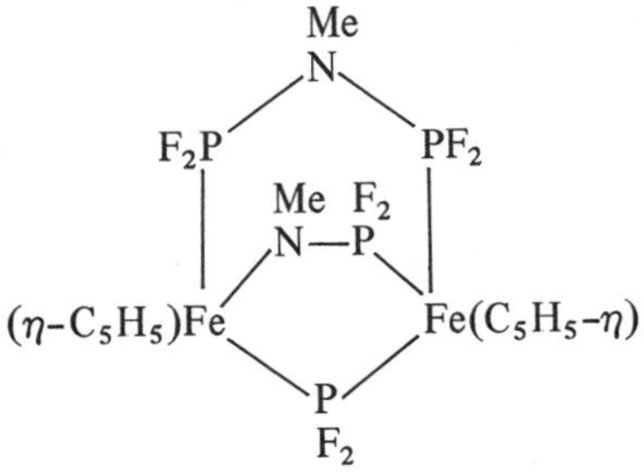 **(XXI)**

authenticated by X-ray crystallography. It is noteworthy that formation of (XXI) involves P-N cleavage. Ultraviolet irradiation of $[M(CO)_6]$ (M = Cr, Mo, or W) with excess of $MeN(PF_2)_2$ in diethyl ether gives the volatile homoleptic $[M\{MeN(PF_2)_2\}_3]$ [375b], and $[Co_2\{MeN(PF_2)_2\}_5]$ is similarly obtained from $[Co_2(CO)_8]$. The octahedral complexes $[Cr(F_2PNMe_2)_6]$ and $[Cr(F_2PNMe_2)_4\text{-}\{MeN(PF_2)_2\}]$ are obtained from co-condensation of Cr metal and the two PN ligands [149b]. $MeN(PPh_2)_2$ acts as a *P,P'*-bridging ligand in $[MeN\{PPh_2AgBr\}_2]_2$ which contains an Ag_4Br_2 octahedron as central unit (X-ray) [668a]. Other examples of complexes involving F_2PNR_2 (R_2 = Me_2, Et_2, or C_5H_{10}) (=L) or $EtN(PF_2)_2$ (=L′) as ligand are *cis*-$[Mo(CO)_4L_2]$ and *cis*-$[Mo(CO)_4L']$ [55e], and *cis*-$[RuX_2L_2(PPh_3)_2]$ (X = H or Cl), *cis*-$[RuH_2L(PPh_3)_3]$, and *cis*-$[RuH_2L(PF_3)(PPh_3)_2]$ [301a].

If NO in $[Cr\{N(SiMe_3)_2\}_3(NO)]$, (XXII), is taken as NO^+, for which there is X-ray structural evidence [96], then this compound may be regarded as an adduct of the chromium-centred Lewis base $[Cr\{N(SiMe_3)_2\}_3]$; the molecule has C_{3v} heavy atom skeletal symmetry. It is diamagnetic and is obtained [98] from its paramagnetic factors. Some analogues have been made [98] and trends in NO stretching modes suggest the following decreasing order of π-donor facility ($\ddot{X} \curvearrowright M$) of the ligand X = NPr^i_2 > 2,6-dimethylpiperidino > $N(SiMe_3)_2$ > OBu^t > OPr^i. The alkoxides were obtained by alcoholysis from (XXII). See also Chapter 8 for further discussion.

D TABLES OF REACTIONS

Table 1 Reactions as Lewis Bases, with Nitrogen as Ligating Atom(s)

Reactants		Products	References
$NaN(SiMe_3)_2$	MeHgCl	$NaN(SiMe_3)_2.MeHgCl$	[669]
$NaN(SiMe_3)_2$	Me_3Al	$NaN(SiMe_3)_2.AlMe_3$	[749]
$Be(NMe_2)_2$	$Me_3N.Me_{3-n}AlH_n$	$(HBeNMe_2)_2.(Me_{3-n}AlH_{n-1}NMe_2)$	[571]
$Be(NMe_2)_2$	Me_3Al	$Be(NMe_2)_2.2AlMe_3$	[571]
$B_2(NMe_2)_4$	Me_3Al	$3B_2(NMe_2)_4.6AlMe_2$	[28]
$B_2(NMe_2)_4$	$[Re(CO)_5]^-$	$B_2(NMe_2)_4.2Re(CO)_3$	[328]
$B(NMe_2)_3$	$HgCl_2$	$B(NMe_2)_3.HgCl_2$	[435]
$B(NR_2)_3$ (NR_2 = NMe_2, $NHBu^t$)	$SnCl_4$	$B(NR_2)_3.SnCl_4$	[435]
$B(NMe_2)_3$	$VOCl_3$	$B(NMe_2)_3.VOCl_3$	[435]
$B(NMe_2)_3.SnCl_4$	$2NC_5H_5$	$B(NMe_2)_3$ + $SnCl_4.NC_5H_5$	[435]
$B(NMe_2)_3$	$[W(CO)_6]$	$[W(CO)_3\{B(NMe_2)_3\}]$	[545]
$XB(NMe_2)_2$	$[Pd(NCPh)_2X_2]$	$[PdX_2\{XB(NMe_2)_2\}]$ (X = Me, Cl) or $[PdX_2(\{(Me_2N)_2B\}_2NX_2CPh)]$ (X = Br, I)	[668]
$PhB(NR_2)_2$	$SnCl_4$	$PhB(NR_2)_2.SnCl_4$	[435]
$ClB(NMe_2)_2$	$SnCl_4$	$ClB(NMe_2)_2.SnCl_4$	[435]
$XB(NMe_2)_2$ [X = Cl, Br, $B(NMe_2)_2$; X′ = Cl, Br]	TiX'_4	$2XB(NMe_2)_2.3TiX'_4$	[417]
$XB(NMe_2)_2$ (X = Cl, Br)	$TiCl_4$	$XB(NMe_2)_2.TiCl_4$	[544]
$MeB(NMe_2)_2$	$[FeBr(\eta\text{-}C_5H_5)(CO)_3]$	$[Fe(CO)_3\{MeB(NMe_2)_2\}]$	[667]

Table 1 *(C'ntd.)*

Reactants		Products	References
$XYBNMe_2$ (X, Y = NMe_2, Me; CH_2=CH, Br)	$[Fe_2(CO)_9]$	$[Fe(CO)_3\{XYB(NMe_2)\}]$	[667]
Me_2BNMe_2	$AlCl_3$	$Me_2BNMe_2.AlCl_3$	[539]
Me_2BNMe_2	$[Ni(\eta\text{-}C_5H_5)_2]$	$[Ni(\eta\text{-}C_5H_5)(Me_2BNMe_2)_2]$	[667]
$R_2BNR'_2$ (R, NR'_2 = Me, NMe_2; Me, $\overline{N(CH_2)_5}$; Pr, NMe_2)	I_2	$R_2BNR'_2.I_2$	[232]
$(MeBNMe)_3$	$TiCl_4$	$(MeBNMe)_3.TiCl_4$	[544]
$(RBNR')_3$ (R, R' = Me, Et, Pr, Pr^i; M = Cr, Mo)	$[M(CO)_3(NCMe)_3]$	$[M(CO)_3(RBNR')_3]$	[204, 325a, 599, 684, 684a]
$\overline{MeBNBu^tCH{=}CHNBu^t}$	$[Cr(CO)_3(NCMe)_3]$	$[Cr(CO)_3\{\overline{MeBNBu^tCH{=}CHNBu^t}\}]$	[667b]
Me_3N_4B (x = 1, and MCl_n = BCl_3 or $SbCl_5$; x = 2 and MCl_n = $SnCl_4$)	MCl_n	$Me_3N_4B.xMCl_n$	[305]
Me_2N_4BX (X = Me, Ph, Cl)	$TiCl_4$	$3Me_2N_4BX.4TiCl_4$	[305]
Br_2AlNMe_2	$[Fe_3(CO)_{12}]$	$[\{Fe(AlNMe_2)Br_2(CO)_3\}_2]$	[582]
$Si(NMe_2)_4$	BH_3	$Si(NMe_2)_4.BH_3$	[52]
$HSi(NMe_2)_3$	BH_3	$HSi(NMe_2)_3.2BH_3$	[52]
$H_2Si(NMe_2)_2$ (R = H, n = 1, 2; R = Me, n = 1)	R_3B	$H_2Si(NMe_2)_2.nBR_3$	[52]
$Me_2Si(NMe_2)_2$ (M = Sn, Ti, Zr)	MCl_4	$Me_2Si(NMe_2)_2.MCl_4$	[435]
$Me_2Si(NMe_2)_2$	$VOCl_3$	$Me_2Si(NMe_2)_2.VOCl_3$	[435]
$Me_2Si(NEt_2)_2$	$SnCl_4$	$Me_2Si(NEt_2)_2.SnCl_4$	[435]
H_3SiNMe_2 (R = H, Me)	R_3B	$H_3SiNMe_2.BR_3$	[707]
Me_3SiNMe_2	$AlBr_3$	$2Me_3SiNMe_2.AlBr_3$	[321]
Me_3SiNMe_2	$SnCl_4$	$2Me_3SiNMe_2.SnCl_4$	[435]
HMe_2SiNEt_2	$AlCl_3$	$HMe_2SiNEt_2.AlCl_3$	[506]
$HN(SiMe_3)_2$ (A = $AlCl_3$, $SnCl_4$, $VOCl_3$, Me_2AlI)	A	$(Me_3Si)_2NH.A$	[435, 505, 745]
$(Me_2SiNH)_n$ (n = 3, 4)	$SnCl_4$	$(Me_2SiNH)_n.SnCl_4$	[435]
$HN(SiHMe_2)_2$	$AlCl_3$	$(Me_2HSi)_2NH.AlCl_3$	[506]
$N(SiMe_3)_3$	$AlCl_3$	$(Me_3Si)_3N.AlCl_3$	[755]
$R_2S(NMMe_3)_2$ (R = Me, Et; M = Si, Ge)	Me_3Al	$R_2S(NMMe_3)_2.AlMe_3$	[759]
$R_2S(NGeMe_3)_2$ (R = Me, Et)	Me_3Al	$R_2S(NGeMe_3)_2.2AlMe_3$ (unstable → $GeMe_4$ ↑)	[759]

Table 1 *(C'ntd.)*

Reactants		Products	References
Me_3SnNMe_2	$[\{PtCl_2(PBu_3)\}_2]$	$[PtCl_2(Me_3SnNMe_2)PBu_3]$	[134]
$Sn(NMe_2)_4$	$VOCl_3$	$Sn(NMe_2)_4.VOCl_3$	[435]
Me_3SnNMe_2	$B_{10}H_{14}$	$Me_3SnNMe_2.B_{10}H_{14}$	[147]
$N(SnMe_3)_3$	Me_3SnBr	$[(Me_3Sn)_4N]Br$	[6]
$Ti(NMe_2)_4$	$B_{10}H_{14}$	$Ti(NMe_2)_4.2B_{10}H_{14}$	[147]
$Ti(NMe_2)_4$	$SnCl_4$	$Ti(NMe_2)_4.SnCl_4$	[435]
$[Ti(\eta-C_5H_5)(NMe_2)_3]$ (M = Cr, Mo, W)	$[M(CO)_6]$	$[M(CO)_3\{Ti(\eta-C_5H_5)(NMe_2)_3\}]$	[97]
$M(NMe_2)_4$ (M = Ti, Zr, Hf)	$[Mo(CO)_6]$	$M(NMe_2)_4.2Mo(CO)_6$*	[90]
$M(NMe_2)_4$ (M = Ti, Zr, Hf)	$[Fe(CO)_5]$	$M(NMe_2)_4.2Fe(CO)_5$	[90]
$M(NMe_2)_4$ (M = Ti, Zr, Hf)	$[Ni(CO)_4]$	$M(NMe_2)_4.2Ni(CO)_4$	[90]

*See also Chapter 10.

Table 2 Reactions as Lewis Bases, with Metal as Ligating Atom

Reactants		Products	References
Br_2BNEt_2	$[Fe_2(CO)_9]$	$[Fe(CO)_4(BNEt_2)]$	[667a]
Br_2AlNMe_2	$[Fe_3(CO)_{12}]$	$[\{FeBr_2(CO)_3(AlNMe_2)\}_2]$	[582]
Br_2AlNMe_2	$[Fe_3(CO)_{12}]$, T.H.F.	$[FeBr_2(CO)_3\{Al(NMe_2)(T.H.F.)\}]$	[582]
$Ge[N(SiMe_3)_2]_2$	$[Cr(CO)_6]$, $h\nu$	$[Cr(CO)_5(Ge\{N(SiMe_3)_2\}_2)]$	[594a]
$Ge[N(SiMe_3)_2]_2$	$[Mo(CO)_6]$, $h\nu$	$[Mo(CO)_5(Ge\{N(SiMe_3)_2\}_2)]$	[594a]
$Ge[N(SiMe_3)_2]_2$	$[W(CO)_6]$, $h\nu$	$[W(CO)_5(Ge\{N(SiMe_3)_2\}_2)]$	[424b]
$Sn[N(SiMe_3)_2]_2$	$[Cr(CO)_6]$, $h\nu$	$[Cr(CO)_5(Sn\{N(SiMe_3)_2\}_2)]$	[594a]
$Sn[N(SiMe_3)_2]_2$	$[Mo(CO)_6]$, $h\nu$	$[Mo(CO)_5(Sn\{N(SiMe_3)_2\}_2)]$	[594a]
$Sn[N(SiMe_3)_2]_2$	$[W(CO)_6]$, $h\nu$	$[W(CO)_5(Sn\{N(SiMe_3)_2\}_2)]$	[428]
$M'[N(SiMe_3)_2]_2$ (M = Mo, W; M' = Ge, Sn)	$[M(CO)_4(NCMe)_2]$ or $[M(CO)_4$-(norbornadiene)]	*trans*-$[M(CO)_4(M'\{N(SiMe_3)_2\}_2)_2]$	[424b, 594a]
$M[N(SiMe_3)_2]_2$ (M = Sn, Pb)	$[\{Pd(\eta-C_3H_5)Cl\}_2]$	$[Pd(\eta-C_3H_5)Cl(M\{N(SiMe_3)_2\}_2)]$	[424b, 428]
$Sn[N(SiMe_3)_2]_2$	$[\{Sc(\eta-C_5H_5)_2Me\}_2]$	$[Sc(\eta-C_5H_5)_2Me(Sn\{N(SiMe_3)_2\}_2)]$	[594a]
$As(NMe_2)_3$	$[Ni(CO)_4]$	$[Ni(CO)_3\{As(NMe_2)_3\}]$	[667a]
$[Cr(NR_2)_3]$ [R = Pr^i, $SiMe_3$; NR_2 = $\overline{N(CH_2)_6}$]	NO	$[Cr(NR_2)_3(NO)]$	[96, 98]

17

Metal Amides as Polymerisation Initiators

A INTRODUCTION

The ability of a metal amide to initiate the polymerisation of a vinyl monomer was demonstrated around 1950 with alkali metal derivatives [109a, 309, 646a, 761a]. Since then, other metal amides which have been shown to initiate the oligomerisation or polymerisation of certain monomer(s) include amides of Li for 1,3-dienes [370a, 388, 499, 523a]; Be and Mg for aldehydes [775]; Mg for methacrylonitrile [350, 501, 502]; Al for aldehydes [775], dienes [168, 476a], methyl methacrylate [484], 2-ethylhexyl methacrylate, or isobutyl vinyl ether [564]; Sn for CH_2=CHCHO [145] or formaldehyde [330]; Ti for acrylonitrile [72–75, 145, 263, 324, 346, 348, 570], methyl methacrylate, styrene, or α-cyanostyrene [324, 570], methacrylonitrile or ethylene sulphide [346, 348], Zr for ethylene [454]; Zr or Hf for acrylonitrile or methacrylonitrile [346, 348]; and Zn for propylene oxide [563]. The use of the amides of Al, Si, P, Ti, V, and Cr in Ziegler-Natta catalyst systems, to polymerise olefins or dienes, has also been claimed in several patents [138, 139, 141, 189, 190, 246, 262, 322, 342, 389, 393, 415, 652, 713, 762, 764–767, 776, 777]; the stereospecific polymerisation of isoprene is catalysed by $(HAlNR)_m[TiCl_4]$ (R = alkyl or Ph) and yields a polymer high in 1,4-*cis* content [476a]. Aminosilanes have been used to remove water from polyester polymers [318].

B THE GROUP 4 TRANSITION METAL AMIDES

The group 4 transition metal amides are the most effective initiators towards acrylonitrile and the system $[Ti(NMe_2)_4]/CH_2$=CHCN has been studied in detail [72–75, 346, 348]. Unlike the Ge or Sn amides R_3MNMe_2 (M = Ge or Sn), which add across the C=C bond of acrylonitrile to yield the 1:1 adduct (Chapter 10), tetrakis(dialkylamido)-titanium [145, 324, 346, 348, 570], -zirconium [346, 348], or -hafnium initiate polymerisation to produce high molecular weight polyacrylonitrile even at −78° C. This difference between the group 4a and 4b metal amides does not lie in the different chosen functionalities, i.e., Me_3MNMe_2 versus $M(NMe_2)_4$, since $Sn(NMe_2)_4$ affords only a 1:4 adduct with acrylonitrile [149]. A rather complicated mechanism, involving either radicals or anions according to the choice of monomer, solvent, or temperature was initially suggested for the polymerisation of acrylonitrile, methyl methacrylate, or styrene initiated by $[Ti(NR_2)_4]$ [324, 570]. The proposed anionic mechanism for acrylonitrile did not take account of the possibility [145] of a multistep insertion process encouraged by the lability of the Ti–C bond.

A detailed mechanistic study of the system $[Ti(NMe_2)_4]/CH_2{=}CHCN$ has recently been made, using various conditions of temperature, time, and reactant concentrations [72–75, 346, 348]. In general, (i) the average molecular weight (or degree of polymerisation) decreases with percentage conversion of the monomer, and (ii) percentage conversion increases with time, temperature, monomer concentration, and catalyst concentration. Kinetic investigations of the reaction at 20° C showed that the polymerisation is essentially second-order with respect to monomer, although there is a significant induction period (even with rigorous precautions [72] to minimise decomposition of catalyst by adventitious impurities, especially moisture). This has been attributed to a two-stage initiation, in which $[Ti(NMe_2)_4]$, (T), rapidly reacts with the monomer (M) to yield an adduct (A), which, in a slow step, forms the true initiator X_1. The reaction scheme [Eqs. (1)-(4)] thereafter involves a sequence of propagation reactions and finally a termination step [Eq. (4)], wherein the propagating chain X_m (containing $m + 1$ monomer units) is converted to the product P_m. Termination is first-order with respect to the growing chains. Because in the system $[Ti(NMe_2)_4]/CH_2{=}CHCN$, the polymerisation rate is unaffected by the presence of the radical scavenger 1,1-diphenyl-2-picrylhydrazyl, the presence of free radical intermediates is unlikely and therefore termination [Eq. (4)] probably involves a single chain carrier; furthermore NEt_3 is not a (radical) chain transfer agent because the amine has no effect on the average molecular weight of the polymer [348].

The formation of the 1:1 adduct may involve either (i) 1,2-addition of the fragments $Ti(NMe_2)_3$ and NMe_2 of $[Ti(NMe_2)_4]$, T, across the C≡N group of the monomer [Eq. (5a)], or (ii) 1:4-addition [Eq. (5b)] [348]. The metallo-amidine, (I), or its isomer, (II), thus formed, reacts further with the monomer producing a species (III) (X_1), having a Ti–C bond [Eq. (6)], which behaves as the effective initiator, whereafter propagation [Eq. (7)] ensues [348].

$$\left.\begin{array}{l} T + M \longrightarrow A\ (\text{fast}) \\ A + M \longrightarrow X_1\ (\text{slow}) \end{array}\right\}\quad \text{Initiation} \qquad \begin{array}{r}(1)\\(2)\end{array}$$

$$X_n + M \longrightarrow X_{n+1} \qquad \text{Propagation} \qquad (3)$$

$$X_m \longrightarrow P_m \qquad \text{Termination} \qquad (4)$$

$$\geqslant Ti{-}NMe_2 + CH_2{=}CHCN \begin{cases} \xrightarrow{(a)} \geqslant Ti{-}N{=}\underset{\displaystyle NMe_2}{\underset{|}{C}}{-}CH{=}CH_2 \quad \textbf{(I)} \\ \xrightarrow{(b)} \geqslant Ti{-}N{=}C{=}CHCH_2NMe_2 \quad \textbf{(II)} \end{cases} \tag{5}$$

$$\geqslant Ti{-}N{=}C\lessgtr + CH_2{=}CHCN \longrightarrow \geqslant Ti{-}\underset{\displaystyle CN}{\underset{|}{C}}H{-}CH_2{-}N{=}C\lessgtr \quad \textbf{(III)} \tag{6}$$

$$\geqslant Ti{-}\underset{\displaystyle CN}{\underset{|}{C}}H{-}CH_2{-}N{=}C\lessgtr + n\underset{\displaystyle CN}{\underset{|}{C}}H{=}CH_2 \longrightarrow \geqslant Ti(\underset{\displaystyle CN}{\underset{|}{C}}H{-}CH_2)_n{-}\underset{\displaystyle CN}{\underset{|}{C}}H{-}CH_2{-}N{=}C\lessgtr \tag{7}$$

Evidence for Eq. (5a) derives from the observation that $[M(NMe_2)_4]$ (M = Ti, Zr, or Hf) readily adds across the nitrile group of benzonitrile or toluonitrile; the addition product, e.g., $[Zr(N{=}\underset{\displaystyle NMe_2}{\underset{|}{C}}{-}C_6H_4{-}Me)_4]$ is an efficient initiator for polymerisation of acrylonitrile [146]. In general, saturated nitriles, such as EtCN, react slowly whereas unsaturated nitriles without protic α-H (e.g., PhCN or PhCH=CHCN) react rapidly [71a].

More direct evidence that the adduct A is a metallo-amidine (I) [Eq. (5a)], is based on i.r. spectra. Thus, when $[Ti(NMe_2)_4]$ (T_4) is stirred (18 h) with an excess of propionitrile in methylcyclohexane a dark brown viscous liquid is obtained which does not show $\nu(C{\equiv}N)$ at 2220 cm^{-1} but has a strong absorption at 1600 cm^{-1}. The product, which is probably a mixture of compounds of the form $[Ti\{N{=}C(NMe_2)Et\}_n(NMe_2)_{4-n}]$ (the overall composition corresponds to approximately 33% of ligand as the amidine) is a very effective initiator for acrylonitrile polymerisation, similar in behaviour under ambient conditions with that of T_4 at high temperature. Moreover, there is no polymerisation induction period [72]. The kinetics of the insertion of EtCN may be monitored by ^1H n.m.r. spectroscopy, and shows a classic sequence of pseudo-first-order reactions of the form $T_4 \rightarrow T_3 \rightarrow T_2 \rightarrow T_1 \rightarrow T_0$, where T_n represents the transition metal complex with n amide groups [71a]. The reaction of T_4 with a styrene/acrylonitrile random copolymer affords a product in which the acrylonitrile sites have been converted into $-CH_2{-}CH\underset{\displaystyle NMe_2}{\underset{|}{C}}{=}N{-}Ti\equiv$ type centres; after careful washing to remove excess T_4, this material is able to polymerise acrylonitrile, grafting polyacrylonitrile links on to the original backbone. This provides further support for the proposed two step initiation reaction [Eqs. (1) and (2)].

However, a study of initiator consumption and polymer molecular weight distribution has shown that over a wide conversion range a significant fraction of the available Ti–N bonds react without initiating polymerisation, although at high conversions more than one polymer chain is produced per Ti atom reacted [73]. Attempts to isolate and characterise the intermediate A in the system $[Ti(NMe_2)_4]/CH_2{=}CHCN$, even at very low temperatures, have been unsuccessful. A single-stage initiation mechanism in which the components of the Ti–N bond add across the C=C of the monomer (slow) and subsequent (fast) successive insertion (propagation) of monomer into the resulting Ti–C bond cannot be wholly ruled out.

A 6-membered transition state was the initial proposal regarding the termination step [Eq. (8)] [348].

$$\gtrdot Ti \cdots N{\equiv}C \cdots CH(CN){-}CH_2 \cdots CH{-}CH_2{-}(CH(CN){-}CH_2)_n N{=}C< \longrightarrow \gtrdot Ti{-}N{=}C{=}CH{-}CH_2(CH(CN){-}CH_2)_n N{=}C< + CH(CN){=}CH_2 \quad (8)$$

However, by comparison with other organometallic initiator systems, it seems more probable that the chain-termination mechanism is a β-hydrogen elimination, and that very few, if any, of the Ti–H species so produced act as initiators; these conclusions are supported by labelling experiments using $[Ti\{N(^{14}CH_3)_2\}_4]$ [71a].

Using tritiated isopropanol in order to terminate the polymerisation at different time intervals and then determining the radioactivity of the polymer (this provides a measure of the number of Ti to polymer bonds at the instant of quenching), it has been shown [74] that 97% of the polymerisation reaction is terminated by a reaction leading to detachment of polymer chain from the metal centre; this is further evidence in favour of β-elimination.

Surprisingly, the rate of polymerisation appears to be at its maximum near 25° C and decreases both with increasing and decreasing temperature. Moreover, the second-order behaviour is not strictly followed at other temperatures [72]. These observations suggest that the reaction scheme presented above is a simplification of the actual mechanism.

Despite the high reactivity of $[Ti(NMe_2)_4]$ towards acrylonitrile, very little tendency to initiate the polymerisation of other potential monomers was apparent under comparable conditions. Methacrylonitrile or ethylene sulphide were only polymerised at ca. 60° C, and vinylidine chloride, styrene, or PhCH=C(CN)Ph proved unreactive in refluxing ether [146, 346, 348]. The heavier group 4 metal homologues are more reactive, both towards acrylonitrile and methacrylonitrile; the hafnium compound readily polymerises methacrylonitrile, even below room temperature. Substitution of cyclopentadienyl- for dimethylamido- groups reduces reactivity with respect to $CH_2{=}CHCN$ polymerisation. On the other hand, substitution of aziridino- for dimethylamido- groups as in $Ti[\overline{N(CH_2)_2}]_4$ [347] causes enhancement of reactivity [75], becoming essentially instantaneous and 'dead-step' in nature. Relative reactivities may be represented by the following inequalities [75, 346, 348]:

(i) $Zr \approx Hf > Ti$
(ii) $[M(NMe_2)_4] > [M(\eta\text{-}C_5H_5)_2(NMe_2)_2]$
(iii) $[Ti\{\overline{N(CH_2)_2}\}_4] \gg [Ti(NMe_2)_4]$
(iv) $CH_2{=}CHCN > CH_2{=}C(Me)CN$

As for (iii), the sequence has been extended for other complexes $[Ti(NR_2)_4]$: $NR_2 = \overline{N(CH_2)_2} \gg NMe_2 > \overline{N(CH_2)_5} > NEt_2 > NPh_2$ [71a]. This, as well as (i)-(iv), indicates that steric effects are dominant. Once again (see Chapter 16), although metal amides are strong bases towards the proton, they may be weak nucleophiles towards carbon. Electronic effects are probably significant only in that the incipient monomer, e.g., acrylonitrile, must be susceptible to anionic, rather than cationic, polymerisation; nevertheless the group 4 metal amide is not believed to behave as an ionic species.

Attempts to copolymerise acrylonitrile with methacrylonitrile, vinylidine chloride, styrene, or ethylene sulphide, using $[Ti(NMe_2)_4]$ invariably afforded polyacrylonitrile with no detectable copolymer [348]. The failure to produce a copolymer of even closely related monomers like $CH_2{=}CHCN$ and $CH_2{=}CMeCN$ (which are known to copolymerise in the presence of typical anionic initiators such as PhMgBr or $NaNH_2$ [203]) supports the above proposal that Ti^{IV} amide initiation involves a molecular insertion rather than an anionic process, and is susceptible to steric effects. A somewhat similar structural selectivity toward acrylic and methacrylic esters has been observed for Et_3SnNMe_2; the former affords $Me_2NCH_2CH(SnEt_3)CO_2Me$ (insertion) and the latter yields (metathesis) $CH_2{=}CMeCONMe_2$ [144, 257] (see Chapters 10 and 13).

It is possible to prepare a supported catalyst, using $[Ti(NMe_2)_4]$ on a silica or alumina surface [71a], and the procedure is convenient in providing a measure of the number of surface hydroxyl groups. However, the supported catalyst has a decreased activity towards polymerisation of acrylonitrile, in contrast to the case for $[TiR_4]$ (R = $PhCH_2$ or Me_3SiCH_2) and ethylene polymerisation [198].

18

Miscellaneous Reactions of Metal or Metalloid Amides: Molecular Rearrangements, Oxidative Addition, Reductive Elimination, Disproportionation, M-N Homolysis, or Elimination

A INTRODUCTION

The reactions discussed in this chapter often do not directly involve the M–N bond of the metal or metalloid amide (Sections B, C, and D). However, certain disproportionations (Section D), e.g., formation of (II), may require M–N bond-breaking and -making in the reaction pathway in order to provide a mechanism of electron transfer, as in Eq. (1) [170a, 286b, 322a, 428], in which there may, in the limit, be M–N homolysis, cf. Eq. (1b).

$$\underset{\textbf{(I)}}{2Sn[N(SiMe_3)_2]_2} \xrightarrow{(a)} \underset{\textbf{(II)}}{\dot{S}n[N(SiMe_3)_2]_3} + \underset{\text{presumed}}{\frac{1}{n}[SnN(SiMe_3)_2]_n}$$

$$2Sn[N(SiMe_3)_2]_2 \xrightarrow{(b)} \underset{\text{presumed}}{\frac{1}{n}[SnN(SiMe_3)_2]_n} + \dot{N}(SiMe_3)_2 \xrightarrow[(b)]{\textbf{(I)}} \textbf{(II)} \quad (1)$$

We deal separately (Section E) with those processes which unambiguously lead to aminyl radicals, $\dot{N}R_2$. Eliminations (Section F) generally do not require M–N bond scission in the sense that the central metal or metalloid M retains the same number of nearest neighbour nitrogen atoms in the product(s) as in the substrate, e.g., Eqs. (2) [50, 417a], (3) [520b, 660a], or (4) [518b]. A special case, however, is that of β-elimination, a method of thermal decomposition only recently recognised for metal amides [207a] (but better known for the isoelectronic metal alkyls [198]), which is illustrated in Eq. (5a) ($R = Me$) [207a]; this should be contrasted, Eq. (5b) for the case of $R = SiMe_3$ [141d], when the Rh^{I} compound (III) might well be formed via a Rh^{III} precursor $[\overline{Rh(C_6H_4PPh_2\text{-}o)}H\{N(SiMe_3)_2\}(PPh_3)_2]$.

$$2B(NHR)_3 \xrightarrow[-H_2NR]{\text{heat}} (RHN)_2B{-}N(R){-}B(NHR)_2 \xrightarrow{\text{heat}} B(NHR)_3 + \tfrac{1}{3}\,[RHNB{-}NR]_3 \quad (2)$$

$$P(Cl)[N(SiMe_3)_2]_2 \xrightarrow{\text{heat}} Me_3SiN{=}PN(SiMe_3)_2 + Me_3SiCl \quad (3)$$

$$As(F)[N(SiMe_3)_2]_2 \xrightarrow{\text{heat}} Me_3SiN\langle As(N(SiMe_3)_2)\rangle_2 NSiMe_3 + Me_3SiF \quad (4)$$

$$[Rh(Cl)(PPh_3)_3] \xrightarrow{LiNR_2} \begin{cases} \xrightarrow[(a)]{R = Me} [Rh(NR_2)(PPh_3)_{3\text{ or }4}] \xrightarrow{(a)} [Rh(H)(PPh_3)_{3\text{ or }4}] + MeN{=}CH_2 \\ \xrightarrow[(b)]{R = SiMe_3} [Rh\{N(SiMe_3)_2\}(PPh_3)_2] \end{cases}$$

$$\xrightarrow[\substack{PPh_3 \\ (b)}]{\text{heat}} [(Ph_3P)_2\overline{Rh(C_6H_4PPh_2\text{-}o)}] \;(\mathbf{III}) + HN(SiMe_3)_2 \quad (5)$$

There are certain other reactions in which the role of the amide ligand is purely incidental. These are excluded, but illustration is provided by Eq. (6) [308], by the use of $HAl[\overline{N(CH_2)_2CH(Me)(CH_2)_2}]_2$ as a reagent for reducing a carboxylic acid to an aldehyde [499b], and by the reaction of $Bu^tN{=}PN(SiMe_3)_2$ with MeOH to give $MeOPN(SiMe_3)_2(NHBu^t)$ [660b].

$$[\{Fe(CO)_3NH\}_2] + 4NO \longrightarrow [\{Fe(NO)_2NH\}_2] + 6CO \quad (6)$$

B MOLECULAR REARRANGEMENTS

Only a small number of molecular rearrangements has been reported in metal or metalloid amide chemistry. These involve the migration of an R_3Si group from nitrogen to carbon (see Chapter 5, Section B.9, for the reverse situation) [104a], oxygen, or sulphur, or another nitrogen. An interesting case of the N→C shift is in the reaction of $[Me_3PN(SiMe_3)_2]I$ with LiBu which yields $Me_3SiN{=}P(CH_2SiMe_3)Me_2$ and not $Me_2P(CH_2)N(SiMe_3)_2$ [756a]. The N→O transformation is illustrated by the reaction of $LiN(SiMe_3)_2$ with $ClP(O)X_2$ to give $Me_3SiN{=}P(OSiMe_3)X_2$ (X = F, Cl, or Ph) [126a, 507b], rather than the isomeric [193] $X_2P(O)N(SiMe_3)_2$. A similar N→S shift is observed in the reaction of $P(S)Cl_3$ with $LiN(SiMe_3)_2$, which in diethyl ether at 25° C yields $Me_3SiN{=}P(SSiMe_3)Cl_2$ and not the isomeric $Cl_2P(S)N(SiMe_3)_2$ [429].

Other examples of N→S shifts are in the PhNCS insertion into $(Me_3Si)_2NMe$ yielding not only $Me_3SiN(Me)C(NPh)SSiMe_3$, but also $Me_3SiN(Me)C(S)N(Ph)SiMe_3$ [337], and similar instances of formal N→O shifts are found in some other insertion reactions (see Chapter 10, Section B.2). An N→As shift is found in the reaction of $Bu^tAsN(Li)SiMe_3$ with MeCl which affords $Me_3SiN{=}As(Me)Bu^t_2$ [660b].

Fluxional behaviour, as in $[\{Ti(NMe_2)_3\}_2]$ where there is rapid (on the n.m.r. time scale) site exchange between terminal and bridging amido ligands [25, 431], may be regarded as a molecular rearrangement. Similar spectroscopic studies have also been carried out on the methyl group exchange in

$$Me_2Al\overset{Ph_2N}{\underset{Me}{\langle\rangle}}AlMe_2$$ [614b];

the first order dependence in dimer suggests that the exchange process does not involve predissociation into monomer, but that only the methyl bridge opens and is followed by an Al–N bond rotation.

C OXIDATIVE ADDITION REACTIONS

These involve the addition of a molecule AB to the subvalent metal or metalloid amide substrate affording an oxidised product, Eq. (7). Some [Eq. (7a)] may alternatively be regarded as insertions of the amide into the A–B bond, but are quite distinct from the insertions discussed in Chapter 10 in which the M–N bond is cleaved, Eq. (8).

$$L_nM{-}NR_2 \xrightarrow{AB} \begin{cases} \xrightarrow[\text{(a)}]{\text{e.g., } AB = Cl_2 \text{ or MeI}} L_m\overset{A}{\underset{B}{\overset{|}{\underset{|}{M}}}}{-}NR_2 \\ \xrightarrow[\text{(b)}]{\text{e.g., } AB = O_2} L_m\overset{A}{\overset{\|}{M}}{-}NR_2 \end{cases} \quad (7)$$

$$L_nM{-}NR_2 + AB \longrightarrow L_mM{-}A{-}B{-}NR_2 \quad (8)$$

Oxidative addition reactions are well-known in transition metal chemistry. A common example is that in which an alkyl halide or a halogen molecule is the addendum and a co-ordinatively unsaturated low oxidation state transition metal complex is the substrate; the latter is frequently a d^6 (Cr^O, Mo^O, or W^O), d^7 (Co^{II}), d^8 (Fe^O, Ru^O, Os^O, Rh^I, Ir^I, or Pt^{II}), or d^{10} (Ni^O, Pd^O, Pt^O, or Au^I) complex, which becomes oxidised to a d^4, d^6, d^6, or d^8 product, respectively. In a similar manner a bivalent Group IVB metal amide (s^2) may undergo oxidative addition to give a d^{10} M^{IV} adduct.

Although the synthesis of a phosphonium salt $[PR_3X]Y$ from a tertiary phosphine, PR_3, and a hydrogen or alkyl halide or a dihalogen, XY, is not often regarded as an oxidation, in the present context it clearly is an oxidative addition ($P^{III} \rightarrow P^V$). However, following convention rather than strict logic, such reactions of P^{III} amides are not considered here but in Chapter 16 in the context of their behaviour as bases [but see Eqs. (18), (20), (25), and (26)].

The amide $Sn[N(SiMe_3)_2]_2$ readily undergoes a wide variety of facile oxidative additions. Thus, insertion reactions into C-Cl, C-Br, C-I, N-Br, Fe-F, Fe-I, Fe-Me, Mn-Br, or Pt-Cl bonds have been observed [286, 286a, 295, 428, 594a], Eq. (7a) in which $L_nM = L_mM = Sn[N(SiMe_3)_2]_2$. Specific examples are shown in Eqs. (9) [286, 428], (10) [594a], and (11) [295]. Metal-nitrogen bonds are occasionally formed as a result of oxidative addition, as in the $Sn[N(SiMe_3)_2]_2$-$BrN(SiMe_3)_2$ [594a] or the $[Pt(PPh_3)_4]$-$\overline{HNC(O)CH_2CH_2CO}$ [637a], or the MR_3-$Br\overline{NC(O)YCO}$ (M = As, Sb, or Bi; R = Me or Ph; X = Cl or Br; Y = $-CH_2CH_2-$ or o-OC_6H_4O) [194a] systems.

$$Sn[N(SiMe_3)_2]_2 + PrBr \longrightarrow SnBr[N(SiMe_3)_2]_2Pr \quad (9)$$

$$Sn[N(SiMe_3)_2]_2 + BrN(SiMe_3)_2 \longrightarrow SnBr[N(SiMe_3)_2]_3 \quad (10)$$

$$Sn[N(SiMe_3)_2]_2 + [Fe(\eta\text{-}C_5H_5)(CO)_2Me] \longrightarrow [Fe(\eta\text{-}C_5H_5)(CO)_2\{Sn(Me)[N(SiMe_3)_2]_2\}] \quad (11)$$

A further example, involving the germanium or lead analogue, is in Eq. (12) (M = Ge or Pb) [424b, 594a].

$$cis\text{-}[\{PtCl_2(PEt_3)\}_2] + 2M[N(SiMe_3)_2]_2 \longrightarrow 2\,cis\text{-}[(Pt(\mu\text{-}Cl)(PEt_3)\{M(Cl)[N(SiMe_3)_2]_2\})_2] \quad (12)$$

The mechanism of the addition of a halogeno-alkane or -arene to a Sn^{II} amide, as in Eq. (9), has been studied in some detail together with parallel investigations in which the isoelectronic Sn^{II} alkyl $Sn[CH(SiMe_3)_2]_2$ was the substrate [286, 286a]. The following observations were made: (i) the reaction, even when an aryl halide is the addendum, proceeds in good yield and under mild conditions [286]; (ii) relative reactivities follow the sequences $Sn[CH(SiMe_3)_2]_2 > Sn[N(SiMe_3)_2]_2$ and RI > RBr > RCl; (iii) in the $Sn[CH(SiMe_3)_2]_2$–RBr (R = Et or Pr) system, e.s.r. spectroscopy shows formation of $\dot{R}$ (by spin-trapping) and probably $\dot{S}nR'_2Br$; (iv) in the Sn^{II} alkyl-(+)-n-C_6H_{13}(Me)CHCl system the Sn^{IV} adduct is the racemate, and if only a trace of the Sn^{II} alkyl is used the optical activity of the (+)-*s*-octyl chloride is unaffected; (v) the proportion of 1:1 adduct to dihalide is solvent dependent, T.H.F. favouring the latter at the expense of the former [286a], Eq. (13) (the 9:1 ratio refers to molar proportions); and (vi) the addition of PhBr to $Sn[N(SiMe_3)_2]_2$ yielding $SnBr[N(SiMe_3)_2]_2Ph$ is catalysed by a trace of the more reactive halide EtBr. Point (vi) is further illustrated by kinetic data using an eight-fold excess of PhBr over the tin(II) amide in benzene (0.4 *M* in amide) at 32° C when $k = 7 \times 10^{-5}$ s^{-1}, but in the presence of EtBr (2.5×10^{-3} per 1 mol amide) $k = 1.8 \times 10^{-4}$ s^{-1} [286a]. Observations (i)–(v) are satisfactorily accommodated by Scheme 1(a) [X = $N(SiMe_3)_2$ or $CH(SiMe_3)_2$] [286]. The formation of the Sn^{IV} dibromide in T.H.F., Eq. (13), is attributed to the rapid abstraction of a hydrogen atom from solvent to form the radical (IV)

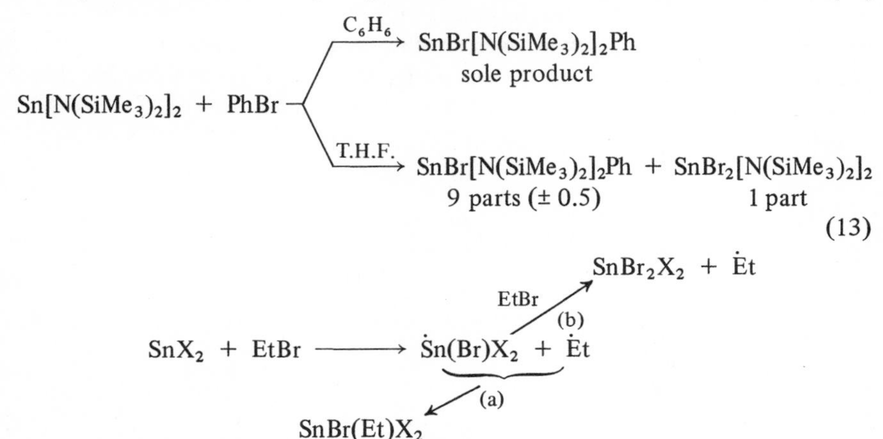

Scheme 1 – The mechanism of addition of an alkyl halide, such as EtBr, to $Sn[N(SiMe_3)_2]_2$ or a related SnX_2 molecule.

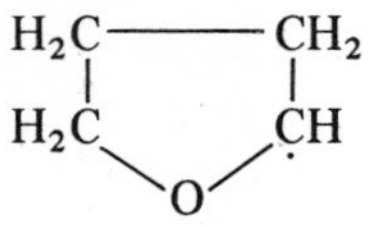

$$\begin{array}{ccc} H_2C & \text{---} & CH_2 \\ | & & | \\ H_2C & & \dot{C}H \\ & \diagdown O \diagup & \end{array}$$

(IV)

allowing reaction (b) of Scheme 1 to be competitive with the geminate radical pair combination (a) [286a]. The catalytic effect, item (vi), is interpreted by the radical chain process of Scheme 2 [$X = N(SiMe_3)_2$ or $CH(SiMe_3)_2$].

$$\left.\begin{array}{l} SnX_2 + EtBr \longrightarrow \dot{S}n(Br)X_2 + \dot{E}t \\ \dot{S}n(Br)X_2 + PhBr \longrightarrow SnBr_2X_2 + \dot{P}h \end{array}\right\} \text{initiation}$$

$$\left.\begin{array}{l} SnX_2 + \dot{P}h \longrightarrow \dot{S}n(Ph)X_2 \\ \dot{S}n(Ph)X_2 + PhBr \longrightarrow SnBr(Ph)X_2 + \dot{P}h \end{array}\right\} \text{propagation}$$

Scheme 2 – The mechanism of EtBr-catalysed addition of PhBr to $Sn[N(SiMe_3)_2]_2$ or a related SnX_2 molecule [286a].

An interesting type of oxidative addition reaction promises to have some synthetic utility. This relates to a Ge^{II} or Sn^{II} amide as substrate and a carboxylic acid chloride as reagent, exemplified by Eq. (14) (M = Ge or Sn, R = Bu^t or Ph) [424c]. Acyl derivatives of main group metals are notoriously difficult to obtain, partly because acyl analogues of Grignard reagents are not available. Acyl-silanes and -germanes have attracted attention in recent years because of their colour and their facile molecular rearrangement into siloxy- or germoxy-carbenes, (V) (M = Si or Ge), while acylstannanes are extremely rare. However, by regarding the bulky amido ligands of Eq. (14) as protecting groups, a convenient procedure to a wide variety of acyl-germanes or -stannanes becomes feasible.

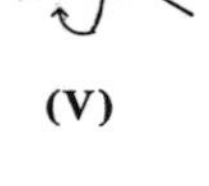

$$RC(=O)\text{—}M\lessgtr$$

(V)

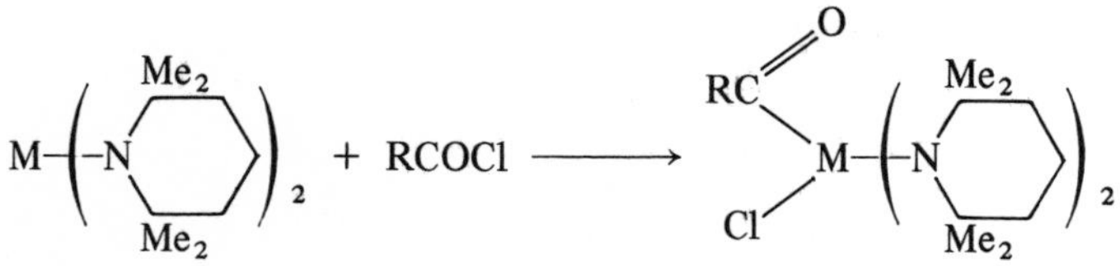

$$M\left(N\text{(2,2,6,6-}Me_4C_5H_6\text{)}\right)_2 + RCOCl \longrightarrow RC(=O)(Cl)M\left(N\text{(2,2,6,6-}Me_4C_5H_6\text{)}\right)_2 \quad (14)$$

There has been significant progress recently by the Toulouse school in the chemistry of 'germylenes', bivalent germanium compounds which are generated as reaction intermediates. Relevant examples are in Eqs. (15) [649] and (16) [619].

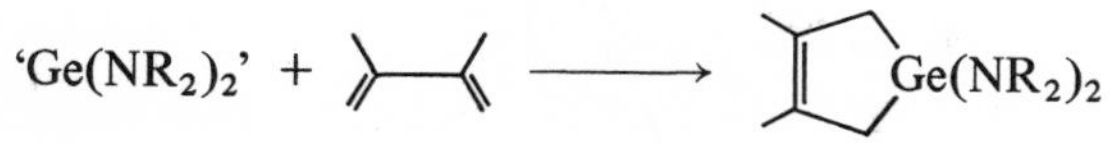

$$\text{'Ge(NR}_2)_2\text{'} + \text{(2,3-dimethylbutadiene)} \longrightarrow \text{(3,4-dimethylgermacyclopent-3-ene)Ge(NR}_2)_2 \qquad (15)$$

$$\text{'Ge(NMe}_2\text{)Ph'} + \text{Me}_3\text{GeX} \longrightarrow \text{Me}_3\text{GeGe(NMe}_2\text{)(Ph)X} \qquad (16)$$

Reactions which fall broadly into addition according to Eq. (7b) are illustrated in the context of Group IV metal chemistry in Eq. (17) [721b].

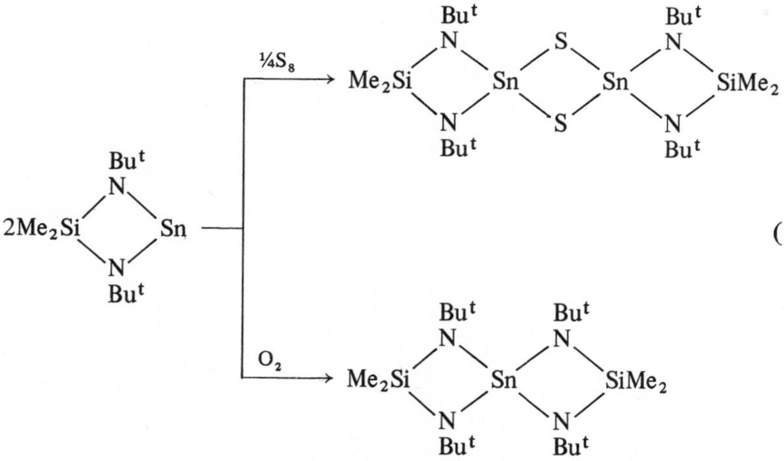

$$2\,\text{Me}_2\text{Si}(\mu\text{-NBu}^t)_2\text{Sn} \xrightarrow{\frac{1}{4}\text{S}_8} \text{Me}_2\text{Si}(\mu\text{-NBu}^t)_2\text{Sn}(\mu\text{-S})_2\text{Sn}(\mu\text{-NBu}^t)_2\text{SiMe}_2$$

$$2\,\text{Me}_2\text{Si}(\mu\text{-NBu}^t)_2\text{Sn} \xrightarrow{\text{O}_2} \text{Me}_2\text{Si}(\mu\text{-NBu}^t)_2\text{Sn}(\mu\text{-NBu}^t)_2\text{SiMe}_2 \qquad (17)$$

As stated in the early part of this Section, amides of the Main Group V elements may behave as substrates for oxidative addition, $L_nM^{III}(NR_2)_{3-n}$ being usually oxidised from M^{III} to M^V. The product is generally the $[L_nM^V(NR_2)]^+$ salt and numerous examples are known for phosphorus, e.g., Eqs. (18), (19) ($R'X = F_2$, Cl_2, Br_2, I_2, CCl_4, PrBr, or EtI), or (20) [274a, 518, 662a]. Related systems are those of Eqs. (21) [48, 48a] and (22) [48c]. However, complications may arise if an *N*-chlorodi(silyl)amine is the reagent, as in Eq. (23) (R = Bu, Ph, OEt, OPr, or OBu) [588a], where the initial 1:1-adduct readily loses Me_3SiCl (for similar eliminations, see Section G).

$$\text{P(NMe}_2)_3 + \text{MeI} \longrightarrow [\text{P(NMe}_2)_3\text{Me}]\text{I} \qquad (18)$$

$$\text{RN=P—NR}_2 + \text{R'X} \longrightarrow \text{RN=P(X)(R')—NR}_2 \qquad (19)$$

$$\mathrm{MeP(\mu\text{-}NBu^t)_2PCl \;+\; MeI \longrightarrow Me_2\overset{+}{P}(\mu\text{-}NBu^t)_2\overset{-}{P}(Cl)I} \qquad (20)$$

$$\mathrm{F_2PNH_2 \;+\; NH_3 \longrightarrow F_2\overset{H}{P}(NH_2)_2} \qquad (21)$$

$$\mathrm{P(NMe_2)_3 \;+\; (NH\text{-macrocycle: } N\text{-}H,\ NH,\ HN,\ H\text{-}N) \longrightarrow (N)_4P\text{-}H} \qquad (22)$$

$$\mathrm{R_3P \;+\; ClN(SiMe_3)_2 \longrightarrow Me_3SiCl \;+\; R_3P{=}NSiMe_3} \qquad (23)$$

The reaction between an alkyl halide and an arsenic amide does not usually lead to oxidative addition. For instance, Eq. (24) [718a] shows a metathetical NR_2/I exchange (see Chapter 13). Rare examples of oxidative additions of As^{III} amides are provided by Eqs. (25) and (26) [723].

$$\mathrm{Et_2AsNR_2 \;+\; 2MeI \longrightarrow Et_2AsI \;+\; [NMe_2Et_2]I} \qquad (24)$$

$$\mathrm{As(NMe_2)_3 \;+\; Br_2 \longrightarrow [AsBr(NMe_2)_3]Br} \qquad (25)$$

$$\mathrm{As(NMe_2)_3 \;+\; ClNMe_2 \longrightarrow [As(NMe_2)_4]Cl} \qquad (26)$$

The Staudinger reaction (see Chapter 6), effectively a nitrene ($R\ddot{N}$) addition, leading from a phosphine and azide to a phosphine-imine is an oxidation as in Eq. (7b), as illustrated by Eqs. (27) and (28) [702a], and (29) [43a, 660a]. In a similar system, Eq. (30) ($R = Pr^i$ or $SiMe_3$), there is, however, no change in the oxidation state of phosphorus [590b].

$$\mathrm{P(NMe_2)_3 \;+\; PhN_3 \longrightarrow PhN{=}P(NMe_2)_3 \;+\; N_2} \qquad (27)$$

$$\mathrm{P(NMe_2)_3 \;+\; RSO_2N_3 \longrightarrow R(O)_2SN{=}P(NMe_2)_3 \;+\; N_2} \qquad (28)$$

$$\mathrm{Me_3SiN{=}PN(SiMe_3)_2 \;+\; Me_3SiN_3 \longrightarrow (Me_3SiN{=})_2PN(SiMe_3)_2 \;+\; N_2} \qquad (29)$$

$$\mathrm{Bu^tN{=}PNMe_2 \;+\; Bu^tN_3 \longrightarrow \underset{\textbf{(VI)}}{N(Bu^t)\text{-}N{=}N\text{-}N(Bu^t)\text{-}PNMe_2}\ (\text{ring})} \qquad (30)$$

This may provide a clue to the pathway of the Staudinger reaction when an iminophosphine is substrate: initial 1,3-dipolar adduct formation [e.g., (VI)], and its subsequent fragmentation with elimination of N_2. A related oxidation is that of Eq. (31) to give either compound (VII) ($R = Me$, Et, Pr^i, or Bu^t) or (VIII) ($R = SiMe_3$) [520c], while Eq. (32) shows oxidative cycloadditions [$Y = S = X$, $R = SiMe_3$; or $Y = (CF_3)_2CN_2$, $X = NSiMe_3$, $R = (CF_3)_2CN$] [43a, 519a].

$$RN{=}PNR_2 + R'(Me)CN_2 \longrightarrow (RN{=})(R'(Me)C{=})PNR_2 \quad \textbf{(VII)}$$

$$\textbf{(VII)} \xrightarrow{R'(Me)CN_2} \left[(RN{=})(NR_2)P(\bar{C}(Me)R')(N{=}\overset{+}{N}{=}C(Me)R') \right] \longrightarrow (RN{=})(NR_2)P(C(R'){=}CH_2)(N(H){-}N{=}C(Me)R') \quad \textbf{(VIII)} \qquad (31)$$

$$Me_3SiN{=}PN(SiMe_3)_2 + Me_3SiN{=}SMe_2 \longrightarrow Me_2S\langle N(SiMe_3) \rangle_2 P({=}NSiMe_3)N(SiMe_3)_2 \qquad (32)$$

$$Me_3SiN{=}PN(SiMe_3)_2 \xrightarrow{Y} \begin{array}{l} Me_3SiN{-}P({=}X)N(SiMe_3)_2 \\ \;\;| \qquad\quad | \\ (Me_3Si)_2NP{-}NR \end{array}$$

Clear examples of Eq. (7b) in the context of phosphorus amide chemistry are illustrated by Eqs. (33) [269c], (34) [83b, 269c, 366a, 732] (X = O, S, or Se), or (35) [293a] [X' = -S-, -S-S-, -S(O)-, $-S(O)_2-$, or $-NC(O)CH_2C(O)CH_2-$, R'' = Me or Et, or $P(NR''_2)_3$ = compound (IX)], the last of which may have considerable synthetic potential in a method of desulphurisation of organic sulphur compounds, as in Eq. (35).

$$P(NR_2)_3 + \frac{1}{n}X_n \longrightarrow XP(NR_2)_3 \qquad (33)$$

$$(OCN)_2PNR_2 + \frac{1}{n}X_n \longrightarrow XP(NR_2)(NCO)_2 \qquad (34)$$

$$P(NR''_2)_3 + RX'SR' \longrightarrow RX'R' + SP(NR''_2)_3 \qquad (35)$$

$$P{-}\left(N\text{(1,2,3,4-tetrahydroisoquinolin-2-yl)}\right)_3 \quad \textbf{(IX)}$$

Amides of sulphur are outside our scope; nevertheless they often behave similarly to those of phosphorus, as in the conversion of $S^{II}(NMe_2)_2$ into $[ClS^{IV}(NMe_2)_2][SbCl_6]$ by means of Cl_2-$SbCl_5$ [739]. There are no clear cut cases as yet of oxidative additions involving an amide of a transition metal, but the formation of a blue colour [24], suggestive of Cr^{IV}, upon exposing $[\{Cr(NEt_2)_3\}_n]$ to oxygen may be a case in point; reaction (36) is an oxidation if NO is regarded as a uninegative ligand [96, 98] (see Chapter 16 for an alternative view).

$$[Cr\{N(SiMe_3)_2\}_3] + NO \longrightarrow [Cr\{N(SiMe_3)_2\}_3(NO)] \qquad (36)$$

D REDUCTIVE ELIMINATION REACTIONS

This is the converse of oxidative addition and, like the latter, has ample precedents in inorganic and organometallic chemistry outside that of metal or metalloid amides. Accordingly, we may define such a reaction as a fragmentation, classified generally by the reverse of Eq. (7), wherein a high oxidation state metal or metalloid (M) amide is transformed into a reduced complex, usually, but not invariably, with decrease in the co-ordination number of M. As we are dealing here and in Section B with redox processes, it is obviously possible to have a simple electron transfer without a change in the number or type of ligand; only a single example of this, Eq. (37), appears to have been reported [252a].

$$[V(NPh_2)_4] + K[C_{10}H_8] \longrightarrow K[V(NPh_2)_4] \qquad (37)$$

Reactions which fall within the scope of the reverse of Eq. (7a) include some dehydrohalogenations which have more conveniently been discussed in Chapter 14.

The pyrolysis of a homoleptic amide $M(NR_2)_n$ of B, Si, Sn, Ti, Zr, Nb, or Ta at 300–500° C has been stated ultimately to yield the metal (B, Si, or Sn) or the nitride [706c]. For $Si(NHR)_4$, however, spirocyclic and branched organosilazanes are said to form upon pyrolysis [40a]. Photolysis and other reactions which unambiguously proceed by M–N homolysis are discussed in Section E.

Wurtz-type condensations of the kind shown in Eq. (38) [303a] are formally reductions of $B^{III} \rightarrow B^{II}$. Similar treatment of $Sn[CH(SiMe_3)_2]_3Cl$ with Na in liquid

$$\begin{array}{l} 2ClB(NMe_2)_2 + 2K \longrightarrow B_2(NMe_2)_4 + 2KCl \\ \quad \Big\downarrow {\scriptstyle Cl_2BNMe_2,\ 4K} \\ Me_2NB[B(NMe_2)_2]_2 + 4KCl \end{array} \qquad (38)$$

ammonia affords the white solid $[Sn\{CH(SiMe_3)_2\}_3]_2$ which dissociates in hydrocarbon solvent to the persistent radical $\dot{S}n[CH(SiMe_3)_2]_3$ [285c]. By contrast, the corresponding reaction of the isoelectronic $SnBr[N(SiMe_3)_2]_3$ yields a transient e.s.r. signal attributable to $\dot{S}n[N(SiMe_3)_2]_3$, and then its fragmentation products, Eq. (39) [594a]; attempts to reduce the compound $Sn[N(SiMe_3)_2]_2(R)X$ (X = Cl, Br, or I; R = Me, Et, Pr^i, or Bu) by irradiation in the presence of an

$$SnBr[N(SiMe_3)_2]_3 \xrightarrow[OEt_2,\ 25°\ C]{Na/Hg} \dot{S}n[N(SiMe_3)_2]_3 \longrightarrow N(SiMe_3)_3 + Sn + HN(SiMe_3)_2 \quad (39)$$

electron-rich olefin were inconclusive. We have already described the synthesis of kinetically-stabilised phosphinyl and arsinyl radicals $\dot{M}(NR_2)_2$ (M = P or As) (Chapters 6 and 7), a procedure [273a, 285a, 285b] which is related to those of Eqs. (38) and (39), and in its most generalised form is illustrated in Eq. (40) {M′ = P or As, X and/or Y is a bulky amino or alkyl group, and e.r.o. = an electron-rich olefin such as $C_2(NMe_2)_4$ or $\mp\overline{CN(Et)CH_2CH_2N}Et]_2\}$.

$$ClM\begin{matrix}X\\Y\end{matrix} \xrightarrow[\text{e.r.o. in PhMe, } h\nu]{\text{Na in PhMe; or}} \dot{M}\begin{matrix}X\\Y\end{matrix} \quad (40)$$

Reductive processes involving arsenic, Eq. (41) [588b]; sulphur $[CF_3SF_4Cl \xrightarrow{Me_3SiNMe_2} CF_3(Cl)S(NMe_2)_2]$ [375e]; chromium, Eqs. (42) [56a] and (43) [151d]; and molybdenum, Eq. (44) [155e] have been demonstrated. Crystallographic data confirm structures (X) and (XI) [151d]. The reductive protic reaction of a higher oxidation state metal amide with cyclopentadiene to give a metallocene was first demonstrated for $[Fe(\eta\text{-}C_5H_5)_2]$ [352, 354] and has also been observed for Nb [347]. Reactions (42)-(44) (see also Section D) may be contrasted with the non-reductive protic cleavage of the M-NR_2 bond

$$Bu^tN{=}AsCl_3 \xrightarrow{heat} \begin{matrix}Bu^tN & \text{—} & AsCl\\ | & & |\\ ClAs & \text{—} & NBu^t\end{matrix} + Cl_2 \quad (41)$$

$$[Cr(NEt_2)_4] \xrightarrow{C_5H_6} [Cr(\eta\text{-}C_5H_5)_2] \quad (42)$$

$$[Cr(NEt_2)_4] + 4CO_2 \longrightarrow \underset{\textbf{(X)}}{[Cr^{III}_2(O_2CNEt_2)_4(\mu\text{-}NEt_2)_2]} + \underset{\textbf{(XI)}}{[Cr^{II}_2(O_2CNEt_2)_4(HNEt_2)_2]} \quad (43)$$

$$[Mo_2Et_2(NMe_2)_4] + 4CO_2 \longrightarrow [Mo_2(O_2CNMe_2)_4] + C_2H_4 + C_2H_6 \quad (44)$$

(Chapter 11) or simple insertion into the M–NR_2 bond (Chapter 10). The oxidative coupling of $RN{=}CPh_2$, using for instance potassium in benzene, to yield $R\bar{N}CPh_2\text{-}CPh_2\bar{N}R$ or $R\bar{N}\bar{C}Ph_2$ ($R = Et_2Al$) [314b] may be classified as a reductive elimination.

A common decomposition mode for metal alkyls is that of β-elimination, as exemplified by Eq. (45) [198]. Clear evidence has only recently emerged for

$$L_nM\text{—}\overset{\alpha}{C}H_2\text{—}\overset{\beta}{C}H_3 \longrightarrow L_nM\text{—}H + CH_2{=}CH_2 \tag{45}$$

such a pathway for the isoelectronic metal amides and comes from experiments on Rh^I complexes. Thus, treatment of Wilkinson's catalyst (XII) with $LiNMe_2$ yields a Rh^I hydride and $MeN{=}CH_2$ [207a]; whereas from $LiN(SiMe_3)_2$, which has no β-H, the amide (XIII) is obtained [141d], Eq. (5). The latter, like

$[RhCl(PPh_3)_3]$ **(XII)** $[Rh\{N(SiMe_3)_2\}(PPh_3)_2]$ **(XIII)**

$[RuH\{N(SiMe_3)_2\}(PPh_3)_2]$ loses $HN(SiMe_3)_2$ upon heating, to yield the ortho-metallated product [(III) from the Rh amide (XIII)] [141d]. This might be regarded as a simple elimination (cf. Section G), but it is likely that such reactions proceed via initial tautomerisation. Taking this view, the intermediate in the conversion of (XIII) into (III) is the Rh^{III} complex (XIV). In the $[RhCl(PPh_3)_3]$-$Li[N(CH_3)CD_3]$ system a large isotope effect ($k_H/k_D \approx 6$) indicates that C–H bond-breaking is rate-limiting [207a].

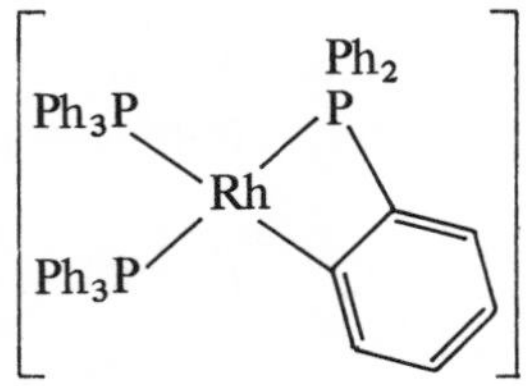

(III)

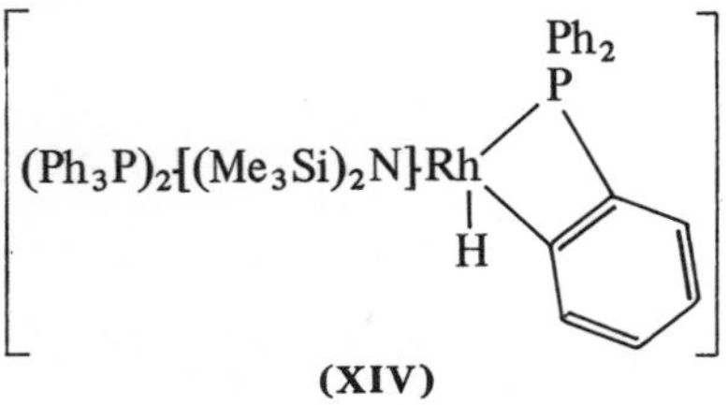

(XIV)

From $[RuCl_2(PPh_3)_4]$ and $LiNMe_2$, $[RuH_2(PPh_3)_4]$ or $[RuH(Cl)(PPh_3)_3]$ is obtained, while $[PdCl_2(PPh_3)_2]$ and the same amide yield Pd or, in presence of PPh_3, $[Pd(PPh_3)_4]$.

E DISPROPORTIONATION REACTIONS

The disproportionation of a metal amide involves a change in oxidation state and possibly of co-ordination number at the metal atom. Stoicheiometrically it is a bimolecular internal redox reaction, yielding both an oxidised and a reduced (often not fully characterised) form of the original complex. The process commonly takes place upon heating or distilling a lower oxidation state metal amide and is particularly prevalent among early transition elements, e.g., Eqs. (46) ($X = NMe_2$, NEt_2, NPr^i_2, or Cl) [431], (47) [25], and (48) [56a]. It appears that the reaction is an equilibrium, being volatility controlled, and does not take place until well above boiling point of the most volatile product.

$$[\{Ti(NMe_2)_2X\}_2] \longrightarrow [Ti(NMe_2)_4] + \frac{1}{n}[TiX_2]_n \qquad (46)$$

$$VCl_3 \xrightarrow{3LiNEt_2} \underset{\text{not isolated}}{V(NEt_2)_3} \longrightarrow [V(NEt_2)_4] + \frac{1}{n}[V(NEt_2)_2]_n \qquad (47)$$

$$CrCl_3 \xrightarrow{3LiNEt_2} \underset{\text{not isolated}}{Cr(NEt_2)_3} \longrightarrow [Cr(NEt_2)_4] + \frac{1}{n}[Cr(NEt_2)_2]_n \qquad (48)$$

Disproportionation has recently been observed for certain bivalent amides of germanium or tin, which upon photolysis yield the corresponding persistent trivalent amide [170a, 286b, 322a, 428]. As there is no deposition of metal, it is inferred that there is concomitant formation of a univalent amide complex; which, being diamagnetic (e.s.r.), must be an oligomer or telomer: thus, we have the system of Eq. (49). Details are in Section F, because it is possible that there may be initial M–N homolysis (see also Table 3 and Chapter 5).

$$2M^{II} \longrightarrow M^{III} + \frac{1}{n}(M^{I})_n \qquad (49)$$

Other instances of disproportionation may arise upon the addition of an unsaturated or protic compound to the metal amide (see also Section C). The clearest example is that of Eq. (50) [25, 432].

$$[\{Ti(NEt_2)_3\}_2] + 6CS_2 \longrightarrow [Ti(S_2CNEt_2)_4] + \frac{1}{n}[Ti(S_2CNEt_2)_2]_n \qquad (50)$$

F M–N HOMOLYSIS

Evidence for M-N bond homolysis rests partly upon plausible mechanistic interpretations of reaction products, especially of photochemical transformations. Recently, however, appropriate spectroscopic techniques have been brought to bear on the problem and serve to identify radicals or radical pairs.

The photolysis of tetrakis(dimethylamido)diboron in a chlorinated solvent yields products most readily interpreted in terms of competing B-B and B-N homolysis [286d], Eq. (51). E.s.r. spectroscopy identifies the siloxynitroxide

$$B_2(NMe_2)_4 \xrightarrow{h\nu} \begin{cases} \xrightarrow{CCl_4} B_2Cl(NMe_2)_3 + B(NMe_2)_3 + CHCl_3 + CH_2(NMe_2)_2 \\ \xrightarrow{CHCl_3} B_2Cl(NMe_2)_3 + HB(NMe_2)_2 + B(NMe_2)_3 + CH_2Cl_2 \\ \xrightarrow{CH_2Cl_2} ClB(NMe_2)_2 + B(NMe_2)_3 \end{cases} \quad (51)$$

radical (XV) during the photolysis of the amide (XVI) in presence of the spin-trap ArNO [614a]; this is suggestive of initial SiN homolysis of (XVI) yielding $\dot{S}iMe_3$.

$$Ar{-}\dot{N}(O)(OSiMe_3)$$

(XV)

$$Me_3Si{-}N\langle C_5H_4(Me)(H)\rangle$$

(XVI)

Photolysis at ca. 20° C of a hexane solution of a bulky Ge^{II} or Sn^{II} amide $M(NR_2)_2$ leads to the kinetically-stabilised M^{III} amide, as shown by e.s.r., Eq. (51) [M = Ge or Sn and $NR_2 = N(Bu^t)SiMe_3$, $N(SiMe_3)_2$, or $N(GeMe_3)_2$; or M = Sn and R = $GeMe_3$] [170a, 286b, 322a, 428] (Chapter 5); however, in the case of an even more bulky metal(II) amide, the aminyl radical is formed, Eq. (52) [M = Ge or Sn and $NR_2 = N(SiEt_3)_2$, NBu^t_2, or N(2,2,6,6-tetramethylpiperidino) ring with Me_2 groups] [286b, 428, 429a, 433a]. The bis(di-t-butylamido)metal(II) compounds decompose in this fashion even in the dark, on gentle heating, or slowly at ambient temperature [433a], as does $Pb[N(M'Me_3)SiMe_3]_2$ (M′ = C or Si) [322a].

$$2M(NR_2)_2 \longrightarrow \underset{\textbf{(XVII)}}{\dot{M}(NR_2)_3} + \underset{\text{not identified}}{\frac{1}{n}(MNR_2)_n} \quad (51)$$

$$M(NR_2)_2 \longrightarrow \dot{N}R_2 + \text{other products} \quad (52)$$

Alternative mechanisms have been proposed to account for the formation of Group IVB metal-centred radicals (XVII), [322a]. Either initial M–N homolysis is followed by spin-trapping of the resulting aminyl radical by a further molecule of the metal(II) amide, Eq. (53); or a bimolecular ligand transfer, i.e. disproportionation (Section D), occurs which may possibly implicate an excited-state metal(II) amide molecule, Eq. (54) (see also ref. [418]).

$$M(NR_2)_2 \longrightarrow \frac{1}{n}(MNR_2)_n + \dot{N}R_2 \xrightarrow{M(NR_2)_2} \underset{\textbf{(XVII)}}{\dot{M}(NR_2)_3} \qquad (53)$$

$$M(NR_2)_2 \longrightarrow M(NR_2)_2{}^* \xrightarrow{M(NR_2)_2} \underset{\textbf{(XVII)}}{\dot{M}(NR_2)_3} + \frac{1}{n}(MNR_2)_n \qquad (54)$$

The technique of chemically-induced dynamic nuclear polarisation, CIDNP, has been applied to help elucidate the mechanism of photolysis of Me_3SnNEt_2, Eq. (55) [444a]; this shows the intermediacy of $\dot{S}nMe_3$ and $\dot{N}Et_2$ radicals.

$$Me_3SnNEt_2 \longrightarrow (Me_3Sn)_2 + HNEt_2 + MeCH{=}NEt \qquad (55)$$

The nature of the organic products obtained upon photolysis [298a] or thermolysis [772a] of a Hg^{II} amide suggests that aminyl radical intermediates are formed, Eq. (56).

$$(CF_3)_2NC(F){=}NCF_3 \xleftarrow{\text{heat}} Hg[N(CF_3)_2]_2 \xrightarrow{h\nu} (CF_3)_2NN(CF_3)_2 + CF_3N{=}CF_2$$
$$+ FN(CF_3)_2 + (CF_3)_2NN(CF_3)CF_2N(CF_3)_2 + \begin{array}{l} CF_2 - CFN(CF_3)_2 \\ \;|\qquad\quad | \\ CF_2 - CFN(CF_3)_2 \end{array} \qquad (56)$$

G ELIMINATION REACTIONS

Here we describe reactions in which a stable molecule is eliminated from the metal amide, generally to afford a compound which may either contain a multiple M=N bond [e.g., Eq. (3)] or be the product of its oligomerisation or polymerisation, e.g., Eqs. (2) or (4). Eliminations of this type are well-established as synthetic routes to amides of B, Si, Ge, Sn, P, or As. They are reviewed in ref. [657]. The driving force includes the relatively-favoured facile removal of ammonia, a primary amine, or a trimethylhalogenosilane, the high standard free energies of formation of these compounds, and the considerable heats of autocomplexation by hydrogen bonding or by μ-X_2-bridging. Because many such reactions relating to main group metals have been described in Chapters 3–7, we mention only a small selection here and in Table 5. However, we note that reactions are generally associative rather than dissociative in character, because bimolecular species are often isolable. Thus, by reference to Eq. (2), the formation of the borazine $[(RHN)BNR]_3$ from $B(NHR)_3$ proceeds via the isolated diborylamine $RN[B(NHR)_2]_2$ [50, 417a], from which the unstable borazyne (RHN)B=NR may be presumed to be formed by a 1,3-shift mechanism. In confirmation, pyrolysis of $B(NHBu^t)_3$ is a poor procedure for preparing the boretane (XVIII), possibly because steric effects make an associative elimination energetically

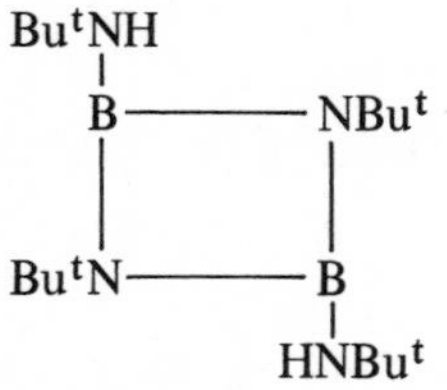

Bu^tNH–B, NBu^t, Bu^tN, B–$HNBu^t$

(XVIII)

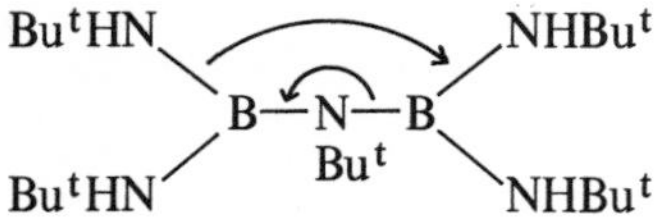

$(Bu^tHN)_2B$–$N(Bu^t)$–$B(NHBu^t)_2$

(XIX)

unfavourable; whereas independently prepared diborylamine (XIX) is a good thermal source of (XVIII) [423b]. Similarly, heating $EtN[B(OBu)_2]_2$ yields $B(OBu)_3$ and $[(BuO)BNEt]_3$ [50, 417a].

Some elimination reactions which involve extrusion of a small molecule other than NH_3, H_2NR, or Me_3SiX (or its Ge or Sn analogues) are illustrated in Eqs. (57) [734a], (58) [314a], (59) [721b], and (60) [734a]. Eq. (58) is particularly interesting as an example of an orthometallation, cf. Eq. (5).

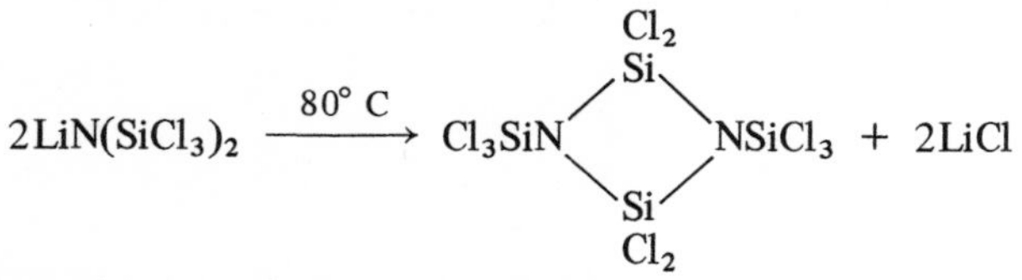

$$2LiN(SiCl_3)_2 \xrightarrow{80^\circ C} Cl_3SiN(SiCl_2)_2NSiCl_3 + 2LiCl \quad (57)$$

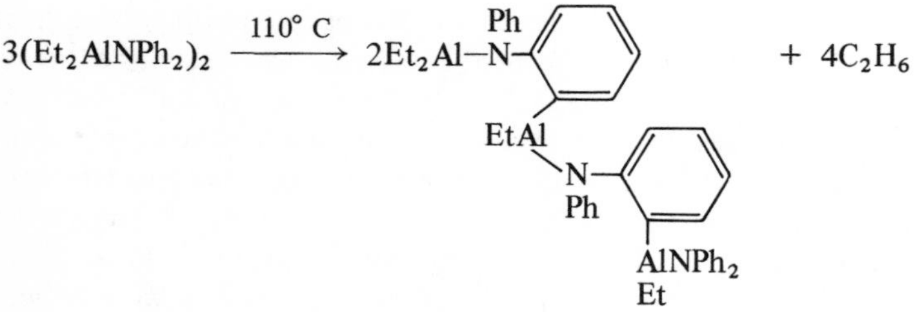

$$3(Et_2AlNPh_2)_2 \xrightarrow{110^\circ C} 2Et_2Al{-}N(Ph){-}C_6H_4{-}Al(Et){-}N(Ph){-}C_6H_4{-}Al(Et)NPh_2 + 4C_2H_6 \quad (58)$$

$$Me_2Si(\mu\text{-}NBu^t)_2Sn(\mu\text{-}S)_2Sn(\mu\text{-}NBu^t)_2SiMe_2 \xrightarrow{\text{heat}} Me_2Si(\mu\text{-}NBu^t)_2Sn(\mu\text{-}NBu^t)_2SiMe_2 + SnS_2 \qquad (59)$$

$$2MeCdN(SiCl_3)_2 \xrightarrow{20^\circ\,C} Cl_3SiN(\mu\text{-}SiCl_2)_2NSiCl_3 + 2MeCdCl \qquad (60)$$

Trimethylchlorosilane elimination has also been reported for some transition metals, e.g., Eq. (61) [20a, 21], (Chapter 8). However, in one case, Eq. (62) [66a],

$$[TiCl_4] + N(SiMe_3)_3 \longrightarrow [Ti(Cl)_3N(SiMe_3)_2] + Me_3SiCl \;\downarrow\; \frac{1}{n}[Cl_2TiNSiMe_3]_n + Me_3SiCl \qquad (61)$$

$$[Ti(\eta\text{-}C_5H_5)_2Cl_2] \xrightarrow{LiN(SiMe_3)_2} [Ti(\eta\text{-}C_5H_5)_2(Cl)N(SiMe_3)_2] \;\downarrow\; [(\eta\text{-}C_5H_5)_2Ti(N(SiMe_3))(CH_2)SiMe_2] \qquad (62)$$

cleavage of a C–H bond and elimination of HCl takes place in preference to loss of Me_3SiCl. A further example of such a process is in the reaction between $SiCl_4$ and $NaN(SiMe_3)_2$ at 200° C, Eq. (63) [732a]. This, however, is not really an elimination but a metathetical exchange (cf., Chapter 13), and other examples

$$NaN(SiMe_3)_2 + SiCl_4 \longrightarrow Me_2Si(N(SiMe_3))(C(SiMe_3)H)Si(N(SiMe_3))_2SiMe_2 \qquad (63)$$

are the loss of Me_3SiCl in the $HN(SiMe_3)_2$–Ph_3CCCl_3 [443a] and $LiN(SiMe_3)_2$-NOCl [754] systems to yield $PhSi(NHSiMe_3)_3$ and $(Me_3Si)_2NNO$, respectively, or the reaction of NSF_3 with $LiN(SiMe_3)_2$ to give $Me_3SiN{=}S(F)_2{=}NSiMe_3$ with Me_3SiF, LiF, and $S(NSiMe_3)_3$ [268].

The formation of the phosphonitrilic compound (XX), as shown in Eq. (64), may involve an amine elimination and a subsequent prototropic shift [672a]; however, the proposed intermediates (XXI) and (XXII) are speculative.

$$ClP(NMe_2)_2 + NH_3 \longrightarrow \underset{\textbf{(XXI)}}{P(NMe_2)_2NH_2} \longrightarrow \underset{\textbf{(XXII)}}{Me_2NP{=}NH} \longrightarrow \tfrac{1}{3}\,[\text{cyclo-}(HP(NMe_2){=}N)_3]\ \textbf{(XX)} \qquad (64)$$

Amine elimination has been reported in the reaction between $CoCl_2$ and $LiNEt_2$ [88a]. The formation of the nitrogen analogue of a β-diketonate (XXIII) [88a], Eq. (65), may result from initial β-elimination (see Section C) to afford a Co^{II} hydride and the imine MeCH=NEt which couples to form the ligand $EtNCHCH_2CH(Me)NHEt$.

$$7CoCl_2 + 14LiNEt_2 \longrightarrow \underset{\textbf{(XXIII)}}{[Co(EtN{=}C(Me)CH{=}CHNEt)_2]} + 14LiCl + 10HNEt_2 + 6Co \qquad (65)$$

The synthesis, (a) in Eq. (66), of 4-membered heterocycles from amino-iminophosphanes and metal halides is limited to a small number of elements [520a]; this is due to α-elimination of the Lewis acid EX_{n-2}, (e.g., $TiCl_2$), (b) in Eq. (66) ($R = SiMe_3$) [520b], which is also an oxidative addition at phosphorus.

$$RX + \text{cyclo-}[{-}N{-}P(X){-}N{-}E(X_{n-2}){-}] \xleftarrow{(a)} {-}N(R){-}P{=}N{-} + EX_n \xrightarrow{(b)} EX_{n-2} + {-}N(R){-}P(X)_2{=}N{-} \qquad (66)$$

H TABLES OF REACTIONS

Table 1 Oxidative Addition Reactions of Metal Amides

Substrate	Reactant	Product	References
$Ge[N(SiMe_3)_2]_2$	*trans*-$[\{PtCl_2(PEt_3)\}_2]$	$[(Pt(\mu\text{-}Cl\{Ge(Cl)[N(SiMe_3)_2]_2\}(PEt_3))_2]$	[594a]
$\frac{1}{n}[Ge(NR_2)_2]_n$	2,3-dimethylbutadiene	cyclic 3,4-dimethyl-germacyclopent-3-ene $Ge(NR_2)_2$	[649]
$\frac{1}{n}[Ge(NMe_2)Ph]_n$	Me_3GeX	$Me_3GeGe(NMe_2)(Ph)X$	[619]
$Ge(NC_5H_6Me_4)_2$ (2,2,6,6-Me_4-piperidino)	PhCOCl	$Ge(COPh)(NC_5H_6Me_4)_2Cl$	[424c]
$Sn[N(SiMe_3)_2]_2$	$[Fe(\eta\text{-}C_5H_5)(CO)_2X]$ (X = F, I, Me)	$[Fe(\eta\text{-}C_5H_5)(CO)_2\text{-}\{Sn(X)[N(SiMe_3)_2]_2\}]$	[295, 594a]
$Sn[N(SiMe_3)_2]_2$	$[Mn(Br)(CO)_5]$	$[Mn(CO)_5\{Sn(Br)\text{-}[N(SiMe_3)_2]_2\}]$	[594a]
$Sn[N(SiMe_3)_2]_2$	*trans*-$[\{PtCl_2(PEt_3)\}_2]$	*trans*-$[(Pt(\mu\text{-}Cl)(PEt_3)\text{-}\{Sn(Cl)[N(SiMe_3)_2]_2\})_2]$	[428, 594a]
$Sn[N(SiMe_3)_2]_2$	$[Pt(COD)Cl_2]$ (COD = 1,5-cyclo-octadiene)	$[Pt(COD)\{Sn(Cl)\text{-}[N(SiMe_3)_2]_2\}_2]$	[594a]
$Sn[N(SiMe_3)_2]_2$	RX [X = Cl, R = Bu; X = Br, R = Me, Et, $N(SiMe_3)_2$, Pr, Bu^t, Ph; X = I, R = Me, Et, Pr^i, Bu, Ph)	$Sn(R)[N(SiMe_3)_2]_2X$ {+ $SnBr_2[N(SiMe_3)_2]_2$ when RX = PhBr, in THF}	[286, 286a, 428, 594a] [286a]
$Sn[N(SiMe_3)_2]_2$	O_2	$\frac{1}{n}\{OSn[N(SiMe_3)_2]_2\}_n$	[655]
$Sn[N(SiMe_3)_2]_2$	Bu^tCOCl	$Sn(COBu^t)[N(SiMe_3)_2]_2Cl$	[424c]
$Sn(NC_5H_6Me_4)_2$ (2,2,6,6-Me_4-piperidino)	RCOCl (R = Bu^t, Ph)	$Sn(COR)(NC_5H_6Me_4)_2Cl$	[424c]
$Sn(NBu^t_2)_2$	PhCOCl	$Sn(COPh)(NBu^t_2)_2Cl$	[424c]
$Me_2Si(NBu^t)_2Sn$ (cyclic)	O_2	$Me_2Si(NBu^t)_2Sn(NBu^t)_2SiMe_2$ (spiro)	[721b]
$Me_2Si(NBu^t)_2Sn$ (cyclic)	¼S_8	½$Me_2Si(NBu^t)_2Sn(\mu\text{-}S)_2Sn(NBu^t)_2SiMe_2$	[721b]
$Pb[N(SiMe_3)_2]_2$	*trans*-$[\{PtCl_2(PEt_3)\}_2]$	*trans*-$[(Pt(\mu\text{-}Cl)(PEt_3)\text{-}\{Pb(Cl)[N(SiMe_3)_2]_2\}_2]$	[424b, 594a]
$As(NMe_2)_3$	Br_2	$[As(Br)(NMe_2)_3]Br$	[723]
$As(NMe_2)_3$	$ClNMe_2$	$[As(NMe_2)_4]Cl$	[723]
$\frac{1}{n}[Cr(NEt_2)_3]_n$	O_2	a Cr^{IV} product	[24]

Table 2 Reductive Elimination Reactions of Metal Amides

Substrate	Reagent	Product	References
$BCl(NMe_2)_2$	K	$B_2(NMe_2)_4$, $(Me_2N)_2BB(NMe_2)B(NMe_2)_2$	[303a]
$SnBr[N(SiMe_3)_2]_3$	Na/Hg	$N(SiMe_3)_3$ + Sn + others	[594a]
$AsCl[N(SiMe_3)_2]_2$	(tetra-N-ethyl bi(imidazolidinylidene)-type olefin: two rings each with N(Et), N(Et), joined by C=C), $h\nu$	$\dot{A}s[N(SiMe_3)_2]_2$	[285a, 285b]
$As(NBu^t)Cl_3$	heat	$ClAs$—NBu^t / Bu^tN—$AsCl$ (four-membered ring)	[588b]
$[V(NPh_2)_4]$	$K[C_{10}H_8]$	$K[V(NPh_2)_4]$	[252a]
$[Nb(NMe_2)_5]$	CS_2	$[Nb(S_2CNMe_2)_4]$	[93]
$[Nb(NMe_2)_5]$	C_5H_6	$[Nb(\eta\text{-}C_5H_5)(NMe_2)_3]$	[347]
$[Cr(NEt_2)_4]$	C_5H_6	$[Cr(\eta\text{-}C_5H_5)_2]$	[56a]
$[Cr(NEt_2)_4]$	CS_2	$[Cr(S_2CNEt_2)_3]$	[56a]
$[Cr(NEt_2)_4]$	RR′CHOH	$[Cr(OCHRR')_3]$	[56a]
$[Cr(NEt_2)_4]$	$4CO_2$	$[Cr_2(O_2CNEt_2)_4(\mu\text{-}NEt_2)_2]$ + $[Cr_2(O_2CNEt_2)_4(HNEt_2)_2]$	[151a]
$[Mo_2Et_2(NMe_2)_4]$	$4CO_2$	$[Mo(O_2CNMe_2)_4]$ + C_2H_4 + C_2H_6	[155e]
$[W(NMe_2)_6]$	MeSH	$\frac{1}{n}[W(SMe)_3]_n$	[90a]
$[W(NMeR)_6]$ (R = Me, Et)	CS_2	$[W(S_2CNMeR)_4]$	[90a, 155]
$\frac{1}{n}[Fe(NMe_2)_3]_n$	C_5H_6	$[Fe(\eta\text{-}C_5H_5)_2]$	[352, 354]
$\frac{1}{n}[M(NR_2)_n]$ (M = B, Si, Sn^{IV}, Ti^{IV}, Zr^{IV}, Nb^{V}, Ta^{V})	300–350°	metal (B, Si, or Sn) or metal nitride	[706a]

Table 3 Disproportionation Reactions of Metal Amides

Substrate	Reaction conditions	Products	References
$Ge(NR_2)_2$ [R = $SiMe_3$, $GeMe_3$; R_2 = $Bu^t(SiMe_3)$]	$h\nu$, n–C_6H_{14}	$\dot{G}e(NR_2)_3$ (+ a Ge^I compound?)	[286b, 322a, 428, 770a]
$Sn(NR_2)_2$ [R = $SiMe_3$, $GeMe_3$, $GeEt_3$; R_2 = $Bu^t(SiMe_3)$]	$h\nu$, n–C_6H_{14}	$\dot{S}n(NR_2)_3$ (+ a Sn^I compound?)	[170a, 286b, 322a, 428]
$[\{Ti(NEt_2)_3\}_2]$	CS_2	$[Ti(S_2CNEt_2)_4]$ + $\frac{1}{n}[Ti(S_2CNEt_2)_2]_n$	[25, 432]
$[\{Ti(NMe_2)_2(NR_2)\}_2]$ (R = Me, Et, Pr^i)	distillation	$[Ti(NMe_2)_4]$ + $\frac{1}{n}[Ti(NR_2)_2]_n$	[431]
$\frac{1}{n}[V(NR_2)_3]_n$	distillation	$[V(NR_2)_4]$ + V^{II}?	[25]
$\frac{1}{n}[Cr(NR_2)_3]_n$ (R = Et, Pr, but not Me)	distillation	$[Cr(NEt_2)_4]$ + $\frac{1}{n}[Cr(NR_2)_2]_n$	[56a]

Table 4 M–N Homolysis Reactions of Metal Amides

Substrate	Conditions	Products	References
$B_2(NMe_2)_4$	hν, CCl_4	$B_2(NMe_2)_3Cl + CHCl_3$	[286d]
$B_2(NMe_2)_4$	hν, $CHCl_3$	$B_2(NMe_2)_3Cl + HB(NMe_2)_2 + B(NMe_2)_3 + CH_2Cl_2$	[286d]
$B_2(NMe_2)_4$	hν, CH_2Cl_2	$B(NMe_2)_3 + ClB(NMe_2)_2$	[286d]
Me_3SiN(ring)(Me)(H)	ArNO, hν	$Ar(Me_3SiO)\dot{N}O + 2C_5H_4N$-Me-4	[614a]
$Ge(NR_2)_2$	hν, n-C_6H_{14}	$\dot{N}R_2$	
(R = $SiEt_3$, $GeEt_3$; NR_2 = N(ring, Me_2, Me_2))			[429a, 594a]
(R = Bu^t)			[433a]
$Sn(NR_2)_2$	hν, n-C_6H_{14}	$\dot{N}R_2$	
(R = $SiEt_3$; NR_2 = N(ring, Me_2, Me_2))			[429a, 594a]
(R = Bu^t)			[433a]
$SnMe_3(NEt_2)$	hν	$(Me_3Sn)_2 + HNEt_2 + MeCH{=}NEt$	[444a]
$Pb[N(MMe_3)SiMe_3]_2$ (M = C, Si)	hν, n-C_6H_{14}	Pb	[286b, 322a]
$Hg[N(CF_3)_2]_2$	hν,	$(CF_3)_2N{=}N(CF_3)_2$, $CF_3N{=}CF_2$, $(CF_3)_2NF$, $(CF_3)_2N{-}N(CF_3)CF_2N(CF_3)_2$, $CF_2{-}CF[N(CF_3)_2]$ / $CF_2{-}CF[N(CF_3)_2]$ (cyclic)	[298a]
$Hg[N(CF_3)_2]_2$	170° C	$(CF_3)_2NC(F){=}NCF_3$	[772a]

Table 5 Selected Elimination Reactions of Metal Amides

Substrate	Conditions	Products	References
$LiN(SiCl_3)_2$	heat	$\frac{1}{2}[(Cl_3Si)NSiCl_2]_2$ + LiCl	[734a]
$B(NHEt)_3$	heat	$\frac{1}{3}[\{Et(H)N\}BNEt]_3$ + polyborazene	[417a]
$B(NHBu^t)_3$	heat	$\frac{1}{2}[\{Bu^t(H)N\}BNBu^t]_2$ (low yield)	[423b]
$Bu^tN[B(NHBu^t)_2]_2$	heat	$[\{Bu^t(H)N\}BNBu^t]_2$	[423b]
$BCl[N(R)SiMe_3]Ph$	heat	$\frac{1}{3}(PhBNR)_3 + Me_3SiCl$	[507a]
$(Et_2AlNPh_2)_2$	heat	$\frac{2}{3}Et_2AlN(Ph)$–C_6H_4–EtAlN(Ph)–C_6H_4–$AlNPh_2$ + $2C_2H_6$	[314a]
$Me_2Si(NBu^t)_2Sn(\mu\text{-}S)_2Sn(NBu^t)_2SiMe_2$	heat	$Me_2Si(NBu^t)_2Sn(NBu^t)_2SiMe_2 + SnS_2$	[721b]
$ClP[N(SiMe_3)_2]_2$	heat	$Me_3SiN{=}PN(SiMe_3)_2 + Me_3SiCl$	[520b]
$ClP(CH_2SiMe_3)[N(SiMe_3)_2]$	heat	$Me_3SiN{=}PCH_2SiMe_3 + Me_3SiCl$	[594a]
$As(F)[N(SiMe_3)_2]_2$	heat	$\frac{1}{2}[(Me_3Si)NAs\{N(SiMe_3)_2\}_2]_2$	[518b]
$[TiCl_3\{N(SiMe_3)_2\}]$*		$\frac{1}{n}[TiCl_2(NSiMe_3)]_n + Me_3SiCl$	[20a, 21]
$[Ti(\eta\text{-}C_5H_5)_2Cl\{N(SiMe_3)_2\}]$*		$[Ti(\eta\text{-}C_5H_5)_2N(SiMe_3)SiMe_2CH_2]$ (ring) + Me_3SiCl	[66a]
$[Rh(NMe_2)(PPh_3)_{3\text{ or }4}]$†		$[Rh(H)(PPh_3)_{3\text{ or }4}] + MeN{=}CH_2$	[207a]
$[Rh\{N(SiMe_3)_2\}(PPh_3)_2]$	heat	$[Rh(\overline{Ph_2P C_6H_4})(PPh_3)_2] + (Me_3Si)_2NH$	[141d]
$MeCdN(SiCl_3)_2$	20° C	$\frac{1}{2}[(Cl_3Si)NSiCl_2]_2$ + MeCdCl	[734a]

*These complexes are presumed to be transiently formed from $TiCl_4/N(SiMe_3)_3$ or $[Ti(\eta\text{-}C_5H_5)_2Cl_2]/LiN(SiMe_3)_2$, respectively.

†This compound is presumed to be transiently formed from $[Rh(Cl)(PPh_3)_3]/LiNMe_2$.

I REFERENCES FOR CHAPTERS 10–18

[1] Abedini, M., and MacDiarmid, A. G., *Inorg. Chem.*, 1963, **2**, 608.
[2] Abel, E. W., *J. Chem. Soc.*, 1960, 4406.
[3] Abel, E. W., *J. Chem. Soc.*, 1961, 4933.
[4] Abel, E. W., and Armitage, D. A., *J. Chem. Soc.*, 1964, 3122.
[5] Abel, E. W., and Armitage, D. A., *J. Organometallic Chem.*, 1966, **5**, 326.
[6] Abel, E. W., Armitage, D. A., and Brady, D. B., *Trans. Faraday Soc.*, 1966, **62**, 3459.
[7] Abel, E. W., Armitage, D. A., and Brady, D. B., *J. Organometallic Chem.*, 1966, **5**, 130.
[8] Abel, E. W., Armitage, D. A., and Bush, R. P., *J. Chem. Soc.*, 1964, 5584.
[9] Abel, E. W., Armitage, D. A., Bush, R. P., and Willey, G. R., *J. Chem. Soc.*, 1965, 62.
[10] Abel, E. W., Armitage, D. A., and Willey, G. R., *Trans. Faraday Soc.*, 1964, **60**, 1257.
[11] Abel, E. W., Armitage, D. A., and Willey, G. R., *J. Chem. Soc.*, 1965, 57.
[12] Abel, E. W., Brady, D. B., and Lerwill, B. R., *Chem. Ind.*, 1962, 1333.
[13] Abel, E. W., and Bush, R. P., *J. Organometallic Chem.*, 1965, **3**, 245.
[14] Abel, E. W., and Crow, J. P., *J. Chem. Soc. (A)*, 1968, 1361.
[15] Abel, E. W., and Dunster, M. O., *J. Organometallic Chem.*, 1973, **49**, 435.
[15a] Abel, E. W., Keppie, S. A., Lappert, M. F., and Moorhouse, S., *J. Organometallic Chem.*, 1970, **22**, C31.
[16] Abel, E. W., and Singh, A., *J. Chem. Soc.*, 1959, 690.
[17] Abel, E. W., and Willey, G. R., *J. Chem. Soc.*, 1964, 1528.
[18] Abel, E. W., and Willey, G. R., *Proc. Chem. Soc.*, 1962, 308.
[18a] Abeler, G., Nöth, H., and Schick, H., *Chem. Ber.*, 1968, **101**, 3981.
[19] Abeler, G., Bayrhuber, H., and Nöth, H., *Chem. Ber.*, 1969, **102**, 2249.
[20] Adler, O., and Kober, F., *J. Organometallic Chem.*, 1974, **72**, 351; *Chem. Z.*, 1976, **100**, 235.
[20a] Akiyama, M., Chisholm, M. H., Cotton, F. A., Extine, M. W., and Murillo, C. A., *Inorg. Chem.*, 1977, **16**, 2407.
[20b] Alcock, N. W., Pierce-Butler, M., and Willey, G. R., *J.C.S. Chem. Comm.*, 1974, 627.
[21] Alcock, N. W., Pierce-Butler, M., and Willey, G. R., *J.C.S. Dalton*, 1976, 707.
[22] Allmann, R., Flatau, K., and Musso, H., *Chem. Ber.*, 1972, **105**, 3067.
[23] Alsobrook, A. L., Collins, A. L., and Wells, R. L., *Inorg. Chem.*, 1965, **4**, 253.
[24] Alyea, E. C., Basi, J. S., Bradley, D. C., and Chisholm, M. H., *Chem. Comm.*, 1968, 495.
[25] Alyea, E. C., Bradley, D. C., Lappert, M. F., and Sanger, A. R., *Chem. Comm.*, 1969, 1064.
[26] Amberger, E., and Kula, M. R., *Chem. Ber.*, 1963, **96**, 2562.

[27] Amberger, E., Kula, M. R., and Lorberth, J., *Angew. Chem. Internat. Edn.*, 1964, **3**, 138.
[28] Amero, B. A., and Schram, E. P., *Inorg. Chem.*, 1976, **15**, 2842.
[29] Anderson, H. H., *J. Amer. Chem. Soc.*, 1951, **73**, 5802.
[30] Anderson, H. H., *J. Amer. Chem. Soc.*, 1952, **74**, 1421.
[31] Anderson, H. H., *J. Amer. Chem. Soc.*, 1953, **75**, 1576.
[32] Anderson, J. W., Barker, G. K., Drake, J. E., and Rodger, M., *J.C.S. Dalton,* 1973, 1716.
[33] Andersen, R. A., and Coates, G. E., *J.C.S. Dalton,* 1974, 1729.
[34] Anderson, R. H., and Cragg, R. H., *Chem. Comm.*, 1970, 425.
[34a] Anderson, R. H., and Cragg, R. H., Unpublished Observations (1970).
[35] Anderson, J. W., and Drake, J. E., *J. Inorg. Nucl. Chem.*, 1973, **35**, 1032.
[36] Ando, F., Mayashi, T., Ohashi, K., and Kobetsu, J., *J. Inorg. Nucl. Chem.*, 1975, **37**, 2011.
[37] Andrianov, K. A., Astakhin, V. V., and Kochkin, D. A., *Bull. Acad. Sci. U.S.S.R.*, 1962, **10**, 1757.
[38] Andrianov, K. A., Dzhashiashvili, T. K., and Astakhin, V. V., *Zh. Obshch. Khim.*, 1966, **36**, 2012.
[39] Andrianov, K. A., Dzhashiashvili, T. K., Astakhin, V. V., and Shumakova, G. N., *Zh. Obshch. Khim.*, 1967, **37**, 928; *Izv. Akad. Nauk S.S.S.R., Ser. Khim.*, 1966, 2229.
[40] Andrianov, K. A., Emel'yanov, V. N., and Rudman, E. V., *Dokl. Akad. Nauk S.S.S.R.*, 1973, **212**, 872; *Chem. Abstr.*, 1974, **80**, 48436.
[40a] Andrianov, K. A., Il'in, M. M., Talanov, V. N., and Khananashvili, L. M., U.S.S.R. Patent, 1971, 304,275; *Chem. Abstr.*, 1972, **76**, 4324.
[41] Andrianov, K. A., Ismailov, B. A., Kononov, A. M., and Kotrelev, G. V., *J. Organometallic Chem.*, 1965, **3**, 129.
[42] Andrianov, K. A., Klement'ev, I. Yu., Kartsev, G. N., and Tikhnov, V. S., *Zh. Obshch. Khim.*, 1972, **42**, 2342.
[43] Ang, H. G., *J. Chem. Soc. (A),* 1968, 2734.
[43a] Appel, R., and Halstenberg, M., *J. Organometallic Chem.*, 1975, **99**, C25.
[44] Appel, R., and Montenarh, M., *Chem. Ber.*, 1975, **108**, 1442.
[45] Appel, R., and Montenarh, M., *Chem. Ber.*, 1975, **108**, 2340.
[46] Armitage, D. A., Clark, M. J., and Kinsey, A. C., *J. Chem. Soc. (C).*, 1971, 3867.
[47] Armitage, D. A., Clark, M. J., Sinden, A. W., Wingfield, J. N., Abel, E. W., and Lonis, E. J., *Inorg. Synth.*, 1974, **15**, 207.
[47a] Arnold, D. E. J., Ebsworth, E. A. V., and Rankin, D. W. H., *J.C.S. Dalton,* 1976, 823.
[48] Arnold, D. E. J., and Rankin, D. W. H., *J.C.S. Dalton,* 1975, 889.
[48a] Arnold, D. E. J., Rankin, D. W. H., and Robinet, G., *J.C.S. Dalton,* 1977, 585.

[48b] Ashby, E. C., and Lin, J. J., *Tetrahedron Letters,* 1976, 3865.
[48c] Atkins, T. J., and Richman, J. E., *Tetrahedron Letters,* 1978, 5149.
[49] Atwood, J. L., and Stucky, G. D., *J. Amer. Chem. Soc.,* 1969, **91**, 4426.
[50] Aubrey, D. W., and Lappert, M. F., *Proc. Chem. Soc.,* 1960, 148.
[51] Aubrey, D. W., Lappert, M. F., and Majumdar, M. K., *J. Chem. Soc.,* 1962, 4088.
[52] Aylett, B. J., and Peterson, L. K., *J. Chem. Soc.,* 1965, 4043.
[53] Aylett, B. J., Ellis, I. A., and Richmond, J. R., *J.C.S. Dalton,* 1973, 981.
[53a] Bachhuber, H., Fischer, G., and Wiberg, N., *Angew. Chem. Internat. Edn.,* 1972, **11**, 829.
[53b] Bagnall, K. W., and Yanir, E., *J. Inorg. Nucl. Chem.,* 1974, **36**, 777.
[54] Baldwin, J. C., Lappert, M. F., Pedley, J. B., and Poland, J. S., *J. Chem. Soc. (A),* 1972, 1943.
[54a] Balulescu, C. R., and Keller, P. C., *Inorg. Chem.,* 1978, **17**, 3707.
[54b] Baney, R. H., and Shindorf, R. J., *J. Organometallic Chem.,* 1966, **6**, 660.
[54c] Banford, L., and Coates, G. E., *J. Chem. Soc.,* 1964, 3564.
[55] Barashenkov, G. C., and Derkach, N. Ya., *J. Gen. Chem. U.S.S.R.,* 1978, **48**, 1012.
[55a] Barlos, K., Christl, H., and Nöth, H., *Ann.,* 1976, 2272.
[55b] Barlos, K., Nölle, D., and Nöth, H., *Z. Naturforsch.,* 1977, **32b**, 1095.
[55c] Barlos, K., and Nöth, H., *Chem. Ber.,* 1977, **110**, 2790.
[55d] Barlos, K., and Nöth, H., *Chem. Ber.,* 1977, **110**, 3460.
[55e] Barlow, C. G., Nixon, J. F., and Swain, J. R., *J. Chem. Soc. (A),* 1968, 2692; 1969, 1082; Barlow, C. G., Nixon, J. F., and Webster, M., *ibid.,* 1968, 2216; Johnson, T. R., and Nixon, J. F., *ibid.,* 1969, 2518.
[55f] Basha, A., Lipton, M., and Weinreb, S. M., *Tetrahedron Letters,* 1977, 4171.
[56] Basi, J. S., and Bradley, D. C., *Proc. Chem. Soc.,* 1963, 305.
[56a] Basi, J. S., Bradley, D. C., and Chisholm, M. H., *J. Chem. Soc. (A),* 1971, 1433.
[57] Batyeva, E. S., Al'fonsov, V. A., Kaufman, M. Z., and Pudovik, A. N., *Izv. Akad. Nauk Ser. Khim. S.S.S.R.,* 1976, **5**, 1193; *Chem. Abstr.,* 1976, **85**, 143200.
[58] Batyeva, E. S., Al'fonsov, V. A., and Pudovik, A. N., *Izv. Akad. Nauk S.S.S.R., Ser. Khim.,* 1976, **2**, 463; *Chem. Abstr.,* 1976, **84**, 180333.
[59] Baukov, Yu. I., Burlachenko, G. S., Kostyuk, A. S., and Lutsenko, I. F., *Zh. Obshch. Khim.,* 1966, **36**, 1859.
[60] Beachley, O. T., and Durkin, T. R., *Inorg. Chem.,* 1974, **13**, 1768.
[61] Becher, H. J., *Z. Anorg. Chem.,* 1956, **288**, 235.
[62] Becke-Goehring, M., and Krill, H., *Chem. Ber.,* 1961, **94**, 1059.
[63] Becke-Goehring, M., and Wunsch, G., *Ann.,* 1958, **618**, 43.
[64] Becke-Goehring, M., and Wunsch, G., *Chem. Ber.,* 1960, **93**, 326.
[65] Becker, F., *J. Organometallic Chem.,* 1973, **51**, C9.

[66] Bellut, H., Miller, C. D., and Köster, R., *Synth. Inorg. Metal-Org. Chem.*, 1971, **1**, 83.

[66a] Bennett, C. R., and Bradley, D. C., *J.C.S. Chem. Comm.*, 1974, 29.

[67] Benzing, E. P., and Kornicker, W. A., *Chem. Ber.*, 1961, **94**, 2263.

[68] Bernard, D., and Burgada, R., *Tetrahedron Letters*, 1973, 3455.

[69] Berry. A. D., Bergerud, J. R., Highsmith, R. E., MacDiarmid, A. G., and Nasta, M. A., 4th International Conference on Organometallic Chemistry, Bristol, 1969, Abstracts paper A 4.

[70] Beyer, H., Dawson. J. W., Jenne, H., and Niedenzu, K., *J. Chem. Soc.*, 1964, 2115.

[71] Beyer, H., Niedenzu, K., and Dawson, J. W., Proc. South Eastern Reg. Meeting Amer. Chem. Soc., Gatlingburg, 1962.

[71a] Billingham, N. C., personal communication (1977).

[72] Billingham, N. C., Boxall, L. M., and Jenkins, A. D., *Europ. Polym. J.*, 1972, **8**, 1045.

[73] Billingham, N. C., Boxall, L. M., Jenkins, A. D., and Lees, P. D., *Europ. Polym. J.*, 1974, **10**, 981.

[74] Billingham, N. C., Boxall, L. M., Jenkins, A. D., and Lees, P. D., *Europ. Polym. J.*, 1974, **10**, 991.

[75] Billingham, N. C., and Jenkins, A. D., *Appl. Polym. Symposium* No. 26, 1975, 13.

[76] Binder, H., and Fischer, R., *Chem. Ber.*, 1974, **107**, 205.

[77] Birkofer, L., and Dickopp, H., *Angew. Chem.*, 1964, **76**, 648.

[78] Birkofer, L., and Ritter, A., *Chem. Ber.*, 1960, **93**, 424.

[79] Birkofer, L., and Schmidtberg, G., *Chem. Ber.*, 1971, **104**, 3831.

[80] Blaschette, A., Schirawski, G., and Wannagat, U., *Inorg. Nuclear Chem. Letters*, 1969, **5**, 707.

[81] Bloodworth, A. J., and Davies, A. G., *J. Chem. Soc. (C)*, 1966, 299.

[82] Bochkarev, M. N., Maiorova, L. P., Bochkarev, L. N., and Vyazankin, N. S., *Izv. Akad. Nauk S.S.S.R., Ser. Khim.*, 1971, 2353; *Chem. Abstr.*, 1972, **76**, 46262j.

[83] Bochkarev, M. N., Maiorova, L. P., and Vyazankin, N. S., *J. Organometallic Chem.*, 1973, **55**, 89.

[83a] Bock, H., and Fuss, W., *Chem. Ber.*, 1971, **104**, 1687.

[83b] Boden, G., Grosskreutz, W., Kessler, G., and Scheler, H., *Z. Chem.*, 1972, **12**, 299.

[84] Boe, B., *J. Organometallic Chem.*, 1976, **107**, 139.

[85] Boenig, I. A., Conway, W. R., and Niedenzu, K., *Synth. Inorg. Metal-Org. Chem.*, 1975, **5**, 1.

[86] Boersma, J., and Noltes, J. G., *J. Organometallic Chem.*, 1970, **21**, P32.

[87] Boersma, J., and Noltes, J. G., *J. Organometallic Chem.*, 1969, **17**, 1.

[88] Boersma, J., Spek, A. L., and Noltes, J. G., *J. Organometallic Chem.*, 1974, **81**, 7.

[88a] Bonnett, R., Bradley, D. C., Fisher, K. J., and Rendall, I. F., *J. Chem. Soc. (A)*, 1971, 1622.
[89] Boone, J. L., and Willcockson, G.W., Abstr. 142nd Meeting, Amer. Chem. Soc., Atlantic City, 1962, 6N.
[89a] Bradley, D. C., *Adv. Inorg. Chem., Radiochem.*, 1972, **15**, 259.
[90] Bradley, D. C., Charalambous, J., and Jain, S., *Chem. Ind.*, 1965, 1730.
[90a] Bradley, D. C., Chisholm, M. H., Extine, M. W., and Stager, M. E., *Inorg. Chem.*, 1977, **16**, 1794.
[91] Bradley, D. C., and Ganorkar, M. C., *Chem. Ind.*, 1968, 1521.
[92] Bradley, D. C., Ghotra, J. S., Hart, F. A., Hursthouse, M. B., and Raithby, P. R., *J.C.S. Chem. Comm.*, 1974, 40.
[93] Bradley, D. C., and Gitlitz, M. H., *Chem. Comm.*, 1965, 289; *J. Chem. Soc (A)*, 1969, 1152.
[94] Bradley, D. C., and Hammersley, P. A., *J. Chem. Soc. (A)*, 1967, 1894.
[95] Bradley, D. C., and Hillyer, M. J., *Trans. Faraday Soc.*, 1966, **62**, 2374.
[96] Bradley, D. C., Hursthouse, M. B., Newing, C. W., and Welch, A. J., *J.C.S. Chem. Comm.*, 1972, 567.
[96a] Bradley, D. C., Hursthouse, M. B., Smallwood, R. J., and Welch, A. J., *J.C.S. Chem. Comm.*, 1972, 872.
[97] Bradley, D. C., and Kasenally, A. S., *Chem. Comm.*, 1968, 1430.
[98] Bradley, D. C., and Newing, C. W., *Chem. Comm.*, 1970, 219.
[99] Breed, L. W., and Elliott, R. L., *J. Organometallic Chem.*, 1968, **11**, 447; Breed, L. W., Budde, W. L., and Elliott, R. L., *J. Organometallic Chem.*, 1966, **6**, 676.
[100] Breederveld, H., *Rec. Trav. Chim.*, 1960, **79**, 1126.
[101] Breederveld, H., *Rec. Trav. Chim.*, 1962, **81**, 276.
[102] Brinckman, F. E., Cooper, J., and Coyle, T. D., 153rd Meeting Amer. Chem. Soc., Miami, 1967, L112.
[103] Brocas, J. M., De, B. J., and Pommier, J. C., *J. Organometallic Chem.*, 1976, **120**, 217.
[104] Brocas, J. M., and Pommier, J. C., *J. Organometallic Chem.*, 1976, **121**, 45.
[104a] Brook, A. G., and Duff, J. M., *J. Amer. Chem. Soc.*, 1974, **96**, 4692; and results cited in West, R., *Adv. Organometallic Chem.*, 1977, **16**, 1.
[105] Brotherton, R. J., McCloskey, A. L., Boone, J. L., and Manasevit, H. M., *J. Amer. Chem. Soc.*, 1960, **82**, 6245.
[106] Brotherton, R. J., Petterson, L. L., and Boone, J. L., *J. Org. Chem.*, 1961, **26**, 3030.
[107] Brotherton, R. J., McCloskey, A. L., Petterson, L. L., and Steinberg, H., *J. Amer. Chem. Soc.*, 1960, **82**, 6242.
[108] Brunner, H., *Chem. Ber.*, 1969, **102**, 305.
[109] Buchanan, M. J., Cragg, R. H., and Steltner, A., *J. Organometallic Chem.*, 1976, **120**, 189.
[109a] Bullitt, O. H., U.S. Patent 2,608,555; *Chem. Abstr.*, 1953, **47**, 1430.

[110] Bulloch, G., Keat, R., and Tennent, N. H., *J.C.S. Dalton,* 1974, 2329.
[110a] Bulten, E. J., and Budding, H. A., *J. Organometallic Chem.,* 1978, **157**, C3.
[111] Burg, A. B., and Banus, J., *J. Amer. Chem. Soc.,* 1954, **76**, 3903.
[112] Burg, A. B., and Kuljian, E. S., *J. Amer. Chem. Soc.,* 1950, **72**, 3103.
[113] Burg, A. B., and Randolph, C. L., *J. Amer. Chem. Soc.,* 1951, **73**, 953.
[114] Burg, A. B., and Randolph, C. L., *J. Amer. Chem. Soc.,* 1949, **71**, 3451.
[115] Burgada, R., *Ann. Chim.,* 1963, **8**, 547.
[116] Bürger, H., and Dämmgen, U., *J. Organometallic Chem.,* 1975, **101**, 295.
[117] Bürger, H., and Neese, H.-J., *Z. Anorg. Chem.,* 1969, **365**, 243; 1969, **370**, 275.
[118] Bürger, H., Sawodny, W., and Wannagat, U., *J. Organometallic Chem.,* 1965, **3**, 113.
[119] Bürger, H., Smrekar, O., and Wannagat, U., *Monatsh.,* 1964, **95**, 292.
[120] Bürger, H., and Wannagat, U., *Monatsh.,* 1963, **94**, 761.
[121] Bürger, H., and Wannagat, U., *Monatsh.,* 1963, **94**, 1007.
[122] Bürger, H., and Wannagat, U., *Monatsh.,* 1964, **95**, 1099.
[123] Bürger, H., and Wiegel, K., *J. Organometallic Chem.,* 1977, **124**, 279.
[124] Bürger, H., and Wiegel, K., *Z. Anorg. Chem.,* 1973, **398**, 257.
[125] Bush, R. P., Lloyd, N. C., and Pearce, C. A., *Chem. Comm.,* 1967, 1269, 1270.
[126] Buss, W., Krannich, H. J., and Sundermeyer, W., *Z. Naturforsch.,* 1975, **30b**, 842.
[126a] Butvinik, J., and Neilson, R. H., *Inorg. Nucl. Chem. Letters,* 1978, 497.
[127] Byrne, J. E., and Russ, C. R., *J. Organometallic Chem.,* 1970, **22**, 357.
[127a] Cadogan, J. I. G., Stewart, N. J., and Tweddle, N. J., *J.C.S. Chem. Comm.,* 1978, 182.
[127b] Calderazzo, F., Dell'Amico, G., Netti, R., and Pasquali, M., *Inorg. Chem.,* 1978, **17**, 471.
[127c] Calderazzo, F., Dell'Amico, G., Pasquali, M., and Perego, G., *Inorg. Chem.,* 1978, **17**, 478.
[128] Campbell, K. N., and Campbell, B. K., *Proc. Indiana Acad. Sci.,* 1940, **50**, 123.
[128a] Cardin, C. J., Cardin, D. J., and Lappert, M. F., *J.C.S. Dalton,* 1977, 767, and refs. therein.
[129] Cardin, D. J., Keppie, S. A., Kingston, B. M., and Lappert, M. F., *Chem. Comm.,* 1967, 1035.
[130] Cardin, D. J., Keppie, S. A., and Lappert, M. F., *Inorg. Nucl. Chem. Letters,* 1968, **4**, 365.
[131] Cardin, D. J., Keppie, S. A., and Lappert, M. F., *J. Chem. Soc. (A),* 1970, 2594.
[132] Cardin, D. J., and Lappert, M. F., *Chem. Comm.,* 1966, 506.
[133] Cardin, D. J., and Lappert, M. F., *Chem. Comm.,* 1967, 1034.

[134] Cardin, D. J., and Lappert, M. F., Unpublished Observations (1966).
[135] Cason, J., Sumrell, G., and Mitchell, R. S., *J. Org. Chem.*, 1950, **15**, 850.
[136] Cass, R., and Coates, G. E., *J. Chem. Soc.*, 1952, 2347.
[137] Castro, B., and Dormoy, J. R., *Tetrahedron Letters*, 1973, 3243.
[138] Caunt, A. D., Ger. Patent, 1973, 2,234,506; *Chem. Abstr.*, 1973, **78**, 125196.
[139] Caunt, A. D., Ger. Patent, 1972, 2,135,511; *Chem. Abstr.*, 1972, **77**, 6104.
[140] Cavell, R. G., *J. Chem. Soc.*, 1964, 1992; Reetz, T., and Katlafsky, B., *J. Amer. Chem. Soc.*, 1960, **82**, 5036; Clemens, D. F., Sisler, H. H., and Brey, W. S., *Inorg. Chem.*, 1966, **5**, 427; *Idem, J. Amer. Chem. Soc.*, 1966, **5**, 527; Spangenberg, S. F., and Sisler, H. H., *Inorg. Chem.*, 1969, **8**, 1004; Brown, D. H., Crosbie, K. D., Fraser, G.W., and Sharp, D.W. A., *J. Chem. Soc. (A)*, 1969, 551; Paine, R. T., *Inorg. Chem.*, 1977, **16**, 2996.
[141] Cesca, S., Marconi, W., and Santostasi, M. L., *Chim. Ind. (Milan)*, 1969, **51**, 1226; *Chem. Abstr.*, 1970, **72**, 55945d.
[141a] Çetinkaya, B., Gümrükçü, I., and Lappert, M. F., Unpublished Observations (1978).
[141b] Çetinkaya, B., Lappert, M. F., and McMeeking, J., *J.C.S. Dalton*, 1973, 1975.
[141c] Çetinkaya, B., Lappert, M. F., and Sehri, H., Unpublished Observations (1978).
[141d] Çetinkaya, B., Lappert, M. F., and Torroni, S., *J.C.S. Chem. Comm.*, 1979, in the press.
[142] Chambers, O. R., Rycroft, D. S., Sharp, D. W. A., and Winfield, J. M., *Inorg. Nucl. Chem. Letters*, 1976, **12**, 559.
[143] Chan, L. H., and Rochow, E. G., *J. Organometallic Chem.*, 1967, **9**, 231.
[144] Chandra, G., George, T. A., and Lappert, M. F., *J. Chem. Soc. (C)*, 1969, 2565.
[145] Chandra, G., George, T. A., and Lappert, M. F., *Chem. Comm.*, 1967, 116.
[146] Chandra, G., Jenkins, A. D., Lappert, M. F., and Srivastava, R. C., *J. Chem. Soc. (A)*, 1970, 2550.
[147] Chandra, G., and Lappert, M. F., *J. Chem. Soc. (A)*, 1968, 1940.
[148] Chandra, G., and Lappert, M. F., *Inorg. Nucl. Chem. Letters*, 1965, **1**, 83.
[149] Chandra, G., Lappert, M. F., and Lynch, J., Unpublished Observations (1967).
[149a] Chang, M., King, R. B., and Newton, M. G., *J. Amer. Chem. Soc.*, 1978, **100**, 998.
[149b] Chang, M., Newton, M. G., and King, R. B., *Inorg. Chim. Acta*, 1978, **30**, L341.
[149c] Chang, M., Newton, M. G., King, R. B., and Lotz, T. J., *Inorg. Chim. Acta*, 1978, **28**, L153.

[150] Chechetkin, S. A., Blagoveshchenskii, V. S., and Nifant'ev, E. E., *Zh. Vses. Khim. Ova,* 1975, **20,** 596; *Chem. Abstr.,* 1976, **84,** 58825.
[151] Chernyshev, E. A., Bugerenko, E. F., Akat'ev, A. S., and Naumov, A. D., *Zh. Obshch. Khim.,* 1975, **45,** 242; *Chem. Abstr.,* 1975, **82,** 98065.
[151a] Chisholm, M. H., Cotton, F. A., Extine, M. W., and Stults, B. R., *J. Amer. Chem. Soc.,* 1976, **98,** 4477.
[151b] Chisholm, M. H., Cotton, F. A., Extine, M. W., and Stults, B. R., *Inorg. Chem.,* 1977, **16,** 603.
[151c] Chisholm, M. H., Cotton, F. A., and Extine, M. W., *Inorg. Chem.,* 1978, **17,** 2000.
[151d] Chisholm, M. H., Cotton, F. A., Extine, M. W., and Rideout, D. C., *Inorg. Chem.,* 1978, **17,** 3536.
[152] Chisholm, M. H., and Extine, M. W., *J. Amer. Chem. Soc.,* 1974, **96,** 6214.
[153] Chisholm, M. H., and Extine, M. W., *J. Amer. Chem. Soc.,* 1975, **97,** 5625.
[154] Chisholm, M. H., and Extine, M. W., *J. Amer. Chem. Soc.,* 1975, **97,** 1623.
[154a] Chisholm, M. H., and Extine, M. W., *J. Amer. Chem. Soc.,* 1977, **99,** 782.
[154b] Chisholm, M. H., and Extine, M. W., *J. Amer. Chem. Soc.,* 1977, **99,** 792.
[155] Chisholm, M. H., Extine, M. W., Cotton, F. A., and Stults, B. R., *J. Amer. Chem. Soc.,* 1976, **98,** 4683.
[155a] Chisholm, M. H., Extine, M. W., and Reichert, W. W., *Adv. Chem. Ser.,* 1976, **150,** 273.
[155b] Chisholm, M. H., and Reichert, W. W., *J. Amer. Chem. Soc.,* 1974, **96,** 1249.
[155c] Chisholm, M. H., and Reichert, W. W., *Inorg. Chem.,* 1978, **17,** 767.
[155d] Chisholm, M. H., Reichert, W. W., and Thornton, P., *J. Amer. Chem. Soc.,* 1978, **100,** 2744.
[155e] Chisholm, M. H., Haitko, D. A., and Murillo, C. A., *J. Amer. Chem. Soc.,* 1978, **100,** 6262.
[156] Chivers, T., and David, B., *J. Organometallic Chem.,* 1967, **10,** P35.
[157] Chivers, T., and David, B., *J. Organometallic Chem.,* 1968, **13,** 177.
[157a] Choukroun, R., and Gervais, D., *Inorg. Chim. Acta,* 1978, **27,** 163.
[157b] Choukroun, R., and Gervais, D., *Synth. Inorg. Metal-Org. Chem.,* 1978, **8,** 137.
[157c] Clare, P., and Glockling, F., *Inorg. Chim. Acta,* 1976, **17,** 229.
[158] Clare, P., Millington, D., and Sowerby, D. B., *J.C.S. Chem. Comm.,* 1972, 324.
[159] Coates, G. E., and Fishwick, A. H., *J. Chem. Soc. (A),* 1967, 1199.
[160] Coates, G. E., Glockling, F., and Huck, N. D., *J. Chem. Soc.,* 1952, 4512.
[161] Coates, G. E., and Ridley, D., *J. Chem. Soc.,* 1965, 1870.
[162] Coates, G. E., and Ridley, D., *J. Chem. Soc. (A),* 1966, 1064.

[163] Coles, B. F., and Walton, D. R. M., *Synthesis,* 1975, 390.
[164] Collier, M. R., Lappert, M. F., and McMeeking, J., *Inorg. Nucl. Chem. Letters,* 1971, **7,** 689.
[165] Collins, A. L., and Wells, R. L., *Inorg. Nucl. Chem. Letters,* 1966, **2,** 201.
[166] Contreras, R., Wolf, R., and Sanchez, M., *Synth. Inorg. Metal-Org. Chem.,* 1973, **3,** 37.
[167] Cook, W. L., and Niedenzu, K., *Synth. Inorg. Metal-Org. Chem.,* 1974, **4,** 53.
[168] Corbellini, M., and Balducci, A., Ger. Patent. 1976, 2,529,318; *Chem. Abstr.,* 1976, **84,** 106360.
[169] Corey, E. J., and Chaykovsky, M., *J. Amer. Chem. Soc.,* 1965, **87,** 1345.
[170] Costes, J. P., Cros, G., and Laurent, J. P., *Compt. Rend.,* 1975, **280C,** 665; *J. Inorg. Nucl. Chem.,* 1978, **40,** 829.
[170a] Cotton, J. D., Cundy, C. S., Harris, D. H., Hudson, A., Lappert, M. F., and Lednor, P. W., *J.C.S. Chem. Comm.,* 1974, 651.
[171] Cowley, A. H., and Dierdrof, D. S., *J. Amer. Chem. Soc.,* 1969, **91,** 6609.
[172] Cowley, A. H., Fairbrother, F., and Scott, N., *J. Chem. Soc.,* 1959, 717; Voronkov, M. G., Dolgov, B. N., and Dmitrieva, N. A., *Dokl. Akad. Nauk S.S.S.R.,* 1952, **84,** 959; *Chem. Abstr.,* 1953, **47,** 3228; and Orlov, N. F., *Dokl. Akad. Nauk S.S.S.R.,* 1957, **114,** 1033; *Chem. Abstr.,* 1958, **52,** 2742.
[173] Cradock, S., Ebsworth, E. A. V., and Jessep, H. F., *J.C.S. Dalton,* 1972, 359.
[174] Cragg, R. H., *J. Inorg. Nucl. Chem.,* 1968, **30,** 395.
[175] Cragg, R. H., and Husband, J. P. N., Unpublished Observations (1970).
[176] Cragg, R. H., and Lappert, M. F., *Adv. Chem. Ser.,* 1964, **42,** 220.
[177] Cragg, R. H., and Lappert, M. F., *J. Chem. Soc. (A),* 1966, 82.
[178] Cragg, R. H., and Lappert, M. F., Unpublished Observations (1966).
[179] Cragg, R. H., Lappert, M. F., Nöth, H., Schweizer, P., and Tilley, B. P., *Chem. Ber.,* 1967, **100,** 2377.
[180] Cragg, R. H., Lappert, M. F., and Tilley, B. P., *J. Chem. Soc.,* 1964, 2108.
[181] Cragg, R. H., Lappert, M. F., and Tilley, B. P., *J. Chem. Soc. (A),* 1967, 947.
[181a] Cragg, R. H., and Miller, T. J., *J. Organometallic Chem.,* 1978, **154,** C3.
[181b] Cragg, R. H., and Nazery, M., *J. Organometallic Chem.,* 1976, **120,** 35.
[182] Cragg, R. H., and Weston, A. F., *J.C.S. Dalton,* 1975, 93.
[183] Creemers, H. M. J. C., and Noltes, J. G., *Rec. Trav. Chim.,* 1965, **84,** 382.
[184] Creemers, H. M. J. C., and Noltes, J. G., *Rec. Trav. Chim.,* 1965, **84,** 590.
[185] Creemers, H. M. J. C., and Noltes, J. G., *J. Organometallic Chem.,* 1967, **7,** 237.
[186] Creemers, H. M. J. C., Noltes, J. G., and van der Kerk, G. J. M., *Rec. Trav. Chim.,* 1964, **83,** 1284.

[187] Creemers, H. M. J. C., Noltes, J. G., and Verbeek, F., *J. Organometallic Chem.*, 1968, **14**, 125.

[188] Creemers, H. M. J. C., Verbeek, F., and Noltes, J. G., *J. Organometallic Chem.*, 1967, **8**, 469.

[189] Cucinella, S., Giuliani, G., Bruzzone, M., and Mazzei, A., Ger. Patent, 1974, 2,348,489; *Chem. Abstr.*, 1974, **81**, 64326.

[190] Cucinella, S., and Mazzei, A., Ger. Patent, 1974, 2,529,299; *Chem. Abstr.*, 1976, **84**, 136316.

[191] Cullen, W. R., and Emeléus, H. J., *J. Chem. Soc.*, 1959, 372.

[192] Cuvigny, T., and Normant, H., *Compt. Rend.*, 1969, **269C**, 1398.

[193] Czieslik, G., Flaskerud, G., Hoefer, R., and Glemser, O., *Chem. Ber.*, 1973, **106**, 399.

[194] Dahl, L. F., Costello, W. R., and King, R. B., *J. Amer. Chem. Soc.*, 1968, **90**, 5422.

[194a] Dahlmann, J., and Winsel, K., East Ger. Patent, 1971, 83,134; *Chem. Abstr.*, 1973, **78**, 43710j.

[195] Dalton, R. F., and Jones, K., *J. Chem. Soc.(A)*, 1970, 590.

[196] Davidson, N., and Brown, H. C., *J. Amer. Chem. Soc.*, 1942, **64**, 316.

[197] Davidson, P. J., Harris, D. H., and Lappert, M. F., *J.C.S. Dalton*, 1976, 2268.

[198] Davidson, P. J., Lappert, M. F., and Pearce, R., *Accounts Chem. Res.*, 1974, **7**, 209; *Chem. Rev.*, 1976, **76**, 219.

[199] Davies, A. G., *N.Y. Acad. Sci.*, 1964, **26**, 923.

[200] Davies, A. G., and Harrison, P. G., *J. Chem. Soc. (C)*, 1967, 1313.

[201] Davies, A. R., Dronsfield, A. T., Haszeldine, R. N., and Taylor, D. R., *J.C.S., Perkin I*, 1973, 379.

[202] Davies, A. G., and Mitchell, T. N., *J. Organometallic Chem.*, 1966, **6**, 568.

[203] Dawans, F., and Smets, G., *Makromol. Chem.*, 1963, **59**, 163.

[204] Deckelmann, K., and Werner, H., *Helv. Chim. Acta*, 1971, **54**, 2189.

[204a] Dehnicke, K., Lorberth, J., Thiel, W., and Weller, F., *Z. Anorg. Chem.*, 1971, **381**, 57.

[205] Demitras, G. C., Kent, R. A., and MacDiarmid, A. G., *Chem. Ind.*, 1964, 1712; *Inorg. Chem.*, 1967, **6**, 1903.

[205a] Dergunov, Yu. I., Gerega, V. F., and D'Yachkovskaya, O. S., *Russ. Chem. Eng.*, 1977, **46**, 1132.

[206] Dergunov, Yu. I., Gordetsov, A. S., Vostokov, I. A., and Boitsov, E. N., *Zh. Obshch. Khim.*, 1976, **46**, 1653.

[207] Dergunov, Yu. I., Vostokov, I. A., Gordetsov, A. S., and Gal'perin, V. A., *Zh. Obshch. Chim.*, 1976, **46**, 1573.

[207a] Diamond, S. E., and Mares, F., *J. Organometallic Chem.*, 1977, **142**, C55.

[208] Dietz, E. A., and Martin, D. R., *Inorg. Chem.*, 1972, **12**, 241.

[209] Dimroth, K., Hettche, A., Kanter, H., and Staede, W., *Tetrahedron Letters*, 1972, 835.

[210] Dobbie, R. C., and Emeléus, H. J., *J. Chem. Soc. (A)*, 1966, 367.
[211] Dobinson, G. C., Mason, R., Robertson, G. B., Ugo, R., Conti, F., Morelli, D., Cenini, S., and Bonati, F., *Chem. Comm.*, 1967, 739.
[212] Dorokhov, V. A., Boldyreva, O. G., and Mikhailov, B. M., *Zh. Obshch. Khim.*, 1973, **43**, 1955.
[213] Dorokhov, V. A., and Mikhailov, B. M., *Dokl. Akad. Nauk, S.S.S.R.*, 1969, **187**, 1300.
[214] Dorokhov, V. A., and Mikhailov, B. M., *Izv. Akad. Nauk, S.S.S.R., Ser. Khim.*, 1972, 1895; *Chem. Abstr.*, 1972, **77**, 164794.
[215] Douglas, W. M., and Ruff, J. K., *J. Chem. Soc. (A)*, 1971, 3558.
[216] Douglas, W. M., and Ruff, J. K., *Synth. Inorg. Metal-Org. Chem.*, 1972, **2**, 151.
[217] Dousse, G., Satgé, J., and Rivière-Baudet, M., *Synth. Inorg. Metal-Org. Chem.*, 1975, **3**, 11.
[218] Druce, P. M., Kingston, B. M., Lappert, M. F., Spalding, T. R., and Srivastava, R. C., *J. Chem. Soc. (A)*, 1969, 2106.
[219] Druce, P. M., Lappert, M. F., and Riley, P. N. K., *Chem. Comm.*, 1967, 486.
[220] Duffaut, N., and Dupin, J.-P., *Bull. Soc. Chim. France*, 1966, 3205.
[220a] Dupreez, J. G. H., Rohwer, H. E., Dewet, J. F., and Caira, M. R., *Inorg. Chim. Acta*, 1978, **28**, L59.
[221] Dustmukhamedov, T. T., Yusupov, M. M., Rozhkova, N. K., and Tulyaganov, S. R., *Zh. Obshch. Khim.*, 1976, **46**, 300; *Chem. Abstr.*, 1976, **84**, 164947.
[221a] Eaborn, C., Pidcock, A., and Steele, B. R., *J.C.S. Dalton*, 1976, 767.
[222] Eaborn, C., Thompson, A. R., and Walton, D. R. M., *Chem. Comm.*, 1968, 1051.
[223] Ebsworth, E. A. V., and Emeléus, H. J., *J. Chem. Soc.*, 1958, 2150.
[224] Ebsworth, E. A. V., Rocktäschel, G., and Thompson, J. C., *J. Chem. Soc. (A)*, 1967, 362.
[225] Efimova, V. D., Kharrasova, F. M., Kel'bedina, Z. A., and Abolonina, I. B., *Zh. Obshch. Khim.*, 1976, **46**, 2208; *Chem. Abstr.*, 1977, **86**, 16748.
[226] Eliseenkov, V. N., Samatova, N. A., Anoshina, N. P., and Pudovik, A. N., *Zh. Obshch. Khim.*, 1976, **46**, 23.
[227] Elter, G., Glemser, O., and Herzog. W., *Inorg. Nucl. Chem. Letters*, 1972, **8**, 191.
[228] Elter, G., Glemser, O., and Herzog, W., *J. Organometallic Chem.*, 1972, **36**, 257.
[229] Elter, G., Glemser, O., and Herzog, W., *Cher. Ber.*, 1972, **105**, 115.
[230] Emeléus, H. J., and Miller, N., *Nature*, 1938, **142**, 996; *J. Chem. Soc.*, 1939, 819.

[230a] Emeléus, H. J., and Tattershall, B. W., *J. Inorg. Nucl. Chem.*, 1966, **28**, 1823.

[231] Emeléus, H. J., and Tattershall, B. W., *J. Chem. Soc.*, 1964, 5892.

[231a] Emsley, J., and Hall, D., *The Chemistry of Phosphorus,* Harper and Row, 1976.

[232] Eubanks, I. D., and Lagowski, J. J., *J. Amer. Chem. Soc.*, 1966, **88**, 2425.

[232a] Fedotov, N. S., Kozyukov, V. P., Golen, G. E., and Mironov, V. F., *Zh. Obshch. Khim.*, 1972, **42**, 358.

[233] Feser, M. F., Glemser, O., Von Halasz, S. P., and Saran, H., *Inorg. Nucl. Chem. Letters,* 1972, **8**, 321.

[234] Fessenden, R., and Fessenden, J. S., *Chem. Rev.*, 1961, **61**, 361.

[234a] Fieser, L. F., and Fieser, M., *Reagents for Organic Synthesis,* Vols. 1–6, Wiley, 1967–1976.

[235] Fild, M., and Stankiewiez, T., *Z. Anorg. Chem.*, 1974, **406**, 115.

[236] Findeiss, W., Davidsohn, W., and Henry, M. C., *J. Organometallic Chem.*, 1967, **9**, 435.

[237] Fink, W., *Helv. Chim. Acta,* 1972, **55**, 1901.

[238] Fink, W., *Chem. Ber.*, 1964, **97**, 1424.

[239] Fink, W., *Chem. Ber.*, 1964, **97**, 1433.

[240] Fisher, K. J., *Inorg. Nucl. Chem. Letters,* 1973, **9**, 921.

[241] Fischer, E. O., and Kollmeier, H. J., *Angew. Chem. Internat. Edn.*, 1970, **9**, 309.

[242] Fischer, E. O., Winkler, E., Kreiter, C. G., Huttner, G., and Krieg, B., *Angew. Chem. Internat. Edn.*, 1971, **10**, 922.

[242a] Fischer, H., and Rewicki, D., *Progr. Org. Chem.*, 1968, **7**, 116.

[243] Fishwick, M., Smith, C. A., and Wallbridge, M. G. H., *J. Organometallic Chem.*, 1970, **21**, P9.

[243a] Fleming, S., Lupton, M. K., and Jekot, K., *Inorg. Chem.*, 1972, **11**, 2534.

[244] Fleming, S., and Parry, R. W., *Inorg. Chem.*, 1972, **11**, 1.

[245] Fluck, E., and Jakobson, G., *Z. Anorg. Chem.*, 1969, **369**, 178.

[246] Fodor, L. M., U.S. Patent, 1973, 3,734,899; *Chem. Abstr.*, 1973, **79**, 43080.

[246a] Foley, P., and Zeldin, M., *Inorg. Chem.*, 1975, **14**, 2264.

[247] Foss, V. L., Semenenko, N. M., Sorokin, N. M., and Lutsenko, I. F., *Zh. Obshch. Khim.*, 1973, **43**, 1264.

[248] Foss, V. L., Semenenko, N. M., Sorokin, N. M., and Lutsenko, I. F., *J. Organometallic Chem.*, 1974, **78**, 107, 115.

[249] Frainnet, E., and Duboudin, F., *Compt. Rend.*, 1966, **262C**, 1663.

[250] Frainnet, E., and Duboudin, F., Internat. Symp. Organosilicon Chem., Sci. Commun., Prague, 1965, 342.

[251] Frainnet, E., Duboudin, F., Jarry, C., and Dabescate, F., *Compt. Rend.*, 1970, **270C**, 240.

[252] Fritz, P., Niedenzu, K., and Dawson, J. W., *Inorg. Chem.*, 1964, **3**, 778.

[252a] Frölich, H.-O., and Märkisch, U., *Z. Chem.*, 1975, **15**, 276.
[253] Fukui, M., Ishii, Y., and Itoh, K., *Tetrahedron Letters*, 1968, 3867.
[254] Fussstetter, H., Nöth, H., and Winterstein, W., *Chem. Ber.*, 1977, **110**, 1931.
[255] George, T. A., Jones, K., and Lappert, M. F., *J. Chem. Soc.*, 1965, 2157.
[256] George, T. A., and Lappert, M. F., *Chem. Comm.*, 1966, 463, *J. Chem. Soc. (A)*, 1969, 992.
[257] George, T. A., and Lappert, M. F., *J. Organometallic Chem.*, 1968, **14**, 327.
[258] Gerrard, W., Lappert, M. F., and Pearce, C. A., *J. Chem. Soc.*, 1957, 381.
[259] Gerrard, W., Lappert, M. F., and Pearce, C. A., *Chem. Ind.*, 1958, 292.
[260] Gerwarth, U. W., and Müller, K.-D., *J. Organometallic Chem.*, 1975, **96**, C33; 1976, **110**, 15.
[261] Ghose, B. N., and Walton, D. R. M., *Synthesis,* 1974, 890.
[262] Giannini, U., Longi, P., Deluca, D., and Pivotto, B., Ital. Patent, 1970, 867,243; *Chem. Abstr.*, 1973, **79**, 19427.
[263] Giannini, U., Longi, P., Deluca, D., and Pricca, A., Ger. Patent, 1972, 2,137,872; *Chem. Abstr.*, 1972, **76**, 141538.
[263a] Gibbins, S. G., Lappert, M. F., Pedley, J. B., and Sharp, G. J., *J.C.S. Dalton,* 1975, 72.
[264] Gibson, J. A., Marat, R. K., and Janzen, A. F., *Can. J. Chem.*, 1975, **53**, 3044.
[265] Gilman, H., and Haiduc, I., 3rd Internat. Symp. Organometallic Chem. Abstr., 1967, 54.
[266] Giurgiu, D., Roman, L., Popescu, I., and Ciobanu, A., *Rev. Roum. Chim.*, 1971, **16**, 1217; *Chem. Abstr.*, 1972, **76**, 25350.
[267] Glemser, O., and Wegener, J., *Angew. Chem. Internat. Edn.*, 1970, **9**, 309.
[268] Glemser, O., Wegener, J., and Hoefer, R., *Chem. Ber.*, 1972, **105**, 474.
[269] Glidewell, C., and Rankin, D. W. H., *J. Chem. Soc. (A)*, 1970, 279.
[269a] Glockling, F., and Mahale, V. B., *Inorg. Chim. Acta,* 1977, **25**, L117.
[269b] Glockling, F., and Sweeney, J. J., *J. Chem. Res.*, 1977, (S)35, (M)615.
[269c] Gloe, K., Kessler, G., and Scheler, H., *Z. Chem.*, 1972, **12**, 337.
[270] Glore, J. D., Hall, R. E., and Schram, E. P., *Inorg. Chem.*, 1972, **11**, 550.
[271] Goetze, R., and Nöth, H., *Chem. Ber.*, 1976, **109**, 3247.
[271a] Goetze, R., and Nöth, H., *J. Organometallic Chem.*, 1978, **145**, 151.
[272] Goldberg, D. E., Harris, D. H., Lappert, M. F., and Thomas, K. M., *J.C.S. Chem. Comm.*, 1976, 261.
[273] Gol'din, G. S., Baturina, L. S., and Gavrilova, T. N., *Zh. Obshch. Khim.*, 1975, **45**, 2189.
[273a] Goldwhite, H., Gynane, M. J. S., Lappert, M. F., Power, P. P., and El-Soueni, A., Unpublished Observations (1976).
[274] Goodrich-Maines, R., and Gilje, J. W., *Inorg. Chem.*, 1976, **15**, 470.

[274a] Gorbatenko. I., and Feshenko, N. G., *J. Gen. Chem. U.S.S.R.*, 1977, **47**, 1752.
[275] Goubeau, J., and Keller, H., *Z. Anorg. Chem.*, 1951, **267**, 1.
[276] Gragg, B. R., Handshoe, R. E., and Niedenzu, K., *J. Organometallic Chem.*, 1976, **116**, 135.
[277] Gragg, B. R., and Niedenzu, K., *J. Organometallic Chem.*, 1976, **117**, 1.
[278] Graves, G. E., McKennon, D. W., and Lustig, M., *Inorg. Chem.*, 1971, **10**, 2083.
[279] Green, B., and Sowerby, D. B., *J. Chem. Soc. (A)*, 1970, 987; Green, B., Sowerby, D. B., and Clare, P., *J. Chem. Soc. (A)*, 1971, 3487.
[280] Green, M. C., Lappert, M. F., and Lynch, J., Unpublished Observations (1969).
[281] Grobe, J., and Möller, U., *J. Organometallic Chem.*, 1969, **17**, 263.
[282] Grosse-Ruyken, H., and Schaarschmidt, K., *Chem. Tech. (Berlin)*, 1959, **11**, 451; *Chem. Abstr.*, 1960, **54**, 4359.
[283] Gunderloy, F. C., and Erickson, C. E., *Inorg. Chem.*, 1962, **1**, 349.
[283a] Gupta, S. K., *J. Organometallic Chem.*, 1978, **156**, 95.
[284] Grüning, R., and Lorberth, J., *J. Organometallic Chem.*, 1974, **78**, 221.
[284a] Gutmann, V., *Coordination Chemistry in Non-aqueous Solutions*, Springer, 1968.
[285] Gverdsiteli, I. M., Baramidze, L. V., and Dzheliya, M. I., *Zh. Obshch. Khim.*, 1972, **42**, 2019; *Chem. Abstr.*, 1973, **78**, 43566.
[285a] Gynane, M. J. S., Hudson, A., Lappert, M. F., Power, P. P., and Goldwhite, H., *J.C.S. Chem. Comm.*, 1976, 623.
[285b] Gynane, M. J. S., Hudson, A., Lappert, M. F., Power, P. P., and Goldwhite, H., *J.C.S. Dalton*, 1980, in the press.
[285c] Gynane, M. J. S., Lappert, M. F., and Miles, S. J., Unpublished Observations (1977).
[286] Gynane, M. J. S., Lappert, M. F., Miles, S. J., and Power, P. P., *J.C.S. Chem. Comm.*, 1976, 256.
[286a] Gynane, M. J. S., Lappert, M. F., Miles, S. J., and Power, P. P., *J.C.S. Chem. Comm.*, 1978, 192.
[286b] Gynane, M. J. S., Harris, D. H., Lappert, M. F., Power, P. P., Rivière, P., and Rivière-Baudet, M., *J.C.S. Dalton*, 1977, 2004.
[286c] Haasemann, P., and Goubeau, J., *Z. Anorg. Chem.*, 1974, **408**, 293.
[286d] Hancock, K. G., Uriarte, A. K., and Dickinson, D. A., *J. Amer. Chem. Soc.*, 1973, **95**, 6980.
[287] Hänssgen, D., and Appel, R., *Chem. Ber.*, 1972, **105**, 3271.
[287a] Hänssgen, D., Kuna, J., and Ross, B., *Chem. Ber.*, 1976, **109**, 1797.
[288] Hänssgen, D., and Pohl, I., *Angew. Chem.*, 1974, **86**, 676.
[289] Hänssgen, D., and Roelle, M.W., *J. Organometallic Chem.*, 1973, **56**, C14.
[290] Halasz, S. P. V., and Glemser, O., *Chem. Ber.*, 1970, **103**, 594.
[291] Hall, R. E., and Schram, E. P., *Inorg. Chem.*, 1969, **8**, 270.

[292] Hammell, M., and Levine, R., *J. Org. Chem.*, 1950, **15**, 162.

[293] Harman, M., Sharp, D. W. A., and Winfield, J. M., *Inorg. Nucl. Chem. Letters*, 1974, **10**, 183.

[293a] Harpp, D. N., Adams, J., Gleason, J. G., Mullins, D., and Steliou, K., *Tetrahedron Letters*, 1978, 3989.

[294] Harris, D. H., D.Phil. Thesis, Sussex, 1975.

[294a] Harris, D. H., Keppie, S. A., and Lappert, M. F., *J.C.S. Dalton*, 1973, 1653.

[295] Harris, D. H., and Lappert, M. F., *J.C.S. Chem. Comm.*, 1974, 895.

[295a] Harris, D. H., and Lappert, M. F., *J. Organometallic Chem. Library*, 1976, **2**, 13.

[295b] Harris, D. H., Lappert, M. F., Pedley, J. B., and Sharp, G. J., *J.C.S. Dalton*, 1976, 945.

[296] Harrison, P. G., *J.C.S. Perkin I*, 1972, 130.

[297] Harrison, P. G., and Zuckerman, J. J., *Inorg. Chem.*, 1970, **9**, 175.

[298] Harrison, P. G., and Richards, J. A., *J. Organometallic Chem.*, 1976, **108**, 61.

[298a] Haszeldine, R. N., and Tipping, A. E., *J. Chem. Soc. (C)*, 1967, 1241.

[299] Haubold, W., and Kraatz, U., *Z. Anorg. Chem.*, 1976, **421**, 105.

[299a] Hausen, P. J., Bordner, J., and Schreiner, A. F., *Inorg. Chem.*, 1973, **12**, 1347.

[300] Hauser, C. R., and Puterbaugh, W. H., *J. Amer. Chem. Soc.*, 1953, **75**, 1068.

[301] Havlicek, M. D., and Gilje, J. W., *Inorg. Chem.*, 1972, **11**, 1624.

[301a] Head, R. A., and Nixon, J. F., *J.C.S. Dalton*, 1978, 913.

[302] Hedaya, E., Kawakami, J. H., and Kwaikowski, G. T., Ger. Patent, 1975, 2,435,385; *Chem. Abstr.*, 1975, **82**, 171723.

[303] Heider, W., Klingebiel, U., and Lin, T., *Chem. Ber.*, 1974, **107**, 592.

[303a] Hermannsdörfer, K. H., Matejčikova, E., and Nöth, H., *Chem. Ber.*, 1970, **103**, 516.

[304] Hesse, G., Witte, H., and Gulden, W., *Tetrahedron Letters*, 1966, 2707.

[305] Hessett, B., Morris, J. H., and Perkins, P. G., *J. Chem. Soc. (A)*, 1971, 2056, 2466.

[306] Hester, R. E., and Jones, K., *Chem. Comm.*, 1966, 317.

[307] Heying, T. L., and Smith, H. D., *Adv. Chem. Ser.*, 1964, **42**, 201.

[308] Hieber, W., and Beutner, H., *Z. Naturforsch.*, 1960, **15b**, 324; *Z. Anorg. Chem.*, 1962, **317**, 63.

[309] Higginson, W. C. E., and Wooding, N. S., *J. Chem. Soc.*, 1952, 760.

[310] Highsmith, R. E., and Sisler, H. H., *Inorg. Chem.*, 1969, **8**, 1029.

[311] Hils, J., Graubaum, H. J., and Rühlmann, K., *J. Prakt. Chem.*, 1966, **32**, 37.

[312] Hirabayashi, T., Imaeda, H., Itoh, K., Sakai, S., and Ishii, Y., *J. Organometallic Chem.*, 1969, **19**, 299.

[313] Hirabayashi, T., Itoh, K., Sakai, S., and Ishii, Y., *J. Organometallic Chem.*, 1970, **21**, 273.
[314] Hirsch, H. V., *Chem. Ber.*, 1967, **100**, 1289.
[314a] Hoberg, H., *Ann.*, 1972, **766**, 142.
[314b] Hoberg, H., and Griebsch, U., *Ann.*, 1977, 1516.
[315] Hoberg, H., and Mur, J. B., *J. Organometallic Chem.*, 1969, **17**, P28.
[316] Hoberg, H., and Mur, J. B., *J. Organometallic Chem.*, 1969, **17**, P30.
[317] Hoefer, R., and Glemser, O., *Z. Naturforsch.*, 1975, **30b**, 460.
[318] Hoffman, E. R., *Anal. Chem.*, 1976, **48**, 445.
[319] Hofmeister, H. K., and Van Wazer, J. R., *J. Inorg. Nucl. Chem.*, 1964, **26**, 1209.
[320] Horder, J. R., and Lappert, M. F., *Chem. Comm.*, 1967, 485; *J. Chem. Soc. (A)*, 1969, 173.
[321] Horder, J. R., and Lappert, M. F., *J. Chem. Soc. (A)*, 1968, 1167.
[322] Horvath, B., U.S. Patent, 3,646,000; *Chem. Abstr.*, 1972, **77**, 6106.
[322a] Hudson, A., Lappert, M. F., and Lednor, P. W., *J.C.S. Dalton*, 1976, 2369.
[323] Hudson, R. F., and Mancuso, A., *Chem. Comm.*, 1969, 522.
[323a] Hudson, R. F., and Searle, R. J. G., *J. Chem. Soc. (B)*, 1968, 1349.
[324] Hughes, L. J., and Perry, E., *J. Polymer Sci. (A)*, 1965, **3**, 1527.
[325] Hundeck, J., *Angew. Chem.*, 1965, **77**, 729.
[325a] Huttner, G., and Krieg, B., *Angew. Chem. Internat. Edn.*, 1971, **10**, 512.
[325b] Huttner, B., and Müller, H.-D., *Angew. Chem. Internat. Edn.*, 1975, **14**, 571.
[326] Inglis, T., Keable, H. R., Kilner, M., and Robertson, E. E., Chem. Uses Molybdenum, Proc. Conf., 1973; *Chem. Abstr.*, 1975, **82**, 57882.
[327] Inoue, S., and Yokoo, Y., *Bull. Chem. Soc. Jap.*, 1972, **45**, 3651.
[327a] Ioffe, S. L., Kashuting, M. V., Shitkin, V. M., Levin, A. A., and Tartakovskii, V. A., *Zh. Obshch. Khim.*, 1973, **9**, 896.
[328] Iqbal, M. J., *Proc. Pak. Akad. Sci.*, 1973, **10**, 57; *Chem. Abstr.*, 1975, **82**, 43568.
[329] Irvine, I., and Keat, R., *J.C.S. Dalton*, 1972, 17.
[330] Ishida, S., Ooshima, N., and Kurita, K., Jap. Patent, 1973, 7,222,332; *Chem. Abstr.*, 1973, **78**, 4776.
[331] Ishii, Y., Sakai, S., and Hattori, T., Jap. Patent, 1976, 7,530,060; *Chem. Abstr.*, 1976, **85**, 108760.
[332] Ishii, Y., Itoh, K., Katsuura, T., and Matsuda, I., *J. Organometallic Chem.*, 1969, **19**, 347.
[333] Issleib, K., and Lischewski, M., *Synth. Inorg. Metal-Org. Chem.*, 1973, **3**, 255.
[334] Itoh, K., Ishii, Y., and Fukumoto, Y., *Tetrahedron Letters*, 1968, 3199.
[335] Itoh, K., Kato, N., and Ishii, Y., *J. Organometallic Chem.*, 1970, **22**, 49.
[336] Itoh, K., Kobayashi, S., Sakai, S., and Ishii, Y., *J. Organometallic Chem.*, 1967, **10**, 451.

[337] Itoh, K., Lee, I. K., Matsuda, I., Sakai, S., and Ishii, Y., *Tetrahedron Letters,* 1967, 2667.
[338] Itoh, K., Matsuda, I., and Ishii, Y., *J. Chem. Soc. (C),* 1971, 1870.
[338a] Itoh, K., Matsuda, I., and Ishii, Y., *Tetrahedron Letters,* 1969, 2675.
[338b] Itoh, K., Matsuda, I., Katsuura, T., and Ishii, Y., *J. Organometallic Chem.,* 1969, **19**, 347.
[339] Itoh, K., Sakai, S., and Ishii, Y., *Tetrahedron Letters,* 1966, 4941.
[340] Itoh, K., Sakai, S., and Ishii, Y., *J. Org. Chem.,* 1966, **31**, 3948.
[341] Itoh, K., Sakai, S., and Ishii, Y., *J. Org. Chem.,* 1967, **32**, 2210.
[342] Iwasaki, K., Yamaguchi, K., Matsuura, J., Hasuo, M., and Kojima, K., Jap. Patent, 1972, 72, 19406; *Chem. Abstr.,* 1972, **77**, 127256.
[342a] Jamerson, J. D., and Takats, J., *J. Organometallic Chem.,* 1974, **78**, C23.
[343] Janssen, M. J., Luijten, J. G. A., and van der Kerk, G. J. M., *J. Organometallic Chem.,* 1964, **1**, 286.
[343a] Jarvis, B. B., and Tong, W. P., *Synthesis,* 1975, 102.
[343b] Jefferson, R., and Lappert, M. F., *Intra-Science Chem. Rept.,* 1973, **7**, 123.
[344] Jefferson, R., Lappert. M. F., Prokai, B., and Tilley, B. P., *J. Chem. Soc. (A),* 1966, 1584.
[345] Jefferson, R., Nixon, J. F., Painter, T. M., Keat, R., and Stobbs, L., *J.C.S. Dalton,* 1973, 1414.
[346] Jenkins, A. D., Lappert, M. F., and Srivastava, R. C., *J. Polymer Sci. (B),* 1968, **6**, 865.
[347] Jenkins, A. D., Lappert, M. F., and Srivastava, R. C., *J. Organometallic Chem.,* 1970, **23**, 165.
[348] Jenkins, A. D., Lappert, M. F., and Srivastava, R. C., *Europ. Polym. J.,* 1971, **7**, 289.
[349] Jenne, H., and Niedenzu, K., *Inorg. Chem.,* 1964, **3**, 68.
[350] Joh, Y., Yoshihara, T., Kotake, Y., Imai, Y., and Kurihara, S., *J. Polymer Sci. (A-I),* 1967, 2503.
[350a] Jonás, K., Nöth, H., and Storch, W., *Chem. Ber.,* 1977, **110**, 2783.
[350b] Jones, K., and Hester, R. E., *Chem. Comm.,* 1966, 317.
[351] Jones, K., and Lappert, M. F., *Proc. Chem. Soc.,* 1962, 358.
[352] Jones, K., and Lappert, M. F., *Proc. Chem. Soc.,* 1964, 22.
[353] Jones, K., and Lappert, M. F., *J. Chem. Soc.,* 1965, 1944.
[354] Jones, K., and Lappert, M. F., *J. Organometallic Chem.,* 1965, **3**, 295.
[355] Jones, K., and Lappert, M. F., *Organometallic Chem. Rev.,* 1966, **1**, 67.
[356] Jones, K., and Lappert, M. F., in *Organotin Compounds,* ed. A. K. Sawyer, Marcel Dekker, 1971.
[357] Jones, R. G., Karmas, G., Martin, G. A., and Gilman, H., *J. Amer. Chem. Soc.,* 1956, **78**, 4285.
[357a] Kalinina, G. S., Petrov, B. I., Kruglaya, O. A., and Vyazankin, N. S., *Zh. Obshch. Khim.,* 1972, **42**, 148.

[358] Kannengiesser, G., and Damm, F., *Bull. Soc. Chim. Fr.*, 1967, 2492.
[359] Kaufmann, J., and Kober, F., *Z. Anorg. Chem.*, 1976, **420**, 177.
[360] Kaufmann, J., and Kober, F., *J. Organometallic Chem.*, 1974, **71**, 49.
[361] Kaufmann, J., and Kober, F., *J. Organometallic Chem.*, 1974, **82**, 333.
[362] Kaufmann, J., and Kober, F., *J. Organometallic Chem.*, 1974, **81**, 59.
[363] Kaufmann, J., and Kober, F., *J. Organometallic Chem.*, 1975, **96**, 243.
[364] Kazantsev, A. V., Zhubekova, M. N., and Zhakharkin, L. I., *Zh. Obshch. Khim.*, 1972, **42**, 1570; *Chem. Abstr.*, 1972, **77**, 12752.
[365] Keat, R., *J.C.S. Dalton,* 1974, 876; *Phosphorus,* 1972, **1**, 253.
[366] Keat, R., Abstr. Joint Annual Meetings, Chemical Society, Nottingham, 1969, p. 7, 35.
[366a] Keat, R., *J.C.S. Dalton,* 1972, 2189.
[367] Keller, P. C., *Inorg. Chem.*, 1972, **11**, 256.
[368] Keller, P. C., *J. Amer. Chem. Soc.*, 1974, **96**, 3073.
[369] Keller, P. C., *J. Amer. Chem. Soc.*, 1974, **96**, 3078.
[370] Keppie, S. A., and Lappert, M. F., Unpublished Observations (1970).
[370a] Kibler, R. W., Bozzacco, F., and Forman, L. E., U.S. Patent, 2,849,432; *Chem. Abstr.*, 1959, **53**, 10844i.
[371] Kiesel, R., and Schram, E. P., *Inorg. Chem.*, 1973, **12**, 1090.
[372] King. F. D., and Walton, D. R. M., *Synthesis,* 1975, 738.
[373] King, F. D., and Walton, D. R. M., *J.C.S. Chem. Comm.*, 1974, 256.
[374] King, R. B., *Inorg. Chem.*, 1967, **6**, 25.
[375] King, R. B., *Inorg. Chem.*, 1963, **2**, 936; Barlow, C. G., Nixon, J. F., and Swain, J. R., *J. Chem. Soc. (A),* 1969, 1082; Douglas, W. M., and Ruff, J. K., *J. Chem. Soc. (A),* 1971, 3558; Atkinson, L. K., and Smith, D. C., *J. Organometallic Chem.*, 1971, **33**, 189; Bennett, M. A., and Turney, T. W., *Austr. J. Chem.*, 1973, **26**, 2321.
[375a] King, R. B., and Gimeno, J., *J.C.S. Chem. Comm.*, 1977, 142.
[375b] King, R. B., and Gimeno, J., *Inorg. Chem.*, 1978, **17**, 2390.
[375c] King, R. B., and Gimeno, J., *Inorg. Chem.*, 1978, **17**, 2396.
[375d] King, R. B., Gimeno, J., and Lotz, T. J., *Inorg. Chem.*, 1978, **17**, 2401.
[375e] Kitazume, T., and Shreeve, J. M., *J. Amer. Chem. Soc.*, 1977, **99**, 3690.
[376] Klebe, J. F., *J. Polymer Sci. (B),* 1964, **2**, 1079.
[377] Klebe, J. F., Bush, J. B., and Lyons, J. E., *J. Amer. Chem. Soc.*, 1964, **86**, 4400.
[377a] Klingebiel, U., *Chem. Ber.*, 1978, **111**, 2735.
[378] Klyuchnikov, N. G., Karabadzhak, F. I., and Losev, V. B., *J. Gen. Chem. U.S.S.R.*, 1971, **41**, 166.
[379] Kober, F., *Z. Anorg. Chem.*, 1973, **401**, 243.
[380] Kober, F., *Z. Anorg. Chem.*, 1973, **397**, 97.
[381] Kober, F., *J. Organometallic Chem.*, 1975, **94**, 393.
[382] Kober, F., *Chem. Z.*, 1976, **100**, 197; *Chem. Abstr.*, 1976, **85**, 143209.
[383] Kober, F., *Z. Anorg. Chem.*, 1975, **412**, 202.

[384] Kober, F., *Z. Anorg. Chem.*, 1973, **401**, 243.
[385] Kober, F., *Z. Naturforsch.*, 1974, **29b**, 358.
[386] Kober, F., and Ruehl, W. J., *Z. Anorg. Chem.*, 1976, **420**, 74.
[387] Kober, F., and Ruehl, W. J., *Z. Anorg. Chem.*, 1974, **406**, 52.
[388] Kodomari, M., Sawa, S., Morozumi, K., and Ohkita, T., *Nippon Kagaku Kaishi*, 1976, **2**, 301; *Chem. Abstr.*, 1976, **95**, 78423.
[389] Koide, N., Iimura, K., and Takeda, M., Ionic Polym. Unsolved Probl., Jpn., U.S. Semin. Polym. Synth., 1st, 1974 (pub. 1976) 321; *Chem. Abstr.*, 1976, **85**, 160628.
[390] Koketsu, J., Sakai, S., and Ishii, Y., *Kogyo Kagaku Zasshi*, 1969, **72**, 2503; *Chem. Abstr.*, 1970, **72**, 79165a.
[391] Kolodyazhnyi, Yu. V., Gruntfest, M. G., Baturina, L. S., Morgunova, M. M., Sizova, N. I., Gol'din, G. S., and Osipov, O. A., *Zh. Obshch. Khim.*, 1975, **45**, 1046.
[391a] Komarov, N. V., and Ol'khovskaya, L. I., *J. Gen. Chem. U.S.S.R.*, 1977, **47**, 1099.
[392] Kopp, R. W., Bond, A. C., and Parry, R. W., *Inorg. Chem.*, 1976, **15**, 3042; Schultz, C. W., and Parry, R. W., *Inorg. Chem.*, 1976, **15**, 3046.
[393] Kormer, V. A., Yufa, T. L., Babitskii, B. D., Poletaeva, I. A., Simanova, N. P., Lapuk, I. M., Markova, V. V., Lobach, M. I., Kovalev, N. F., and Kholodnitskaya, G. V., U.S.S.R. Patent, 427,002; *Chem. Abstr.*, 1976, **84**, 60265.
[394] Köster, R., and Iwasaki, K., *Adv. Chem. Ser.*, 1964, **42**, 148.
[395] Kostyuk, A. S., Baukov, Yu. I., and Lutsenko, I. F., *Zh. Obshch. Khim.*, 1973, **43**, 603.
[396] Kostyuk, A. S., Lutsenko, I. F., Baukov, Yu. I., Savel'eva, N. I., and Krysina, V. K., *J. Organometallic Chem.*, 1969, **17**, 241.
[397] Kostyuk, A. S., Ruderfer, I. B., Baukov, Yu. I., and Lutsenko, I. F., *Zh. Obshch. Khim.*, 1975, **45**, 819.
[398] Kowalski, J., and Lasocki, Z., *J. Organometallic Chem.*, 1976, **116**, 75.
[398a] Kozyukov, V. P., Mironov, V. F., Petrovskaya, L. I., and Sheludyakova, S. V., *Khim. Geterotsikl. Soedin, Akad. Nauk Latv. S.S.S.R.*, 1967, 185.
[399] Kozyukov, V. P., Sheludyakova, S. V., and Mironov, V. F., *Zh. Obshch. Khim.*, 1976, **46**, 442; *Chem. Abstr.*, 1976, **84**, 164922.
[400] Krapf, H., Lorberth, J., and Nöth, H., *Chem. Ber.*, 1967, **100**, 3511.
[401] Kricheldorf, H. R., and Leppert, F., *Synthesis*, 1975, 49.
[401a] Krohmer, P., and Goubeau, J., *Chem. Ber.*, 1971, **104**, 1347.
[402] Krommes, P., and Lorberth, J., *J. Organometallic Chem.*, 1976, **110**, 195.
[403] Krommes, P., and Lorberth, J., *J. Organometallic Chem.*, 1975, **97**, 59.
[404] Krommes, P., and Lorberth, J., *J. Organometallic Chem.*, 1975, **88**, 329.
[405] Krommes, P., and Lorberth, J., *J. Organometallic Chem.*, 1975, **93**, 339.
[406] Krommes, P., and Lorberth, J., *J. Organometallic Chem.*, 1976, **120**, 131.
[407] Krommes, P., and Lorberth, J., *J. Organometallic Chem.*, 1977, **127**, 19.

[408] Kronawitter, I., and Nöth, H., *Chem. Ber.*, 1972, **105**, 2423.
[408a] Kroshefsky, R. D., and Verkade, J. G., *Inorg. Chem.*, 1975, **14**, 3090.
[409] Krück, T., Mocueler, G. M., and Schmidgem, G., *Z. Anorg. Chem.*, 1975, **412**, 239.
[410] Kruger, C., *J. Organometallic Chem.*, 1967, **9**, 125.
[411] Kruger, C., and Rochow, E. G., *Angew. Chem.*, 1963, **75**, 617.
[412] Kruger, C., Rochow, E. G., and Wannagat, U., *Chem. Ber.*, 1963, **96**, 2132, 2138.
[413] Krutskii, L. N., Zykova, T. V., Salakhutdinov, R. A., and Tsinunin, V. S., *Zh. Obshch. Khim.*, 1972, **42**, 1493.
[414] Kula, M.-R., Lorberth, J., and Amberger, E., *Chem. Ber.*, 1964, **97**, 2087.
[415] Kunioka, E., and Tajima, Y., *J. Polymer Sci. (A-I)*, 1968, **6**, 241.
[416] Kuramshin, I. Ya., Bhashkirov, Sh. Sh., Muratova, A. A., Manapov, R. A., Khramov, A. S., and Pudovik, A. N., *Zh. Obshch. Khim.*, 1975, **45**, 701; *Chem. Abstr.*, 1975, **83**, 36925.
[417] Kyker, G. S., and Schram, E. P., *J. Amer. Chem. Soc.*, 1968, **90**, 3672, 3678.
[417a] Lappert, M. F., *Proc. Chem. Soc.*, 1959, 59.
[418] Lappert, M. F., and Lednor, P. W., *Adv. Organometallic Chem.*, 1976, **14**, 345.
[419] Lappert, M. F., and Levitt, T. E., Unpublished Observations (1970).
[420] Lappert, M. F., Litzow, M. R., Pedley, J. B., Riley, P. N. K., Spalding, T. R., and Tweedale, A., *J. Chem. Soc. (A)*, 1970, 2320.
[421] Lappert, M. F., and Lorberth, J., *Chem. Comm.*, 1967, 836.
[421a] Lappert, M. F., Lorberth, J., and Poland, J. S., *J. Chem. Soc. (A)*, 1970, 2954; Lorberth, J., *J. Organometallic Chem.*, 1971, **27**, 303.
[422] Lappert, M. F., Lorberth, J., and Poland, J. S., *J. Chem. Soc. (A)*, 1970, 2954.
[423] Lappert, M. F., and Lynch, J., *Chem. Comm.*, 1968, 750.
[423a] Lappert, M. F., Majumdar, M. K., and Tilley, B. P., *J. Chem. Soc. (A)*, 1966, 1590.
[423b] Lappert, M. F., and Majumdar, M. K., *Proc. Chem. Soc.*, 1963, 88; *Adv. Chem. Ser.*, 1964, **42**, 208.
[424] Lappert, M. F., and Majumdar, M. K., *J. Organometallic Chem.*, 1966, **6**, 316.
[424a] Lappert, M. F., McMeeking, J., and Palmer, D. E., *J.C.S. Dalton*, 1973, 151.
[424b] Lappert, M. F., Miles, S. J., Power, P. P., Carty, A. J., and Taylor, N. J., *J.C.S. Chem. Comm.*, 1977, 458.
[424c] Lappert, M. F., Onyszchuk, M., and Slade, M. J., Unpublished Observations (1978).
[425] Lappert, M. F., and Poland, J. S., *Chem. Comm.*, 1969, 156.
[426] Lappert, M. F., and Poland, J. S., *Chem. Comm.*, 1969, 1061.

[427] Lappert, M. F., and Poland, J. S., *Adv. Organometallic Chem.*, 1971, **9**, 397.
[428] Lappert, M. F., and Power, P. P., *Adv. Chem. Ser.*, 1976, **157**, 70.
[429] Lappert, M. F., and Power, P. P., Unpublished Observations (1977).
[429a] Lappert, M. F., Power, P. P., and Slade, M. J., Unpublished Observations (1977).
[430] Lappert, M. F., and Prokai, B., *Adv. Organometallic Chem.*, 1967, **5**, 225.
[431] Lappert, M. F., and Sanger, A. R., *J. Chem. Soc. (A)*, 1971, 874.
[432] Lappert, M. F., and Sanger, A. R., *J. Chem. Soc. (A)*, 1971, 1314.
[433] Lappert, M. F., and Sanger, A. R., Unpublished Observations (1969).
[433a] Lappert, M. F., and Slade, M. J., Unpublished Observations (1978).
[434] Lappert, M. F., and Smith, J. K., *J. Chem. Soc.*, 1961, 3224.
[435] Lappert, M. F., and Srivastava, G., *Inorg. Nucl. Chem. Letters*, 1965, **1**, 53.
[435a] Lappert, M. F., and Srivastava, G., Unpublished Observations (1965).
[436] Lappert, M. F., and Srivastava, G., *Proc. Chem. Soc.*, 1964, 120.
[437] Lappert, M. F., and Srivastava, G., *J. Chem. Soc. (A)*, 1967, 602.
[438] Lappert, M. F., and Tilley, B. P., Unpublished Observations (1966).
[439] Larsson, E., and Marin, R. E. I., Swed. Patent, 1954, 138,357; *Chem. Abstr.*, 1954, **48**, 2761.
[440] Lasocki, Z., *Synth. Inorg. Metal-Org. Chem.*, 1973, **3**, 29.
[441] Lebedev, E. P., Baburina, V. A., and Fridland, D. V., *Zh. Obshch. Khim.*, 1975, **45**, 1647.
[442] Lebedev, E. P., Fridland, D. V., Reikhsfel'd, V. O., and Korol, E. N., *Zh. Obshch. Khim.*, 1976, **46**, 315; *Chem. Abstr.*, 1976, **85**, 5741.
[442a] Lebedev, E. P., and Valimukhametova, R. G., *J. Gen. Chem. U.S.S.R.*, 1977, **47**, 978.
[443] Lebedev, E. P., Valimukhametova, R. G., Korol, E. N., Reikhsfel'd, V. O., *Zh. Obshch. Khim.*, 1974, **44**, 1941.
[444] LeBlond, R. D., and DesMarteau, D. D., *J.C.S. Chem. Comm.*, 1974, 555; DesMarteau, D. D., *J. Amer. Chem. Soc.*, 1978, **100**, 6270.
[444a] Lehnig, M., *Tetrahedron Letters*, 1974, 3323.
[445] Lemire, A. E., and Thompson, J. C., *Can. J. Chem.*, 1972, **50**, 1386.
[446] Lesbre, M., Satgé, J., and Baudet, M., *Compt. Rend.*, 1964, **259C**, 4733.
[447] Levitt, T. E., and Pelter, A., *Nature*, 1966, **211**, 299.
[448] Lidy, W., and Sundermeyer, W., *Chem. Ber.*, 1976, **109**, 2542.
[449] Lidy, W., and Sundermeyer, W., *Z. Naturforsch.*, 1974, **29b**, 276.
[450] Lidy, W., Sundermeyer, W., and Verbeek, W., *Z. Anorg. Chem.*, 1974, **406**, 228.
[451] Limburg, W. W., and Post, H. W., *Rec. Trav. Chim.*, 1962, **81**, 430.
[452] Liston, T. V., U.S. Patent, 1971, 3,631,084; *Chem. Abstr.*, 1972, **77**, 19795.

[453] Lochmann, L., and Šorm, M., *Coll. Czech. Chem. Commun.*, 1973, **38**, 3449.
[453a] Lockhart, J. C., *Redistribution Reactions*, Academic Press, 1970.
[454] Longi, P., Greco, F., and Ressi, U., Ital. Patent, 1970, 871,096; *Chem. Abstr.*, 1973, **78**, 148474.
[455] Lorberth, J., Dissertation, München, 1965.
[456] Lorberth, J., *J. Organometallic Chem.*, 1968, **15**, 251.
[457] Lorberth, J., *J. Organometallic Chem.*, 1969, **16**, 235.
[458] Lorberth, J., *J. Organometallic Chem.*, 1969, **16**, 327.
[459] Lorberth, J., Unpublished Observations (1969).
[460] Lorberth, J., *Chem. Ber.*, 1965, **98**, 1201.
[461] Lorberth, J., *J. Organometallic Chem.*, 1971, **27**, 303; 1974, **71**, 159.
[462] Lorberth, J., and Lange, G., *J. Organometallic Chem.*, 1973, **54**, 165.
[463] Lorberth, J., and Weller, F., *J. Organometallic Chem.*, 1971, **32**, 145.
[464] Lukevics, E., and Liberts, L., *Latv. PSR Zinal. Akad. Vestis, Khim. Ser.*, 1972, 203; *Chem. Abstr.*, 1972, **77**, 75256.
[465] Malavaud, C., and Barrans, J., *Tetrahedron Letters,* 1975, 3077.
[466] Malavaud, C., Charbonnel, Y., and Barrans, J., *Tetrahedron Letters,* 1975, 497.
[467] Malhortra, S. C., *Inorg. Chem.*, 1964, **3**, 862.
[468] Manoussakis, G. E., *J. Inorg. Nucl. Chem.*, 1968, **30**, 3100.
[469] Manoussakis, G. E., and Tossidis, J. A., *J. Inorg. Nucl. Chem.*, 1972, **34**, 2449.
[470] Manoussakis, G. E., and Tossidis, J. A., *Prakt. Panelleniou Chem. Synedriou, 4th,* 1970, **1**, 86; *Chem. Abstr.*, 1976, **85**, 93724.
[470a] Maringgele, W., and Meller, A., *Z. Anorg. Chem.*, 1977, **436**, 173.
[471] Martel, B., and Aly, E., *J. Organometallic Chem.*, 1971, **29**, 61.
[472] Martel, B., and Hiriart, J. M., *Tetrahedron Letters,* 1971, 2737.
[472a] Matsuda, I., Itoh, K., and Ishii, Y., *J. Chem. Soc. (C),* 1969, 701.
[473] Matsuda, I., Itoh, K., and Ishii, Y., *J. Chem. Soc. (C),* 1969, 2675; *J. Organometallic Chem.*, 1974, **69**, 353.
[474] Matsuda, I., Itoh, K., and Ishii, Y., *J.C.S. Perkin, I,* 1972, 1678.
[475] Matsuda, I., Itoh, K., and Ishii, Y., *Tetrahedron Letters,* 1969, 2675.
[476] Matsuda, I., Ishii, Y., and Itoh, K., *J. Organometallic Chem.*, 1969, **19**, 339.
[476a] Mazzei, A., Cucinella, S., and Marconi, W., *Makromol. Chem.*, 1969, **122**, 168.
[476b] Meller, A., Maringgele, W., and Hirninger, F. J., *J. Organometallic Chem.*, 1977, **136**, 289.
[476c] Meller, A., and Ossico, A., *Monatsh.*, 1972, **103**, 577.
[477] Mews, R., and Glemser, O., *Inorg. Chem.*, 1972, **11**, 2521.
[478] Middleton, W. J., U.S. Patent, 1976, 3,914,265; *Chem. Abstr.*, 1976, **84**, 42635.

[479] Mikhailov, B. M., and Dorokhov, V. A., *J. Gen. Chem., U.S.S.R.*, 1961, **31**, 3504; *Bull. Acad. Sci. U.S.S.R., Div. Chem. Sci.*, 1961, 1082.
[480] Mikhailov, B. M., Ter-Sarkisyan, G. S., Govorov, N. N., Nikolaeva, N. A., and Kiselev, V. G., *Izv. Akad. Nauk S.S.S.R., Ser. Khim.*, 1976, **4**, 870; *Chem. Abstr.*, 1976, **85**, 63112.
[481] Mikhailov, B. M., Ter-Sarkisyan, G. S., and Nikolaeva, N. A., *Izv. Akad. Nauk S.S.S.R., Ser. Khim.*, 1972, 2372; *Chem. Abstr.*, 1973, **78**, 43561.
[481a] Mikhailov, B. M., Ter-Sarkisyan, G. S., Nikolaeva, N. A., and Kiselev, V. G., *Zh. Obshch. Khim.*, 1973, **43**, 857.
[482] Miller, J. M., and Wilson, G. L., *J. Fluorine Chem.*, 1974, **4**, 207.
[483] Millington, D., and Sowerby, D. B., *J.C.S. Dalton*, 1974, 1070.
[484] Milovskaya, E. B., Makarycher, M. N., and Skvortscvich, E. P., *Vysokomol. Soedin, Ser. A.*, 1975, **17**, 1217; *Chem. Abstr.*, 1975, **83**, 115051.
[485] Mironov, V. F., Kozyukov, V. P., and Bulatov, V. P., *Zh. Obshch. Khim.*, 1973, **43**, 2089; *Chem. Abstr.*, 1974, **80**, 37216.
[486] Mironov, V. F., Pachurina, S. Ya., Grigos, V. I., Zhigach, A. F., and Siryatakay, V. I., *Dokl. Akad. Nauk, S.S.S.R.*, 1972, **202**, 1345; *Chem. Abstr.*, 1972, **76**, 153833.
[487] Mironov, V. F., Sheludyakov, V. D., and Rodionov, E. S., *Zh. Obshch. Khim.*, 1974, **44**, 1502.
[488] Mironov, V. F., Sheludyakov, V. D., Rodionov, E. S., and Popov, A. I., *Zh. Obshch. Khim.*, 1972, **42**, 1651; *Chem. Abstr.*, 1972, **77**, 140214.
[489] Mitchell, T. N., *J. Organometallic Chem.*, 1975, **92**, 311.
[490] Mitchell, T. N., and Neumann, W. P., *J. Organometallic Chem.*, 1970, **22**, C25.
[490a] Moedritzer, K., *Adv. Organometallic Chem.*, 1968, **6**, 171.
[491] Moles, A., Sacco, A., and Ugo, R., *Coord. Chem. Rev.*, 1966, **1**, 67.
[492] Montemayor, R. G., and Parry, R. W., *Inorg. Chem.*, 1973, **12**, 2482.
[492a] Montemayor, R. G., Sauer, D. T., Fleming, S., Bennett, D. W., Thomas, M. G., and Parry, R. W., *J. Amer. Chem. Soc.*, 1978, **100**, 2231.
[493] Montenarh, M., *Z. Naturforsch.*, 1976, **318b**, 993.
[494] Moore, W. S., and Yoder, C. H., *J. Organometallic Chem.*, 1975, **87**, 389.
[495] Moretto, H. H., Schmidt, P., and Wannagat, U., *Z. Anorg. Chem.*, 1972, **394**, 125.
[496] Morgunova, M. M., Popkov, K. K., Zhinkin, D. Ya., and Andrianov, K. A., *Dokl. Akad. Nauk S.S.S.R.*, 1964, **159**, 641.
[497] Morris, J. H., *J. Inorg. Nucl. Chem.*, 1974, **36**, 2439.
[498] Morse, J. G., and Morse, K. W., *Inorg. Chem.*, 1973, **12**, 2119.
[498a] Muckle, G., Nöth, H., and Storch, W., *Chem. Ber.*, 1976, **109**, 2572.
[499] Mueller, E., and Marwede, G., Ger. Patent, 1975, 2,355,941; *Chem. Abstr.*, 1975, **83**, 133050.
[499a] Müller, K.-D., Komorowski, L., and Niedenzu, K., *Synth. Inorg. Metal-Org. Chem.*, 1978, **8**, 149.

[499b] Muraki, M., and Mukakyama, T., *Chem. Letters,* 1974, 1447.

[500] Murray, M., Schirawski, G., and Wannagat, U., *J.C.S. Dalton,* 1972, 911.

[501] Nakatsuka, K., Ide, F., Shiro, Y., Jo, Y., Kotake, Y., Kotake, K., and Furutake, K., Jap. Patent, 1969, 69 27,745; *Chem. Abstr.,* 1970, **72,** 56063b, 67458k.

[502] Nakatsuka, K., Ide, F., Shiro, Y., Jo, Y., Kotake, Y., Kotake, K., and Furutake, K., Jap. Patent, 1970, 6,927,745; *Chem. Abstr.,* 1970, **72,** 67459m.

[503] Nametkin, N. S., Batalova, L. G., Perchenko, V. N., and Ter-Asaturova, N. I., *Khim. Geterotsikl. Soedin, Akad. Nauk Latv. S.S.S.R.,* 1967, 106; *Chem. Abstr.,* 1967, **67,** 21960m.

[504] Nametkin, N. S., Ledina, L. E., Perchenko, V. N., and Batalova, L. G., *Dokl. Akad. Nauk S.S.S.R.,* 1974, **219,** 1130.

[505] Nametkin, N. S., Vdovin, V. M., Karel'skii, V. N., Babich, E. D., and Kacharmin, B. V., *Izv. Akad. Nauk S.S.S.R., Ser. Khim.,* 1973, 1364; *Chem. Abstr.,* 1973, **79,** 105338.

[506] Nametkin, N. S., Vdovin, V. M., Karel'skii, V. N., Kacharmin, B. V., and Babich, E. D., *Izv. Akad. Nauk S.S.S.R., Ser. Khim.,* 1973, 2149; *Chem. Abstr.,* 1974, **80,** 22239.

[507] Nast, R., and Dilly, P., *Angew. Chem. Internat. Edn.,* 1967, **6,** 357.

[507a] Neilson, R. H., and Wells, R. L., *Inorg. Chem.,* 1974, **13,** 480.

[507b] Neilson, R. H., Jacobs, R. D., Scheirman, R. W., and Wilburn, J. C., *Inorg. Chem.,* 1978, **17,** 1880.

[508] Nelson, P., and Pelter, A., *J. Chem. Soc.,* 1965, 5142.

[509] Nesterov, L. V., Krepysheva, N. E., and Sabirova, R. A., *Izv. Akad. Nauk S.S.S.R., Ser. Khim.,* 1975, 1908; *Chem. Abstr.,* 1975, **83,** 206371.

[510] Neumann, W. P., *Ann.,* 1963, **667,** 1.

[511] Neumann, W. P., *Ann.,* 1963, **667,** 12.

[512] Neumann, W. P., *Die Organische Chemie des Zinns,* Ferdinand Elke, 1967; *The Organic Chemistry of Tin,* Wiley, 1970.

[513] Neumann, W. P., and Heymann, E., *Angew. Chem. Internat. Edn.,* 1963, **2,** 100.

[514] Neumann, W. P., and Kühlein, K., *Tetrahedron Letters,* 1966, 3423.

[515] Neumann, W. P., and Kühlein, K., *Tetrahedron Letters,* 1966, 3419.

[516] Neumann, W. P., and Schneider, B., *Angew. Chem. Internat. Edn.,* 1964, **3,** 751.

[517] Neumann, W. P., Schneider, B., and Sommer, R., *Ann.,* 1966, **692,** 1.

[517a] Newton, M. G., King, R. B., Chang, M., and Gimeno, J., *J. Amer. Chem. Soc.,* 1978, **100,** 326.

[517b] Newton, M. G., King, R. B., Chang, M., and Gimeno, J., *J. Amer. Chem. Soc.,* 1978, **100,** 1635.

[517c] Newton, M. G., King, R. B., Chang, M., Pantaleo, N. S., and Gimeno, J., *J.C.S. Chem. Comm.,* 1977, 531.

[517d] Newton, M. G., Pantaleo, N. S., King, R. B., and Lotz, T. J., *J.C.S. Chem. Comm.*, 1978, 514.
[518] Niecke, E., and Bitter, W., *Chem. Ber.*, 1976, **109**, 415.
[518a] Niecke, E., and Bitter, W., *Angew. Chem. Internat. Edn.*, 1975, **14**, 56.
[518b] Niecke, E., and Bitter, W., *Synth. Inorg. Metal-Org. Chem.*, 1975, **5**, 231.
[519] Niecke, E., and Bitter, W., *Inorg. Nucl. Chem. Letters*, 1973, **9**, 127.
[519a] Niecke, E., and Flick, W., *J. Organometallic Chem.*, 1976, **104**, C23.
[520] Niecke, E., and Flick, W., *Angew. Chem. Internat. Edn.*, 1973, **12**, 585.
[520a] Niecke, E., Kröher, R., and Pohl, S., *Angew. Chem. Internat. Edn.*, 1977, **16**, 864.
[520b] Niecke, E., and Scherer, O. J., *Nachr. Chem. Tech.*, 1975, **23**, 395.
[520c] Niecke, E., and Wildbredt, D. A., *Angew. Chem. Internat. Edn.*, 1978, **17**, 199.
[521] Niedenzu, K., Beyer, H., Dawson, J. W., and Jenne, H., *Chem. Ber.*, 1963, **96**, 2653.
[522] Nifant'ev, E. E., and Shilov, I. V., *Zh. Obshch. Khim.*, 1975, **45**, 1264; *Chem. Abstr.*, 1975, **83**, 147546.
[523] Nifant'ev, E. E., and Shilov, I. V., *Zh. Obshch. Khim.*, 1971, **41**, 2372, *Chem. Abstr.*, 1972, **76**, 113314.
[523a] Nikolaev, N. I., Geller, N. M., Dolgoplask, B. A., Zgonnik, V. N., and Kropachev, V. A., *Vysokomol. Soedin.*, 1963, **5**, 811; *Chem. Abstr.*, 1965, **59**, 103240.
[523b] Nixon, J. F., *Adv. Inorg. Chem. Radiochem.*, 1970, **13**, 363.
[524] Nogaideli, A. I., Tkeshelashvili, R. Sh., and Makharashvili, N. P., *Zh. Obshch. Khim.*, 1974, **44**, 119; *Chem. Abstr.*, 1974, **80**, 96083.
[525] Noltes, J. G., *Rec. Trav. Chim.*, 1964, **83**, 515.
[526] Noltes, J. G., *Rec. Trav. Chim.*, 1965, **84**, 126.
[527] Noltes, J. G., and Boersma, J., *J. Organometallic Chem.*, 1969, **16**, 345.
[528] Noskov, V. G., Kirpichnikova, A. A., Sokal'skii, M. A., and Englin, M. A., *Zh. Obshch. Khim.*, 1973, **43**, 2090.
[529] Nöth, H., *Z. Naturforsch.*, 1961, **16b**, 618.
[530] Nöth, H., 142nd Meeting Amer. Chem. Soc., New York, 1962, Abstracts of papers, p. 29N.
[531] Nöth, H., *Angew. Chem.*, 1962, **74**, 506.
[532] Nöth, H., unpublished results, in Brotherton, R. J., and Steinberg, H., (Eds.) *Progress in Boron Chemistry,* Vol. III, Pergamon Press, 1970.
[533] Nöth, H., *Progr. Boron Chem.*, 1970, **3**, 211.
[533a] Nöth, H., *Chem. Ber.*, 1971, **104**, 558.
[534] Nöth, H., Dorokhov, V. A., Fritz, P., and Pfab, F., *Z. Anorg. Chem.*, 1962, **318**, 293.
[535] Nöth, H., and Fritz, P., *Z. Anorg. Chem.*, 1963, **322**, 297.
[536] Nöth, H., and Hollerer, G., *Angew. Chem.*, 1962, **74**, 718.
[537] Nöth, H., and Hollerer, G., *Angew. Chem. Internat. Edn.*, 1962, **1**, 551.

[538] Nöth, H., and Konrad, P., *Z. Naturforsch.*, 1975, **30b**, 681.
[539] Nöth, H., and Konrad, P., Unpublished Observations (1967).
[540] Nöth, H., and Lukas, S., *Chem. Ber.*, 1962, **95**, 1505.
[540a] Nöth, H., and Regnet, W., *Chem. Ber.*, 1969, **102**, 167.
[540b] Nöth, H., Regnet, W., Rihl, H., and Standfest, R., *Chem. Ber.*, 1971, **104**, 722.
[540c] Nöth, H., Reichenbach, W., and Winterstein, W., *Chem. Ber.*, 1977, **110**, 2158.
[541] Nöth, H., Reiner, D., and Storch, W., *Chem. Ber.*, 1973, **106**, 1508.
[541a] Nöth, H., Reith, H., and Thorn, V., *J. Organometallic Chem.*, 1978, **159**, 165.
[542] Nöth, H., Schick, H., and Meister, W., *J. Organometallic Chem.*, 1964, **1**, 401.
[543] Nöth, H., and Schmid, G., *J. Organometallic Chem.*, 1966, **5**, 109.
[544] Nöth, H., Schmid. G., and Chung, Y., Proc. 8th Internat. Conf. Co-ord. Chem., 1964, p. 180, Springer.
[545] Nöth, H., Schmid, G., and Deberitz, J., *Angew. Chem. Internat. Edn.*, 1968, **7**, 293.
[545a] Nöth, H., and Schuchardt, U., *Z. Anorg. Chem.*, 1975, **418**, 97.
[546] Nöth, H., and Schweizer, P., *Chem. Ber.*, 1964, **97**, 1464.
[547] Nöth, H., and Sprague, M. J., *J. Organometallic Chem.*, 1970, **22**, 11.
[547a] Nöth, H., and Sprague, M. J., *J. Organometallic Chem.*, 1970, **23**, 323.
[548] Nöth, H., and Storch, W., *Chem. Ber.*, 1974, **107**, 1028.
[549] Nöth, H., and Storch, W., *Chem. Ber.*, 1976, **109**, 884.
[549a] Nöth, H., and Storch, W., *Chem. Ber.*, 1977, **110**, 2607.
[550] Nöth, H., and Suchy, H., *Chem. Ber.*, 1971, **104**, 549.
[551] Nöth, H., and Tinhof, W., *Chem. Ber.*, 1974, **107**, 3806.
[551a] Nöth, H., and Tinhof, W., *Chem. Ber.*, 1975, **108**, 3109.
[552] Nöth, H., Tinhof, W., and Taeger, T., *Chem. Ber.*, 1974, **107**, 3113.
[553] Nöth, H., and Ullmann, R., *Chem. Ber.*, 1976, **109**, 1942.
[554] Nöth, H., and Vahrenkamp, H., *J. Organometallic Chem.*, 1969, **16**, 357.
[555] Nöth, H., and Vetter, H.-J., *Chem. Ber.*, 1963, **96**, 1298; Holmes, R. R., and Wagner, R. P., *J. Amer. Chem. Soc.*, 1962, **84**, 357.
[556] Nöth, H., and Vetter, H.-J., *Chem. Ber.*, 1963, **96**, 1298.
[556a] Nugent, W. A., and Harlow, R. L., *J.C.S. Chem. Comm.*, 1978, 579.
[557] Oediger, H., Kabbe, H. J., Moller, F., and Eiter, K., *Chem. Ber.*, 1966, **99**, 2012.
[558] Oertel, G., Holtschmidt, H., and Malz, H., Ger. Patent, 1964, 1,170,393.
[559] Oertel, G., Holtschmidt, H., and Malz, H., Ger. Patent, 1963, 1,157,226.
[560] Oertel, G., Holtschmidt, H., and Malz, H., *Chem. Ber.*, 1964, **97**, 891.
[561] Oertel, G., Malz, H., Holtschmidt, H., and Degener, E., Ger. Patent, 1963, 1,154,475.

[562] Oertel, G., Malz, H., Holtschmidt, H., and Degener, E., Ger. Patent, 1963, 1,155,433.
[563] Oguni, N., Tani, H., Araki, T., and Ueyama, N., *J. Amer. Chem. Soc.*, 1967, **89**, 173.
[564] Oguni, N., Sano, S., Fujimura, M., and Tani, H., *Polym. J.*, 1973, **4**, 607.
[564a] Olofson, R. A., Hoskin, D. H., and Lotts, K. D., *Tetrahedron Letters*, 1978, 1667.
[564b] Onan, K. D., McPhail, A. T., Yoder, C. H., and Hillyard, R. W., *J.C.S. Chem. Comm.*, 1978, 209.
[565] Osthoff, R. C., and Brown, C. A., *J. Amer. Chem. Soc.*, 1952, **74**, 2378.
[566] Overberger, C. G., and Ishida, S., *J. Polymer Sci. (B)*, 1965, **3**, 789.
[566a] Paine, R. T., *J. Amer. Chem. Soc.*, 1977, **99**, 3884.
[567] Patterson, W. J., and Bilow, N., *J. Polymer Sci. (A-I)*, 1969, **7**, 1089.
[568] Pelter, A., and Levitt, T. E., *Tetrahedron*, 1970, **26**, 1545.
[569] Pelter, A., Levitt, T. E., and Nelson, P., *Tetrahedron*, 1970, **26**, 1539.
[570] Perry, E., *Makromol. Chem.*, 1963, **65**, 145.
[571] Peters, F. M., and Fetter, N. R., *J. Organometallic Chem.*, 1965, **4**, 181.
[572] Peterson, W. R., and Sollott, G. P., *J. Amer. Chem. Soc.*, 1967, **89**, 5054, 6783.
[573] Peterson, L. K., and The, K. I., *Can. J. Chem.*, 1972, **50**, 562.
[574] Peterson, L. K., and The, K. I., *Can. J. Chem.*, 1972, **50**, 553.
[575] Peterson, L. K., and Wilson, G. L., *Can. J. Chem.*, 1971, **49**, 3171.
[576] Petrov, A. A., and Kormer, V. A., *Zh. Obshch. Khim.*, 1964, **34**, 1868.
[577] Petrovskaya, L. I., Fedin, E. I., Sheludyakov, V. D., and Kozyukov, V. P., *Zh. Strukt. Khim.*, 1967, **8**, 51.
[578] Petz, W., *J. Organometallic Chem.*, 1973, **55**, C42.
[579] Petz, W., *J. Organometallic Chem.*, 1974, **72**, 369.
[579a] Petz, W., *J. Organometallic Chem.*, 1975, **90**, 223.
[580] Petz, W., and Jonas, A., *J. Organometallic Chem.*, 1976, **120**, 423.
[581] Petz, W., and Schmid, G., *Angew. Chem. Internat. Edn.*, 1972, **11**, 934.
[582] Petz, W., and Schmid, G., *J. Organometallic Chem.*, 1972, **35**, 321.
[583] Pike, R. M., *J. Org. Chem.*, 1961, **26**, 232.
[584] Pike, R. M., *J. Polymer Sci.*, 1961, **50**, 151.
[585] Pike, R. M., *Rec. Trav. Chim.*, 1961, **80**, 819.
[586] Pike, R. M., *Rec. Trav. Chim.*, 1961, **80**, 885.
[587] Pike, R. M., and Weaver, S. J., *Rec. Trav. Chim.*, 1967, **86**, 606.
[587a] Pinchuk, A. M., Filonenko, L. P., and Kirsanov, A. V., *Khim. Elementoorg. Soedin*, 1976, 98; *Chem. Abstr.*, 1977, **86**, 43777.
[588] Pinchuk, A. M., Filonenko, L. P., and Suleimanova, M. G., *Zh. Obshch. Khim.*, 1972, **42**, 2116.
[588a] Pinchuk, A. M., Filonenko, L. P., and Suleimanova, M. G., *Zh. Obshch. Khim.*, 1972, **42**, 2115.

[588b] Pinchuk, A. M., Khranovskii, V. A., Kuplennik, Z. I., and Filonenko, L. P., *J. Gen. Chem. U.S.S.R.*, 1977, **47**, 49.
[589] Pittman, C. U., Patterson, W. J., and McManus, S. P., *J. Polymer Sci. A-I,* 1971, **9**, 3187.
[590] Pittman, C. U., Patterson, W. J., and McManus, S. P., *J. Polymer Sci.*, 1976, **14**, 1715.
[590a] Pohl, S., *J. Organometallic Chem.*, 1977, **142**, 185.
[590b] Pohl, S., Niecke, E., and Schäfer, H.-G., *Angew. Chem. Internat. Edn.*, 1978, **17**, 136.
[591] Pommier, J. C., and Roubineau, A., *J. Organometallic Chem.*, 1969, **16**, P23.
[592] Pommier, J. C., and Roubineau, A., *J. Organometallic Chem.*, 1969, **17**, P25.
[593] Pommier, J. C., and Roubineau, A., *J. Organometallic Chem.*, 1973, **50**, 101.
[594] Ponomarev, S. V., Zakharova, O. A., Lebedev, S. A., and Lutsenko, I. F., *Zh. Obshch. Khim.*, 1975, **45**, 2680.
[594a] Power, P. P., D.Phil. Thesis, Sussex, 1977.
[594b] Prade, W. D. La., and Nordman, C. E., *Inorg. Chem.*, 1969, **8**, 1669.
[595] Pratt, J. R., and Thames, S. F., *Synthesis,* 1973, 223.
[596] Pressl, K., and Schmidt, A., *Chem. Ber.*, 1972, **105**, 3518.
[597] Pressl, K., and Schmidt, A., *Chem. Ber.*, 1973, **106**, 2217.
[598] Pribytkova, I. M., Kisin, A. V., Luzikov, Yu. N., Makoveyeva, N. P., Torocheshnikov, V. N., and Ustynyuk, Yu. A., *J. Organometallic Chem.*, 1971, **30**, C57.
[599] Prinz, R., and Werner, H., *Angew. Chem. Internat. Edn.*, 1967, **6**, 91.
[600] Prons, V. N., Grinblat, M. P., and Klebanski, A. L., *Zh. Obshch. Khim.*, 1975, **45**, 2423.
[601] Pudovik, A. N., Batyeva, E. S., and Ofitserov, E. N., *Zh. Obshch. Khim.*, 1976, **46**, 1441.
[602] Pudovik, A. N., Batyeva, E. S., Ofitserov, E. N., and Al'fonsov, V. A., *Zh. Obshch. Khim.*, 1975, **45**, 2338; *Chem. Abstr.*, 1976, **84**, 42805.
[603] Pudovik, M. A., Medvedeva, M. D., Kibardina, L. K., and Pudovik, A. N., *Zh. Obshch. Khim.*, 1975, **45**, 941; *Chem. Abstr.*, 1975, **83**, 28336.
[604] Pudovik, M. A., Medvedeva, M. D., and Pudovik, A. N., *Zh. Obshch. Khim.*, 1975, **45**, 700; *Chem. Abstr.*, 1975, **83**, 10260.
[605] Pudovik, M. A., Medvedeva, M. D., and Pudovik, A. N., *Zh. Obshch. Khim.*, 1976, **46**, 773.
[606] Pudovik, A. N., Pudovik, M. A., Terent'eva, S. A., and Gol'dfrab, E. I., *Zh. Obshch. Khim.*, 1972, **42**, 1901.
[607] Pudovik, M. A., Terent'eva, S. A., Nebogatikova, I. V., and Pudovik, A. N., *Zh. Obshch. Khim.*, 1974, **44**, 1020.
[608] Quane, D., and Roberts, S. D., *J. Organometallic Chem.*, 1976, **108**, 27.

[609] Raedler, K. P., Benndorf, G., Horn, V., and Fuehrling, W., *J. Prakt. Chem.*, 1976, **318**, 697.
[610] Rogozhin, S. V., Davidovich, Yu. A., Andreev, S. M., Yurtanov, A. I., Mironova, N. V., and Samoilova, N. A., Tezisy Dokl. -Vses. Simp. Khim. Pept. Belkov, 3rd, 1974, 126; *Chem. Abstr.*, 1976, **85**, 193061.
[611] Raigorodskii, I. M., Bakhaeva, G. P., and Savin, V. A., U.S.S.R. Patent, 1976, 525,714; *Chem. Abstr.*, 1977, **86**, 5967.
[612] Razumov, A. I., Liorber, B. G., Sokolov, M. P., and Zykova, T. V., *Zh. Obshch. Khim.*, 1971, **41**, 2106.
[613] Razuvaev, G. A., Dodonov, V. A., Mysin, N. I., and Starvostina, T. I., *Zh. Obshch. Khim.*, 1972, **42**, 152.
[614] Razuvaev, G. A., Latyaeva, V. N., Gladyshev, E. N., Lineva, A. N., and Krasil'nikova, E. V., *Dokl. Akad. Nauk S.S.S.R.*, 1975, **223**, 1144; *Chem. Abstr.*, 1976, **84**, 5102.
[614a] Reuter, K., and Neumann, W. P., *Tetrahedron Letters*, 1978, 5235.
[614b] Rie, J. E., and Oliver, J. P., *J. Organometallic Chem.*, 1977, **133**, 147; and Oliver, J. P., *Adv. Organometallic Chem.*, 1977, **16**, 111.
[615] Rijkens, F., Janssen, M. J., and van der Kerk, G. J. M., *Rec. Trav. Chim.*, 1965, **84**, 1597.
[616] Rilsel, L., and Somieski, R., *Z. Anorg. Chem.*, 1975, **411**, 148.
[616a] Rivière, P., Cazes, A., Castel, A., Rivière-Baudet, M., and Satgé, J., *J. Organometallic Chem.*, 1978, **155**, C58.
[617] Rivière, P., Rivière-Baudet, M., Couret, C., and Satgé, J., *Synth. Inorg. Metal-Org. Chem.*, 1974, **4**, 295.
[618] Rivière, P., and Satgé, J., *Bull. Soc. Chim. Fr.*, 1971, 3221.
[619] Rivière, P., Satgé, J., Dousse, G., Rivière-Baudet, M., and Couret, C., *J. Organometallic Chem.*, 1974, **72**, 339.
[620] Rivière, P., Satgé, J., and Soula, D., *J. Organometallic Chem.*, 1974, **72**, 329.
[621] Rivière-Baudet, M., and Rivière, P., *J. Organometallic Chem.*, 1976, **116**, C49.
[621a] Rivière-Baudet, M., Rivière, P., and Satgé, J., *J. Organometallic Chem.*, 1978, **154**, C23.
[622] Rivière-Baudet, M., and Satgé, J., *J. Organometallic Chem.*, 1973, **56**, 159.
[623] Rivière-Baudet, M., and Satgé, J., *Synth. Inorg. Metal-Org. Chem.*, 1971, **1**, 249.
[624] Rivière-Baudet, M., and Satgé, J., *Synth. Inorg. Metal-Org. Chem.*, 1971, **1**, 257.
[625] Roesky, H. W., *Chem. Ber.*, 1972, **105**, 1439.
[626] Roesky, H. W., and Dietl, M., *Angew. Chem.*, 1973, **85**, 453.
[627] Roesky, H. W., and Janssen, E., *Chem. Z.*, 1974, **98**, 260; *Chem. Abstr.*, 1974, **81**, 32713.

[628] Roesky, H. W., and Kloker, W., *Z. Naturforsch.*, 1972, **27b**, 486.
[628a] Roesky, H. W., and Mehrotra, S. K., *Angew. Chem. Internat. Edn.*, 1978, **17**, 599.
[629] Roesky, H. W., and Peterson, O., *Angew. Chem.*, 1973, **85**, 413.
[630] Roesky, H. W., Schaper, W., Gross-Boewing, W., and Dietl, M., *Z. Anorg. Chem.*, 1975, **416**, 306.
[630a] Roesky, H. W., and Sidiropoulos, G., *Z. Naturforsch.*, 1977, **32b**, 628.
[631] Roesky, H. W., and Sidiropoulos, G., *Angew. Chem.*, 1976, **88**, 759.
[631a] Roesky, H. W., and Wiezer, H., *Angew. Chem.*, 1973, **85**, 722.
[632] Roesky, H. W., and Wiezer, H., *Chem. Ber.*, 1973, **106**, 280.
[633] Roesky, H. W., and Wiezer, H., *Chem. Ber.*, 1974, **107**, 3186.
[634] Roesky. H. W., and Wiezer, H., *Chem. Ber.*, 1974, **107**, 1153.
[635] Roesky, H. W., and Wiezer, H., *J. Inorg. Nucl. Chem.*, (H. H. Herbert Memorial Vol.), 1976, 45.
[636] Rosnati, L., *Gazz. Chim. Ital.*, 1948, **78**, 516; *Chem. Abstr.*, 1949, **43**, 1005.
[637] Roubineau, A., and Pommier, J. C., *J. Organometallic Chem.*, 1976, **107**, 63.
[637a] Roundhill, D. M., *Inorg. Chem.*, 1970, **9**, 254.
[638] Ruehl, W. J., and Kober, F., *Z. Naturforsch.*, 1976, **31b**, 190.
[639] Ruehl, W. J., and Kober, F., *Z. Naturforsch.*, 1976, **31b**, 307.
[640] Ruff, J. K., *J. Org. Chem.*, 1962, **27**, 1020.
[641] Ruff, J. K., *J. Amer. Chem. Soc.*, 1961, **83**, 2835.
[642] Ruppert, I., Bastian, V., and Appel, R., *Chem. Ber.*, 1974, **107**, 3426.
[643] Russ, C. R., and MacDiarmid, A. G., *Angew. Chem. Internat. Edn.*, 1964, **3**, 509.
[644] Saegusa, T., Ito, Y., Kobayashi, S., and Hirota, K., *J. Amer. Chem. Soc.*, 1967, **89**, 2240.
[645] Sagan, L. S., Zingaro, R. A., and Irgolic, K. J., *J. Organometallic Chem.*, 1972, **39**, 301.
[646] Sakakibara, T., Hirabayashi, T., and Ishii, Y., *J. Organometallic Chem.*, 1972, **46**, 231.
[646a] Sanderson, J. J., and Hauser, C. R., *J. Amer. Chem. Soc.*, 1949, **71**, 1595.
[647] Sanger, A. R., *Inorg. Nucl. Chem. Letters*, 1973, **9**, 351.
[648] Santini-Scampucci, C., and Wilkinson, G., *J.C.S. Dalton*, 1976, 807.
[648a] Sarraje, I., and Lorberth, J., *J. Organometallic Chem.*, 1978, **146**, 113.
[649] Satgé, J., and Dousse, G., *J. Organometallic Chem.*, 1973, **61**, C26.
[650] Satgé, J., and Dousse, G., *Helv. Chim. Acta*, 1972, **55**, 2406.
[651] Satgé, J., and Lesbre, M., *Bull. Soc. Chim. Fr.*, 1965, 2578.
[651a] Satgé, J., and Rivière-Baudet, M., *Rec. Trav. Chim.*, 1975, **94**, 22.
[652] Sato, T., Iwasaki, K., Kakogawa, G., Kasahara, K., and Okano, T., Jap. Patent, 1971, 71 34,970; *Chem. Abstr.*, 1972, **76**, 86440.

[653] Savel'eva, N. I., Baukov, Yu. I., Kostyuk, A. S., and Lutsenko, I. F., *Zh. Obshch. Khim.*, 1974, **44**, 1753; *Chem. Abstr.*, 1975, **82**, 4352.
[654] Scantlin, W. M., and Norman, A. D., *Inorg. Chem.*, 1972, **11**, 3082.
[655] Schaeffer, C. D., and Zuckerman, J. J., *J. Amer. Chem. Soc.*, 1974, **96**, 7160.
[656] Sheludyakov, V. D., Rodionov, E. S., Khatuntsev, G. D., and Mironov, V. F., *Zh. Obshch. Khim.*, 1972, **42**, 367.
[657] Scherer, O. J., *J. Organometallic Chem., Rev. (A)*, 1968, **3**, 281.
[658] Scherer, O. J., and Hornig, P., *Angew. Chem.*, 1966, **78**, 776.
[659] Scherer, O. J., and Hornig, P., *Angew. Chem. Internat. Edn.*, 1967, **6**, 89.
[659a] Scherer, O. J., and Hornig, P., *J. Organometallic Chem.*, 1967, **8**, 465.
[660] Scherer, O. J., and Janssen, W., *J. Organometallic Chem.*, 1969, **16**, P69.
[660a] Scherer, O. J., and Kuhn, N., *Chem. Ber.*, 1974, **107**, 2123.
[660b] Scherer, O. J., and Kuhn, N., *J. Organometallic Chem.*, 1974, **82**, C3.
[661] Scherer, O. J., and Schmidt, M., *Angew. Chem.*, 1965, **77**, 456.
[662] Scherer, O. J., and Schmidt, M., *Chem. Ber.*, 1965, **98**, 2243.
[662a] Scherer, O. J., and Schnabl, G., *Chem. Ber.*, 1976, **109**, 2996.
[663] Scherer, O. J., and Schnabl, G., *J. Organometallic Chem.*, 1973, **52**, C18.
[664] Scherer, O. J., and Wolmershaeuser, G., *Z. Naturforsch.*, 1974, **29b**, 444.
[665] Schlak, O., Stadelmann, W., Stelzer, O., and Schmutzler, R., *Z. Anorg. Chem.*, 1976, **419**, 275.
[665a] Schlientz, W. J., and Ruff, J. K., *Inorg. Chem.*, 1972, **11**, 2265, and refs. therein.
[666] Schmid, G., *Angew. Chem. Internat. Edn.*, 1972, **11**, 1034.
[667] Schmid, G., *Chem. Ber.*, 1970, **103**, 528.
[667a] Schmid, G., Petz, W., and Nöth, H., *Inorg. Chim. Acta*, 1970, **4**, 423.
[667b] Schmid, G., and Schulze, J., *Angew. Chem. Internat. Edn.*, 1977, **16**, 249.
[668] Schmid, G., and Weber, L., *Z. Naturforsch.*, 1970, **25b**, 1083; 1971, **26b**, 994.
[668a] Schmidbaur, H., Aly, A. A. M., and Schubert, U., *Angew. Chem. Internat. Edn.*, 1978, **17**, 846.
[669] Schmidbaur, H., and Rathlein, K. H., *Chem. Ber.*, 1974, **107**, 102.
[669a] Schmidbaur, H., Scharf, W., and Füller, H.-J., *Z. Naturforsch.*, 1977, **32b**, 858.
[670] Schmidbaur, H., and Schmidt, M., *Angew. Chem. Internat. Edn.*, 1962, **1**, 327.
[671] Schmidbaur, H., and Schmidt, M., *Angew. Chem.*, 1958, **70**, 657.
[671a] Schmidbaur, H., and Shiotani, A., *J. Amer. Chem. Soc.*, 1970, **92**, 7003.
[672] Schmidbaur, H., and Wolfsberger, W., *Synth. Inorg. Metal-Org. Chem.*, 1971, **1**, 111.
[672a] Schmidpeter, A., and Rossknecht, H., *Chem. Ber.*, 1974, **107**, 3146.
[673] Schmidpeter, A., and Weinmaier, J. H., *Angew. Chem.*, 1975, **87**, 517.
[674] Schmitz-Dumont, O., Chem. Soc. (London), Spec. Publ., 1961, **15**, 100.

[675] Schmitz-Dumont, O., and Schulte, H., *Z. Anorg. Chem.*, 1955, **282**, 253.
[676] Schmutzler, R., *J.C.S. Dalton,* 1973, 2687.
[676a] Schmutzler, R., *Inorg. Chem.*, 1964, **3**, 415.
[677] Schmutzler, R., *Z. Naturforsch.*, 1964, **19b**, 1101.
[678] Schmutzler, R., *Angew. Chem. Internat. Edn.*, 1964, **3**, 753.
[679] Schmutzler, R., *Chem. Comm.*, 1965, 19; *Inorg. Chem.*, 1968, **7**, 1327.
[680] Schöllkopf, U., Banhidai, B., and Scholz, H. U., *Ann.*, 1972, **761**, 137.
[681] Schram, E. P., *Inorg. Chem.*, 1966, **5**, 1291.
[682] Schumann, H., and Roth, A., *J. Organometallic Chem.*, 1968, **11**, 125.
[682a] Schumann, H., Du Mont, W. W., and Kroth, H. J., *Chem. Ber.*, 1976, **109**, 237.
[682b] Schumann, H., and Roth, A., *Chem. Ber.*, 1969, **102**, 3731.
[682c] Schuster, K., and Wiberg, E., *Z. Anorg. Chem.*, 1933, **213**, 77.
[683] Schwartz, L. D., and Keller, P. C., *J. Amer. Chem. Soc.*, 1972, **94**, 3015.
[684] Scotti, M., and Werner, H., *Helv. Chim. Acta,* 1974, **57**, 1234.
[684a] Scotti, M., Werner, M., Brown, D. L. S., Cavell, S., Connor, J. A., and Skinner, H. A., *Inorg. Chim. Acta,* 1977, **25**, 261.
[685] Seppelt, K., *Inorg. Chem.*, 1973, **12**, 2837.
[686] Sevast'yanova, I. V., Ponamarev, A. I., Klebanskii, A. L., and Kozlov, V. T., *Vysokomol. Soedin. Ser. B.*, 1973, **15**, 282; *Chem. Abstr.*, 1973, **79**. 66846.
[687] Seyffert, H., and Wannagat, U., *Angew. Chem. Internat. Edn.*, 1965, **4**, 438.
[687a] Shatenshtein, A. I., and Shapiro, I. O., *Russ. Chem. Rev.*, 1968, **37**, 845.
[687b] Shea, K. J., Gobeille, R., Bramblett, J., and Thompson, E., *J. Amer. Chem. Soc.*, 1978, **100**, 1611.
[688] Sheludyakov, V. D., Kozyukov, V. P., Petrovskaya, L. I., and Mironov, V. F., *Khim. Geterotsikl, Soedin., Akad. Nauk Latv. S.S.S.R.*, 1967, 185; *Chem. Abstr.*, 1967, **67**, 43858m.
[689] Sheludyakov, V. D., Kirilin, A. D., and Mironov, V. F., *Zh. Obshch. Khim.*, 1975, **45**, 479; *Chem. Abstr.*, 1975, **82**, 156445.
[690] Sheludyakov, V. D., Kirilin, A. D., and Mironov, V. F., *Zh. Obshch. Khim.*, 1975, **45**, 707; *Chem. Abstr.*, 1975, **83**, 10261.
[691] Sheludyakov, V. D., Rodionov, E. S., Bochkarev, V. N., Polivanov, A. N., Strelenko, Yu. A., and Mironov, V. F., *Zh. Obshch. Khim.*, 1974, **44**, 1506.
[691a] Sheludyakov, V. D., Tkachev, A. S., Sheludyakova, S. V., and Mironov, V. F., *J. Gen. Chem. U.S.S.R.*, 1977, **47**, 974.
[692] Shibata, Y., Kondo, S., Fujimura, Y., and Koketsu, J., *Nippon Kagaku Kaishi,* 1975, 1868; *Chem. Abstr.*, 1976, **84**, 30049.
[693] Simonyan, L. A., Avetisyan, E. A., and Gambaryan, N. P., *Izv. Akad. Nauk S.S.S.R. Ser. Khim.*, 1972, 2352; *Chem. Abstr.*, 1973, **78**, 43574.

[694] Singer, R. J., Eisenhut, M., and Schmutzler, R., *J. Fluorine Chem.*, 1971, **1**, 193.
[695] Singh, A., and Mehrotra, R. C., *Indian J. Chem.*, 1975, **13**, 1197.
[696] Singh, A., Rai, A. K., and Mehrotra, R. C., *J.C.S. Dalton,* 1972, 1911.
[697] Sisido, K., and Kozima, S., *J. Org. Chem.*, 1962, **27**, 4051.
[698] Sisido, K., and Kozima, S., *J. Org. Chem.*, 1964, **29**, 907.
[699] Sollott, G. P., and Peterson, W. R., U.S. Patent, 1973, 3,759,976; *Chem. Abstr.*, 1973, **79**, 126623.
[700] Sommer, R., Neumann, W. P., and Schneider, B., *Tetrahedron Letters,* 1964, 3875.
[701] Sommer, R., Schneider, B., and Neumann, W. P., *Ann.*, 1966, **692**, 12.
[701a] Srivastava, G., *J.C.S. Perkin I,* 1974, 916.
[701b] Srivastava, G., *J. Organometallic Chem.*, 1978, **152**, 39.
[702] Stadnichuk, T. V., Kormer, V. A., and Petrov, A. A., *Zh. Obshch. Khim.*, 1964, **34**, 3284.
[702a] Staudinger, H., and Meyer, J., *Helv. Chim. Acta,* 1919, **2**, 635.
[703] Steinberg, H., and Brotherton, R. J., *Organoboron Chemistry,* 1966, Vol. 2, Interscience, London.
[704] Steinberger, H., and Kuchen, W., *Z. Naturforsch.*, 1973, **28b**, 44.
[704a] Still, W. C., *J. Amer. Chem. Soc.*, 1978, **100**, 1481.
[705] Stober, M. R., Michael, K. W., and Speier, J. L., *J. Org. Chem.*, 1967, **32**, 2740.
[705a] Storch, W., Jackstiess, W., Nöth, H., and Winter, G., *Angew. Chem.*, 1977, **89**, 494.
[706] Storch, W., and Nöth, H., *Angew. Chem.*, 1976, **88**, 231.
[706a] Storch, W., and Nöth, H., *Chem. Ber.*, 1977, **110**, 1636.
[706b] Streitwieser, A., and Scannon, P. J., *J. Amer. Chem. Soc.*, 1973, **95**, 6273; Streitwieser, A., Berke, C. M., and Robbers, K., *J. Amer. Chem. Soc.*, 1978, **100**, 8271.
[706c] Sugyama, K., Takahashi, Y., Pac, S., and Motoyima, S., Proc. Conf. Chem. Vap. Deposition Int. Conf., 1975, 140; *Chem. Abstr.*, 1976, **84**, 46655.
[707] Sujishi, S., and Witz, S., *J. Amer. Chem. Soc.*, 1954, **76**, 4631.
[708] Sujishi, S., and Witz, S., *J. Amer. Chem. Soc.*, 1957, **79**, 2447.
[709] Sujishi, S., and Witz, S., U.S., Department Comm., P.B. Report 143, p. 572; *Chem. Abstr.*, 1961, **55**, 17333.
[710] Sujishi, S., and Witz, S., *J. Amer. Chem. Soc.*, 1957, **79**, 2447; Grosse-Ruyken, H., and Kleesaat, R., *Z. Anorg. Chem.*, 1961, **308**, 122.
[711] Suliman, M. R., and Schram, E. P., *Inorg. Chem.*, 1973, **12**, 920.
[712] Suliman, M. R., and Schram, E. P., *Inorg. Chem.*, 1973, **12**, 923.
[713] Suzuki, K., and Tanaka, Y., Jap. Patent, 1972, 72 31,701; *Chem. Abstr.*, 1973, **78**, 30545.
[713a] Tada, H., and Okawara, R., *J. Org. Chem.*, 1970, **35**, 1666.
[714] The, K. I., and Cavell, R. G., *Inorg. Chem.*, 1976, **15**, 2518.

[714a] The, K. I., Vande Griend, L., Whitla, W. A., and Cavell, R. G., *J. Amer. Chem. Soc.*, 1977, **99**, 7379.

[715] Thomas, I. M., *Can. J. Chem.*, 1961, **39**, 1386.

[716] Thomas, M. G., Kopp, R. W., Schultz, C. W., and Parry, R. W., *J. Amer. Chem. Soc.*, 1974, **96**, 2646.

[716a] Thomas, M. G., Schultz, C. W., and Parry, R. W., *Inorg. Chem.*, 1977, **16**, 994.

[717] Toyo Rayon Co. Ltd., Fr. Patent, 1,386,994; *Chem. Abstr.*, 1965, **62**, 13063.

[718] Tzschach, A., and Reiss, E., *J. Organometallic Chem.*, 1967, **8**, 255.

[718a] Tzschach, A., and Lange, W., *Z. Anorg. Chem.*, 1964, **326**, 280.

[719] Uhle, K., and Hahnfeld, K., *Z. Chem.*, 1973, **13**, 376; *Chem. Abstr.*, 1974, **80**, 96077.

[720] Uhle, K., Kinting, A., and Geissler, W., *J. Organometallic Chem.*, 1975, **99**, 53.

[721] Urata, K., Itoh, K., and Ishii, Y., *J. Organometallic Chem.*, 1974, **66**, 229.

[721a] Van Wazer, J. R., and Moedritzer, K., *Inorg. Chem.*, 1964, **3**, 268.

[721b] Veith, M., Recktenwald, O., and Humpfer, E., *Z. Naturforsch.*, 1978, **33b**, 14.

[722] Vetter, H.-J., and Nöth, H., *Z. Naturforsch.*, 1964, **19b**, 167.

[723] Vetter, H.-J., and Nöth, H., *Z. Anorg. Chem.*, 1964, **330**, 233.

[724] Vetter, H.-J., Strametz, H., and Nöth, H., *Angew. Chem.*, 1963, **75**, 417.

[725] Voronkov, M. G., Marmur, L. Z., Dolgov, O. N., Pestunovich, V. A., Pokrovskii, E. I., and Popelis, J., *Zh. Obshch. Khim.*, 1971, **41**, 1987.

[726] Voronkov, M. G., Mirskov, R. G., Yarosh, O. G., Ishchenko, O. S., and Tsetlina, E. O., *Zh. Obshch. Khim.*, 1973, **43**, 2425; *Chem. Abstr.*, 1974, **80**, 59992.

[727] Voronkov, M. G., and Yarosh, O. G., *Zh. Obshch. Khim.*, 1972, **42**, 2030; *Chem. Abstr.*, 1973, **78**, 43570.

[728] Wada, M., Suda, T., and Okawara, R., *J. Organometallic Chem.*, 1974, **65**, 335.

[729] Walther, B., Mahrwald, R., Jahn, C., and Klar, W., *Z. Anorg. Chem.*, 1976, **423**, 144.

[730] Walton, D. R. M., Brit. Patent, 1970, 1,153,132; *Chem. Abstr.*, 1970, **72**, 79212p.

[731] Wannagat, U., *Pure and Appl. Chem. Rev.*, 1969, **19**, 329.

[732] Wannagat, U., and Autzen, H., *Z. Anorg. Chem.*, 1976, **420**, 119.

[732a] Wannagat, U., Autzen, H., Kuckertz, H., and Wismar, H. J., *Z. Anorg. Chem.*, 1972, **394**, 254.

[732b] Wannagat, U., Autzen, H., and Schlingmann, M., *Z. Anorg. Chem.*, 1976, **419**, 41.

[732c] Wannagat, U., Herzig, J., and Bürger, H., *J. Organometallic Chem.*, 1970, **23**, 373.

[733] Wannagat, U., Herzig, J., Schmidt, P., and Schulze, M., *Monatsh.*, 1971, **102**, 1817.
[734] Wannagat, U., Kuckertz, H., Kruger, C., and Pump, J., *Z. Anorg. Chem.*, 1964, **333**, 54.
[734a] Wannagat, U., Moretto, H. H., and Schmidt, P., *Z. Anorg. Chem.*, 1971, **385**, 164.
[735] Wannagat, U., Pump, J., and Bürger, H., *Monatsh.*, 1963, **94**, 1013.
[736] Wannagat, U., and Seyffert, H., *Angew. Chem. Internat. Edn.*, 1965, **4**, 438.
[737] Wannagat, U., Schmidt, P., and Schulze, M., *Angew. Chem. Internat. Edn.*, 1967, **6**, 447.
[738] Warthmann, W., and Schmidt, A., *Z. Anorg. Chem.*, 1975, **418**, 57, 61.
[739] Warthmann, W., and Schmidt, A., *Z. Anorg. Chem.*, 1975, **418**, 145.
[740] Washburne, S. S., and Peterson, W. R., *J. Organometallic Chem.*, 1970, **21**, 59.
[740a] Weingarten, H., *Chem. Comm.*, 1966, 293.
[741] Weingarten, H., and Miles, M. G., *J. Org. Chem.*, 1968, **33**, 1506.
[742] Weingarten, H., Miles, M. G., Byrn, S. R., and Hobbs, C. F., *J. Amer. Chem. Soc.*, 1967, **89**, 1968.
[743] Weingarten, H., Miles, M. G., and Edelmann, N. K., *Inorg. Chem.*, 1968, **7**, 879.
[744] Weingarten, H., and Van Wazer, J. R., *J. Amer. Chem. Soc.*, 1965, **87**, 724.
[745] Weingarten, H., and Van Wazer, J. R., *J. Amer. Chem. Soc.*, 1966, **88**, 2700.
[746] Weingarten, H., and White, W. A., *J. Amer. Chem. Soc.*, 1966, **88**, 850; *J. Org. Chem.*, 1966, **31**, 2874.
[747] Weingarten, H., and White, W. A., *J. Org. Chem.*, 1966, **31**, 4041.
[748] Wells, R. L., and Neilson, R. H., *Synth. Inorg. Metal-Org. Chem.*, 1973, **3**, 137.
[748a] White, W. A., and Weingarten, H., *J. Org. Chem.*, 1967, **32**, 213.
[749] Wiberg, E., and Sturm, W., *Z. Naturforsch.*, 1953, **8b**, 529; 1955, **10b**, 108, 111.
[750] Wiberg, E., and Sturm. W., *Z. Naturforsch.*, 1955, **10b**, 109.
[751] Wiberg, N., Baumeister, W., and Zahn, P., *J. Organometallic Chem.*, 1972, **36**, 267.
[752] Wiberg, N., and Joo, W. C., *Z. Naturforsch.*, 1966, **21b**, 1234.
[753] Wiberg, N., Pracht, H. J., *Chem. Ber.*, 1972, **105**, 1377, 1392, 1399; *J. Organometallic Chem.*, 1972, **40**, 289.
[754] Wiberg, N., and Schmid, K. H., *Z. Naturforsch.*, 1966, **21b**, 1107.
[755] Wiberg, N., and Schmid, K. H., *Z. Anorg. Chem.*, 1966, **345**, 93.
[755a] Wiberg, N., Vasischt, S. K., Fischer, G., and Weinberg, E., *Chem. Ber.*, 1976, **109**, 710.

[756] Wiberg, N., and Veith, M., *Chem. Ber.*, 1971, **104**, 3191.
[756a] Wilburn, J. C., and Neilson, R. H., *J.C.S. Chem. Comm.*, 1977, 308.
[756b] Wilson, R. D., and Bau, R., *J. Amer. Chem. Soc.*, 1974, **96**, 7601.
[757] Winters, L. J., and Hill, D. T., *Inorg. Chem.*, 1965, **4**, 1433.
[758] Witke, K., Reich, P., and Kriegsmann, H., *J. Organometallic Chem.*, 1968, **15**, 37.
[759] Wolfsberger, W., and Forsterling, H., *J. Organometallic Chem.*, 1976, **122**, 13.
[760] Wolfsberger, W., Pickel, H. H., and Schmidbaur, H., *Z. Naturforsch.*, 1971, **26b**, 979.
[761] Wolfsberger, W., and Pickel, H. H., *J. Organometallic Chem.*, 1973, **54**, C8.
[761a] Wooding, N. S., and Higginson, W. C. E., *J. Chem. Soc.*, 1952, 1178.
[762] Yamaguchi, K., Hasuo, M., and Maruyama, Y., Jap. Patent, 1974, 74 80,187; *Chem. Abstr.*, 1975, **83**, 115411.
[763] Yamamoto, Y., and Kimura, H., *Chem. Pharm. Bull.*, 1976, **24**, 1236; *Chem. Abstr.*, 1977, **86**, 43543.
[764] Yamasaki, Y., and Yokoto, M., Jap. Patent, 1969, 69 21,748; *Chem. Abstr.*, 1970, **72**, 56075g.
[765] Yamasaki, Y., and Yokoto, M., Jap. Patent, 1969, 69 21,747; *Chem. Abstr.*, 1970, **72**, 56067r.
[766] Yamazaki, I., Tooyama, Y., Hagino, I., and Hirota, K., Jap. Patent, 1972, 72 14,447; *Chem. Abstr.*, 1972, **77**, 115117.
[767] Yamazaki, I., Tooyama, Y., and Hirota, K., Jap. Patent, 1971, 71 34,847; *Chem. Abstr.*, 1972, **76**, 113928.
[768] Yambushev, F. D., Gatilov, Yu. F., Tenisheva, N. Kh., and Savin, V. I., *Zh. Obshch. Khim.*, 1974, **44**, 2205; *Chem. Abstr.*, 1975, **82**, 57838.
[768a] Yambushev, F. D., Tenisheva, N. Kh., and Khusainov, Kh. Z., *J. Gen. Chem. U.S.S.R.*, 1977, **47**, 386.
[769] Yoder, C. H., Komoriya, A., Kochanowski, J. E., and Suydam, F. H., *J. Amer. Chem. Soc.*, 1971, **93**, 6515.
[770] Yoder, C. H., and Zuckerman, J. J., *J. Amer. Chem. Soc.*, 1966, **88**, 2170.
[770a] Yoshida, Z., Yoneda, S., Yato, T., Hagama, M., *Tetrahedron Letters*, 1973, 873.
[771] Young, J. A., Durrell, W. S., and Dresdner, R. D., *J. Amer. Chem. Soc.*, 1962, **84**, 2105.
[772] Young, J. A., Tsoukalas, S. N., and Dresdner, R. D., *J. Amer. Chem. Soc.*, 1958, **80**, 3604.
[772a] Young, J. A., Tsoukalas, S. N., and Dresdner, R. D., *J. Amer. Chem. Soc.*, 1960, **82**, 396.
[773] Zelles, G. R., and Lewis, R., *J. Org. Chem.*, 1948, **13**, 160.
[774] Zhigach, A. F., Sobolev, E. S., Svitsyn, R. A., and Nikitin, V. S., *Zh. Obshch. Khim.*, 1973, **43**, 1966.

[775] Swiss Patent, 1967, 434,749; *Organometallic Compd.*, 1967, **11**, 214.
[776] Fr. Patent 1,358,503; *Chem. Abstr.*, 1964, **61**, 4548; Belg. Patent 637,825; *Chem. Abstr.*, 1965, **62**, 7893; Brit. Patent 969,074; *Chem. Abstr.*, 1965, **62**, 11930; Fr. Patent 1,386,994; *Chem. Abstr.*, 1965, **62**, 13063; Neth. Patent 284,092; *Chem. Abstr.*, 1965, **63**, 701; Fr. Patent 1,369,866; *Chem. Abstr.*, 1965, **63**, 1896; Jap. Patent 1966, 5379; *Chem. Abstr.*, 1966, **65**, 4714; Belg. Patent 664,699; *Chem. Abstr.*, 1966, **65**, 15537.
[777] U.S. Patent 1977, 4,042,610.

Author Index

Numbers in parentheses are reference numbers, which are preceded by the page number on which the author's work is cited.

Subject Index

(The reader is also referred to the detailed lists of contents which preface each chapter)